IBM Software Systems Integration

With IBM MQ Series for JMS, IBM FileNet Case Manager, and IBM Business Automation Workflow

Alan S. Bluck

apress®

IBM Software Systems Integration: With IBM MQ Series for JMS, IBM FileNet Case Manager, and IBM Business Automation Workflow

Alan S. Bluck
Ashley Heath, Hampshire, UK

ISBN-13 (pbk): 978-1-4842-8860-3   ISBN-13 (electronic): 978-1-4842-8861-0
https://doi.org/10.1007/978-1-4842-8861-0

Copyright © 2023 by Alan S. Bluck

This work is subject to copyright. All rights are reserved by the Publisher, whether the whole or part of the material is concerned, specifically the rights of translation, reprinting, reuse of illustrations, recitation, broadcasting, reproduction on microfilms or in any other physical way, and transmission or information storage and retrieval, electronic adaptation, computer software, or by similar or dissimilar methodology now known or hereafter developed.

Trademarked names, logos, and images may appear in this book. Rather than use a trademark symbol with every occurrence of a trademarked name, logo, or image we use the names, logos, and images only in an editorial fashion and to the benefit of the trademark owner, with no intention of infringement of the trademark.

The use in this publication of trade names, trademarks, service marks, and similar terms, even if they are not identified as such, is not to be taken as an expression of opinion as to whether or not they are subject to proprietary rights.

While the advice and information in this book are believed to be true and accurate at the date of publication, neither the authors nor the editors nor the publisher can accept any legal responsibility for any errors or omissions that may be made. The publisher makes no warranty, express or implied, with respect to the material contained herein.

Managing Director, Apress Media LLC: Welmoed Spahr
Acquisitions Editor: Mirashi Aditee
Development Editor: James Markham
Coordinating Editor: Mirashi Aditee

Cover image designed by eStudioCalamar

Distributed to the book trade worldwide by Springer Science+Business Media New York, 1 New York Plaza, Suite 4600, New York, NY 10004-1562, USA. Phone 1-800-SPRINGER, fax (201) 348-4505, e-mail orders-ny@springer-sbm.com, or visit www.springeronline.com. Apress Media, LLC is a California LLC and the sole member (owner) is Springer Science + Business Media Finance Inc (SSBM Finance Inc). SSBM Finance Inc is a **Delaware** corporation.

For information on translations, please e-mail booktranslations@springernature.com; for reprint, paperback, or audio rights, please e-mail bookpermissions@springernature.com.

Apress titles may be purchased in bulk for academic, corporate, or promotional use. eBook versions and licenses are also available for most titles. For more information, reference our Print and eBook Bulk Sales web page at http://www.apress.com/bulk-sales.

Any source code or other supplementary material referenced by the author in this book is available to readers on GitHub via the book's product page, located at www.apress.com/. For more detailed information, please visit http://www.apress.com/source-code.

Printed on acid-free paper

*Dedicated to
my beloved wife, Jennifer,
and
my children, Julie and Rosalie,
and
their families*

Table of Contents

About the Author vii
About the Technical Reviewer ix
Acknowledgments xi
Introduction xiii

Chapter 1: IBM FileNet Case Manager 5.3.3 Case Builder Solution Development Steps for the Audit System 1
    Chapter Organization 1
    Introduction 2
    Part 1 – Bill of Materials 3
    Starting the IBM Case Manager System 6
    Part 2 – IBM Case Manager Solution Metadata 11
    Adding a New Solution 11
    Adding Properties and Business Objects 13
    Adding Roles 15
    Adding Properties to the In-Baskets for Each Role 21
    Adding Document Class 23
    Adding Business Objects 27
    Adding Business Object Types As Case Properties 43
    Part 3 – Adding IBM Case Manager Case Types 44
    Adding Case Types 44
    Adding Business Objects to Be Searched 54
    Part 4 – Testing and Administration of the Audit Solution 70
    Exporting and Production Deployment of the Audit Master Solution 98
    Debugging the Case Manager Client 114
    Checked FileNet Version 118
    Checked FileNet Health 119
    Check FileNet Object Store Upgrade Status 120
    Fix/Workaround for Error on Case Search 123
    Fix (Recommended) 131
    Chapter 1 Exercises 139

Chapter 2: Configuring Java Custom Components 143
    Chapter Organization 144
    Part 1 – Supporting Documents 144
    IBM FileNet Java API Call Examples 144
    Java Code Development 145
    Part 2 – Configuring Java Components for Content Engine Events 146
    Custom Code Module Java JAR API Call Development 179
    Creating a Custom Event Object 188
    Configuring Workflow Subscriptions 193
    Creating the fn_eventshandler.jar 199
    Part 3 – Testing the Add Document to Folder Event 205
    Part 4 – The DbExecute Workflow Step for Calling Database Stored Procedures 251
    The IBM Auto-Claims Case Manager Solution Example 251
    Methods of Displaying a Case Type Task Workflow 252
    The Case Builder Workflow Step Designer Application 255
    The Process Designer Plugin Applet in IBM Content Navigator 258
    Configuring the DbExecute Connection 271
    Part 5 – Standalone Process Designer Installation 276
    Fixing the Process Designer Shell Script Environment Variables 290
    The DbExecute Workflow Step Addition 298
    The DbExecute Stored Procedure SQL 304
    Configuring the Workflow for the DbExecute Step 307
    Stored Procedure Parameter Limitations 311
    Example SQL Table Creation for the DB2 Database 312
    Testing on Different Database Platforms 329
    Oracle Version of the Stored Procedure SQL Code 330
    MS SQL Server Version of the Stored Procedure SQL Code 331
    Chapter 2 Exercises 332

TABLE OF CONTENTS

Chapter 3: IBM JMS Interface Development IBM FileNet 5.5.x Workflow 335
Chapter Organization 336
Part 1 – Bill of Materials 337
Part 2 – Custom Operations Component Development 386
Part 3 – Creating a Non-privileged MQ User for a Client Application Connection 458
Procedure 458
Listener Authorities 478
Stopping and Starting the Queue Manager 486
Creating a Client Channel for Messaging 491
Setting Up the Client on Linux 493
Sending a Message from a Client to a Server 495
Testing for Errors 495
Part 4 – Second Run of the AUDOperationsTest.java JUnit for JMS Messaging 558
Code Now Updated As Follows for AUDOperations.java 559
AUDOperations Rebuild and Deploy .jar: Final Prebuild Test 570
Part 5 – Building the AUDOperations.jar 575
Rebuilding the AUDOperations.jar File 583
FileNet Workflow System Component AUDOperations.jar Deployment 584
Editing the Parameters 604
IBM Process Designer Component Queue Configuration 607
Part 6 – Transferring Workflow and Setting Up Workflow Subscriptions 623
Creating a Workflow Subscription on a New Audit Report Document Event 623
Transferring the JMS Test Workflow to the Target Object Store 626
Testing the Updated Audit Master Solution 635
Chapter 3 Exercises 644

Chapter 4: A Replication Java Program for IBM FileNet Object Stores 647
Chapter Organization 648
Part 1 – Bill of Materials 648
Part 2 – The Replication Program Introduction 650
Nonfunctional Requirements 652
Development Tools Used 653
The P8 5.5.5 Environment 654
Unit Test Data 656
Creator Property 657
Setting Object Store Access Rights 658
Linux Directory Paths Required 663
Jars Required for Content Engine Client 667
Static Constants for Property Types 682
Event Setup 684
Unit Test Phases 687
Creating the Java Projects 698
Part 3 – Code Developed for the Replication Program 699
CEReplicate – 2957 Lines of Java Code – replication.jar 699
Project Creation 704
Part 4 – Code Listing – CEReplicateConfig 760
CEReplicateConfig – 502 Lines of Java Code –fn_common.jar 862
Part 5 – Code Listing – CEMigrateConfig 878
CEMigrateConfig – 425 Lines of Java Code – fn_common.jar 878
Part 6 – Code Listing – Supporting Utility Code 891
PropsUtil – 68 Lines of Java Code – fn_utils.jar 891
CEConnect – 67 Lines of Java Code – fn_connect.jar 893
CEConnection – 119 Lines of Java Code – fn_connection.jar 895
Chapter 4 Exercises 899

Chapter 5: IBM Cognos Analytics Custom Development 903
Bill of Materials 903
Chapter 5 Exercises 989

Chapter 6: PDF Document Creation Using Java 991
Chapter Organization 991
Part 1 – Bill of Materials 993
Imports Used with the Test Audit Report Stub from the iText jar Library 993
Example 1 – A Simple Audit PDF Document 994
Example 2 – Test Code to Add the Audit Master Logo 1000
Part 2 – Example 3 – An Audit Report from the Audit Master 1002
Part 3 – Supporting Java Classes for the Main Audit Report Program 1152
Part 4 – Example 4 – Create an Auditor Calendar Table in a PDF Document 1263
Chapter 6 Exercises 1267

Index 1271

About the Author

 **Alan S. Bluck** has over 45 years of IT experience. He has been a Solutions Architect for IBM for over 10 years. Elected as an IBM Champion (2022), he is now the director and owner of ASB Software Development Limited, an IBM PartnerWorld partner, and a consultancy providing systems architecture for a broad range of services. He is a Member of the British Computer Society (MBCS, CITP).

About the Technical Reviewer

Giles Metcalf has been a software architect in the process, content, and case management arena for more than 20 years. With extensive experience in the Financial Services market, he is keenly interested in integrating modern solutions with legacy systems, as well as enhancing the flow of information through organizations. He has worked with most of the high-street banks, as well as many other institutions. He has recently moved into the Identity and Access Management sphere.

Acknowledgments

There are a few people whom I want to thank for the continued and ongoing support that they gave me while I wrote this book. First and foremost, I would like to thank my wife, Jenny, for continuously encouraging me to write the book – I could never have completed this book without her support.

I am grateful to the excellent online courses and materials provided by Red Hat (www.redhat.com) and IBM RedBooks (www.redbooks.ibm.com).

I would also like to extend my gratitude to the team at Apress for their support in the editing and publication of the book, and the technical reviewer, Giles Metcalf, for his helpful suggestions and diligence.

Introduction

This book describes the design and build of an Audit Master System for ISO9000/BS5750 internal quality system audits, implemented using the IBM Case Manager Builder web application.

This system is then extended from the initial working IBM Case Manager solution to demonstrate a number of Java-based integrations to show the possibilities of integration with other systems, using the IBM FileNet Content Manager API, IBM MQ Series, JMS Java API calls, and the IBM Process Designer Workflow application.

The fully working integrations are described step by step, including

- The replication and synchronization of documents and the Case Manager folder system to another Object Store.

- The transfer of the Case metadata to an IBM MQ Series queue using JMS Java calls. This is demonstrated with the use of an IBM Process Designer Workflow, Java library Operations Step method to call the JMS (Java Messaging Service) with the Case data.

- The creation and deployment of a Custom Java EventHandler Code Module to automatically link the separate Case Audit Report documents to a central folder.

- The implementation of a workflow step to call a database Stored Procedure to store the Case Audit metadata to a separate database system.

- The creation of graphical analytical summary data using the IBM Cognos Analytics system and the extension of the IBM Case Analyzer OLAP Event table database to use custom Views for date range and multilevel Time Dimensions using the Cognos links to a Microsoft SQL Server Analysis Manager system.

- The use of Java iText library methods to generate pdf Audit Report questions from the Case Management data, which are then stored against a Department Case Document folder.

INTRODUCTION

This book is designed for use by IT consultants, Systems and Solution Architects with basic general IT knowledge to provide working templates for the development of solutions using IBM software products to integrate diverse corporate systems, and for developers of document management, workflow, and case management solutions for implementation in manufacturing organizations, banks, and insurance companies.

CHAPTER 1

IBM FileNet Case Manager 5.3.3 Case Builder Solution Development Steps for the Audit System

This chapter covers the step-by-step procedure to use the IBM FileNet Case Manager 5.3.3 Case Builder Solution Development steps and the similar IBM system, IBM Business Automation Workflow, to create an Audit system.

See:

`www.ibm.com/docs/en/case-manager/5.3.3?topic=overview-whats-new-in-case-manager-v533`

An overview of the IBM Case Manager system can be found in the following link:

`www.ibm.com/docs/en/case-manager/5.3.3?topic=documentation-case-management-overview`

Chapter Organization

Note that most chapters in this book are divided into Parts featuring the core topics being covered, further organized by sections within.

CHAPTER 1   IBM FILENET CASE MANAGER 5.3.3 CASE BUILDER SOLUTION DEVELOPMENT STEPS
            FOR THE AUDIT SYSTEM

This chapter contains the following four Parts:

Part 1 – Bill of Materials. This Part lists the prerequisite system and IBM Software components required to develop the IBM Case Builder solution used in this book as a base for the system integrations which are described in later chapters.

Part 2 – IBM Case Manager Solution Metadata. This Part covers the definition of the global elements of a Solution, Properties, Security Roles, Document Classes, and Business Objects.

Part 3 – Adding IBM Case Manager Case Types. In this Part, the Audit Master Solution Case Type creation procedure is described. Case types are used to specify the Case Solution tasks. This specification includes the required document classes attached to the task workflow, the task workflow steps, and the In-baskets and roles that are used for each of the Workflow steps.

Part 4 – Testing and Administration of the Audit Solution. This Part describes the development and administration of the Audit Solution after the initial deployment. This includes the deployment to a production system from a development environment.

Introduction

IBM Case Manager is now available as part of IBM Business Automation Workflow as described in the following.

Note   IBM Case Manager is still available as a standalone IBM FileNet product for upgrades.

For viewing an IBM Business Process Manager (BPM) process from Case Manager Builder after creating a case task based on an existing workflow process, the process is visible in IBM Process Designer. After you create a task from an existing process, you can click the Open Web Process Designer icon to edit the process in Process Designer.

To change an existing process or add a new process, you must open the process from the IBM Process Center, update or create the process, create a snapshot, and then make that snapshot the default. To view any new processes in the Add Task dialog box, click **Refresh** on the **Select Process** page.

The following video tutorial shows an example of this procedure:
www.youtube.com/watch?v=VtobZqwBh9g

CHAPTER 1   IBM FILENET CASE MANAGER 5.3.3 CASE BUILDER SOLUTION DEVELOPMENT STEPS FOR THE AUDIT SYSTEM

(IBM BPM BAW – Moving to Web based Process Designer from Desktop Process Designer)

Access to the IBM Business Process Manager Work Dashboard is also available from the Case Manager Client.

The IBM Process Portal Work Dashboard is accessed from the Case Manager Client by adding the IBM Business Automation Workflow plugin to the IBM Content Navigator desktop.

The Work Dashboard tasks can then be viewed in a Content Navigator desktop which has the same functionality as the Business Process Management Process Portal.

When the IBM Business Automation Workflow plugin and IBM BPM V8.6.0 CF2018.03 are installed, the Workflow ➤ Launch Process action is added to the Action menu and the pop-up menu in the Browse view (configurable in the IBM Case Manager desktop).

This action can be used to launch an IBM BPM process from a document.

The Launch Process action can also be added to the menu in the Documents view for the IBM Case Manager Information widget.

The values for a property that is defined as type Business Object are now listed in the IBM case package solution PDF file along with the other case properties.

The Audit system is based on an article in the journal, Quality Forum, Volume 19 No. 3, September 1993. *The Journal of the Institute of Quality Assurance*. Pages 116–126. The Quality Audit Process, by Alan S. Bluck.

This article covers the department audits based on the ISO9000/BS5750 Quality Standard. Downloadable from

```
www.researchgate.net/publication/334094923_Case_Manager_Installation_04-05-2019
```

Part 1 – Bill of Materials

The installation steps required for IBM Case Manager 5.3.3 are covered step by step in the following free ResearchGate documents, downloaded using the DOI URL as follows:

```
https://doi.org/10.13140/RG.2.2.21708.16001
```

entitled *"Case Manager 5.3.3 Installation on RHEL 8.0 with Content Navigator 3.0.6"*

Click the **Download file PDF** command button in Figure 1-1.

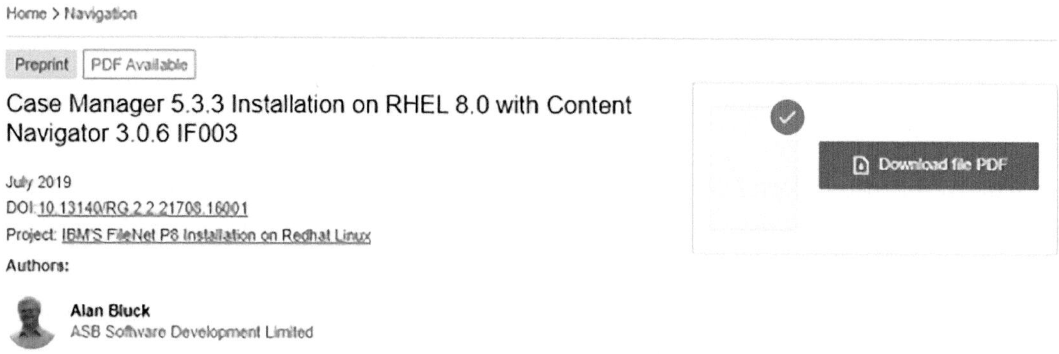

CHAPTER 1

IBM FILENET CASE MANAGER 5.3.3 CASE BUILDER SOLUTION DEVELOPMENT STEPS FOR THE AUDIT SYSTEM

***Figure 1-1.** Download the IBM Case Manager Installation document*

This gives the downloaded document, entitled *CaseManagerInstallationonRHEL8.0_V3.docx*.

Note As part of this installation, the requirement for the prerequisite IBM FileNet Content Engine Object Store **Workplace Base Extensions** for use with the IBM Case Manager installation is covered.

The step-by-step procedure for the installation of IBM Business Automation Workflow is available as a free download from the ResearchGate website, downloaded using the DOI URL as follows:

https://doi.org/10.13140/RG.2.2.27345.89440

entitled *"IBM BAW 18.0 Installation phase1 preprint with install of IBM Workflow Center 8.6.1.19002"*

Click the **Download file PDF** command button in Figure 1-2.

CHAPTER 1

IBM FILENET CASE MANAGER 5.3.3 CASE BUILDER SOLUTION DEVELOPMENT STEPS FOR THE AUDIT SYSTEM

Home > Econ > Multinational Corporations > International Economics > IBM

Preprint | PDF Available

IBM BAW 18.0 Installation phase1 preprint with install of IBM Workflow Center 8.6.1.19002

August 2019
DOI: 10.13140/RG.2.2.27345.89440
Project: IBM'S FileNet P8 Installation on Redhat Linux
Authors:

Alan Bluck
ASB Software Development Limited

Figure 1-2. *Download the IBM Business Automation Workflow Installation document*

This gives the downloaded document, entitled
IBMBAW18.0Installation_phase1_preprint (1).docx

The IBM Case Manager Audit system components used are listed as follows with the IBM part numbers shown in brackets. These part numbers can then be searched and downloaded from IBM's Software Access Catalog.

IBM Case Foundation V5.3.0 Linux Multilingual (CNPF8ML)

IBM Case Foundation V5.3.0 Windows Multilingual (CNPG0ML)

IBM FileNet Content Manager V5.5.0 for IBM Case Foundation V5.3.0 Multiplatform Multilingual eAssembly (CJ2VNML)

FileNet Content Manager V5.5.0 Quick Start Guide Multiplatform Multilingual (CNP8XML)

IBM FileNet Content Platform Engine V5.5.0 Linux Multilingual (CNP8ZML)

IBM FileNet Content Platform Engine V5.5.0 Windows Multilingual (CNP90ML)

IBM FileNet Content Platform Engine Client V5.5.0 Linux English (CNP93EN)

IBM FileNet Content Platform Engine Client V5.5.0 Windows English (CNP94EN)

IBM FileNet Content Search Services V5.5.0 Windows Multilingual (CNP9BML)

IBM FileNet Content Search Services V5.5.0 Linux64 Multilingual (CNP9CML)

IBM DB2 Enterprise Server Edition Restricted Use Quick Start and Activation V11.1 for Linux, UNIX and Windows Multilingual (CNB25ML)

Quick Start Guide for IBM WebSphere Application Server V9.0 (CNA8LML)

IBM WebSphere Application Server V9.0 (CND1AML)

IBM WebSphere Application Server V9.0 Supplements – Application Client (CND1CML)

CHAPTER 1 IBM FILENET CASE MANAGER 5.3.3 CASE BUILDER SOLUTION DEVELOPMENT STEPS
FOR THE AUDIT SYSTEM

IBM WebSphere Application Server V9.0 Supplements – IBM HTTP Server (CND1DML)

IBM WebSphere Application Server V9.0 Supplements – Web Server Plugins (CND1EML)

IBM WebSphere Application Server V9.0 Supplements – WebSphere Customization Toolkit (CND1FML)

IBM WebSphere Application Server Liberty (IBM Installation Manager install) (CND1GML)

IBM Installation Manager V1.8.5 for Linux x86_64 (CND0ZML)

IBM SDK, Java (TM) Technology Edition, Version 8 for Windows (CND15ML)

IBM SDK, Java (TM) Technology Edition, Version 8 for Linux (CND18ML)

Note WebSphere 8.5.5 Fix pack 15 automatically installs and defaults the JDK to 1.8.

Starting the IBM Case Manager System

The ecmukdemo6 VMware system described in the IBM Case Manager installation earlier can be started after booting as follows, logging in as the root user.

Start the LDAP Servers

Find the desktop icon for the Instance Administration tool.

Figure 1-3. Click the Linux desktop Instance Administration tool for LDAP

Click to open the LDAP, IBM Security Directory Server Instance Administration Tool:

CHAPTER 1 IBM FILENET CASE MANAGER 5.3.3 CASE BUILDER SOLUTION DEVELOPMENT STEPS
          FOR THE AUDIT SYSTEM

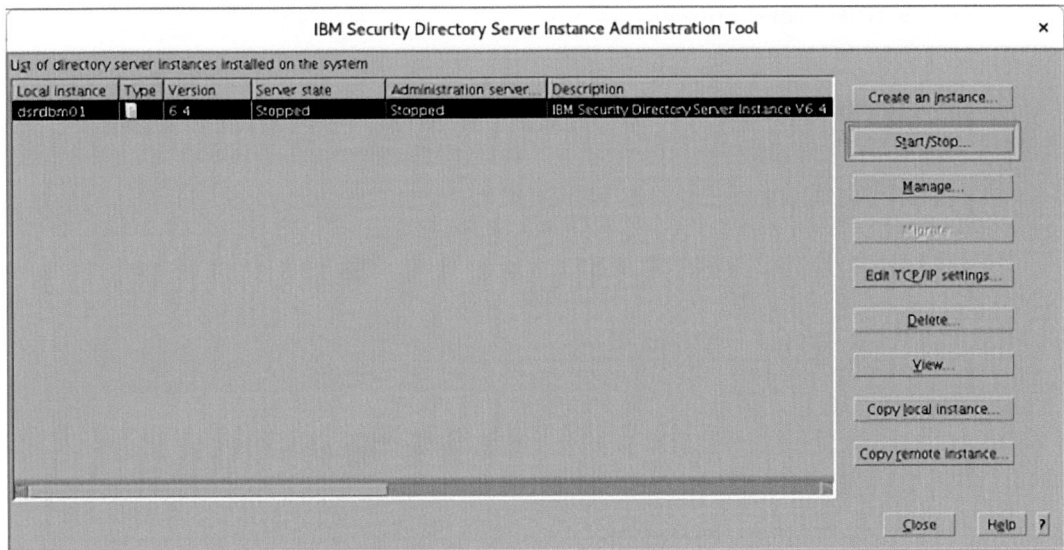

Figure 1-4. *Select the **Start/Stop** command to start the LDAP servers*

Click the **Start/Stop** command.

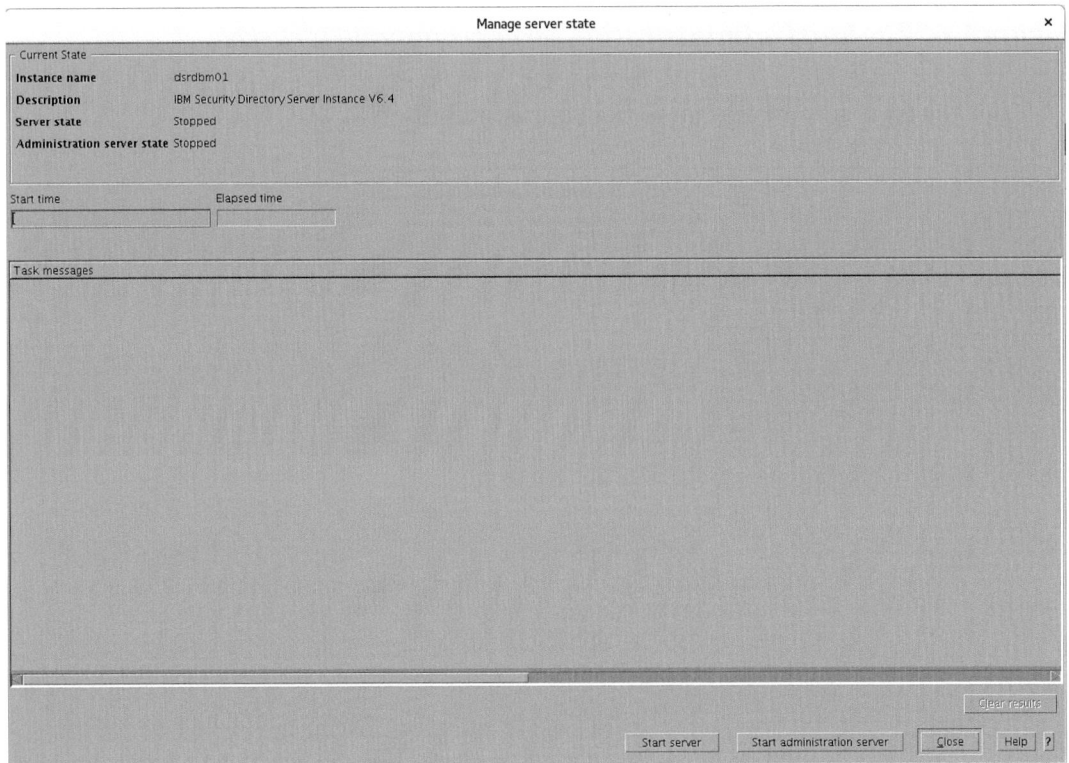

Figure 1-5. *The **Start server** command is clicked to start the LDAP server*

CHAPTER 1   IBM FILENET CASE MANAGER 5.3.3 CASE BUILDER SOLUTION DEVELOPMENT STEPS
            FOR THE AUDIT SYSTEM

Click the **Start server** command button.

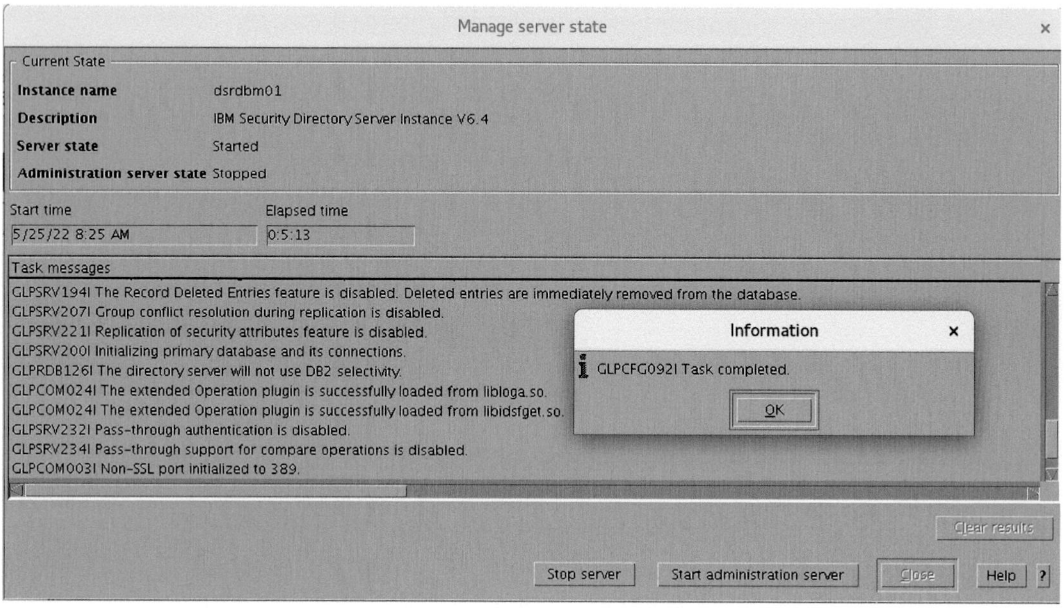

*Figure 1-6.   The **Start administration server** command is clicked*

Click the Start administration server command button.

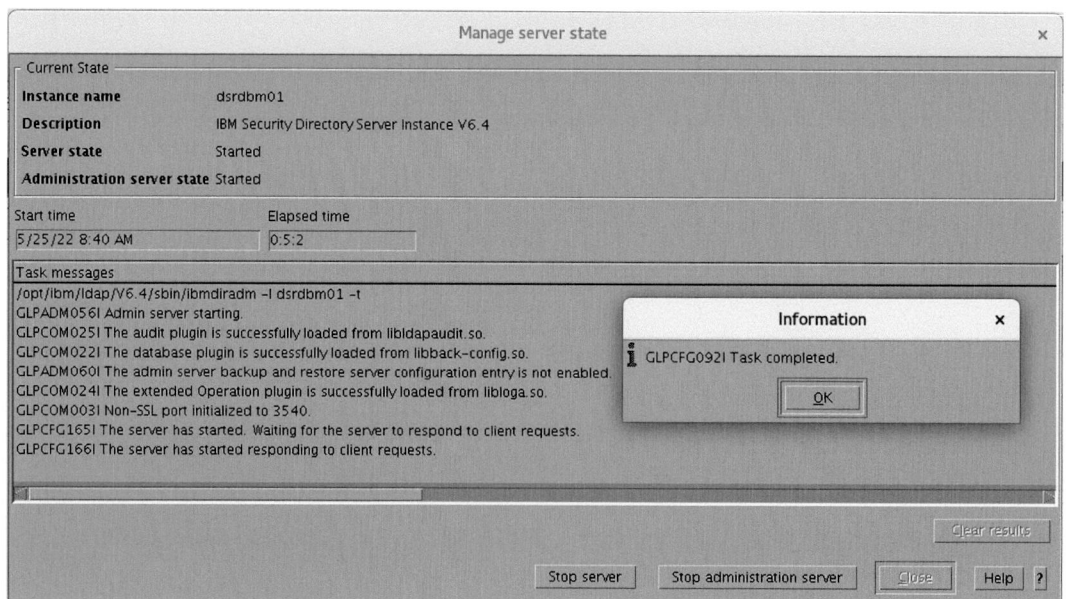

*Figure 1-7.   The **Close** command is disabled until the servers have started*

CHAPTER 1 IBM FILENET CASE MANAGER 5.3.3 CASE BUILDER SOLUTION DEVELOPMENT STEPS
FOR THE AUDIT SYSTEM

After around six minutes, the **Start complete** pop-up window is displayed, then you can click the **Close** command button.

Start the IBM FileNet Case Manager Database

Type the commands as follows:

su - db2inst1
db2start
exit

From the command window in Linux:

```
[root@ECMUKDEMO6 IBM]# pwd
/opt/IBM
[root@ECMUKDEMO6 IBM]# su - db2inst1
Last login: Mon Mar 14 04:56:45 PDT 2022 on pts/0
[db2inst1@ECMUKDEMO6 ~]$ db2start
05/10/2022 08:47:16     0   0  SQL1063N  DB2START processing was successful.
SQL1063N  DB2START processing was successful.
[db2inst1@ECMUKDEMO6 ~]$ exit
logout
[root@ECMUKDEMO6 IBM]#
```

Figure 1-8. *The IBM FileNet and IBM Case Manager databases are started*

Start the IBM FileNet Content Engine Web Service

From the command window in Linux, type the commands as follows:

cd /opt/IBM/WebSphere/AppServer/profiles/AppSrv01/bin
./startServer.sh server1

```
[root@ECMUKDEMO6 IBM]# cd /opt/IBM/WebSphere/AppServer/profiles/AppSrv01/bin
[root@ECMUKDEMO6 bin]# ./startServer.sh server1
ADMU0116I: Tool information is being logged in file
           /opt/IBM/WebSphere/AppServer/profiles/AppSrv01/logs/server1/startServer.log
ADMU0128I: Starting tool with the AppSrv01 profile
ADMU3100I: Reading configuration for server: server1
ADMU3200I: Server launched. Waiting for initialization status.
ADMU3000I: Server server1 open for e-business; process id is 9468
[root@ECMUKDEMO6 bin]#
```

Figure 1-9. *The IBM FileNet Content Engine web application is started*

CHAPTER 1   IBM FILENET CASE MANAGER 5.3.3 CASE BUILDER SOLUTION DEVELOPMENT STEPS
              FOR THE AUDIT SYSTEM

Start the IBM Case Manager Web Application

From the command window in Linux, type the commands as follows:

cd /opt/IBM/WebSphere/AppServer/profiles/AppSrv02/bin

./startServer.sh server1

```
[root@ECMUKDEMO6 bin]# cd /opt/IBM/WebSphere/AppServer/profiles/AppSrv02/bin
[root@ECMUKDEMO6 bin]# ./startServer.sh server1
ADMU0116I: Tool information is being logged in file
           /opt/IBM/WebSphere/AppServer/profiles/AppSrv02/logs/server1/startServer.log
ADMU0128I: Starting tool with the AppSrv02 profile
ADMU3100I: Reading configuration for server: server1
ADMU3200I: Server launched. Waiting for initialization status.
ADMU3000I: Server server1 open for e-business; process id is 10151
[root@ECMUKDEMO6 bin]#
```

Figure 1-10. *The IBM Case Manager web application server is started*

Load the IBM Case Manager Builder Web Application

Using the URL as follows, we can launch the installed IBM Case Builder development tool:

   http://ecmukdemo6:9081/CaseBuilder/designer/DesignerHome.jsp

The version of the IBM Case Manager and the available development IBM FileNet Object stores can be viewed using the **About** menu option in the IBM Case Builder web application as shown in Figure 1-11.

Figure 1-11. *The **?** ➤ **About** menu is selected to display the version details*

CHAPTER 1   IBM FILENET CASE MANAGER 5.3.3 CASE BUILDER SOLUTION DEVELOPMENT STEPS
            FOR THE AUDIT SYSTEM

The referenced IBM version is displayed in the About pop-up window as shown Figure 1-12.

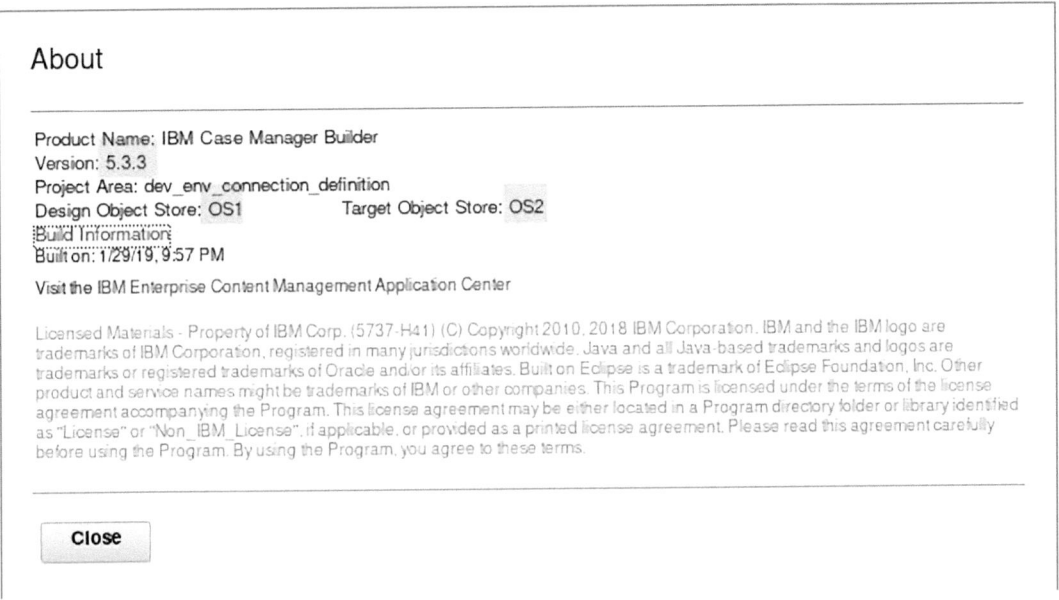

Figure 1-12.   The highlighted details show the Version and Object Store names

Part 2 – IBM Case Manager Solution Metadata

Adding a New Solution

A new solution for IBM Case Manager is created using the + icon on the top right of the logged in Case Builder screen.

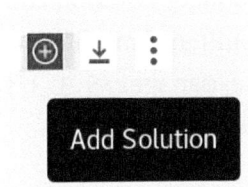

Figure 1-13.   The Add solution (+) icon is selected to create a new Case Solution

CHAPTER 1   IBM FILENET CASE MANAGER 5.3.3 CASE BUILDER SOLUTION DEVELOPMENT STEPS
            FOR THE AUDIT SYSTEM

We will call this new solution *Audit Master*.

This is based on the article "The Quality Audit Process," Quality Forum, *The Journal of the Institute of Quality Assurance*, Volume 19, Number 3, dated September 1993.

See Appendix A ©1992, Alan S. Bluck (`https://www.researchgate.net/publication/334095565_Case_Manager_Solution_Development`).

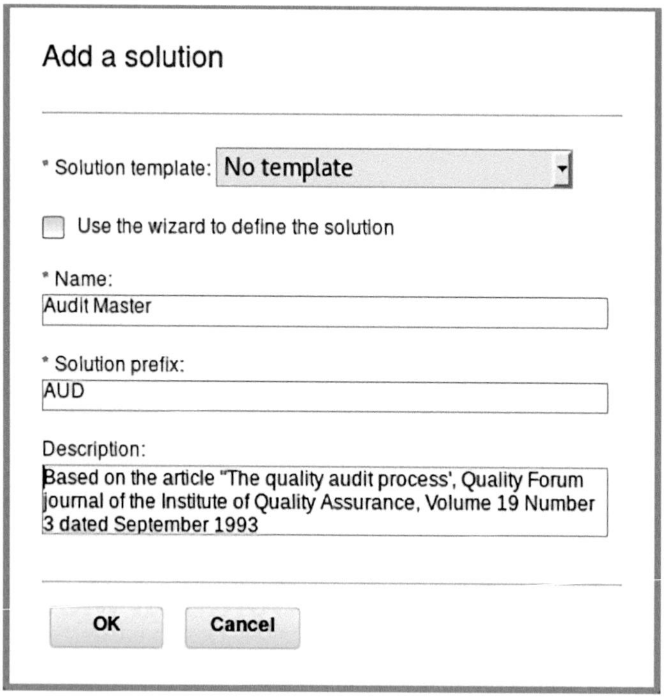

Figure 1-14. The Solution Name is entered and OK is clicked

The Solution can be based on a similar **Solution template** in the drop-down shown earlier. (Any deployed IBM Case Manager Solution can be saved as a template so that different Solution versions can be built and modified as required).

See `www.ibm.com/docs/no/case-manager/5.3.3?topic=application-creating-distributing-case-manager-solution-templates`.

CHAPTER 1 IBM FILENET CASE MANAGER 5.3.3 CASE BUILDER SOLUTION DEVELOPMENT STEPS
        FOR THE AUDIT SYSTEM

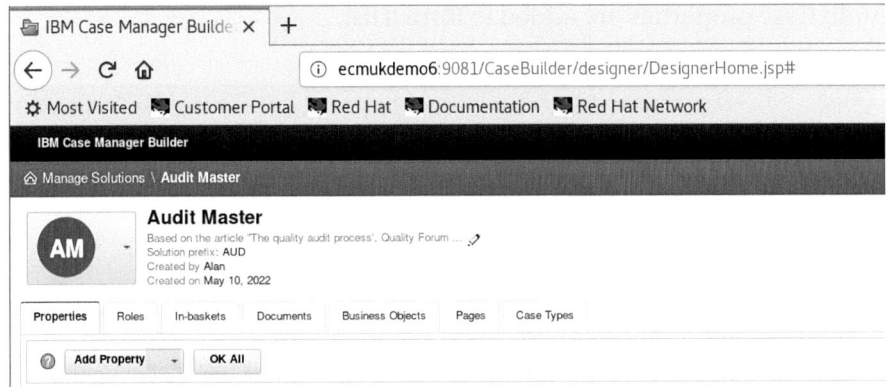

Figure 1-15. *The Add Property command is selected to add Global Solution properties*

The solution is created by entering the required solution criteria using the tabs as shown earlier, working from left to right.

The first tab adds the globally referenced solution properties, which are used to create the required IBM FileNet Content Manager custom object attributes, which are generated from the Solution Template structure for each case created.

Adding Properties and Business Objects

Properties can be added or reused. In this solution, the **New** option is selected as shown in Figure 1-16.

Figure 1-16. *The New menu item is selected to create a new Case Solution property*

CHAPTER 1   IBM FILENET CASE MANAGER 5.3.3 CASE BUILDER SOLUTION DEVELOPMENT STEPS FOR THE AUDIT SYSTEM

The Audit Case properties are added to form a list.

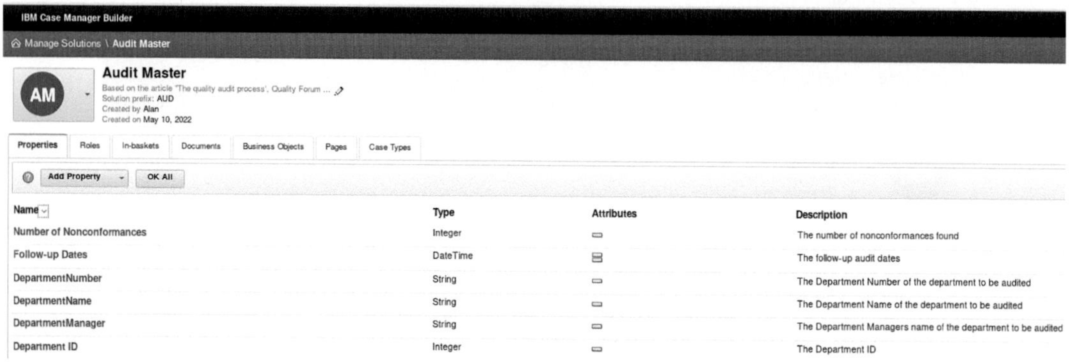

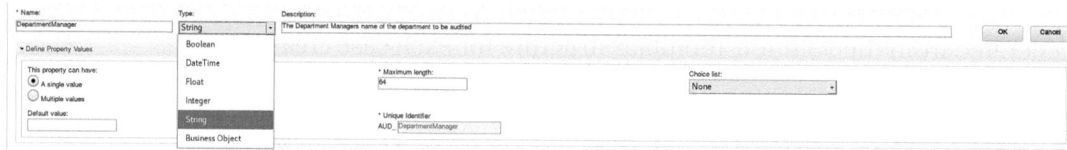

Figure 1-17. *The list of Audit global properties*

The Add Property command button is clicked to add the Audit system department properties for the ISO9000/BS5750 auditing of an organization's department.

Figure 1-18. *The property type is selected from a drop-down list*

For each property added, as shown earlier for the **DepartmentManager** property, the type of the property can be selected from the usual IBM FileNet Content Manager property types, including the new **Business Object** property type.

A choice list of drop-down options can also be selected by first creating them, using the **Manage Choice Lists** command.

CHAPTER 1   IBM FILENET CASE MANAGER 5.3.3 CASE BUILDER SOLUTION DEVELOPMENT STEPS
            FOR THE AUDIT SYSTEM

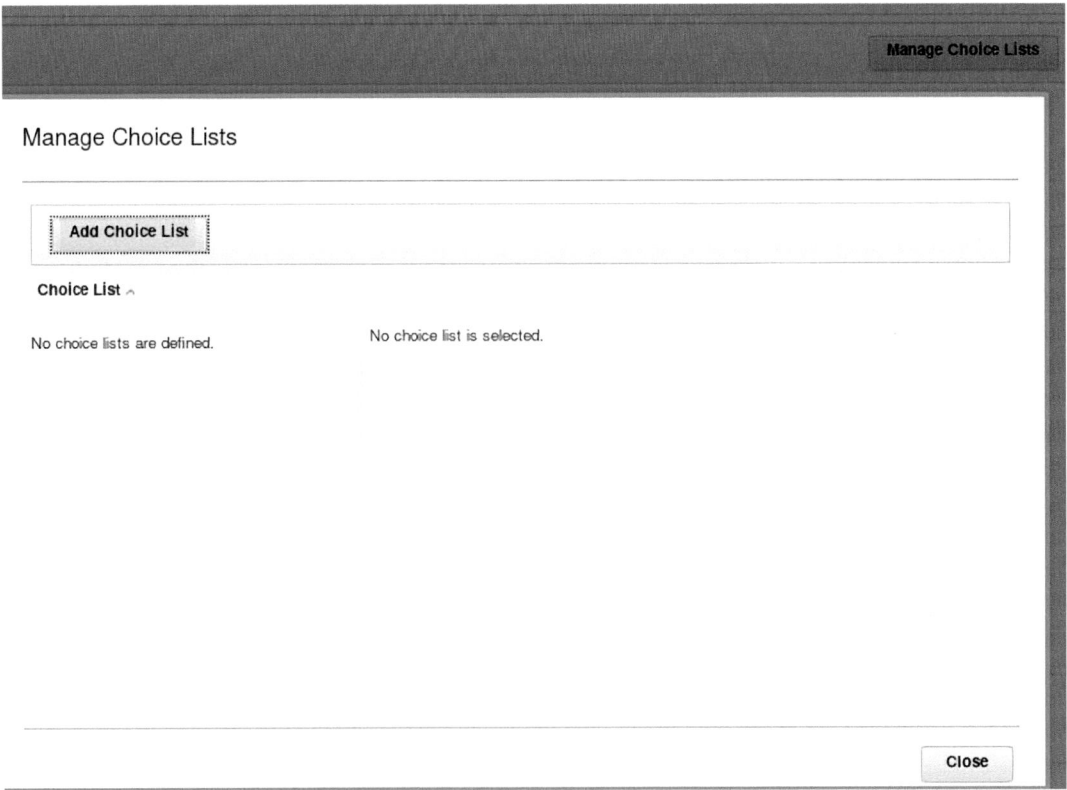

Figure 1-19. *The Manage Choice Lists command*

A Choice List consists of a displayed name and a separately held internal value. This allows a visible name, for example, **New York**, to be displayed, but the state code of **NY** can be stored.

The allowed length and, for numbers, valid range can also be entered for the property here. Also, a default value can be entered to facilitate the ease of entry.

The property can be selected to be a single value, or it can be defined to allow multiple values to be entered.

Adding Roles

The second tab allows the Solution Roles to be added. The security system uses the concept of a Role to define a group of users with the same access level to the Case Manager project. These roles are attached to the Case In-baskets used and also define functions available to a user with a specific Role. (A user can be allocated more than one Role, which extends and enhances the possible security matrix which can be defined.)

CHAPTER 1 IBM FILENET CASE MANAGER 5.3.3 CASE BUILDER SOLUTION DEVELOPMENT STEPS
        FOR THE AUDIT SYSTEM

The Audit system **Auditor** Role is added with the option **Personal** selected.

The options to move work into the user's Personal In-basket and the ability to reassign work to other users are selected using the tick boxes against each option.

Click the **Roles** tab.

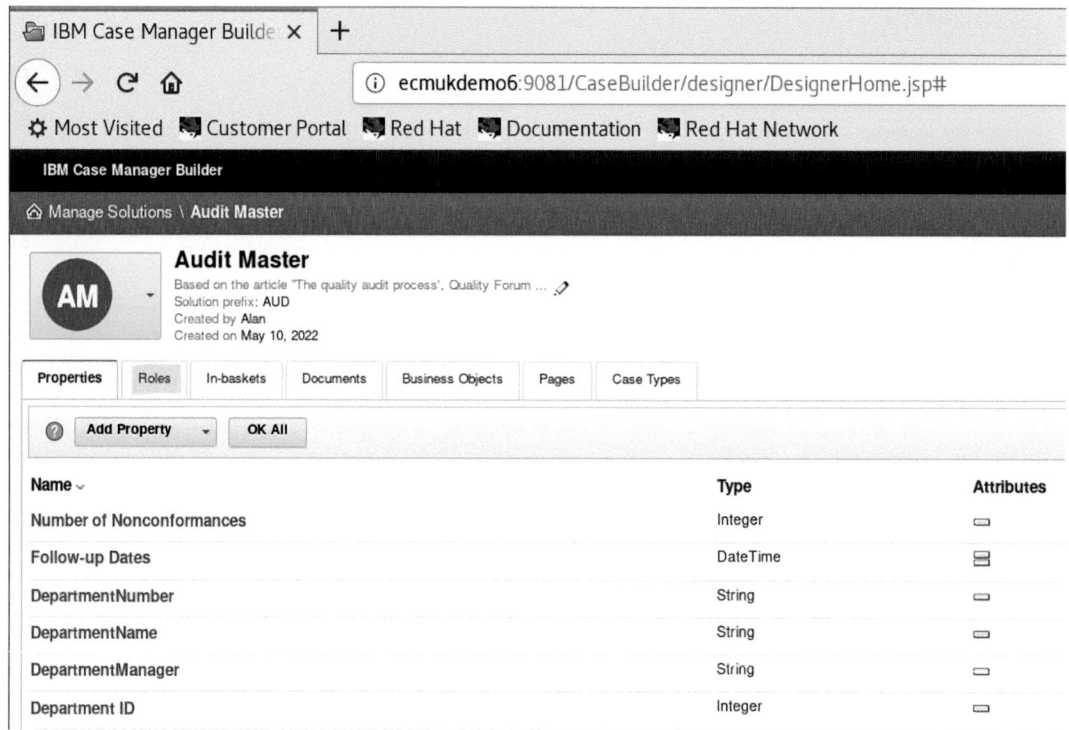

Figure 1-20. *The **Roles** tab is selected next to create the Case Security*

The **Auditor** role is added with the access options shown in Figure 1-21.

CHAPTER 1   IBM FILENET CASE MANAGER 5.3.3 CASE BUILDER SOLUTION DEVELOPMENT STEPS
            FOR THE AUDIT SYSTEM

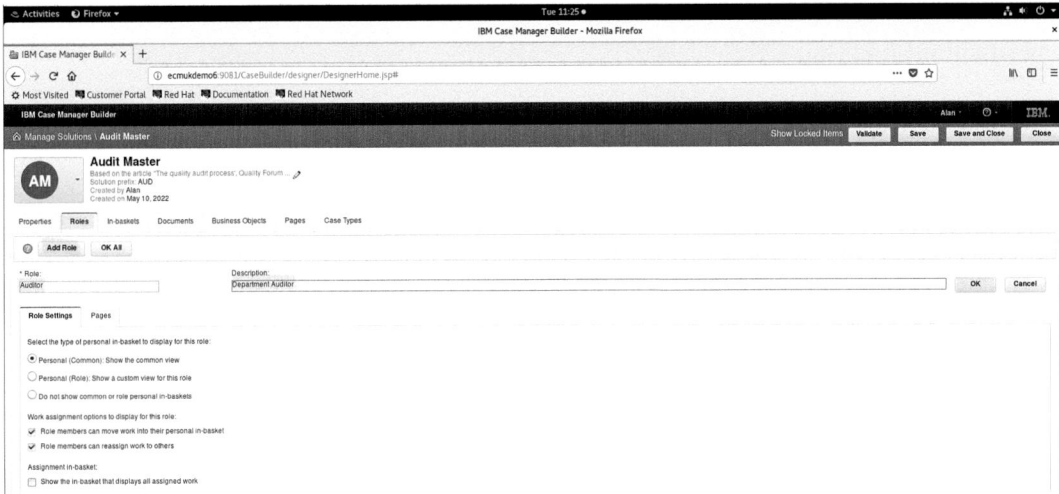

Figure 1-21. *The **Auditor** role is added with the options selected as shown*

Note   It is good practice to click the Save command button at regular intervals to ensure you do not lose any edits during the Case Solution build.

The **Pages** tab is then selected to show the Assigned pages for the Role.

At this point, custom pages could be added (see later in this chapter for the details of the entry for the **Pages** tab).

CHAPTER 1 IBM FILENET CASE MANAGER 5.3.3 CASE BUILDER SOLUTION DEVELOPMENT STEPS FOR THE AUDIT SYSTEM

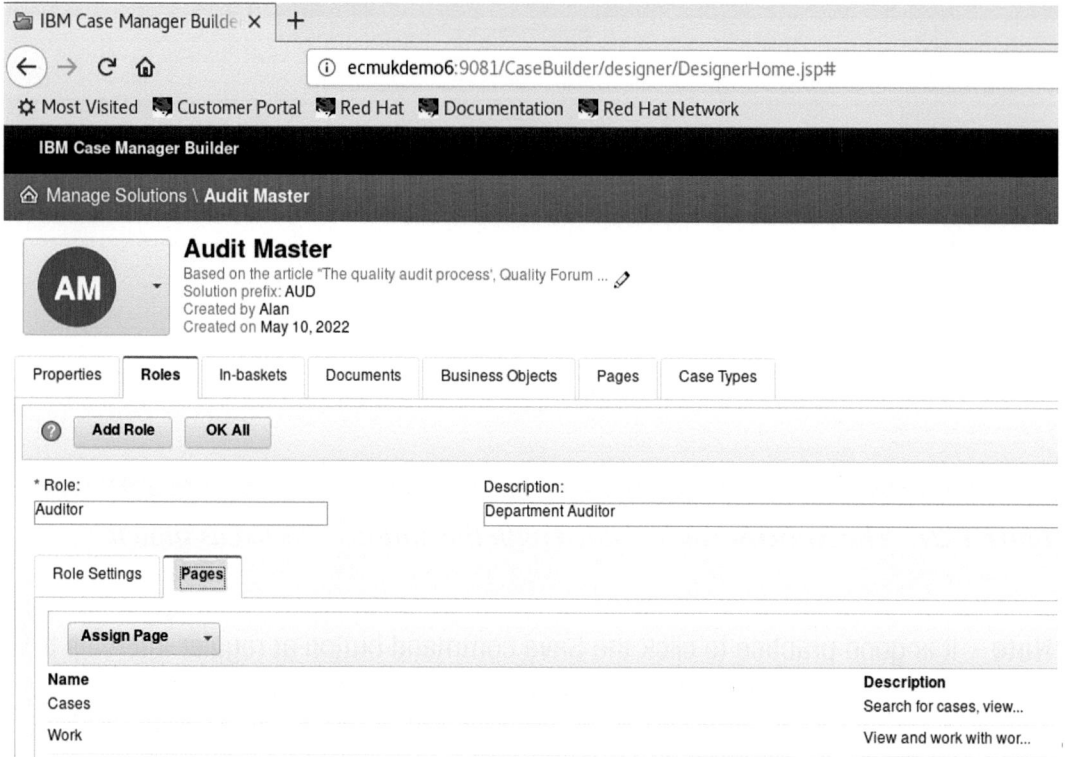

Figure 1-22. *The **Pages** tab is left with the displayed defaults*

The **Pages** tab is next selected as shown in Figure 1-22.

This tab allows the Role to be linked with standard and custom created web widget pages.

(The prebuilt **Work** and **Cases** pages are shown.)

The **Auditor's Manager** Role is added next to define a supervisor role for the **Auditor** Role users.

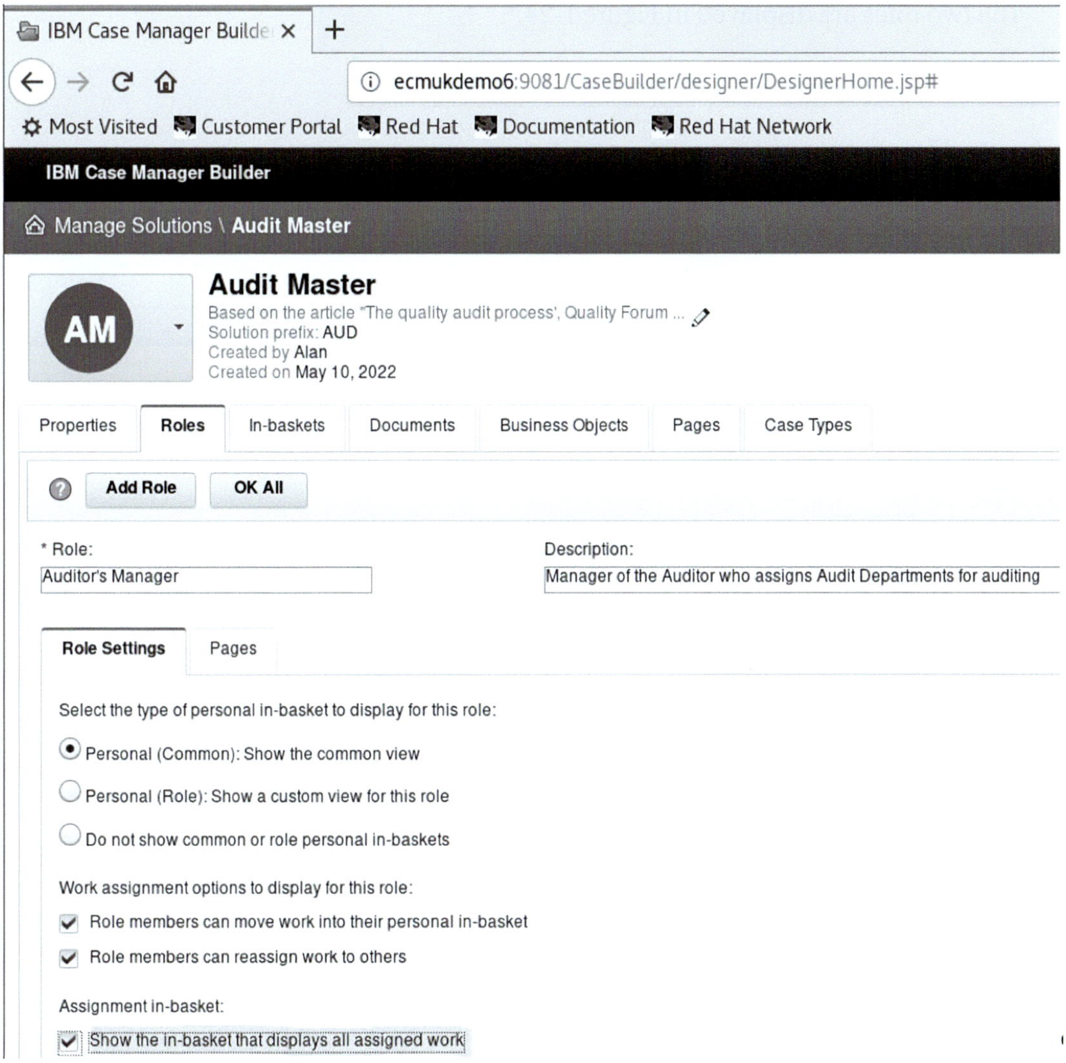

CHAPTER 1 IBM FILENET CASE MANAGER 5.3.3 CASE BUILDER SOLUTION DEVELOPMENT STEPS
FOR THE AUDIT SYSTEM

Figure 1-23. *The Auditor's Manager role is added with the assigned work in-basket*

CHAPTER 1   IBM FILENET CASE MANAGER 5.3.3 CASE BUILDER SOLUTION DEVELOPMENT STEPS
            FOR THE AUDIT SYSTEM

The two roles are displayed in Figure 1-24.

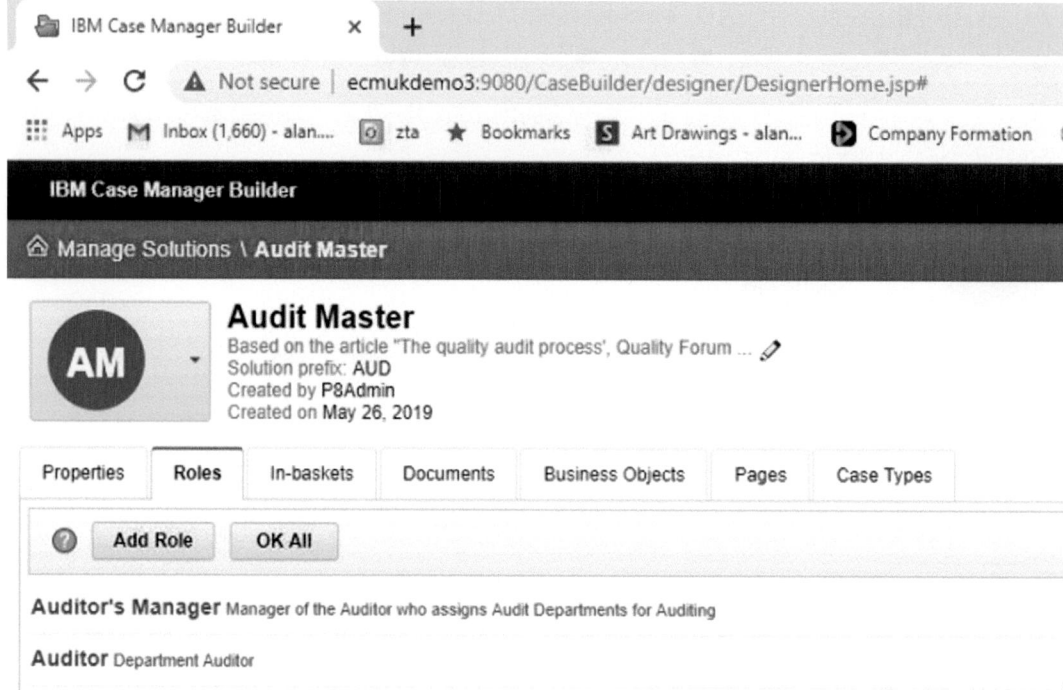

Figure 1-24.   The list of Roles defined is shown

The next step is to define the In-baskets of the generated Audit cases.

CHAPTER 1 IBM FILENET CASE MANAGER 5.3.3 CASE BUILDER SOLUTION DEVELOPMENT STEPS FOR THE AUDIT SYSTEM

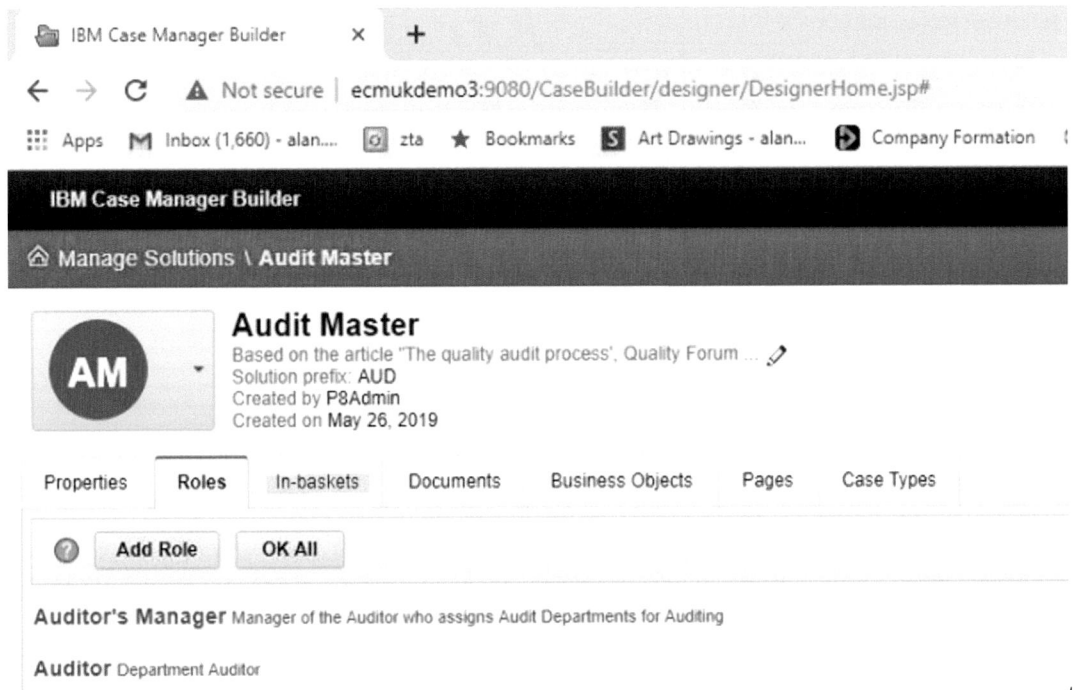

Figure 1-25. *The **In-baskets** tab (highlighted) is clicked to define In-baskets*

The **In-baskets** tab is now selected.

Adding Properties to the In-Baskets for Each Role

The properties we defined earlier can now be selected to be added as attributes of the Case, visible in the In-baskets, as required.

CHAPTER 1　IBM FILENET CASE MANAGER 5.3.3 CASE BUILDER SOLUTION DEVELOPMENT STEPS FOR THE AUDIT SYSTEM

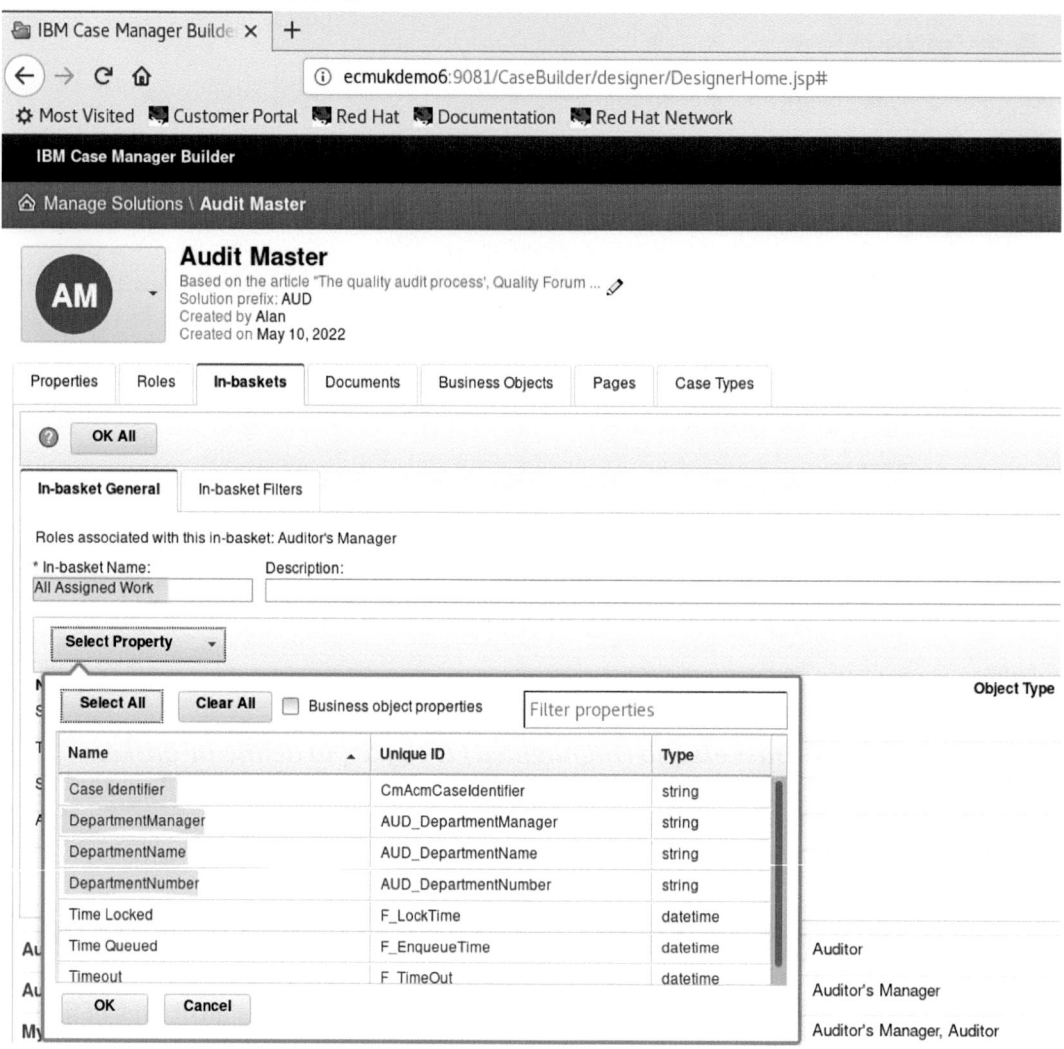

Figure 1-26. *The Global properties are selected as required for **All Assigned Work***

The required properties are selected by holding down the **CTRL key** and clicking using the left mouse button. Then select the **OK** command button.

CHAPTER 1   IBM FILENET CASE MANAGER 5.3.3 CASE BUILDER SOLUTION DEVELOPMENT STEPS FOR THE AUDIT SYSTEM

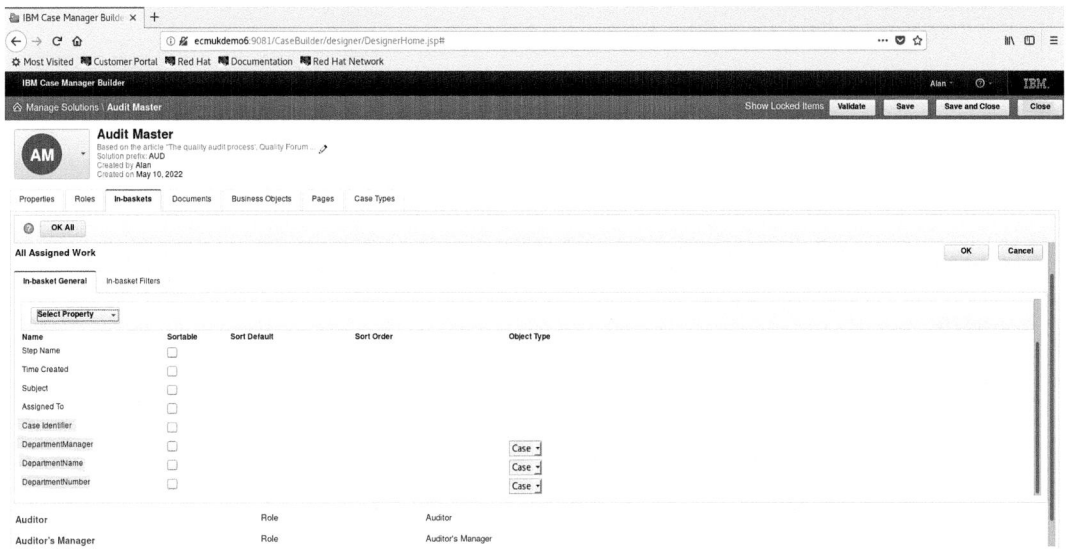

Figure 1-27. *The new In-basket properties are displayed after selection (highlighted)*

In addition, the columns which can be sorted can be selected by setting a tick in the **Sortable** column boxes against the required property to be sortable.

Note   The In-basket Filters tab allows a property to be selected to be used to filter a subset of the Cases in the In-basket.

Repeat the preceding configuration for the other In-baskets.

Adding Document Class

The attached IBM FileNet Content Engine document class can be created using the IBM Case Manager Builder as shown in Figure 1-28.

CHAPTER 1 IBM FILENET CASE MANAGER 5.3.3 CASE BUILDER SOLUTION DEVELOPMENT STEPS
            FOR THE AUDIT SYSTEM

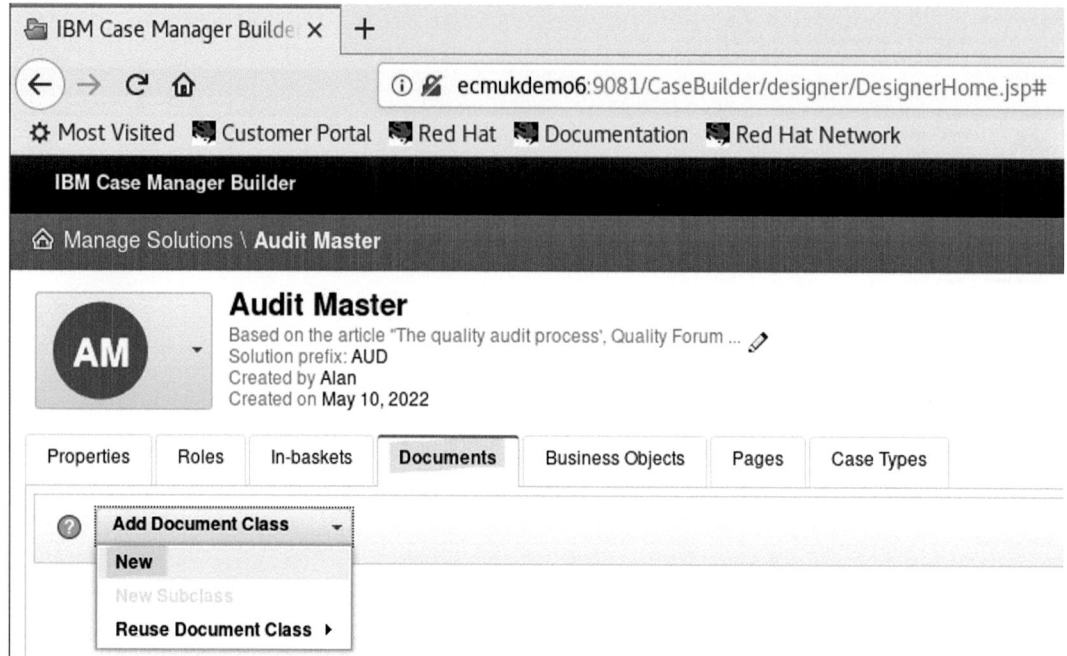

Figure 1-28. *The Documents tab is selected to add IBM FileNet document classes*

The Audit Report document class is created as shown in Figure 1-29.

CHAPTER 1   IBM FILENET CASE MANAGER 5.3.3 CASE BUILDER SOLUTION DEVELOPMENT STEPS FOR THE AUDIT SYSTEM

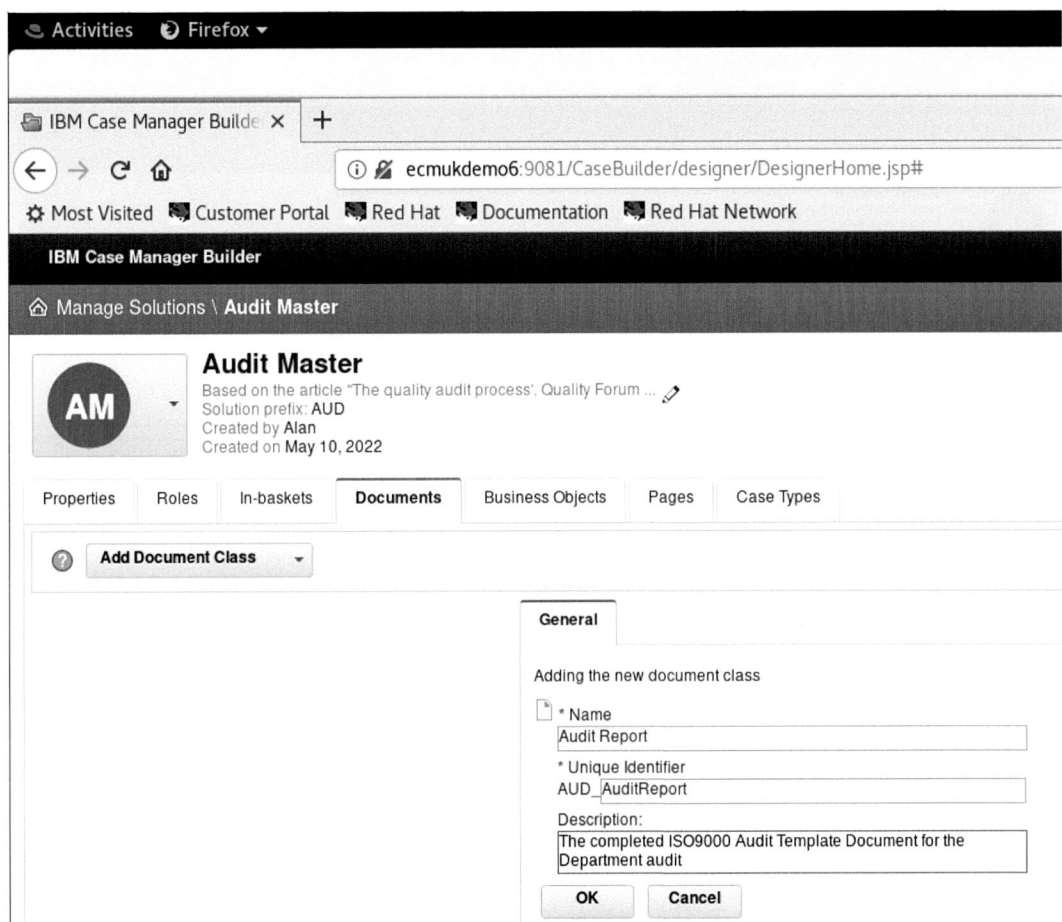

*Figure 1-29.   The new **Audit Report** document class details are added*

Click **OK** to generate the new Document Class and then select the properties required.

CHAPTER 1   IBM FILENET CASE MANAGER 5.3.3 CASE BUILDER SOLUTION DEVELOPMENT STEPS FOR THE AUDIT SYSTEM

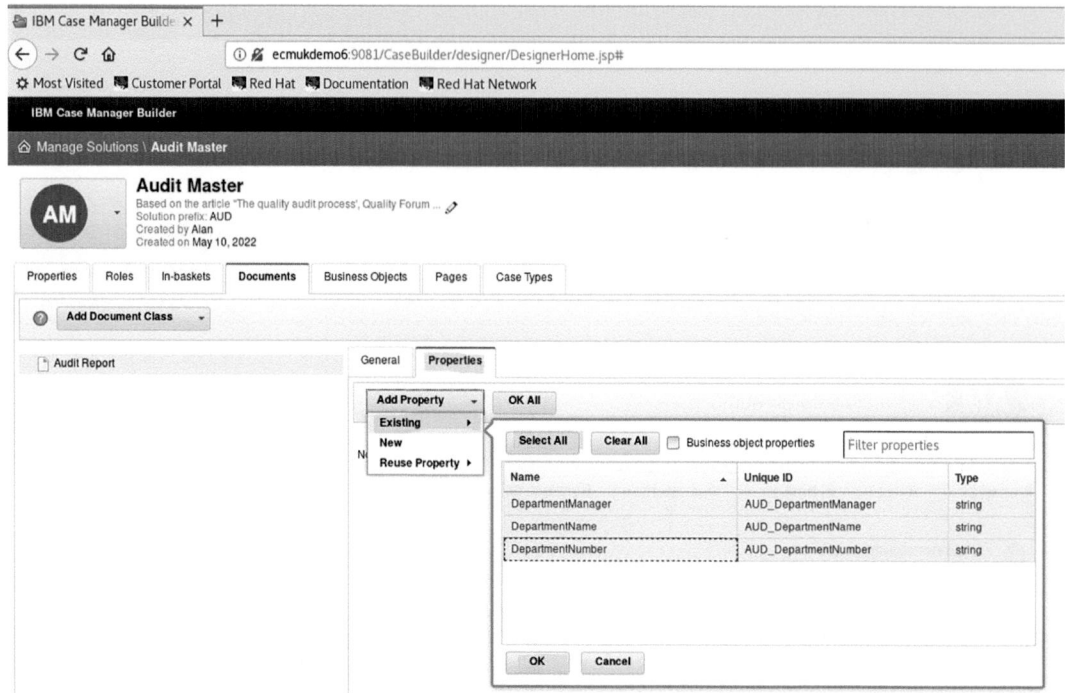

Figure 1-30. *The existing global properties created earlier are added*

Click **OK**.

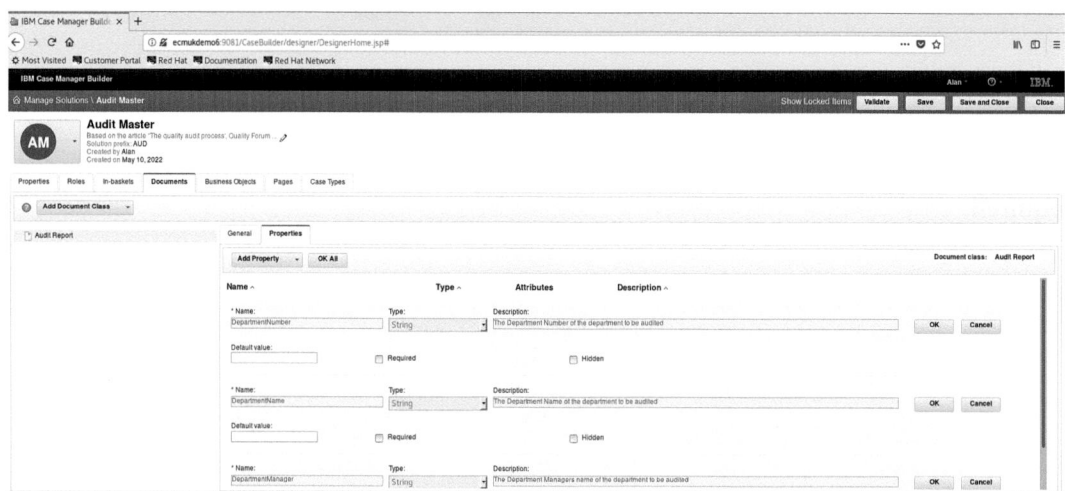

Figure 1-31. *On selecting the properties, each one is displayed with options to be added*

CHAPTER 1   IBM FILENET CASE MANAGER 5.3.3 CASE BUILDER SOLUTION DEVELOPMENT STEPS
             FOR THE AUDIT SYSTEM

If required, for each property the attributes of **Default value**, **Required**, or **Hidden** can be selected, and then the **OK** command button can be clicked for each of the properties.

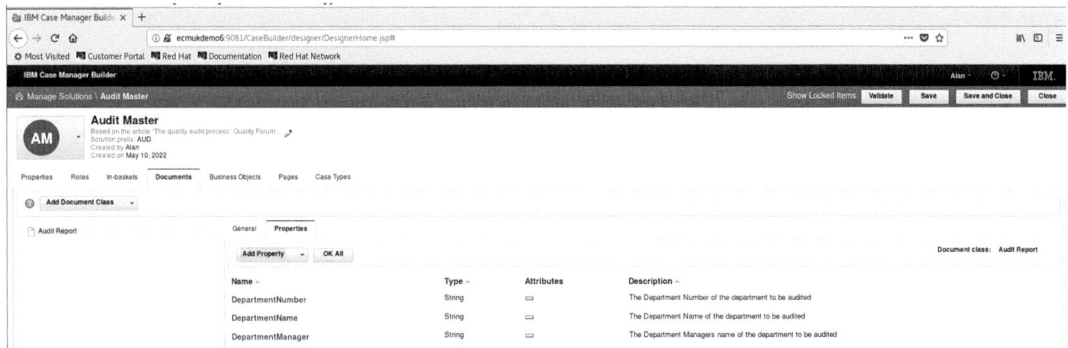

*Figure 1-32. The updated property attributes are completed after each **OK** is clicked*

Adding Business Objects

The **Business Objects** are a fairly recent IBM FileNet Content Engine property type which provides a compound property consisting of a collection of different properties which are tied together as one entity, that is, a structured data type that contains a collection of properties.

CHAPTER 1   IBM FILENET CASE MANAGER 5.3.3 CASE BUILDER SOLUTION DEVELOPMENT STEPS
            FOR THE AUDIT SYSTEM

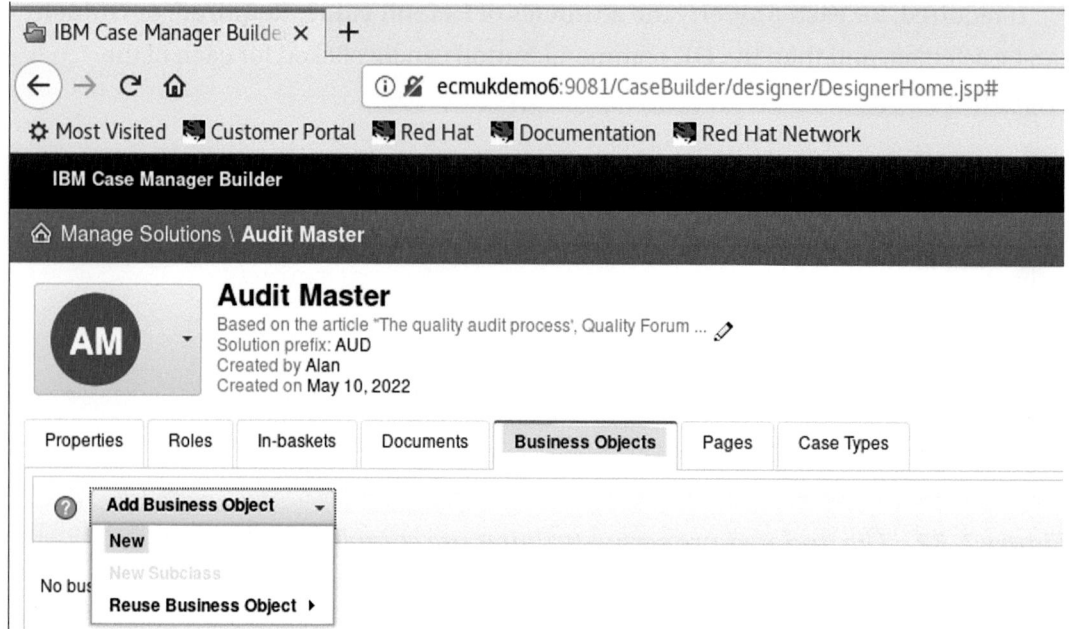

Figure 1-33. *The* **Business Objects** *tab is selected and a* **New** *Object is created*

The **Business Objects** tab is selected as shown in Figure 1-33, and then select the **New** menu option from the **Add Business Object** drop-down.

CHAPTER 1 IBM FILENET CASE MANAGER 5.3.3 CASE BUILDER SOLUTION DEVELOPMENT STEPS
            FOR THE AUDIT SYSTEM

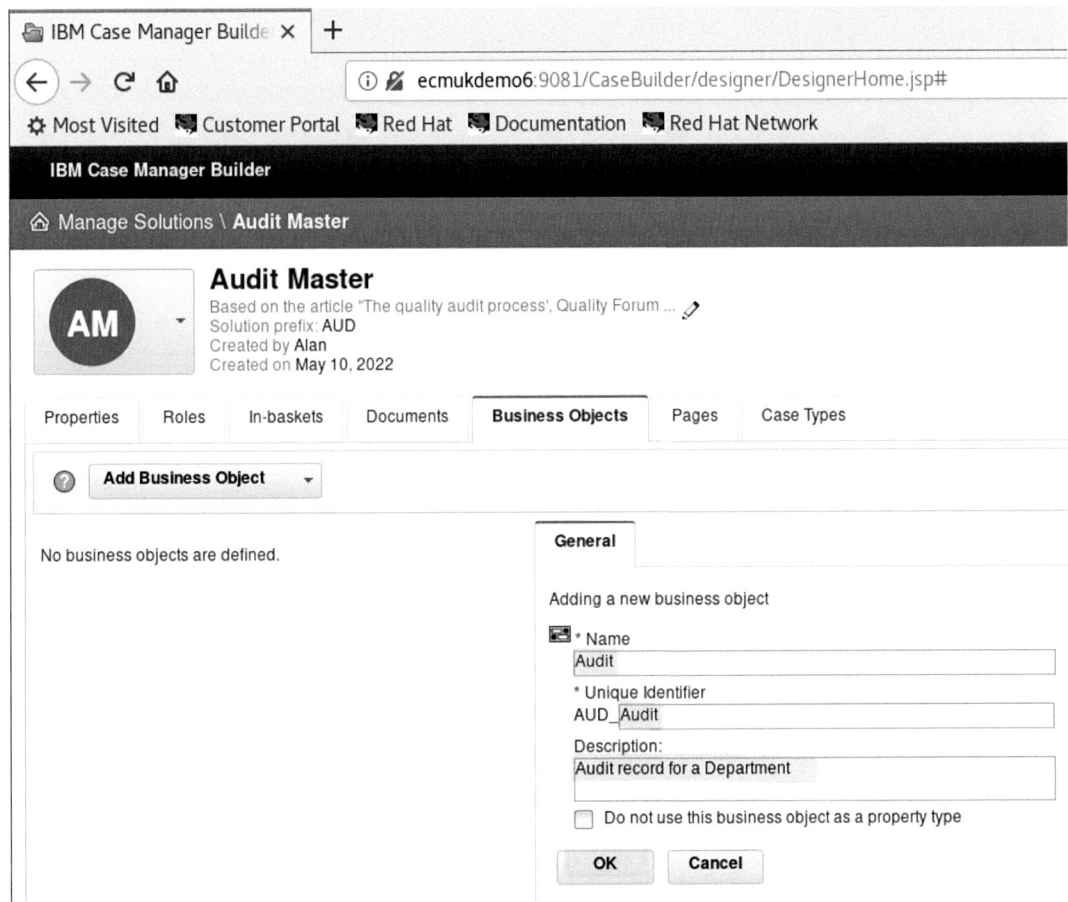

*Figure 1-34. The new **Audit** Business Object is added*

Enter the Business Object **Name** and enter a **Description** and click **OK** as highlighted in Figure 1-34.

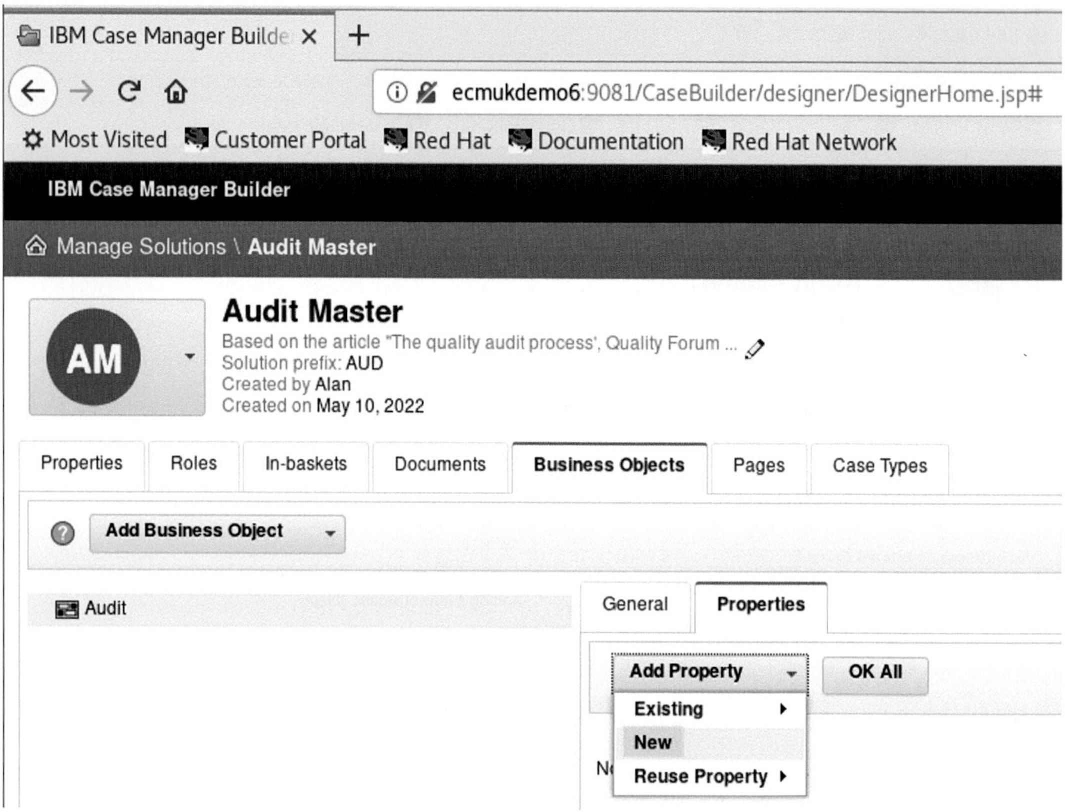

CHAPTER 1   IBM FILENET CASE MANAGER 5.3.3 CASE BUILDER SOLUTION DEVELOPMENT STEPS
              FOR THE AUDIT SYSTEM

*Figure 1-35.   The **New** menu item is selected for properties for the **Audit** Business Object*

Click **New** to add a new property to the **Audit** Business Object.

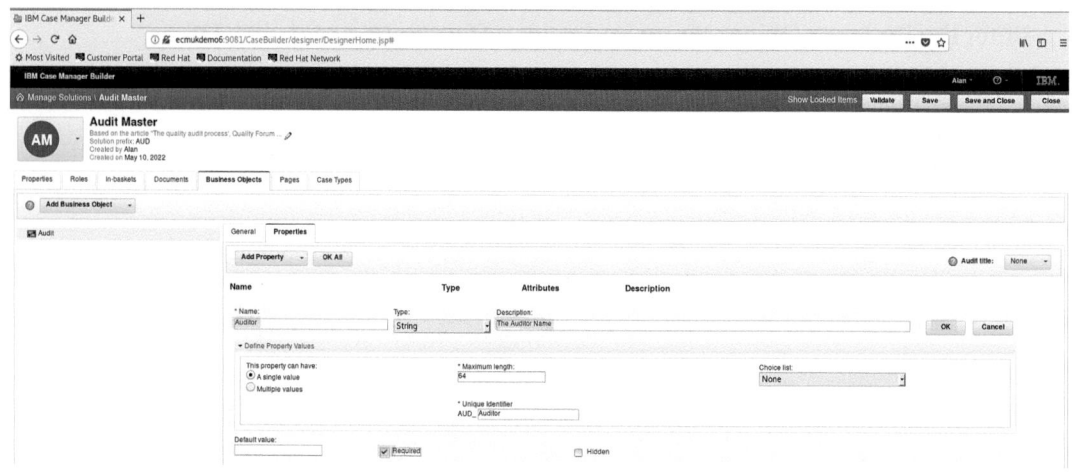

*Figure 1-36.   The **Auditor Name** property details are entered*

30

CHAPTER 1 IBM FILENET CASE MANAGER 5.3.3 CASE BUILDER SOLUTION DEVELOPMENT STEPS
         FOR THE AUDIT SYSTEM

The next property is a **DateTime** type. Notice that each property is held more uniquely for the Case Solution by using the Case Solution prefix (set to **AUD** for the Audit Master solution) for the internal **Unique identifier** value. This enables separate properties to be stored in the IBM FileNet Content Store for different Case solutions where the visible name may be coincidentally set the same, but the Solution property attributes are required to be set differently.

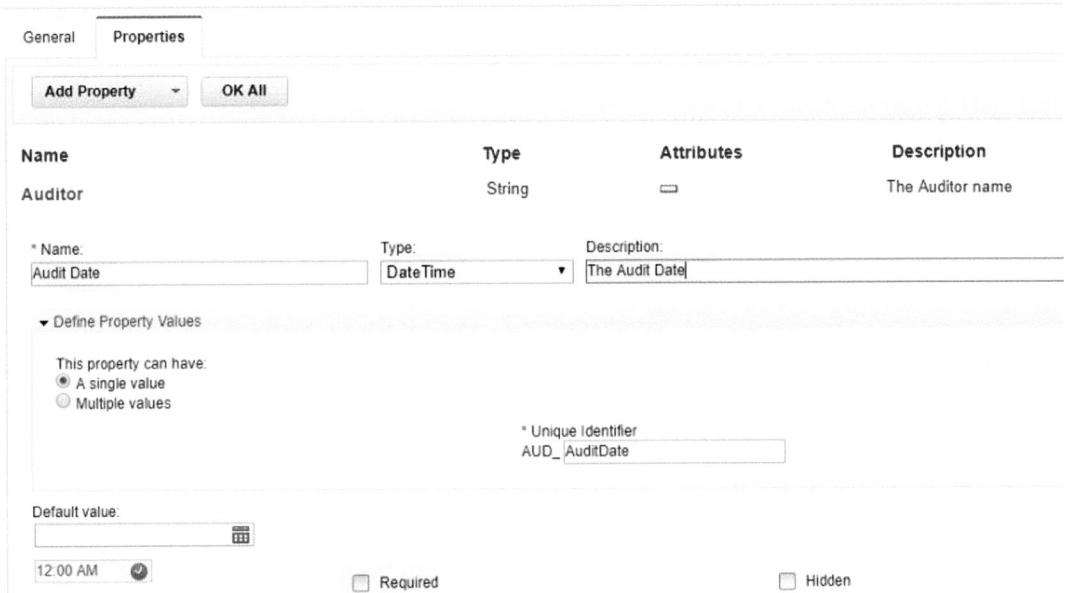

Figure 1-37. The Audit Date property is added and flagged as Required

The **Audit Date** property is added to the **Audit** Business Object.

CHAPTER 1 IBM FILENET CASE MANAGER 5.3.3 CASE BUILDER SOLUTION DEVELOPMENT STEPS FOR THE AUDIT SYSTEM

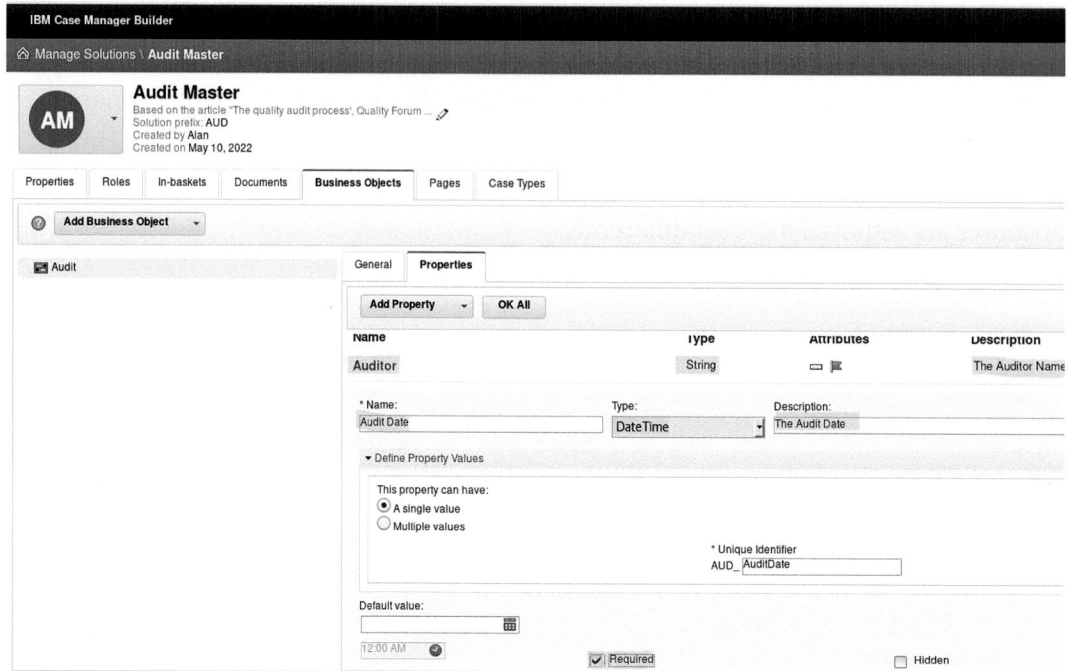

Figure 1-38. *The **Auditor** property is added to the **Audit** Business Object*

The **Comments** multi-value property is added to the **Audit** Business Object.

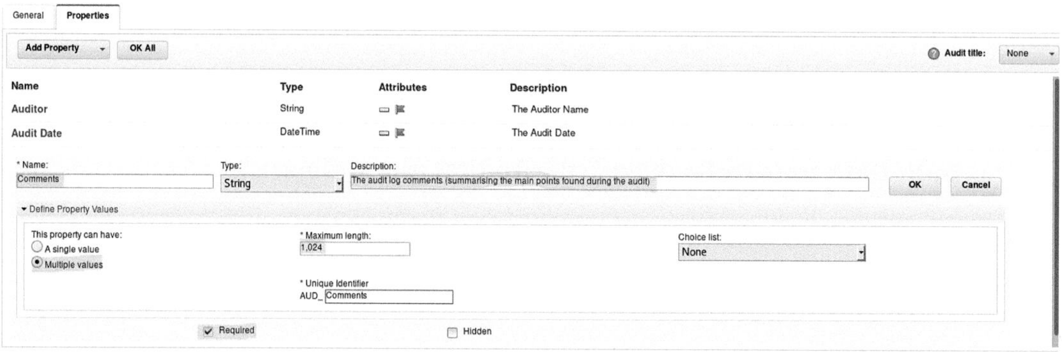

Figure 1-39. *The **Comments** multi-value property is added to the **Audit** Business Object*

The complete list of properties added to the **Audit** Business Object is shown in Figure 1-40.

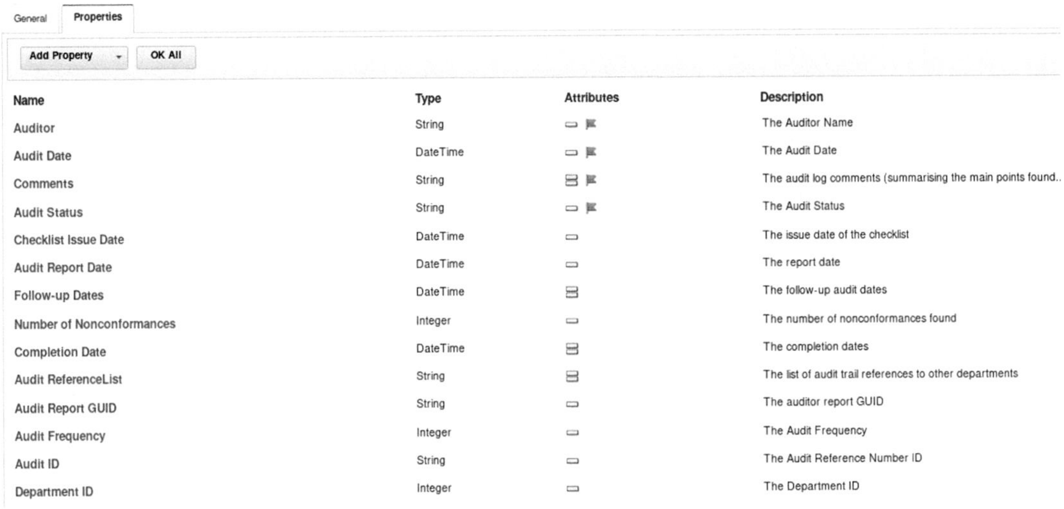

CHAPTER 1 IBM FILENET CASE MANAGER 5.3.3 CASE BUILDER SOLUTION DEVELOPMENT STEPS FOR THE AUDIT SYSTEM

Name	Type	Attributes	Description
Auditor	String		The Auditor Name
Audit Date	DateTime		The Audit Date
Comments	String		The audit log comments (summarising the main points found..
Audit Status	String		The Audit Status
Checklist Issue Date	DateTime		The issue date of the checklist
Audit Report Date	DateTime		The report date
Follow-up Dates	DateTime		The follow-up audit dates
Number of Nonconformances	Integer		The number of nonconformances found
Completion Date	DateTime		The completion dates
Audit ReferenceList	String		The list of audit trail references to other departments
Audit Report GUID	String		The auditor report GUID
Audit Frequency	Integer		The Audit Frequency
Audit ID	String		The Audit Reference Number ID
Department ID	Integer		The Department ID

Figure 1-40. *The completed list of properties added to the **Audit** Business Object*

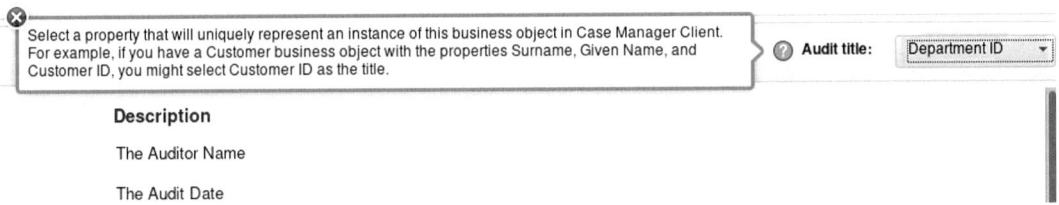

Figure 1-41. *The **Audit** Business Object is linked to the **Department ID** property*

Finally, the **Department ID** property is selected to represent a unique instance of the **Audit** Business Object.

Next, the **Auditor** Business Object is defined as shown in Figure 1-42.

CHAPTER 1 IBM FILENET CASE MANAGER 5.3.3 CASE BUILDER SOLUTION DEVELOPMENT STEPS FOR THE AUDIT SYSTEM

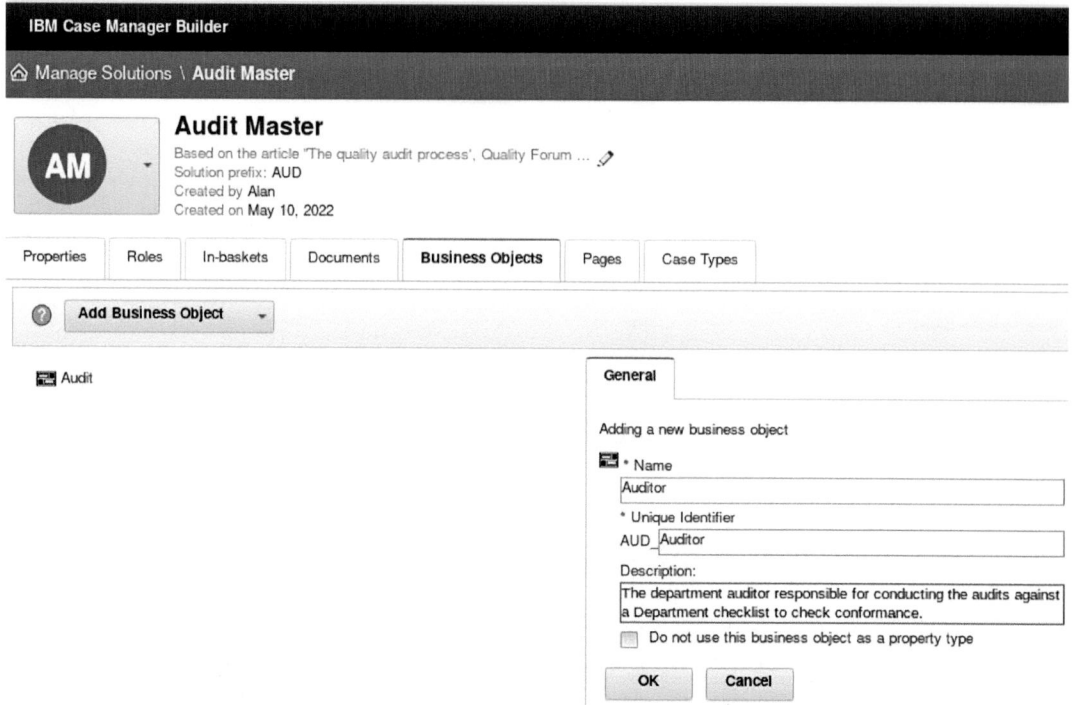

*Figure 1-42. The **Auditor** Business Object details are entered*

The properties are added from the Global property list (defaulting the attributes).

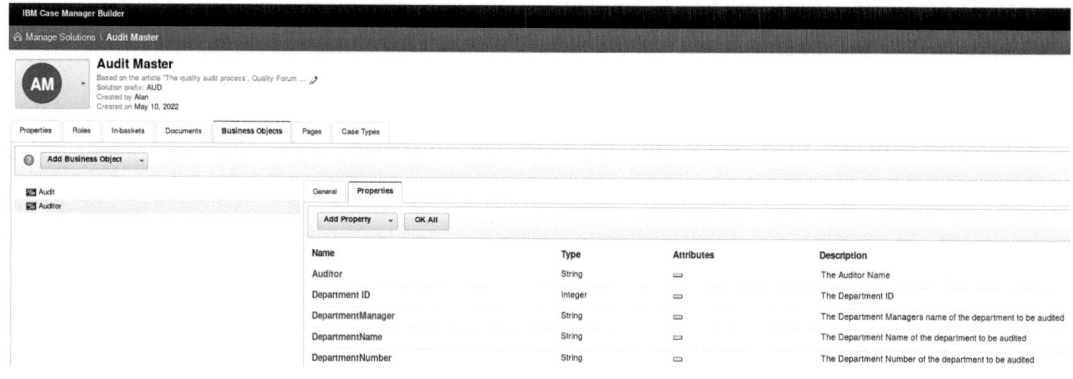

*Figure 1-43. The **Auditor** Business Object properties are selected and entered from the Global list*

CHAPTER 1 IBM FILENET CASE MANAGER 5.3.3 CASE BUILDER SOLUTION DEVELOPMENT STEPS
          FOR THE AUDIT SYSTEM

The **Department** Business Object is created next, with the Name and Description defined as shown in Figure 1-44.

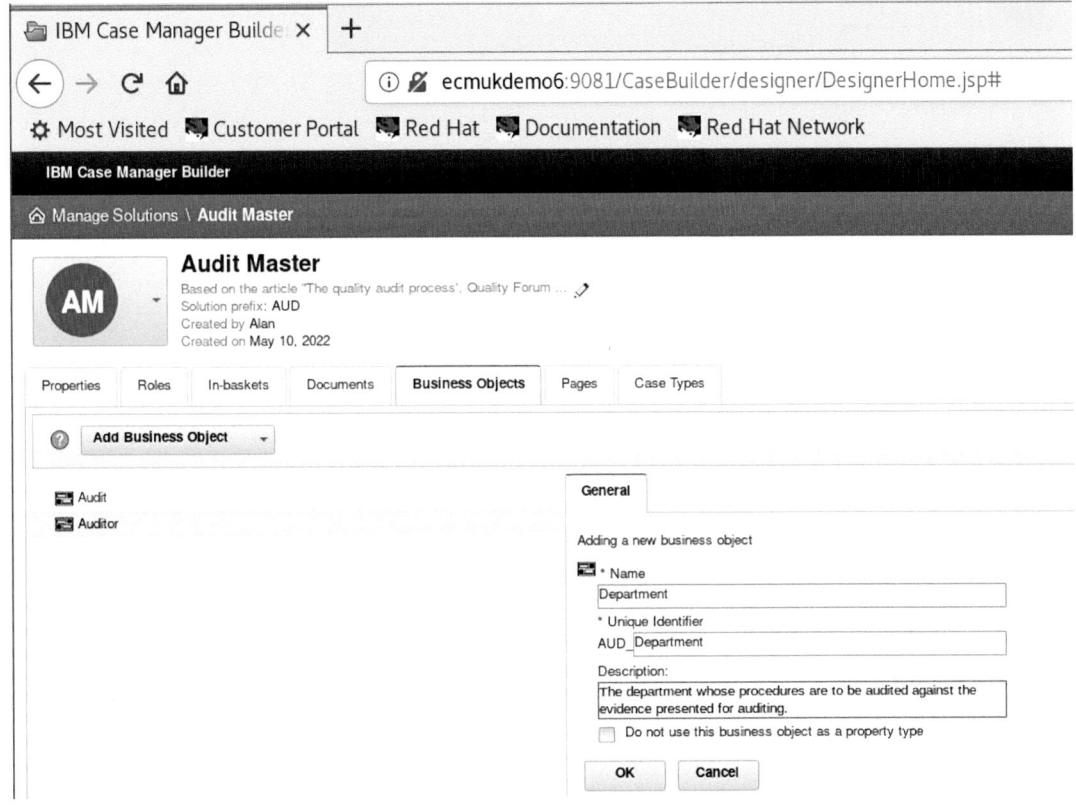

Figure 1-44. *The Department Name and Description are entered*

The **Department** Business Object properties are selected from the Existing Global property list as shown in Figure 1-45.

35

CHAPTER 1   IBM FILENET CASE MANAGER 5.3.3 CASE BUILDER SOLUTION DEVELOPMENT STEPS
            FOR THE AUDIT SYSTEM

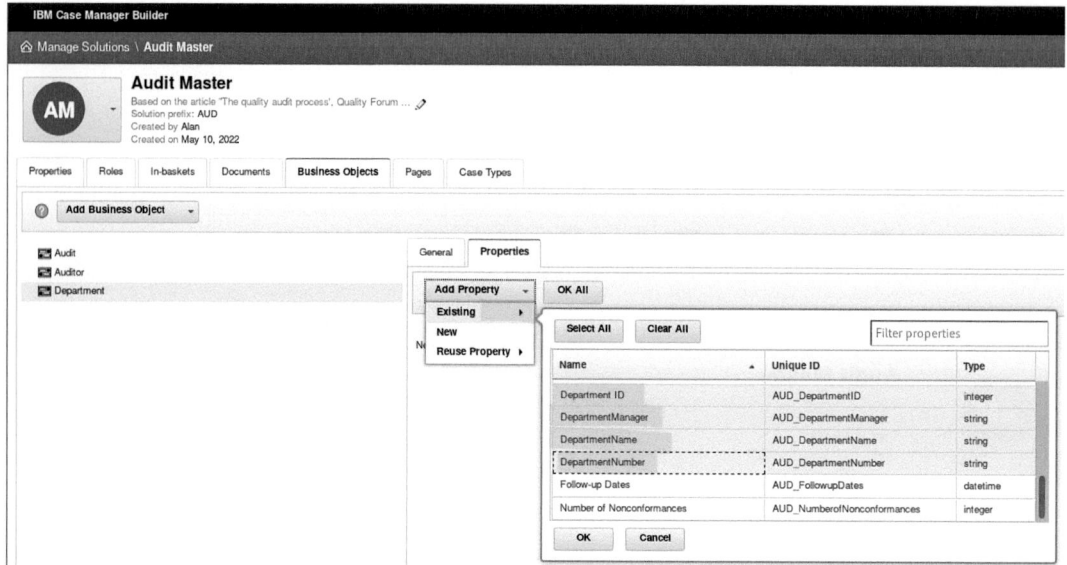

Figure 1-45.   The Department Business Object Global properties are selected

New properties are then also added as follows:

Department Emails (String, multiple values)

Department Procedure IDs (String, multiple values)

Department Procedure Names (String, multiple values)

(Note the red flag fields denote that the **Required** box is ticked for the property.)

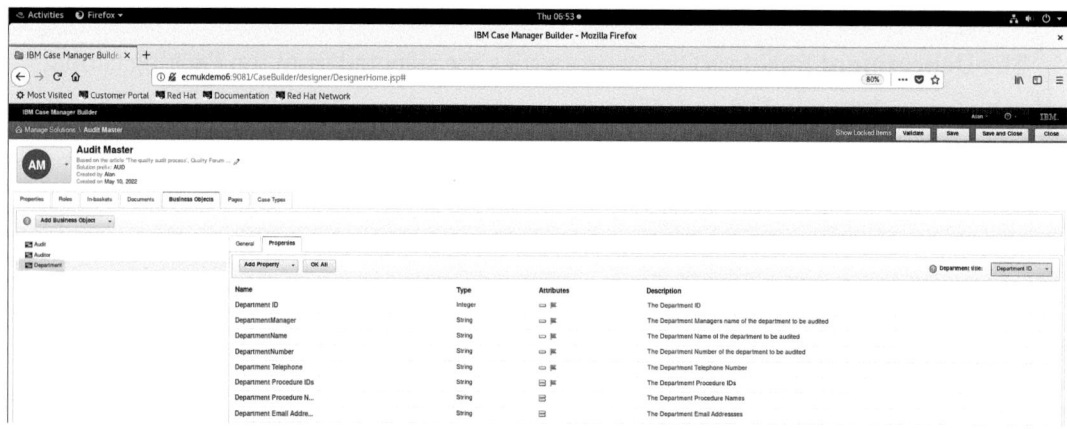

Figure 1-46.   The Department ID is also selected for the Department Business Object unique ID

36

CHAPTER 1 IBM FILENET CASE MANAGER 5.3.3 CASE BUILDER SOLUTION DEVELOPMENT STEPS
FOR THE AUDIT SYSTEM

The details of the properties for the Department Business Object are shown in Figure 1-47.

Name	Type	Attributes	Description
Department ID	Integer		The Department ID
DepartmentManager	String		The Department Managers name of the department to be audited
DepartmentName	String		The Department Name of the department to be audited
DepartmentNumber	String		The Department Number of the department to be audited
Department Telephone	String		The Department Telephone Number
Department Procedure IDs	String		The Departmemt Procedure IDs
Department Procedure N...	String		The Department Procedure Names
Department Email Addre...	String		The Department Email Addressses

Figure 1-47. *The Department Business Object property details*

The **Procedure** Business Object is created next, with the Name and Description defined as shown in Figure 1-48.

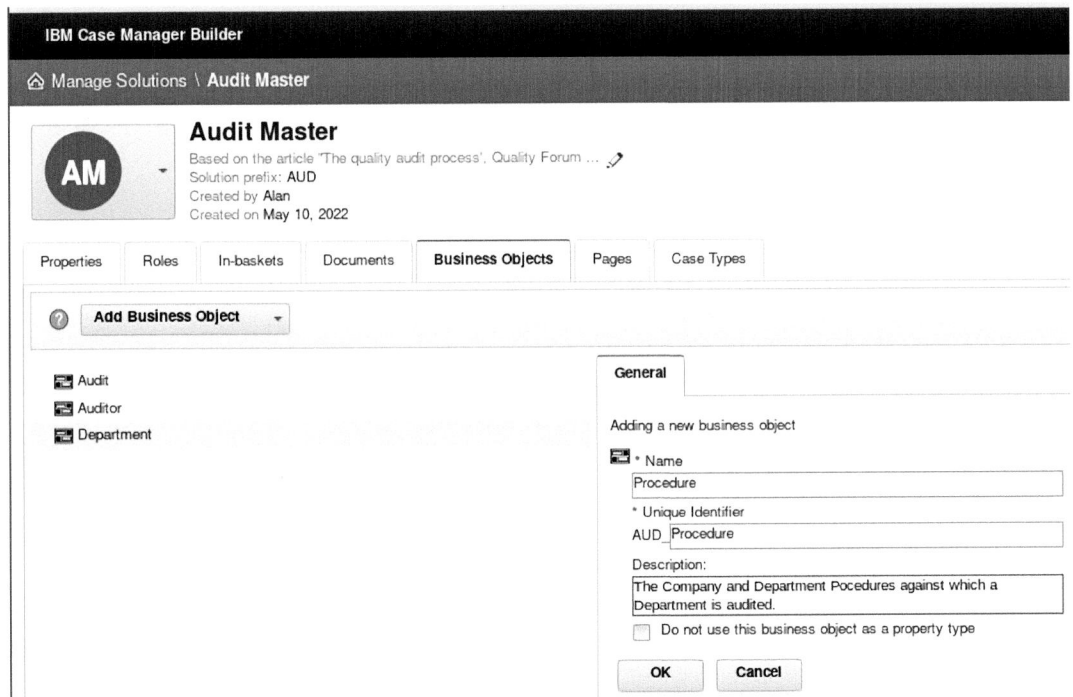

Figure 1-48. *The Procedure Name and Description are entered*

37

CHAPTER 1   IBM FILENET CASE MANAGER 5.3.3 CASE BUILDER SOLUTION DEVELOPMENT STEPS
            FOR THE AUDIT SYSTEM

The **Procedure** Business Object properties are selected as **NEW** properties as shown in the following and in Figure 1-49 (using the **Add Property** drop-down command button):

> **Procedure ID** (Required) The Procedure ID
>
> **Procedure Publication Date** (Required) The Procedure publication date
>
> **Procedure Version** (Required) The issue/version number of the current version of the Procedure
>
> **Procedure Main Authors** (Multiple values) (Required) The main authors of the procedure
>
> **Procedure Main Reference** The procedure reference
>
> **Procedure Signatories** (Multiple values) (Required) The authorization signatories
>
> **Procedure Reviewers** (Multiple values) (Required) The main reviewers
>
> **Procedure Location** The master copy location
>
> **Distribution List** (Multiple values) The Owner email
> **Distribution list**
>
> **Procedure Name** (Required) The Procedure Name

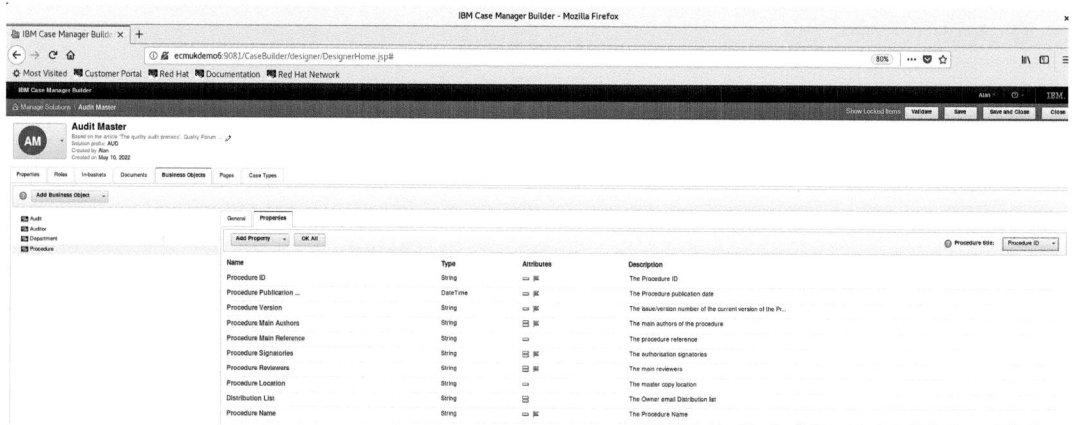

Figure 1-49. *The New properties added to the **Procedure** Business Object*

The **Checklist Item** Business Object is added next to the list of Business Objects.

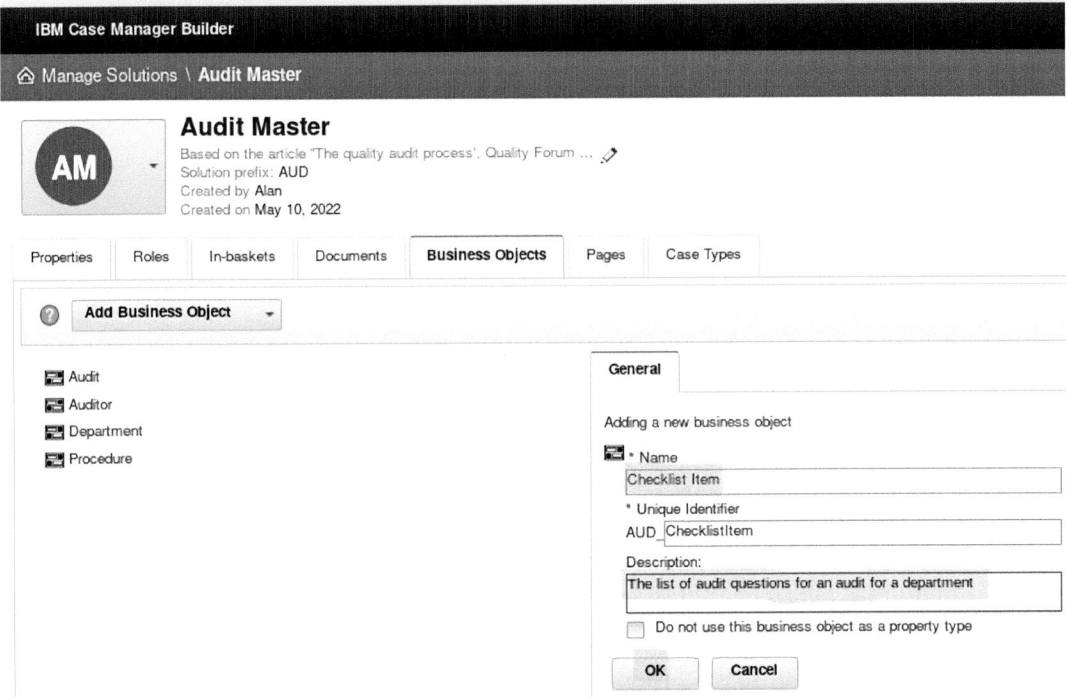

Figure 1-50. The New properties added to the Checklist Item Business Object

The **Checklist Item** Business Object properties are selected as **NEW** properties as shown in the following and in Figure 1-51 (using the **Add Property** drop-down command button):

Checklist Item Ref (Required) The Checklist Item Reference

Checklist Item Audit Date Time (Required) The Checklist Item Audit Date Time

Checklist Item Department (Required) The Checklist Item Audit Date

Checklist Item Description (Required) The Checklist Item Description

(Maximum 1024 Characters Long)

CHAPTER 1   IBM FILENET CASE MANAGER 5.3.3 CASE BUILDER SOLUTION DEVELOPMENT STEPS
            FOR THE AUDIT SYSTEM

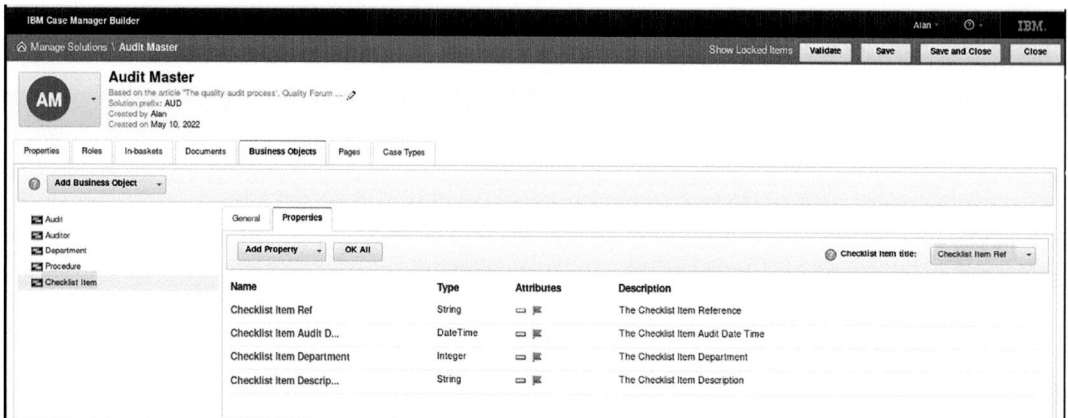

Figure 1-51. *The New properties added to the **Checklist Item** Business Object*

In the preceding property list, we have changed the default **String** length for the **Checklist Item Description** from 64 characters to 1024 characters.

> **Note** In an IBM FileNet Content Store, Content Store Object String properties (for Object Classes such as Document and Folder) are limited to a maximum of 4000 characters for DB2 and SQL Server–based database systems and 1333 characters for an Oracle-based database system (as of IBM FileNet Version 5.5.x).

The details of the properties for the **Checklist Item** Business Object are shown in Figure 1-52.

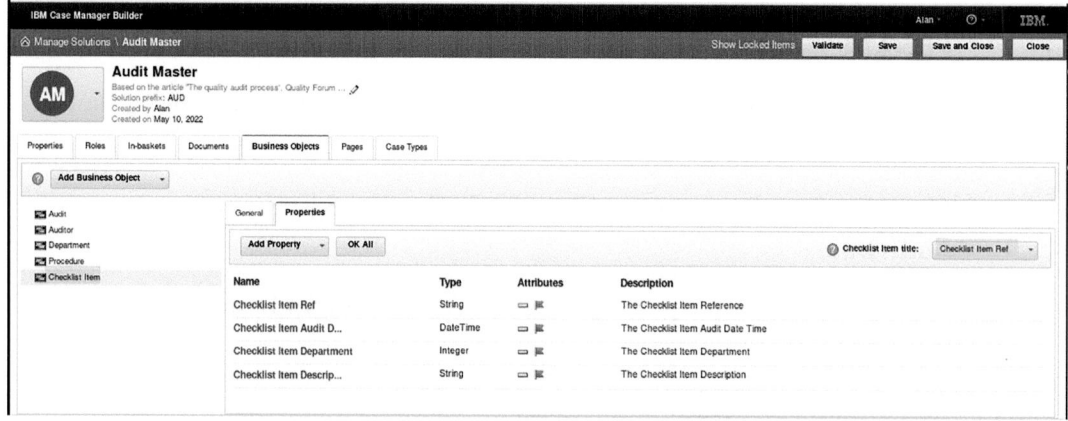

Figure 1-52. *The details of the properties for the **Checklist Item** Business Object*

CHAPTER 1　IBM FILENET CASE MANAGER 5.3.3 CASE BUILDER SOLUTION DEVELOPMENT STEPS FOR THE AUDIT SYSTEM

The summary of the properties for each of the Business Object and Solution properties is shown in Table 1-1, based on the Appendix A System Design (https://www.researchgate.net/publication/334095565_Case_Manager_Solution_Development).

Table 1-1. *The list of* **Audit Master** *Solution properties*

Global Solution Properties		
Property Name	**Property Type**	**Description**
Department Number	String 64	The Department Number of the department to be audited
Department Name	String 64	The Department Name of the department to be audited
Department Manager	String 64	The Department Manager of the department to be audited
Audit Business Object		
Auditor	String 64	The Auditor name
Audit Date	Date	The Audit Date
Audit Status	String 64	The Audit Status
Checklist Issue Date	Date	The Issue date of the checklist
Audit Report Date	Date	The report date
Follow-up Dates	Multi Date	The follow-up audit dates
Completion Date	Multi Date	The completion dates
Number of Nonconformances	Number	The number of nonconformances found
Audit References	Multi String 64	The list of audit trail references to other departments
Comments	Multi String 1024	The audit log comments (summarizing the main points found during the audit)
Audit Report GUID	String	The auditor report
Audit Frequency	Number	The Audit Frequency
Audit ID	Number	The Audit Reference Number ID
Department ID	Number	The Department ID

(*continued*)

CHAPTER 1   IBM FILENET CASE MANAGER 5.3.3 CASE BUILDER SOLUTION DEVELOPMENT STEPS
FOR THE AUDIT SYSTEM

Table 1-1. (*continued*)

Global Solution Properties		
Property Name	**Property Type**	**Description**
Auditor Business Object		
Auditor ID	String 64	The Auditor's Unique ID
Auditor Name	String 64	The Auditor's Name
Auditor Department	String 64	The Auditor's Department Number
Auditor Status	String 64	The Auditor's status
Auditor Manager	String 64	The Auditor's manager
Auditor Skills	Multi String 128	The list of the Auditor's skills
Department Business Object		
Department ID	Number	The Department ID
Department Telephone	String 64	The Department Telephone Number
Department Emails	Multi String 64	The Department email addresses
Department Procedure IDs	Multi String 64	The Department Procedure IDs
Department Procedure Names	Multi String 128	The Department Procedure Names
Department Manager	String 64	The Department Manager
Procedure Business Object		
Procedure ID	String 64	The Procedure ID
Procedure Name	String 64	The Procedure Name
Procedure Publication Date	Date	The Procedure publication date
Procedure Version	String 64	The issue/version number of the current version of the Procedure
Procedure Main Authors	Multi String 64	The main authors of the procedure
Procedure Main Reference	String 64	The procedure references
Procedure Signatories	Multi String 64	The authorization signatories
Procedure Reviewers	Multi String 64	The main reviewers

(*continued*)

CHAPTER 1   IBM FILENET CASE MANAGER 5.3.3 CASE BUILDER SOLUTION DEVELOPMENT STEPS
            FOR THE AUDIT SYSTEM

Table 1-1. (*continued*)

Global Solution Properties		
Property Name	**Property Type**	**Description**
Procedure Location	String 64	The master copy location
Distribution List	Multi String 64	The Owner Distribution list
Checklist Item Business Object		
Checklist Item Ref	String 64	The Checklist Item Reference
Checklist Item Audit Date	Date	The Checklist Item Audit Date
Checklist Item Department	Number	The Checklist Item Department number
Checklist Item Description	String 1024	The Checklist Item Description

Adding Business Object Types As Case Properties

The Business Object types we created earlier have to be added back into the Global properties section as Case properties of the Solution as follows in this section.

See "Adding and modifying business objects":

www.ibm.com/docs/en/case-manager/5.3.3?topic=solution-adding-modifying-business-objects

In the properties tab of Case Builder, select the Business Object type.

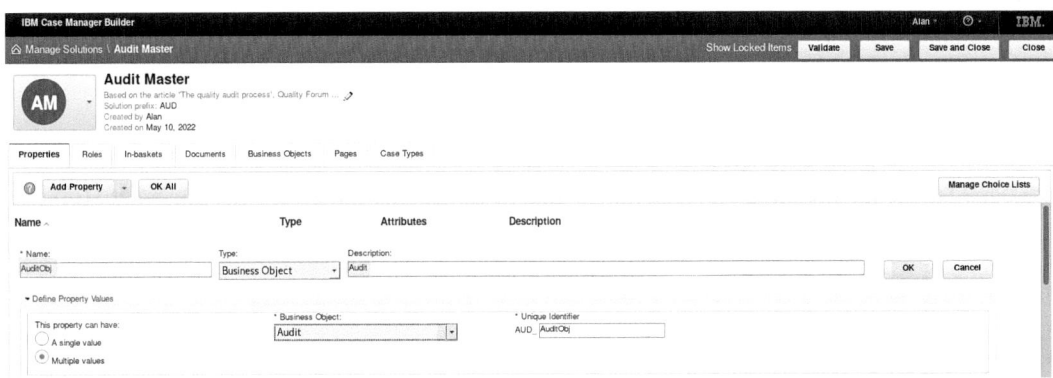

Figure 1-53. The **AuditObj** property of type Business Object is added to the Solution

CHAPTER 1   IBM FILENET CASE MANAGER 5.3.3 CASE BUILDER SOLUTION DEVELOPMENT STEPS
                FOR THE AUDIT SYSTEM

The **AuditObj** property is automatically assigned as a multi-value object and is selected as an **Audit** Business Object type from the drop-down list of Business Objects which we created earlier.

Next, we will add the other Business Object types, similarly, as Case properties as shown in the list in Figure 1-54.

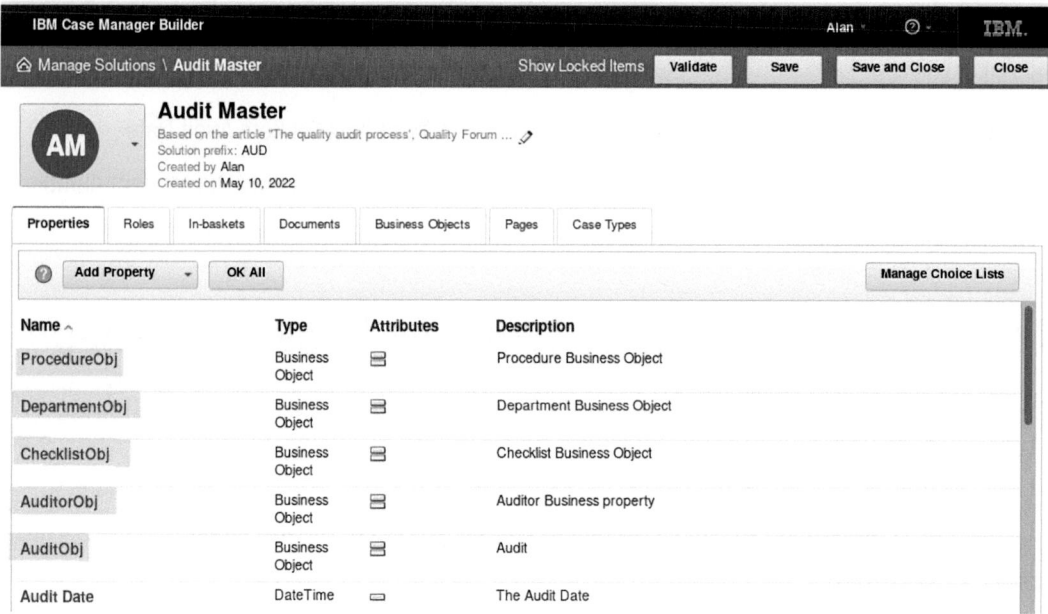

Figure 1-54. *The list of the Case Solution Business Object properties is highlighted*

Part 3 – Adding IBM Case Manager Case Types

There can be any number of Case Types added to an IBM Case Manager Solution. These are used to define the specific Workflows and their associated steps and for each step the In-baskets where the individual Cases can be viewed and opened.

Adding Case Types

In this next step, we can now create the Audit Master Solution Case Types.

Case types are used to specify the Case Solution tasks. This specification includes the required document classes attached to the task workflow, the task workflow steps, and the In-baskets and roles that are used for each of the Workflow steps.

For each Case Type, the following attribute elements can be created/edited:

- The name and description for each case type, including the unique task identifier.

- The document class which, when stored in the IBM FileNet Object Store, can initiate the task workflow.

- The option for a user to start additional custom tasks for the case.

- The page types to use. The default layout for the page can be selected for the Add Case, Split Case, and Case Details function pages.

- Additional properties can also be assigned to the case type.

- Case Folders can be added for a specific Case Type to store the associated documents.

See `www.ibm.com/docs/en/case-manager/5.3.3?topic=types-case` for the full description of the usage of the IBM Case Manager Case Types.

The Case Type tab is the last tab which is navigated, in order, from the top to the bottom of a Tree menu, to add the required Case Type properties, Views, Case Folders (if required), Stages, Rules, and Tasks.

Auditing Department Task

The Audit Department Task is added as follows.

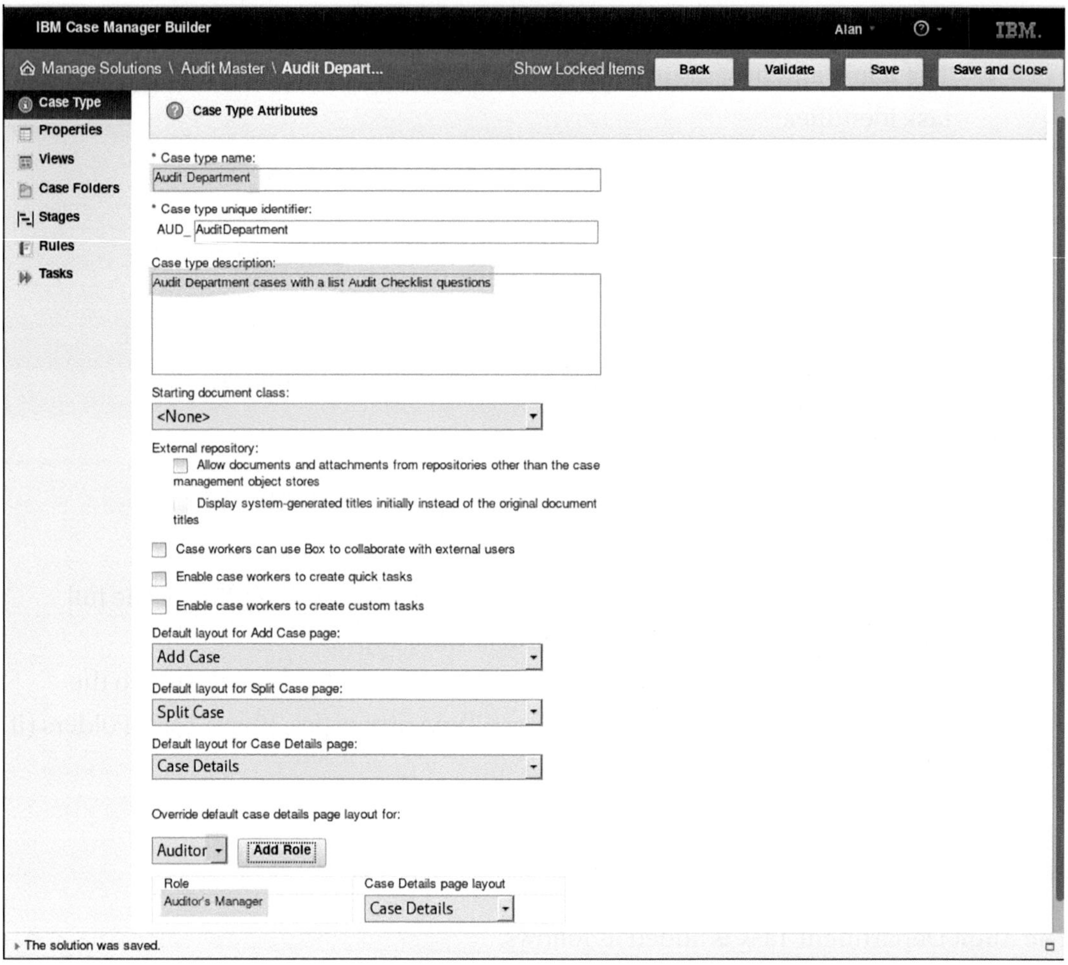

CHAPTER 1   IBM FILENET CASE MANAGER 5.3.3 CASE BUILDER SOLUTION DEVELOPMENT STEPS
            FOR THE AUDIT SYSTEM

*Figure 1-55. The highlighted attributes for the **Audit Department** case type are added*

The Case Type **Properties** Tree menu item is selected to add the required properties for the Task.

CHAPTER 1   IBM FILENET CASE MANAGER 5.3.3 CASE BUILDER SOLUTION DEVELOPMENT STEPS
            FOR THE AUDIT SYSTEM

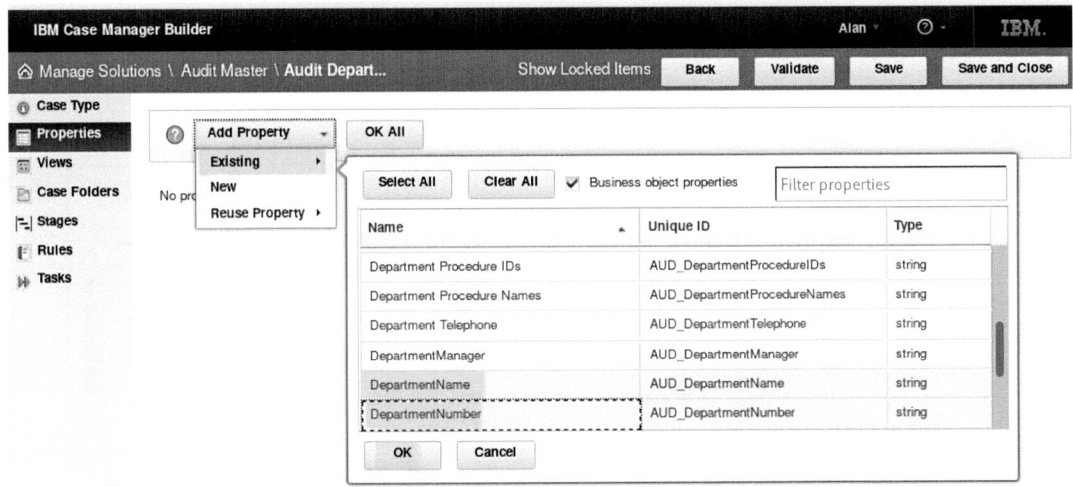

*Figure 1-56. The **DepartmentName** and **DepartmentNumber** properties are added*

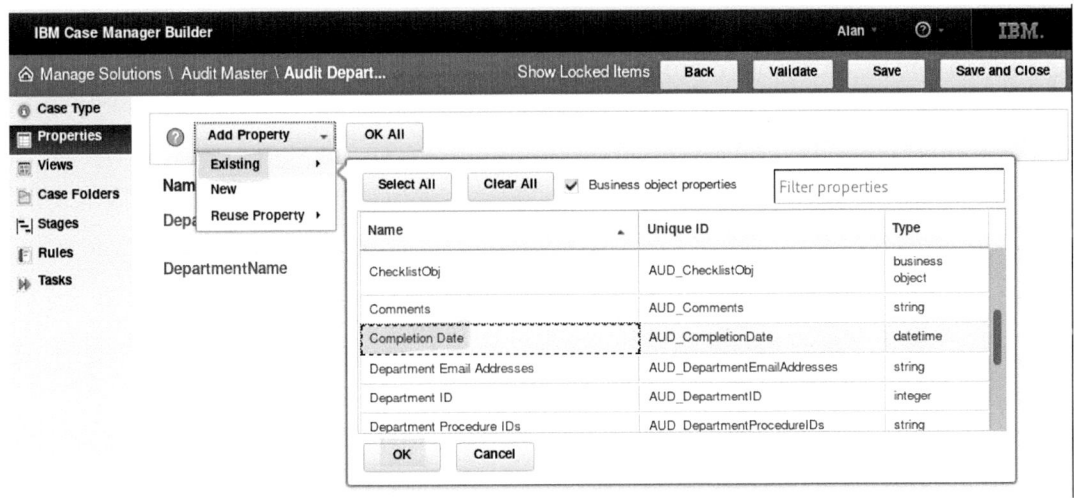

*Figure 1-57. The existing **Completion Date** property is added*

47

CHAPTER 1   IBM FILENET CASE MANAGER 5.3.3 CASE BUILDER SOLUTION DEVELOPMENT STEPS
            FOR THE AUDIT SYSTEM

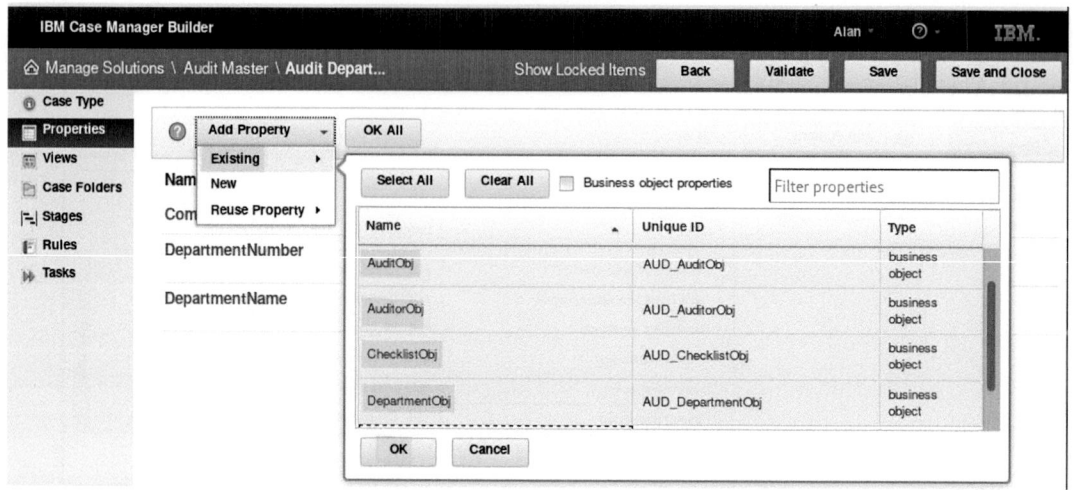

Figure 1-58. *The Business Object type properties are selected for the Audit Case Type*

Add the required Business Objects to the Audit Department Case Type. In this case, they are all set with the **Required** option. The **OK All** command button can then be selected to update all the open property attribute values.

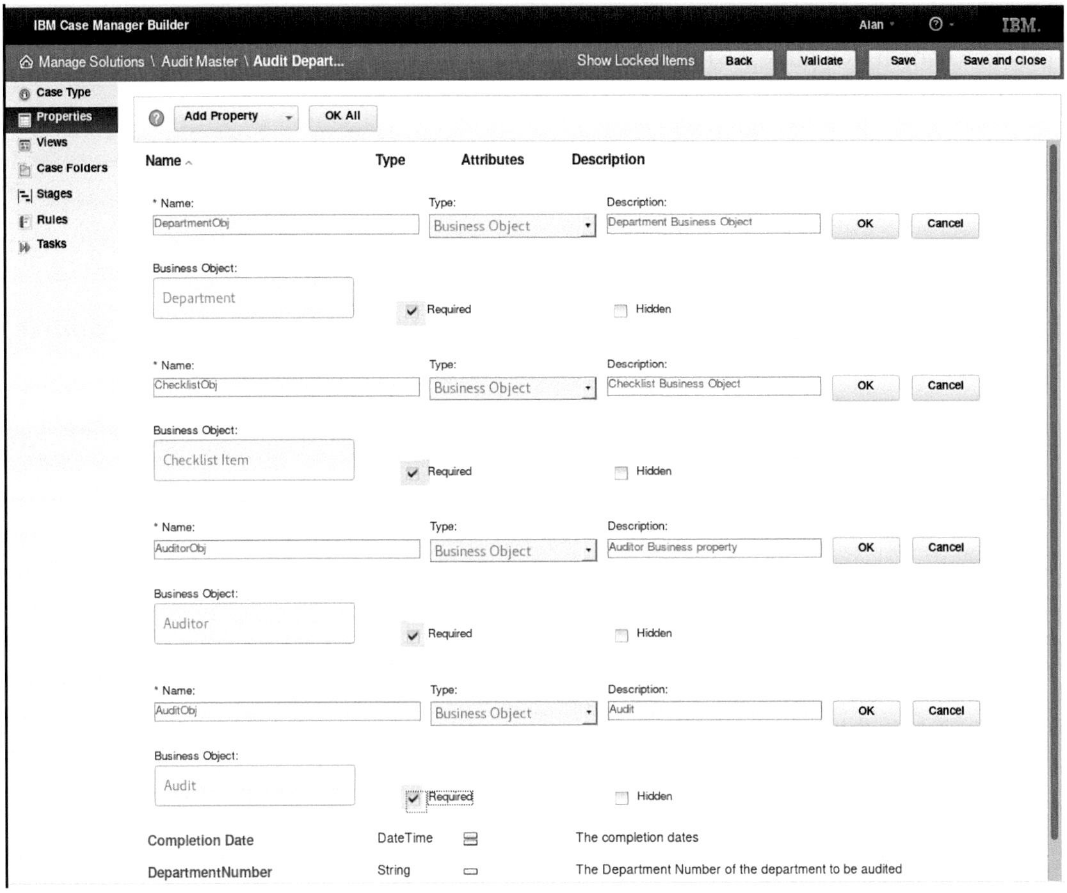

CHAPTER 1   IBM FILENET CASE MANAGER 5.3.3 CASE BUILDER SOLUTION DEVELOPMENT STEPS
            FOR THE AUDIT SYSTEM

Figure 1-59. *The Required Option is selected for all the Business Object properties*

The Save command can be used to save all the Solution build updates at this point.

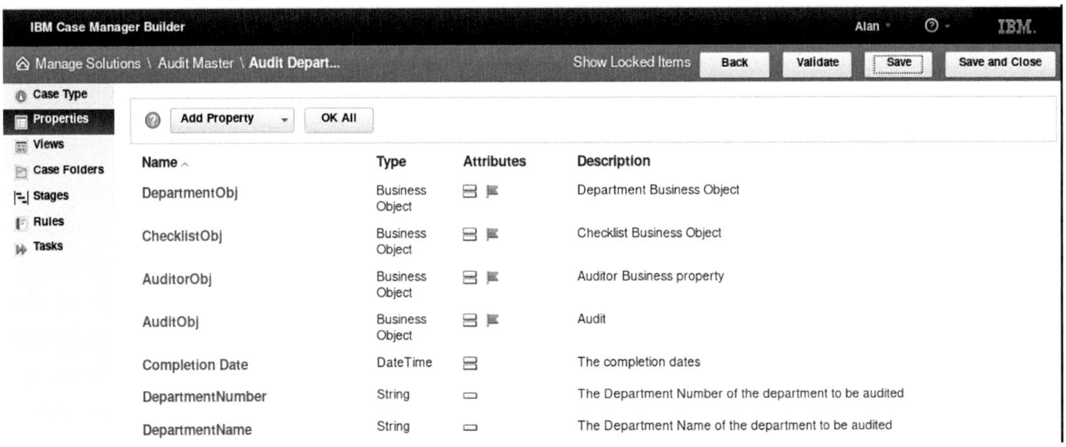

Figure 1-60. *The completed list of Audit Department Case Type properties*

49

CHAPTER 1   IBM FILENET CASE MANAGER 5.3.3 CASE BUILDER SOLUTION DEVELOPMENT STEPS
            FOR THE AUDIT SYSTEM

The **Views** tab is selected, and the related Case summary properties are added from the list as shown in Figure 1-61.

Figure 1-61. *A selected Case Property (the internal Case Identifier in this example) is added*

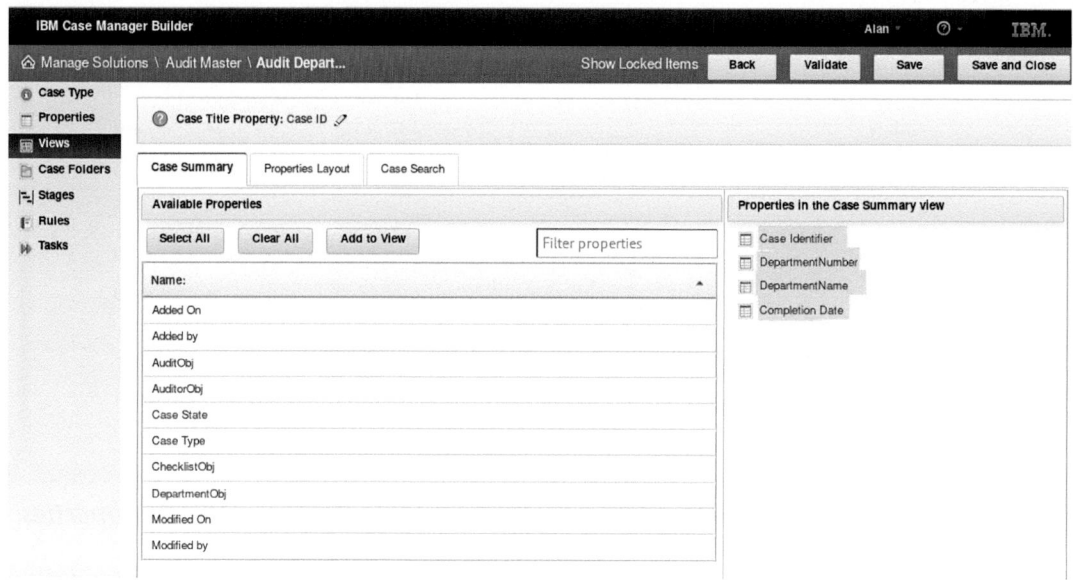

Figure 1-62. *The list of selected Summary Case properties to be viewed*

The Audit Business Object properties are added next, to the Case Summary view.

CHAPTER 1   IBM FILENET CASE MANAGER 5.3.3 CASE BUILDER SOLUTION DEVELOPMENT STEPS
             FOR THE AUDIT SYSTEM

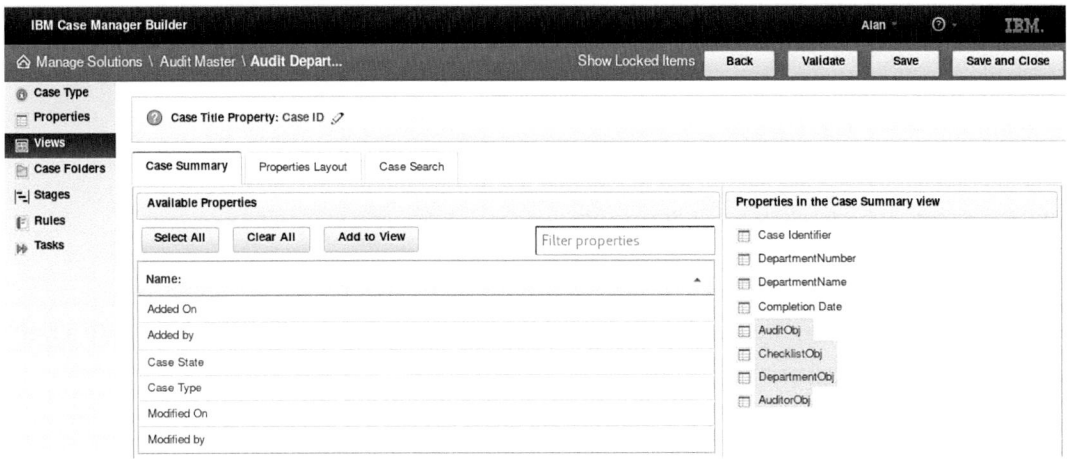

Figure 1-63. The highlighted Audit Business Objects are selected for the summary view

Next, we select the **Properties Layout** tab to define the layout of the displayed property values.

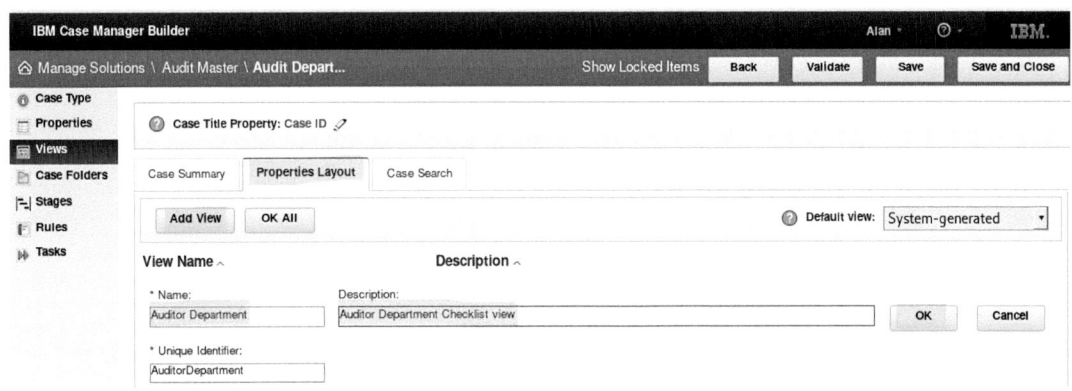

Figure 1-64. The new View Auditor Department is created

On clicking the OK command in Figure 1-64, the Audit Department entry line for the new View is shown, and the icon highlighted in Figure 1-65 appears on using the mouse-over on this new entry line.

51

CHAPTER 1 IBM FILENET CASE MANAGER 5.3.3 CASE BUILDER SOLUTION DEVELOPMENT STEPS
           FOR THE AUDIT SYSTEM

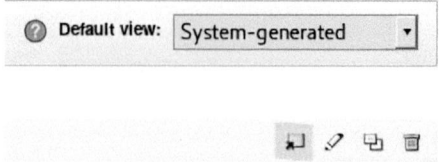

Figure 1-65. The highlighted "Open Properties View Designer" icon is clicked to open

The View editor can be used to drag and drop the required properties to create a new **View Auditor Department** layout page.

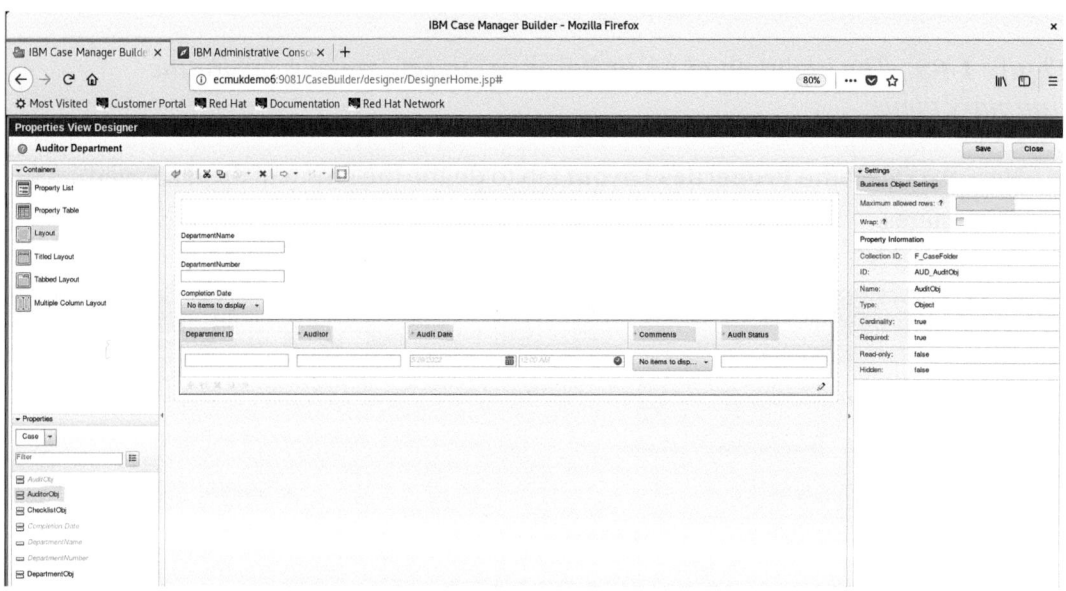

Figure 1-66. The required properties for the View are dragged and dropped to the page

First, the **Layout** Container in the list on the left in Figure 1-66 is dragged to the central panel of the **Properties View Designer**.

Then the properties are dragged from the Case **Properties** list on the left to the **Layout** Container area as shown in Figure 1-66.

CHAPTER 1   IBM FILENET CASE MANAGER 5.3.3 CASE BUILDER SOLUTION DEVELOPMENT STEPS
FOR THE AUDIT SYSTEM

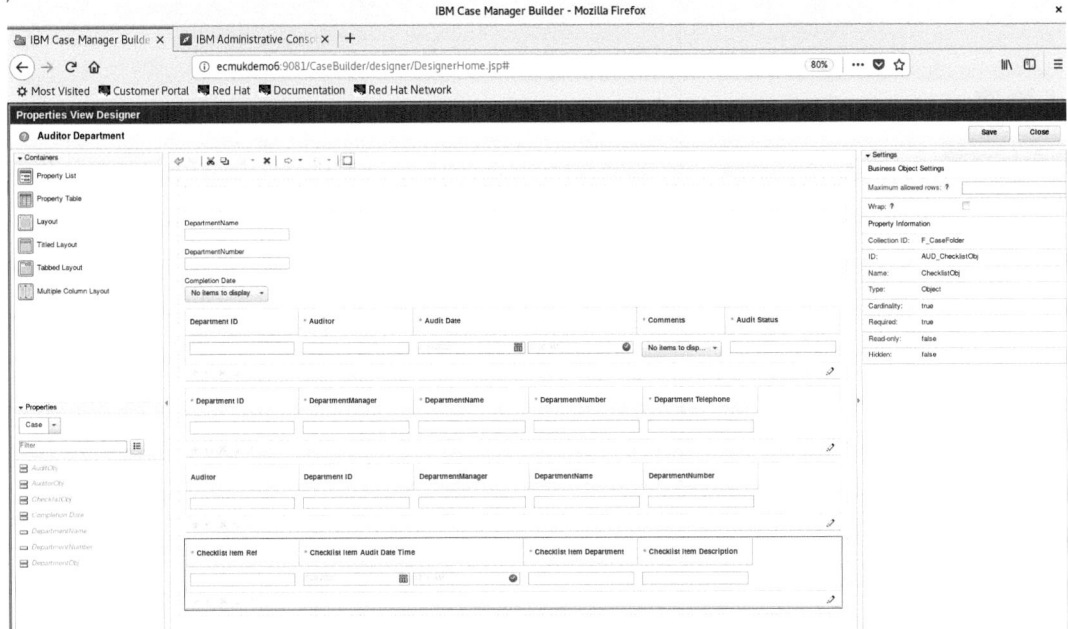

Figure 1-67. *The completed properties layout, including Business Objects*

Finally, the Case Search tab is selected, and the required Search Properties are added to the Case Search view as shown in Figure 1-68.

CHAPTER 1 IBM FILENET CASE MANAGER 5.3.3 CASE BUILDER SOLUTION DEVELOPMENT STEPS FOR THE AUDIT SYSTEM

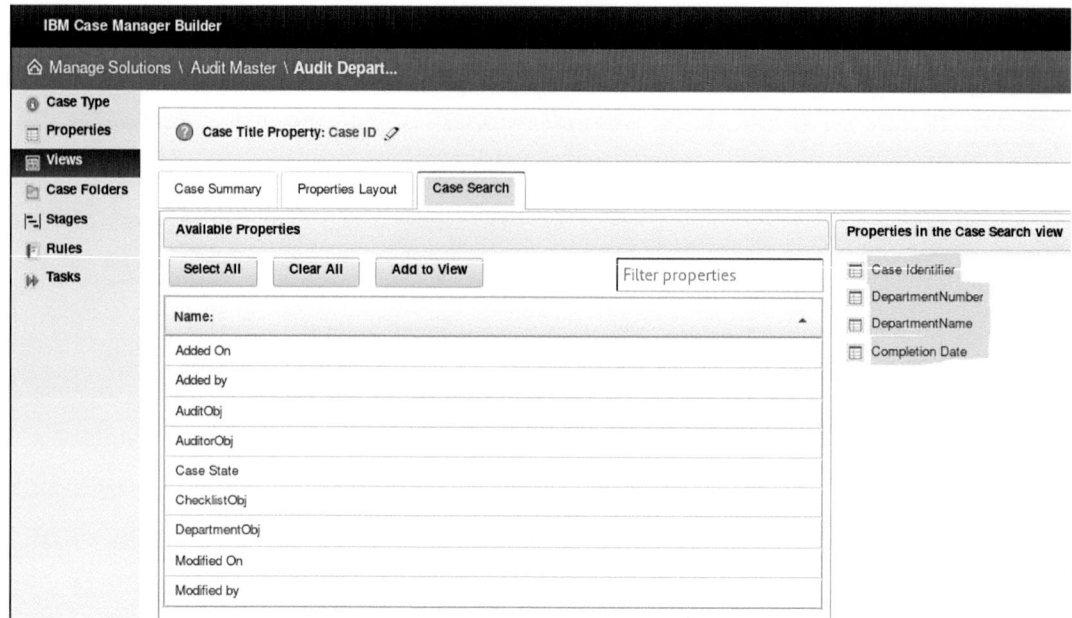

Figure 1-68. The Case Properties to be searched are added to the Case Search view

Adding Business Objects to Be Searched

The Business Objects to be searched are also added to the Case Search view.

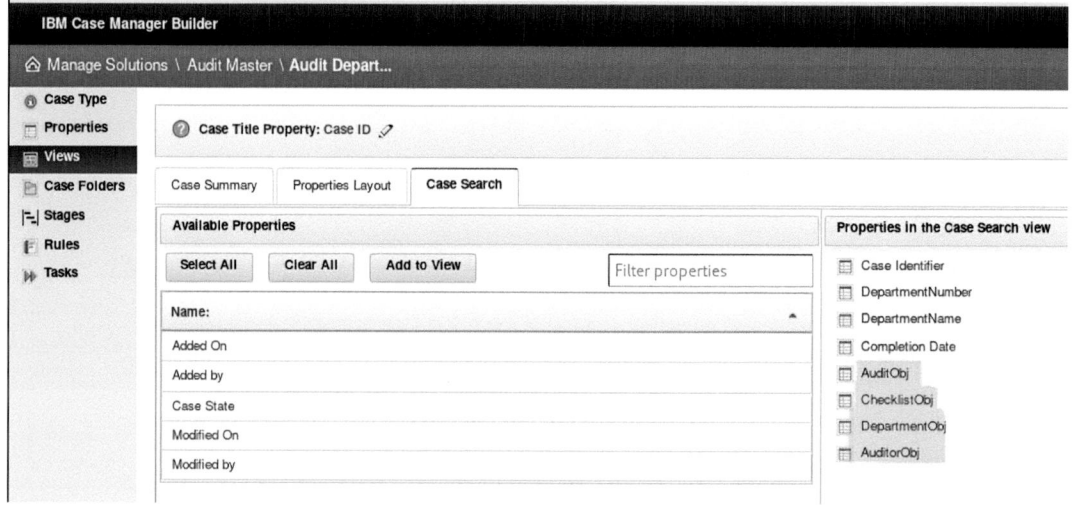

Figure 1-69. The Case Business Objects to be searched are added to the Case Search view

CHAPTER 1   IBM FILENET CASE MANAGER 5.3.3 CASE BUILDER SOLUTION DEVELOPMENT STEPS FOR THE AUDIT SYSTEM

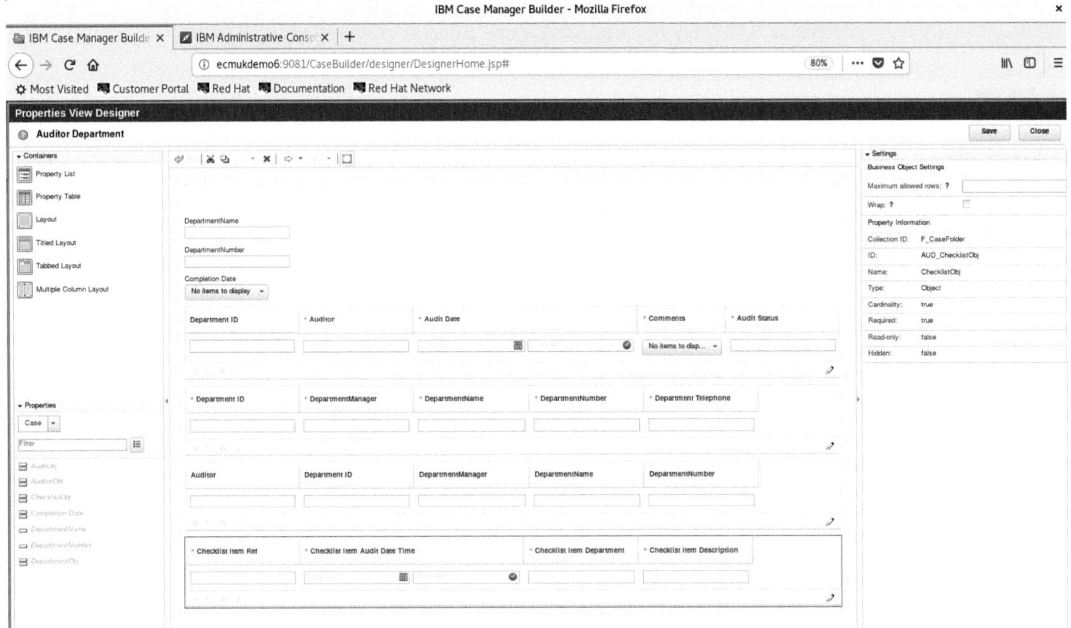

Figure 1-67. *The completed properties layout, including Business Objects*

Finally, the Case Search tab is selected, and the required Search Properties are added to the Case Search view as shown in Figure 1-68.

53

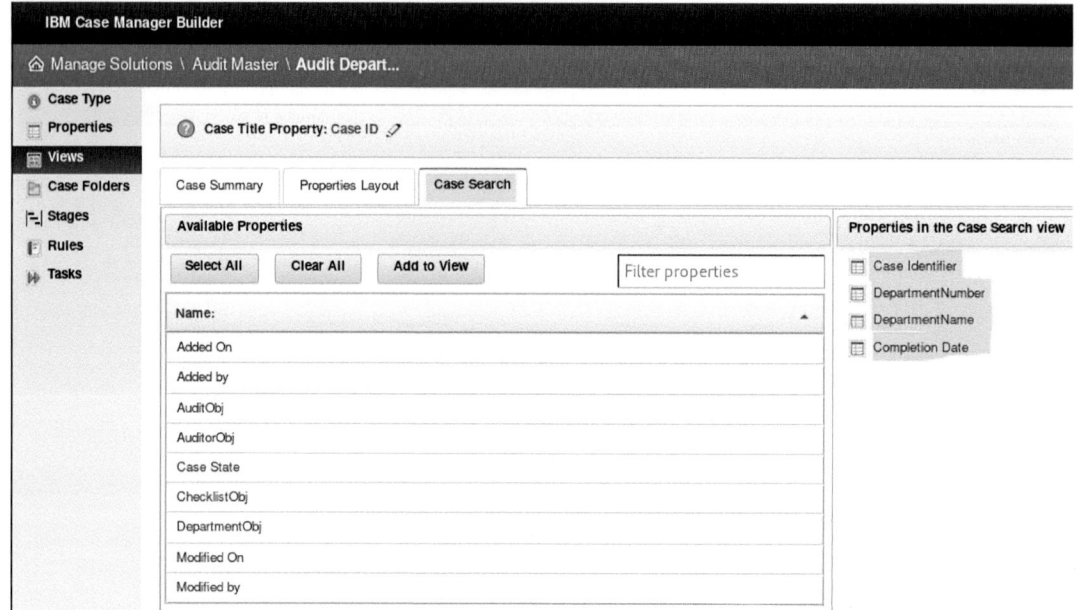

Figure 1-68. The Case Properties to be searched are added to the Case Search view

Adding Business Objects to Be Searched

The Business Objects to be searched are also added to the Case Search view.

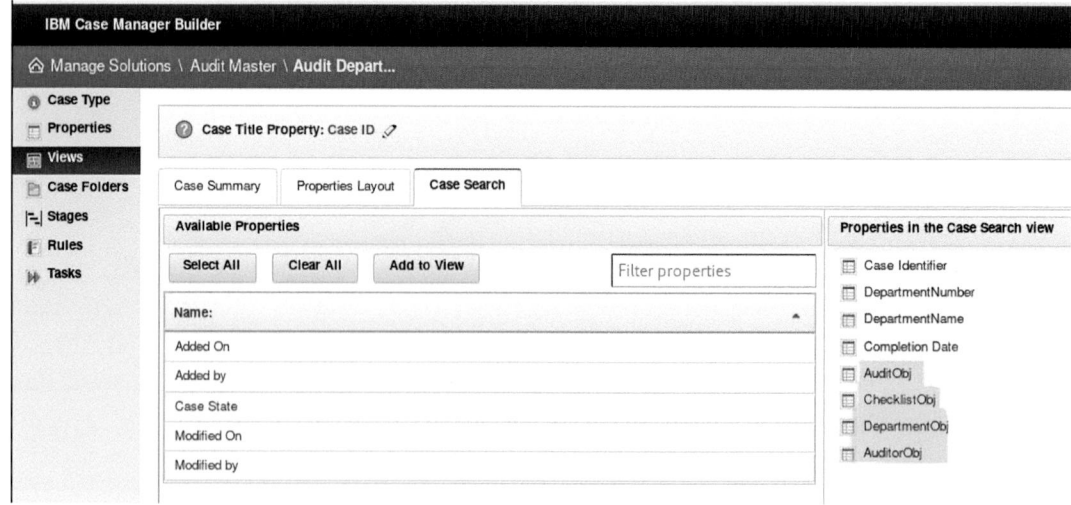

Figure 1-69. The Case Business Objects to be searched are added to the Case Search view

CHAPTER 1 IBM FILENET CASE MANAGER 5.3.3 CASE BUILDER SOLUTION DEVELOPMENT STEPS
FOR THE AUDIT SYSTEM

The default root Case Folder (already created by the system) of **Audit Department** is left for the **Case Folders** item as shown in Figure 1-70. (A set of subfolders could be created to provide a different location for different document class types.)

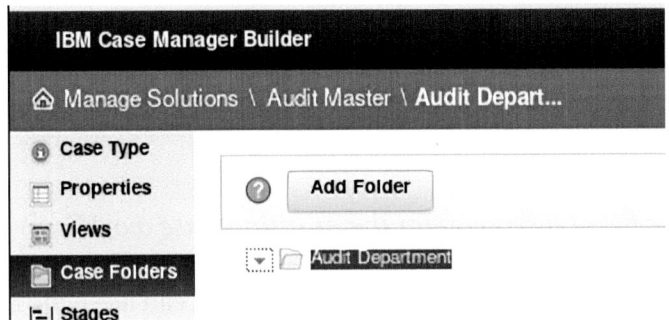

*Figure 1-70. The default **Audit Department** Case Folder is left "as is."*

The **Stages** section is used to add stages. An **Audit Check List Creation** and an **Audit Checklist Review** stage are added, as shown in Figure 1-71.

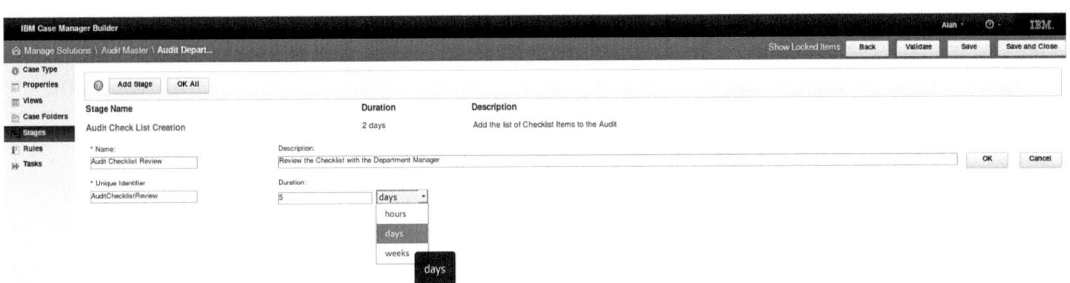

Figure 1-71. The Stages are added to the Audit Department Case workflow steps

Additional Stages are added as shown in Figure 1-72.

CHAPTER 1　IBM FILENET CASE MANAGER 5.3.3 CASE BUILDER SOLUTION DEVELOPMENT STEPS FOR THE AUDIT SYSTEM

Stage Name	Duration	Description
Audit Check List Creation	2 days	Add the list of Checklist Items to the Audit
Audit Checklist Review	5 days	Review the Checklist with the Department Manager
Audit	3 hours	Audit
Audit Follow Up	5 days	Follow up after audit on Nonconformances
Audit Report	1 days	Audit Report Creation

Figure 1-72. The list of the stages for the Auditing Case workflow

The Business rules are not set on this Solution, so we can leave the Rules step empty.

A Case Type can have Business rules which determine the actions to take if particular conditions are met. After creation, the business rules in a task can be used to affect the route of the Task Workflow process or update the case properties.

Rules can be created which include the user-defined case properties and case system properties.

For a full description and usage, see the following link:

www.ibm.com/docs/en/case-manager/5.3.3?topic=solution-business-rules

The Case Workflow Tasks can be added with the attributes shown as follows using the Case Type **Tasks** section.

There are four possible options for the Workflow Task types as follows:

- **Task with New FileNet P8 Process**

  (Creates a task that uses a FileNet P8 Workflow Process to define one or more steps that a case worker completes)

- **To-do Task**

  (Creates a to-do task that can be used to record information and can be completed without any workflow steps)

- **Container Task**

  (Creates a container workflow task that contains subtasks that start after the Container workflow task changes to a working state)

- **Task with Existing FileNet P8 Process**

  (Creates a Workflow task that reuses an existing FileNet P8 Workflow Process)

CHAPTER 1   IBM FILENET CASE MANAGER 5.3.3 CASE BUILDER SOLUTION DEVELOPMENT STEPS
            FOR THE AUDIT SYSTEM

The first type of Task is selected (**Task with New FileNet P8 Process**).

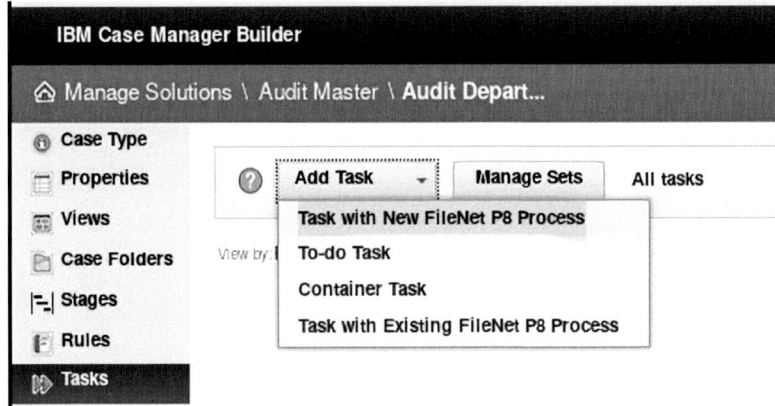

Figure 1-73. *A new Case Type Workflow Task is selected*

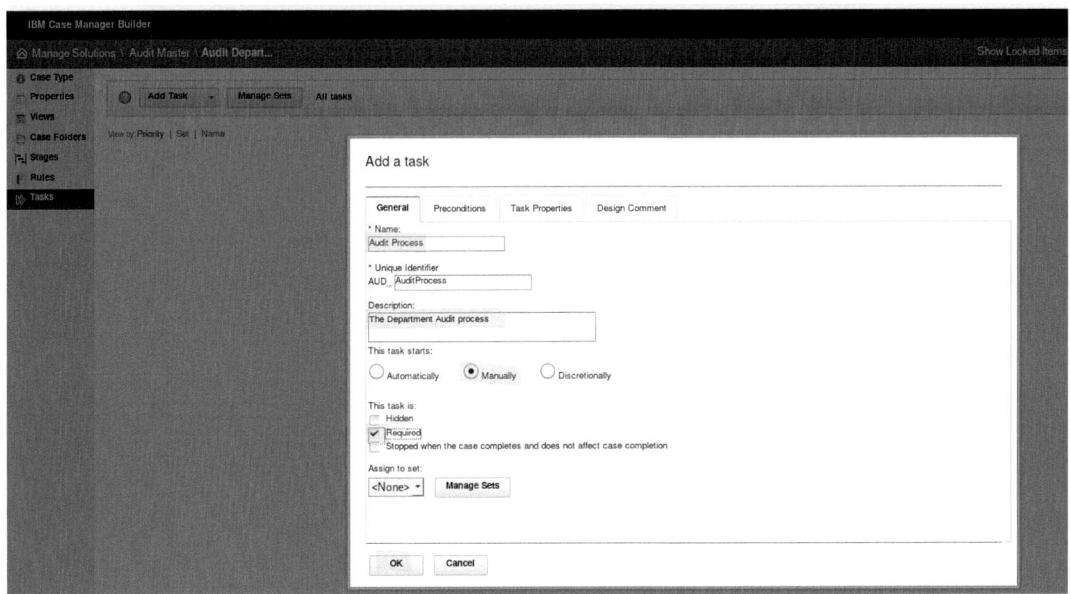

Figure 1-74. *The Audit Process workflow task attributes are entered*

The Task Properties tab is entered next. (The Preconditions tab allows for preconditions that must be met before the task can start. Tasks can start automatically after all preconditions are met or manually by a user after all the preconditions are met. The Case property values are usually used to determine the preconditions.

CHAPTER 1 IBM FILENET CASE MANAGER 5.3.3 CASE BUILDER SOLUTION DEVELOPMENT STEPS
           FOR THE AUDIT SYSTEM

See the full details here:

www.ibm.com/docs/en/case-manager/5.3.3?topic=tasks-task-preconditions

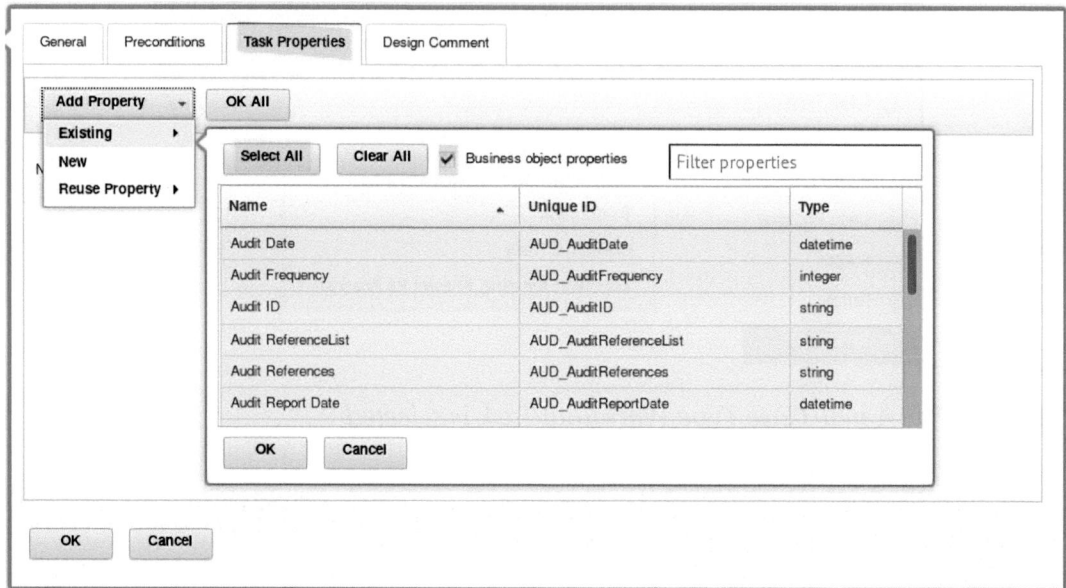

Figure 1-75. Select All is clicked to reference all the Audit Solution properties

Add a Design Comment. (This appears on the Solution reports.)

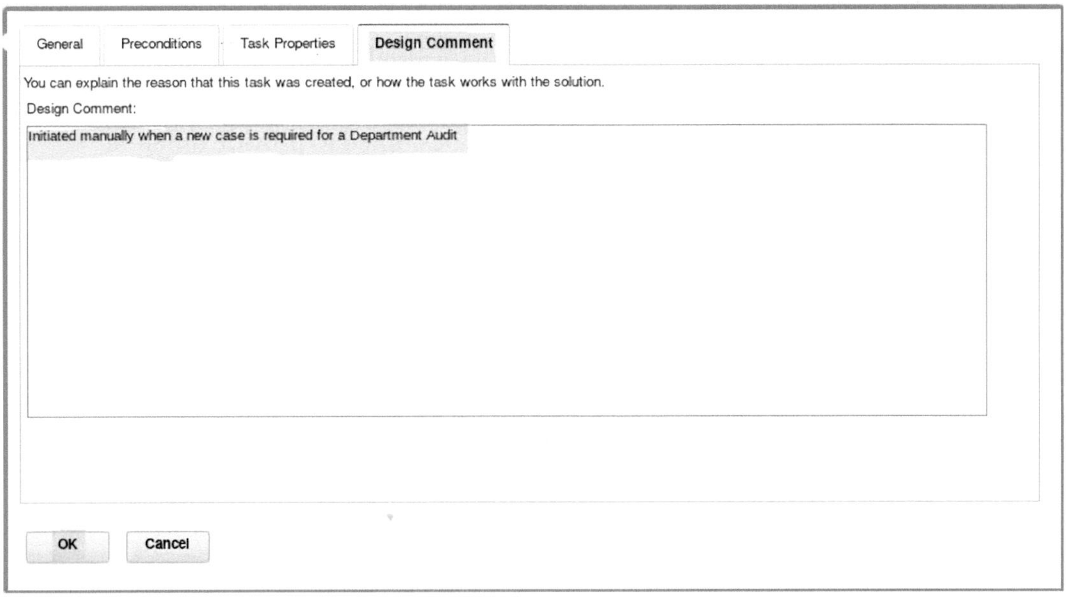

CHAPTER 1   IBM FILENET CASE MANAGER 5.3.3 CASE BUILDER SOLUTION DEVELOPMENT STEPS
              FOR THE AUDIT SYSTEM

Figure 1-76. The Design Comments are added to the Audit Master Audit Process Task

The **Pages** tab of the main solution is selected to copy the default Case Details page to add the custom property's view we created.

Copy and edit the Case Details page and add the custom property's view created.

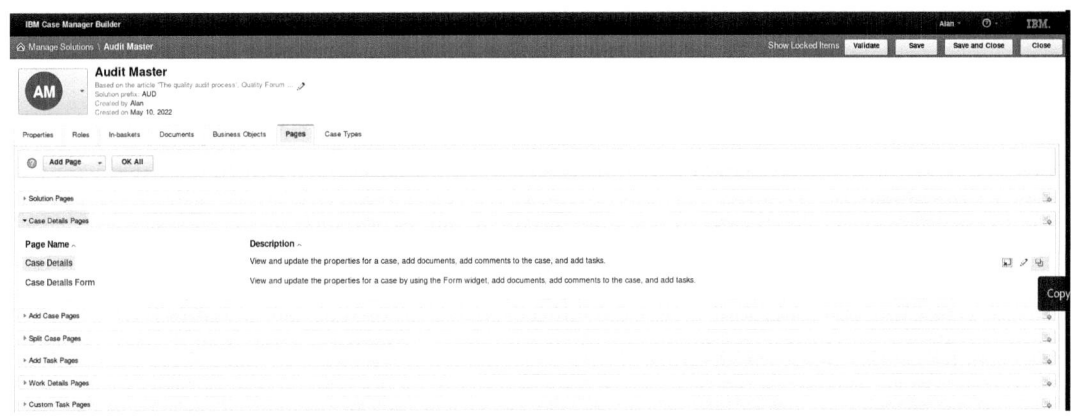

*Figure 1-77. The default **Case Details** page is copied to customize the view*

CHAPTER 1 IBM FILENET CASE MANAGER 5.3.3 CASE BUILDER SOLUTION DEVELOPMENT STEPS
            FOR THE AUDIT SYSTEM

It is good practice to use the copy facility of the **IBM Case Builder** for customizing pages since this provides a ready-made backup of the original default **Case Details** page which can be reverted back to, if there are any issues with the customization. In addition, a working customized **Case Details** page can itself be copied to provide different functions for different **Roles** within a Solution.

Figure 1-78. *The highlighted Copy icon is selected to copy the Case Details page*

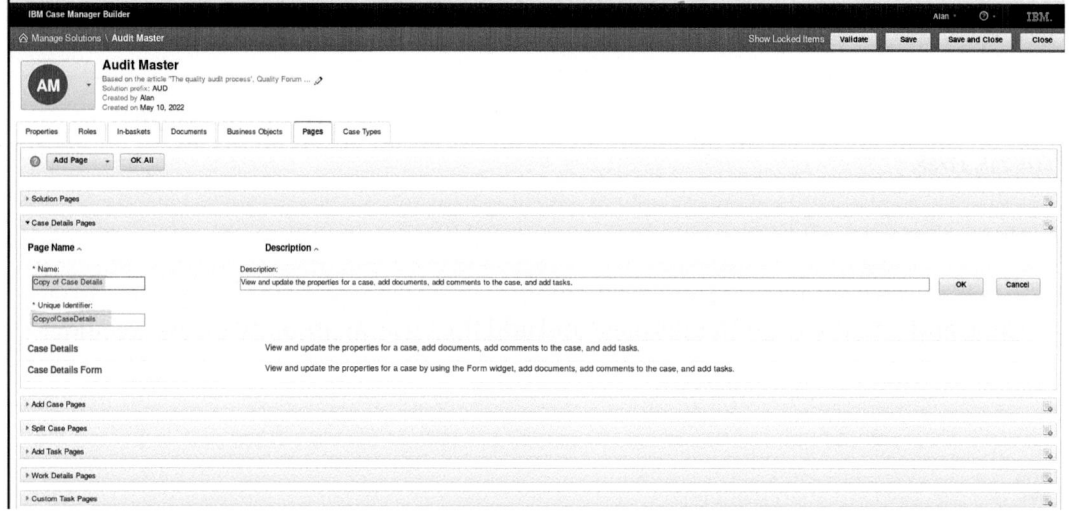

Figure 1-79. *The copied **Case Details** page name can be altered from the highlighted default*

The **Case Details** page can now be updated with the name **Case Details with Business Objects** as shown in Figure 1-80.

CHAPTER 1   IBM FILENET CASE MANAGER 5.3.3 CASE BUILDER SOLUTION DEVELOPMENT STEPS
            FOR THE AUDIT SYSTEM

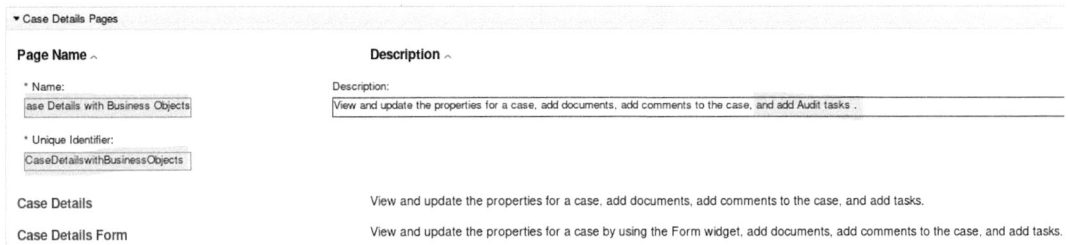

*Figure 1-80.   The **Name** and **Description** fields are updated for the Custom **Case Details** page*

The **Page Designer** icon is clicked to edit the new **Case Details** page.

Case Pages are set up with predefined templates containing web widgets which have built-in functionality to support the display of Case details in the In-baskets for a specifically selected Case. It is possible to modify the layout of these template pages in the IBM Case Builder Page Designer as shown in Figure 1-81.

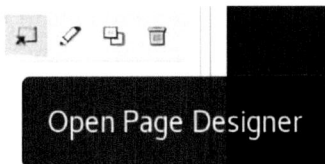

*Figure 1-81.   The **Open Page Designer** icon is selected using the mouse-over on the new **Case Details** page line*

The **Edit Settings** icon is selected on the **Properties** section of the Page Designer.

CHAPTER 1 IBM FILENET CASE MANAGER 5.3.3 CASE BUILDER SOLUTION DEVELOPMENT STEPS
          FOR THE AUDIT SYSTEM

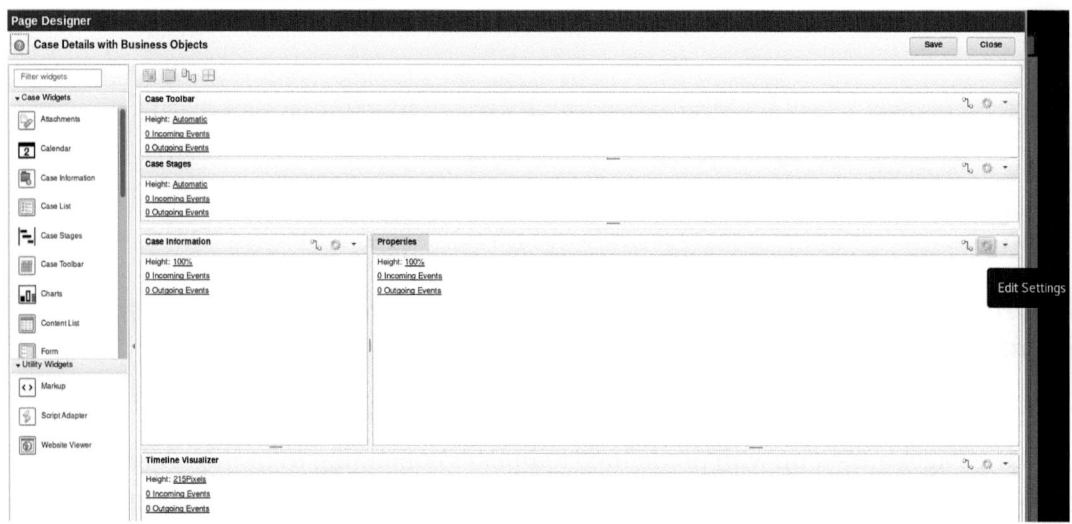

Figure 1-82. *The **Properties** panel is edited using the **Edit Settings** icon*

The **Case Type** we created can now be associated with the **Case Details** page properties by clicking the **Add** icon highlighted in Figure 1-83.

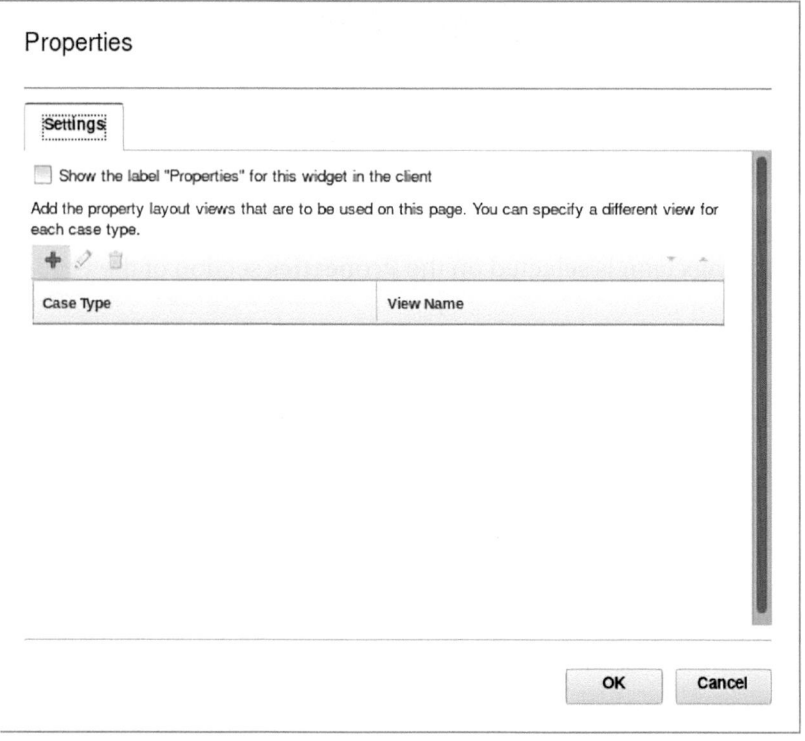

Figure 1-83. *The **Add (+)** icon is clicked to add the **Audit Department** Case Type*

62

CHAPTER 1   IBM FILENET CASE MANAGER 5.3.3 CASE BUILDER SOLUTION DEVELOPMENT STEPS
            FOR THE AUDIT SYSTEM

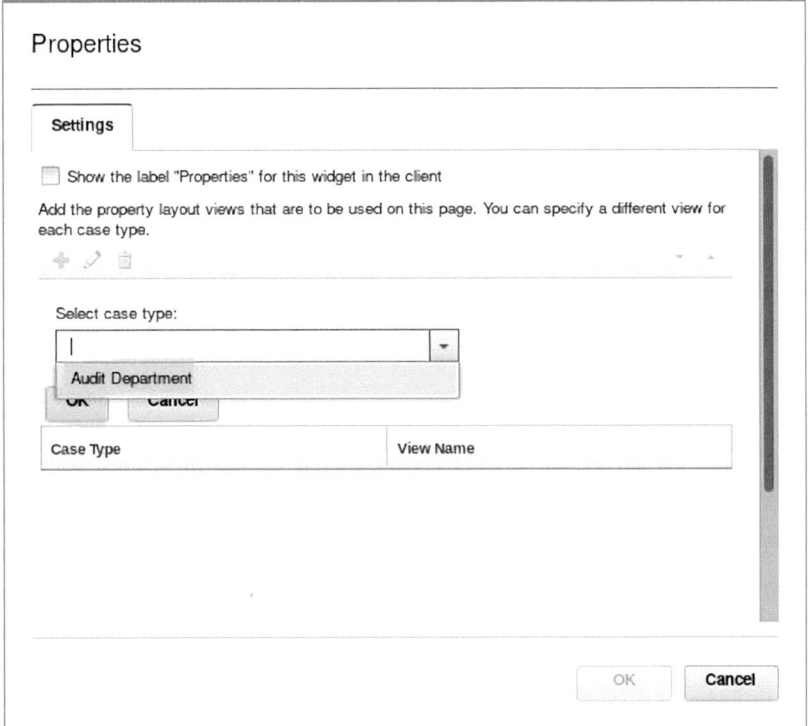

Figure 1-84. *The **Audit Department** Case Type we created earlier is selected from the list*

The selected Case Type will then give access to one or more Views created under the Case Type.

CHAPTER 1 IBM FILENET CASE MANAGER 5.3.3 CASE BUILDER SOLUTION DEVELOPMENT STEPS FOR THE AUDIT SYSTEM

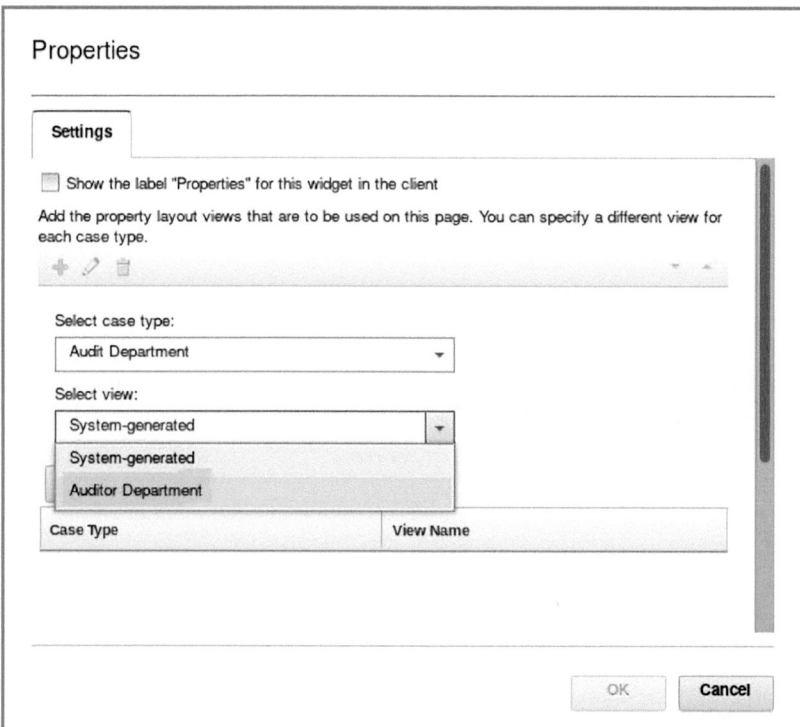

Figure 1-85. *The Auditor Department view, created earlier, is selected*

The Add button is now grayed out. If there were additional Views created for a second Case Type, then this could be added and then associated with a relevant properties frame view.

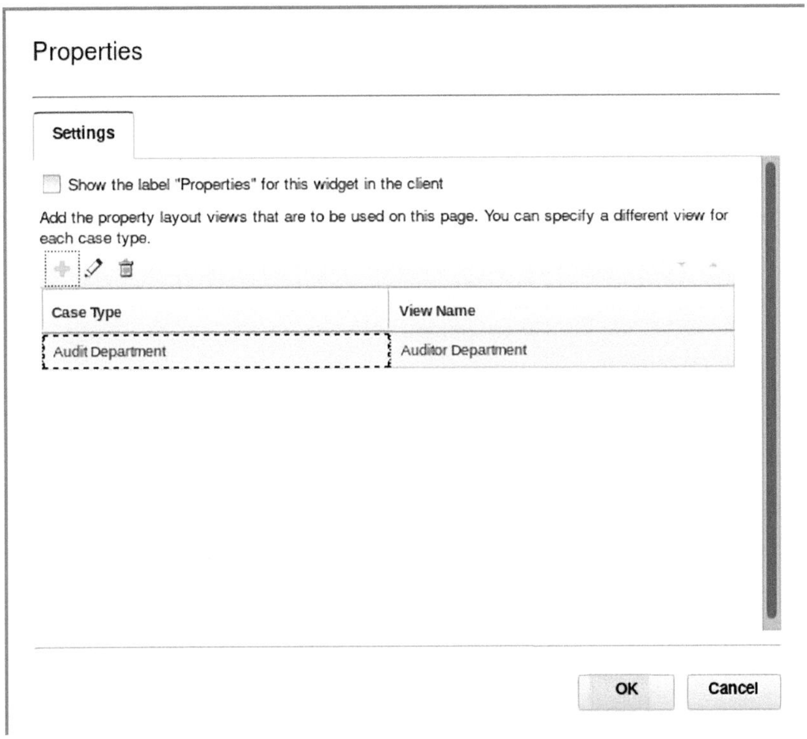

CHAPTER 1 IBM FILENET CASE MANAGER 5.3.3 CASE BUILDER SOLUTION DEVELOPMENT STEPS
FOR THE AUDIT SYSTEM

Figure 1-86. The OK command button is clicked to save the settings

The **Audit Department** Case Type can now be updated to associate the **Case Details with Business Objects** page, which we copied from the default **Case Details** page earlier, with the **Auditor's Manager** role.

The **Case Details with Business Objects** page allows a **Business Object** (which consists of a number of global properties) to be treated as a single multi-property entity. The page widget has additional display features such as a + icon to add a complete business object in a pop-up window.

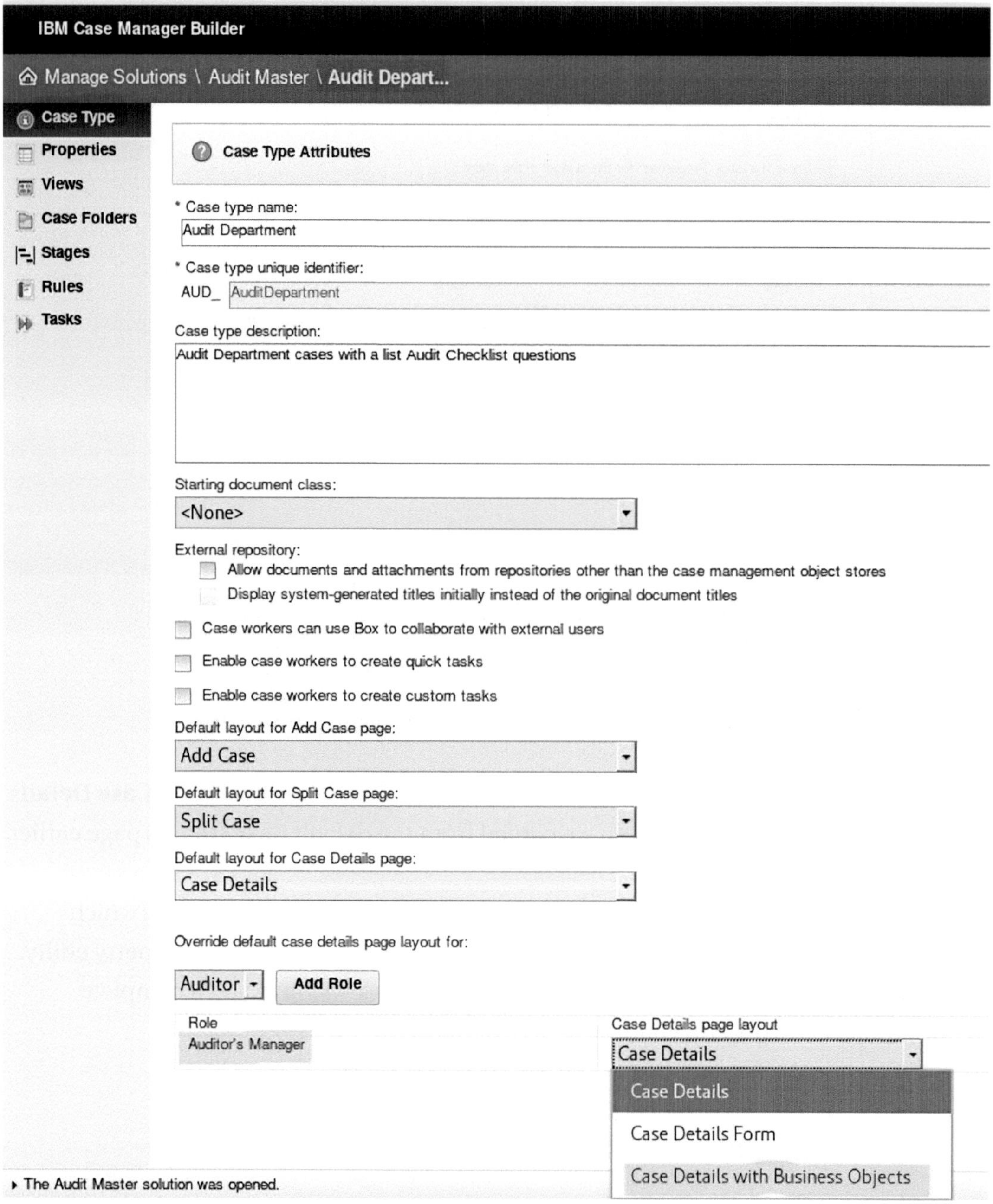

CHAPTER 1   IBM FILENET CASE MANAGER 5.3.3 CASE BUILDER SOLUTION DEVELOPMENT STEPS
            FOR THE AUDIT SYSTEM

*Figure 1-87.   The **Auditor's Manager** role is associated with the custom **Case Details** page layout*

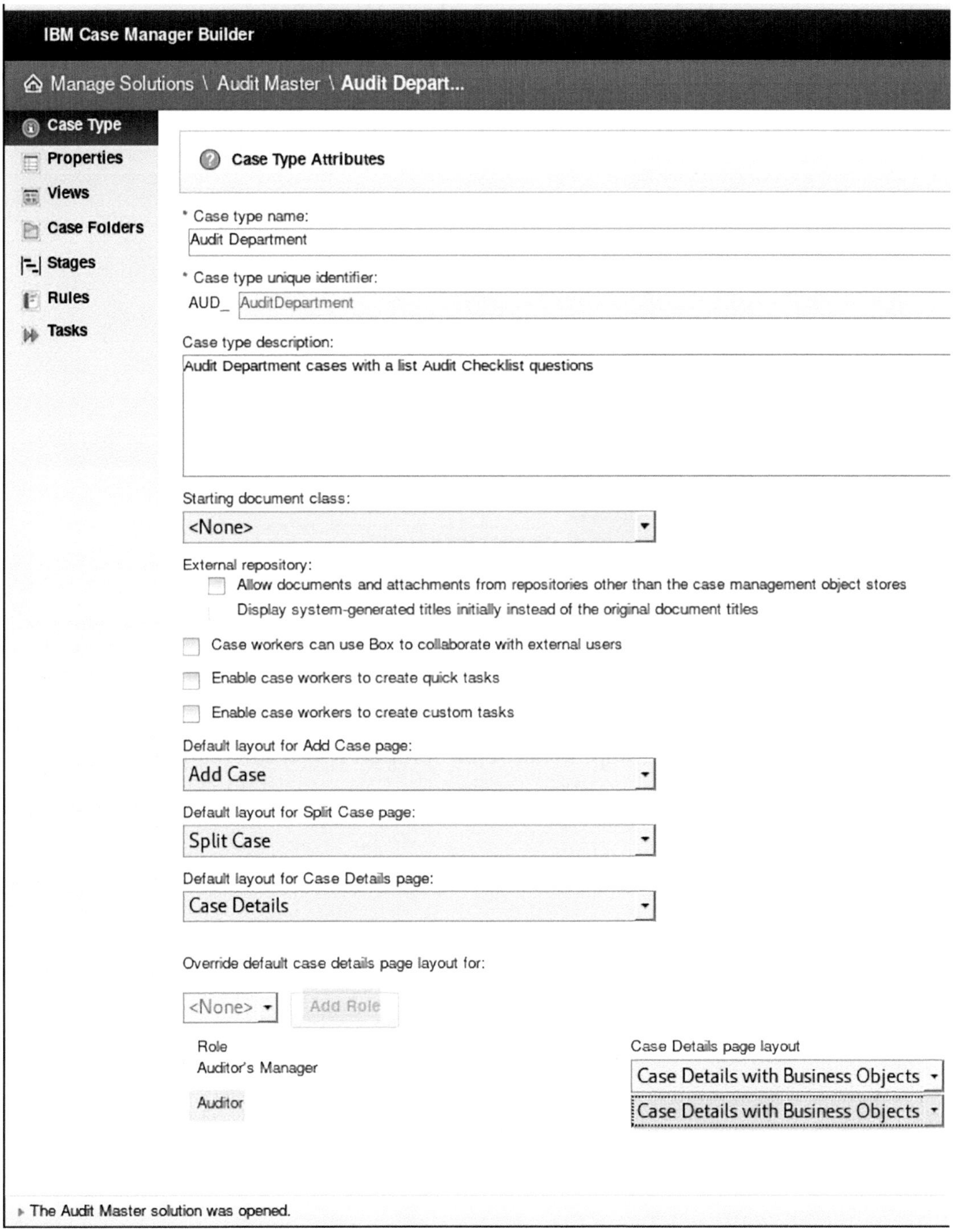

CHAPTER 1　IBM FILENET CASE MANAGER 5.3.3 CASE BUILDER SOLUTION DEVELOPMENT STEPS FOR THE AUDIT SYSTEM

*Figure 1-88.　The **Auditor** role is also associated with the custom **Case Details** page layout*

CHAPTER 1   IBM FILENET CASE MANAGER 5.3.3 CASE BUILDER SOLUTION DEVELOPMENT STEPS
            FOR THE AUDIT SYSTEM

We can now save, deploy, and then test the Audit Master Solution.

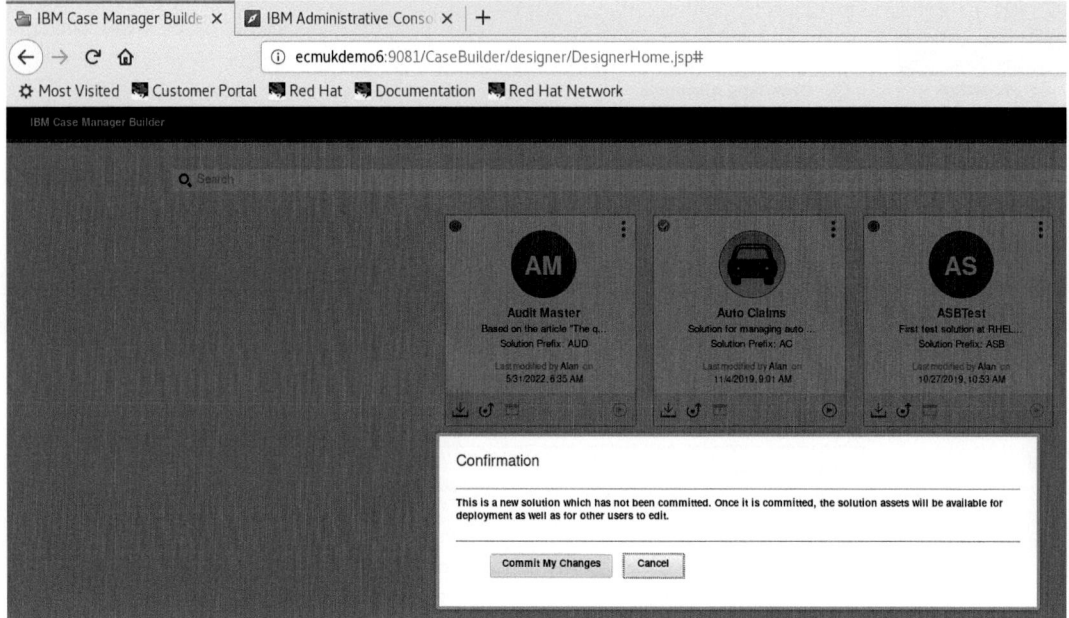

Figure 1-89. *The "Commit all the components of the Solution" icon (↓) is clicked*

If there are some locked components of the Solution which still require saving, you might get the message as shown in Figure 1-90.

Figure 1-90. *The "Commit all the components of the Solution" icon (↓) message is shown if items are locked for a Solution*

CHAPTER 1 IBM FILENET CASE MANAGER 5.3.3 CASE BUILDER SOLUTION DEVELOPMENT STEPS
          FOR THE AUDIT SYSTEM

Next, we have to deploy the Audit Master Solution to the Target Object Store in order to test the Case system.

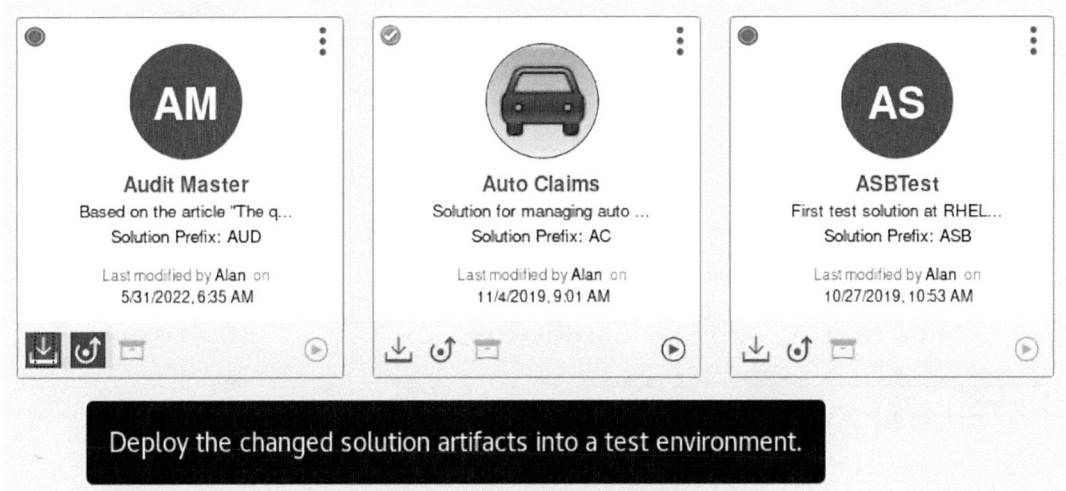

Figure 1-91. The "Deploy the changed solution artifacts into a test environment" icon is clicked

The deployment takes around two minutes and is confirmed with a green tick if all is successful or a red "x" if there are errors requiring remediation.

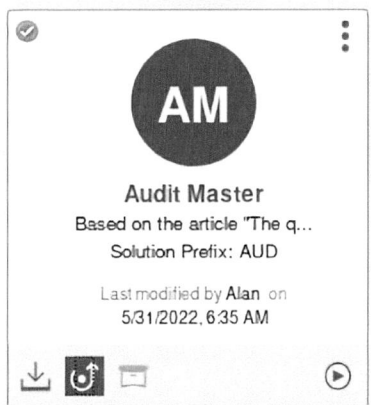

Figure 1-92. The Audit Master solution is successfully deployed for testing

CHAPTER 1   IBM FILENET CASE MANAGER 5.3.3 CASE BUILDER SOLUTION DEVELOPMENT STEPS
            FOR THE AUDIT SYSTEM

In the preceding example, the Solution is deployed with no deployment errors.

We can now click the icon as shown in Figure 1-93 for a first test of the deployed **Audit Master** Solution.

Figure 1-93. The "Start the IBM Case Manager Client to test the solution" icon is clicked

Part 4 – Testing and Administration of the Audit Solution

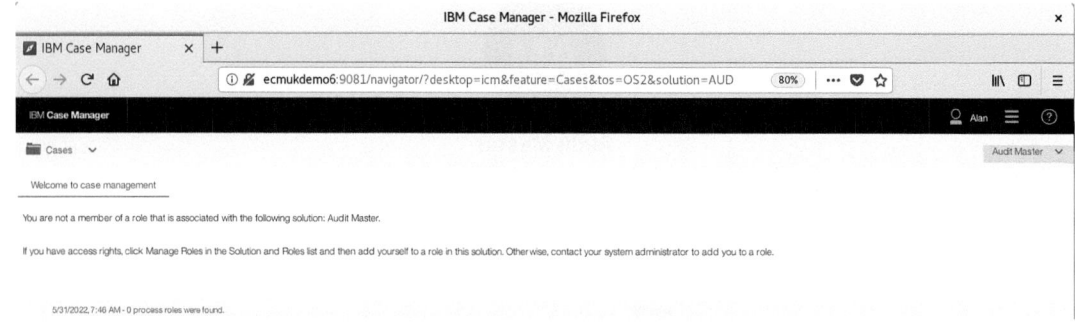

Figure 1-94. We need to be a member of a Solution **Role** associated with **Audit Master**

Add the Solution role for user Alan.

CHAPTER 1 IBM FILENET CASE MANAGER 5.3.3 CASE BUILDER SOLUTION DEVELOPMENT STEPS
           FOR THE AUDIT SYSTEM

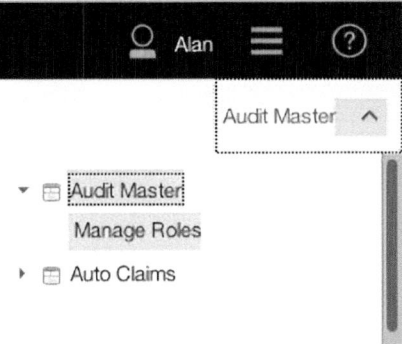

Figure 1-95. *The Manage Roles menu is selected under the Audit Master Solution*

The **Add Users and Groups** command button is clicked on the first role which is selected, by default, as the **Auditor's Manager** role.

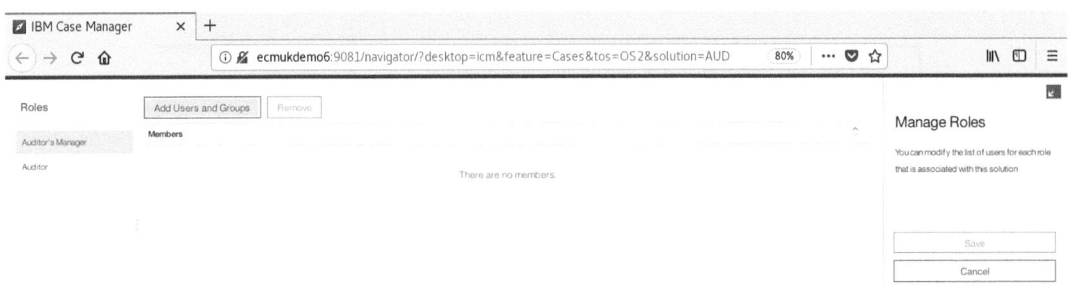

Figure 1-96. *The **Add Users and Groups** command button is clicked to add users*

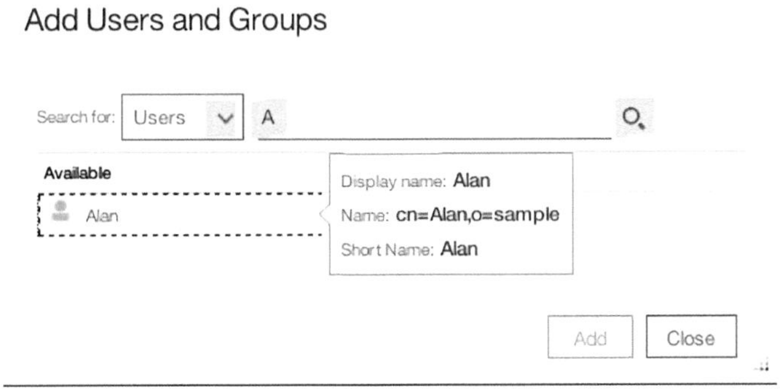

Figure 1-97. *The User Names beginning with A are searched, and Alan is selected*

71

CHAPTER 1 IBM FILENET CASE MANAGER 5.3.3 CASE BUILDER SOLUTION DEVELOPMENT STEPS
           FOR THE AUDIT SYSTEM

After a list of Auditor's Manager role users is added, the Auditor role users can be added. (A user can be added to more than one role in an IBM Case Manager Solution.)

Figure 1-98. *The selected users are saved, and the Auditor role users are added next*

The Auditor role users are added next.

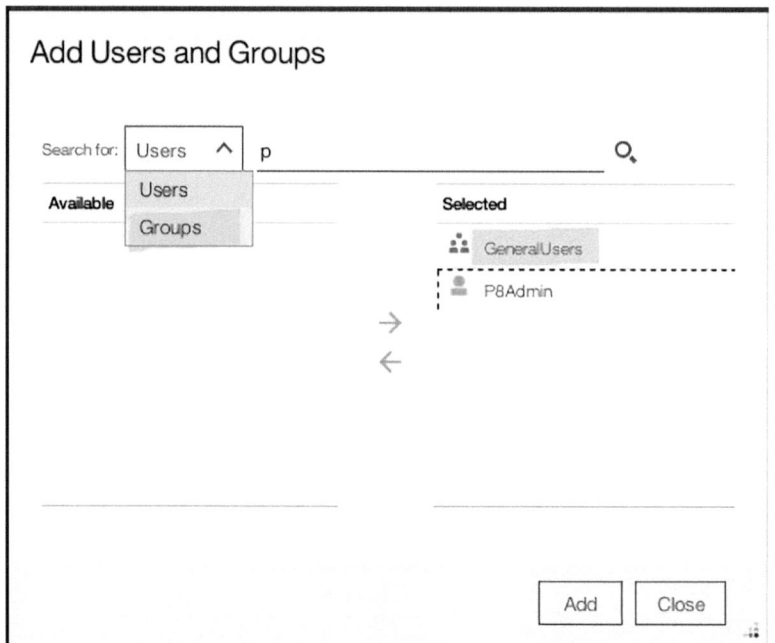

Figure 1-99. *A Group can be added to a Solution role*

Now we can see the **Add Case** button. Also, **Roles** can now be managed to add more users, using the **Manage Roles** command.

CHAPTER 1 IBM FILENET CASE MANAGER 5.3.3 CASE BUILDER SOLUTION DEVELOPMENT STEPS FOR THE AUDIT SYSTEM

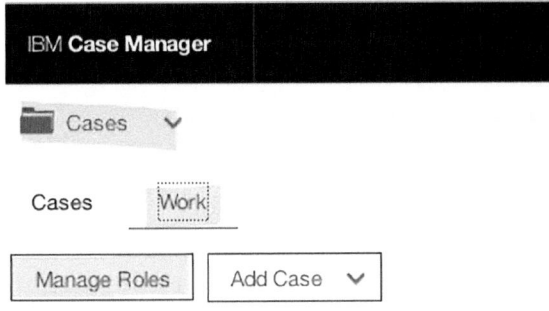

Figure 1-100. *The Manage Roles command can now be used to add new users*

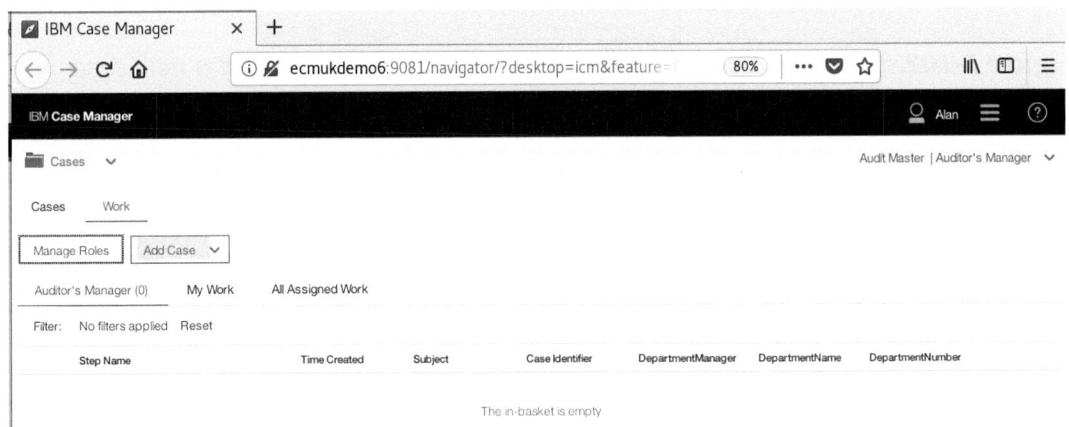

Figure 1-101. *The **Add Case** command button is clicked to add a new Audit Department*

The Audit Department Case Type is selected from the drop-down (an IBM Case Manager Solution can have multiple Case Types).

73

CHAPTER 1   IBM FILENET CASE MANAGER 5.3.3 CASE BUILDER SOLUTION DEVELOPMENT STEPS
            FOR THE AUDIT SYSTEM

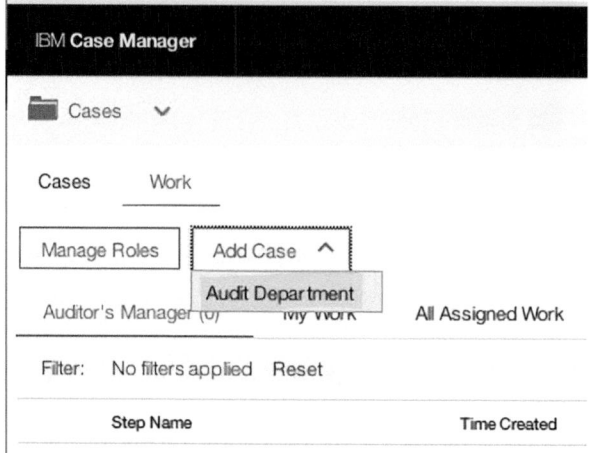

Figure 1-102. *The new **Audit Department** case is selected to start a new case*

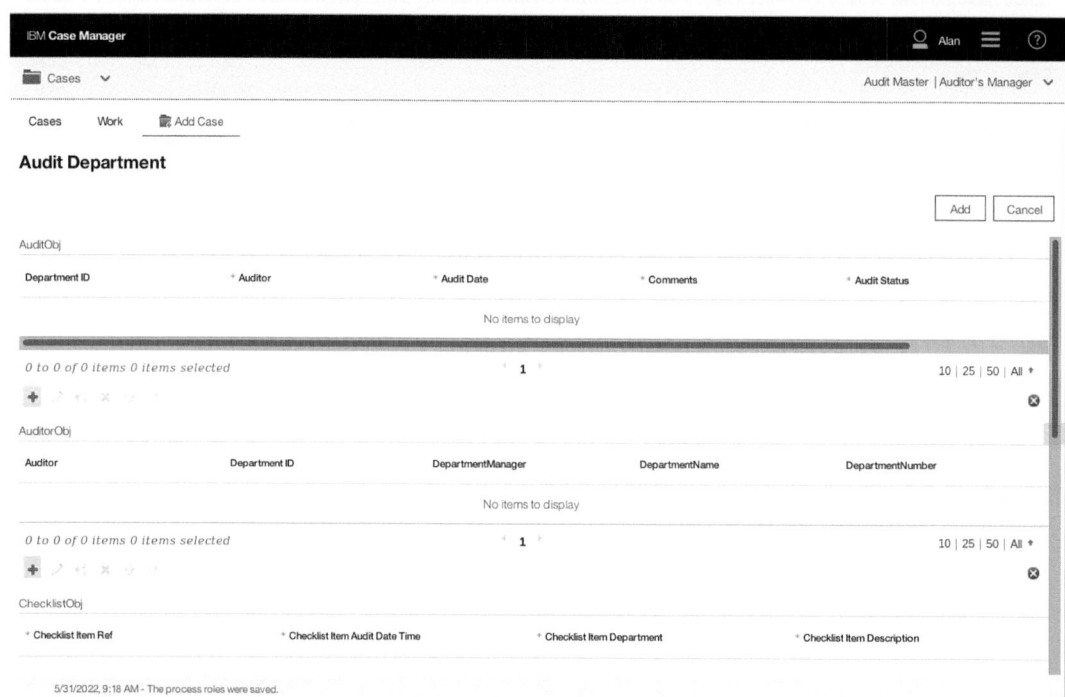

Figure 1-103. *The required Business Objects are edited using the + icons (highlighted)*

CHAPTER 1 IBM FILENET CASE MANAGER 5.3.3 CASE BUILDER SOLUTION DEVELOPMENT STEPS
          FOR THE AUDIT SYSTEM

The + icons are selected in turn, and the Case entry form is scrolled down as shown in Figure 1-103.

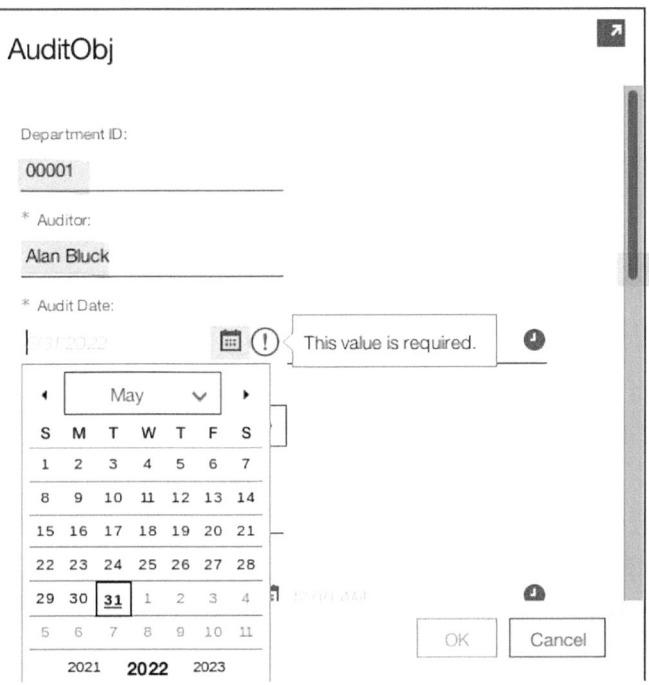

Figure 1-104. The Audit Date/Time field is a required entry field

The Date picker shows the current date, but we need to select an Audit Date in the future to allow the Checklist questions to be reviewed by the Department Manager.

75

CHAPTER 1   IBM FILENET CASE MANAGER 5.3.3 CASE BUILDER SOLUTION DEVELOPMENT STEPS FOR THE AUDIT SYSTEM

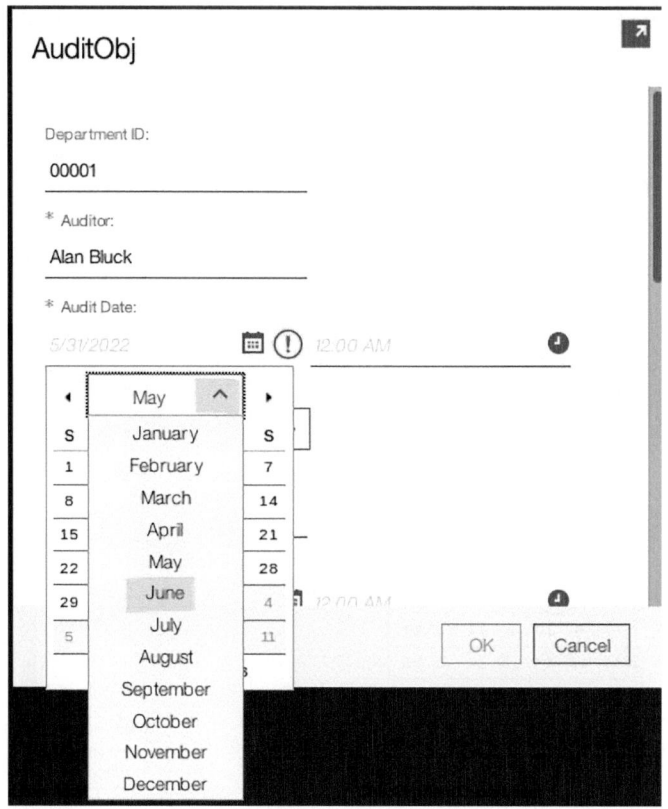

Figure 1-105. *June is selected using the Date picker drop-down for the month*

CHAPTER 1 IBM FILENET CASE MANAGER 5.3.3 CASE BUILDER SOLUTION DEVELOPMENT STEPS
FOR THE AUDIT SYSTEM

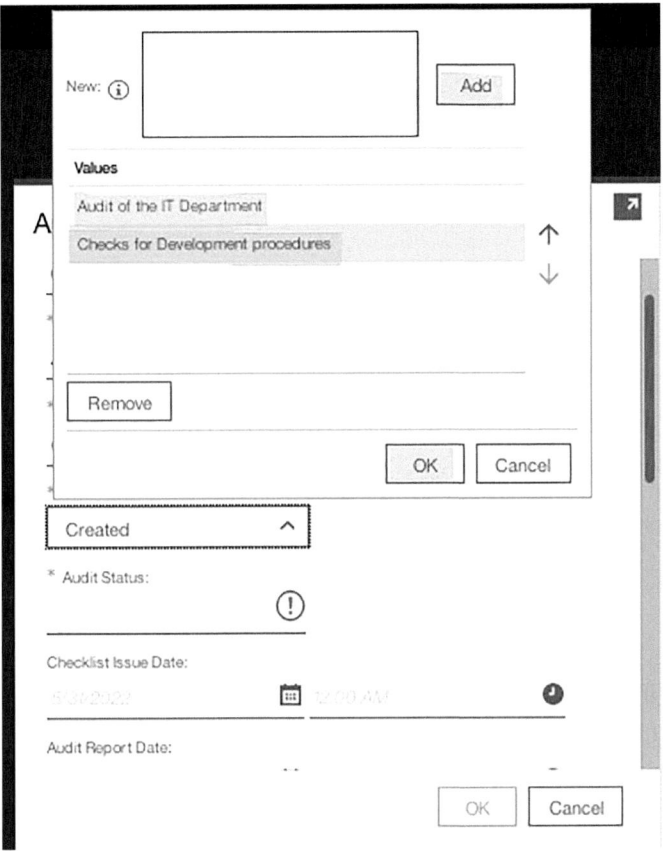

Figure 1-106. *The required fields are entered as shown (red *) for required fields*

The Audit Business Object is entered and updated using the OK command.

CHAPTER 1   IBM FILENET CASE MANAGER 5.3.3 CASE BUILDER SOLUTION DEVELOPMENT STEPS FOR THE AUDIT SYSTEM

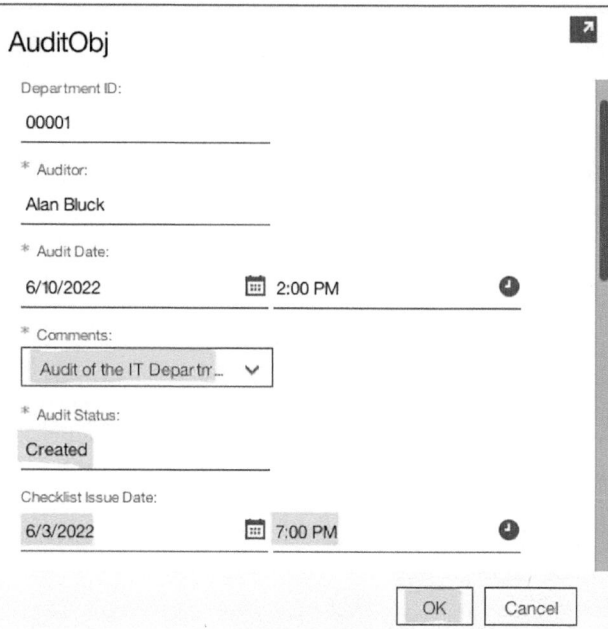

Figure 1-107. *The Comments (multi-value) are entered and the Audit information fields*

Figure 1-108. *The Auditor Business Object fields are entered*

CHAPTER 1　IBM FILENET CASE MANAGER 5.3.3 CASE BUILDER SOLUTION DEVELOPMENT STEPS FOR THE AUDIT SYSTEM

One or more Checklist Business Objects are entered to confirm the Department procedures are followed correctly.

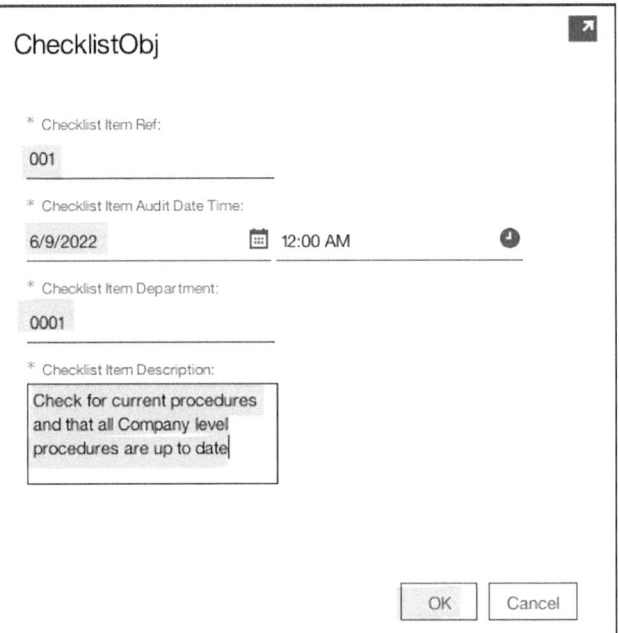

Figure 1-109. *A Checklist item is entered*

Next, the Department details are entered using the Department Business Object.

79

CHAPTER 1   IBM FILENET CASE MANAGER 5.3.3 CASE BUILDER SOLUTION DEVELOPMENT STEPS FOR THE AUDIT SYSTEM

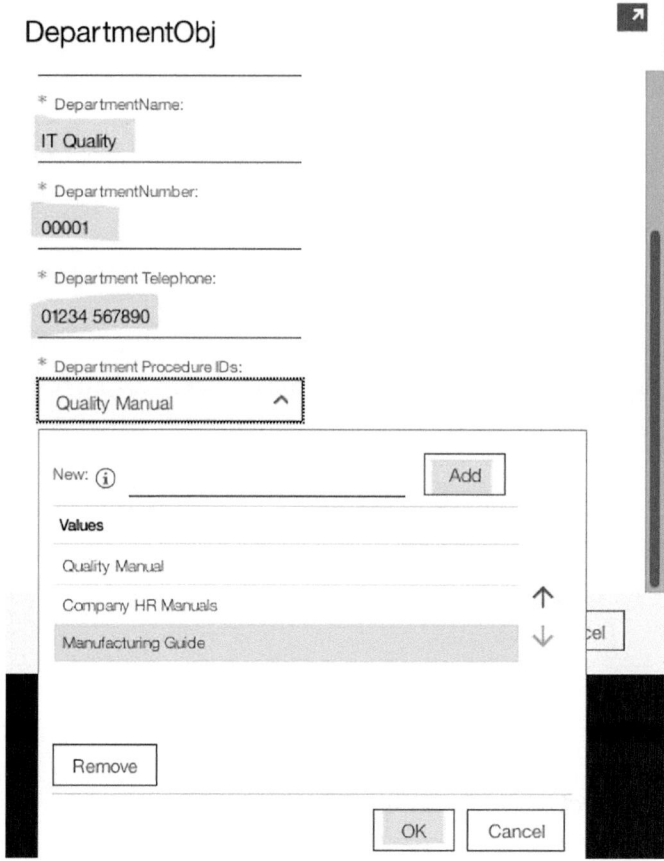

Figure 1-110. The list of the Department procedure documents to be audited is entered

The list of department emails for user contacts for the audit is entered.

CHAPTER 1 IBM FILENET CASE MANAGER 5.3.3 CASE BUILDER SOLUTION DEVELOPMENT STEPS FOR THE AUDIT SYSTEM

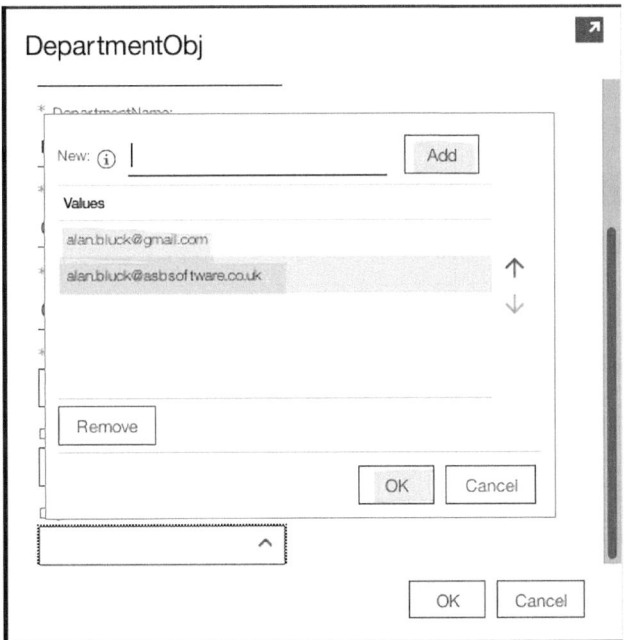

Figure 1-111. *The multi-value list of emails for the audit is entered*

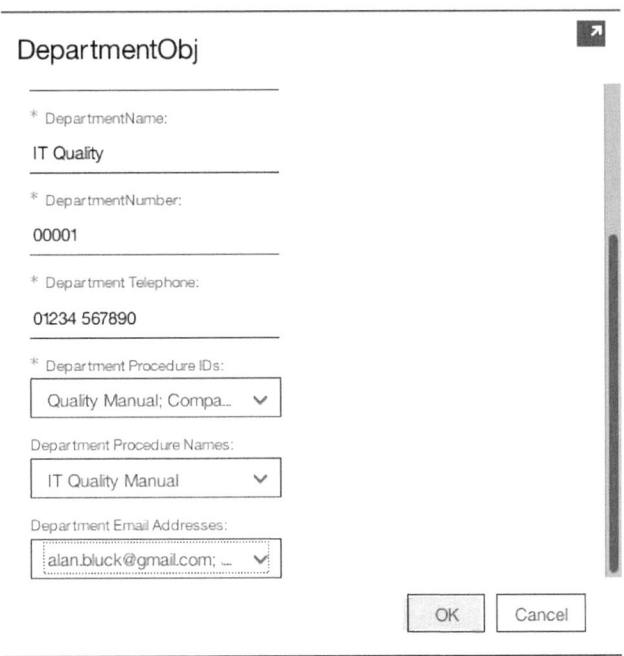

Figure 1-112. *The OK command is clicked to signify that all the Department fields are completed*

CHAPTER 1   IBM FILENET CASE MANAGER 5.3.3 CASE BUILDER SOLUTION DEVELOPMENT STEPS
            FOR THE AUDIT SYSTEM

The completed Case details are reviewed, and after checking, the **Add** command button is clicked to launch the "live" **Audit Master** Case into the **IBM Case Manager** system.

Figure 1-113. *The Case details are reviewed, and the Case is launched by clicking the Add command button*

The Case is visible in the Cases tab of the Case Manager Client, after selecting the **Search** command button.

CHAPTER 1 IBM FILENET CASE MANAGER 5.3.3 CASE BUILDER SOLUTION DEVELOPMENT STEPS
        FOR THE AUDIT SYSTEM

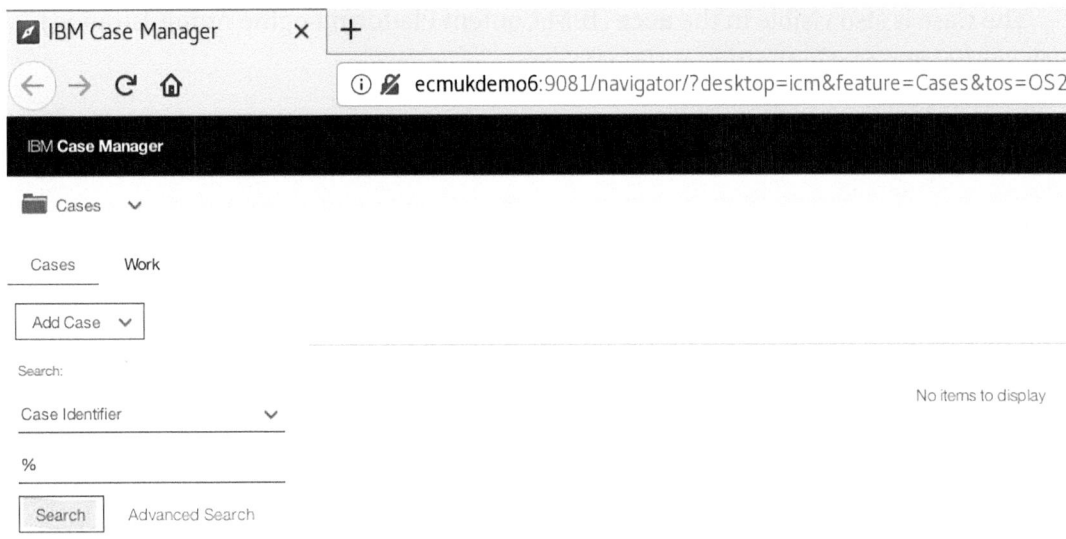

Figure 1-114. *The Search command button is used to refresh the Cases In-basket view*

The case we created is automatically provided with a unique Case ID.

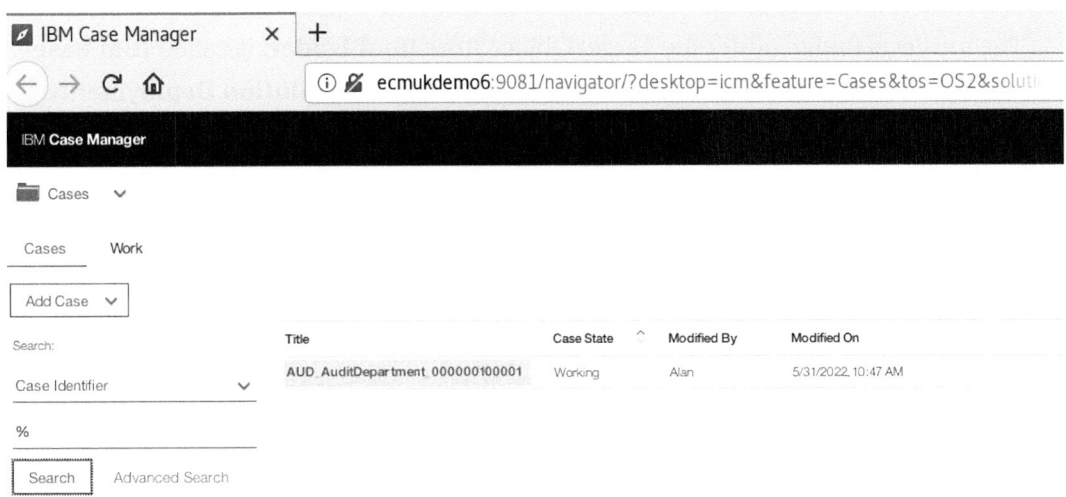

Figure 1-115. *The refreshed Cases In-basket displays the new Audit Case we created*

CHAPTER 1   IBM FILENET CASE MANAGER 5.3.3 CASE BUILDER SOLUTION DEVELOPMENT STEPS
           FOR THE AUDIT SYSTEM

The Case is also visible in the **acce** (IBM Content Platform Engine Administration) web application tool in the Target Object Store.

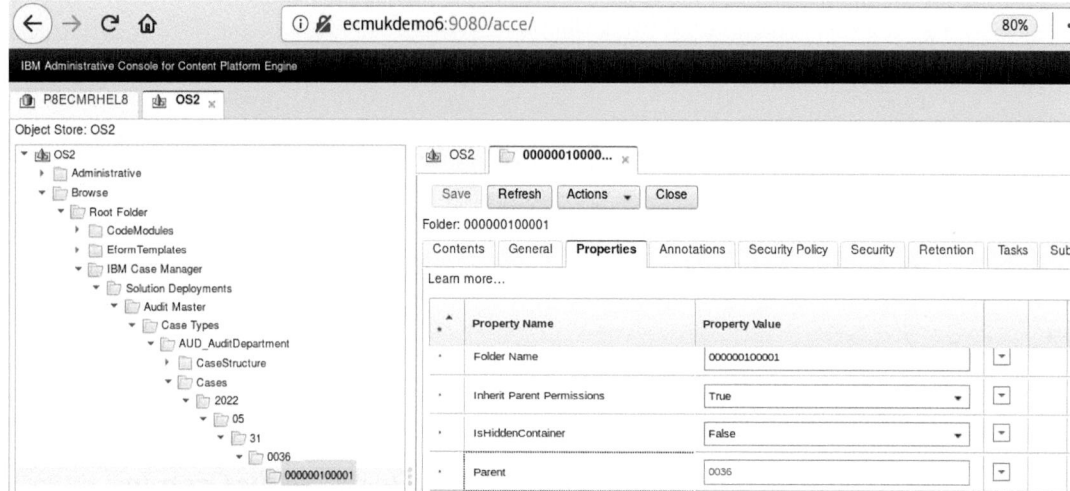

Figure 1-116.   *The Case structure created in the IBM FileNet Content Object Store OS2*

The Case Folder structure is held in the IBM FileNet Case Manager Target Object Store, OS2. The structure of the nested folders is shown in Figure 1-117.

The top-level folder, under the Target Object Store **Root Folder**, is called **IBM Case Manager**. Under this folder, each solution is stored under the **Solution Deployments** folder. (The **Audit Master** solution folder can be seen to be linked to this folder.)

Each solution **Case Type** for **Audit Master** is then held under the **Case Types** folder. The next subfolder is the **AUD_AuditDepartment** case type for the **Audit Master** solution. The live Audit Master Solution cases are then stored under the **Cases** folder, which is one of the subfolders under the **AUD_AuditDepartment** Case Type parent folder.

This top-level **Cases** folder for the **AUD_AuditDepartment** Case Type is split by date subfolders organized by nested folder levels of year (**2022**), month (**5**), day (**31**), the four-digit parent folder number, a randomly generated number to identify the parent folder (**0036**), and under this a unique Case ID sequence value starting from **0000000100001**.

The Case ID and sequence value and customization are described in detail on the URL page:

www.ibm.com/docs/en/case-manager/5.3.3?topic=system-customizing-case-unique-identifier-prefix

CHAPTER 1   IBM FILENET CASE MANAGER 5.3.3 CASE BUILDER SOLUTION DEVELOPMENT STEPS
            FOR THE AUDIT SYSTEM

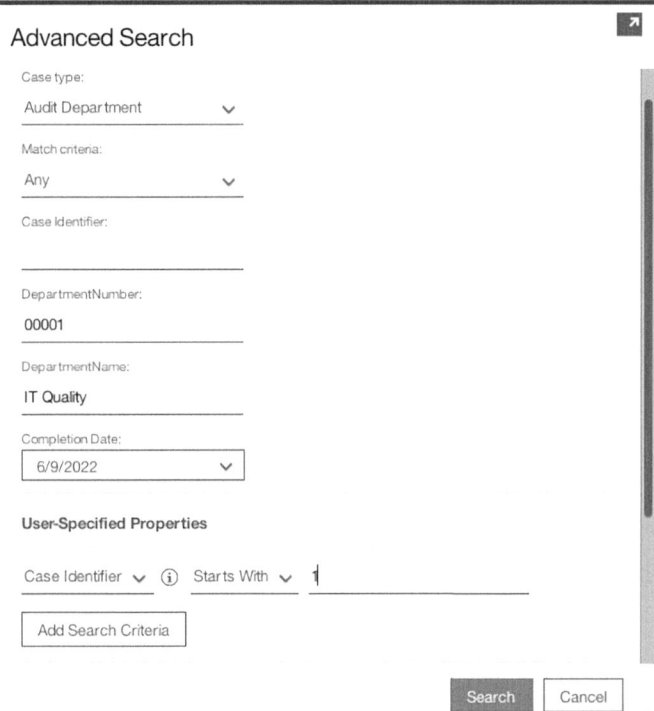

Figure 1-117.   The Case Advanced Search criteria entry window

The **Advanced Search** command button on the main IBM Case Manager **Cases** tab is clicked to display the search criteria entry window shown in Figure 1-117.

Figure 1-118.   The Advanced Search results are shown

CHAPTER 1 IBM FILENET CASE MANAGER 5.3.3 CASE BUILDER SOLUTION DEVELOPMENT STEPS
         FOR THE AUDIT SYSTEM

The returned row can be clicked on the **DepartmentNumber** property column to show the **Summary** tab details in the right panel, as displayed in Figure 1-119.

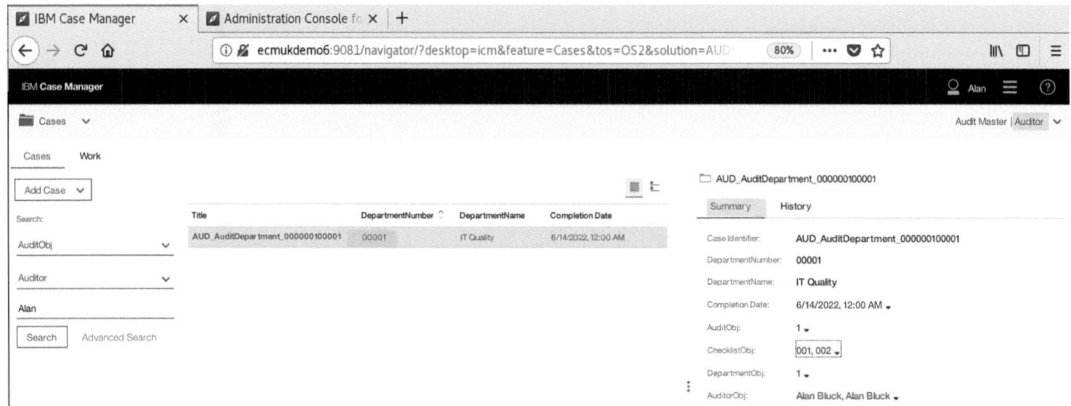

Figure 1-119. The Summary tab details are shown for the Case

On clicking the **History** tab, the recorded events for the Case and Task audit history are displayed as shown in Figure 1-120, with the latest events shown at the top of the list.

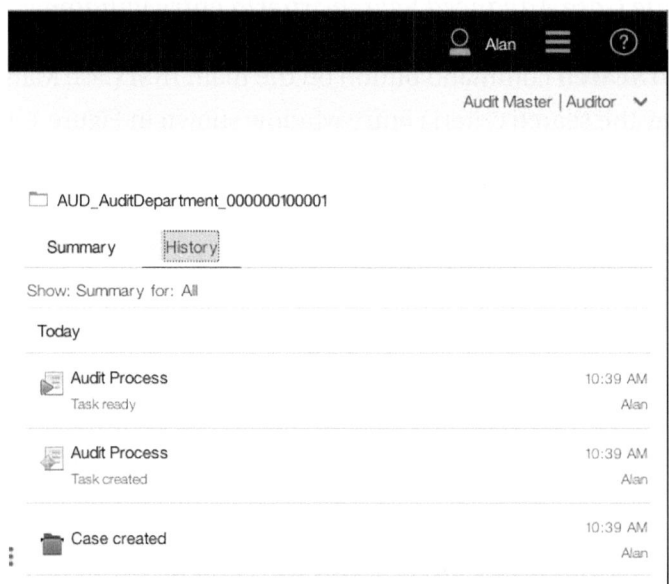

Figure 1-120. The Case History tab shows the events generated for the Case and task

CHAPTER 1   IBM FILENET CASE MANAGER 5.3.3 CASE BUILDER SOLUTION DEVELOPMENT STEPS
            FOR THE AUDIT SYSTEM

The layout of the Case details panel can be changed from a summary **Details view** to the **Magazine view** of each property using the icon highlighted in Figure 1-121.

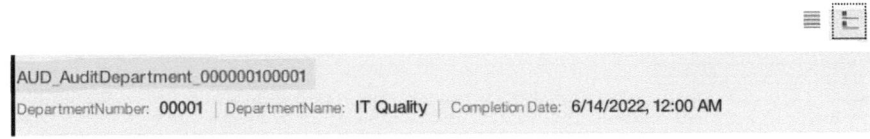

Figure 1-121.   The Magazine view of the Case properties

On clicking the **AUD_AuditDepartment_000000100001** Case link in Figure 1-121, the Case Details with the Business Object property entry view, **Case Details with Business Objects** (the page view we created earlier), is displayed.

*Figure 1-122.   The Business Object property entry page view, **Case Details with Business Objects***

The + icon against each multi-value Business Object property allows new entries to be inserted, as shown for the **ChecklistObj** property lines in Figure 1-123.

CHAPTER 1   IBM FILENET CASE MANAGER 5.3.3 CASE BUILDER SOLUTION DEVELOPMENT STEPS
            FOR THE AUDIT SYSTEM

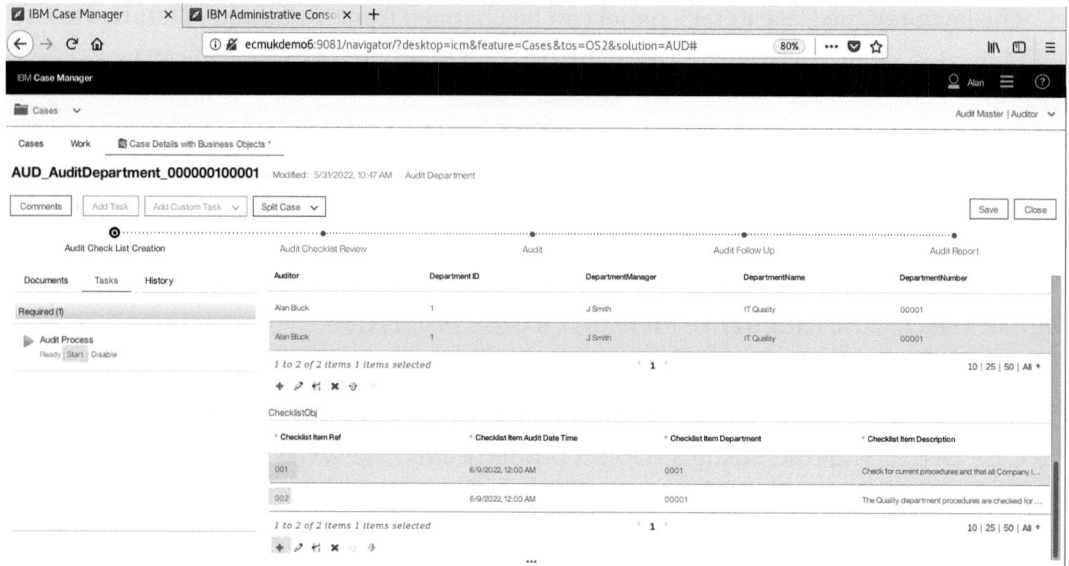

Figure 1-123. *Checklist items are added to the Case using the + icon (highlighted)*

In Figure 1-123, the Audit Process **Start** command can be clicked to move the case onto the next stage of the process (from the **Audit Check List Creation** to the **Audit Checklist Review** stage). This shows the pop-up window shown in Figure 1-124.

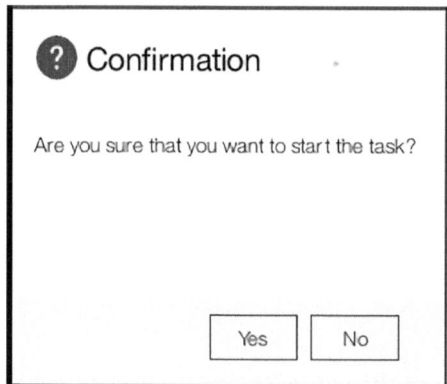

Figure 1-124. *The Yes command is used to start the Audit Process*

CHAPTER 1   IBM FILENET CASE MANAGER 5.3.3 CASE BUILDER SOLUTION DEVELOPMENT STEPS
              FOR THE AUDIT SYSTEM

The Audit Process Task status is shown with a status of Completed with the date of completion.

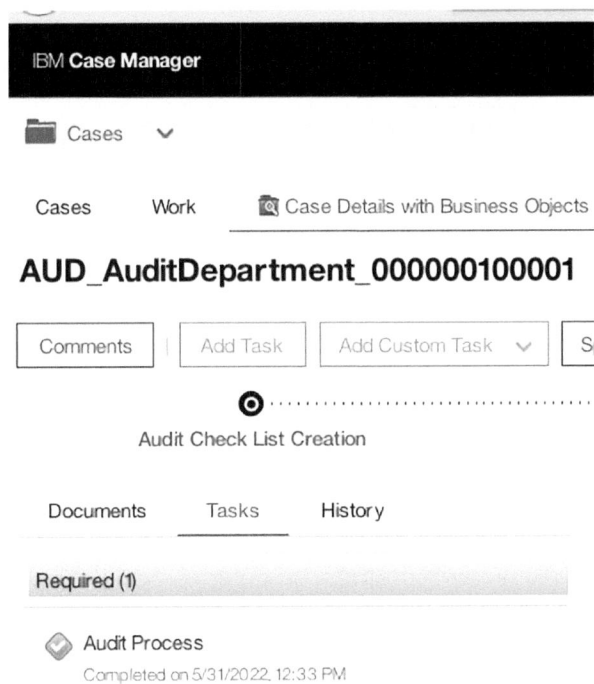

Figure 1-125. *The Audit Process Task status is shown with a status of **Completed***

CHAPTER 1 IBM FILENET CASE MANAGER 5.3.3 CASE BUILDER SOLUTION DEVELOPMENT STEPS
FOR THE AUDIT SYSTEM

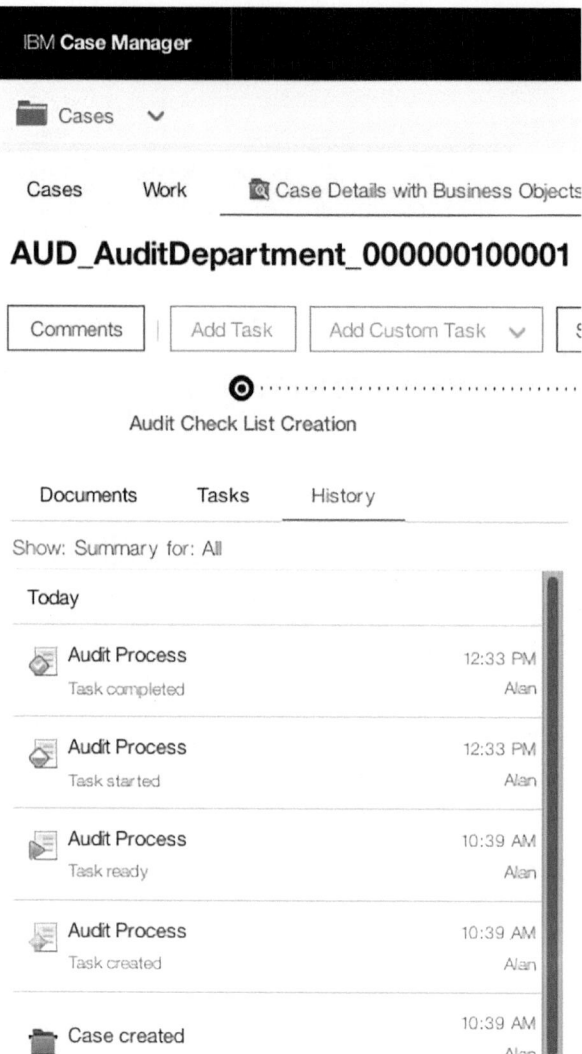

Figure 1-126. *The History tab shows the latest Audit Process event*

Each of the Audit Case Stages shown in Figure 1-127 can be clicked to show the Stage details entered (defined during the IBM Case Builder development, as shown in Figures 1-71 and 1-72).

Figure 1-127. *The Stages defined for the Audit Master solution process*

CHAPTER 1 IBM FILENET CASE MANAGER 5.3.3 CASE BUILDER SOLUTION DEVELOPMENT STEPS
FOR THE AUDIT SYSTEM

On clicking the **Audit Checklist Review** stage link (highlighted in Figure 1-127), we get the pop-up window shown in Figure 1-128.

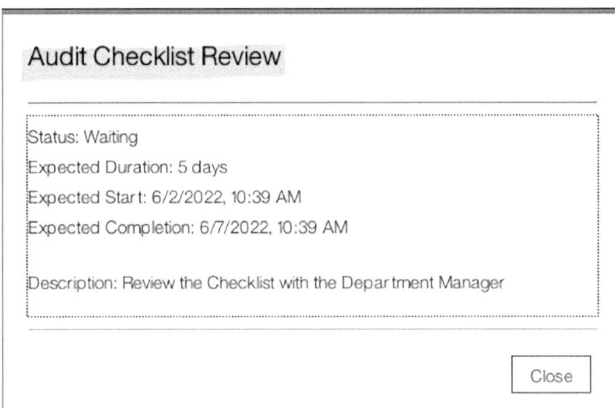

Figure 1-128. *The **Audit Checklist Review** stage details are displayed*

It is possible to configure the state of the case stage at runtime by changing the case stage state by

a) Adding the following actions to a menu or toolbar in the Case List widget or Case Toolbar widget:

- Complete Stage

- Restart Stage

- Toggle Stage

b) Adding the stage steps to the System Lane of the workflow for a task to run the following case operations:

- completeCurrentCaseStage

- placeCurrentCaseStageOnHold

- releaseCurrentOnHoldCaseStage

- restartPreviousCaseStage

CHAPTER 1　IBM FILENET CASE MANAGER 5.3.3 CASE BUILDER SOLUTION DEVELOPMENT STEPS FOR THE AUDIT SYSTEM

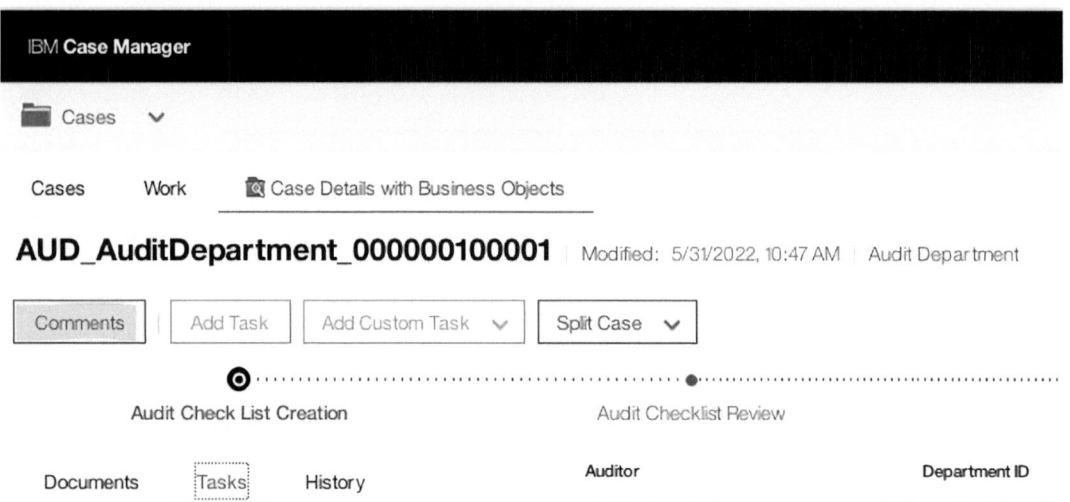

Figure 1-129. Comments can be added at any time for an Audit Case by clicking the Comments command button (highlighted)

One or more comments can be added to the Audit Case, as shown in Figure 1-130, using the **Add** command button. The **Close** command is then used to complete the addition of comments.

CHAPTER 1   IBM FILENET CASE MANAGER 5.3.3 CASE BUILDER SOLUTION DEVELOPMENT STEPS
            FOR THE AUDIT SYSTEM

Figure 1-130. A new comment is added to the Case

An Excel or MS Word Office document can be added to the Audit case as shown in Figure 1-131.

CHAPTER 1 IBM FILENET CASE MANAGER 5.3.3 CASE BUILDER SOLUTION DEVELOPMENT STEPS FOR THE AUDIT SYSTEM

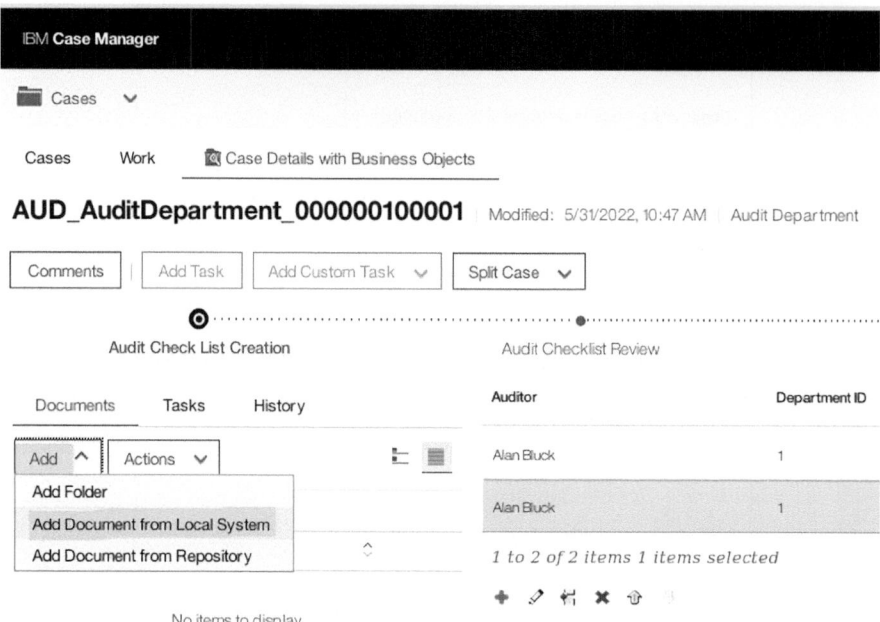

Figure 1-131. *The Documents tab is selected, and the **Add** drop-down is used to Add a Document from the Local System*

The **Add ➤ Add Document from Local System** menu option on the **Documents** tab allows a document to be browsed to from a pop-up window, as shown on the Linux system window in Figure 1-132.

CHAPTER 1 IBM FILENET CASE MANAGER 5.3.3 CASE BUILDER SOLUTION DEVELOPMENT STEPS
         FOR THE AUDIT SYSTEM

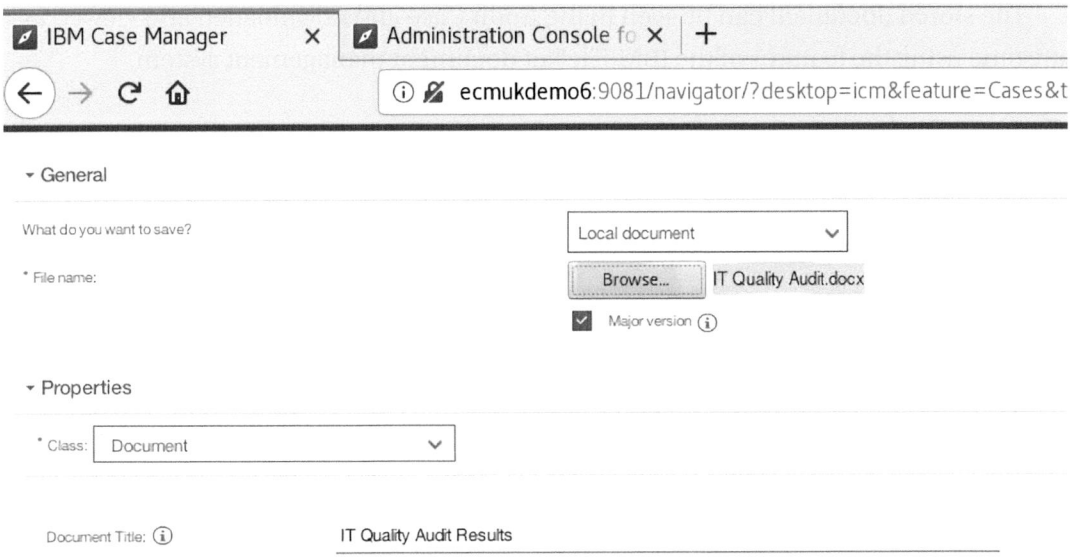

Figure 1-132. The **Browse** command is used to navigate to an Audit results Word document

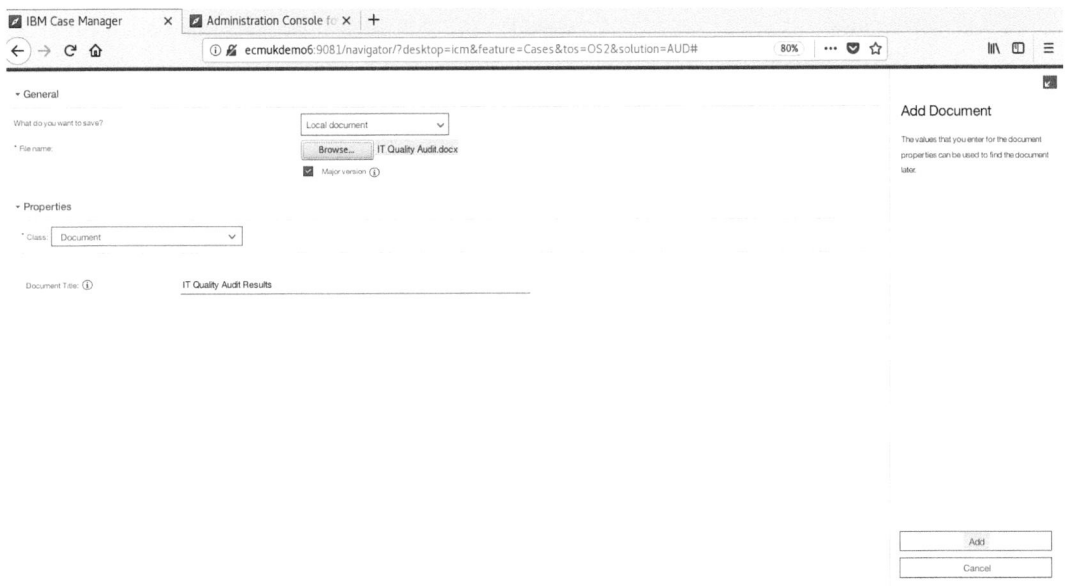

Figure 1-133. The Add command is used to store the selected Word document in the IBM FileNet Content Engine Object Store

CHAPTER 1   IBM FILENET CASE MANAGER 5.3.3 CASE BUILDER SOLUTION DEVELOPMENT STEPS
            FOR THE AUDIT SYSTEM

The stored document can be seen in the Audit Case and downloaded and viewed at any time using the features of the IBM FileNet document management system.

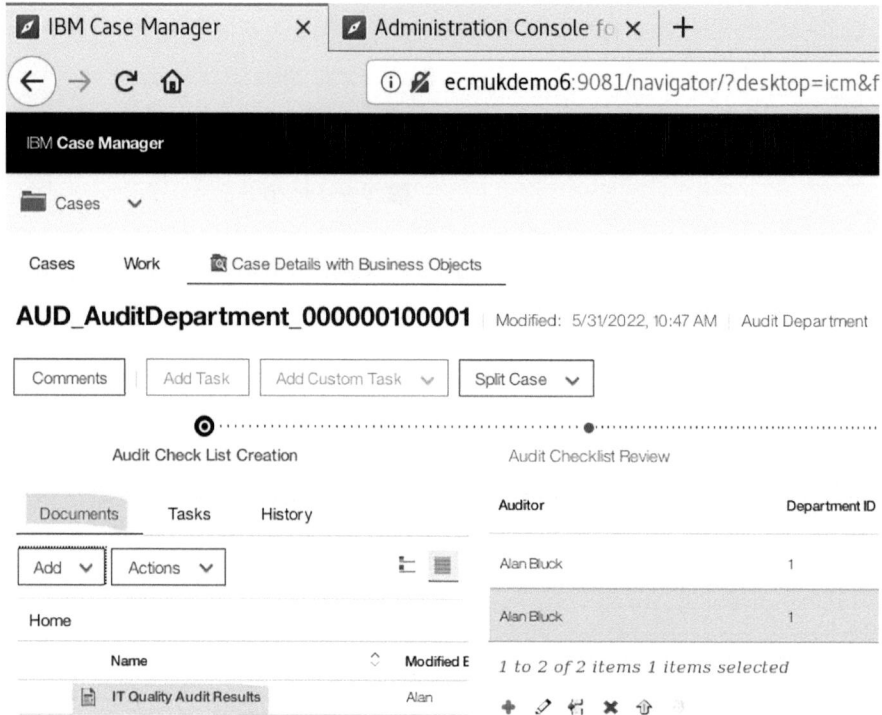

*Figure 1-134. The uploaded **IT Quality Audit Results** document is available for viewing*

The **History** tab of the Case now shows the uploaded Document we added.

CHAPTER 1 IBM FILENET CASE MANAGER 5.3.3 CASE BUILDER SOLUTION DEVELOPMENT STEPS
FOR THE AUDIT SYSTEM

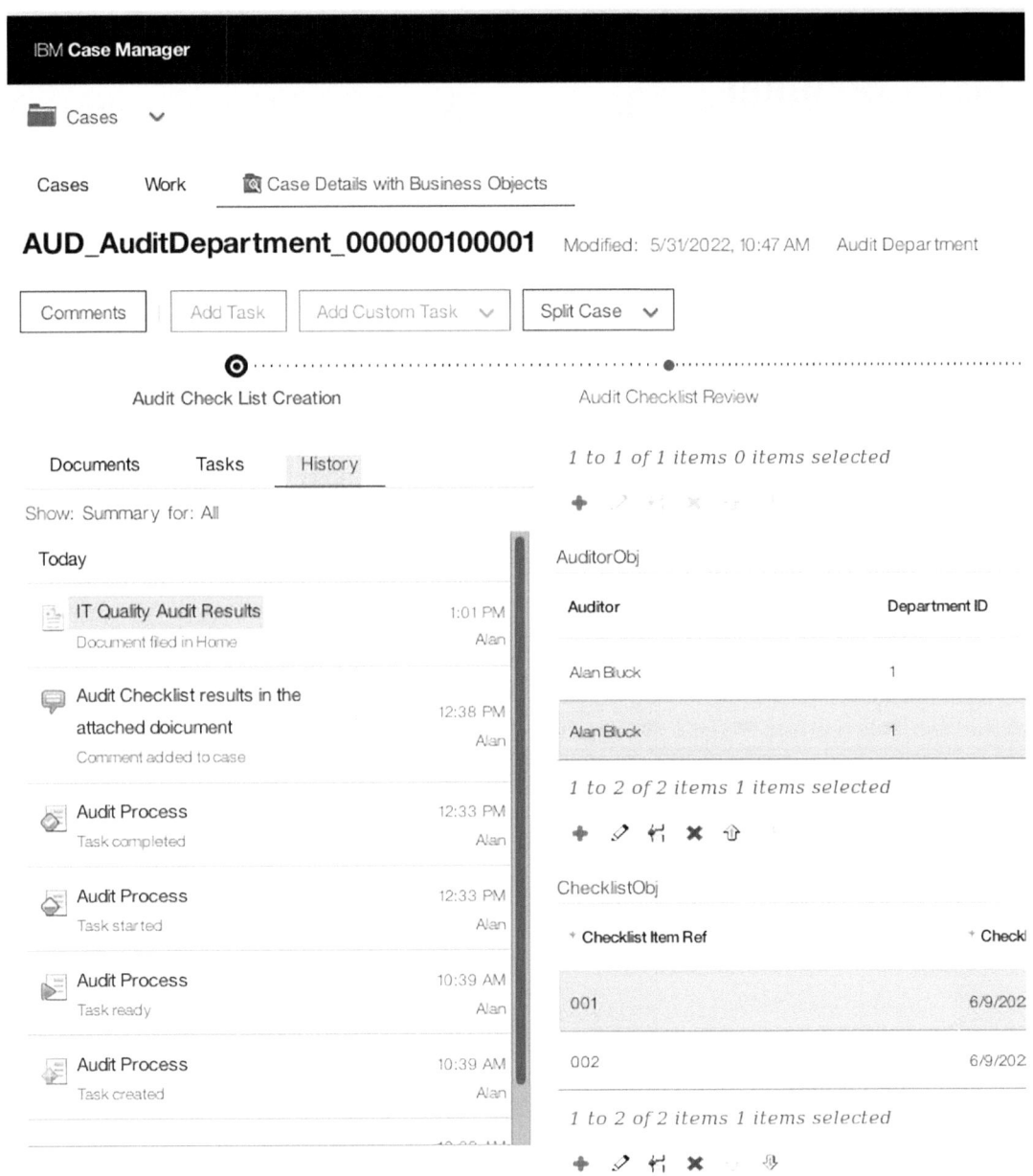

Figure 1-135. *The **History** tab shows the uploaded Audit results Word document event*

CHAPTER 1 IBM FILENET CASE MANAGER 5.3.3 CASE BUILDER SOLUTION DEVELOPMENT STEPS
          FOR THE AUDIT SYSTEM

Exporting and Production Deployment of the Audit Master Solution

Usually, the IBM Case Manager Solution will be developed in a development environment and will then be transferred to one or more test environments and finally imported to a production system. IBM Case Manager supports this development life cycle by providing an export/import of the complete Solution, packaged as a single zip file, as shown in this section.

The IBM Case Administration Content Navigator desktop application is run using the URL as follows:

http://ecmukdemo6:9081/navigator/?desktop=icmadmin

where **ecmukdemo6** is the web application server and **9081** is the WebSphere application server port for the IBM Case Manager applications.

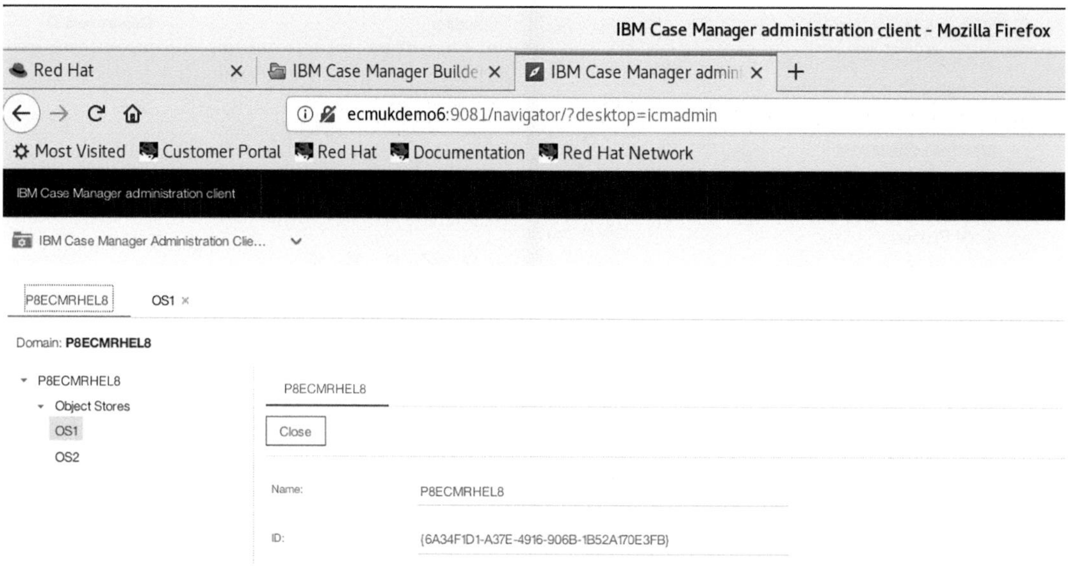

Figure 1-136. *The IBM Case Manager Administration Desktop is launched*

The Design Object store, **OS1**, in Figure 1-136, is clicked to show the Solutions drop-down as illustrated in Figure 1-137.

CHAPTER 1　IBM FILENET CASE MANAGER 5.3.3 CASE BUILDER SOLUTION DEVELOPMENT STEPS FOR THE AUDIT SYSTEM

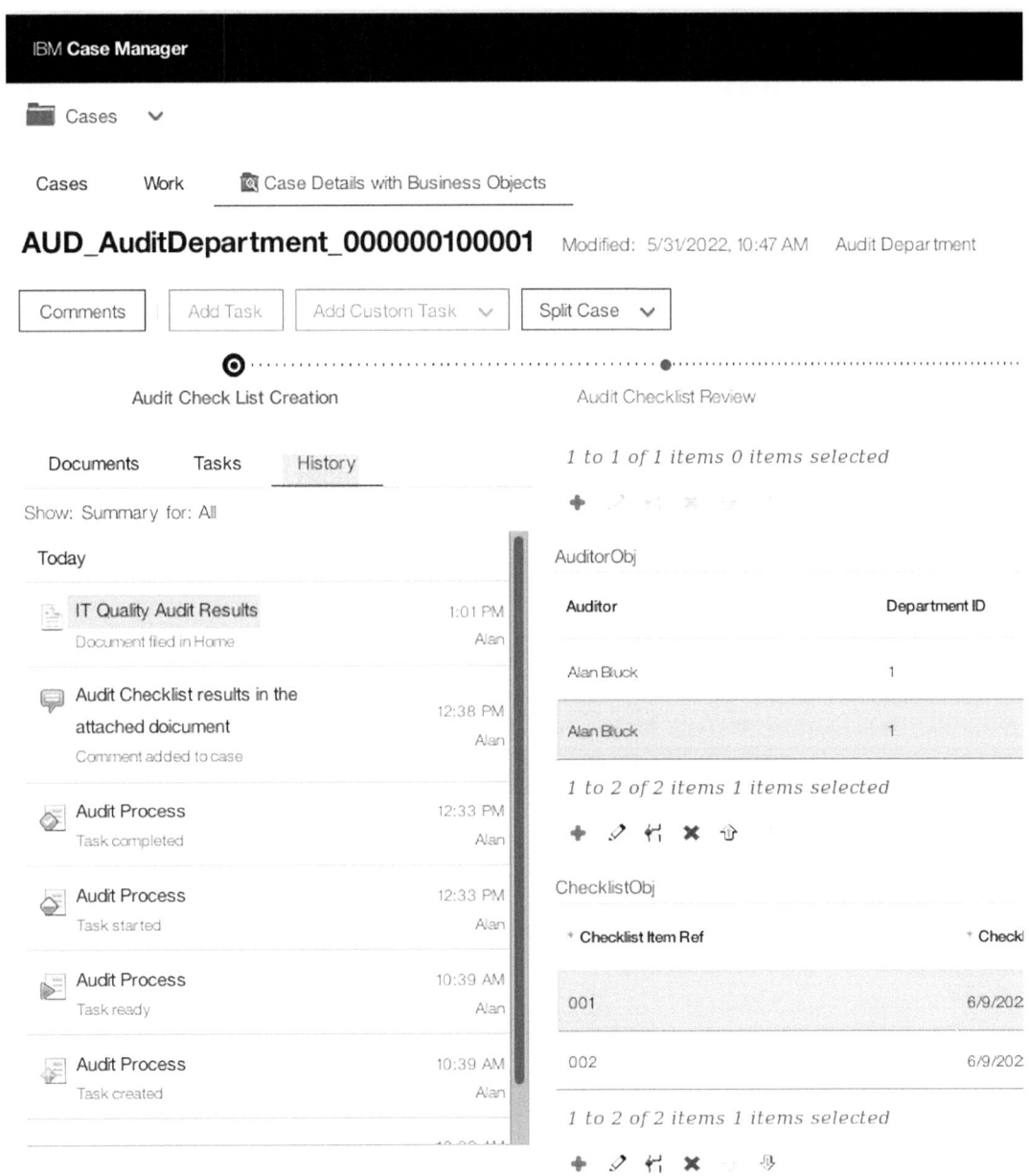

Figure 1-135. *The **History** tab shows the uploaded Audit results Word document event*

CHAPTER 1   IBM FILENET CASE MANAGER 5.3.3 CASE BUILDER SOLUTION DEVELOPMENT STEPS
            FOR THE AUDIT SYSTEM

Exporting and Production Deployment of the Audit Master Solution

Usually, the IBM Case Manager Solution will be developed in a development environment and will then be transferred to one or more test environments and finally imported to a production system. IBM Case Manager supports this development life cycle by providing an export/import of the complete Solution, packaged as a single zip file, as shown in this section.

The IBM Case Administration Content Navigator desktop application is run using the URL as follows:

http://ecmukdemo6:9081/navigator/?desktop=icmadmin

where **ecmukdemo6** is the web application server and **9081** is the WebSphere application server port for the IBM Case Manager applications.

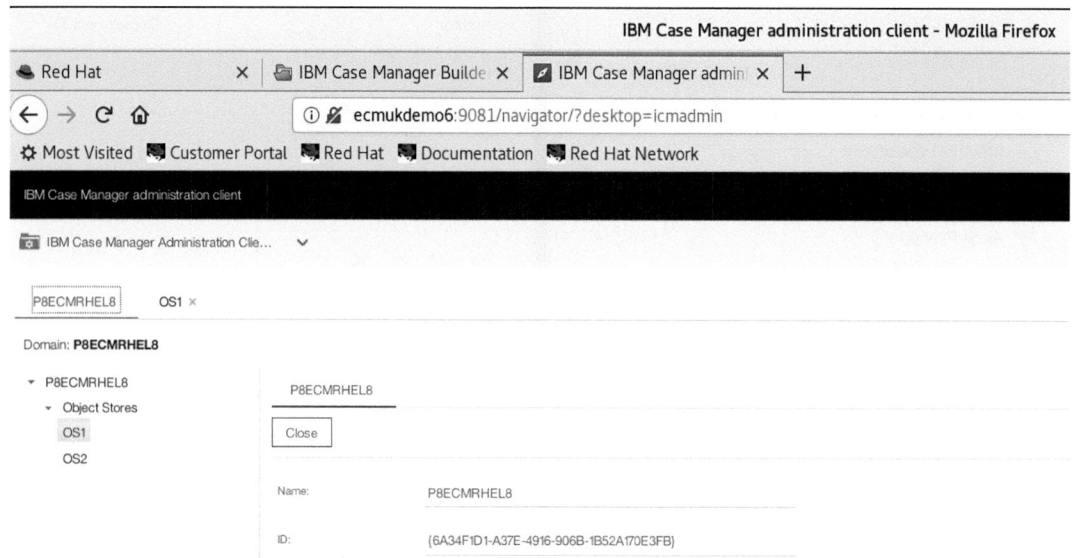

Figure 1-136. *The IBM Case Manager Administration Desktop is launched*

The Design Object store, **OS1**, in Figure 1-136, is clicked to show the Solutions drop-down as illustrated in Figure 1-137.

CHAPTER 1   IBM FILENET CASE MANAGER 5.3.3 CASE BUILDER SOLUTION DEVELOPMENT STEPS FOR THE AUDIT SYSTEM

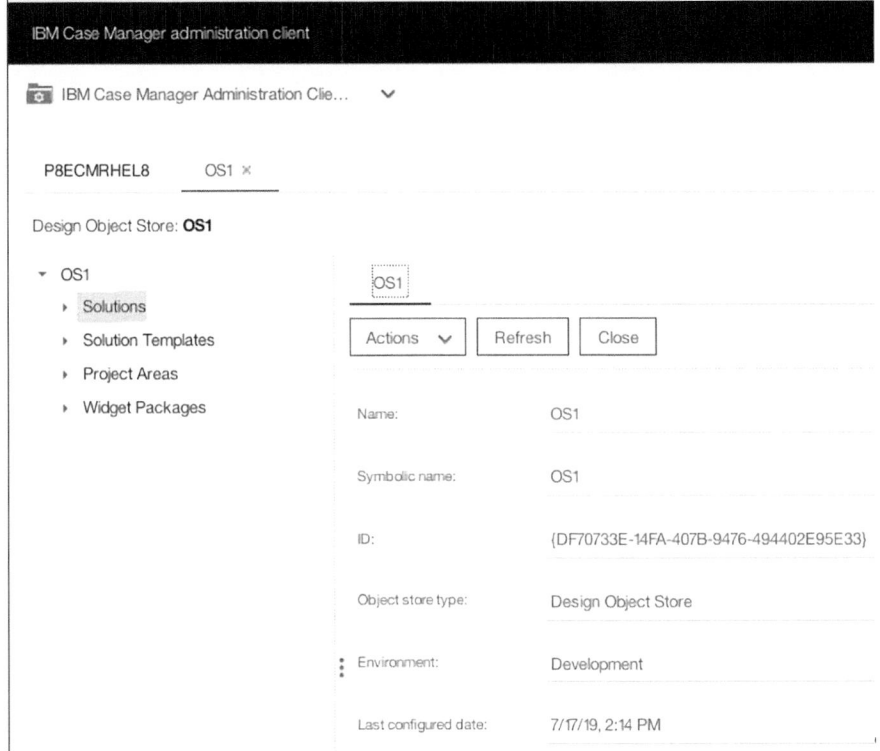

*Figure 1-137. The Design Object Store, **OS1**, menu options are displayed*

The drop-down **Solutions** is then clicked to show the list of IBM Case Manager Solutions, from which we can export the **Audit Master** Solution file we require.

Figure 1-138. The list of Solutions in the Design Object store

CHAPTER 1   IBM FILENET CASE MANAGER 5.3.3 CASE BUILDER SOLUTION DEVELOPMENT STEPS
            FOR THE AUDIT SYSTEM

The **Audit Master** solution is selected on the **Solutions** tab shown in Figure 1-138, and the **Actions** drop-down command button is clicked to select the **Export** option.

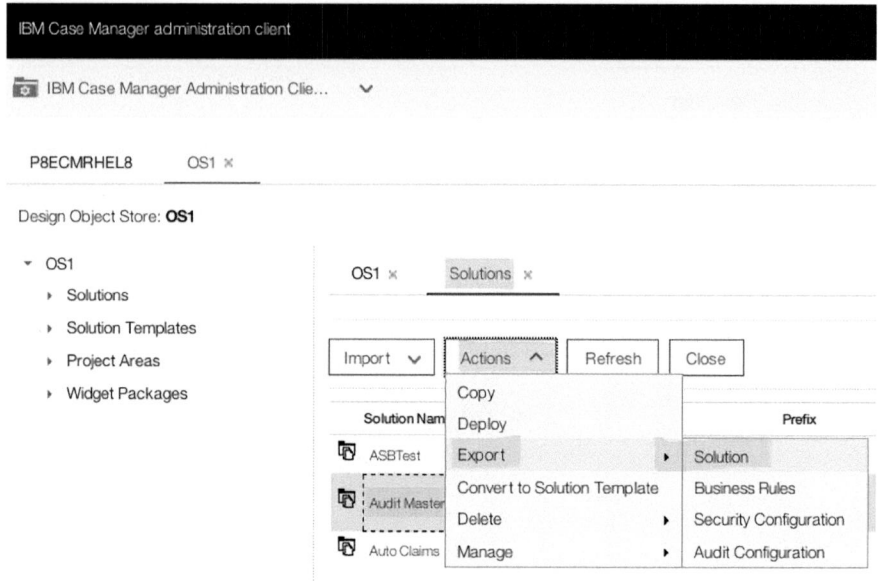

Figure 1-139. *The Audit Master Solution is selected for Export*

The Solution name and prefix values can (optionally) be changed for export as shown in the next screen in Figure 1-140.

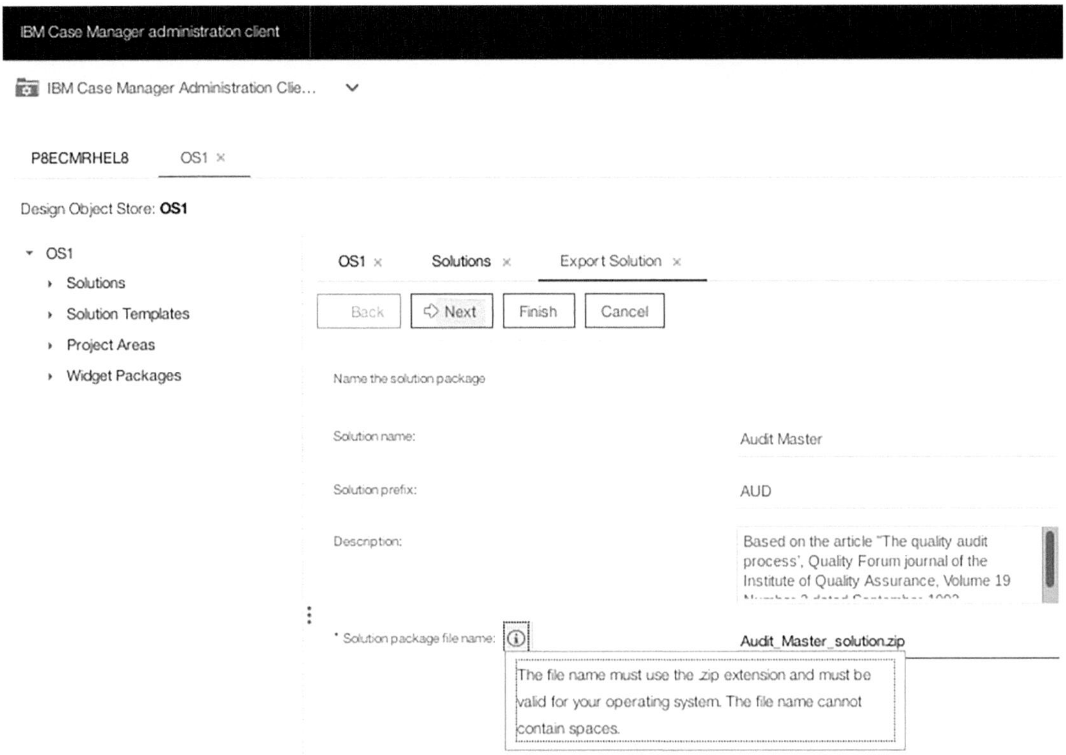

CHAPTER 1　IBM FILENET CASE MANAGER 5.3.3 CASE BUILDER SOLUTION DEVELOPMENT STEPS FOR THE AUDIT SYSTEM

Figure 1-140. The Next command is clicked after any required changes are made

The Next command is clicked after verifying the Export details are correct.

CHAPTER 1   IBM FILENET CASE MANAGER 5.3.3 CASE BUILDER SOLUTION DEVELOPMENT STEPS
             FOR THE AUDIT SYSTEM

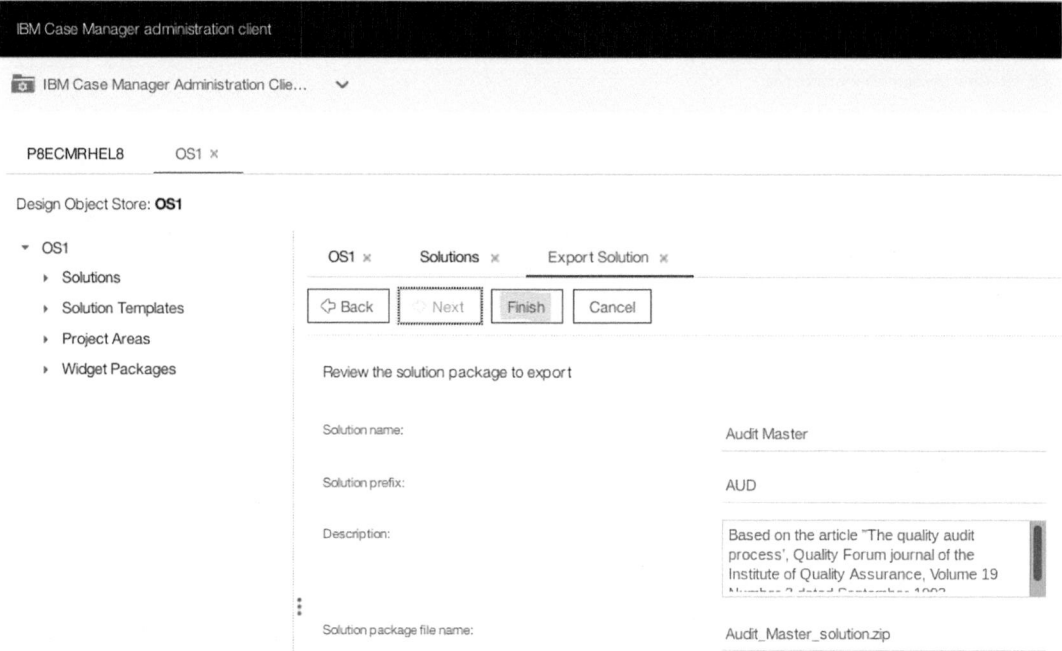

Figure 1-141. *The Finish command is clicked to start the Export to the Solution zip file*

When the **Finish** command is clicked, the **Audit_Master_solution.zip** file is created with the exported Solution parameters. This zip file can be used as a backup of the Solution during design and development, so that any changes which require backing out can easily be removed, since the saved .zip file can be used to import the backed up Solution and reset the Design Object store.

(In practice, the Design/Target Object Stores may need to be reinitialized to clear out some property templates which are not easy to remove programmatically before loading the backed up Solution.)

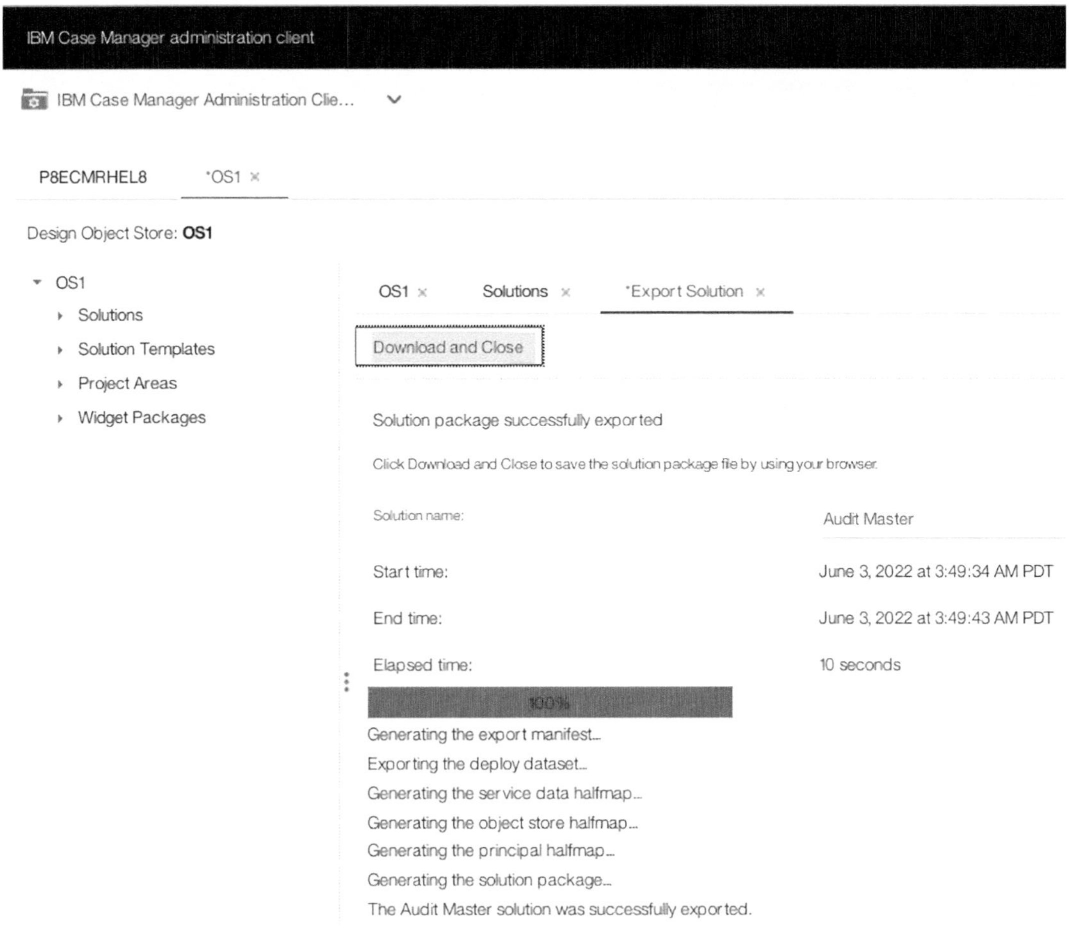

CHAPTER 1 IBM FILENET CASE MANAGER 5.3.3 CASE BUILDER SOLUTION DEVELOPMENT STEPS
           FOR THE AUDIT SYSTEM

Figure 1-142. The **Download and Close** command is executed after the solution is exported

The web browser download option is used to retrieve the exported Solution .zip file.

CHAPTER 1 IBM FILENET CASE MANAGER 5.3.3 CASE BUILDER SOLUTION DEVELOPMENT STEPS
FOR THE AUDIT SYSTEM

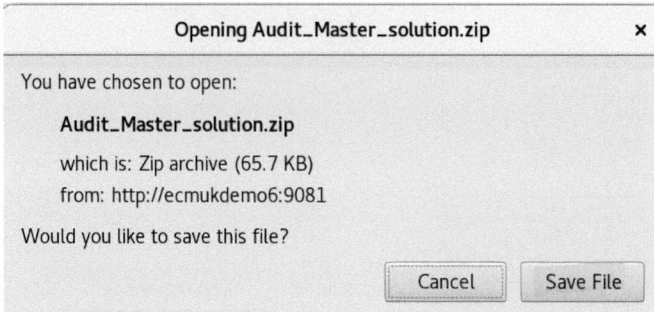

Figure 1-143. *The exported solution .zip file is saved to the Downloads area*

The exported **Audit_Master_solution.zip** file can now be copied to another test or production system to be imported.

Figure 1-144. *The exported zip file can be copied to a production system*

CHAPTER 1   IBM FILENET CASE MANAGER 5.3.3 CASE BUILDER SOLUTION DEVELOPMENT STEPS
               FOR THE AUDIT SYSTEM

Importing the Zipped Audit Master Solution to a Production System

The copied **Audit_Master_solution.zip** file is also imported using the IBM Case Manager administration web application.

Copy the exported Audit_Master_solution.zip file to a new folder /opt/AuditMaster.

```
(base) [root@ECMUKDEMO6 Downloads]# cd /opt/
(base) [root@ECMUKDEMO6 opt]# mkdir AuditMaster
(base) [root@ECMUKDEMO6 opt]# cp /mnt/hgfs/Installs/Audit_Master_solution.zip .
(base) [root@ECMUKDEMO6 opt]# mv Audit_Master_solution.zip ./Audit
AuditMaster/              Audit_Master_solution.zip
(base) [root@ECMUKDEMO6 opt]# mv Audit_Master_solution.zip ./AuditMaster/
(base) [root@ECMUKDEMO6 opt]# cd AuditMaster/
(base) [root@ECMUKDEMO6 AuditMaster]# ls
Audit_Master_solution.zip
(base) [root@ECMUKDEMO6 AuditMaster]#
```

Figure 1-145.   The exported zip file is copied to the production system

The zipped solution can be loaded into the Design Object store as shown in Figure 1-146.

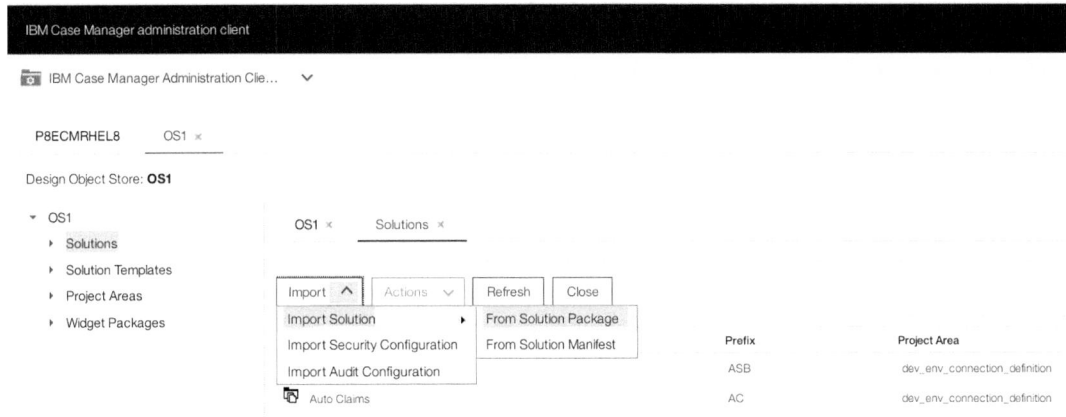

Figure 1-146.   The Import ➤ Import Solution ➤ From Solution Package option is selected

105

CHAPTER 1   IBM FILENET CASE MANAGER 5.3.3 CASE BUILDER SOLUTION DEVELOPMENT STEPS
            FOR THE AUDIT SYSTEM

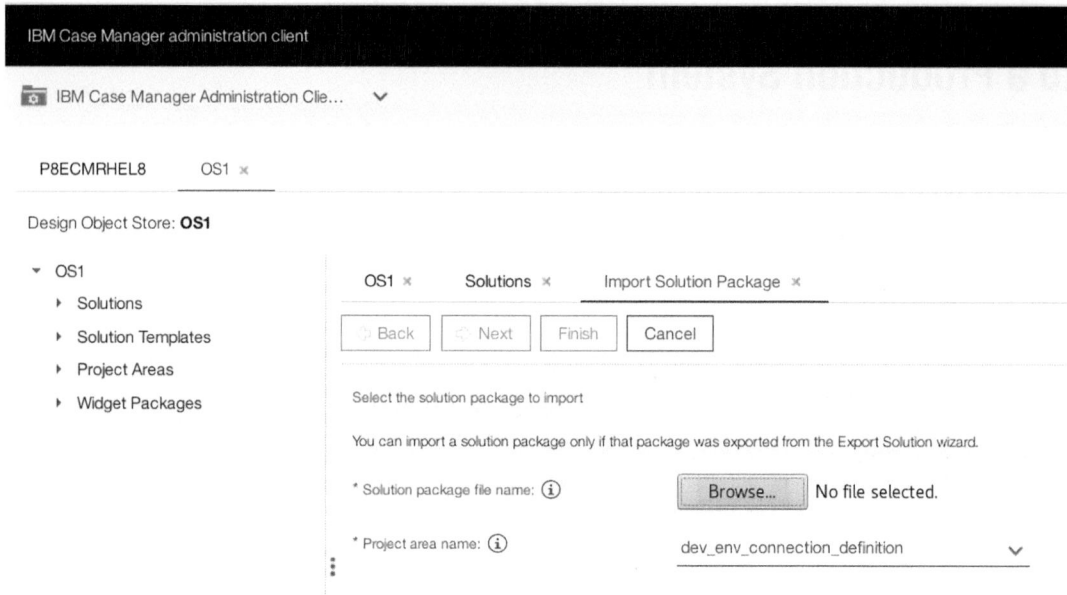

Figure 1-147. *The exported Audit Master Solution zip file is browsed to*

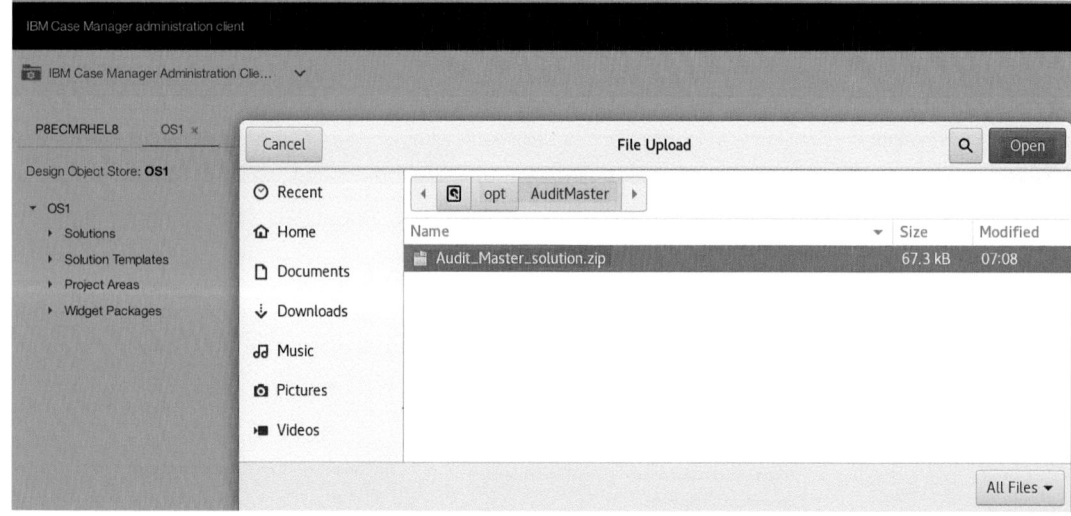

Figure 1-148. *The Audit_Master_solution.zip is selected to open*

CHAPTER 1 IBM FILENET CASE MANAGER 5.3.3 CASE BUILDER SOLUTION DEVELOPMENT STEPS
            FOR THE AUDIT SYSTEM

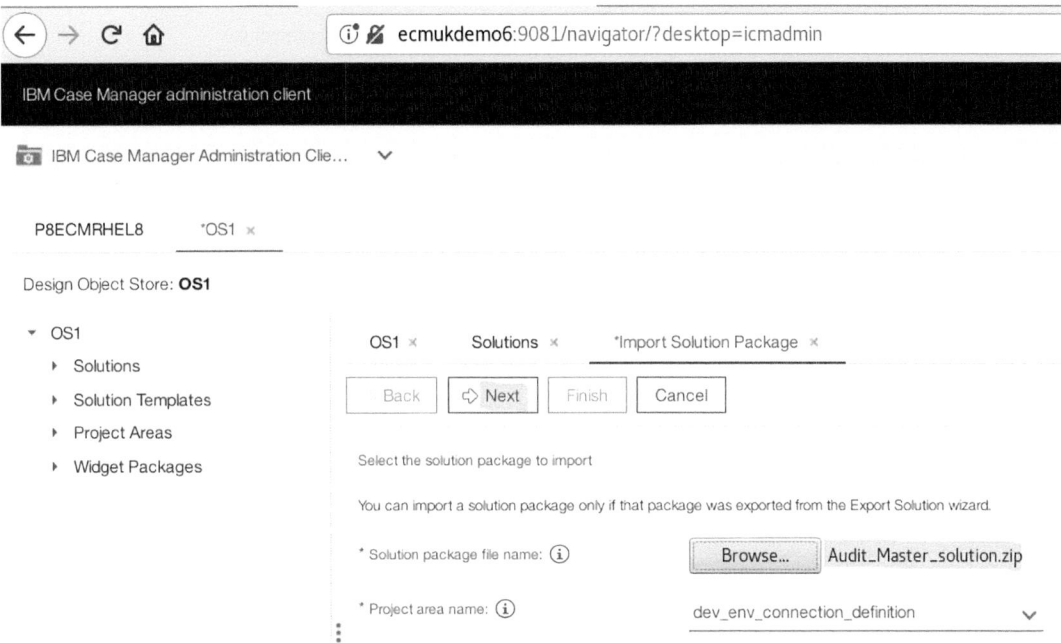

Figure 1-149. *The Next command is used to display the Audit Master solution attributes*

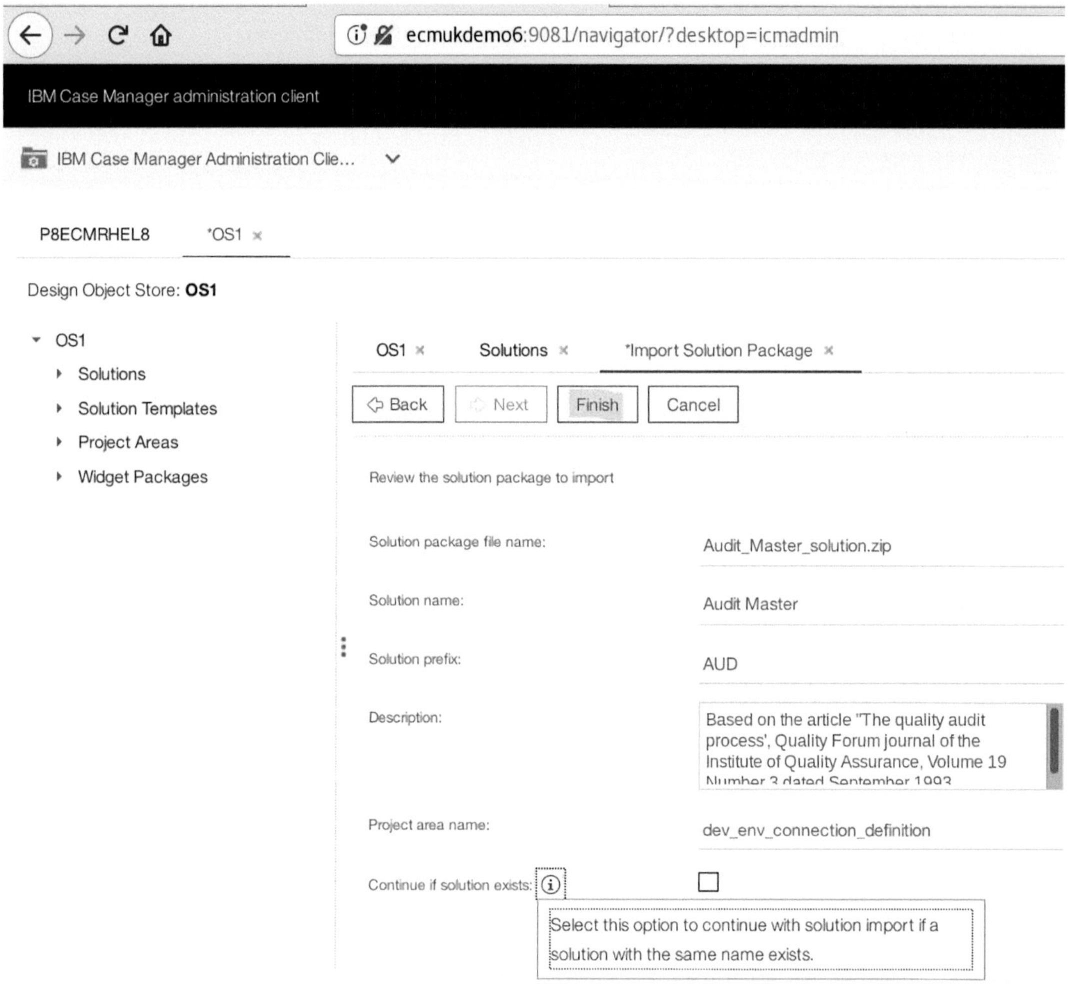

CHAPTER 1   IBM FILENET CASE MANAGER 5.3.3 CASE BUILDER SOLUTION DEVELOPMENT STEPS
            FOR THE AUDIT SYSTEM

Figure 1-150. The Audit Master zipped Solution attributes are displayed

CHAPTER 1　IBM FILENET CASE MANAGER 5.3.3 CASE BUILDER SOLUTION DEVELOPMENT STEPS
　　　　　　FOR THE AUDIT SYSTEM

The **Finish** command entered to start the Import process for Audit Master.

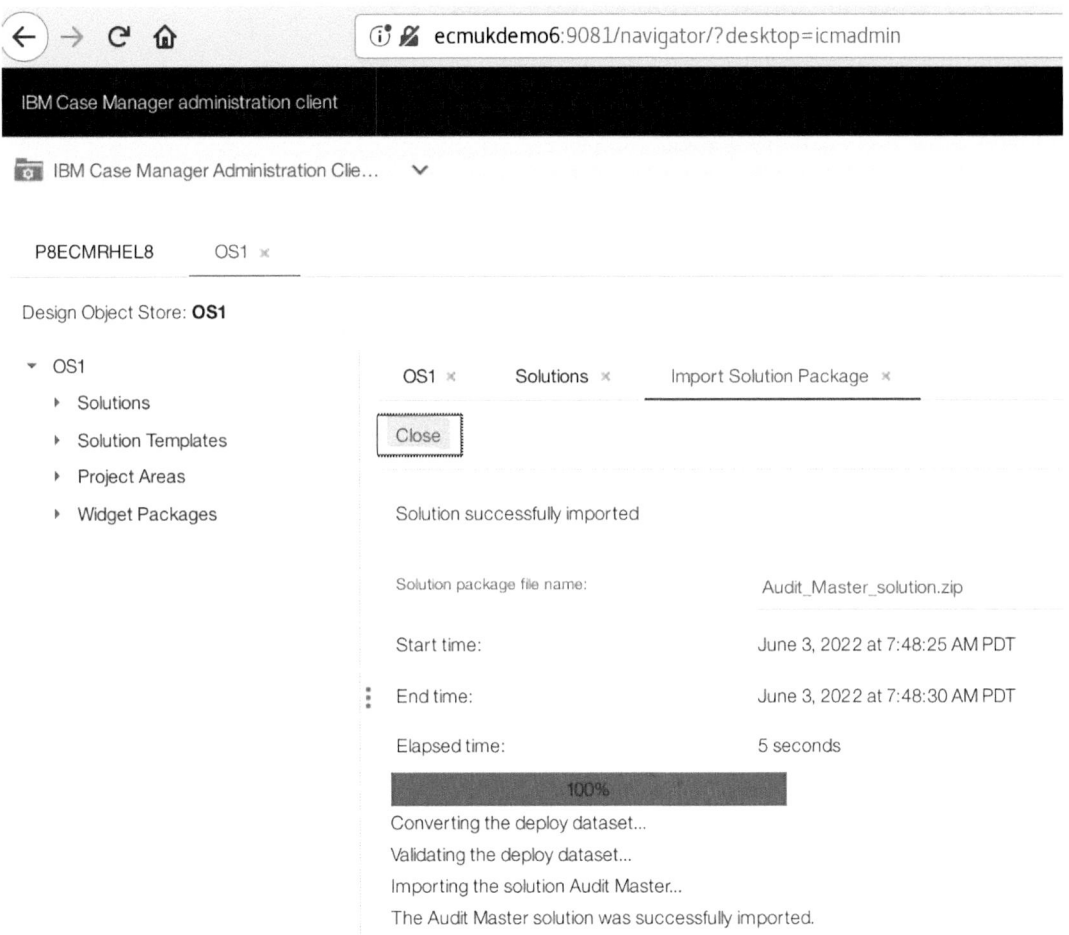

Figure 1-151. *The Audit Master Solution is successfully imported to the production system*

CHAPTER 1   IBM FILENET CASE MANAGER 5.3.3 CASE BUILDER SOLUTION DEVELOPMENT STEPS
            FOR THE AUDIT SYSTEM

The **Close** command button can now be clicked on the Figure 1-151 page.

The Audit Master Solution is now added to the **Design** Object Store, **OS1**, but now we need to deploy it to the **Target** Object Store, **OS2**.

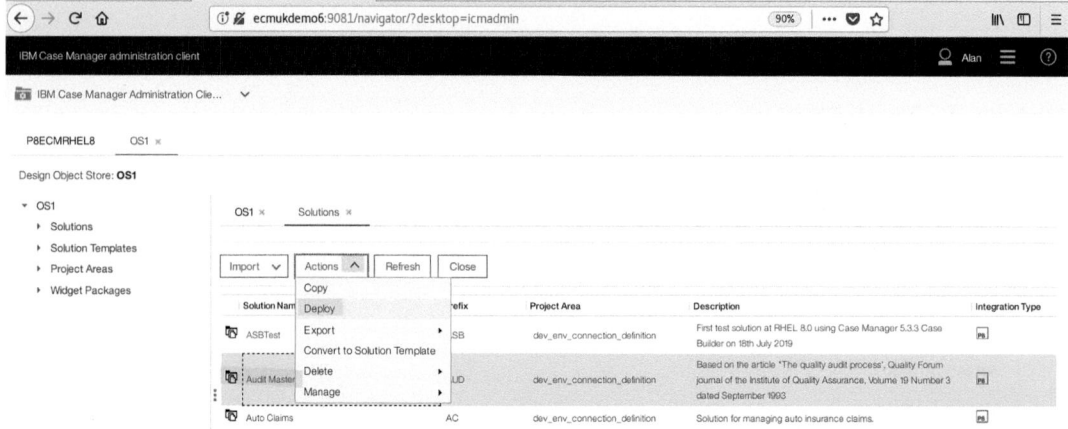

Figure 1-152. The Audit Master solution is deployed to the Target Object Store

CHAPTER 1   IBM FILENET CASE MANAGER 5.3.3 CASE BUILDER SOLUTION DEVELOPMENT STEPS
           FOR THE AUDIT SYSTEM

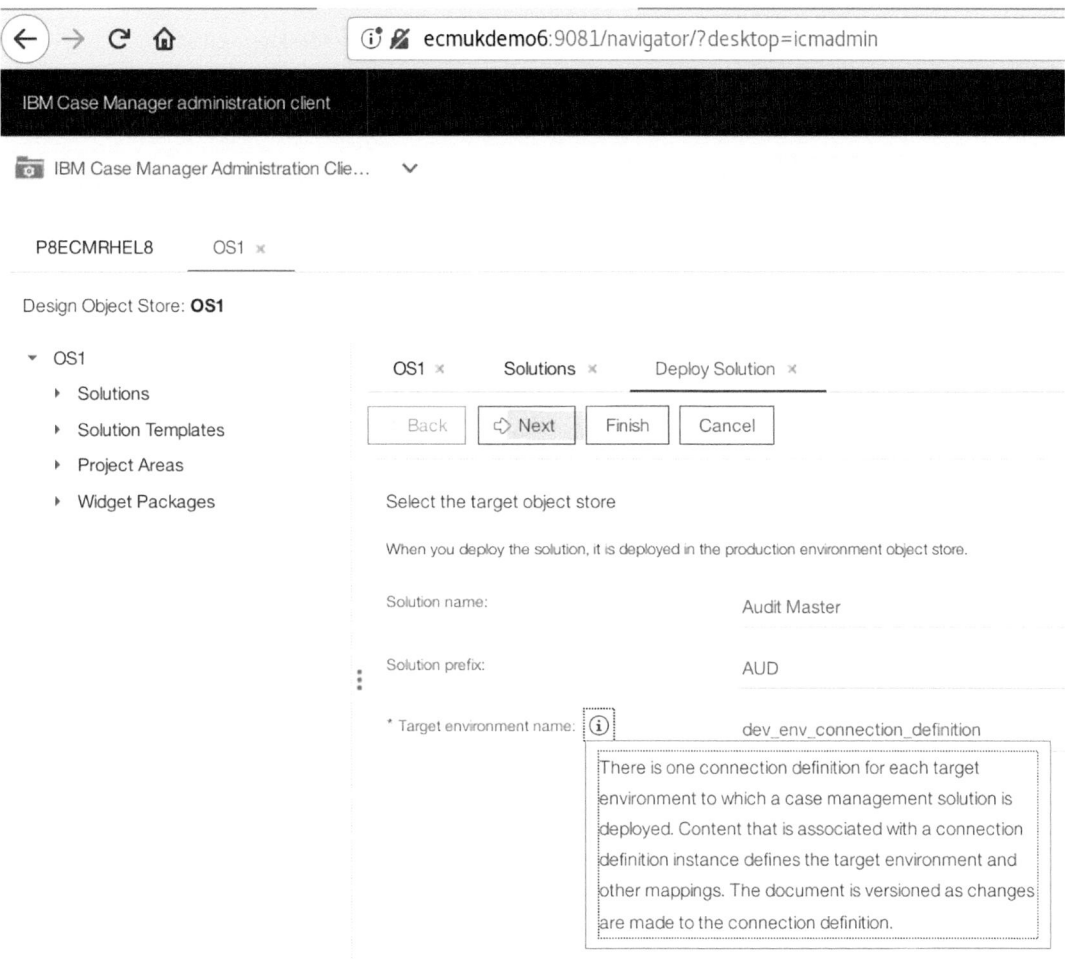

Figure 1-153. *The Next command is used to view the deployment attributes*

CHAPTER 1   IBM FILENET CASE MANAGER 5.3.3 CASE BUILDER SOLUTION DEVELOPMENT STEPS
            FOR THE AUDIT SYSTEM

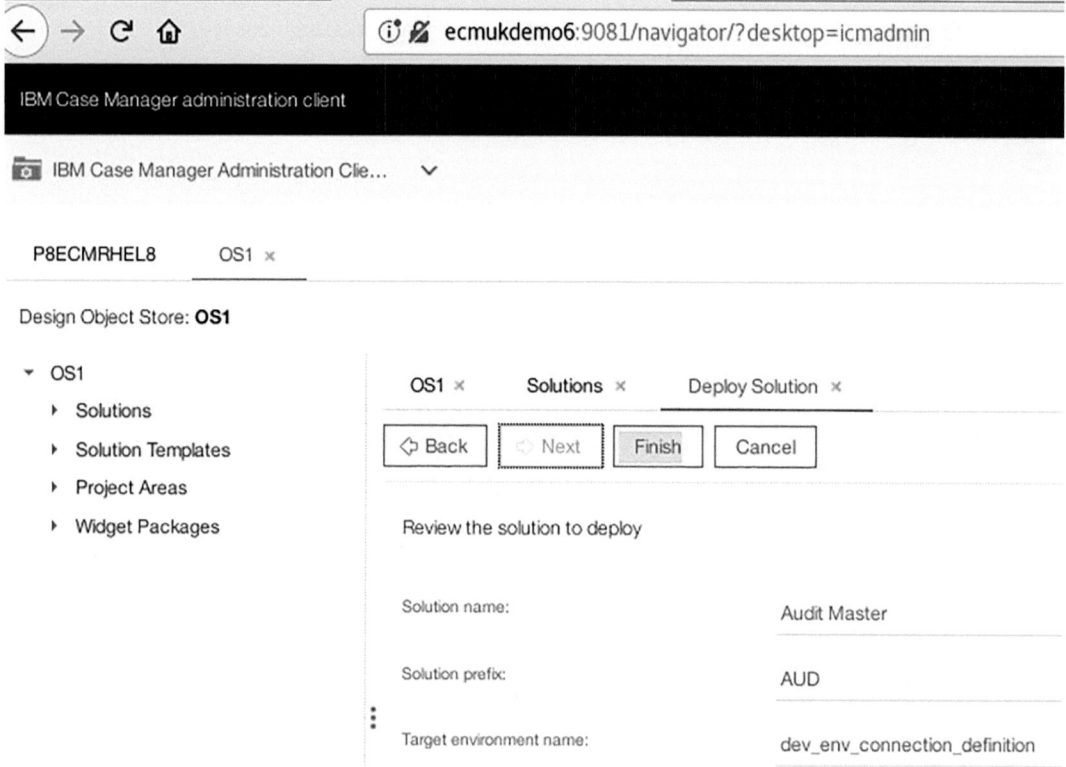

Figure 1-154. *The Finish command is clicked to deploy the imported Solution*

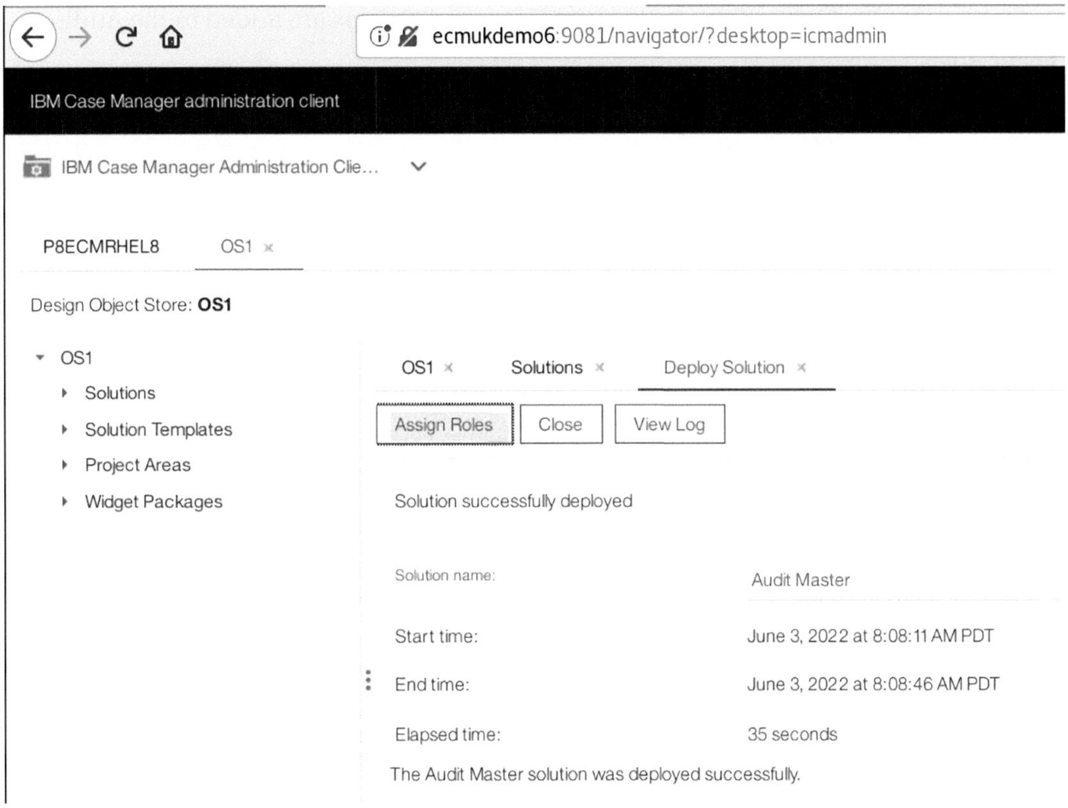

CHAPTER 1   IBM FILENET CASE MANAGER 5.3.3 CASE BUILDER SOLUTION DEVELOPMENT STEPS
            FOR THE AUDIT SYSTEM

Figure 1-155. The Audit Master solution is successfully deployed

After a successful deployment to the Production Target Object Store, the Assign Roles command button launches a separate IBM Case Manager Client window to allow the Production Role security to be configured.

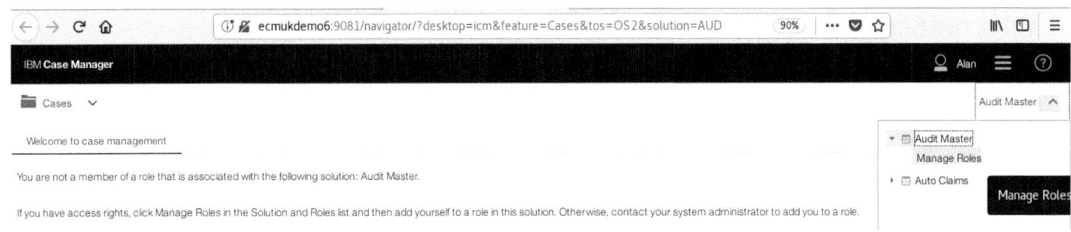

Figure 1-156. The IBM Case Manager Client Manage Roles is selected

113

CHAPTER 1   IBM FILENET CASE MANAGER 5.3.3 CASE BUILDER SOLUTION DEVELOPMENT STEPS
            FOR THE AUDIT SYSTEM

Finally, the Production System Security Users and Groups are added to the Audit Master Roles.

Figure 1-157.   The Production Users and Groups are added to the Audit Master solution roles

Debugging the Case Manager Client

This section covers an earlier issue which was fixed by adding a missing component for the IBM FileNet Content Server Target Object Store, used by IBM Case Manager.

REF: www.ibm.com/support/knowledgecenter/en/SSCTJ4_5.3.3/com.ibm.casemgmt.design.doc/acmta044.htm

To configure logging for **IBM Case Manager** web applications, you must enable debugging on the **IBM Content Navigator** server and then restart the **IBM Content Navigator** server.

Procedure

To configure logging settings in IBM Content Navigator:

Log in to the IBM Content Navigator administration desktop.

Click the Settings tab and then click the Logging subtab.

CHAPTER 1   IBM FILENET CASE MANAGER 5.3.3 CASE BUILDER SOLUTION DEVELOPMENT STEPS
            FOR THE AUDIT SYSTEM

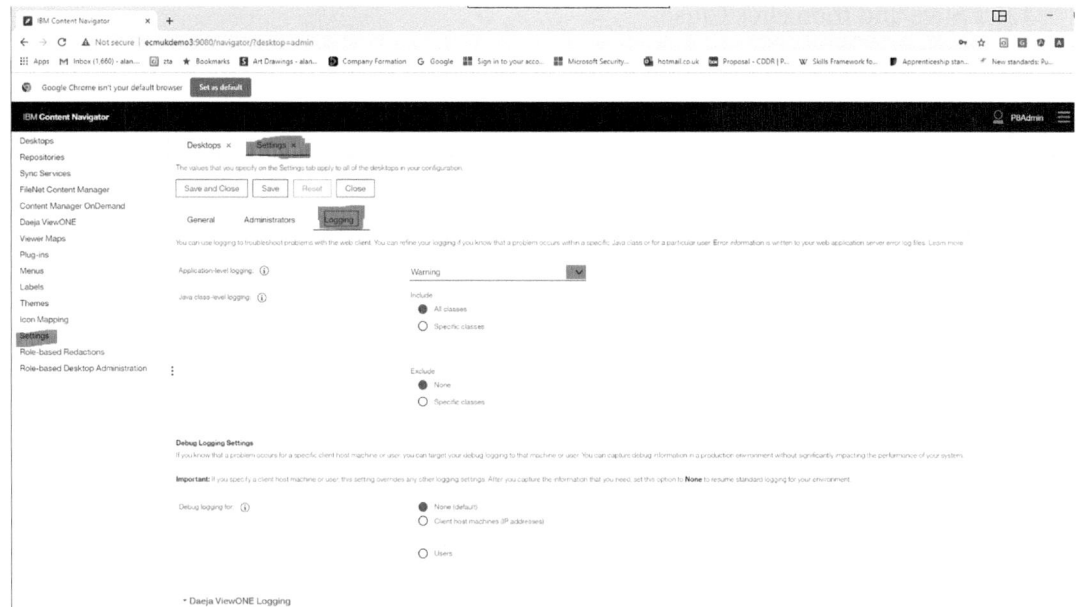

*Figure 1-158. The **Logging** tab of the **Settings** menu in the admin desktop is selected*

Set the logging level for the applications to Debug.

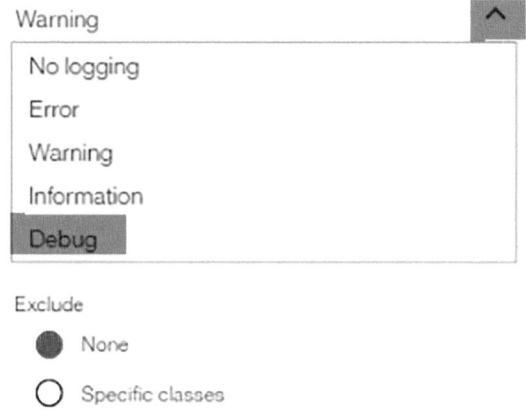

*Figure 1-159. The **Debug** level option is selected to replace the **Warning** level*

CHAPTER 1   IBM FILENET CASE MANAGER 5.3.3 CASE BUILDER SOLUTION DEVELOPMENT STEPS
             FOR THE AUDIT SYSTEM

Click Save and then click Close.

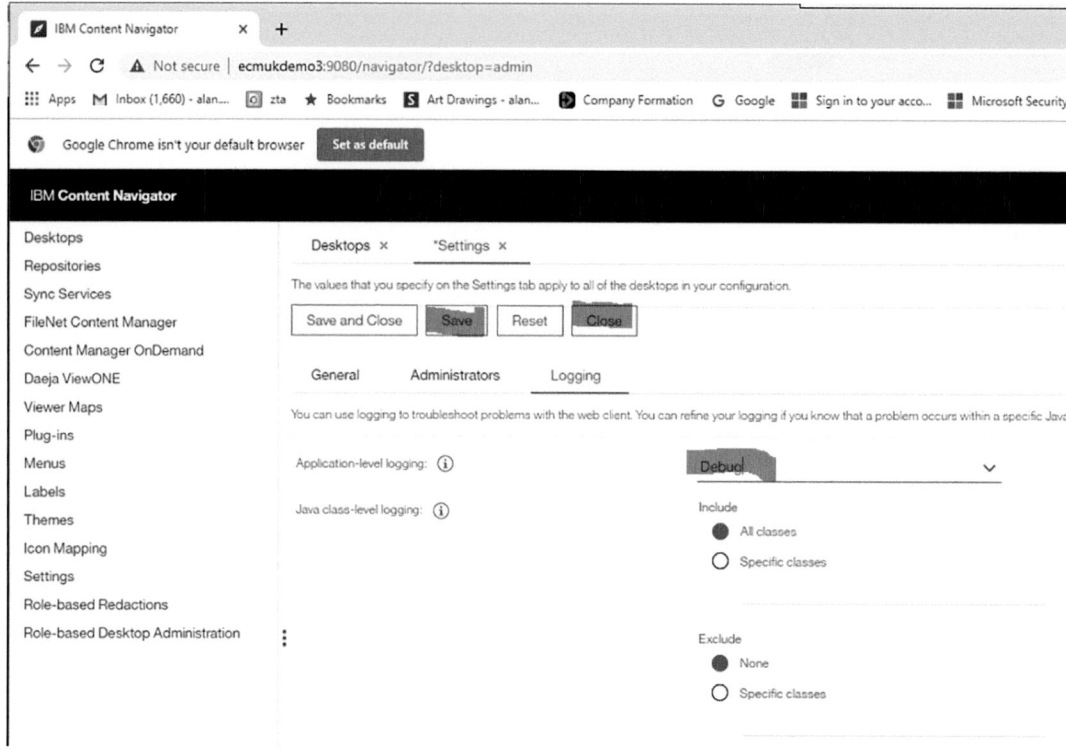

*Figure 1-160.   The **Logging** level is now displayed as **Debug***

Log out of the IBM Content Navigator administration desktop.
Restart the application server instance where IBM Content Navigator is deployed.
In the SystemOut.log file, I get

```
[REQUEST 19] com.ibm.ecm.struts.actions.p8.P8SearchAction.executeAction()
```
**com.filenet.api.exception.EngineRuntimeException: FNRCR0081E: RETRIEVE_
PROPERTY_NOT_DEFINED: The property ContainerType is not defined.**
```
ObjectStore: "OS2", SQL: "SELECT t.[FolderName], t.[CmAcmHealthIndicator],
t.[LastModifier], t.[DateLastModified], t.[CmAcmCaseTypeFolder],
t.[CmAcmCaseState], t.[CmAcmCaseIdentifier], t.[DateCreated], t.[Creator],
t.[Id], t.[ASB_TestString], t.[ClassDescription],
```
t.[ContainerType],
```
t.[LockToken], t.[LockTimeout] FROM [ASB_Test] t WHERE t.[ASB_TestString]
LIKE 'ASB%%' and t.[CmAcmCaseState] > 1 ORDER BY t.[CmAcmCaseIdentifier]
OPTIONS ( COUNT_LIMIT 2147483647 )" errorStack={
```

CHAPTER 1   IBM FILENET CASE MANAGER 5.3.3 CASE BUILDER SOLUTION DEVELOPMENT STEPS
            FOR THE AUDIT SYSTEM

Analysis with acce showed:

[REQUEST 19] com.ibm.ecm.struts.actions.p8.P8SearchAction.executeAction()
com.filenet.api.exception.EngineRuntimeException: FNRCR0081E: RETRIEVE_
PROPERTY_NOT_DEFINED:
The property **ContainerType** is not defined. ObjectStore: "OS2", SQL:

```
                                        Exists
"SELECT t.[FolderName],                 Yes
        t.[CmAcmHealthIndicator],       Yes
        t.[LastModifier],               Yes
        t.[DateLastModified],           Yes
        t.[CmAcmCaseTypeFolder],        Yes
        t.[CmAcmCaseState],             Yes
        t.[CmAcmCaseIdentifier],        Yes
        t.[DateCreated],                Yes
        t.[Creator],                    Yes
        t.[Id],                         Yes
        t.[ASB_TestString],             Yes
        t.[ClassDescription],           Yes
        t.[ContainerType],              No
        t.[LockToken],                  Yes
        t.[LockTimeout]                 Yes

FROM [ASB_Test] t
WHERE t.[ASB_TestString] LIKE 'ASB%%'
and t.[CmAcmCaseState] > 1
ORDER BY t.[CmAcmCaseIdentifier] OPTIONS ( COUNT_LIMIT 2147483647 )"
errorStack={
```

CHAPTER 1   IBM FILENET CASE MANAGER 5.3.3 CASE BUILDER SOLUTION DEVELOPMENT STEPS
            FOR THE AUDIT SYSTEM

Checked FileNet Version

http://ecmukdemo6:9080/FileNet/Engine

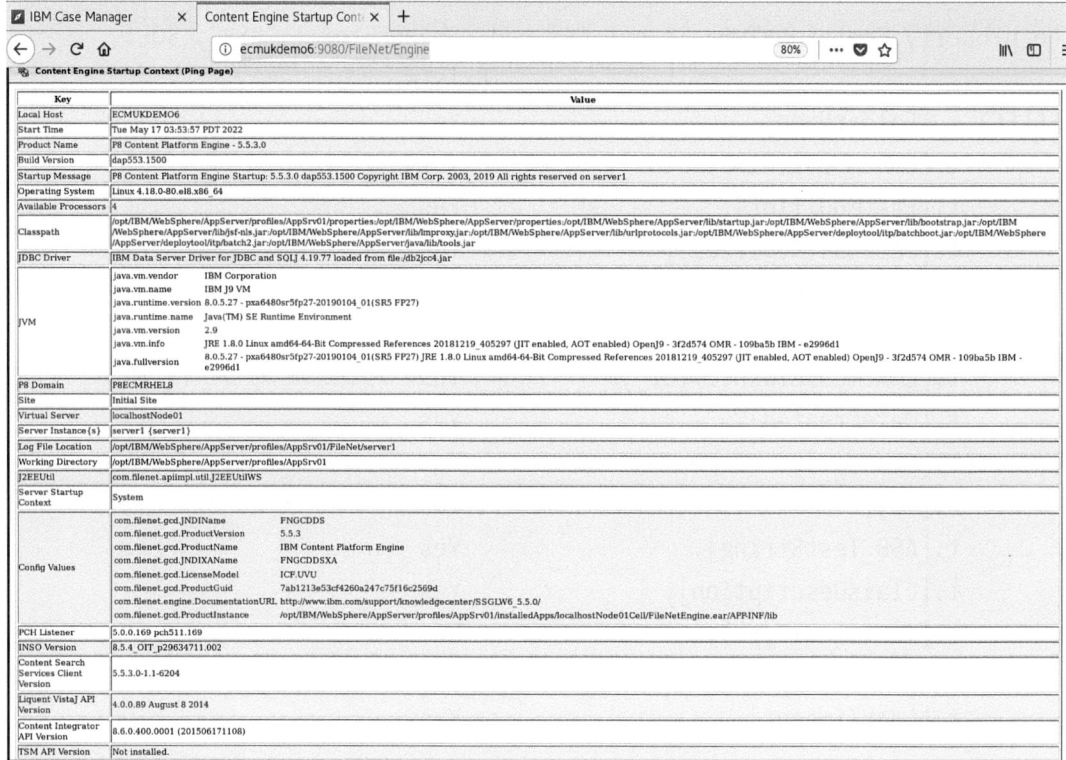

Figure 1-161.   *The FileNet version details are checked*

CHAPTER 1   IBM FILENET CASE MANAGER 5.3.3 CASE BUILDER SOLUTION DEVELOPMENT STEPS
            FOR THE AUDIT SYSTEM

Checked FileNet Health

http://ecmukdemo6:9080/P8CE/Health

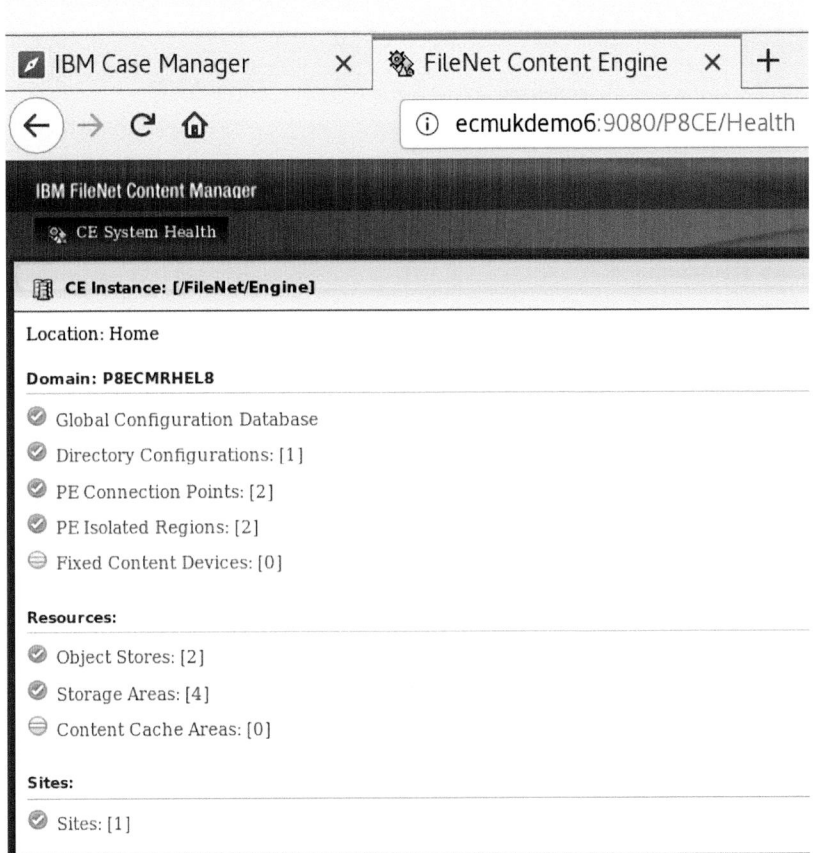

Figure 1-162.   The FileNet Health status is checked and validated by the system

CHAPTER 1　IBM FILENET CASE MANAGER 5.3.3 CASE BUILDER SOLUTION DEVELOPMENT STEPS FOR THE AUDIT SYSTEM

Check FileNet Object Store Upgrade Status

http://ecmukdemo6:9080/FileNet/AutomaticUpgradeStatus

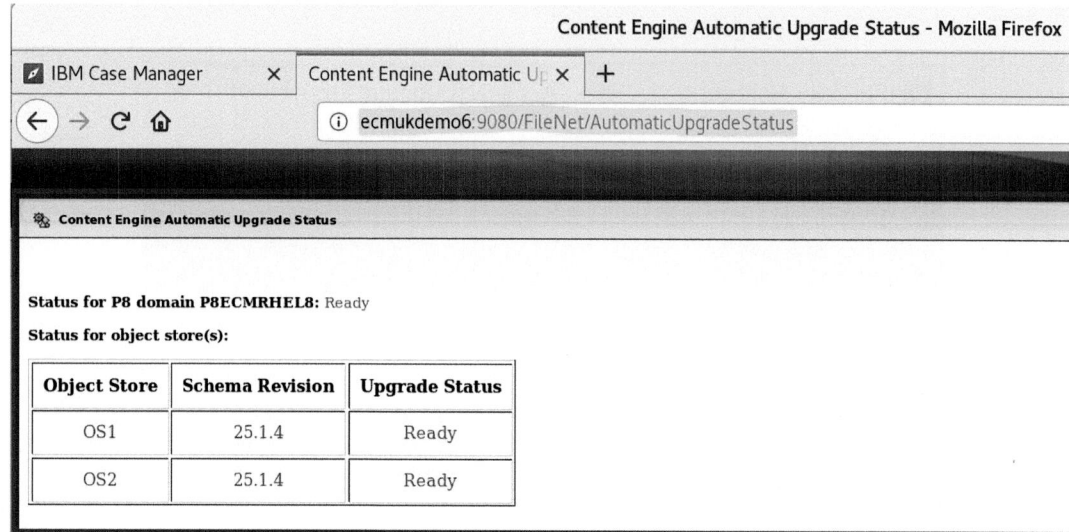

Figure 1-163. The IBM FileNet Object Store Upgrade status is checked

Upgrading Status Meaning

Visit the following URL to see the description of the status values shown:
　　www.ibm.com/support/knowledgecenter/en/SSGLW6_5.5.0/com.ibm.p8.install.doc/p8pup323.htm

Check FileNet Workflow Upgrade Status

http://ecmukdemo6:9080/peengine/IOR/ping
　　NB: You will be prompted to log in using the administration user (or other workflow Administrator accounts).

CHAPTER 1   IBM FILENET CASE MANAGER 5.3.3 CASE BUILDER SOLUTION DEVELOPMENT STEPS FOR THE AUDIT SYSTEM

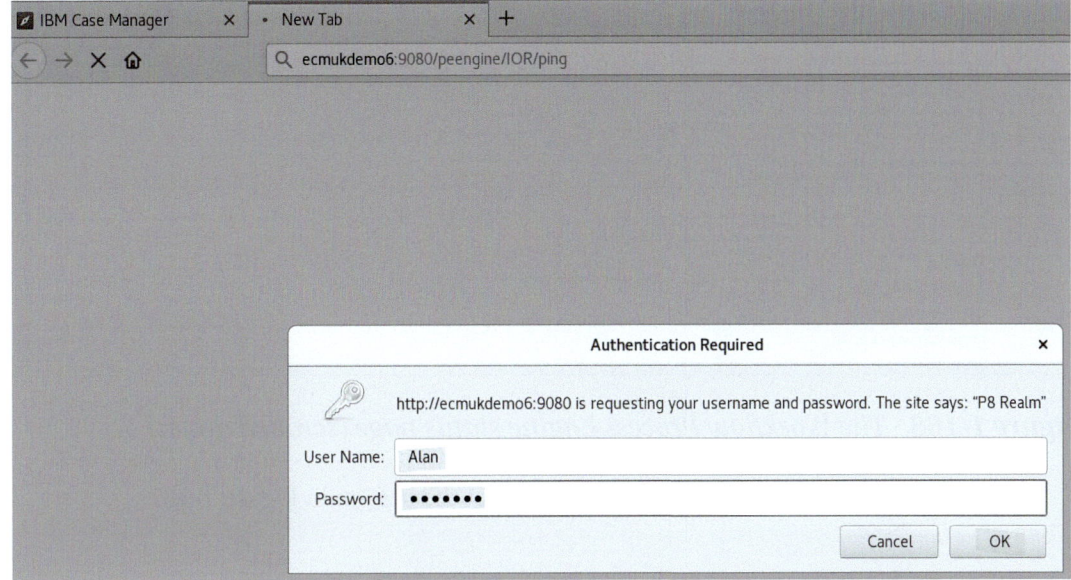

Figure 1-164. *Log in to the Workflow Process Engine status page*

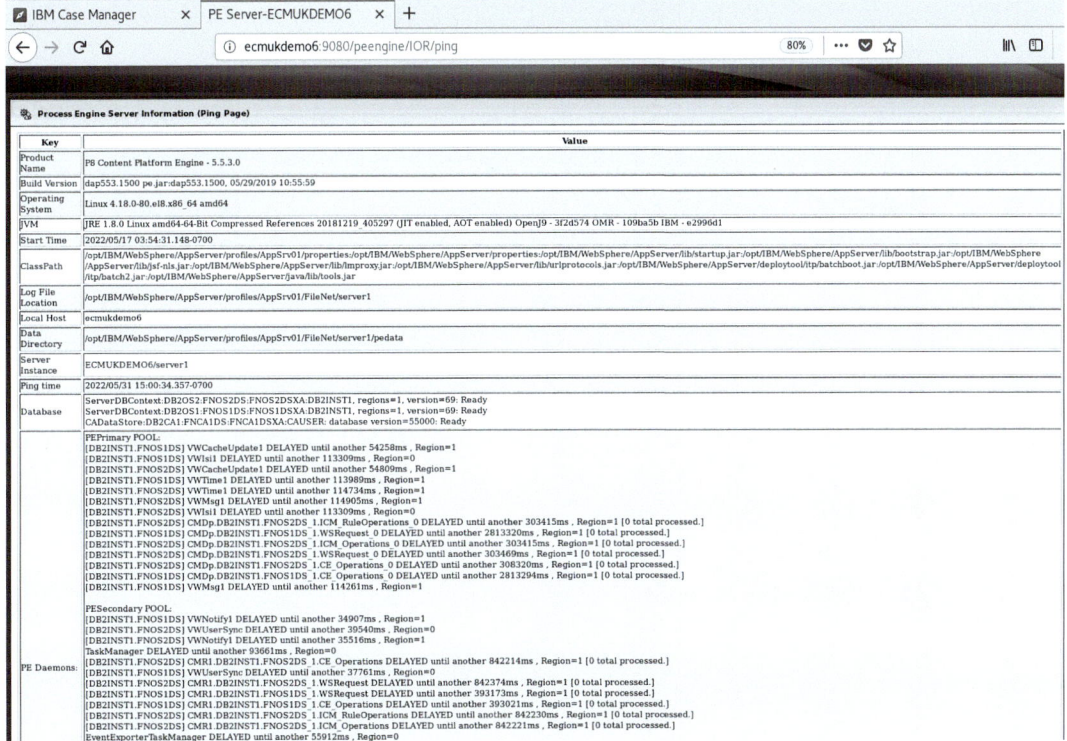

Figure 1-165. *The Workflow Process Engine status page*

121

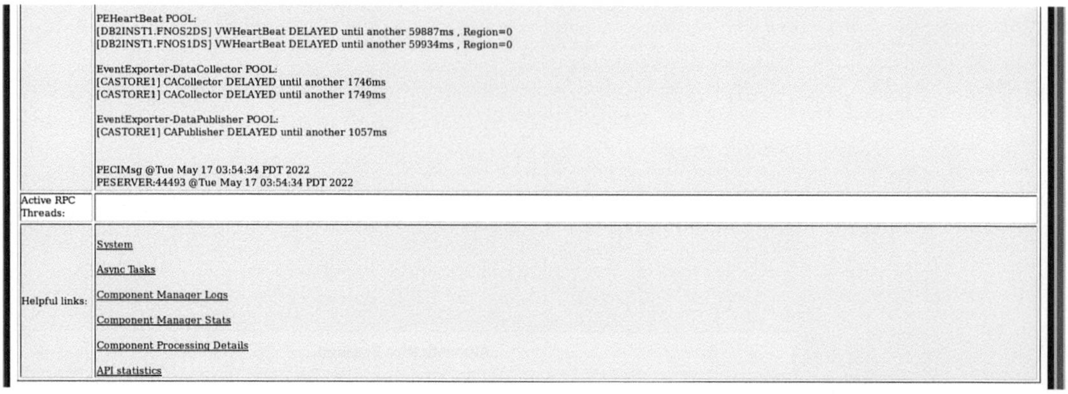

CHAPTER 1 IBM FILENET CASE MANAGER 5.3.3 CASE BUILDER SOLUTION DEVELOPMENT STEPS FOR THE AUDIT SYSTEM

```
PEHeartBeat POOL:
[DB2INST1.FNOS2DS] VWHeartBeat DELAYED until another 59887ms , Region=0
[DB2INST1.FNOS1DS] VWHeartBeat DELAYED until another 59934ms , Region=0

EventExporter-DataCollector POOL:
[CASTORE1] CACollector DELAYED until another 1746ms
[CASTORE1] CACollector DELAYED until another 1749ms

EventExporter-DataPublisher POOL:
[CASTORE1] CAPublisher DELAYED until another 1057ms

PECIMsg @Tue May 17 03:54:34 PDT 2022
PESERVER:44493 @Tue May 17 03:54:34 PDT 2022
```

Active RPC Threads:

Helpful links:
- System
- Async Tasks
- Component Manager Logs
- Component Manager Stats
- Component Processing Details
- API statistics

Figure 1-166. *The Workflow Process Engine status page (scrolled down)*

Clicking the **System** link in Figure 1-166, we get the screen in Figure 1-167.

CHAPTER 1   IBM FILENET CASE MANAGER 5.3.3 CASE BUILDER SOLUTION DEVELOPMENT STEPS
            FOR THE AUDIT SYSTEM

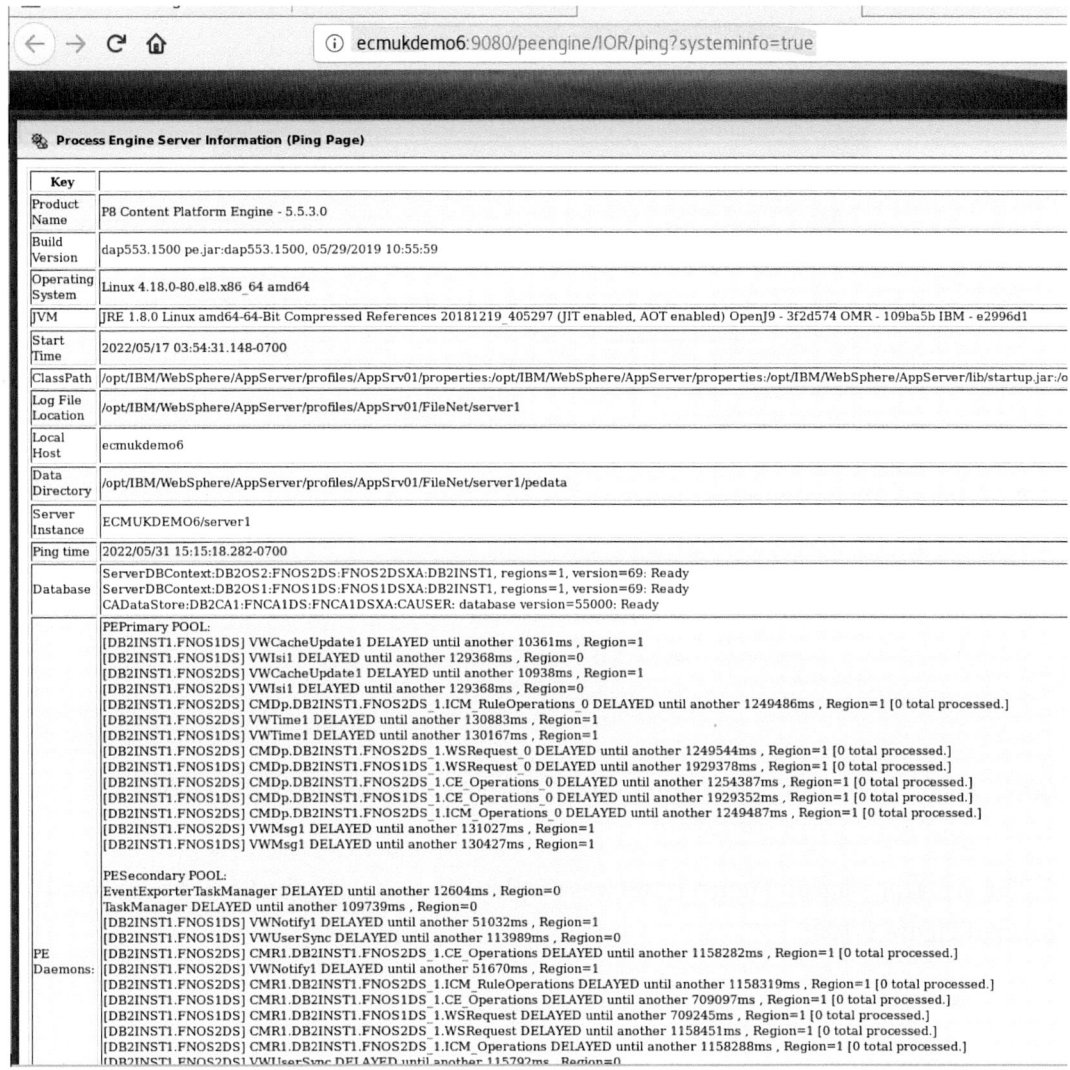

Figure 1-167. The Process Engine status System Information details

Fix/Workaround for Error on Case Search

Issue Found in Case Manager Client Search in the Solution

Initially, I get the error in Figure 1-168 trying to search a Case that appears to be created correctly.

CHAPTER 1   IBM FILENET CASE MANAGER 5.3.3 CASE BUILDER SOLUTION DEVELOPMENT STEPS FOR THE AUDIT SYSTEM

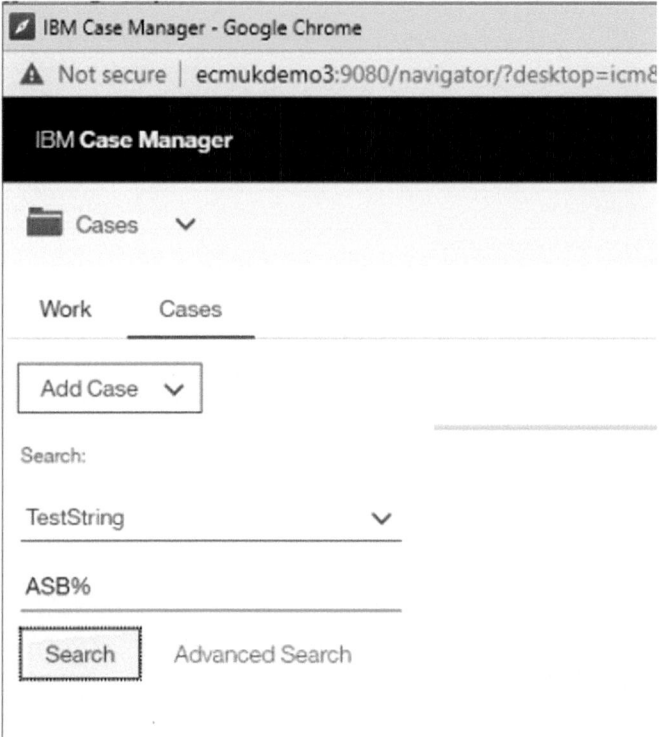

Figure 1-168. *The initial version of the deployed solution was searched*

The error message in Figure 1-169 was displayed on clicking the Search command shown in Figure 1-168.

CHAPTER 1 IBM FILENET CASE MANAGER 5.3.3 CASE BUILDER SOLUTION DEVELOPMENT STEPS
FOR THE AUDIT SYSTEM

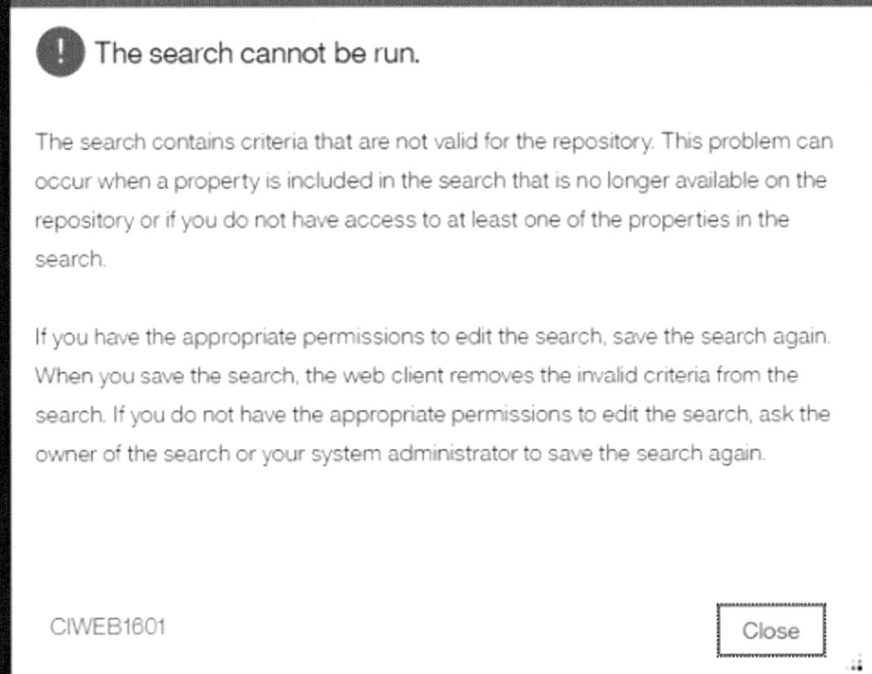

Figure 1-169. The error window was shown on using the Search command

Initial Workaround

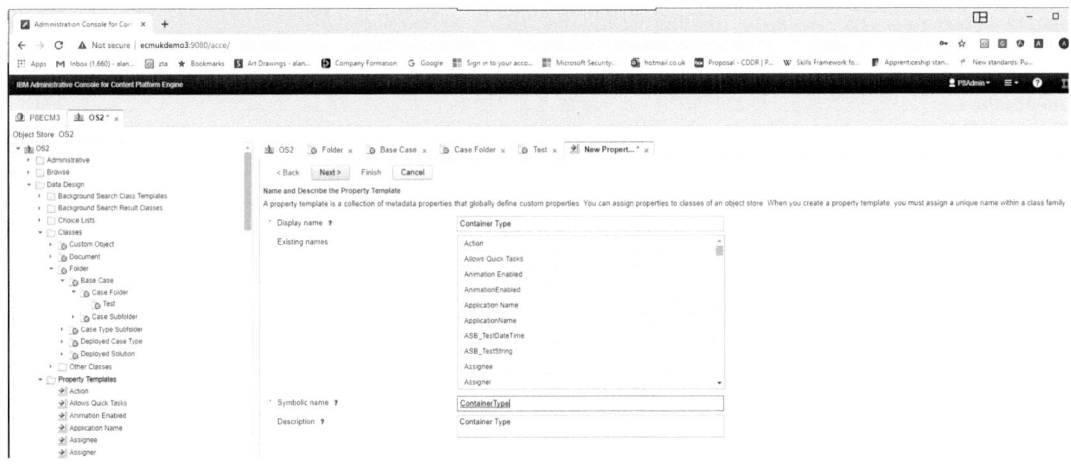

*Figure 1-170. The **ContainerType** Property Template was added to the Object Store*

CHAPTER 1 IBM FILENET CASE MANAGER 5.3.3 CASE BUILDER SOLUTION DEVELOPMENT STEPS
         FOR THE AUDIT SYSTEM

Add a Property Template with a Symbolic Name of **ContainerType**.

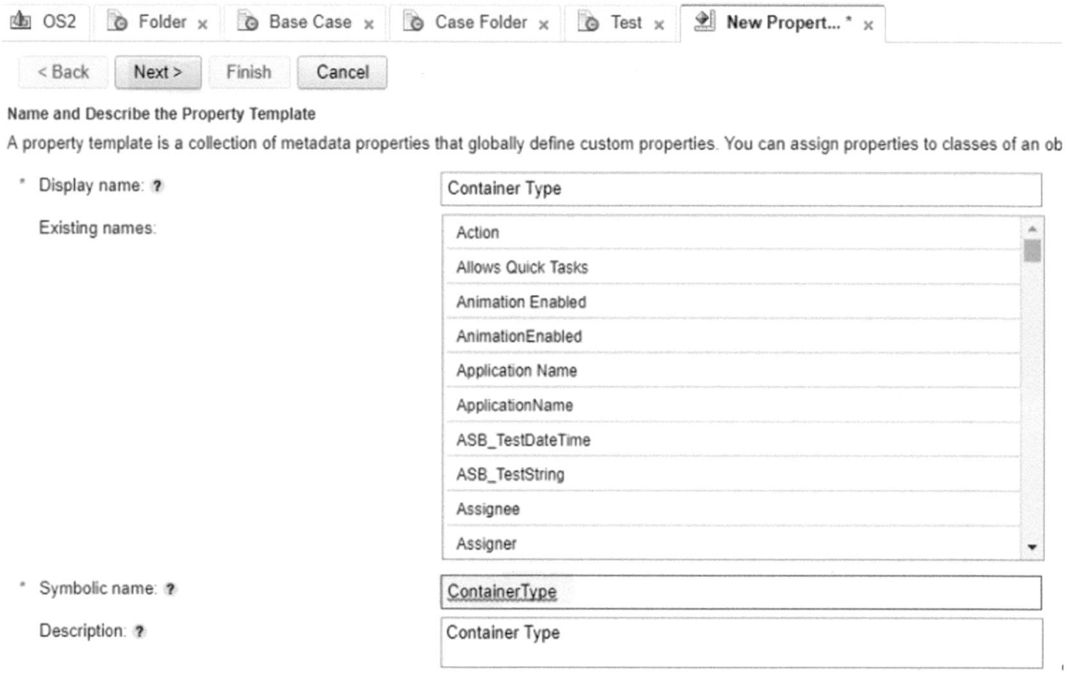

*Figure 1-171. A Property Template with a Symbolic Name of **ContainerType** was added*

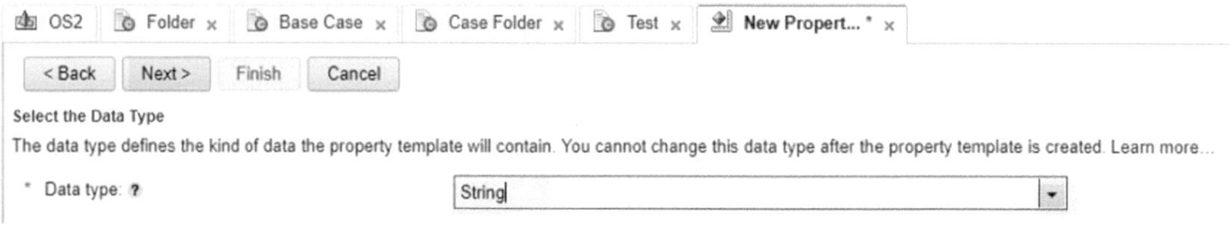

Figure 1-172. The Data type was selected as String

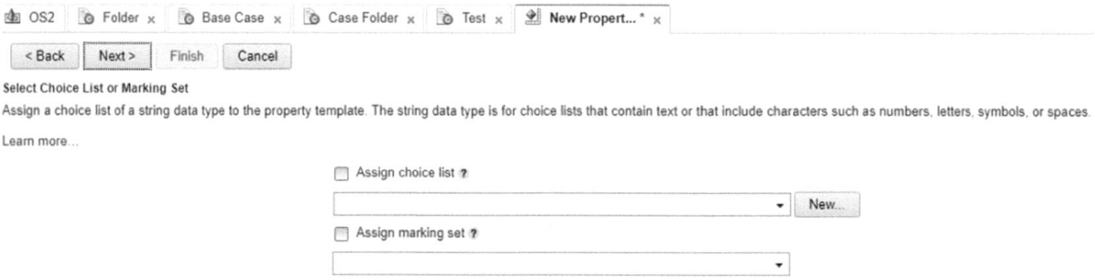

Figure 1-173. The Choice list values were left empty

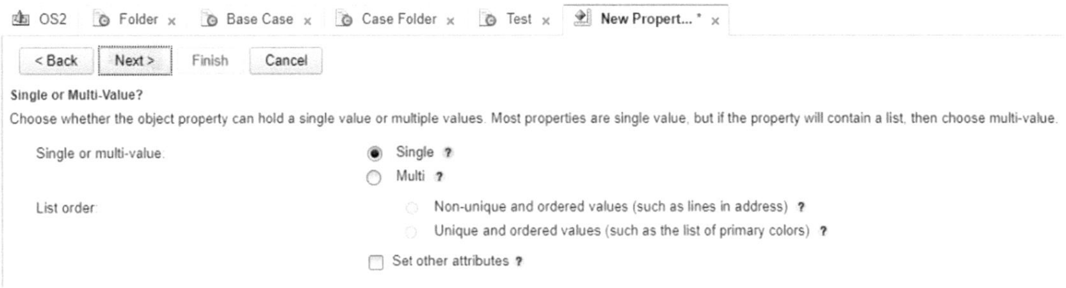

CHAPTER 1　IBM FILENET CASE MANAGER 5.3.3 CASE BUILDER SOLUTION DEVELOPMENT STEPS FOR THE AUDIT SYSTEM

*Figure 1-174.　The **Single** option default was set for the value type*

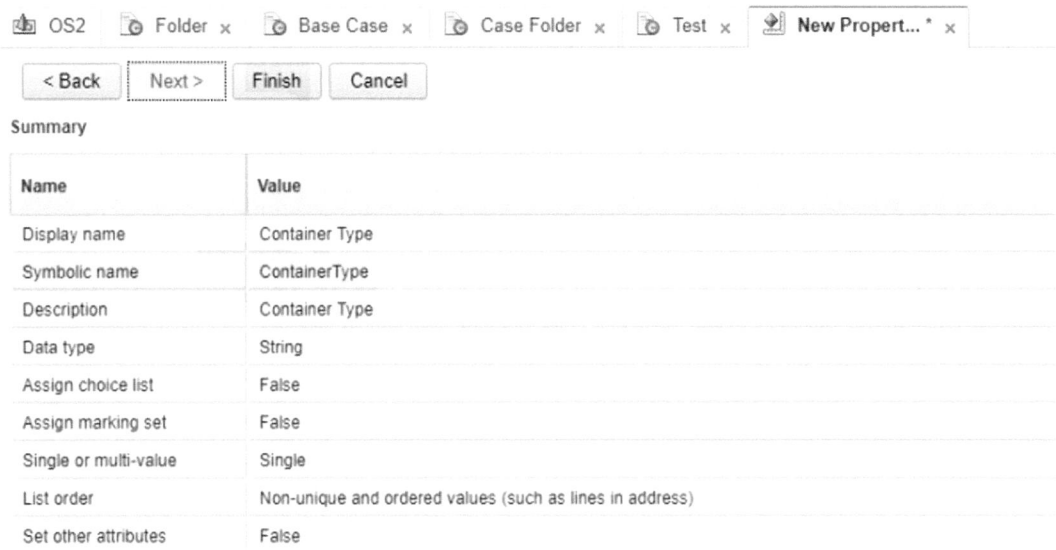

*Figure 1-175.　The reviewed properties for the **ContainerType** Property Template*

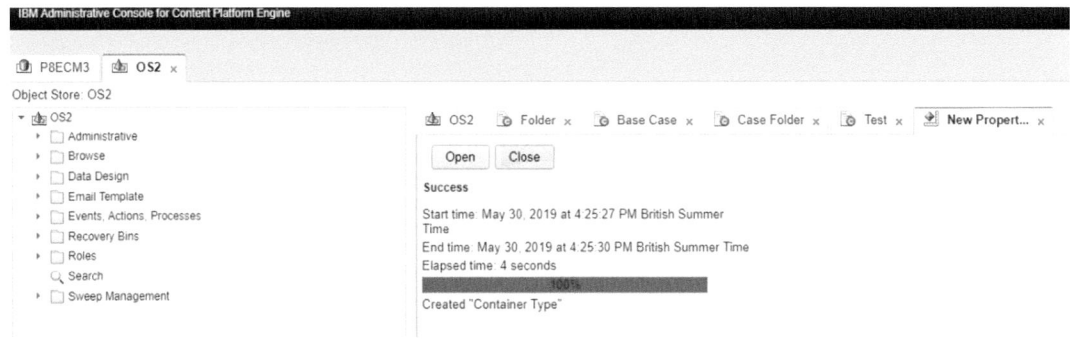

Figure 1-176.　The ContainerType Property Template was successfully created

CHAPTER 1   IBM FILENET CASE MANAGER 5.3.3 CASE BUILDER SOLUTION DEVELOPMENT STEPS
            FOR THE AUDIT SYSTEM

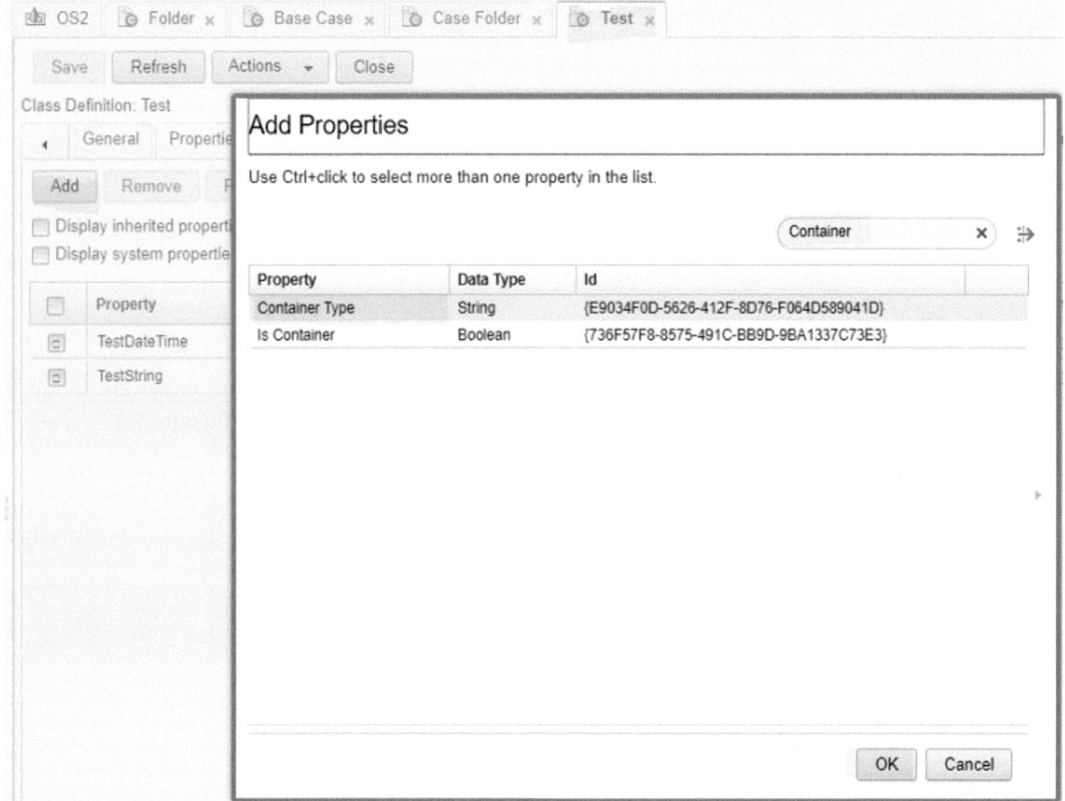

Figure 1-177. *The ContainerType property template was added to the Test Case Object class*

Select the "Container Type" Property Template Definition we just created to add to the Test Case Object class.

CHAPTER 1 IBM FILENET CASE MANAGER 5.3.3 CASE BUILDER SOLUTION DEVELOPMENT STEPS
           FOR THE AUDIT SYSTEM

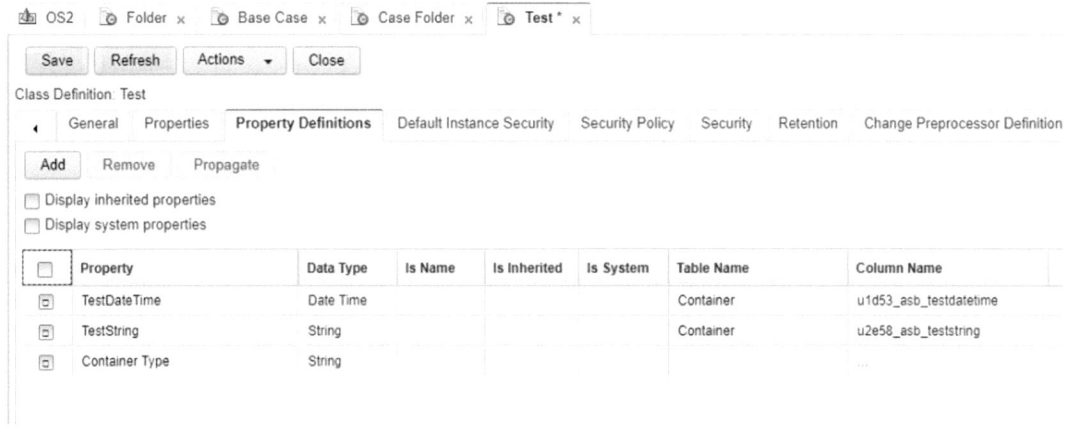

Figure 1-178. *The Container Type property is now visible in the Test Object class definitions*

Save and refresh.

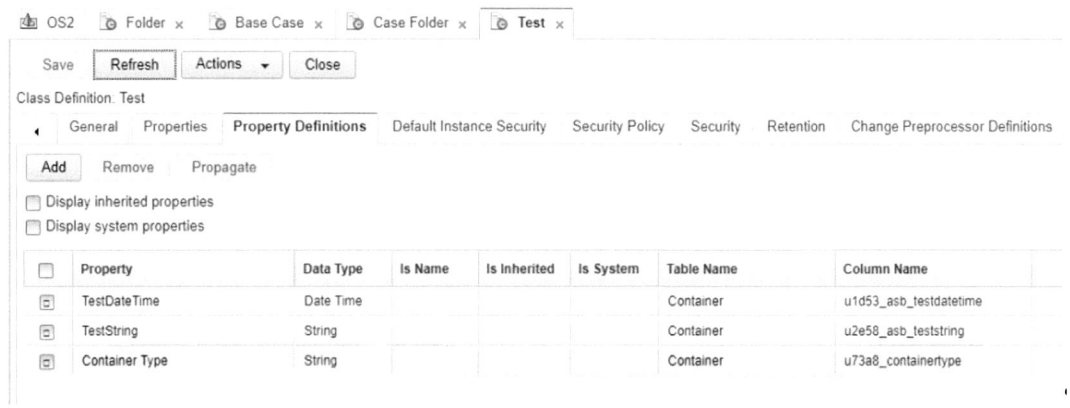

Figure 1-179. *The Container Type property is now saved and ready for use*

CHAPTER 1   IBM FILENET CASE MANAGER 5.3.3 CASE BUILDER SOLUTION DEVELOPMENT STEPS
            FOR THE AUDIT SYSTEM

Tested and working in the acce tool, search the SQL as follows:

SELECT [This], t.[FolderName], t.[CmAcmHealthIndicator], t.[LastModifier], t.[DateLastModified], t.[CmAcmCaseTypeFolder], t.[CmAcmCaseState], t.[CmAcmCaseIdentifier], t.[DateCreated], t.[Creator], t.[Id], t.[ASB_TestString], t.[ClassDescription], t.[ContainerType], t.[LockToken], t.[LockTimeout] FROM [ASB_Test] t WHERE [ASB_TestString] like '%ASB%'

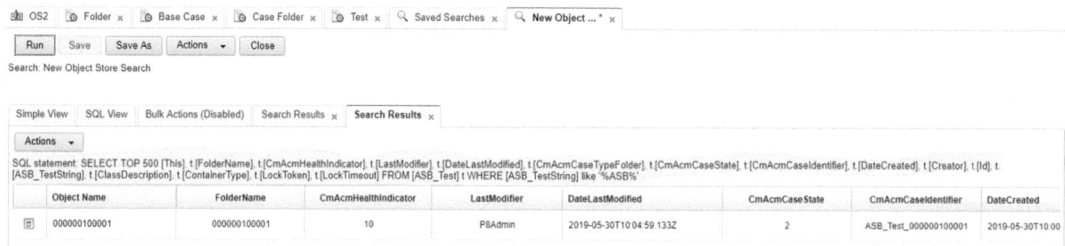

Figure 1-180. The SQL is run and now works correctly returning the first Case details

Case Search is now working correctly.

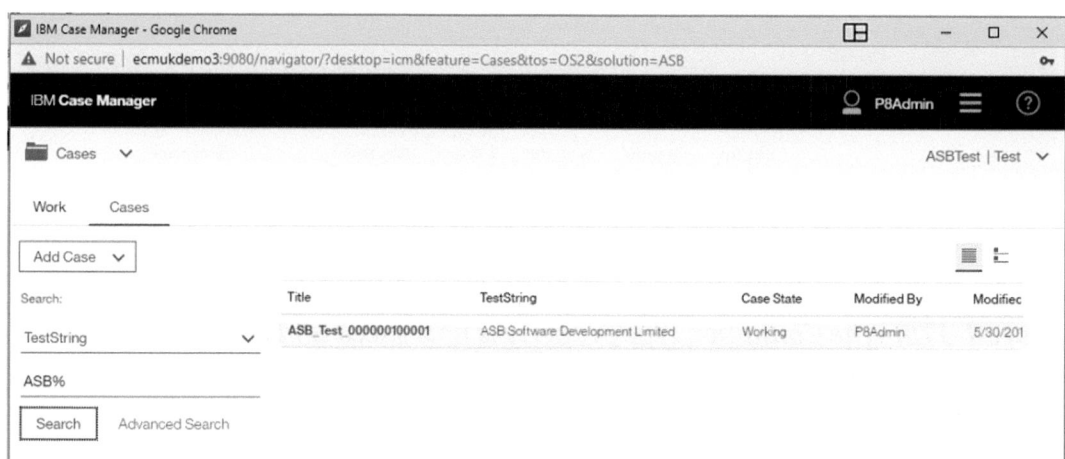

Figure 1-181. The IBM Case Manager shows the retrieved Test case

But see other issues!!

130

CHAPTER 1   IBM FILENET CASE MANAGER 5.3.3 CASE BUILDER SOLUTION DEVELOPMENT STEPS FOR THE AUDIT SYSTEM

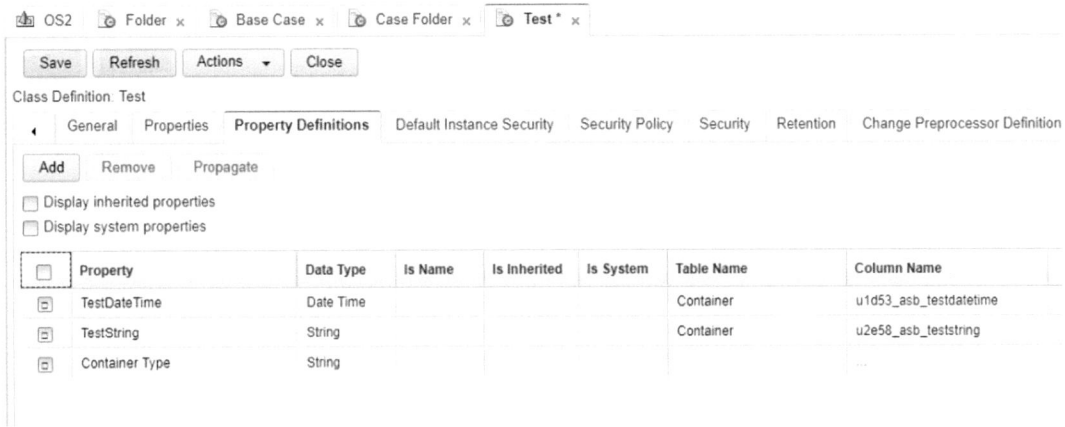

Figure 1-178. *The Container Type property is now visible in the Test Object class definitions*

Save and refresh.

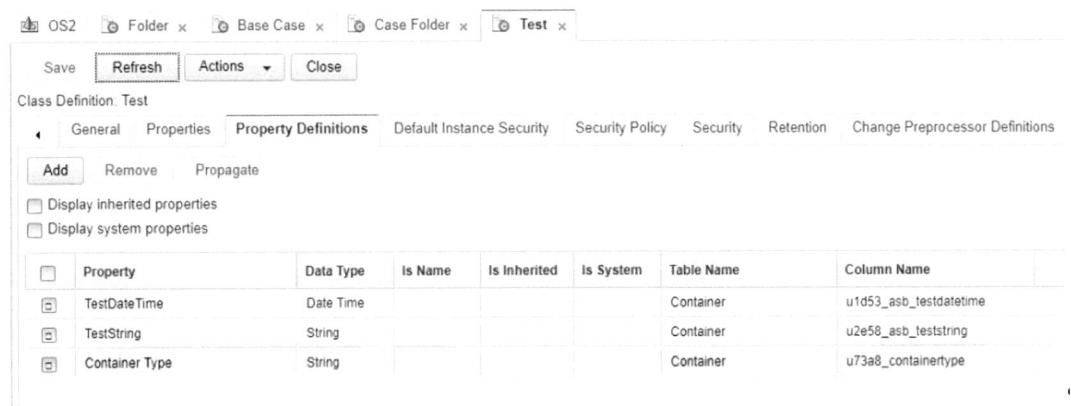

Figure 1-179. *The Container Type property is now saved and ready for use*

CHAPTER 1   IBM FILENET CASE MANAGER 5.3.3 CASE BUILDER SOLUTION DEVELOPMENT STEPS
              FOR THE AUDIT SYSTEM

Tested and working in the acce tool, search the SQL as follows:

SELECT [This], t.[FolderName], t.[CmAcmHealthIndicator], t.[LastModifier], t.[DateLastModified], t.[CmAcmCaseTypeFolder], t.[CmAcmCaseState], t.[CmAcmCaseIdentifier], t.[DateCreated], t.[Creator], t.[Id], t.[ASB_TestString], t.[ClassDescription], t.[ContainerType], t.[LockToken], t.[LockTimeout] FROM [ASB_Test] t WHERE [ASB_TestString] like '%ASB%'

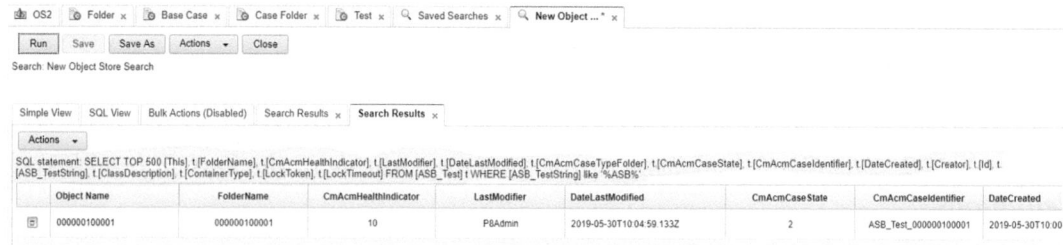

Figure 1-180. The SQL is run and now works correctly returning the first Case details

Case Search is now working correctly.

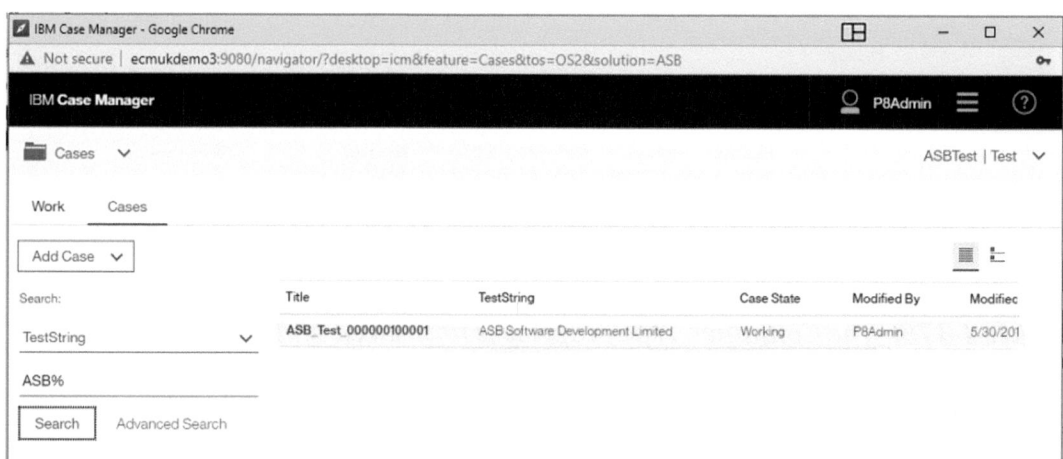

Figure 1-181. The IBM Case Manager shows the retrieved Test case

But see other issues!!

CHAPTER 1   IBM FILENET CASE MANAGER 5.3.3 CASE BUILDER SOLUTION DEVELOPMENT STEPS
             FOR THE AUDIT SYSTEM

Fix (Recommended)

After searching the Internet with Google using

What is the Container Type for the Folder Class in FileNet P8?

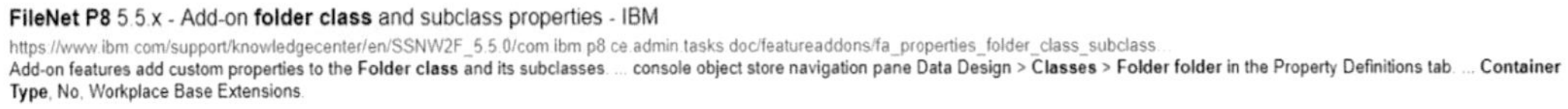

FileNet P8 5.5.x - Add-on **folder class** and subclass properties - IBM
https://www.ibm.com/support/knowledgecenter/en/SSNW2F_5.5.0/com.ibm.p8.ce.admin.tasks.doc/featureaddons/fa_properties_folder_class_subclass
Add-on features add custom properties to the **Folder class** and its subclasses ... console object store navigation pane Data Design > **Classes** > **Folder** folder in the Property Definitions tab. ... **Container Type**, No, Workplace Base Extensions

Figure 1-182.   *The IBM Knowledge base article on the Container Type property*

REF:
www.ibm.com/support/knowledgecenter/en/SSNW2F_5.5.0/com.ibm.p8.ce.admin.tasks.doc/featureaddons/fa_properties_folder_class_subclasses.htm

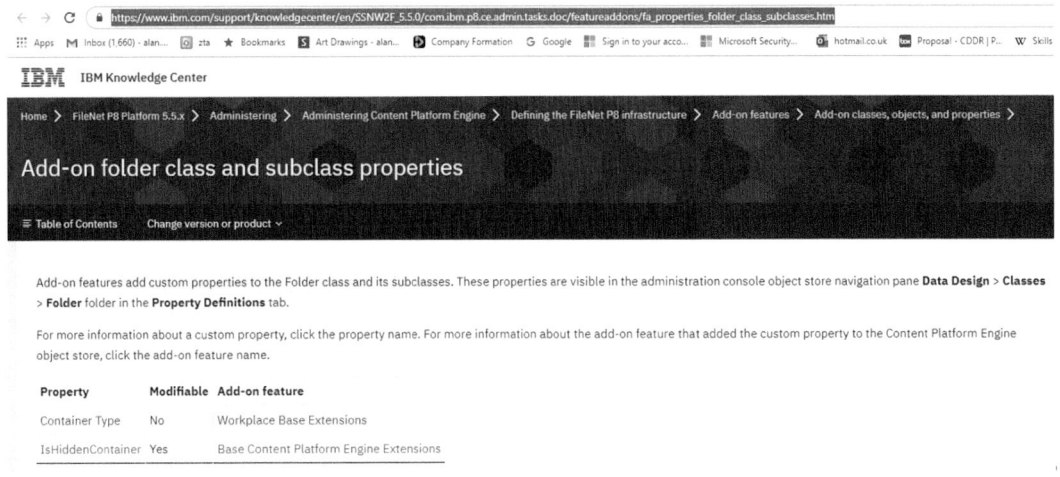

Figure 1-183.   *The Container Type is installed using the Workplace Base Extensions*

So we just need to add a Workplace Base Extensions Add-On to the OS2 Target Object store.

First, reverse out our initial workaround by removing the property definition we added:

a)   From the Test Folder

CHAPTER 1   IBM FILENET CASE MANAGER 5.3.3 CASE BUILDER SOLUTION DEVELOPMENT STEPS
             FOR THE AUDIT SYSTEM

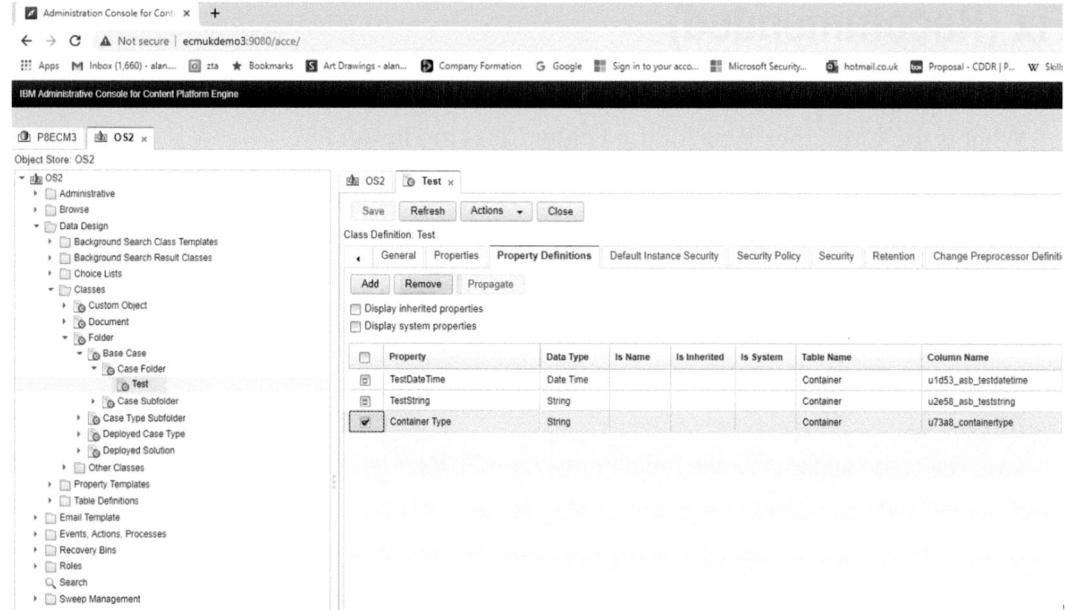

Figure 1-184.   The manually added Container Type property is removed from the Test class

Figure 1-185.   The Save and Refresh command shows the Container Type has been removed

CHAPTER 1   IBM FILENET CASE MANAGER 5.3.3 CASE BUILDER SOLUTION DEVELOPMENT STEPS
           FOR THE AUDIT SYSTEM

b) And then from the property definition list

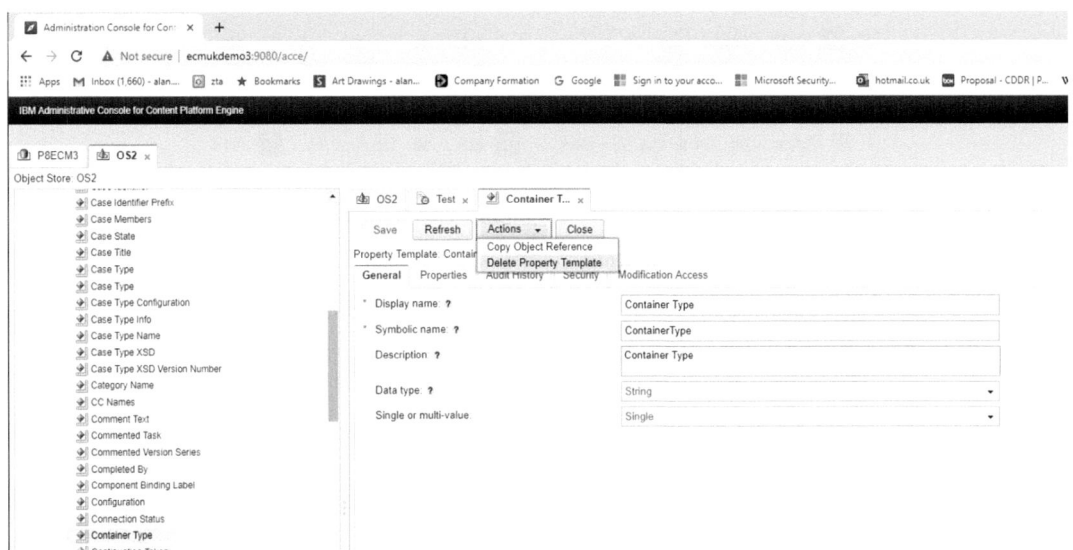

Figure 1-186. *The Container Type property template is also removed*

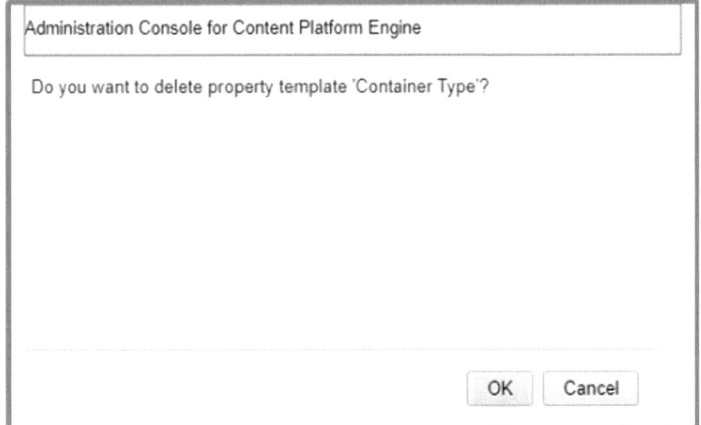

Figure 1-187. *Click OK to confirm the deletion of the Container Type template property*

133

CHAPTER 1   IBM FILENET CASE MANAGER 5.3.3 CASE BUILDER SOLUTION DEVELOPMENT STEPS
            FOR THE AUDIT SYSTEM

Now install the Workplace Base add-on feature.

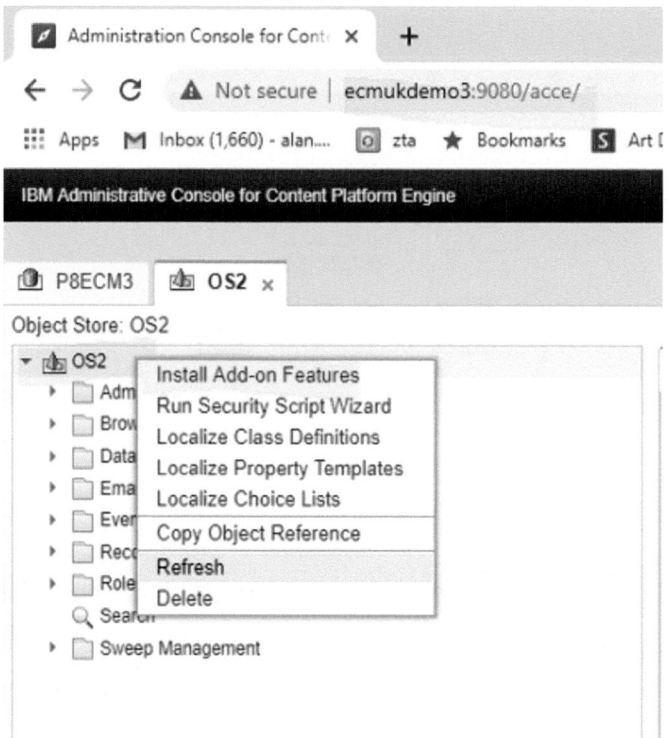

Figure 1-188. *The Install Add-on Features menu option is selected by right-clicking the **OS2** Object Store node*

CHAPTER 1   IBM FILENET CASE MANAGER 5.3.3 CASE BUILDER SOLUTION DEVELOPMENT STEPS
            FOR THE AUDIT SYSTEM

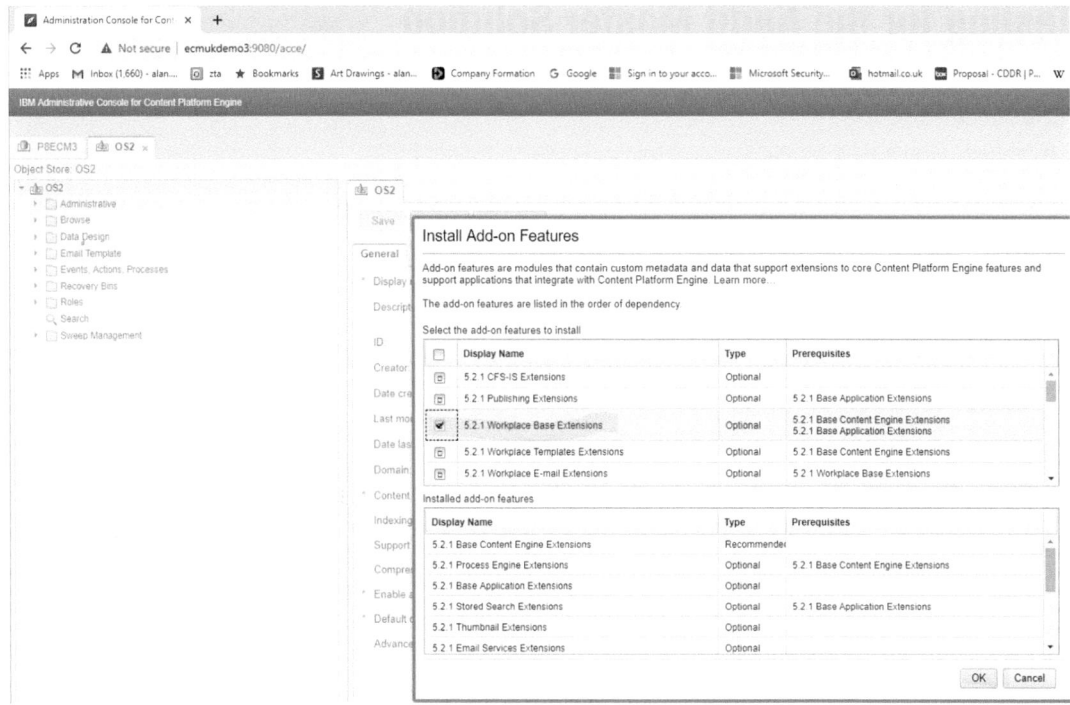

Figure 1-189. *The 5.2.1 Workplace Base Extensions add-on feature is selected for installation*

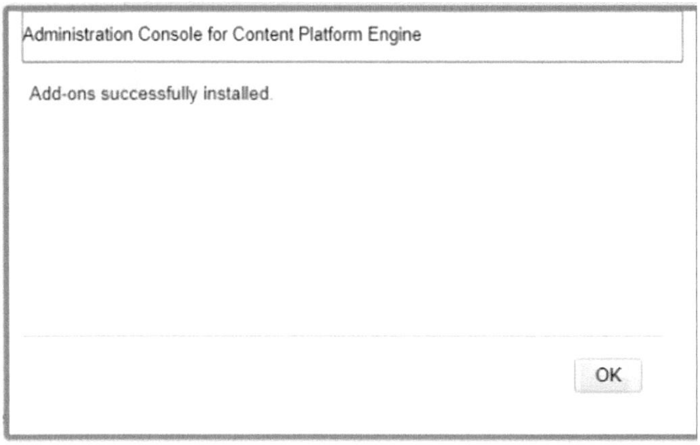

Figure 1-190. *The Add-on feature status is displayed*

And then restart the WebSphere application server.

CHAPTER 1　IBM FILENET CASE MANAGER 5.3.3 CASE BUILDER SOLUTION DEVELOPMENT STEPS FOR THE AUDIT SYSTEM

Testing for the Audit Master Solution

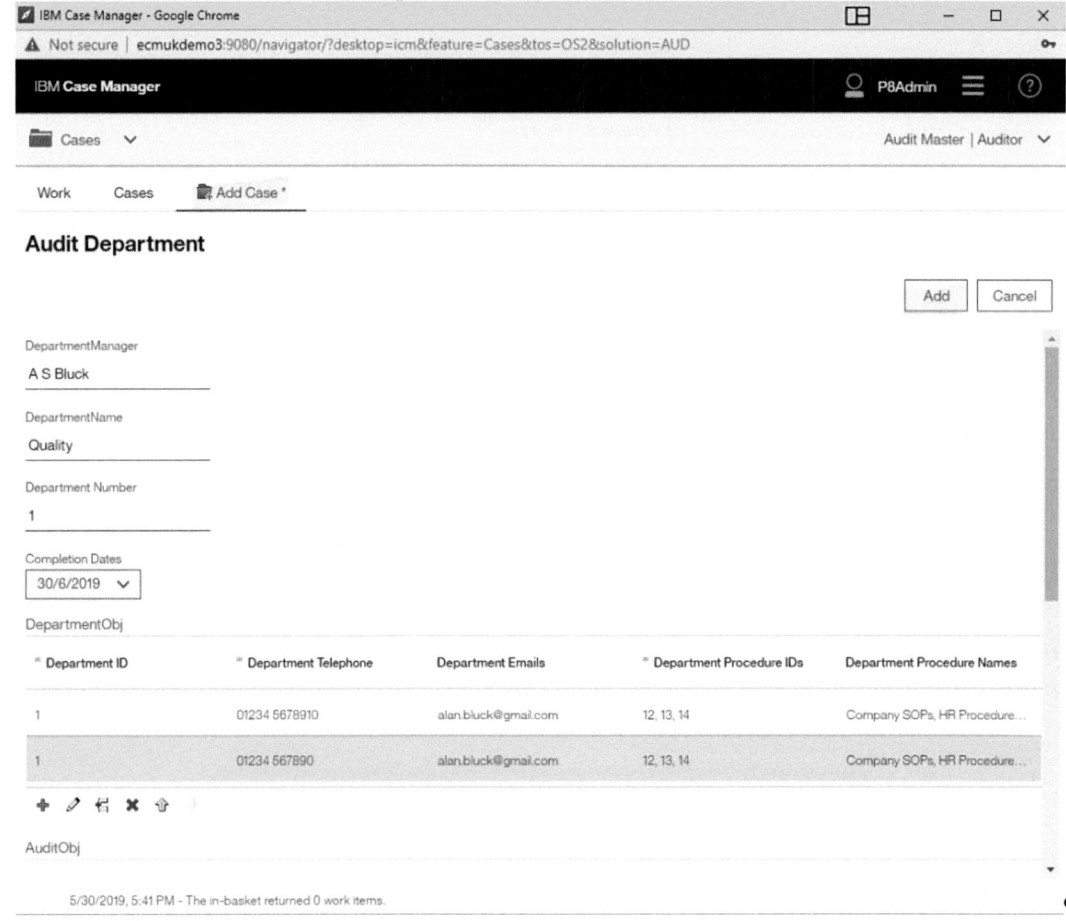

Figure 1-191. *The Test Case is added using the IBM Case Manager desktop*

CHAPTER 1   IBM FILENET CASE MANAGER 5.3.3 CASE BUILDER SOLUTION DEVELOPMENT STEPS
            FOR THE AUDIT SYSTEM

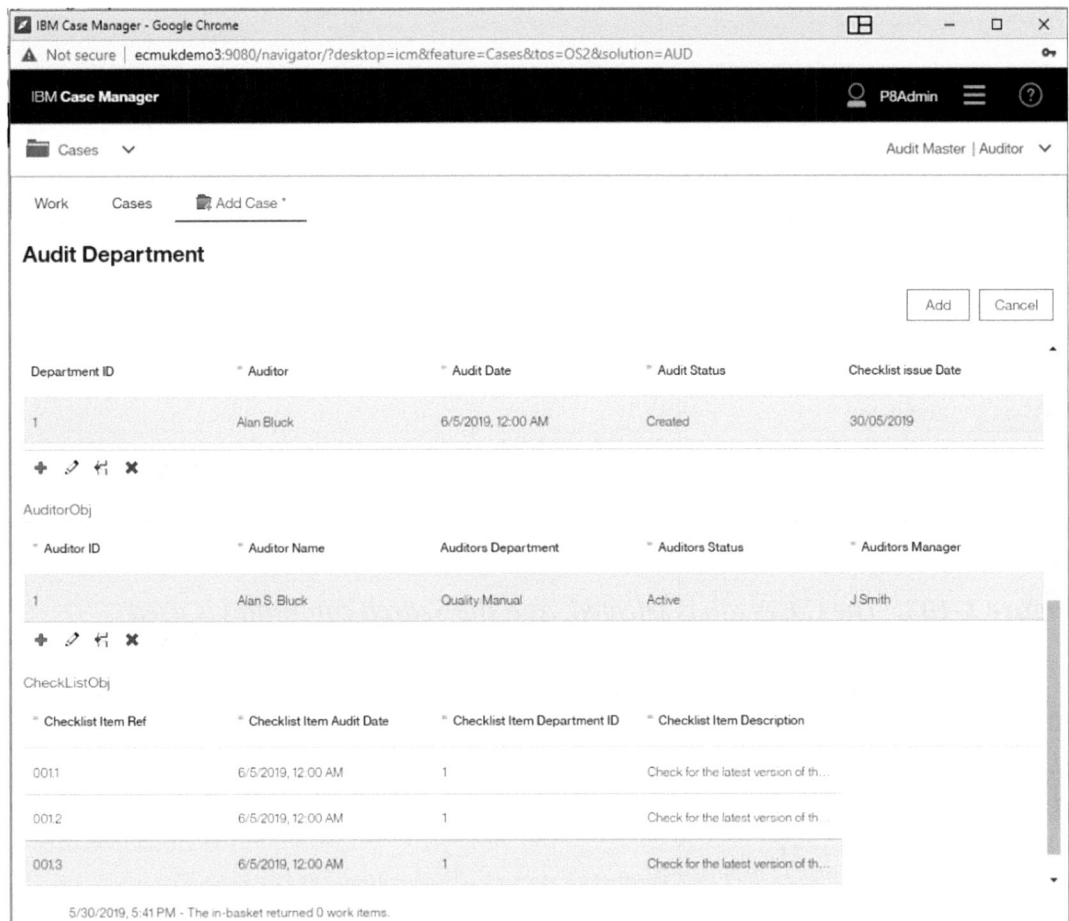

Figure 1-192.   The example Audit Master case is added

The **Search** command is tested against the added Case in Figure 1-192.

CHAPTER 1   IBM FILENET CASE MANAGER 5.3.3 CASE BUILDER SOLUTION DEVELOPMENT STEPS
            FOR THE AUDIT SYSTEM

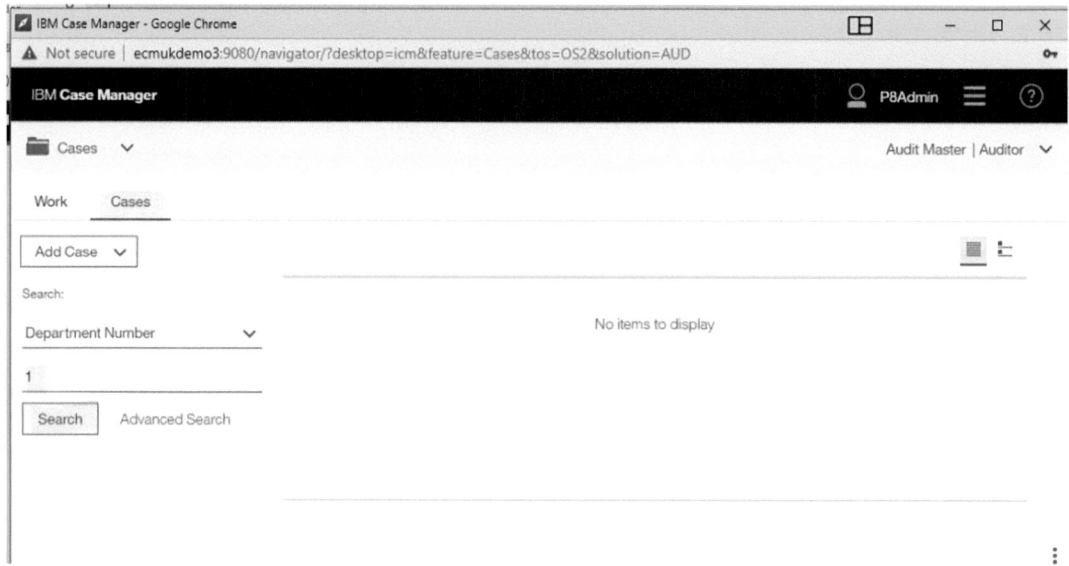

Figure 1-193. The Cases tab is selected, and the Search command is used

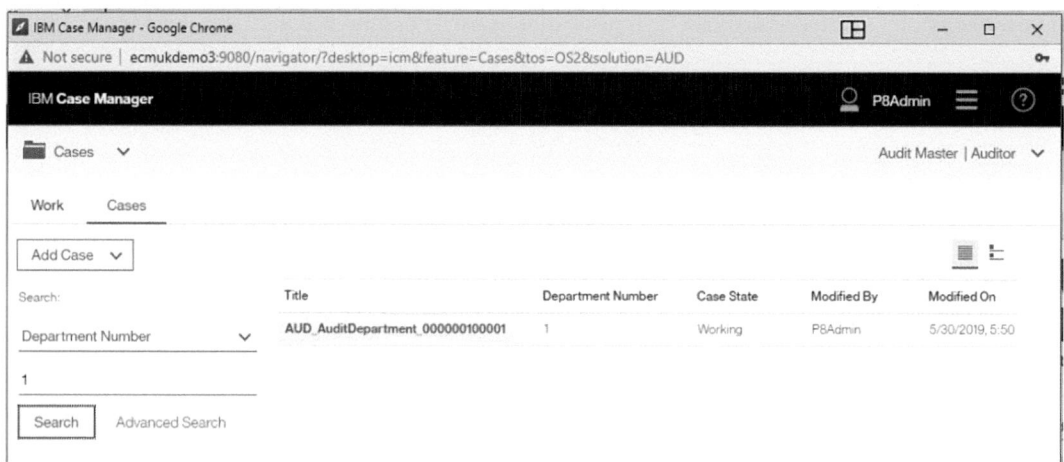

Figure 1-194. The Search command now works correctly and returns the case

Double-click the Case Title.
Search is now working correctly!

CHAPTER 1   IBM FILENET CASE MANAGER 5.3.3 CASE BUILDER SOLUTION DEVELOPMENT STEPS
            FOR THE AUDIT SYSTEM

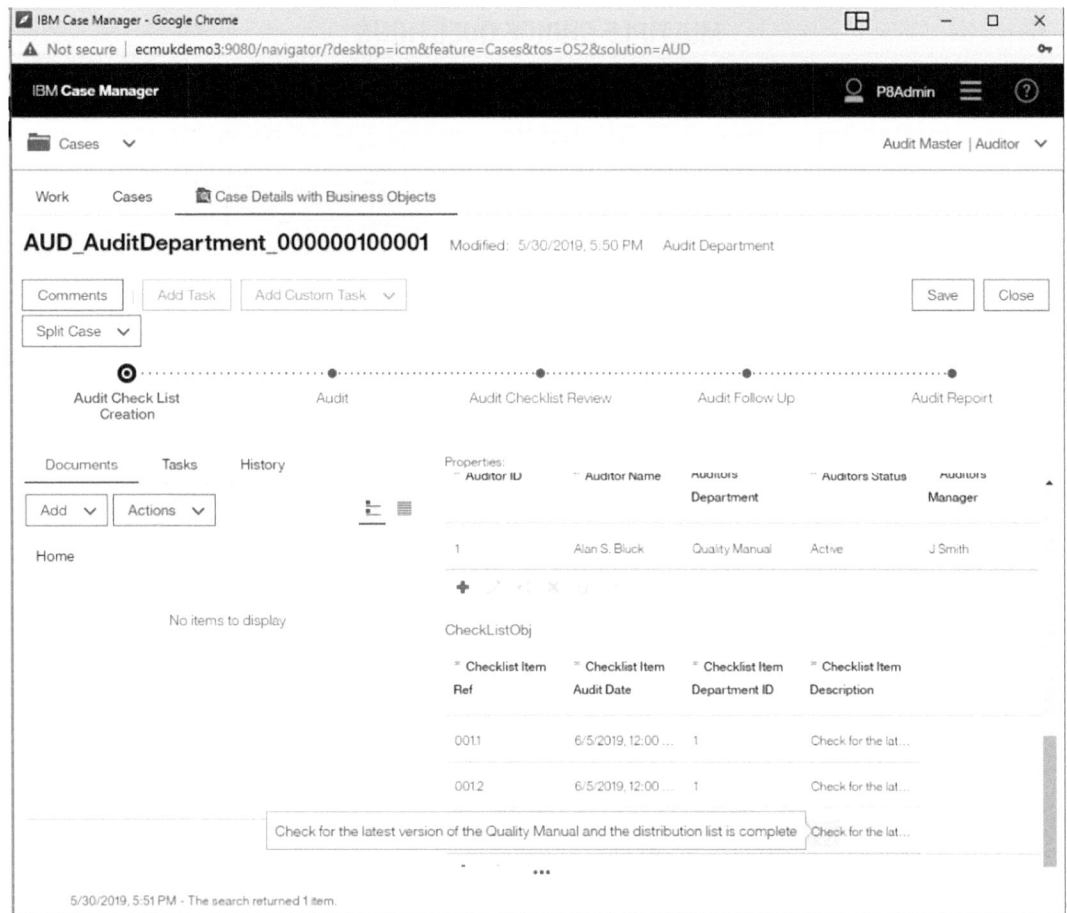

Figure 1-195. The Audit case details are now all displayed correctly

Chapter 1 Exercises

The following questions cover the functions using the IBM Case Builder which we covered in this chapter.

CHAPTER 1 IBM FILENET CASE MANAGER 5.3.3 CASE BUILDER SOLUTION DEVELOPMENT STEPS
FOR THE AUDIT SYSTEM

MULTIPLE CHOICE QUESTIONS

1. The Red Flag icon against the IBM Case Property Attributes column defines

    a) A multiple-value property

    b) A Hidden attribute property

    c) A Required attribute property

    d) A New property is added

2. An IBM Case String property type for the Oracle Database system is limited to

    a) 64 characters

    b) 1333 characters

    c) 4000 characters

    d) 024 characters

3. The In-basket Filters tab is used:

    a) To filter a subset of the Cases

    b) To alter the sort order of the Cases

    c) To define the Object Type of the Case properties

    d) To define the list of Hidden Case properties

4. Business Object Types as Case properties are

    a) Multi-value String properties

    b) Multi-property Business Objects

    c) Multi-value Choice lists

    d) Multiple Case Types

CHAPTER 1   IBM FILENET CASE MANAGER 5.3.3 CASE BUILDER SOLUTION DEVELOPMENT STEPS
             FOR THE AUDIT SYSTEM

MULTIPLE CHOICE ANSWERS

1. c) A Required attribute property
2. b) 1333 characters
3. a) To filter a subset of the Cases
4. b) Multi-property Business Objects

QUESTIONS

1. Describe how you would implement an IBM Case Manager Solution to support moving it from a development environment to a production system in particular, describing how the development system security is kept separate from the security of the production system.

2. What databases does IBM Case Manager support and what impact does this have on how String Case properties are stored?

3. What methods can you use to track issues in the deployment and testing of an IBM Case Manager Solution?.

In Chapter 2, we will cover the use of the Java language to customize Workflow Component Events and the **DBEXECUTE** workflow step for calling Database Stored Procedures.

CHAPTER 2

Configuring Java Custom Components

This chapter covers the creation and configuration of Java customizations required for processing **Events** triggered by the **IBM FileNet Content Engine**. It describes the development of custom Java code and the deployment of a **Code Module** which is invoked. The use of **Workflow Subscriptions** which enable a workflow to be launched on the storage of a document in the **Content Engine Object Store** is described with a step-by-step example.

The full **Content Engine Java API** documentation can be found with the following link:

```
www.ibm.com/docs/en/filenet-p8-platform/5.5.x?topic=development-content-engine-java-api-reference
```

Note IBM recommends that Content Engine Web Services can be used where a system is not compatible with the use of the IBM FileNet Java or .NET API. For further Content Engine Web Service code examples, see

www.ibm.com/docs/en/filenet-p8-platform/5.5.x?topic=development-content-engine-web-service-developers-guide

For the application of a Code Module for IBM Business Process Automation, see

www.ibm.com/docs/en/baw/20.x?topic=events-using-event-handler-filenet-content-manager

---

CHAPTER 2 CONFIGURING JAVA CUSTOM COMPONENTS

Chapter Organization

This chapter contains the following five Parts:

Part 1 – Supporting Documents. This Part provides references to freely downloadable documents which provide additional reading material supporting the chapter.

Part 2 – Configuring Java Components for Content Engine Events. In this Part, a Java project is built, step by step, to create and deploy the example fn_eventhandlers.jar file, which we then show how to deploy, for use with the Audit Master solution covered in Chapter 1.

Part 3 – Testing the Add Document to Folder Event. In this Part, we test the Add Document to Folder Event. The TestHarness code in the Eclipse project requires a valid Document Object to be used to link to an existing AUDIT_TEST folder which is created and used as a test setup for the EventHandler code.

Part 4 – The DbExecute Workflow Step for Calling Database Stored Procedures. In this Part, we describe the use of the DbExecute workflow step. This allows a Database Stored procedure to be invoked from an IBM Case Manager Workflow step. This is supported using the IBM FileNet P8 Process Designer tool.

Part 5 – Standalone Process Designer Installation. This Part describes the installation and configuration required to install the Windows Client Standalone Process Designer, which is then used to add a DbExecute workflow step to the Audit Master Solution we created in Chapter 1.

Part 1 – Supporting Documents

There are two very useful downloads which provide additional examples and links to downloadable Java code supplied freely by IBM.

IBM FileNet Java API Call Examples

This MS Word document can be downloaded as shown in Figure 2-1. It describes the download of an **IBM DemoJava.zip** file and the configuration of the Java Eclipse IDE which can be configured to run working examples to show the use of the IBM FileNet Content Manager Java API calls to create, store, and retrieve Document and Folder objects in a Content Object Store.

CHAPTER 2   CONFIGURING JAVA CUSTOM COMPONENTS

Figure 2-1. *Download the free IBM FileNet P8 Java Development Document*

This gives the downloaded document, entitled
"IBM FileNet P8 Java Development on ECM Cloud Private Container P8 Examples"

This free download document, **IBMFileNetP8JavaDevelopmentonECMCloudPrivateContainerP8Examples.docx**, can be downloaded using the URL link:

```
https://doi.org/10.13140/RG.2.2.20160.69129
```

Java Code Development

The Java code developed for custom events should be compiled into an appropriately named Java archive, or "jar" file. To install this code, the jar file should be made accessible to a machine that has access to the **acce** (IBM Administrative Console for Content Platform Engine) application.

"IBM FileNet P8 Java Development on ECM Cloud Private Container P8 Examples"

This describes the download and installation of an Eclipse IDE for Java development and the basic configuration of the supporting Java jar files for the IBM FileNet API on a Red Hat RHEL 8.0 Linux system.

For the Audit system, the following Java code in this chapter provides the custom event, **AddDocToFolder**. This Java code will work on Red Hat Linux or MS Windows operating systems which have IBM FileNet Content Engine installed.

In enterprise applications, normally code needs to be included in a "war" file and deployed so that its classes are accessible. The Content Engine Event code, however, uses a Java jar file with the classes required for events which can be contained within the Content Engine Object Store. This makes for easy deployment and maintenance. The jar file is added to the Object Store through the **acce** tool, which has the ability to choose hidden classes.

Part 2 – Configuring Java Components for Content Engine Events

In this Part, we will build a Java project, step by step, to create and deploy the example **fn_eventhandlers.jar** file, which we then show how to deploy, for use with the **Audit Master** solution covered in Chapter 1.

The final Java project structure we build in the **Eclipse** Java IDE will have the main Java components as shown in Figure 2-2.

CHAPTER 2　CONFIGURING JAVA CUSTOM COMPONENTS

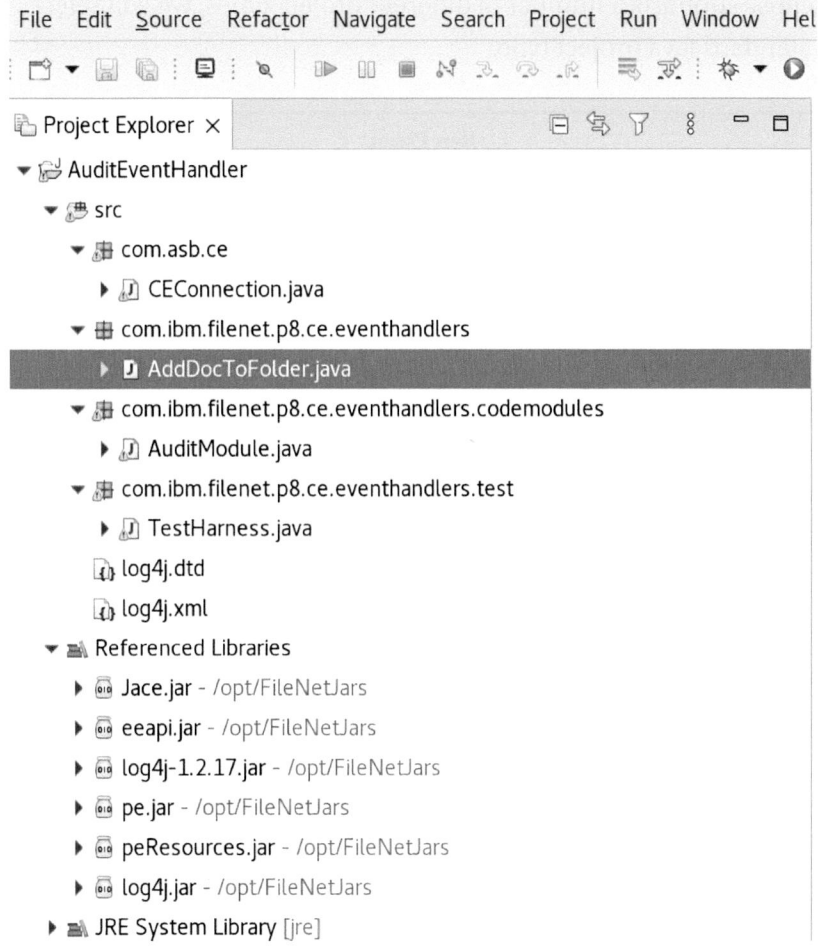

Figure 2-2. *The AuditEventHandler Eclipse Java project structure*

The Java code project is created using the following steps:

1. In the Eclipse Java IDE, select the menu items File ➤ New Project.

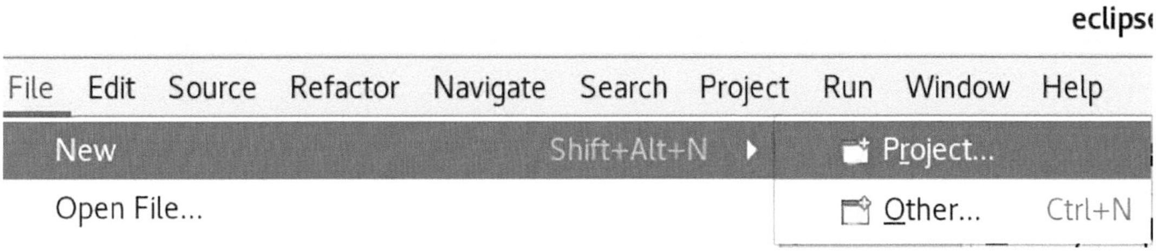

Figure 2-3. *Open a New Project in Eclipse*

CHAPTER 2   CONFIGURING JAVA CUSTOM COMPONENTS

2. Eclipse supports a number of different project types; we will select a standard Java project type.

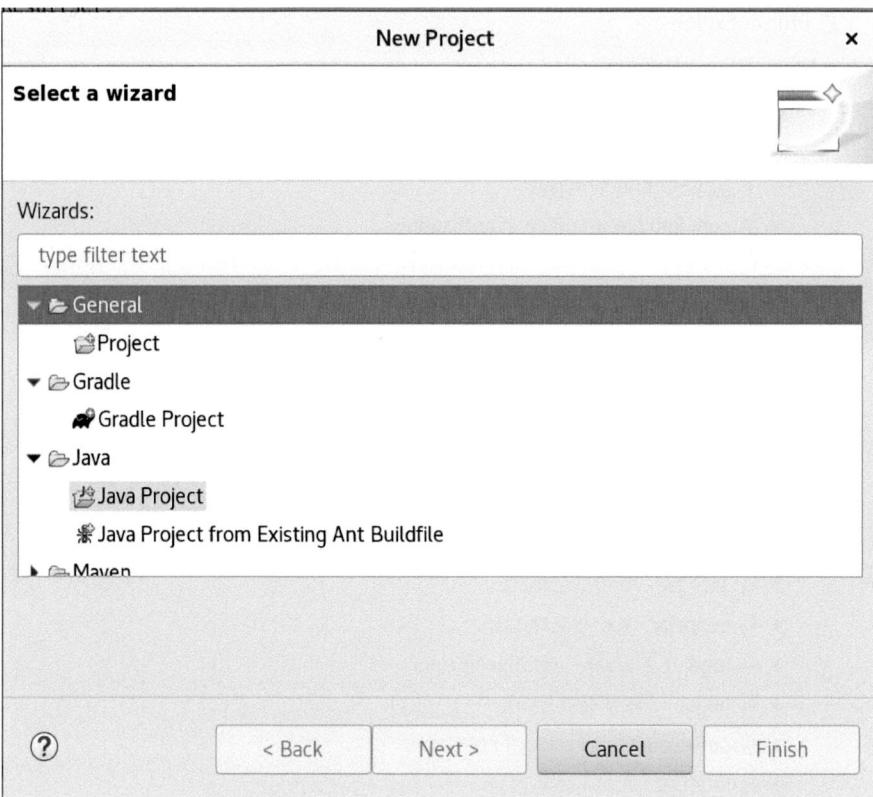

Figure 2-4. *Select the Java Project type from the Project Wizard types*

3. The Next command is clicked to add the Project attributes.

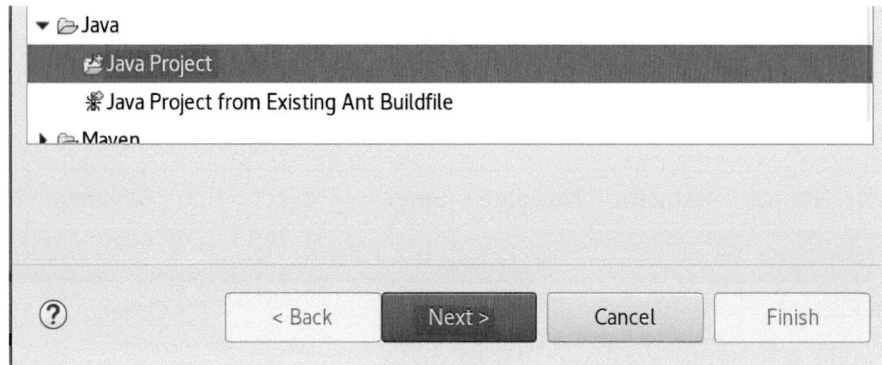

Figure 2-5. *Select the Next command button to add the Project attributes*

CHAPTER 2 CONFIGURING JAVA CUSTOMS COMPONENTS

4. The Project name is entered as **AuditEventHandler**.

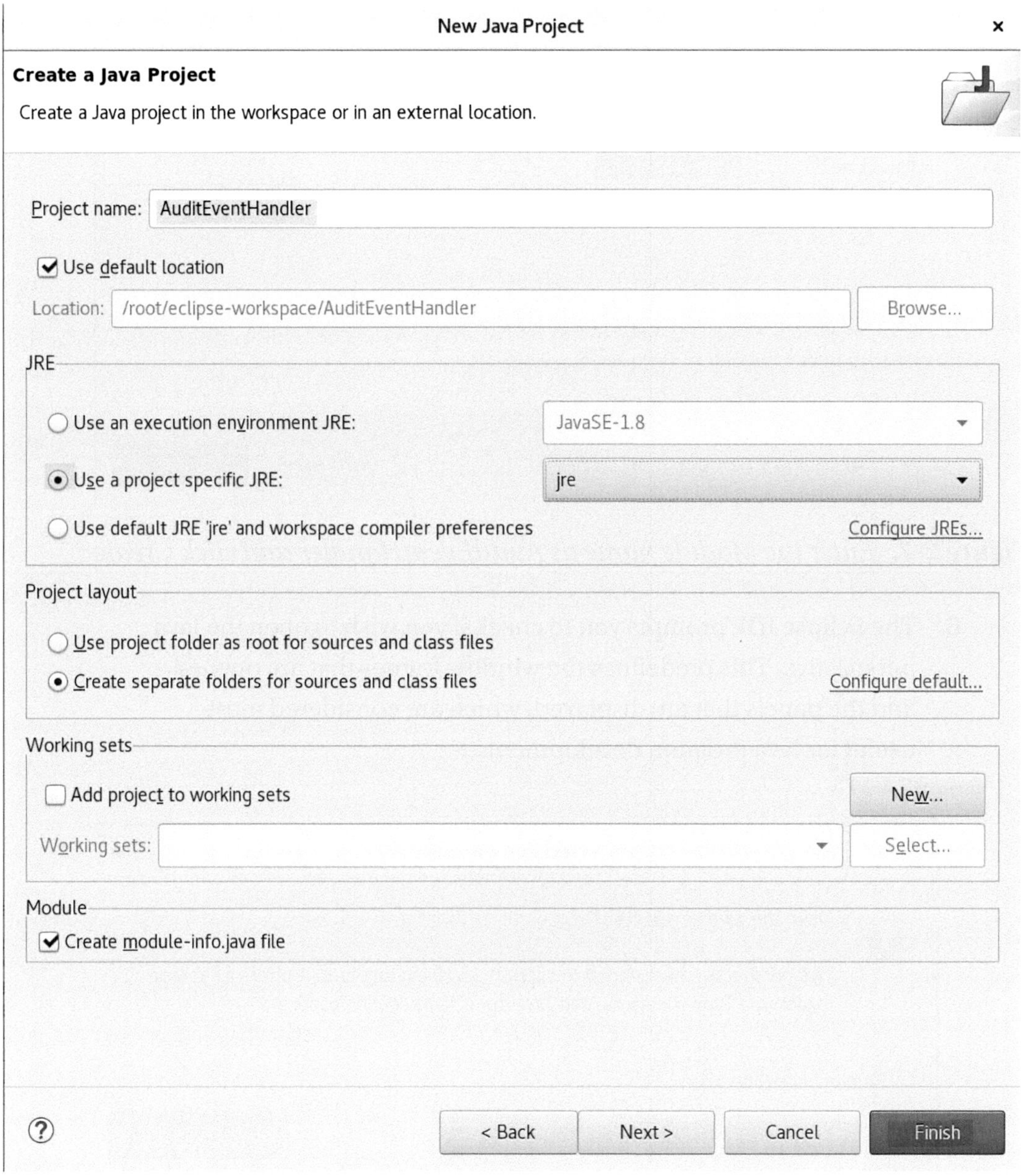

Figure 2-6. Select the Finish command button to create the Project outline

CHAPTER 2   CONFIGURING JAVA CUSTOM COMPONENTS

5. The module-info.java was created as follows.

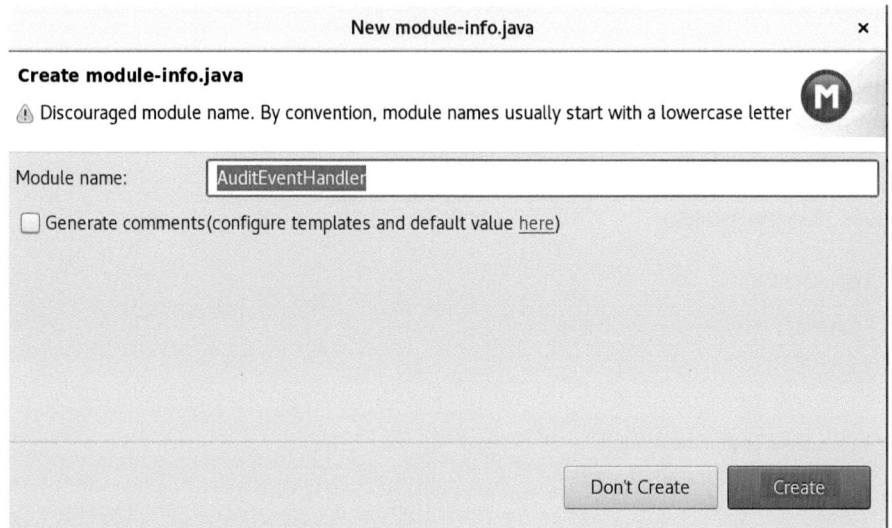

Figure 2-7. *Enter the Module name as AuditEventHandler and click Create*

6. The Eclipse IDE prompts you to check if you wish to open the Java perspective. This predefines the window frames that are opened and the panels that are displayed, which are considered most useful for Java program development.

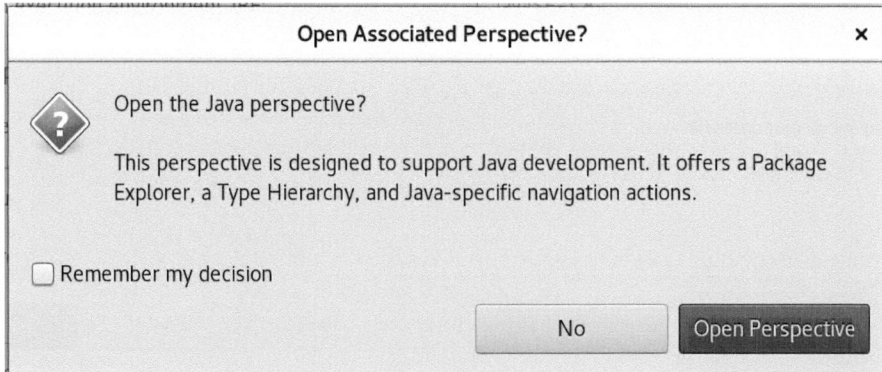

Figure 2-8. *Click the Open Perspective command to view the standard Java layout*

CHAPTER 2 CONFIGURING JAVA CUSTOM COMPONENTS

7. At this point, we have just one empty Java class and the Java JRE System Library defined for the project.

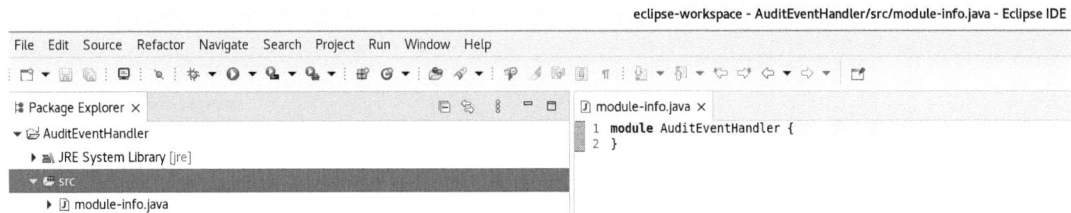

Figure 2-9. The basic Java project shown in the standard Java layout

8. To create a new Java package, ***com.asb.ce***, we need to right-click the ***src*** folder, highlighted in Figure 2-9, and select the ***New ➤ Package*** option from the menu drop-downs. We can then enter the package name as shown in Figure 2-10.

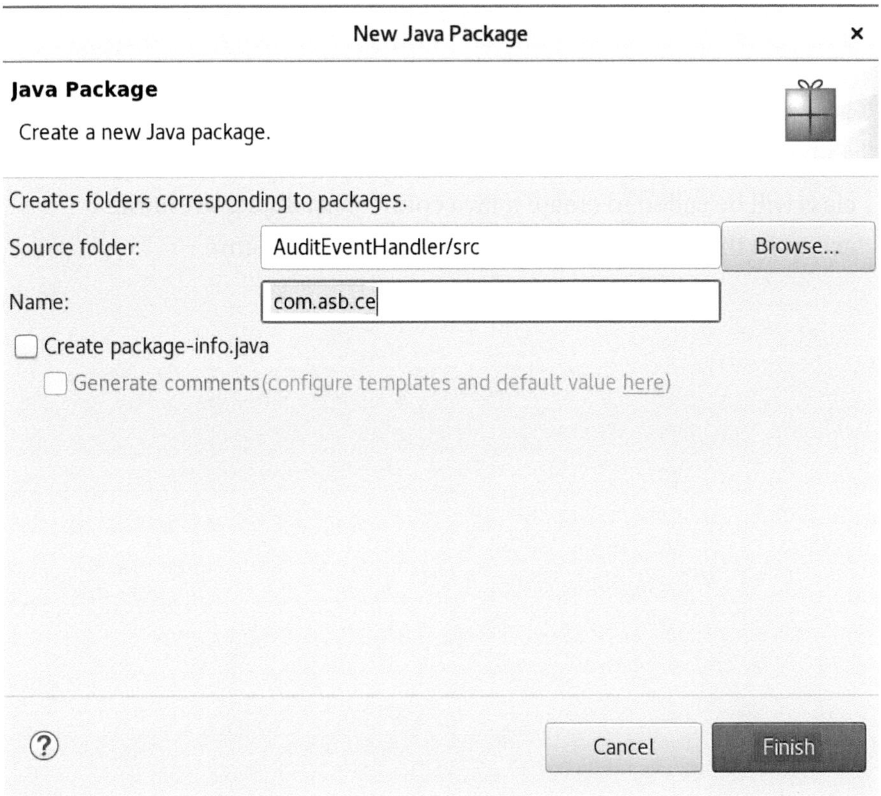

Figure 2-10. The Java package com.asb.ce is created under the Java src folder

151

CHAPTER 2   CONFIGURING JAVA CUSTOM COMPONENTS

9. We can now add Java code as a Java *Class* to the "empty" *com.asb.ce* package by right-clicking the *com.asb.ce* package we created in step 8 and then selecting the menu items, *New ➤ Class* (highlighted in Figure 2-11).

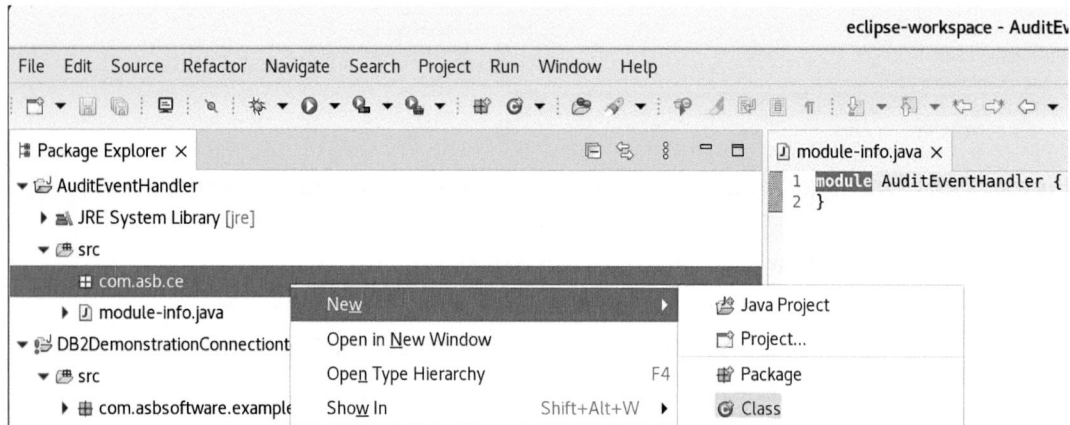

Figure 2-11. *A new Java Class is added to the Java package **com.asb.ce***

10. The attributes required for the Java Class are displayed; we just add the name of the class we want to give as **CEConnection**. This class will be called to create a Java connection object to enable access to the IBM FileNet Content Engine Object Store.

152

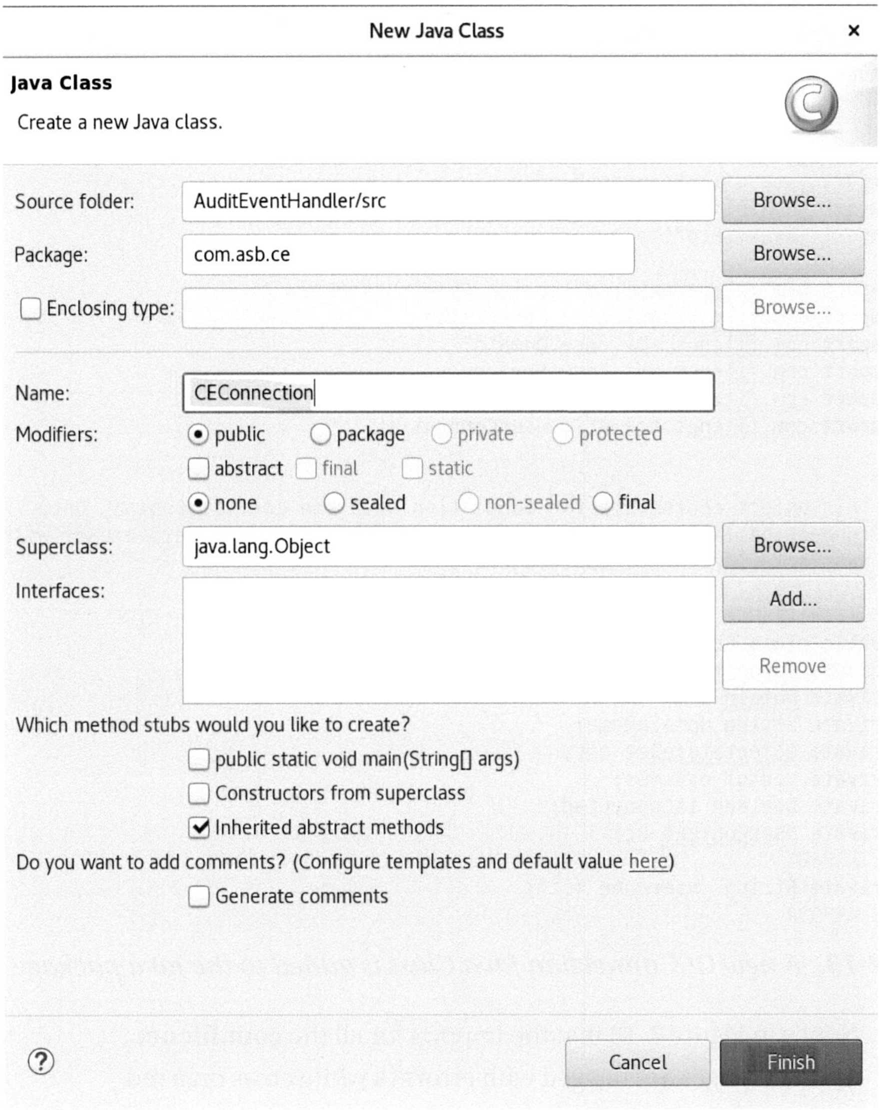

CHAPTER 2 CONFIGURING JAVA CUSTOM COMPONENTS

Figure 2-12. A new CEConnection Java Class is added to the Java package

11. For the **CEConnection.java** Java code, we just cut and paste a Standard IBM code example as shown in Figure 2-13 and listed fully in Listing 2-1 later in this chapter.

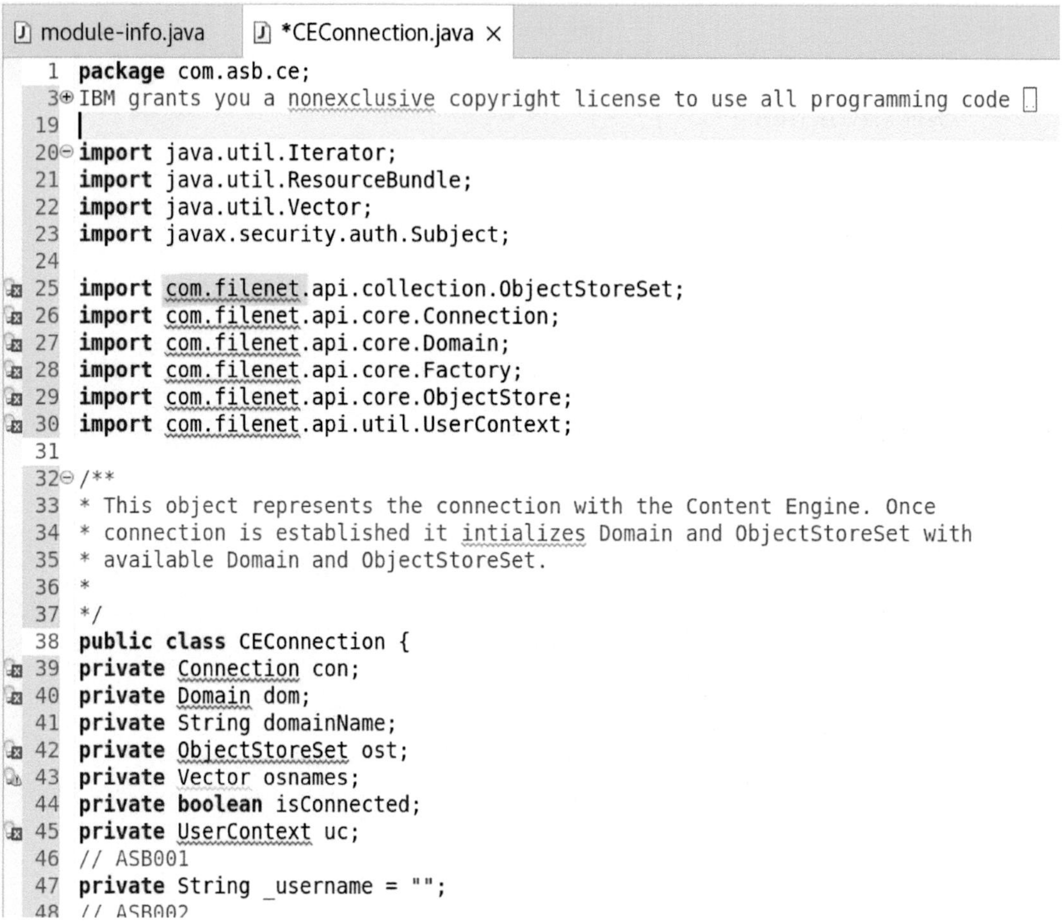

CHAPTER 2   CONFIGURING JAVA CUSTOM COMPONENTS

```java
 module-info.java    *CEConnection.java ×
 1  package com.asb.ce;
 3⊕ IBM grants you a nonexclusive copyright license to use all programming code ▯
19
20⊖ import java.util.Iterator;
21  import java.util.ResourceBundle;
22  import java.util.Vector;
23  import javax.security.auth.Subject;
24
25  import com.filenet.api.collection.ObjectStoreSet;
26  import com.filenet.api.core.Connection;
27  import com.filenet.api.core.Domain;
28  import com.filenet.api.core.Factory;
29  import com.filenet.api.core.ObjectStore;
30  import com.filenet.api.util.UserContext;
31
32⊖ /**
33   * This object represents the connection with the Content Engine. Once
34   * connection is established it intializes Domain and ObjectStoreSet with
35   * available Domain and ObjectStoreSet.
36   *
37   */
38  public class CEConnection {
39  private Connection con;
40  private Domain dom;
41  private String domainName;
42  private ObjectStoreSet ost;
43  private Vector osnames;
44  private boolean isConnected;
45  private UserContext uc;
46  // ASB001
47  private String _username = "";
48  // ASB002
```

Figure 2-13. *A new CEConnection Java Class is added to the Java package*

12. Notice in Figure 2-13 that the imports for all the **com.filenet.api...** packages are flagged with errors (a white cross on a red background). This is because we need to add the IBM FileNet API jar files to the Java project properties, which we will now provide in the next steps.

CHAPTER 2 CONFIGURING JAVA CUSTOM COMPONENTS

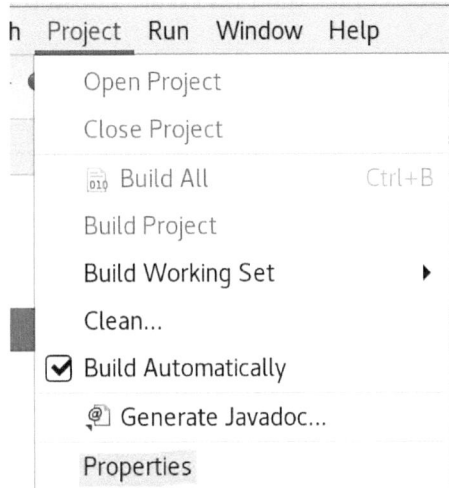

Figure 2-14. *The Project Properties are selected to update the Classpath details*

First, we have to copy the IBM FileNet client jar files from their installed location on the Linux server to the reference classpath directory we create using

mkdir /opt/FileNetJars

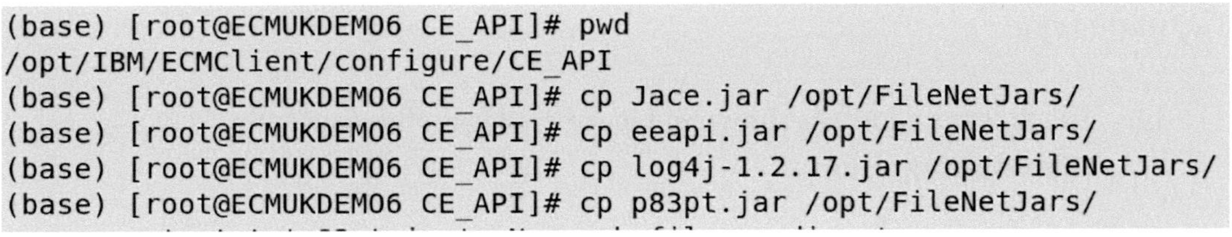

```
(base) [root@ECMUKDEMO6 CE_API]# pwd
/opt/IBM/ECMClient/configure/CE_API
(base) [root@ECMUKDEMO6 CE_API]# cp Jace.jar /opt/FileNetJars/
(base) [root@ECMUKDEMO6 CE_API]# cp eeapi.jar /opt/FileNetJars/
(base) [root@ECMUKDEMO6 CE_API]# cp log4j-1.2.17.jar /opt/FileNetJars/
(base) [root@ECMUKDEMO6 CE_API]# cp p83pt.jar /opt/FileNetJars/
```

Figure 2-15. *The IBM FileNet Client API Content Engine .jar files are copied*

```
(base) [root@ECMUKDEMO6 CE_API]# cp pe3pt.jar /opt/FileNetJars/
(base) [root@ECMUKDEMO6 CE_API]# cp pe.jar /opt/FileNetJars/
(base) [root@ECMUKDEMO6 CE_API]# cp peResources.jar /opt/FileNetJars/
(base) [root@ECMUKDEMO6 CE_API]# cp jaas.conf.WSI /opt/FileNetJars/
(base) [root@ECMUKDEMO6 CE_API]#
```

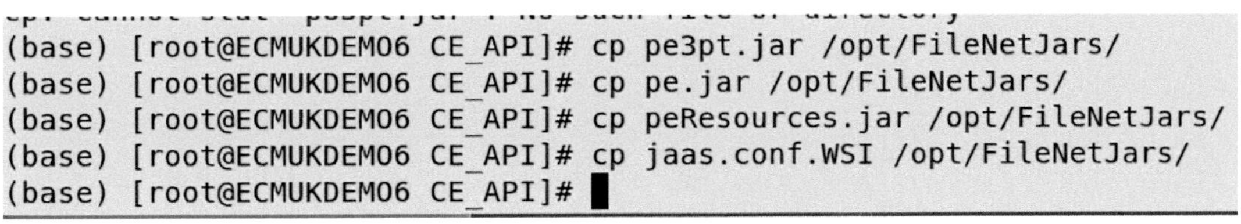

Figure 2-16. *The IBM FileNet Client API Process Engine .jar files are copied*

155

CHAPTER 2   CONFIGURING JAVA CUSTOM COMPONENTS

13. Next, we can add the copied .jar files created in step 12 to the Project Classpath.

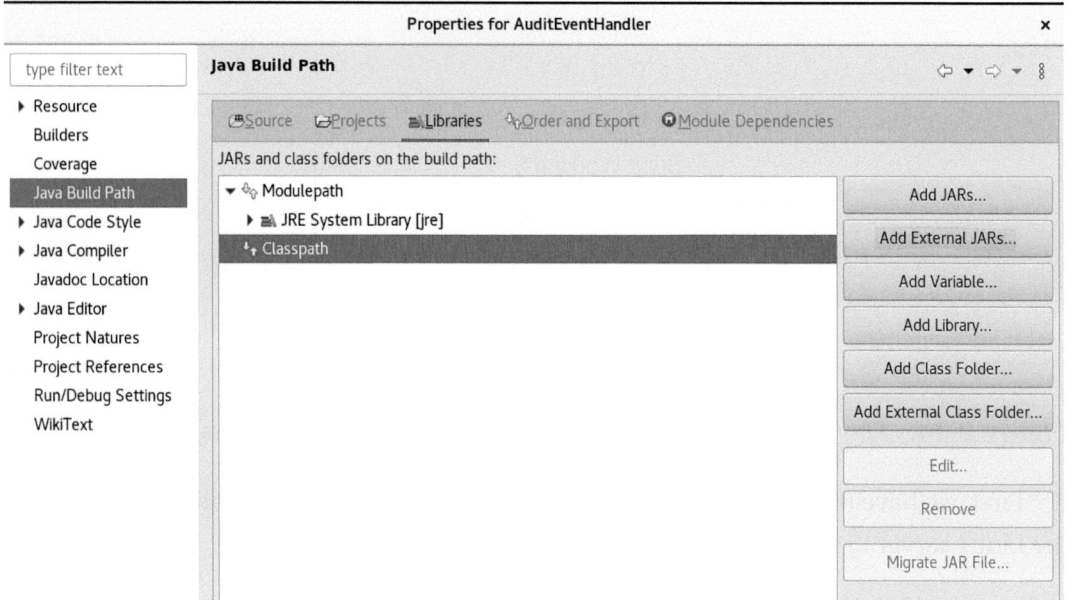

Figure 2-17. The **Add External JARs** command is clicked on the highlighted Classpath node

14. The files are added as shown in Figure 2-18 from the **/opt/FileNetJars** directory.

CHAPTER 2 CONFIGURING JAVA CUSTOM COMPONENTS

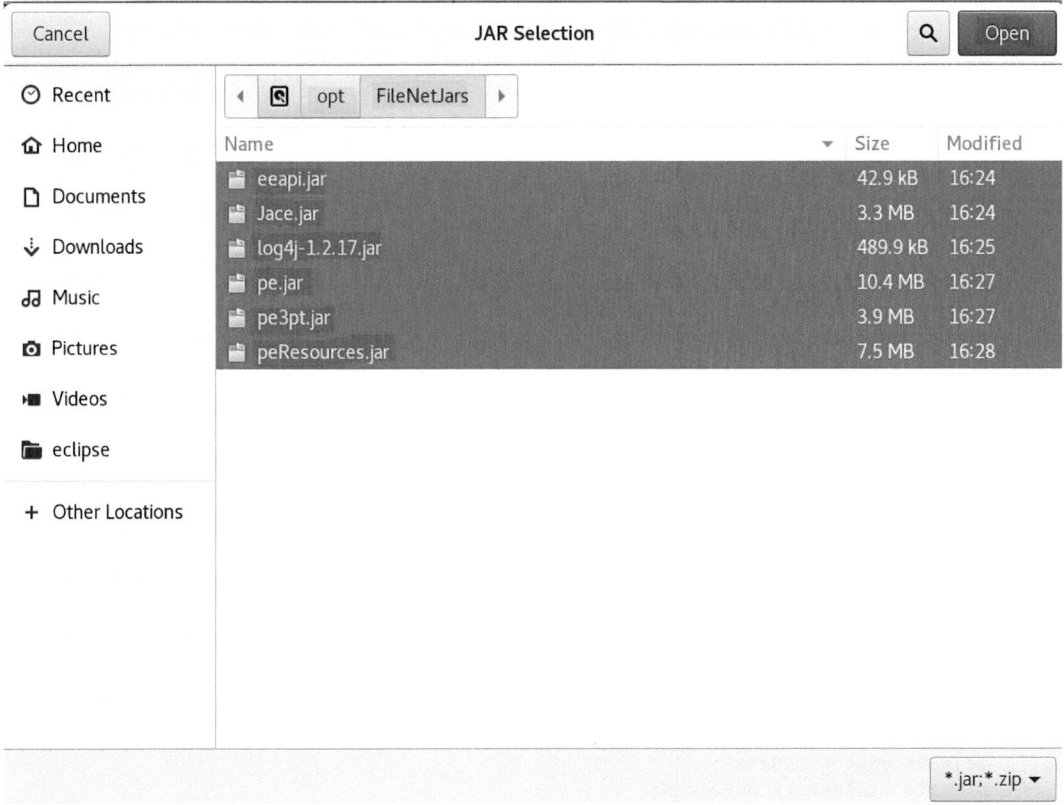

Figure 2-18. *The **Add External JARs** command is clicked on the highlighted Classpath node*

It can be seen that the result of adding the jar files has corrected the errors as highlighted in Figure 2-19. However, there are still some import failures showing on the **java.util...** package imports and the **javax.security** package.

```
module-info.java    CEConnection.java ×
  1  package com.asb.ce;
  3⊕ IBM grants you a nonexclusive copyright license to use all programming code ⌷
 19
 20⊖ import java.util.Iterator;
 21  import java.util.ResourceBundle;
 22  import java.util.Vector;
 23  import javax.security.auth.Subject;
 24
 25  import com.filenet.api.collection.ObjectStoreSet;
 26  import com.filenet.api.core.Connection;
 27  import com.filenet.api.core.Domain;
 28  import com.filenet.api.core.Factory;
 29  import com.filenet.api.core.ObjectStore;
 30  import com.filenet.api.util.UserContext;
 31
 32⊖ /**
 33   * This object represents the connection with the Content Engine. Once
 34   * connection is established it intializes Domain and ObjectStoreSet with
 35   * available Domain and ObjectStoreSet.
 36   *
 37   */
 38  public class CEConnection {
 39      private Connection con;
 40      private Domain dom;
 41      private String domainName;
 42      private ObjectStoreSet ost;
 43      private Vector osnames;
 44      private boolean isConnected;
 45      private UserContext uc;
 46      // ASB001
 47      private String _username = "";
 48      // ASB002
```

Figure 2-19. *The **Add External JARs** command fixes the highlighted imports*

15. To correct the Java package **import** errors, we need to add a JRE library, using the **Add Library** command, to the Project **Classpath** as indicated in Figure 2-20.

CHAPTER 2   CONFIGURING JAVA CUSTOM COMPONENTS

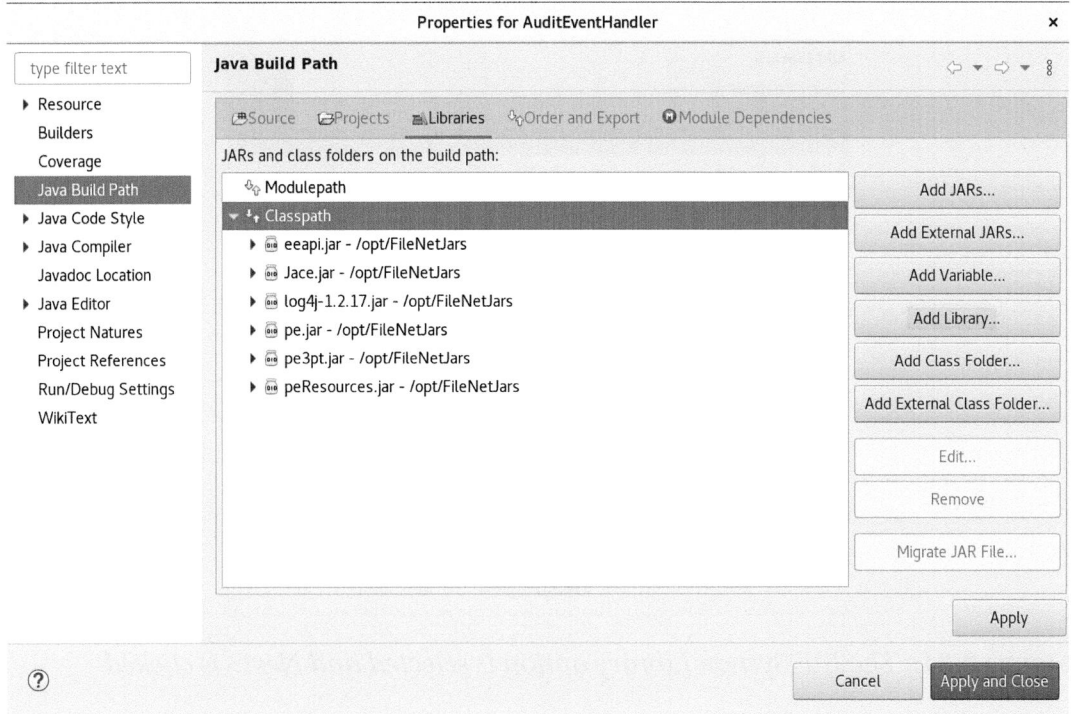

Figure 2-20. *The **Add Library** command fixes the Java package imports*

16. We need to add a **JRE System Library** type which we select and then click the **Next>** command button to show a command which will give a list of the available **JRE** versions.

159

CHAPTER 2   CONFIGURING JAVA CUSTOM COMPONENTS

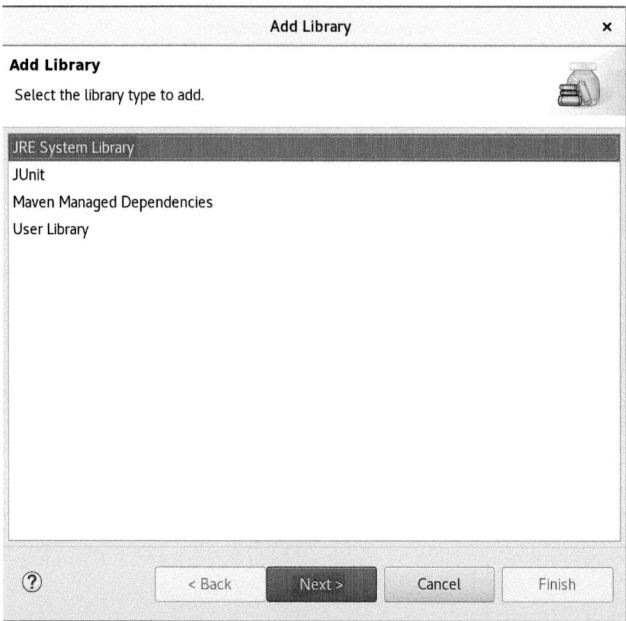

Figure 2-21. *The JRE System Library option is selected and Next> is clicked*

17. The **Installed JREs** command is clicked to allow a choice of installed JRE Java System .jar files.

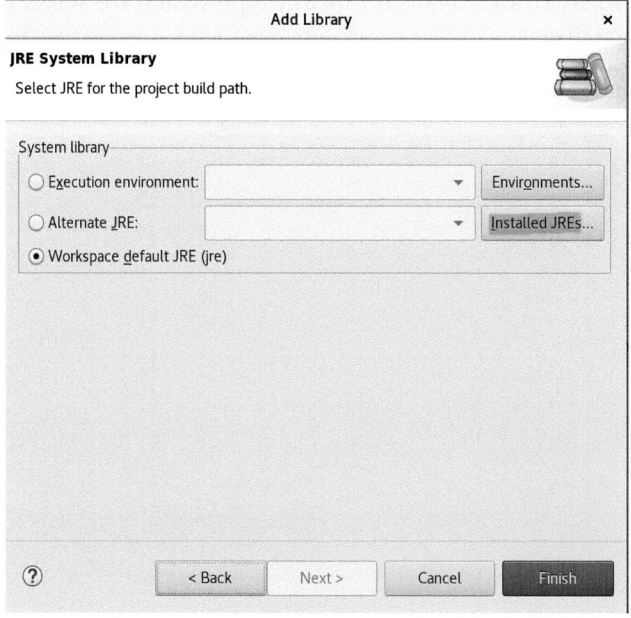

Figure 2-22. *The Installed JREs command button is selected*

160

CHAPTER 2 CONFIGURING JAVA CUSTOM COMPONENTS

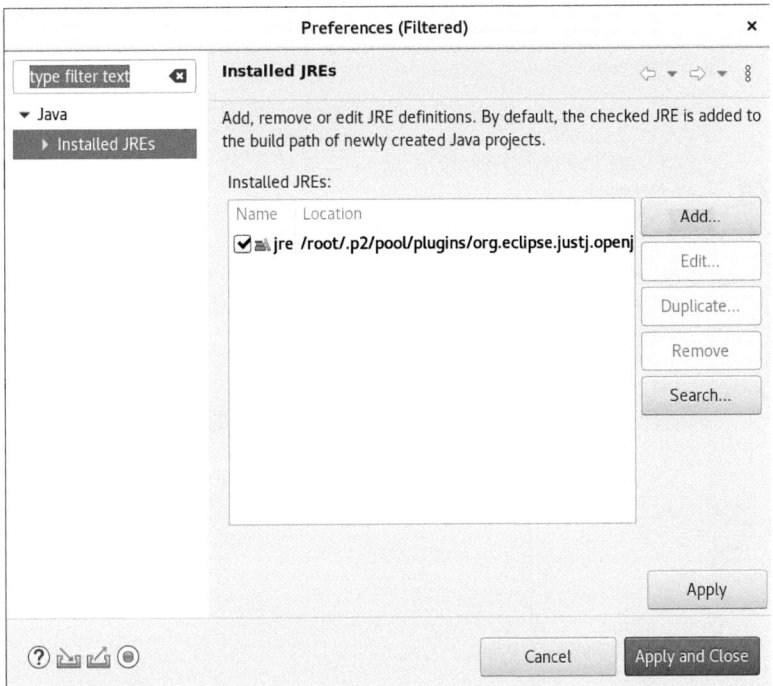

Figure 2-23. *The Installed system JRE is selected, and the **Apply and Close** is clicked*

18. As shown in Figure 2-23, the Add command is used to add the JRE to the **Classpath** for the project, and then the **Apply and Close** command is clicked.

CHAPTER 2   CONFIGURING JAVA CUSTOM COMPONENTS

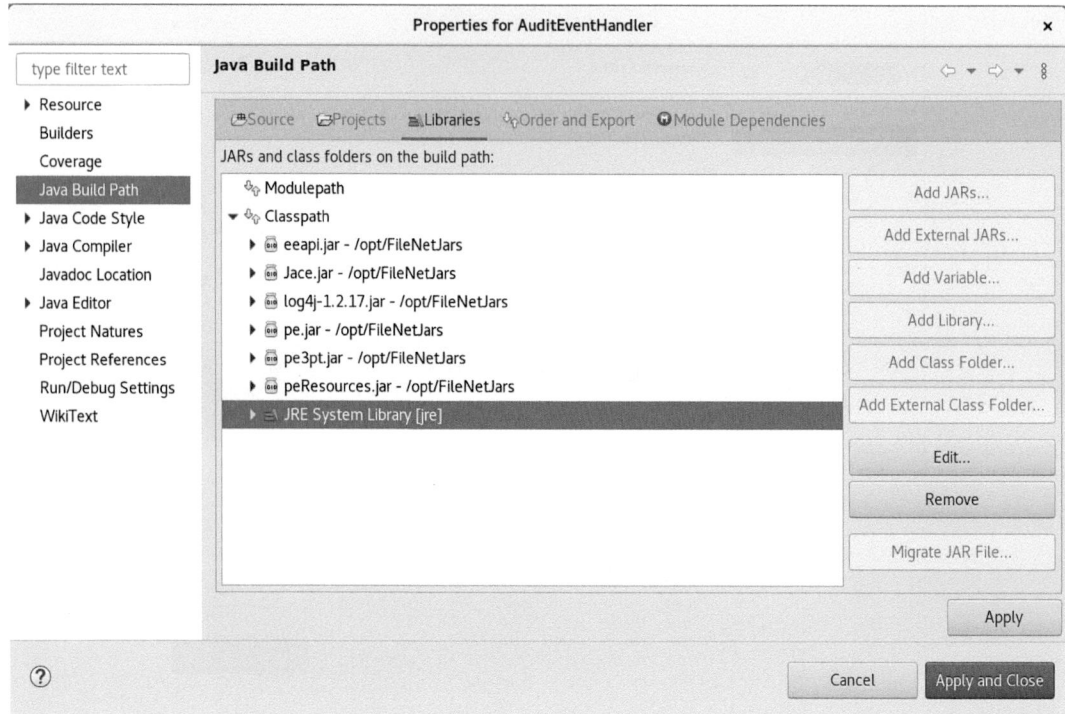

***Figure 2-24.** The JRE System Library appears in the Project Classpath*

The Properties for the **AuditEventHandler** project now show the JRE System Library in the Java Build Classpath.

19. In order to ensure the correct import packages are used to build the project, it is sometimes necessary to use a specific order of .jar files, so that the earlier versions of a package are consistent with the functions required, since it is possible that a Class module with the same name appears in more than one .jar file. The **Classpath** order of the referenced .jar files can be adjusted using the panel shown in Figure 2-25.

CHAPTER 2 CONFIGURING JAVA CUSTOM COMPONENTS

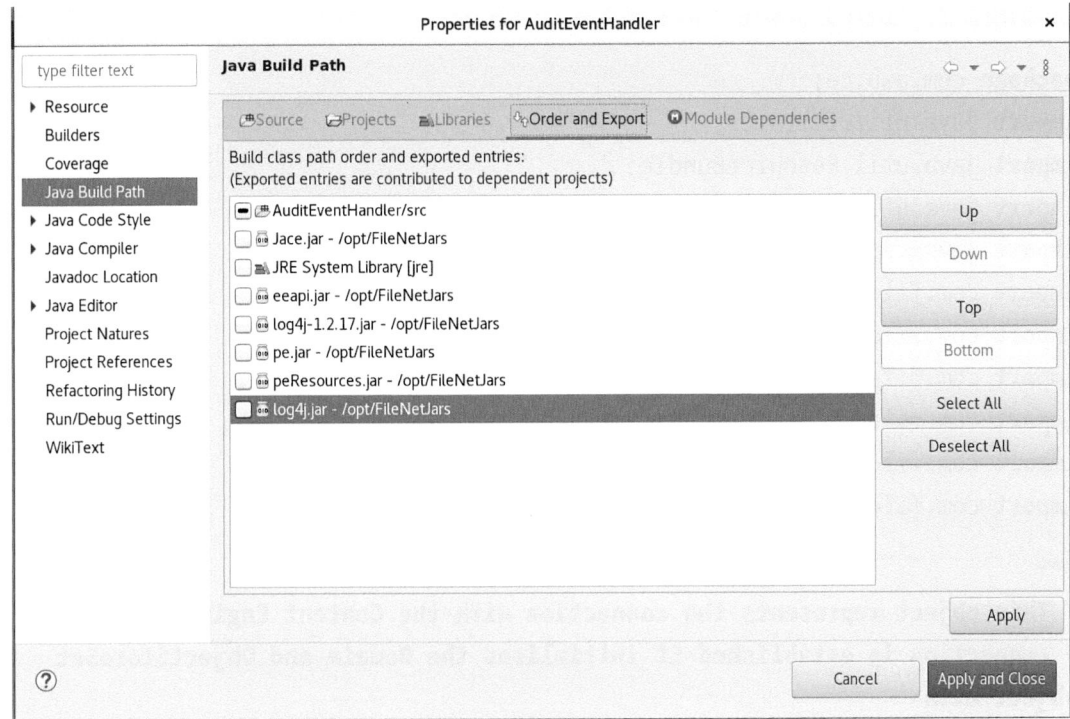

Figure 2-25. *The final working order of the Classpath libraries is set*

20. The CEConnection.java source code is now free of errors.

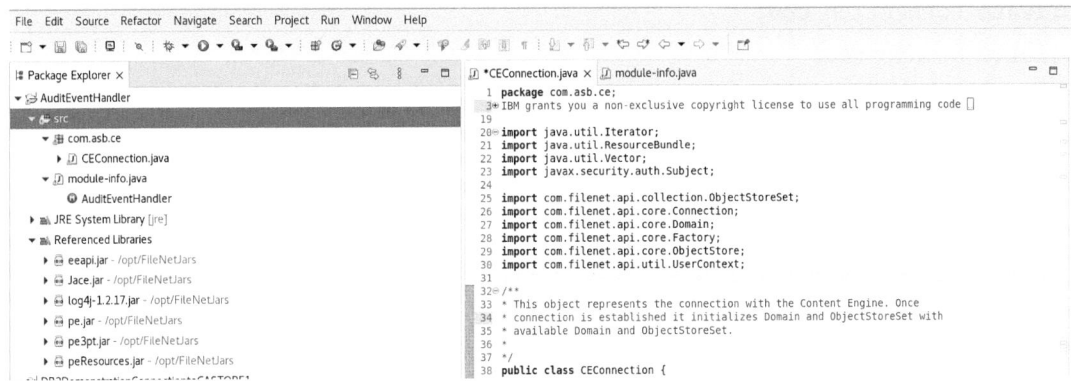

Figure 2-26. *The working CEConnection.java code*

This code is deployed as follows:

Create the Project and add the **CEConnection.java** (Standard IBM code).

Listing 2-1. The CEConnection Java class object

```java
package com.asb.ce;
import java.util.Iterator;
import java.util.ResourceBundle;
import java.util.Vector;
import javax.security.auth.Subject;
import com.filenet.api.collection.ObjectStoreSet;
import com.filenet.api.core.Connection;
import com.filenet.api.core.Domain;
import com.filenet.api.core.Factory;
import com.filenet.api.core.ObjectStore;
import com.filenet.api.util.UserContext;

/**
 * This object represents the connection with the Content Engine. Once
 * connection is established it initialises the Domain and ObjectStoreSet
object with
 * the available Target Store Domain and ObjectStoreSet.
 *
 */
public class CEConnection {
private Connection con;
private Domain dom;
private String domainName;
private ObjectStoreSet ost;
private Vector osnames;
private boolean isConnected;
private UserContext uc;
// ASB001
private String _username = "";
// ASB002
private static final String encryption_delimiter = "\n"; // Newline
/*
 * constructor
 */
```

```java
public CEConnection() {
    con = null;
    uc = UserContext.get();
    dom = null;
    domainName = null;
    ost = null;
    osnames = new Vector();
    isConnected = false;
}

// Object Finalize (called before Garbage collection)
// ASB003.
protected void finalize() {
    this.disconnect();
}
// ASB004.
public void disconnect() {
    if (isConnected) {
        uc.popSubject();
        con = null;
        dom = null;
        isConnected = false;
    }
}

// ASB005
public String getCurrentUser() {
    return _username;
}

// ASB006 Returns the IBM Object Store java Object
public ObjectStore getObjectStore(String ObjectStore)
{
    // This hard coded example would have String variables read from  an encrypted Config.xml or config.properties file
    // in a Production Environment!
```

CHAPTER 2 CONFIGURING JAVA CUSTOM COMPONENTS

```
        //              Values passed are User , password , the stanza jaas.
                        conf.WSI value, MTOM SOAP URL link
    // The jaas.conf.WSI is a file in the path /opt/IBM/ECMClient/
       configure/CE_API/
    // The stanza is as follows in the jaas.conf.WSI file :
    //    FileNetP8WSI {
    //        com.filenet.api.util.WSILoginModule required;
    //    };
    this.establishUnencryptedConnection("Alan", "filenet", "FileNetP8WSI",
    "http://localhost:9080/wsi/FNCEWS40MTOM/");

    return fetchOS(ObjectStore);
}

/*
 * Establishes connection with Content Engine using supplied username,
 * password, JAAS stanza and CE Uri.
 */
public void establishUnencryptedConnection(String userName, String
password, String stanza, String uri) {
    con = Factory.Connection.getConnection(uri);
    Subject sub = UserContext.createSubject(con, userName, password,
    stanza);
    uc.pushSubject(sub);
    dom = fetchDomain();
    domainName = dom.get_Name();
    ost = getOSSet();
    isConnected = true;
    _username = userName;
}

// ASB007 The form of the method called in the CodeModule distribution
public void establishConnection(String userName, String password,
ResourceBundle ResourceBundle) {

    String stanza = ResourceBundle.getString("stanza");
    String uri = ResourceBundle.getString("uri");
```

```java
public CEConnection() {
    con = null;
    uc = UserContext.get();
    dom = null;
    domainName = null;
    ost = null;
    osnames = new Vector();
    isConnected = false;
}

// Object Finalize (called before Garbage collection)
// ASB003.
protected void finalize() {
    this.disconnect();
}
// ASB004.
public void disconnect() {
    if (isConnected) {
        uc.popSubject();
        con = null;
        dom = null;
        isConnected = false;
    }
}

// ASB005
public String getCurrentUser() {
    return _username;
}

// ASB006 Returns the IBM Object Store java Object
public ObjectStore getObjectStore(String ObjectStore)
{
    // This hard coded example would have String variables read from  an
    encrypted Config.xml or config.properties file
    // in a Production Environment!
```

```java
        //                   Values passed are User , password , the stanza jaas.
        //                   conf.WSI value, MTOM SOAP URL link
        // The jaas.conf.WSI is a file in the path /opt/IBM/ECMClient/
           configure/CE_API/
        // The stanza is as follows in the jaas.conf.WSI file :
        //    FileNetP8WSI {
        //         com.filenet.api.util.WSILoginModule required;
        //    };
        this.establishUnencryptedConnection("Alan", "filenet", "FileNetP8WSI",
        "http://localhost:9080/wsi/FNCEWS40MTOM/");

        return fetchOS(ObjectStore);
}

/*
 * Establishes connection with Content Engine using supplied username,
 * password, JAAS stanza and CE Uri.
 */
public void establishUnencryptedConnection(String userName, String
password, String stanza, String uri) {
    con = Factory.Connection.getConnection(uri);
    Subject sub = UserContext.createSubject(con, userName, password,
    stanza);
    uc.pushSubject(sub);
    dom = fetchDomain();
    domainName = dom.get_Name();
    ost = getOSSet();
    isConnected = true;
    _username = userName;
}

// ASB007 The form of the method called in the CodeModule distribution
public void establishConnection(String userName, String password,
ResourceBundle ResourceBundle) {

    String stanza = ResourceBundle.getString("stanza");
    String uri = ResourceBundle.getString("uri");
```

```java
        // Delegate method
        establishUnencryptedConnection(userName, password, stanza, uri);
}

/*
 * Returns Domain object.
 */
public Domain fetchDomain() {
    dom = Factory.Domain.fetchInstance(con, null, null);
    return dom;
}

/*
 * Returns ObjectStoreSet from Domain
 */
public ObjectStoreSet getOSSet() {
    ost = dom.get_ObjectStores();
    return ost;
}
/*
 * Returns vector containing ObjectStore names from object stores available
 * in ObjectStoreSet.
 */
public Vector getOSNames() {
    if (osnames.isEmpty()) {
        Iterator it = ost.iterator();
        while (it.hasNext()) {
            ObjectStore os = (ObjectStore) it.next();
            osnames.add(os.get_DisplayName());
        }
    }
    return osnames;
}
```

```java
/*
 * Boolean method which checks whether the connection has been established
   with the Content Engine or not.
 */
public boolean isConnected() {
    return isConnected;
}

/*
 * Returns the IBM FileNet ObjectStore object for the passed object store
   name string.
 */
public ObjectStore fetchOS(String name) {
    ObjectStore os = Factory.ObjectStore.fetchInstance(dom, name, null);
    return os;
}

/*
 * Getter method which Returns the domain name.
 */
public String getDomainName() {
    return domainName;
}

/*
 * Getter method which Returns the current connection object.
 */
public Connection getCurrentConnection() {
    return con;
}

/*
 * Getter method which Returns the domain object.
 */
public Domain getCurrentDomain() {
    return dom;
}
}
```

CHAPTER 2 CONFIGURING JAVA CUSTOM COMPONENTS

Now we can create another Java Class for the project. This is the Event Handler itself, so we add a new class package, **com.ibm.filenet.p8.ce.eventhandlers.codemodules**, as shown in Figure 2-27.

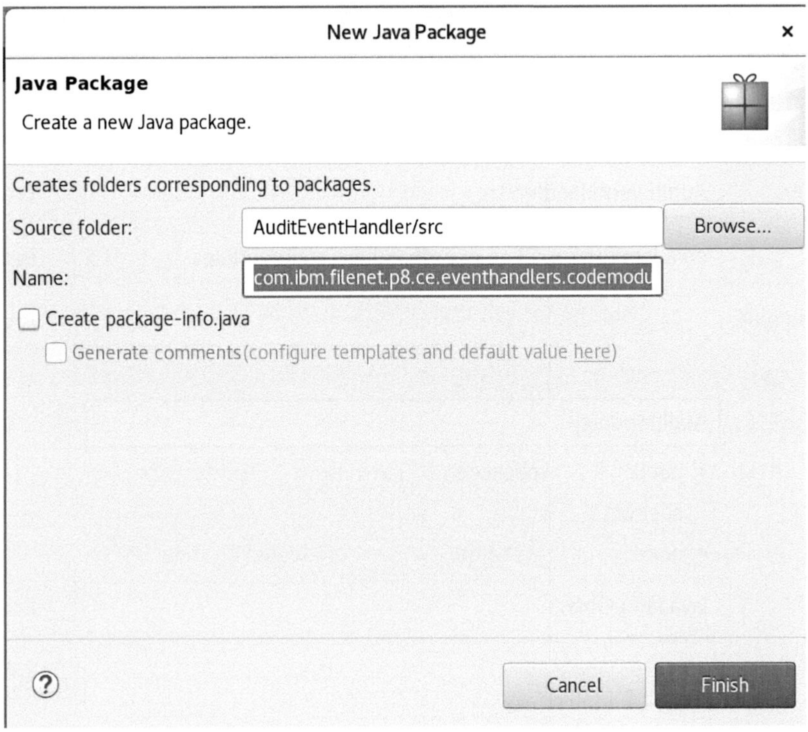

Figure 2-27. The AuditModule Class package is created

CHAPTER 2   CONFIGURING JAVA CUSTOM COMPONENTS

Next, the **AuditModule** Java class is created under this package using the screen in Figure 2-28.

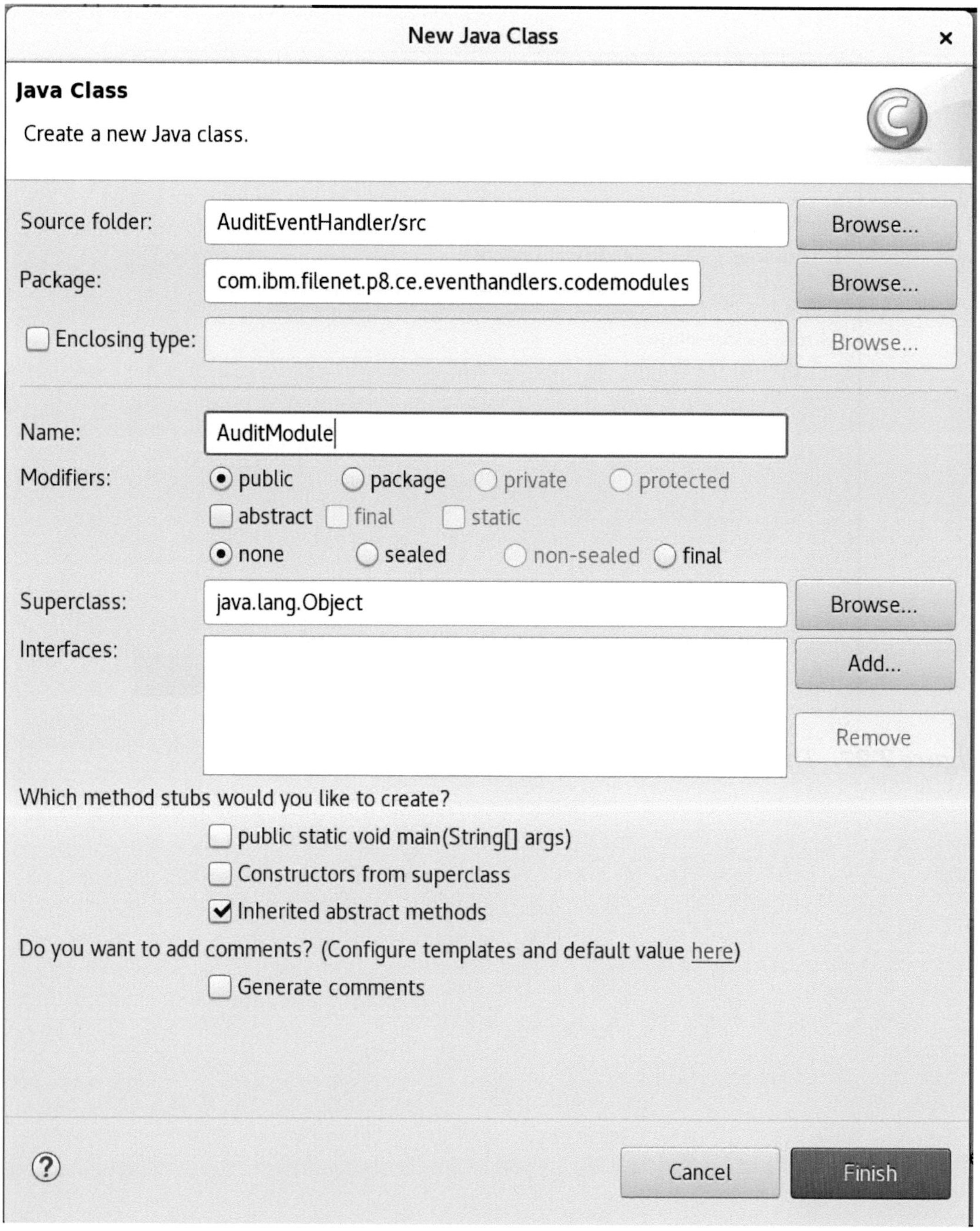

Figure 2-28. The AuditModule Java class is created

CHAPTER 2　CONFIGURING JAVA CUSTOM COMPONENTS

The following code is the main CodeModule class with the Event Handler **AuditModule** class for the **addDocToFolder** Java method.

Listing 2-2. The AuditModule Java class object

```java
package com.ibm.filenet.p8.ce.eventhandlers.codemodules;

import java.util.Iterator;

import org.apache.log4j.Logger;
import com.filenet.api.collection.ContentElementList;
import com.filenet.api.collection.ReferentialContainmentRelationshipSet;
import com.filenet.api.collection.StringList;
import com.filenet.api.constants.AutoUniqueName;
import com.filenet.api.constants.DefineSecurityParentage;
import com.filenet.api.constants.RefreshMode;
import com.filenet.api.core.Annotation;
import com.filenet.api.core.ContentElement;
import com.filenet.api.core.CustomObject;
import com.filenet.api.core.Document;
import com.filenet.api.core.DynamicReferentialContainmentRelationship;
import com.filenet.api.core.Factory;
import com.filenet.api.core.Folder;
import com.filenet.api.core.ObjectStore;
import com.filenet.api.events.ObjectChangeEvent;
import com.filenet.api.query.SearchSQL;
import com.filenet.api.query.SearchScope;
import com.filenet.api.util.Id;

public class AuditModule {

    private static Logger logger;
    static {
        System.setProperty("log4j.configuration", "events_log4j.xml");
        logger = Logger.getLogger(AuditModule.class);
    }
```

```java
public void addDocToFolder(Document d) {
    //We retrieve the folder ID from the Document folder_Id_to_file_
      in property
    //(which is added to the Audit Document Class)
    String propNames[] = {"AUD_folder_Id_to_file_in"};
    d.fetchProperties(propNames);
    //The folderId string contains a GUID as a string as in our test
      example e.g.: {00D14A81-0000-C31E-A7D2-F2B7CF2D4803}
      String folderId = d.getProperties().getStringValue("AUD_folder_
      Id_to_file_in");
    //IBM FileNet Folder object is retrieved from the ObjectStore
      based on the string folderId
      Folder parentFolder = Factory.Folder.fetchInstance(d.
      getObjectStore(), folderId, null);
    //The IBM FileNet drcr Object is a linking IBM FileNet object
      which links a Document Object to a Folder Object
      DynamicReferentialContainmentRelationship drcr
      = (DynamicReferentialContainmentRelationship)
      parentFolder.file(d, AutoUniqueName.AUTO_UNIQUE, null,
      DefineSecurityParentage.DO_NOT_DEFINE_SECURITY_PARENTAGE);
    drcr.save(RefreshMode.NO_REFRESH);
}

}
```

CHAPTER 2 CONFIGURING JAVA CUSTOM COMPONENTS

The next Java class, **TestHarness**, is used to test the Event method(s).

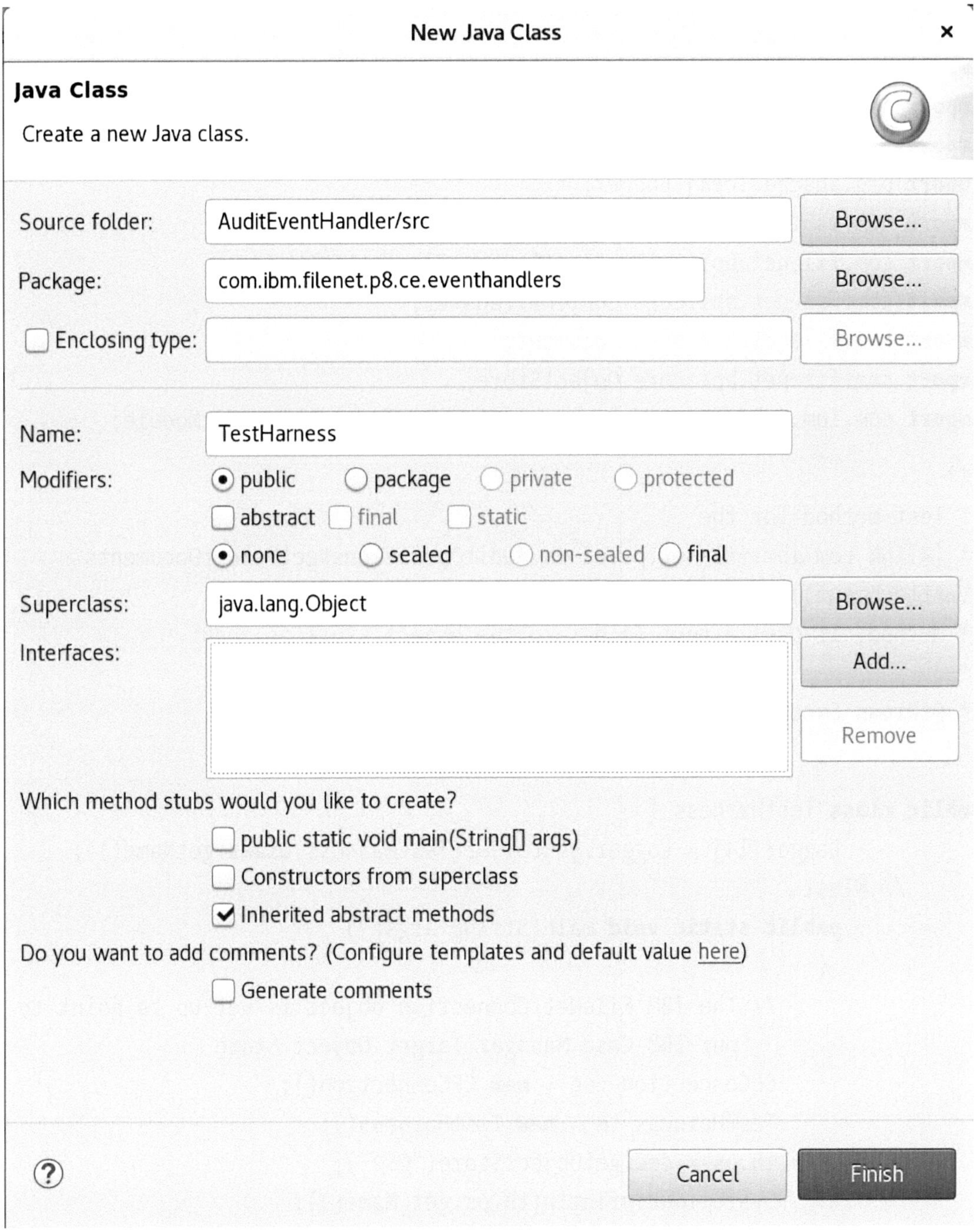

Figure 2-29. The TestHarness Java class is created

CHAPTER 2 CONFIGURING JAVA CUSTOM COMPONENTS

Listing 2-3. The TestHarness Java class object

```java
package com.ibm.filenet.p8.ce.eventhandlers.test;

import java.io.IOException;
import java.io.InputStream;
import java.util.Iterator;
import org.apache.log4j.Logger;
import com.asb.ce.CEConnection;
import com.filenet.api.collection.ContentElementList;
import com.filenet.api.core.ContentTransfer;
import com.filenet.api.core.Document;
import com.filenet.api.core.ObjectStore;
import com.ibm.filenet.p8.ce.eventhandlers.codemodules.AuditModule;

/**
 * Test method for the
 * {@link com.ibm.filenet.ps.ciops.AuditOperations#getFolderDocuments
 (VWAttachment)}
 * method. It uses a test folder in the object store as input.
 *
 * @throws Exception
 */
public class TestHarness {
        Logger l4j = Logger.getLogger(TestHarness.class.getName());
    // @Test
        public static void main(String args[])
         {
            // The IBM FileNet Connection object is set up to point to
               our IBM Case Manager Target Object Store
            CEConnection cec = new CEConnection();
            TestHarness th = new TestHarness();
            th.os = cec.getObjectStore("OS2");
            System.out.println(th.os.get_Name());
        //
```

```java
            try {
                th.test_AddDocToFolder();
            } catch (Exception e) {
                // TODO Auto-generated catch block
                e.printStackTrace();
            }

        }
        //The Test Document ID for AC.pdf file is {00D14A81-0000-C31E-
        A7D2-F2B7CF2D4803}
        private void test_AddDocToFolder() throws Exception {
            l4j.debug("test_AddDocToFolder");
            Document d = (Document) os.getObject("Document",
            "{00D14A81-0000-C31E-A7D2-F2B7CF2D4803}");
            am.addDocToFolder(d);
        }
// Test using the AUDIT_TEST folder Id = {896F3369-7AF8-405E-85A7-3E18F90
ACE43}
    private ObjectStore os;
    private AuditModule am = new AuditModule();
    public static InputStream getDocContentStream(com.filenet.api.core.
    Document doc) {
        ContentElementList docContentList = doc.get_ContentElements();
        Iterator iter = docContentList.iterator();
        InputStream stream = null;
        while (iter.hasNext()) {
            ContentTransfer ct = (ContentTransfer) iter.next();
            ct.set_RetrievalName("AuditDocTest"); // ASB Test
            Document name

            int docLen = ct.get_ContentSize().intValue();
            byte[] buf = new byte[docLen];
            stream = ct.accessContentStream();
            try {
                stream.read(buf, 0, docLen);
                String readStr = new String(buf);
                System.out.println("Content:\n " + readStr);
```

CHAPTER 2 CONFIGURING JAVA CUSTOM COMPONENTS

```
            } catch (IOException ioe) {
                ioe.printStackTrace();
            }
            System.out.println("\nElement Sequence number: " + ct.get_
            ElementSequenceNumber().intValue() + "\n"
                    + "Content type: " + ct.get_ContentType() + "\n");
        }
        return stream;
    }
}
```

The integration with the IBM FileNet Event system is interfaced using the **AddDocToFolder** Event Java Class. This implements the abstract IBM FileNet Content Engine class, **EventActionHandler**, using the code listed as follows.

Listing 2-4. The AddDocToFolder Event Java Class object code

```
package com.ibm.filenet.p8.ce.eventhandlers;

import org.apache.log4j.Logger;

import com.filenet.api.core.Document;
import com.filenet.api.engine.EventActionHandler;
import com.filenet.api.events.ObjectChangeEvent;
import com.filenet.api.exception.EngineRuntimeException;
import com.filenet.api.util.Id;
import com.ibm.filenet.p8.ce.eventhandlers.codemodules.AuditModule;

public class AddDocToFolder implements EventActionHandler {

    private static Logger logger = Logger.getLogger(AddDocToFolder.class.
    getName());
    private static final String M_NAME = "AddDocToFolder";
    private AuditModule AMModule = new AuditModule(); //Construct a new
    Audit Master event handler

    public AddDocToFolder(){
        logger.debug(M_NAME + " Constructor");
    }
```

```java
    public void onEvent(ObjectChangeEvent event, Id arg1) throws
EngineRuntimeException {
        logger.debug(M_NAME + " Started");
        try {
            Document d = AMModule.getDocument(event);
            AMModule.addDocToFolder(d);
        }
        catch (Exception e){
            logger.error(M_NAME + " failed - " + e.getMessage(),e);
            EngineRuntimeException ere = new EngineRuntimeException();
            ere.setStackTrace(e.getStackTrace());
            throw ere;
        }
        finally {
            logger.debug(M_NAME + " Finished");
        }
    }
}
```

Add the FileNet 5.5.x jar libraries to the project **AuditEventHandler**, **TestHarness class module run configurations**:

Modulepath

Jace.jar
eeapi.jar
log4j-1.2.17.jar
pe.jar
peResources.jar
log4j.jar

Classpath

AuditEventHandler
JRE System Library[jre]
peResources.jar
pe.jar

CHAPTER 2   CONFIGURING JAVA CUSTOM COMPONENTS

```
log4j-1.2.17.jar
log4j.jar
Jace.jar
eapi.jar
```

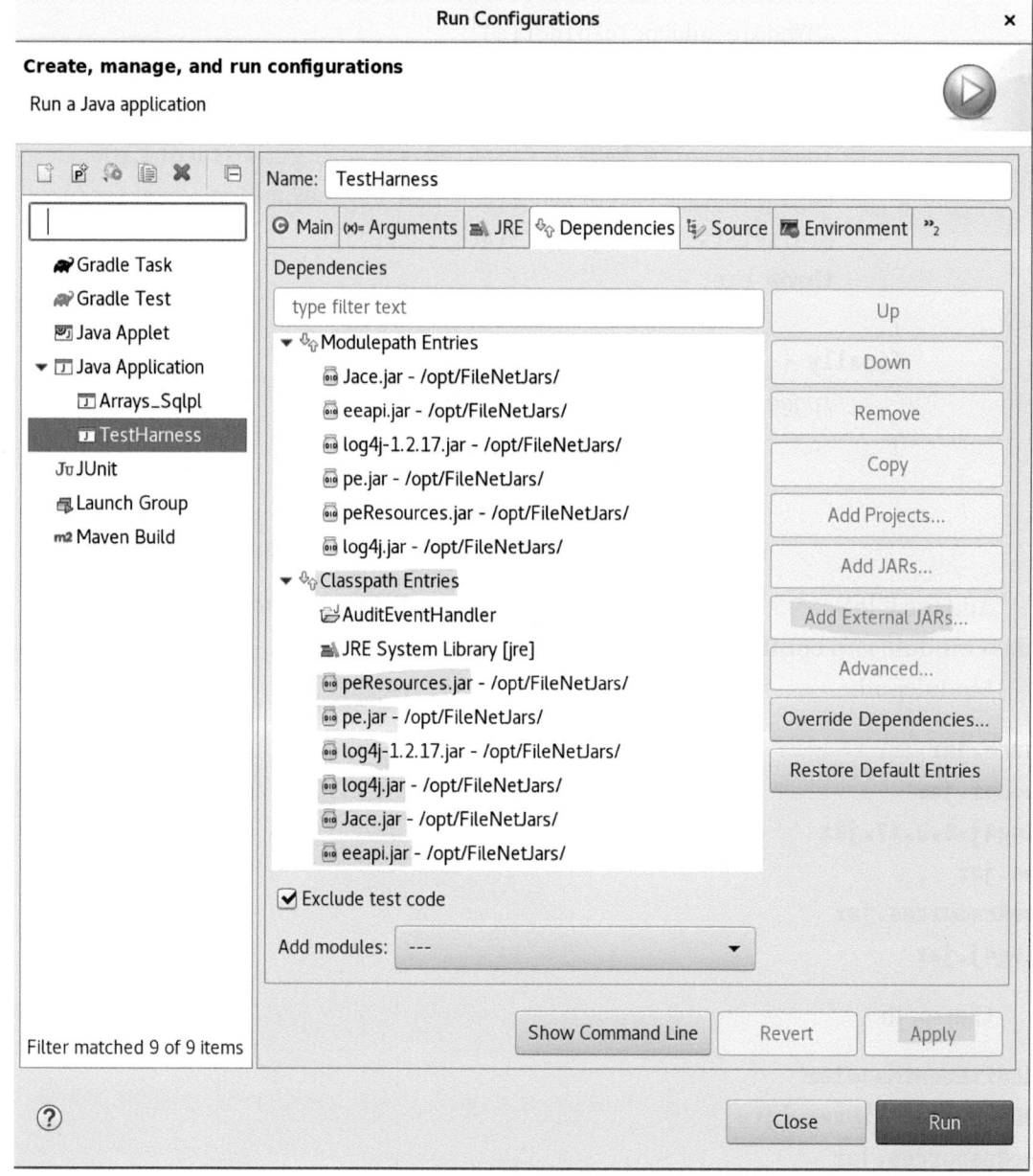

Figure 2-30. The TestHarness Run Configurations for the Classpath

```java
    public void onEvent(ObjectChangeEvent event, Id arg1) throws
EngineRuntimeException {

        logger.debug(M_NAME + " Started");
        try {
            Document d = AMModule.getDocument(event);
            AMModule.addDocToFolder(d);
        }
        catch (Exception e){
            logger.error(M_NAME + " failed - " + e.getMessage(),e);
            EngineRuntimeException ere = new EngineRuntimeException();
            ere.setStackTrace(e.getStackTrace());
            throw ere;
        }
        finally {
            logger.debug(M_NAME + " Finished");
        }

    }
}
```

Add the FileNet 5.5.x jar libraries to the project **AuditEventHandler**, **TestHarness** **class module run configurations**:

Modulepath

Jace.jar
eeapi.jar
log4j-1.2.17.jar
pe.jar
peResources.jar
log4j.jar

Classpath

AuditEventHandler
JRE System Library[jre]
peResources.jar
pe.jar

CHAPTER 2   CONFIGURING JAVA CUSTOM COMPONENTS

```
log4j-1.2.17.jar
log4j.jar
Jace.jar
eapi.jar
```

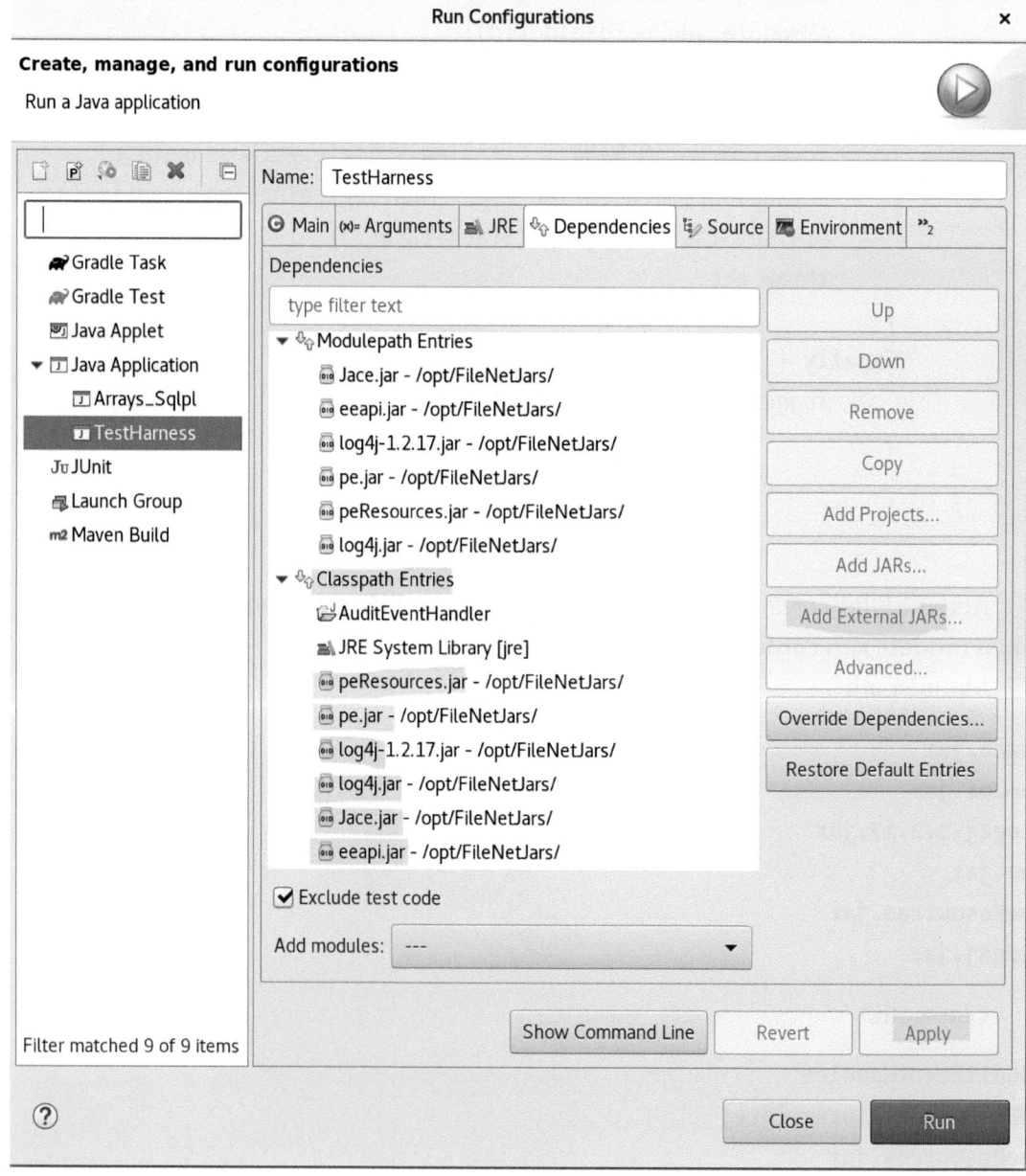

Figure 2-30.   The TestHarness Run Configurations for the Classpath

CHAPTER 2 CONFIGURING JAVA CUSTOM COMPONENTS

```java
    public void onEvent(ObjectChangeEvent event, Id arg1) throws
    EngineRuntimeException {

        logger.debug(M_NAME + " Started");
        try {
            Document d = AMModule.getDocument(event);
            AMModule.addDocToFolder(d);
        }
        catch (Exception e){
            logger.error(M_NAME + " failed - " + e.getMessage(),e);
            EngineRuntimeException ere = new EngineRuntimeException();
            ere.setStackTrace(e.getStackTrace());
            throw ere;
        }
        finally {
            logger.debug(M_NAME + " Finished");
        }

    }
}
```

Add the FileNet 5.5.x jar libraries to the project **AuditEventHandler**, **TestHarness class module run configurations**:

Modulepath

Jace.jar
eeapi.jar
log4j-1.2.17.jar
pe.jar
peResources.jar
log4j.jar

Classpath

AuditEventHandler
JRE System Library[jre]
peResources.jar
pe.jar

CHAPTER 2 CONFIGURING JAVA CUSTOM COMPONENTS

```
log4j-1.2.17.jar
log4j.jar
Jace.jar
eapi.jar
```

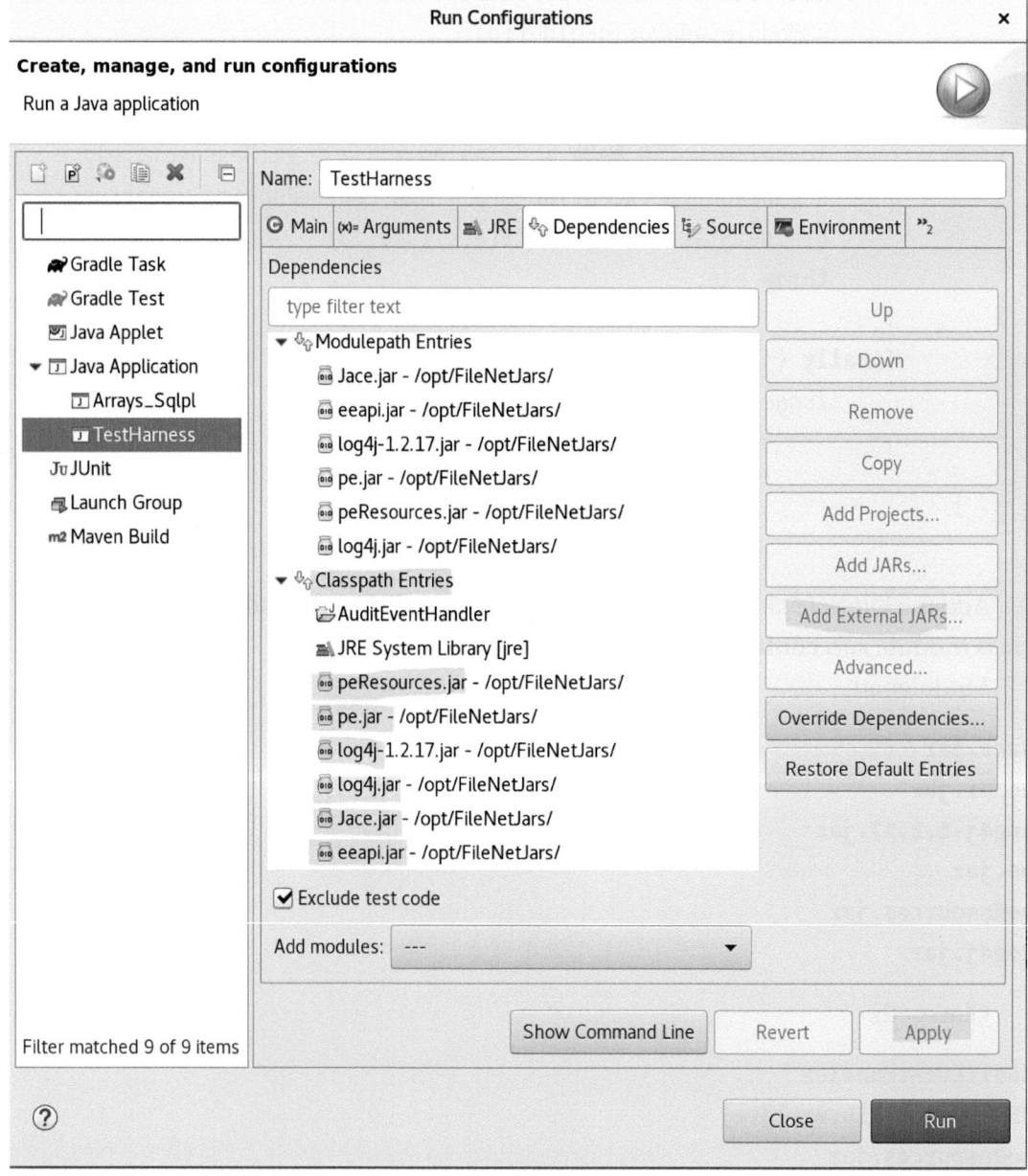

Figure 2-30. *The TestHarness Run Configurations for the Classpath*

This **AuditModule** Code Module contains one Event Handler class, **addDocToFolder**. This is passed the Case Audit Report document object to be filed in an **AUDIT_TEST** common Folder for ease of access to audits using the IBM Case Manager Audit Master solution.

The code loads the **Audit Report** class Document Objects with the String property, **AUD_folder_Id_to_file_in,** which contains the GUID of the Case Target Object Store's **AUDIT_TEST** folder. The **addDocToFolder** Event Handler is triggered and causes a new **DRCR** (Dynamic Referential Containment Relationship Object) **Folder** link object to be created, which is the standard linking method to allow an IBM FileNet Folder to link in a many-to-many relationship with the FileNet Audit Report Document objects.

Custom Code Module Java JAR API Call Development

1. Confirm that the "**CodeModules**" folder exists. This is a root-level folder of class "Folder." It only has one special attribute. It is a "**hidden**" folder and so is not visible in IBM Content Navigator. If this folder does not exist, create it as you would create any other folder, and then change its "**isHiddenContainer**" attribute as shown in Figure 2-31.

CHAPTER 2 CONFIGURING JAVA CUSTOM COMPONENTS

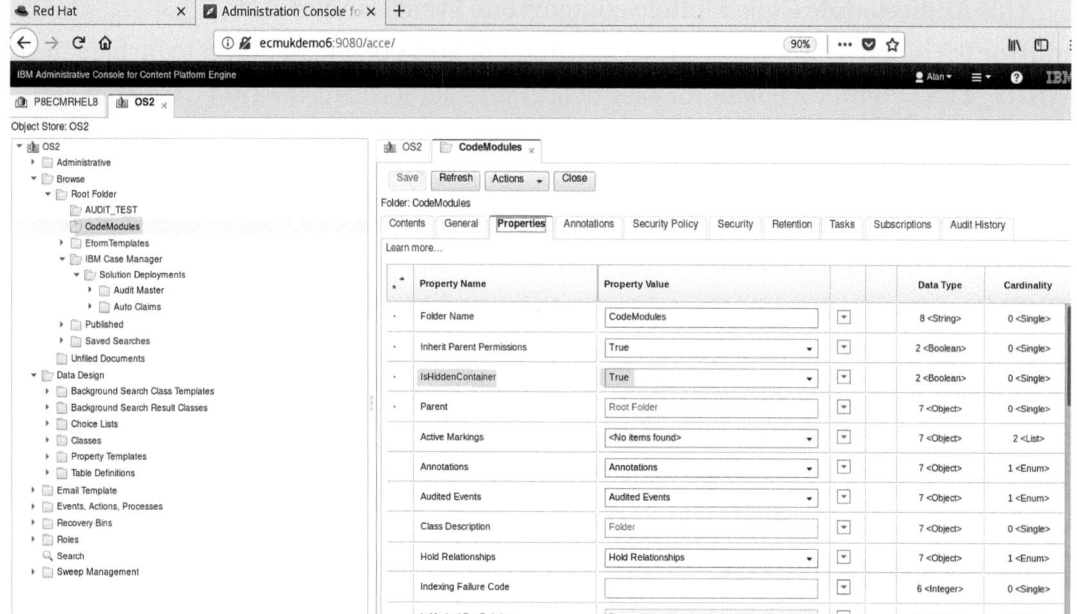

Figure 2-31. *The CodeModules Folder isHiddenContainer property is set to True*

Adding a Code Module

2. Create a new Document in this folder. Right-click the "**CodeModules**" folder in the tree view, and select "**New Document.**"

CHAPTER 2 CONFIGURING JAVA CUSTOM COMPONENTS

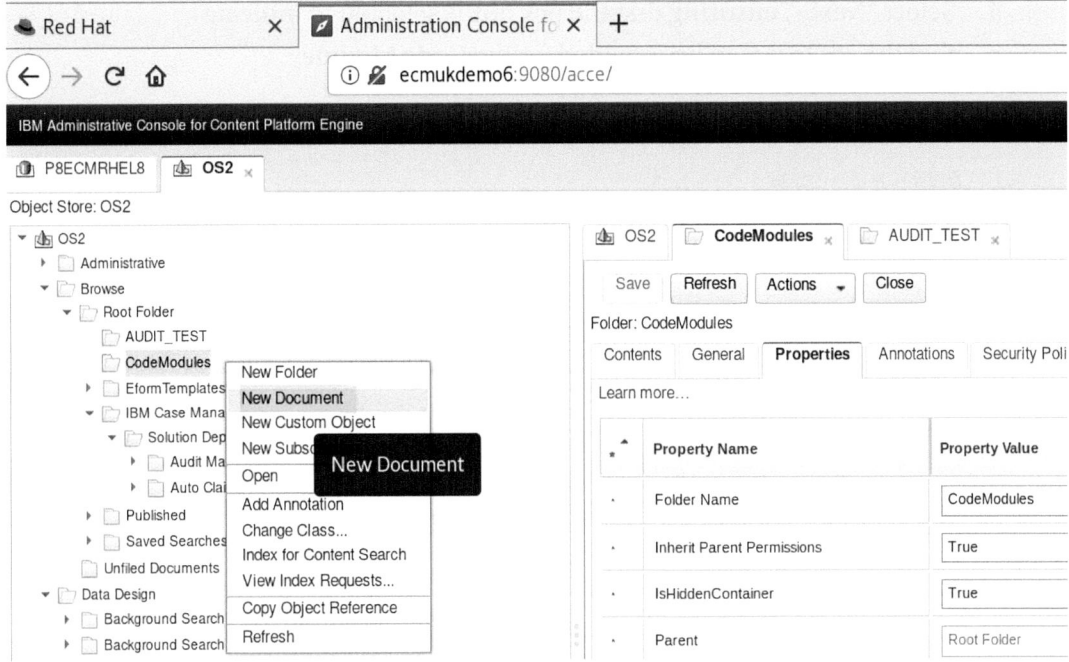

Figure 2-32. *The CodeModules folder New Document menu item is selected*

3. You will be presented with the **AddDocument** wizard. Enter a name (**DocumentTitle**) for the Document. We will see this later when we come to configure the Events. This will show as the name of the code module.

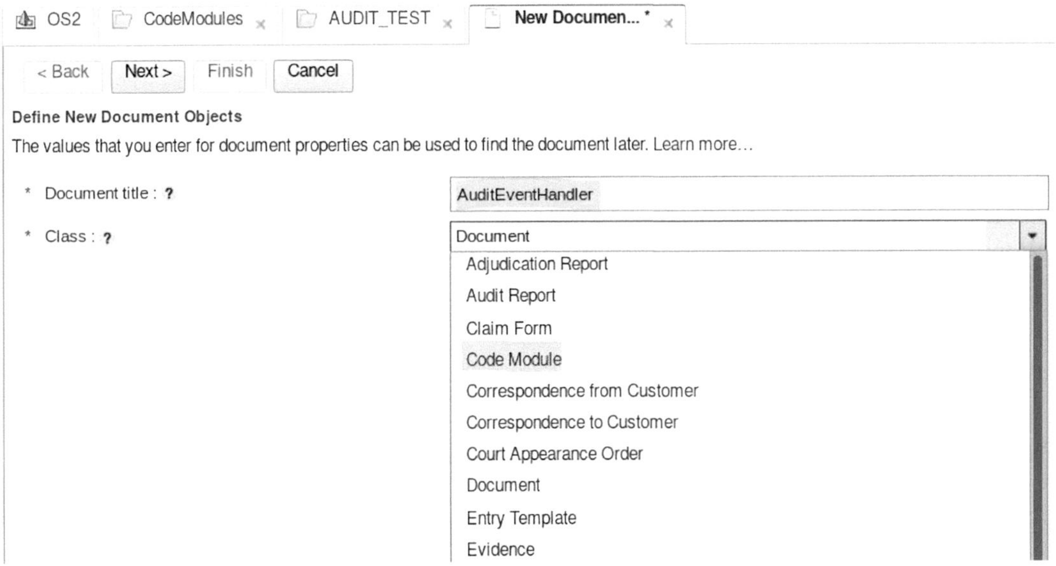

Figure 2-33. *The Document title and Class of Code Module is selected*

181

CHAPTER 2   CONFIGURING JAVA CUSTOM COMPONENTS

4. Select "Next>", ensuring that the tick box is selected to indicate that the content is to be provided for the Code Module.

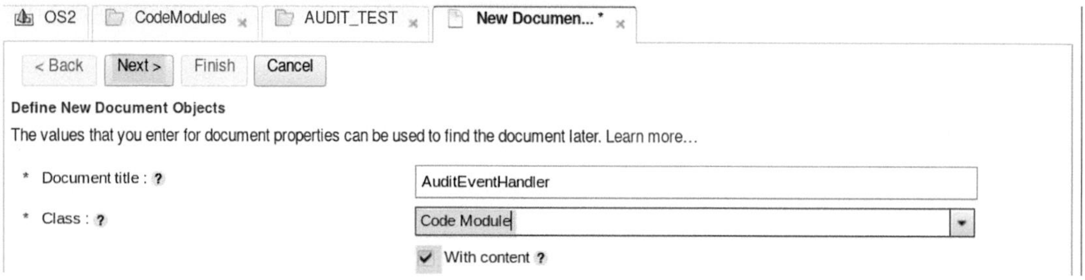

Figure 2-34. The Next command is selected with the With content tick box ticked

5. On the next wizard page, use the "**Add**" button to select your jar file (**fn_eventhandlers.jar**).

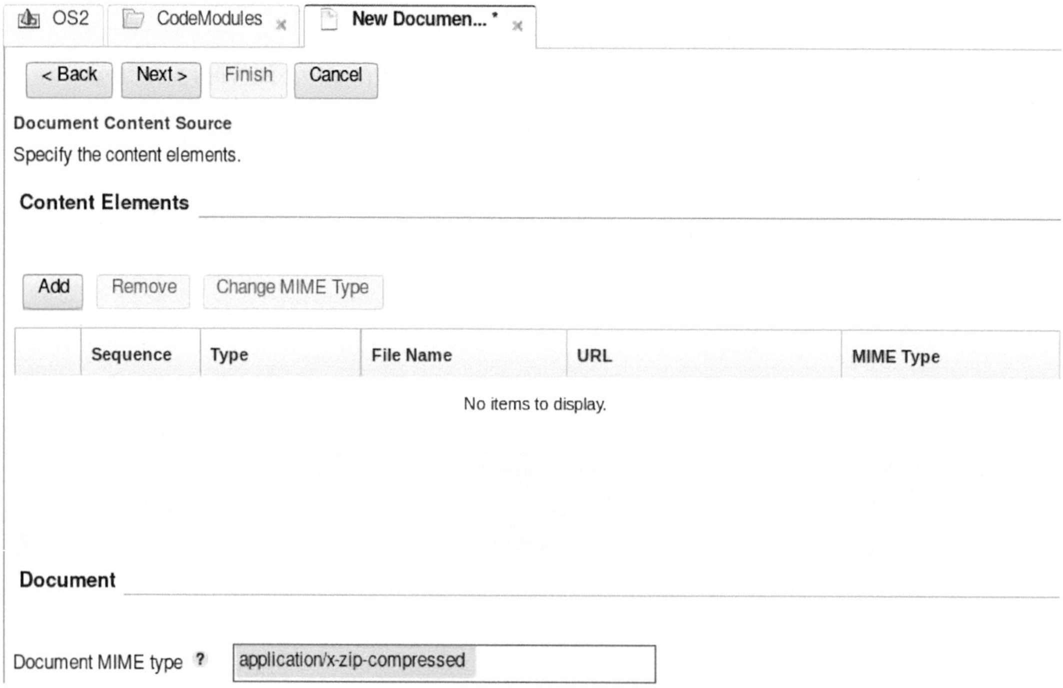

Figure 2-35. The mime type is changed to application/x-zip-compressed

6. Change the MIME type **before adding the File**.

You will need to enter this as **application/x-zip-compressed**.

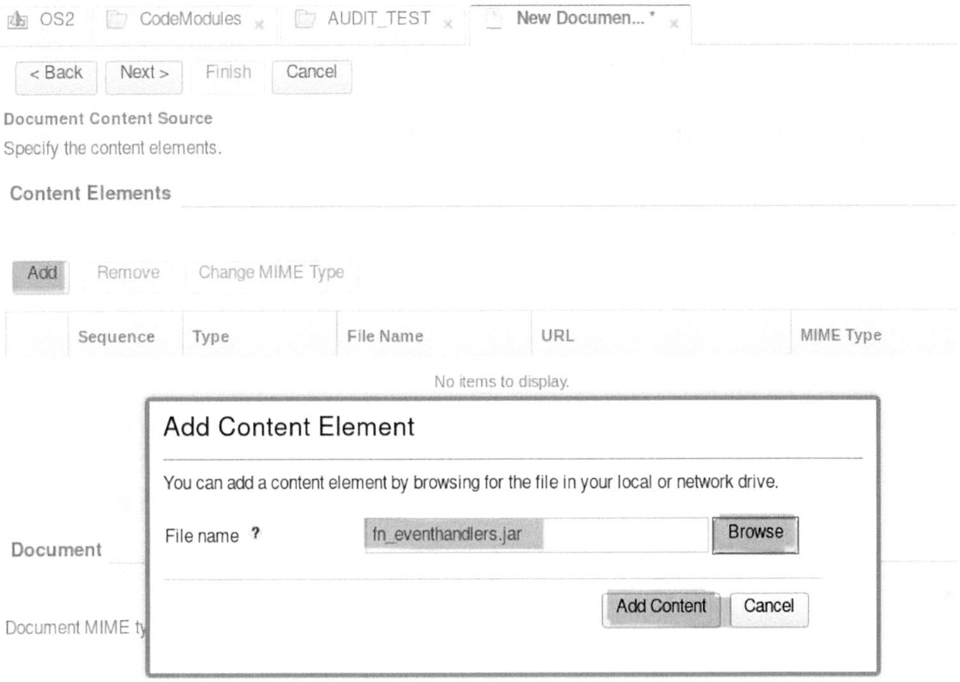

CHAPTER 2 CONFIGURING JAVA CUSTOM COMPONENTS

Figure 2-36. Click the Add button to add the compiled fn_eventhandlers.jar file

The **Browse** button is used to find the compiled **fn_eventhandlers.jar** file. On our Linux system, this is in the path **/root/eclipse/java-2021-12/eclipse**, but this is system dependent.

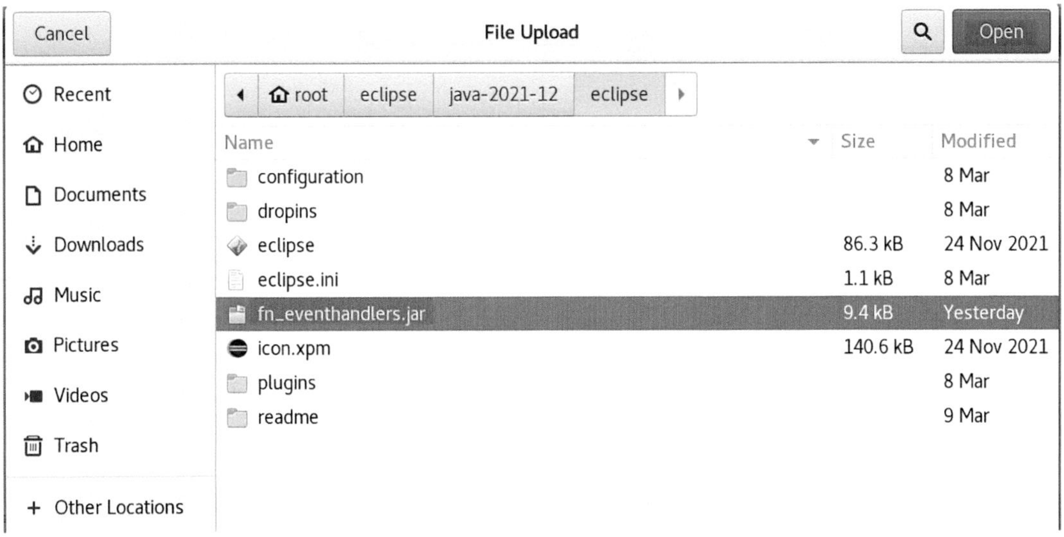

Figure 2-37. The Browse button shows the fn_eventhandlers.jar which is selected

CHAPTER 2 CONFIGURING JAVA CUSTOM COMPONENTS

The **Open** command button is then used to select the **fn_eventhandlers.jar** file for the **Add Content** button in Figure 2-36 to add into the Object store.

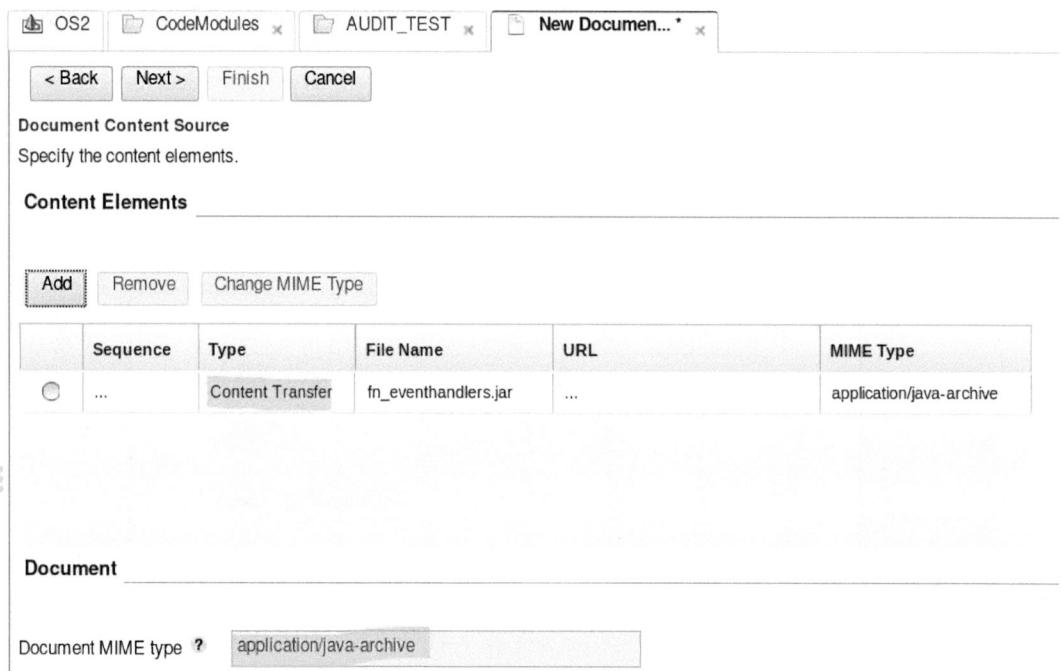

Figure 2-38. *The system identifies the Mime type as **applicationjava-archive***

On clicking the **Next>** command button, we see Figure 2-39.

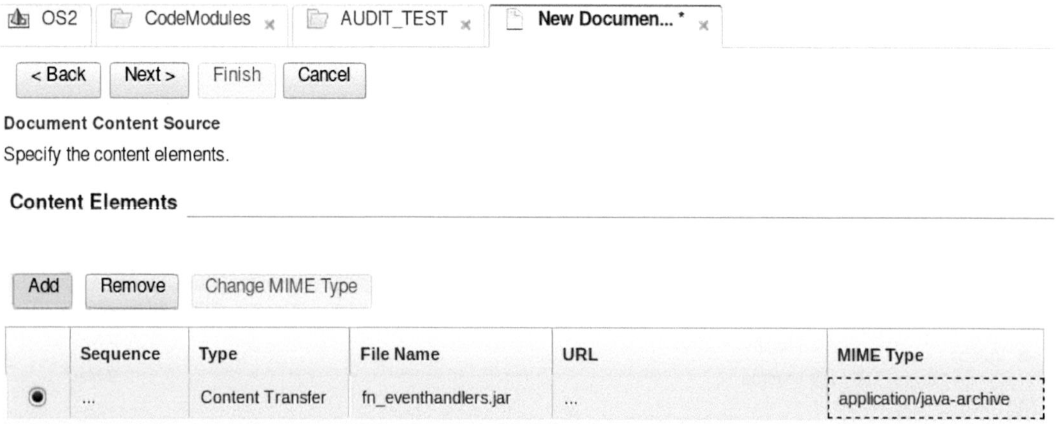

Figure 2-39. *The Next button is clicked again*

CHAPTER 2 CONFIGURING JAVA CUSTOM COMPONENTS

The Property values for the **fn_eventhandlers** Code Module are displayed; we just click the **Next>** command again (highlighted in Figure 2-40).

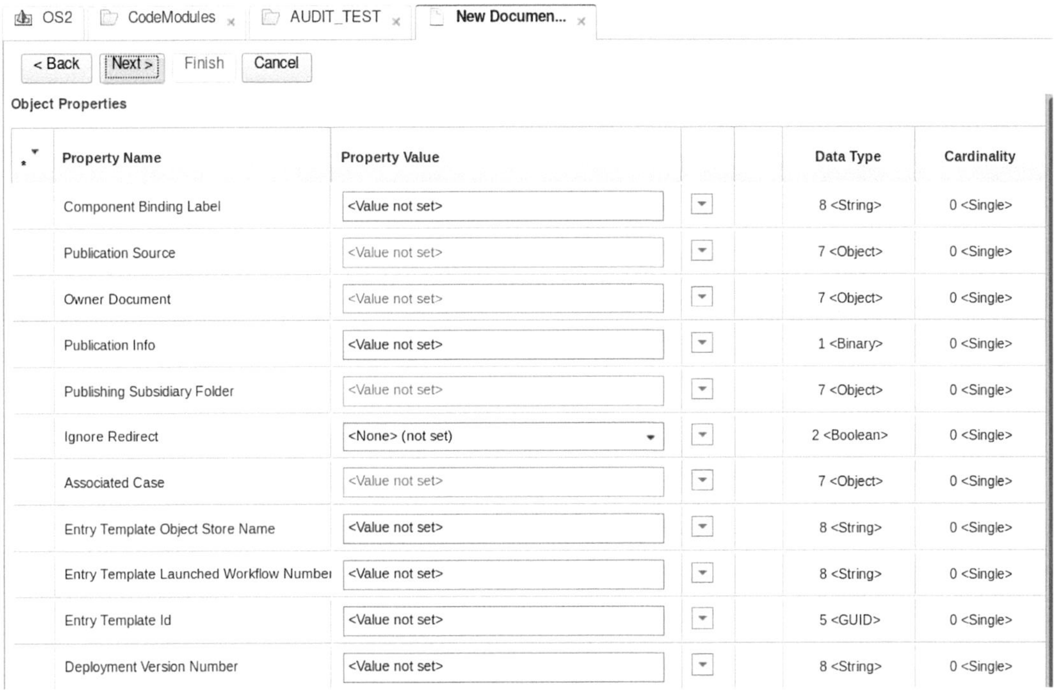

Figure 2-40. The Next> command is clicked (no changes are required here)

The next screen shows the **Version** property options for the document; in this case, we use the default options as shown in Figure 2-41.

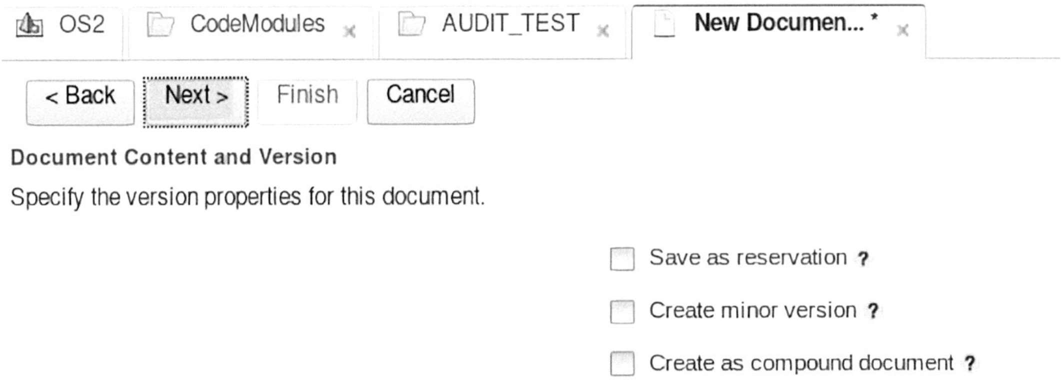

Figure 2-41. The default Version property is used here

CHAPTER 2   CONFIGURING JAVA CUSTOM COMPONENTS

The **Next>** command shows the **Retention** time options for the document, which can be set to allow automatic removal after a selected time interval.

For this functionality, we use the default, which never deletes the document.

Figure 2-42. The Next command is clicked to use the defaults

The next screen shows the Code Modules Storage area. We set this to the default using the drop-down highlighted in Figure 2-43 and then click the Next> command button.

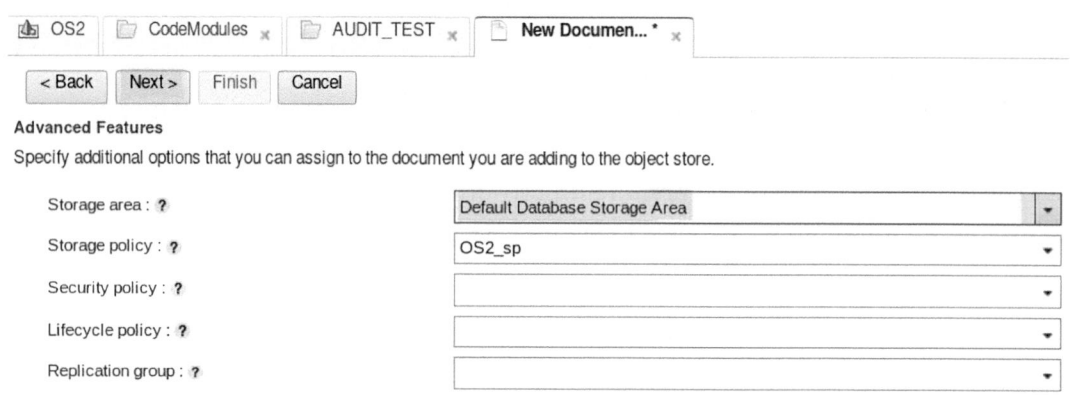

Figure 2-43. The Default Database Storage Area is selected

Finally, we can create the AuditEventHandler Code Module by clicking the **Finish** command highlighted in Figure 2-44.

CHAPTER 2 CONFIGURING JAVA CUSTOM COMPONENTS

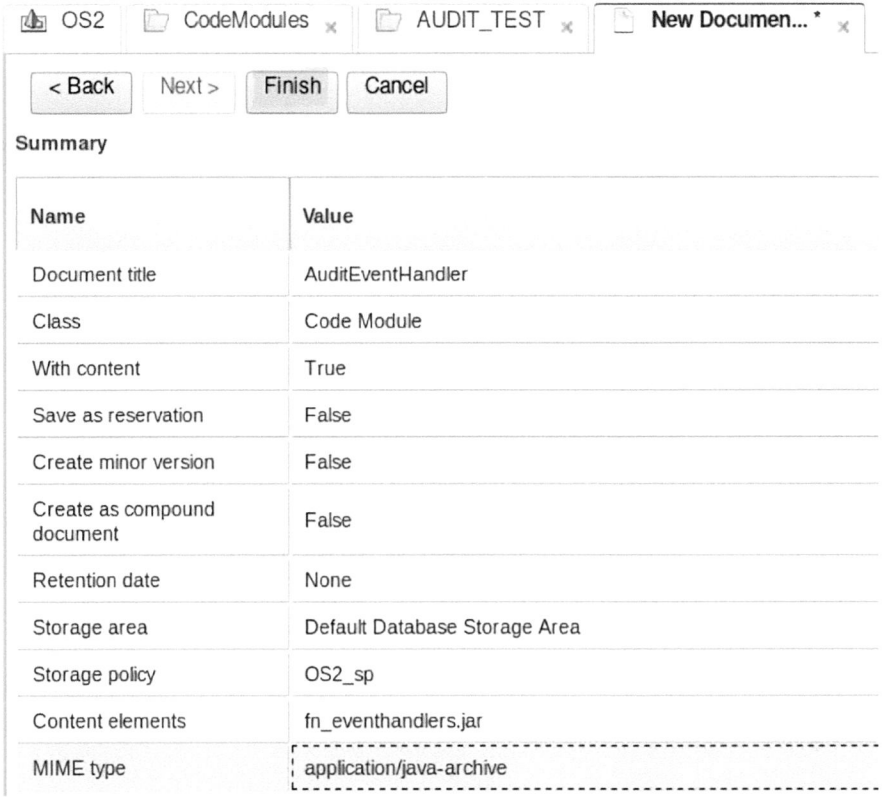

Figure 2-44. *The Finish command creates the Code Module document*

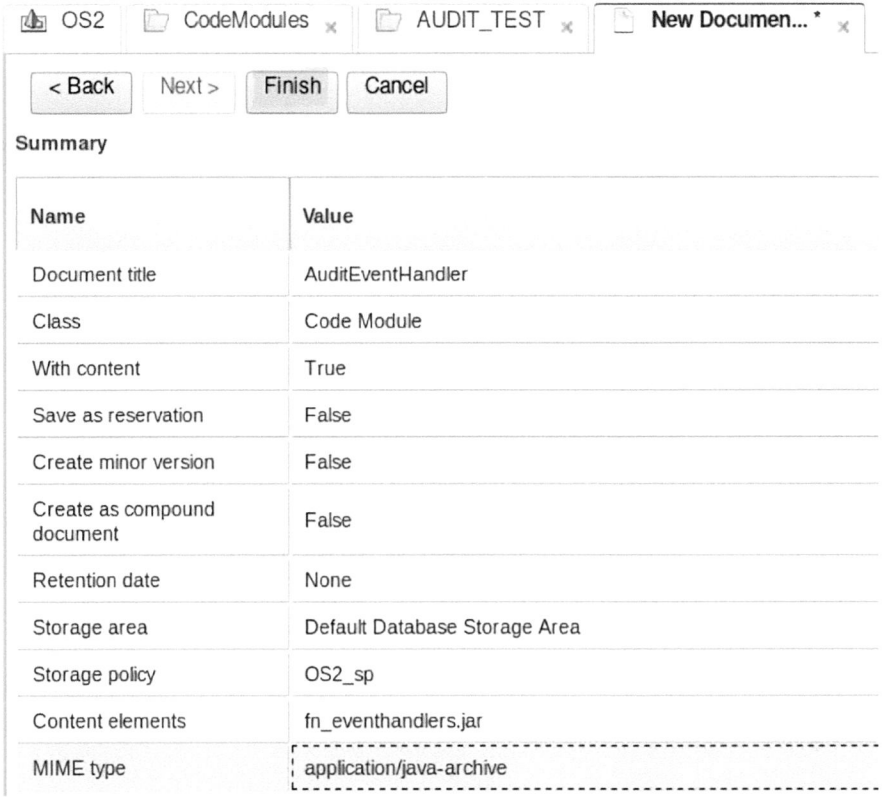

Figure 2-45. *The status of the Document creation is displayed*

187

CHAPTER 2 CONFIGURING JAVA CUSTOM COMPONENTS

The Code Module can now be seen in the CodeModules folder.

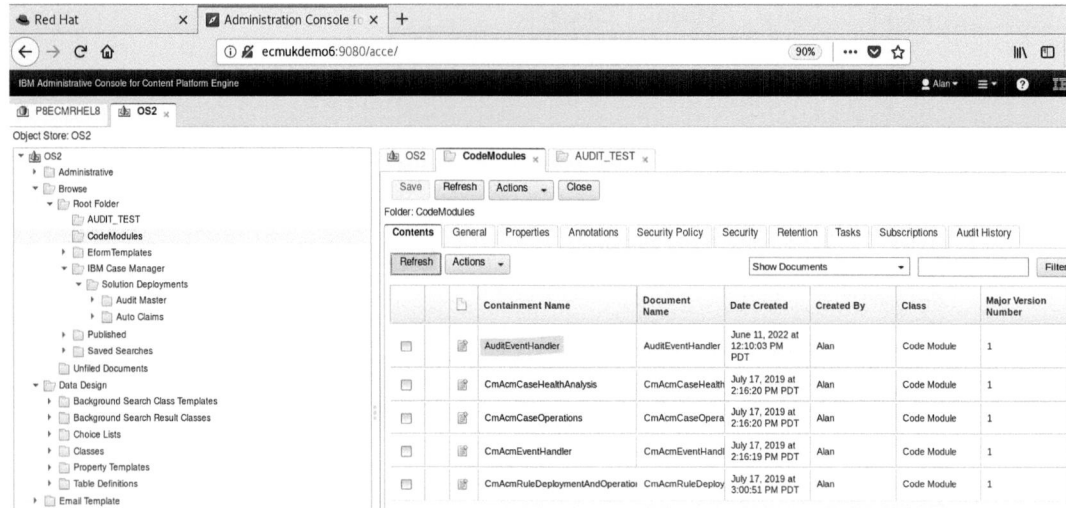

Figure 2-46. *The AuditEventHandler Code Module is now ready to be used*

Creating a Custom Event Object

Before the AuditEventHandler Code Module can be integrated as a new Custom Event in the OS2 Object Store for IBM Case Manager, it has to be linked to other Object Store objects. The first of these is the Event Object.

We will now create a new Event Object called **AddDocToFolder**.

CHAPTER 2 CONFIGURING JAVA CUSTOM COMPONENTS

You will need to perform the following steps for the Event:

1. In the **acce** tree view, expand "**Events, Actions, Processes**" and select "**Event Actions.**" Right-click "**Event Actions**" and select "**New Event Action.**" You will be presented with the "**Name and Describe the Event Action**" wizard.

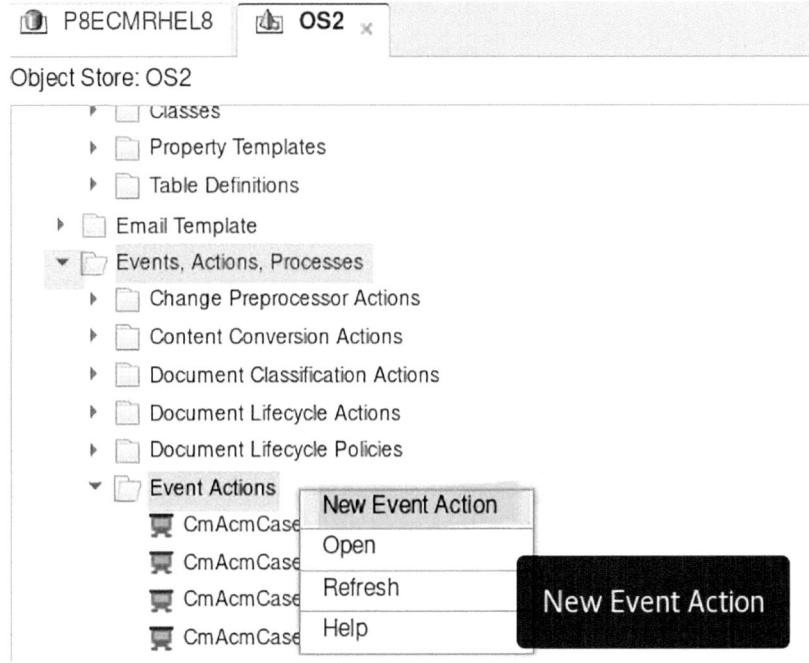

Figure 2-47. *The New Event Action menu is clicked*

The **AddDocToFolder** Display name and Description boxes for the event are entered as shown in Figure 2-48.

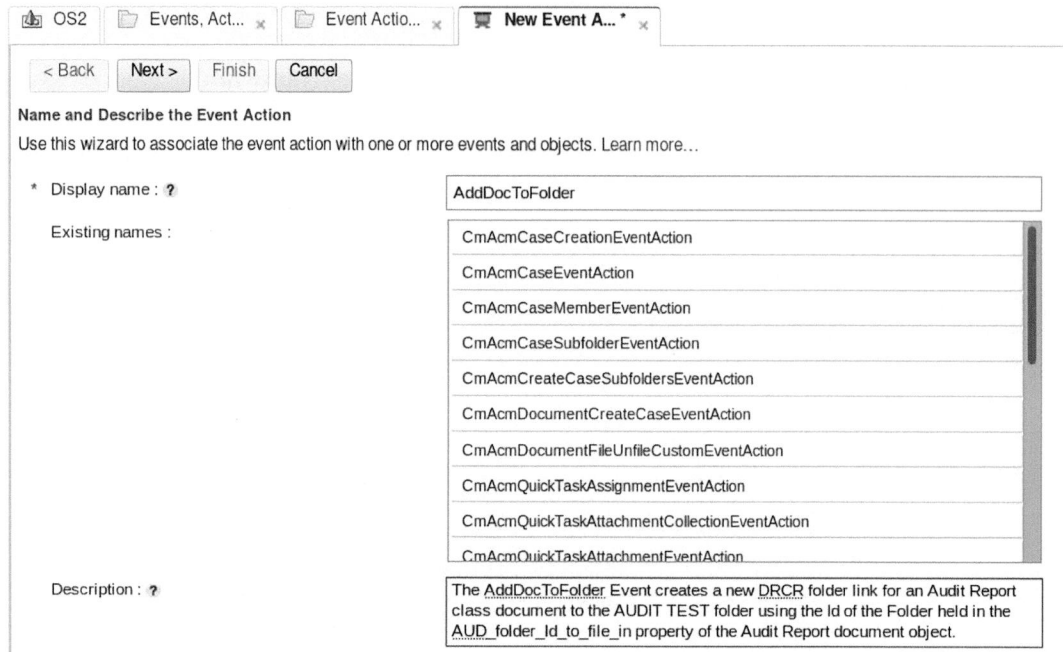

Figure 2-48. The "Name and Describe the Event Action" wizard

2. Select "Next." In the "**Name and Describe the Event Action**" wizard page, enter "**AddDocToFolder**" in the name. The description is optional and defaults to the same as the name value. In this case, we select a slightly more detailed description.

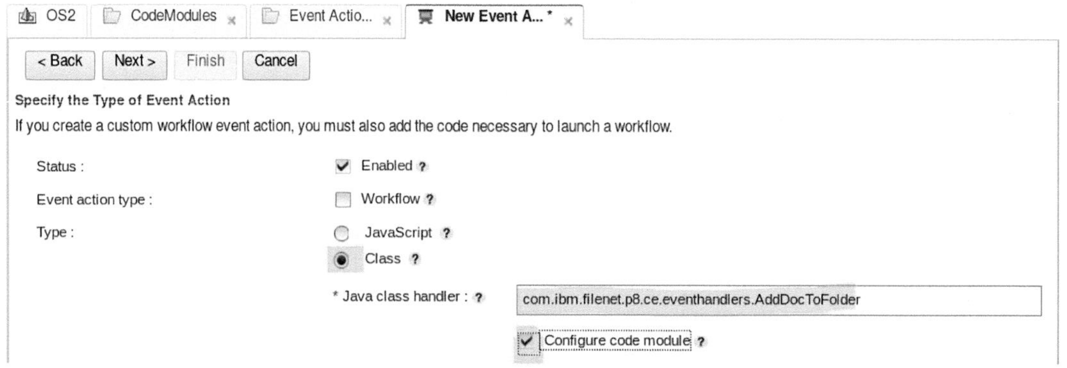

Figure 2-49. The Java class handler full package name and class are entered

CHAPTER 2　CONFIGURING JAVA CUSTOM COMPONENTS

3. Select "Next." On the next wizard page, enter the "Event Action Handler Java Class Name." This value is as follows for the event:

   com.ibm.filenet.p8.ce.eventhandlers.AddDocToFolder

   and select the "Configure Code Module" checkbox shown in Figure 2-49.

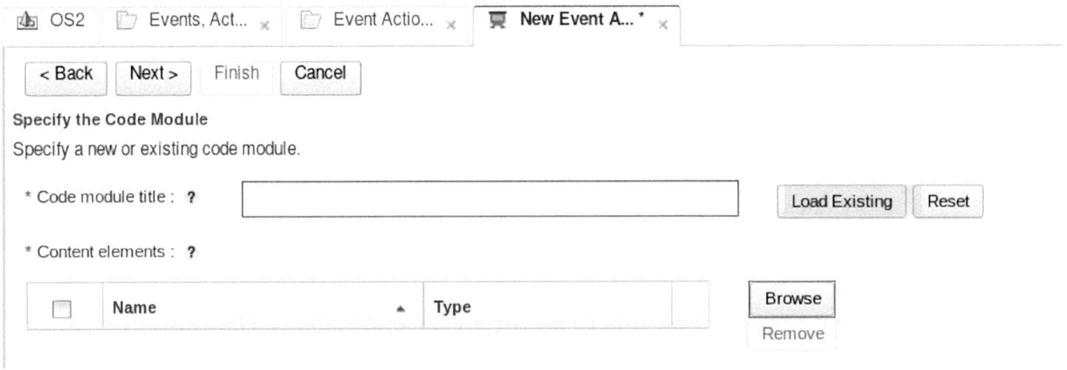

Figure 2-50. The Load Existing command button is clicked

4. Click "Next." On the next wizard page, click the "Load Existing" button; you should see a list with the code module name we created in the "Create Code Module Step 2" earlier. Select this item.

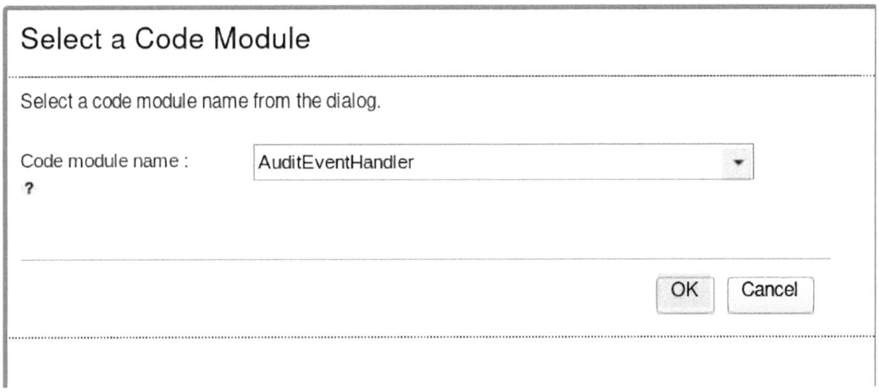

Figure 2-51. The AuditEventHandler code module we created is selected to be used

CHAPTER 2   CONFIGURING JAVA CUSTOM COMPONENTS

5. Click OK.

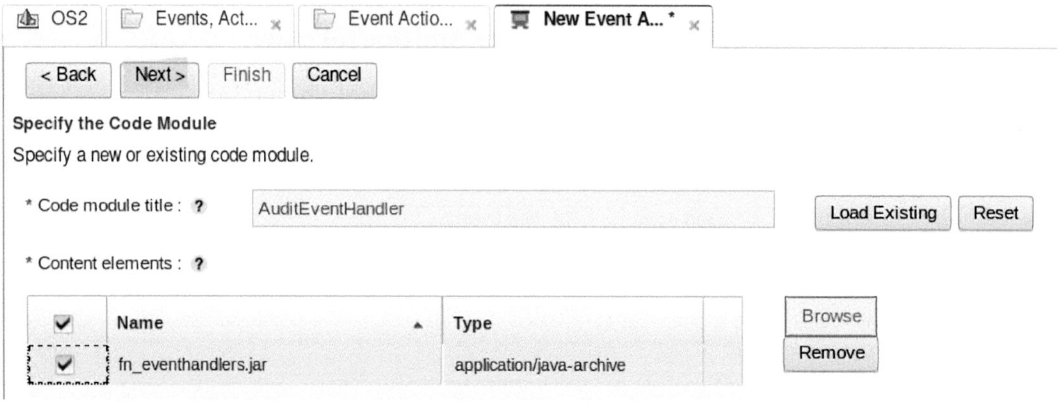

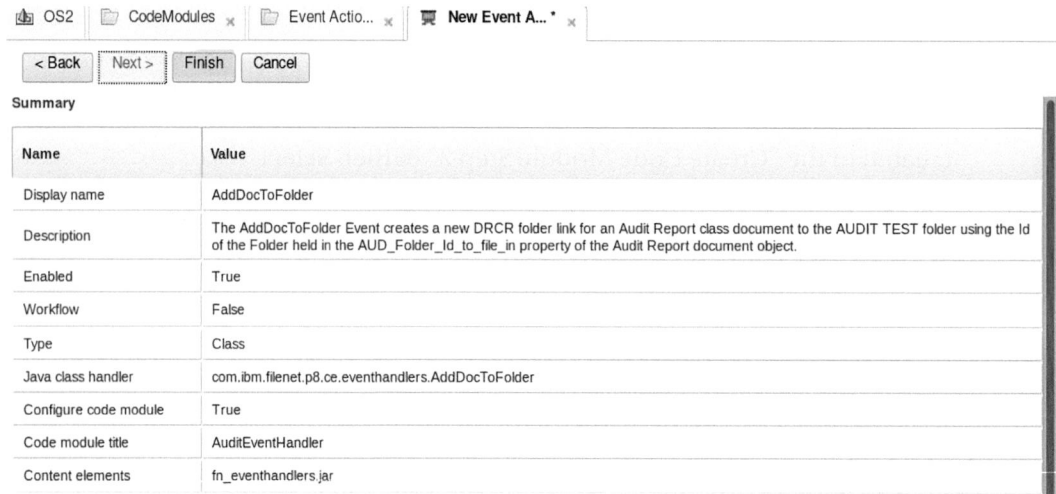

Figure 2-52.   *The Next command is selected*

6. Click "Next." The parameters to be run to create an Event Handler are displayed.

Name	Value
Display name	AddDocToFolder
Description	The AddDocToFolder Event creates a new DRCR folder link for an Audit Report class document to the AUDIT TEST folder using the Id of the Folder held in the AUD_Folder_Id_to_file_in property of the Audit Report document object.
Enabled	True
Workflow	False
Type	Class
Java class handler	com.ibm.filenet.p8.ce.eventhandlers.AddDocToFolder
Configure code module	True
Code module title	AuditEventHandler
Content elements	fn_eventhandlers.jar

Figure 2-53.   *Click Finish to build the Event Handler object and links*

CHAPTER 2 CONFIGURING JAVA CUSTOM COMPONENTS

7. Click "Finish" to complete the wizard.

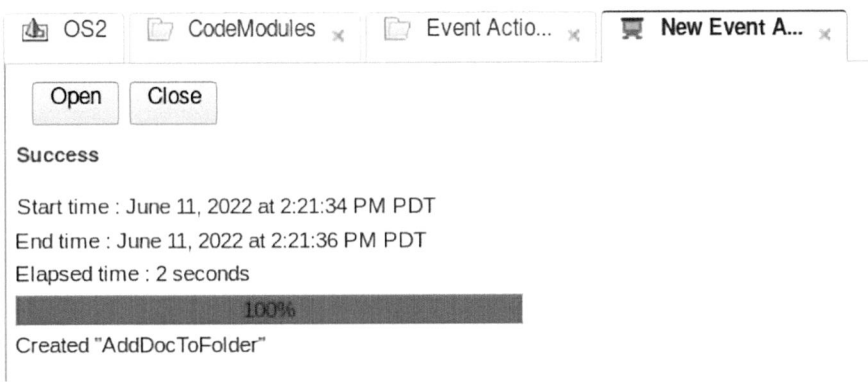

Figure 2-54. The status of the New Event Handler creation process is shown

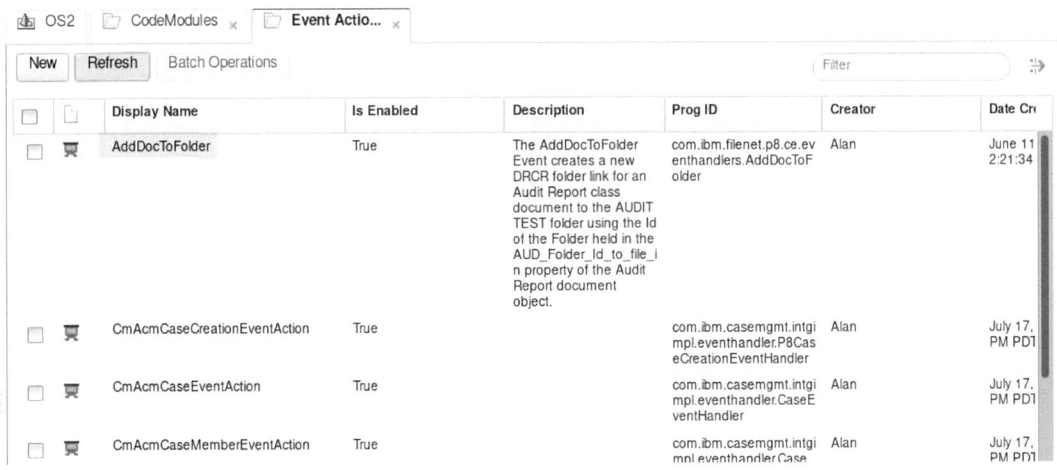

Figure 2-55. The new AddDocToFolder Event Action is created in the OS2 Target Object Store

Configuring Workflow Subscriptions

Content Engine Objects, such as Folders and Documents, can subscribe to certain events and have the preceding "event action" code fire on these events. We will now describe how to subscribe the creation event of a Document of class "**Audit Report**" to the **AddDocToFolder** event action.

CHAPTER 2   CONFIGURING JAVA CUSTOM COMPONENTS

Note   The AddDocToFolder event code creates a link of the created Audit Report Document to the AUDIT_TEST folder that is defined.

1. In acce, select Classes ➤ Document ➤ Audit Report from the tree view.

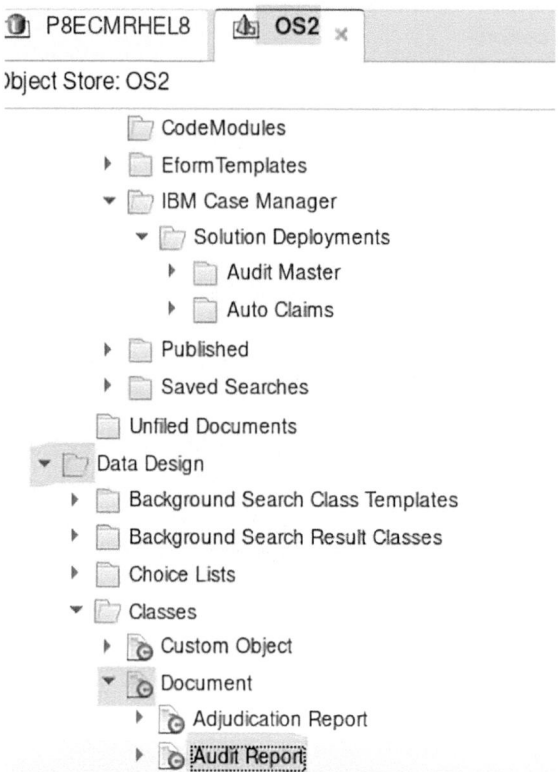

Figure 2-56.   *The Audit Report class is navigated down to*

2. Right-click the Audit Report class and select "New Subscription" from the context menu.

194

CHAPTER 2　CONFIGURING JAVA CUSTOM COMPONENTS

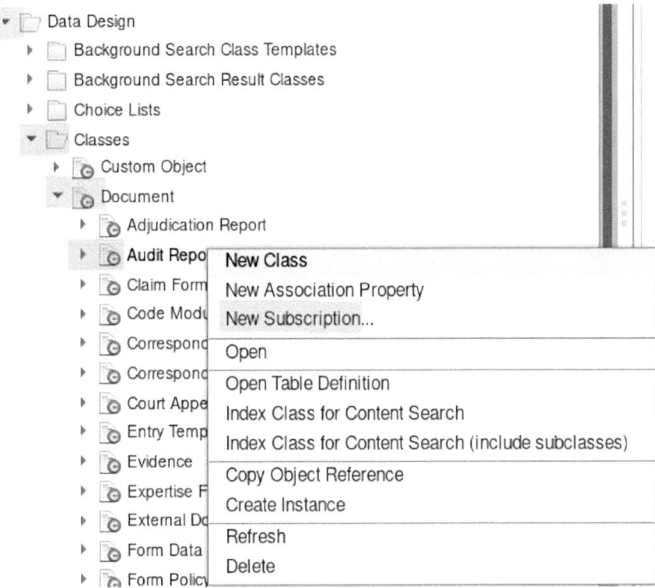

Figure 2-57. *Select "New Subscription" from the context menu for the Audit Report class*

3. Enter the Subscription Name as **AUD_AddDocumentToFolder** and click "Next" on the "Create Subscription" screen.

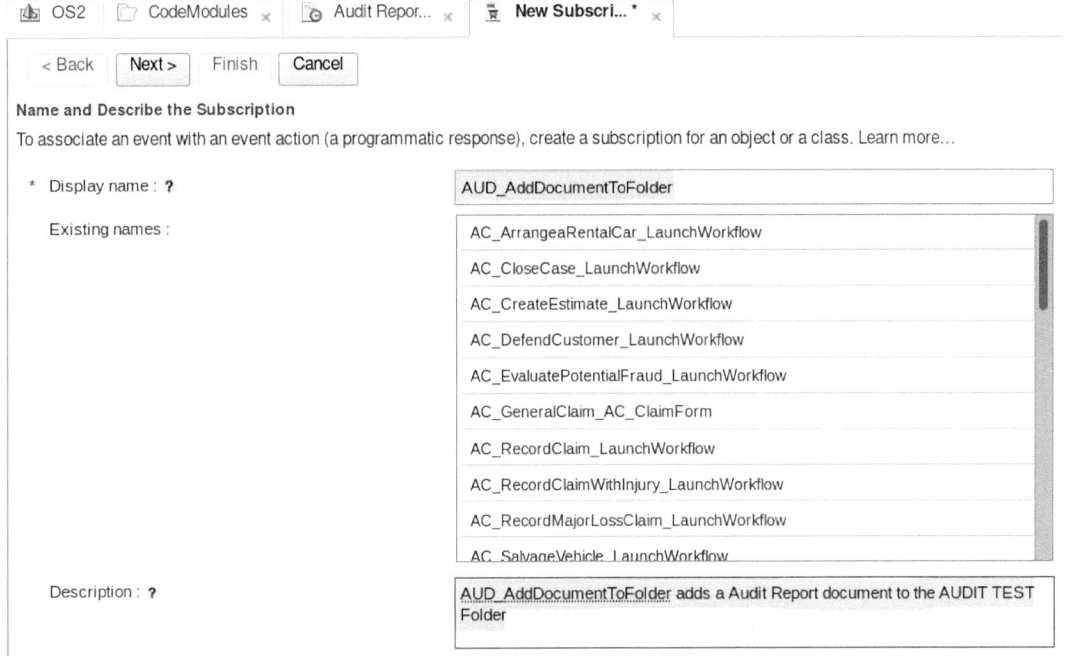

Figure 2-58. *Enter the Subscription Name as **AUD_AddDocumentToFolder***

4. Select a name for the subscription that describes it for this type of object. This is because Event Actions can be used for multiple object types and for multiple events (Create, Update, etc.), and so well-named subscriptions are easier to understand and maintain.

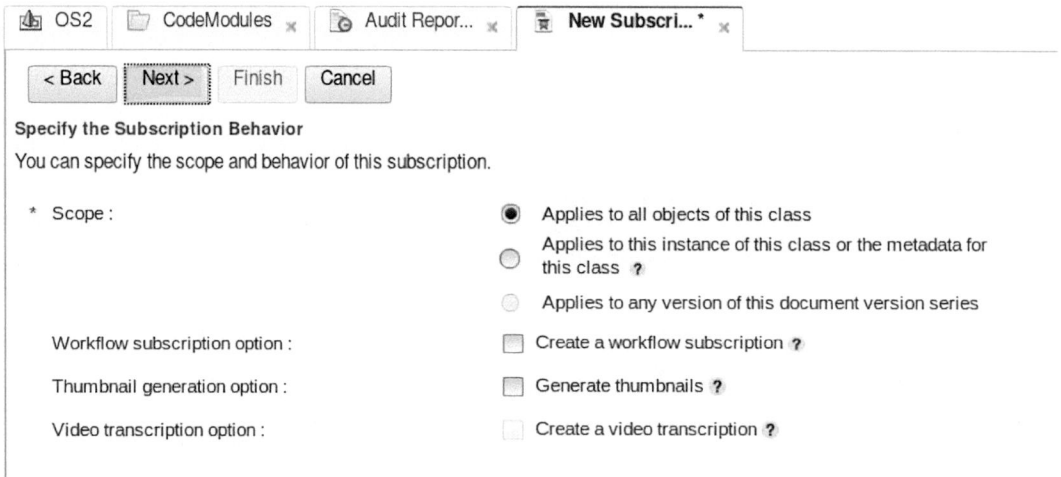

Figure 2-59. *The default of **Applies to all Objects of this class** is left; **Next>** is clicked*

At this point, you can see that there is an option to Create a workflow subscription.

5. Click "Next." The one or more system events (or custom events) can now be selected using tick boxes. We are just going to select the Document Object Creation Event.

CHAPTER 2　CONFIGURING JAVA CUSTOM COMPONENTS

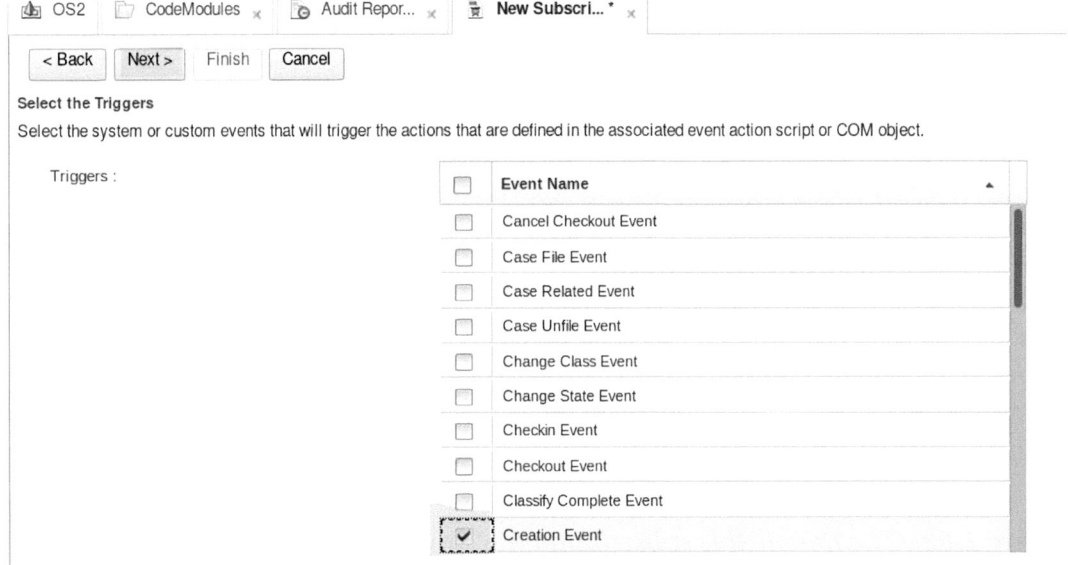

Figure 2-60. *The Creation Event is selected for triggering the Event Handler*

6. On the "Select the Triggers" screen, select "Creation Event" from the list and click "Next."

Figure 2-61. *The AddDocToFolder Event Handler action we created is selected*

197

CHAPTER 2 CONFIGURING JAVA CUSTOM COMPONENTS

7. Click "**Next**." On the "**Specify an Event Action**" page, select "**AddDocToFolder**" from the drop-down.

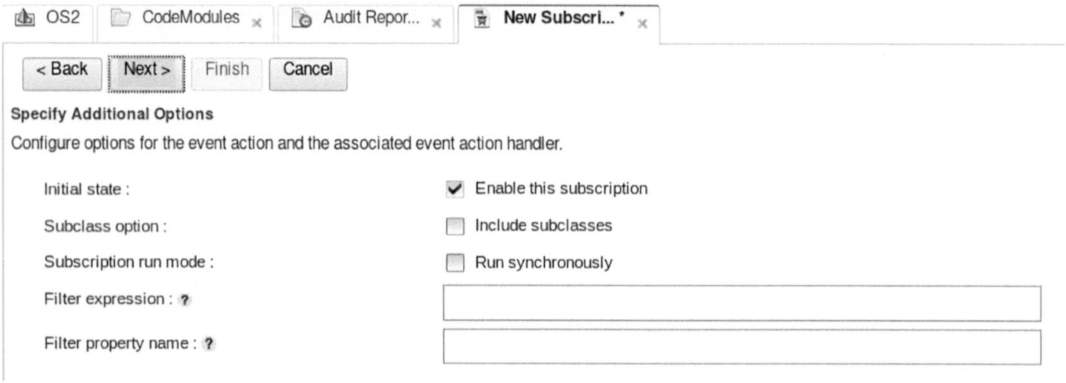

Figure 2-62. *The Enable this subscription tick box is selected, and Next> is clicked*

8. Select "Next."

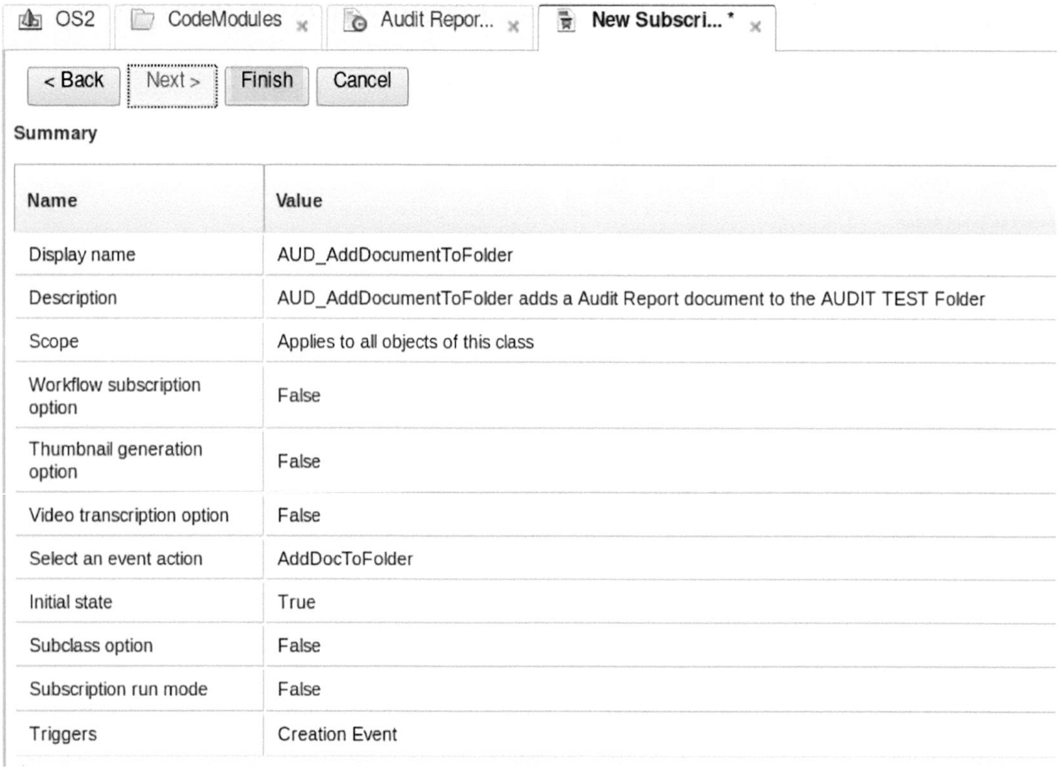

Figure 2-63. *The Finish command is clicked to build the Event Action Subscription*

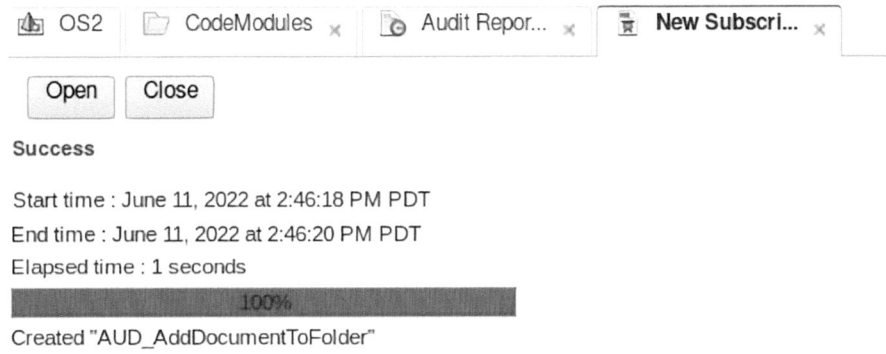

Figure 2-64. *The AUD_AddDocumentToFolder Subscription is created*

Click Finish. The preceding screen is displayed showing the status of the AUD_AddDocumentToFolder New subscription creation.

This subscription can now be tested by creating a document of class "Audit Report." Once created, you should see the Document linked to the **AUDIT_TEST** folder.

Creating the fn_eventshandler.jar

The following steps were used to create the **fn_eventshandler.jar**, Events Handler jar file, which we used in the previous section, from the Eclipse Project.

CHAPTER 2　CONFIGURING JAVA CUSTOM COMPONENTS

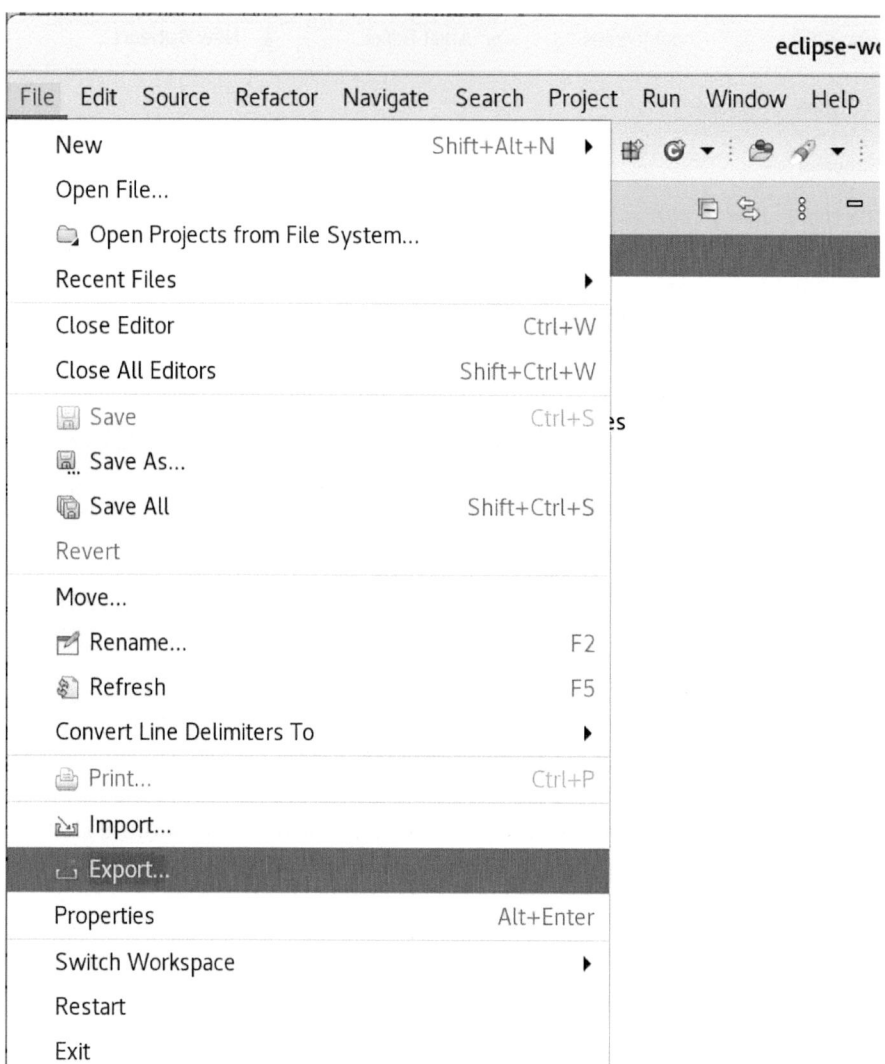

Figure 2-65. *The Eclipse **File ➤ Export** menu option is selected for the AuditEventHandler Java project*

In the Eclipse Java IDE tool, we can export the Project classes and supporting files to the **fn_eventshandler.jar** file we need for integration with IBM Case Manager.

CHAPTER 2 CONFIGURING JAVA CUSTOM COMPONENTS

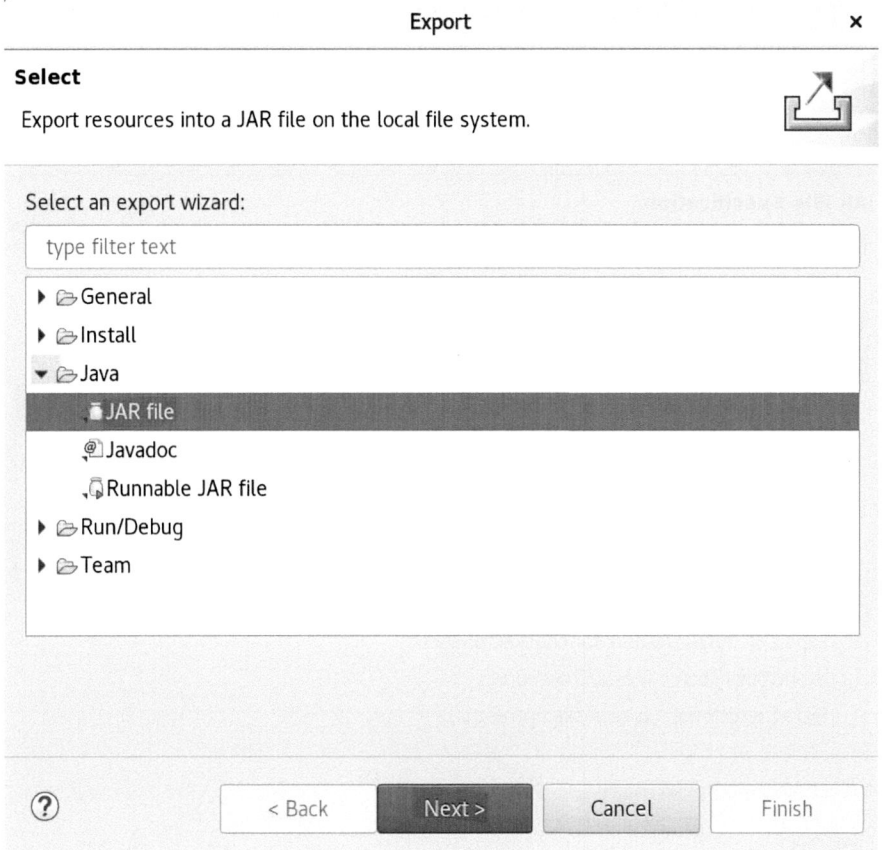

Figure 2-66. *The Java ➤ JAR file export wizard option is selected, and the Next> command is clicked*

CHAPTER 2   CONFIGURING JAVA CUSTOM COMPONENTS

The Eclipse Java JAR file export wizard is selected and the Next> command is clicked as shown in Figure 2-66.

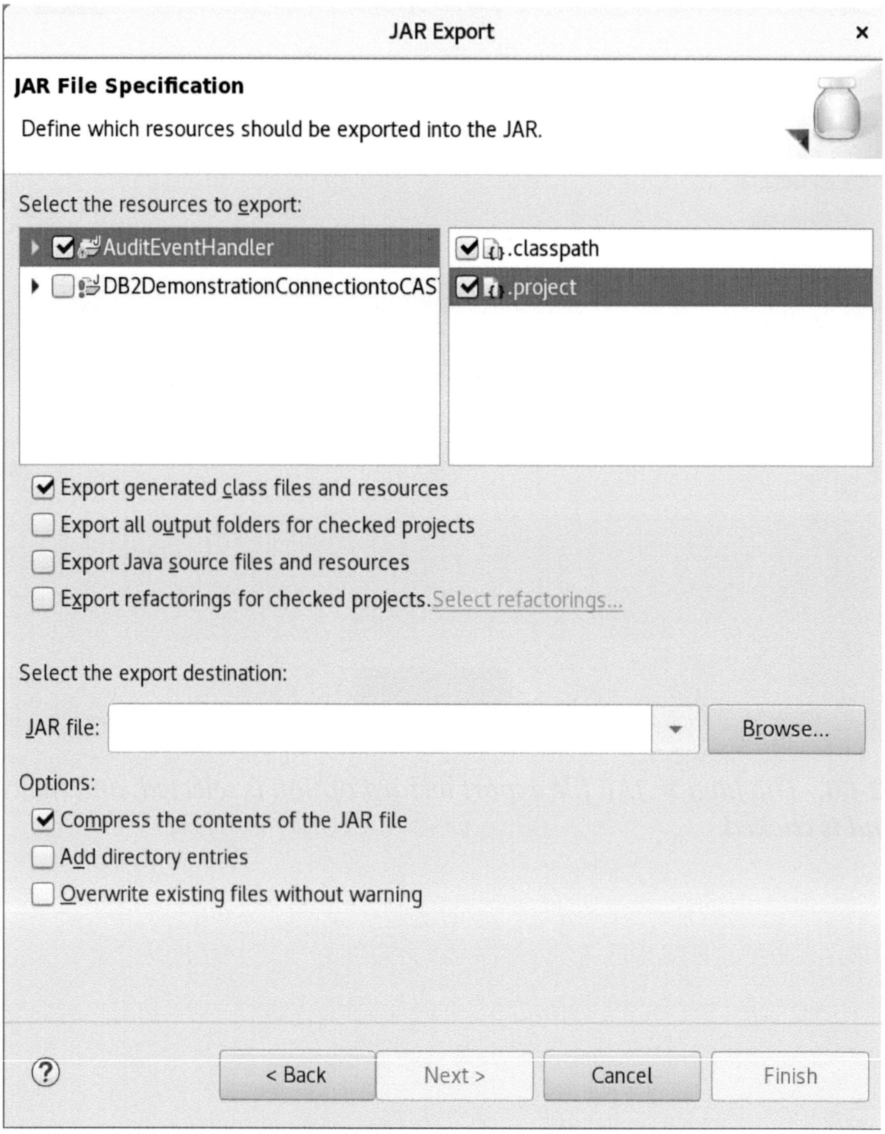

Figure 2-67. *The AuditEventHandler project is selected with the classpath and project attributes included*

CHAPTER 2 CONFIGURING JAVA CUSTOM COMPONENTS

The defaults shown in Figure 2-67 are selected for the **AuditEventHandler** project. Then the **Browse** button is used to choose the exported JAR file destination path.

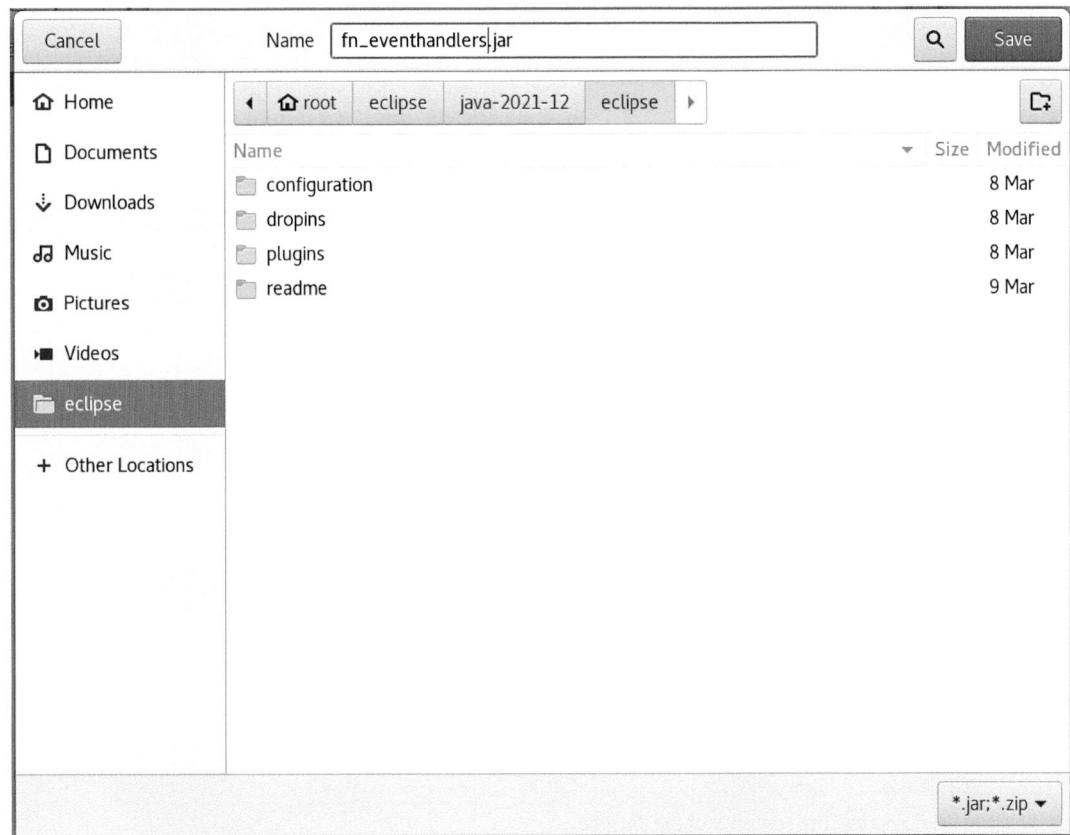

Figure 2-68. *The Browse button allows the Folder Path and the export JAR file name to be entered*

CHAPTER 2   CONFIGURING JAVA CUSTOM COMPONENTS

The default Eclipse path shown in Figure 2-68 was used with the file name entered as **fn_eventhandlers.jar**.

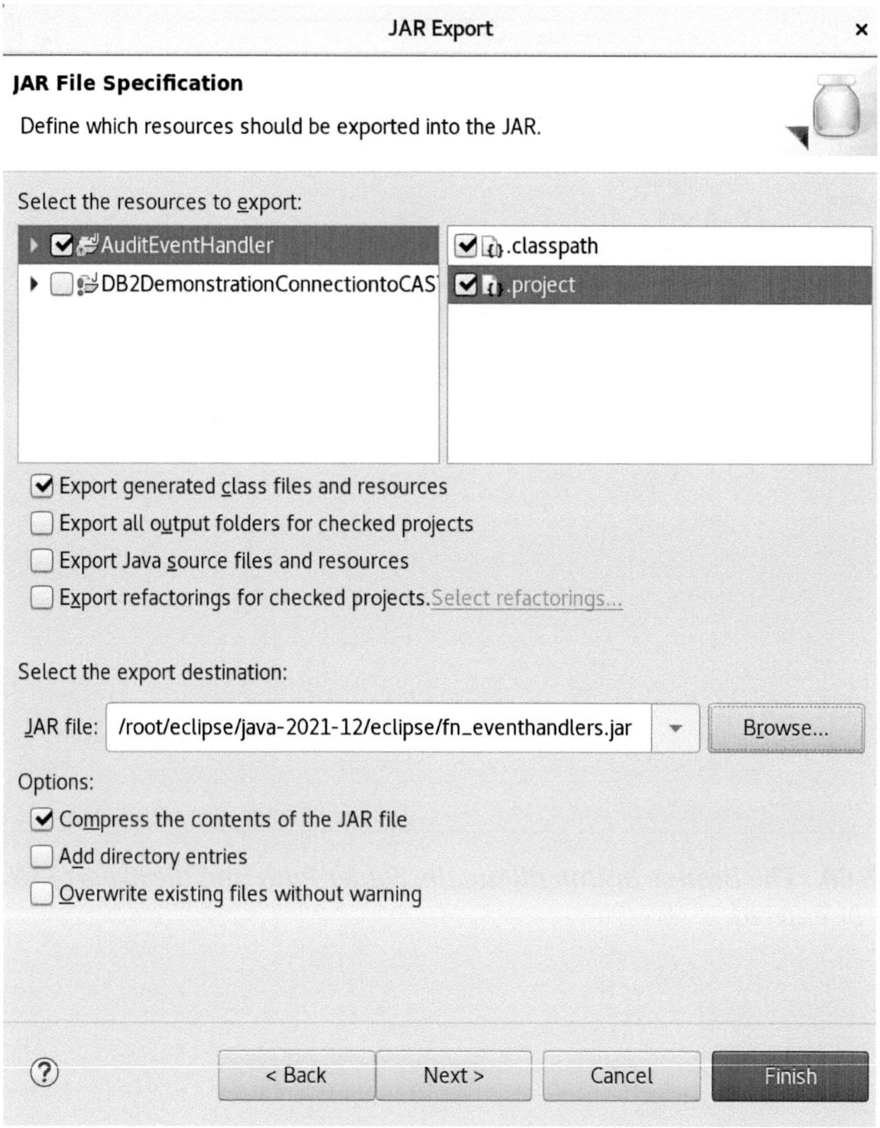

Figure 2-69.   The Finish command button is clicked to create the fn_eventhandlers.jar file

CHAPTER 2   CONFIGURING JAVA CUSTOM COMPONENTS

Part 3 – Testing the Add Document to Folder Event

The TestHarness code in the Eclipse project requires a valid Document Object to be used to link to an existing **AUDIT_TEST** folder which we need to create, and also we need to note the Object Id property values which identify them. These Id property values uniquely identify Folder and Document objects in the IBM FileNet Target Object Store which is used by the IBM Case Manager system.

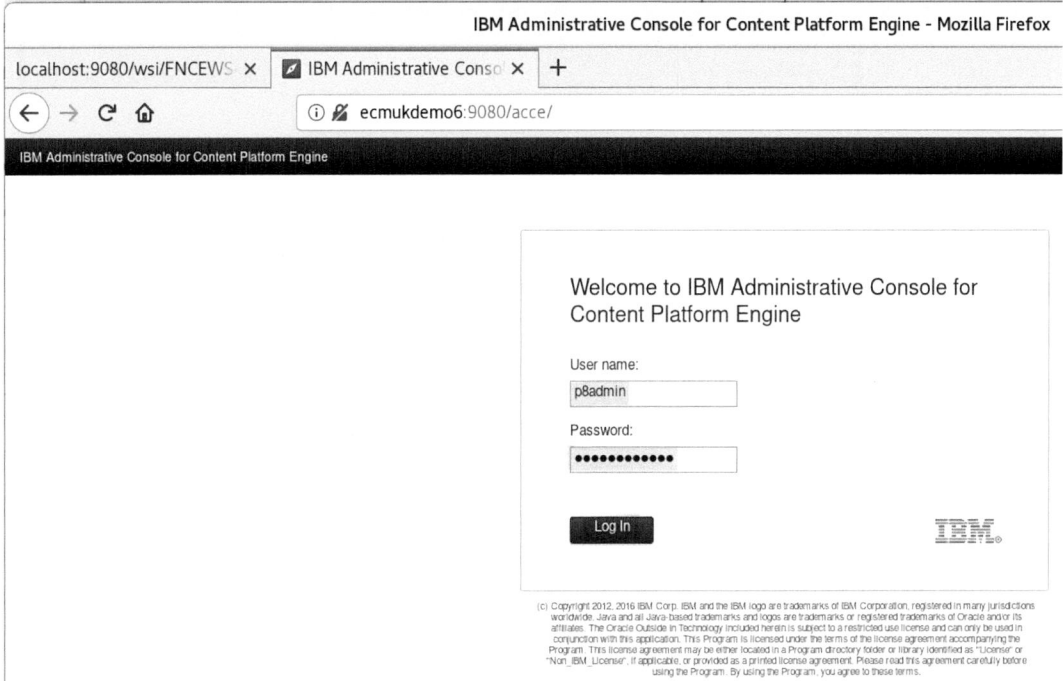

Figure 2-70. *Log in to the IBM FileNet acce administration web application tool*

In **acce**, the IBM Administrative Console for Content Platform Engine, we can create Folder objects, import and view Document and property template classes, and identify the Document Id values for a specific Object Store class object.

CHAPTER 2   CONFIGURING JAVA CUSTOM COMPONENTS

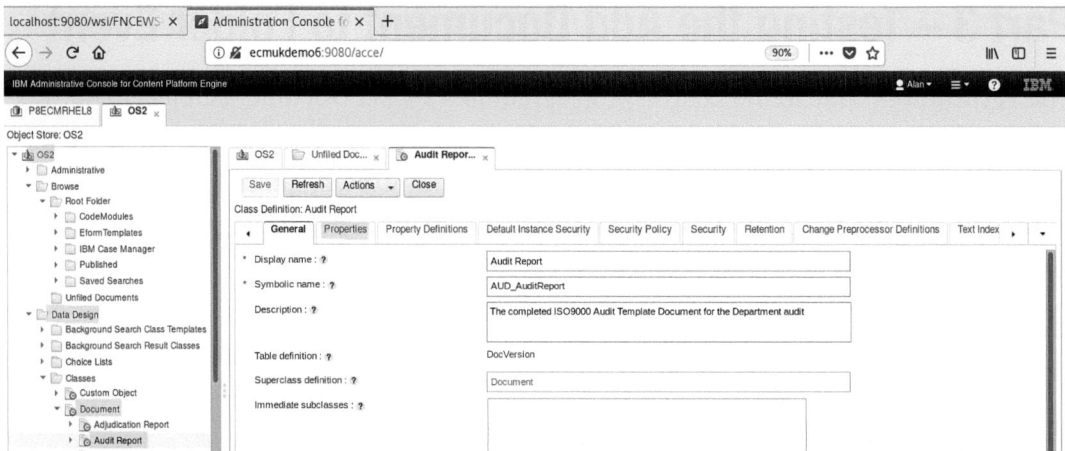

Figure 2-71. *The Audit Report document class Properties tab is highlighted in acce*

The IBM Case Builder web application of the IBM Case Manager system can also be used to create new Document Classes and their related properties. We will create the new property with the symbolic name, **AUD_folder_Id_to_file_in**.

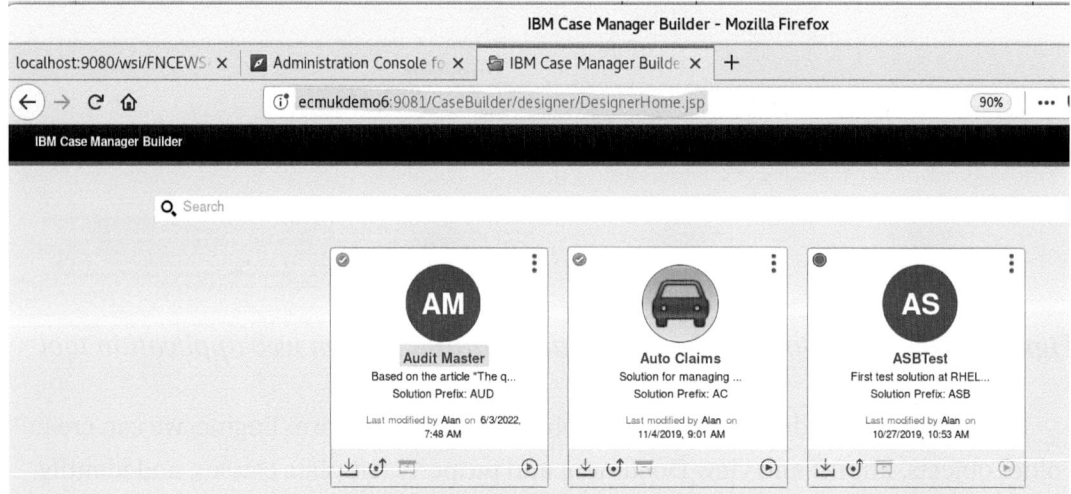

Figure 2-72. *The Audit Master solution created in Chapter 1 is edited again*

206

CHAPTER 2   CONFIGURING JAVA CUSTOM COMPONENTS

We are going to add the property as a Global Solution property template as follows.

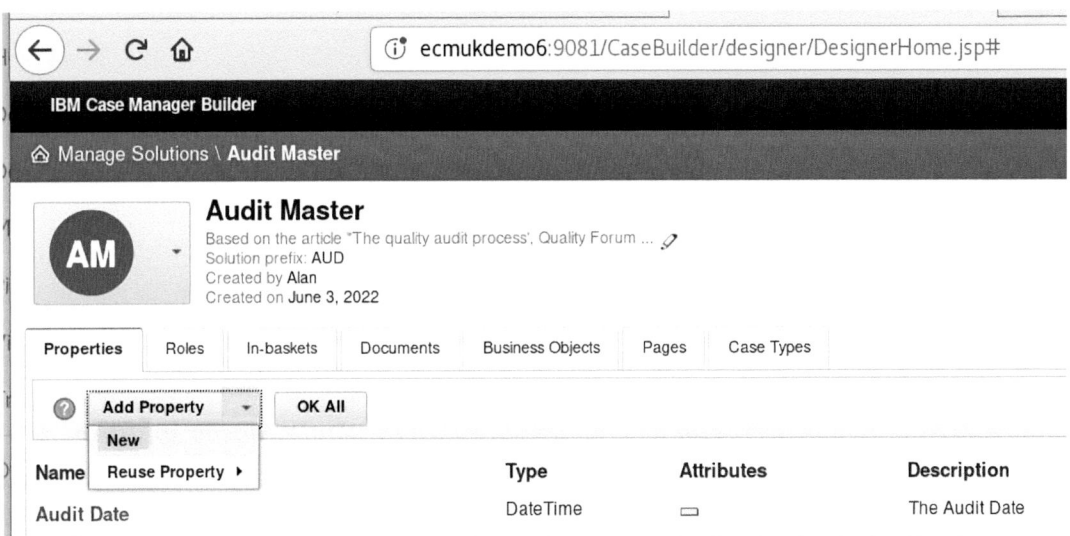

Figure 2-73. The Add Property ➤ New menu item is selected on the Solution Properties tab

We select to add a **New** property and enter the property details shown in the screen in Figure 2-74.

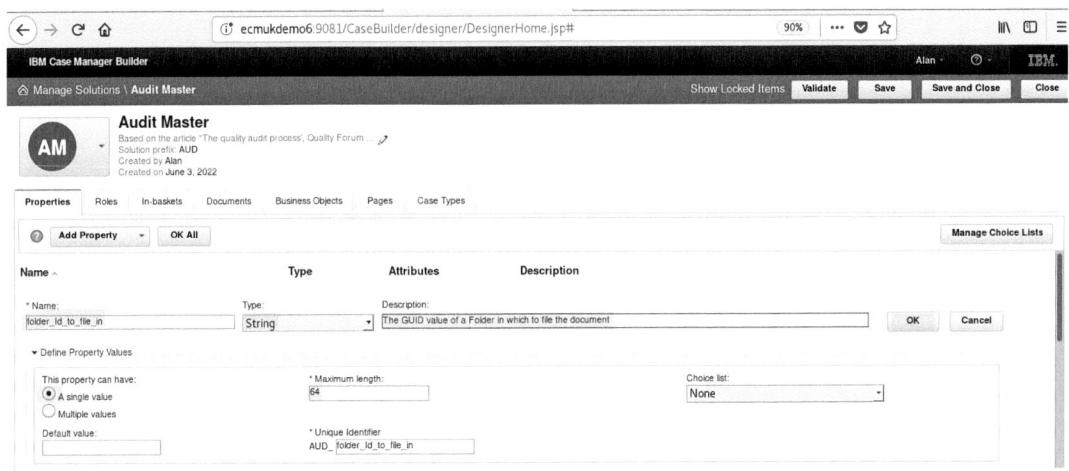

*Figure 2-74. The property with the **SymbolicName AUD_folder_Id_to_file_in** is created*

207

CHAPTER 2   CONFIGURING JAVA CUSTOM COMPONENTS

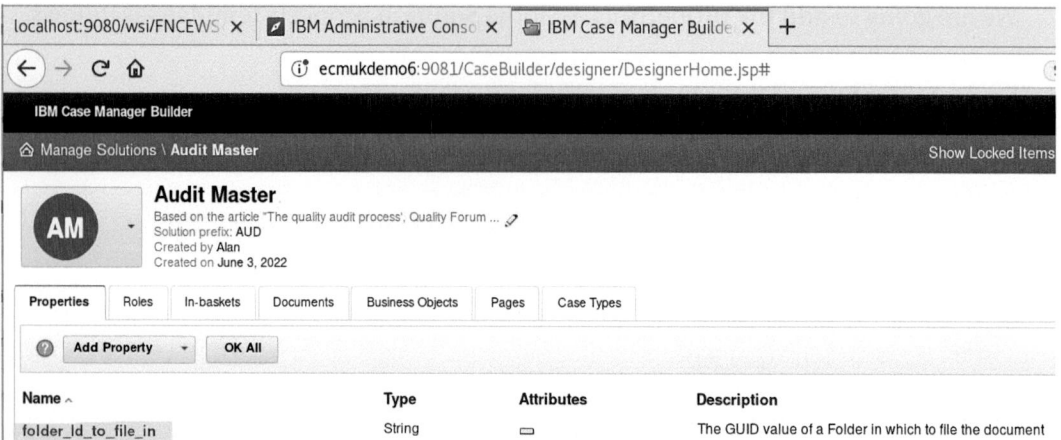

Figure 2-75. *The property can be seen displayed in the Audit Master solution*

Next, we need to add this new property to the **Audit Report** Document class so we can programmatically read the Folder Id string property from it and so identify the Folder to link the Audit Report class documents to. (This is the function of the Java Event program we have developed, which is now available in the Target Object Store.)

CHAPTER 2 CONFIGURING JAVA CUSTOM COMPONENTS

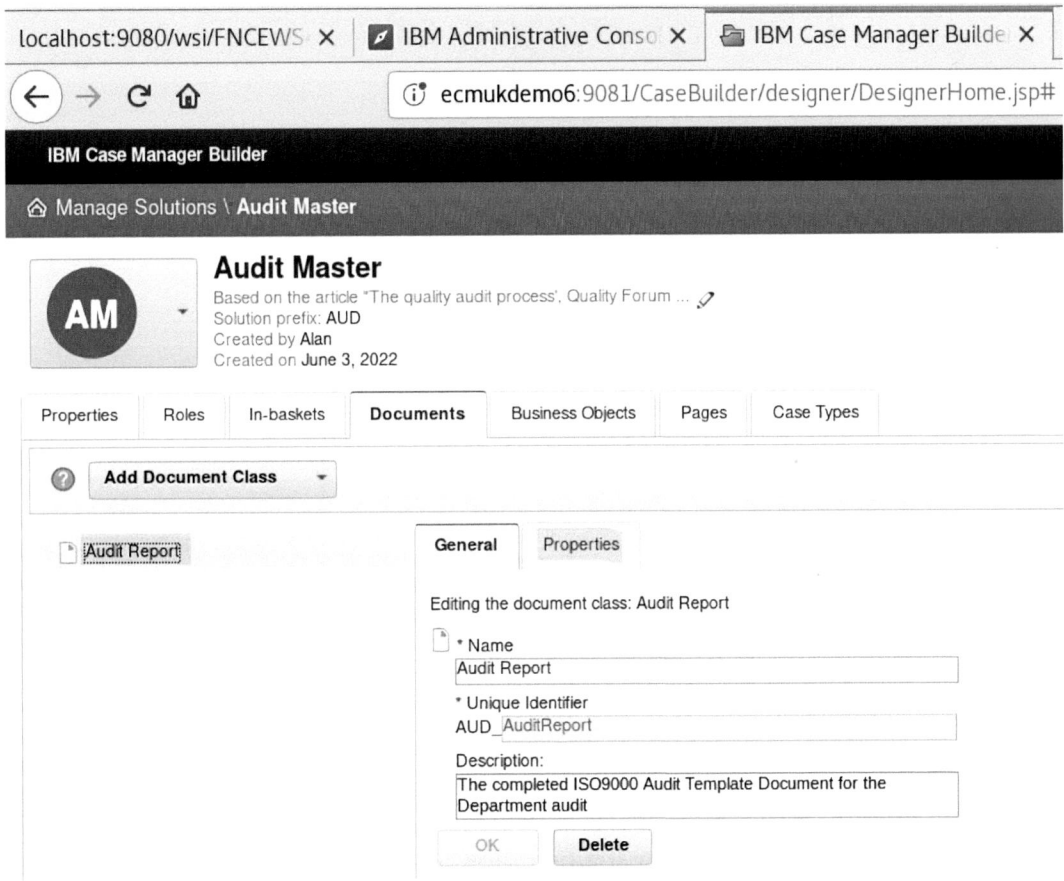

Figure 2-76. *The **Properties** tab of the **Audit Report** Document class is selected*

CHAPTER 2   CONFIGURING JAVA CUSTOM COMPONENTS

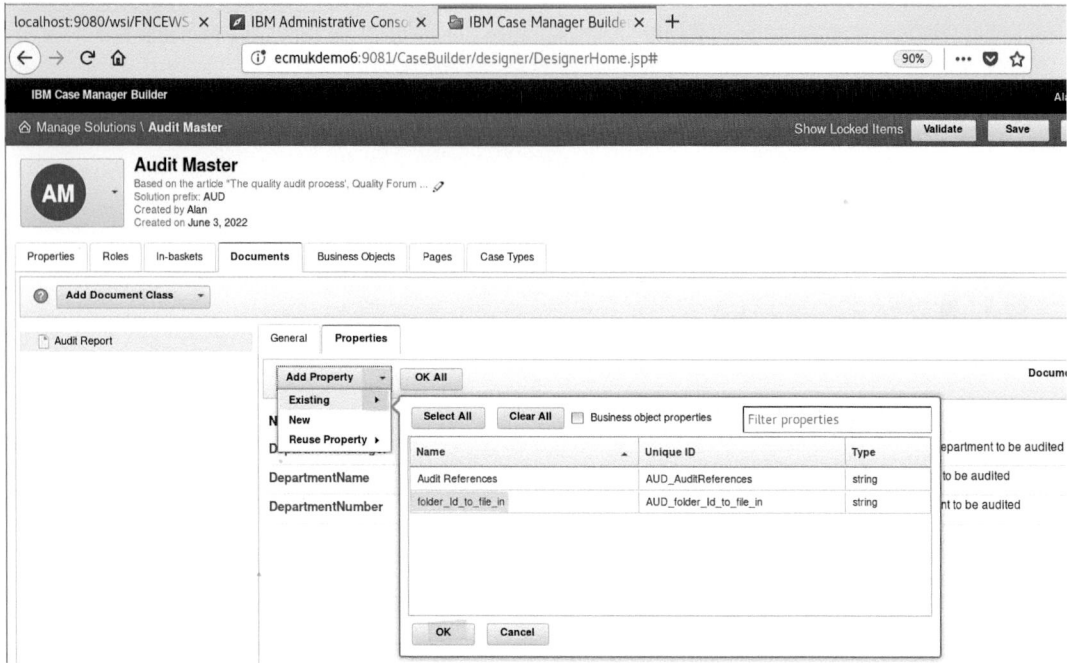

Figure 2-77.   *The Add Property drop-down is used to select the Existing **folder_id_to_file_in** property we created*

The new **Audit Report** Document class, **folder_id_to_file_in** class property, is now defined in the updated **Audit Master** solution. Now we need to redeploy the **Audit Master** Solution (stored in the IBM Case Manager Design Object Store, **OS1**) to the working Target Object Store (**OS2**). This process will update the existing **Audit Report** document class with the new property we have created.

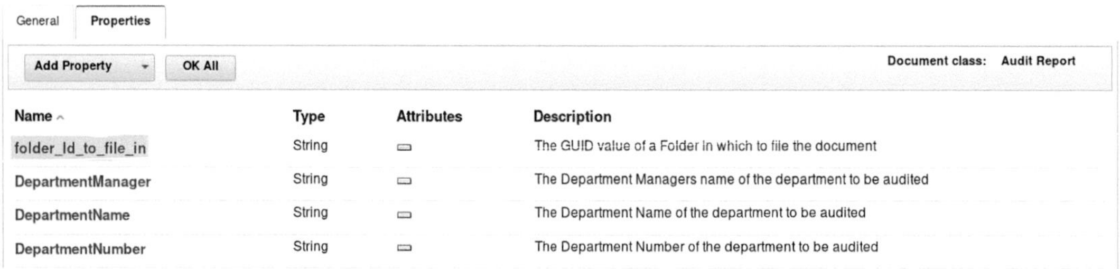

Figure 2-78.   *The **Audit Report** Document class can be seen to have the new **folder_id_to_file_in** property*

CHAPTER 2   CONFIGURING JAVA CUSTOM COMPONENTS

After we save and close the changes to the **Audit Master** solution, we use the **Redeploy All** menu option shown in Figure 2-79.

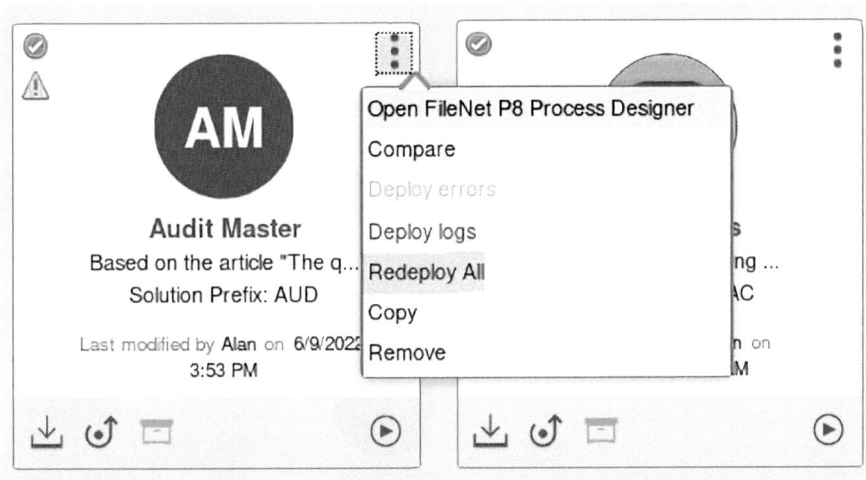

Figure 2-79.*   The **Redeploy All** menu option is used in the IBM Case Manager Builder to deploy the updated solution*

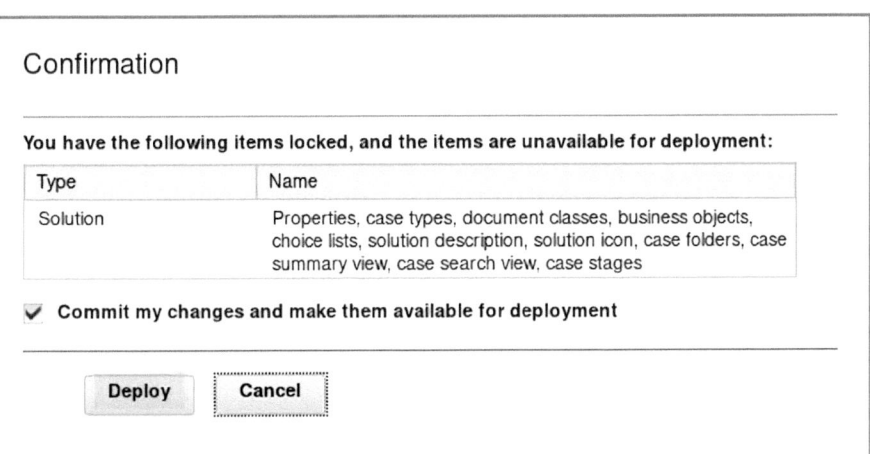

Figure 2-80.*   The Deploy command button is clicked to start the Audit Master Solution redeployment*

211

CHAPTER 2　CONFIGURING JAVA CUSTOM COMPONENTS

If there are changes edited into the Audit Master solution, and these are not deployed, a yellow warning triangle appears below the green tick mark icon.

This yellow warning triangle disappears after a successful redeployment.

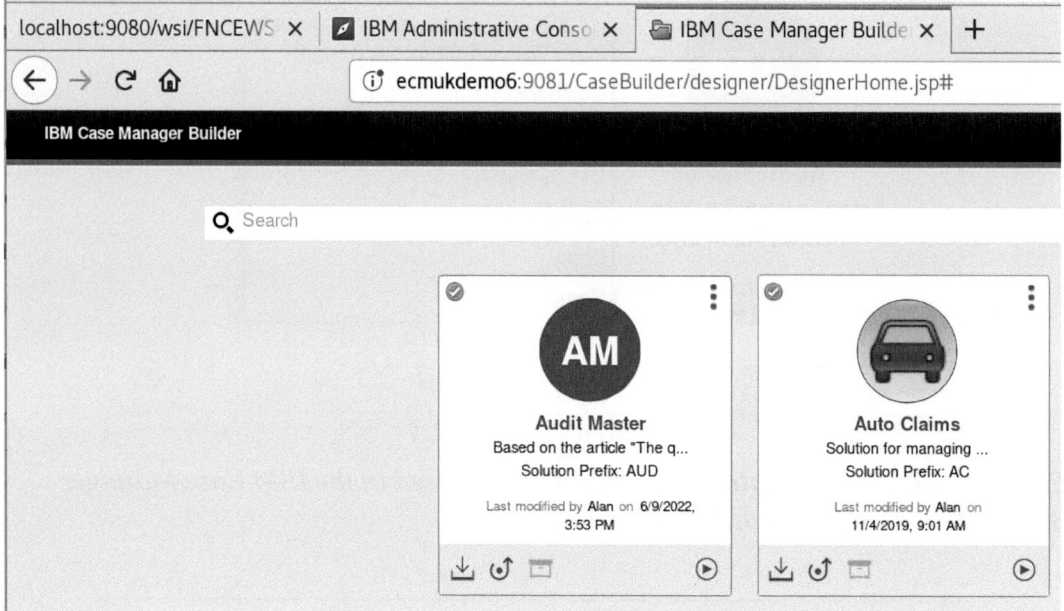

Figure 2-81. *The Audit Master solution is successfully redeployed to the OS2 Target Object store*

Now we create the **AUDIT_TEST** folder in the **OS2** Target Object store using the IBM FileNet **acce** web application tool. We right-click the **Root Folder** node and select the **New Folder** menu option to show a **New Folder** tab as shown in Figure 2-82.

212

CHAPTER 2 CONFIGURING JAVA CUSTOM COMPONENTS

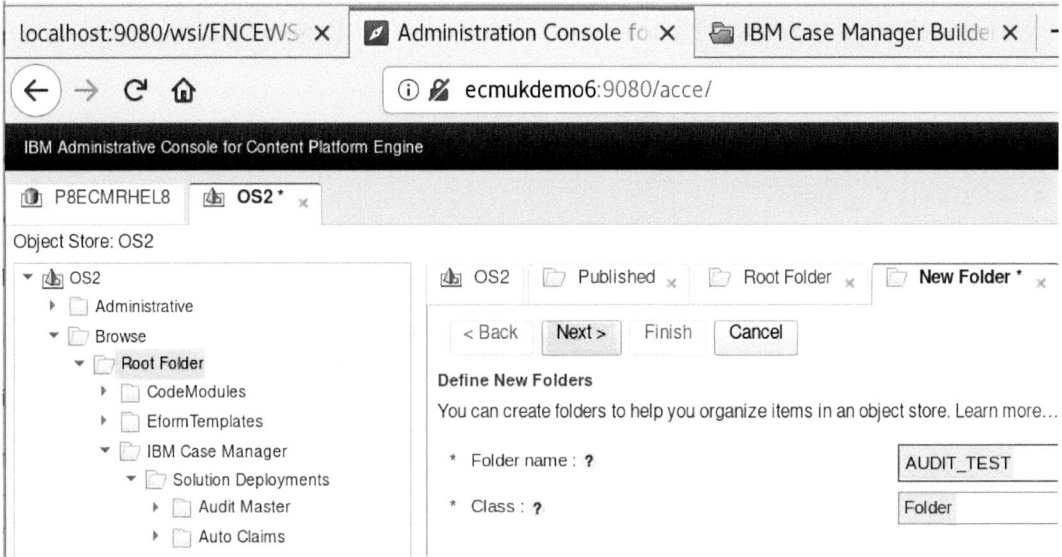

*Figure 2-82. The **AUDIT_TEST** folder is created in the OS2 Target object store under the Root Folder*

*Figure 2-83. The **Next>** command is used to leave the default Retention options*

The **Finish** command is used, as shown in Figure 2-84, to create the Folder **AUDIT_TEST**, which we will use to hold links to all the **Audit Report** case Documents of the **Audit Master** solution.

213

CHAPTER 2 CONFIGURING JAVA CUSTOM COMPONENTS

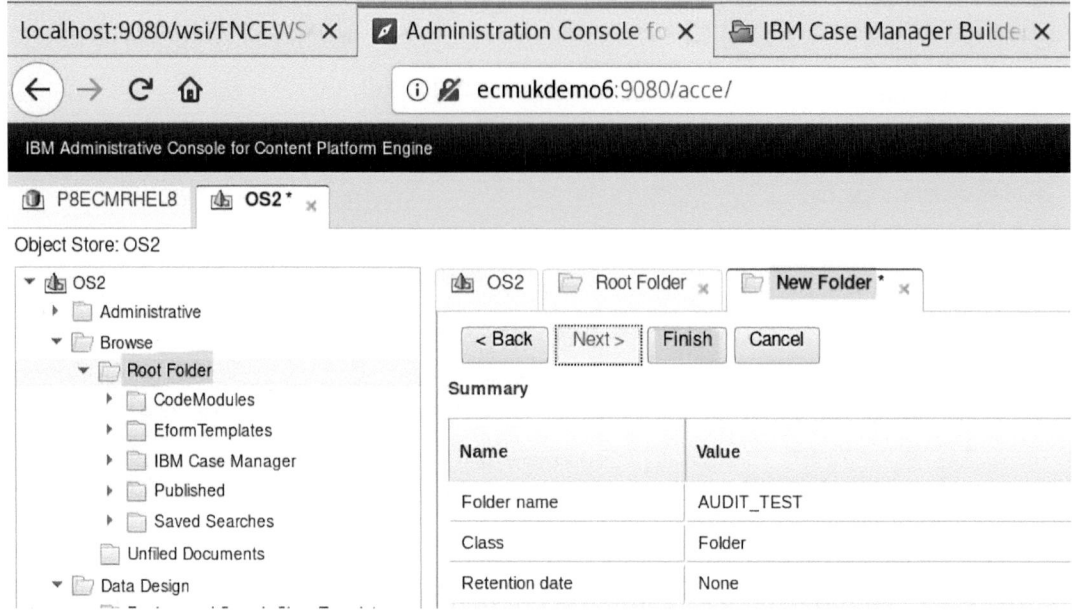

Figure 2-84. The Finish command is used to create the AUDIT_TEST folder

The status window shows the Date and Time of the folder creation and the time taken.

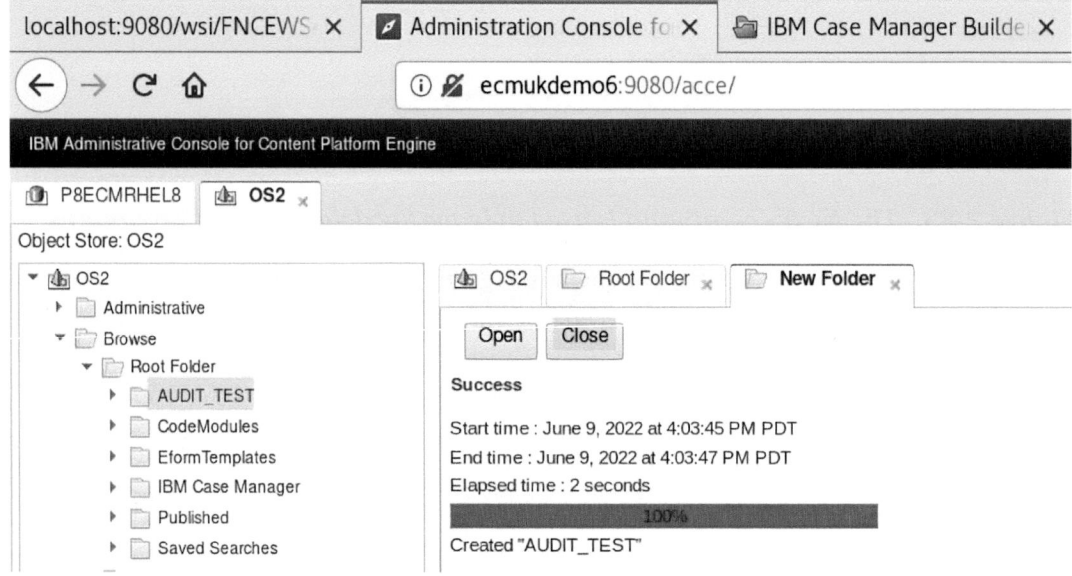

Figure 2-85. The status of the New Folder creation of AUDIT_TEST is shown

CHAPTER 2 CONFIGURING JAVA CUSTOM COMPONENTS

We now need to find and display the unique Id of the **AUDIT_TEST** folder object we created and use this in our **TestHarness** Java program to test the functionality of our **AddDocToFolder** Event Java Class.

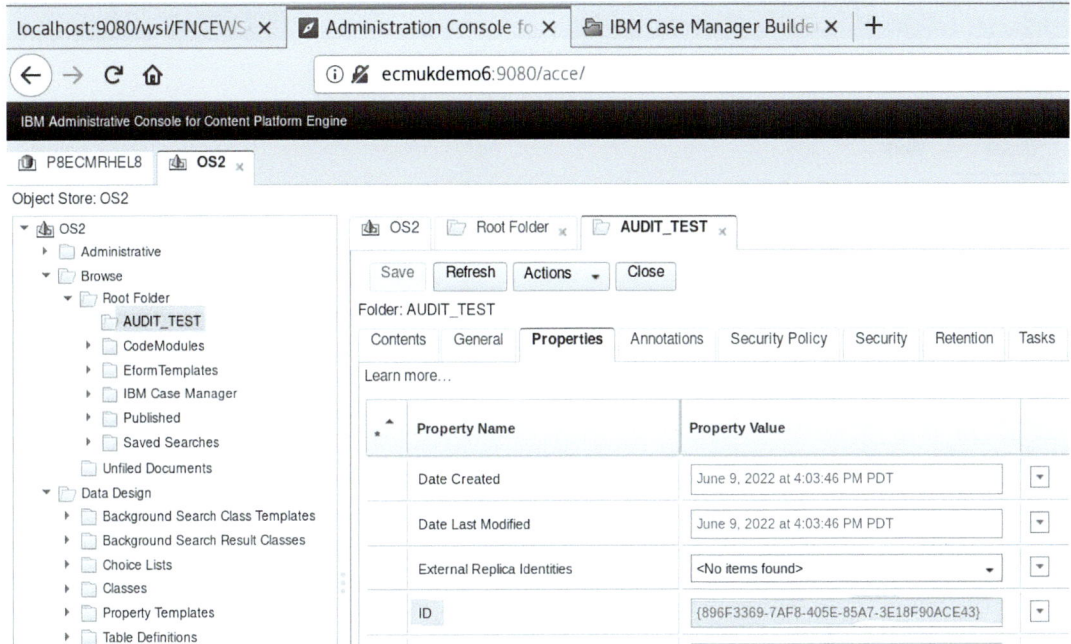

Figure 2-86. *The ID property of the AUDIT_TEST folder is displayed*

The **AUDIT_TEST** folder ID string value is found to be **{896F3369-7AF8-405E-85A7-3E18F90ACE43}** for the Object Store I am using and can be copied and pasted to the **TestHarness** Java program to run a test in the Eclipse Java IDE. First, we need to create an **Audit Department** Case to use for the Tests.

CHAPTER 2   CONFIGURING JAVA CUSTOM COMPONENTS

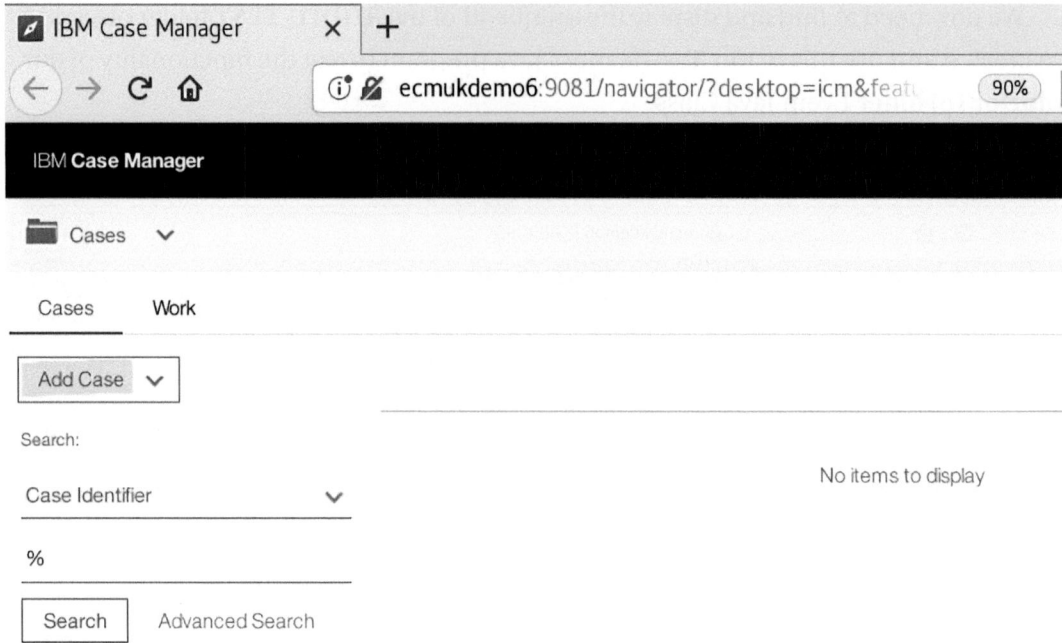

Figure 2-87.   A new Case is created using the Add Case drop-down

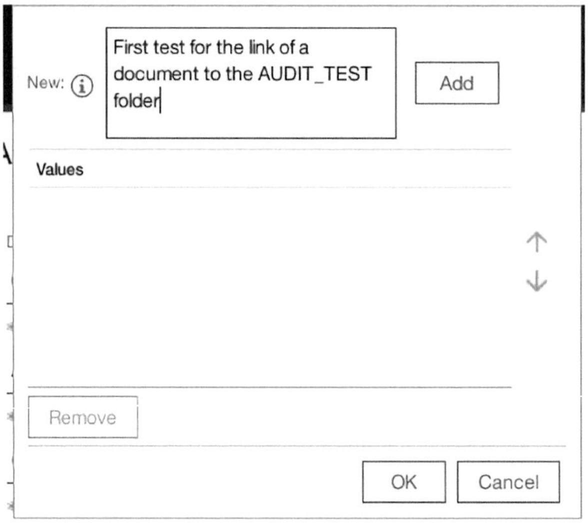

Figure 2-88.   The Case description is added

CHAPTER 2 CONFIGURING JAVA CUSTOM COMPONENTS

The case is searched for using the % wildcard for the Case Identifier as shown in Figure 2-89.

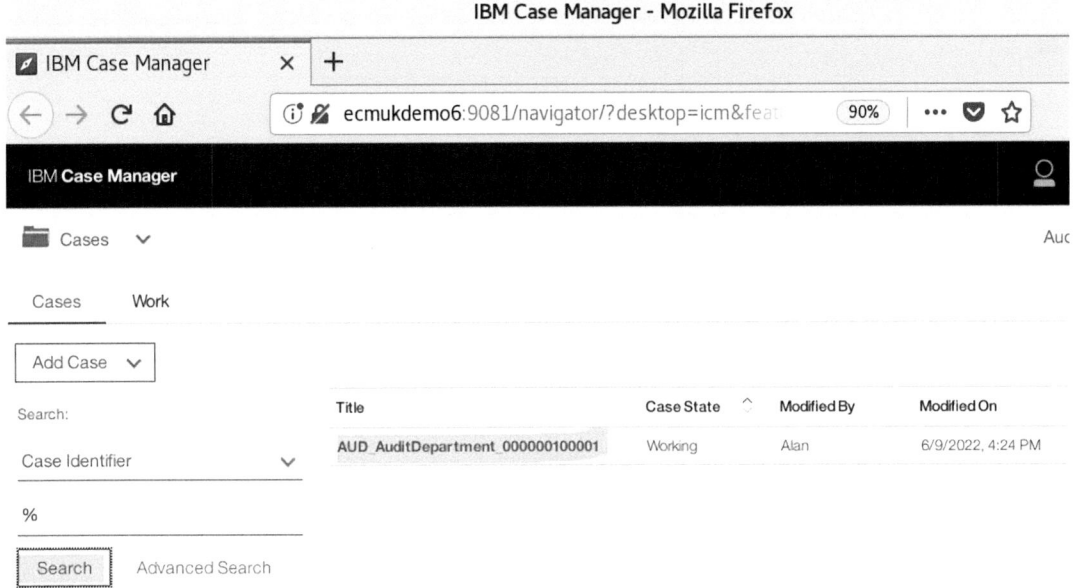

Figure 2-89. *A **Search** shows that this case is the first Case entered in the system*

Since we have redeployed the Solution, this case is now the first Case entered in the system.

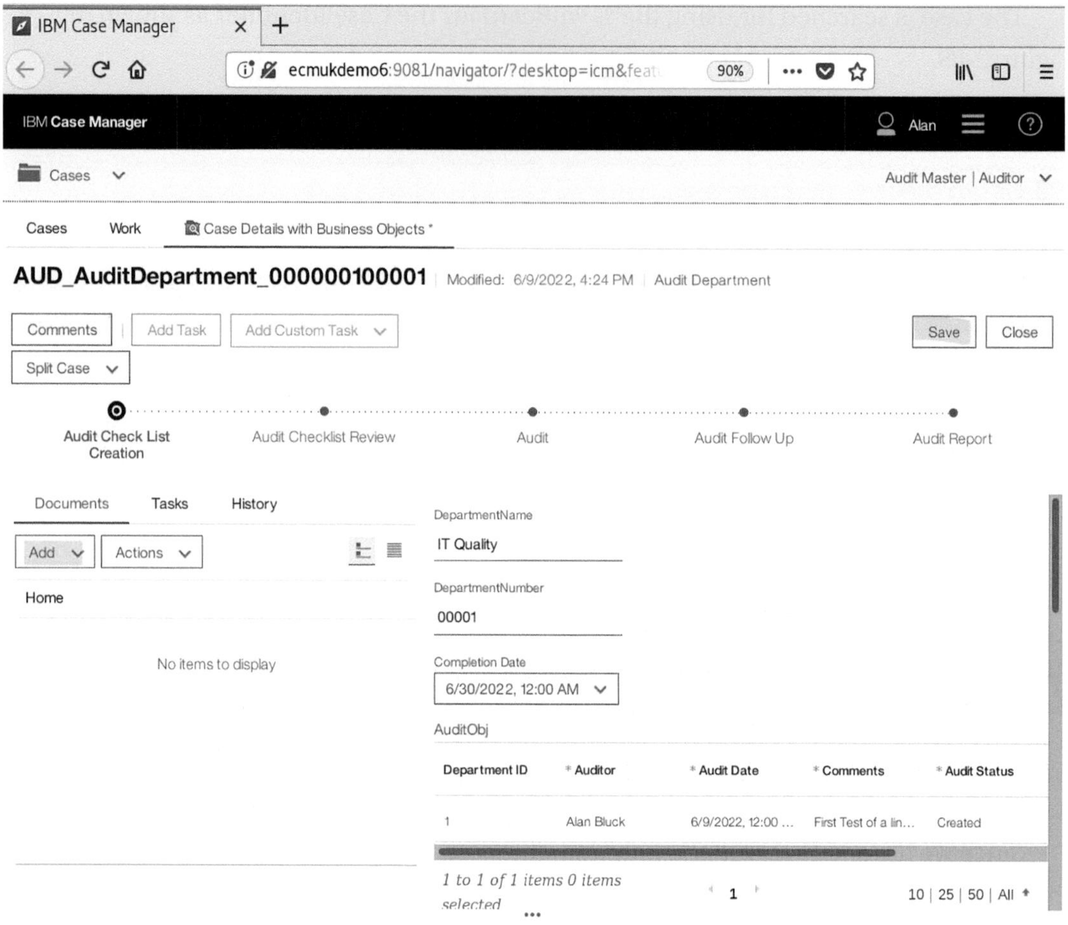

CHAPTER 2   CONFIGURING JAVA CUSTOM COMPONENTS

Figure 2-90. *The Audit Department Audit details are entered for the Case*

To test that an Audit Report Document is created, we use the Add ➤ Add Document from Local System option shown in Figure 2-91.

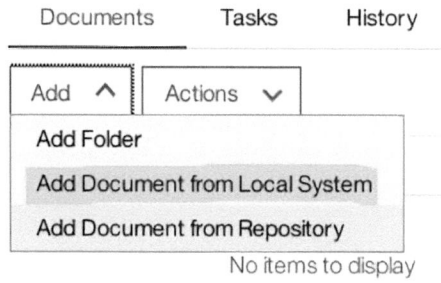

Figure 2-91. *The **Add Document from Local System** option is selected for the Case **Audit Report** document*

CHAPTER 2 CONFIGURING JAVA CUSTOM COMPONENTS

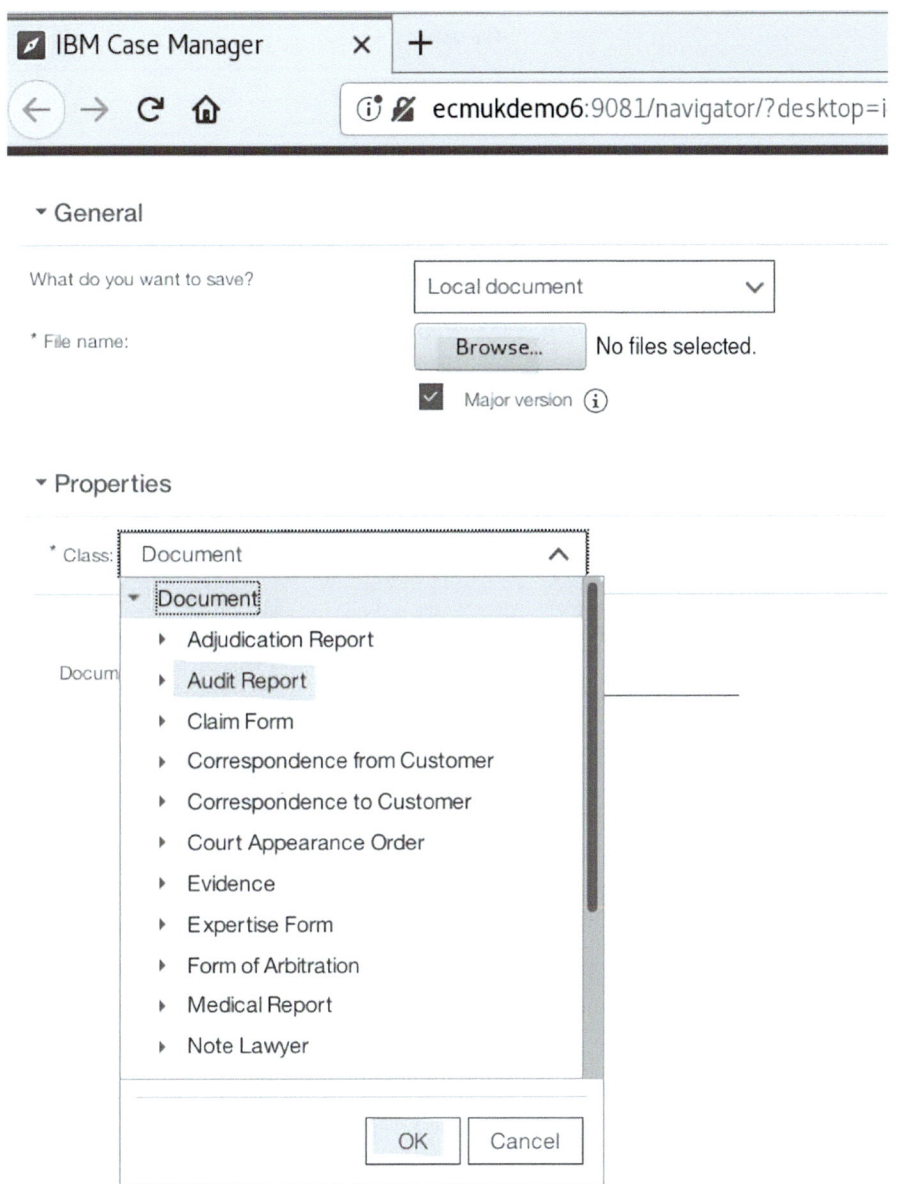

*Figure 2-92. The highlighted **Audit Report** class of Document is selected from the drop-down choice*

The Audit Report Document class is selected as shown for our test.

CHAPTER 2   CONFIGURING JAVA CUSTOM COMPONENTS

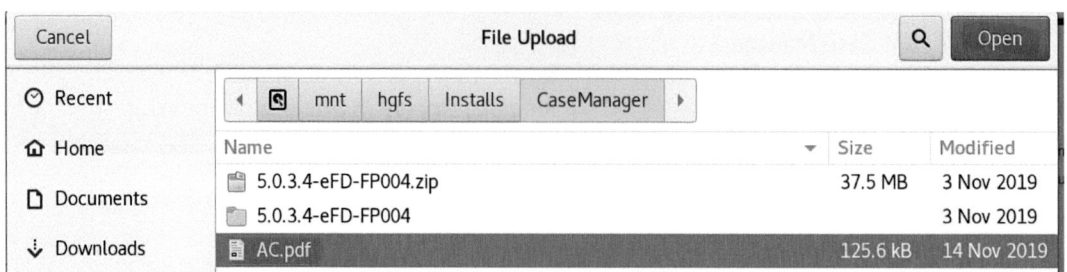

Figure 2-93. *A pdf document called **AC.pdf** is selected for the Document content using the **Browse** command*

We can now enter the Audit Report Document Class properties, which now include our **AUDIT_TEST** folder Id string in the **folder_id_to_file_in** property.

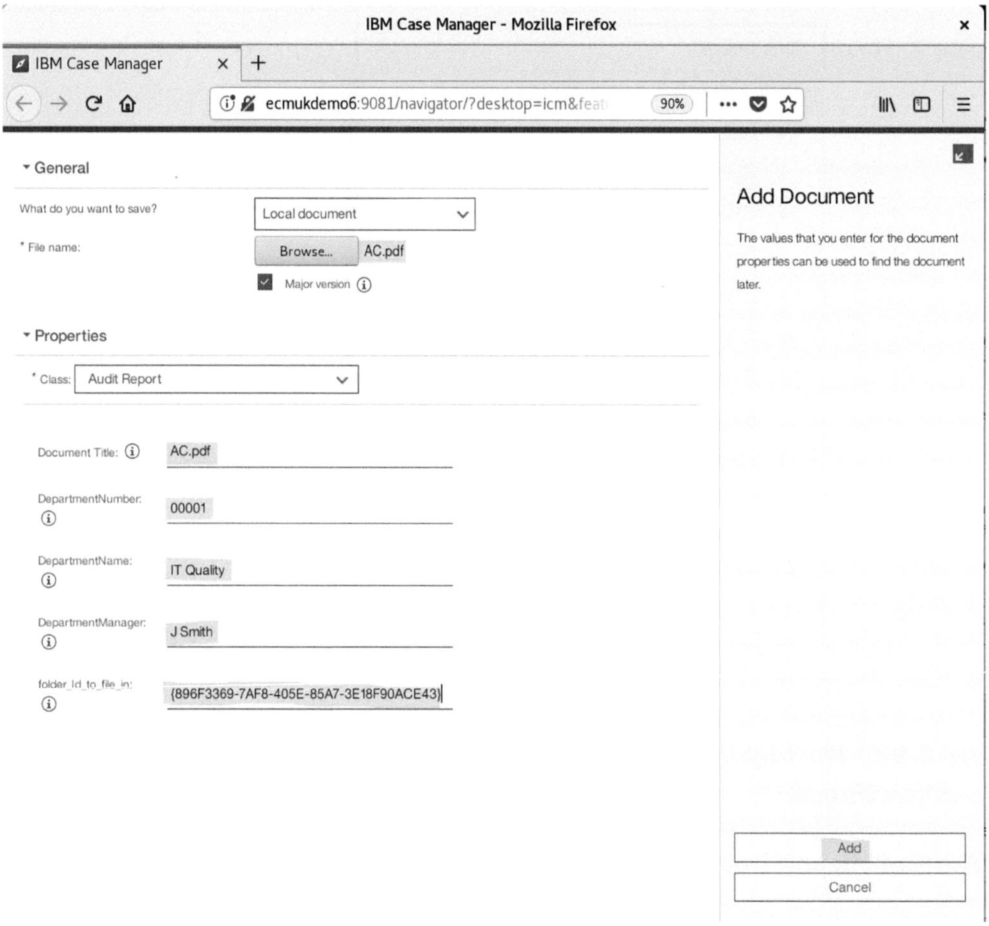

Figure 2-94. *The Audit Report Document class properties are entered and committed using the **Add** command*

220

CHAPTER 2   CONFIGURING JAVA CUSTOM COMPONENTS

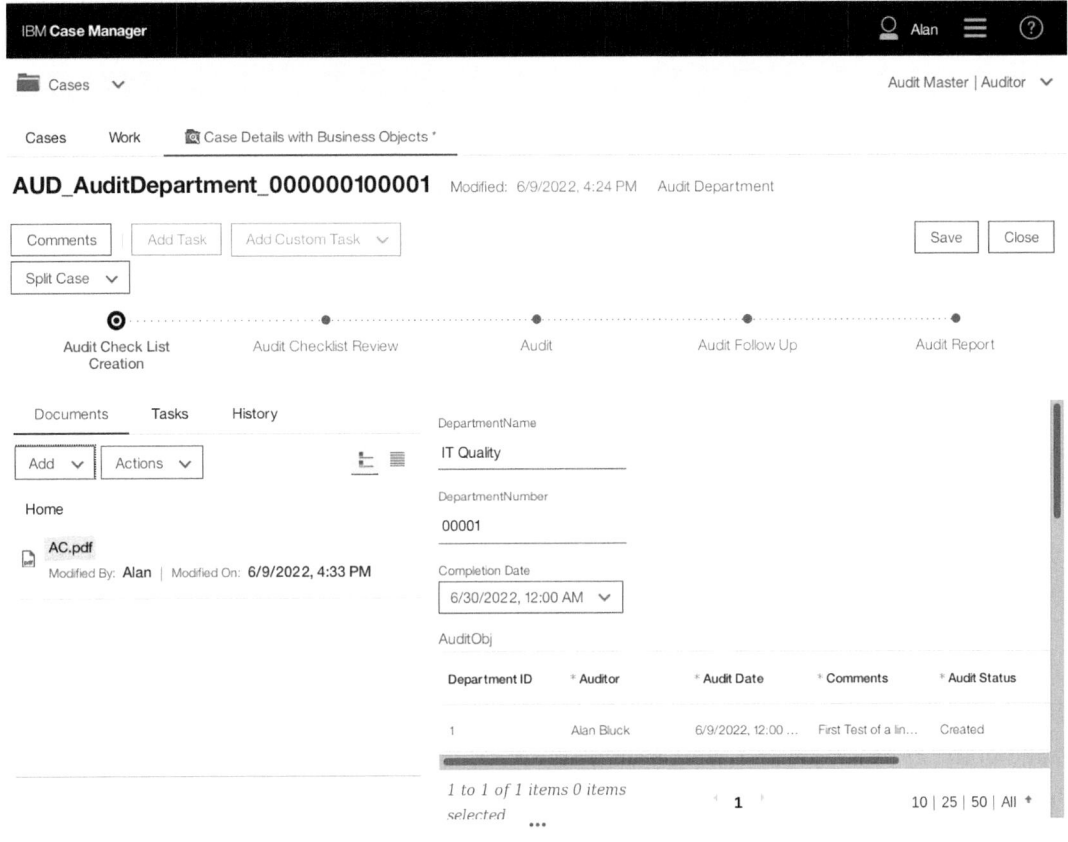

Figure 2-95. *The AC.pdf document (highlighted) is added to the Case*

We can now run the **TestHarness** Java program in the Eclipse IDE and use the unique **SymbolicName** property **AUD_folder_Id_to_file_in** String value, **{896F3369-7AF8-405E-85A7-3E18F90ACE43}**, to link the test document to the **AUDIT_TEST** folder.

221

CHAPTER 2   CONFIGURING JAVA CUSTOM COMPONENTS

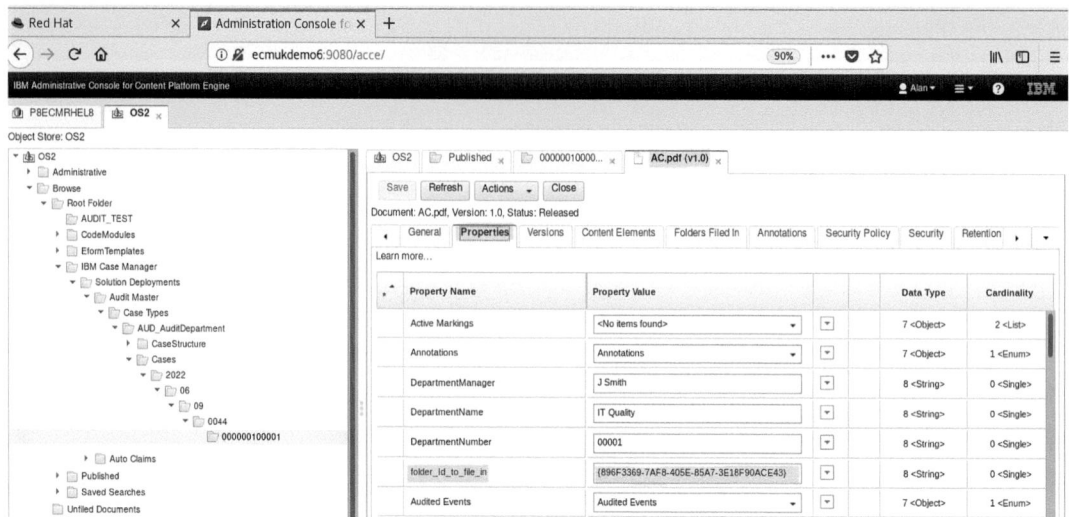

Figure 2-96.   The Folder Id {896F3369-7AF8-405E-85A7-3E18F90ACE43} is in the AC.pdf Properties

We also need to manually add the **AC.pdf** example file's **Audit Report** Document class Id to the **TestHarness** code to allow it to locate and link the document. In the live usage of the Event Handler, the Custom Event subscription automatically passes the Document's Id to the Event Java module.

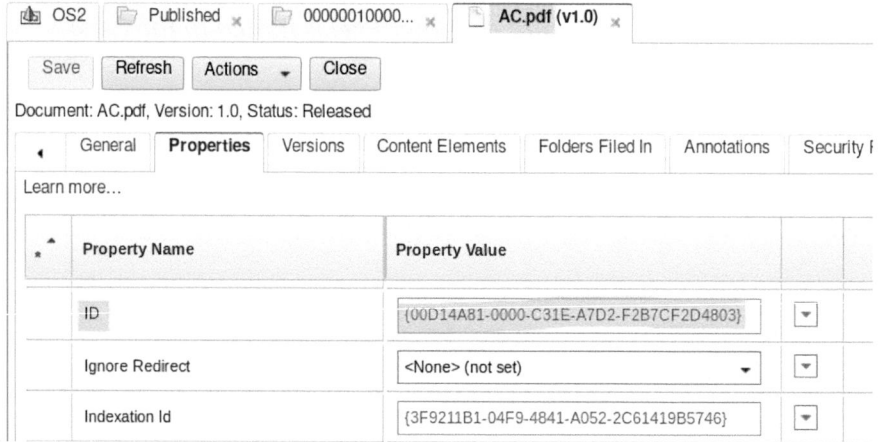

Figure 2-97.   Detail of the AC.pdf Document Property ID is shown highlighted

CHAPTER 2 CONFIGURING JAVA CUSTOM COMPONENTS

The Test Document ID for the AC.pdf file shown earlier is {00D14A81-0000-C31E-A7D2-F2B7CF2D4803}; this is then entered into the TestHarness code.

Listing 2-5. The test_AddDocToFolder Java Class code

```
//The Test Document ID for AC.pdf file is {00D14A81-0000-C31E-
A7D2-F2B7CF2D4803}
private void test_AddDocToFolder() throws Exception {
    l4j.debug("test_AddDocToFolder");
    Document d = (Document) os.getObject("Document",
    "{00D14A81-0000-C31E-A7D2-F2B7CF2D4803}");
    am.addDocToFolder(d);
}
```

In the preceding code, the Document, AC.pdf, is retrieved into the Java Object **d**, which is an Audit Report FileNet Content Engine Document class. The document now has the Id of the AUDIT_TEST folder it is to be linked to.

Figure 2-98. *The Display Value option allows a unique Id to be copied and so can then be pasted*

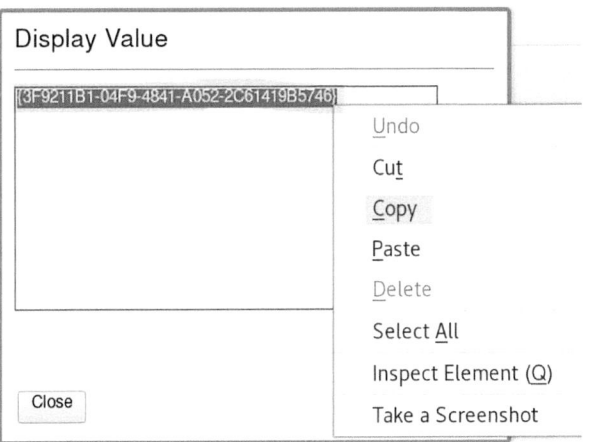

Figure 2-99. *A unique Id can be selected for copy in acce by using the drop-down menu Copy item (highlighted)*

CHAPTER 2   CONFIGURING JAVA CUSTOM COMPONENTS

The TestHarness.java code is shown in Figure 2-100.

```java
/**
 * Test method for the
 * {@link com.ibm.filenet.ps.ciops.AuditOperations#getFolderDocuments(VWAttachment)}
 * method. It uses a test folder in the object store as input.
 *
 * @throws Exception
 */
// @Test
public static void main(String args[]) {
    CEConnection cec = new CEConnection();
    TestHarness th = new TestHarness();
    th.os = cec.getObjectStore("OS2");
    System.out.println(th.os.get_Name());
    //
    try {
        th.test_AddDocToFolder();
    } catch (Exception e) {
        // TODO Auto-generated catch block
        e.printStackTrace();
    }

}
// The Test Document ID for AC.pdf file is {00D14A81-0000-C31E-A7D2-F2B7CF2D4803}
private void test_AddDocToFolder() throws Exception {
    l4j.debug("test_AddDocToFolder");
    Document d = (Document) os.getObject("Document", "{00D14A81-0000-C31E-A7D2-F2B7CF2D4803}");
    am.addDocToFolder(d);
}
```

Figure 2-100. *The TestHarness.java program*

CHAPTER 2 CONFIGURING JAVA CUSTOM COMPONENTS

For the code block, see Listing 2-3 listed in the section "Configuring Java Components for Content Engine Events."

In order to successfully run the TestHarness.java, we have to ensure the Classpath for the Run configurations option of the TestHarness.java program contain the references to all the .jar files it references.

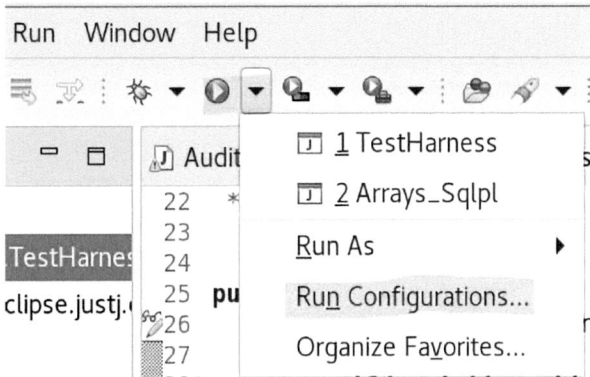

Figure 2-101. *The TestHarness Run Configurations... menu item is selected to set the required Classpath*

In order to run the TestHarness program without "Class Not Found" Java errors, we need to check that all the program dependencies, including the required supporting Java library files, are on the Classpath.

CHAPTER 2　CONFIGURING JAVA CUSTOM COMPONENTS

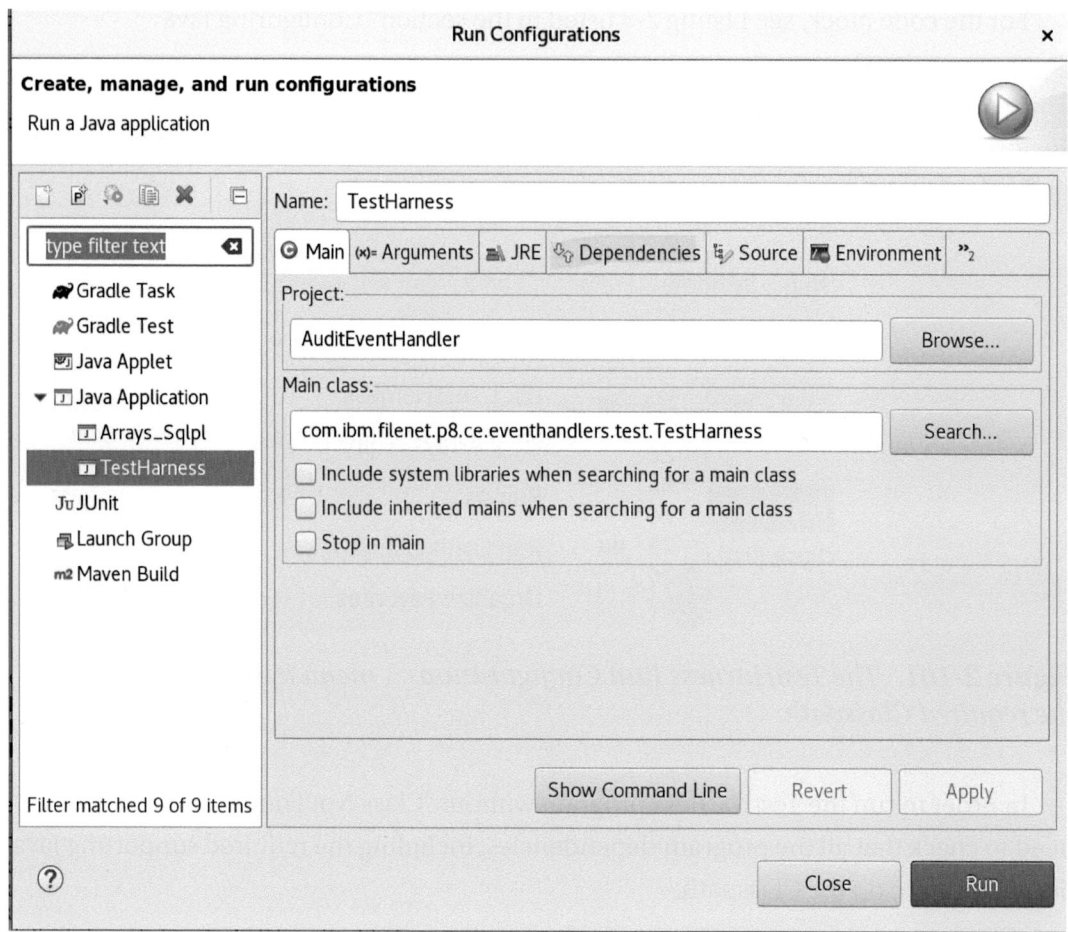

Figure 2-102.　*The Dependencies tab is selected*

CHAPTER 2   CONFIGURING JAVA CUSTOM COMPONENTS

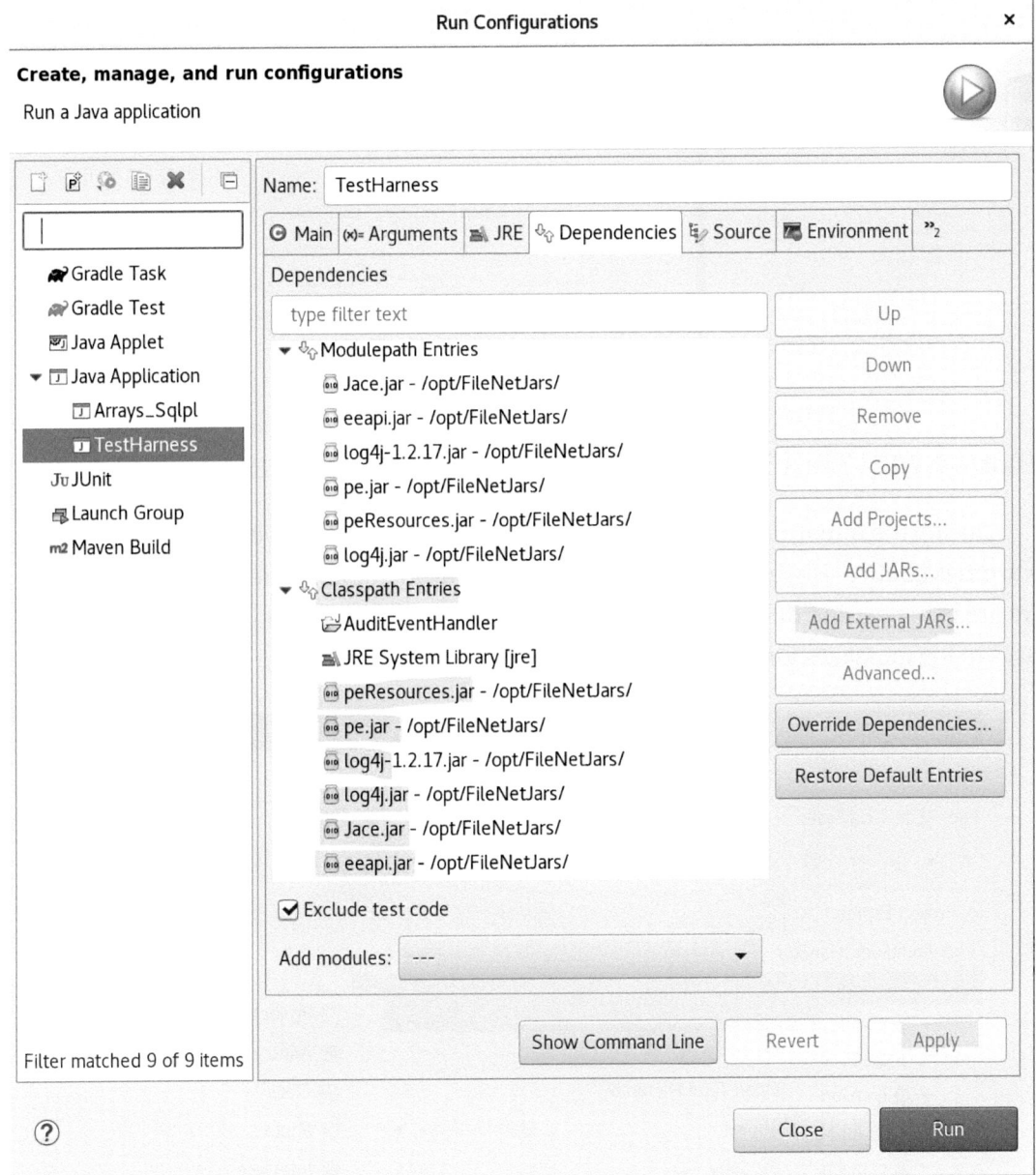

Figure 2-103. *The list of Classpath .jar files we used is shown highlighted*

After running the **TestHarness** program, we can see that the **AC.pdf** file has been successfully linked to the **AUDIT_TEST** folder in the **OS2** target object store Content Store.

CHAPTER 2 CONFIGURING JAVA CUSTOM COMPONENTS

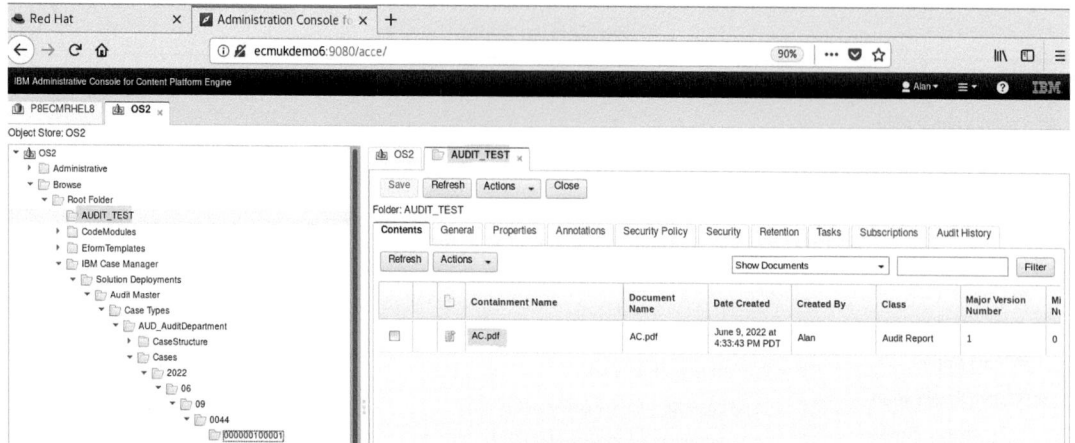

Figure 2-104. *The AC.pdf file appears linked to the AUDIT_TEST folder*

During the initial TestHarness program runs, we noticed that we get Java log4j warnings about a missing appender. This requires us to initialize the log4j system which we are using to send debug output (and which the IBM FileNet API jar file, jace.jar, also uses to log messages for debug).

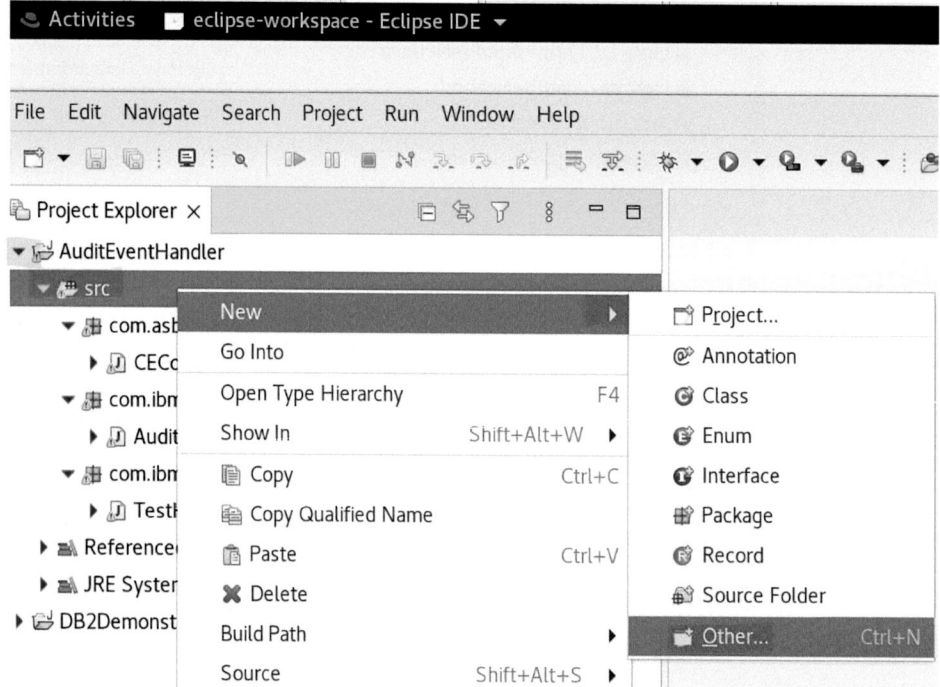

Figure 2-105. *The* **New ▶ Other** *option is used to create a* **log4j.xml** *configuration file and the* **log4j.dtd** *file*

CHAPTER 2 CONFIGURING JAVA CUSTOM COMPONENTS

The src folder needs to be updated with a **log4j.xml** and **log4j.dtd** file for configuring the logger commands.

```
log4j:WARN No appenders could be found for logger (filenet_error.api.com.filenet.apiimpl.util.ConfigValueLookup)
log4j:WARN Please initialize the log4j system properly.
log4j:WARN See http://logging.apache.org/log4j/1.2/faq.html#noconfig for more info.
[Perf Log] No interval found. Auditor disabled.
OS2
```

Figure 2-106. *The logger WARN messages indicate that we need to add a log4j.xml file to define appenders*

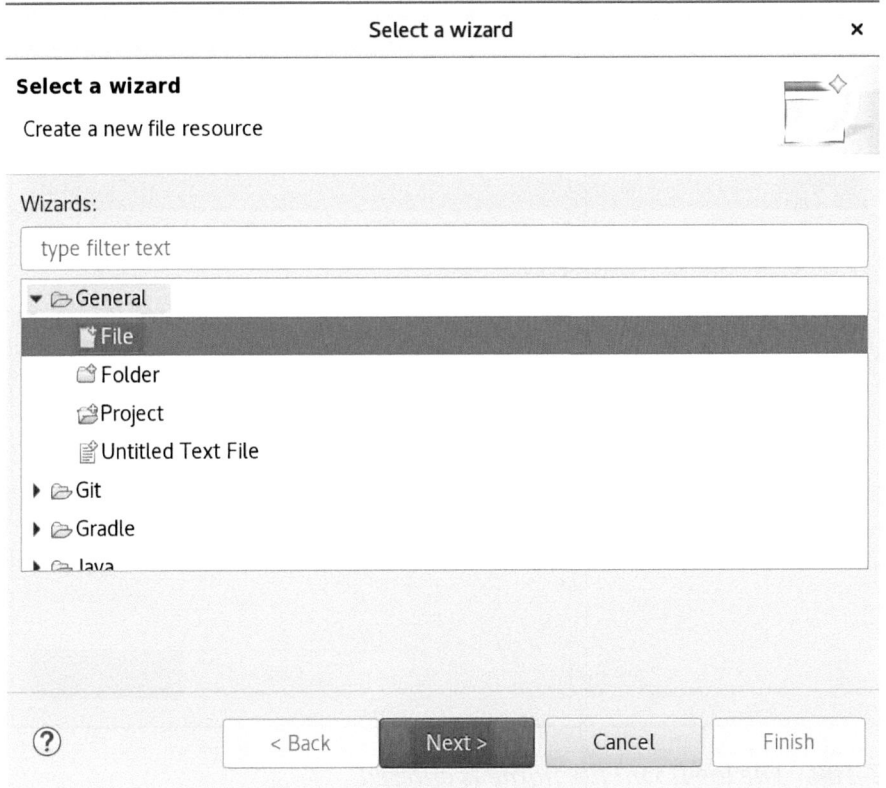

Figure 2-107. *The **General ▶ File** option is selected for adding the log4j.xml file*

CHAPTER 2 CONFIGURING JAVA CUSTOM COMPONENTS

The file is expected to be called **log4j.xml** with an XML format with standard tags to define the **appenders**.

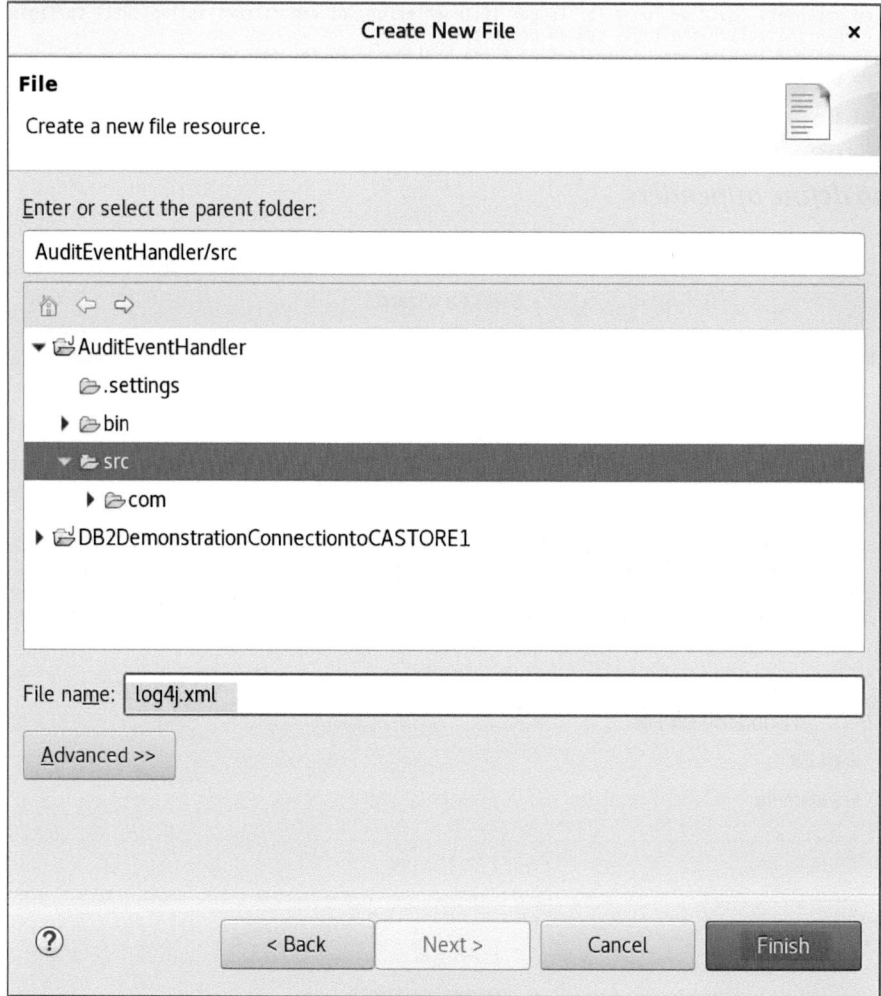

Figure 2-108. *The log4j.xml file name is entered*

CHAPTER 2 CONFIGURING JAVA CUSTOM COMPONENTS

The src folder needs to be updated with a **log4j.xml** and **log4j.dtd** file for configuring the logger commands.

```
log4j:WARN No appenders could be found for logger (filenet_error.api.com.filenet.apiimpl.util.ConfigValueLookup)
log4j:WARN Please initialize the log4j system properly.
log4j:WARN See http://logging.apache.org/log4j/1.2/faq.html#noconfig for more info.
[Perf Log] No interval found. Auditor disabled.
OS2
```

Figure 2-106. *The logger WARN messages indicate that we need to add a log4j.xml file to define appenders*

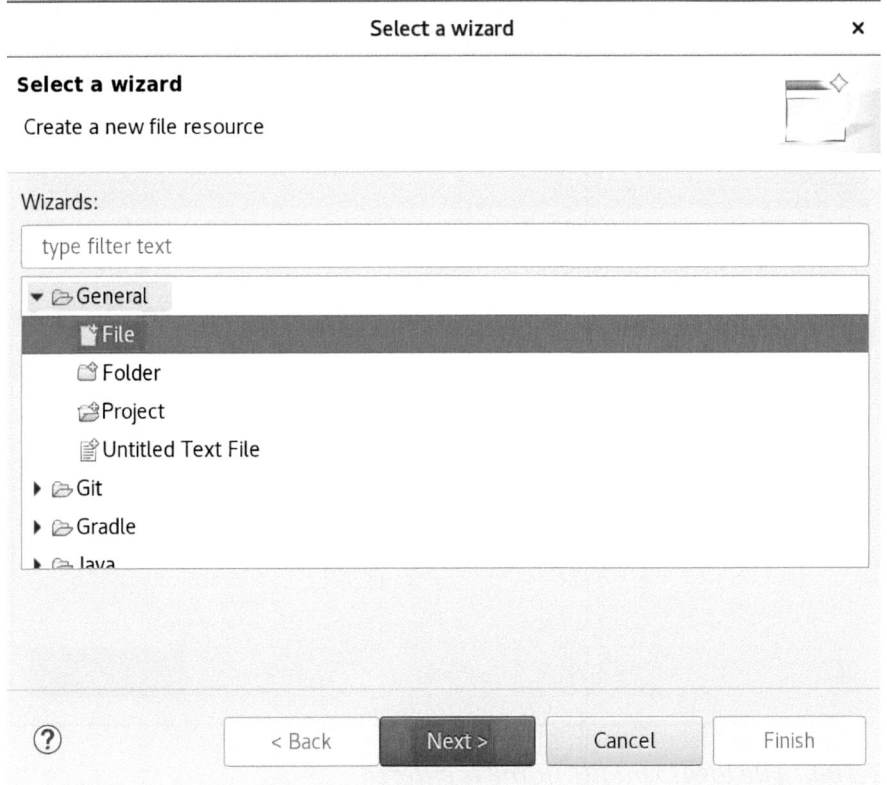

Figure 2-107. *The **General ▶ File** option is selected for adding the log4j.xml file*

229

CHAPTER 2 CONFIGURING JAVA CUSTOM COMPONENTS

The file is expected to be called **log4j.xml** with an XML format with standard tags to define the **appenders**.

Figure 2-108. *The log4j.xml file name is entered*

CHAPTER 2 CONFIGURING JAVA CUSTOM COMPONENTS

```xml
 1  <?xml version="1.0" encoding="UTF-8"?>
 2  <!DOCTYPE log4j:configuration SYSTEM "log4j.dtd" >
 3  <log4j:configuration>
 4      <appender name="stdout" class="org.apache.log4j.ConsoleAppender">
 5          <layout class="org.apache.log4j.PatternLayout">
 6              <param name="ConversionPattern" value="%m%n" />
 7          </layout>
 8          <filter class="org.apache.log4j.varia.LevelRangeFilter">
 9              <param name="levelMin" value="DEBUG" />
10              <param name="levelMax" value="INFO" />
11          </filter>
12      </appender>
13      <appender name="rfa" class="org.apache.log4j.RollingFileAppender">
14          <param name="file" value="/opt/filenet_app.log"/>
15          <param name="MaxFileSize" value="100KB"/>
16          <!-- Keep some backup files -->
17          <param name="MaxBackupIndex" value="10"/>
18          <layout class="org.apache.log4j.PatternLayout">
19              <param name="ConversionPattern" value="%p %t %c - %m%n"/>
20          </layout>
21          <filter class="org.apache.log4j.varia.LevelRangeFilter">
22              <param name="levelMin" value="WARN" />
23              <param name="levelMax" value="FATAL" />
24          </filter>
25      </appender>
26
27      <logger name="com.ibm.filenet.ce" additivity="false">
28          <level value="debug" />
29          <appender-ref ref="rfa" />
30          <appender-ref ref="stdout" />
31      </logger>
32  </log4j:configuration>
```

***Figure 2-109.** The XML for the standard log4j.xml file content is shown*

***Listing 2-6.** The log4j.xml file content*

```xml
<?xml version="1.0" encoding="UTF-8"?>
<!DOCTYPE log4j:configuration SYSTEM "log4j.dtd" >
 <log4j:configuration>
     <appender name="stdout" class="org.apache.log4j.ConsoleAppender">
         <layout class="org.apache.log4j.PatternLayout">
             <param name="ConversionPattern" value="%m%n" />
         </layout>
```

CHAPTER 2 CONFIGURING JAVA CUSTOM COMPONENTS

```xml
        <filter class="org.apache.log4j.varia.LevelRangeFilter">
            <param name="levelMin" value="DEBUG" />
            <param name="levelMax" value="INFO" />
    </filter>
  </appender>
<appender name="rfa" class="org.apache.log4j.RollingFileAppender">
    <param name="file" value="/opt/filenet_app.log"/>
    <param name="MaxFileSize" value="100KB"/>
    <!-- Keep some backup files -->
    <param name="MaxBackupIndex" value="10"/>
    <layout class="org.apache.log4j.PatternLayout">
            <param name="ConversionPattern" value="%p %t %c - %m%n"/>
    </layout>
    <filter class="org.apache.log4j.varia.LevelRangeFilter">
        <param name="levelMin" value="WARN" />
        <param name="levelMax" value="FATAL" />
    </filter>
  </appender>
<logger name="com.ibm.filenet.ce" additivity="false">
    <level value="debug" />
    <appender-ref ref="rfa" />
    <appender-ref ref="stdout" />
</logger>
<logger name="filenet_error.api.com.filenet.apiimpl.util.
ConfigValueLookup" additivity="false">
    <level value="debug" />
    <appender-ref ref="rfa" />
    <appender-ref ref="stdout" />
</logger>
<logger name="filenet_tracing.api.detail.moderate.summary.com.filenet.
apiimpl.util.ConfigValueLookup" additivity="false">
```

```
        <level value="debug" />
        <appender-ref ref="rfa" />
        <appender-ref ref="stdout" />
    </logger>

</log4j:configuration>
```

We have entered the content for the **log4j.xml** and also a standard **log4j.dtd** from the **apache** download site.

Save Resource
Save 'log4j.xml'?
Don't Save
Cancel
Save

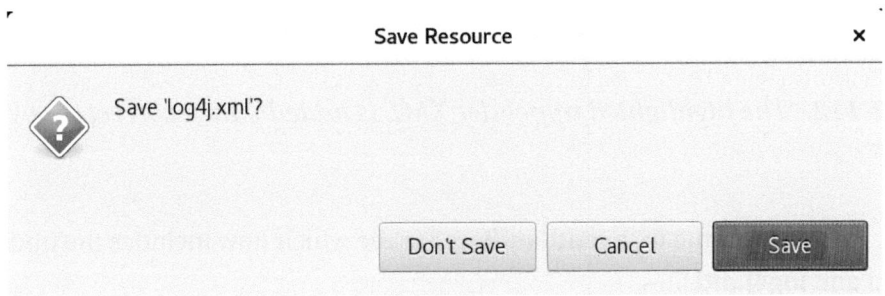

Figure 2-110. *The log4j.xml file and contents are saved, using the Eclipse file Save command*

But, after saving the log4j.xml and rerunning the TestHarness program, we still get the following warnings.

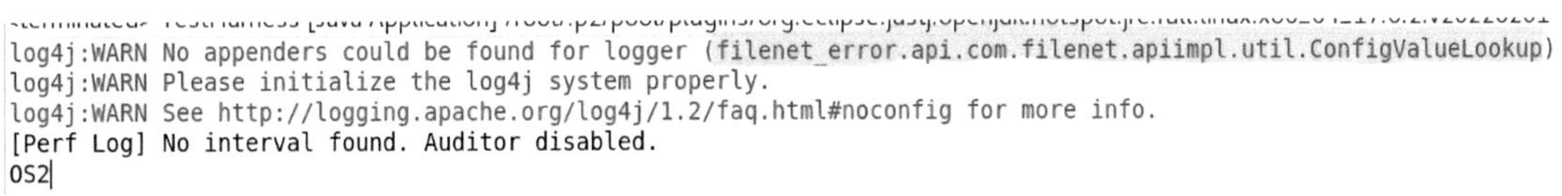

log4j:WARN No appenders could be found for logger (filenet_error.api.com.filenet.apiimpl.util.ConfigValueLookup)
log4j:WARN Please initialize the log4j system properly.
log4j:WARN See http://logging.apache.org/log4j/1.2/faq.html#noconfig for more info.
[Perf Log] No interval found. Auditor disabled.
OS2

Figure 2-111. *The logger appender is still requested and displays the highlighted class*

CHAPTER 2 CONFIGURING JAVA CUSTOM COMPONENTS

As highlighted in Figure 2-111, we need to add the appender for the **filenet_error.api.com.filenet.apiimpl.util.ConfigValueLookup** class to the **log4j.xml** file.

```xml
26  <logger name="com.ibm.filenet.ce" additivity="false">
27      <level value="debug" />
28      <appender-ref ref="rfa" />
29      <appender-ref ref="stdout" />
30  </logger>
31  <logger name="filenet_error.api.com.filenet.apiimpl.util.ConfigValueLookup" additivity="false">
32      <level value="debug" />
33      <appender-ref ref="rfa" />
34      <appender-ref ref="stdout" />
35  </logger>
36  <logger name="filenet_tracing.api.detail.moderate.summary.com.filenet.apiimpl.util.ConfigValueLookup" additivity="false">
37      <level value="debug" />
38      <appender-ref ref="rfa" />
39      <appender-ref ref="stdout" />
40  </logger>
41
42  </log4j:configuration>
```

Figure 2-112. *The highlighted appender XML is added which corrects the WARN messages*

Next, we re-export the **fn_eventhandlers.jar** file which now includes the updated **log4j.xml** and **log4j.dtd** files.

CHAPTER 2　CONFIGURING JAVA CUSTOM COMPONENTS

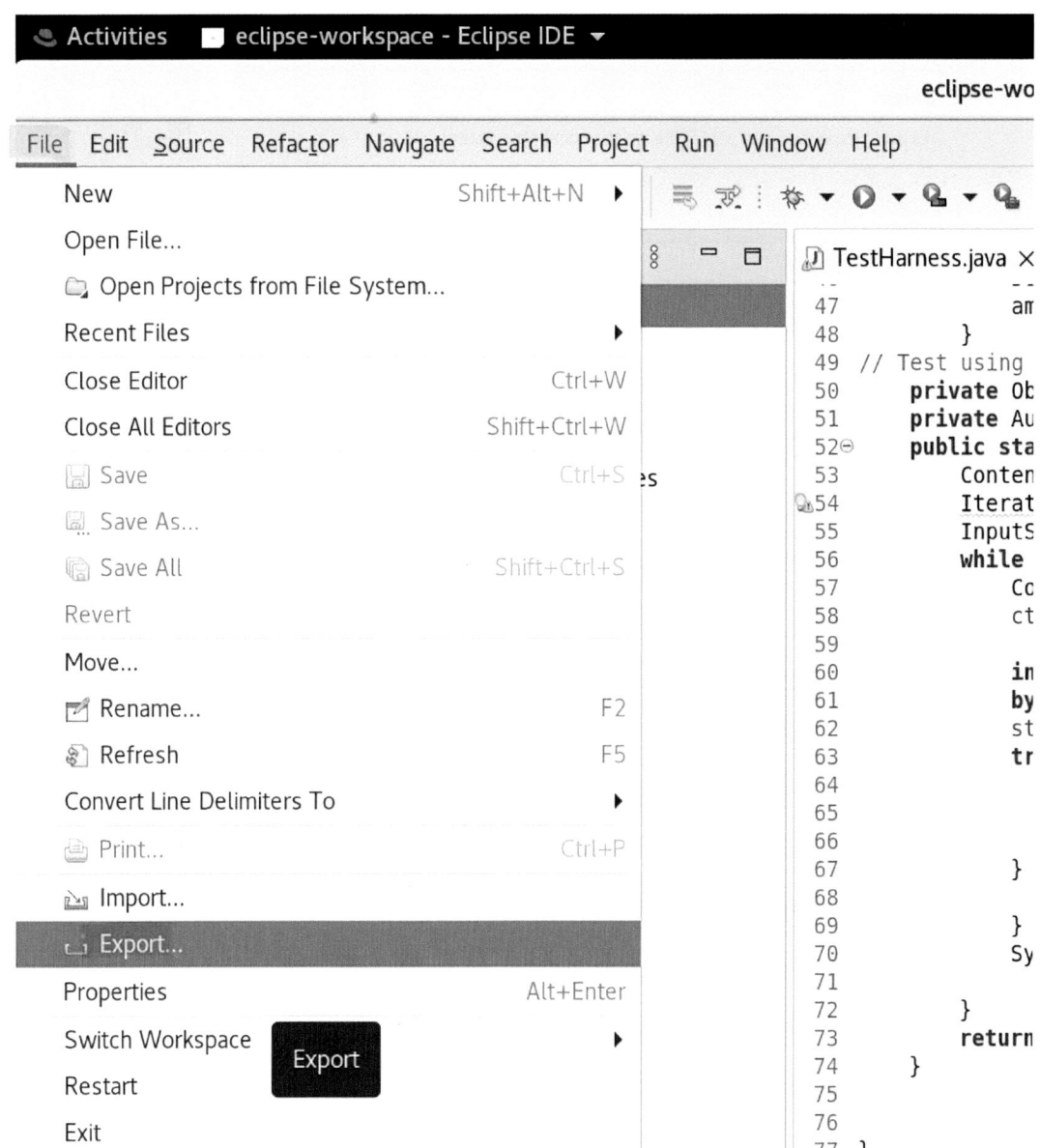

Figure 2-113. *The* **File ▶ Export** *command is used to create an updated* *fn_eventhandlers.jar* *file*

CHAPTER 2　CONFIGURING JAVA CUSTOM COMPONENTS

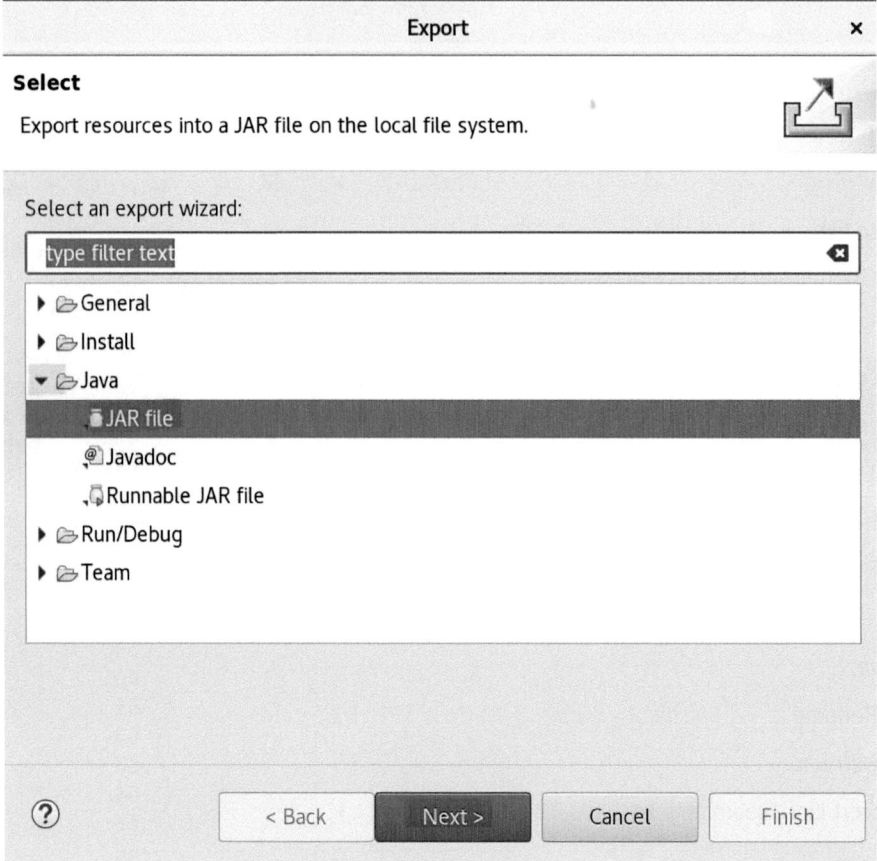

Figure 2-114. *The **Java ➤ JAR** file export wizard is selected, and the **Next>** command is clicked*

CHAPTER 2 CONFIGURING JAVA CUSTOM COMPONENTS

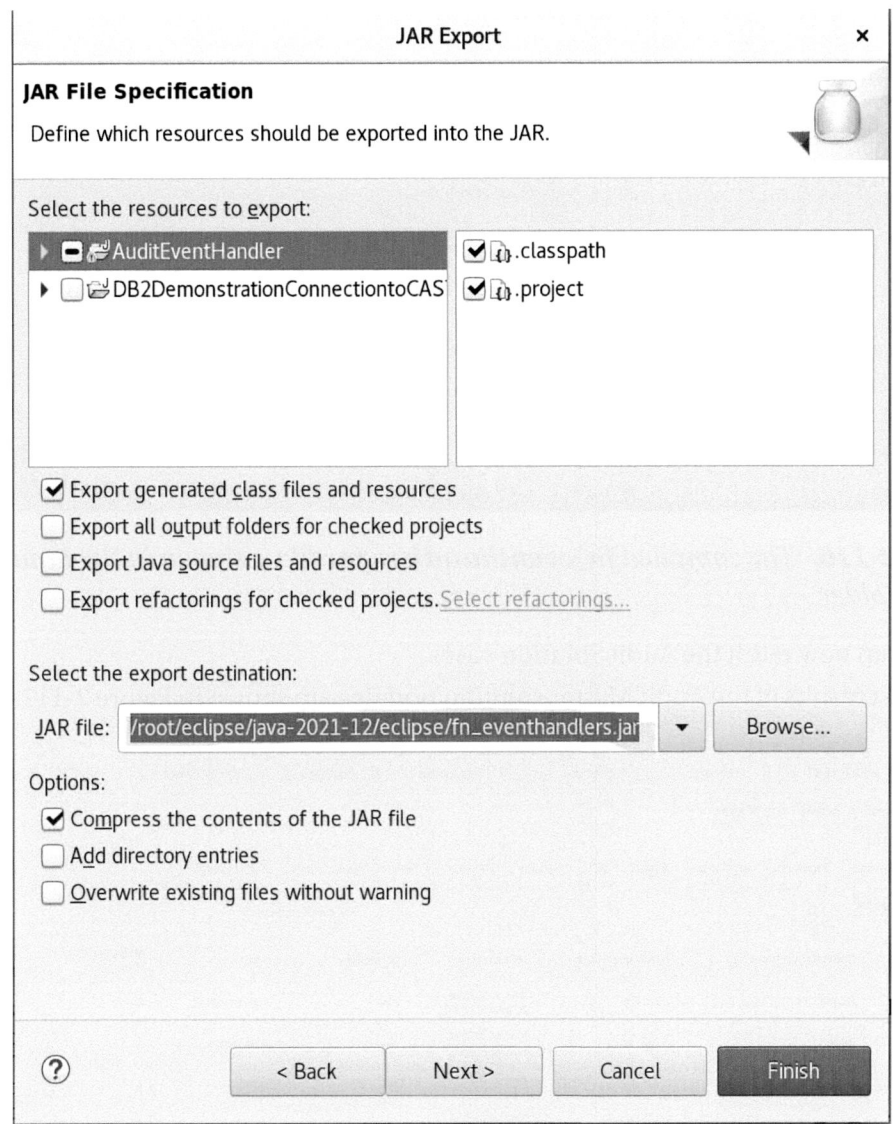

Figure 2-115. *The full path to the fn_eventhandlers.jar file is selected, and Finish is clicked*

The updated **fn_eventhandlers.jar** file created earlier is shown in the Linux folder path in Figure 2-116.

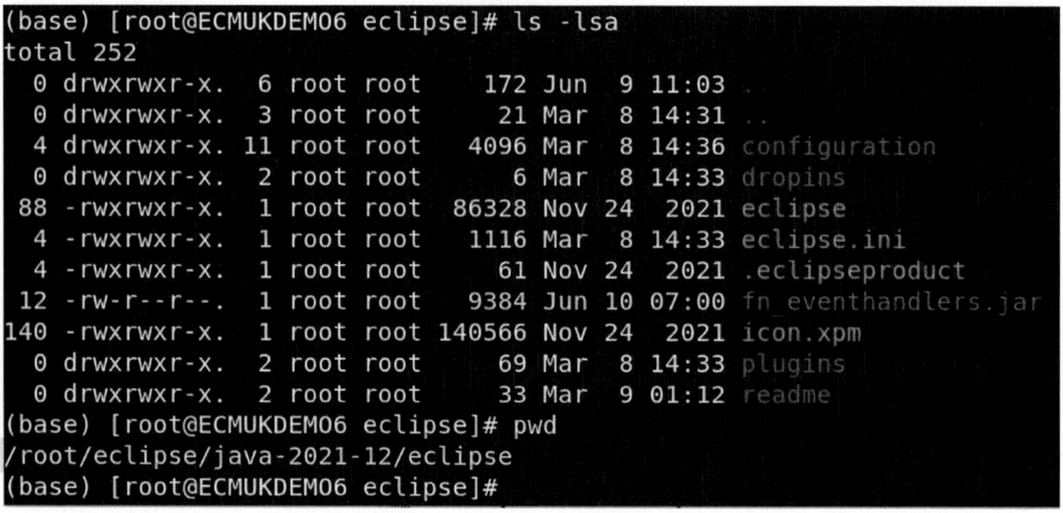

CHAPTER 2　CONFIGURING JAVA CUSTOM COMPONENTS

Figure 2-116. *The compiled **fn_eventhandlers.jar** file is shown in the Linux Eclipse folder*

We can now retest the Audit Solution cases.

Retest results of the Audit Master solution updates are shown in Figure 2-117.

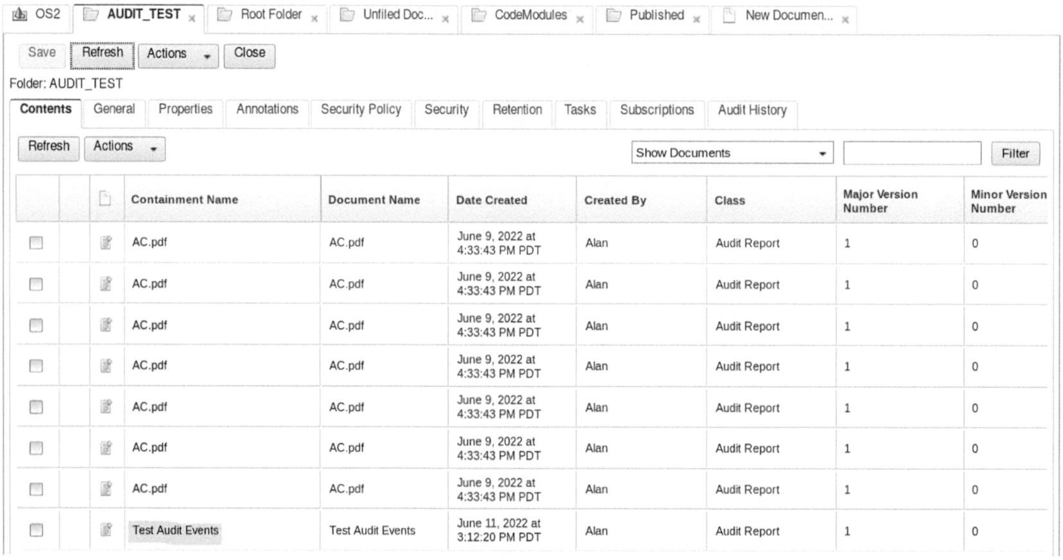

Figure 2-117. *The Test Audit Events document is linked to the AUDIT_TEST folder by the Event call*

238

CHAPTER 2　CONFIGURING JAVA CUSTOM COMPONENTS

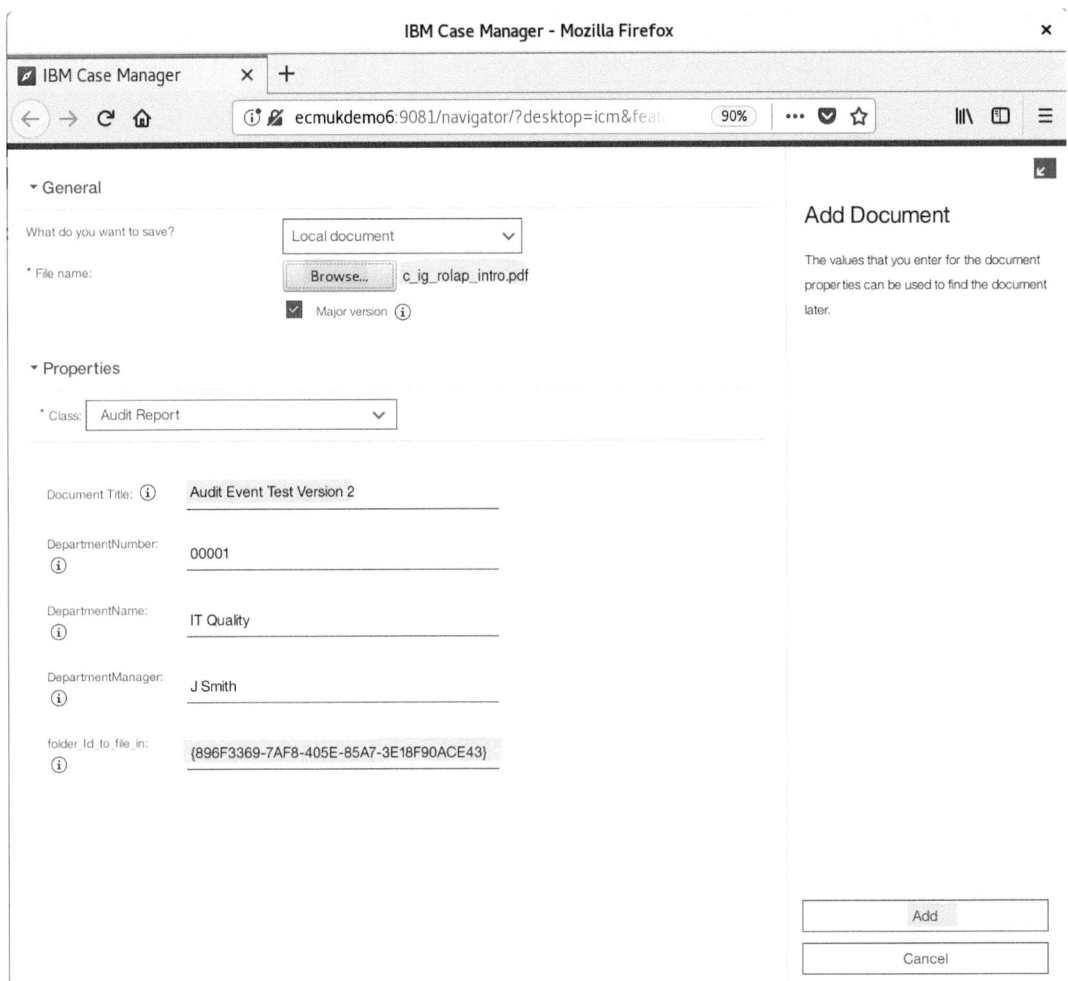

Figure 2-118. *A second test Audit Report class Case document is added to the Case*

We can add multiple documents of type **Audit Report** to the same **Audit Department** case, and they should also appear in the **AUDIT_TEST** folder.

CHAPTER 2   CONFIGURING JAVA CUSTOM COMPONENTS

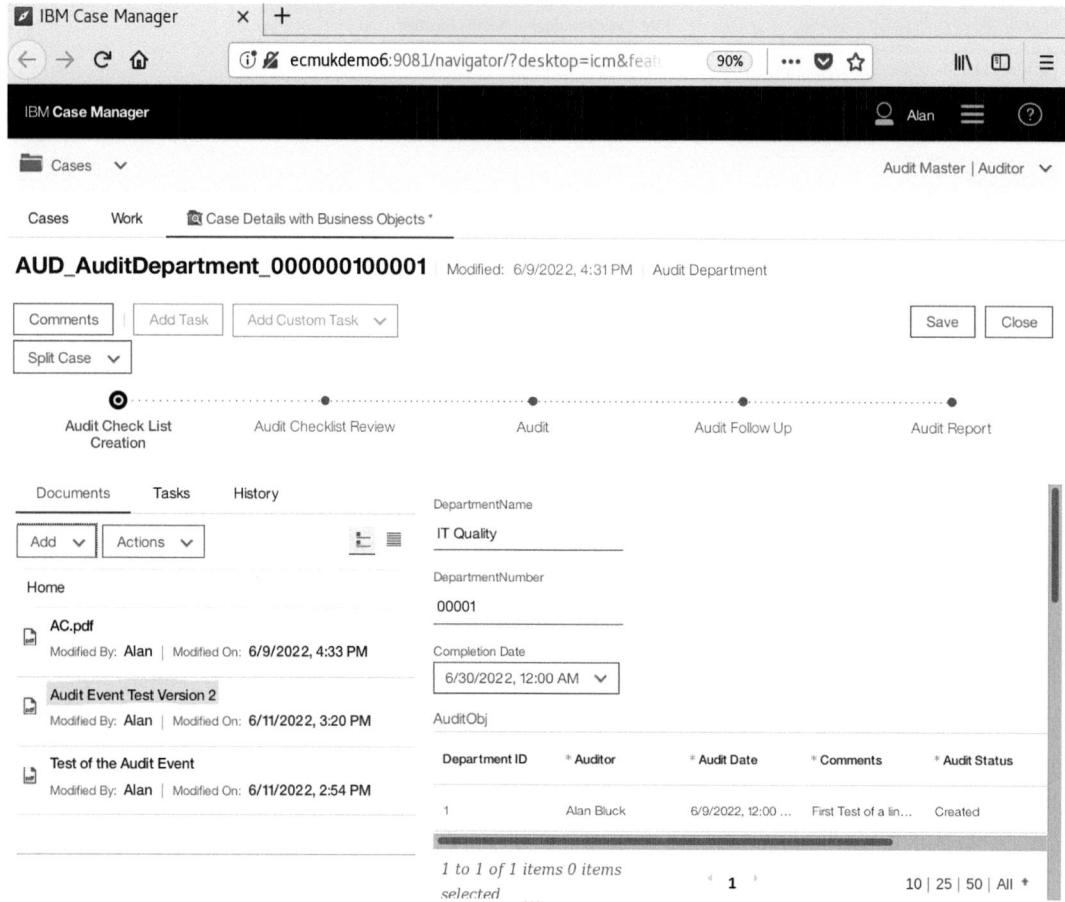

Figure 2-119. *The highlighted document is shown saved in the case*

In the **AUDIT_TEST** folder, we now also see the same highlighted document appear linked. We should emphasize here that this is a single document version copy with links in multiple folders.

CHAPTER 2   CONFIGURING JAVA CUSTOM COMPONENTS

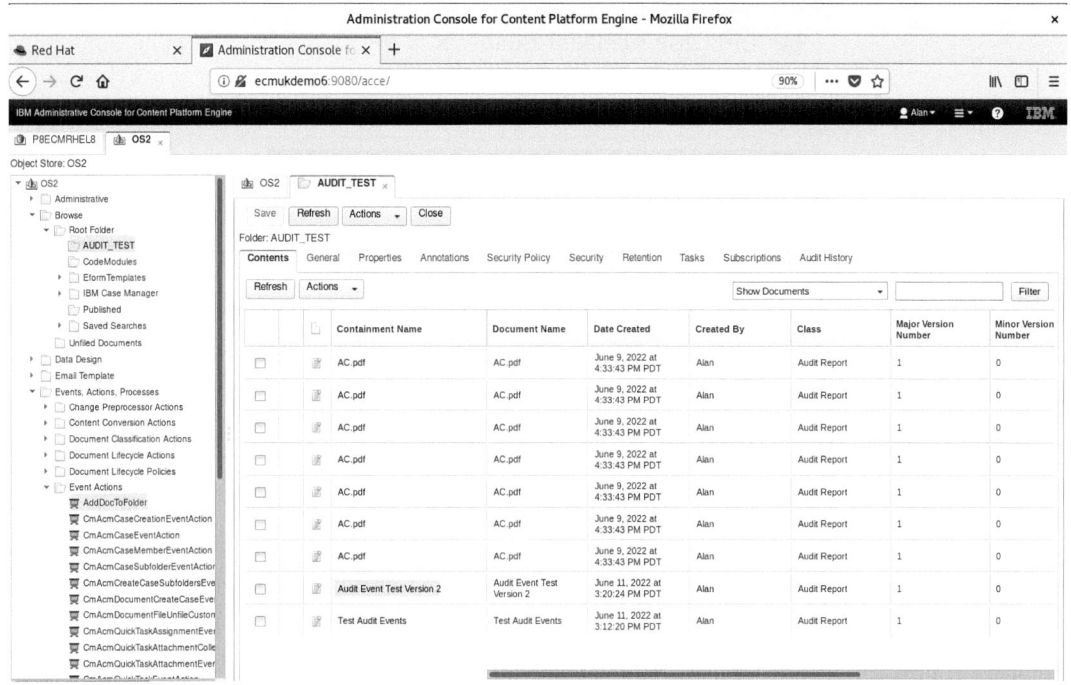

Figure 2-120. *The test successfully shows the Audit Report document highlighted appearing*

In the preceding examples we tested, we have manually added the Folder Id, for **AUDIT_TEST** folder, which is the string GUID, **{896F3369-7AF8-405E-85A7-3E18F90A CE43}**, to the **Audit Report Document class String** property, **folder_Id_to_file_in** (with the full unique Id, **AUD_folder_Id_to_file_in**).

However, we have the option to set this Folder Id as a default value for the **folder_Id_to_file_in** property of the **Audit Report** document class by editing it into the **Audit Master** solution. What is more interesting is that the default value stored in this way could be set differently for different Document classes, which would open up the possibility of automatically filing different types of documents in different folders, using the same Java code module we have created. We will now show this just for the Audit Report document class at the moment.

241

CHAPTER 2 CONFIGURING JAVA CUSTOM COMPONENTS

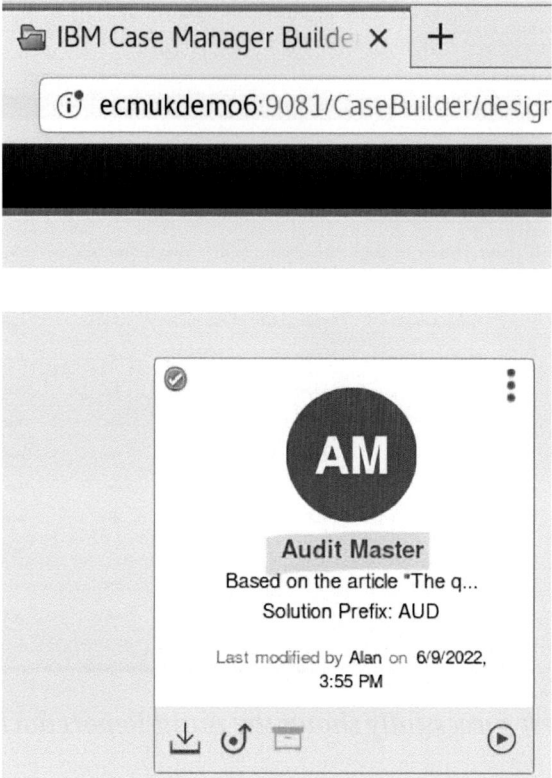

Figure 2-121. *Click the Audit Master solution highlighted to open the IBM Case Builder editor*

The IBM Case Builder web application can now be used to modify our initial Audit Master solution again.

CHAPTER 2   CONFIGURING JAVA CUSTOM COMPONENTS

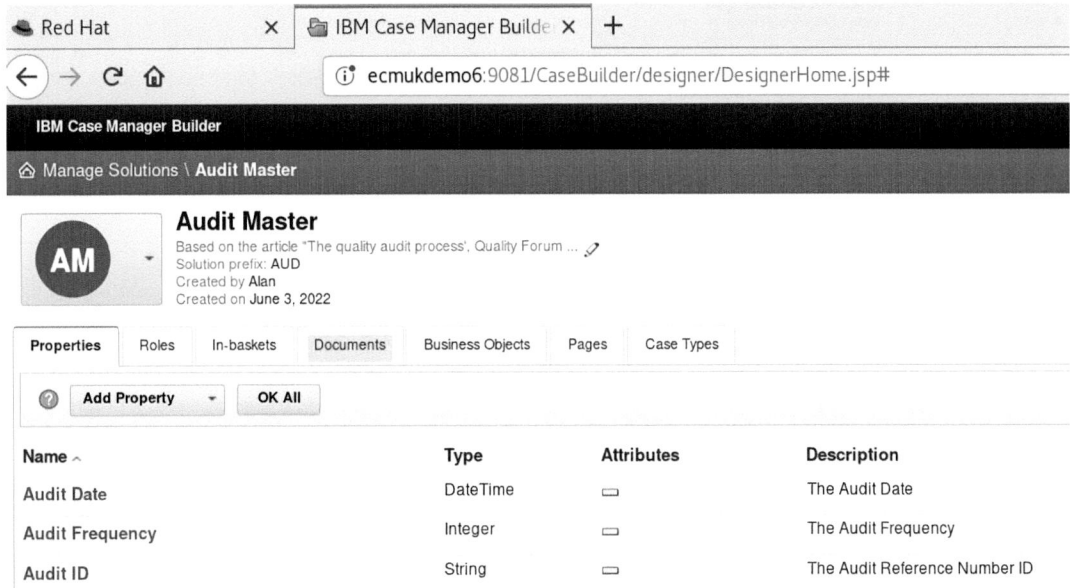

Figure 2-122.   Select the Documents tab of the Audit Master Solution editor

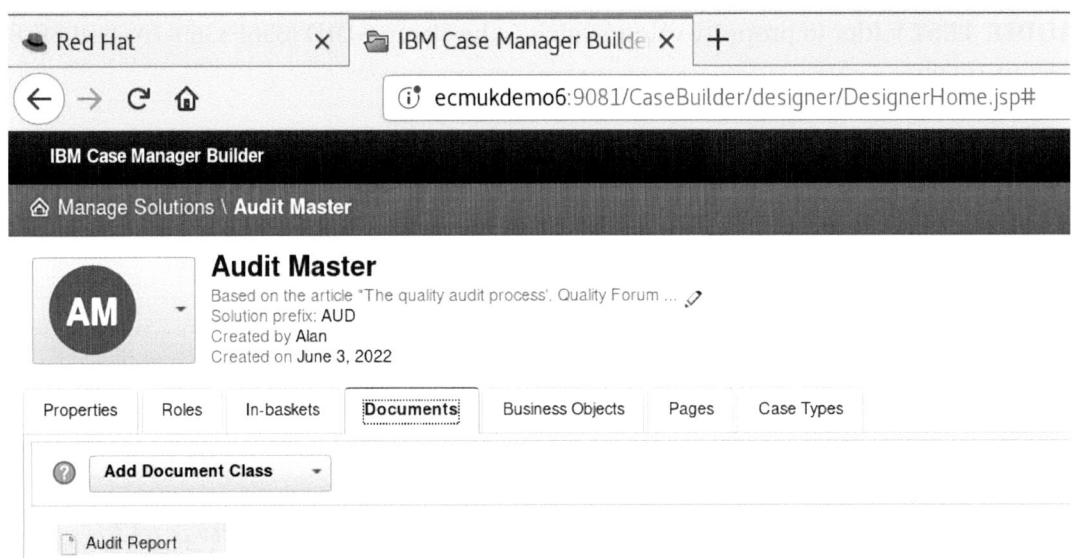

Figure 2-123.   Select the Audit Report Document Class to update the properties

We can now update the **folder_id_to_file_in** property of the **Audit Report** Document class by selecting the **Properties** tab, and using the mouse-over, we can see the **Edit** icon (highlighted in Figure 2-124).

243

CHAPTER 2   CONFIGURING JAVA CUSTOM COMPONENTS

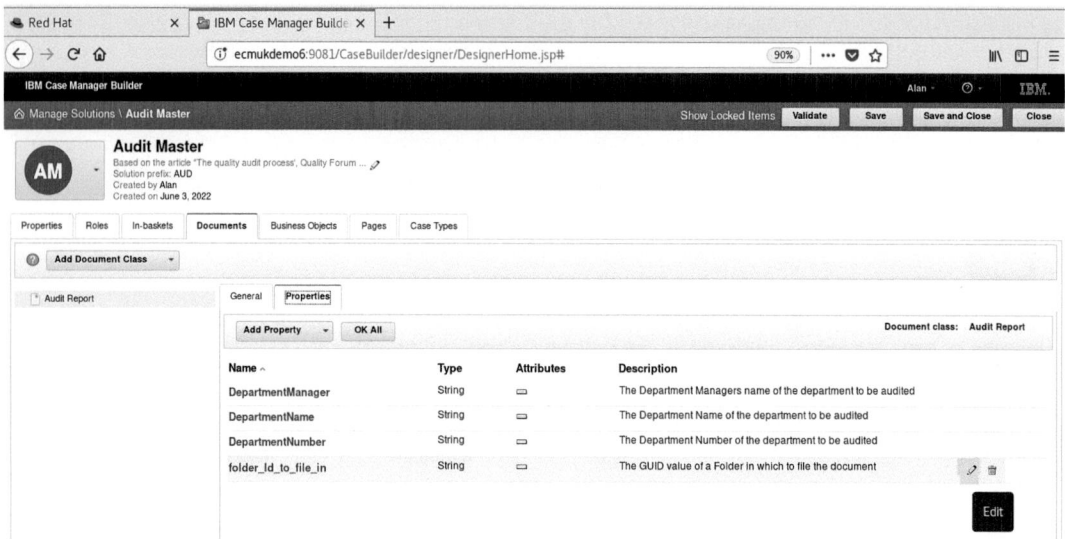

Figure 2-124. *The* ***folder_Id_to_file_in*** *property of the Audit Report Document class is edited*

Now, we can set the Default value for the **folder_Id_to_file_in** property to the **AUDIT_TEST** folder Id property value (which is the string GUID {896F3369-7AF8-405E-85A7-3E18F90ACE43}).

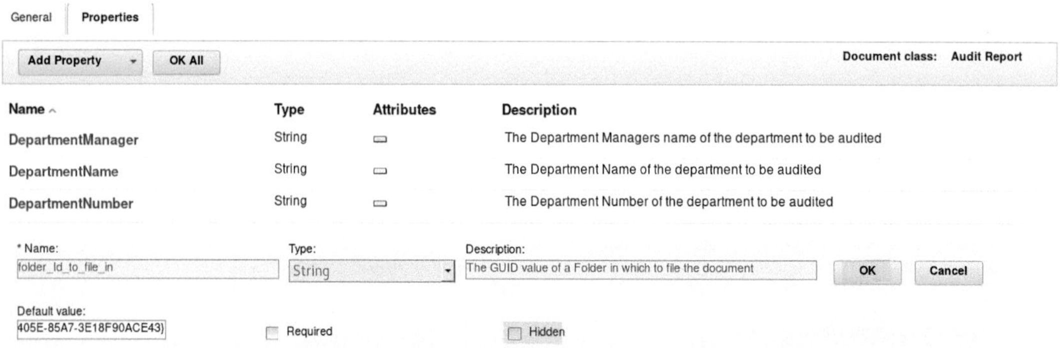

Figure 2-125. *The Default value of the folder_id_to_file_in property is set to {896F3369-7AF8-405E-85A7-3E18F90ACE43}*

In addition, notice that there is a **Hidden** boolean option available (highlighted) for properties, which we could utilize if required, so that a normal user would not even be aware that this property existed. The **Audit Reports** would appear "as if by magic" to this user in the **AUDIT_TEST** folder!

244

CHAPTER 2   CONFIGURING JAVA CUSTOM COMPONENTS

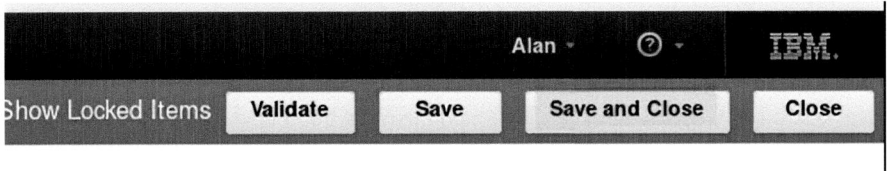

Figure 2-126. *The Audit Master Solution is updated with the **Save and Close** command button*

Having updated the Audit Master solution again, we can now deploy the changed solution as shown in Figure 2-127.

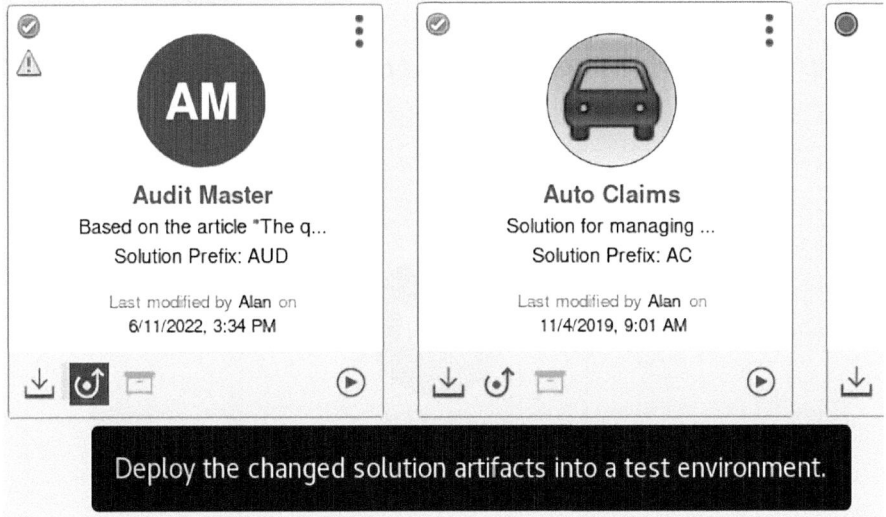

Figure 2-127. *The Deploy icon can be clicked to redeploy the solution we updated*

A Confirmation pop-up dialog window is then displayed as shown in Figure 2-128.

245

CHAPTER 2　CONFIGURING JAVA CUSTOM COMPONENTS

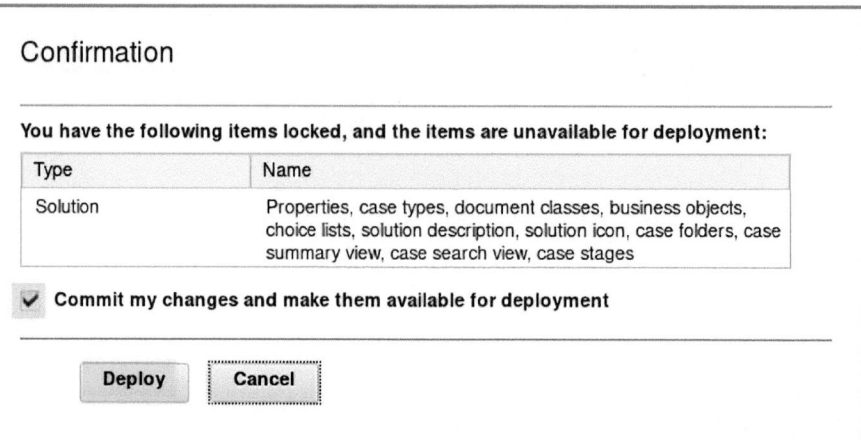

Figure 2-128. *The **Deploy** command is used to redeploy the edited Audit Master solution*

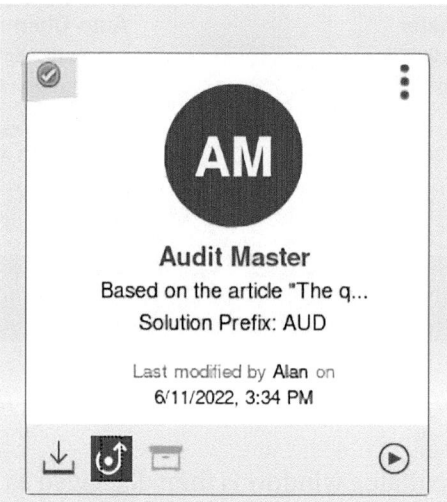

Figure 2-129. *The green tick icon highlighted shows the Audit Master solution is successfully deployed*

On adding an **Audit Report** document class to a new Case now, we automatically have the **folder_Id_to_file_in** property filled with the **AUDIT_TEST** folder **Id** value.

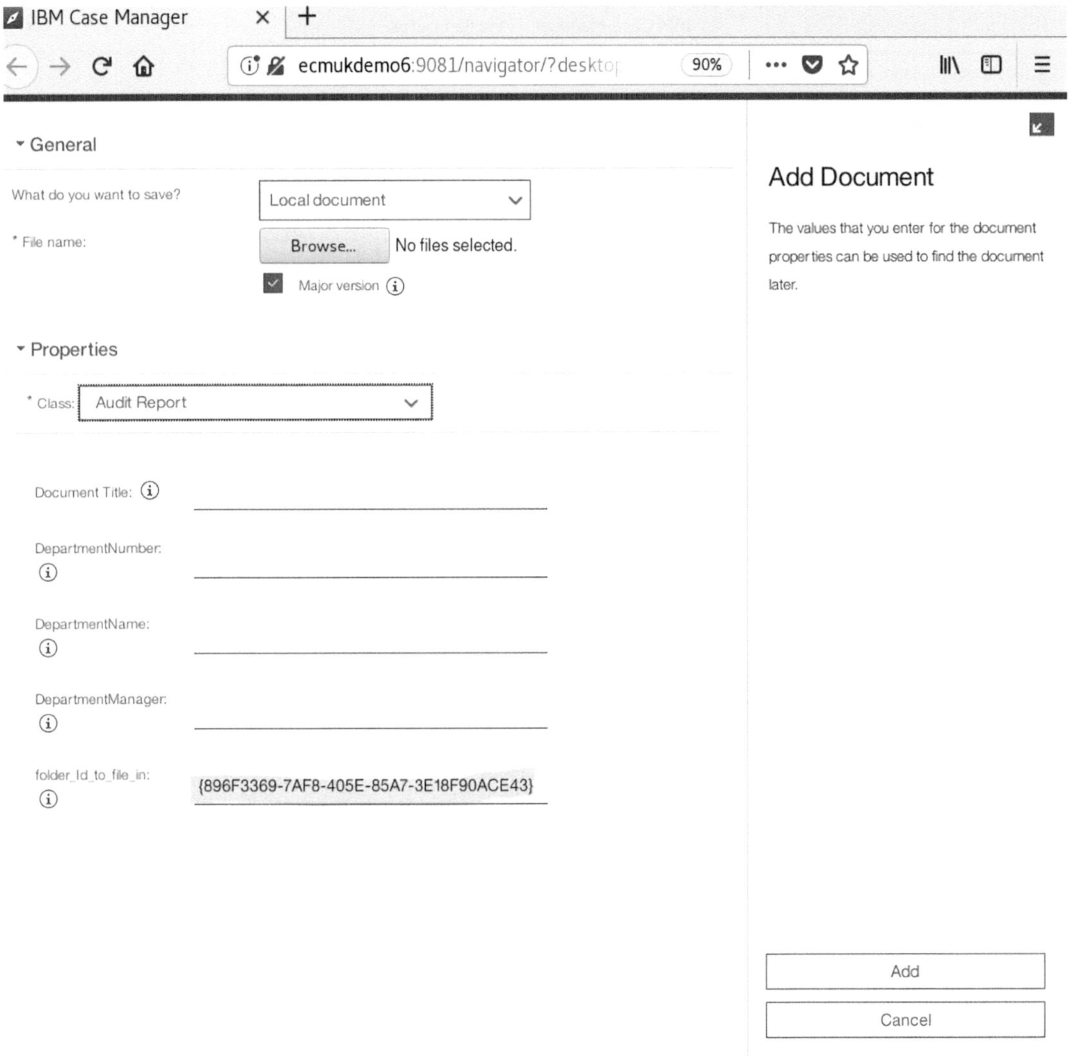

*Figure 2-130. The **folder_Id_to_file_in** property is now filled automatically with the default value*

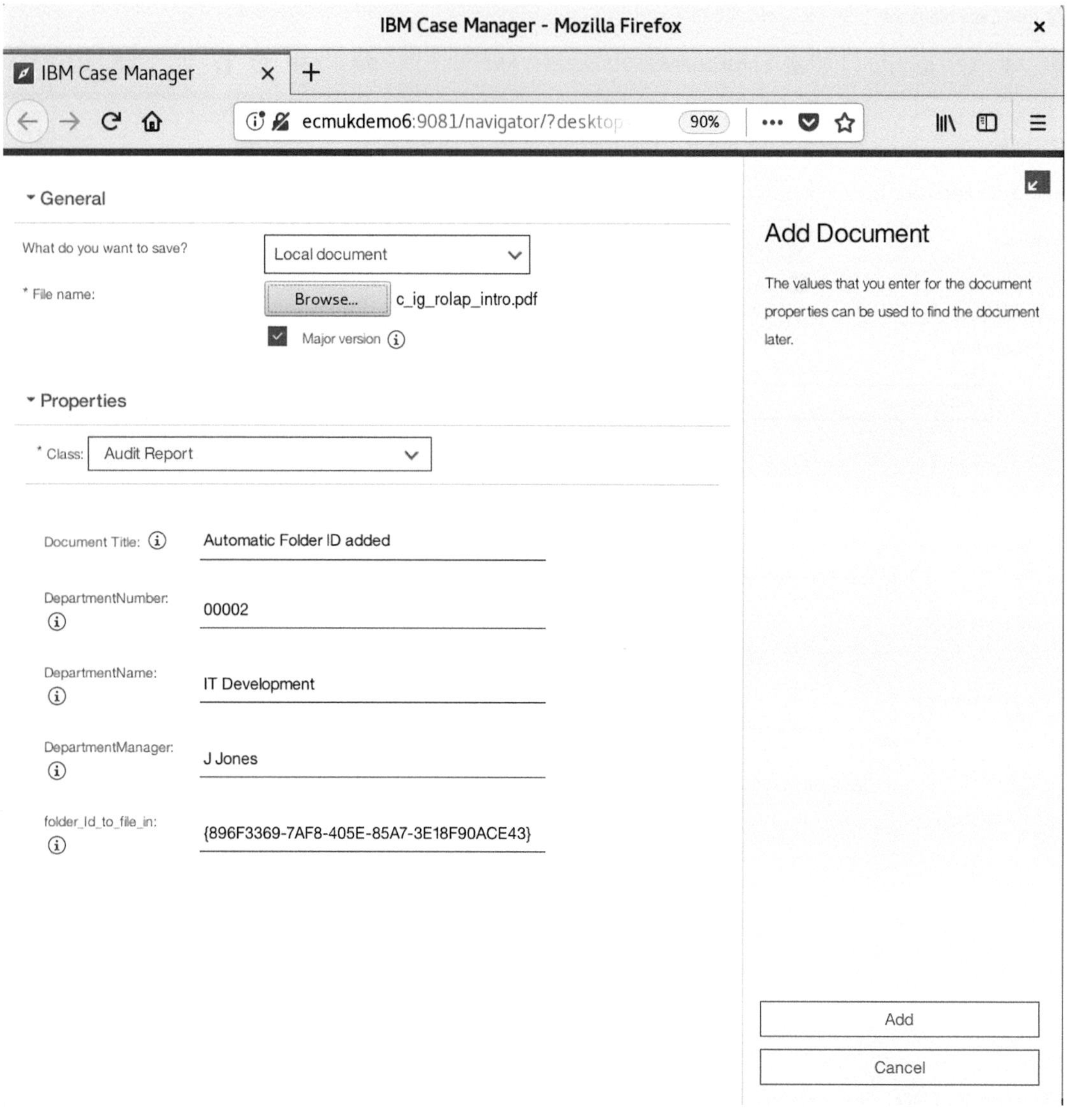

CHAPTER 2   CONFIGURING JAVA CUSTOM COMPONENTS

*Figure 2-131.   A new **Automatic Folder ID added** Audit Report Document is added to test the Event Handler*

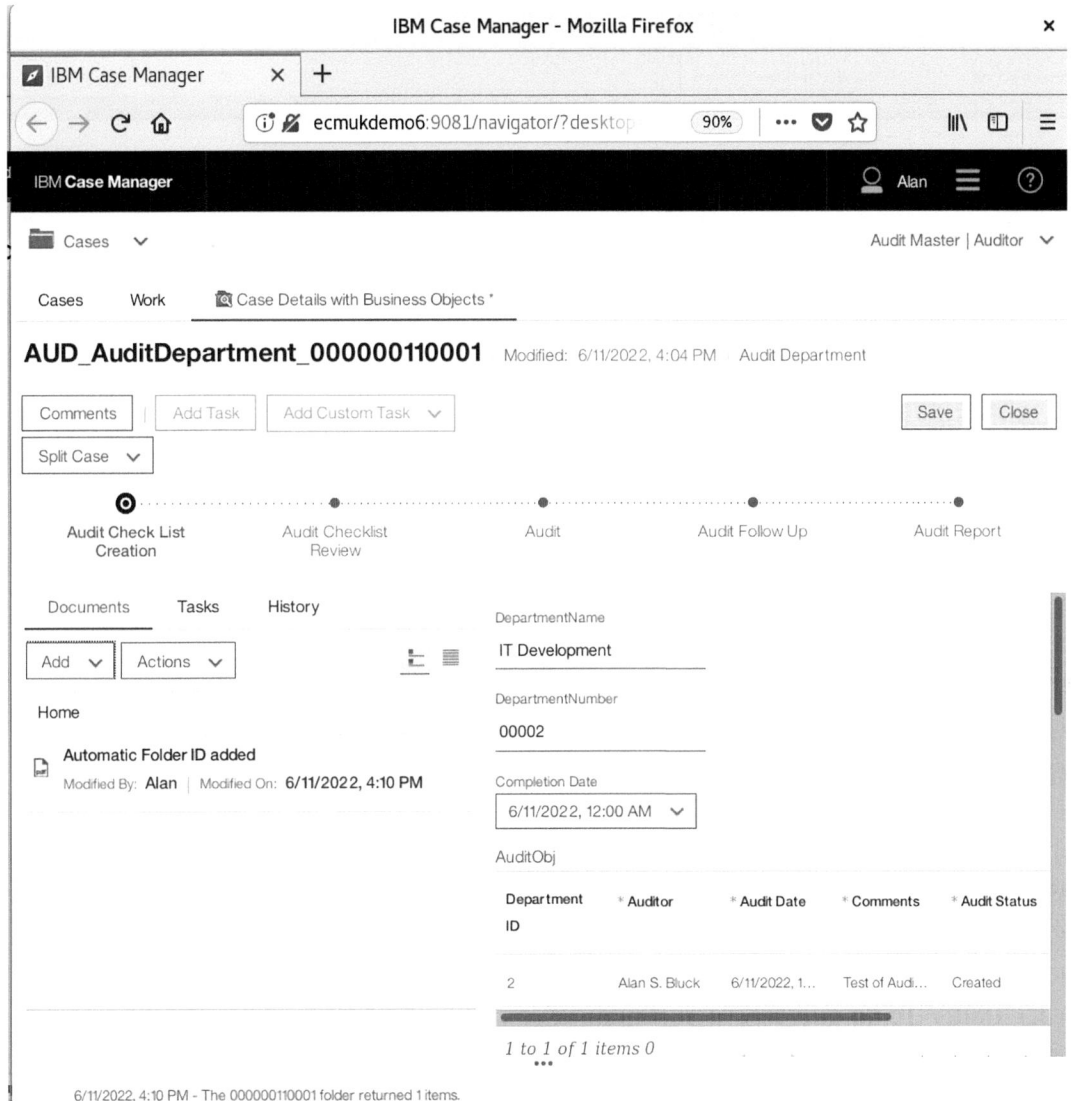

***Figure 2-132.** The saved document appears in the Audit Master case*

In the preceding screen, it should be noted that the Auditor can be still actively using the Case, in the IBM Case Manager application, because as soon as the system detects the **Create Event** for the **Audit Report** document class, it will appear (on selecting the **Refresh** command) in the **acce** web application shown in Figure 2-133.

CHAPTER 2 CONFIGURING JAVA CUSTOM COMPONENTS

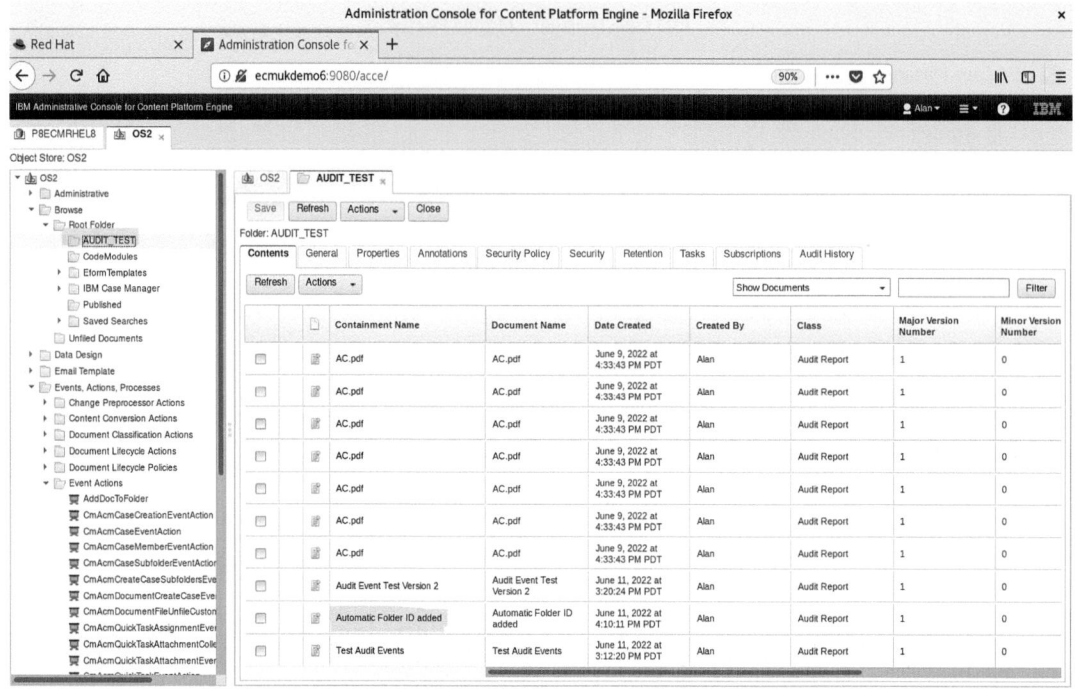

Figure 2-133. The Audit Report document is also found linked to the AUDIT_TEST folder

In summary, the **Audit Report** Document class Creation Event triggers the **AddDocToFolder** Java **EventHandler,** a **Class** in the **fn_eventhandlers.jar** (which is registered as a **Code Module** in the **OS2** Object Store). This links the new **Audit Report** document of class type **AUD_AuditReport** to the **AUDIT_TEST** folder using the property **AUD_folder_Id_to_file_in** which contains the unique ID (**GUID**) of the **AUDIT_TEST** folder.

This functionality can then be used to automatically provide a second link to all the Audit Report case documents, which can then be viewed centrally in one Folder location.

CHAPTER 2 CONFIGURING JAVA CUSTOM COMPONENTS

Part 4 – The DbExecute Workflow Step for Calling Database Stored Procedures

One other integration we cover in this chapter is the DbExecute workflow step. This allows a Database Stored procedure to be invoked from an IBM Case Manager Workflow step. This is supported using the IBM FileNet P8 Process Designer tool. This tool is available as a .jar applet and also a standalone Java program version running on Windows as a client application linked to the IBM Case Manager server.

The IBM Auto-Claims Case Manager Solution Example

The following free download covers the example DB2 Stored Procedure for an Insurance Company, IBM Case Manager **Auto Claims** Solution:

https://doi.org/10.13140/RG.2.2.16987.52008

"Importing Case Manager Solution Auto Claims example into Case Manager 5.3.3 on RHEL 8.0"

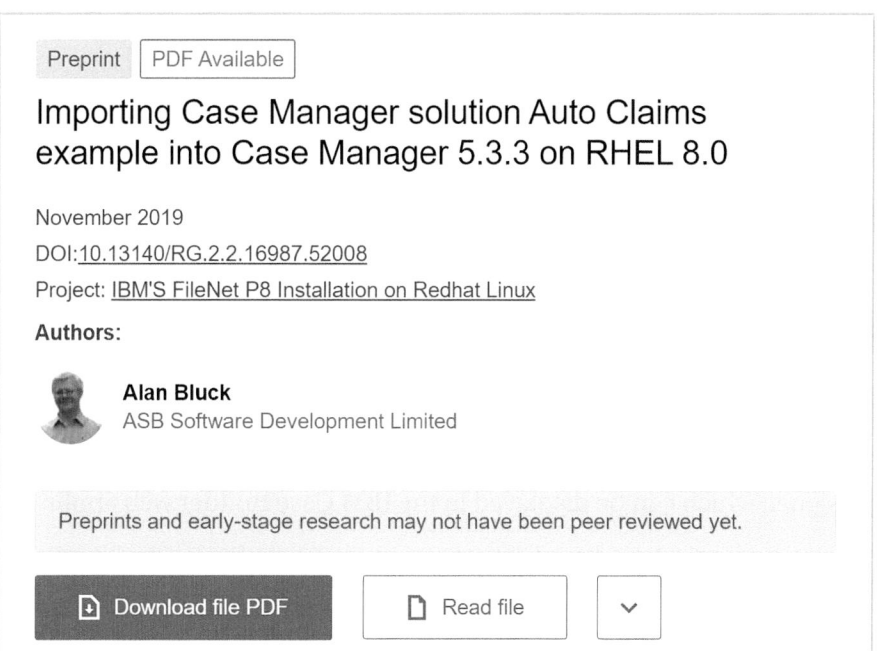

Figure 2-134. The free download Document web page for details of a DB2 Stored Procedure development

CHAPTER 2　CONFIGURING JAVA CUSTOM COMPONENTS

There are several points of integration with the IBM FileNet P8 Process Designer for building an IBM Case Manager workflow, shown in this section.

Methods of Displaying a Case Type Task Workflow

Figure 2-136 highlights the web-based Process Designer applet menu, which can be launched for the development of the Solution Case Type Workflow tasks used to process and display active Cases to Auditors and Case Administrators.

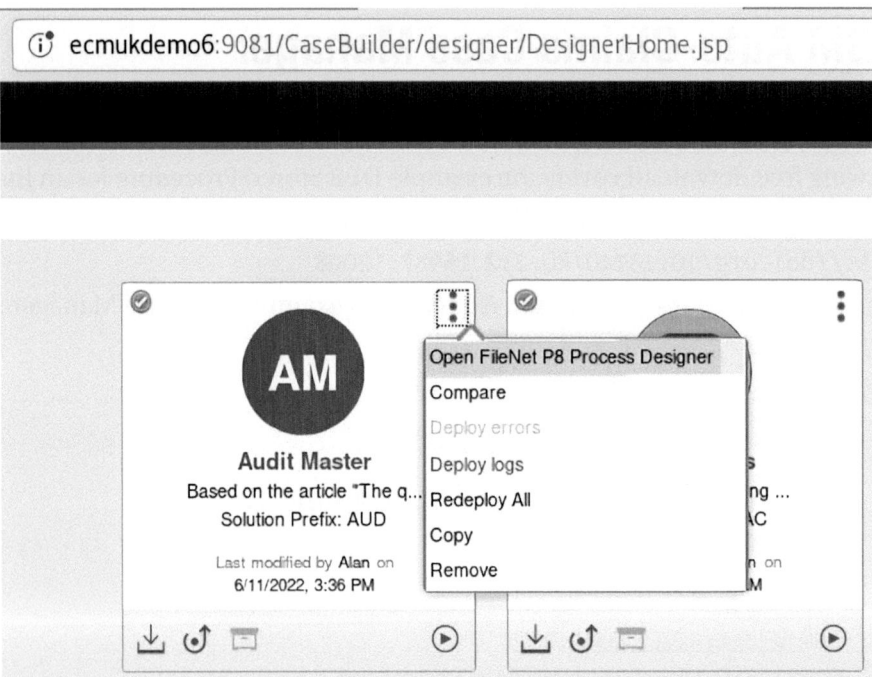

Figure 2-135.　*The IBM Case Builder Solution display web page, with the link for FileNet P8 Process Designer*

There is also a graphical view available of the same workflow as used in FileNet P8 Process Designer, which can be displayed in the IBM Case Builder web application tool as follows.

First, we load the Audit Master solution by clicking the blue **Audit Master** link shown in Figure 2-135.

CHAPTER 2   CONFIGURING JAVA CUSTOM COMPONENTS

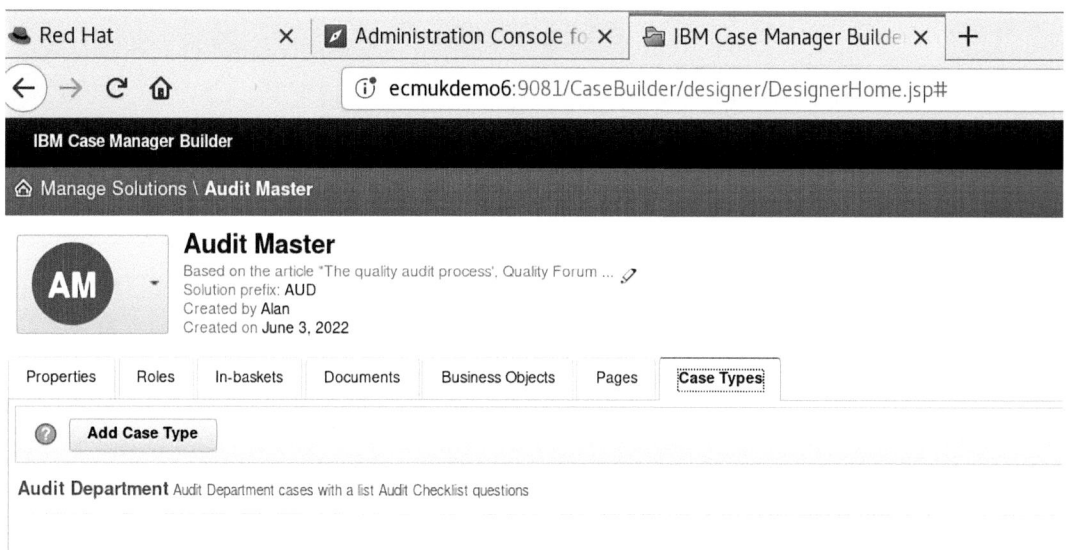

Figure 2-136. *The **Case Types** tab is selected from the Audit Master solution IBM Case Builder view*

1. We then click the **Audit Department** Case type shown earlier to display the attributes of the Case Type as shown in Figure 2-137.

2. From here, we select the **Tasks** option from **Audit Department** Case Type menu options.

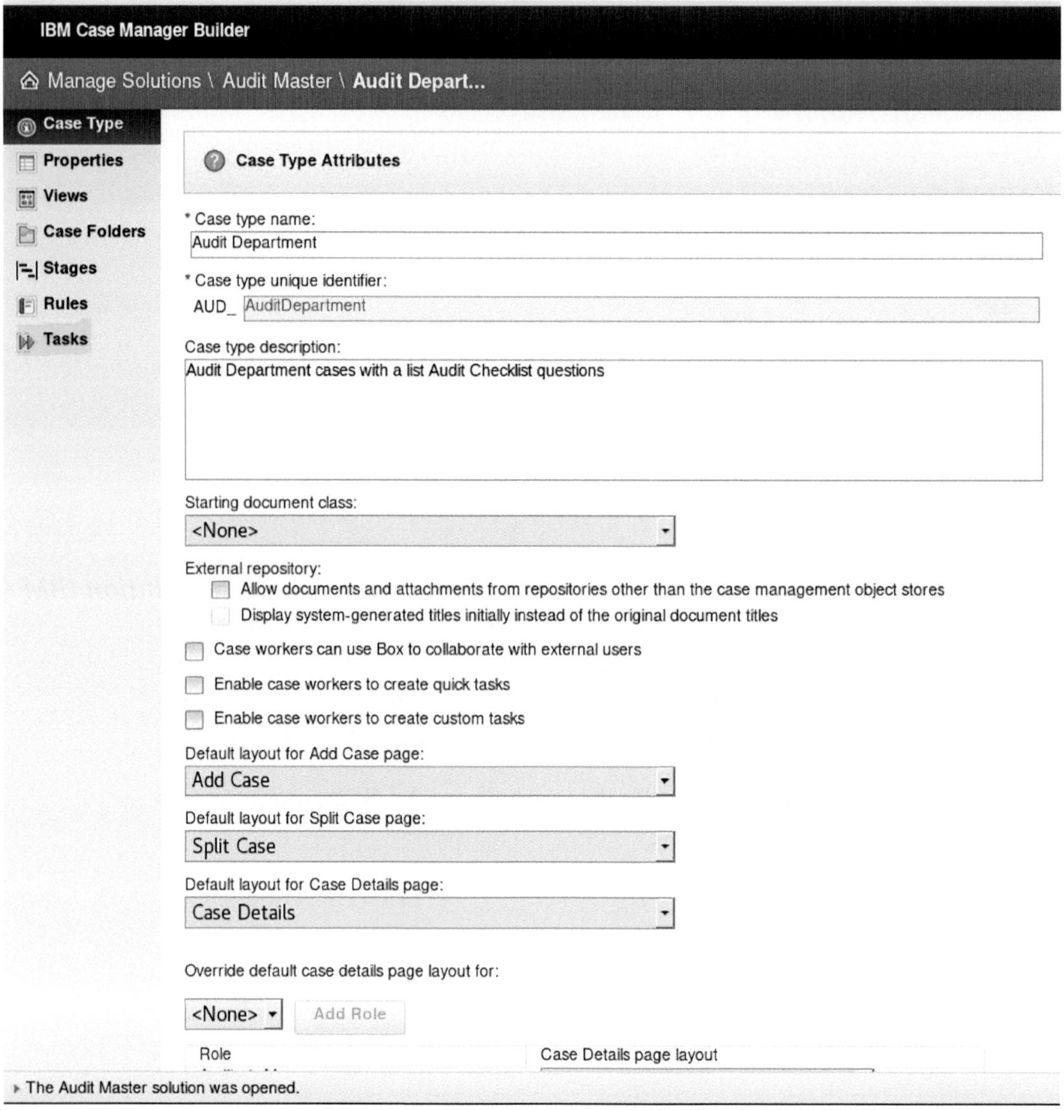

CHAPTER 2 CONFIGURING JAVA CUSTOM COMPONENTS

Figure 2-137. *The Audit Department type is double-clicked to access the Case Type attributes*

The Figure 2-137 screen has a drop-down which allows a document loaded into the IBM **OS2** Target Object Store to automatically invoke the Audit Master solution Audit Department task workflow, if we select a **Starting document class** from the drop-down shown earlier. For the moment, we will leave the default option, which is to start the Audit Department Task manually from each Case, when this is required.

CHAPTER 2　CONFIGURING JAVA CUSTOM COMPONENTS

The **Tasks** option Audit Process page is shown in Figure 2-138.

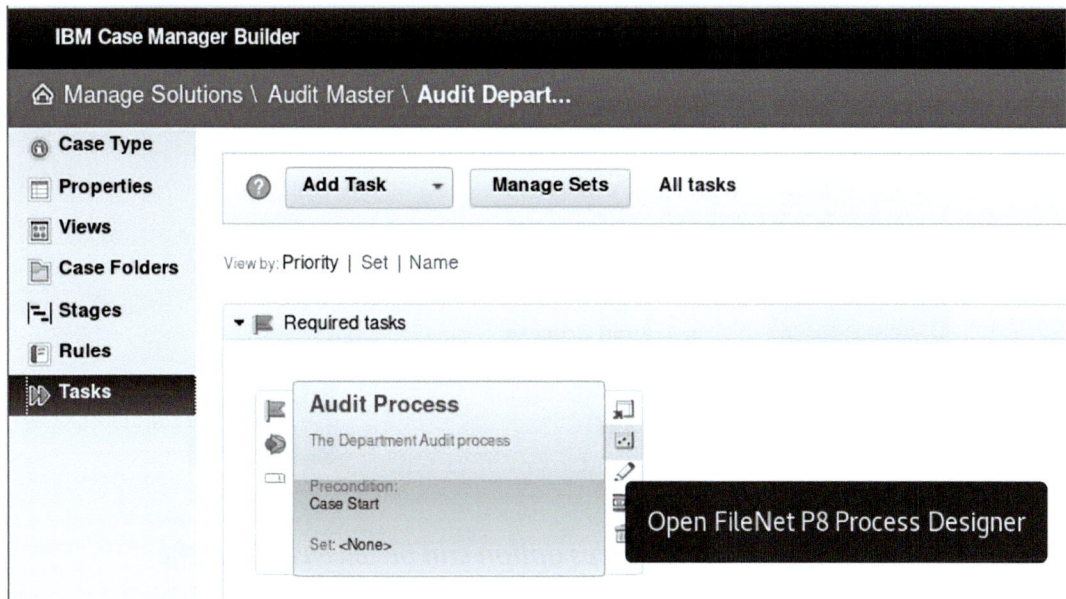

Figure 2-138. *The highlighted icon can be selected to launch the IBM FileNet P8 Process Designer applet*

Now, we have two options to view the Workflow here. The first option, shown in Figure 2-138, launches the IBM FileNet P8 Process Designer as an embedded Java Applet. (But see later why this is rarely able to be used!)

The Case Builder Workflow Step Designer Application

The second option is to use the Step Designer, shown using the highlighted icon in Figure 2-139.

255

CHAPTER 2   CONFIGURING JAVA CUSTOM COMPONENTS

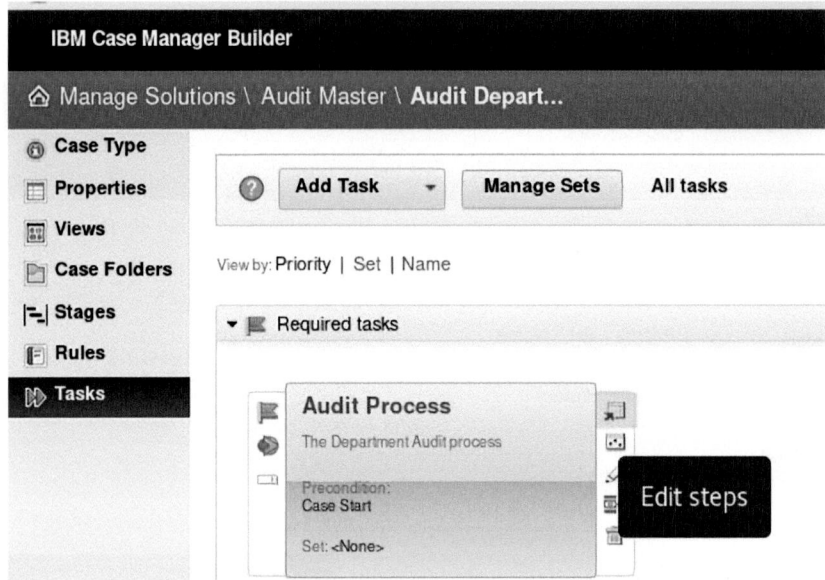

Figure 2-139. *The Edit Workflow steps option can be selected using the highlighted icon*

Clicking the icon highlighted in Figure 2-139 gives the Task Workflow, Step Designer, a high-level Workflow design page for the IBM Case Builder, shown in Figure 2-140.

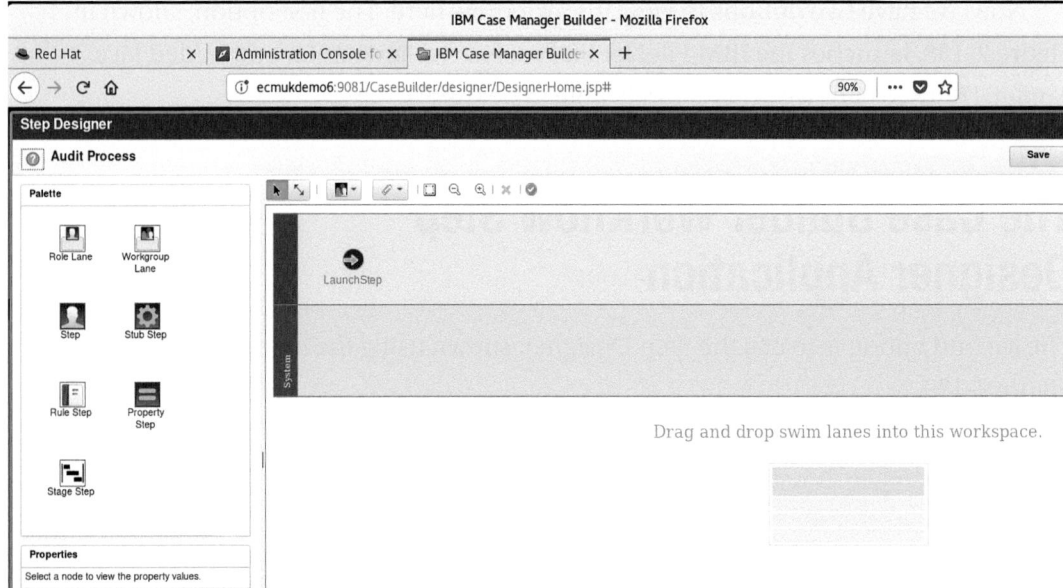

Figure 2-140. *The Task Workflow Step Designer page is displayed using the Edit Steps icon*

CHAPTER 2 CONFIGURING JAVA CUSTOM COMPONENTS

The **Palette** on the left-hand side of the screen, shown in Figure 2-140, of the Workflow Step Designer contains different types of Workflow step and Role Lanes which can be dragged and dropped using the mouse to the right-hand pane in the dragged lanes which are then defined for each Security Role.

Unfortunately, in our requirement, for which we wish to use the **DbExecute** step (which is a Stub Step dragged into the **System Lane**), the details of the **DbExecute** step have to be entered using the **IBM FileNet P8 Process Designer** Java program.

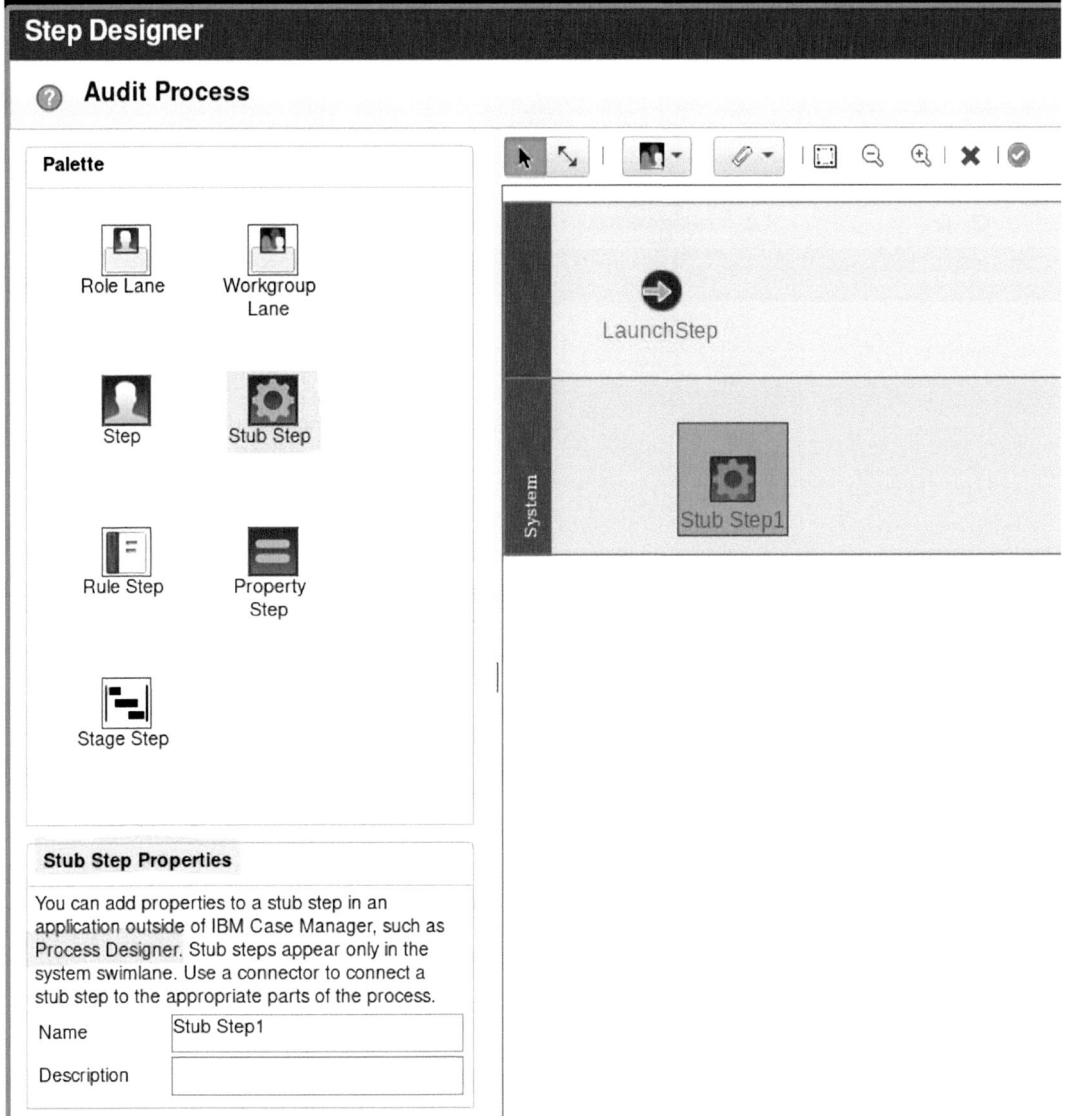

Figure 2-141. The Stub Step1 is dragged into the System Lane

CHAPTER 2   CONFIGURING JAVA CUSTOM COMPONENTS

The Process Designer Plugin Applet in IBM Content Navigator

The next option for accessing the IBM FileNet P8 Process Designer is to configure the access from the IBM FileNet Content Navigator web application, using the IBM FileNet P8 Process Designer plugin applet.

This is accessed using the link from our IBM Content Navigator web application server, ecmukdemo6, using the URL:

http://ecmukdemo6:9081/navigator/?desktop=admin

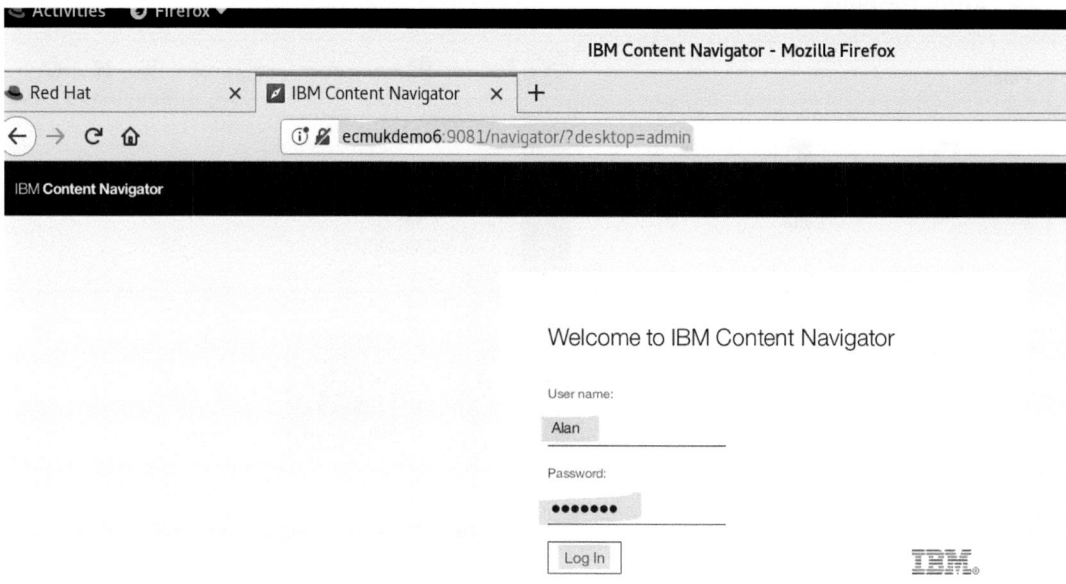

Figure 2-142. The login screen for the IBM Content Navigator web application

CHAPTER 2 CONFIGURING JAVA CUSTOM COMPONENTS

After logging in to the IBM Content Navigator web application tool shown in Figure 2-142, we can select the Plug-ins menu item as follows.

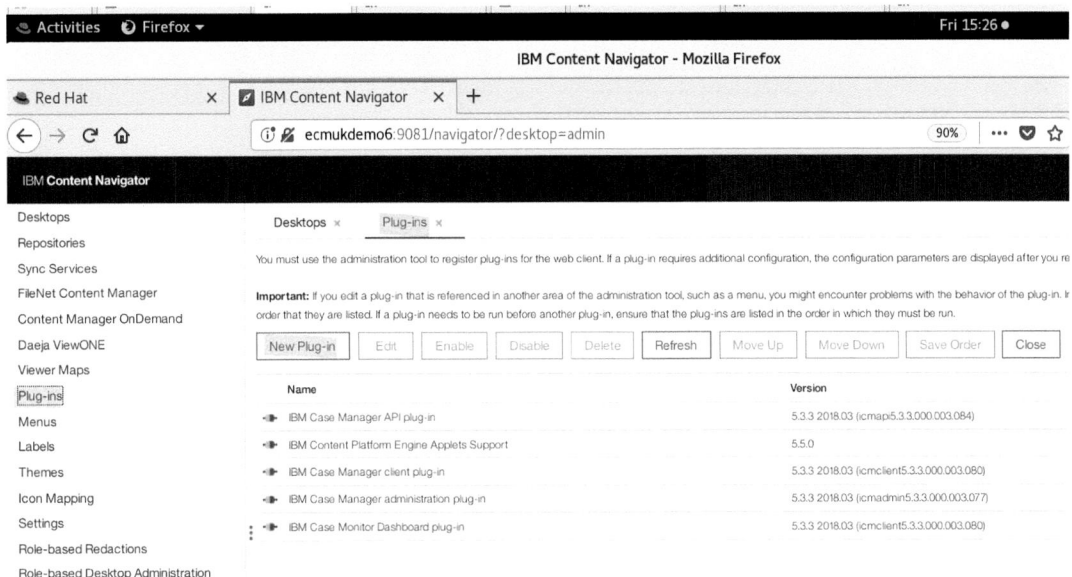

Figure 2-143. *The New Plug-in command is selected from the Plug-ins administration menu*

CHAPTER 2   CONFIGURING JAVA CUSTOM COMPONENTS

The Content Platform Engine **CPEAppletsPlugin.jar** file is available by downloading it from the installed Content Platform Engine server (ecmukdemo6) using the URL:

http://ecmukdemo6:9080/peengine/plugins/CPEAppletsPlugin.jar

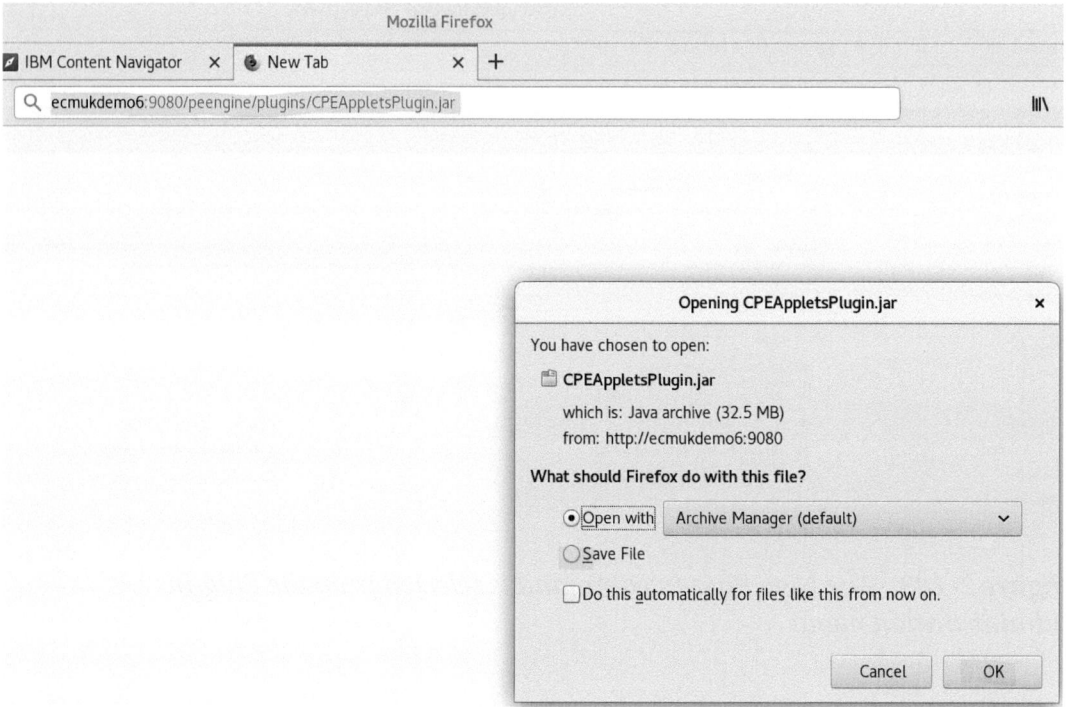

Figure 2-144.   The CPEAppletsPlugin.jar file is downloaded for access by the IBM Content Navigator admin

The Content Platform Engine **CPEAppletsPlugin.jar** file is then available from the /**root/Downloads** Linux directory path for use.

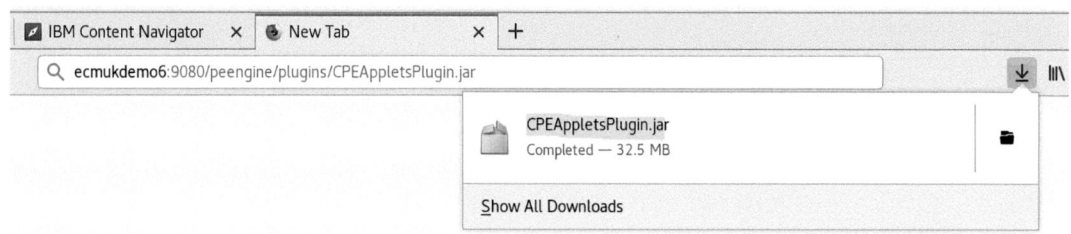

Figure 2-145.   The Content Platform Engine CPEAppletsPlugin.jar file

CHAPTER 2 CONFIGURING JAVA CUSTOM COMPONENTS

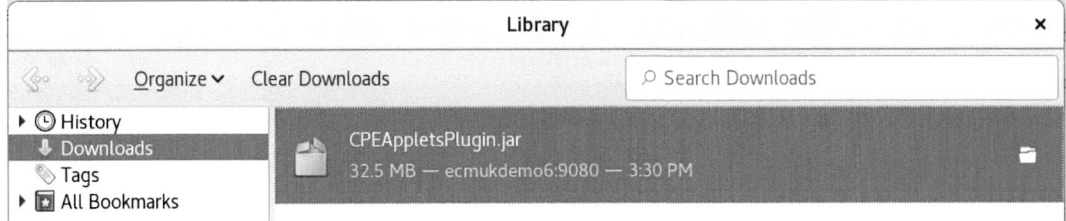

Figure 2-146. *The CPEAppletsPlugin.jar file is downloaded to the /root/Downloads folder*

The downloaded **CPEAppletsPlugin.jar** file name and path have to be manually typed into the selected admin **JAR File path** as **/root/Downloads/CPEAppletsPlugin.jar** to **Load** it from the Linux directory as shown in the highlighted screenshot in Figure 2-147.

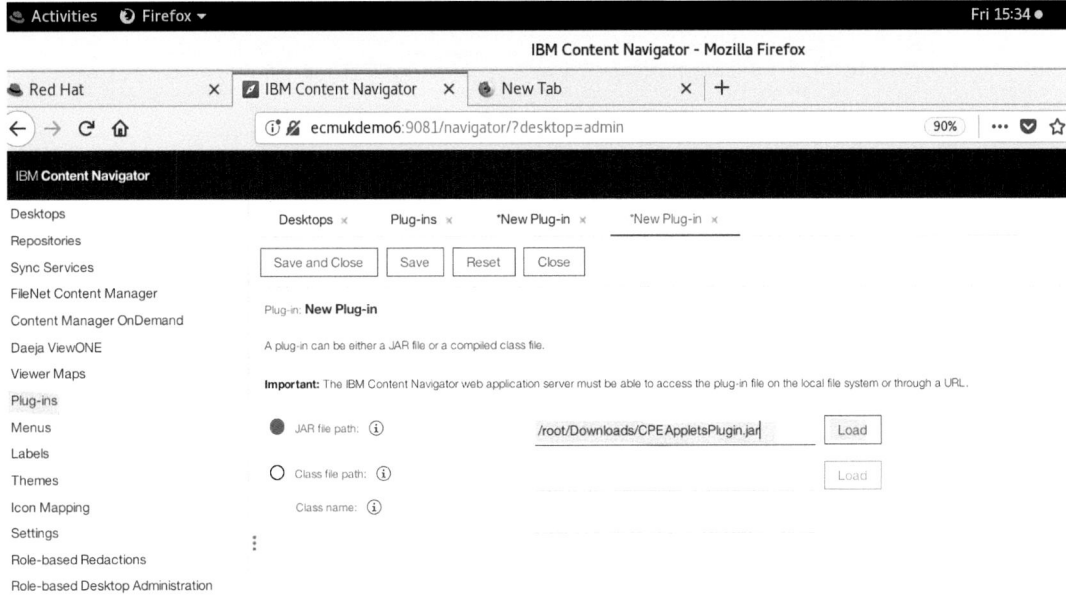

Figure 2-147. *The Load command is clicked to import the /root/Downloads/CPEAppletsPlugin.jar file*

The loaded **CPEAppletsPlugin.jar** file displays the Version details and the contained Actions available.

CHAPTER 2   CONFIGURING JAVA CUSTOM COMPONENTS

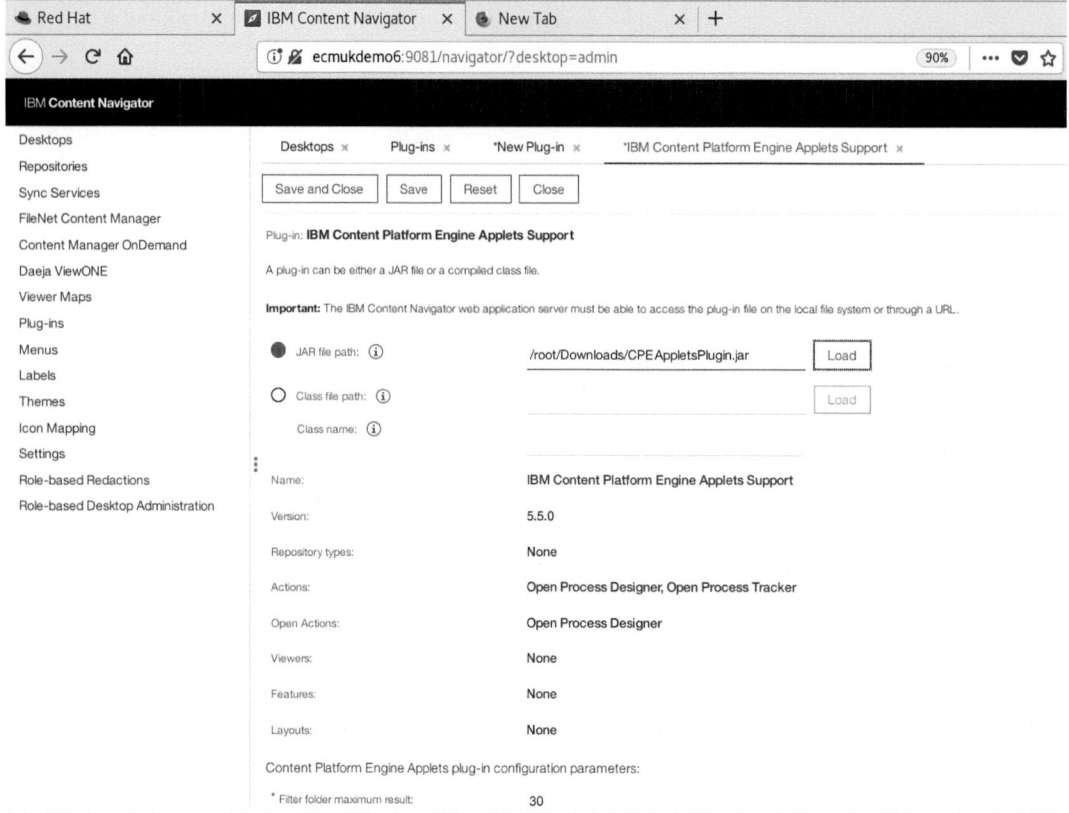

Figure 2-148. The details of the CPEAppletsPlugin.jar are displayed after loading

The **CPEAppletsPlugin.jar** security must be changed before loading as follows.
We can use the Linux command:

chmod 775 CPEAppletsPlugin.jar

Without this, there will be Load errors in the Content Navigator web application.

```
(base) [root@ECMUKDEMO6 Downloads]# chmod 775 CPEAppletsPlugin.jar
(base) [root@ECMUKDEMO6 Downloads]# ls -lsa CPEAppletsPlugin.jar
33268 -rwxrwxr-x. 1 root grrdbm01 34066245 Jun 17 15:28 CPEAppletsPlugin.jar
(base) [root@ECMUKDEMO6 Downloads]#
```

Figure 2-149. The security is changed using the Linux command cmod 775 CPEAppletsPlugin.jar

CHAPTER 2 CONFIGURING JAVA CUSTOM COMPONENTS

On refreshing the desktop, the **Plug-ins** tab page should now have the highlighted IBM FileNet P8 Process Designer and Process Tracker Actions plugin:

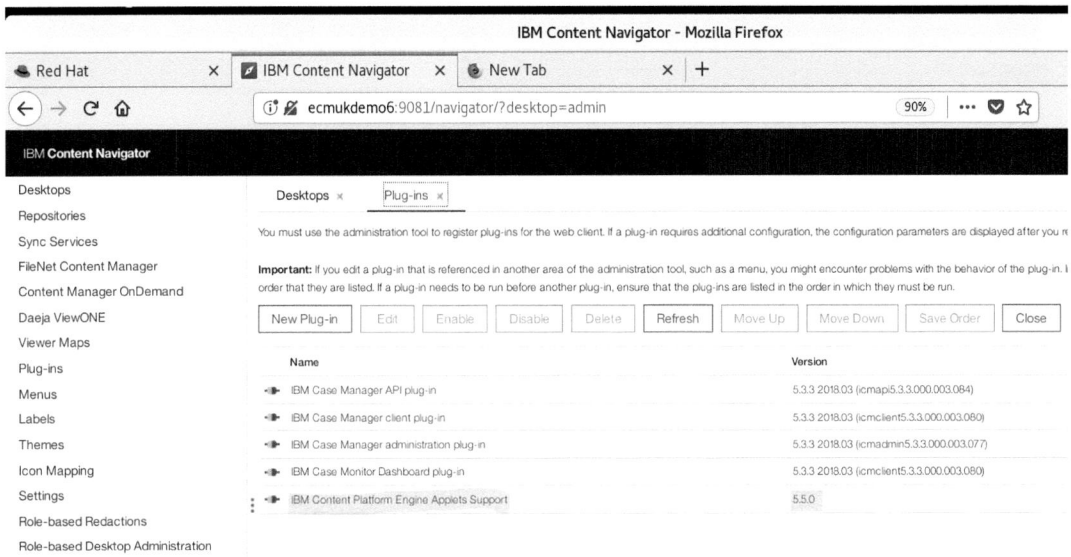

Figure 2-150. *The IBM FileNet P8 Process Designer and Process Tracker Actions plugin*

The next step is to select the IBM Content Navigator Repositories menu highlighted as follows, in order to create new connections to the IBM Case Manager Design (OS1) and Target (OS2) Object stores.

CHAPTER 2   CONFIGURING JAVA CUSTOM COMPONENTS

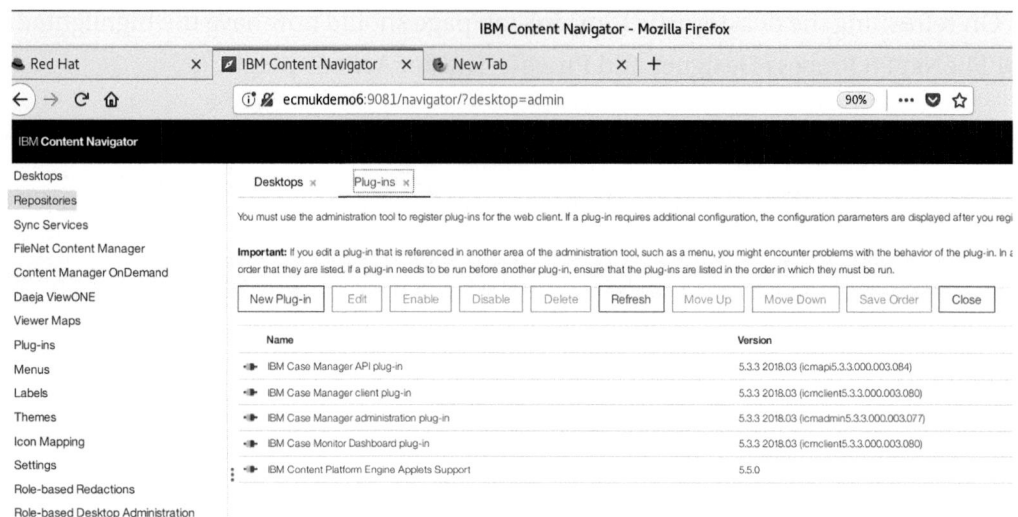

Figure 2-151.   The Repositories (IBM FileNet Object Store) menu is selected

A **New Repository** connection can be added using the **FileNet Content Manager** drop-down option.

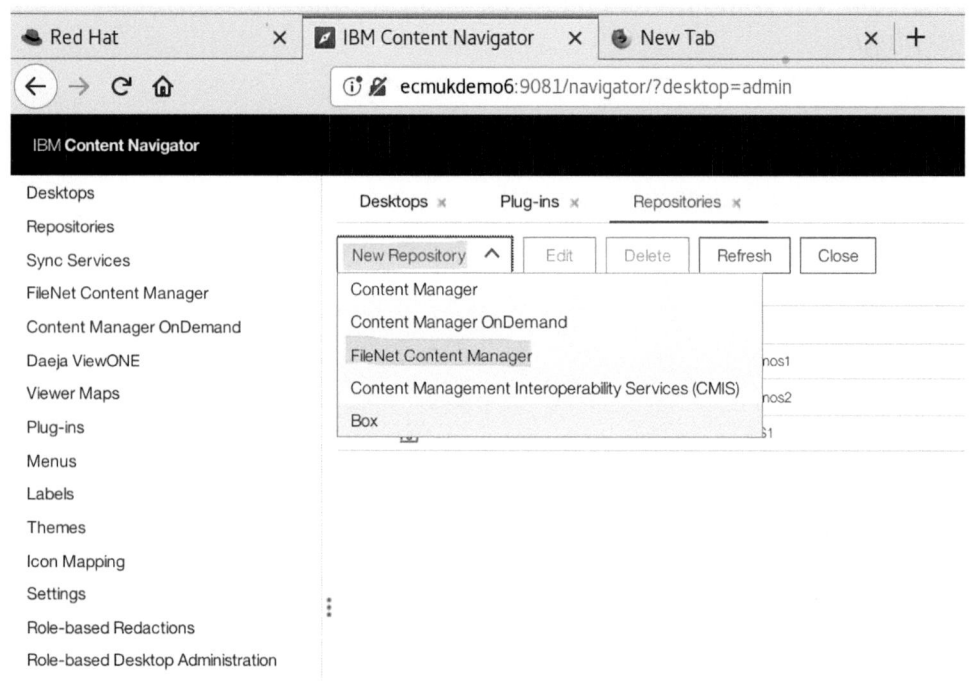

*Figure 2-152.   The **New Repository** ➤ **FileNet Content Manager** menu option is selected*

CHAPTER 2 CONFIGURING JAVA CUSTOM COMPONENTS

The connection parameters are then filled out as highlighted in Figure 2-153.

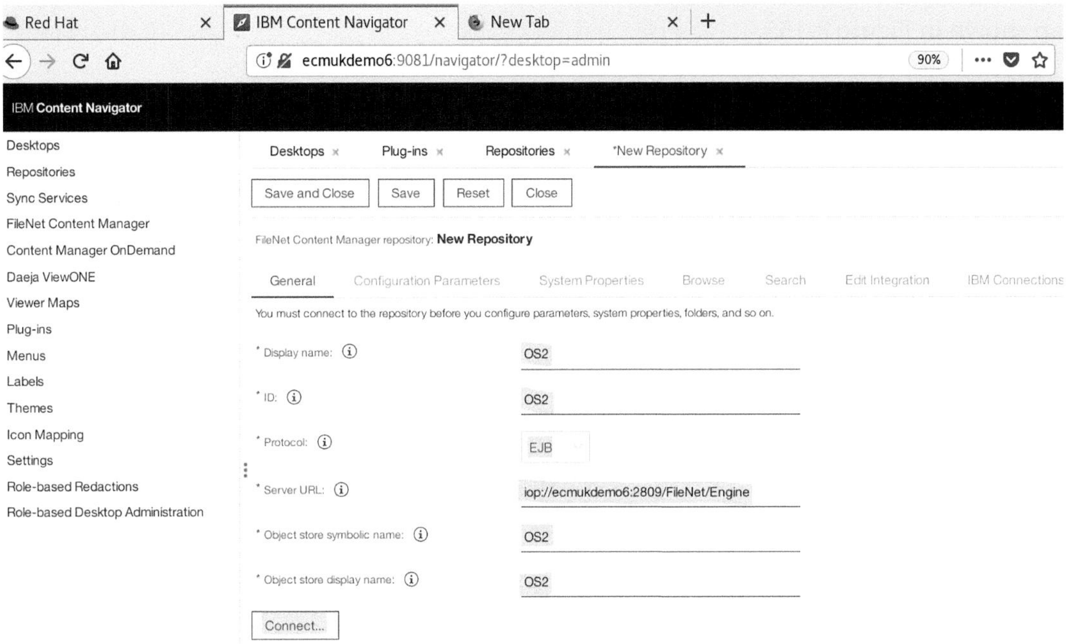

Figure 2-153. The new Repository connection parameters are filled out

The EJB connection string is used, iop://ecmukdemo6:2809/FileNet/Engine.

The user login credentials for access to the IBM FileNet P8 OS2 Content Object store are prompted for.

Figure 2-154. The login to OS2 is prompted for

CHAPTER 2   CONFIGURING JAVA CUSTOM COMPONENTS

On acceptance of the credentials by the IBM FileNet Content Manager system, the list of tabs is enabled for configuration. We need to select the **Configuration Parameters** tab shown in Figure 2-155.

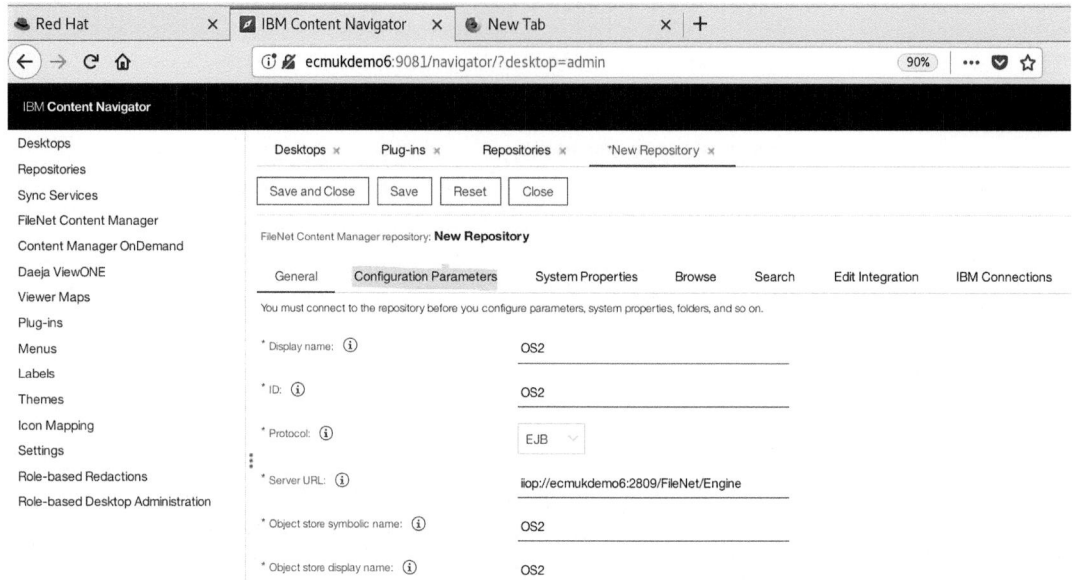

Figure 2-155. *The tabs are all enabled, and we can now select the Configuration Parameters tab*

We now have to enable the **Display workflow definition class** option shown on the next screen.

CHAPTER 2   CONFIGURING JAVA CUSTOM COMPONENTS

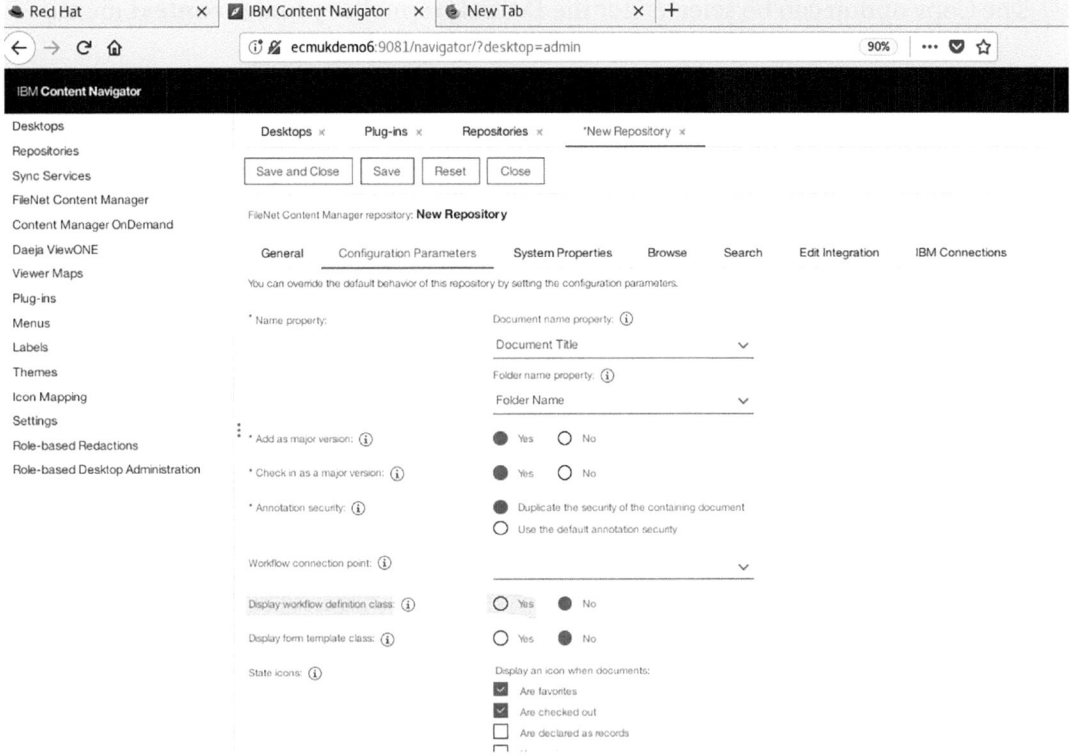

*Figure 2-156.   The highlighted **Display workflow definition class** option is selected as Yes*

The preceding procedure is repeated to add the OS2 repository.

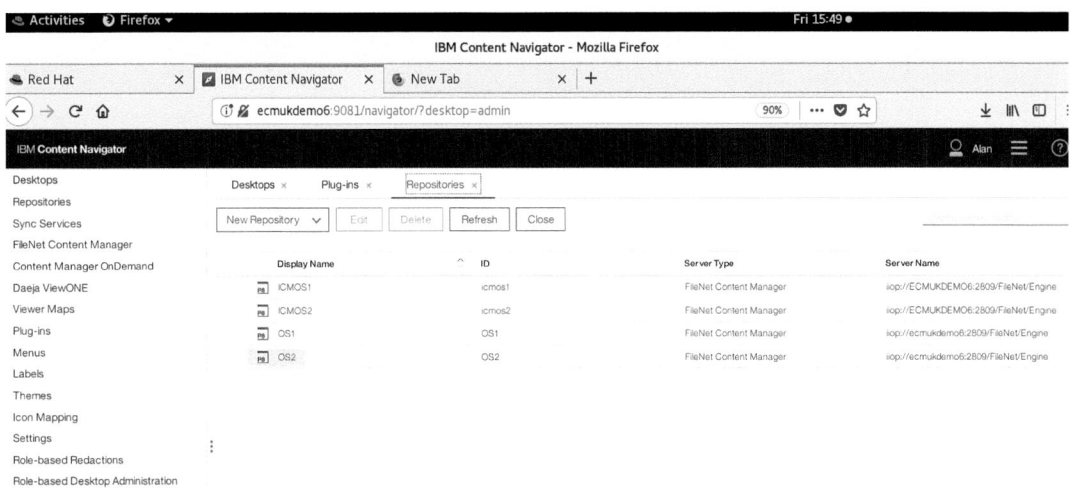

Figure 2-157.   The OS2 repository is configured

267

CHAPTER 2   CONFIGURING JAVA CUSTOM COMPONENTS

The Copy option can be selected for the **Default repository folder context** menu.

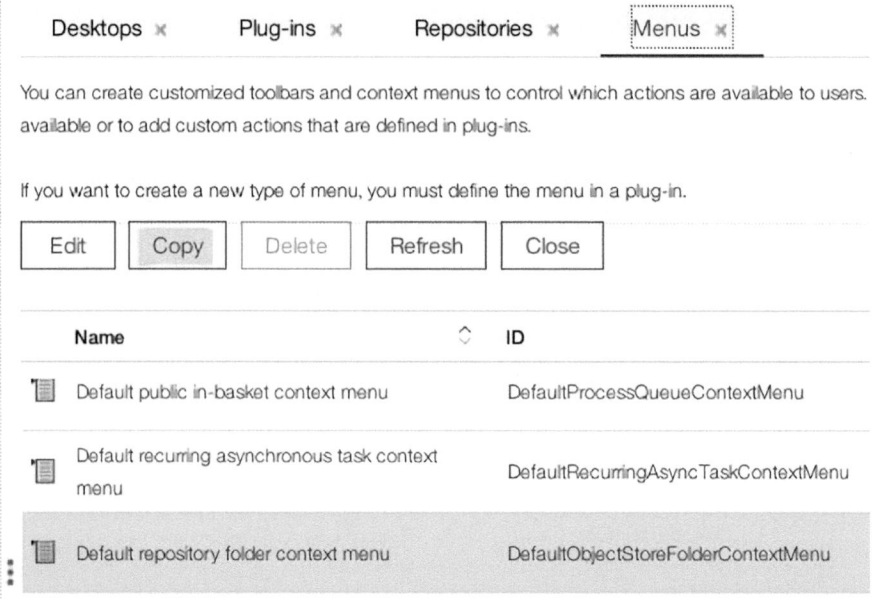

*Figure 2-158.   The menu option named **Default repository folder context menu** is copied*

CHAPTER 2 CONFIGURING JAVA CUSTOM COMPONENTS

This is a template menu which is found by scrolling down to the highlighted position shown in Figure 2-159.

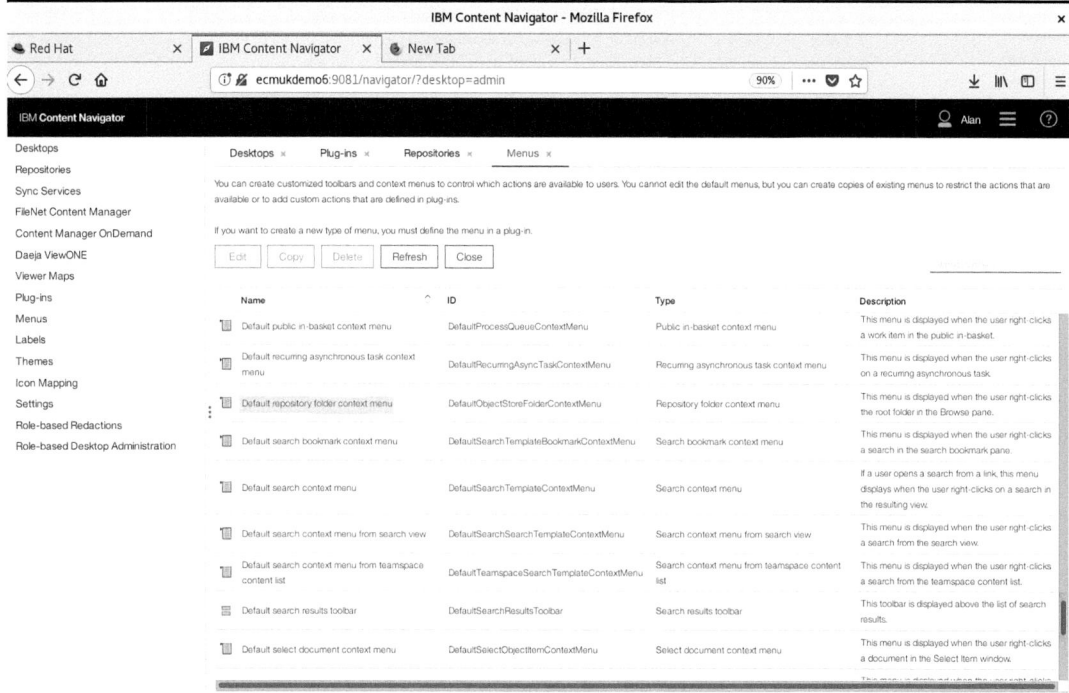

Figure 2-159. *The highlighted menu template is copied for configuration*

The copied menu now shows the **Open Process Designer** and **Open Process Tracker** menu actions.

CHAPTER 2　CONFIGURING JAVA CUSTOM COMPONENTS

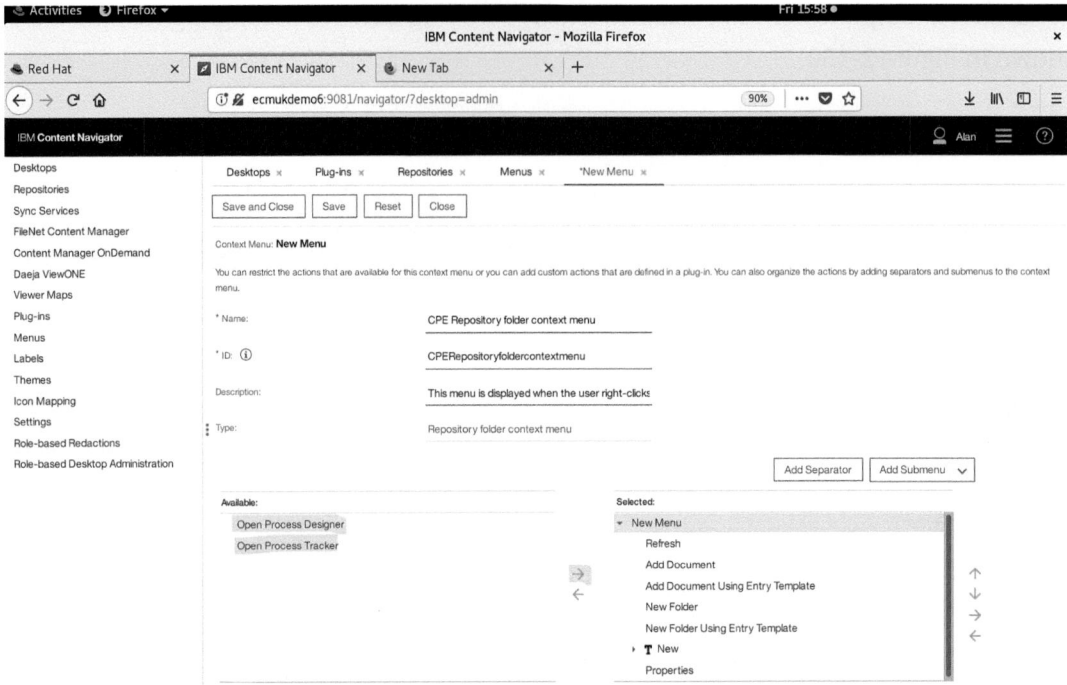

Figure 2-160. The new Open Process Designer and Open Process Tracker menu items are added

The highlighted arrow key in Figure 2-160 is used to copy the selected **Open Process Designer** and **Open Process Tracker** options, and then their positions can be edited using the up/down arrow keys on the right of the panel to change the menu item order.

CHAPTER 2 CONFIGURING JAVA CUSTOM COMPONENTS

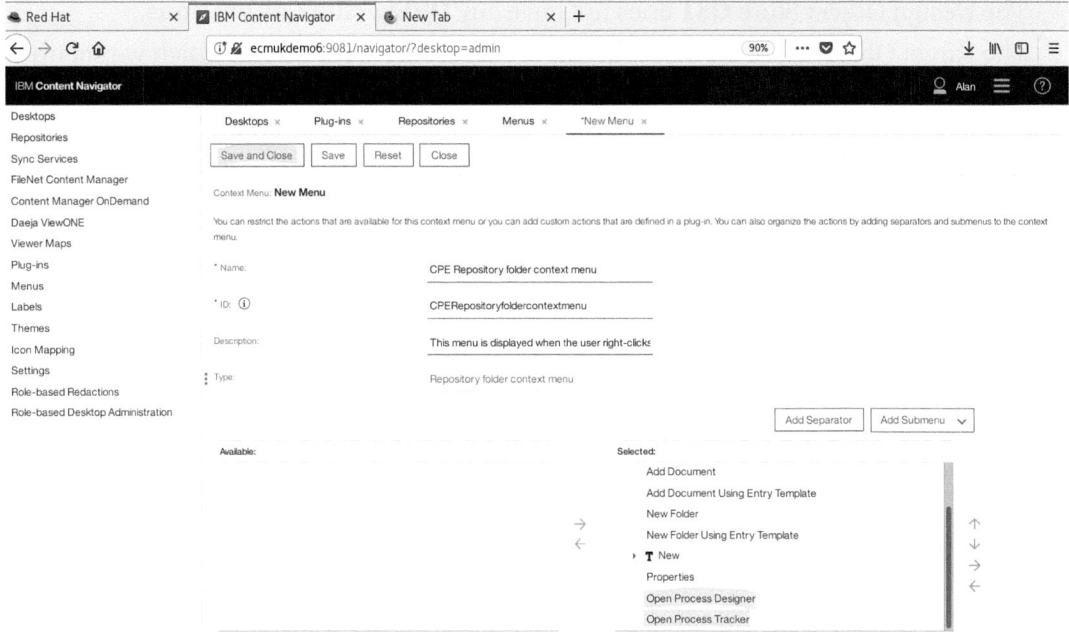

Figure 2-161. *The menu items are shown transferred to the copied menu item list*

Configuring the DbExecute Connection

In order to call the DbExecute step from a Task Workflow, calling a database stored procedure, we have to log in to the IBM FileNet Content Manager **acce** administration tool and create a DbExecute connection using the highlighted tab in Figure 2-162.

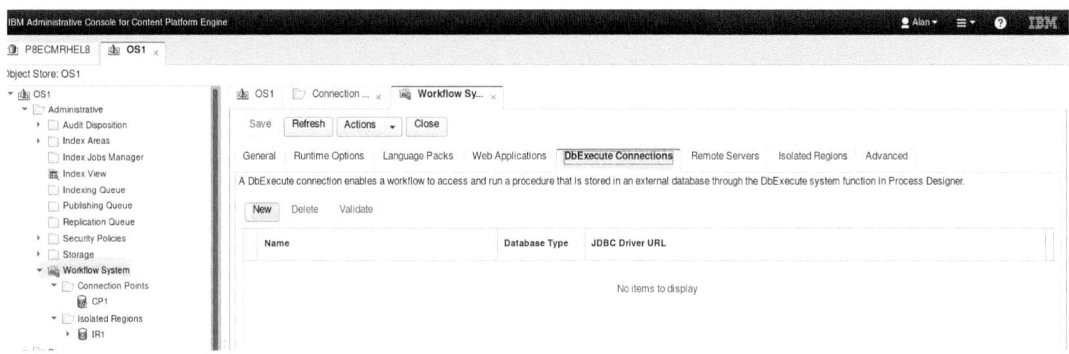

Figure 2-162. *The **DbExecute Connections** tab on **OS1** is selected, and the **New** command button is clicked*

271

CHAPTER 2 CONFIGURING JAVA CUSTOM COMPONENTS

The Design Object Store **OS1** does not have any **DbExecute Connections**.

The Target Object Store **OS2** already has a **DbExecute connection** which we set up for the standard **IBM Auto Insurance Claims** example solution.

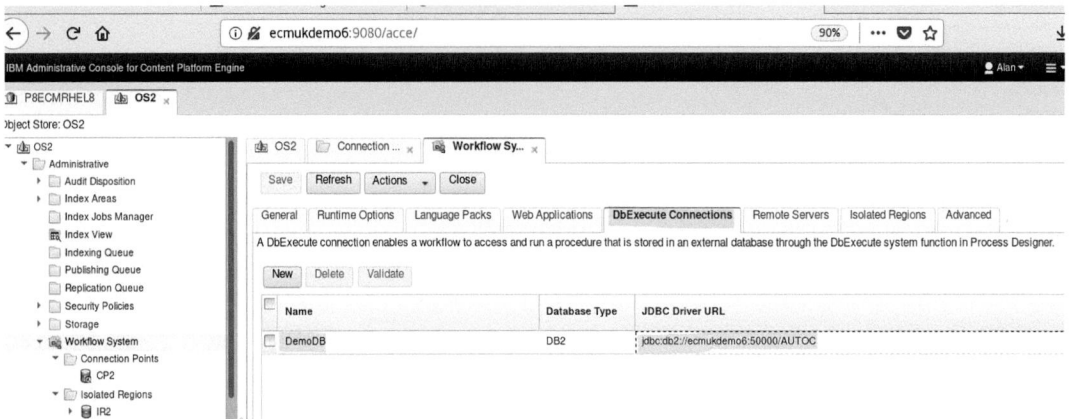

Figure 2-163. *The IBM Auto Insurance Claims example solution DbExecute connection*

The Edit command shows the example connection parameters as shown in Figure 2-164.

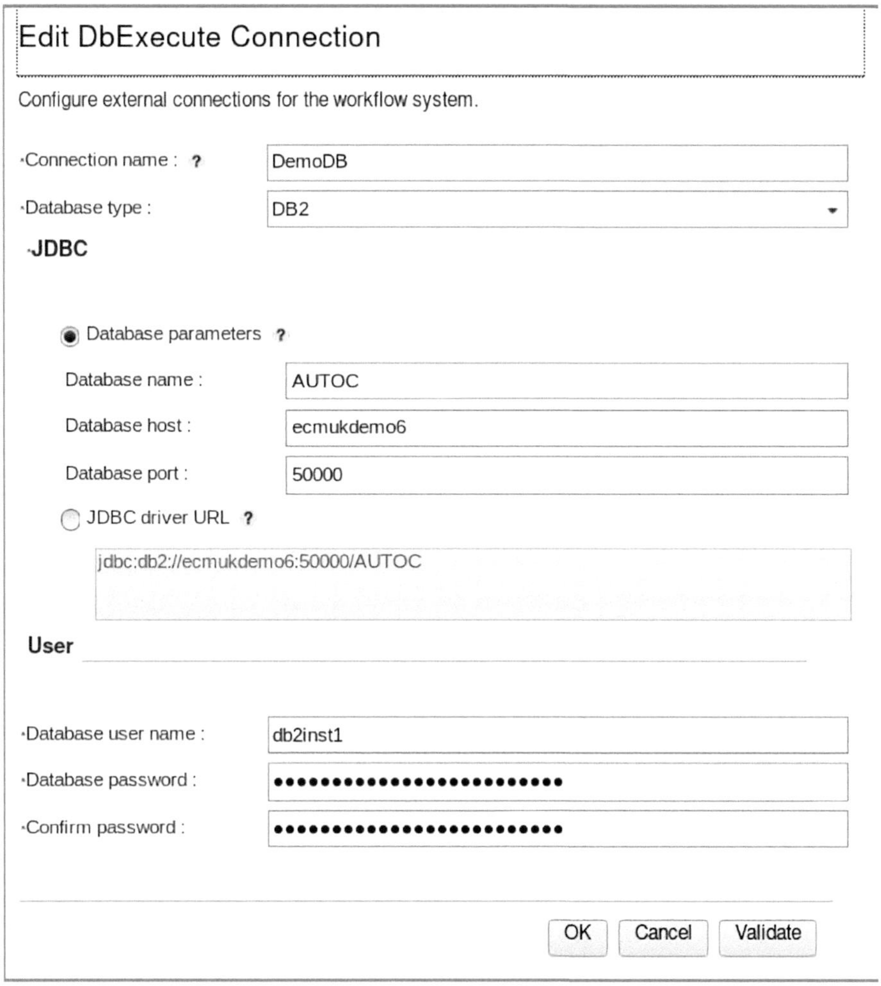

CHAPTER 2 CONFIGURING JAVA CUSTOM COMPONENTS

Figure 2-164. The Edit DbExecute Connection for the Auto Claims Solution is displayed

The Additional Desktop Components for IBM Content Navigator are also changed to enable the display of the Global toolbar.

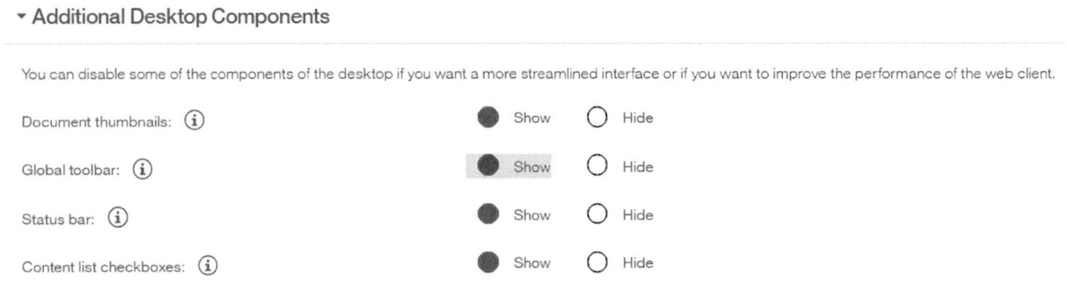

CHAPTER 2 CONFIGURING JAVA CUSTOM COMPONENTS

Figure 2-165. *The Global toolbar settings are changed to show the Process Designer plugin menu actions*

In the Advanced settings of the IBM Case Manager administration client desktop, we must also select the tick box to "Enable the desktop for FileNet P8 workflow email notifications," as shown in Figure 2-166.

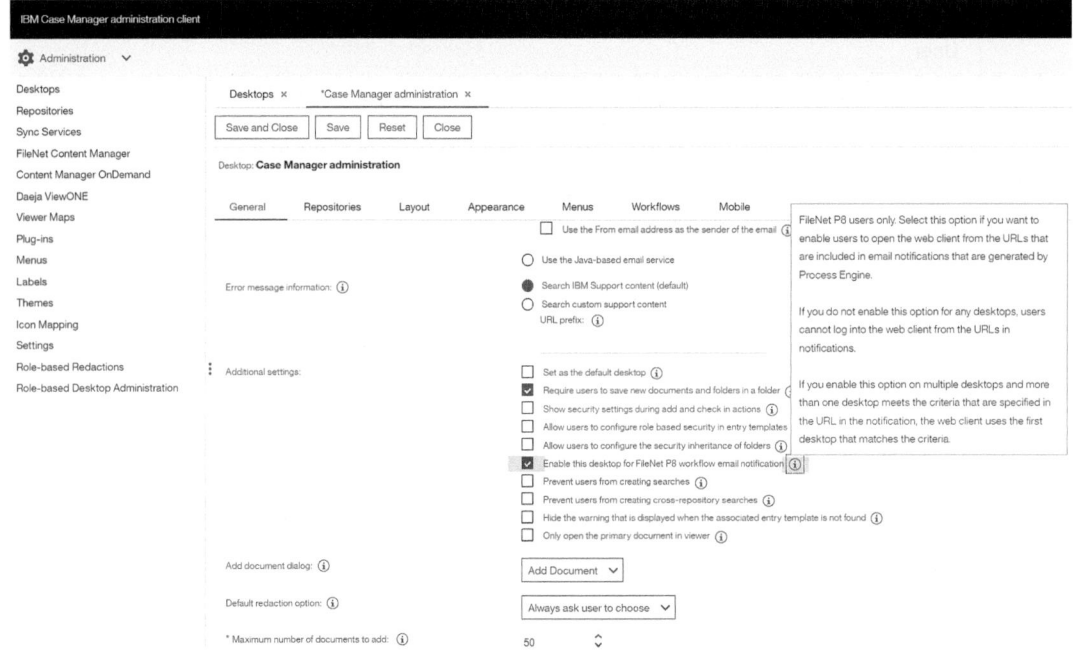

Figure 2-166. *The option to enable the IBM Case Manager administration client desktop to display emails*

Figure 2-167 is a zoom in to show the mouse-over information against this option.

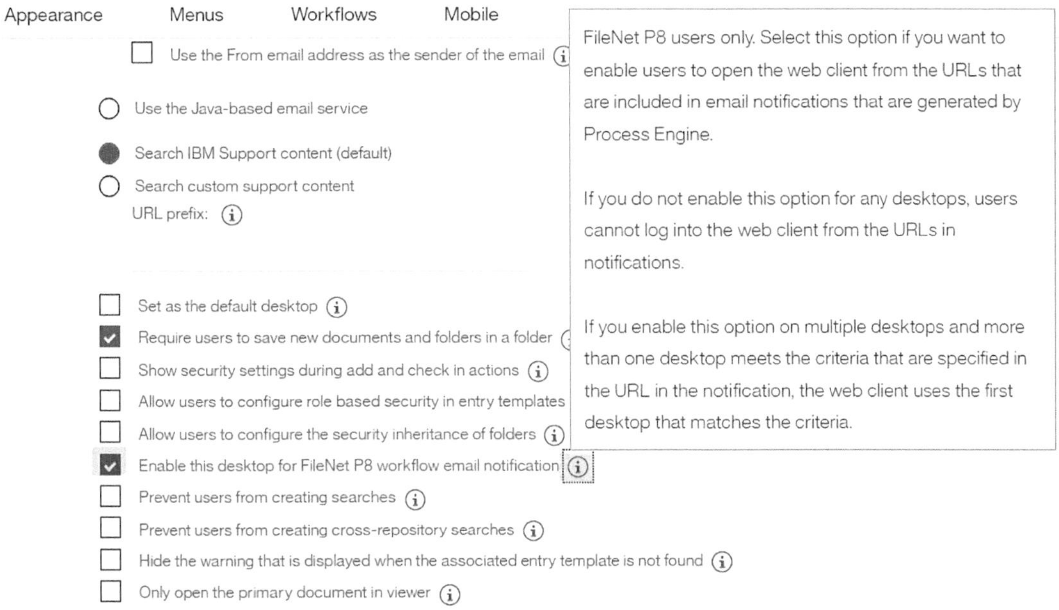

CHAPTER 2　CONFIGURING JAVA CUSTOM COMPONENTS

Appearance　Menus　Workflows　Mobile

☐ Use the From email address as the sender of the email ⓘ

○ Use the Java-based email service
● Search IBM Support content (default)
○ Search custom support content
　URL prefix: ⓘ

☐ Set as the default desktop ⓘ
☑ Require users to save new documents and folders in a folder ⓘ
☐ Show security settings during add and check in actions ⓘ
☐ Allow users to configure role based security in entry templates ⓘ
☐ Allow users to configure the security inheritance of folders ⓘ
☑ Enable this desktop for FileNet P8 workflow email notification ⓘ
☐ Prevent users from creating searches ⓘ
☐ Prevent users from creating cross-repository searches ⓘ
☐ Hide the warning that is displayed when the associated entry template is not found ⓘ
☐ Only open the primary document in viewer ⓘ

> FileNet P8 users only. Select this option if you want to enable users to open the web client from the URLs that are included in email notifications that are generated by Process Engine.
>
> If you do not enable this option for any desktops, users cannot log into the web client from the URLs in notifications.
>
> If you enable this option on multiple desktops and more than one desktop meets the criteria that are specified in the URL in the notification, the web client uses the first desktop that matches the criteria.

Figure 2-167. *The details of the Information icon help for the option are displayed on mouse-over*

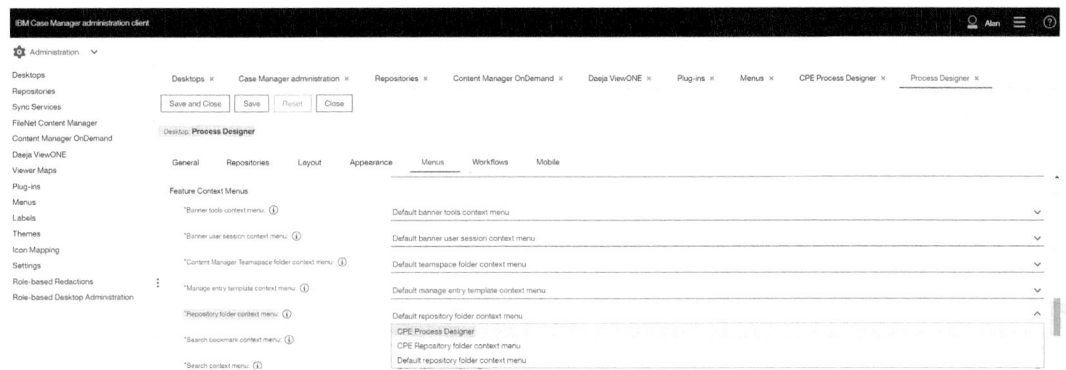

Figure 2-168. *The Process Designer Desktop tab shows the CPE Process Designer menu we created*

In Figure 2-168, the copy of the Repository folder context menu we made can be seen in the drop-down. We can now select our copy of this menu type for our IBM Content Navigator Desktop, which we named **CPE Process Designer**.

275

CHAPTER 2   CONFIGURING JAVA CUSTOM COMPONENTS

See the following web pages for details of running the Workflow Process Designer Applet in IBM Content Navigator:

www.ibm.com/support/pages/do-you-know-now-you-can-run-process-designer-and-process-tracker-ibm-content-navigator-icn

For setting up IBM Case Manager for using the Process Designer:

www.ibm.com/docs/en/case-manager/5.3.3?topic=cdecnd-configuring-case-manager-use-stand-alone-process-designer

Also, this has useful information:

www.ibm.com/docs/en/filenet-p8-platform/5.5.x?topic=workflows-running-filenet-process-designer

To register the IBM Content Navigator plugins, see

www.ibm.com/docs/en/filenet-p8-platform/5.5.x?topic=navigator-registering-process-applets-plug-in

Followed by this web page section:

www.ibm.com/docs/en/filenet-p8-platform/5.5.x?topic=navigator-configuring-process-designer

Part 5 – Standalone Process Designer Installation

In the current release of browsers, the embedded applet jar file used by the IBM FileNet Workflow Process Designer is often disabled by the browser (including on my systems!). So, this Part describes the installation and configuration required to install the Windows Client Standalone Process Designer, which is then used to add a DbExecute workflow step to the Audit Master Solution we created in Chapter 1.

References are as follows:

www.ibm.com/docs/en/case-manager/5.3.3?topic=cdecnd-configuring-case-manager-use-stand-alone-process-designer

www.ibm.com/docs/en/filenet-p8-platform/5.2.1?topic=system-setting-dbexecute-connections

For viewing the error logs in FileNet, the following link is useful:

www.ibm.com/docs/en/filenet-p8-platform/5.5.x?topic=logging-viewing-filenet-p8-logs

CHAPTER 2　CONFIGURING JAVA CUSTOM COMPONENTS

We will install the **IBM FileNet Process Engine Client tools** using the Windows client installer for IBM Content Foundation, Content Platform Engine for Windows system, which can be downloaded from the IBM Software Catalog as a zip file, **ICF_5.5.0_WINDOWS_ML.zip**.

We first unzip the **ICF_5.5.0_WINDOWS_ML.zip** software package to subfolders as in the example shown in Figure 2-169.

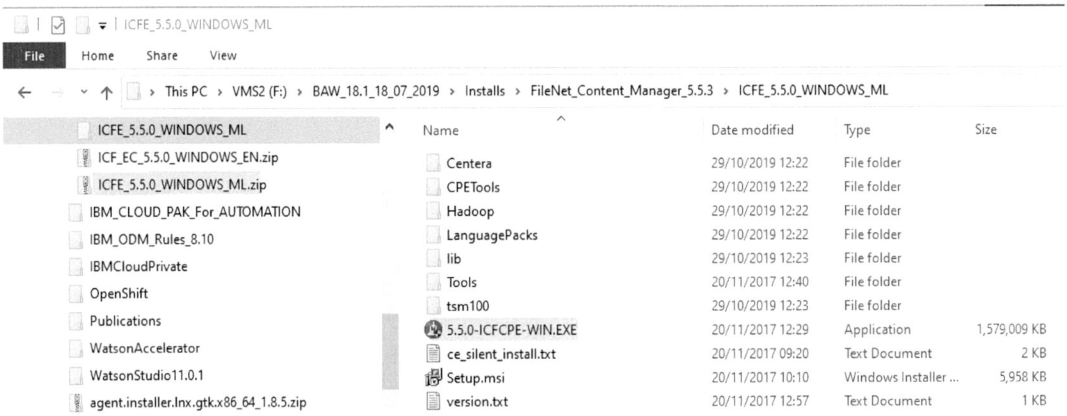

Figure 2-169.　The unpacked ICF_5.5.0_WINDOWS_ML.zip software package

Next, we run the **5.5.0-ICFCPE-WIN.EXE** program by right-clicking the program and selecting Run as administrator as shown in Figure 2-170.

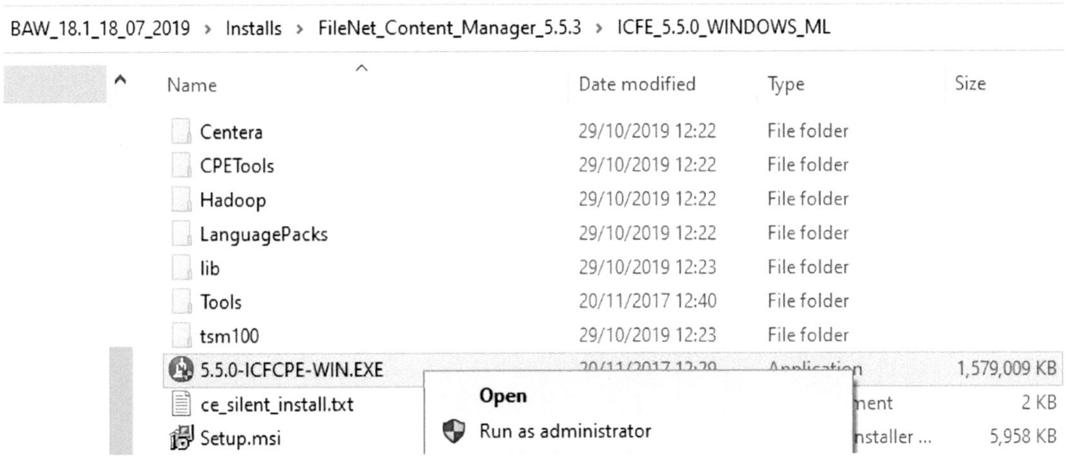

Figure 2-170.　The 5.5.0-ICFCPE-WIN.EXE program is run as the Windows Administrator

277

CHAPTER 2 CONFIGURING JAVA CUSTOM COMPONENTS

The installation program checks for existing deployments and ensures the required prerequisites are installed.

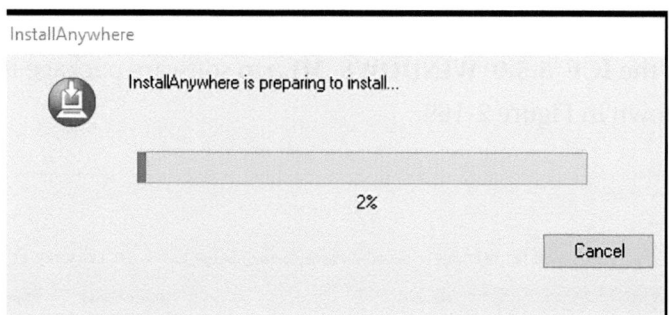

Figure 2-171. *The progress status bar of the installation. It can take several minutes to prepare the installation*

After accepting the windows pop-up for the installer to use the Windows Administrator, we see the progress status bar shown in Figure 2-171.

After several minutes, the following initial installation window is displayed.

CHAPTER 2 CONFIGURING JAVA CUSTOM COMPONENTS

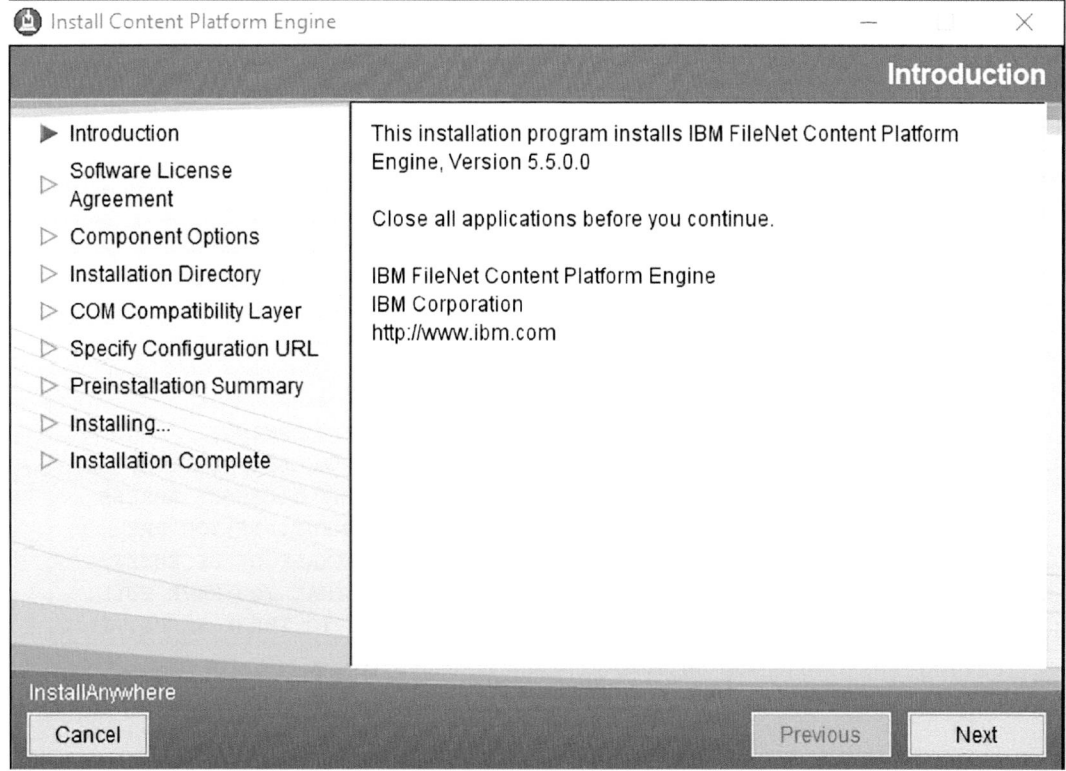

Figure 2-172. *The initial installation window shows the version of the IBM Content Platform Engine*

CHAPTER 2   CONFIGURING JAVA CUSTOM COMPONENTS

We click the **Next** command button to show the IBM License Acceptance splash screen, with a scrollable license text window.

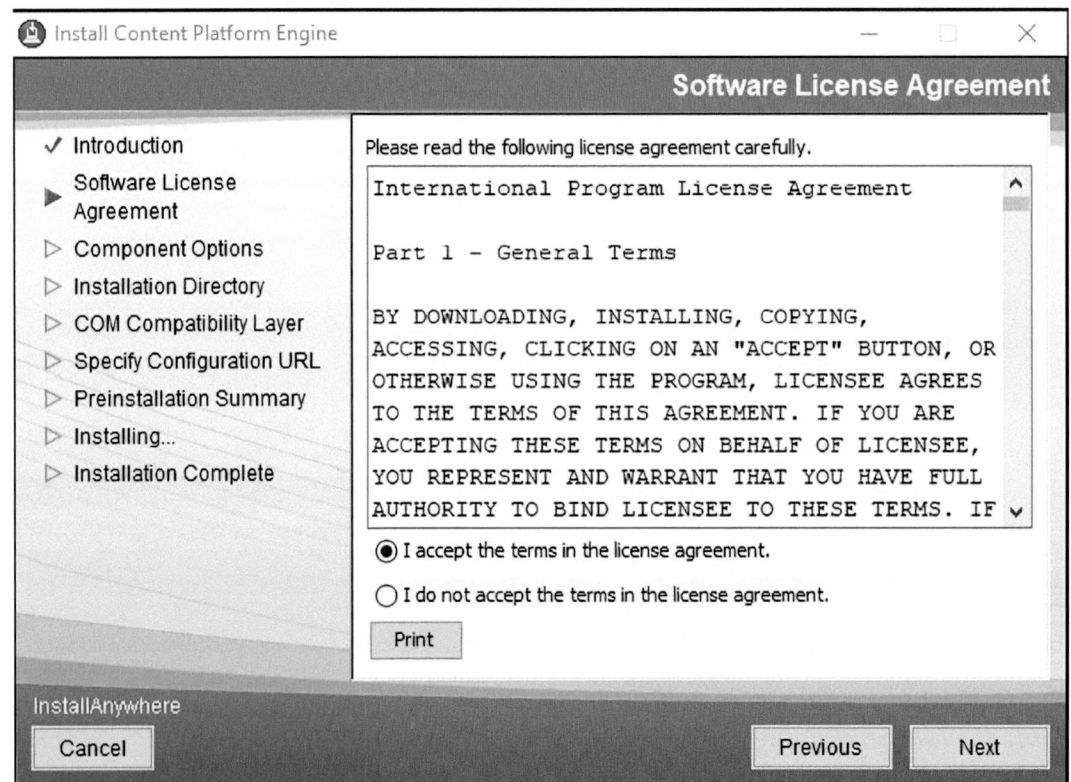

Figure 2-173. *The I Accept... option is selected, and click the Next command button*

The installation options are chosen on the next screen.

CHAPTER 2　CONFIGURING JAVA CUSTOM COMPONENTS

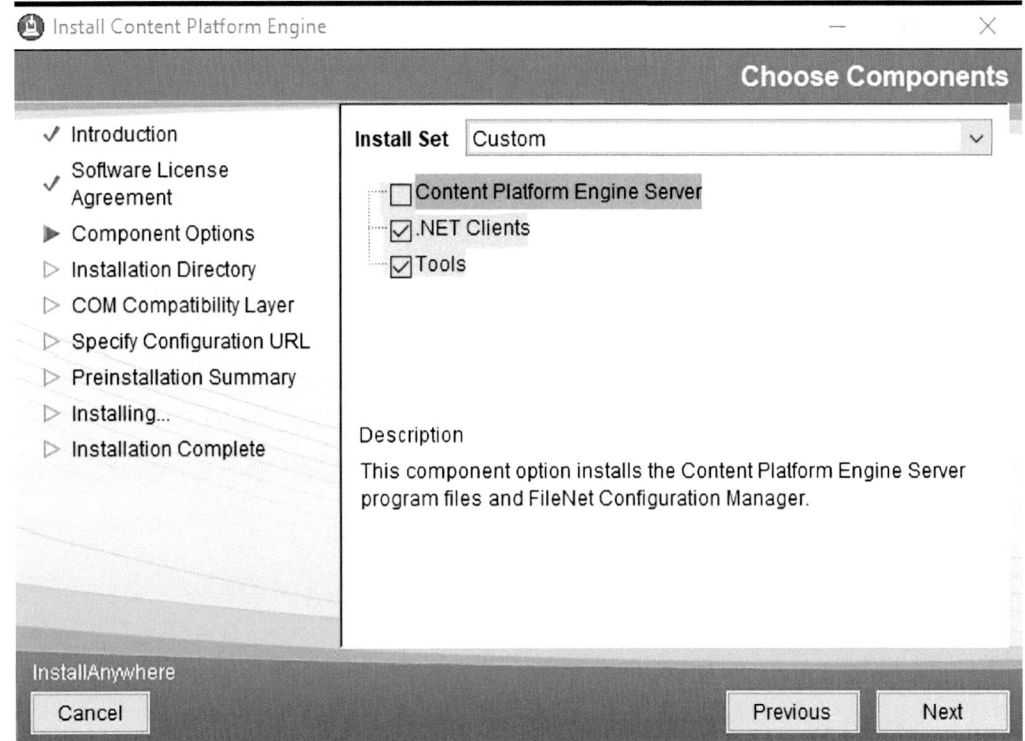

Figure 2-174. *The installation option for the Content Platform Engine Server is deselected*

We just need to install the Client tools for the IBM Content Platform Engine, so we just select for the .NET Clients and the Tools.

CHAPTER 2   CONFIGURING JAVA CUSTOM COMPONENTS

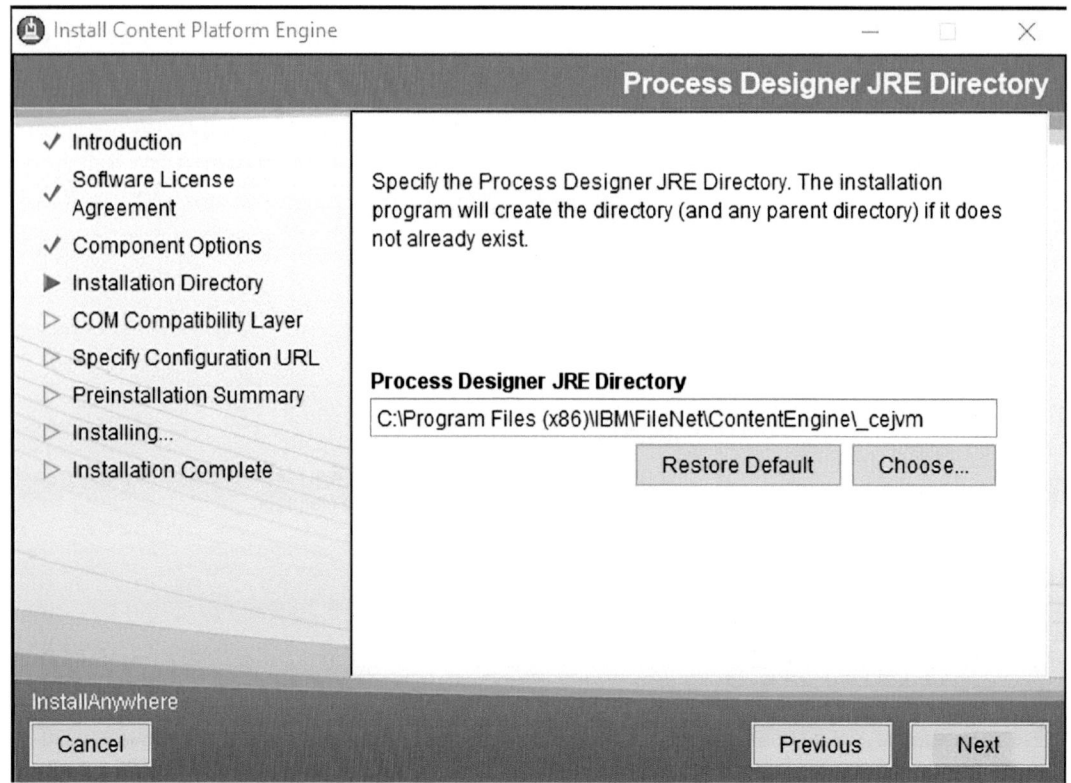

Figure 2-175. *The Process Designer JRE Directory was left as the default (but see the following text!)*

Note   The preceding default path does not exist on our MS Windows client system, so we had to post-edit the Process Designer environment batch file later. Also, if a valid path is entered, you must ensure that it does not contain any spaces (so, use the DOS short form for a directory, e.g., "C:\PROGRA~1\..." rather than the "C:\Program Files..." directory name).

CHAPTER 2 CONFIGURING JAVA CUSTOM COMPONENTS

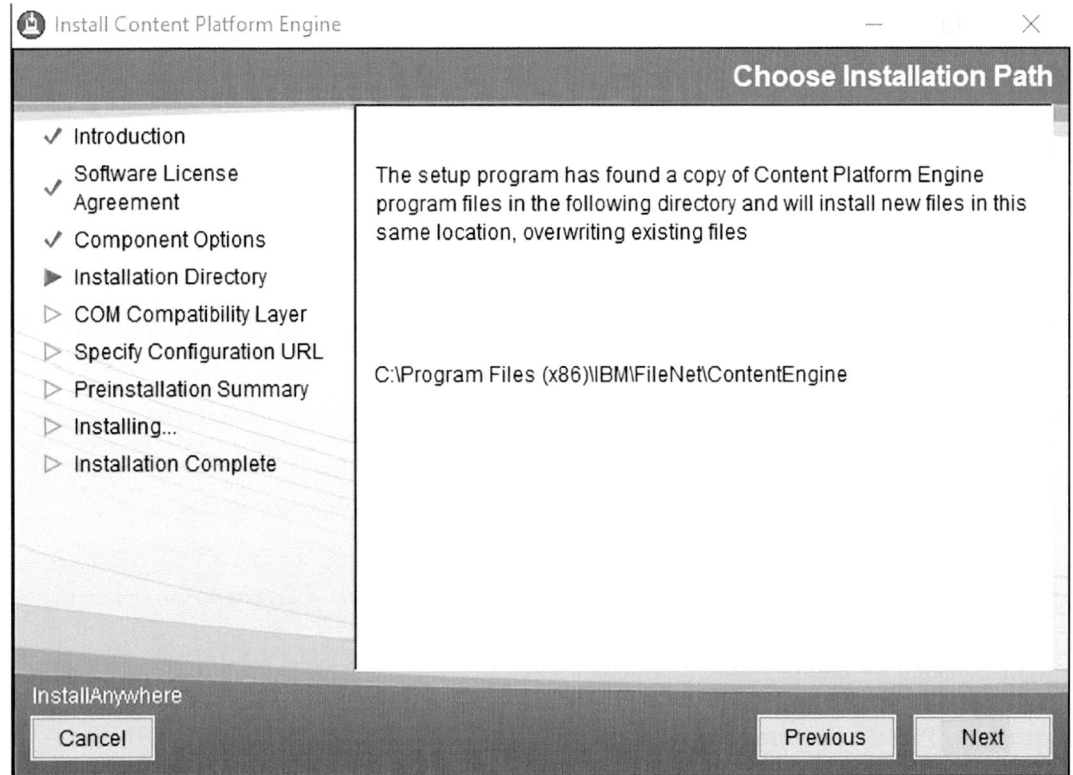

Figure 2-176. *The **Next** command is used after reading the information that we have an older version*

We will be asked in the next installation screen to enter the IBM FileNet P8 Content Engine Service Definition SOAP WSDL URL link. This can be tested in a browser which can connect to the Linux ecmukdemo6 server (the Windows hosts file should have an entry for the IP address).

This SOAP URL is `http://ecmukdemo6:9080/wsi/FNCEWS40MTOM`.

CHAPTER 2 CONFIGURING JAVA CUSTOM COMPONENTS

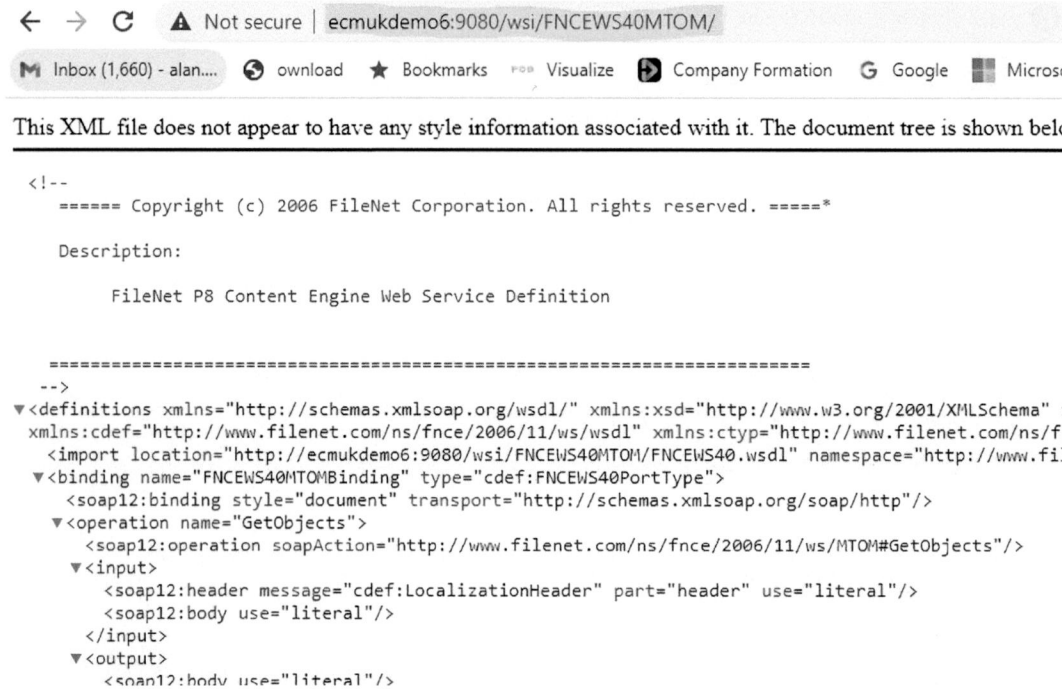

Figure 2-177. *The output expected from a correctly configured SOAP URL link*

CHAPTER 2 CONFIGURING JAVA CUSTOM COMPONENTS

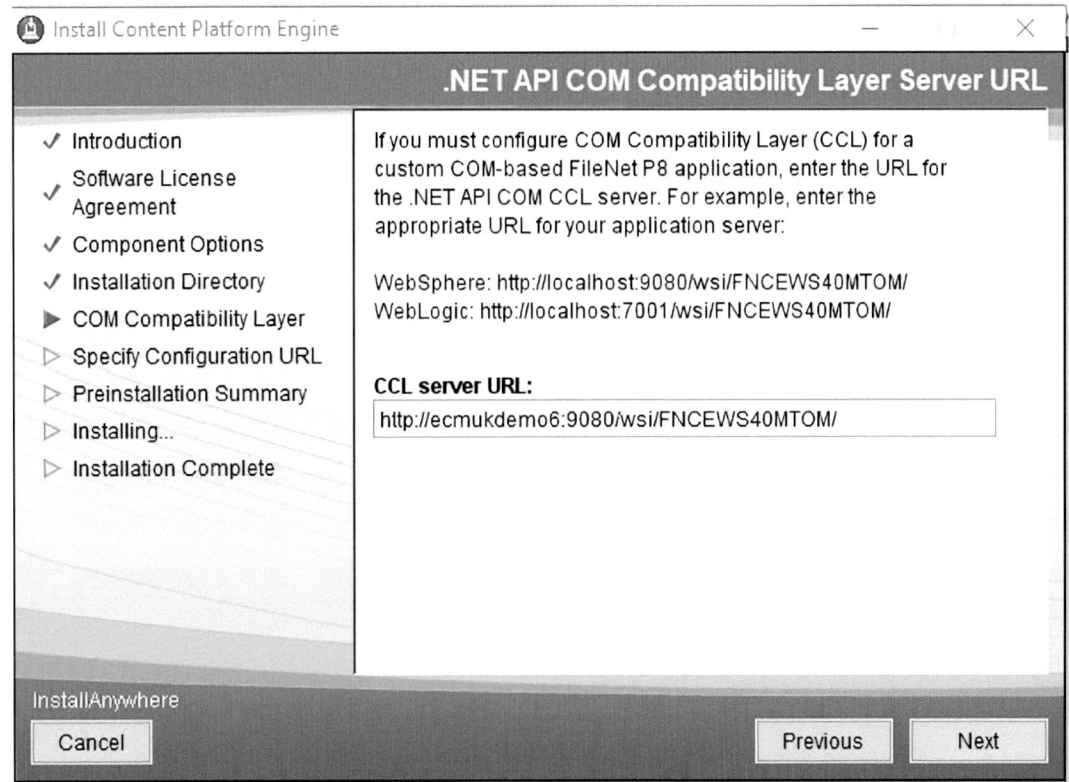

Figure 2-178. *The SOAP URL link used for the COM CCL API layer*

The tested URL is entered in the screen in Figure 2-178.

This URL link is also used by the Java API in a properties file called **WcmApiConfig. properties**, so this is also entered in the next screen shown as follows.

CHAPTER 2   CONFIGURING JAVA CUSTOM COMPONENTS

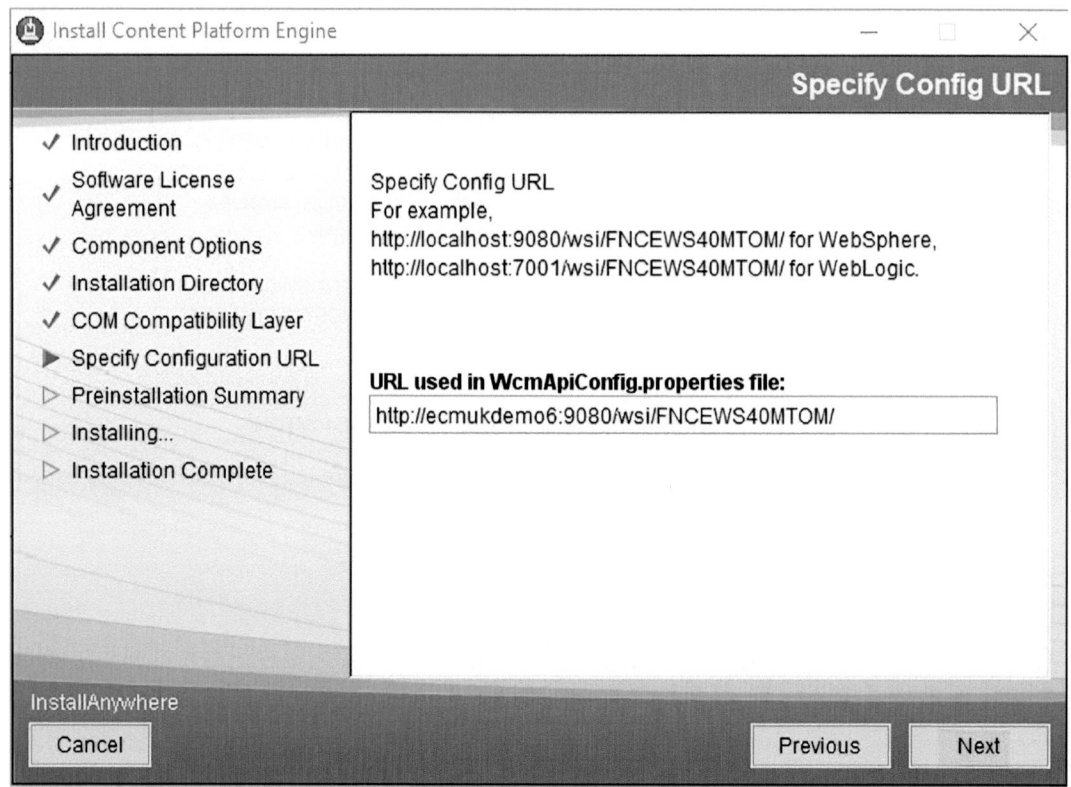

Figure 2-179.   The SOAP link is entered to be used in the WcmApiConfig. properties file

The following screen, shown on clicking the **Next** command button in Figure 2-179, shows a summary of the selected installation options which can be initiated by clicking the **Install** command button.

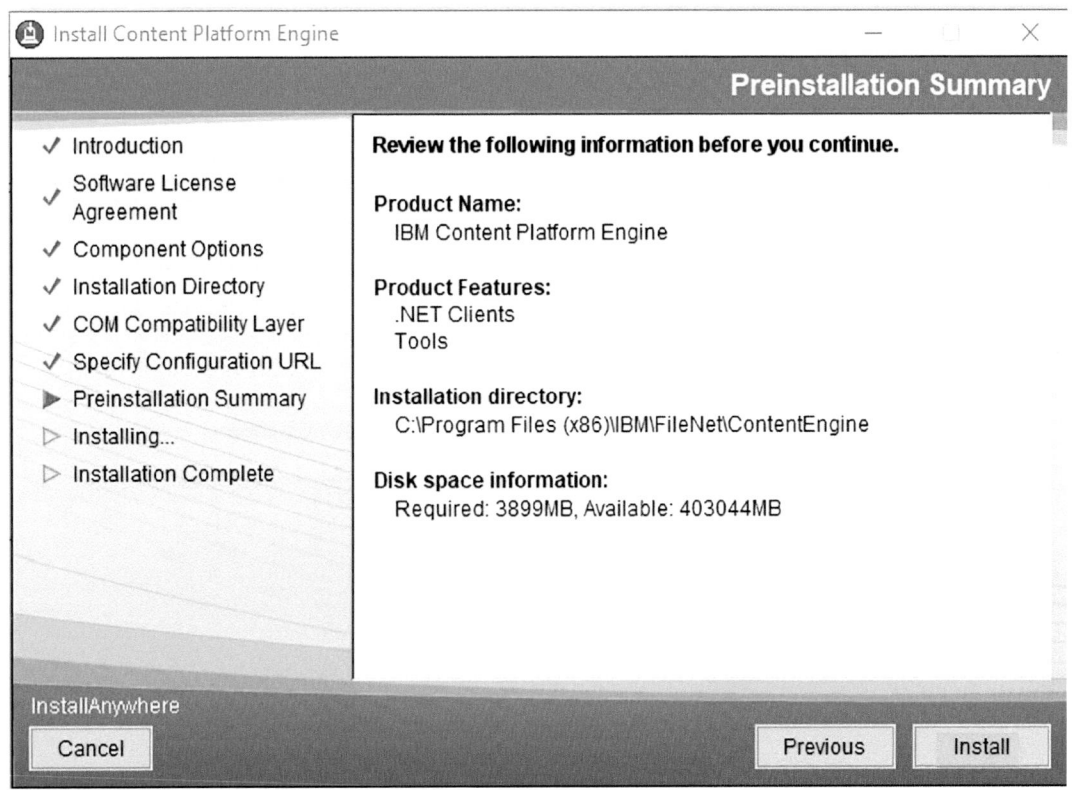

Figure 2-180. *The **Install** command button is clicked to start the installation*

CHAPTER 2   CONFIGURING JAVA CUSTOM COMPONENTS

The installation status is shown in the next Install screen in Figure 2-181.

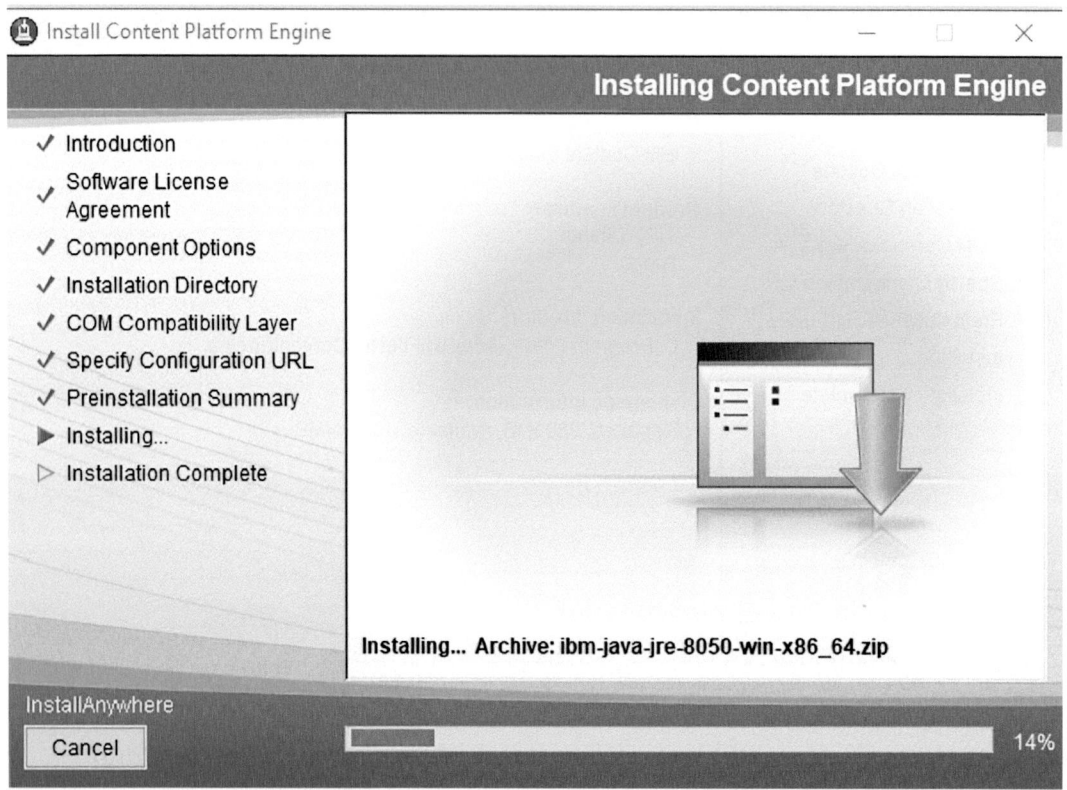

Figure 2-181. The installation progress status bar is shown

After the installation is completed, there is an option to launch the configuration manager with a tick box, which in this instance we don't require, since we are using the Windows Client for the Standalone Process Designer Java application. (The Configuration Manager is used to prepare and install a new IBM Content Platform Engine server system.)

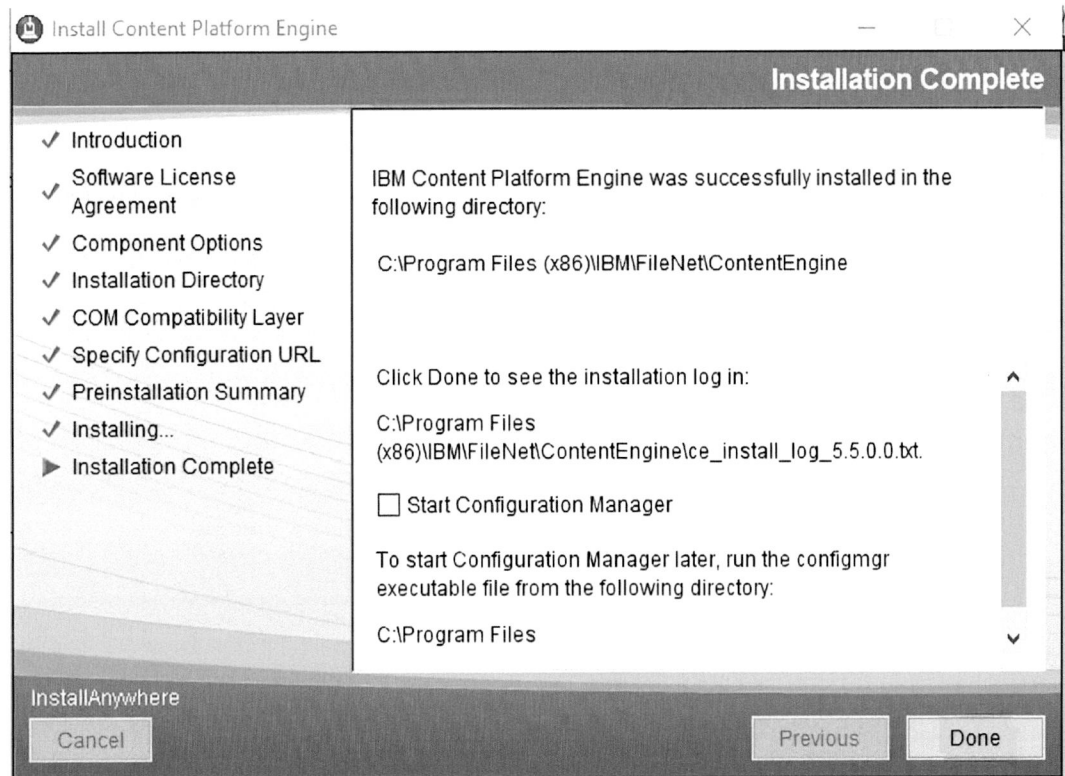

Figure 2-182. *The Done command button is clicked to exit from the final installation screen*

CHAPTER 2   CONFIGURING JAVA CUSTOM COMPONENTS

Fixing the Process Designer Shell Script Environment Variables

The Windows DOS **cmd.exe** command window (run as the Windows Administrator) can now be used to run the **pedesigner.bat** Windows shell script file. However, as will be seen later, we have to do some editing on the environment variable initiation shell scripting to obtain a working **JAVA_HOME** and **CLASSPATH** environment configuration!

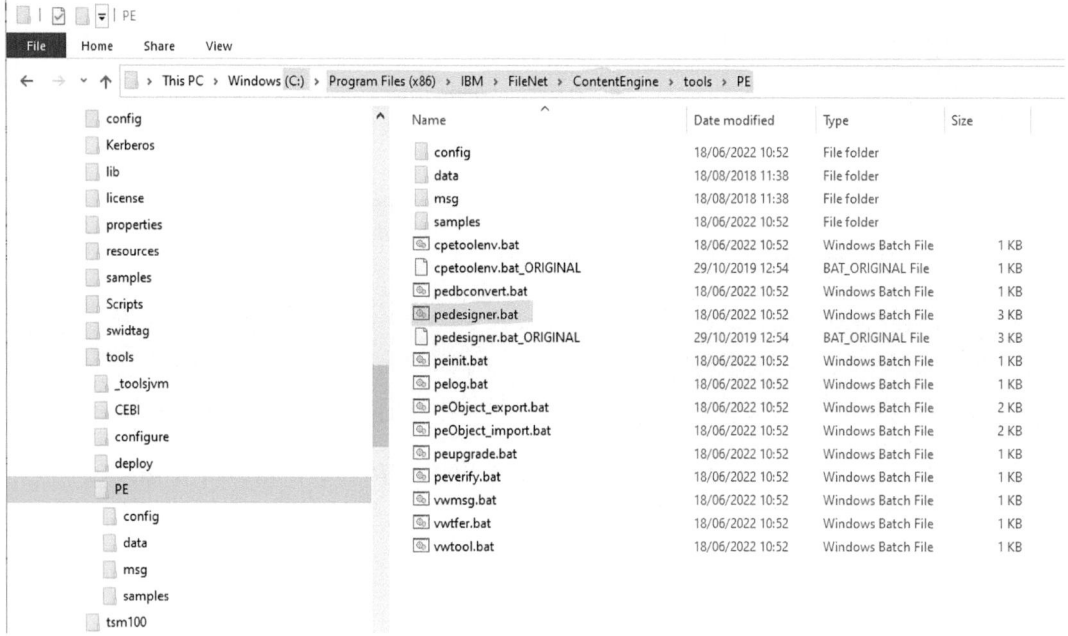

Figure 2-183. *The location of the **pedesigner.bat** file which will launch the Standalone Process Designer*

CHAPTER 2   CONFIGURING JAVA CUSTOM COMPONENTS

```
C:\WINDOWS\system32\cmd.exe
c:\PROGRA~2\IBM\FileNet\ContentEngine\tools\PE>dir
 Volume in drive C is Windows
 Volume Serial Number is 2690-9B6E

 Directory of c:\PROGRA~2\IBM\FileNet\ContentEngine\tools\PE

18/06/2022  10:52    <DIR>          .
18/06/2022  10:52    <DIR>          ..
18/06/2022  10:52    <DIR>          config
18/06/2022  10:52               115 cpetoolenv.bat
29/10/2019  13:54               115 cpetoolenv.bat_ORIGINAL
18/08/2018  11:38    <DIR>          data
18/08/2018  11:38    <DIR>          msg
18/06/2022  10:52               636 pedbconvert.bat
18/06/2022  10:52             2,344 pedesigner.bat
29/10/2019  13:54             2,344 pedesigner.bat_ORIGINAL
18/06/2022  10:52               821 peinit.bat
18/06/2022  10:52               715 pelog.bat
18/06/2022  10:52             1,108 peObject_export.bat
18/06/2022  10:52             1,110 peObject_import.bat
18/06/2022  10:52               827 peupgrade.bat
18/06/2022  10:52               697 peverify.bat
18/06/2022  10:52    <DIR>          samples
18/06/2022  10:52               624 vwmsg.bat
18/06/2022  10:52               689 vwtfer.bat
18/06/2022  10:52               689 vwtool.bat
              14 File(s)         12,834 bytes
               6 Dir(s)  424,049,319,936 bytes free

c:\PROGRA~2\IBM\FileNet\ContentEngine\tools\PE>pedesigner.bat
```

Figure 2-184. *The launch of the pedesigner.bat shell script to load the Standalone Process Designer*

We have to include the Process Engine Connection point for the Designer Object Store which has the name **CP1**.

```
c:\PROGRA~2\IBM\FileNet\ContentEngine\tools\PE>pedesigner.bat
Usage:  pedesigner connection_point_name

c:\PROGRA~2\IBM\FileNet\ContentEngine\tools\PE>pedesigner.bat CP1
```

Figure 2-185. *The correct command line including the Connection Point, CP1*

Unfortunately, we have the **JAVA_HOME** Windows environment variable pointing at a **JRE** Classpath which does not exist.

However, we also have a Folder path with spaces, which does not work for the shell script!

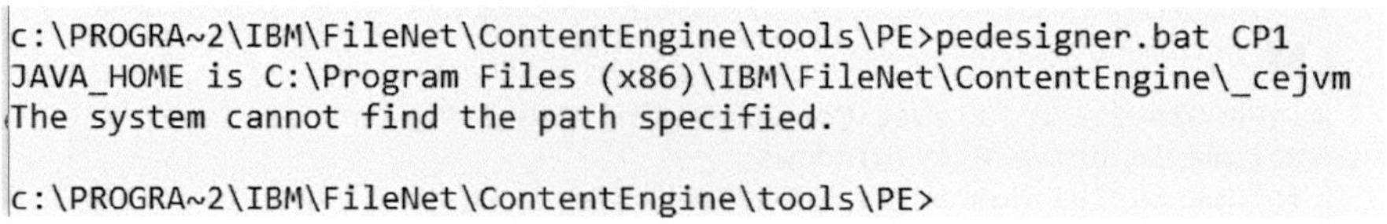

CHAPTER 2 CONFIGURING JAVA CUSTOM COMPONENTS

```
c:\PROGRA~2\IBM\FileNet\ContentEngine\tools\PE>pedesigner.bat CP1
JAVA_HOME is C:\Program Files (x86)\IBM\FileNet\ContentEngine\_cejvm
The system cannot find the path specified.

c:\PROGRA~2\IBM\FileNet\ContentEngine\tools\PE>
```

Figure 2-186. *The system error message appears*

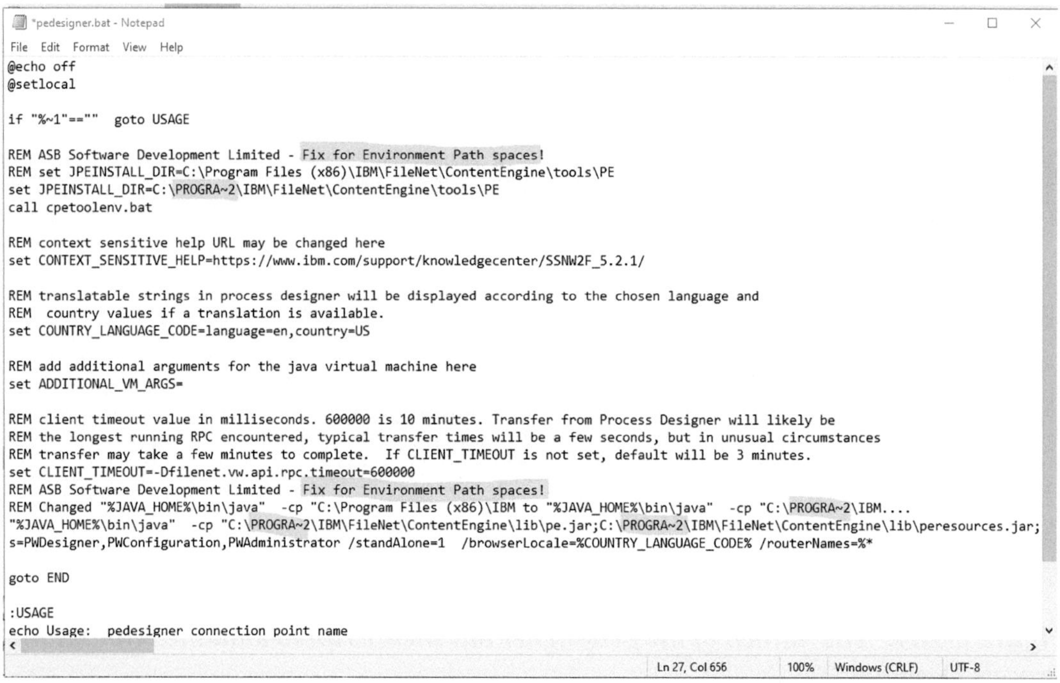

```
*pedesigner.bat - Notepad                                          —  □  ×
File  Edit  Format  View  Help
@echo off
@setlocal

if "%~1"=="" goto USAGE

REM ASB Software Development Limited - Fix for Environment Path spaces!
REM set JPEINSTALL_DIR=C:\Program Files (x86)\IBM\FileNet\ContentEngine\tools\PE
set JPEINSTALL_DIR=C:\PROGRA~2\IBM\FileNet\ContentEngine\tools\PE
call cpetoolenv.bat

REM context sensitive help URL may be changed here
set CONTEXT_SENSITIVE_HELP=https://www.ibm.com/support/knowledgecenter/SSNW2F_5.2.1/

REM translatable strings in process designer will be displayed according to the chosen language and
REM    country values if a translation is available.
set COUNTRY_LANGUAGE_CODE=language=en,country=US

REM add additional arguments for the java virtual machine here
set ADDITIONAL_VM_ARGS=

REM client timeout value in milliseconds. 600000 is 10 minutes. Transfer from Process Designer will likely be
REM the longest running RPC encountered, typical transfer times will be a few seconds, but in unusual circumstances
REM transfer may take a few minutes to complete.  If CLIENT_TIMEOUT is not set, default will be 3 minutes.
set CLIENT_TIMEOUT=-Dfilenet.vw.api.rpc.timeout=600000
REM ASB Software Development Limited - Fix for Environment Path spaces!
REM Changed "%JAVA_HOME%\bin\java"  -cp "C:\Program Files (x86)\IBM to "%JAVA_HOME%\bin\java"  -cp "C:\PROGRA~2\IBM....
"%JAVA_HOME%\bin\java"   -cp "C:\PROGRA~2\IBM\FileNet\ContentEngine\lib\pe.jar;C:\PROGRA~2\IBM\FileNet\ContentEngine\lib\peresources.jar;
s=PWDesigner,PWConfiguration,PWAdministrator /standAlone=1  /browserLocale=%COUNTRY_LANGUAGE_CODE% /routerNames=%*

goto END

:USAGE
echo Usage:   pedesigner connection point name
                                                        Ln 27, Col 656   100%  Windows (CRLF)  UTF-8
```

Figure 2-187. *The highlighted changes are made in the pedesigner.bat shell script*

In Figure 2-187, we make changes to ensure there are no spaces in the classpath which we can achieve by using the **DOS** short form of the directory path for the top level **c:\Program Files (x86)** directory, which has a DOS short form of **c:\PROGRA~2**.

Then, we also have to edit the cpetoolenv.bat shell script file, which is called by the main pedesigner.bat shell script file, as shown in Figure 2-188.

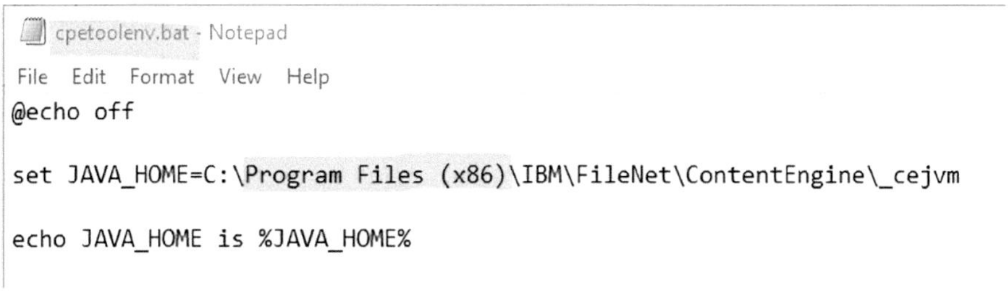

CHAPTER 2   CONFIGURING JAVA CUSTOM COMPONENTS

```
cpetoolenv.bat - Notepad
File  Edit  Format  View  Help
@echo off

set JAVA_HOME=C:\Program Files (x86)\IBM\FileNet\ContentEngine\_cejvm

echo JAVA_HOME is %JAVA_HOME%
```

Figure 2-188. *The highlighted change is made in the **cpetoolenv.bat** shell script*

This results in the updated shell script file in Figure 2-189.

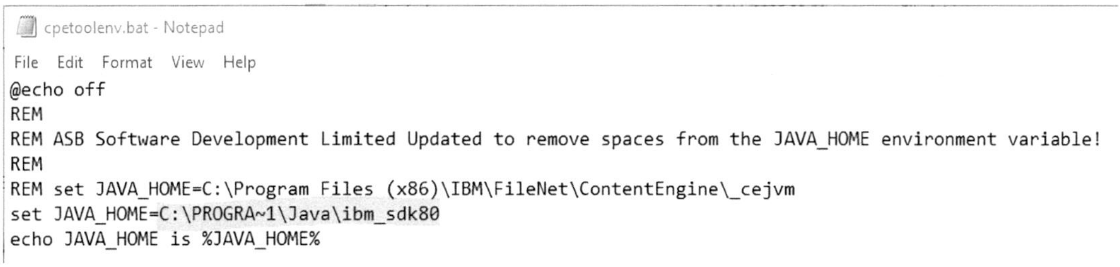

```
cpetoolenv.bat - Notepad
File  Edit  Format  View  Help
@echo off
REM
REM ASB Software Development Limited Updated to remove spaces from the JAVA_HOME environment variable!
REM
REM set JAVA_HOME=C:\Program Files (x86)\IBM\FileNet\ContentEngine\_cejvm
set JAVA_HOME=C:\PROGRA~1\Java\ibm_sdk80
echo JAVA_HOME is %JAVA_HOME%
```

Figure 2-189. *The updated cpetoolenv.bat file*

We can now launch the Standalone Process Designer Java program using the Windows cmd.exe DOS window, with the commands as follows:

```
cd c:\PROGRA~2\IBM\FileNet\ContentEngine\PE
pedesigner.bat CP1
```

This loads a pop-up window prompting us for a Process Engine administrator login.

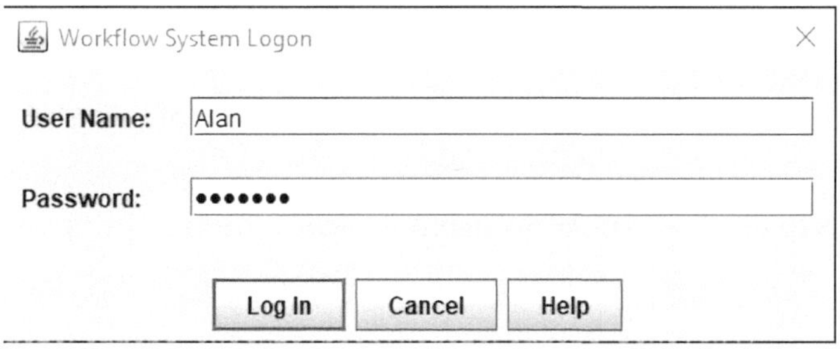

```
Workflow System Logon
User Name:  Alan
Password:   •••••••
            [ Log In ]  [ Cancel ]  [ Help ]
```

Figure 2-190. *The User Name and Password are entered, and the Log In command button is clicked*

CHAPTER 2   CONFIGURING JAVA CUSTOM COMPONENTS

We now see the first screen of the IBM Standalone Java program, **Process Designer**, which we can use for the editing of IBM Case Manager solution Task workflows.

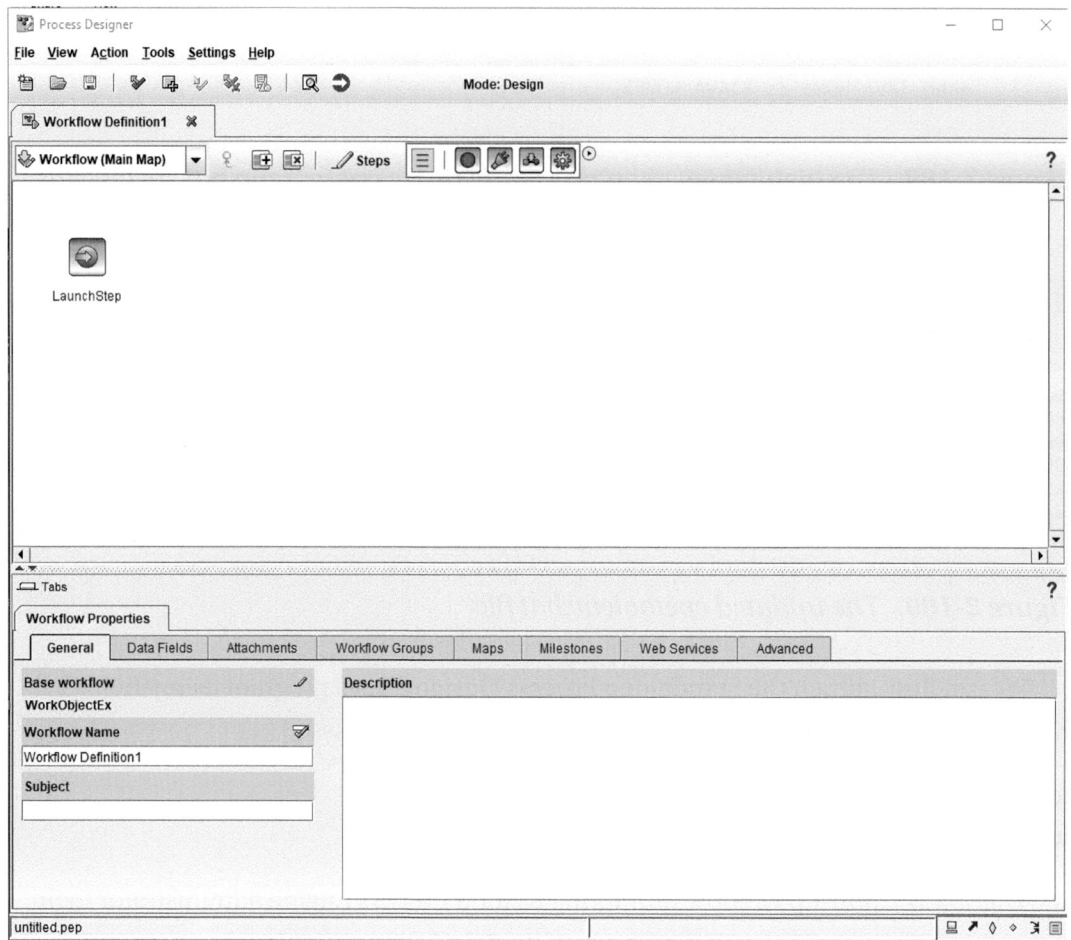

Figure 2-191. *The first edit screen of the Process Designer Java program*

We can now load the **Audit Department** Task Workflow of our **Audit Master** Case Manager Solution.

The first step is to use the **File ➤ Solution ➤ Edit...** menu option as shown in Figure 192.

CHAPTER 2   CONFIGURING JAVA CUSTOM COMPONENTS

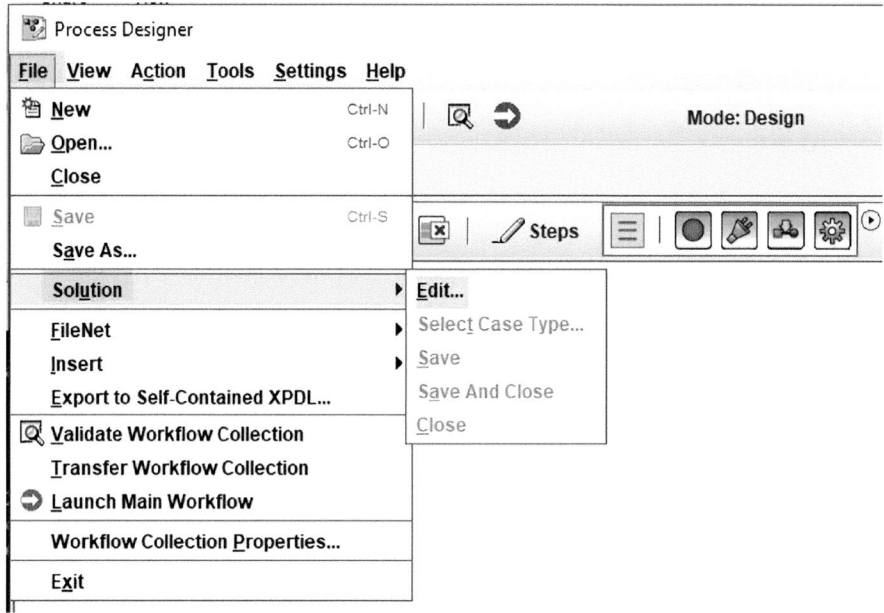

Figure 2-192. *The menu option is selected to navigate to the Audit Master solution to edit the Task Workflow*

We now see the option to select from the available IBM FileNet Object Stores, used by Case Manager for the **Design** (**OS1**) and **Target** (**OS2**) object stores, where our solution is stored. We select the **OS1** Design Object store where the **Audit Master** solution is developed.

CHAPTER 2　CONFIGURING JAVA CUSTOM COMPONENTS

*Figure 2-193.　The Design Object Store, **OS1**, is selected*

From the **OS1** Object Store, we navigate down to the **Audit Master** solution following the Object Store folder path **OS1 ➤ IBM Case Manager ➤ Solutions ➤ Audit Master**, as shown in Figure 2-194.

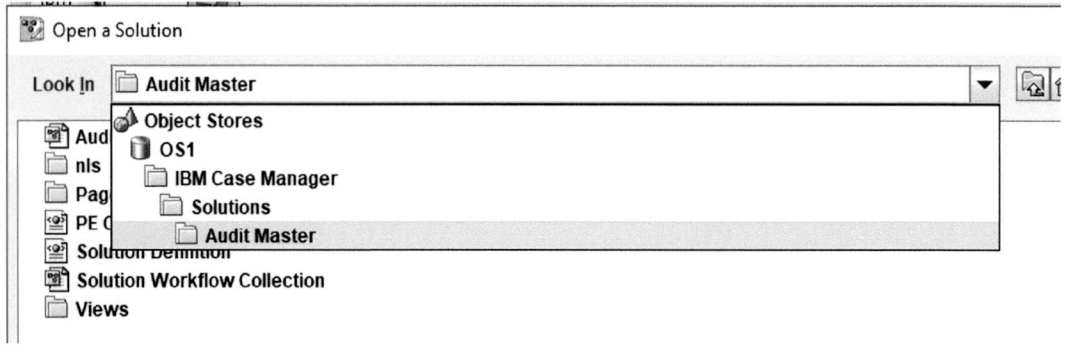

Figure 2-194.　The path to the Audit Master solution is selected and opened

We are now presented with a list of IBM Case Manager Solution components. For the Workflows, we select the **Solution Definition** file shown in Figure 2-195.

CHAPTER 2 CONFIGURING JAVA CUSTOM COMPONENTS

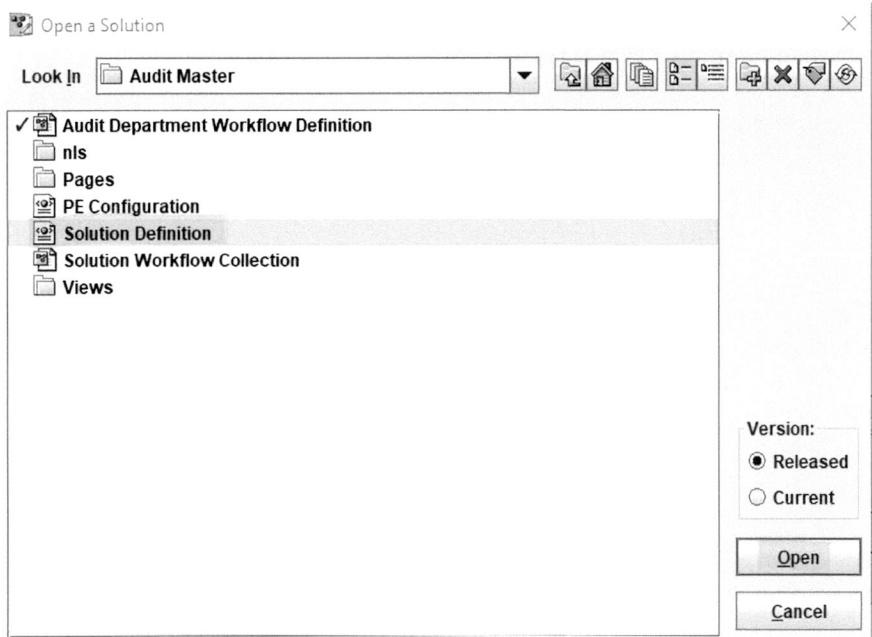

Figure 2-195. *The **Solution Definition** file for the **Audit Master** solution is selected and **Open** clicked*

We are now shown a pop-up window with a drop-down which shows the Case Types available. (For the Audit Master solution, we only have one Case Type at the moment, **Audit Department**... there can be many!)

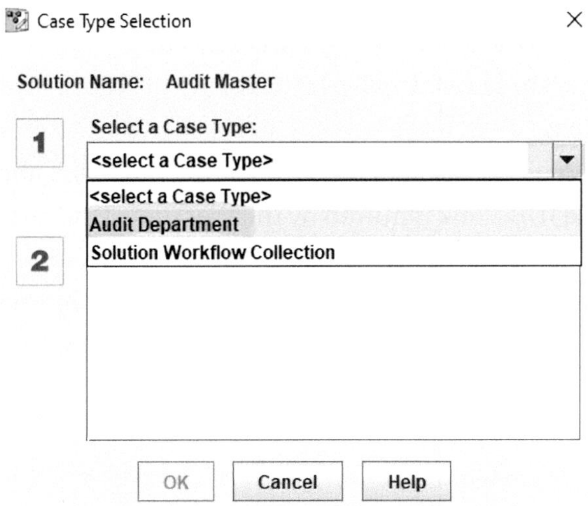

Figure 2-196. *The Audit Department Case Type is selected*

297

The selected Case Type then shows the **AUD_AuditProcess** Task Workflow to be used. (There can be many Case Type workflows!)

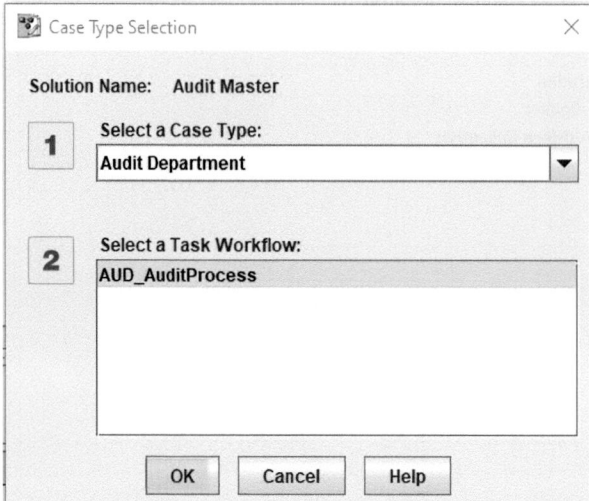

Figure 2-197. *The AUD_AuditProcess Task Workflow for the Audit Department Case Type is selected*

We can now add the DbExecute Workflow step.

The DbExecute Workflow Step Addition

The following URL covers the system setup for DbExecute:
    www.ibm.com/docs/en/filenet-p8-platform/5.2.1?topic=system-setting-dbexecute-connections

The following procedure is used in the IBM FileNet **acce** web application to configure the database JDBC link required by the Workflow DbExecute step. It defines the JDBC URL which allows the FileNet system to connect to a server database, by an administrator specifying a server name and port number and entering the required database name and logon details to access the Stored Procedure to be used.

CHAPTER 2 CONFIGURING JAVA CUSTOM COMPONENTS

Procedure to Configure a DbExecute Connection in the Content Engine Store

To set a DbExecute connection:

1. Access the **DbExecute Connections** tab in the administration console:

    a. In the domain navigation pane, select the object store.

    b. In the object store navigation pane, click **Administrative ➤ Workflow System**.

    c. In the details pane, click the **DbExecute Connections** tab.

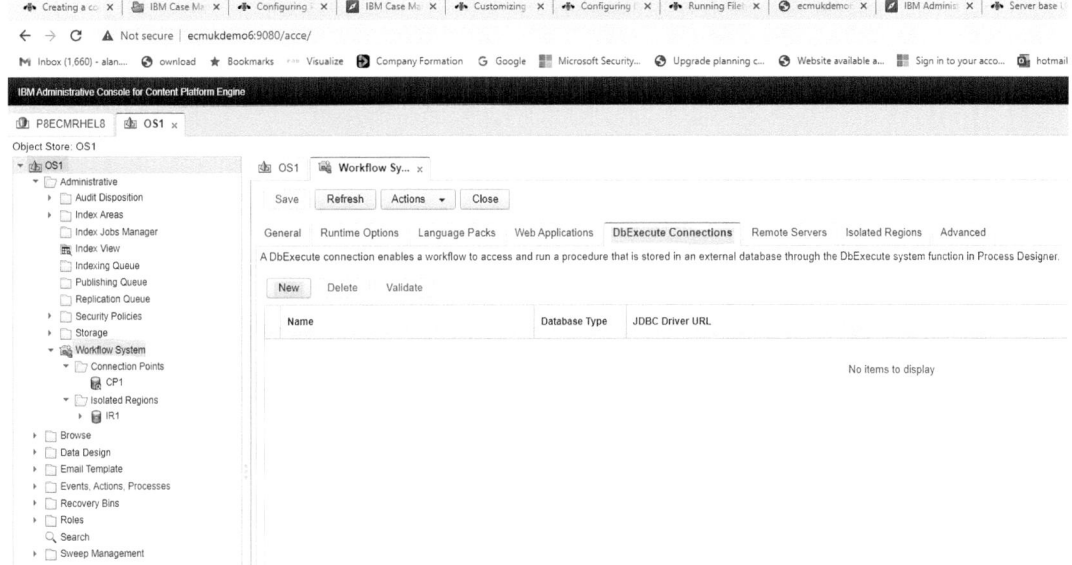

Figure 2-198. *The DbExecute Connections in the Design Object Store*

2. Click **New**.

3. Enter the connection information and database parameters.

299

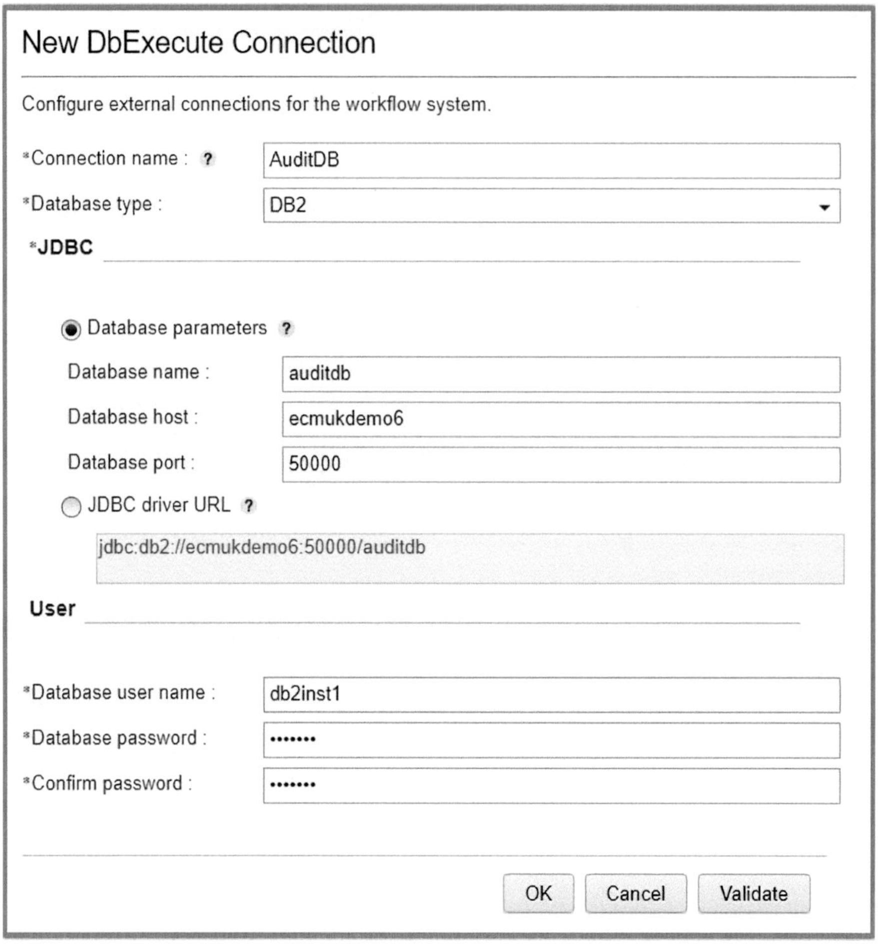

CHAPTER 2   CONFIGURING JAVA CUSTOM COMPONENTS

*Figure 2-199. The **AuditDB** connection parameters are entered*

Click Validate to test the database connection.

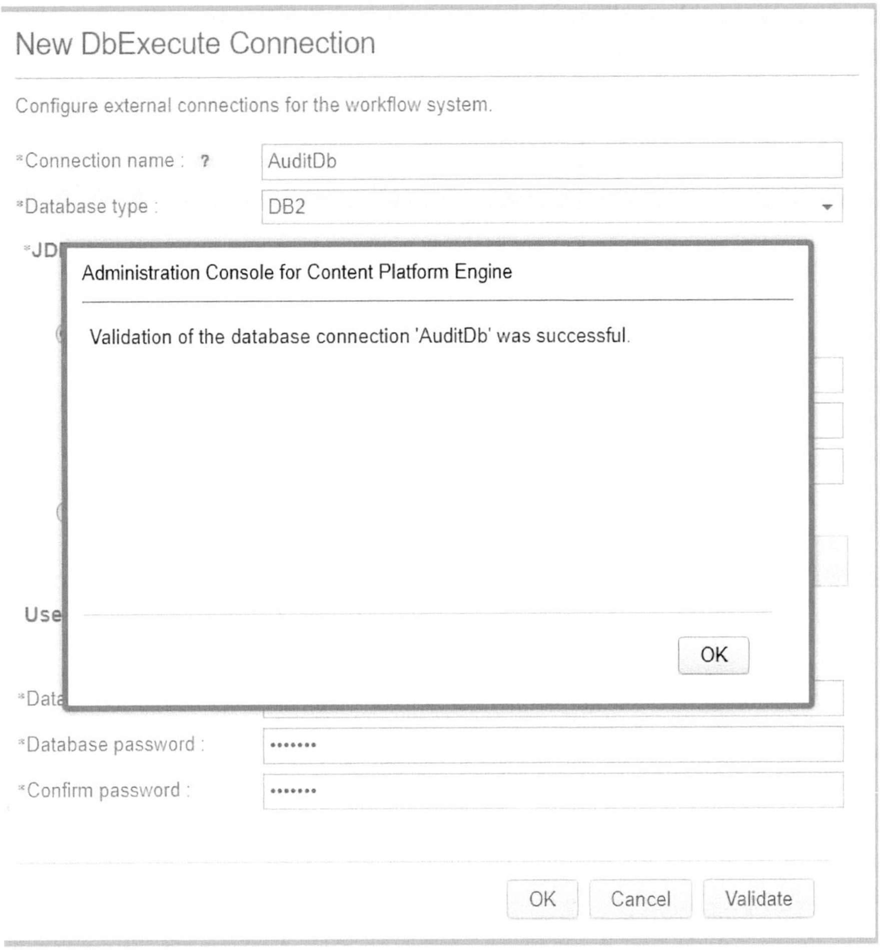

CHAPTER 2 CONFIGURING JAVA CUSTOM COMPONENTS

*Figure 2-200. The status of the **AuditDb** connection is shown*

CHAPTER 2   CONFIGURING JAVA CUSTOM COMPONENTS

We can now see the **AuditDb** object we have successfully created in the **DbExecute Connections** tab.

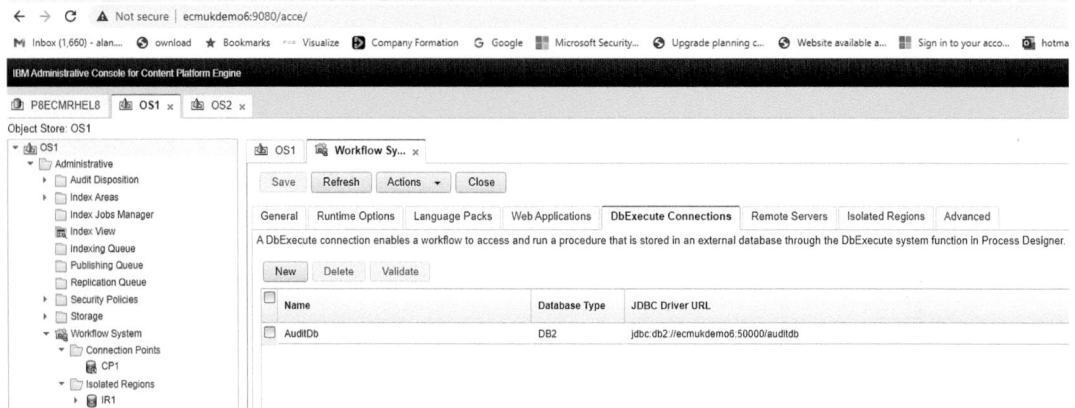

Figure 2-201.   *The **AuditDb** object is successfully created in the **DbExecute Connections** tab of **OS1***

The preceding steps were repeated for the OS2 Target Object Store.

CHAPTER 2   CONFIGURING JAVA CUSTOM COMPONENTS

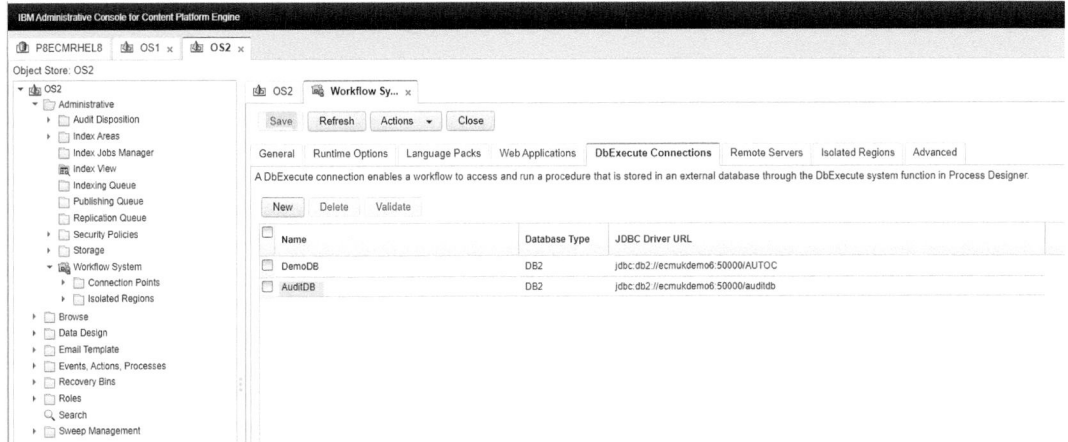

Figure 2-202. *The* **AuditDB** *object is successfully created in the* **DbExecute Connections** *tab of* **OS2**

We can now insert test rows into the **AUDIT_DEPT** table which is in the DB2 **auditdb** database.

```
▸ Connection: ecmukdemo6 - db2inst1 - auditdb [db2inst1]

    INSERT INTO
        AUDIT_DEPT (AUDITOR_NAME,DEPTNO,AUDIT_DATE,COMMENTS,AUDIT_STATUS)
    VALUES
        ('Alan',00001,'2022-06-18','First Audit','Working');
```

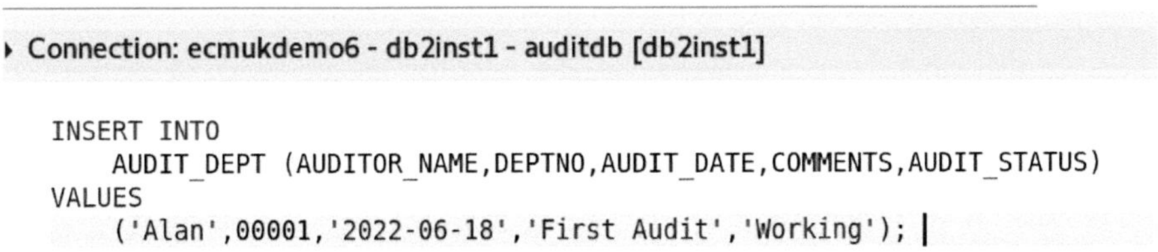

Figure 2-203. *The IBM Data Studio is used to view and add test rows in the DB2 auditdb database*

303

CHAPTER 2　CONFIGURING JAVA CUSTOM COMPONENTS

The DbExecute Stored Procedure SQL

```
                                          db2inst1@ECMUKDEMO6:~
 File  Edit  View  Search  Terminal  Help
 CREATE OR REPLACE PROCEDURE ADD_AUDIT_DEPT (INOUT VARRESULT INT,
                                            INOUT AUDITOR VARCHAR(10),
                                            INOUT DEPTNO INT,
                                            INOUT  AUDITDATE VARCHAR(20),
                                            INOUT COMMENTS VARCHAR(200),
                                            INOUT   AUDITSTATUS VARCHAR(10))

 LANGUAGE SQL
 MODIFIES SQL DATA
 P1: BEGIN
 DECLARE VARRESULT INT;
 DECLARE AUDITORVAR VARCHAR(10);
 DECLARE DEPTNOVAR INT;
 DECLARE AUDITDATEVAR VARCHAR(20);
 DECLARE COMMENTSVAR VARCHAR(20);
 DECLARE AUDIT_STATUSVAR     VARCHAR(256);
         -- ######################################################################
         -- # PROCEDURE ADD_AUDIT_DEPT.
         -- #   Alan S. Bluck, Director, ASB Software Development Limited
         -- #
         -- # This stored procedure is writes a row of Audit Department
         -- # data to the DB2INST1.AUDIT_DEPT table in the auditdb database
         -- #
         -- # NB Any date fields which are set empty on the initial Add of the CASE
         -- #    may cause insert failure!
         -- #    These need setting as the Fields are set NOT NULL so give
         -- #    them 2022-01-01 defaults
         -- ######################################################################
 SET AUDITORVAR=AUDITOR;
 SET DEPTNOVAR = DEPTNO;
 SET AUDITDATEVAR = AUDITDATE;
 SET COMMENTSVAR = COMMENTS;
 SET AUDIT_STATUSVAR = AUDITSTATUS;
 INSERT INTO
         DB2INST1.AUDIT_DEPT (AUDITOR_NAME,DEPTNO,AUDIT_DATE,COMMENTS,AUDIT_STATUS)
 VALUES(AUDITORVAR, DEPTNOVAR, AUDITDATEVAR, COMMENTSVAR, AUDIT_STATUSVAR);
 SET VARRESULT=1;
 END
 @
```

Figure 2-204.　*The Stored Procedure **ADD_AUDIT_DEPT** is edited in the file SetAddAudit_Stored_ProcedureV1.sql*

CHAPTER 2 CONFIGURING JAVA CUSTOM COMPONENTS

The Stored Procedure, **ADD_AUDIT_DEPT**, listed in Figure 2-204, which we have developed, can be called by the **DbExecute** step to add a row to the **DB2INST1.AUDIT_DEPT** table in the **auditdb** database, using parameters passed for each column of the table.

```
[db2inst1@ECMUKDEMO6 ~]$ db2 connect to auditdb

   Database Connection Information

 Database server        = DB2/LINUXX8664 11.5.0.0
 SQL authorization ID   = DB2INST1
 Local database alias   = AUDITDB

[db2inst1@ECMUKDEMO6 ~]$ db2 -td@ -vf SetAddAudit_Stored_ProcedureV1.sql
```

Figure 2-205. *The stored procedure is created in the auditdb database, using the preceding db2 commands*

```
                                                    db2inst1@ECMUKDEMO6:~
File  Edit  View  Search  Terminal  Help
[db2inst1@ECMUKDEMO6 ~]$ vi SetAddAudit_Stored_ProcedureV1.sql
[db2inst1@ECMUKDEMO6 ~]$ db2 -td@ -vf SetAddAudit_Stored_ProcedureV1.sql
CREATE OR REPLACE PROCEDURE ADD_AUDIT_DEPT (INOUT VARRESULT INT,
                                            INOUT AUDITOR VARCHAR(10),
                                            INOUT DEPTNO INT,
                                            INOUT   AUDITDATE VARCHAR(20),
                                            INOUT COMMENTS VARCHAR(200),
                                            INOUT    AUDITSTATUS VARCHAR(10))

LANGUAGE SQL
MODIFIES SQL DATA
P1: BEGIN
DECLARE VARRESULT INT;
DECLARE AUDITORVAR VARCHAR(10);
DECLARE DEPTNOVAR INT;
DECLARE AUDITDATEVAR VARCHAR(20);
DECLARE COMMENTSVAR VARCHAR(20);
DECLARE AUDIT_STATUSVAR     VARCHAR(256);
        -- ####################################################################
        -- # PROCEDURE ADD_AUDIT_DEPT.
        -- #   Alan S. Bluck, Director, ASB Software Development Limited
        -- #
        -- # This stored procedure is writes a row of Audit Department
        -- # data to the DB2INST1.AUDIT_DEPT table in the auditdb database
        -- #
        -- # NB Any date fields which are set empty on the initial Add of the CASE
        -- #    may cause insert failure!
        -- #    These need setting as the Fields are set NOT NULL so give
        -- #    them 2022-01-01 defaults
        -- ####################################################################
SET AUDITORVAR=AUDITOR;
SET DEPTNOVAR = DEPTNO;
SET AUDITDATEVAR = AUDITDATE;
SET COMMENTSVAR = COMMENTS;
SET AUDIT_STATUSVAR = AUDITSTATUS;
INSERT INTO
       DB2INST1.AUDIT_DEPT (AUDITOR_NAME,DEPTNO,AUDIT_DATE,COMMENTS,AUDIT_STATUS)
VALUES(AUDITORVAR, DEPTNOVAR, AUDITDATEVAR, COMMENTSVAR, AUDIT_STATUSVAR);
SET VARRESULT=1;
END

DB20000I  The SQL command completed successfully.

[db2inst1@ECMUKDEMO6 ~]$
```

Figure 2-206. *The install of the **DB2INST1.AUDIT_DEPT** stored procedure is successfully loaded into **auditdb***

See the following web page for the full DbExecute procedure steps:
www.ibm.com/docs/en/filenet-p8-platform/5.5.x?topic=activity-dbexecute-system-function

Configuring the Workflow for the DbExecute Step

We can now define the **DbExecute** system step for the workflow system using the installed Standalone **Process Designer**.

Note To create a new DbExecute connection, an XA and non-XA JDBC Datasource will need to be set up for the IBM FileNet WebSphere application server to enable access for the database.

See the link for a full description of an alternative JDBC DbExecute connection creation (on Page 51 of the document):

https://doi.org/10.13140/RG.2.2.16987.52008

"Importing Case Manager solution Auto Claims example into Case Manager 5.3.3 on RHEL 8.0" Document download:

ImportingCaseManagersolutionAutoClaimsexampleintoCaseManager5.3.3on RHEL8.0.docx

CHAPTER 2   CONFIGURING JAVA CUSTOM COMPONENTS

Procedure to Add the DbExecute System Step to the Task Workflow

To specify a new DbExecute system step:

1. From the **General System Palette**, use the mouse to drag a **DbExecute** step onto the workflow map.

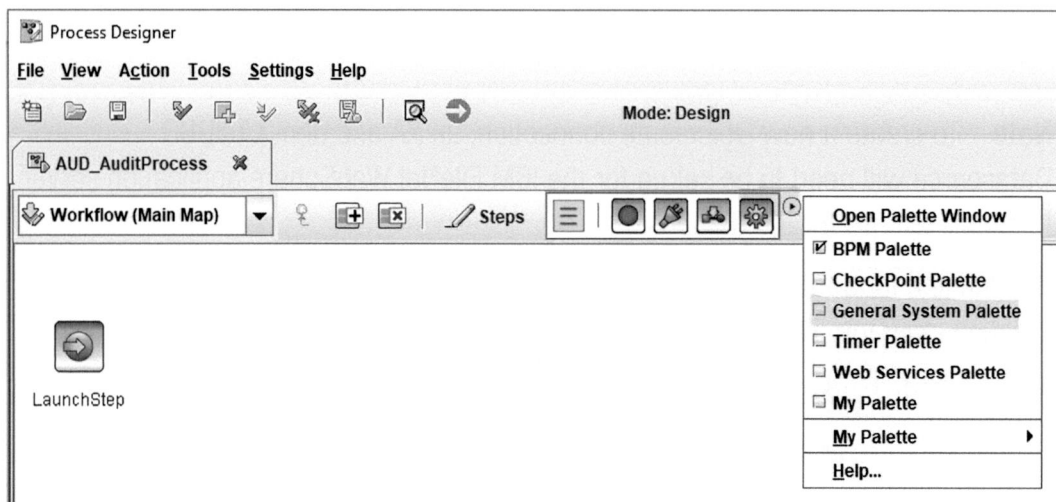

Figure 2-207. *The General System Palette window is ticked in Process Designer*

After the **General System Palette** select box is ticked, the icons open up as shown in Figure 2-208. Then the highlighted DbExecute step can be dragged and dropped into the Process Designer work area.

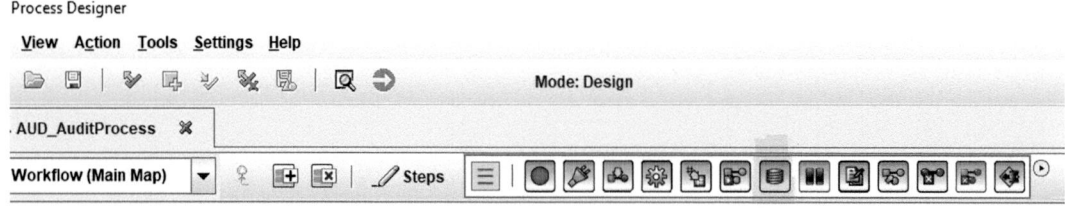

Figure 2-208. *The **DbExecute** workflow step icon is as highlighted*

CHAPTER 2   CONFIGURING JAVA CUSTOM COMPONENTS

2. Enter the database name in the parameter window shown in Figure 2-209.

Note   The name that is used to create a stored procedure is used to call it in the Workflow. Process Designer does not allow a Stored Procedure name to be entered with quotes for the DbExecute system step. See the link as follows for the full details:

www.ibm.com/docs/en/filenet-p8-platform/5.5.x?topic=activity-dbexecute-system-function

3. Enter the name of the stored procedure to execute (**ADD_AUDIT_DEPT**).

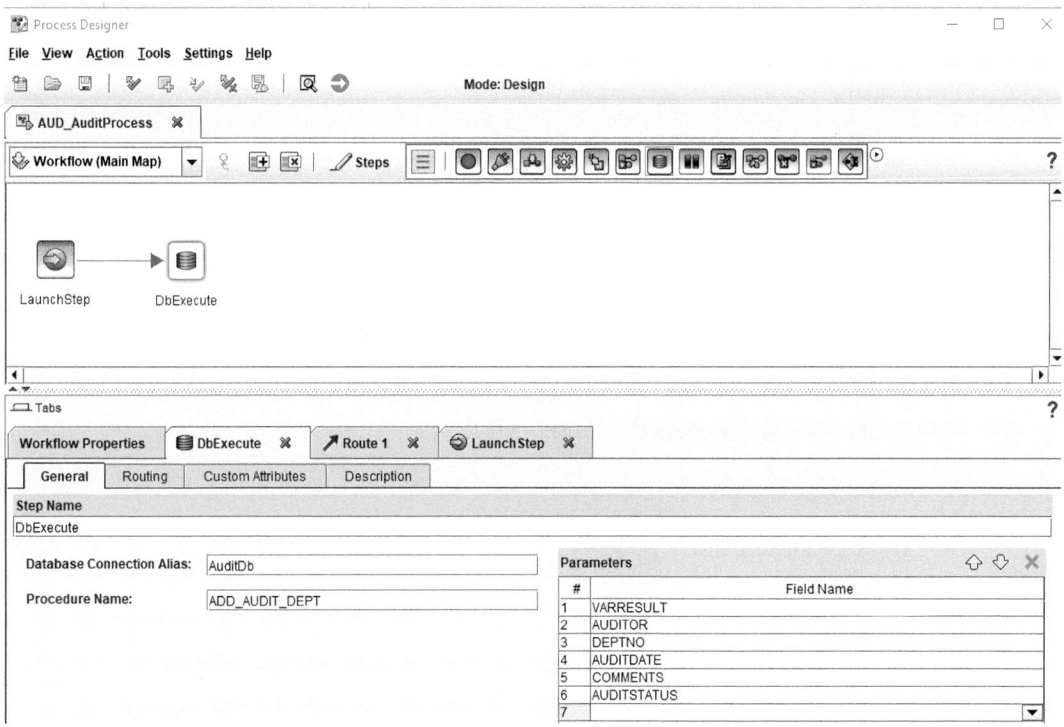

Figure 2-209. *The parameters for the ADD_AUDIT_DEPT Stored Procedure are entered*

309

CHAPTER 2   CONFIGURING JAVA CUSTOM COMPONENTS

In Figure 2-209, we have entered the required DbExecute step parameters which require the Database Connection Alias we set up.

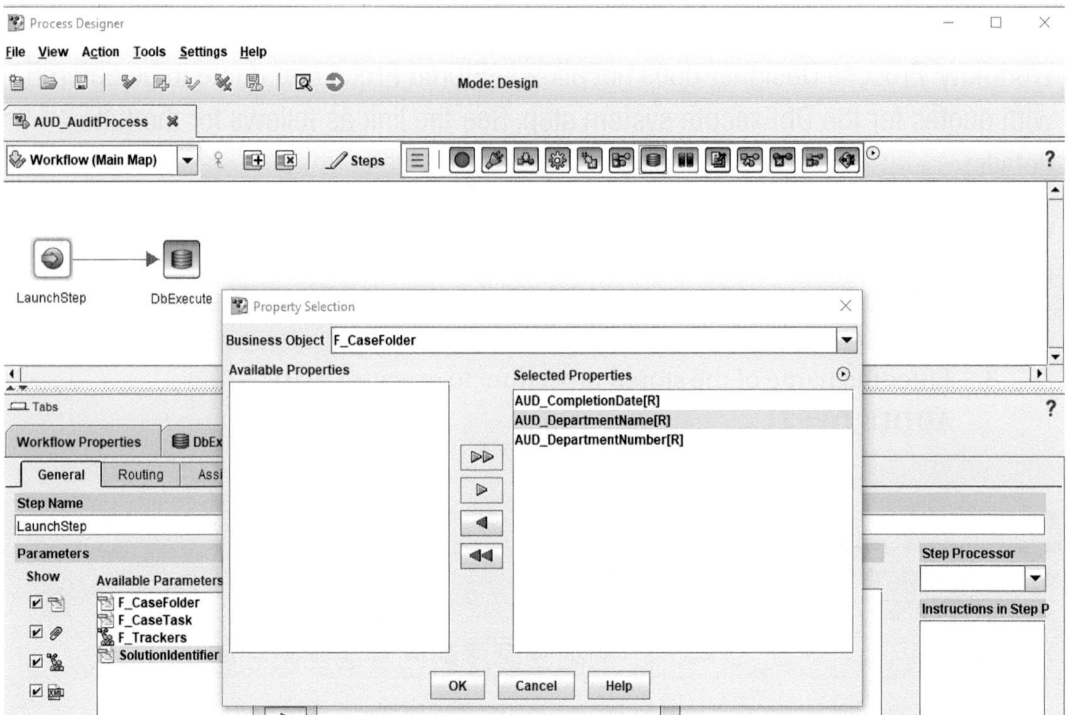

Figure 2-210. *The Case field properties are now selected for assigning to the Stored Procedure*

4. The different supported databases for the DbExecute calls must have all of its parameters declared as follows:

- out – SQL Server
- in out – Oracle
- inout – DB2

The preceding parameter types are set to allow the values to be maintained bidirectionally. This allows the updated values to reflect the stored values passed in as parameters to the stored procedure.

CHAPTER 2 CONFIGURING JAVA CUSTOM COMPONENTS

The parameter data types passed in must match the types that are specified in the stored procedure SQL and can be of the following types:

Designer Param Type	Oracle Param Type	SQL Server Param Type	DB2 Param Type
String	varchar	Varchar	varchar
Integer	number	Int	int
Boolean	number	Bit	number
Float	number	Float	float
Time	date	Datetime	timestamp

Stored Procedure Parameter Limitations

Arrays are not supported for Stored Procedure parameters.

Note If a Date/time, Integer, Boolean, or Float parameter is set to null by a stored procedure, an exception will be thrown.

	Oracle	SQL Server	DB2
Maximum number of parameters in procedure	1024	1024	1024
String parameter maximum characters	4000	4000	4000

CHAPTER 2 CONFIGURING JAVA CUSTOM COMPONENTS

Example SQL Table Creation for the DB2 Database

The table in the auditdb database we are interested in has the following DDL.

Listing 2-7. The Create Audit_Dept table SQL script

```
CREATE TABLE "DB2INST1"."AUDIT_DEPT" (
        "AUDITOR_NAME" CHAR(40 OCTETS) NOT NULL,
        "DEPTNO" INTEGER NOT NULL,
        "AUDIT_DATE" DATE,
        "COMMENTS" CHAR(200 OCTETS) NOT NULL,
        "AUDIT_STATUS" CHAR(10 OCTETS) NOT NULL
    )
    ORGANIZE BY ROW
    DATA CAPTURE NONE
    IN "USERSPACE1"
    COMPRESS NO;
```

DbExecute Stored Procedure Installation

The following db2 commands are used to install the Stored Procedure SQL code. The code was edited into a **SetAddAudit_Stored_ProcedureV1.sql** file containing the SQL.

db2 connect to auditdb
db2 -td@ -vf SetAddAudit_Stored_ProcedureV1.sql

The SQL code we enter for the Stored Procedure is as follows.

Listing 2-8. The ADD_AUDIT_DEPT Stored Procedure DB2 SQL script

```
CREATE OR REPLACE PROCEDURE ADD_AUDIT_DEPT (INOUT VARRESULT INT,
                                            INOUT AUDITOR VARCHAR(10),
                                            INOUT DEPTNO INT,
                                            INOUT  AUDITDATE VARCHAR(20),
                                            INOUT COMMENTS VARCHAR(200),
                                            INOUT   AUDITSTATUS VARCHAR(10))
```

```
LANGUAGE SQL
MODIFIES SQL DATA
P1: BEGIN
DECLARE VARRESULT INT;
DECLARE AUDITORVAR VARCHAR(10);
DECLARE DEPTNOVAR INT;
DECLARE AUDITDATEVAR VARCHAR(20);
DECLARE COMMENTSVAR VARCHAR(20);
DECLARE AUDIT_STATUSVAR    VARCHAR(256);
        -- ############################################################
##########
-- # PROCEDURE ADD_AUDIT_DEPT.
-- #  Alan S. Bluck, Director, ASB Software Development Limited
-- #
-- # This stored procedure writes a row of Audit Department
-- # data to the DB2INST1.AUDIT_DEPT table in the auditdb database
-- #
-- # NB Any date fields which are set empty on initial Add of a CASE
-- #    may cause insert failure!
####################################################################
SET AUDITORVAR=AUDITOR;
SET DEPTNOVAR = DEPTNO;
SET AUDITDATEVAR = AUDITDATE;
SET COMMENTSVAR = COMMENTS;
SET AUDIT_STATUSVAR = AUDITSTATUS;
INSERT INTO
        DB2INST1.AUDIT_DEPT (AUDITOR_NAME,DEPTNO,AUDIT_
        DATE,COMMENTS,AUDIT_STATUS)
VALUES(AUDITORVAR, DEPTNOVAR, AUDITDATEVAR, COMMENTSVAR, AUDIT_STATUSVAR);
SET VARRESULT=1;
END
@
```

CHAPTER 2   CONFIGURING JAVA CUSTOM COMPONENTS

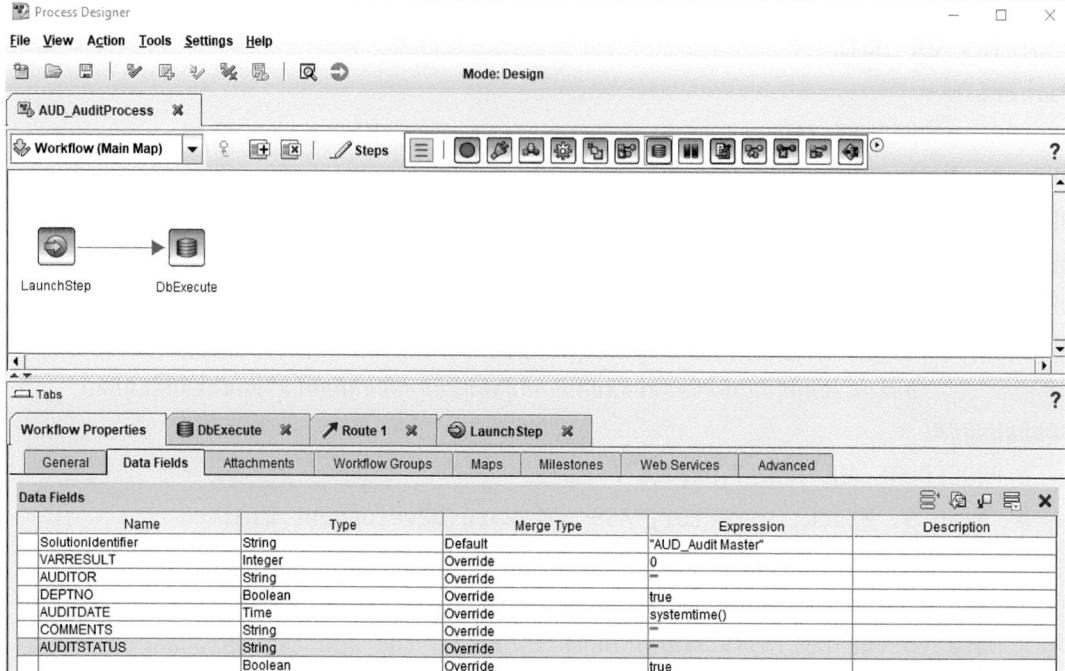

Figure 2-211.   The Data Fields are set up in the Workflow Properties tab for the AUD_AuditProcess Workflow

To support the transfer of the IBM Case Manager **Audit Master** solution **Audit Department** case data to the separate **auditdb** database server table **AUDIT_DEPT**, we need to define some Workflow Data Fields to hold the Case Data. These are set up with initial default values in the Workflow Properties, Data Fields tab shown in Figure 2-211.

CHAPTER 2 CONFIGURING JAVA CUSTOM COMPONENTS

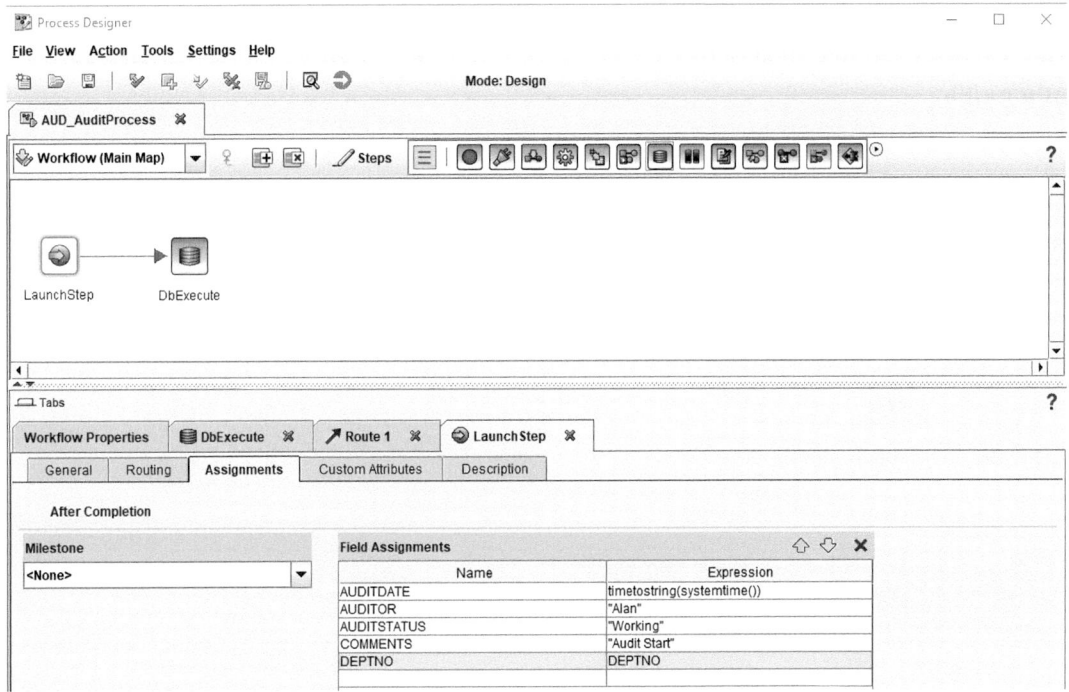

Figure 2-212. *The Workflow LaunchStep parameters are first set up with test data using local String constants*

The **LaunchStep** Field Assignment parameters are first set up with String values to test the Stored Procedure and Workflow during "debugging" as shown in Figure 2-212.

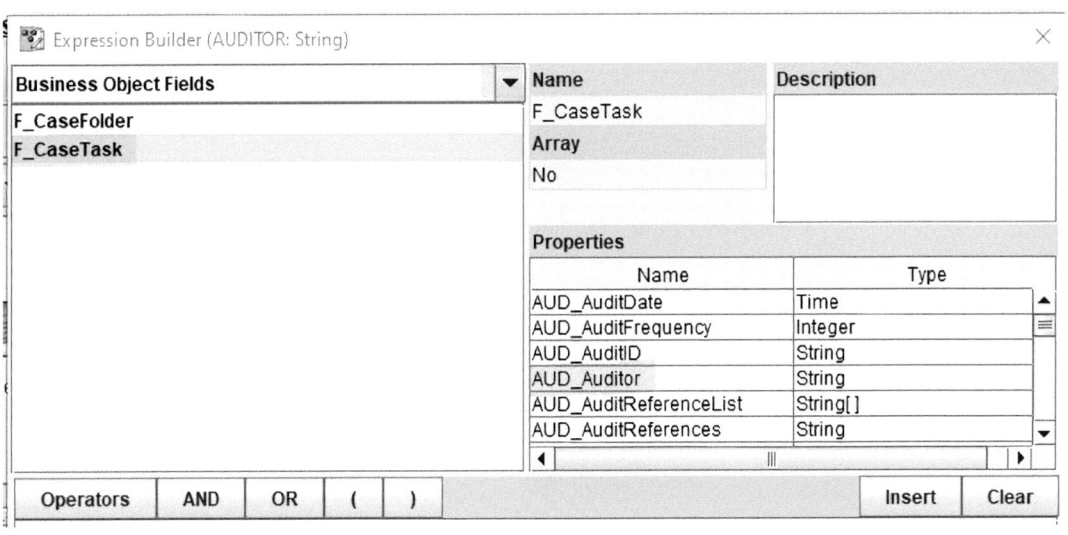

Figure 2-213. *The Case Task Fields are selected for the Field Assignments*

315

CHAPTER 2 CONFIGURING JAVA CUSTOM COMPONENTS

The Case Task Fields can be added next, after testing the Stored Procedure with fixed values for a "live" test of triggering the **DbExecute** step to populate the **AUDIT_DEPT** table.

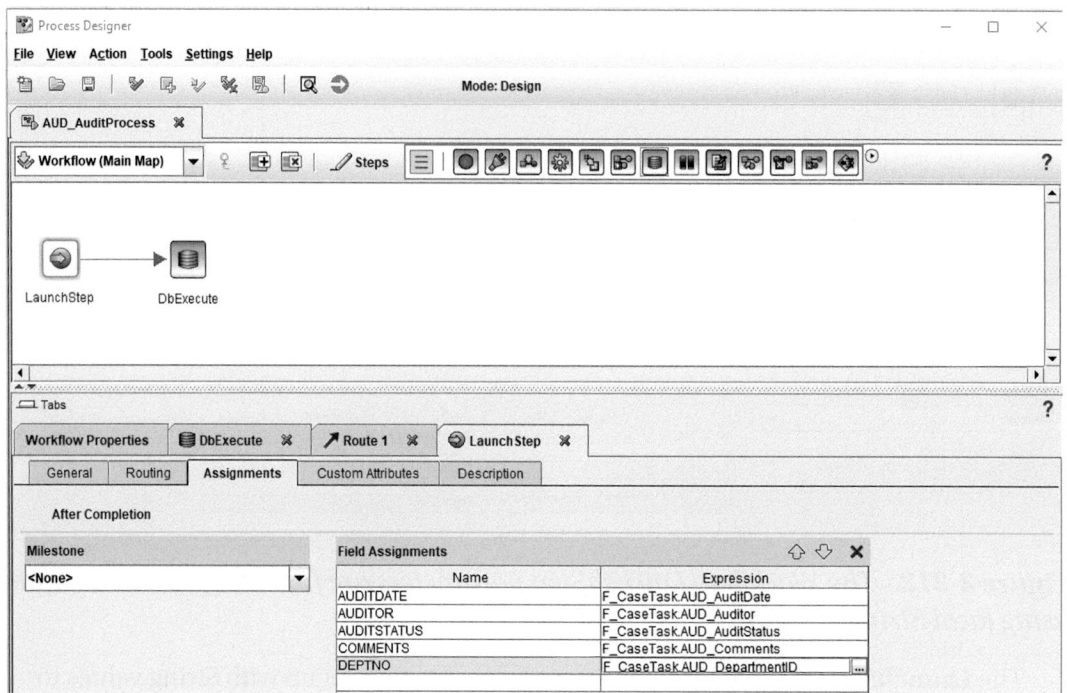

Figure 2-214. *The Field Assignments are now populated from the Case Task Business Object values*

CHAPTER 2 CONFIGURING JAVA CUSTOM COMPONENTS

In Figure 2-214, we now have the Field Assignments set with Case Task values for the Audit Case.

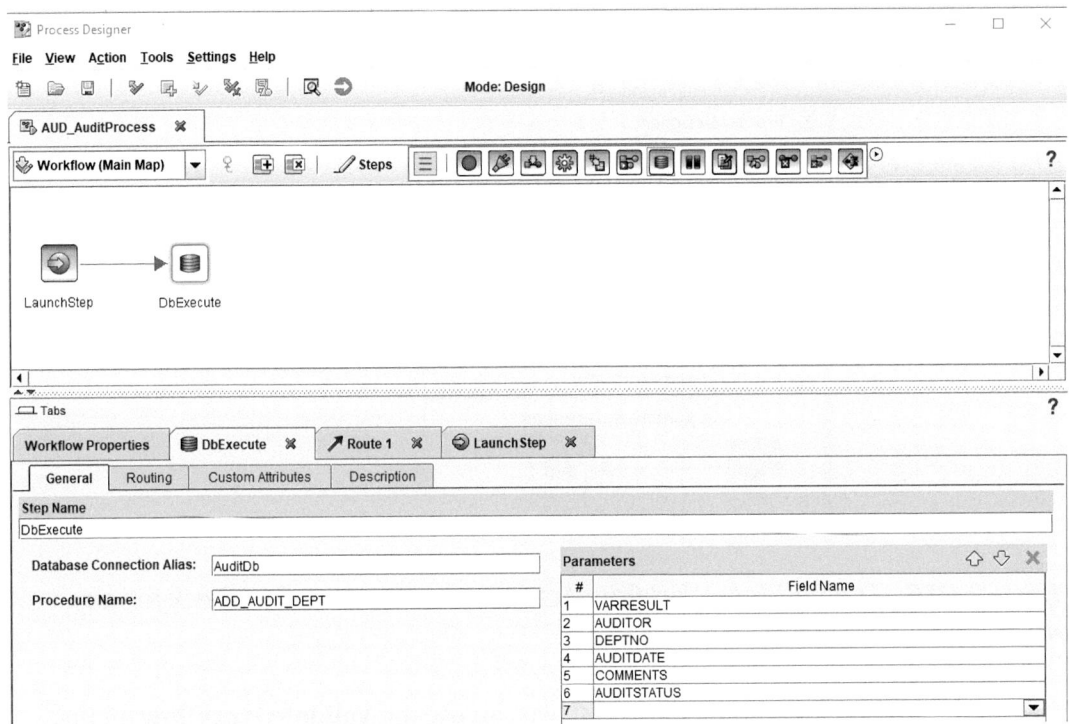

Figure 2-215. *The six DbExecute Stored Procedure parameters are mapped to the Workflow Data Fields*

317

After the **LaunchStep** Workflow Data Field Assignments are configured to the Case Fields, we can now select the **DbExecute** step and choose the Workflow Data Fields, in the order they are required to be used as **Parameters** for the Stored Procedure, as shown in Figure 2-215.

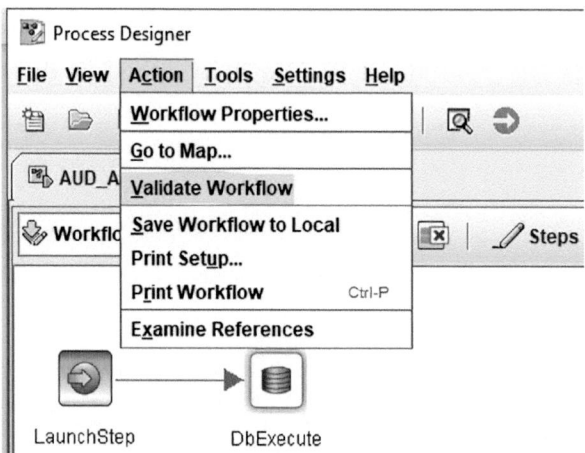

Figure 2-216. *The Validate Workflow Action menu is selected to check for any syntax errors*

Now the **DbExecute** step is completed, we can use the **Validate Workflow** menu action to ensure the Workflow "compiles" with no syntax errors (the workflow can either be stored as the industry standard **XPDL**, or there is an IBM FileNet P8 proprietary workflow **.pep** file structure which can be used).

CHAPTER 2 CONFIGURING JAVA CUSTOM COMPONENTS

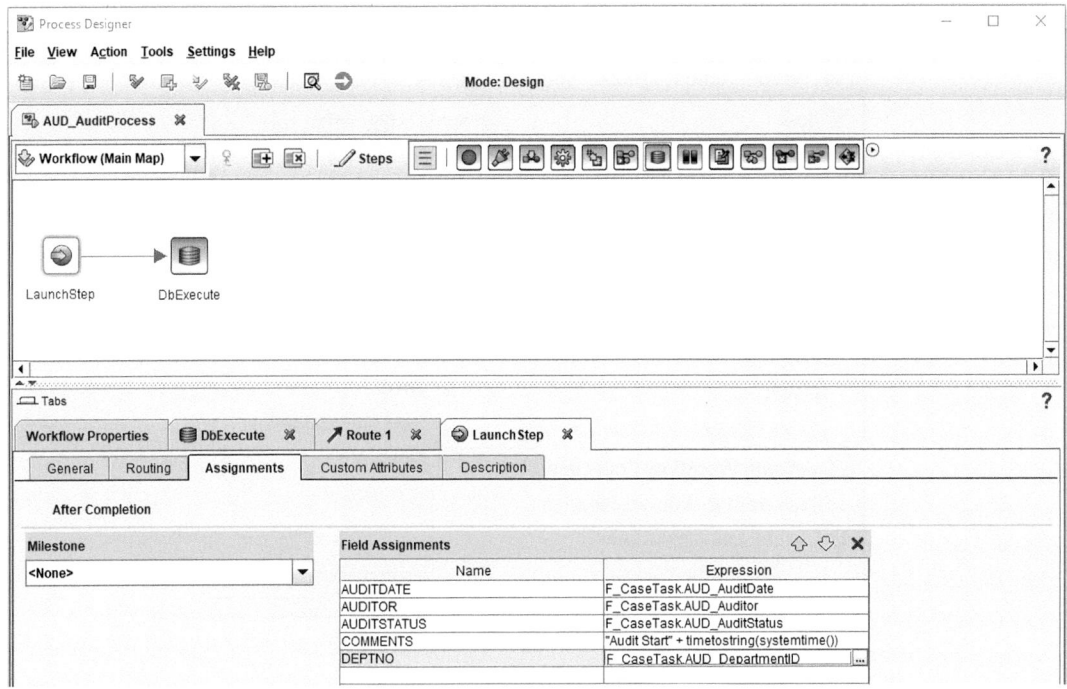

Figure 2-217. *The Field Assignments are updated to convert the Comments to add a date/time string*

A few minor changes were made to the Comments field to use one of the built-in Functions available in the Workflow design system, as shown in Figure 2-217.

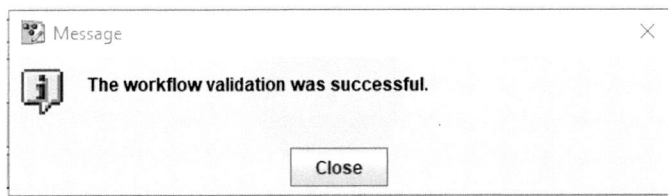

Figure 2-218. *The Workflow validation test was completed successfully*

319

CHAPTER 2   CONFIGURING JAVA CUSTOM COMPONENTS

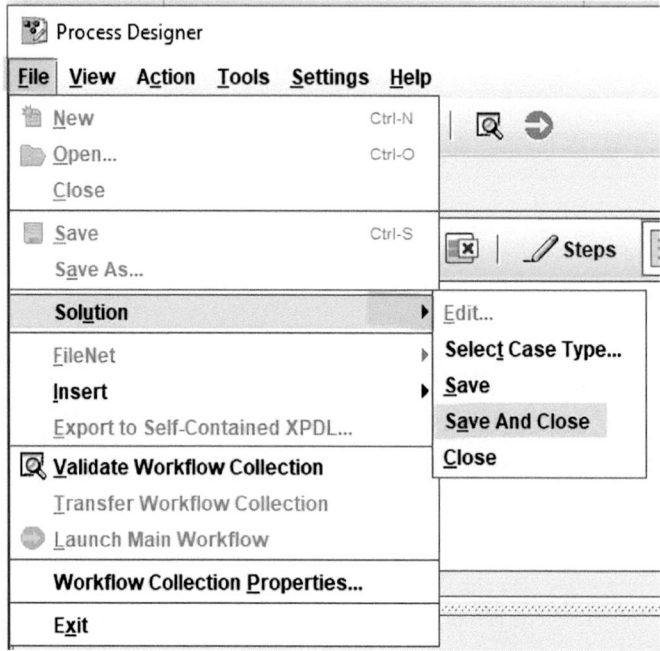

Figure 2-219. *The Workflow(s) for the Solution are saved for testing*

After the Workflow DbExecute step has been entered and validated, we can **Save and Close** the Audit Master Solution Task workflow we have created and redeploy the **Audit Master** solution using the IBM Case Builder web application as shown in Figure 2-220.

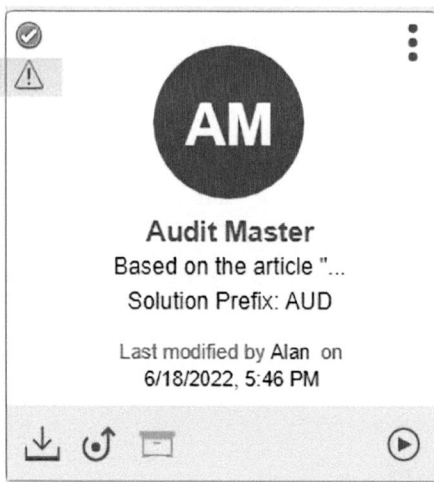

Figure 2-220. *The IBM Case Builder has detected that the Audit Master solution has new changes*

CHAPTER 2 CONFIGURING JAVA CUSTOM COMPONENTS

In the **IBM Case Builder**, we should see a **Warning** triangle (if a refresh of the Case Builder application has been conducted). This also shows the User and Date and Time that the solution was last modified.

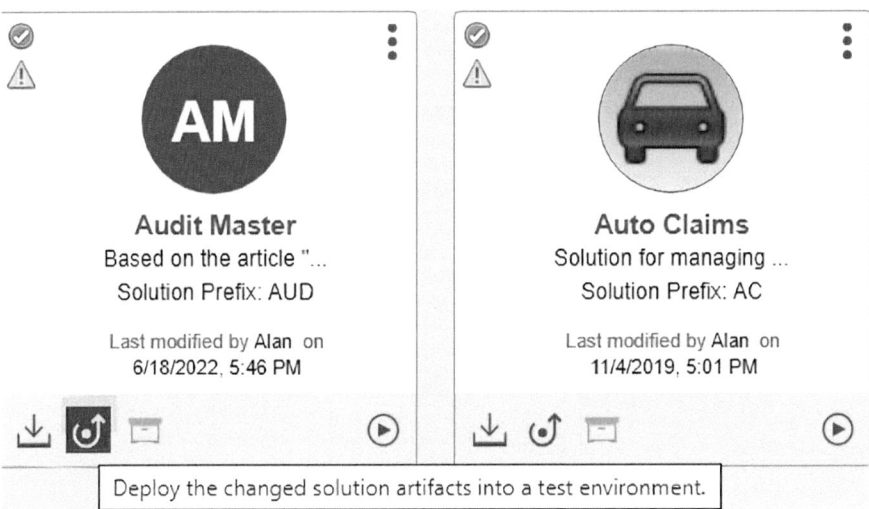

Figure 2-221. *The Deploy the changed solution icon is clicked*

We can now deploy the updated Audit Master solution, which now has a Task workflow with our DbExecute workflow step.

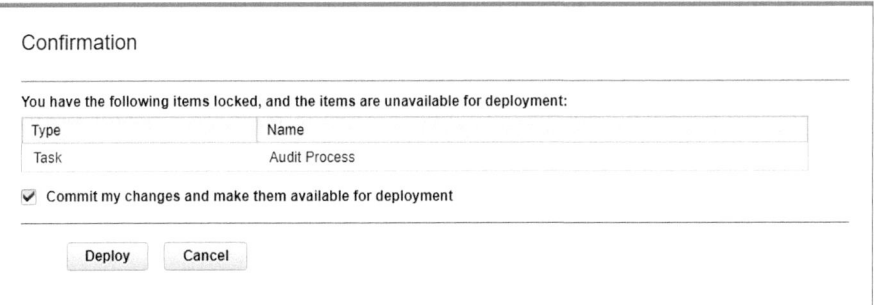

Figure 2-222. *The Deploy pop-up window shows the Audit Process Task has changes*

321

CHAPTER 2　CONFIGURING JAVA CUSTOM COMPONENTS

We now get a pop-up window which lists the detected changes which require deployment. Notice that the system locks the items as a warning, because the IBM Case Builder supports multiple Solution editors for the same Solution, as long as they are working on different areas of a Solution. In this case, we have a lock from the Standalone Process Designer.

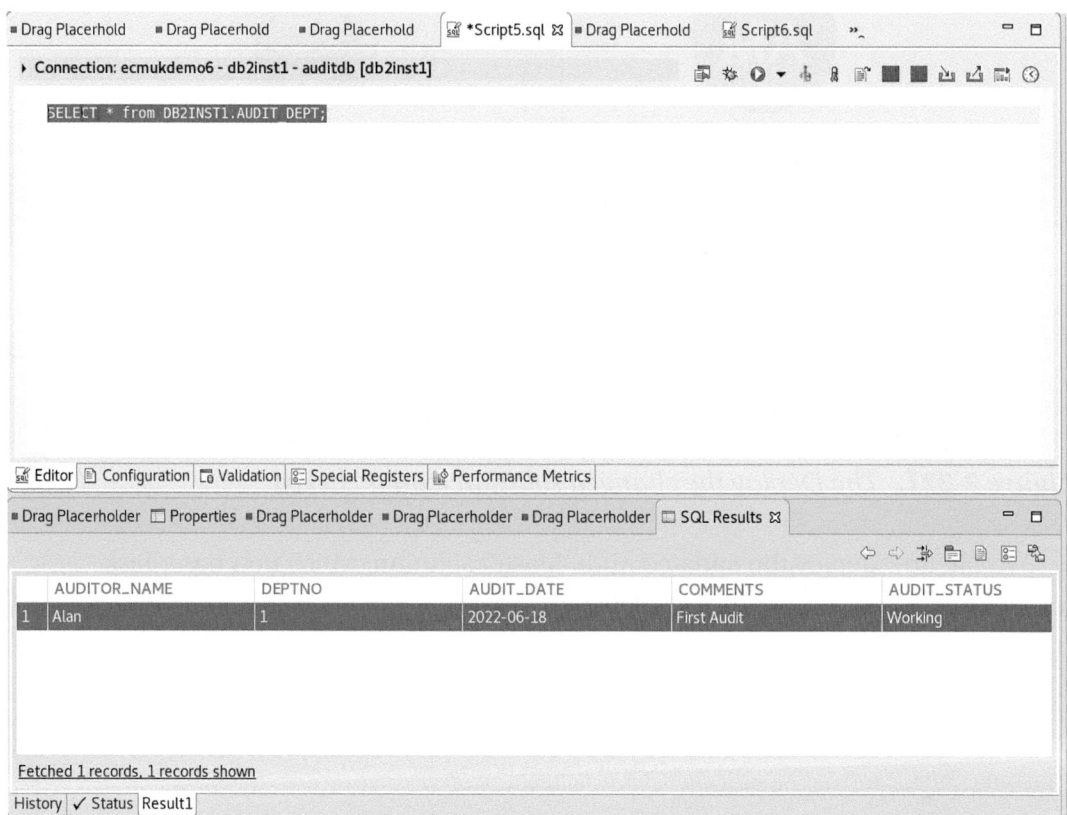

Figure 2-223. *The query from IBM Data Studio on the Linux server shows that there is a row in the table*

CHAPTER 2 CONFIGURING JAVA CUSTOM COMPONENTS

In the **IBM Case Builder**, we should see a **Warning** triangle (if a refresh of the Case Builder application has been conducted). This also shows the User and Date and Time that the solution was last modified.

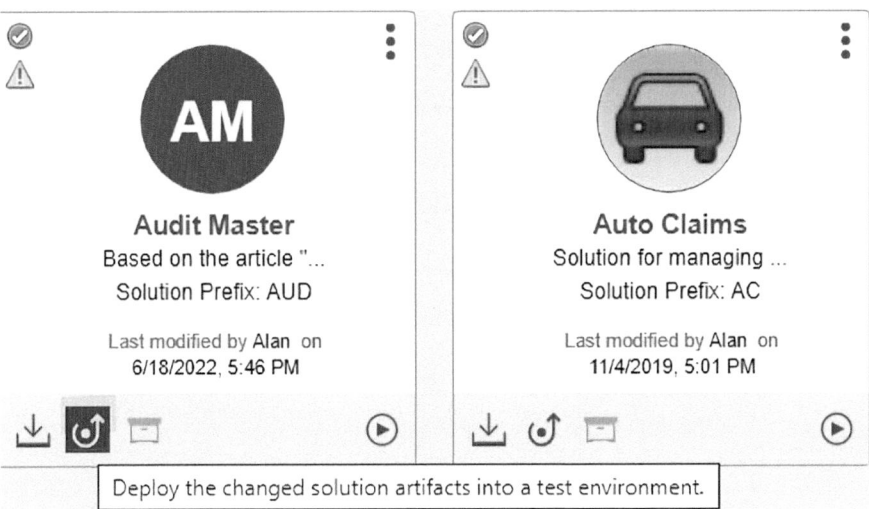

Figure 2-221. *The Deploy the changed solution icon is clicked*

We can now deploy the updated Audit Master solution, which now has a Task workflow with our DbExecute workflow step.

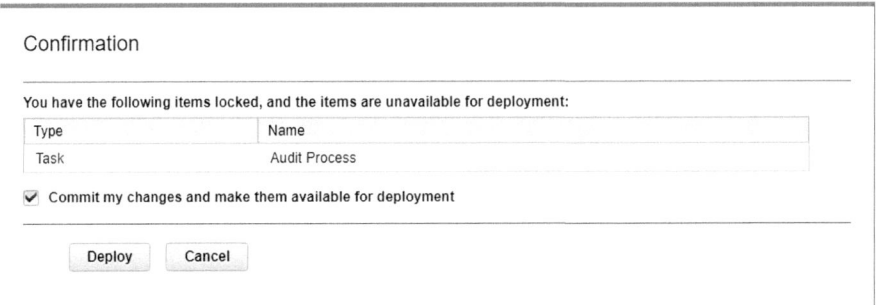

Figure 2-222. *The Deploy pop-up window shows the Audit Process Task has changes*

CHAPTER 2   CONFIGURING JAVA CUSTOM COMPONENTS

We now get a pop-up window which lists the detected changes which require deployment. Notice that the system locks the items as a warning, because the IBM Case Builder supports multiple Solution editors for the same Solution, as long as they are working on different areas of a Solution. In this case, we have a lock from the Standalone Process Designer.

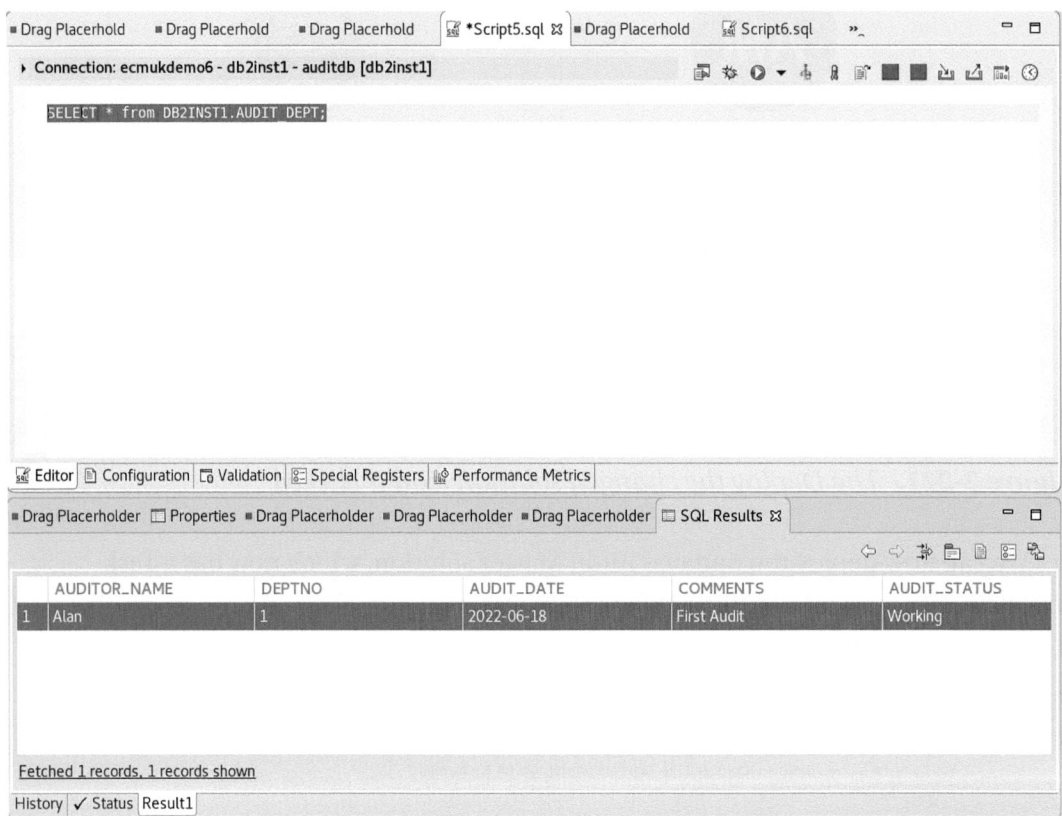

Figure 2-223. *The query from IBM Data Studio on the Linux server shows that there is a row in the table*

CHAPTER 2 CONFIGURING JAVA CUSTOM COMPONENTS

We can check the result of adding records into the **AUDIT_DEPT** table using the IBM Data Studio application. Initially, we have the record from the Test Insert SQL we ran earlier in the development.

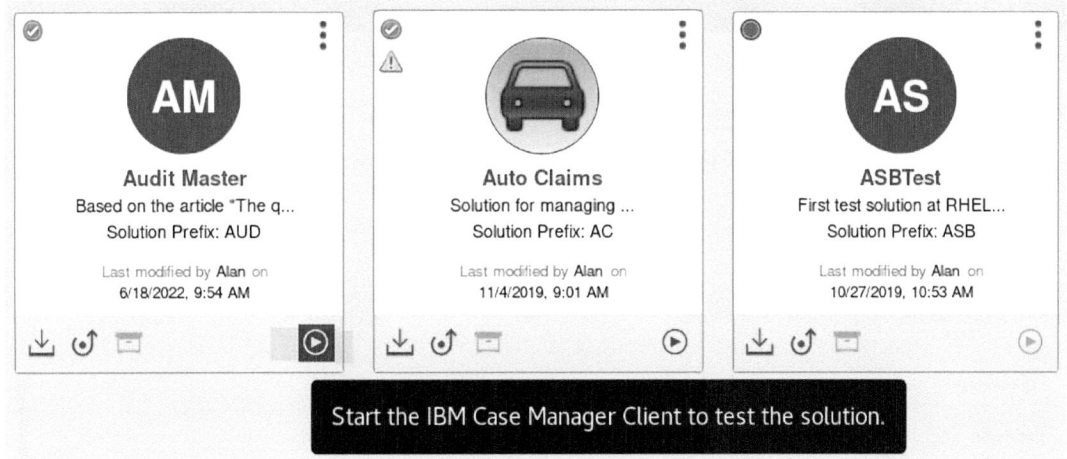

Figure 2-224. *The icon to start the IBM Case Manager Client is selected from the IBM Case Builder*

We can now create a new **Audit Department** Case from the deployed **Audit Master** solution which is now deployed to the Target Object Store (**OS2**). This can then be accessed to launch our Task workflow.

CHAPTER 2   CONFIGURING JAVA CUSTOM COMPONENTS

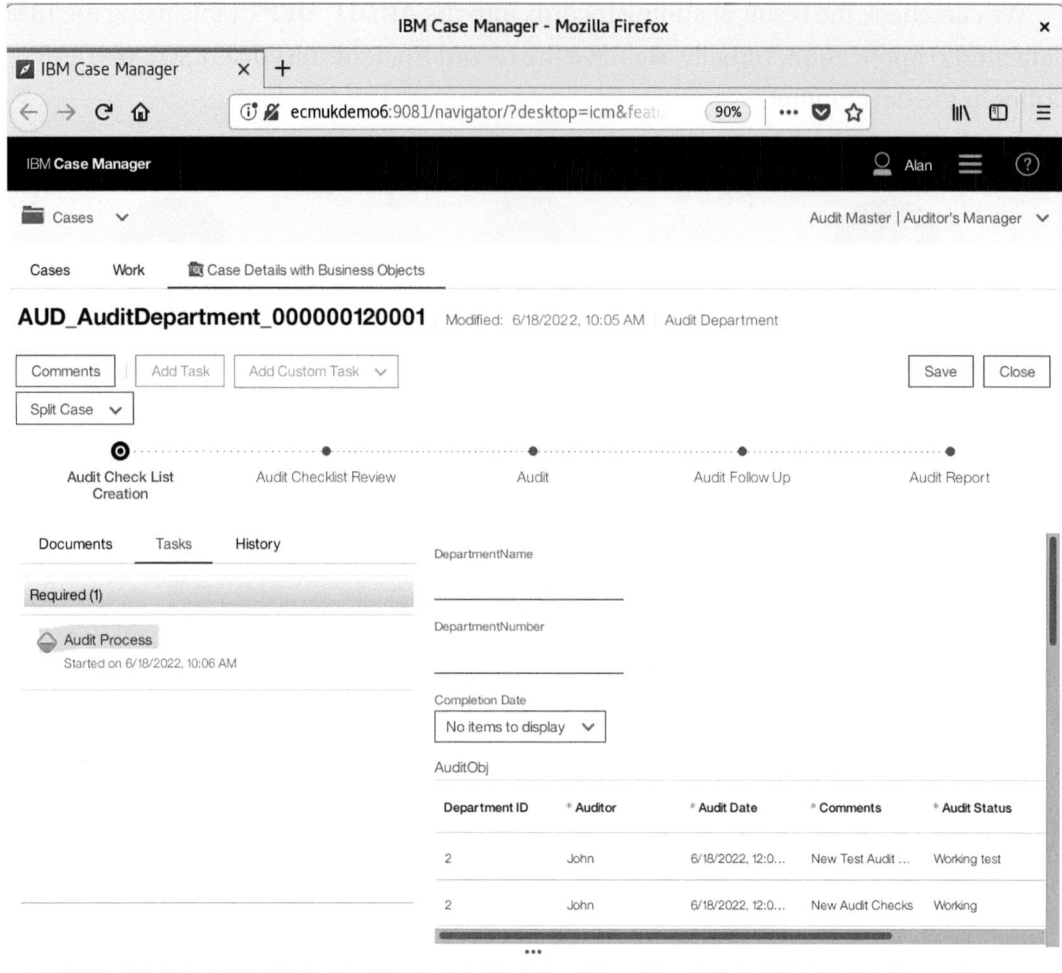

Figure 2-225. *The Audit Process Task is shown Started after selecting the manual Start link*

We can now start the Audit Process task as shown in Figure 2-225.

CHAPTER 2   CONFIGURING JAVA CUSTOM COMPONENTS

```
GRANT EXECUTE ON PROCEDURE ADD_AUDIT_DEPT TO PUBLIC


Query execution time => 4 ms

Script: file:/opt/IBM/DS4.1.3/Script6.sql
Database Name: auditdb
Authorization Id (Database): db2inst1
System/IP Address : ECMUKDEMO6/10.10.10.90
User Id (System) : root
```

Figure 2-226. *The Stored Procedure is required to have a GRANT on it to allow it to execute*

The first test showed that we need to grant security access on the ADD_AUDIT_DEPT stored procedure to give it execute permissions.

```
[db2inst1@ECMUKDEMO6 ~]$ db2 "call ADD_AUDIT_DEPT (1,'A S Bluck',00003,'2022-06-18','Test of the Procedure','Working')"

  Value of output parameters
  --------------------------
  Parameter Name  : VARRESULT
  Parameter Value : 1

  Parameter Name  : AUDITOR
  Parameter Value : A S Bluck

  Parameter Name  : DEPTNO
  Parameter Value : 00003

  Parameter Name  : AUDITDATE
  Parameter Value : 2022-06-18

  Parameter Name  : COMMENTS
  Parameter Value : Test of the Procedure

  Parameter Name  : AUDITSTATUS
  Parameter Value : Working

  Return Status = 0
[db2inst1@ECMUKDEMO6 ~]$
```

Figure 2-227. *The Stored Procedure was tested using a db2 call to show that it would function correctly*

After the Security update, we used the **db2** command line to launch the **ADD_AUDIT_DEPT** stored procedure to check that it was functioning correctly. This is shown using the db2 command in Figure 2-227. Notice that on the Linux system, we have to wrap the **call** command in double quotes, as follows:

db2 "call ADD_AUDIT_DEPT (1, 'A S Bluck',00003,'2022-06-18','Test of the Procedure','Working')"

CHAPTER 2   CONFIGURING JAVA CUSTOM COMPONENTS

DB2INST1.AUDIT_DEPT				
AUDITOR_NAME [CHAR(40 OCTETS)]	DEPTNO [INTEGER]	AUDIT_DATE [DATE]	COMMENTS [CHAR(200 OCTETS)]	AUDIT_STATUS [CHAR(1
1 Alan	1	2022-06-18	First Audit	Working
2 A S Bluck DB2INST1	3	2022-06-18	Test of the Procedure	Working

Figure 2-228. *The row is added from the db2 call test for the ADD_AUDIT_DEPT stored procedure*

We can now check that the Table row has been inserted into the **AUDIT_DEPT** table, which can be seen in Figure 2-228.

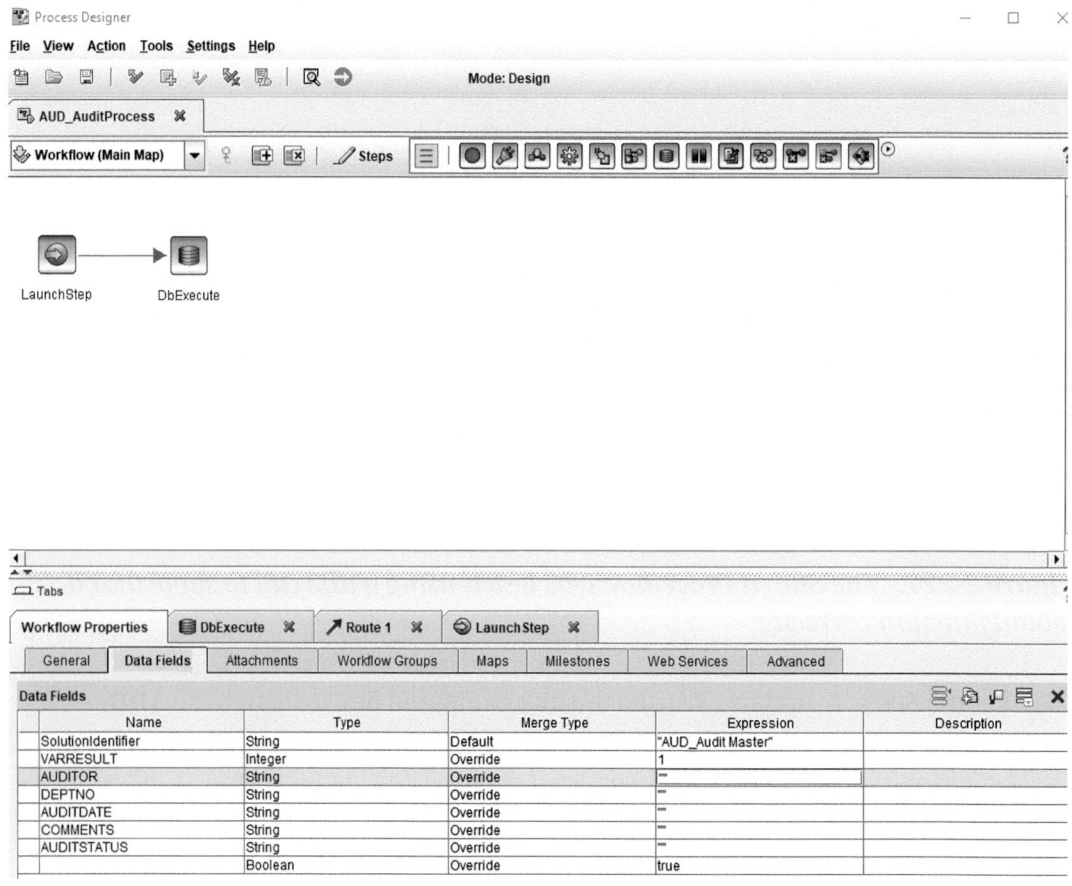

Figure 2-229. *The Data Fields are configured in the Workflow*

The Data Fields are all initialized now as String values, except for the VARRESULT status parameter, which is left as an Integer.

326

CHAPTER 2 CONFIGURING JAVA CUSTOM COMPONENTS

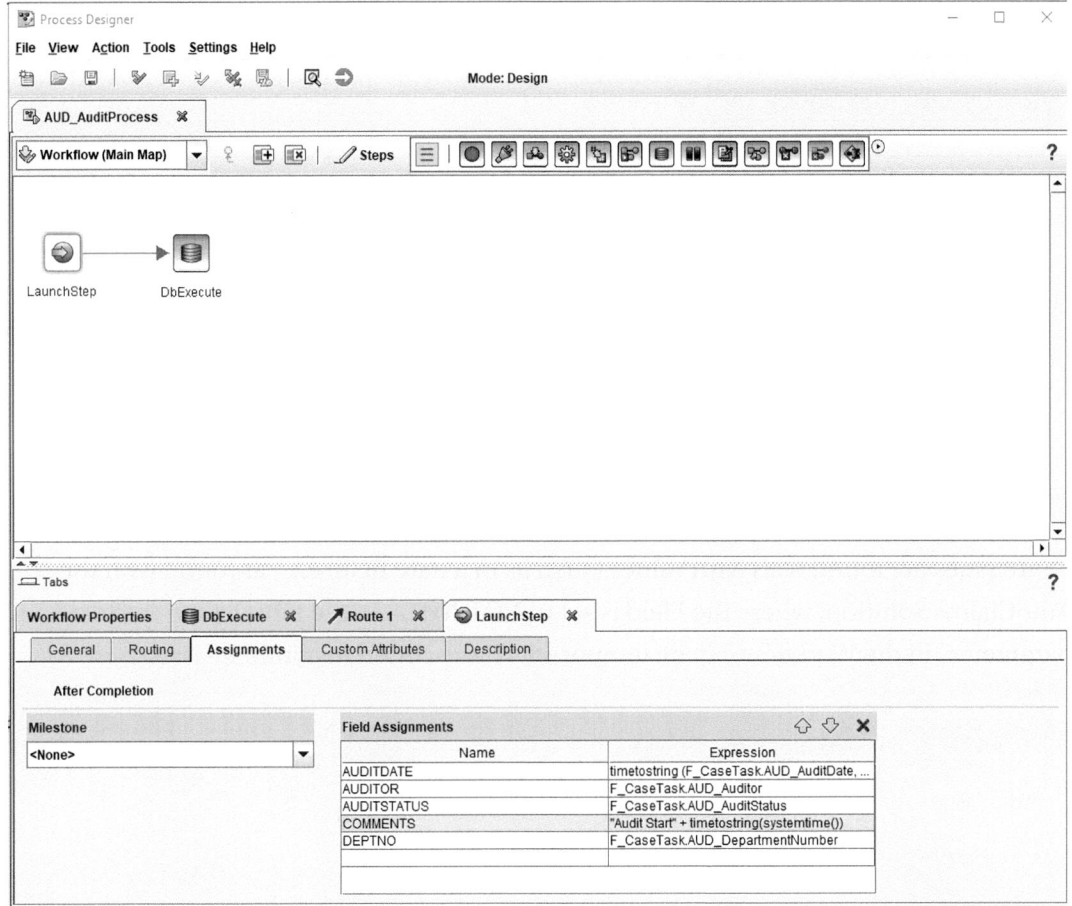

Figure 2-230. *The Data Field Assignments from the LaunchStep are now set*

We can run a first test using the local String variables to eliminate any Null field issues.

CHAPTER 2 CONFIGURING JAVA CUSTOM COMPONENTS

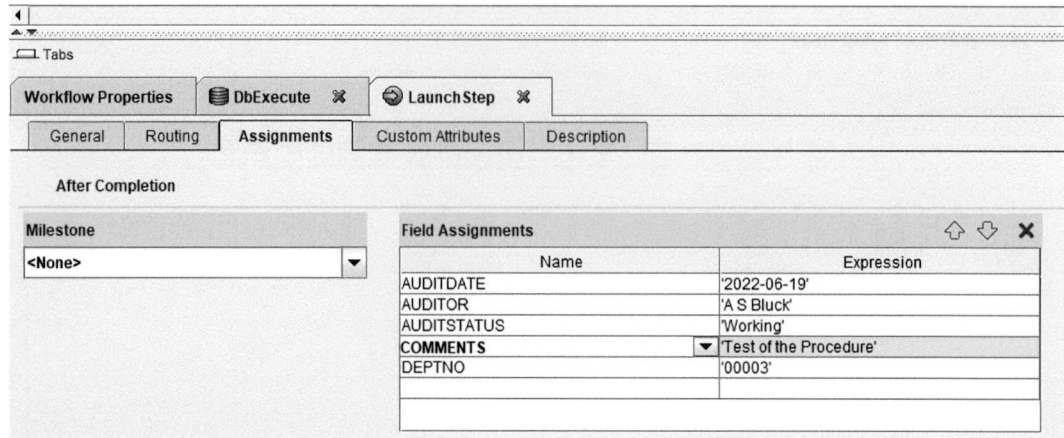

Figure 2-231. *The first test using hard-coded values for the parameters*

If issues are found with **Null** values (such as in a date field, e.g., as found with the AutoClaims Solution, where the Field is set to **NOT NULL** in the table), then the test parameters in the Workflow can be temporarily hard-coded to eliminate issues.

```
[db2inst1@ECMUKDEMO6 ~]$ db2 call ADD_AUDIT_DEPT (1,'A S Bluck',00099,'2022-06-19','Test 2 of the Procedure','Issues')
bash: syntax error near unexpected token `('
[db2inst1@ECMUKDEMO6 ~]$ db2 "call ADD_AUDIT_DEPT (1,'A S Bluck',00099,'2022-06-19','Test 2 of the Procedure','Issues')"

  Value of output parameters
  --------------------------
  Parameter Name  : VARRESULT
  Parameter Value : 1

  Parameter Name  : AUDITOR
  Parameter Value : A S Bluck

  Parameter Name  : DEPTNO
  Parameter Value : 00099

  Parameter Name  : AUDITDATE
  Parameter Value : 2022-06-19

  Parameter Name  : COMMENTS
  Parameter Value : Test 2 of the Procedure

  Parameter Name  : AUDITSTATUS
  Parameter Value : Issues

  Return Status = 0
[db2inst1@ECMUKDEMO6 ~]$
```

Figure 2-232. *The Stored Procedure call is retested after minor changes*

Note In the preceding db2 call on the Linux server, the Stored Procedure call has to be surrounded by double quotes to be executed with the db2 command-line processor.

CHAPTER 2 CONFIGURING JAVA CUSTOM COMPONENTS

Testing on Different Database Platforms

If you want to compare SQL statements in different databases, a website called SQL Fiddle allows a developer to switch between different database providers.

It supports a drop-down for **MySQL**, **PostgreSQL**, **MS SQL Server**, and **SQLite**, which you can run test some SQL queries and the syntax against.

This is very useful if a database platform would take time to install and configure and you have time constraints. Using SQL Fiddle, you just create your database and can run the queries in the web-based application!

See the following link:

`http://sqlfiddle.com`

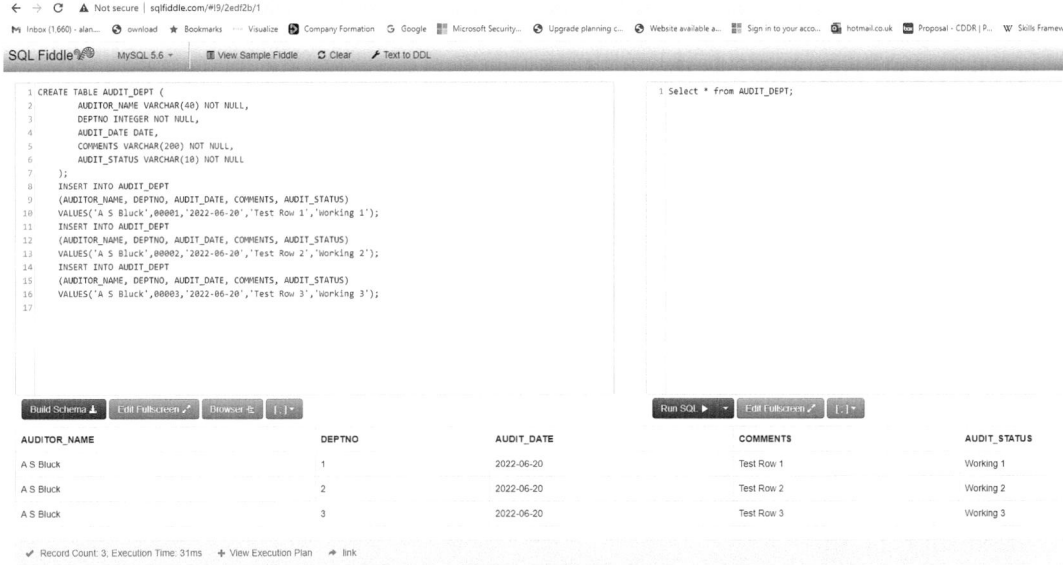

***Figure 2-233.** The sqlfiddle.com free web application SQL test environment*

CHAPTER 2 CONFIGURING JAVA CUSTOM COMPONENTS

Oracle Version of the Stored Procedure SQL Code

Generate the Stored Procedure for DbExecute using Oracle.

Listing 2-9. The Oracle example stored procedure

```
CREATE OR REPLACE PROCEDURE OS2.ADD_AUDIT_DEPT (
VARRESULT IN OUT INT,
AUDITOR IN OUT VARCHAR2,
DEPTNO IN OUT INT,
AUDITDATE IN OUT VARCHAR2,
COMMENTS IN OUT VARCHAR2,
AUDIT_STATUS IN OUT VARCHAR2
)
-- # PROCEDURE ADD_AUDIT_DEPT.
-- #  Alan S. Bluck, Director, ASB Software Development Limited
-- #
-- # This stored procedure writes a row of Audit Department
-- # data to the OS2.AUDIT_DEPT table in the OS2 database
-- #
-- # NB Any date fields which are set empty on initial Add of a CASE
-- #    may cause insert failure!
IS
   v_STATUS   INT;
BEGIN
   VARRESULT := 0;
   INSERT INTO OS2.AUDIT_DEPT
 (AUDITOR_NAME,DEPTNO,AUDIT_DATE,COMMENTS,AUDIT_STATUS)
   VALUES (AUDITOR, DEPTNO, AUDITDATE, COMMENTS, AUDIT_STATUS
);
   VARRESULT := 1;
END;
/
Test run the procedure above using:
SQL> variable CALLSTATUS NUMBER;
```

```
SQL> execute OS2.ADD_AUDIT_DEPT ('A S Bluck','00001','2022-06-19', 'Test
Stored Procedure','Working',:CALLSTATUS);
PL/SQL procedure successfully completed.
SQL> print CALLSTATUS;
```
CALLSTATUS

    1

MS SQL Server Version of the Stored Procedure SQL Code

Generate the Stored Procedure for DbExecute using MS SQL Server.

Listing 2-10. The MS SQL Server example stored procedure

```
SET ANSI_NULLS ON
GO
SET QUOTED_IDENTIFIER ON
GO
IF EXISTS (SELECT * FROM sys.objects WHERE object_id = OBJECT_ID(N'ADD_
AUDIT_DEPT') AND type in (N'P', N'PC')) DROP PROCEDURE ADD_AUDIT_DEPT;
GO
CREATE PROCEDURE ADD_AUDIT_DEPT
     @VARRESULT BIGINT,
    @AUDITOR nvarchar(10),
    @DEPTNO nvarchar BIGINT,
    @AUDITDATE nvarchar(20),
    @COMMENTS nvarchar(20),
    @AUDIT_STATUS nvarchar(256)
-- # PROCEDURE ADD_AUDIT_DEPT.
-- #  Alan S. Bluck, Director, ASB Software Development Limited
-- #
-- # This stored procedure writes a row of Audit Department
-- # data to the DBO.AUDIT_DEPT table in the auditdb database
-- #
```

CHAPTER 2 CONFIGURING JAVA CUSTOM COMPONENTS

```
-- # NB Any date fields which are set empty on initial Add of a CASE
-- #    may cause insert failure!
AS
BEGIN
    --DECLARE THE VARIABLES
    DECLARE @VARRESULT BIGINT
    ,@AUDITORVAR VARCHAR(10)
    ,@DEPTNOVAR BIGINT
    ,@AUDITDATEVAR VARCHAR(20)
    ,@COMMENTSVAR VARCHAR(20)
    ,@AUDIT_STATUSVAR VARCHAR(256)

    SET @AUDITORVAR = @AUDITOR
     SET @DEPTNOVAR = @DEPTNO
     SET @AUDITDATEVAR = @AUDITDATE
     SET @COMMENTSVAR = @COMMENTS
     SET @AUDIT_STATUSVAR = @AUDIT_STATUS
   BEGIN
        BEGIN TRANSACTION
Insert into dbo.AUDIT_DEPT
(AUDITOR_NAME,DEPTNO,AUDIT_DATE,COMMENTS,AUDIT_STATUS)
 VALUES (@AUDITORVAR, @DEPTNOVAR, @AUDITDATEVAR, @COMMENTSVAR, @AUDIT_STATUSVAR);
        COMMIT TRANSACTION
END;
```

Test run the preceding procedure using

```
EXEC dbo.ADD_AUDIT_DEPT 1,N'A S Bluck', N'2022-06-19',N'Test Run',N'Working';
```

Chapter 2 Exercises

The following questions cover the functions using Code Module Events, IBM Process Designer, and DbExecute, which we covered in this chapter.

CHAPTER 2   CONFIGURING JAVA CUSTOM COMPONENTS

MULTIPLE CHOICE QUESTIONS

1. What IBM application would you use to confirm that the "CodeModules" folder exists?

   a) The IBM Content Navigator

   b) The IBM acce Administrative Console for Content Platform Engine

   c) The IBM Case Builder

   d) The IBM Process Designer

2. An IBM Stored Procedure String parameter for the Oracle Database system is limited to

   a) 64 characters

   b) 1333 characters

   c) 4000 characters

   d) 1024 characters

3. The IBM Standalone Process Designer Tool is initially installed using the program

   a) cmd.exe

   b) pedesigner.bat

   c) 5.5.0-ICFCPE-WIN.EXE

   d) cpetoolenv.bat

4. The main purpose of a Workflow Subscription is

   a) To enable the creation of a Folder Object

   b) To enable the creation of a Document Object

   c) To enable a workflow to be launched on the storage of a document

   d) To enable a Case to be created in IBM Case Manager

CHAPTER 2   CONFIGURING JAVA CUSTOM COMPONENTS

MULTIPLE CHOICE ANSWERS

1. b) The IBM acce Administrative Console for Content Platform Engine

2. c) 4000 characters

3. c) 5.5.0-ICFCPE-WIN.EXE

4. c) To enable a workflow to be launched on the storage of a document

QUESTIONS

1. Describe what steps you would use to install a new Custom Event, given that you have been given a working **fn_eventshandler.jar** file and have been asked to configure an **AddFolde**r Java Class Event Handler triggered on the creation of a new Folder.

2. What databases does IBM Case Manager Workflow support and what types of Datasource are recommended to be configured to support a connection for use with DbExecute?

3. What methods are used for resolving issues in the deployment of an IBM Case Manager Solution?

In Chapter 3, we will cover the use of the Java language to customize Workflow Components with the example development of a Java jar file for the IBM Java Messaging Service calls and its deployment for use in an IBM Case Manager Workflow.

CHAPTER 2   CONFIGURING JAVA CUSTOM COMPONENTS

MULTIPLE CHOICE QUESTIONS

1. What IBM application would you use to confirm that the "CodeModules" folder exists?

   a) The IBM Content Navigator

   b) The IBM acce Administrative Console for Content Platform Engine

   c) The IBM Case Builder

   d) The IBM Process Designer

2. An IBM Stored Procedure String parameter for the Oracle Database system is limited to

   a) 64 characters

   b) 1333 characters

   c) 4000 characters

   d) 1024 characters

3. The IBM Standalone Process Designer Tool is initially installed using the program

   a) cmd.exe

   b) pedesigner.bat

   c) 5.5.0-ICFCPE-WIN.EXE

   d) cpetoolenv.bat

4. The main purpose of a Workflow Subscription is

   a) To enable the creation of a Folder Object

   b) To enable the creation of a Document Object

   c) To enable a workflow to be launched on the storage of a document

   d) To enable a Case to be created in IBM Case Manager

CHAPTER 2   CONFIGURING JAVA CUSTOM COMPONENTS

MULTIPLE CHOICE ANSWERS

1. b) The IBM acce Administrative Console for Content Platform Engine
2. c) 4000 characters
3. c) 5.5.0-ICFCPE-WIN.EXE
4. c) To enable a workflow to be launched on the storage of a document

QUESTIONS

1. Describe what steps you would use to install a new Custom Event, given that you have been given a working **fn_eventshandler.jar** file and have been asked to configure an **AddFolder** Java Class Event Handler triggered on the creation of a new Folder.

2. What databases does IBM Case Manager Workflow support and what types of Datasource are recommended to be configured to support a connection for use with DbExecute?

3. What methods are used for resolving issues in the deployment of an IBM Case Manager Solution?

In Chapter 3, we will cover the use of the Java language to customize Workflow Components with the example development of a Java jar file for the IBM Java Messaging Service calls and its deployment for use in an IBM Case Manager Workflow.

CHAPTER 3

IBM JMS Interface Development IBM FileNet 5.5.x Workflow

This chapter describes an example development of a Java jar file for the IBM Java Messaging Service calls and its deployment for use in an IBM Case Manager Workflow. See https://doi.org/10.13140/RG.2.2.21708.16001.

For the installation steps required for IBM Case Manager 5.3.3, which are covered step by step in the following free ResearchGate documents, downloaded using the DOI URLs as follows:

entitled *"Case Manager 5.3.3 Installation on RHEL 8.0 with Content Navigator 3.0.6"* Click the **Download file PDF** command button in Figure 3-1.

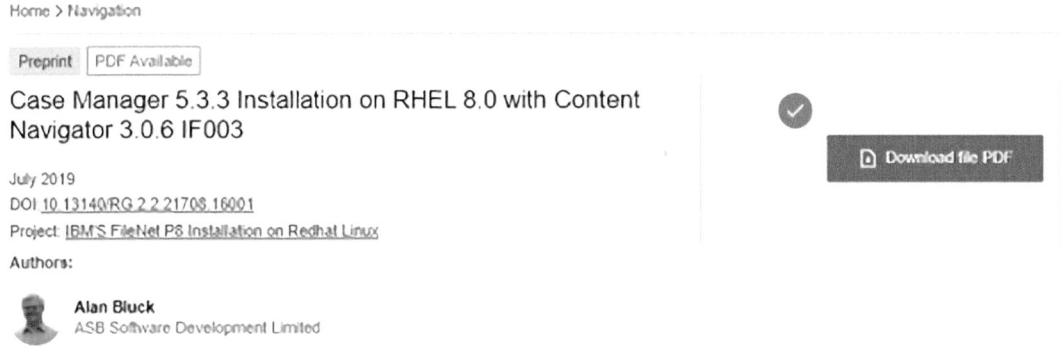

Figure 3-1. *Download the IBM Case Manager Installation Document*

This gives the downloaded document, entitled *CaseManagerInstallationonRHEL8.0_V3.docx*.

CHAPTER 3   IBM JMS INTERFACE DEVELOPMENT IBM FILENET 5.5.X WORKFLOW

Note The Audit system is based on an article in the journal, Quality Forum, Volume 19 No. 3, September 1993. *The Journal of the Institute of Quality Assurance.* Pages 116–126. The Quality Audit Process, by Alan S. Bluck (https://www.researchgate.net/publication/334095565_Case_Manager_Solution_Development#fullTextFileCssontent).

This chapter includes the steps for the development of a prototype system component, with code working with a Test harness including the development of a process for deploying and testing an **AUDOperations.jar** CI (Component Integration) jar file for use in a Workflow system step.

The IBM MQ Series is installed on a Linux VMware and requires the latest Fix Packs. After installation, we describe the setup test queues and the configuration of a JNDI link using a Tivoli Directory Services LDAP for LDAP context lookup.

An IBM MQ server is an installation supporting one or more queue managers that provide message queuing services to one or more clients. All the IBM MQ objects, for example, queues, exist only on the queue manager server. The MQ client does not store any MQ Objects. An IBM MQ MQI client is available which allows an application running on one system server to communicate with a queue manager running on another system server.

Note During development, it was found that the com.ibm.mq.jms.MQQueue cannot be cast to javax.jms.QueueConnectionFactory.

Chapter Organization

This chapter contains the following six Parts:

Part 1 – Bill of Materials. This Part lists the prerequisite IBM Software components, including the IBM product code numbers, required to download the latest IBM MQ Series software components, used to test the Java JMS message service API calls. It also covers the installation of the downloaded software on a Red Hat Linux server.

Part 2 – Custom Operations Component Development. The Custom Operations Component is defined in the subscribed Workflow as a special Component Integrator step, which we will describe in this Part.

Part 3 – Creating a Non-privileged MQ User for a Client Application Connection. This Part describes a procedure to create a non-privileged user ID to be used for a client application which connects to the queue manager. Access is granted for the client application only to be able to use the channel it needs and the queue it needs by use of this user ID.

Part 4 – Second Run of the AUDOperationsTest.java JUnit for JMS Messaging. This Part covers the second run of the JUnit test, which is more successful. The console screen output shown in Figure 3-232 is from the Eclipse IDE JUnit run of the AUDOperationsTest.java Class.

This shows the transmission of a test message to the IBM MQ Connection Factory, AUDCF3 queue, and the retrieval of the ten Audit Documents in the output list from the Audit Master Case Manager solution, Target Object Store.

Part 5 – Building the AUDOperations.jar. This Part describes the method for exporting the Java program in the AUDOperations.jar file with the exposed methods we wish to call from our IBM Process Designer Workflow.

Part 6 – Transferring Workflow and Setting Up Workflow Subscriptions. In this Part, the Target Object store, OS2, is updated with a Workflow Definition document class (AUD_JMSMessage.pep file), which is used to run the Workflow we created, triggered by a new Audit Report document created in the Target Object store from an Audit Master solution Case Manager Audit Department Case. The acce web Administration console is used to create the required Workflow Subscription.

Part 1 – Bill of Materials

The following URL link describes the latest available IBM MQ Series 9.x downloads and provides links to the versions of IBM MQ available for download:

`www.ibm.com/docs/en/ibm-mq/9.2?topic=roadmap-mq-downloads`

The next link displays the available pdf documentation for IBM MQ:

`www.ibm.com/docs/en/ibm-mq/9.2?topic=am-mq-92-pdf-files-product-documentation-program-directories`

CHAPTER 3   IBM JMS INTERFACE DEVELOPMENT IBM FILENET 5.5.X WORKFLOW

The hardware and software requirements for IBM MQ series are described at this URL link:

www.ibm.com/support/pages/node/318077

An overview of the installation procedure for IBM MQ 9.2 can be seen at this link:

www.ibm.com/docs/en/ibm-mq/9.2?topic=uninstalling-installing-mq-linux

The development of a Java JMS message queue for Linux, Windows, and IBM Cloud is described in the following link:

https://developer.ibm.com/tutorials/mq-develop-mq-jms/

The IBM MQ Series Version 9.2 for Continuous Deployment full installation requires 2010 MBytes of disk space. This version is now released as a **Continuous Delivery (CD)** package offer with regular update packages. Each update incrementally delivers new capability and defect fixes (compared with Long-Term Support (LTS) packages which just receive security and defect fixes through fix packs and individual fixes with no incremental functional enhancements).

Product details

File name	IBM_MQ_9.2.4_LINUX_X86-64.tar.gz
Platform(s)	RHEL 7, RHEL 8, SLES 12 LE, SLES 12.4, SLES 12.5, SLES 15, SLES 15.1, SLES 15.2, SLES12.2, SLES12.3
Language(s)	Chinese Simplified, Czech, English U.S., French, German, Italian, Japanese, Korean, Polish, Portuguese Brazilian, Russian, Spanish

Estimated download duration

File size	923,318,779 bytes		
Connection		Download Director	Http transfer
56K Modem		2142 minutes	2142 minutes
256K Cable Modem		191 minutes	479 minutes
T1		32 minutes	80 minutes

Figure 3-2. *Product details of the IBM MQ 9.2.4 Linux version including download times*

The version and product code details for the installations are as follows:
IBM SDK, Java (TM) Technology Edition, Version 8 for Linux (CND18ML)
IBM MQ V9.2.5 Continuous Delivery Release for Windows 64-bit Multilingual (M0458ML)

CHAPTER 3   IBM JMS INTERFACE DEVELOPMENT IBM FILENET 5.5.X WORKFLOW

IBM MQ V9.2.x Continuous Delivery Quick Start Guide (G014JML)

IBM MQ V9.2.5 Continuous Delivery Release for Linux on x86 64-bit Multilingual (M0454ML)

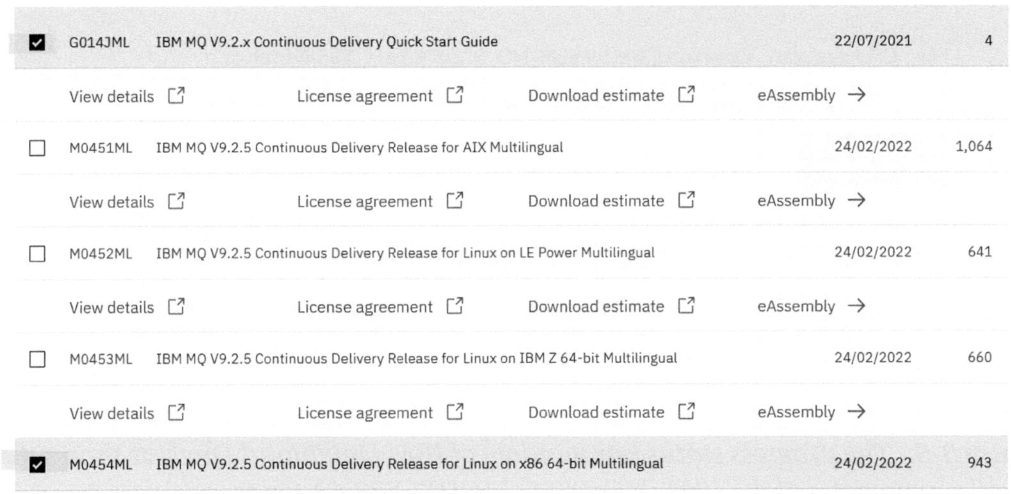

☑	G014JML	IBM MQ V9.2.x Continuous Delivery Quick Start Guide			22/07/2021	4
	View details	License agreement	Download estimate	eAssembly →		
☐	M0451ML	IBM MQ V9.2.5 Continuous Delivery Release for AIX Multilingual			24/02/2022	1,064
	View details	License agreement	Download estimate	eAssembly →		
☐	M0452ML	IBM MQ V9.2.5 Continuous Delivery Release for Linux on LE Power Multilingual			24/02/2022	641
	View details	License agreement	Download estimate	eAssembly →		
☐	M0453ML	IBM MQ V9.2.5 Continuous Delivery Release for Linux on IBM Z 64-bit Multilingual			24/02/2022	660
	View details	License agreement	Download estimate	eAssembly →		
☑	M0454ML	IBM MQ V9.2.5 Continuous Delivery Release for Linux on x86 64-bit Multilingual			24/02/2022	943

Figure 3-3. *Product details for the download of the IBM MQ V9.2.5 installation packages*

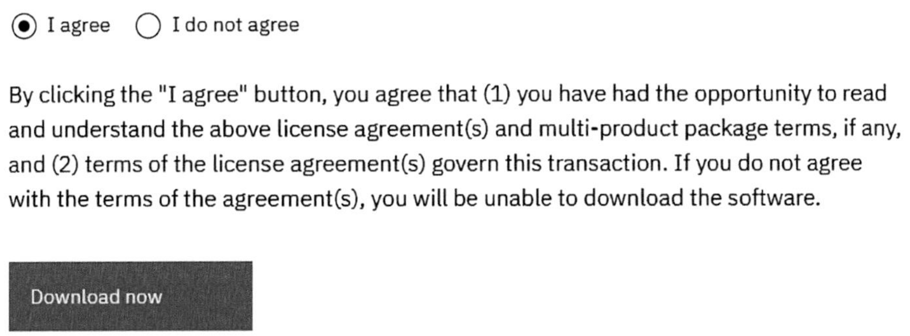

● I agree   ○ I do not agree

By clicking the "I agree" button, you agree that (1) you have had the opportunity to read and understand the above license agreement(s) and multi-product package terms, if any, and (2) terms of the license agreement(s) govern this transaction. If you do not agree with the terms of the agreement(s), you will be unable to download the software.

Download now

Figure 3-4. *The "I agree" radio button option is selected for the packages required*

CHAPTER 3 IBM JMS INTERFACE DEVELOPMENT IBM FILENET 5.5.X WORKFLOW

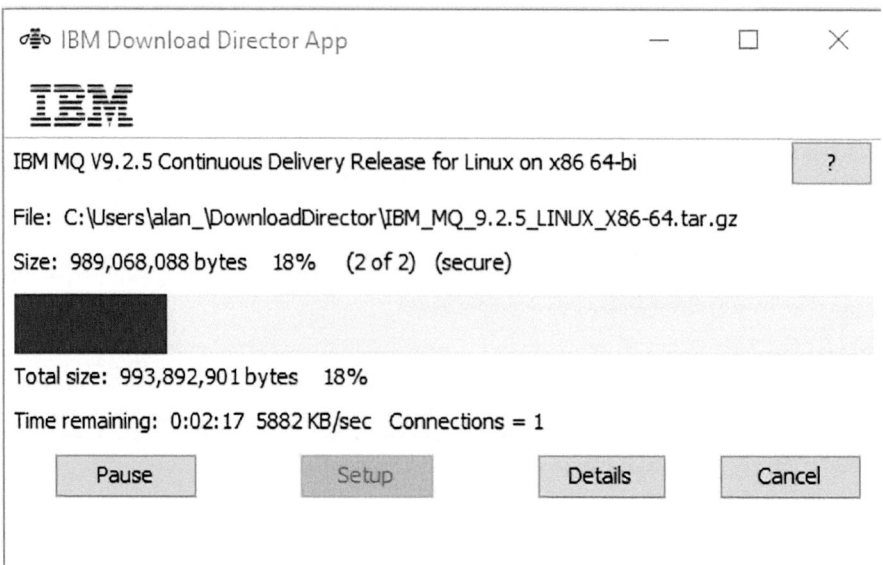

Figure 3-5. *The progress status bar window of IBM Download Director shows the download progress of the IBM_MQ_9.2.5_LINUX_X86-64.tar.gz*

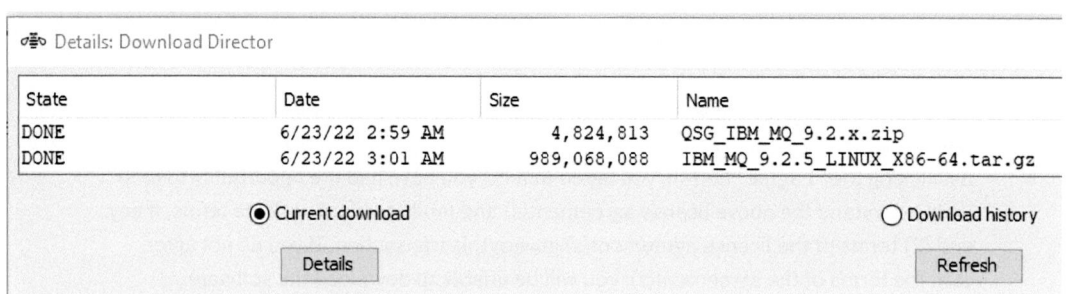

Figure 3-6. *The details of the downloaded packages are displayed*

The preceding downloads are for the Linux version of IBM MQ Version 9.2.5, which we will show how to install and configure later in this chapter.

CHAPTER 3   IBM JMS INTERFACE DEVELOPMENT IBM FILENET 5.5.X WORKFLOW

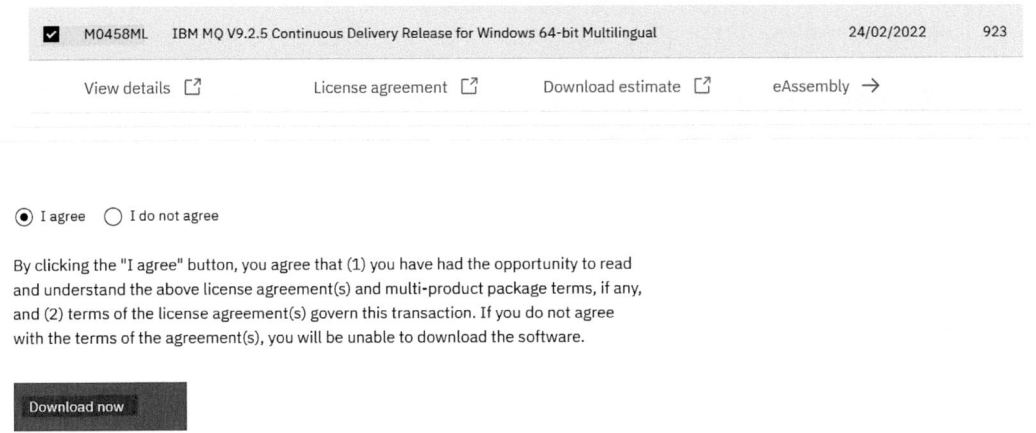

Figure 3-7. *The Windows 64-bit version of IBM MQ V9.2.5 Continuous Delivery Release is selected for download*

The Windows version of IBM MQ V9.2.5 is selected for download.

Figure 3-8. *The Download Director Java applet is run in a Firefox browser*

341

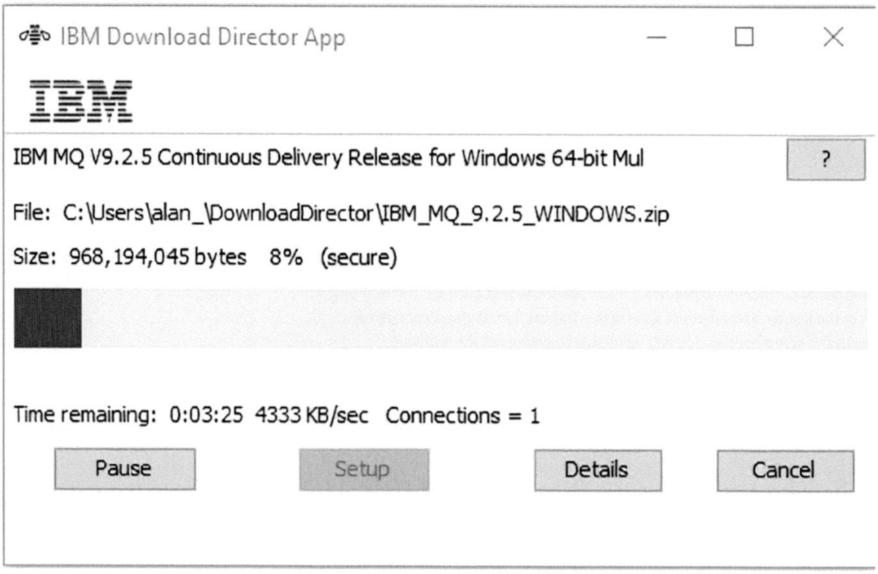

CHAPTER 3   IBM JMS INTERFACE DEVELOPMENT IBM FILENET 5.5.X WORKFLOW

Figure 3-9.   The download status of the IBM_MQ_9.2.5_WINDOWS.zip file is shown

The **Details** command button in Figure 3-9 shows the date and size of the downloaded installation file for IBM MQ version 9.2.5 for Windows.

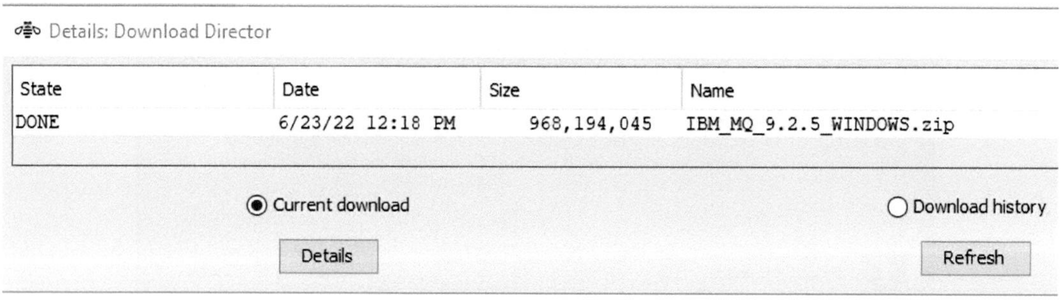

Figure 3-10.   The downloaded IBM_MQ_9.2.5_WINDOWS.zip installation file

1) First, we create the **/home/IBM/MQ** directory logged in as the root user:

    cd /home
    mkdir IBM
    cd IBM
    mkdir MQ
    cd MQ

CHAPTER 3   IBM JMS INTERFACE DEVELOPMENT IBM FILENET 5.5.X WORKFLOW

2) Copy the **IBM_MQ_9.2.5_LINUX_X86-64.tar.gz** file to the /**home/IBM/MQ** folder path created using wasadm on the ecmukdemo6 server:

```
cp /mnt/hgfs/Installs/MQ/IBM_MQ_9.2.5_LINUX_X86-64.tar.gz .
```

3) The IBM MQ 9.2.5 installation is unpacked, and then we change to the MQServer subfolder:

```
tar -zxvf IBM_MQ_9.2.5_LINUX_X86-64.tar.gz
cd /home/IBM/MQM/MQServer
```

4) The license is registered using the following shell script:

```
./mqlicense.sh -text_only
```

Note From IBM MQ 9.2.0, you have the option of accepting the license before or after installing the product. To accept the license before installing, run the mqlicense.sh script first.

Press Enter to continue viewing the license agreement, or enter "1" to accept the agreement, "2" to decline it, "3" to print it, "4" to read non-IBM terms, or "99" to go back to the previous screen.

1

This returns the message:

Agreement accepted: Proceed with install.

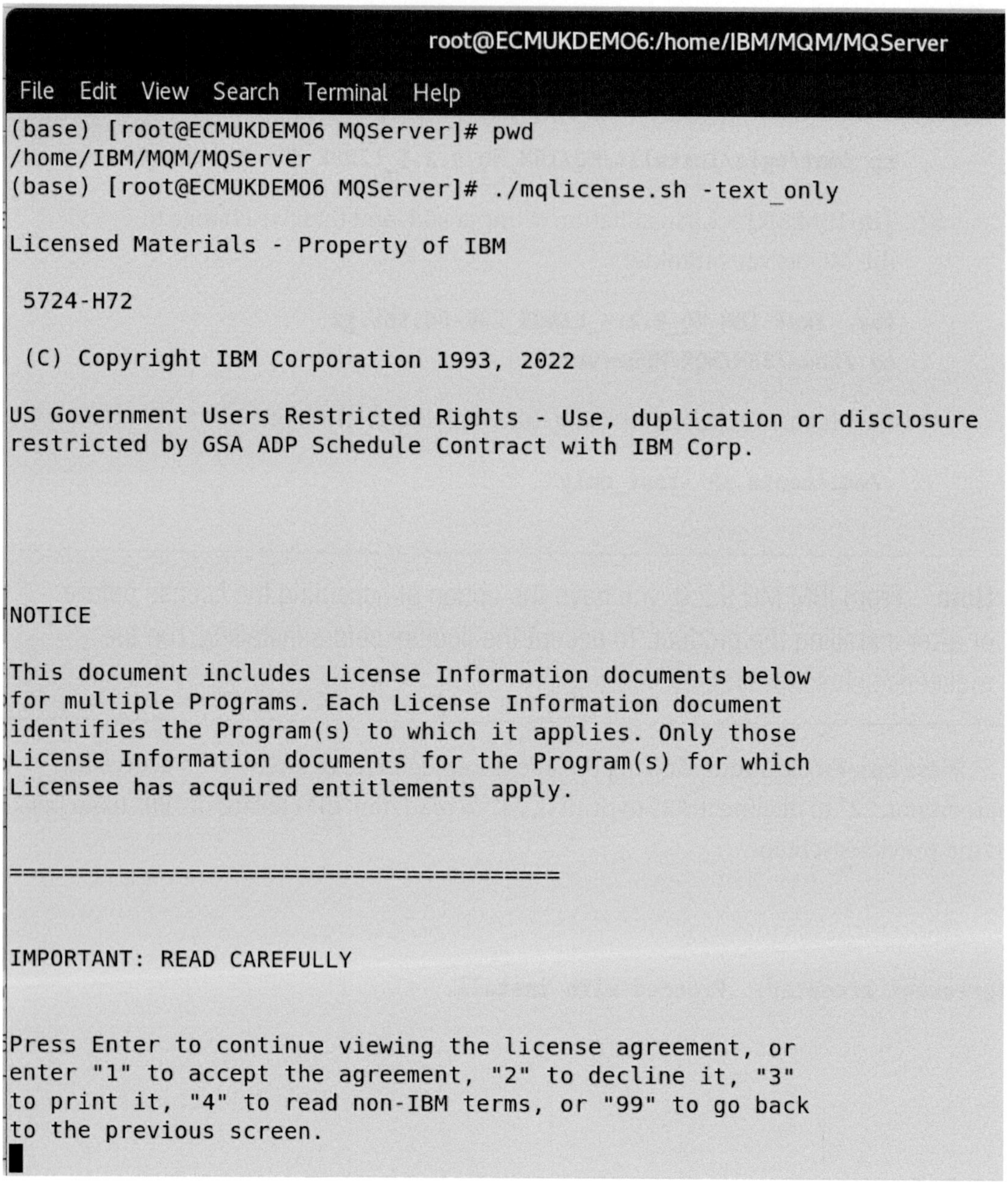

CHAPTER 3   IBM JMS INTERFACE DEVELOPMENT IBM FILENET 5.5.X WORKFLOW

```
(base) [root@ECMUKDEMO6 MQServer]# pwd
/home/IBM/MQM/MQServer
(base) [root@ECMUKDEMO6 MQServer]# ./mqlicense.sh -text_only

Licensed Materials - Property of IBM

 5724-H72

 (C) Copyright IBM Corporation 1993, 2022

US Government Users Restricted Rights - Use, duplication or disclosure
restricted by GSA ADP Schedule Contract with IBM Corp.




NOTICE

This document includes License Information documents below
for multiple Programs. Each License Information document
identifies the Program(s) to which it applies. Only those
License Information documents for the Program(s) for which
Licensee has acquired entitlements apply.



========================================


IMPORTANT: READ CAREFULLY


Press Enter to continue viewing the license agreement, or
enter "1" to accept the agreement, "2" to decline it, "3"
to print it, "4" to read non-IBM terms, or "99" to go back
to the previous screen.
```

Figure 3-11. The License shell script is run

CHAPTER 3   IBM JMS INTERFACE DEVELOPMENT IBM FILENET 5.5.X WORKFLOW

5) The temporary environment variable used for installation is set as follows:

mkdir /opt/IBM/software/
mkdir /opt/IBM/software/tmp
TMPDIR=/opt/IBM/software/tmp

6) The rpm command is used to install the packages:

rpm -ivh MQSeries*.rpm

The preceding command causes the default directory path to be used.

The default location **/opt/mqm** is selected with the **rpm -ivh** command. The preceding command will install all components that are available in the current location on the installation media to the default location.

Listing 3-1. The installation list of IBM MQ Series 9.2.5.0 components

```
(base) [root@ECMUKDEMO6 MQServer]# mkdir /opt/IBM/software/
(base) [root@ECMUKDEMO6 MQServer]# mkdir /opt/IBM/software/tmp
(base) [root@ECMUKDEMO6 MQServer]# TMPDIR=/opt/IBM/software/tmp
(base) [root@ECMUKDEMO6 MQServer]# rpm -ivh MQSeries*.rpm
warning: MQSeriesAMQP-9.2.5-0.x86_64.rpm: Header V3 RSA/SHA256 Signature, key ID 0209b828: NOKEY
Verifying...                          ################################# [100%]
Preparing...                          ################################# [100%]
Creating group mqm
Creating user mqm
Updating / installing...
   1:MQSeriesRuntime-9.2.5-0           ################################# [  3%]
   2:MQSeriesJRE-9.2.5-0               ################################# [  6%]
   3:MQSeriesJava-9.2.5-0              ################################# [  9%]
   4:MQSeriesFTBase-9.2.5-0            ################################# [ 12%]
   5:MQSeriesGSKit-9.2.5-0             ################################# [ 15%]
   6:MQSeriesServer-9.2.5-0            ################################# [ 18%]
Updated PAM configuration in /etc/pam.d/ibmmq
```

CHAPTER 3   IBM JMS INTERFACE DEVELOPMENT IBM FILENET 5.5.X WORKFLOW

WARNING: System settings for this system do not meet recommendations for this product
    See the log file at "/tmp/mqconfig.43911.log" for more information

```
   7:MQSeriesFTAgent-9.2.5-0            ################################## [ 21%]
   8:MQSeriesFTService-9.2.5-0          ################################## [ 24%]
Licensed entitlement 'advanced' set for installation at '/opt/mqm'.
   9:MQSeriesAMQP-9.2.5-0               ################################## [ 26%]
  10:MQSeriesAMS-9.2.5-0                ################################## [ 29%]
Licensed entitlement 'advanced' set for installation at '/opt/mqm'.
  11:MQSeriesFTLogger-9.2.5-0           ################################## [ 32%]
  12:MQSeriesWeb-9.2.5-0                ################################## [ 35%]
  13:MQSeriesXRService-9.2.5-0          ################################## [ 38%]
Licensed entitlement 'advanced' set for installation at '/opt/mqm'.
  14:MQSeriesClient-9.2.5-0             ################################## [ 41%]
  15:MQSeriesFTTools-9.2.5-0            ################################## [ 44%]
  16:MQSeriesBCBridge-9.2.5-0           ################################## [ 47%]
  17:MQSeriesSFBridge-9.2.5-0           ################################## [ 50%]
  18:MQSeriesExplorer-9.2.5-0           ################################## [ 53%]
  19:MQSeriesMan-9.2.5-0                ################################## [ 56%]
  20:MQSeriesMsg_cs-9.2.5-0             ################################## [ 59%]
  21:MQSeriesMsg_de-9.2.5-0             ################################## [ 62%]
  22:MQSeriesMsg_es-9.2.5-0             ################################## [ 65%]
  23:MQSeriesMsg_fr-9.2.5-0             ################################## [ 68%]
  24:MQSeriesMsg_hu-9.2.5-0             ################################## [ 71%]
  25:MQSeriesMsg_it-9.2.5-0             ################################## [ 74%]
  26:MQSeriesMsg_ja-9.2.5-0             ################################## [ 76%]
  27:MQSeriesMsg_ko-9.2.5-0             ################################## [ 79%]
  28:MQSeriesMsg_pl-9.2.5-0             ################################## [ 82%]
  29:MQSeriesMsg_pt-9.2.5-0             ################################## [ 85%]
  30:MQSeriesMsg_ru-9.2.5-0             ################################## [ 88%]
  31:MQSeriesMsg_Zh_CN-9.2.5-0          ################################## [ 91%]
  32:MQSeriesMsg_Zh_TW-9.2.5-0          ################################## [ 94%]
  33:MQSeriesSamples-9.2.5-0            ################################## [ 97%]
  34:MQSeriesSDK-9.2.5-0                ################################## [100%]
(base) [root@ECMUKDEMO6 MQServer]#
```

CHAPTER 3 IBM JMS INTERFACE DEVELOPMENT IBM FILENET 5.5.X WORKFLOW

```
mqconfig: Analyzing Red Hat Enterprise Linux 8.0 (Ootpa) settings for IBM
         MQ V9.2

System V Semaphores
  semmsl     (sem:1)   32000 semaphores                    IBM>=32           PASS
  semmns     (sem:2)   416 of 1024000000 semaphores (0%)   IBM>=4096         PASS
  semopm     (sem:3)   500 operations                      IBM>=32           PASS
  semmni     (sem:4)   340 of 32000 sets           (1%)    IBM>=128          PASS

System V Shared Memory
  shmmax               18446744073692774399 bytes          IBM>=268435456    PASS
  shmmni               64 of 4096 sets              (1%)   IBM>=4096         PASS
  shmall               1485772 of 18446744073692774399 pages (0%)   IBM>=2097152   PASS

System Settings
  file-max             18304 of 1216133 files      (1%)    IBM>=524288       PASS
  pid_max              1680 of 131072 processids   (1%)    IBM>=32768        PASS
  threads-max          1679 of 95378 threads       (1%)    IBM>=32768        PASS

Current User Limits (root)
  nofile     (-Hn)     4096 files                          IBM>=10240        FAIL
  nofile     (-Sn)     1024 files                          IBM>=10240        FAIL
  nproc      (-Hu)     0 of 47689 processes        (0%)    IBM>=4096         PASS
  nproc      (-Su)     0 of 47689 processes        (0%)    IBM>=4096         PASS

mqconfig: Any values listed in the "Current User Limits" section are resource
         limits for the user who ran mqconfig.

         If the user account that is used to invoke this script (root)
         is not the same as the user account that is used to start the
         queue manager, then the assessed values will not be accurate.

         If you normally start your queue managers as the mqm user, you
         should switch to mqm and run mqconfig there.

         If other members of the mqm group also start queue managers, all
         those members should run mqconfig, to ensure that their limits
         are suitable for IBM MQ.

mqconfig: A PASS score means your system meets the minimum IBM
         recommendations but busy systems might need higher limits to run
         production workloads.
```

Figure 3-12. *The current root User file limits are displayed as FAIL*

The current root User limits need to be increased!

nofile (-Hn) 4096 files

and **nofile (-Sn) 1024 files**

These two limits need to be increased to **10240**.

The root user limits can be changed as follows:

vi /root/.bashrc

ulimit -u unlimited

Also, the settings in **/etc/security/limits.conf** need to be edited.

CHAPTER 3   IBM JMS INTERFACE DEVELOPMENT IBM FILENET 5.5.X WORKFLOW

Log in to the computer as the root user or as a user with **sudo** permissions.

Go to the **/etc/security** directory.

Open the **limits.conf** file for editing.

Add the following lines to the file:

@root soft nofile 10240
@root hard nofile 16384
* soft nofile 10240
* hard nofile 16384

Save and close the file.

Restart the computer for the changes to take effect.

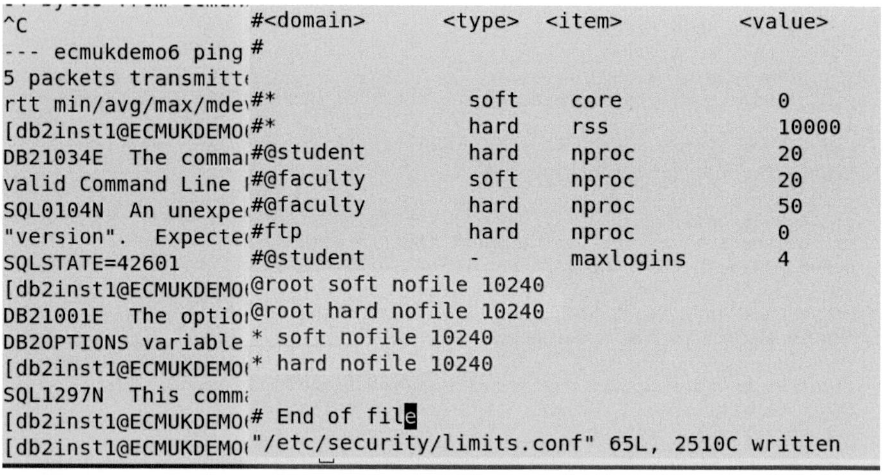

Figure 3-13.   The /etc/security/limits.conf file is edited to fix the Limits FAIL

Note   After changing the limits, the ecmukdemo6 system requires a reboot.

Set the MQ Series as the primary installation:

cd /opt/mqm/bin
/opt/mqm/bin/setmqinst -i -p /opt/mqm

147 of 147 tasks have been completed successfully.

'Installation1' (/opt/mqm) set as the primary installation.

(base) [root@ECMUKDEMO6 bin]#

```
(base) [root@ECMUKDEMO6 bin]# pwd
/opt/mqm/bin
(base) [root@ECMUKDEMO6 bin]# /opt/mqm/bin/setmqinst -i -p /opt/mqm
147 of 147 tasks have been completed successfully.
'Installation1' (/opt/mqm) set as the primary installation.
(base) [root@ECMUKDEMO6 bin]#
```

Figure 3-14. *The /etc/security/limits.conf file is edited to fix the Limits FAIL*

7) **env** gives the following MQ-related environment variables.

Listing 3-2. The MQ-related environment variables

```
(base) [root@ECMUKDEMO6 mqm]# . /opt/mqm/bin/setmqenv -s
(base) [root@ECMUKDEMO6 mqm]# env
MQ_INSTALLATION_NAME=Installation1
HOSTNAME=ECMUKDEMO6
OLDPWD=/opt/mqm/bin
CLASSPATH=/opt/mqm/java/lib/com.ibm.mq.jar:/opt/mqm/java/lib/com.ibm.mqjms.
jar:/opt/mqm/java/lib/com.ibm.mq.allclient.jar:/opt/mqm/samp/wmqjava/
samples:/opt/mqm/samp/jms/samples
MQ_INSTALLATION_PATH=/opt/mqm
PWD=/opt/mqm
HOME=/root
MQ_JAVA_INSTALL_PATH=/opt/mqm/java
MQ_DATA_PATH=/var/mqm
MQ_JAVA_DATA_PATH=/var/mqm
MANPATH=/opt/mqm/man:/usr/share/man
MQ_ENV_MODE=64
MQ_JRE_PATH=/opt/mqm/java/jre64/jre
PATH=/opt/mqm/bin:/root/miniconda3/bin:/root/miniconda3/condabin:/usr/
local/bin:/usr/local/sbin:/usr/bin:/usr/sbin:/opt/GO/go/bin:/root/.dotnet/
tools:/root/bin:/opt/GO/go/bin
MQ_JAVA_LIB_PATH=/opt/mqm/java/lib64
(base) [root@ECMUKDEMO6 mqm]#
```

CHAPTER 3   IBM JMS INTERFACE DEVELOPMENT IBM FILENET 5.5.X WORKFLOW

8) Launch the MQ Explorer.

```
(base) [root@ECMUKDEMO6 bin]# pwd
/opt/mqm/bin
(base) [root@ECMUKDEMO6 bin]# ./MQExplorer
```

Figure 3-15. *The MQ Explorer program is loaded*

9) The MQ Explorer GUI is shown in Figure 3-16.

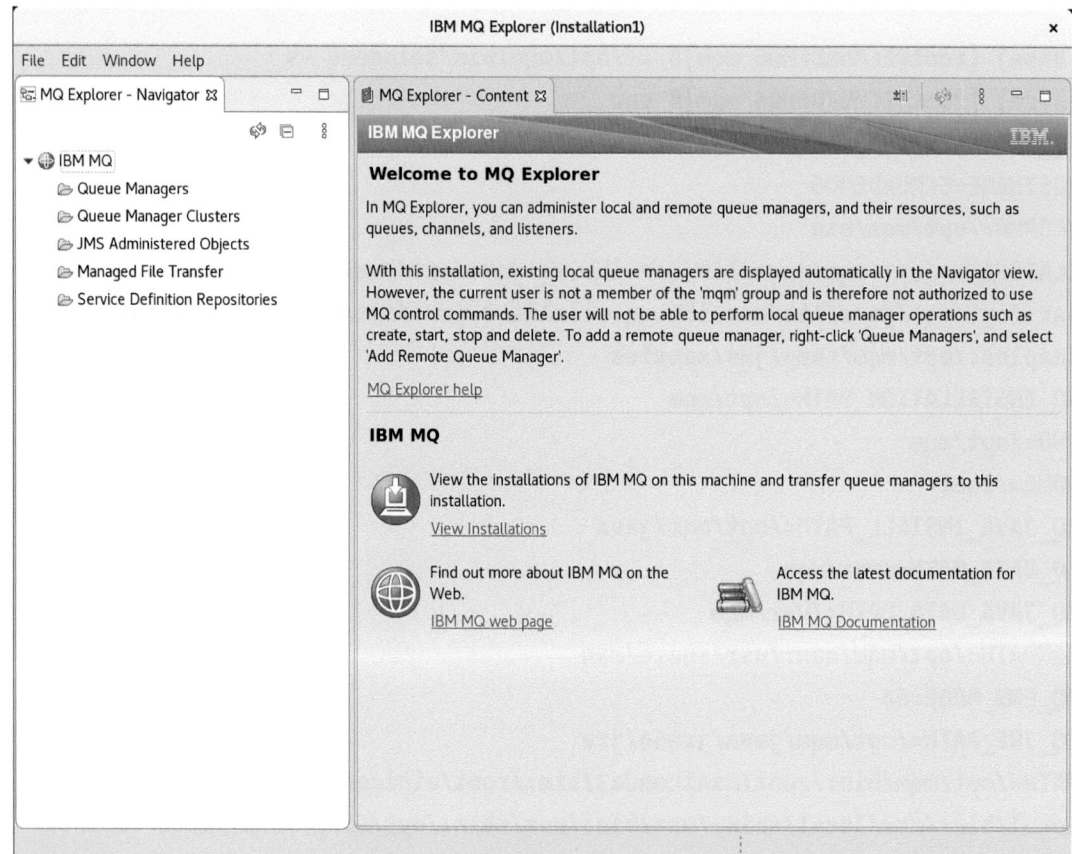

Figure 3-16. *The IBM MQ Explorer GUI is displayed using ./MQExplorer*

Note The message displayed in Figure 3-16:

With this installation, existing local queue managers are displayed automatically in the Navigator view. However, the current user is not a member of the **'mqm'** group and is therefore not authorized to use MQ control commands. The user will not be able to perform local queue manager operations such as create, start, stop and delete. To add a remote queue manager, right-click 'Queue Managers', and select 'Add Remote Queue Manager'.

10) So, we must add **wasadm** as a member of the **mqm** group:

   usermod -a -G mqm wasadm

   Add **mqm** to the mqm group:

   usermod -a -G mqm mqm

Note This must be done for the **mqm** user; otherwise, the **wasadm** Q Manager create will fail with an error message.

11) Log out and then log back in as **wasadm** for the changes to take effect.

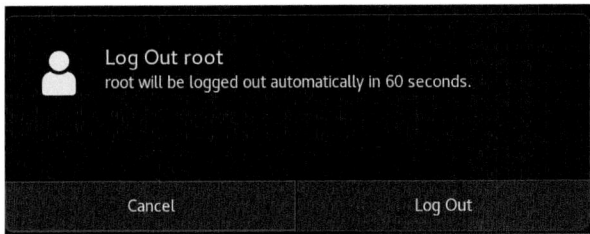

Figure 3-17. *The Linux root user is logged out*

CHAPTER 3 IBM JMS INTERFACE DEVELOPMENT IBM FILENET 5.5.X WORKFLOW

We need to switch to log in to the wasadm Linux user.

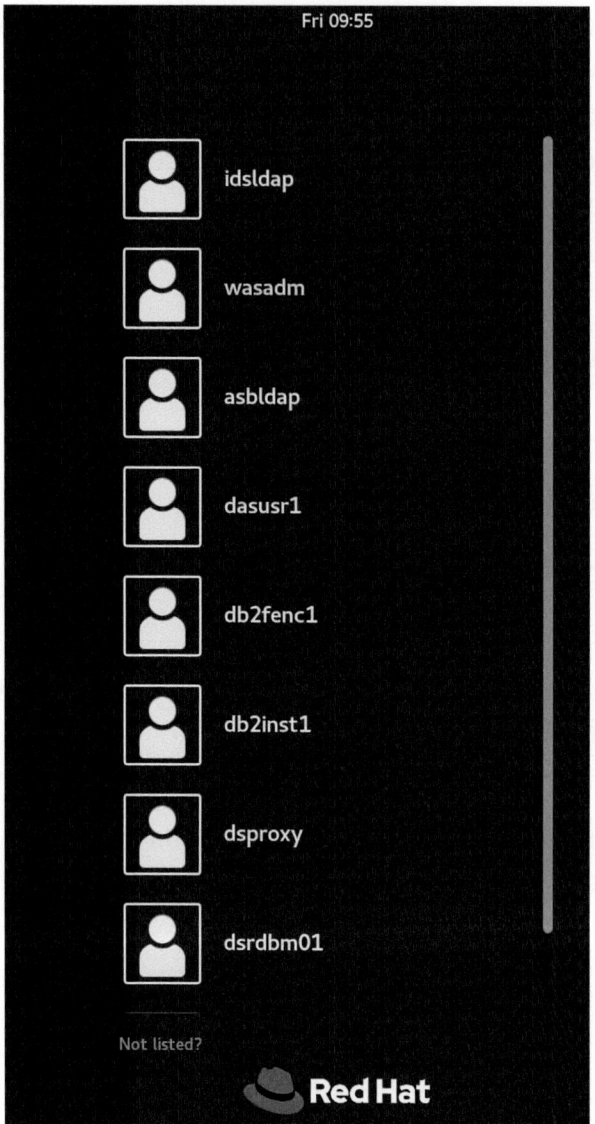

Figure 3-18. *The wasadm Linux user is selected for login*

CHAPTER 3   IBM JMS INTERFACE DEVELOPMENT IBM FILENET 5.5.X WORKFLOW

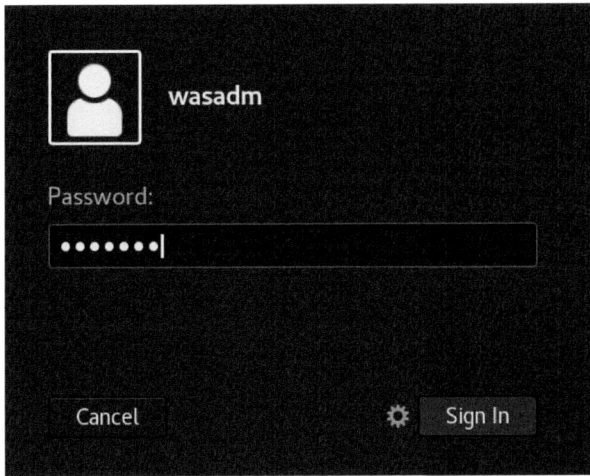

Figure 3-19. *The wasadm password is entered in the Linux Desktop*

After installation of IBM MQ Series 9.2.5, there is a Linux Desktop icon installed which can be searched in the Activities Desktop search, as shown in Figure 3-20, and will then launch the IBM MQ Series 9.2.5 administration application, **IBM MQ Explorer**.

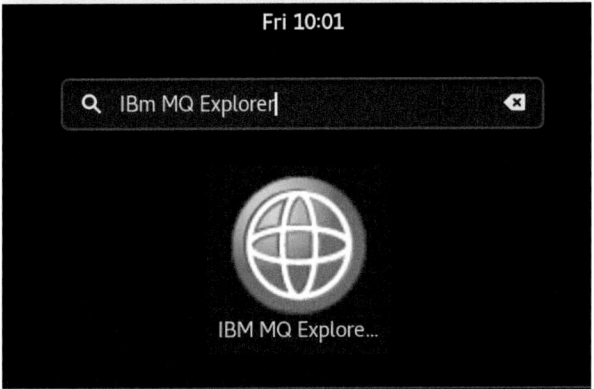

Figure 3-20. *The IBM MQ Explorer 9.2.5 program icon to be clicked*

CHAPTER 3 IBM JMS INTERFACE DEVELOPMENT IBM FILENET 5.5.X WORKFLOW

Figure 3-21. *The splash screen of the IBM MQ Series Version 9.2.5 Explorer*

12) Now displays:

CHAPTER 3 IBM JMS INTERFACE DEVELOPMENT IBM FILENET 5.5.X WORKFLOW

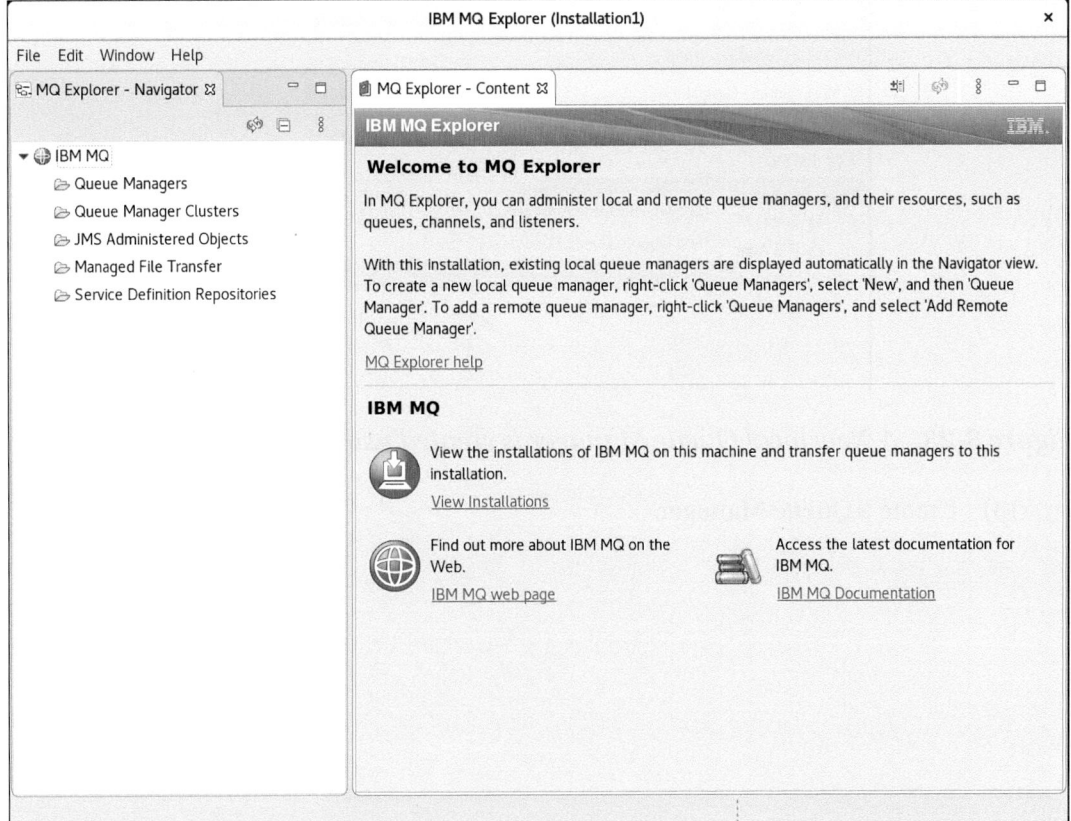

Figure 3-22. *The initial screen of the IBM MQ Series Version 9.2.5 Explorer*

With this installation, existing local queue managers are displayed automatically in the Navigator view. To create a new local queue manager, right-click **Queue Managers**, and select **New ➤ Queue Manager**. To add a remote queue manager, right-click **Queue Managers**, and select **Add Remote Queue Manager**.

CHAPTER 3 IBM JMS INTERFACE DEVELOPMENT IBM FILENET 5.5.X WORKFLOW

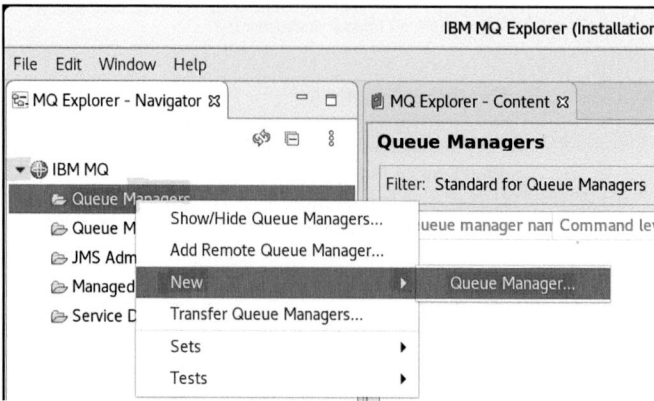

Figure 3-23. *A New local Queue Manager is created using the highlighted menus*

13) Create a Queue Manager.

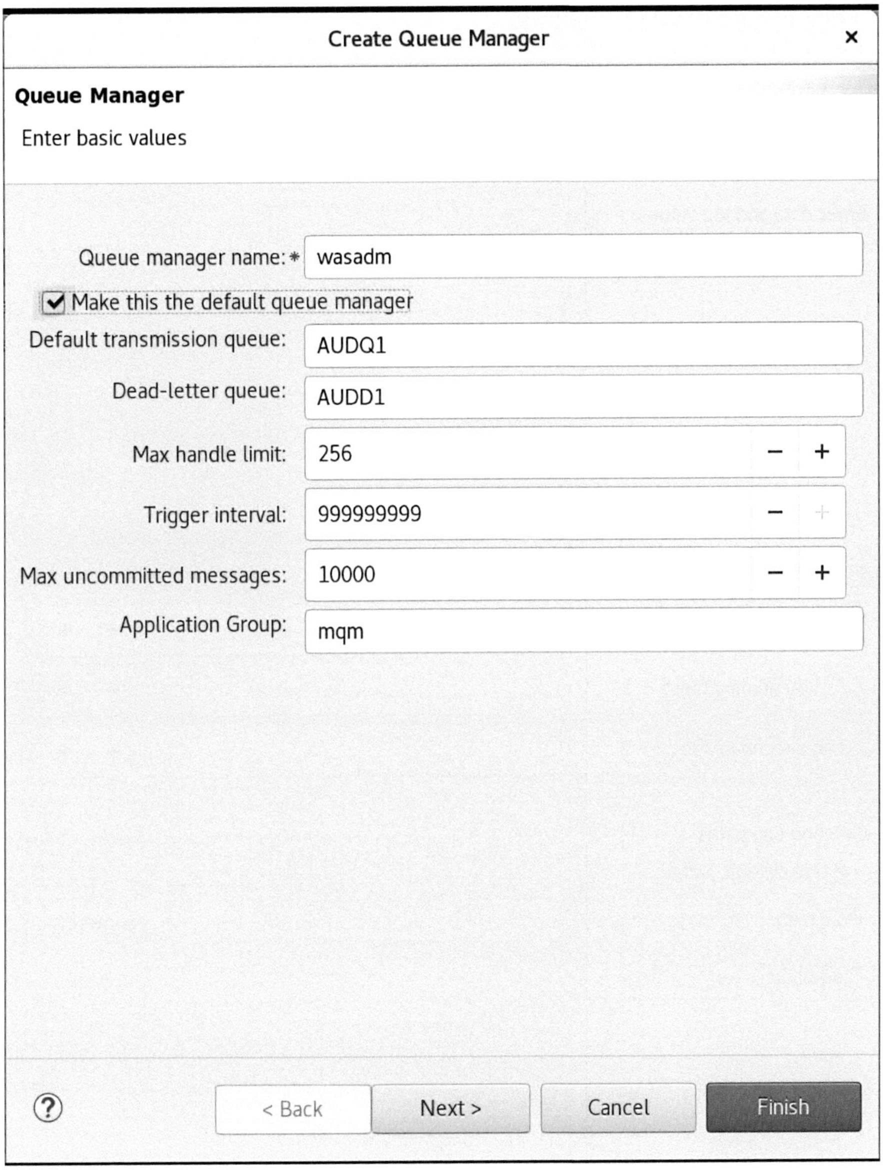

CHAPTER 3   IBM JMS INTERFACE DEVELOPMENT IBM FILENET 5.5.X WORKFLOW

Figure 3-24. *The wasadm Queue Manager is created with Group mqm*

CHAPTER 3 IBM JMS INTERFACE DEVELOPMENT IBM FILENET 5.5.X WORKFLOW

14) Defaults are used for logging.

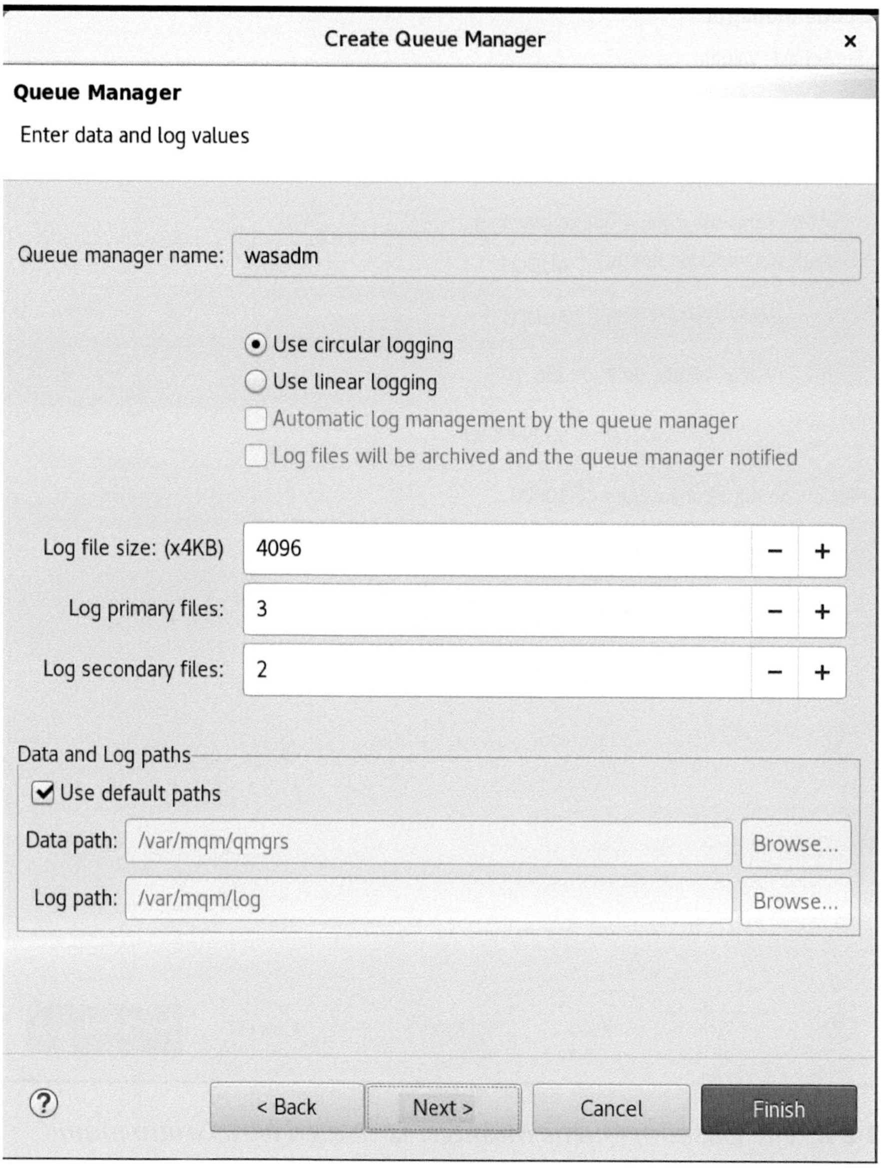

*Figure 3-25. The Queue Manager Logging defaults are used, and **Next>** is clicked*

15) Click **Next>** and select **Permit a standby instance** (tick box) and remote admin by **Create server-connection channel** (tick box) to allow remote.

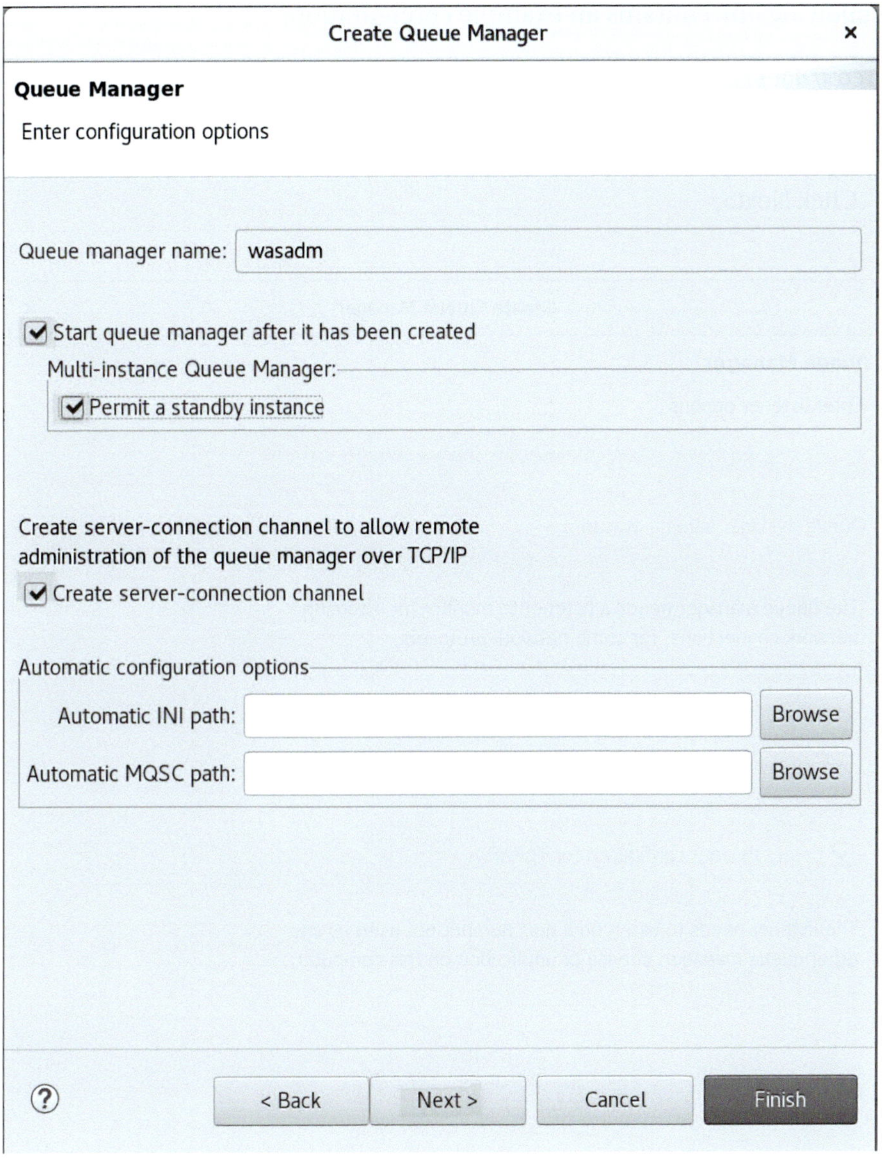

CHAPTER 3 IBM JMS INTERFACE DEVELOPMENT IBM FILENET 5.5.X WORKFLOW

Figure 3-26. The required Queue Manager options are ticked as highlighted

Note From IBM MQ 9.2.0, you can configure your queue manager to automatically apply the contents of an MQSC script, or set of MQSC scripts, on every queue manager start. (This is also available from version 9.1.4 onward.)

(See `www.ibm.com/docs/en/ibm-mq/9.2?topic=commands-automatic-configuration-from-mqsc-script-startup`.)

CHAPTER 3   IBM JMS INTERFACE DEVELOPMENT IBM FILENET 5.5.X WORKFLOW

The following link contains an example configuration:

www.ibm.com/docs/en/ibm-mq/9.2?topic=manager-example-supplying-mqsc-ini-files

16) Click Next>.

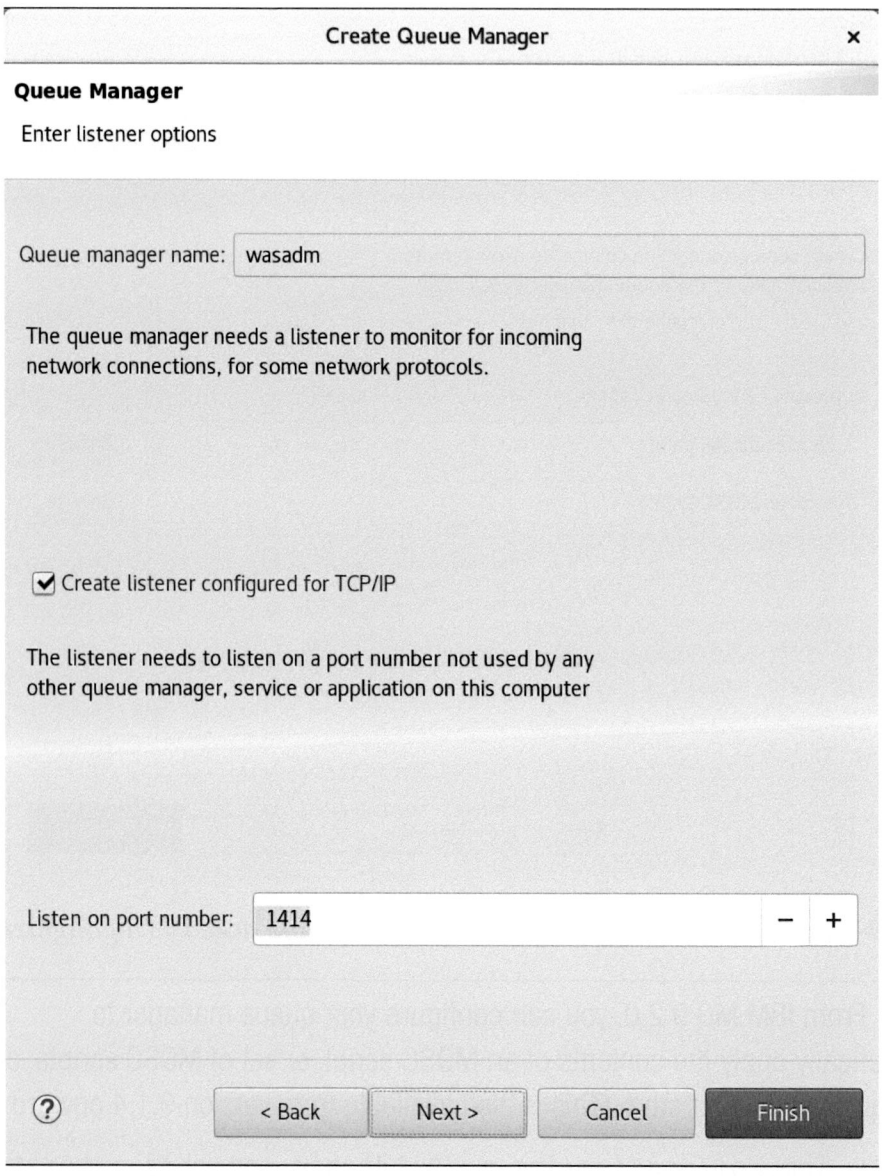

Figure 3-27. *The Create listener configured for TCP/IP is ticked with port 1414*

CHAPTER 3 IBM JMS INTERFACE DEVELOPMENT IBM FILENET 5.5.X WORKFLOW

17) Use the default port 1414 and click Next>.

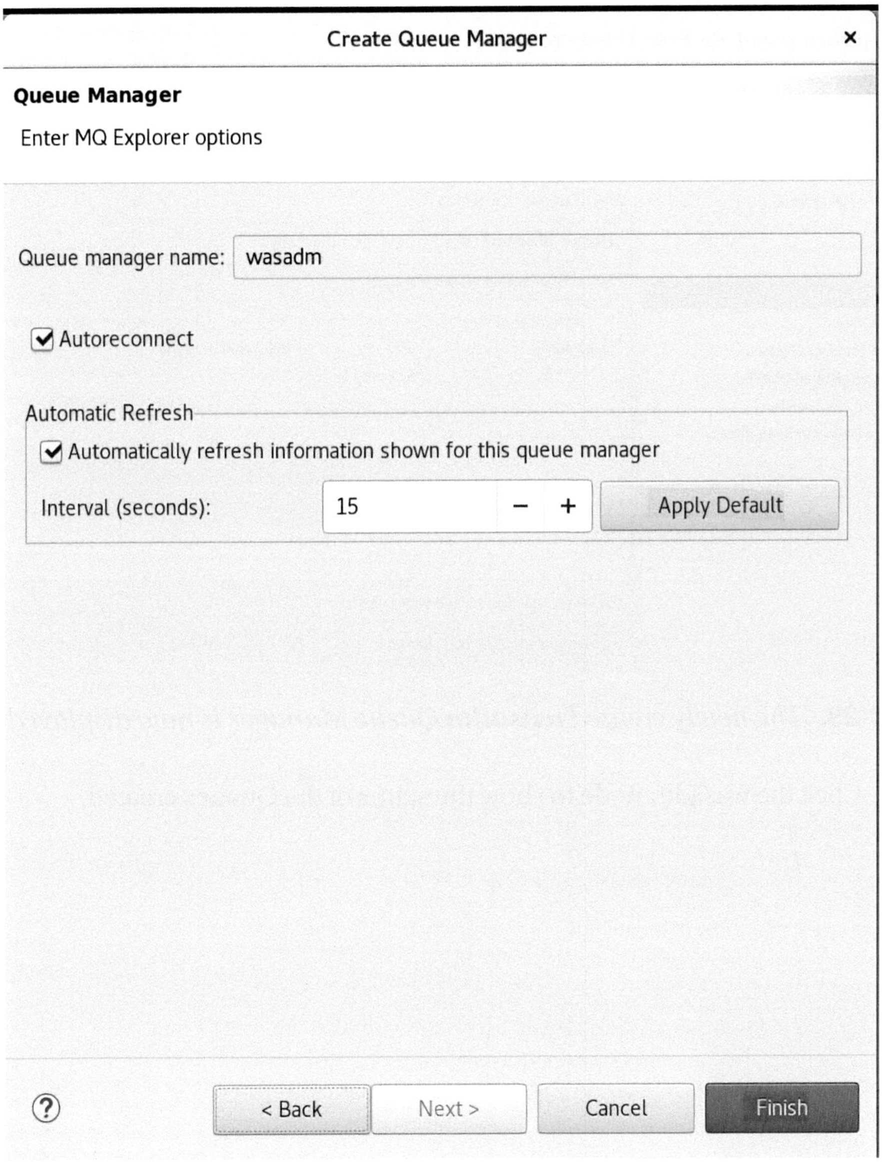

Figure 3-28. The wasadm Queue Manager reconnection settings are entered

CHAPTER 3   IBM JMS INTERFACE DEVELOPMENT IBM FILENET 5.5.X WORKFLOW

18) Use a default 15-second refresh interval, click **Apply Default**, and click **Finish**.

19) Queue Manager is now created.

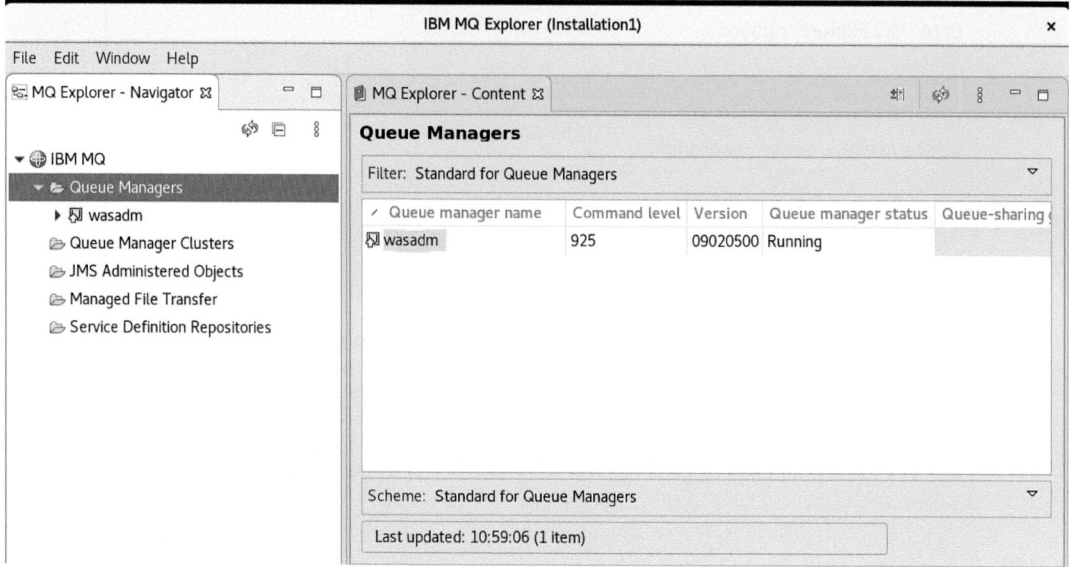

Figure 3-29. *The newly created **wasadm** Queue Manager is now displayed*

20) Click the wasadm node to show the status of the Queues created.

CHAPTER 3 IBM JMS INTERFACE DEVELOPMENT IBM FILENET 5.5.X WORKFLOW

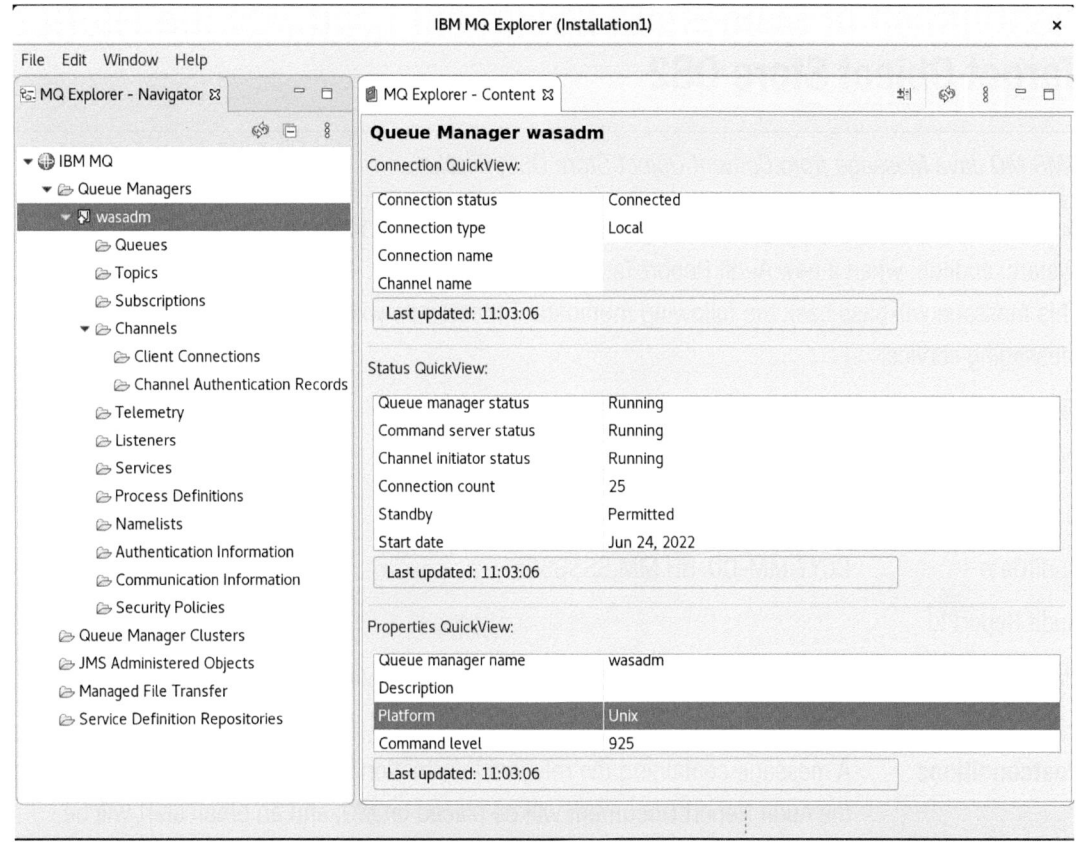

Figure 3-30. *The status of the Queue Manager is displayed at the wasadm node*

CHAPTER 3   IBM JMS INTERFACE DEVELOPMENT IBM FILENET 5.5.X WORKFLOW

Notification of Successful Document Load into the FileNet Target Object Store OS2

JMS MQ Java Message from Content Object Store OS2, FileNet

An event will be created in FileNet to send a real-time MQ message to the Quality Audit Database system, auditdb, when a new Audit Report Task Workflow is created.

This message will pass back the following metadata. Timestamps will be generated as part of the messaging service:

Field ID	Attributes
Case ID	
Department Id	
AuditDate	CCYY-MM-DD-HH.MM.SSSSSS
Audit Report Id	
Preconditions	An Audit Report document has been successfully loaded and stored in the FileNet Target (OS2) Object Store.
Postconditions	A message containing the relevant information to identify the Department and the Audit Report Document will be placed on MQ, and an email alert will be issued from the Workflow operation system step to the Department manager.
Exceptions	If MQ messages fail due to Audit Case processing issues or MQ being unavailable, the load of any new documents may be paused for a limited time period until the root cause of the issues is identified.
	Controls will be implemented where necessary to prevent a flood of MQ messages causing performance issues where a larger than normal volume of documents is processed (i.e., from a backlog following a Document load failure).

Set up point-to-point messaging using queues. Configuring the application and the messaging system requires the following steps:

1. Add initial JMS Context.

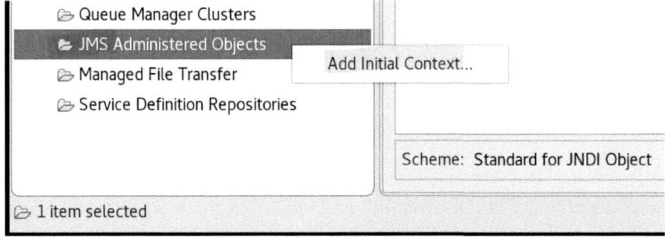

Figure 3-31. *The Initial JMS Content is added*

The Add Initial Context is used to select an LDAP server for the JNDI namespace.

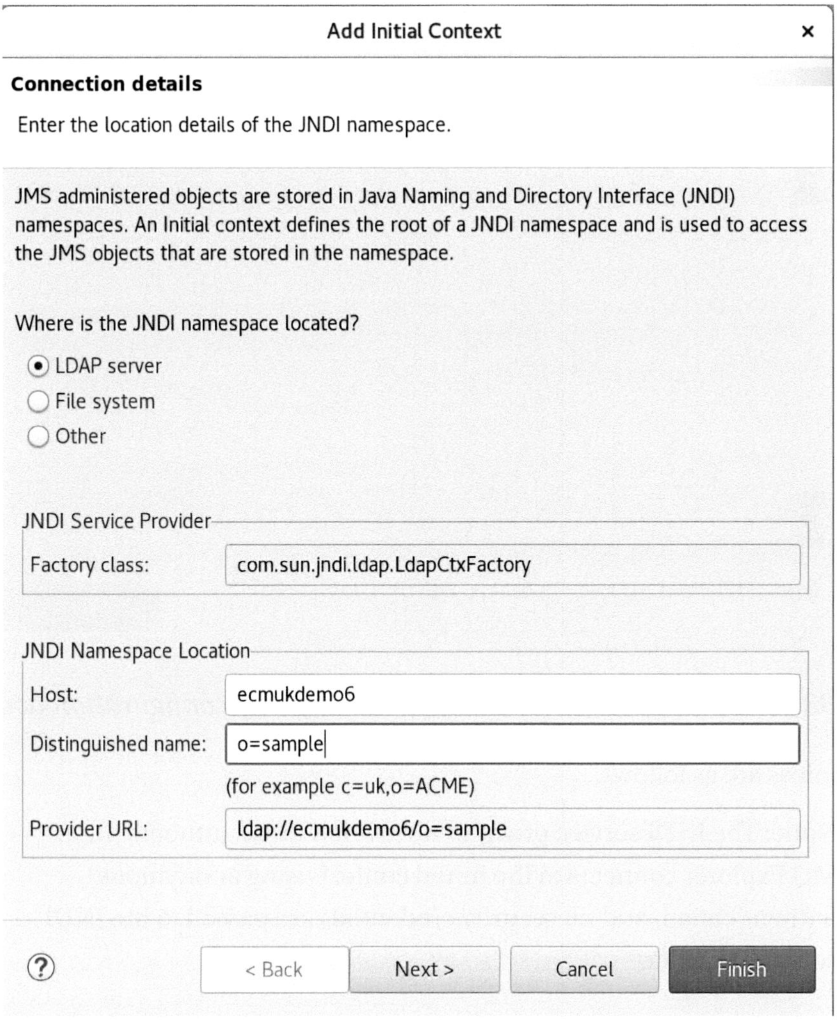

Figure 3-32. *The JNDI namespace is selected to be on the ecmukdemo6 LDAP server*

CHAPTER 3   IBM JMS INTERFACE DEVELOPMENT IBM FILENET 5.5.X WORKFLOW

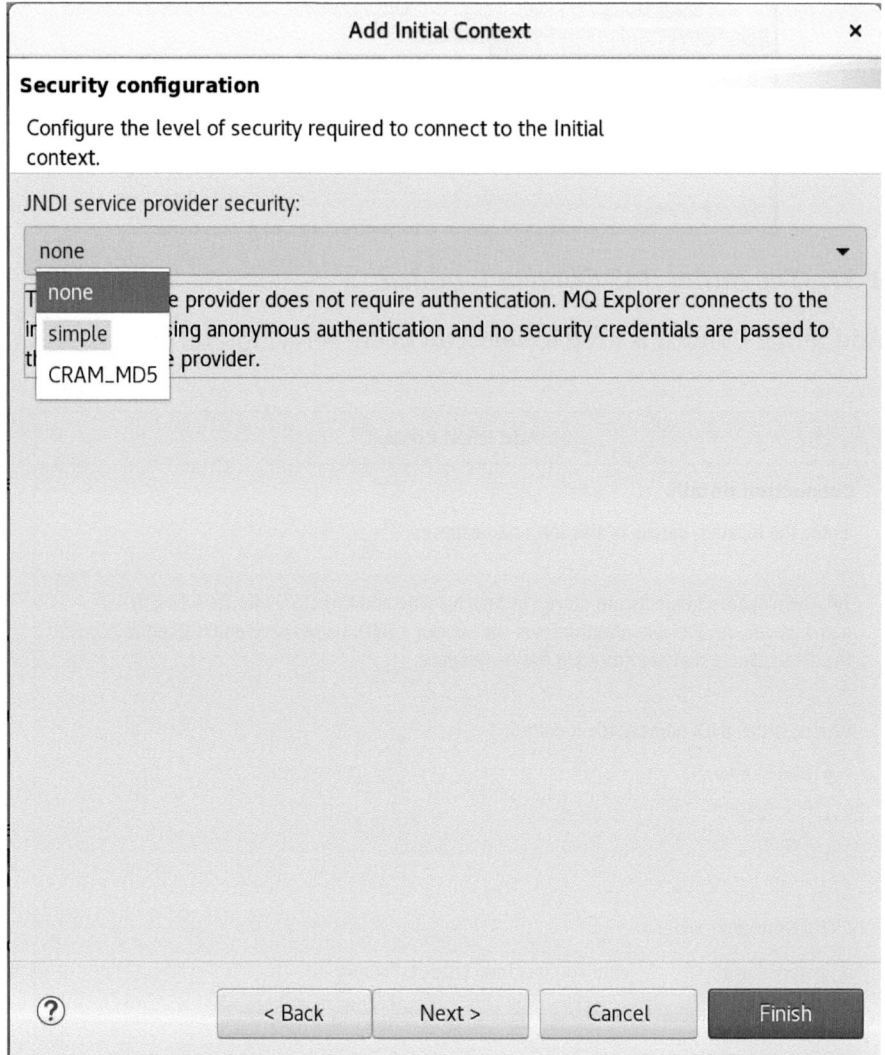

Figure 3-33. *The drop-down shows the types of security configuration available*

The choices are as follows:

- None: The JNDI service provider does not require authentication. MQ Explorer connects to the initial context using anonymous authentication, and no security credentials are passed to the JNDI service provider.

CHAPTER 3　IBM JMS INTERFACE DEVELOPMENT IBM FILENET 5.5.X WORKFLOW

- Simple: The JNDI service provider requires simple authentication. MQ Explorer must pass the user-distinguished name and password to the JNDI service provider to connect to the initial context. (**We select this option.**)

- CRAM MD5: The JNDI service provider requires CRAM-MD5 authentication. MQ Explorer must pass the MD5-encrypted password to the JNDI service provider to connect to the initial context.

Tick the option to Automatically reconnect to context on startup.

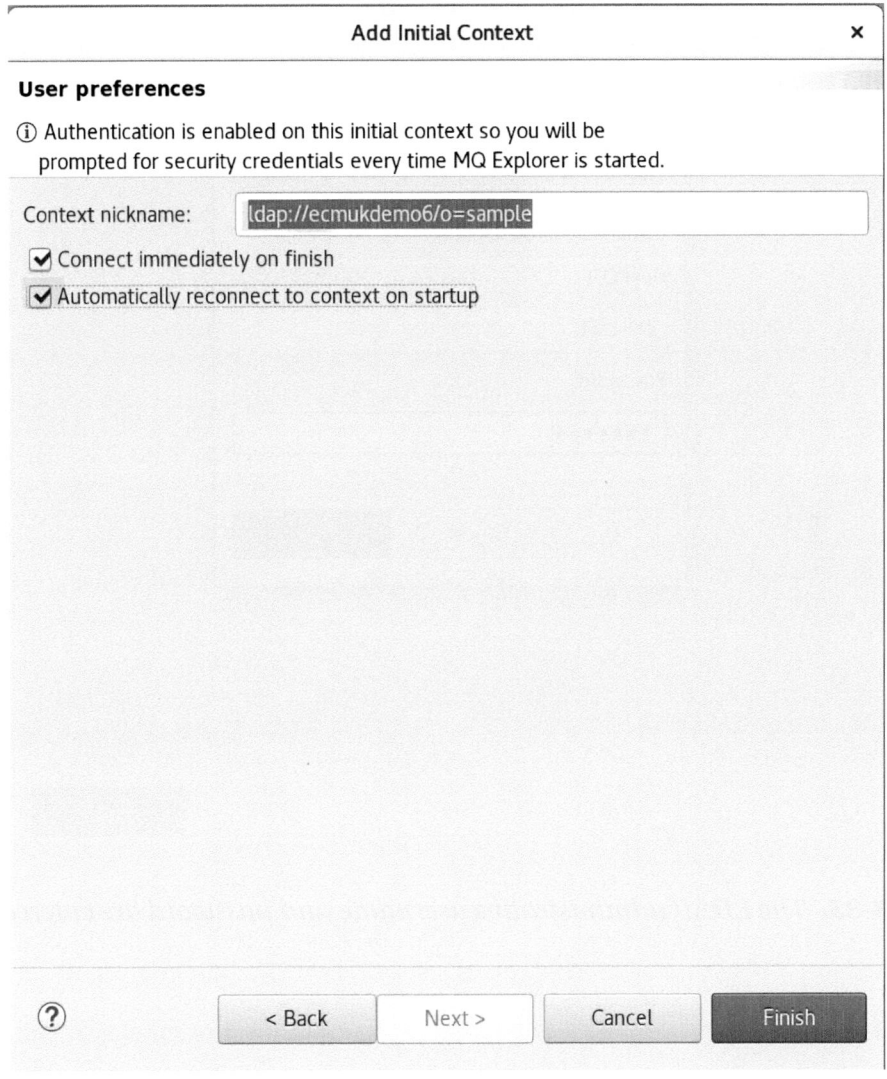

Figure 3-34. *The option to Automatically reconnect to context on startup is ticked*

CHAPTER 3 IBM JMS INTERFACE DEVELOPMENT IBM FILENET 5.5.X WORKFLOW

Click the **Finish** command button.

Enter the user as **cn=root** (password =**filenet**) for logging into the **LDAP** as an administrative user.

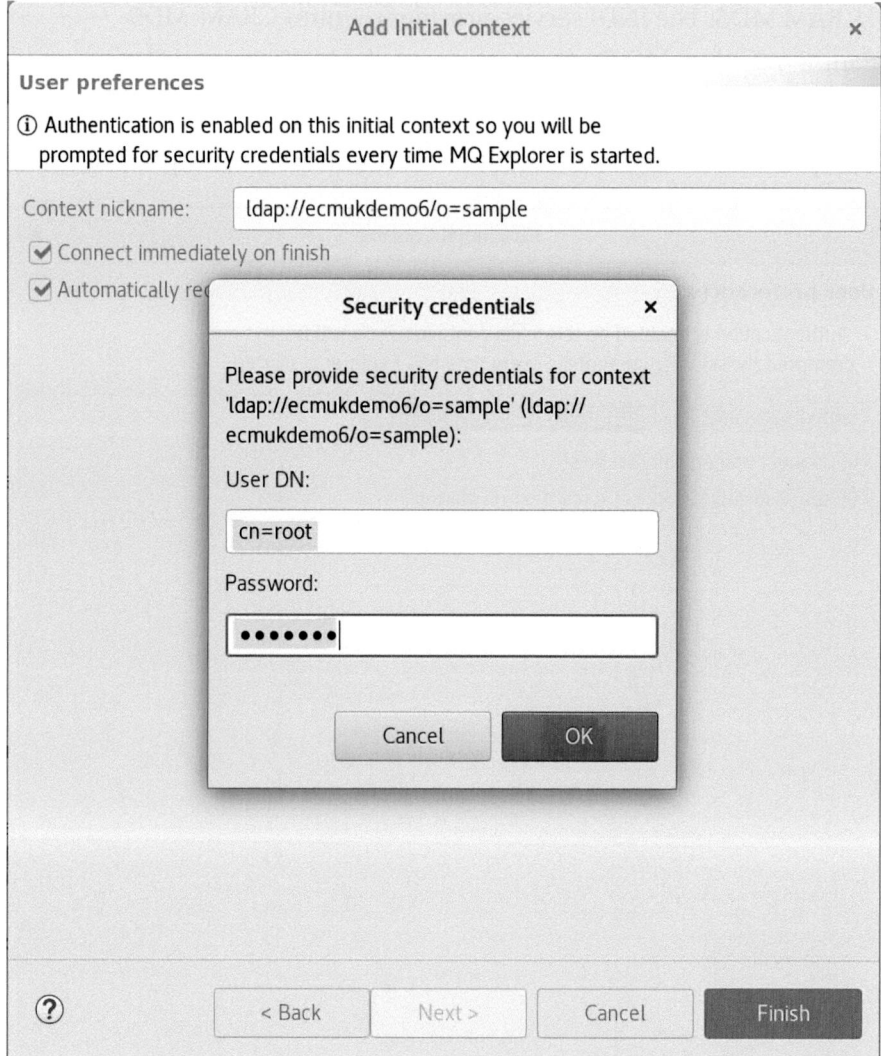

Figure 3-35. *The LDAP administrative username and password are entered*

CHAPTER 3　IBM JMS INTERFACE DEVELOPMENT IBM FILENET 5.5.X WORKFLOW

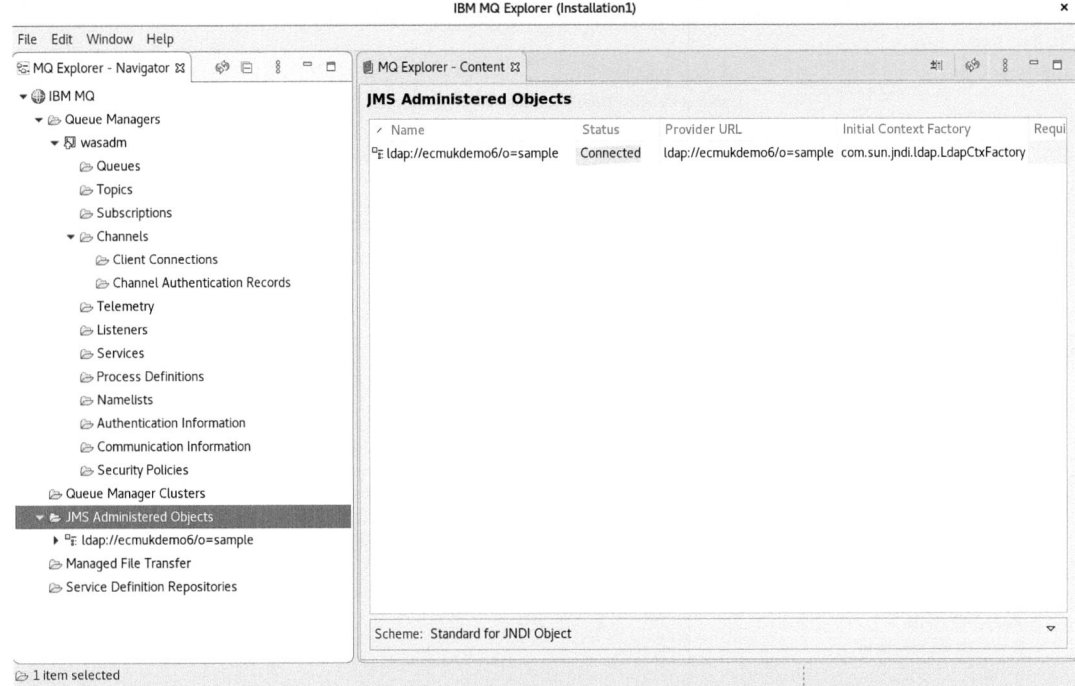

Figure 3-36. *The status of the JMS LDAP Server Context is displayed as Connected*

Clicking the **JMS** Administered Objects ldap node loads the **LDAP** users and groups.

CHAPTER 3   IBM JMS INTERFACE DEVELOPMENT IBM FILENET 5.5.X WORKFLOW

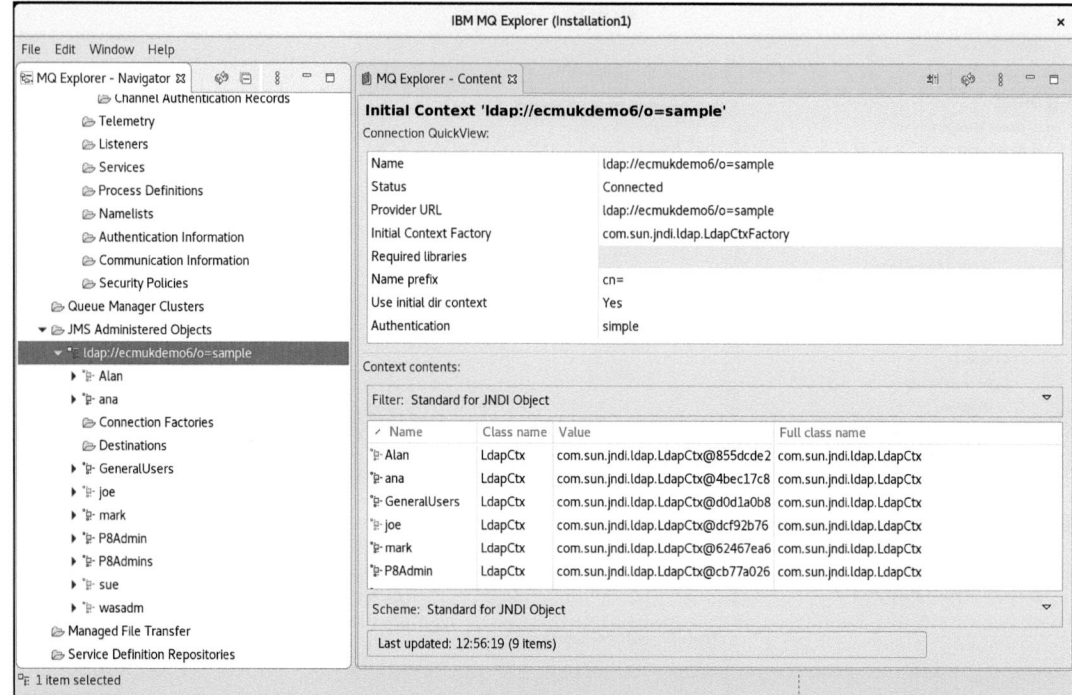

Figure 3-37. *The opener for the LDAP server URL displays the Users and Groups*

Create a new test **Local queue**, **AUDQAR**.

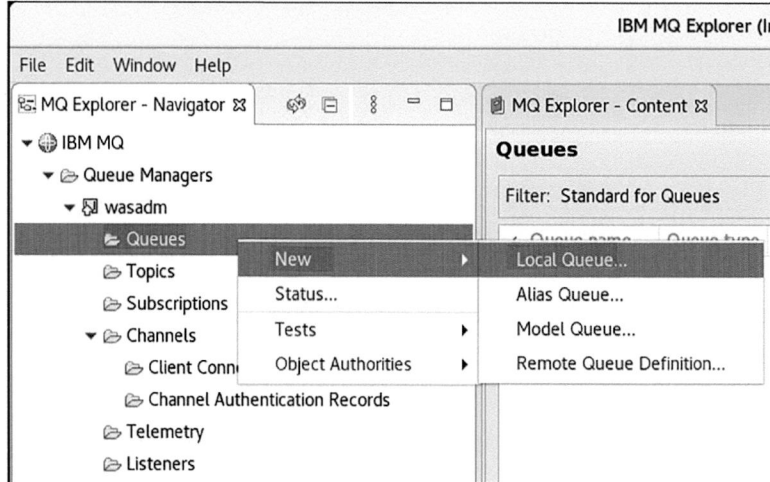

Figure 3-38. *The new AUDQAR Local Queue is created*

CHAPTER 3 IBM JMS INTERFACE DEVELOPMENT IBM FILENET 5.5.X WORKFLOW

The new **Local Queue** is named **AUDQAR**, and the Start wizard option is ticked to launch the **LDAP JMS Queue Context** to match the new local **Queue**.

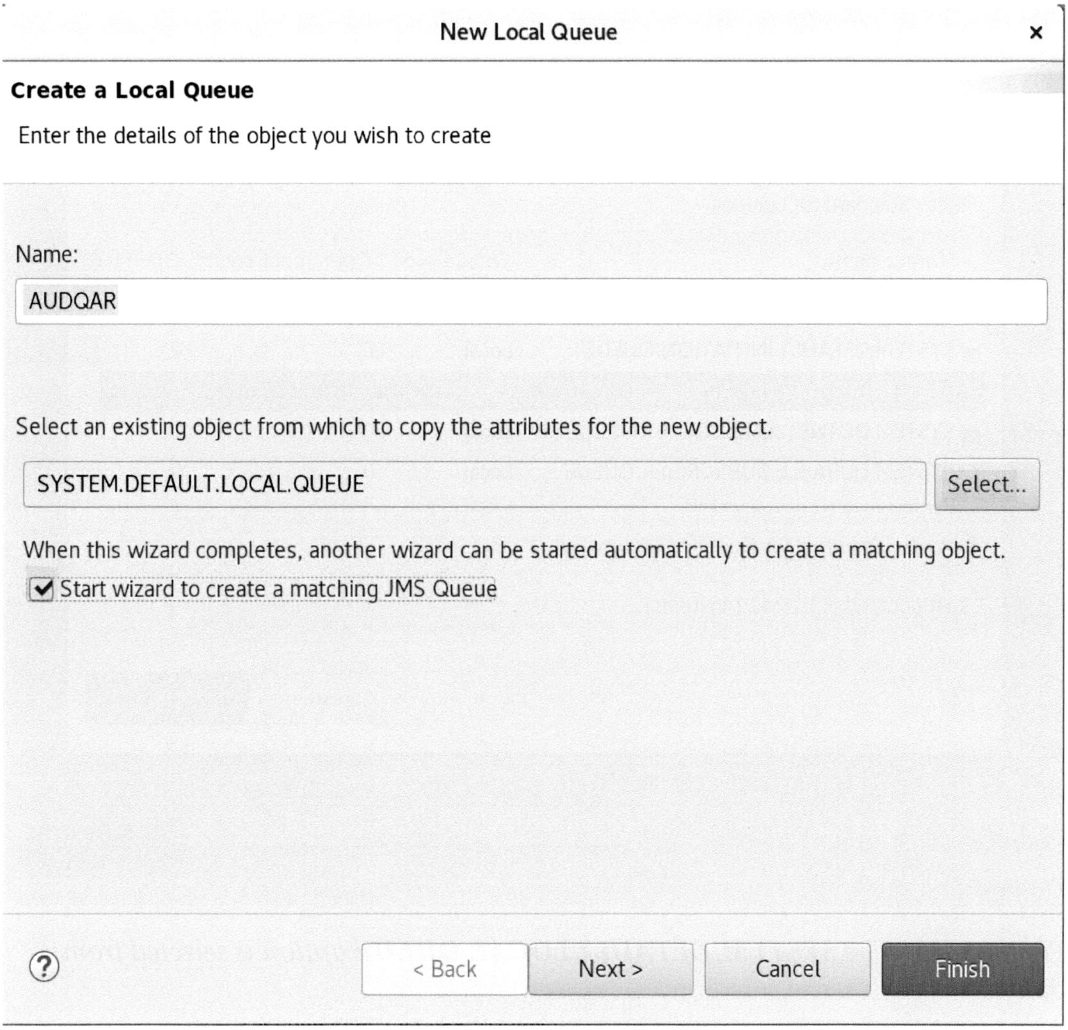

Figure 3-39. The new Local Queue is named AUDQAR and Finish clicked

Click Select.

CHAPTER 3 IBM JMS INTERFACE DEVELOPMENT IBM FILENET 5.5.X WORKFLOW

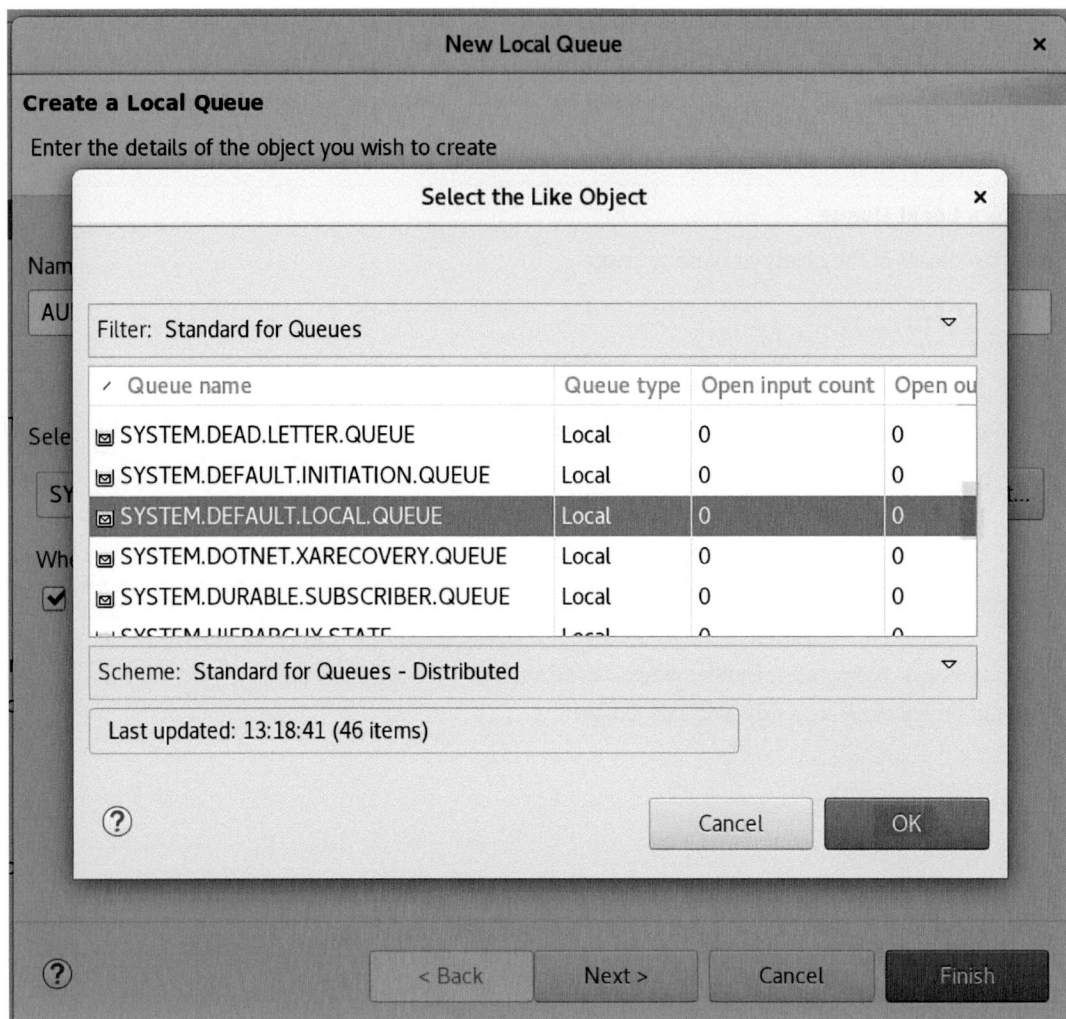

Figure 3-40. The SYSTEM.DEFAULT.LOCAL.QUEUE option is selected from the list

CHAPTER 3 IBM JMS INTERFACE DEVELOPMENT IBM FILENET 5.5.X WORKFLOW

Click OK on **SYSTEM.DEFAULT.LOCAL.QUEUE**.

Tick the "Start wizard to create a matching JMS Queue" and click **Next>**.

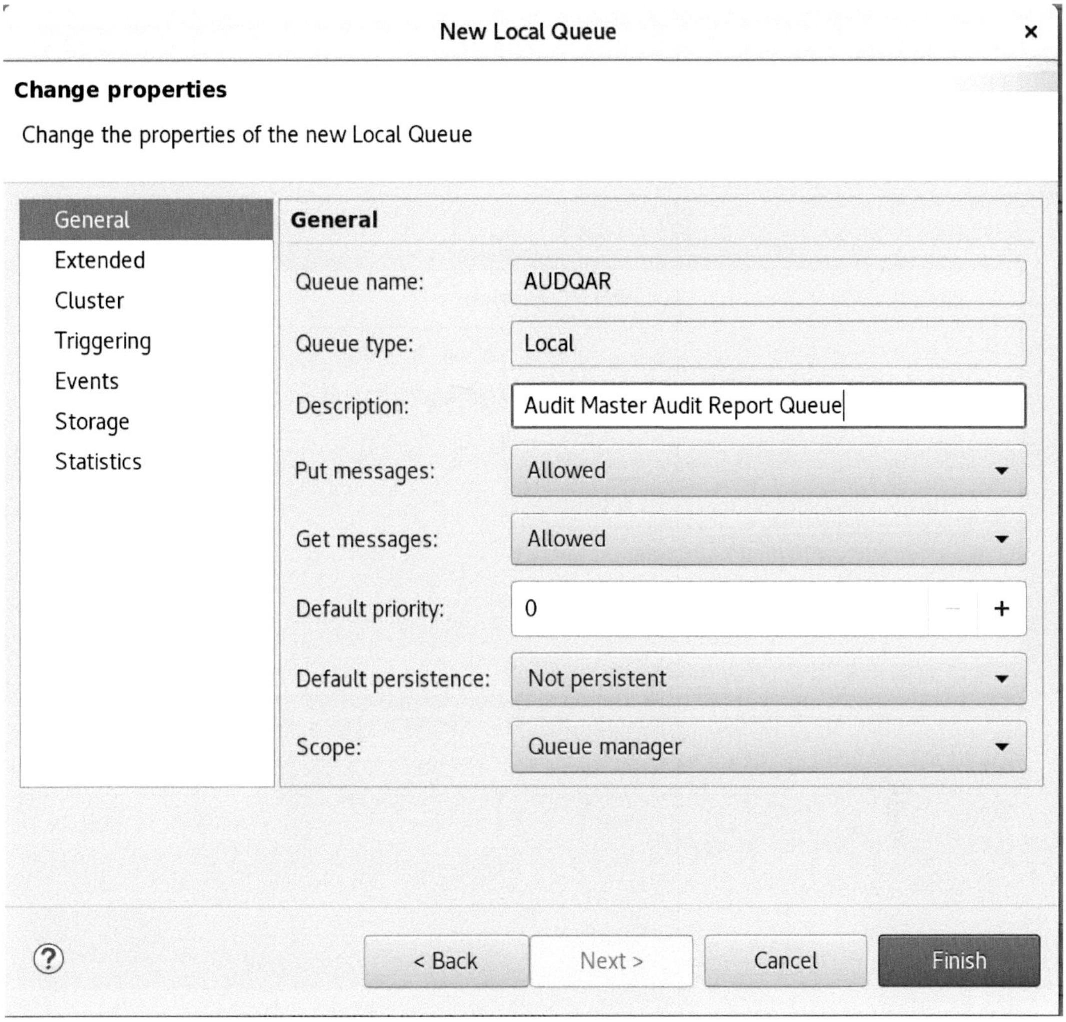

Figure 3-41. Enter the Queue name and click Finish

CHAPTER 3 IBM JMS INTERFACE DEVELOPMENT IBM FILENET 5.5.X WORKFLOW

Click **Finish**.

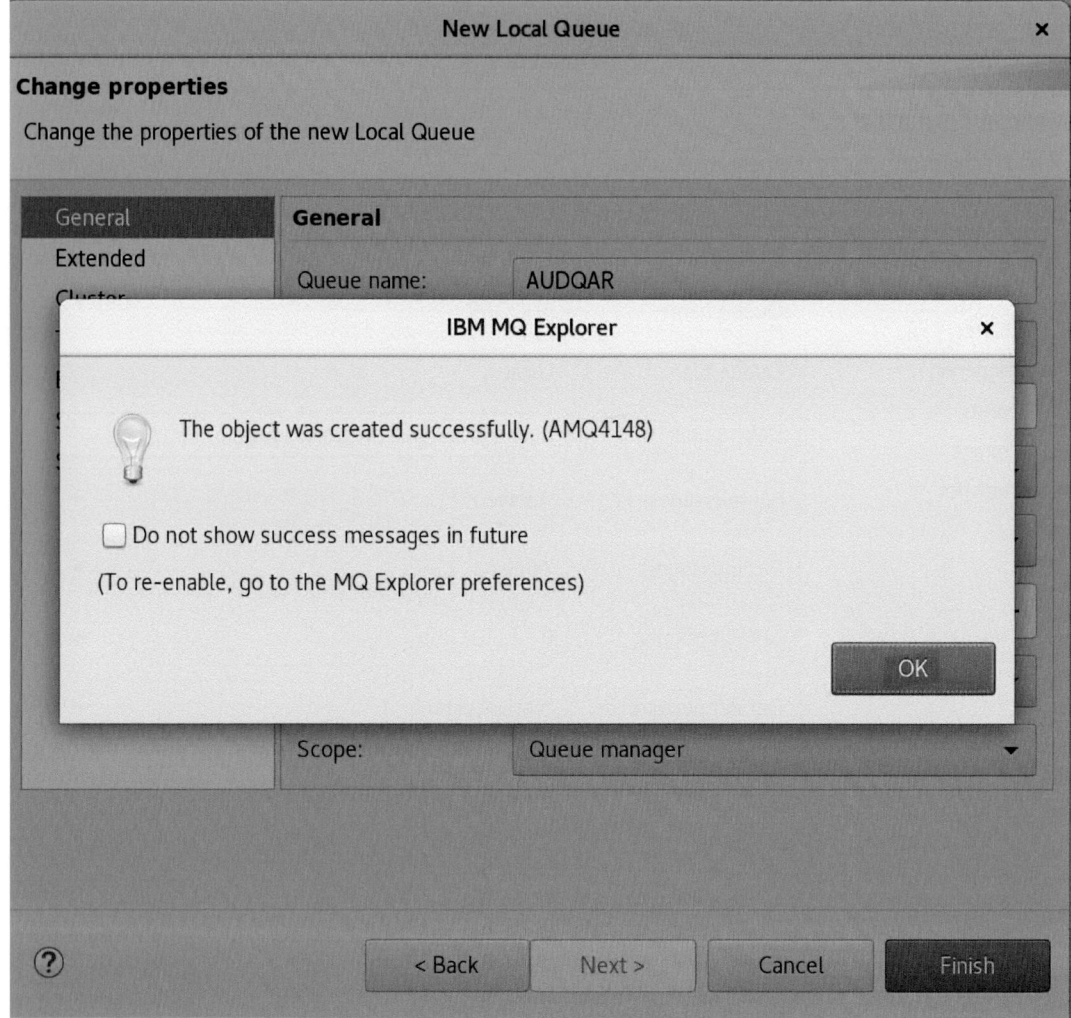

Figure 3-42. *The status of the Queue object creation is displayed*

CHAPTER 3 IBM JMS INTERFACE DEVELOPMENT IBM FILENET 5.5.X WORKFLOW

Click **OK**.

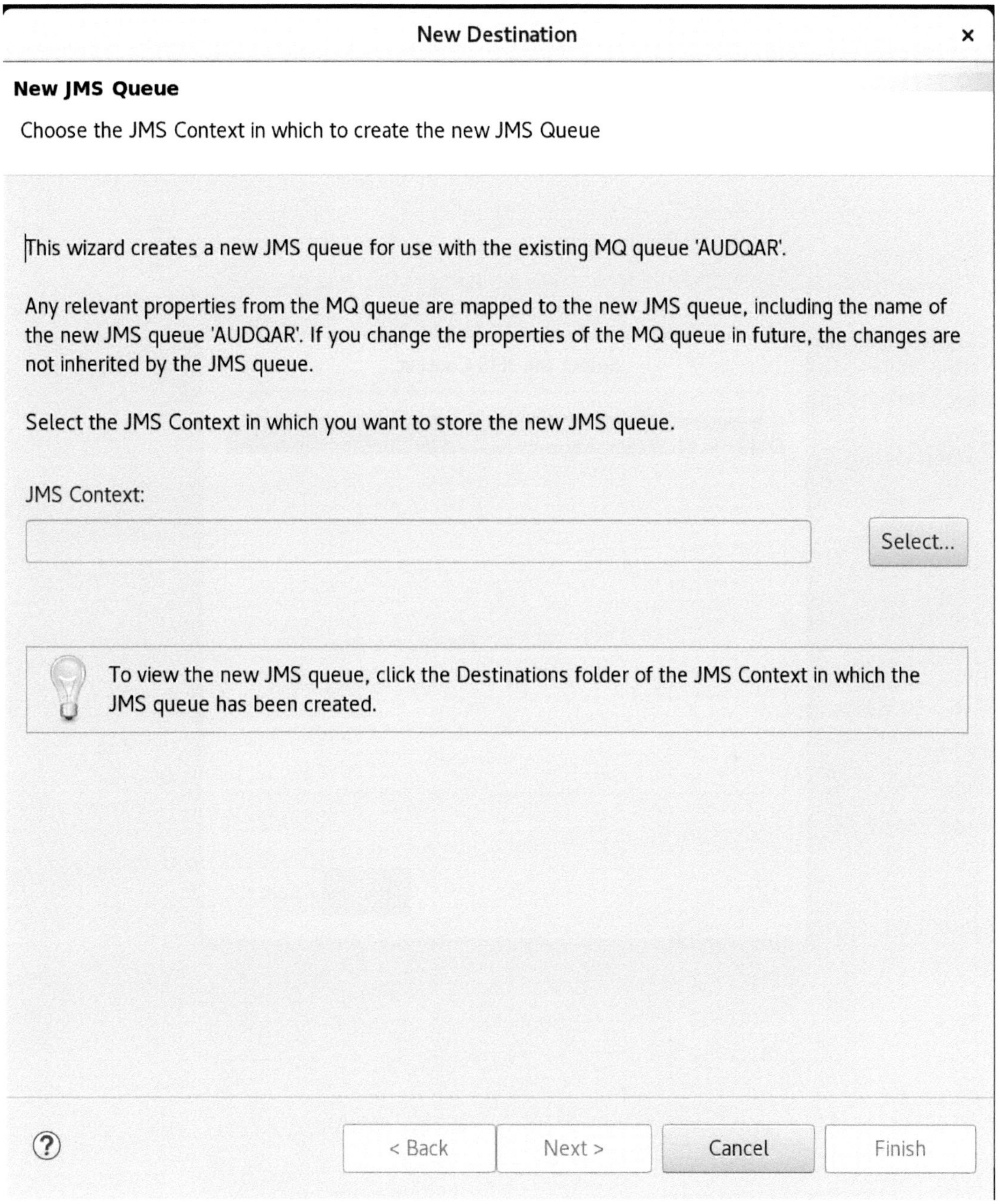

Figure 3-43. The JMS Queue wizard requires the JMS Context to be selected

CHAPTER 3 IBM JMS INTERFACE DEVELOPMENT IBM FILENET 5.5.X WORKFLOW

The JMS Queue wizard is invoked.

Click **Select**.

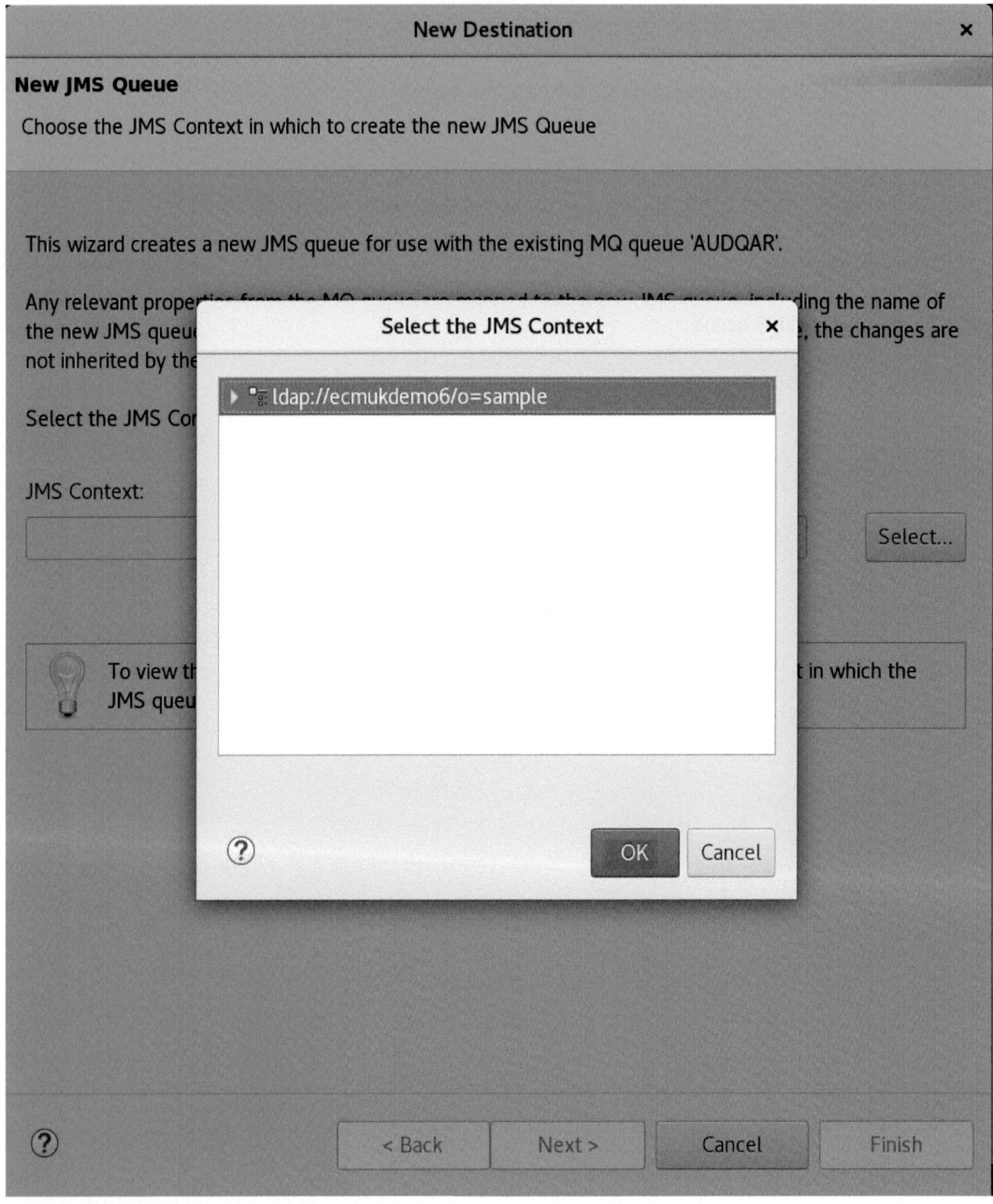

Figure 3-44. *The LDAP server is highlighted; click OK*

CHAPTER 3 IBM JMS INTERFACE DEVELOPMENT IBM FILENET 5.5.X WORKFLOW

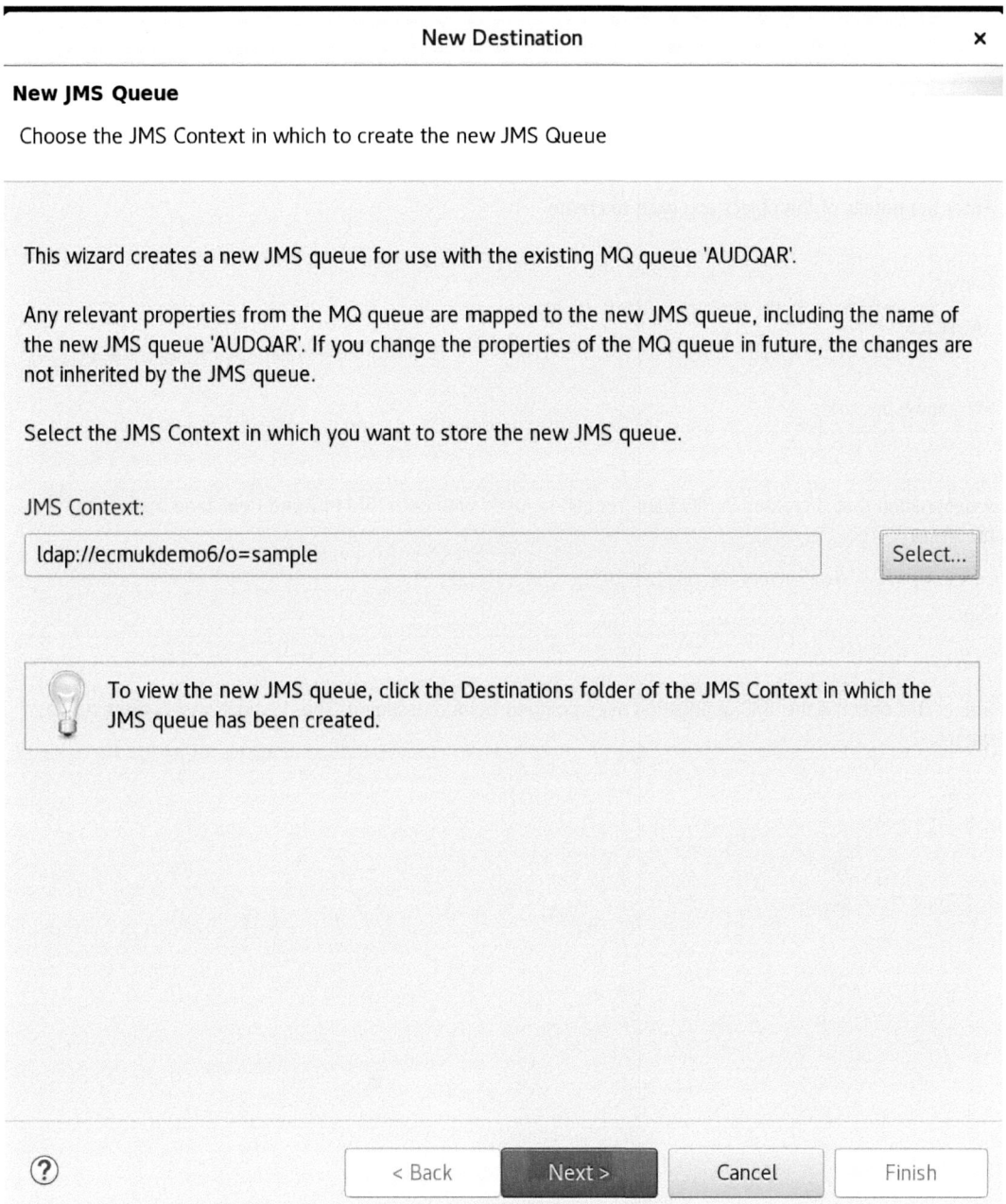

Figure 3-45. The JMS Context shows our LDAP server

CHAPTER 3 IBM JMS INTERFACE DEVELOPMENT IBM FILENET 5.5.X WORKFLOW

Click **Next>**.

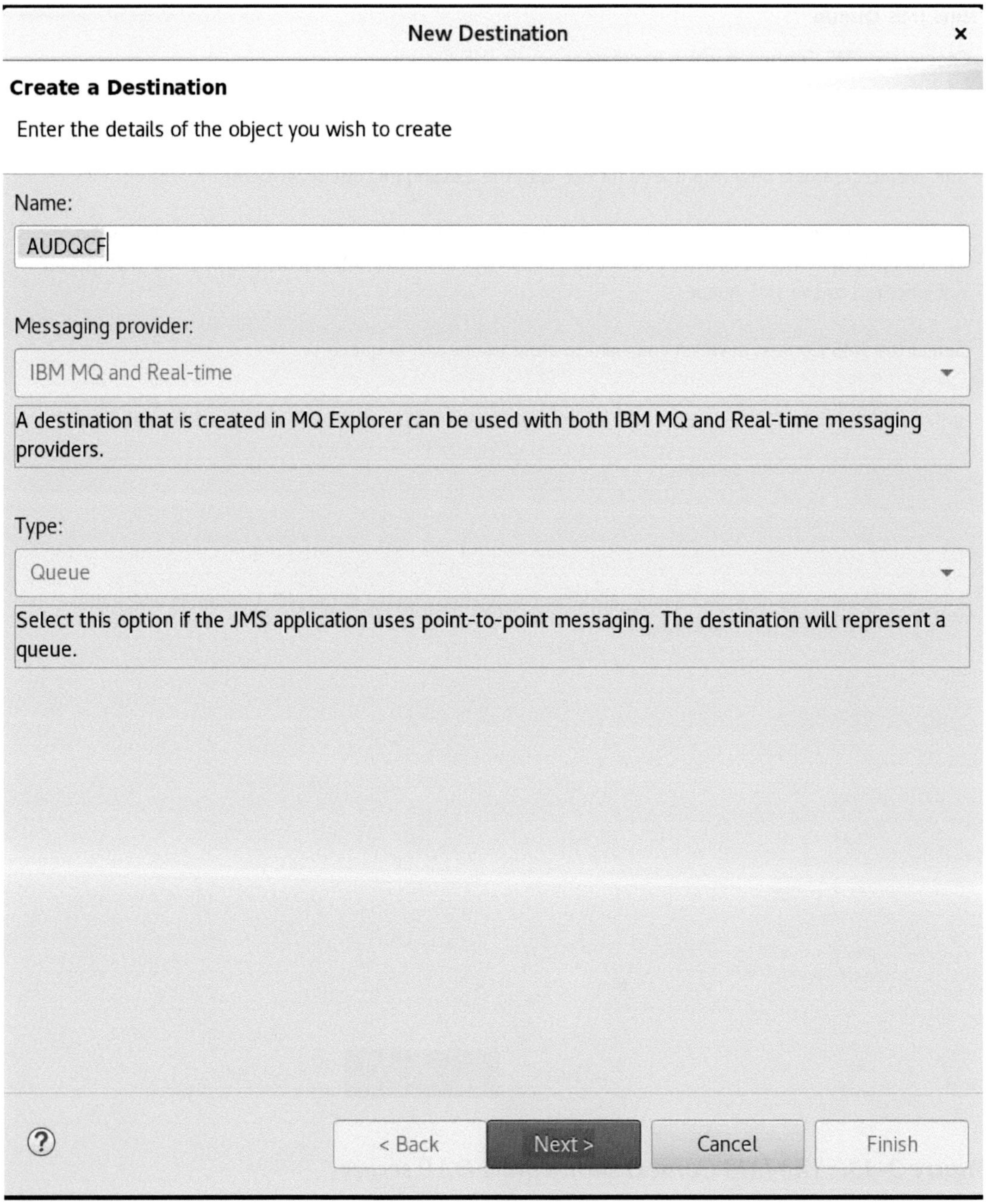

New Destination

Create a Destination

Enter the details of the object you wish to create

Name:

AUDQCF

Messaging provider:

IBM MQ and Real-time

A destination that is created in MQ Explorer can be used with both IBM MQ and Real-time messaging providers.

Type:

Queue

Select this option if the JMS application uses point-to-point messaging. The destination will represent a queue.

< Back | Next > | Cancel | Finish

Figure 3-46. *The Audit Queue Context Factory name is entered as AUDQCF; click Next*

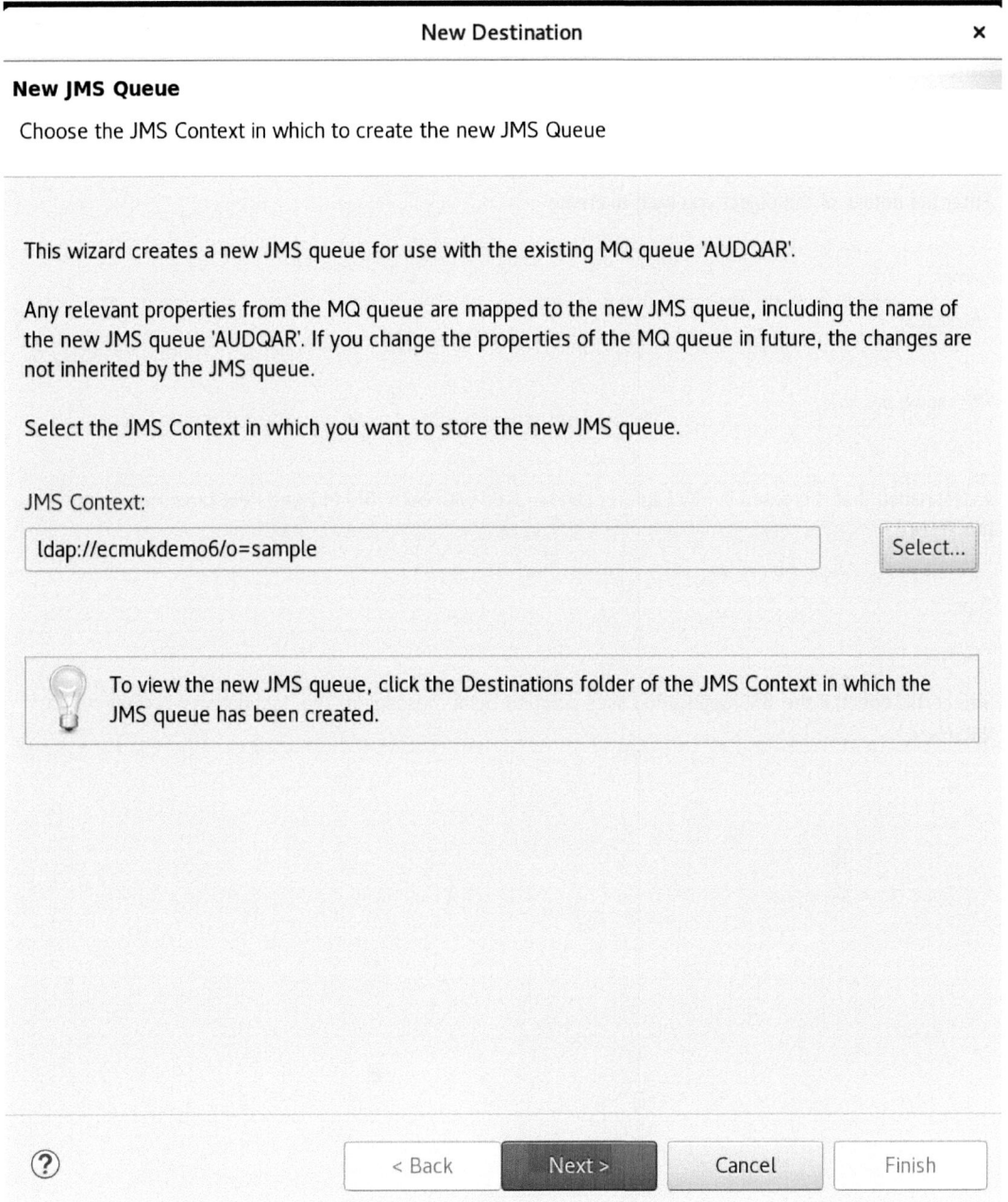

Figure 3-45. *The JMS Context shows our LDAP server*

CHAPTER 3 IBM JMS INTERFACE DEVELOPMENT IBM FILENET 5.5.X WORKFLOW

Click **Next>**.

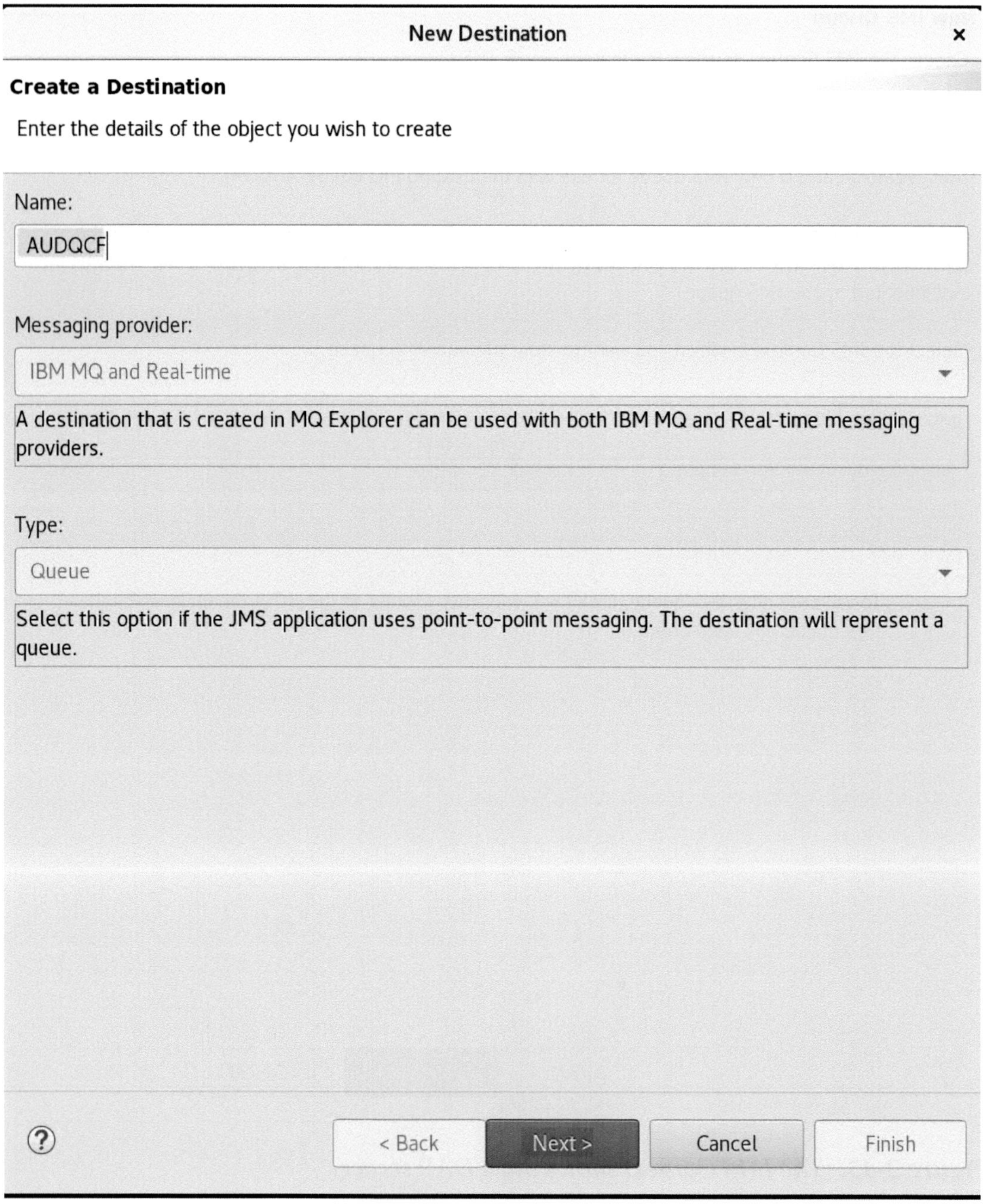

Figure 3-46. The Audit Queue Context Factory name is entered as AUDQCF; click Next

CHAPTER 3   IBM JMS INTERFACE DEVELOPMENT IBM FILENET 5.5.X WORKFLOW

The **AUDQCF** (Audit Queue Context Factory) object name is entered.

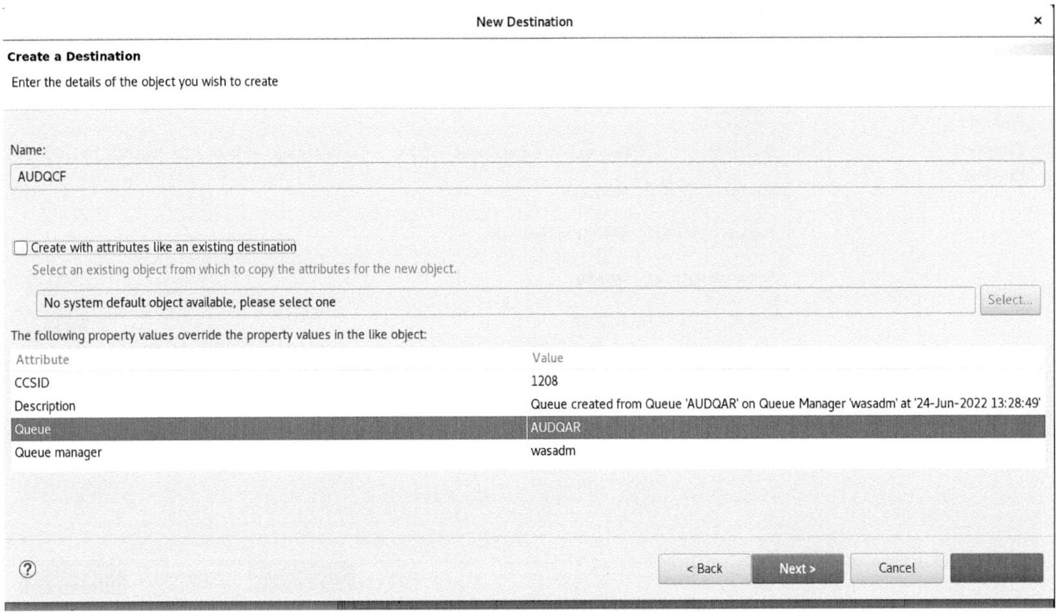

Figure 3-47. The AUDQCF queue name is displayed

Click **Next**.

CHAPTER 3 IBM JMS INTERFACE DEVELOPMENT IBM FILENET 5.5.X WORKFLOW

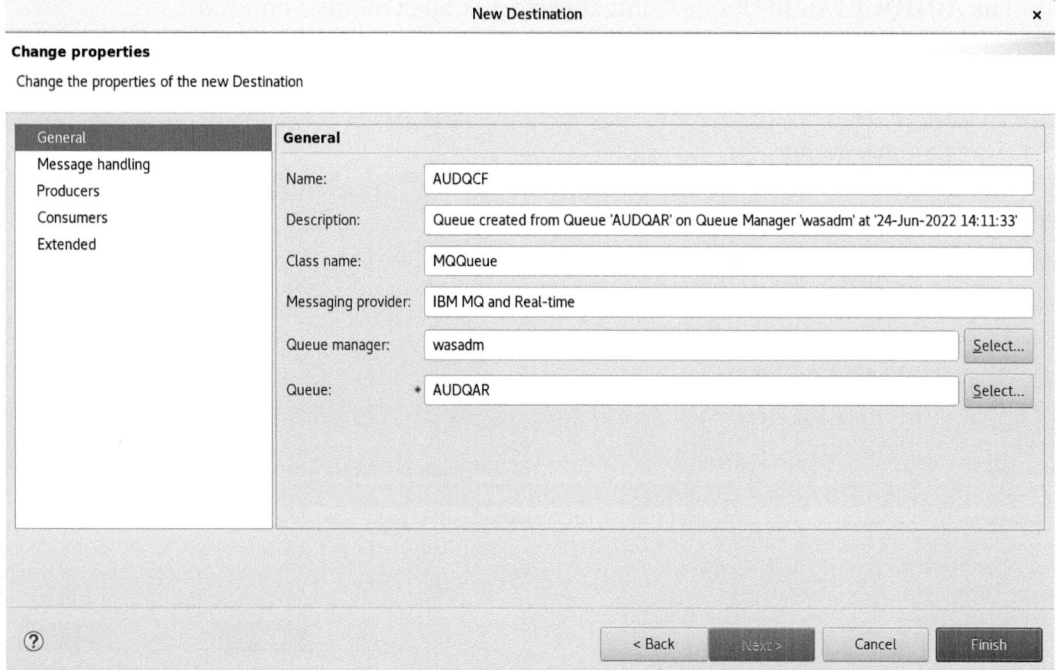

Figure 3-48. *The AUDQCF properties are displayed. Click Finish to create the JMS Queue*

Click Finish.

CHAPTER 3 IBM JMS INTERFACE DEVELOPMENT IBM FILENET 5.5.X WORKFLOW

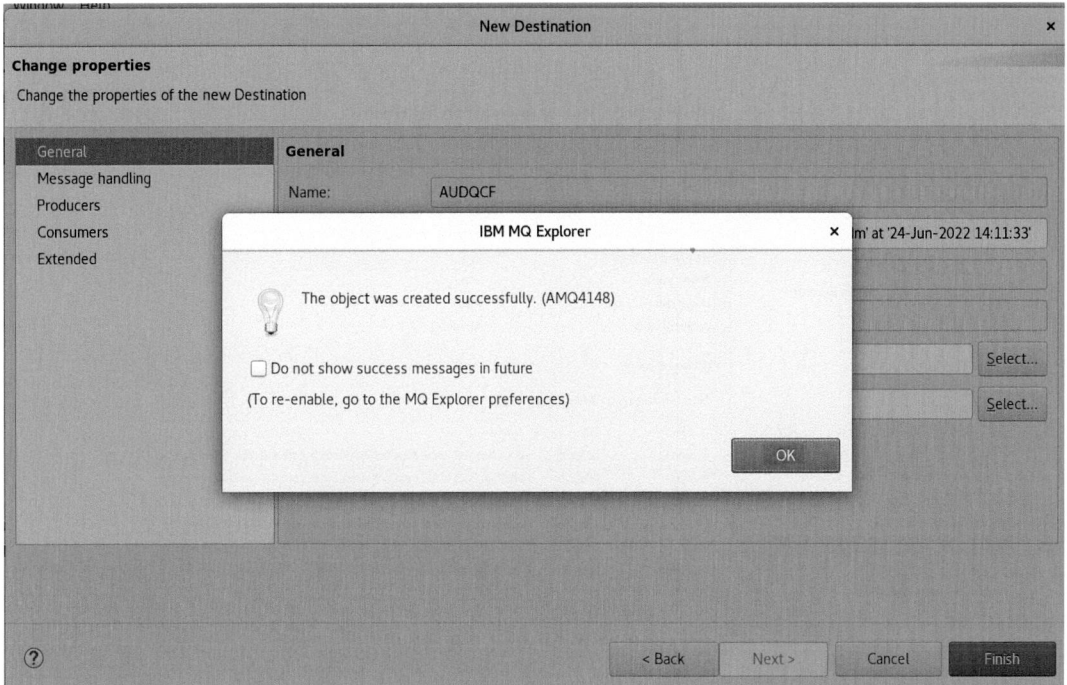

Figure 3-49. *The AUDQCF JMS Queue Context Factory queue status shows it was created*

Click **OK**.

The **JMS** Queue is now created.

CHAPTER 3 IBM JMS INTERFACE DEVELOPMENT IBM FILENET 5.5.X WORKFLOW

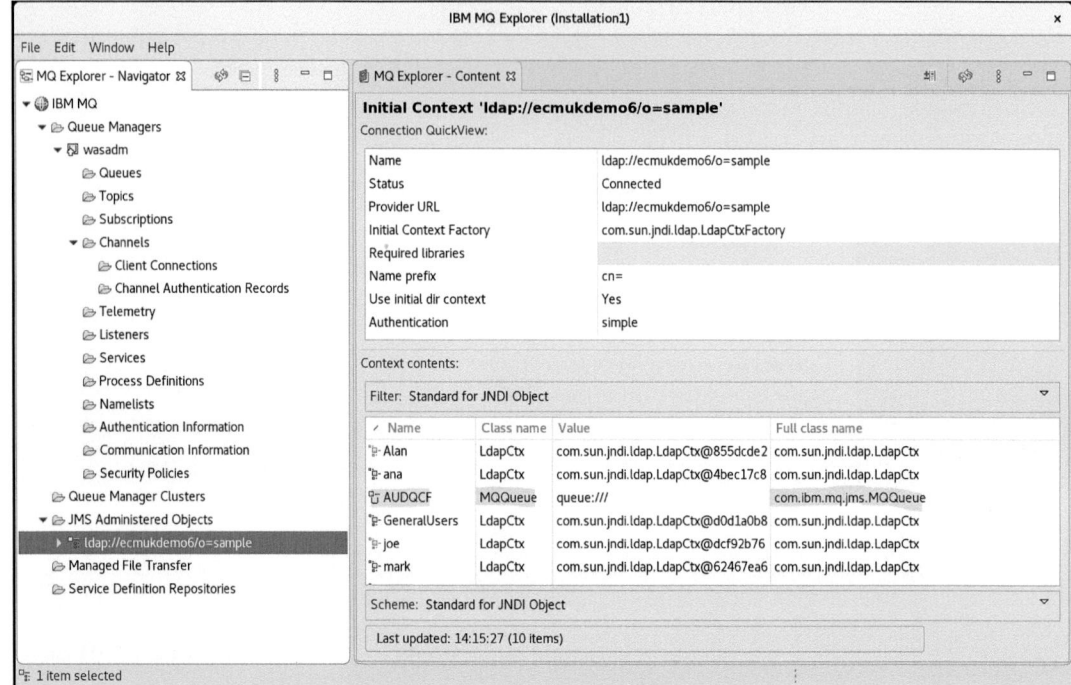

Figure 3-50. *The AUDQCF queue is now visible as a JMS MQQueue class JNDI Object*

Test the **Queue Message** service.

CHAPTER 3   IBM JMS INTERFACE DEVELOPMENT IBM FILENET 5.5.X WORKFLOW

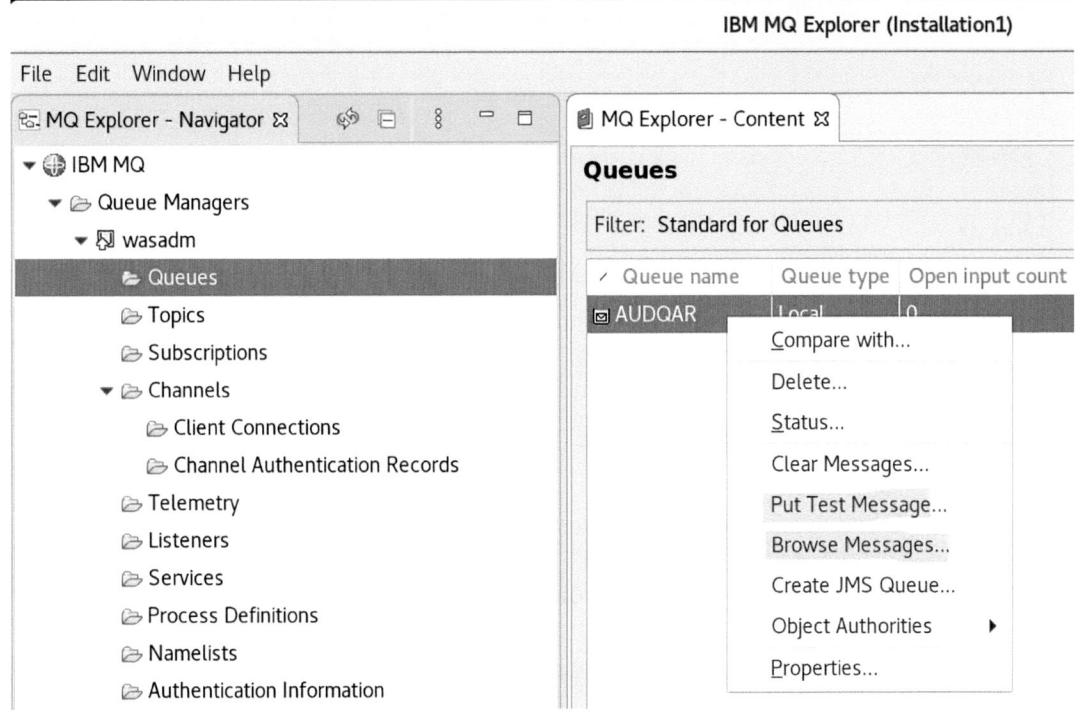

Figure 3-51. *The AUDQAR Queue is tested using the highlighted menu options*

The test message is entered, and the **Put message** command is clicked.

Put test message	
Put message to:	
Queue manager:	
wasadm	
Queue:	
AUDQAR	
Message data:	
This is a test message for the Audit Database auditdb. Audit for Department 00001 and the Audit Report BS5750_0001	
ⓘ The queue which will receive the test message is on this computer. The message will be put directly on the queue.	

Figure 3-52. *The test message is created and the Put message command clicked*

383

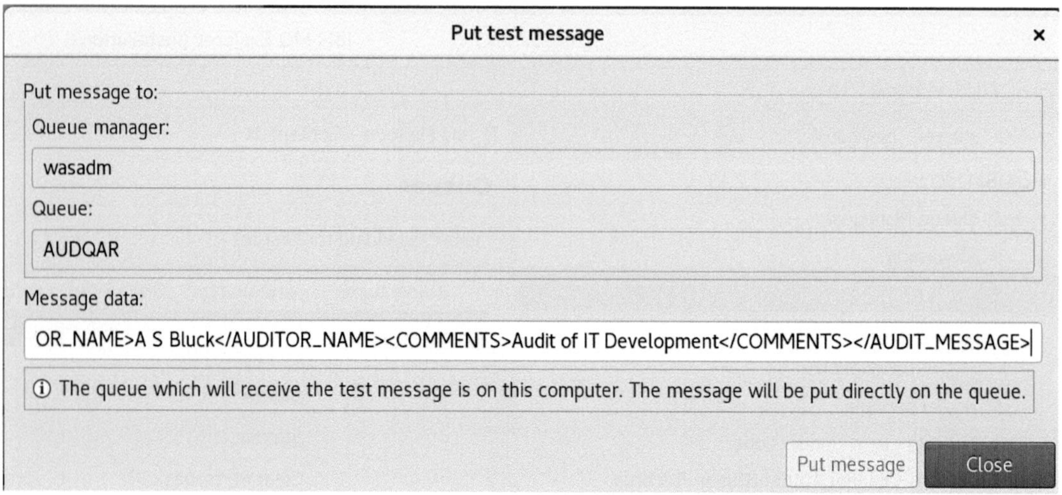

CHAPTER 3 IBM JMS INTERFACE DEVELOPMENT IBM FILENET 5.5.X WORKFLOW

Put test message

Put message to:
Queue manager: wasadm
Queue: AUDQAR
Message data: OR_NAME>A S Bluck</AUDITOR_NAME><COMMENTS>Audit of IT Development</COMMENTS></AUDIT_MESSAGE>

ⓘ The queue which will receive the test message is on this computer. The message will be put directly on the queue.

Figure 3-53. *The browse message shows the transferred test message*

The test message is also sent in **XML** form as shown in Figure 3-53.

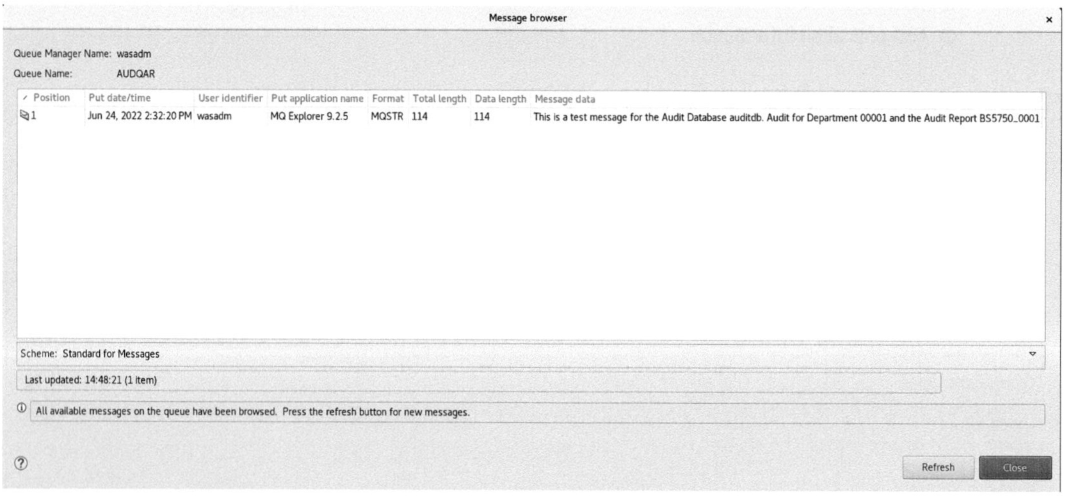

Figure 3-54. *The first test message for the AUDQAR queue*

CHAPTER 3 IBM JMS INTERFACE DEVELOPMENT IBM FILENET 5.5.X WORKFLOW

The message can be seen in the queue using the **Browse Message** function.

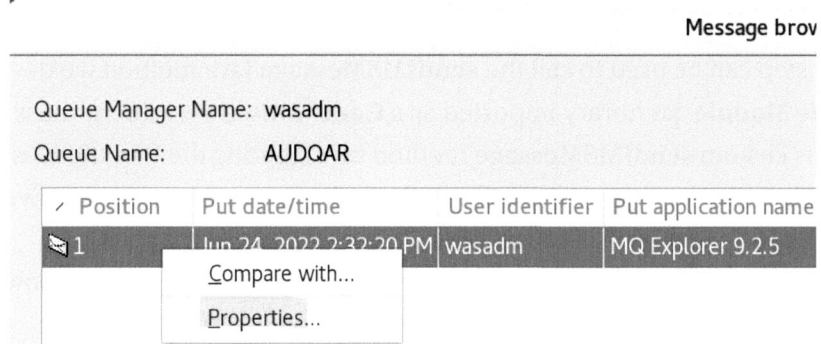

Figure 3-55. *The **Properties** menu item is clicked to display the test message data from the **AUDQAR** queue*

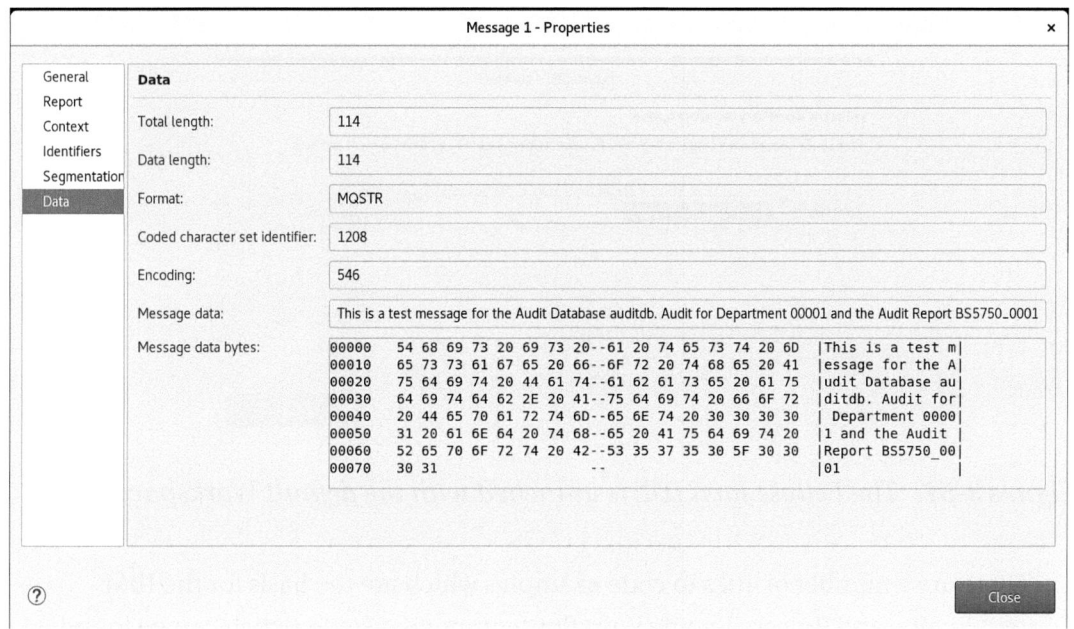

Figure 3-56. *The Data Property of Message 1 displays the expected put message data*

The JMS MQ series queue is now set up ready to be processed.

The following link shows a simplified architecture diagram:

www.ibm.com/docs/en/ibm-mq/9.3?topic=messaging-example-single-queue-manager-publishsubscribe-configuration

Part 2 – Custom Operations Component Development

A Workflow step can be used to call the **sendJMSMessage** Java method we develop, by using a **Code Module** .jar library imported as a **CodeModule** Document Class and then exposing this custom **sendJMSMessage** method by triggering the Workflow using a Workflow **Subscription** object, which can be linked to a Document **Create Event** object reference triggered on the creation of an **Audit Report** Document class.

The Custom Operations Component is defined in the subscribed Workflow as a special **Component Integrator** step, which we will describe in this Part.

First, log out as the wasadm user and log in as root.

cd /root/eclipse/java-2021-12/eclipse
./eclipse

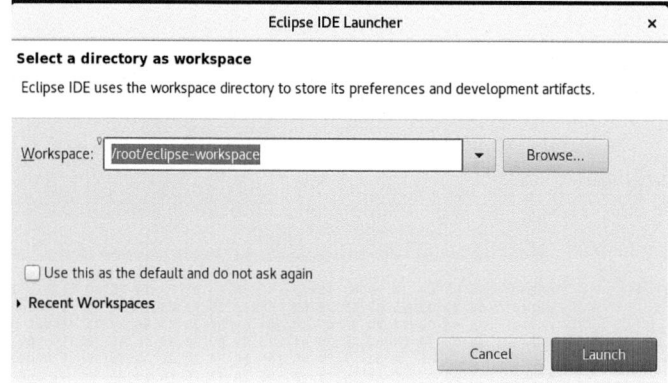

Figure 3-57. *The Eclipse Java IDE is launched with the default Workspace*

There are a number of links to code examples which are the basis for the IBM Content Engine and Process Engine Workflow component steps, which can be found on GitHub:

https://github.com/ibm-ecm/ibm-content-platform-engine-samples

The following IBM License is displayed in the code samples I have used in this book:

https://github.com/ibm-ecm

CHAPTER 3   IBM JMS INTERFACE DEVELOPMENT IBM FILENET 5.5.X WORKFLOW

See the link here for the Java code examples used as a basis for this section.

Licensed Materials - Property of IBM (c) Copyright IBM Corp. 2019 - 2021 All Rights Reserved.

US Government Users Restricted Rights - Use, duplication or disclosure restricted by GSA ADP Schedule Contract with IBM Corp.

DISCLAIMER OF WARRANTIES:

Permission is granted to copy and modify this Sample code, and to distribute modified versions provided that both the copyright notice, and this permission notice and warranty disclaimer appear in all copies and modified versions.

THIS SAMPLE CODE IS LICENSED TO YOU AS-IS. IBM AND ITS SUPPLIERS AND LICENSORS DISCLAIM ALL WARRANTIES, EITHER EXPRESS OR IMPLIED, IN SUCH SAMPLE CODE, INCLUDING THE WARRANTY OF NON-INFRINGEMENT AND THE IMPLIED WARRANTIES OF MERCHANTABILITY OR FITNESS FOR A PARTICULAR PURPOSE. IN NO EVENT WILL IBM OR ITS LICENSORS OR SUPPLIERS BE LIABLE FOR ANY DAMAGES ARISING OUT OF THE USE OF OR INABILITY TO USE THE SAMPLE CODE, DISTRIBUTION OF THE SAMPLE CODE, OR COMBINATION OF THE SAMPLE CODE WITH ANY OTHER CODE. IN NO EVENT SHALL IBM OR ITS LICENSORS AND SUPPLIERS BE LIABLE FOR ANY LOST REVENUE, LOST PROFITS OR DATA, OR FOR DIRECT, INDIRECT, SPECIAL, CONSEQUENTIAL, INCIDENTAL OR PUNITIVE DAMAGES, HOWEVER CAUSED AND REGARDLESS OF THE THEORY OF LIABILITY, EVEN IF IBM OR ITS LICENSORS OR SUPPLIERS HAVE BEEN ADVISED OF THE POSSIBILITY OF SUCH DAMAGES.

The following DemoJava.zip download link contains the code for the core connection Java for links to the IBM MQ Series JMS Queues:

www.ibm.com/support/pages/filenet-content-engine-java-api-demo-sample-code

CHAPTER 3   IBM JMS INTERFACE DEVELOPMENT IBM FILENET 5.5.X WORKFLOW

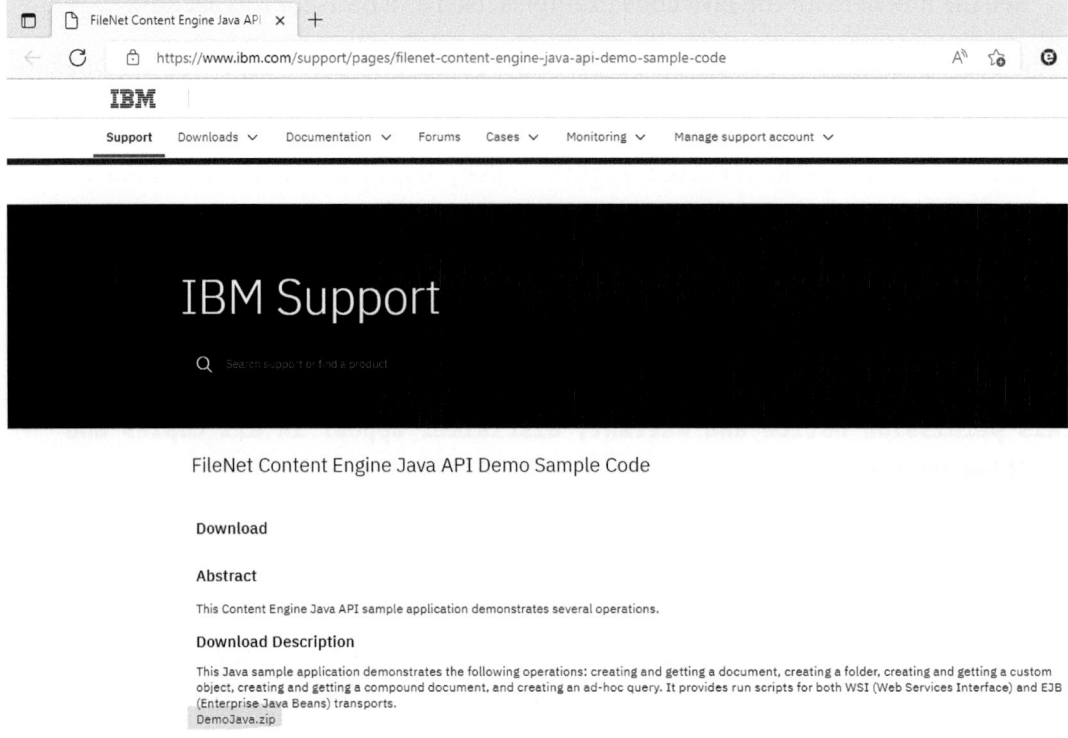

*Figure 3-58. The **DemoJava.zip** link highlighted downloads the sample Java*

The base Java code examples are downloaded from the link in Figure 3-58, and the **DemoJava.zip** file is unpacked to the **DemoJava** folder.

*Figure 3-59. The example Java source code is unpacked to a **DemoJava** subfolder*

In Eclipse, copy the sample Java components, and edit to a new **Java AUD_CIOps** project (this time, we deselected the module name option).

Figure 3-60. *The new AUD_CIOps project is created clicking Finish*

See also Chapter 2, in the start of the section "Configuring Java Components for Content Engine Events," to see the step-by step screenshots and procedure for creating a new Eclipse Java Project.

CHAPTER 3   IBM JMS INTERFACE DEVELOPMENT IBM FILENET 5.5.X WORKFLOW

This Component Integrator code component is described in overview in the IBM Documentation:

www.ibm.com/docs/en/filenet-p8-platform/5.5.x?topic=tools-component-integrator

and in more detail with the following link:
"Developing Component Integrator-Based Work Performers – IBM Documentation"

www.ibm.com/docs/en/filenet-p8-platform/5.5.x?topic=performers-developing-component-integrator-based-work

The following **IBM Technote** has a link to a downloadable pdf document, which describes changes in the deployment for the Component Integrator Java libraries, into the Component Manager, which were introduced in the IBM FileNet P8 5.2 Component Manager version.

See www-01.ibm.com/support/docview.wss?uid=swg27043131.

From this downloaded pdf document, **Migrating to the 5.2 Component Manager.pdf**, it should be noted that the JAR files are added to the object store as a code module, as we described for the Event Handler Java library in Chapter 2. It is also recommended that either the JAR file is stored as a separate content element (since the code module is a versionable object, and JAR files can be added, removed, and updated as required), where the same code module Object can be used by more than one component queue, or, as IBM recommends, that the best practice is to use a separate code module for each component queue.

Creating the AUD_CIOps Java Project

The new Java Project is created as **AUD_CIOps**.
The Package and Class outlines used are as follows.

Listing 3-3. The specification for the AUDOperations Component Integrator code module

```
Package: com.ibm.filenet.ps.ciops
Class AUDOperations (void)
=========================
Variables:
JMSsession: Session
JMSconnection: Connection
```

```
connFactory:ConnectionFactory
messageString: String
dLookup: String
logger: Logger
session: Session
connection: Connection
Methods
=======
AUDOperations() (Constructor)
getConnection():Connection
getVWSession():VWSession
getDatabaseConnection(): Connection
getFolderFromAttachment(VWAttachment):Folder (Private)
getDocumentFromAttachment(VWAttachment): VWAttachment[]:Document (Private)
getObjectStore(String):ObjectStore (Private)
getFolderDocuments(VWAttachment):VWAttachment[] (Public)
getContainedDocuments(Folder):DocumentSet (Private)
getContainedDocumentsPropertyFilter():PropertyFilter (Private)
getAsVWAttachment(Document):VWAttachment (Private)
sendJMSMessage(VWAttachment,String):void (Public)
putMsg(Destination,String):void (Private)
```

The code for the **CELoginModule** Class, working at FileNet P8 5.5.x, is listed at the following link:

www.ibm.com/docs/fr/filenet-p8-platform/5.5.x?topic=performers-celoginmodule-class

A Jar file is created to test the base compilation and functionality.
The Java code created is as follows.

Listing 3-4. The AUDOperations.java code for the Component Integrator

```java
package com.ibm.filenet.ps.ciops;

import java.security.AccessController;
import java.util.ArrayList;
import java.util.Iterator;
import java.util.Set;
```

CHAPTER 3   IBM JMS INTERFACE DEVELOPMENT IBM FILENET 5.5.X WORKFLOW

```java
//JMS packages
import javax.jms.*;
import javax.naming.*;
import javax.naming.directory.*;
import java.util.Hashtable;
import javax.jms.ConnectionFactory;
import javax.jms.Destination;
import javax.jms.JMSException;
import javax.jms.MessageProducer;
import javax.jms.Queue;
import javax.jms.TextMessage;
import javax.jms.Topic;
import javax.mail.Session;
import javax.mail.Transport;
import javax.mail.internet.InternetAddress;
import javax.mail.internet.MimeMessage;
import javax.naming.Context;
import javax.naming.directory.InitialDirContext;
import javax.naming.ldap.InitialLdapContext;
import javax.security.auth.Subject;

import com.filenet.api.collection.DocumentSet;
import com.filenet.api.constants.PropertyNames;
import com.filenet.api.core.Connection;
import com.filenet.api.core.Document;
import com.filenet.api.core.Domain;
import com.filenet.api.core.EntireNetwork;
import com.filenet.api.core.Factory;
import com.filenet.api.core.Folder;
import com.filenet.api.core.ObjectStore;
import com.filenet.api.property.FilterElement;
import com.filenet.api.property.PropertyFilter;
import com.ibm.filenet.ps.ciops.database.DatabasePrincipal;
import com.ibm.mq.MQC;
import com.ibm.mq.MQEnvironment;
import com.ibm.mq.MQMessage;
```

```java
import com.ibm.mq.MQPutMessageOptions;
import com.ibm.mq.MQQueueManager;
import com.ibm.mq.constants.CMQC;
import com.ibm.mq.constants.MQConstants;
import com.ibm.mq.jms.MQConnectionFactory;
import com.ibm.mq.jms.MQQueue;

import filenet.vw.api.VWAttachment;
import filenet.vw.api.VWAttachmentType;
import filenet.vw.api.VWException;
import filenet.vw.api.VWLibraryType;
import filenet.vw.api.VWSession;
import filenet.vw.base.logging.Logger;
//Import JMS related packages

/**
 * Alan S. Bluck - 26th June 2022
 * AUD JMS Message sender Component Integrator code for AUDOperations
 *
 *
 */
public class AUDOperations {
    javax.jms.Session JMSsession = null;
    javax.jms.Connection JMSconnection = null;
    javax.jms.ConnectionFactory  connFactory  = null;
    String messageString = null;
    String dLookup      = "cn=AUDQAR"; // Generic JMS Destination
        private static Logger logger = Logger.getLogger(
        AUDOperations.class );
    //JMS Variables
      static               Session    session    = null; // JMS Session
      static               Connection connection = null; // JMS Connection
      public AUDOperations() {
          //java.sql.Connection databaseConnection =
          getDatabaseConnection();
```

```java
        //if ( databaseConnection != null ) {
        //    logger.debug( databaseConnection.toString() );
        //} else {
            logger.debug( "No database connection" );
        //}
    }

    protected Connection getConnection() {
        String uri = System.getProperty("filenet.pe.bootstrap.ceuri");
        return Factory.Connection.getConnection(uri);
    }

    protected VWSession getVWSession() throws VWException {
        String connectionPoint = System.getProperty("filenet.pe.cm.connectionPoint");
        return new VWSession(connectionPoint);
    }

    protected java.sql.Connection getDatabaseConnection() {
        Subject subject = Subject.getSubject(
        AccessController.getContext() );
        Set<DatabasePrincipal> principals = subject.getPrincipals(
        DatabasePrincipal.class );
        if ( principals != null && ! principals.isEmpty() ) {
            DatabasePrincipal principal = principals.iterator().next();
            return principal.getConnection();
        }
        return null;
    }

    private Folder getFolderFromAttachment(VWAttachment folderAttachment) {
        ObjectStore objectStore = getObjectStore( folderAttachment.getLibraryName() );
        Folder folder = (Folder) objectStore.getObject("Folder",
        folderAttachment.getId() );
        return folder;
    }
```

```java
private Document getDocumentFromAttachment(VWAttachment
documentAttachment) {
    ObjectStore objectStore = getObjectStore( documentAttachment.
    getLibraryName() );
    Document folder = (Document) objectStore.getObject("Document",
    documentAttachment.getId() );
    return folder;
}

private ObjectStore getObjectStore( String objectStoreName ) {
    Connection connection = getConnection();
    EntireNetwork entireNetwork = Factory.EntireNetwork.fetchInstance
    (connection, null);
    Domain domain = entireNetwork.get_LocalDomain();
    return Factory.ObjectStore.getInstance( domain,
    objectStoreName );
}

/**
 * Returns the documents filed in the folder.
 *
 * @param folderAttachment the input folder.
 * @return an array of documents filed in the folder.
 * @throws Exception
 */
public VWAttachment[] getFolderDocuments(VWAttachment folderAttachment
) throws Exception {

    Folder folder = getFolderFromAttachment(folderAttachment);
    DocumentSet containedDocuments = getContainedDocuments(folder);
    Iterator<?> iterator = containedDocuments.iterator();
    ArrayList<VWAttachment> containedDocumentList = new
    ArrayList<VWAttachment>();
```

```java
        while ( iterator.hasNext() ) {
            Document document = (Document) iterator.next();
            VWAttachment documentAttachment =
            getAsVWAttachment(document);
            containedDocumentList.add( documentAttachment );
        }

        return containedDocumentList.toArray( new VWAttachment[0] );
    }

    private DocumentSet getContainedDocuments(Folder folder) {
        PropertyFilter propertyFilter =
        getContainedDocumentsPropertyFilter();
        folder.fetchProperties( propertyFilter );
        DocumentSet containedDocuments = folder.get_ContainedDocuments();
        return containedDocuments;
    }

    private PropertyFilter getContainedDocumentsPropertyFilter() {
        PropertyFilter propertyFilter = new PropertyFilter();
        propertyFilter.addIncludeProperty( new FilterElement( null, null,
        null, PropertyNames.CONTAINED_DOCUMENTS, null ) );
        propertyFilter.addIncludeProperty( new FilterElement( 2, null,
        null, PropertyNames.ID, null ) );
        propertyFilter.addIncludeProperty( new FilterElement( 2, null,
        null, "DocumentTitle", null ) );
        return propertyFilter;
    }

    private VWAttachment getAsVWAttachment(Document document) throws
    VWException {

        VWAttachment documentAttachment = new VWAttachment();

        documentAttachment.setLibraryType( VWLibraryType.LIBRARY_TYPE_
        CONTENT_ENGINE );
        ObjectStore objectStore = document.getObjectStore();
```

```java
        objectStore.fetchProperties( new String[] {
        PropertyNames.NAME } );
        documentAttachment.setLibraryName( objectStore.get_Name() );

        document.fetchProperties( new String[] { PropertyNames.ID,
        PropertyNames.NAME } );
        documentAttachment.setId( document.get_Id().toString() );
        documentAttachment.setAttachmentName( document.get_Name() );
        documentAttachment.setType( VWAttachmentType.ATTACHMENT_TYPE_
        FOLDER );

        return documentAttachment;
    }

    public void sendJMSMessage(VWAttachment docAttachment, String Message
    ) throws Exception {
{
  messageString = Message;

  // A single try block is used here to allow us to focus on the
  JNDI and I/O
  // operations.
// TODO Production code would need to have much more exception handling.
    try {
      //
      //ASB - Start using Tivoli Directory Services LDAP Context
      //create initial context properties
      String url = "ldap://ecmukdemo6/o=sample";
      String icf = "com.sun.jndi.ldap.LdapCtxFactory";
      java.util.Hashtable environment = new java.util.Hashtable();
      environment.put(Context.PROVIDER_URL, url);
      environment.put(Context.INITIAL_CONTEXT_FACTORY, icf);
      Context ctx = new InitialDirContext(environment);
      Destination       myDest       = null;
      String qManager = "AUDQM";
      // and define the name of the Queue
      String qName = "AUDQAR";
```

```java
// Put the message to the queue
System.out.println("Sending a message...");
String     cfLookup    = "cn=AUDQCF";
System.out.println("Lookup connection factory " + cfLookup);
String     JNDITopic   = "cn=AUDTopic";
String     JNDIQueue   = "cn=AUDQAR";
// Class variables
javax.jms.Session       jmssession     = null; // JMS Session
javax.jms.Connection jmsconnection    = null; // JMS Connection
String     dLookup     = "cn=AUDQAR"; // Generic JMS Destination
String     myMode      = null; // Program mode
String     destType    = null; // Destination type
//JMSDEMO

MQEnvironment.hostname = "ecmukdemo6";
MQEnvironment.channel  = "SYSTEM.DEF.CLNT.CONN";
MQEnvironment.port = 1417;
MQEnvironment.properties.put(CMQC.TRANSPORT_PROPERTY, CMQC.TRANSPORT_MQSERIES_CLIENT);
MQMessage hello_world = new MQMessage();
hello_world.writeUTF("Hello World! - First Java Message");
// Create a connection to the queue manager

ConnectionFactory connFactory = (ConnectionFactory) ctx.lookup( cfLookup );
ConnectionFactory factory;
Queue queue;
Topic topic;
queue = (Queue)ctx.lookup(JNDIQueue);
System.out.println("Create and start the connection");
jmsconnection = connFactory.createConnection("wasadm","filenet");
jmsconnection.start();

System.out.println("Create the session");

boolean transacted = true;
```

```java
javax.jms.Session session= jmsconnection.createSession(transacted,
javax.jms.Session.AUTO_ACKNOWLEDGE);
myDest = (Destination)ctx.lookup(dLookup);

//Create the session XML Test Message
String xmlTestMessage = "<?xml version=\"1.0\"
encoding=\"utf-8\"?><GRP.SDR.DOC_ACK.REPLY><corrItemID>12345678</
corrItemID><CIN>CIN1234567</CIN><msgTimeStamp>2012-12-13T12:12:12</ms
gTimeStamp><CustMsgDesc>Disposal Date of Document</CustMsgDesc></GRP.
SDR.DOC_ACK.REPLY>";
putMsg( myDest, xmlTestMessage,session );
 // Clean up session and connection
jmssession.close();
jmssession = null;

jmsconnection.close();
jmsconnection = null;

jmssession.close();
jmssession = null;

JMSconnection.close();
JMSconnection = null;

} catch( JMSException je ) {
  System.out.println("caught JMSException: " + je);
  Exception le = je.getLinkedException();
  if (le != null)  System.out.println("linked exception: "+le);
   //TODO Log exception message

} catch( Exception e ) {
   //TODO Log exception message
      System.out.println("Error : " + e.getMessage());
      System.out.println("Error Stack : " + e.toString());
// A finally block is a good place to ensure that we don't forget
// to close the most important JMS objects
} finally {
```

```java
    if (JMSsession != null) {
      //Closing Session
      JMSsession.close();
    }
    if (JMSconnection != null) {
      //Closing Connection
      JMSconnection.close();
    }
  }
      //if (le.getMessage() != null)
         //TODO Add to logger
   //Finished
      }
    }

  // A single
  private void putMsg( Destination myDest, String outString, javax.jms.
  Session jmssession )
    throws JMSException, Exception
  {
//MessageProducer myProducer = JMSsession.createProducer(myDest);//
REFLECTION ISSUE!
      if (Message.length() > 0) {
      TextMessage outMessage = JMSsession.createTextMessage();
        //myProducer.send(outMessage); //REFLECTION ISSUE!
         JMSsession.commit();
  }
}
```

Adding the Supporting Library JAR Files

a) First, we add the same IBM FileNet P8 API library .jar files to the Classpath as we added in Chapter 2. Additionally, we need to add the JMS Message library, **javax.jms-api-2.0.1.jar**, to the Eclipse IDE classpath.

CHAPTER 3   IBM JMS INTERFACE DEVELOPMENT IBM FILENET 5.5.X WORKFLOW

Create the **MQJars** folder for the jar files we need:

```
(base) [root@ECMUKDEMO6 opt]# mkdir MQJars
(base) [root@ECMUKDEMO6 opt]# cd MQJars
(base) [root@ECMUKDEMO6 MQJars]# pwd
/opt/MQJars
```

b) From the MQJars folder, download the **com.ibm.mq.allclient.jar** file by using curl:

```
(base) [root@ECMUKDEMO6 MQJars]#
 curl -o com.ibm.mq.allclient-9.2.4.0.jar https://repo1.maven.org/maven2/com/ibm/mq/com.ibm.mq.allclient/9.2.4.0/com.ibm.mq.allclient-9.2.4.0.jar
```

c) From the MQJars folder, download the JMS API file by using curl:

```
curl -o javax.jms-api-2.0.1.jar https://repo1.maven.org/maven2/javax/jms/javax.jms-api/2.0.1/javax.jms-api-2.0.1.jar
```

d) From the MQJars folder, download the JSON .jar file by using curl:

```
curl -o json-20211205.jar https://repo1.maven.org/maven2/org/json/json/20211205/json-20211205.jar
```

e) Set the security on the /opt/MQJars and the contained .jar files:

```
chmod 755 -R MQJars/
```

```
(base) [root@ECMUKDEMO6 opt]# ls -ls MQJars/
total 8280
8144 -rwxr-xr-x. 1 root root 8338334 Jun 26 10:31 com.ibm.mq.allclient-9.2.4.0.jar
  64 -rwxr-xr-x. 1 root root   64009 Jun 26 10:31 javax.jms-api-2.0.1.jar
  72 -rwxr-xr-x. 1 root root   70678 Jun 26 10:36 json-20211205.jar
(base) [root@ECMUKDEMO6 opt]#
```

Figure 3-61. *The referenced JMS MQ message .jar files are downloaded*

f) In the Eclipse project Classpath, we now add the preceding downloaded .jar files.

Chapter 3   IBM JMS Interface Development IBM Filenet 5.5.x Workflow

Figure 3-62. *The copied JMS Java and IBM MQ libraries are placed in the Classpath*

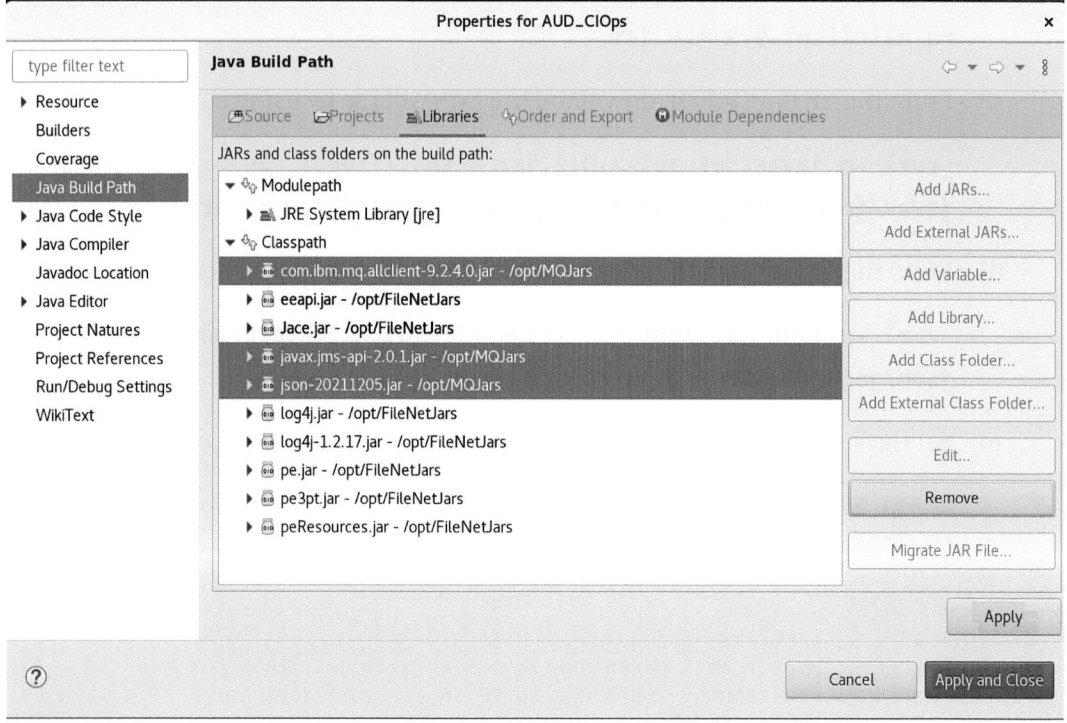

Figure 3-63. *The **Apply** and **Apply and Close** update the Classpath*

Installing the IBM JRE Java SDK 1.8

We also need to install and load the IBM JRE jre1.8.0.261 Java JRE library on Red Hat Linux 8.0, using the download URL:

www.ibm.com/support/pages/java-sdk-downloads-version-80

402

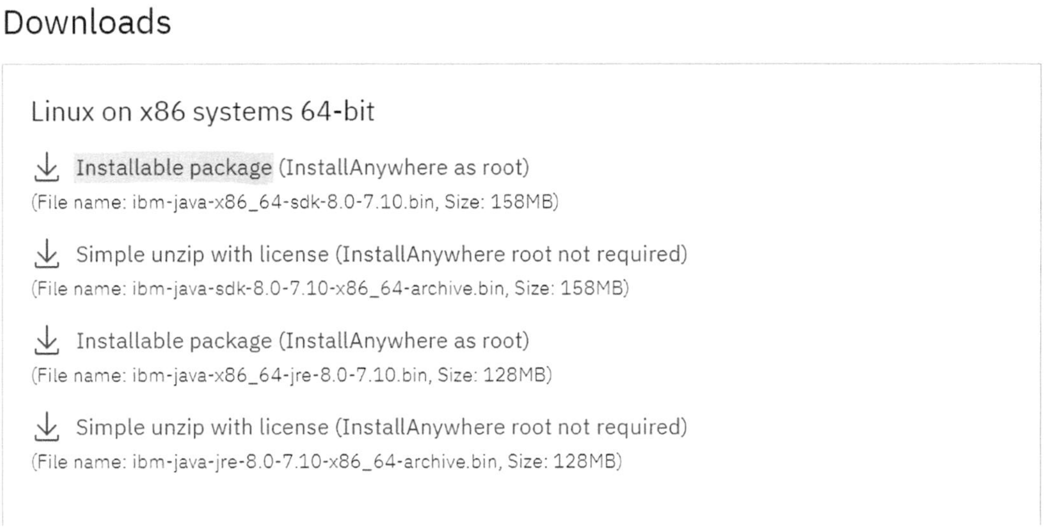

CHAPTER 3　IBM JMS INTERFACE DEVELOPMENT IBM FILENET 5.5.X WORKFLOW

Downloads

Linux on x86 systems 64-bit

↓ Installable package (InstallAnywhere as root)
(File name: ibm-java-x86_64-sdk-8.0-7.10.bin, Size: 158MB)

↓ Simple unzip with license (InstallAnywhere root not required)
(File name: ibm-java-sdk-8.0-7.10-x86_64-archive.bin, Size: 158MB)

↓ Installable package (InstallAnywhere as root)
(File name: ibm-java-x86_64-jre-8.0-7.10.bin, Size: 128MB)

↓ Simple unzip with license (InstallAnywhere root not required)
(File name: ibm-java-jre-8.0-7.10-x86_64-archive.bin, Size: 128MB)

Figure 3-64. *The link to the required IBM JRE 1.8.0.261 library*

Note　For Eclipse on Windows systems, the following URL link can be used.

www.ibm.com/support/pages/java-sdk-downloads-eclipse

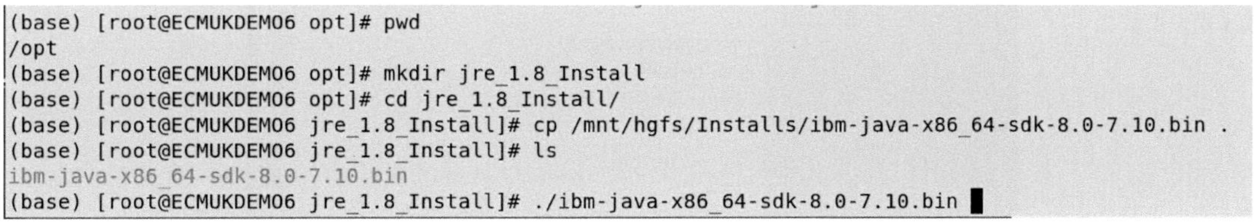

```
(base) [root@ECMUKDEMO6 opt]# pwd
/opt
(base) [root@ECMUKDEMO6 opt]# mkdir jre_1.8_Install
(base) [root@ECMUKDEMO6 opt]# cd jre_1.8_Install/
(base) [root@ECMUKDEMO6 jre_1.8_Install]# cp /mnt/hgfs/Installs/ibm-java-x86_64-sdk-8.0-7.10.bin .
(base) [root@ECMUKDEMO6 jre_1.8_Install]# ls
ibm-java-x86_64-sdk-8.0-7.10.bin
(base) [root@ECMUKDEMO6 jre_1.8_Install]# ./ibm-java-x86_64-sdk-8.0-7.10.bin
```

Figure 3-65. *The downloaded IBM JRE is installed on the Linux VMware*

The first screen is displayed, as shown in Figure 3-66.

CHAPTER 3 IBM JMS INTERFACE DEVELOPMENT IBM FILENET 5.5.X WORKFLOW

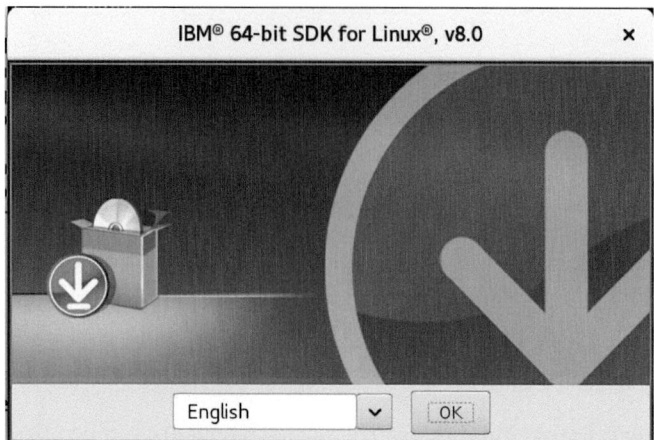

Figure 3-66. *The first install step is used to select the language from the drop-down*

The language required is selected from the drop-down, and the OK command button is clicked.

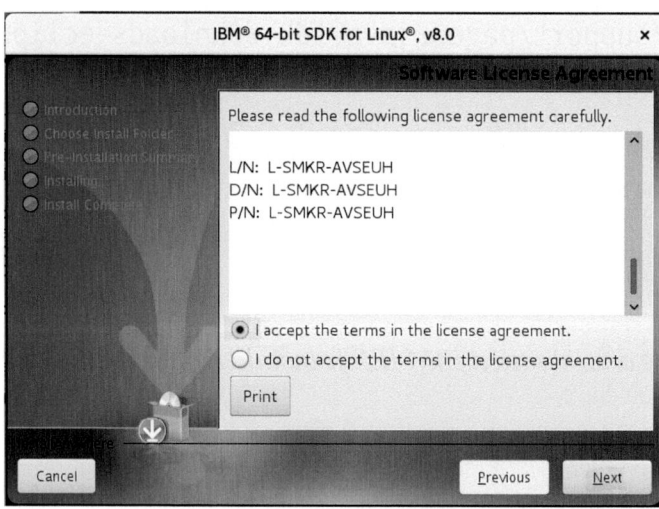

Figure 3-67. *The "I accept the terms in the license agreement" is selected*

CHAPTER 3   IBM JMS INTERFACE DEVELOPMENT IBM FILENET 5.5.X WORKFLOW

The installation is started with the first screen prompt.

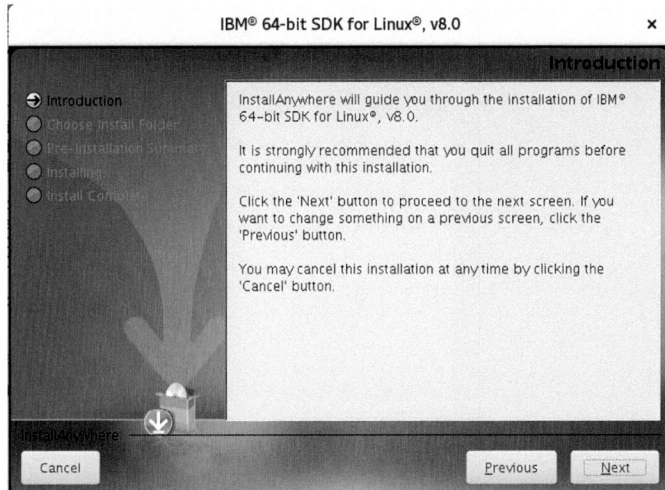

Figure 3-68. *The first installation screen is displayed*

The default folder path displayed, /opt/ibm/java-x86_64-80, is used for the installation.

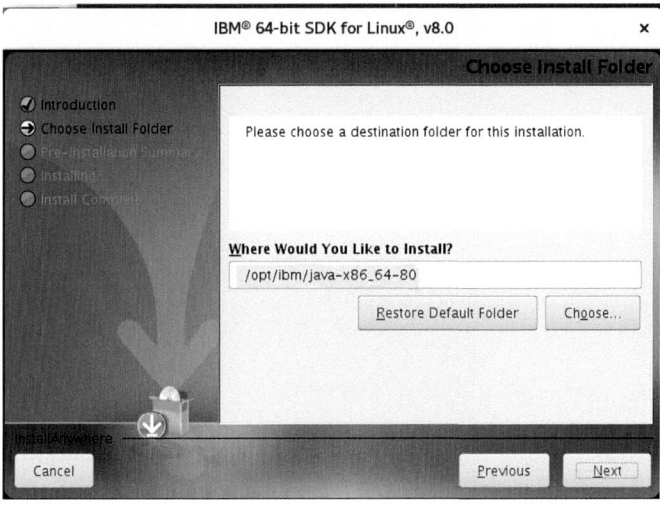

Figure 3-69. *The installation path is selected as highlighted; click Next>*

Chapter 3 IBM JMS Interface Development IBM FileNet 5.5.x Workflow

Click Next on the Installation summary page (or you can cancel at this point). Then the following installation status should be shown.

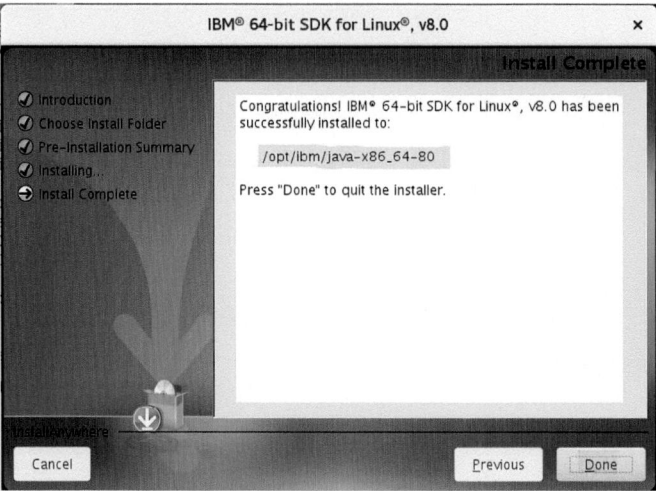

Figure 3-70. *The **Done** command is clicked on the success pop-up window*

This JRE installation was then used in the Classpath for the Java Project in place of the Eclipse default.

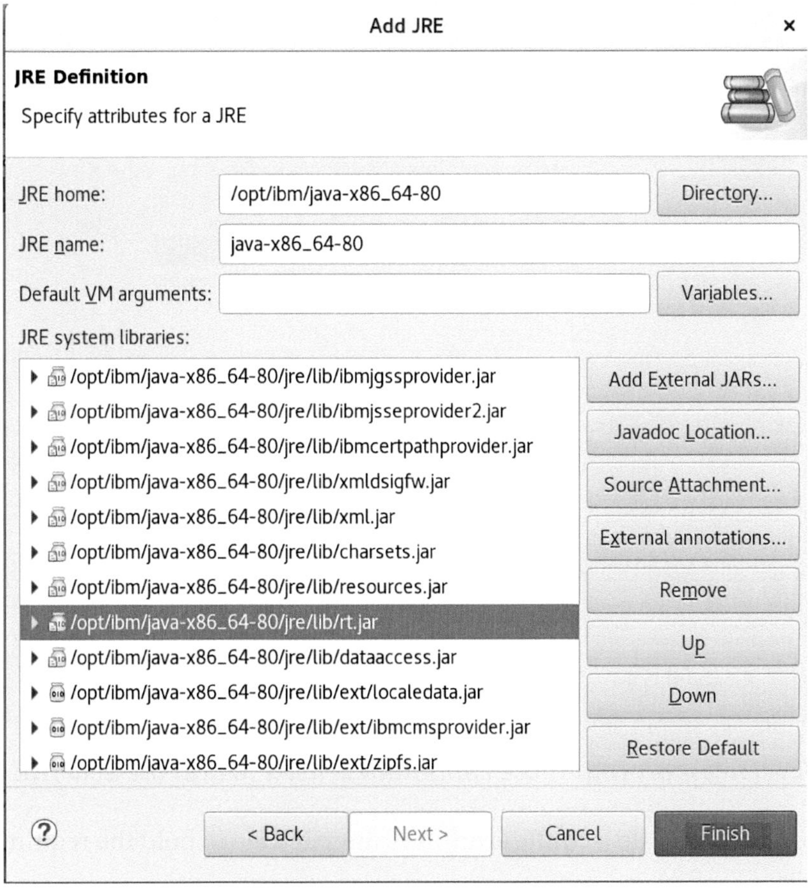

Figure 3-71. The JRE system libraries we installed are confirmed by clicking Finish

The Java Project JRE Libraries are added to the Classpath using the Apply and Close command button.

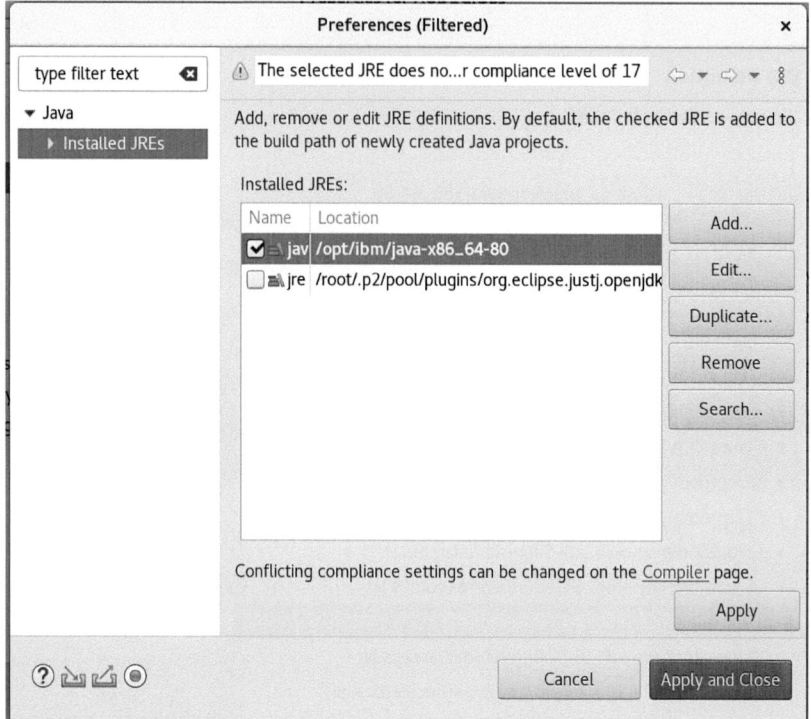

Figure 3-72. *The Apply and Close command is used to add the new JRE library*

Next, we need to create a **configuration.properties** file to hold the required JMS Queue details.

Creating the configuration.properties File

To avoid hard-coding of the many string variables which hold the names of elements, such as the Object Store name and the encrypted IBM FileNet logon credentials, a configuration file is used to hold the string values used by the Java program, as shown in the following section.

*Figure 3-73. The **configuration.properties** file is created*

The **configuration.properties** file text is created as follows.

Listing 3-5. The **configuration.properties** JUnit test file

```
AUDOperationsTest.CEWsiUrl=http://ecmukdemo6:9080/wsi/FNCEWS40MTOM/
AUDOperationsTest.ConnectionPointName=CP1
AUDOperationsTest.EmailFrom=alan.bluck@asbsoftware.co.uk
AUDOperationsTest.EmailTemplatePath=/AUDIT_TEST/AC.pdf
AUDOperationsTest.EmailTo= alan.bluck@asbsoftware.co.uk
AUDOperationsTest.ObjectstoreName=OS2
AUDOperationsTest.Password=filenet
AUDOperationsTest.TestFolderName=/AUDI_TEST
```

CHAPTER 3 IBM JMS INTERFACE DEVELOPMENT IBM FILENET 5.5.X WORKFLOW

```
AUDOperationsTest.Username=Alan
AUDOperationsTest.WaspLocation=/opt/IBM/WebSphere/AppServer/profiles/
AppSrv01/installedApps/localhostNode01Cell/FileNetEngine.ear/cews.war/WEB-
INF/classes/com/filenet/engine/wsi
AUDOperationsTest.WsiJaasConfigFile=/opt/IBM/ECMClient/configure/CE_
API/config/jaas.conf.WSI
```

> **Note** Initially, this **configuration.properties** file was added to the project src folder, but we reference this from a package folder, so it was moved (dragged and dropped) to the **com.ibm.filenet.ps.ciops.test** subfolder under the src folder.

Creating the Test Java Package and Code

The **com.ibm.filenet.ps.ciops.test** Java package subfolder is created as follows.

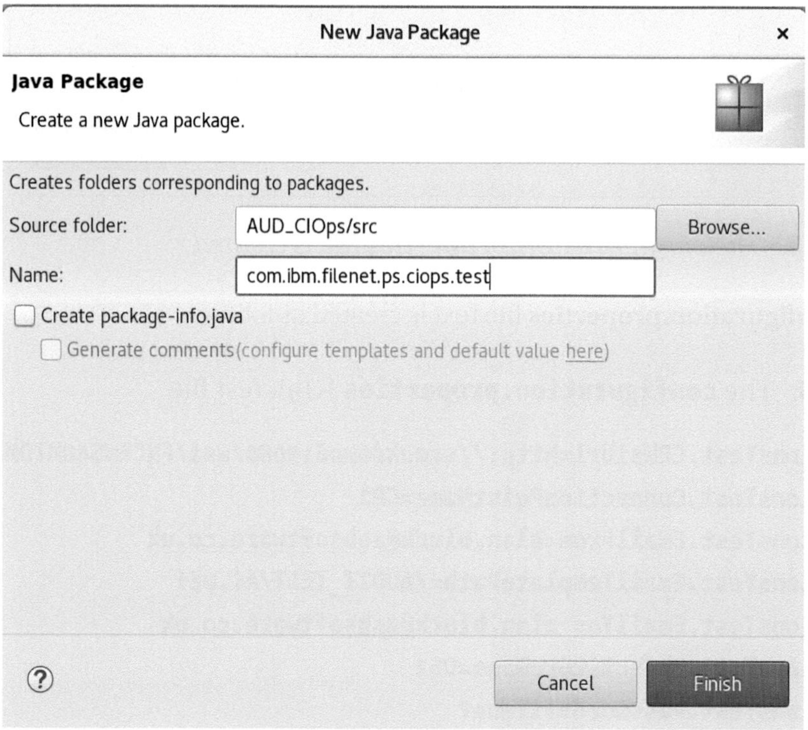

Figure 3-74. The package for testing is created

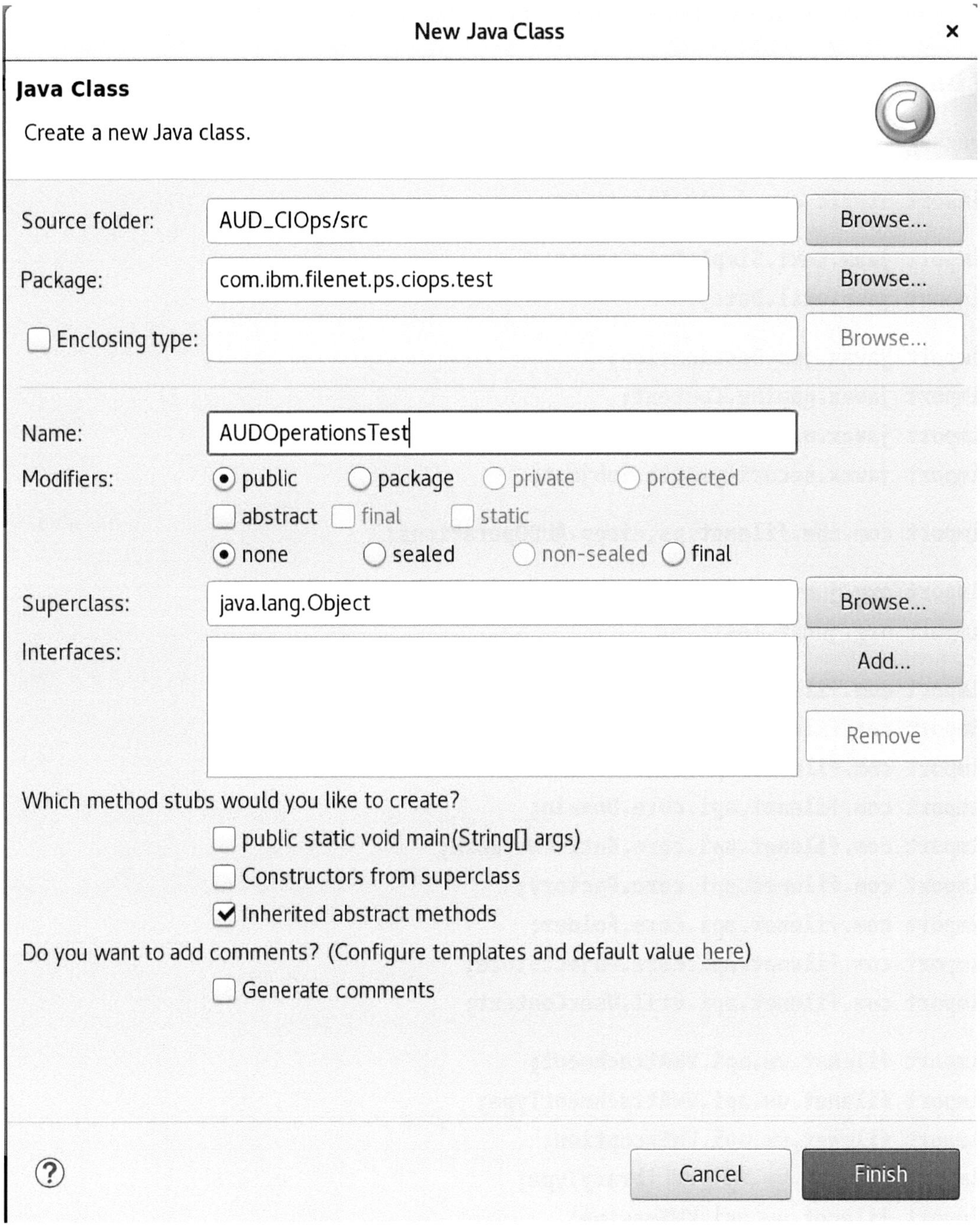

*Figure 3-75. The **AUDOperationsTest** Java class is created*

CHAPTER 3 IBM JMS INTERFACE DEVELOPMENT IBM FILENET 5.5.X WORKFLOW

The code for the **AUDOperationsTest** Java class is entered as follows.

Listing 3-6. The **AUDOperationsTest.java** code used for JUnit testing

```
package com.ibm.filenet.ps.ciops.test;

import static org.junit.Assert.*;

import java.text.SimpleDateFormat;
import java.util.Date;

import javax.jms.Destination;
import javax.naming.Context;
import javax.naming.directory.InitialDirContext;
import javax.security.auth.Subject;

import com.ibm.filenet.ps.ciops.AUDOperations;

import org.junit.BeforeClass;
import org.junit.Test;

import com.filenet.api.constants.PropertyNames;
import com.filenet.api.core.Connection;
import com.filenet.api.core.Document;
import com.filenet.api.core.Domain;
import com.filenet.api.core.EntireNetwork;
import com.filenet.api.core.Factory;
import com.filenet.api.core.Folder;
import com.filenet.api.core.ObjectStore;
import com.filenet.api.util.UserContext;

import filenet.vw.api.VWAttachment;
import filenet.vw.api.VWAttachmentType;
import filenet.vw.api.VWException;
import filenet.vw.api.VWLibraryType;
import filenet.vw.api.VWSession;
```

```java
/**
 * This class contains code for off line testing of the custom component {@
   link com.ibm.filenet.ps.ciops.AUDOperations AUDOperations}. To
 * run this class the different configuration parameters must be set in the
   configuration.properties file.
 *
 */
public class AUDOperationsTest
{
    private static final String EMAIL_TO = Configuration.getParameter("AUD
    OperationsTest.EmailTo"); //$NON-NLS-1$
    private static final String EMAIL_FROM = Configuration.getParameter("A
    UDOperationsTest.EmailFrom"); //$NON-NLS-1$
    private static final String EMAIL_TEMPLATES_PATH = Configuration.getPa
    rameter("AUDOperationsTest.EmailTemplatePath"); //$NON-NLS-1$
    private static final String OBJECT_STORE_NAME = Configuration.getParam
    eter("AUDOperationsTest.ObjectstoreName"); //$NON-NLS-1$
    private static final String TEST_FOLDER_NAME = Configuration.getParame
    ter("AUDOperationsTest.TestFolderName"); //$NON-NLS-1$
    private static final String USERNAME = Configuration.getParameter("AUD
    OperationsTest.Username"); //$NON-NLS-1$
    private static final String PASSWORD = Configuration.getParameter("AUD
    OperationsTest.Password"); //$NON-NLS-1$
    private static final String CONNECTION_POINT_NAME = Configuration.getP
    arameter("AUDOperationsTest.ConnectionPointName"); //$NON-NLS-1$
    private static final String CE_WSI_URL = Configuration.getParameter("A
    UDOperationsTest.CEWsiUrl"); //$NON-NLS-1$
    private static final String WASP_LOCATION = Configuration.getParameter
    ("AUDOperationsTest.WaspLocation"); //$NON-NLS-1$
    private static final String WSI_JAAS_CONFIG_FILE = Configuration.getPa
    rameter("AUDOperationsTest.WsiJaasConfigFile"); //$NON-NLS-1$

    private static Connection connection;
    private static VWSession vwSession;
```

```java
/**
 * This method is run before testing is started. It creates
   connections to the Content Engine and
 * the Process Engine.
 *
 * @throws Exception
 */
@BeforeClass
public static void setUpBeforeClass() throws Exception {

    try {

        System.setProperty("java.security.auth.login.config", WSI_
        JAAS_CONFIG_FILE); //$NON-NLS-1$
        System.setProperty("wasp.location",WASP_LOCATION );
        //$NON-NLS-1$

        String url = CE_WSI_URL;
        String connectionPointName = CONNECTION_POINT_NAME;
        String password = PASSWORD;
        String username = USERNAME;

        createCEConnection(username, password, url);
        createVWSession(username, password, url,
        connectionPointName);

    } catch (Exception e) {
        e.printStackTrace();
        throw e;
    }
}

private static void createCEConnection(String username, String
password, String url) {
    connection = Factory.Connection.getConnection(url);

    Subject subject = UserContext.createSubject(connection, username,
    password, "FileNetP8"); //$NON-NLS-1$
```

```java
        UserContext uc = UserContext.get();
        uc.pushSubject(subject);
    }

    private static void createVWSession(String username, String password,
    String url, String connectionPointName)
            throws VWException {
        vwSession = new VWSession();
        vwSession.setBootstrapCEURI(url);
        vwSession.logon( username, password, connectionPointName );
    }

    private AUDOperations getAUDOperations() {
        return new AUDOperations() {
            @Override
            protected Connection getConnection() {
                return connection;
            }

            @Override
            protected VWSession getVWSession() throws VWException {
                return vwSession;
            }
        };
    }
    /**
     * Test method for the {@link com.ibm.filenet.ps.ciops.AUDOperations#g
       etFolderDocuments(VWAttachment)} method. It
     * uses a test folder in the object store as input.
     *
     * @throws Exception
     */
    @Test
    public void testGetFolderDocuments() throws Exception {
        VWAttachment folderAttachment = getTestFolder();
        AUDOperations AUDOperations = getAUDOperations();
```

CHAPTER 3   IBM JMS INTERFACE DEVELOPMENT IBM FILENET 5.5.X WORKFLOW

```java
        VWAttachment[] folderDocuments = AUDOperations.getFolderDocuments
        (folderAttachment);
        showResults(folderDocuments);
    }
    /**
     * Test method for the {@link com.ibm.filenet.ps.ciops.AUDOperatio
       ns#sendJMSMessage(VWAttachment docAttachment, String Message )
       sendJMSMessage()} method.
     *
     * @throws Exception
     */
    @Test
    public void testJMSMessage() throws Exception {
        ObjectStore objectStore = getTestObjectStore();
        Document document = (Document) objectStore.getObject( "Document",
        EMAIL_TEMPLATES_PATH ); //$NON-NLS-1$
        VWAttachment documentAttachment = getDocumentAsVWAttachment(
        document);
        AUDOperations AUDOperations = getAUDOperations();
        //String Message = "Hello JMS MQ World";
        String Message = "<?xml version=1.0 encoding= utf-8/?><AUDIT_
        REPORT><CaseID>00001</CaseID><COMMENTS>AUD Test on 17th February
        pm</COMMENTS><AUDIT_DATE>2022-06-27</AUDIT_DATE><AUDIT_STATUS>Date
        of Audit Report Document</AUDIT_STATUS></AUDIT_REPORT>";
        System.out.println( Message ); //$NON-NLS-1$
          //create initial context properties
          String url = "ldap://ecmukdemo6/o=sample";
          String icf = "com.sun.jndi.ldap.LdapCtxFactory";
          // define the name of the QueueManager
          String qManager = "AUDQM";
          // and define the name of the Queue
          String qName = "AUDQAR";
          // Note that the generic Connection Factory works for both
          queues & topics
            String     cfLookup    = "cn=AUDQCF";     //ASB
```

```java
            System.out.println("Lookup connection factory " + cfLookup);
            String     JNDITopic    = "cn=AUDTopic";
            String     JNDIQueue    = "cn=AUDQAR";
            String     dLookup      = "cn=AUDQAR";       // LDAP JMS
            Destination
            String     hostName = "ecmukdemo6";
            String     channel  = "SYSTEM.DEF.CLNT.CONN";
            String     MQuser = "wasadm";
            String     MQpassword = "filenet"     ;
            String     MQport = "1417";

        AUDOperations.sendJMSMessage(documentAttachment, Message, url, icf,
        qManager,   qName,
                                            cfLookup,
   JNDITopic,    JNDIQueue, dLookup,
                                                            hostName,
    channel,      MQport, MQuser, MQpassword);
    }

    private void showResults(VWAttachment[] folderDocuments) {
        System.out.println( folderDocuments.length +  " documents found"
); //$NON-NLS-1$
        for (VWAttachment attachment : folderDocuments) {
            System.out.println( attachment.toString() );
        }
    }

    private VWAttachment getTestFolder() throws VWException {
        ObjectStore objectStore = getTestObjectStore();
        Folder folder = (Folder) objectStore.getObject( "Folder", TEST_
        FOLDER_NAME ); //$NON-NLS-1$
        VWAttachment folderAttachment = getFolderAsVWAttachment(folder);
        return folderAttachment;
    }
```

```java
private ObjectStore getTestObjectStore()
{
    EntireNetwork entireNetwork = Factory.EntireNetwork.fetchInstance
    (connection, null);
    Domain domain = entireNetwork.get_LocalDomain();
    ObjectStore objectStore = Factory.ObjectStore.getInstance(
    domain, OBJECT_STORE_NAME );
    return objectStore;
}

private VWAttachment getDocumentAsVWAttachment(Document document)
throws VWException {
    VWAttachment folderAttachment = getCEAttachment(document.
    getObjectStore() );
    document.fetchProperties( new String[] { PropertyNames.ID,
    PropertyNames.NAME } );
    folderAttachment.setId( document.get_Id().toString() );
    folderAttachment.setAttachmentName( document.get_Name() );
    folderAttachment.setType( VWAttachmentType.ATTACHMENT_TYPE_
    FOLDER );
    return folderAttachment;
}

private VWAttachment getFolderAsVWAttachment(Folder folder) throws
VWException {
    VWAttachment folderAttachment = getCEAttachment(folder.
    getObjectStore() );
    folder.fetchProperties( new String[] { PropertyNames.ID,
    PropertyNames.NAME } );
    folderAttachment.setId( folder.get_Id().toString() );
    folderAttachment.setAttachmentName( folder.get_Name() );
    folderAttachment.setType( VWAttachmentType.ATTACHMENT_TYPE_
    FOLDER );
    return folderAttachment;
}
```

```
    private VWAttachment getCEAttachment(ObjectStore objectStore) throws
    VWException {
        VWAttachment ceAttachment = new VWAttachment();
        ceAttachment.setLibraryType( VWLibraryType.LIBRARY_TYPE_CONTENT_
        ENGINE );
        objectStore.fetchProperties( new String[] {
        PropertyNames.NAME } );
        ceAttachment.setLibraryName( objectStore.get_Name() );
        return ceAttachment;
    }

    /**
     * Utility function for time stamping the different things produced by
     the test code. This
     * way the result of different tests can be kept apart.
     *
     * @return a timestamp string.
     */
    private static String getTimestamp() {
        SimpleDateFormat timestampFormatter = new
        SimpleDateFormat("yyyyMMdd HHmmSSS"); //$NON-NLS-1$
        return timestampFormatter.format( new Date() );
    }
}
```

Note The comment **//$NON-NLS-1$** at the end of some statements is the way to communicate to the compiler that UI messages should not be embedded as string literals, but rather sourced from a resource file (so that they can be translated or easily modified, etc.). Consequently, Eclipse can be configured to detect string literals, so that you don't accidentally leave unexternalized UI strings in the code. Also, note that some strings, such as regular expressions, are not externalized.

CHAPTER 3 IBM JMS INTERFACE DEVELOPMENT IBM FILENET 5.5.X WORKFLOW

The next step is to create a **Configuration** Java class.

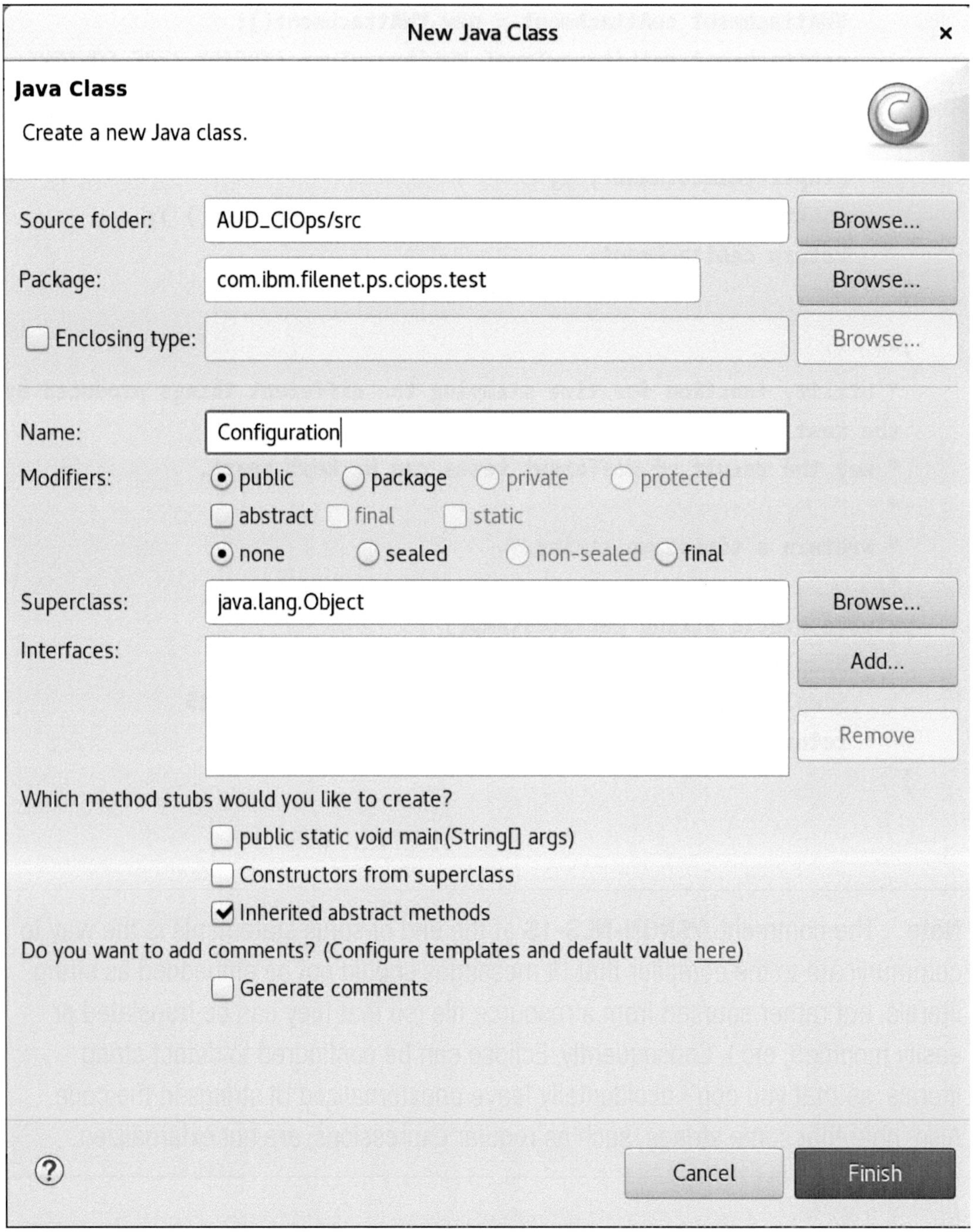

*Figure 3-76. The **Configuration** Java class is created*

The code for the **Configuration** Java class is entered as follows.

Listing 3-7. The Configuration.java code file used to load the Configuration. properties

```java
package com.ibm.filenet.ps.ciops.test;

import java.util.MissingResourceException;
import java.util.ResourceBundle;

public class Configuration {
    //ASB Updated package path for the configuration
    private static final String BUNDLE_NAME = "com.ibm.filenet.ps.ciops.test.configuration"; //$NON-NLS-1$

    private static final ResourceBundle RESOURCE_BUNDLE = ResourceBundle.getBundle(BUNDLE_NAME);

    private Configuration() {
    }

    public static String getParameter(String key) {
        try {
            return RESOURCE_BUNDLE.getString(key);
        } catch (MissingResourceException e) {
            return '!' + key + '!';
        }
    }
}
```

CHAPTER 3   IBM JMS INTERFACE DEVELOPMENT IBM FILENET 5.5.X WORKFLOW

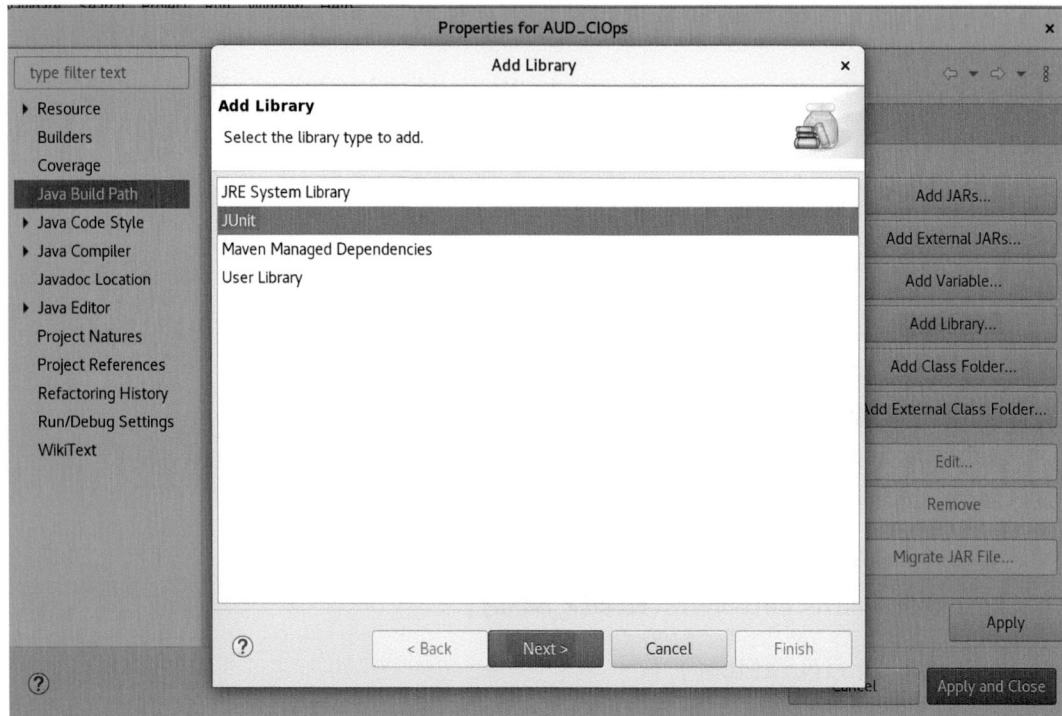

Figure 3-77.   *The JUnit library is added for code testing support*

The JUnit library is selected from a drop-down list as shown in Figure 3-78.

CHAPTER 3 IBM JMS INTERFACE DEVELOPMENT IBM FILENET 5.5.X WORKFLOW

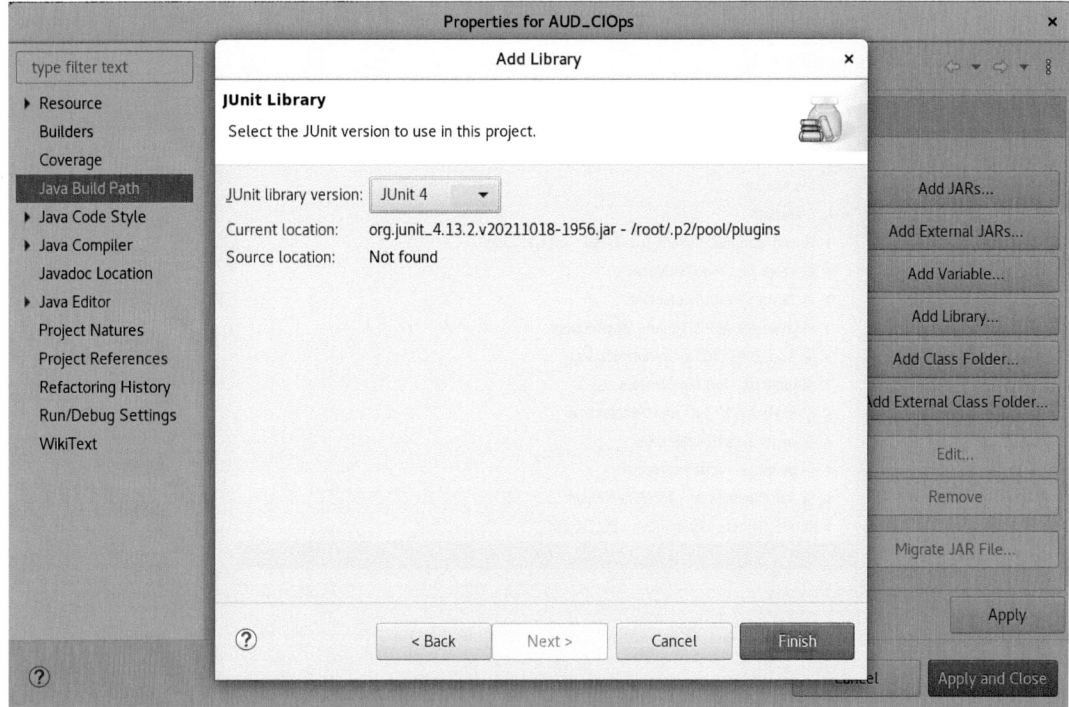

Figure 3-78. *The JUnit 4 library is selected for the Classpath*

The JUnit 4 library version is selected from the drop-down command list.

CHAPTER 3 IBM JMS INTERFACE DEVELOPMENT IBM FILENET 5.5.X WORKFLOW

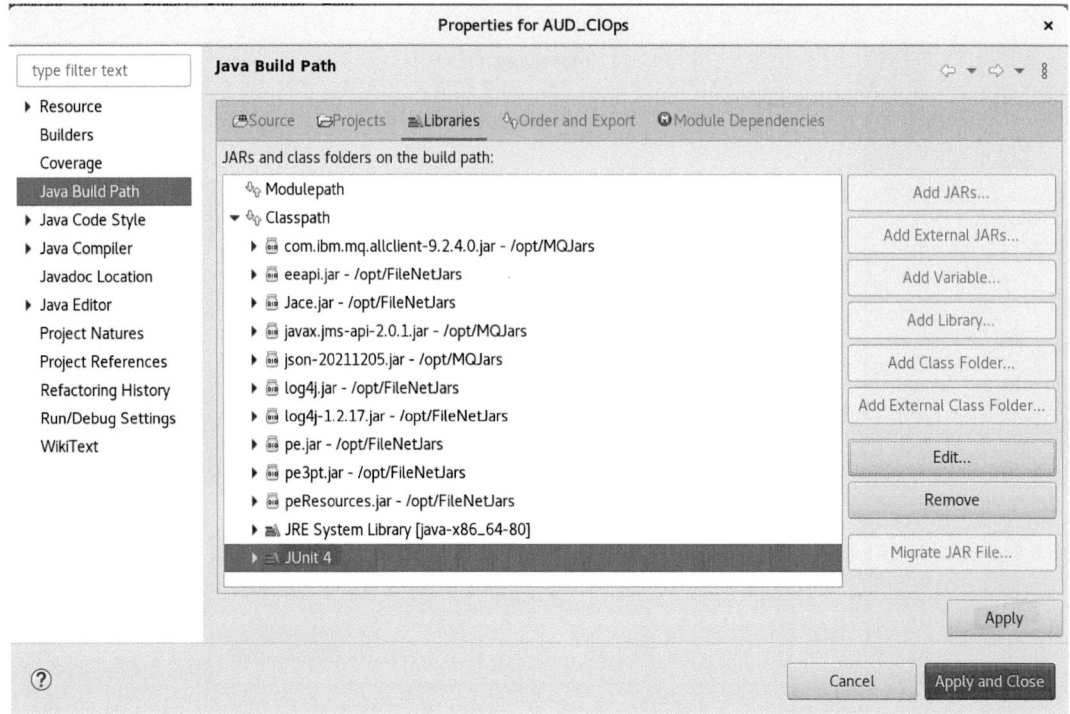

Figure 3-79. *The JUnit 4 library we selected is confirmed with the Apply and Apply and Close command buttons*

Initial Error Fixes

1) First run of the JUnit test class, **AUDOperationsTest**, we get

    JVMCFRE003 bad major version; class=com/ibm/filenet/ps/ciops/test/AUDOperationsTest, offset=6

    indicating that the compiled class JRE is different from the JVM version, so this was changed in the Eclipse project from 1.7 to 1.8.

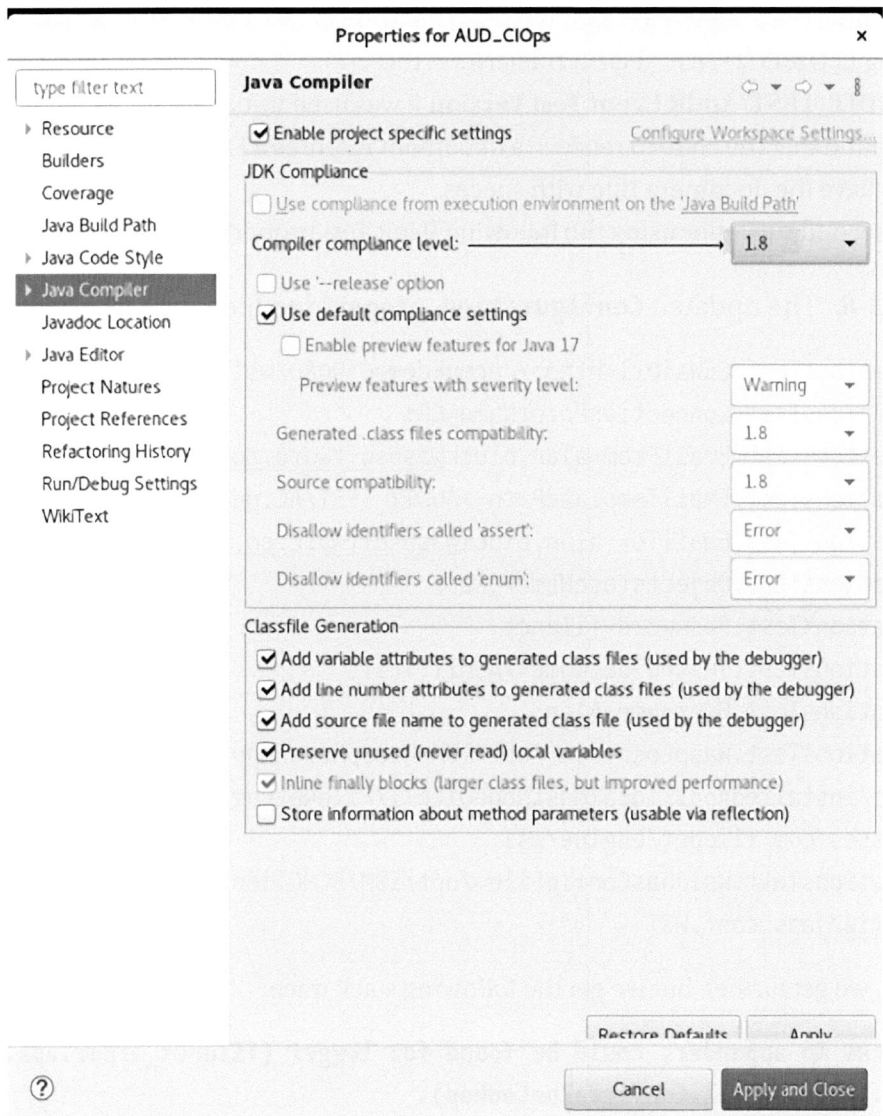

Figure 3-80. *The Java Compiler is changed from Java version 1.7 to 1.8*

The next run of the JUnit test class, **AUDOperationsTest**, gave the output:

com.filenet.api.exception.EngineRuntimeException: FNRCE0051E: **E_OBJECT_NOT_ FOUND: The requested item was not found. Object identity: /AUDIT_TEST/Audit Event Test Version 2. Class name: Versionable. errorStack={**

CHAPTER 3   IBM JMS INTERFACE DEVELOPMENT IBM FILENET 5.5.X WORKFLOW

The initial document we selected had a Title with spaces in the path, which gave the preceding E_OBJECT_NOT_FOUND FileNet error.

(**/AUDIT_TEST/Audit Event Test Version 2** was used initially.)

We found that the code to retrieve a document requires a Document path which does not have the document title with spaces.

The second run, after using the following JUnit Test properties:

Listing 3-8. The updated **Configuration.properties** file

```
AUDOperationsTest.CEWsiUrl=http://ecmukdemo6:9080/wsi/FNCEWS40MTOM/
AUDOperationsTest.ConnectionPointName=CP1
AUDOperationsTest.EmailFrom=alan.bluck@asbsoftware.co.uk
AUDOperationsTest.EmailTemplatePath=/AUDIT_TEST/AC.pdf
AUDOperationsTest.EmailTo= alan.bluck@asbsoftware.co.uk
AUDOperationsTest.ObjectstoreName=OS2
AUDOperationsTest.Password=filenet
AUDOperationsTest.TestFolderName=/AUDIT_TEST
AUDOperationsTest.Username=Alan
AUDOperationsTest.WaspLocation=/opt/IBM/WebSphere/AppServer/profiles/
AppSrv01/installedApps/localhostNode01Cell/FileNetEngine.ear/cews.war/WEB-
INF/classes/com/filenet/engine/wsi
AUDOperationsTest.WsiJaasConfigFile=/opt/IBM/ECMClient/configure/CE_
API/config/jaas.conf.WSI
```

Now, we get further, but we get the following stack trace:

```
log4j:WARN No appenders could be found for logger (filenet_error.api.com.
filenet.apiimpl.util.ConfigValueLookup).
log4j:WARN Please initialize the log4j system properly.
log4j:WARN See http://logging.apache.org/log4j/1.2/faq.html#noconfig for
more info.
[Perf Log] No interval found. Auditor disabled.
<?xml version=1.0 encoding= utf-8/?><AUDIT_REPORT><CaseID>00001</
CaseID><COMMENTS>AUD Test on 17th February pm</COMMENTS><AUDIT_
DATE>2022-06-27</AUDIT_DATE><AUDIT_STATUS>Date of Audit Report Document</
AUDIT_STATUS></AUDIT_REPORT>
Lookup connection factory cn=AUDQCF
```

```
Sending a message...
Lookup connection factory cn=AUDQCF
Error : com.ibm.mq.jms.MQQueue incompatible with javax.jms.
ConnectionFactory
Error Stack : java.lang.ClassCastException: com.ibm.mq.jms.MQQueue
incompatible with javax.jms.ConnectionFactory
```

So, now we get the error indicating that the class **com.ibm.mq.jms.MQQueue** cannot be cast to **javax.jms.ConnectionFactory**.

There is an IBM Technote for this; see the link:

www.ibm.com/support/pages/javalangclasscastexception-error-occurs-during-jndi-lookup-queue-connection-factory

In summary, the ClassCastException occurs when a queue connection factory is defined as a WebSphere MQ Connection Factory instead of a WebSphere MQ **Queue** Connection Factory.

The application code is similar to the following example:

```
javax.jms.QueueConnectionFactory myQCF = null;
InitialContext ic = null;
//... setup InitialContext here ...
try
{
myQCF = (QueueConnectionFactory)ic.lookup("jms/myQCF");
}
catch(Throwable e)
{
e.printStackTrace();
}
```

You can resolve the problem using one of the following methods:

- Define the queue connection factory as a WebSphere MQ Queue Connection Factory.

- Use a **javax.jms.ConnectionFactory** object in the application code rather than a **javax.jms.QueueConnectionFactory** object.

Configuring the Ports and Library Jar Files for IBM MQ Series

Update JMS tcp 1414 port in the server Firewall for ibm-mqseries using the commands:

firewall-cmd --zone=public --permanent --add-port=1414/tcp
firewall-cmd --reload

The server Firewall is now updated to allow the MQ Series port 1414 to be opened. Also, add

firewall-cmd --zone=public --permanent --add-port=1417/tcp
firewall-cmd --reload

We also need to be able to reference the following IBM jar file packages:
com.ibm.mq.jar
com.ibm.mq.jmqi.jar
com.ibm.msg.client.jms.internal.jar
com.ibm.msg.client.jms.jar
com.ibm.msg.client.provider.jar

We also need the WAS_ROOT/runtimes/**com.ibm.ws.ejb.thinclient_8.5.0.jar** classpath, where WAS_ROOT is the installation home of the WAS, for example, /opt/WebSphere80/AppServer/runtimes/**com.ibm.ws.ejb.thinclient_8.5.0.jar**.

```
(base) [root@ECMUKDEMO6 jms]# find /opt -name com.ibm.mq.jar
/opt/IBM/WebSphere/AppServer/installedConnectors/wmq.jmsra.rar/com.ibm.mq.jar
/opt/mqm/java/lib/com.ibm.mq.jar
(base) [root@ECMUKDEMO6 jms]# cd /opt/IBM/WebSphere/AppServer/installedConnectors/wmq.jmsra.rar/
(base) [root@ECMUKDEMO6 wmq.jmsra.rar]# ls
com.ibm.mq.commonservices.jar            com.ibm.msg.client.commonservices.jar
com.ibm.mq.connector.jar                 com.ibm.msg.client.jms.internal.jar
com.ibm.mq.headers.jar                   com.ibm.msg.client.jms.jar
com.ibm.mq.jar                           com.ibm.msg.client.matchspace.jar
com.ibm.mq.jmqi.jar                      com.ibm.msg.client.provider.jar
com.ibm.mq.jmqi.local.jar                com.ibm.msg.client.ref.jar
com.ibm.mq.jmqi.remote.jar               com.ibm.msg.client.wmq.common.jar
com.ibm.mq.jmqi.system.jar               com.ibm.msg.client.wmq.factories.jar
com.ibm.mq.jms.admin.jar                 com.ibm.msg.client.wmq.jar
com.ibm.mqjms.jar                        com.ibm.msg.client.wmq.v6.jar
com.ibm.mq.pcf.jar                       dhbcore.jar
com.ibm.msg.client.commonservices.j2se.jar  META-INF
(base) [root@ECMUKDEMO6 wmq.jmsra.rar]# cp com.ibm.mq.jar /opt/MQJars/
(base) [root@ECMUKDEMO6 wmq.jmsra.rar]# cp com.ibm.mq.jmqi.jar /opt/MQJars/
(base) [root@ECMUKDEMO6 wmq.jmsra.rar]# cp com.ibm.msg.client.jms.internal.jar /opt/MQJars/
(base) [root@ECMUKDEMO6 wmq.jmsra.rar]# cp com.ibm.msg.client.jms.jar /opt/MQJars/
(base) [root@ECMUKDEMO6 wmq.jmsra.rar]# cp com.ibm.msg.client.provider.jar /opt/MQJars/
(base) [root@ECMUKDEMO6 wmq.jmsra.rar]#
```

Figure 3-81. *The required IBM MQ reference .jars are found and copied*

The copied jars are selected for the Eclipse Java project Classpath.

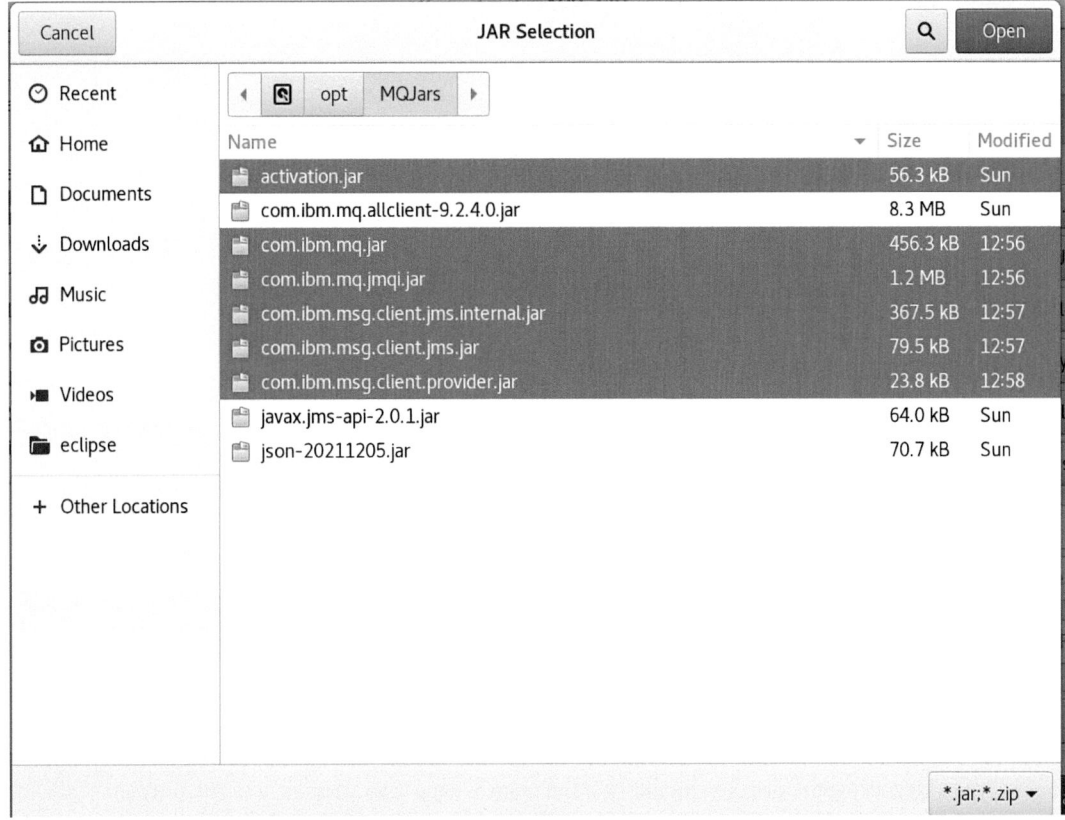

Figure 3-82. *The selected files are referenced*

The Java Build Path now has the following selected library .jar files.

CHAPTER 3   IBM JMS INTERFACE DEVELOPMENT IBM FILENET 5.5.X WORKFLOW

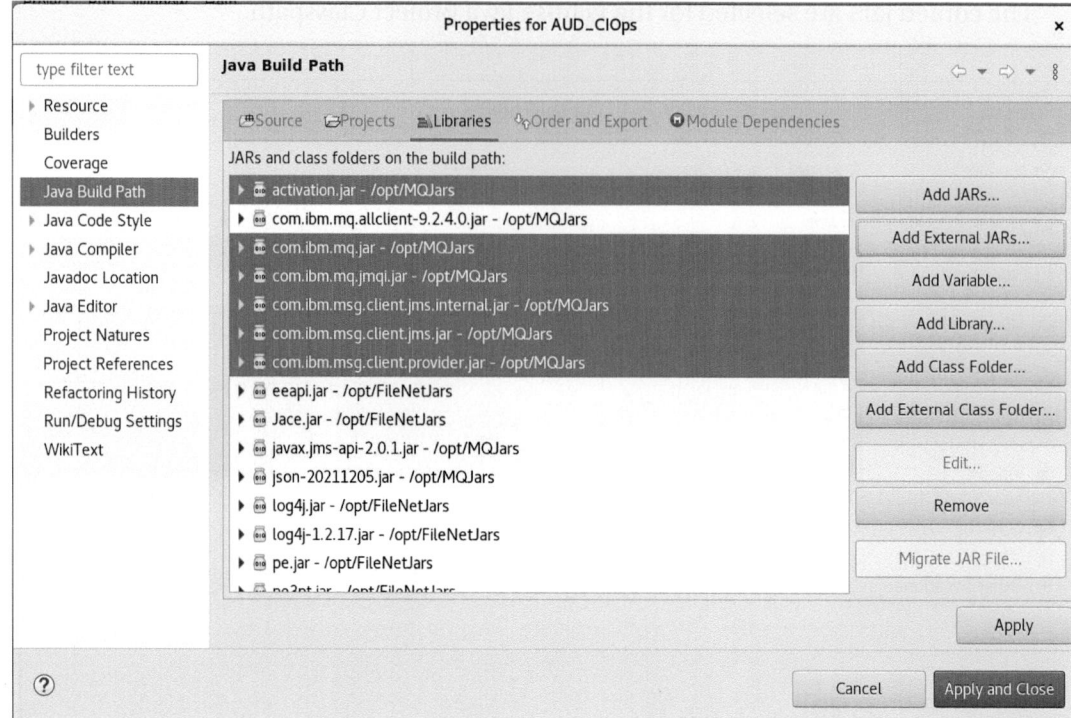

Figure 3-83. *The required IBM MQ jars are selected from the /opt/MQJars folder*

```
(base) [root@ECMUKDEM06 wmq.jmsra.rar]# find /opt -name com.ibm.ws.ejb.thinclient*.jar
/opt/IBM/WebSphere/AppServer/runtimes/com.ibm.ws.ejb.thinclient_8.5.0.jar
(base) [root@ECMUKDEM06 wmq.jmsra.rar]# cd /opt/IBM/WebSphere/AppServer/runtimes/
(base) [root@ECMUKDEM06 runtimes]# cp com.ibm.ws.ejb.thinclient_8.5.0.jar /opt/MQJars/
(base) [root@ECMUKDEM06 runtimes]#
```

Figure 3-84. *The com.ibm.ws.ejb.thinclient_8.5.0.jar is copied to /opt/MQJars*

CHAPTER 3 IBM JMS INTERFACE DEVELOPMENT IBM FILENET 5.5.X WORKFLOW

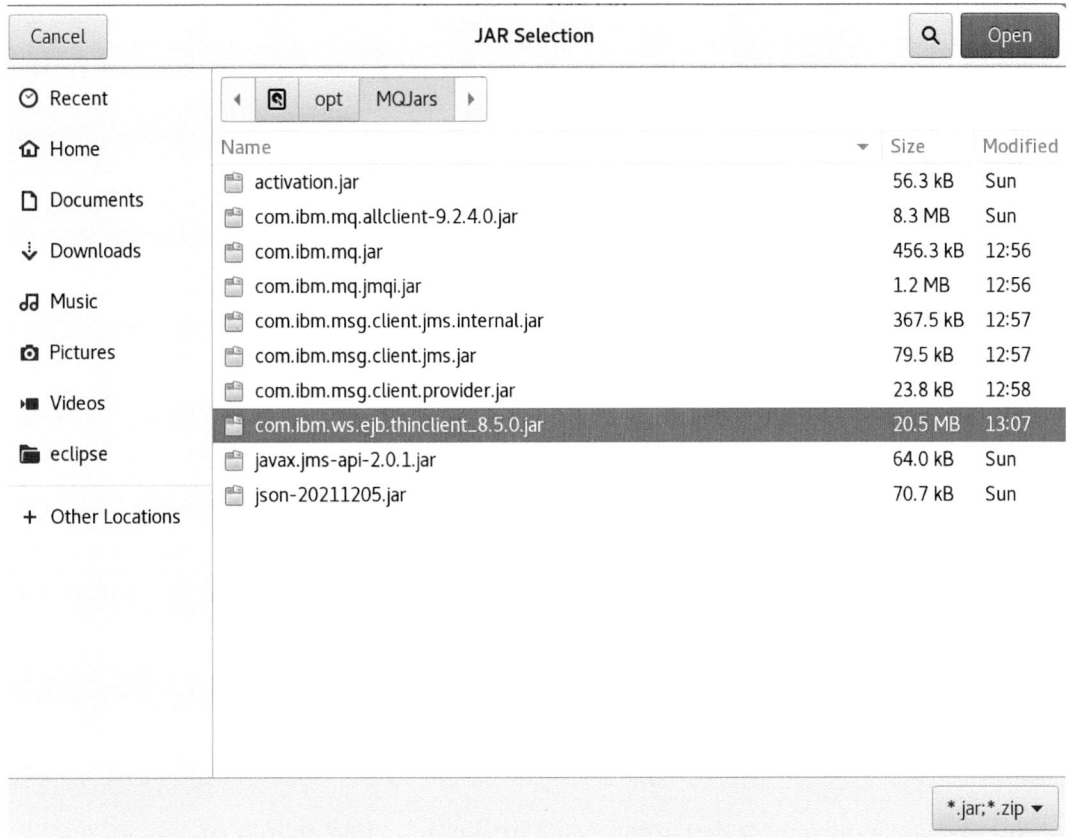

Figure 3-85. *The **com.ibm.ws.ejb.thinclient_8.5.0.jar** is loaded to **Eclipse***

CHAPTER 3   IBM JMS INTERFACE DEVELOPMENT IBM FILENET 5.5.X WORKFLOW

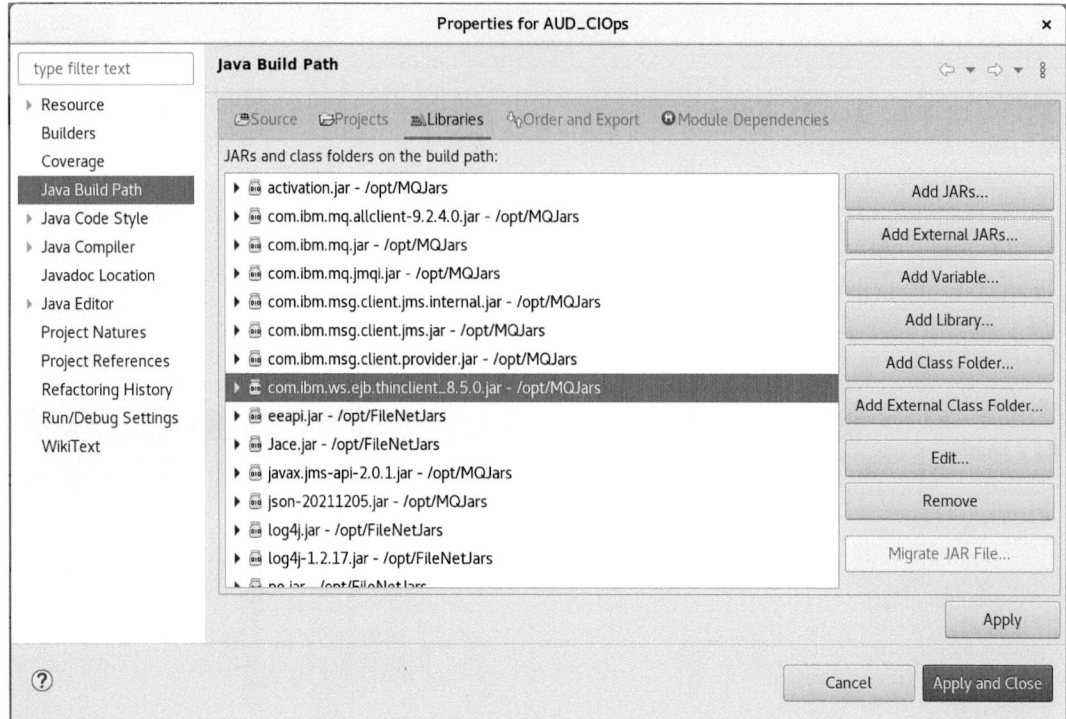

*Figure 3-86.   The **com.ibm.ws.ejb.thinclient_8.5.0.jar** is added to the Classpath*

We also need the following .jar files from <WebSphere Installation Directory>/lib:

j2ee.jar
bootstrap.jar

```
(base) [root@ECMUKDEMO6 lib]# pwd
/opt/IBM/WebSphere/AppServer/lib
(base) [root@ECMUKDEMO6 lib]# cp bootstrap.jar /opt/MQJars/
(base) [root@ECMUKDEMO6 lib]# cp j2ee.jar /opt/MQJars/
(base) [root@ECMUKDEMO6 lib]#
```

Figure 3-87.   The j2ee.jar and bootstrap.jar files are copied

CHAPTER 3   IBM JMS INTERFACE DEVELOPMENT IBM FILENET 5.5.X WORKFLOW

Next, we need the following JARs from <WebSphere Installation Directory>/plugins:

-- **com.ibm.ws.runtime.jar**
-- **com.ibm.ws.emf.jar**
-- **org.eclipse.emf.ecore.jar**
-- **org.eclipse.emf.common.jar**
-- **com.ibm.ffdc.jar**

```
(base) [root@ECMUKDEMO6 plugins]# cd /opt/IBM/WebSphere/AppServer/plugins/
(base) [root@ECMUKDEMO6 plugins]# cp com.ibm.ws.runtime.jar /opt/MQJars/
(base) [root@ECMUKDEMO6 plugins]# cp com.ibm.ws.emf.jar /opt/MQJars/
(base) [root@ECMUKDEMO6 plugins]# cp org.eclipse.emf.ecore.jar /opt/MQJars/
(base) [root@ECMUKDEMO6 plugins]# cp org.eclipse.emf.common.jar /opt/MQJars/
(base) [root@ECMUKDEMO6 plugins]# cp com.ibm.ffdc.jar /opt/MQJars/
(base) [root@ECMUKDEMO6 plugins]# pwd
/opt/IBM/WebSphere/AppServer/plugins
(base) [root@ECMUKDEMO6 plugins]#
```

Figure 3-88. *The listed .jars are also copied*

We also required the **ibmorb.jar**.

```
/opt/mqm/java/jre64/jre/lib/ibmorb.jar
(base) [root@ECMUKDEMO6 plugins]# cp /opt/mqm/java/jre64/jre/lib/ibmorb.jar /opt/MQJars/
(base) [root@ECMUKDEMO6 plugins]#
```

Figure 3-89. *The ibmorb.jar is copied from the /opt/mqm/java/jre64/jre/lib folder*

Added providerutil.jar from the path:
/opt/mqm/java/lib/

```
(base) [root@ECMUKDEMO6 plugins]# find /opt -name providerutil.jar
/opt/mqm/java/lib/providerutil.jar
(base) [root@ECMUKDEMO6 plugins]# cp /opt/mqm/java/lib/providerutil.jar /opt/MQJars/
(base) [root@ECMUKDEMO6 plugins]#
```

Figure 3-90. *The **providerutil.jar** file is added to the **/opt/MQJars** folder*

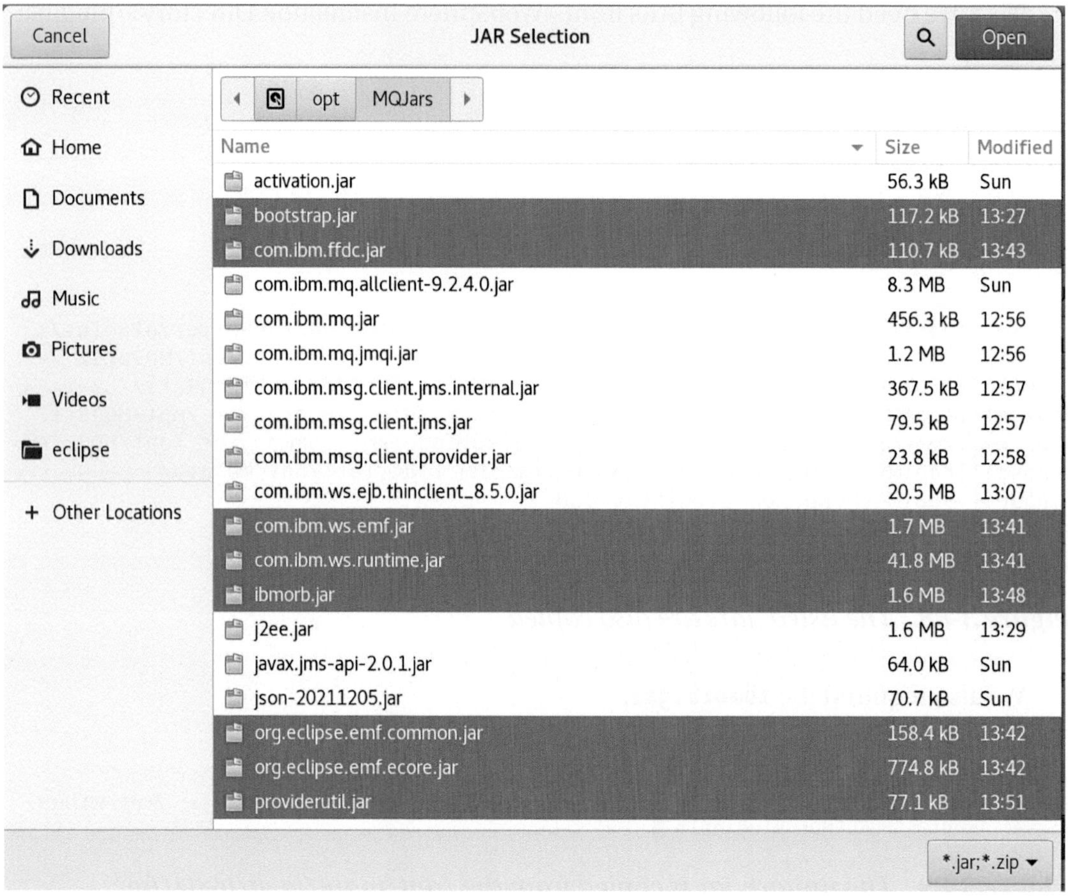

CHAPTER 3 IBM JMS INTERFACE DEVELOPMENT IBM FILENET 5.5.X WORKFLOW

Name	Size	Modified
activation.jar	56.3 kB	Sun
bootstrap.jar	117.2 kB	13:27
com.ibm.ffdc.jar	110.7 kB	13:43
com.ibm.mq.allclient-9.2.4.0.jar	8.3 MB	Sun
com.ibm.mq.jar	456.3 kB	12:56
com.ibm.mq.jmqi.jar	1.2 MB	12:56
com.ibm.msg.client.jms.internal.jar	367.5 kB	12:57
com.ibm.msg.client.jms.jar	79.5 kB	12:57
com.ibm.msg.client.provider.jar	23.8 kB	12:58
com.ibm.ws.ejb.thinclient_8.5.0.jar	20.5 MB	13:07
com.ibm.ws.emf.jar	1.7 MB	13:41
com.ibm.ws.runtime.jar	41.8 MB	13:41
ibmorb.jar	1.6 MB	13:48
j2ee.jar	1.6 MB	13:29
javax.jms-api-2.0.1.jar	64.0 kB	Sun
json-20211205.jar	70.7 kB	Sun
org.eclipse.emf.common.jar	158.4 kB	13:42
org.eclipse.emf.ecore.jar	774.8 kB	13:42
providerutil.jar	77.1 kB	13:51

Figure 3-91. *The selected .jars from the /opt/MQJars are added to the Classpath*

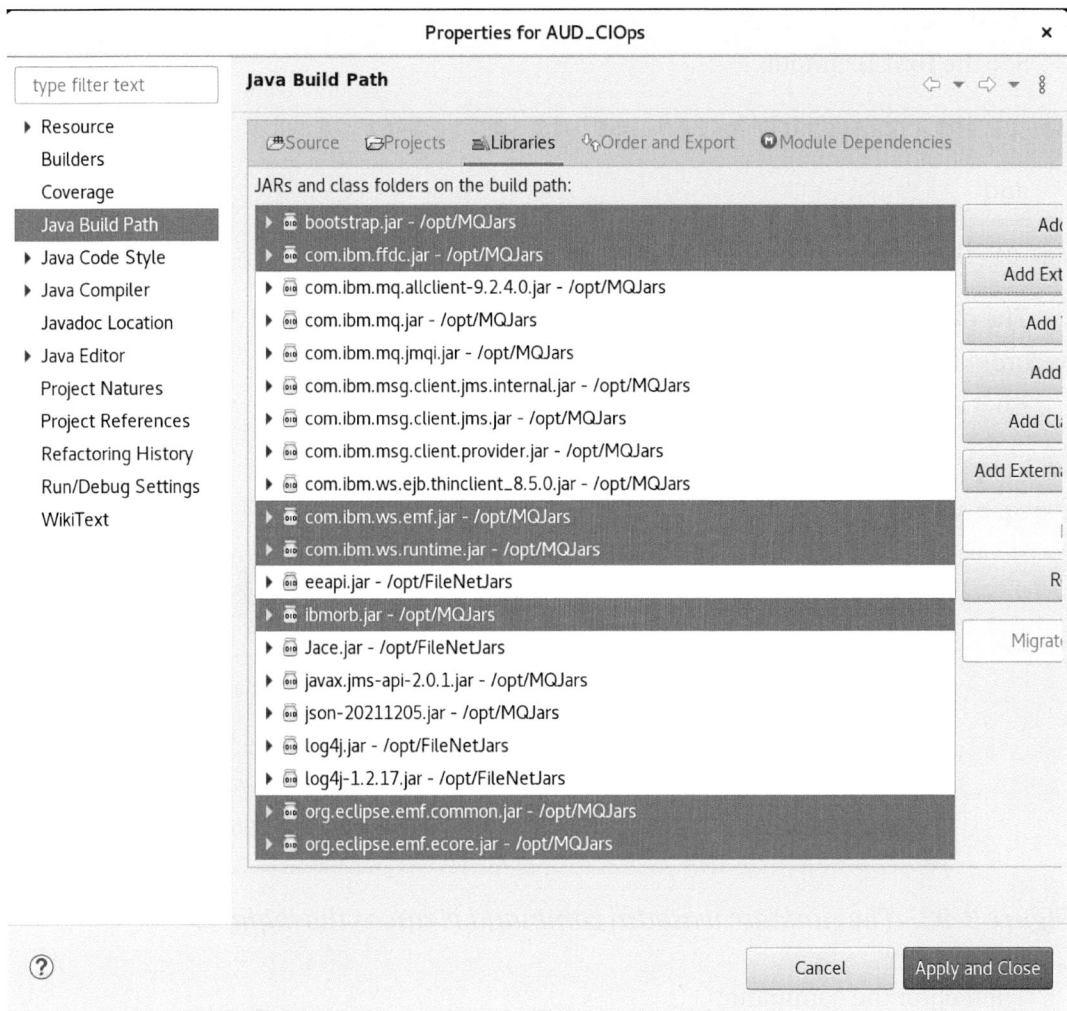

Figure 3-92. *The listed .jars are added using the Apply and Close command*

MQ Series Channel Security Settings

The security is set to provide limited access to the stored messages held in the MQ Series message queues.

Read additional information in

"Configuring and running simple JMS P2P and Pub/Sub applications in MQ 7.0, 7.1, 7.5 and 8.0"

www-01.ibm.com/support/docview.wss?uid=swg27023212&aid=1

(IBM Techdoc: 7023212)

See the IBM Technote:

www-01.ibm.com/support/docview.wss?uid=swg21138961

and

www-01.ibm.com/support/docview.wss?uid=swg21577137

The default value for the new feature introduced in 7.1, "Channel Authentication Records" **(CHLAUTH)**, is **ENABLED**.

Check using the following commands:

su - mqm
runmqsc wasadm

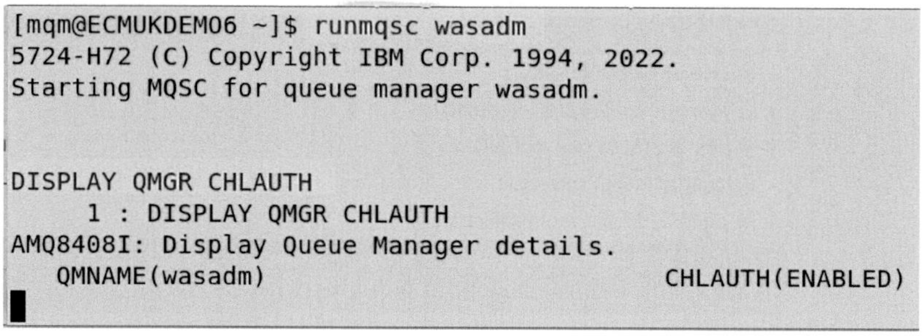

```
[mqm@ECMUKDEM06 ~]$ runmqsc wasadm
5724-H72 (C) Copyright IBM Corp. 1994, 2022.
Starting MQSC for queue manager wasadm.


DISPLAY QMGR CHLAUTH
    1 : DISPLAY QMGR CHLAUTH
AMQ8408I: Display Queue Manager details.
   QMNAME(wasadm)                                    CHLAUTH(ENABLED)
```

Figure 3-93. *The runmqsc wasadm command is run as the mqm user*

Then enter the command:

DISPLAY QMGR CHLAUTH

This gives the following (also, see Figure 3-92):

   1 : DISPLAY QMGR CHLAUTH
AMQ8408I: Display Queue Manager details.
   QMNAME(wasadm)   **CHLAUTH(ENABLED)**

By default, the following three channel authentication records are generated when a new queue manager is created in 7.1 or later using the command:

```
DISPLAY CHLAUTH(*)
AMQ8878: Display channel authentication record details.
   CHLAUTH(SYSTEM.ADMIN.SVRCONN)          TYPE(ADDRESSMAP)
   ADDRESS(*)                             USERSRC(CHANNEL)
AMQ8878: Display channel authentication record details.
   CHLAUTH(SYSTEM.*)                      TYPE(ADDRESSMAP)
   ADDRESS(*)                             USERSRC(NOACCESS)
AMQ8878: Display channel authentication record details.
   CHLAUTH(*)                             TYPE(BLOCKUSER)
   USERLIST(*MQADMIN)
```

```
DISPLAY CHLAUTH(*)
    2 : DISPLAY CHLAUTH(*)
AMQ8878I: Display channel authentication record details.
   CHLAUTH(SYSTEM.ADMIN.SVRCONN)          TYPE(ADDRESSMAP)
   ADDRESS(*)                             USERSRC(CHANNEL)
AMQ8878I: Display channel authentication record details.
   CHLAUTH(SYSTEM.*)                      TYPE(ADDRESSMAP)
   ADDRESS(*)                             USERSRC(NOACCESS)
AMQ8878I: Display channel authentication record details.
   CHLAUTH(*)                             TYPE(BLOCKUSER)
   USERLIST(*MQADMIN)
```

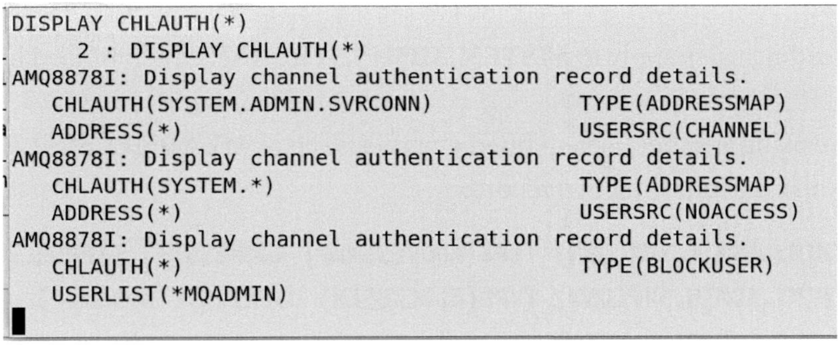

Figure 3-94. *The DISPLAY CHLAUTH(*) command is entered*

The last record blocks all remote channel access to any MQ Administrator. The effect is that non-administrative users can still connect if suitably authorized to do so, but administrative connections and anonymous connections are disallowed regardless of any Object Authority Manager (OAM) authorization settings. This means that new queue managers in V7.1 are much more secure by default than in previous versions, but with the trade-off that administrative access must be explicitly defined.

1) If this is a production queue manager, then you could stop trying to use a userid that is an MQ Administrator and instead use a non-administrator userid to access the queue manager.

2) If you really want the MQ Administrator to be able to access the queue manager via client channels, you could do one of the following actions:

2a) You can add the following two Channel Authentication Records.
User ID blocking

The first rule blocks administrative users and the **MCAUSER "nobody"** (which prevents someone from creating a user ID **"nobody"** and putting it into an authorized group):

```
$ runmqsc QmgrName
SET CHLAUTH(*) TYPE(BLOCKUSER) USERLIST('nobody','*MQADMIN')
```

The second rule provides a reduced blacklist for **SYSTEM.ADMIN** channels that allows administrators to use these. It is assumed here that some other **CHLAUTH** rule such as an **SSLPEERMAP** has validated the administrator's connection or that an exit has done so.

```
SET CHLAUTH(SYSTEM.ADMIN.*) TYPE(BLOCKUSER) USERLIST('nobody')
```

The preceding rules apply to **SYSTEM.ADMIN.SVRCONN** which is used by the MQ Explorer.

If you are using another user-defined channel, such as **MY.ADMIN.SVRCONN**, then you need to add the following two records:

```
SET CHLAUTH(MY.ADMIN.SVRCONN) TYPE(ADDRESSMAP) ADDRESS(*) USERSRC(CHANNEL)
SET CHLAUTH(MY.ADMIN.SVRCONN) TYPE(BLOCKUSER) USERLIST('nobody')
```

Note It is not advisable to use **SYSTEM.DEF.*** channels for active connections. The system default channels are the objects from which all user-defined channels inherit properties. The recommended practice is that **SYSTEM.DEF.*** and **SYSTEM.AUTO.*** channels should **NOT** be configured to be usable.

2b) This is a variation of (2a) but allowing the MQ Administrator to only use a particular host.

The first rule blocks **MCAUSER "nobody"**:

```
SET CHLAUTH(SYSTEM.ADMIN.SVRCONN) TYPE(BLOCKUSER) USERLIST('nobody')
```

The second rule removes all access to **SYSTEM.ADMIN.SVRCONN**:

```
SET CHLAUTH(SYSTEM.ADMIN.SVRCONN) TYPE(ADDRESSMAP) ADDRESS(*) ACTION(REMOVE)
```

and the third rule adds an entry for the server that needs access:

```
SET CHLAUTH(SYSTEM.ADMIN.SVRCONN) TYPE(ADDRESSMAP) ADDRESS(9.27.4x.7y)
USERSRC(CHANNEL)
```

2c) Disable the Channel Authentication Records feature:

```
ALTER QMGR CHLAUTH(DISABLED)
```

Warning Disabling this new feature is not recommended for MQ 7.1 production queue managers due to security implications.

Note Disabling **CHLAUTH** results in a policy that accepts administrative connections by default. The administrative effort to lock down administrative access with **CHLAUTH(DISABLED)** is much greater than to do so with **CHLAUTH(ENABLED)**.

We have run the following just for the test version of the **sendJMSMessage** method used in the Component Integrator queue in this chapter:

```
ALTER QMGR CHLAUTH(DISABLED)
```

Executing **ALTER QMGR CHLAUTH(DISABLED)** displays the following:

```
AMQ8005: WebSphere MQ queue manager changed.
```

Note It is recommended to leave CHLAUTH(ENABLED) and use the other security features of MQ V9.2.x to authenticate administrator connections.

Create a CRL LDAP Authentication

See `www.ibm.com/docs/en/ibm-mq/9.2?topic=manager-accessing-crls-arls-using-mq-explorer`.

Note Certificate Revocation Lists (CRLs) also apply to Authority Revocation Lists (ARLs).

For the overview of the use of the CRL lists used for security, see

`www.ibm.com/docs/en/ibm-mq/9.2?topic=users-working-revoked-certificates`

Use the following procedure to set up an LDAP connection to a CRL:

1. Ensure that you have started your queue manager.
2. Right-click the **Authentication Information** folder and click **New ➤ Authentication Information**. In the property sheet that opens:

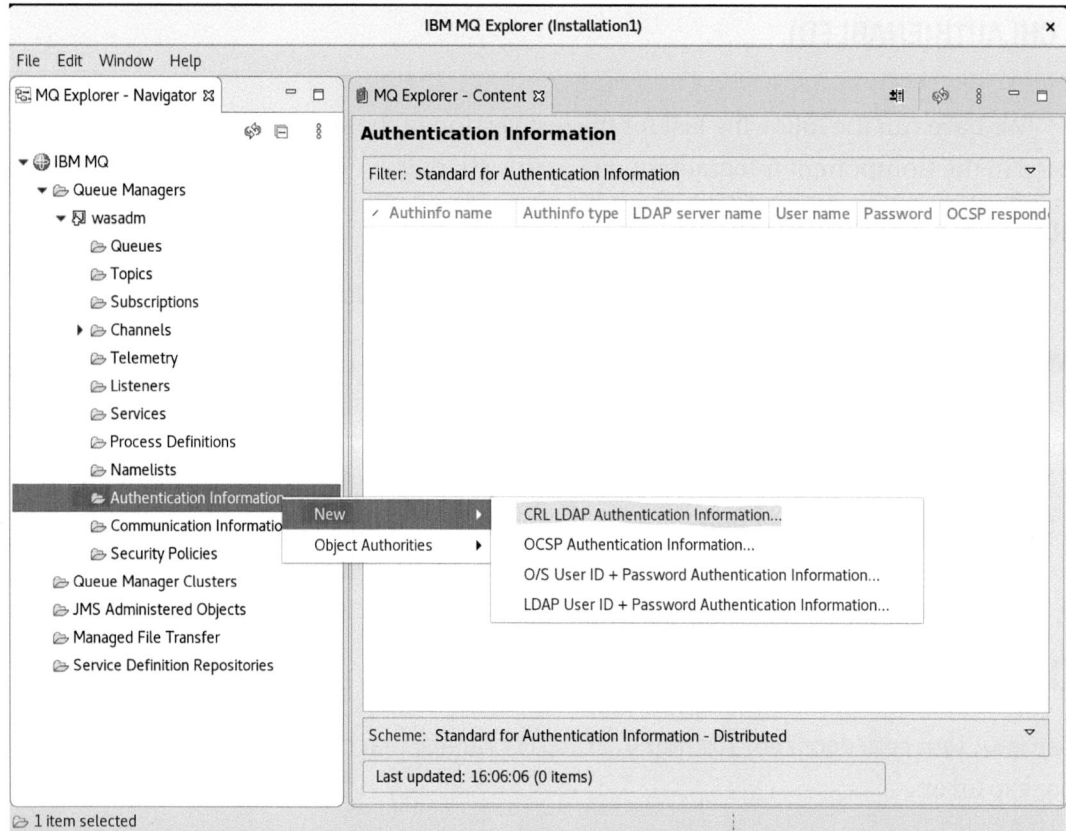

Figure 3-95. *The MQ Explorer CRL LDAP Authentication Information menu is selected*

CHAPTER 3　IBM JMS INTERFACE DEVELOPMENT IBM FILENET 5.5.X WORKFLOW

a. On the first page, Create Authentication Information, enter a name for the CRL (LDAP) object.

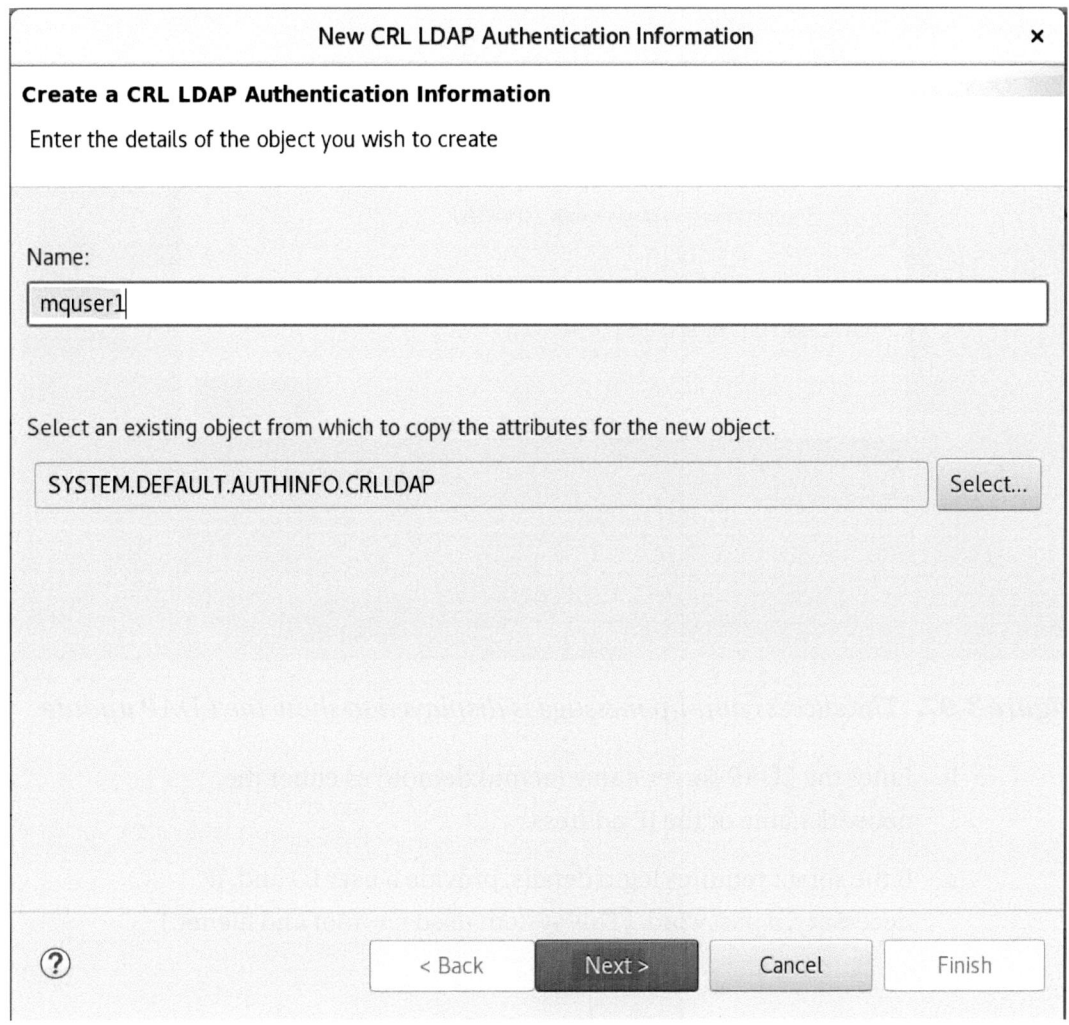

Figure 3-96. *The mquser1 User Name is entered for the CRL LDAP Authentication*

CHAPTER 3 IBM JMS INTERFACE DEVELOPMENT IBM FILENET 5.5.X WORKFLOW

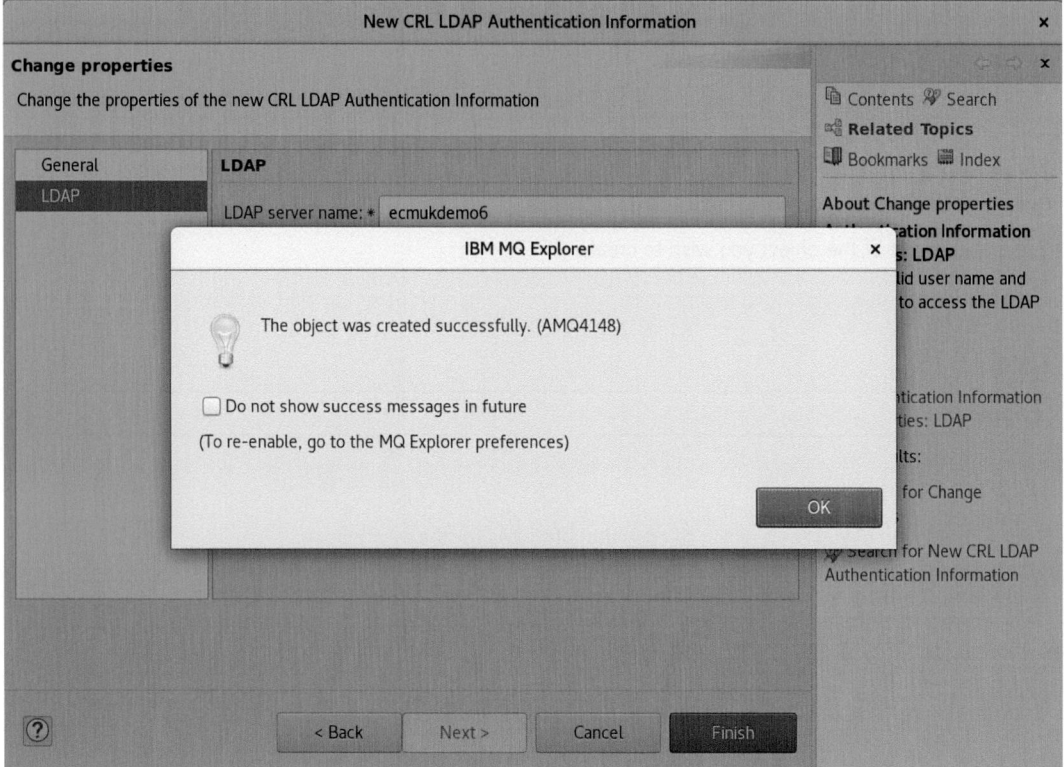

Figure 3-97. *The success pop-up message is displayed to show the LDAP update*

    b. Enter the LDAP server name (ecmukdemo6) as either the network name or the IP address.

    c. If the server requires login details, provide a user ID and, if necessary, a password. (This system used cn=root and filenet.)

CHAPTER 3   IBM JMS INTERFACE DEVELOPMENT IBM FILENET 5.5.X WORKFLOW

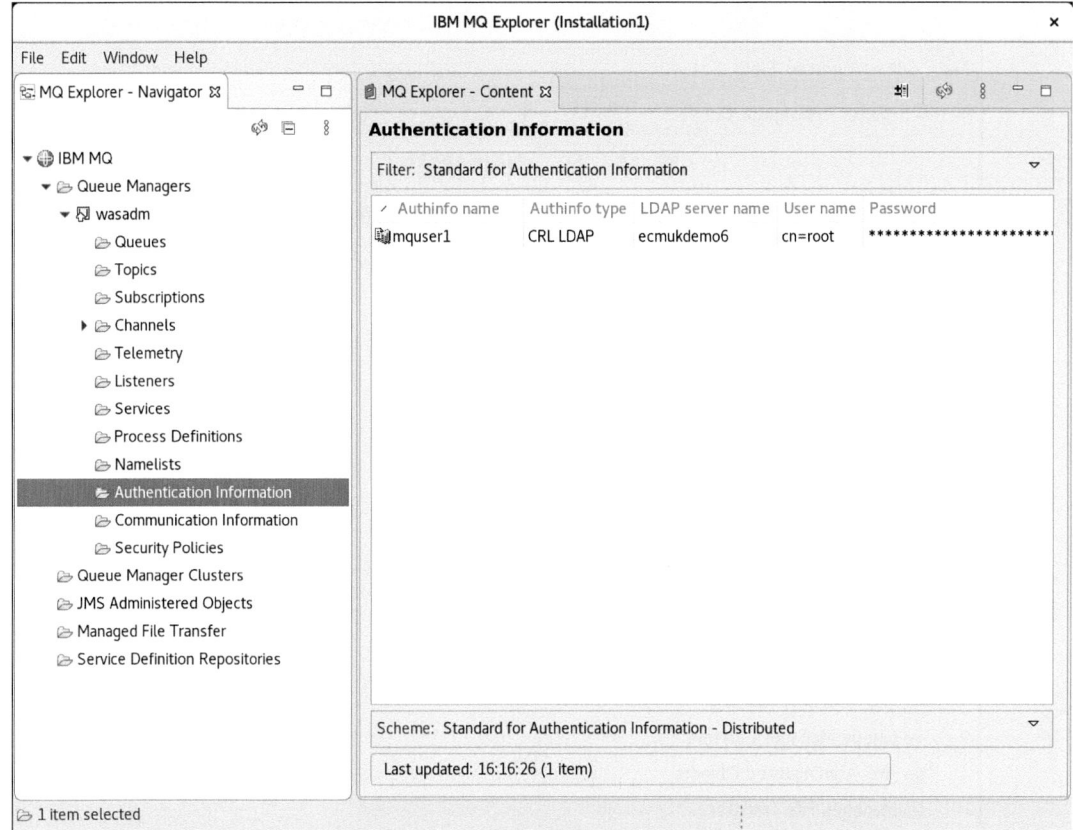

Figure 3-98. *The mquser1 Authentication name is added successfully*

3. Right-click the Namelists folder and click **New ➤ Namelist**. In the property sheet that opens:

CHAPTER 3 IBM JMS INTERFACE DEVELOPMENT IBM FILENET 5.5.X WORKFLOW

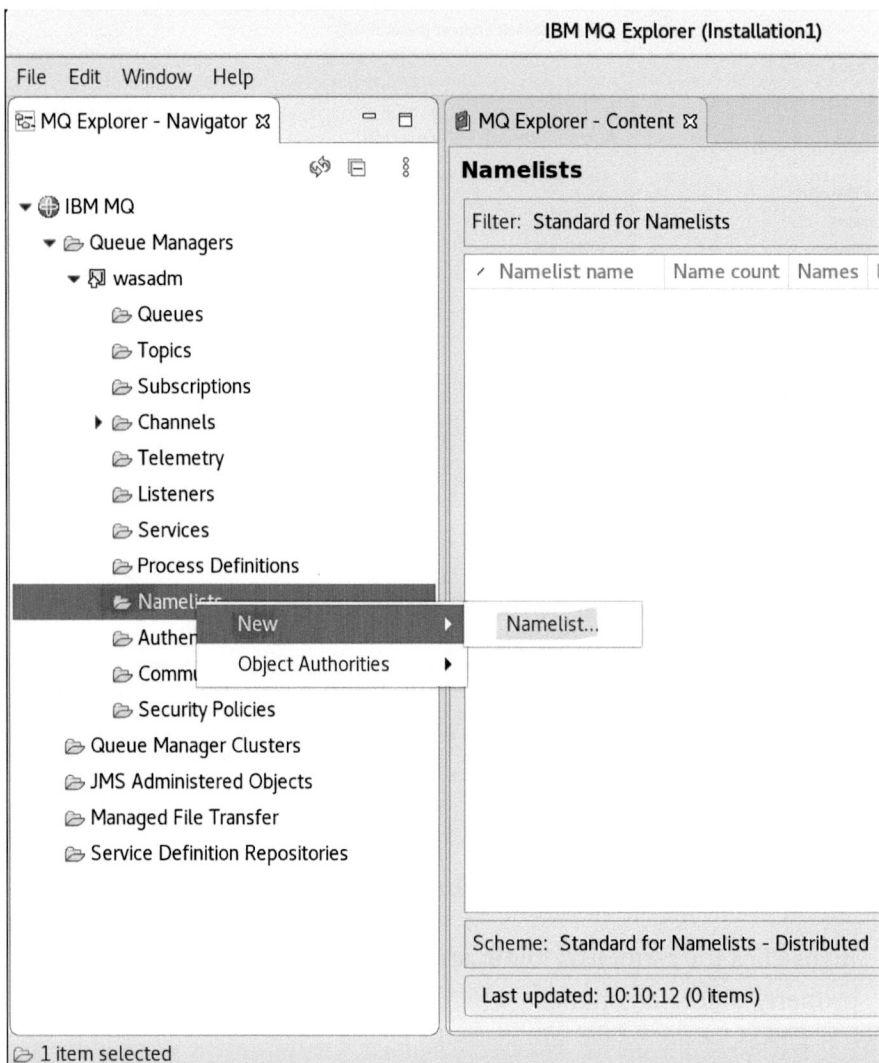

Figure 3-99. *The New Namelist node menu is selected*

CHAPTER 3 IBM JMS INTERFACE DEVELOPMENT IBM FILENET 5.5.X WORKFLOW

a. Type a name for the Namelist (AUDNAMES).

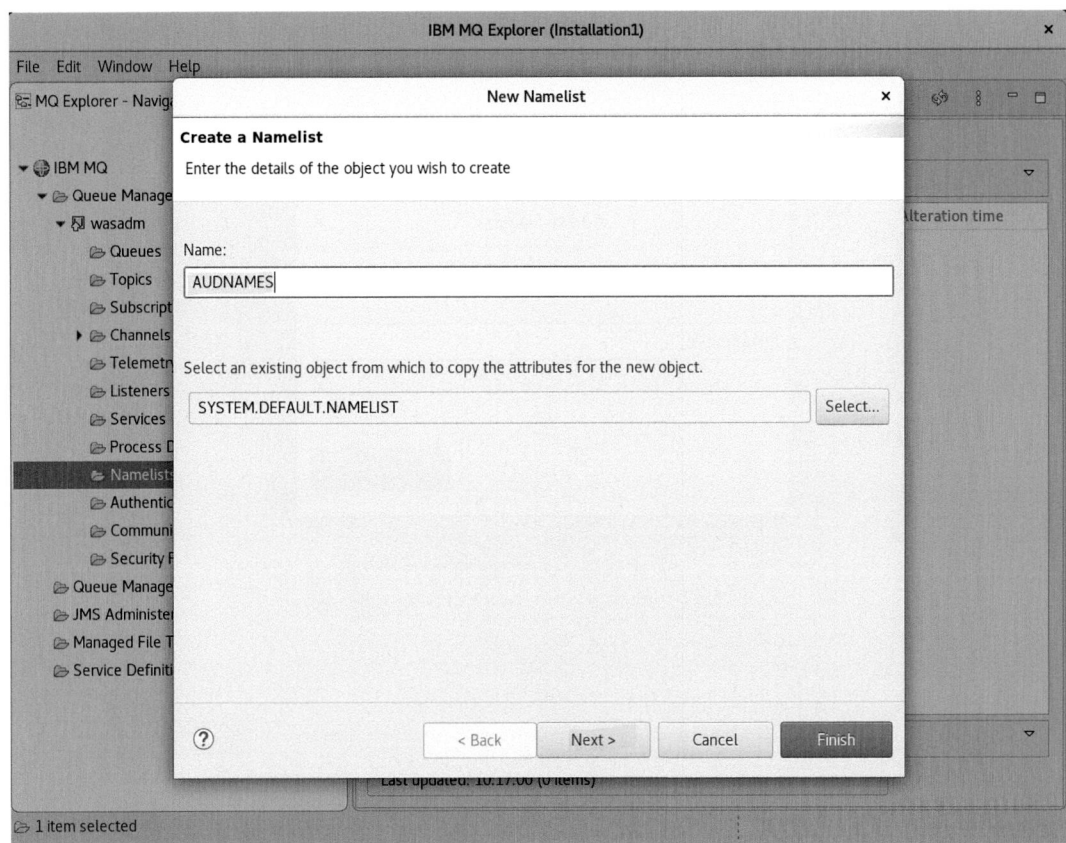

Figure 3-100. The AUDNAMES list is added and Next clicked

b. Add the name of the CRL (LDAP) object (**mquser1**, from step **2a**) to the list.

CHAPTER 3 IBM JMS INTERFACE DEVELOPMENT IBM FILENET 5.5.X WORKFLOW

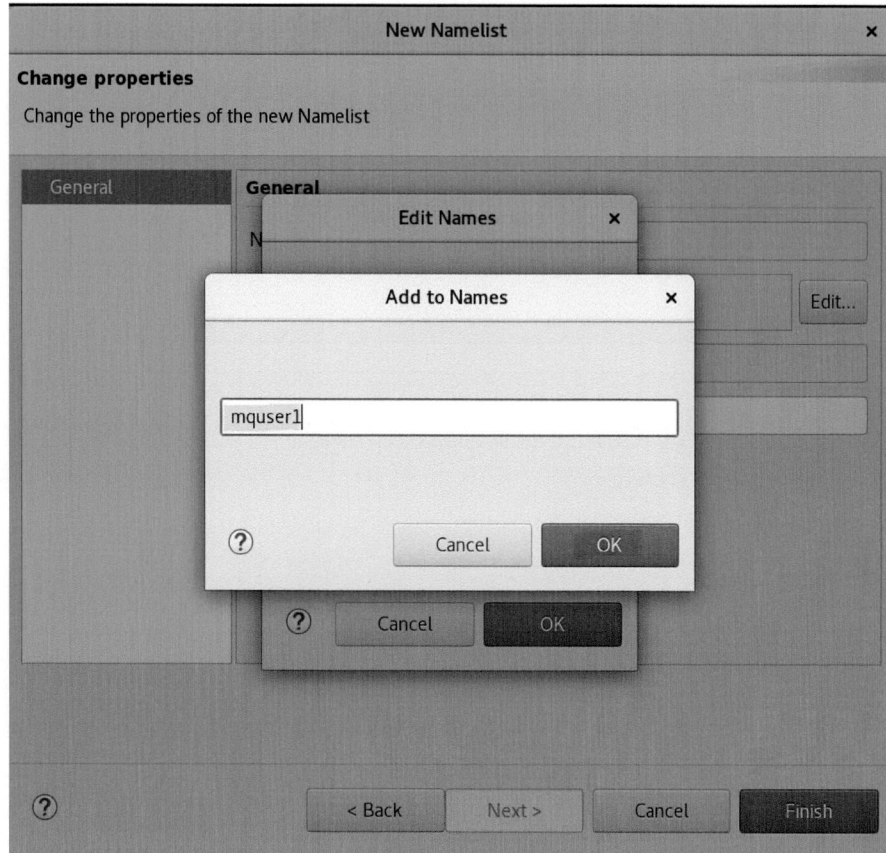

Figure 3-101. *The AUDNAMES Namelist has an entry of* ***mquser1*** *added*

   c. Click **OK**.

CHAPTER 3 IBM JMS INTERFACE DEVELOPMENT IBM FILENET 5.5.X WORKFLOW

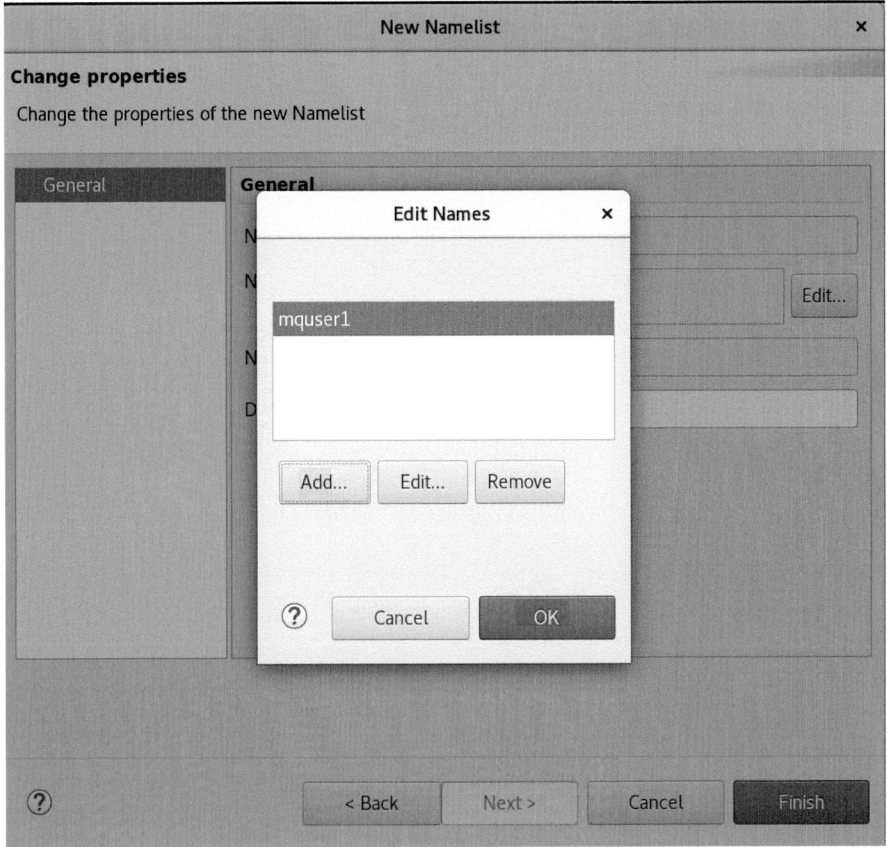

Figure 3-102. *The first name, mquser1, is added and the OK button clicked*

The OK button is then clicked after adding all the required names to the AUDNAMES list.

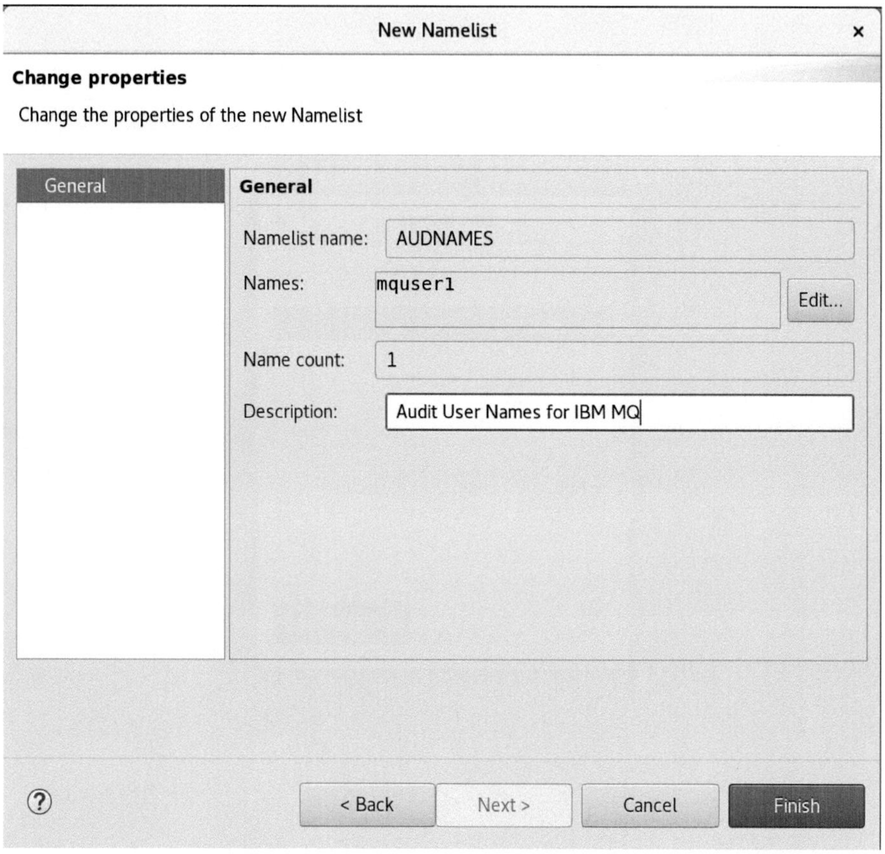

CHAPTER 3   IBM JMS INTERFACE DEVELOPMENT IBM FILENET 5.5.X WORKFLOW

Figure 3-103. The AUDNAMES description is entered and the Finish button clicked

After entering the **AUDNAMES** description and clicking the **Finish** command button, you get a confirmation that the **Namelist** object was successfully created.

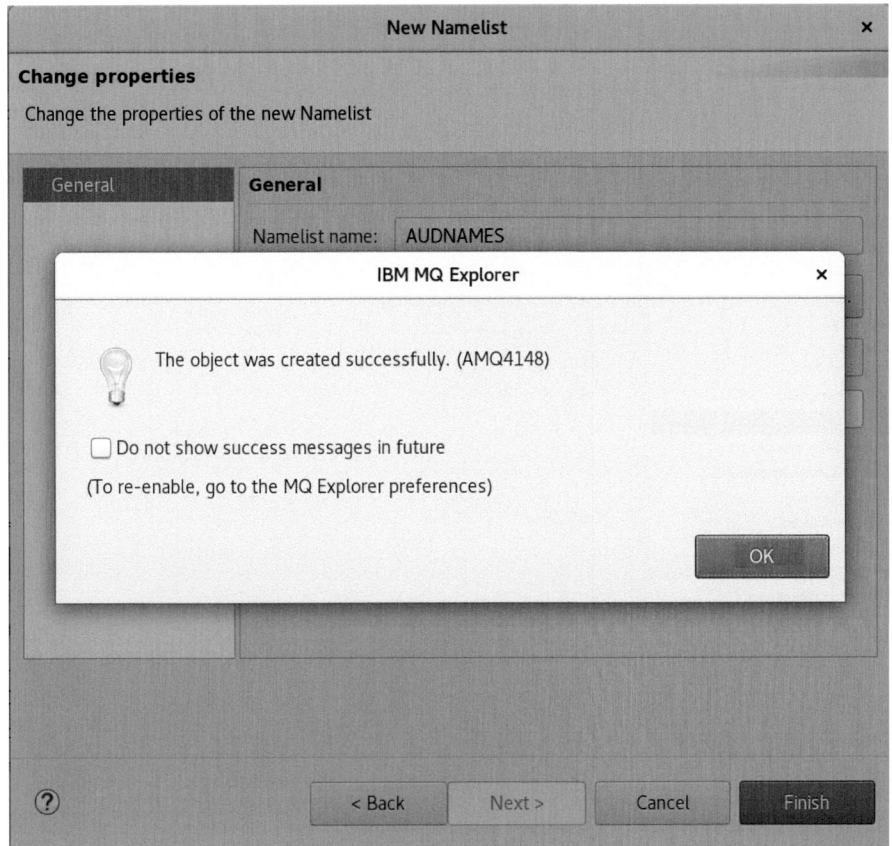

Figure 3-104. *The success message is displayed, indicating the name is added*

On clicking the OK button on the pop-up window, the **mquser1** name in the **AUDNAMES** Namelist is displayed.

CHAPTER 3   IBM JMS INTERFACE DEVELOPMENT IBM FILENET 5.5.X WORKFLOW

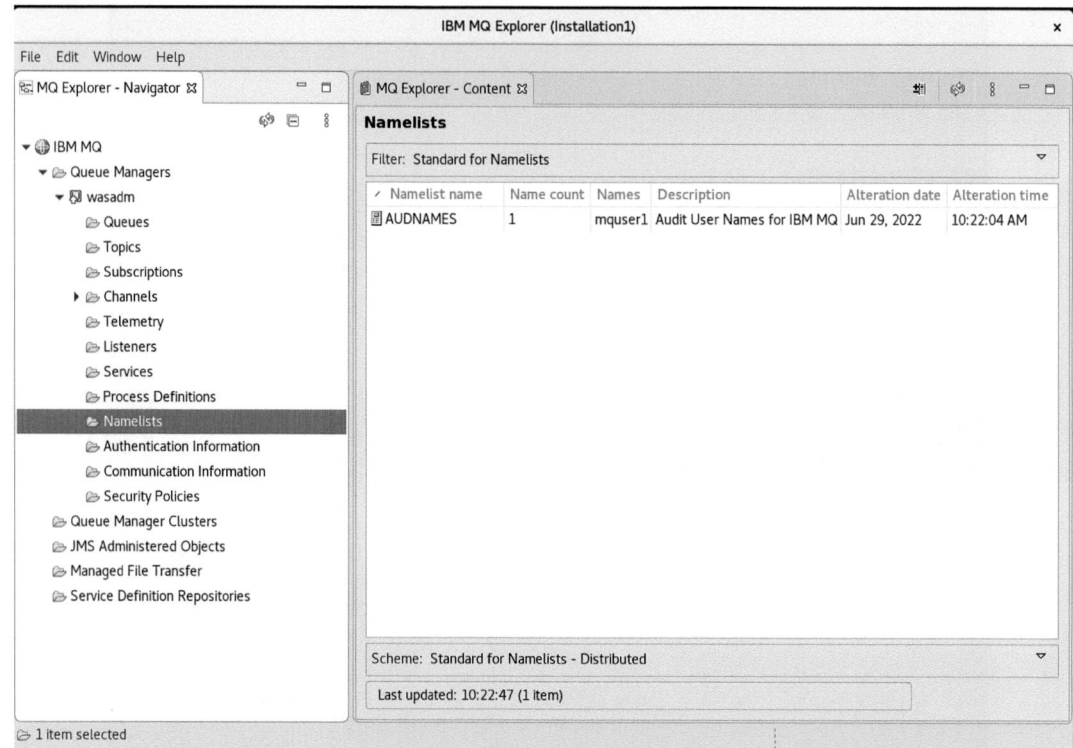

Figure 3-105. *The AUDNAMES Namelist shows the mquser1 name*

    4. Right-click the queue manager, select **Properties**, and select the SSL page.

CHAPTER 3   IBM JMS INTERFACE DEVELOPMENT IBM FILENET 5.5.X WORKFLOW

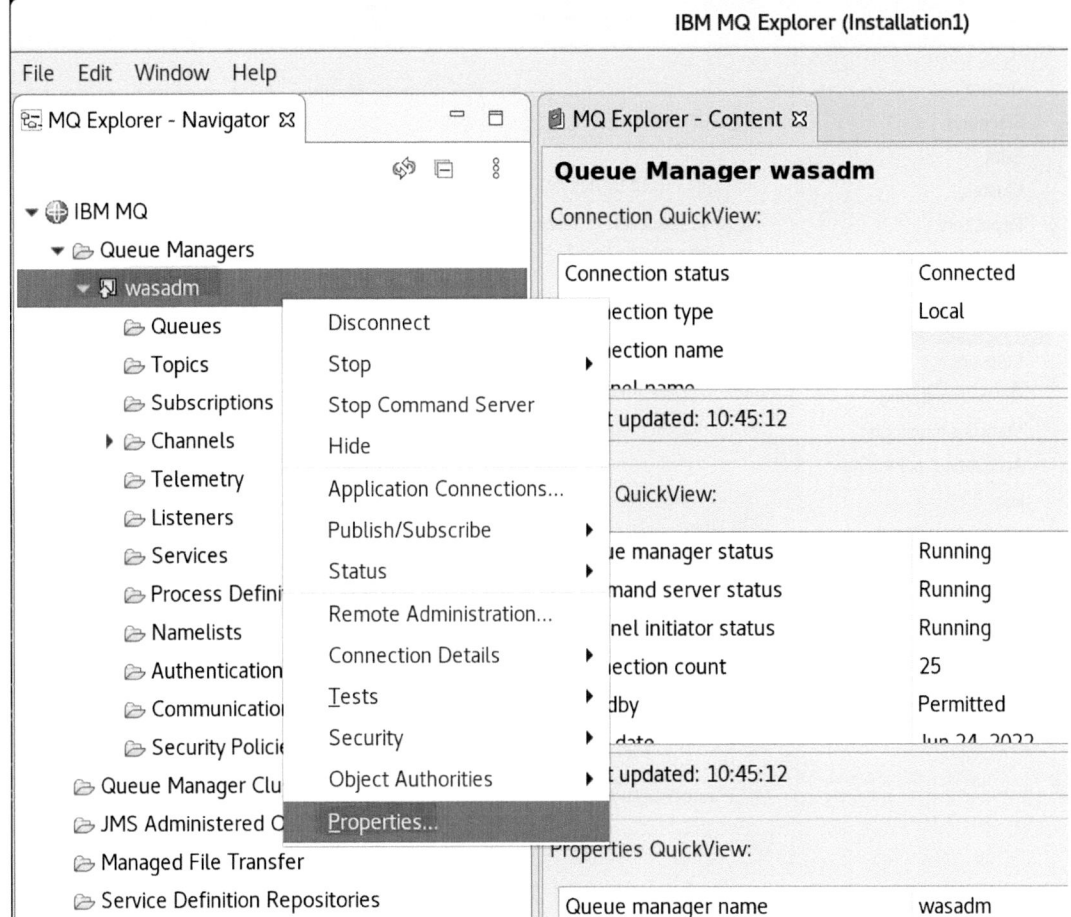

Figure 3-106. *The wasadm queue manager properties are selected*

The SSL page is selected.

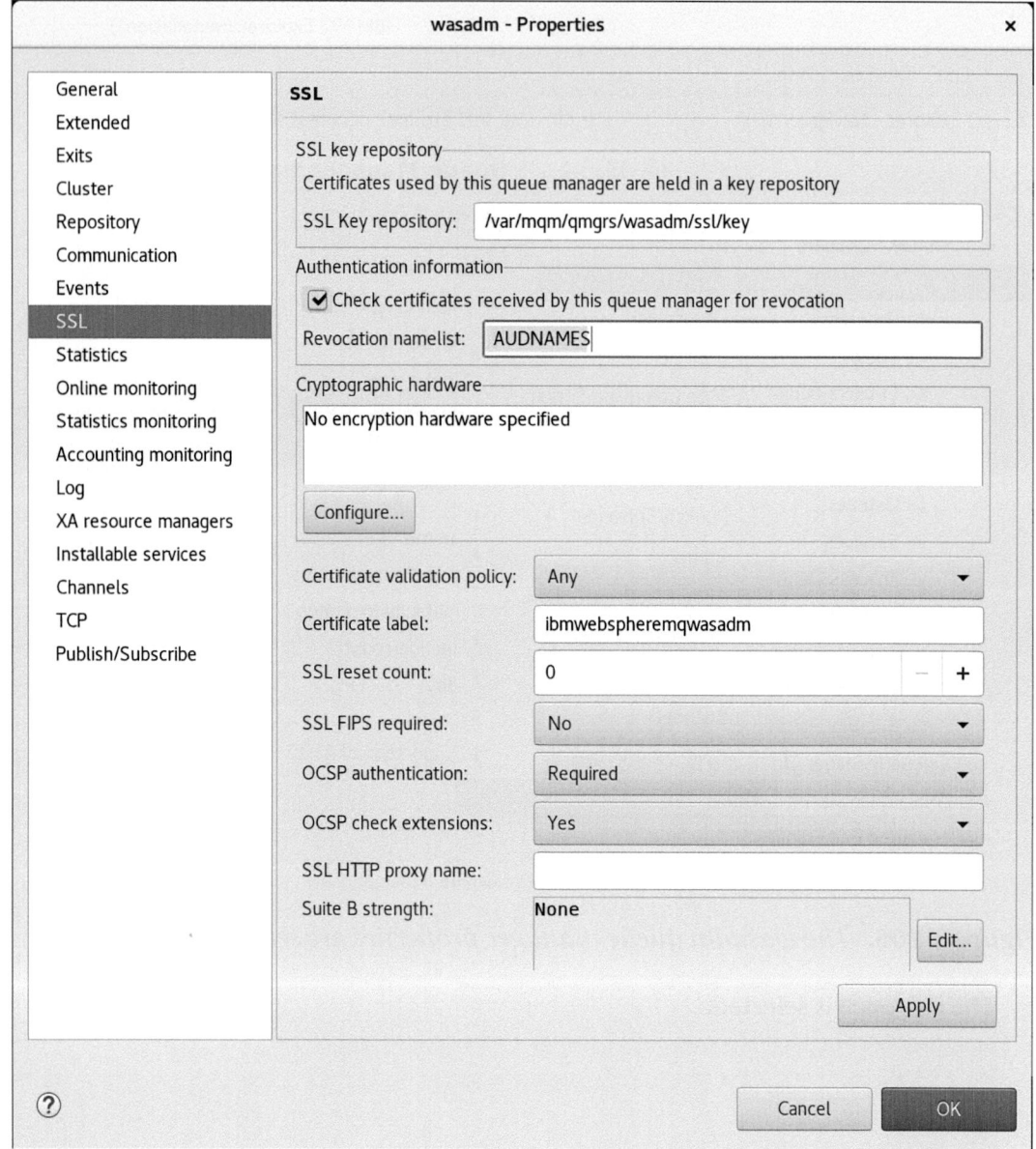

Figure 3-107. *The SSL page is selected, and the check box highlighted is ticked*

   a. Select the Check certificates received by this queue manager against Certification Revocation Lists check box.

   b. Type the name of the namelist (**AUDNAMES** from step **3a**) in the **CRL Namelist** field.

Next, we need to create a New Client connection Channel, **AUD.CLNTCONN**. The New Client Connection

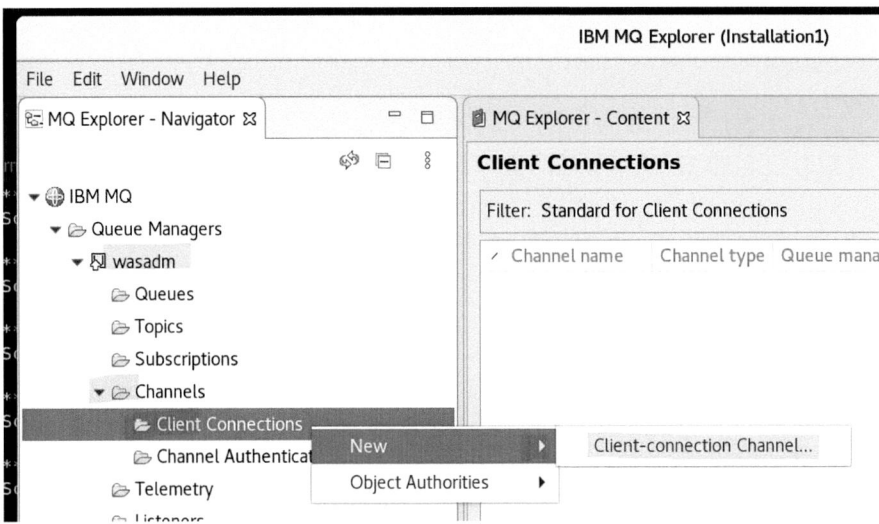

Figure 3-108. *The New Client-connection Channel option is selected*

The **AUD.CLNTCONN** Client-connection Channel is created.

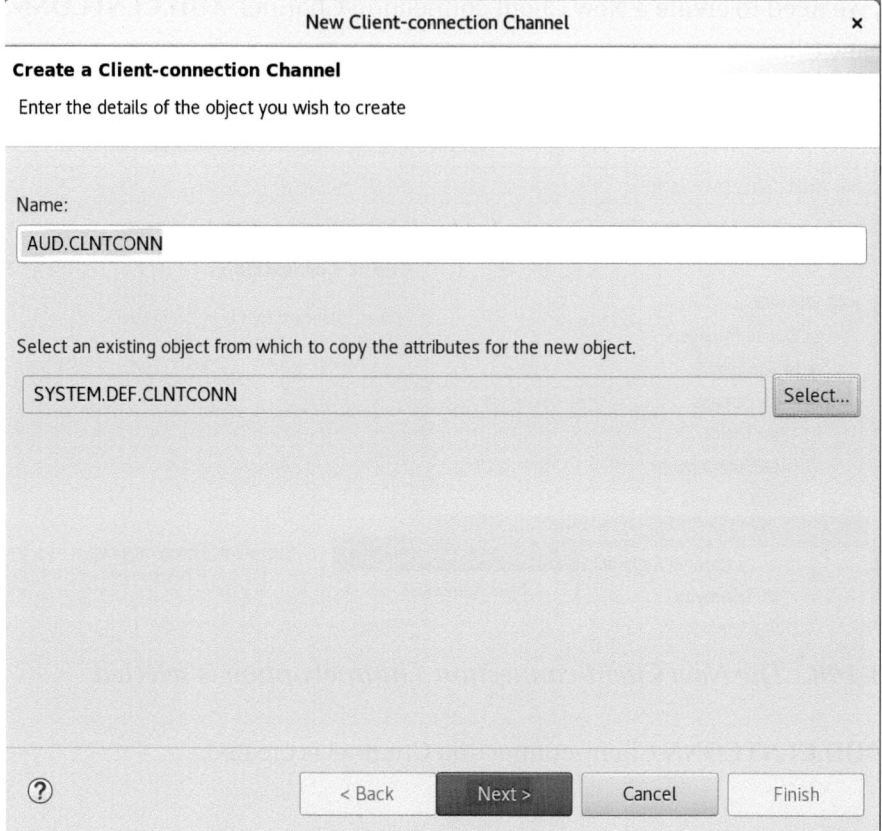

Figure 3-109. *The AUD.CLNTCONN with the SYSTEM.DEF.CLNTCONN is created*

Next is then clicked to display the properties to be entered for using the **AUD.CLNTCONN** with the **SYSTEM.DEF.CLNTCONN** Channel connection.

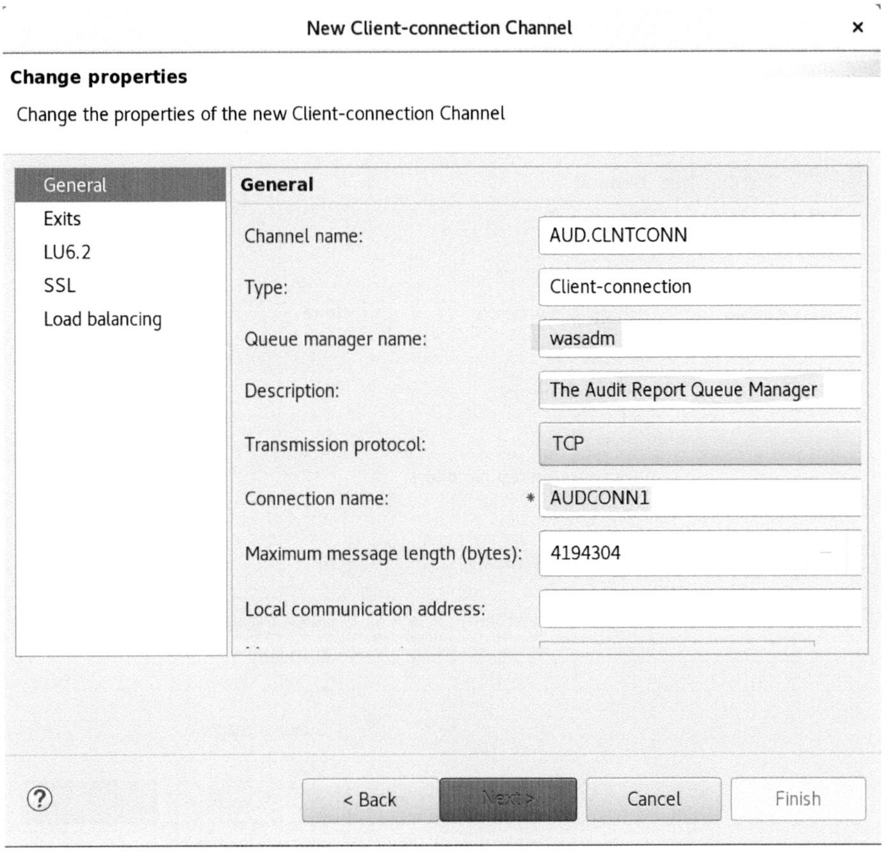

CHAPTER 3 IBM JMS INTERFACE DEVELOPMENT IBM FILENET 5.5.X WORKFLOW

Figure 3-110. *The Queue Manager name and the Connection name are entered*

Then we scroll down to enter the required value for the **Default reconnection** and select this as **Yes**.

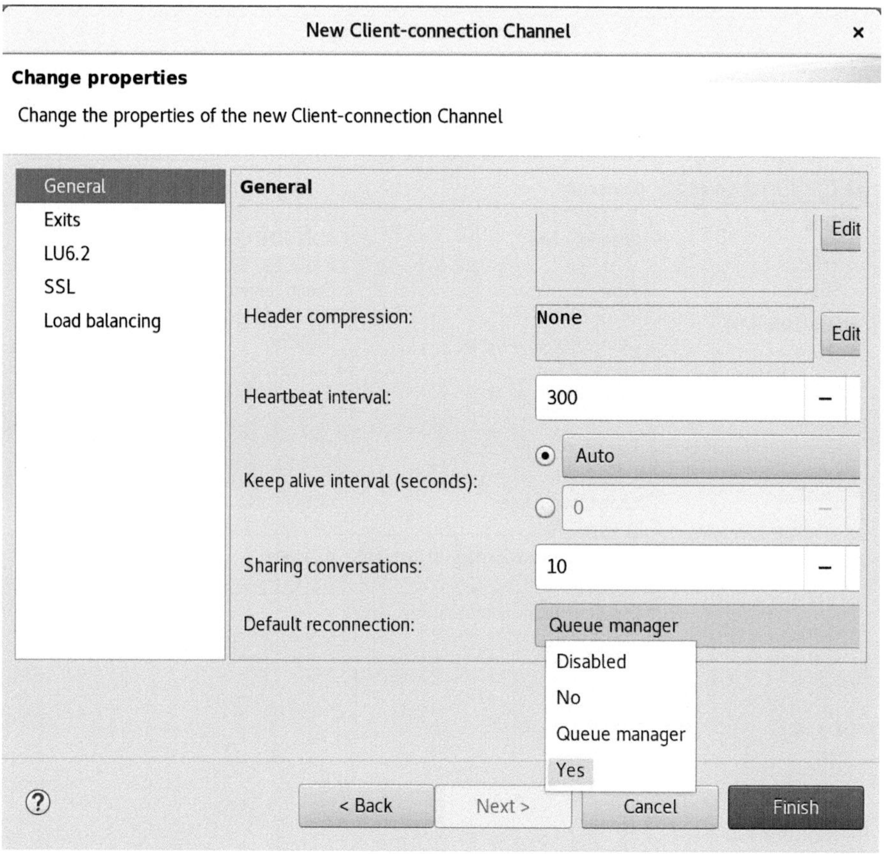

CHAPTER 3 IBM JMS INTERFACE DEVELOPMENT IBM FILENET 5.5.X WORKFLOW

*Figure 3-111. The **Default reconnection** option is set to **Yes** from the drop-down*

CHAPTER 3 IBM JMS INTERFACE DEVELOPMENT IBM FILENET 5.5.X WORKFLOW

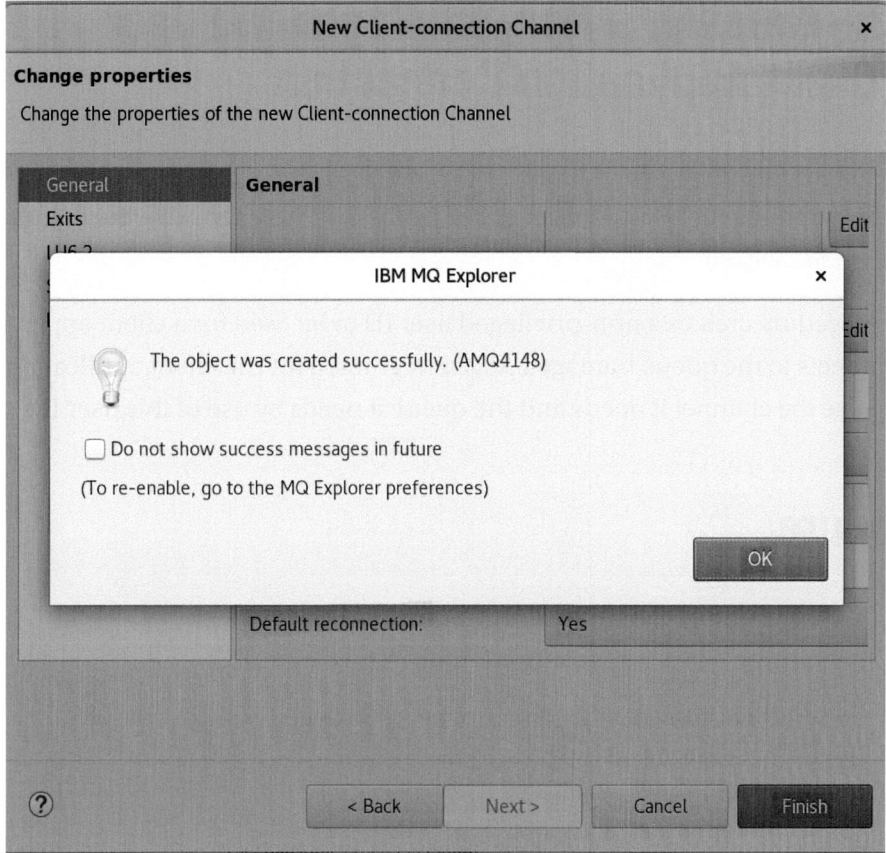

Figure 3-112. *The Client Connection Channel AUD.CLNTCONN is created*

The new Client Connection Channel **AUD.CLNTCONN** is displayed.

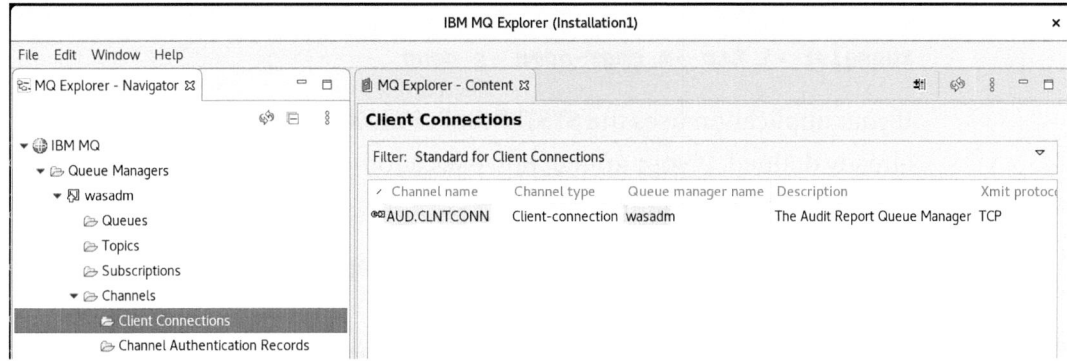

Figure 3-113. *The Client Connection Channel AUD.CLNTCONN*

CHAPTER 3   IBM JMS INTERFACE DEVELOPMENT IBM FILENET 5.5.X WORKFLOW

Part 3 – Creating a Non-privileged MQ User for a Client Application Connection

The instructions from this link were followed:

www-01.ibm.com/support/knowledgecenter/SSFKSJ_8.0.0/com.ibm.mq.dev.doc/q023960_.htm?lang=en

This procedure creates a non-privileged user ID to be used for a client application which connects to the queue manager. Access is granted for the client application only to be able to use the channel it needs and the queue it needs by use of this user ID.

Procedure

1. Obtain a user ID on the system where the queue manager is running on (the server **ecmukdemo6** in this example).

2. For this task, this user ID must not be a privileged administrative user. This user ID will be the authority under which the client connection will run on the queue manager.

3. Start a listener program with the following commands where

    - *qmgr-name* is the name of your queue manager.
    - *nnnn* is your chosen port number.
    - For UNIX and Windows systems:

    runmqlsr -t tcp -m *qmgr-name* **-p** *nnnn*

    If your application uses the **SYSTEM.DEF.SVRCONN**, then this channel is already defined. If your application uses another channel, create it by issuing the MQSC command:

    DEFINE CHANNEL(' *channel-name* **') CHLTYPE(SVRCONN)**
    TRPTYPE(TCP) +
    DESCR('Channel for use by sample programs')

    - *channel-name* is the name of your channel.

4. Create a channel authentication rule allowing only the IP address of your client system to use the channel by issuing the MQSC command:

SET CHLAUTH(' *channel-name* ') TYPE(ADDRESSMAP) ADDRESS(' *client-machine-IP-address* ') +
MCAUSER(' *non-privileged-user-id* ')

In the command window:

su - mqm
runmqsc wasadm
SET CHLAUTH('SYSTEM.ADMIN.SVRCONN') TYPE(ADDRESSMAP)
ADDRESS('10.10.10.90') MCAUSER('db2inst1')

```
[mqm@ECMUKDEMO6 ~]$ runmqsc wasadm
5724-H72 (C) Copyright IBM Corp. 1994, 2022.
Starting MQSC for queue manager wasadm.

SET CHLAUTH('SYSTEM.ADMIN.SVRCONN') TYPE(ADDRESSMAP) ADDRESS('10.10.10.90') MCAUSER('db2inst1')
     1 : SET CHLAUTH('SYSTEM.ADMIN.SVRCONN') TYPE(ADDRESSMAP) ADDRESS('10.10.10.90') MCAUSER('db2inst1')
AMQ8877I: IBM MQ channel authentication record set.
```

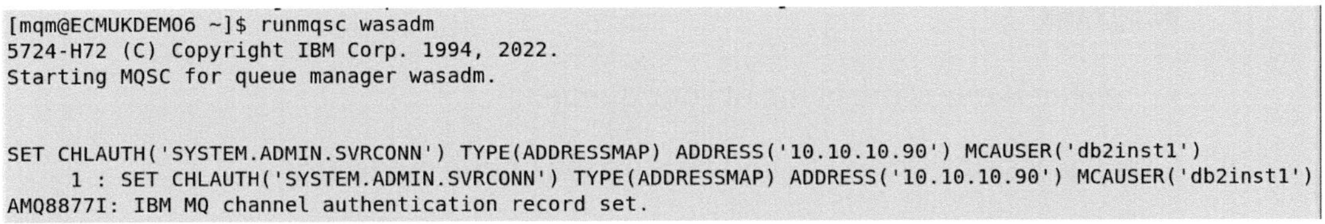

Figure 3-114. The db2inst1 user is set as a non-privileged user id

Exit from the runmqsc command.

```
AMQ8877I: IBM MQ channel authentication record set.
quit
     2 : quit
One MQSC command read.
No commands have a syntax error.
All valid MQSC commands were processed.
[mqm@ECMUKDEMO6 ~]$
```

Figure 3-115. Exit using the quit command

- *channel-name* is the name of your channel.
- *client-machine-IP-address* is the IP address of your client system.

CHAPTER 3   IBM JMS INTERFACE DEVELOPMENT IBM FILENET 5.5.X WORKFLOW

If your sample client application is running on the same machine as the queue manager, then use an IP address of "127.0.0.1" if your application is going to connect using "localhost." If several different client machines are going to connect in, you can use a pattern or a range instead of a single IP address. See www-01.ibm.com/support/knowledgecenter/SSFKSJ_8.0.0/com.ibm.mq.ref.adm.doc/q086080_.htm?lang=en-us (Generic IP addresses) for details.

- *non-privileged-user-id* is the user ID you obtained in step 1.

5. If your application uses the **SYSTEM.DEFAULT.LOCAL.QUEUE**, then this queue is already defined. If your application uses another queue, create it by issuing the **MQSC** command:

DEFINE QLOCAL(' *queue-name* ') DESCR('Queue for use by sample programs')

- *queue-name* is the name of your queue.

6. Grant access to connect to and inquire the queue manager:

- For IBM i, UNIX, and Windows systems, issue the MQSC commands:

SET AUTHREC OBJTYPE(QMGR) PRINCIPAL(' *non-privileged-user-id* ') +
AUTHADD(CONNECT, INQ)

- *non-privileged-user-id* is the user ID you obtained in step 1.

If your application is a point-to-point application, that is, it makes use of queues, grant access to allow inquiring and then putting and getting messages using your queue by the user ID to be used, by issuing the MQSC commands:

- For IBM i, UNIX, and Windows systems, issue the MQSC commands:

SET AUTHREC PROFILE(' *queue-name* ') OBJTYPE(QUEUE) +

PRINCIPAL(' *non-privileged-user-id* ') AUTHADD(PUT, GET, INQ, BROWSE)

- *queue-name* is the name of your queue.
- *non-privileged-user-id* is the user ID you obtained in step 1.

7. If your application is a publish/subscribe application, that is, it makes use of topics, grant access to allow publishing and subscribing using your topic by the user ID to be used, by issuing the MQSC commands:

   - For IBM i, UNIX, and Windows systems, issue the MQSC commands:

   SET AUTHREC PROFILE('SYSTEM.BASE.TOPIC') OBJTYPE(TOPIC) + PRINCIPAL(' *non-privileged-user-id* ') AUTHADD(PUB, SUB)

   - *non-privileged-user-id* is the user ID you obtained in step 1.
   - This will give *non-privileged-user-id* access to any topic in the topic tree; alternatively, you can define a topic object using DEFINE TOPIC and grant accesses only to the part of the topic tree referenced by that topic object. See www-01.ibm.com/support/knowledgecenter/SSFKSJ_8.0.0/com.ibm.mq.sec.doc/q013980_.htm?lang=en-us (**Controlling user access to topics**) for details.

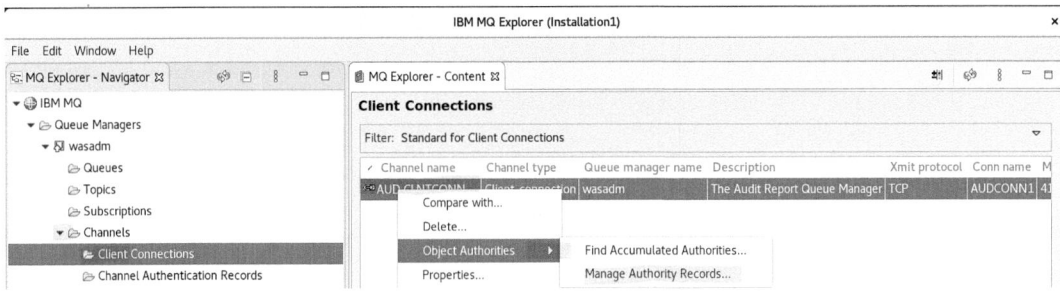

Figure 3-116. *The AUD.CLNTCONN* **Manage Authority Records** *option is selected*

The Manage Authority Records submenu allows the security on a Client Connections object to be modified to add or remove Users and Groups.

CHAPTER 3   IBM JMS INTERFACE DEVELOPMENT IBM FILENET 5.5.X WORKFLOW

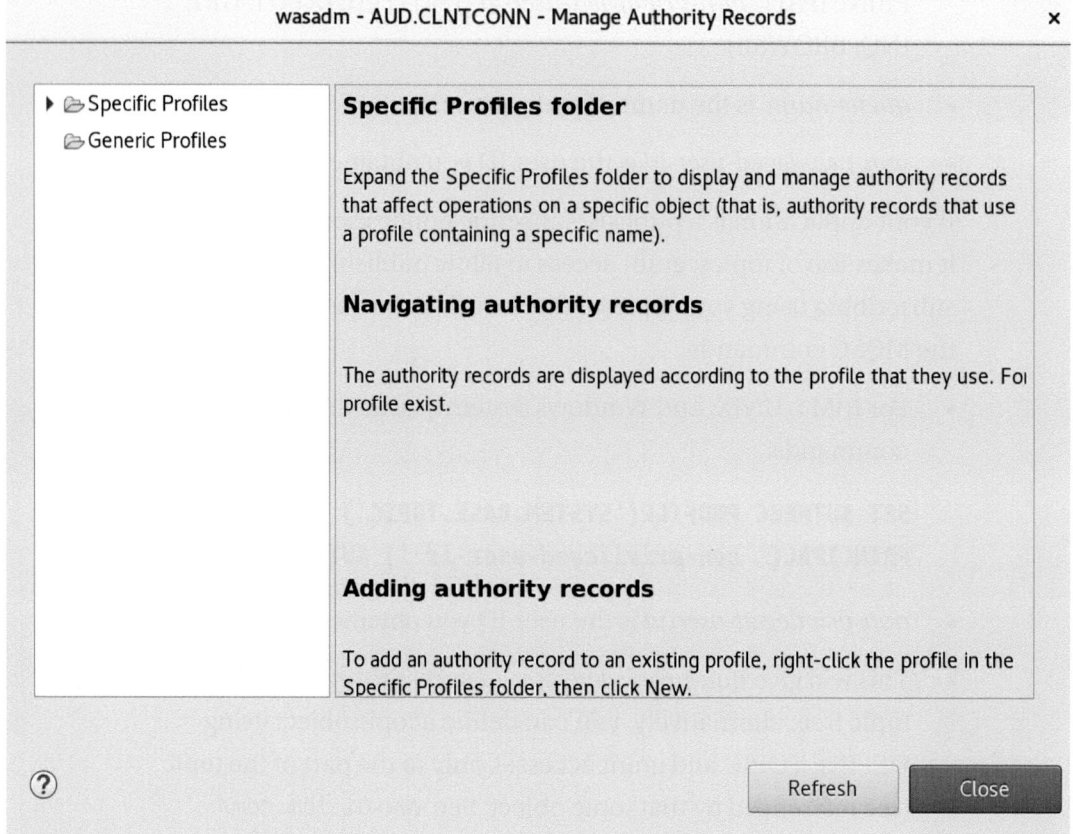

Figure 3-117. *The AUD.CLNTCONN* **Manage Authority Records** *option*

The **Specific Profiles** opener is clicked, and the **New** command button is selected.

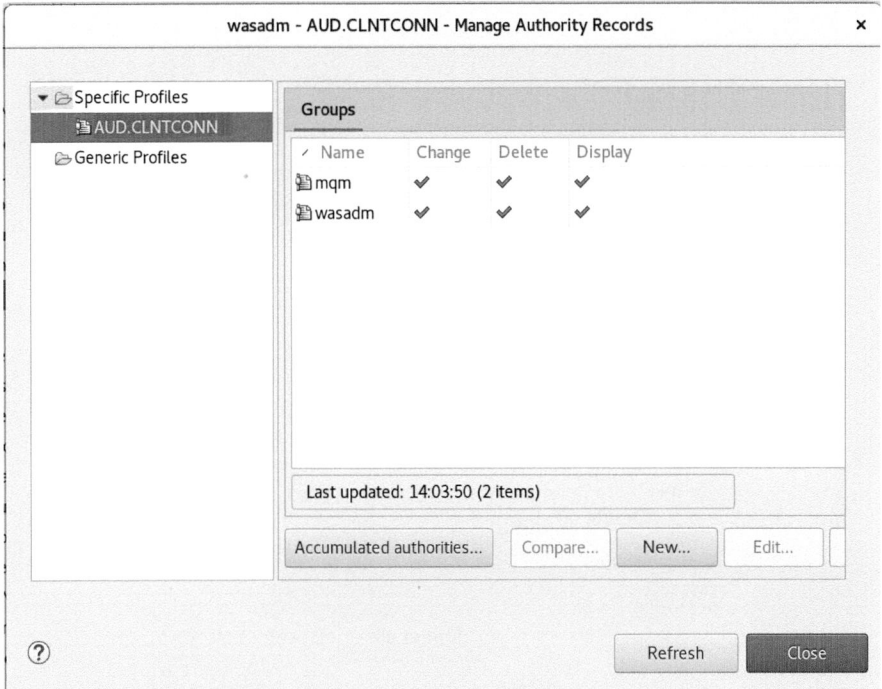

Figure 3-118. *The New command button is clicked*

A search can now be made to look for a Group or User in the connected LDAP server to add to the security list for the Connection Object.

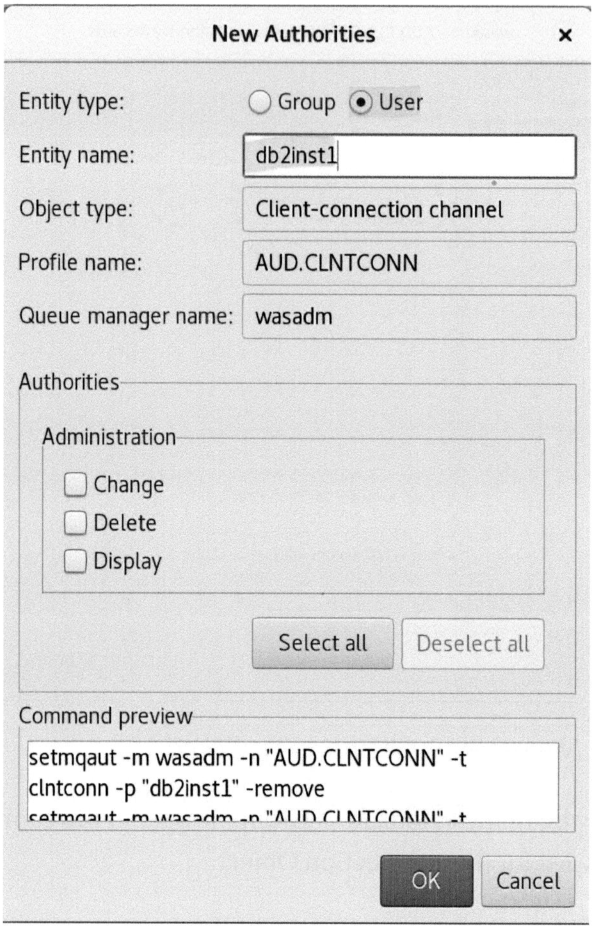

CHAPTER 3   IBM JMS INTERFACE DEVELOPMENT IBM FILENET 5.5.X WORKFLOW

Figure 3-119. *The db2inst1 user is added to the AUD.CLNTCONN profile*

A command-line preview is shown in the **New Authorities** screen.

After the user is selected, and the required Administration security attributes added, if required from the tick box list shown in Figure 3-119 (Change, Delete, and Display), a pop-up box showing the success of the security changes is shown.

CHAPTER 3   IBM JMS INTERFACE DEVELOPMENT IBM FILENET 5.5.X WORKFLOW

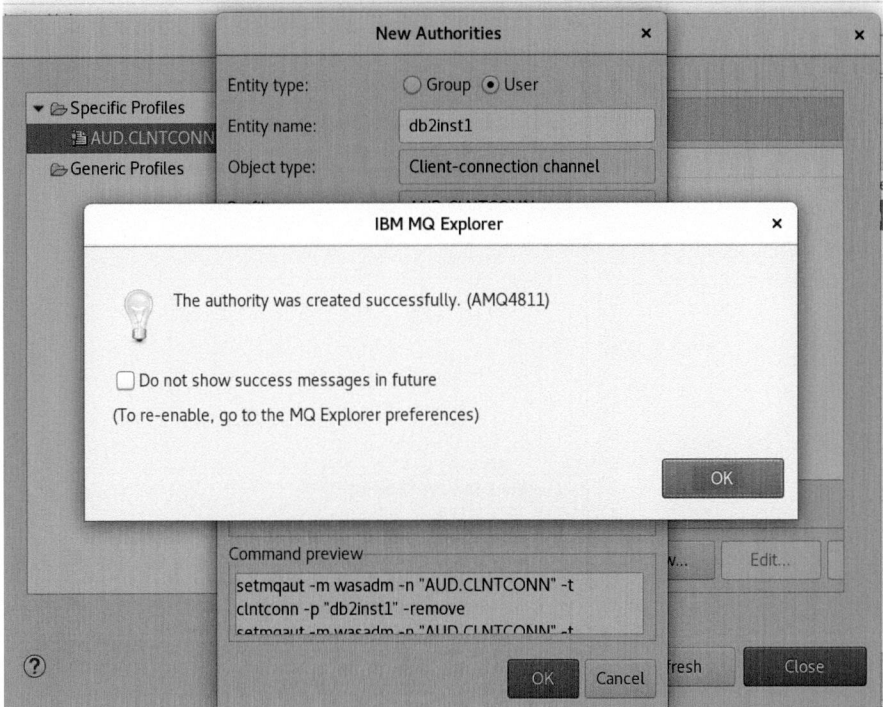

Figure 3-120. *The db2inst1 user is successfully added to the Client-connection channel*

Now we can edit the new db2inst1 user's db2iadm1 group to give the group Change, Delete, and Display security access.

465

CHAPTER 3   IBM JMS INTERFACE DEVELOPMENT IBM FILENET 5.5.X WORKFLOW

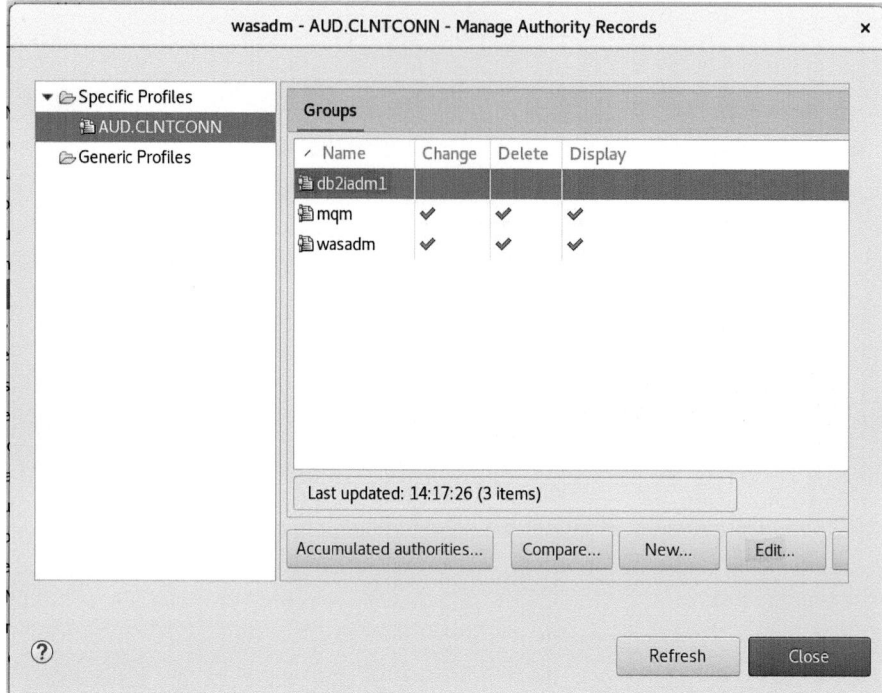

Figure 3-121. *The db2iadm1 group is edited to add the security access required for the Group*

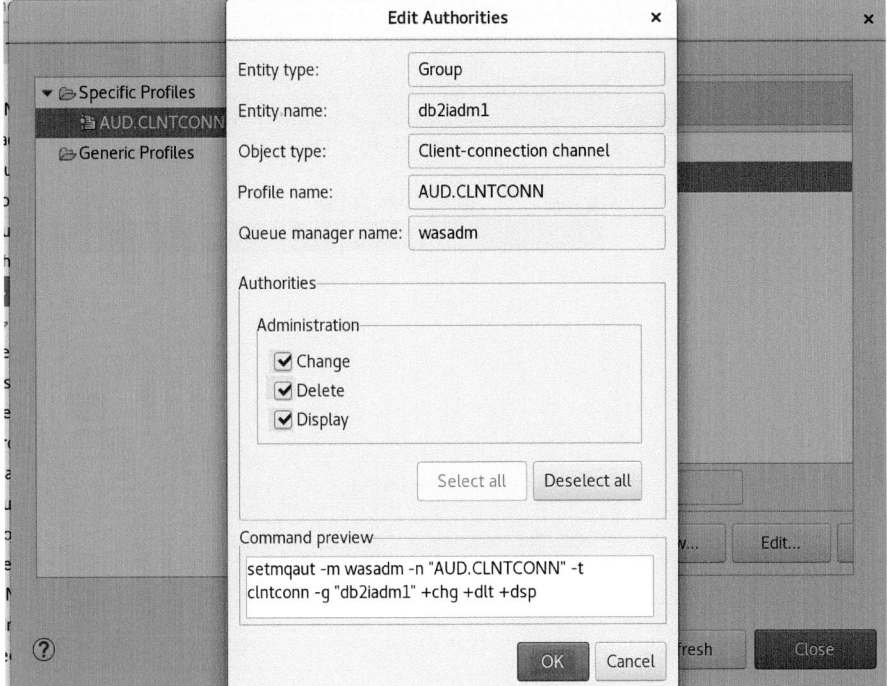

Figure 3-122. *The tick box is selected to give access as required*

The user's Group is also displayed, and the Group security can now be edited to give any Group security access required.

CHAPTER 3   IBM JMS INTERFACE DEVELOPMENT IBM FILENET 5.5.X WORKFLOW

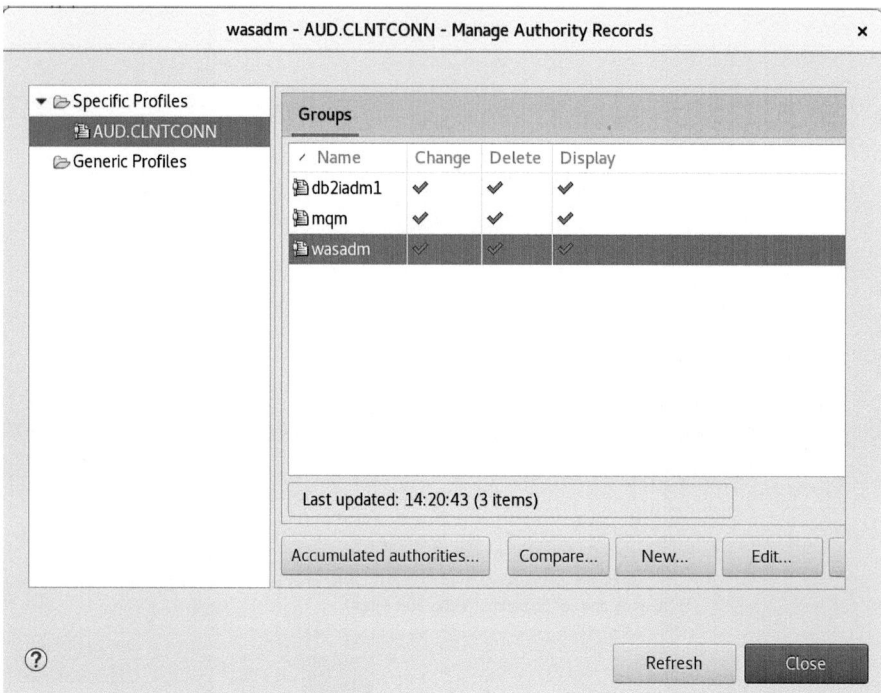

Figure 3-123. *The db2iadm1 Group now has full security access*

In the command window, we can add the user's security access:

```
su - mqm
runmqsc wasadm
SET AUTHREC PROFILE('AUDQCF') OBJTYPE(QUEUE) PRINCIPAL('db2inst1')
AUTHADD(PUT, GET, INQ, BROWSE)
```

```
[mqm@ECMUKDEMO6 ~]$ runmqsc wasadm
5724-H72 (C) Copyright IBM Corp. 1994, 2022.
Starting MQSC for queue manager wasadm.


SET AUTHREC PROFILE('AUDQCF') OBJTYPE(QUEUE) PRINCIPAL('db2inst1') AUTHADD(PUT, GET, INQ, BROWSE)
     1 : SET AUTHREC PROFILE('AUDQCF') OBJTYPE(QUEUE) PRINCIPAL('db2inst1') AUTHADD(PUT, GET, INQ, BROWSE)
AMQ8862I: IBM MQ authority record set.
```

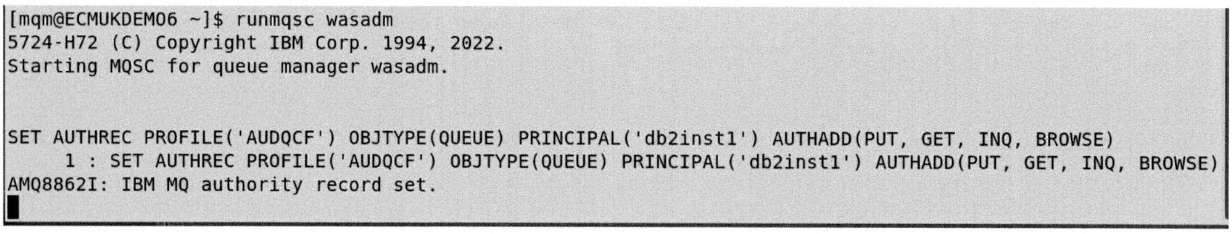

Figure 3-124. *The command is run to give the db2inst1 user access to the AUDQCF Queue*

The Find Accumulated Authorities menu option is selected.

CHAPTER 3   IBM JMS INTERFACE DEVELOPMENT IBM FILENET 5.5.X WORKFLOW

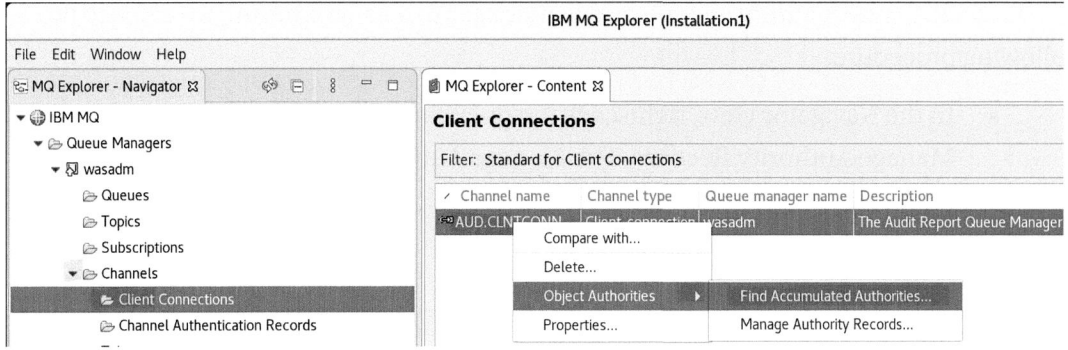

Figure 3-125. *The Object Authorities ➤ Find Accumulated Authorities is selected*

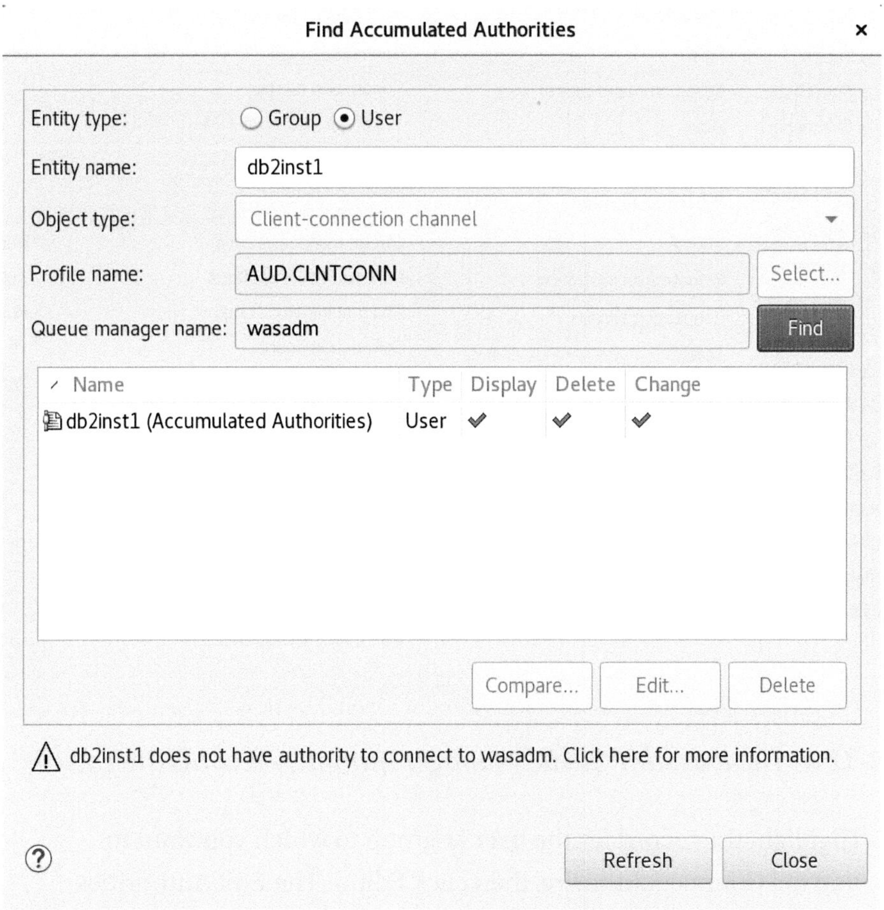

Figure 3-126. *The db2inst1 user has no authority for Queue Manager*

CHAPTER 3 IBM JMS INTERFACE DEVELOPMENT IBM FILENET 5.5.X WORKFLOW

To grant Connect authority for a queue manager to a user or group, we can use the following procedure:

- In the Navigator view, right-click the queue manager, then click Manage Authority Records. The Manage Authority Records dialog opens.

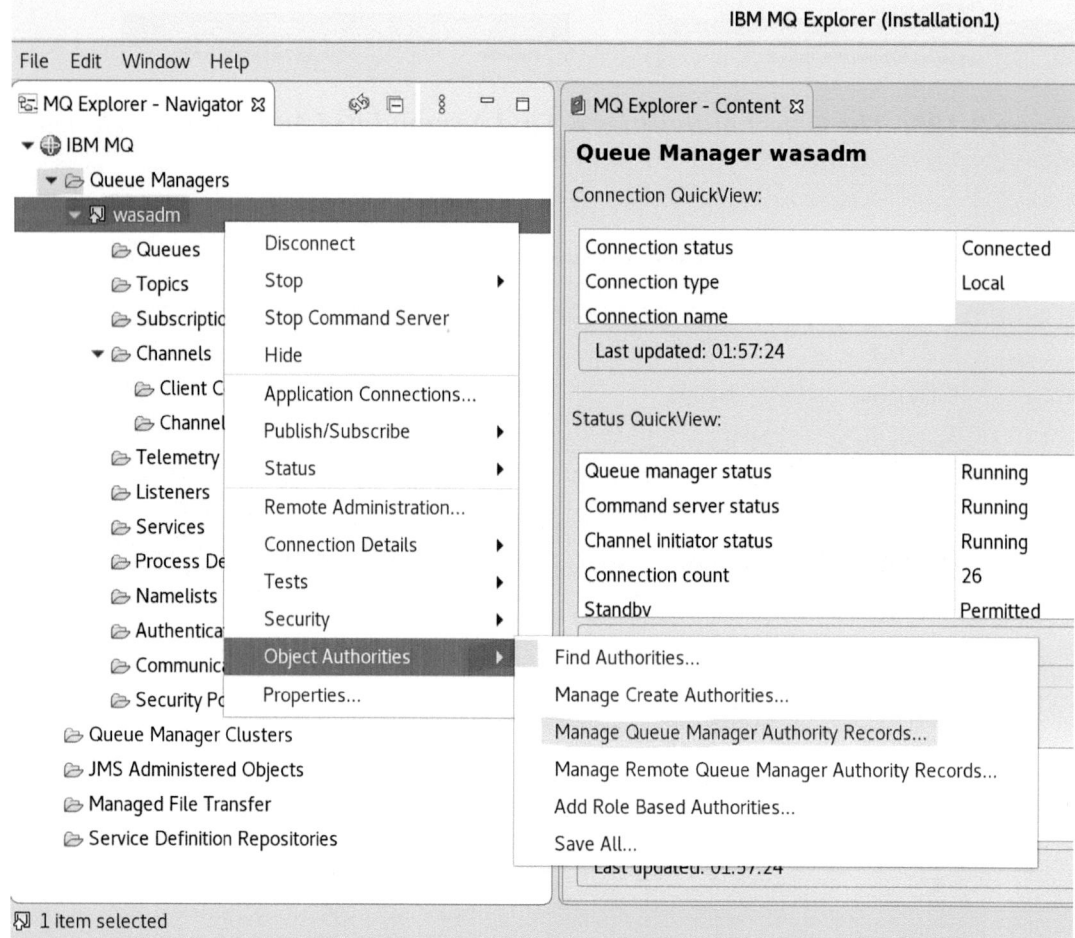

Figure 3-127. *The wasadm Queue Manager Authority records are edited*

- Highlight the record for the user or group to which you want to add the Connect authority, then click Edit…. The Edit Authorities dialog opens.
- Select the Connect check box, then click OK.

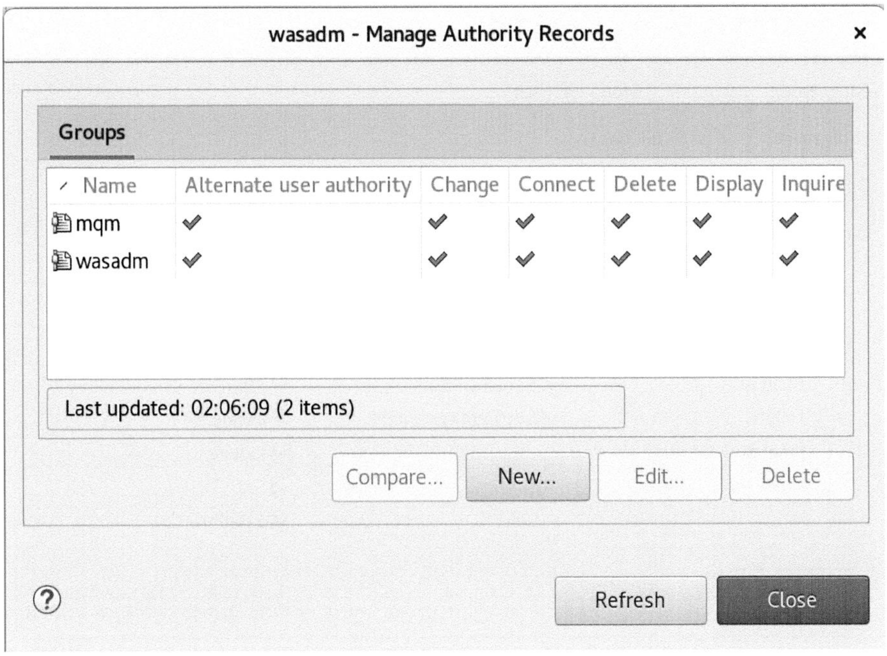

Figure 3-128. *The New command is used to add the db2inst1 user*

The New command is used to add the missing db2inst1 user to the wasadm queue.

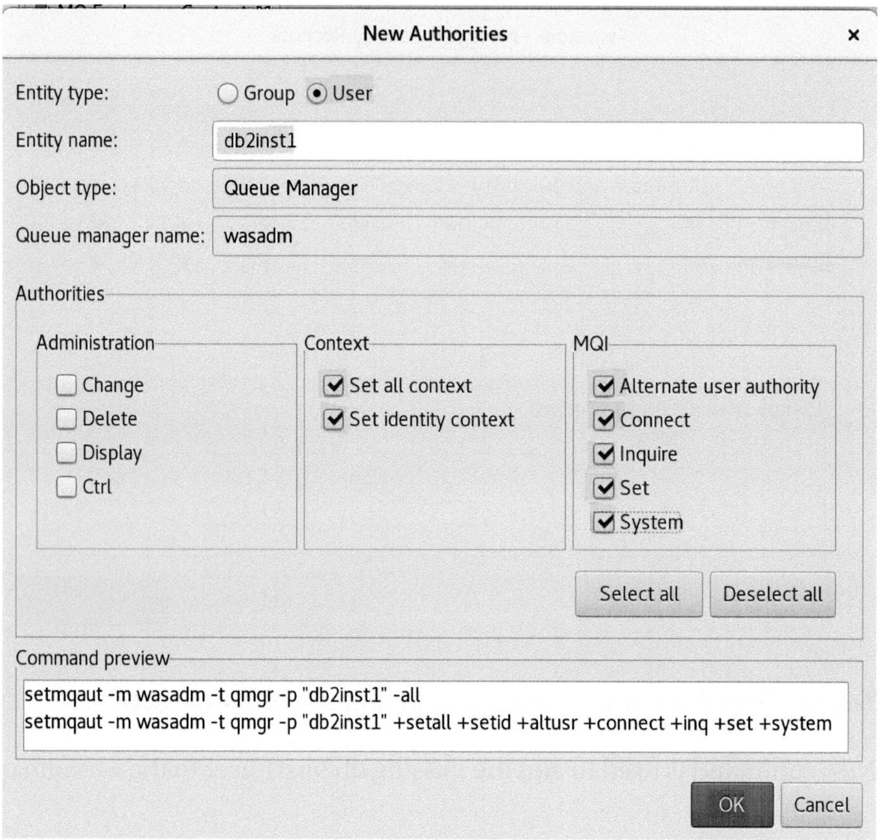

CHAPTER 3   IBM JMS INTERFACE DEVELOPMENT IBM FILENET 5.5.X WORKFLOW

Figure 3-129. The security options for the db2inst1 user are added

The OK command button is clicked after setting the required db2inst1 user access options.

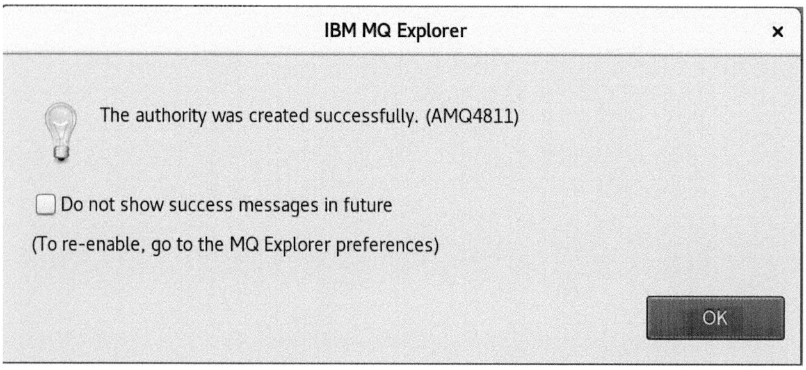

Figure 3-130. The authority for db2inst1 is successfully added

CHAPTER 3   IBM JMS INTERFACE DEVELOPMENT IBM FILENET 5.5.X WORKFLOW

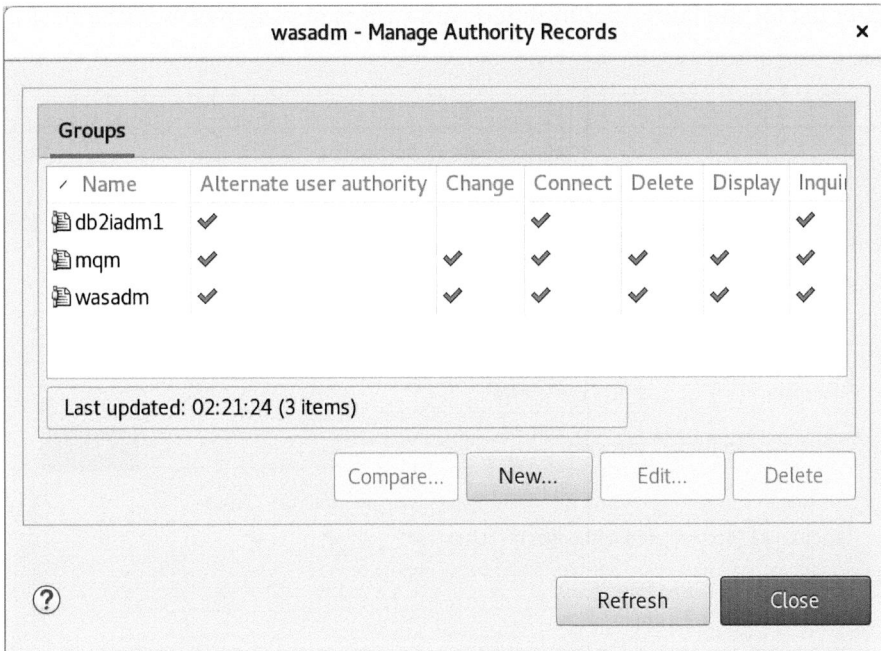

Figure 3-131. *The authority for the db2inst1 user's group, db2iadm, can now be seen*

The Find Accumulated Authorities menu item can be used to show that the db2inst1 user has access.

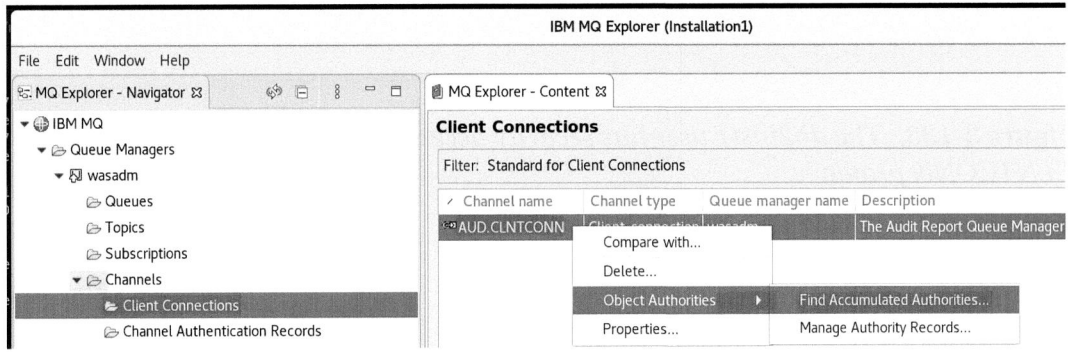

Figure 3-132. *The Find Accumulated Authorities menu item is selected*

473

CHAPTER 3   IBM JMS INTERFACE DEVELOPMENT IBM FILENET 5.5.X WORKFLOW

The user list for the Find Accumulated Authorities now shows that db2inst1 user has access.

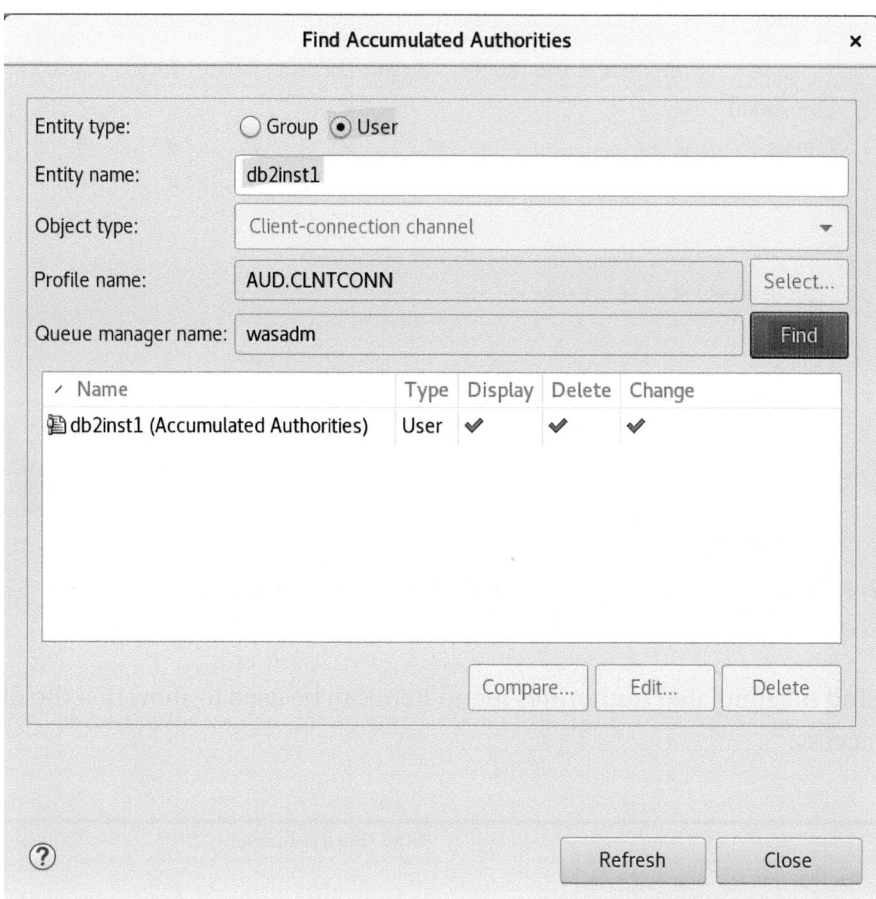

Figure 3-133. *The db2inst1 user has security access on the AUD. CLNTCONN profile*

Reconnect to the JMS LDAP server on ecmukdemo6.

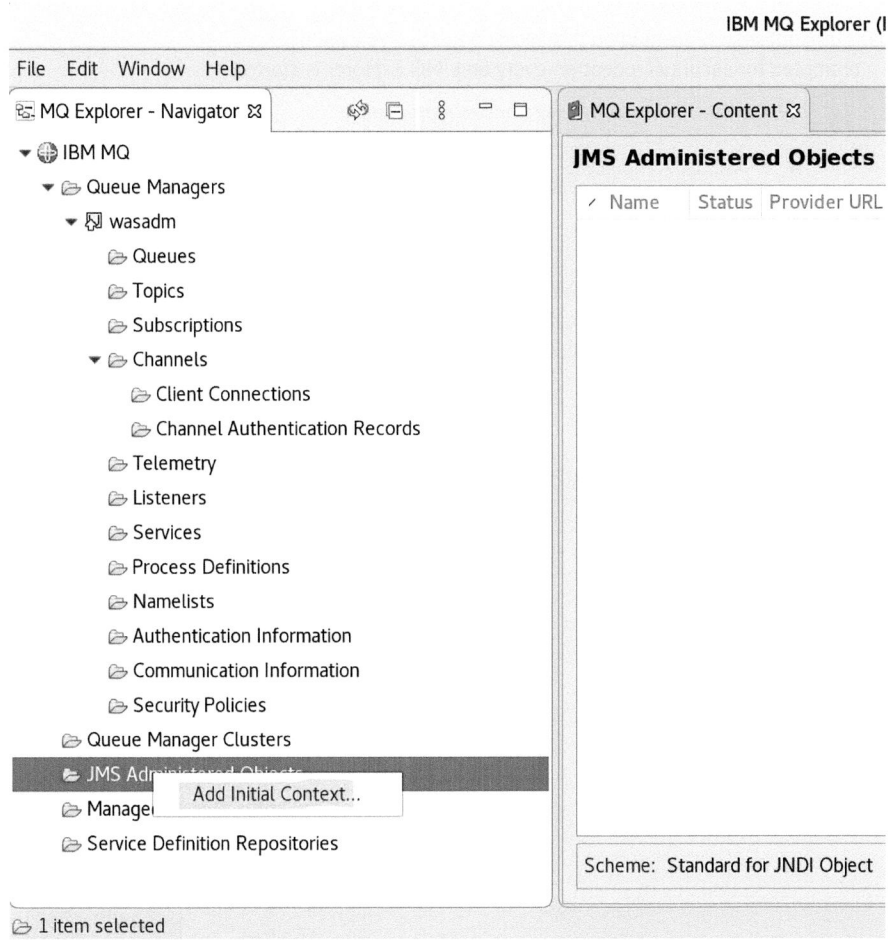

Figure 3-134. *The Add Initial Context is selected on the right-mouse click*

CHAPTER 3 IBM JMS INTERFACE DEVELOPMENT IBM FILENET 5.5.X WORKFLOW

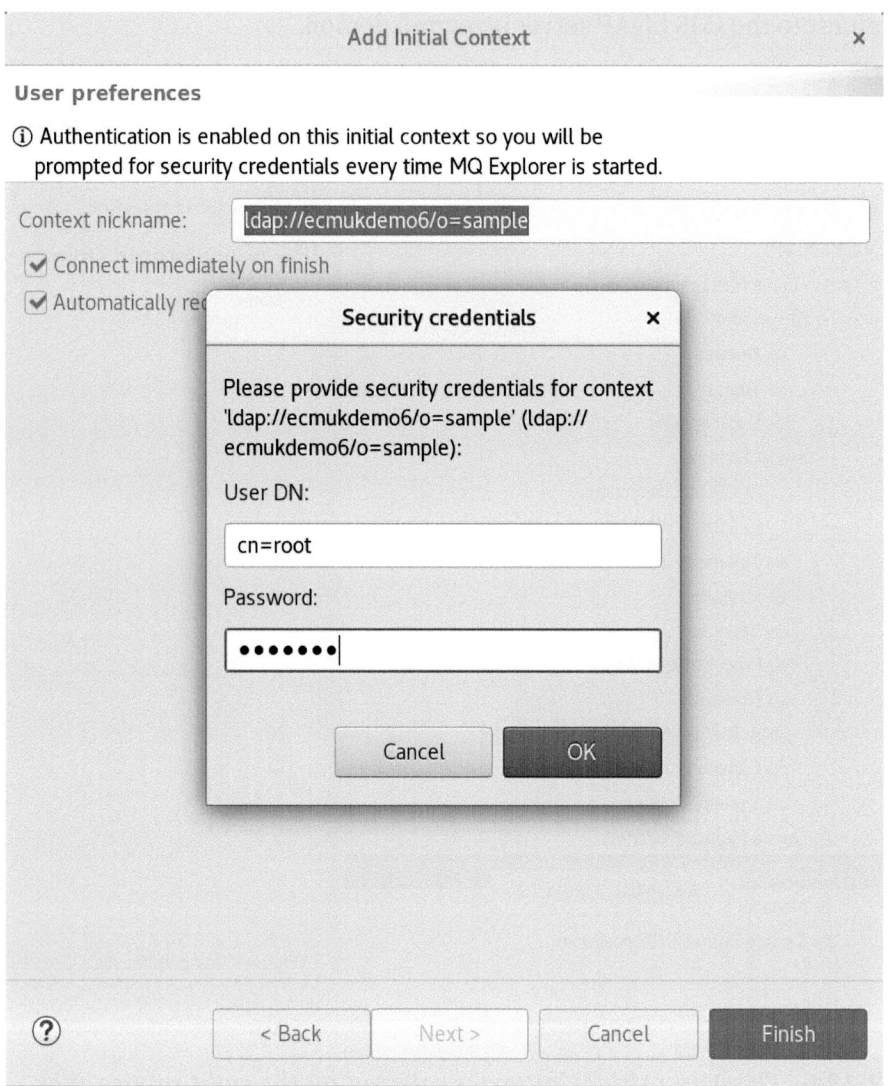

Figure 3-135. *The ecmukdemo6 context is entered with the o=sample LDAP root*

CHAPTER 3   IBM JMS INTERFACE DEVELOPMENT IBM FILENET 5.5.X WORKFLOW

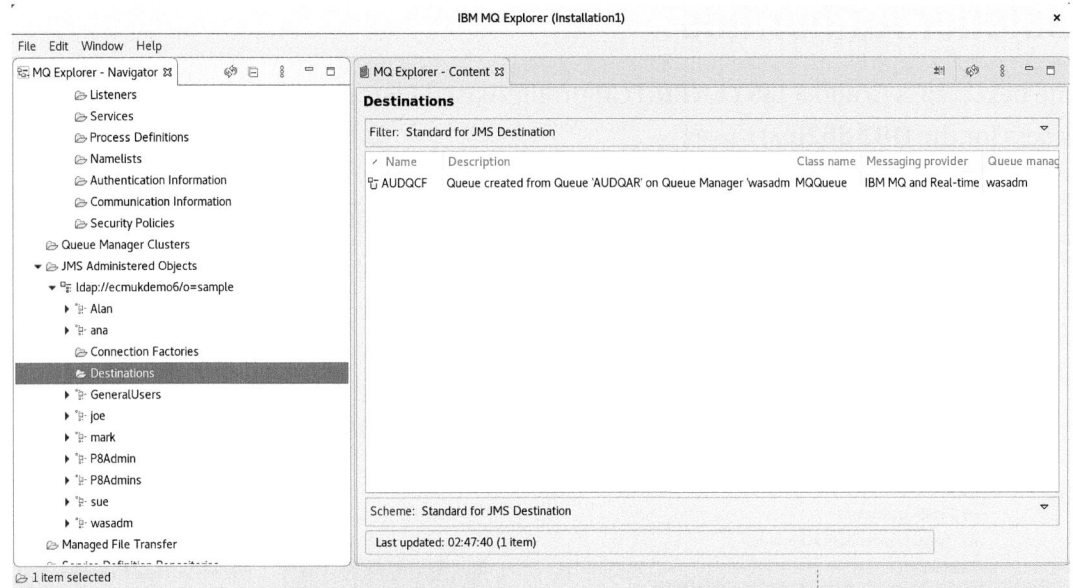

Figure 3-136.   The JMS referenced Queue, AUDQCF, is shown

The TOPIC Profile is set with authorization for db2inst1:

su - mqm
runmqsc wasadm

SET AUTHREC PROFILE('SYSTEM.BASE.TOPIC') OBJTYPE(TOPIC)
PRINCIPAL('db2inst1') AUTHADD(PUB, SUB)

```
SET AUTHREC PROFILE('SYSTEM.BASE.TOPIC') OBJTYPE(TOPIC) PRINCIPAL('db2inst1') AUTHADD(PUB,SUB)
    2 : SET AUTHREC PROFILE('SYSTEM.BASE.TOPIC') OBJTYPE(TOPIC) PRINCIPAL('db2inst1') AUTHADD(PUB,SUB)
AMQ8862I: IBM MQ authority record set.
```

Figure 3-137.   The IBM MQ Authority record is set for db2inst1 to access the Base TOPIC

The user now has Connect access to the queue manager. When the user accesses the queue manager's objects, the authorities that you have granted to the user take effect.

CHAPTER 3　IBM JMS INTERFACE DEVELOPMENT IBM FILENET 5.5.X WORKFLOW

Listener Authorities

The next step is to add a **LISTENER.TCP** profile-based **Listener** object type with User access for the **DB2 db2inst1** user.

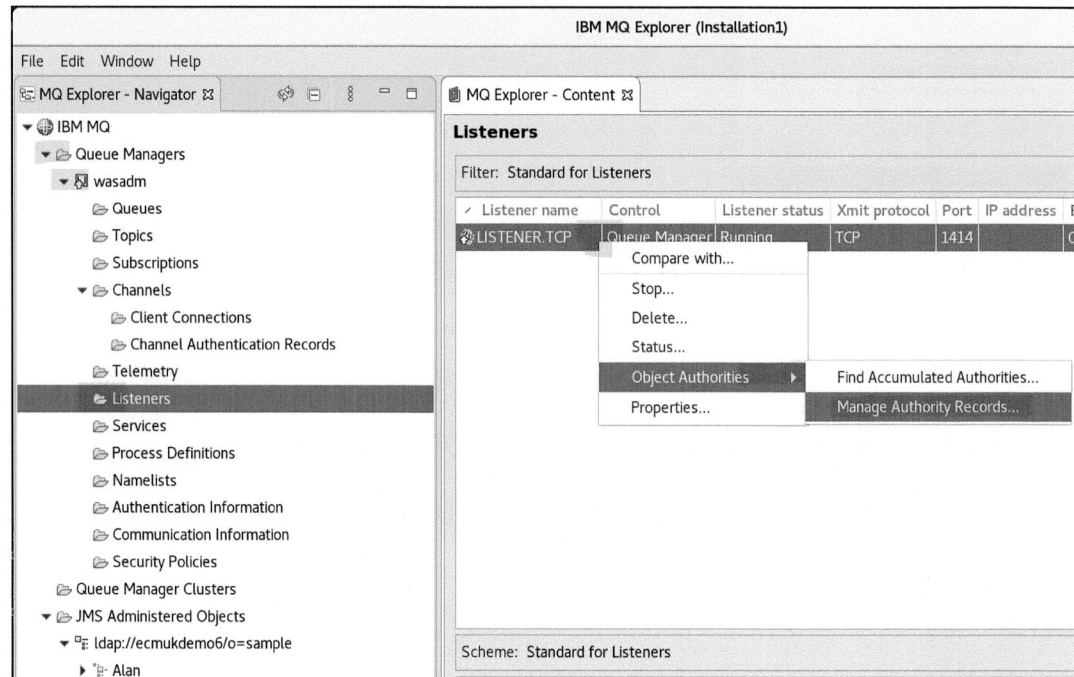

Figure 3-138. *The Listener object security access is updated for the db2inst1 user*

The Listener node is selected, and the LISTENER.TCP security is updated for the db2inst1 Linux DB2 database user, as shown in Figure 3-139.

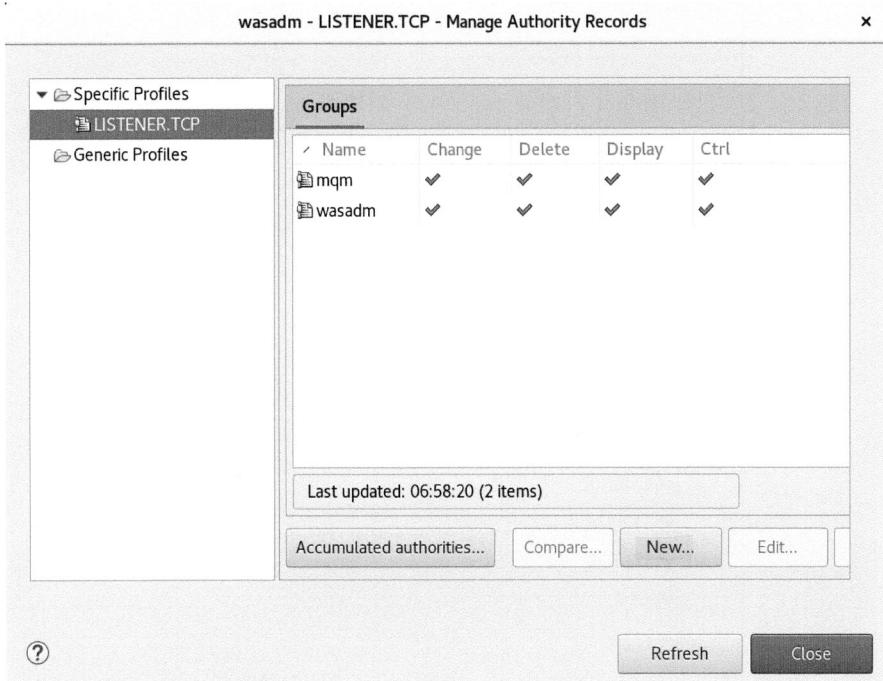

Figure 3-139. *The existing user authorities for the Listener, LISTENER.TCP, profile is displayed*

The **New** command button shown in Figure 3-139 is selected to add the new user, **db2inst1**.

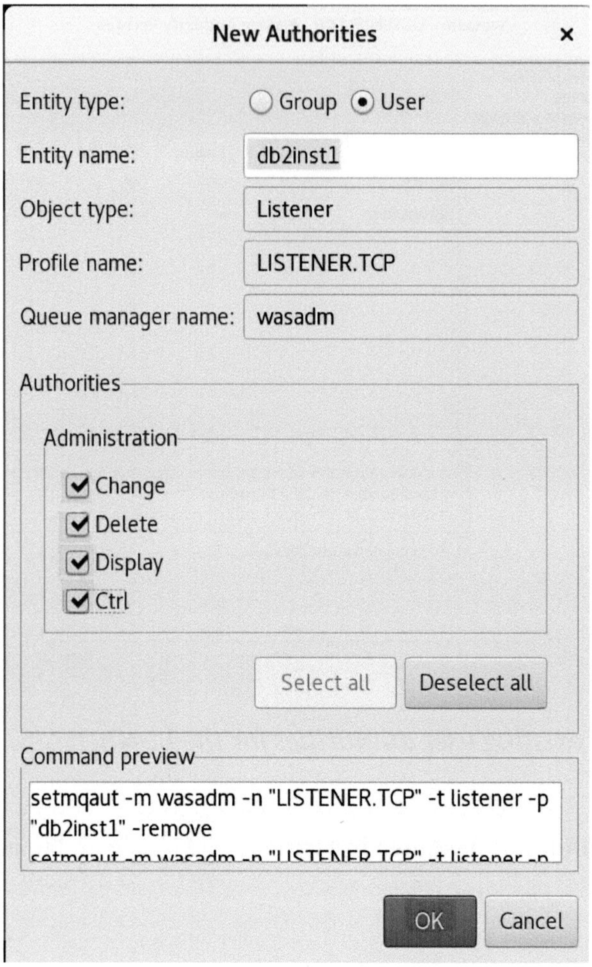

CHAPTER 3   IBM JMS INTERFACE DEVELOPMENT IBM FILENET 5.5.X WORKFLOW

Figure 3-140. *The Listener Authorities are updated for the db2inst1 user*

The OK command button in Figure 3-140 is clicked to add the new user authorities.

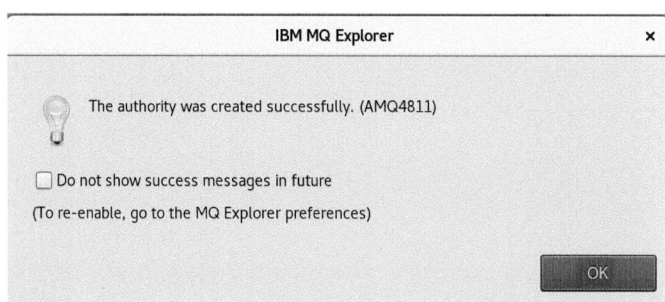

Figure 3-141. *The Listener Authorities are confirmed as updated*

CHAPTER 3　IBM JMS INTERFACE DEVELOPMENT IBM FILENET 5.5.X WORKFLOW

The OK command now shows the updated **Authorities** security list.

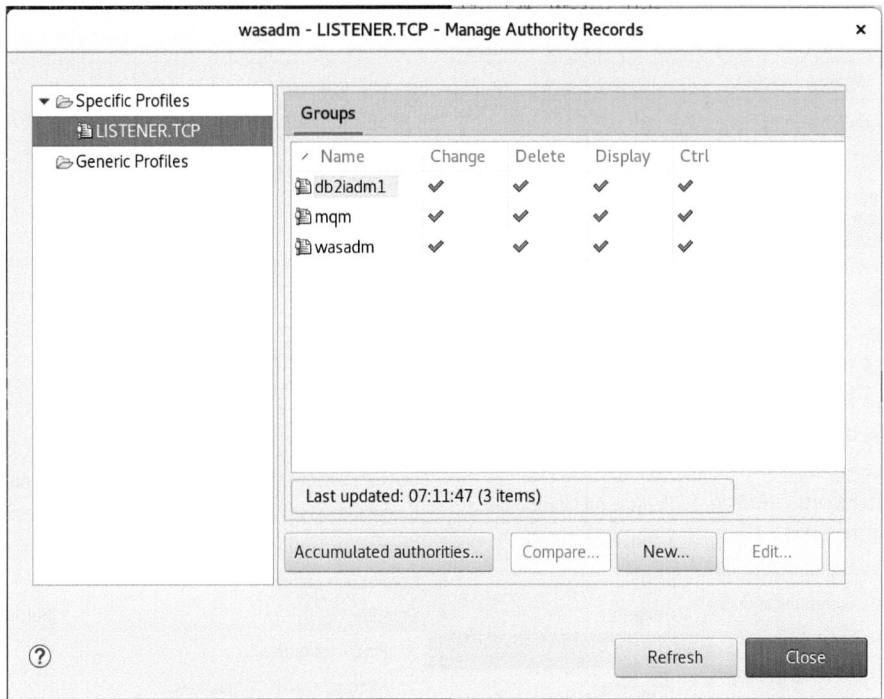

Figure 3-142.　The updated LISTENER.TCP profile shows the new db2inst1 user's group

The command lines are run as follows:

```
su - mqm
runmqsc wasadm
ALTER QMGR CHLAUTH(ENABLED)
```

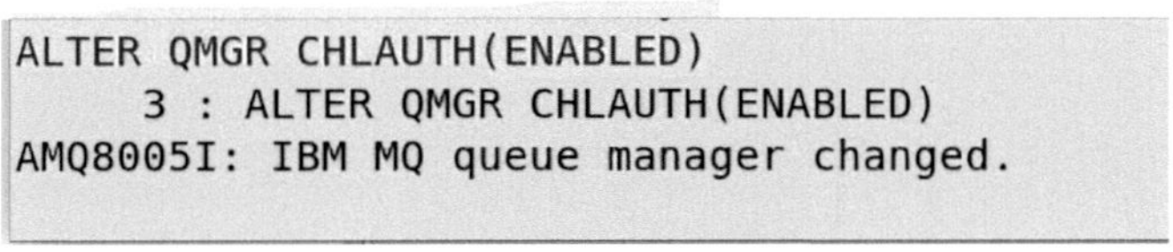

```
ALTER QMGR CHLAUTH(ENABLED)
     3 : ALTER QMGR CHLAUTH(ENABLED)
AMQ8005I: IBM MQ queue manager changed.
```

Figure 3-143.　The Queue Manager Channel Authentication security system is enabled

CHAPTER 3   IBM JMS INTERFACE DEVELOPMENT IBM FILENET 5.5.X WORKFLOW

The **wasadm** Queue Manager Create Authorities are updated for the **db2inst1** user.

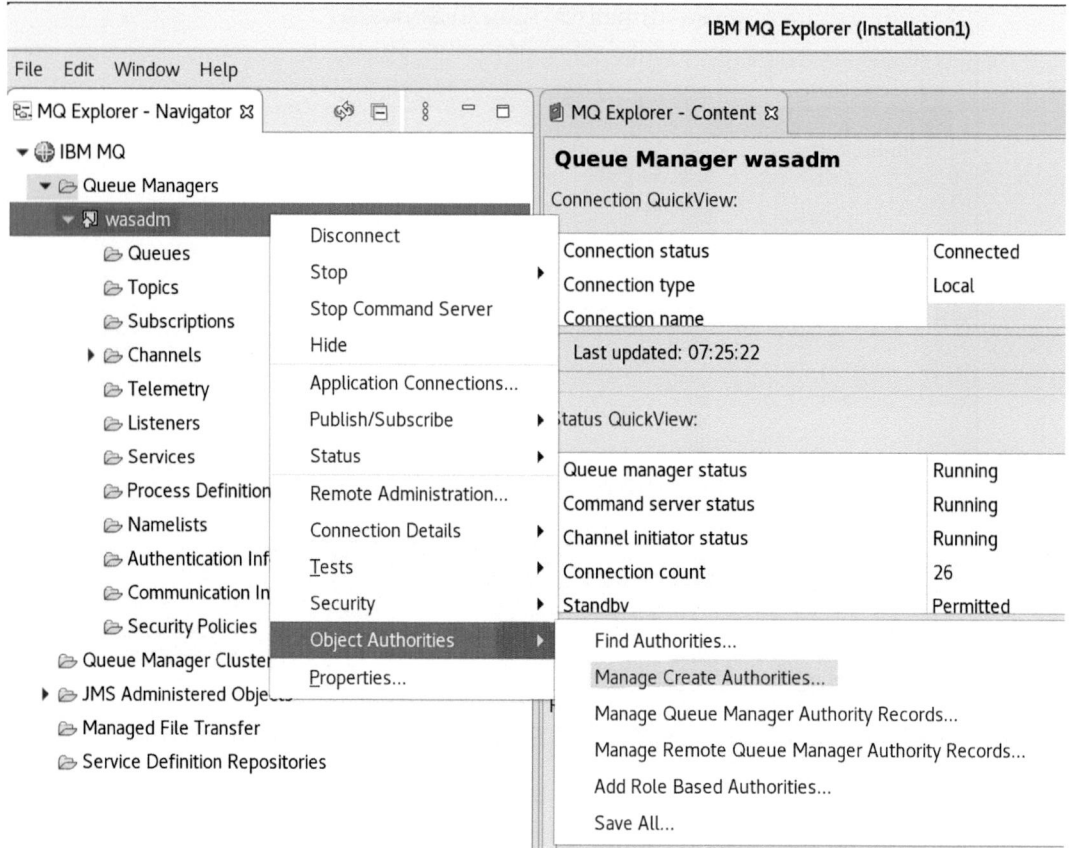

Figure 3-144. *The menu option to **Manage Create Authorities** is selected*

The wasadm Queue Manager node is selected and right-clicked to select the Object Authorities submenu and then the Manage Create Authorities window.

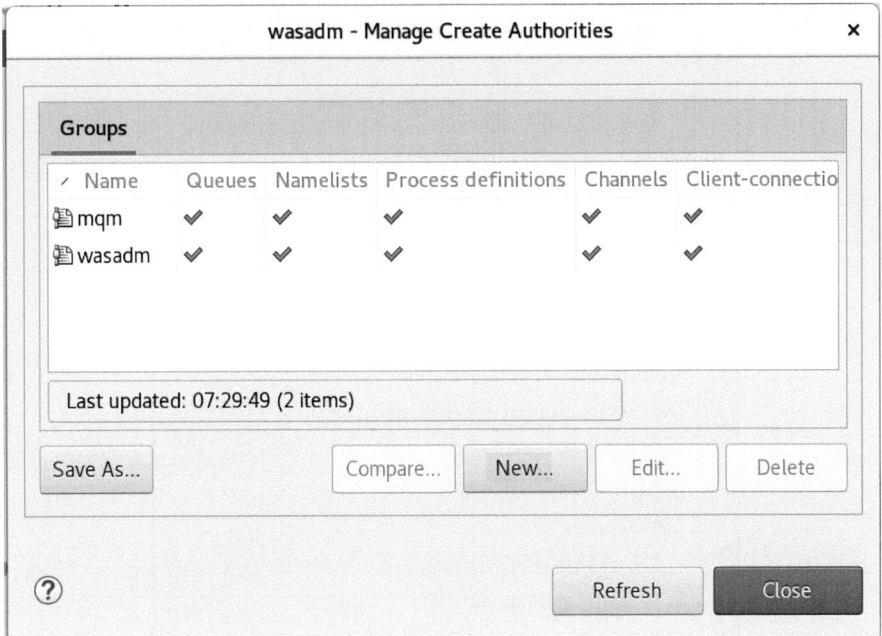

Figure 3-145. *The **New** button is selected to create an entry for the db2inst1 user*

The required entries are entered for the db2inst1 user for the security access required.

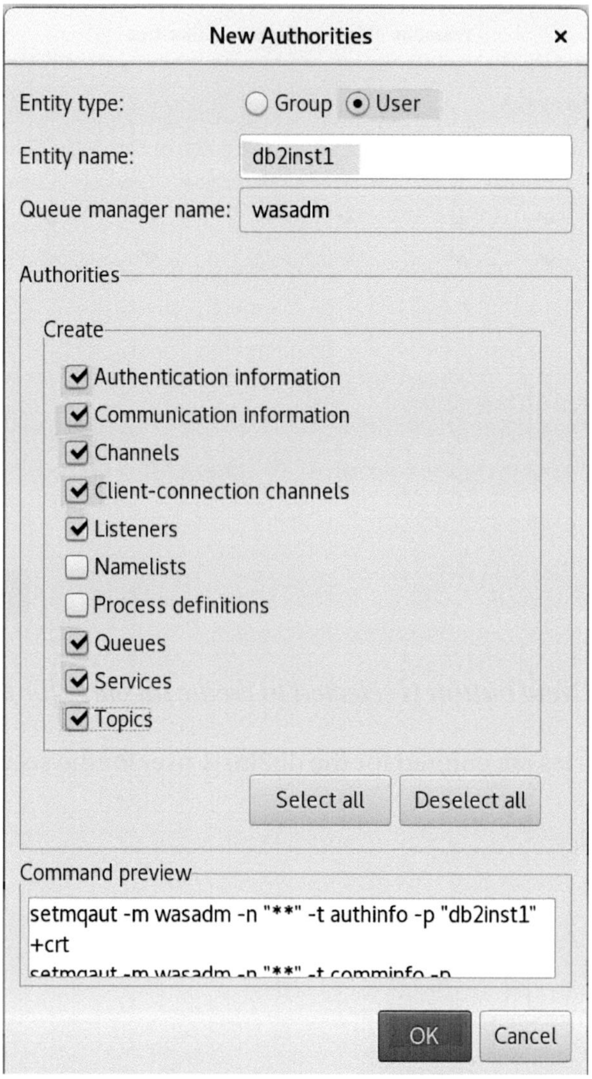

CHAPTER 3 IBM JMS INTERFACE DEVELOPMENT IBM FILENET 5.5.X WORKFLOW

Figure 3-146. *The required entries are entered for the db2inst1 user for the security access*

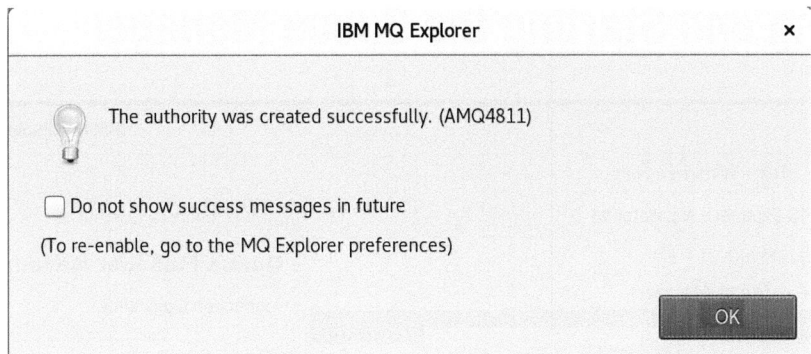

Figure 3-147. *The db2inst1 user security access updates are successfully added*

On clicking OK for the update authorities, the db2inst1 user's group is displayed.

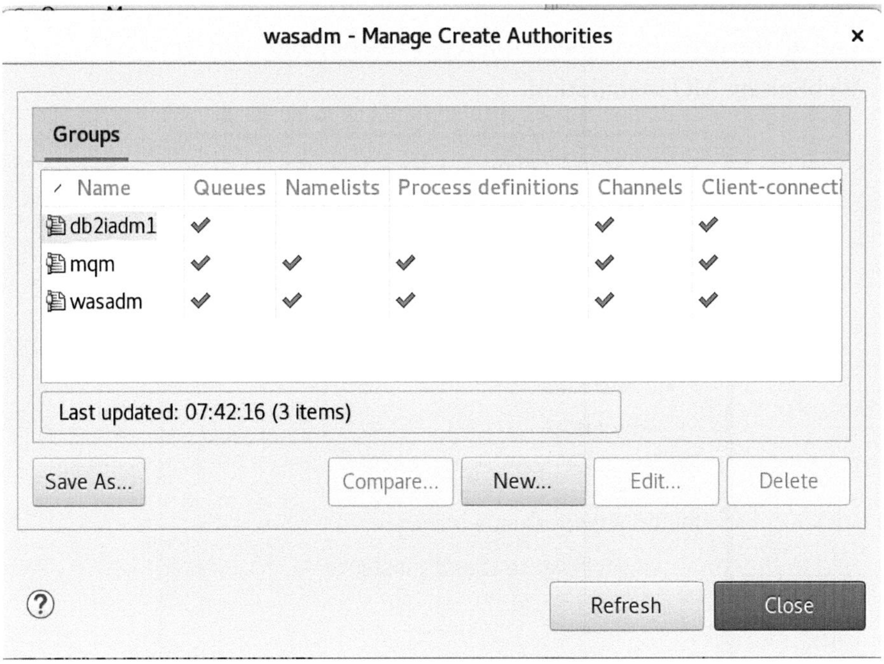

Figure 3-148. *The db2iadm1 Group user is displayed*

Create test for example MQ Main program REF:

www-01.ibm.com/support/knowledgecenter/SSEQTP_8.5.5//com.ibm.websphere.
nd.multiplatform.doc/ae/tmj_pgmng.html#

Chapter 3 IBM JMS Interface Development IBM Filenet 5.5.x Workflow

Stopping and Starting the Queue Manager

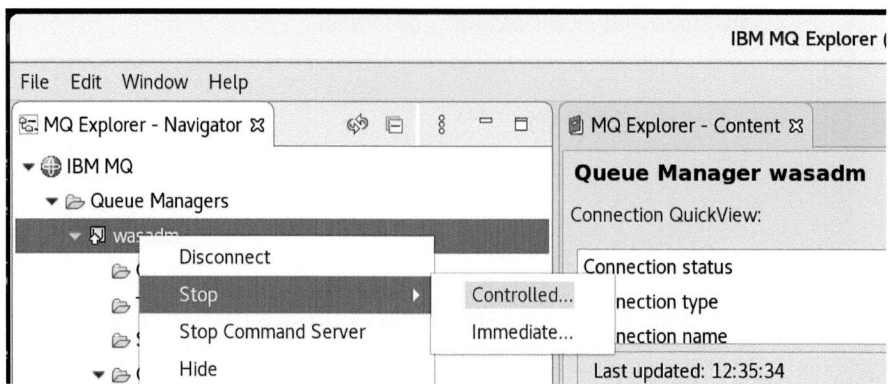

Figure 3-149. *The wasadm Queue Manager is stopped*

1. End all the activity of queue managers associated with the WebSphere MQ installation.

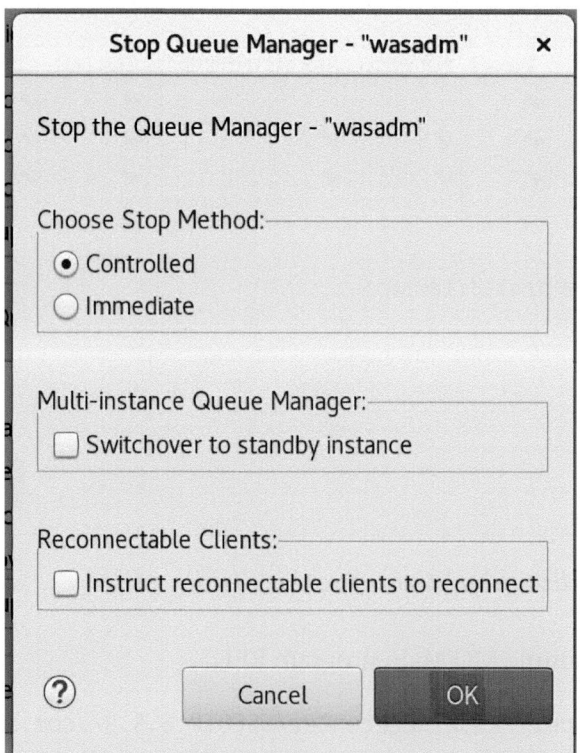

Figure 3-150. *The wasadm Queue manager default option is selected to stop Controlled*

Select the type of Stop required and then click **OK**.

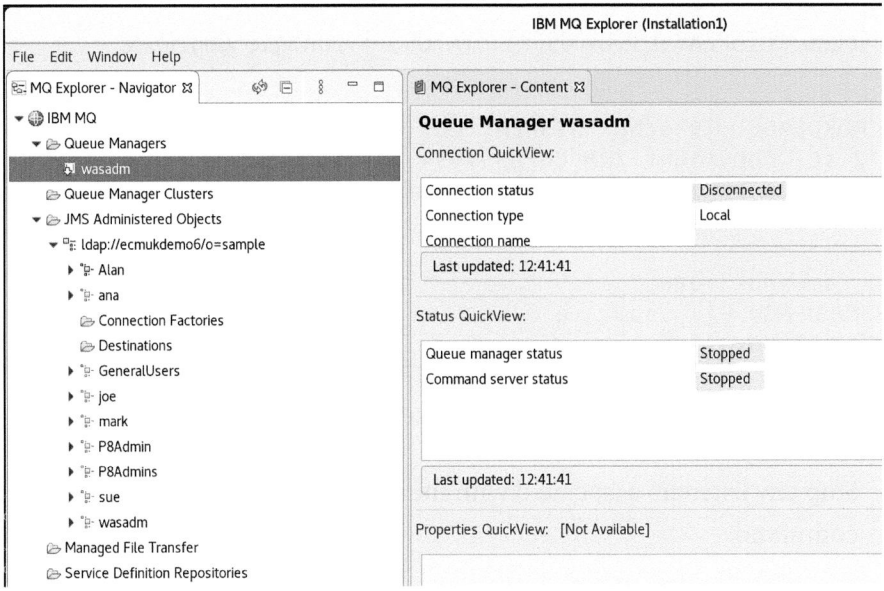

Figure 3-151. *The wasadm Queue Manager status is shown*

a. Run the **dspmq** command to list the state of all the queue managers on the system.

Run either of the following commands from the installation that you are updating:

```
dspmq -o installation -o status
dspmq -a
```

dspmq -o installation -o status displays the installation name and status of all operational queue managers associated with all installations of all WebSphere MQ.

```
[mqm@ECMUKDEMO6 ~]$ dspmq -o installation -o status
QMNAME(wasadm)                                    STATUS(Ended normally) INSTNAME(Installation1) I
NSTPATH(/opt/mqm) INSTVER(9.2.5.0)
[mqm@ECMUKDEMO6 ~]$
```

Figure 3-152. *The dspmq -o installation -o status command lists the state*

dspmq -a displays the status of active queue managers associated with the installation from which the command is run.

CHAPTER 3   IBM JMS INTERFACE DEVELOPMENT IBM FILENET 5.5.X WORKFLOW

b. Run the MQSC command, DISPLAY LSSTATUS(*) STATUS, to list the status of listeners associated with a queue manager:

echo "DISPLAY LSSTATUS(*) STATUS" | runmqsc *QmgrName*

```
[mqm@ECMUKDEM06 ~]$ echo "DISPLAY LSSTATUS(*) STATUS" | runmqsc wasadm
5724-H72 (C) Copyright IBM Corp. 1994, 2022.
Starting MQSC for queue manager wasadm.
AMQ8146E: IBM MQ queue manager not available.

No MQSC commands read.
[mqm@ECMUKDEM06 ~]$ dspmq -a
[mqm@ECMUKDEM06 ~]$
```

Figure 3-153. *The status of the wasadm queue is confirmed*

c. Stop any listeners associated with the queue managers, using the command:

endmqlsr -m QMgrName

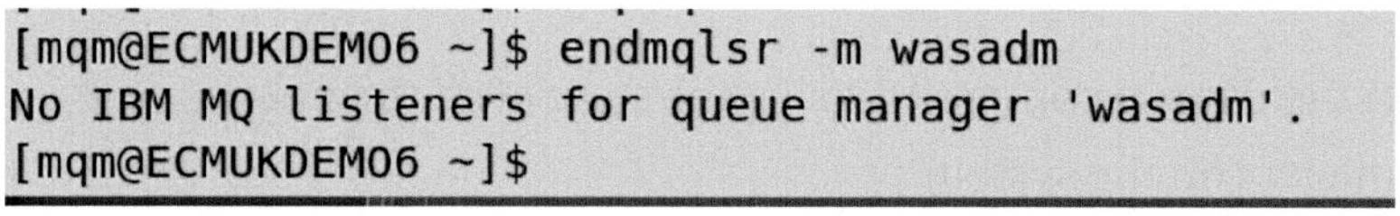

```
[mqm@ECMUKDEM06 ~]$ endmqlsr -m wasadm
No IBM MQ listeners for queue manager 'wasadm'.
[mqm@ECMUKDEM06 ~]$
```

Figure 3-154. *Any listeners for the wasadm Queue Manager are stopped*

/opt/mqm/bin/mqconfig

CHAPTER 3   IBM JMS INTERFACE DEVELOPMENT IBM FILENET 5.5.X WORKFLOW

```
[mqm@ECMUKDEMO6 ~]$ /opt/mqm/bin/mqconfig
mqconfig: Analyzing Red Hat Enterprise Linux 8.0 (Ootpa) settings for IBM
         MQ V9.2
System V Semaphores
  semmsl     (sem:1)   32000 semaphores                    IBM>=32         PASS
  semmns     (sem:2)   370 of 1024000000 semaphores (0%)    IBM>=4096       PASS
  semopm     (sem:3)   500 operations                      IBM>=32         PASS
  semmni     (sem:4)   300 of 32000 sets          (0%)     IBM>=128        PASS
System V Shared Memory
  shmmax               18446744073692774399 bytes          IBM>=268435456  PASS
  shmmni               81 of 4096 sets            (1%)     IBM>=4096       PASS
  shmall               1397628 of 18446744073692774399 pages (0%)   IBM>=2097152   PASS
System Settings
  file-max             26912 of 1216131 files     (2%)     IBM>=524288     PASS
  pid_max              2028 of 131072 processids  (1%)     IBM>=32768      PASS
  threads-max          2028 of 95378 threads      (2%)     IBM>=32768      PASS
Current User Limits (mqm)
  nofile     (-Hn)     10240 files                         IBM>=10240      PASS
  nofile     (-Sn)     10240 files                         IBM>=10240      PASS
  nproc      (-Hu)     158 of 47689 processes     (0%)     IBM>=4096       PASS
  nproc      (-Su)     158 of 47689 processes     (0%)     IBM>=4096       PASS
```

Figure 3-155.   *The configuration settings are checked*

a)   Restart the MQ wasadm Queue.

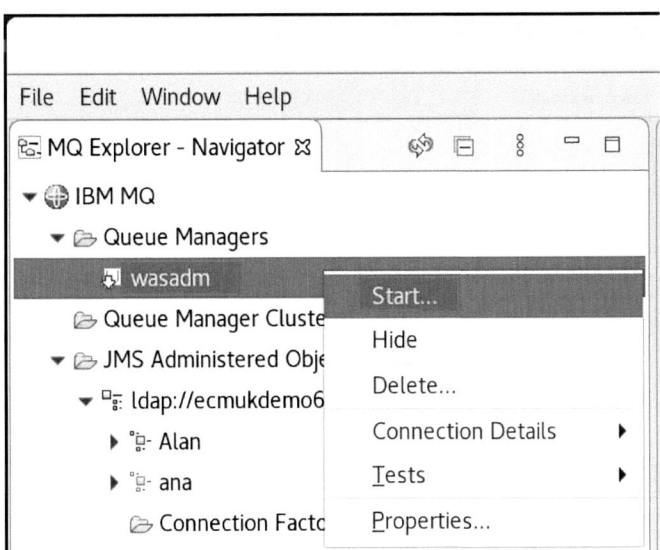

Figure 3-156.   *The wasadm Queue Manager is restarted*

489

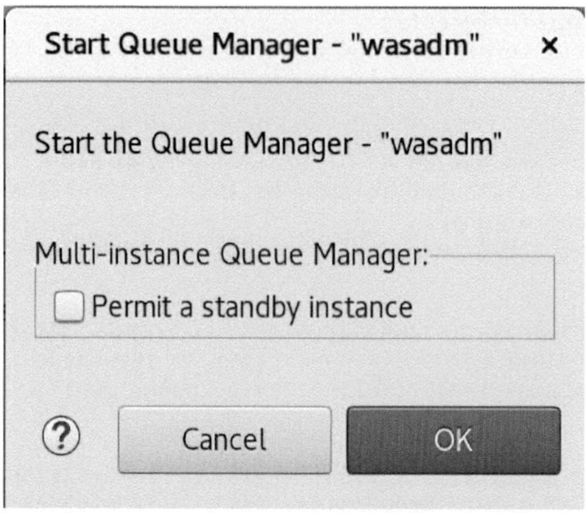

CHAPTER 3   IBM JMS INTERFACE DEVELOPMENT IBM FILENET 5.5.X WORKFLOW

Figure 3-157. *The Start Queue Manager for wasadm is confirmed by clicking OK*

The status of the wasadm Queue Manager can be seen as highlighted in Figure 3-158.

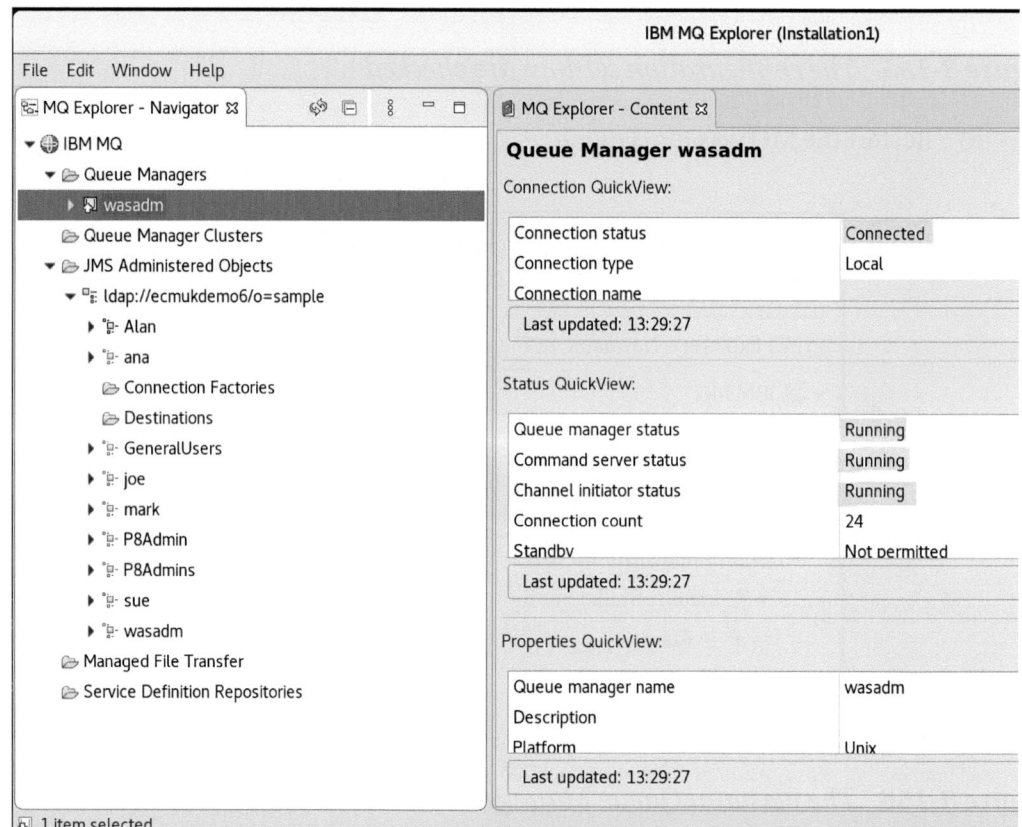

Figure 3-158. *The active status of the restarted wasadm Queue Manager is displayed*

490

CHAPTER 3　IBM JMS INTERFACE DEVELOPMENT IBM FILENET 5.5.X WORKFLOW

Creating a Client Channel for Messaging

A custom Client Channel can be created for messaging based on a standard SYSTEM.
DEF.SERVER template Object type, as shown in this section.

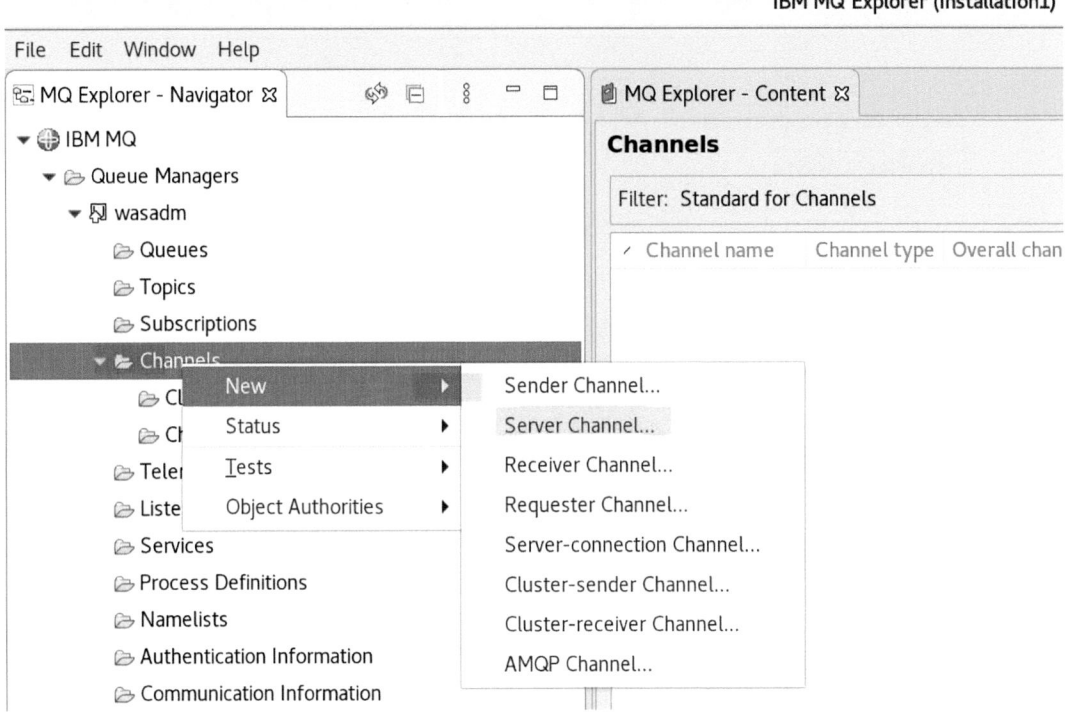

Figure 3-159. *A new Server Channel is created as highlighted*

The **CLIENT.wasadm** Server Channel is created as a **SYSTEM.DEF.SERVER** object type.

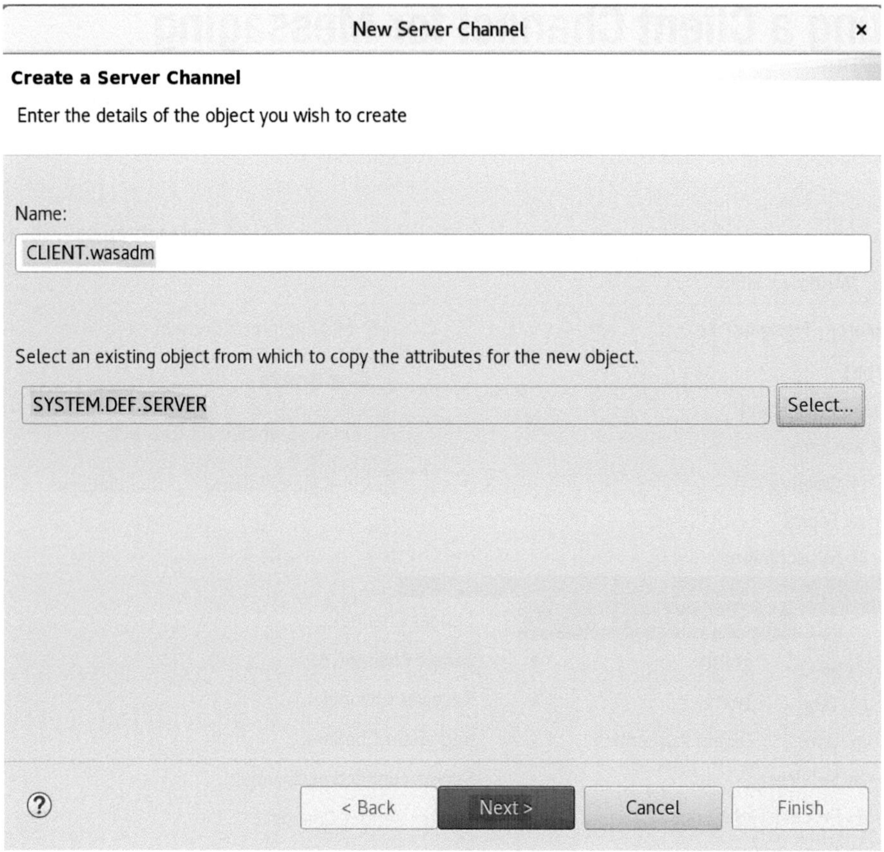

CHAPTER 3 IBM JMS INTERFACE DEVELOPMENT IBM FILENET 5.5.X WORKFLOW

*Figure 3-160. A new **CLIENT.wasadm** Server Channel is created and Next clicked*

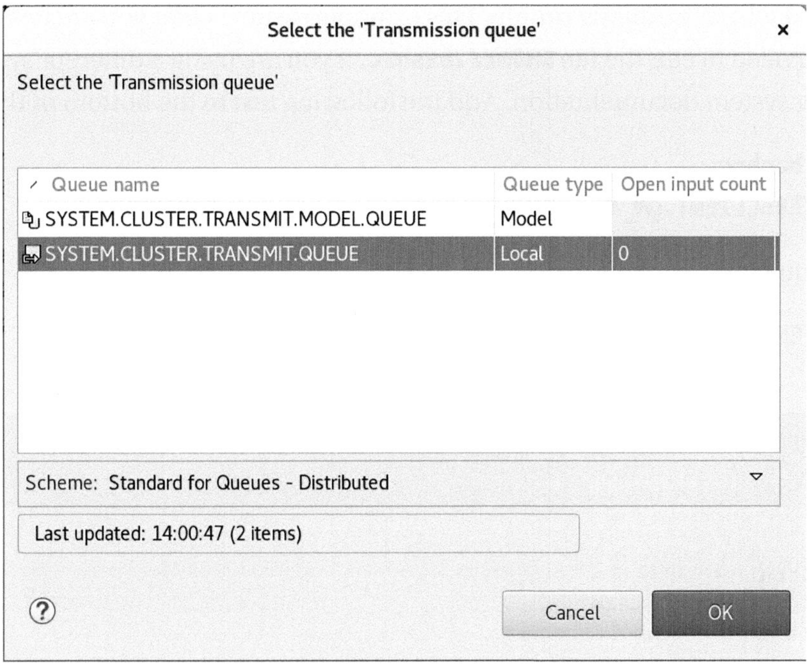

Figure 3-161. *The Local Transmission queue type is selected*

Add the db2inst1 user to the mqm group:

`usermod -a -G mqm db2inst1`

Setting Up the Client on Linux

Set up the client component using the MQSERVER environment variable.

We need to find out the network name of the machine which hosts the queue manager (our example QM is called **wasadm**) from the system administrator.

Log in as the user who will be running the Express File Transfer, who must be a member of the mqm group. (On the ecmukdemo6 Linux server, this is wasadm.)

Open a command prompt.

Type

`cd $HOME`

CHAPTER 3　IBM JMS INTERFACE DEVELOPMENT IBM FILENET 5.5.X WORKFLOW

Use a text editor to edit the profile. This example assumes that you are using the bash shell, so you need to edit the file **$HOME/.bashrc**. If you are using a different system shell, consult your system documentation. Add the following text to the bottom of the file:

vi $HOME/.bashrc

　　MQSERVER=CLIENT.*QM_Name*/TCP/'*hostname*'; export MQSERVER

For example:

　　MQSERVER=CLIENT.wasadm/TCP/'ecmukdemo6'; export MQSERVER

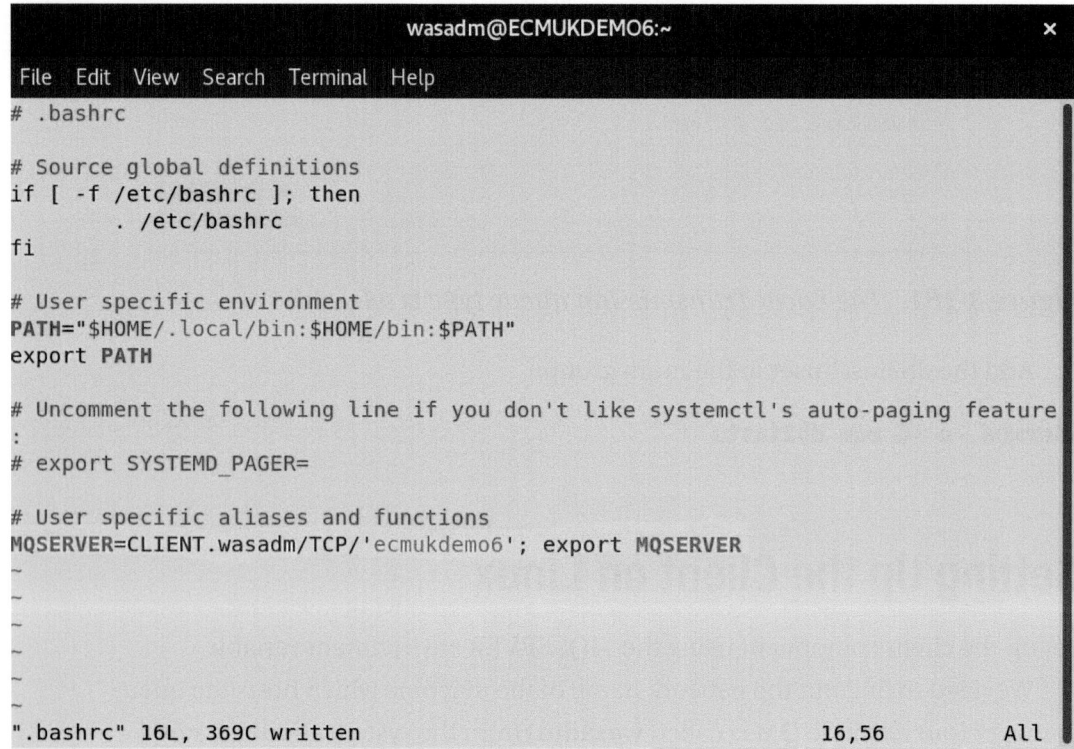

Figure 3-162. *The MQSERVER environment variable is set up*

Replace *hostname* with the name (ecmukdemo6) that identifies the server machine on the network.

Close the command prompt.

Log out and log back in for the change to take effect.

Check as follows.

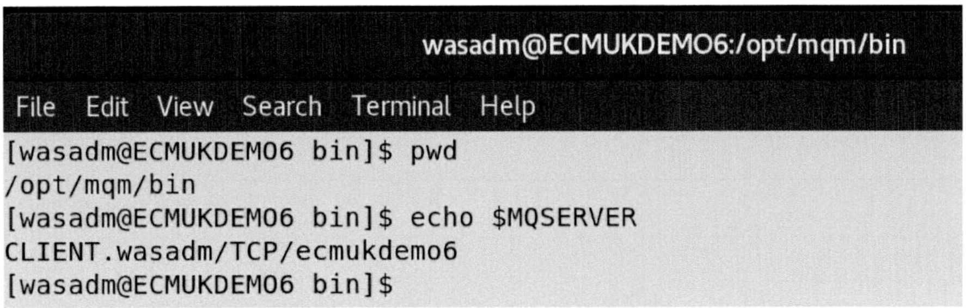

```
                                wasadm@ECMUKDEMO6:/opt/mqm/bin
 File  Edit  View  Search  Terminal  Help
[wasadm@ECMUKDEMO6 bin]$ pwd
/opt/mqm/bin
[wasadm@ECMUKDEMO6 bin]$ echo $MQSERVER
CLIENT.wasadm/TCP/ecmukdemo6
[wasadm@ECMUKDEMO6 bin]$
```

Figure 3-163. *The MQSERVER environment variable is checked*

You have now set up the client and server components needed. The next task is to send a message from the client to the server queue manager **wasadm**.

Sending a Message from a Client to a Server

To send a message from the client to the server queue manager **wasadm**, which uses the remote queue definition **AUDQAR**:

Open a command prompt on the client and follow these steps:

Start the **amqsputc** sample program as follows.

On Linux, change to the **MQ_INSTALLATION_PATH/samp/bin** directory, where **MQ_INSTALLATION_PATH** represents the high-level directory in which WebSphere MQ is installed:

```
[wasadm@ecmukdemo1 Router]$ cd /opt/mqm/samp/bin
```

Type the command:

```
./amqsputc AUDCONN1
```

Testing for Errors

The **error "Dead-letter Queue attribute refers to a queue that does not exist (AUDD1) wasadm Queue Manager / General"**

CHAPTER 3   IBM JMS INTERFACE DEVELOPMENT IBM FILENET 5.5.X WORKFLOW

Remediation for Missing Dead-Letter Queue, AUDD1

Create Dead-Letter Queue AUDD1.

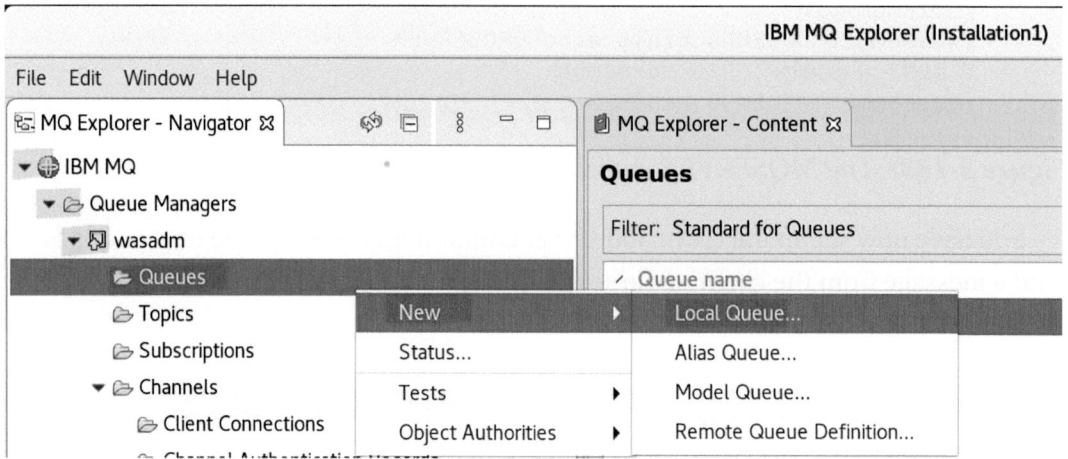

Figure 3-164. *The Queues ➤ New ➤ Local Queue menu*

The menu is selected as shown in Figure 3-164 to create a new AUDD1 Dead-Letter Queue.

CHAPTER 3 IBM JMS INTERFACE DEVELOPMENT IBM FILENET 5.5.X WORKFLOW

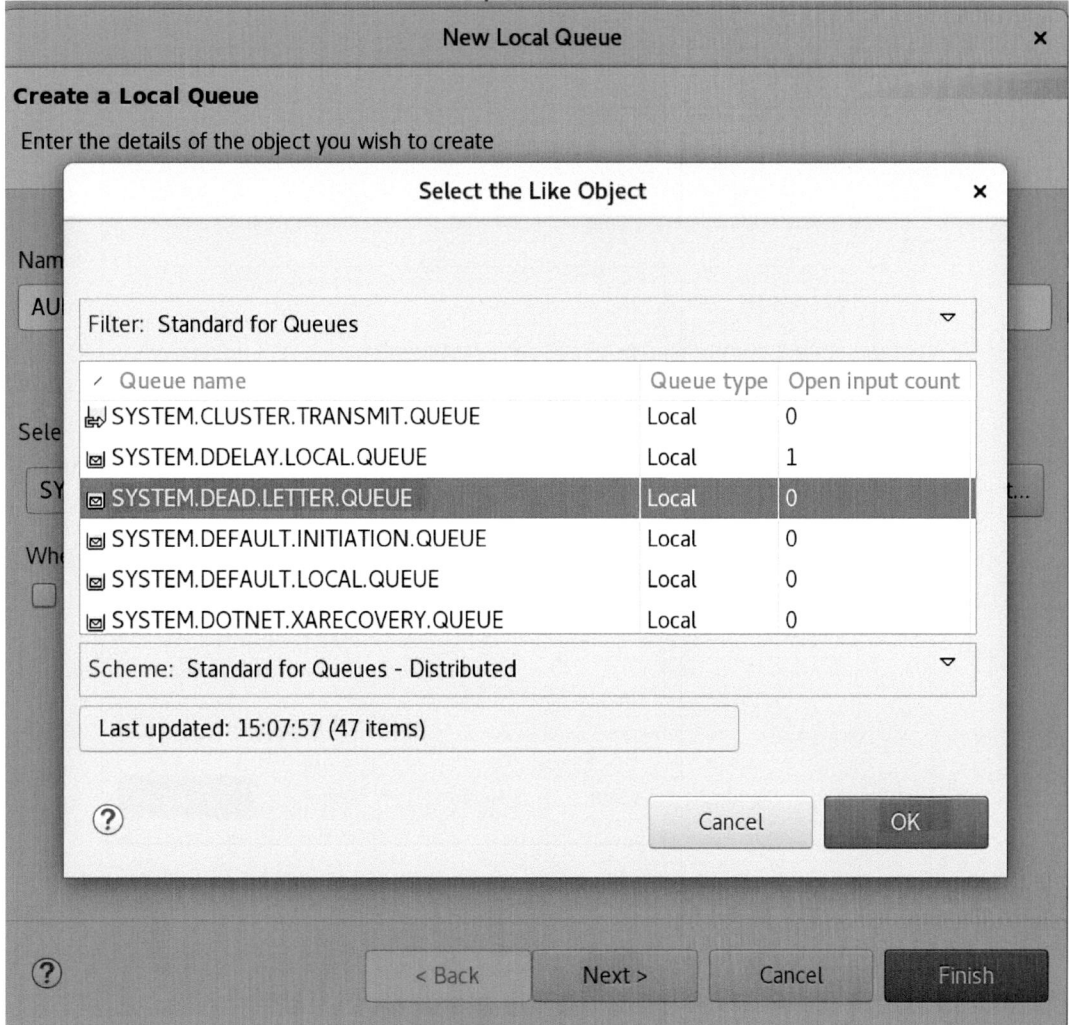

Figure 3-165. *The SYSTEM.DEAD.LETTER.QUEUE template Object is selected*

The Next> command is selected to start the Queue creation process.

CHAPTER 3 IBM JMS INTERFACE DEVELOPMENT IBM FILENET 5.5.X WORKFLOW

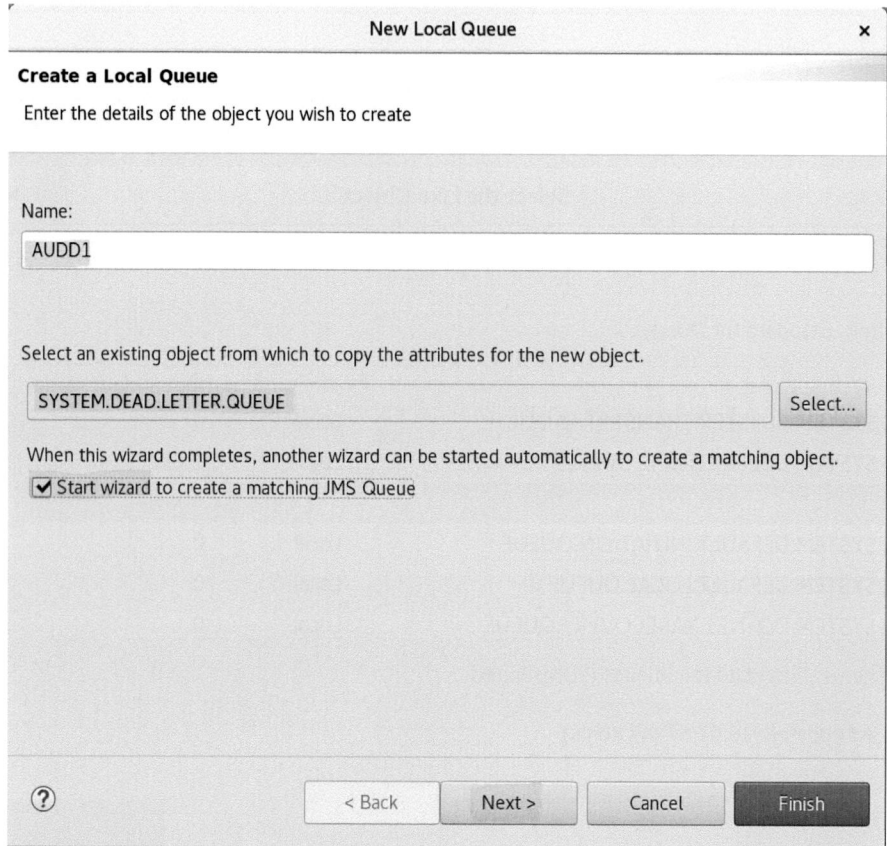

Figure 3-166. *The Start wizard to create a matching JMS Queue tick box is selected*

Next, we have to select the option to create a matching JMS Queue.

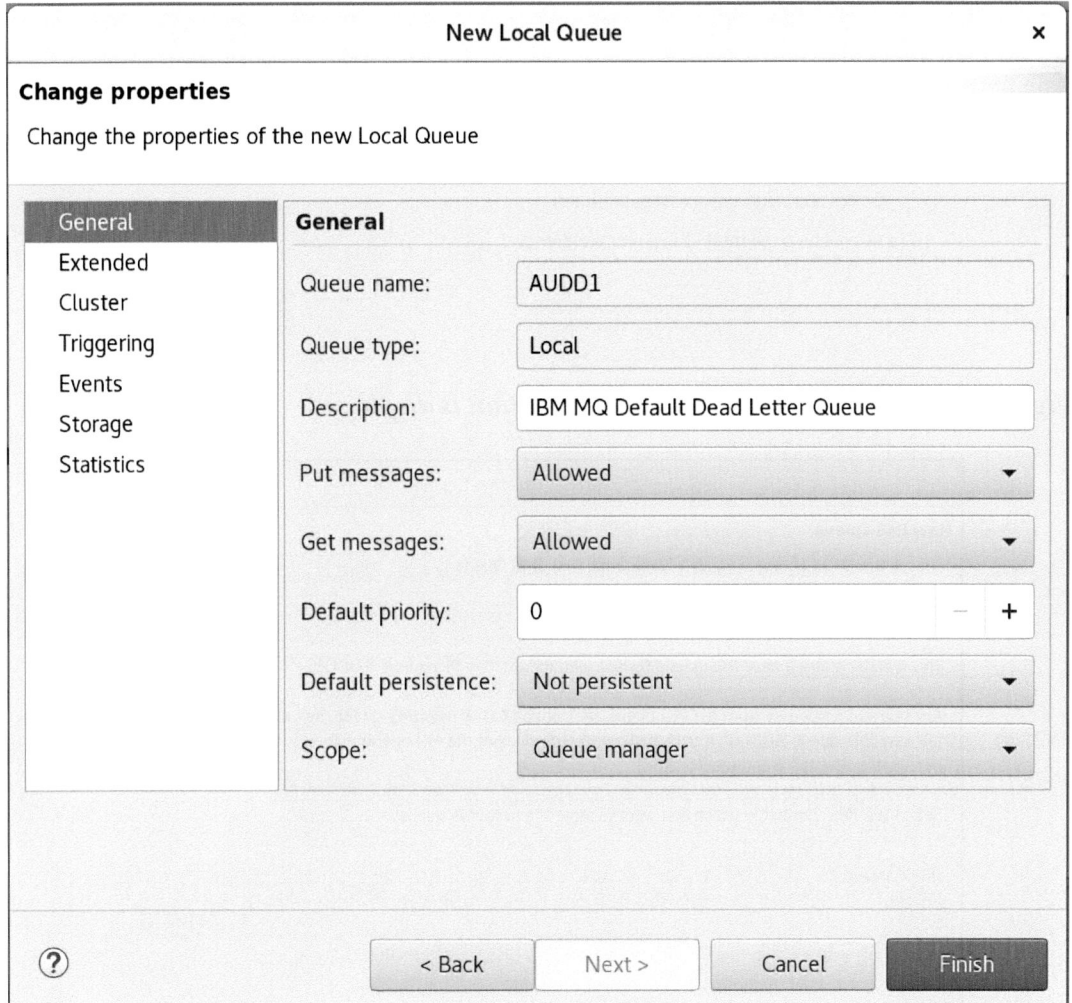

Figure 3-167. *The Finish command button is clicked to create the Queue*

The pop-up window is loaded to display the creation status for the Queue.

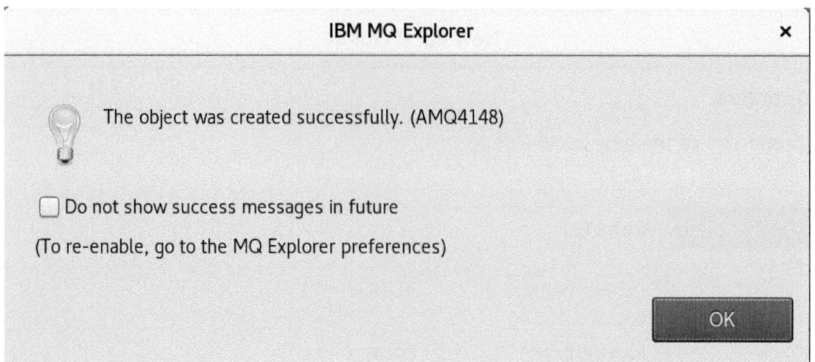

Figure 3-168. *The status of the Queue creation is displayed*

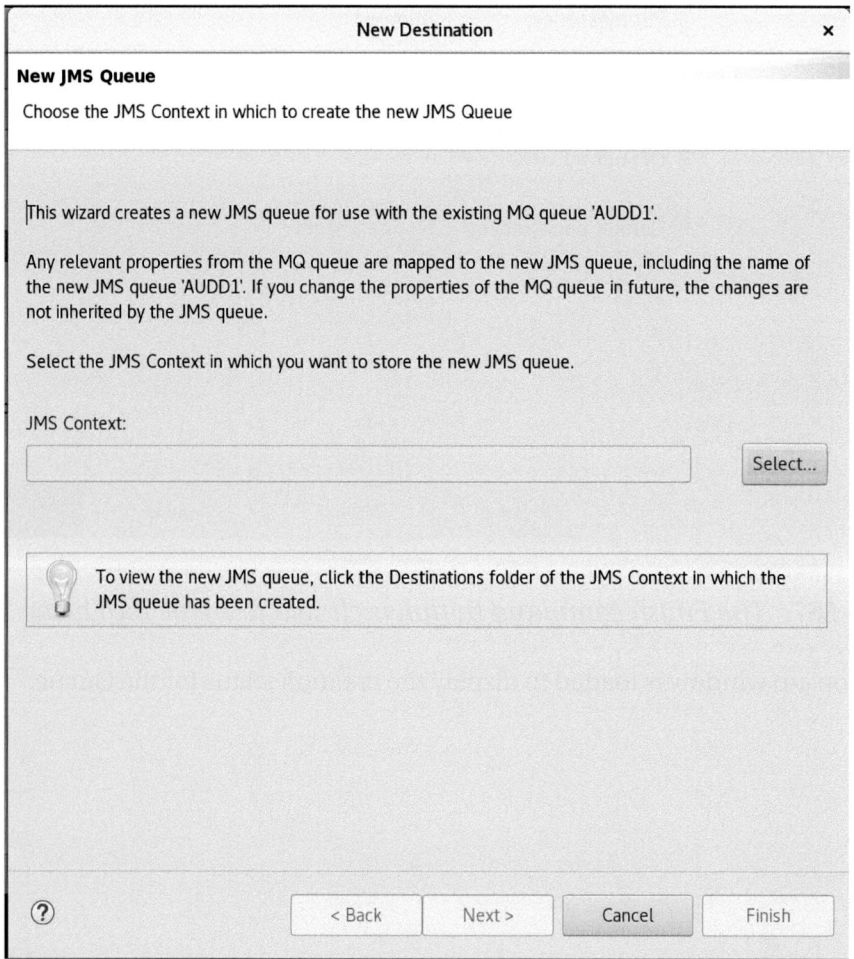

Figure 3-169. *The wizard to create a new JMS queue matching the Dead-Letter Queue is launched*

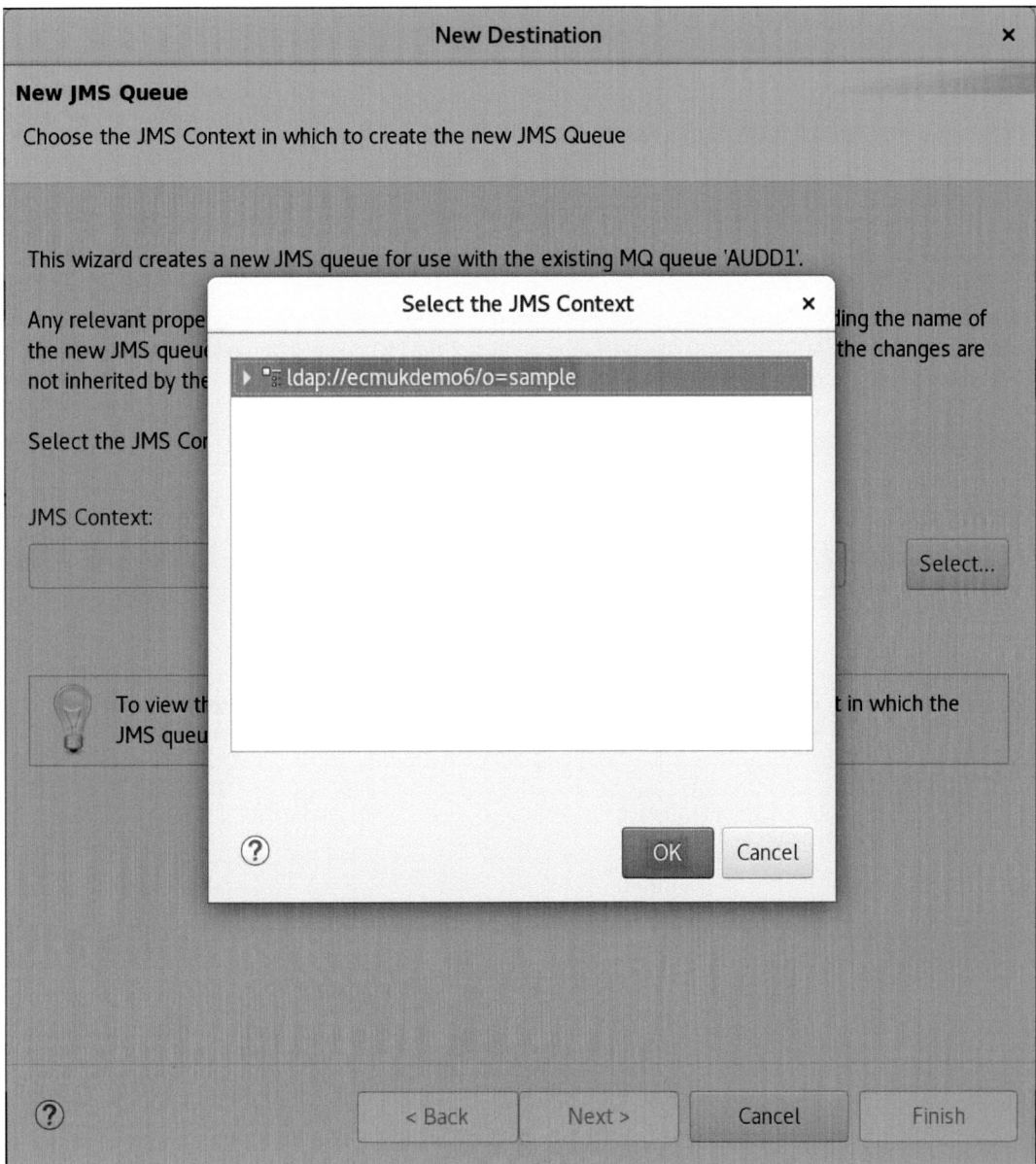

Figure 3-170. The LDAP server JMS Context is displayed

CHAPTER 3 IBM JMS INTERFACE DEVELOPMENT IBM FILENET 5.5.X WORKFLOW

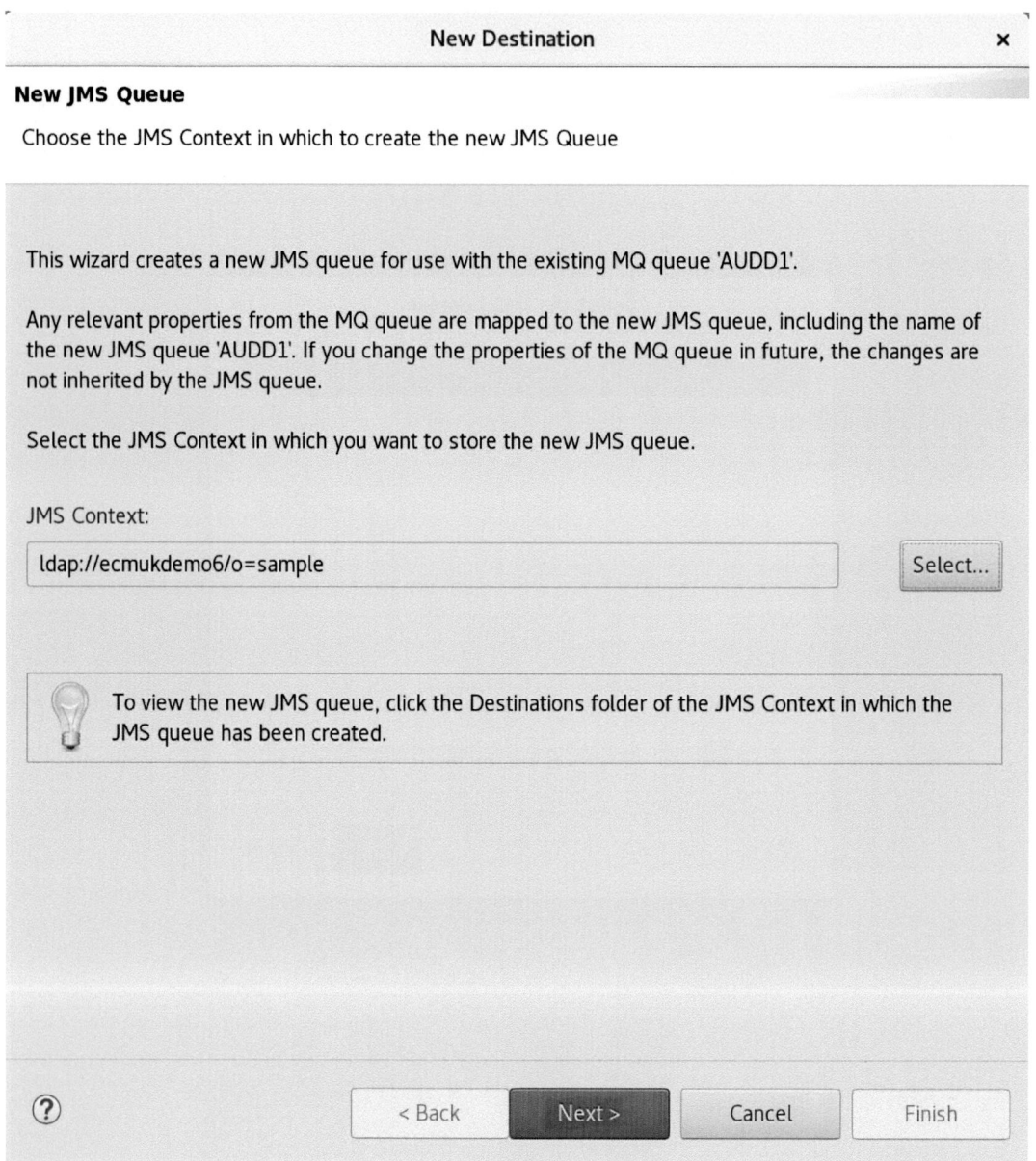

Figure 3-171. *The Next command is clicked to display the Destination JMS object*

CHAPTER 3 IBM JMS INTERFACE DEVELOPMENT IBM FILENET 5.5.X WORKFLOW

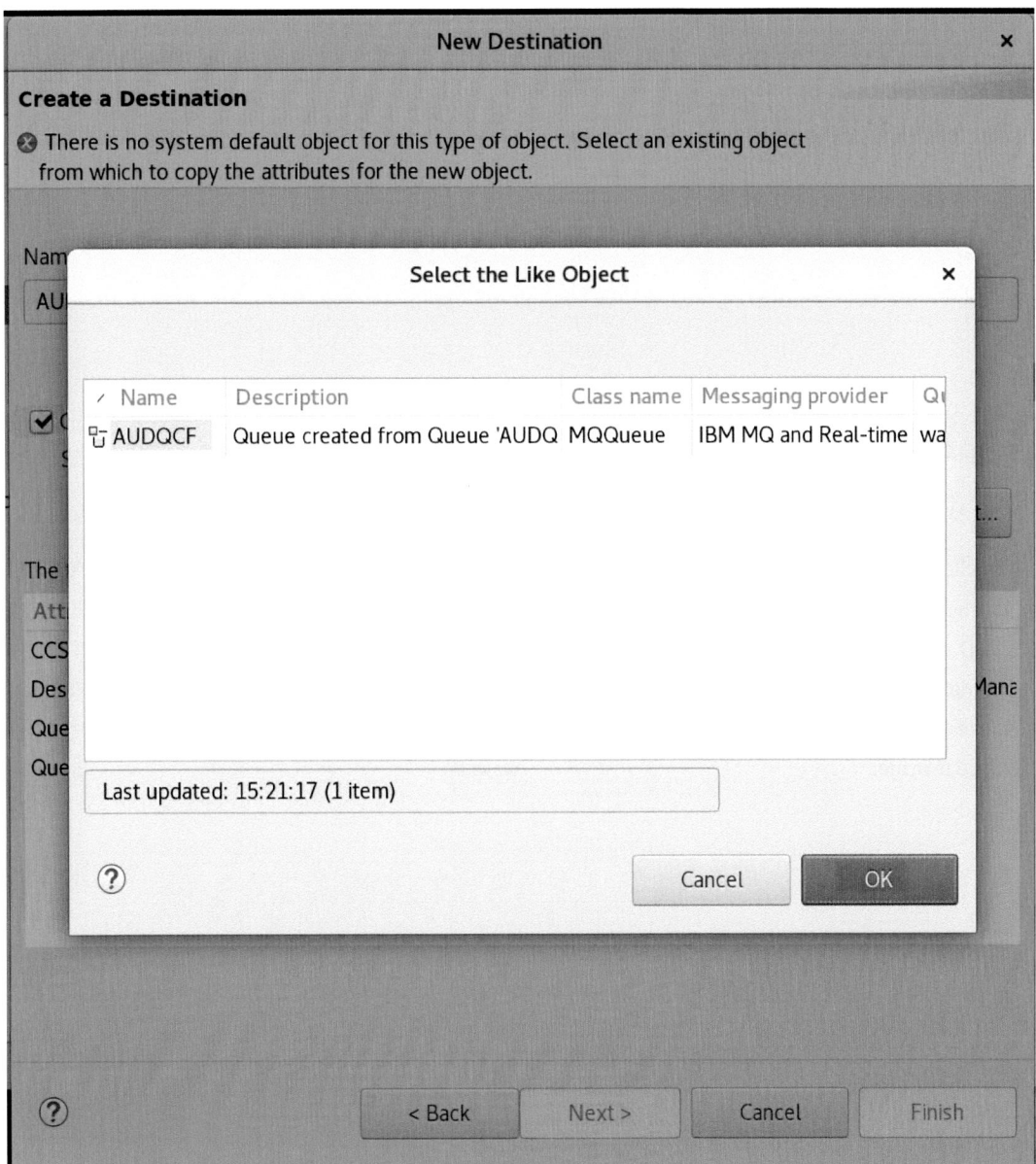

Figure 3-172. *The AUDQCF queue is displayed*

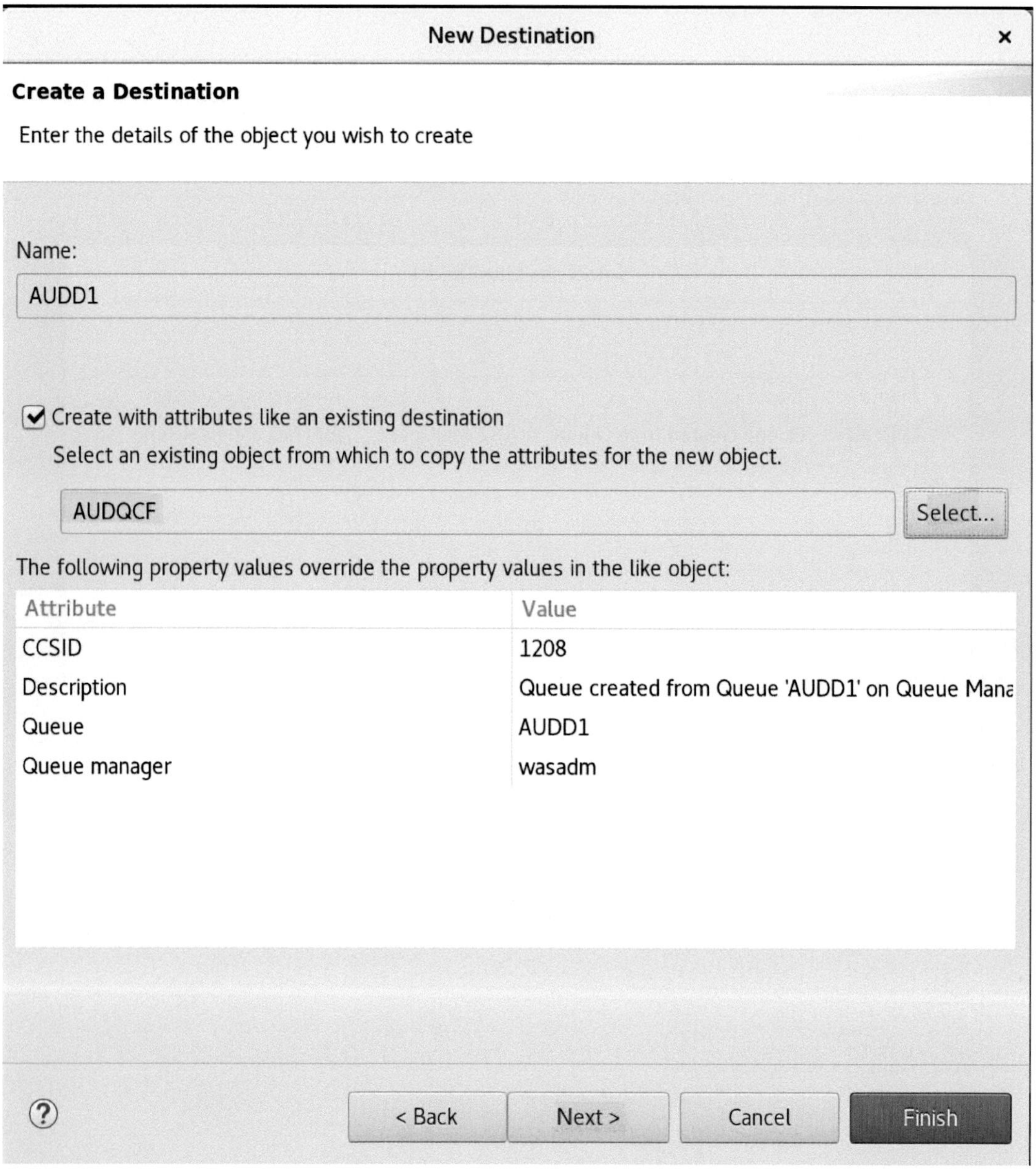

CHAPTER 3 IBM JMS INTERFACE DEVELOPMENT IBM FILENET 5.5.X WORKFLOW

Figure 3-173. The Next> command is selected to match the AUDD1 Dead-Letter Queue

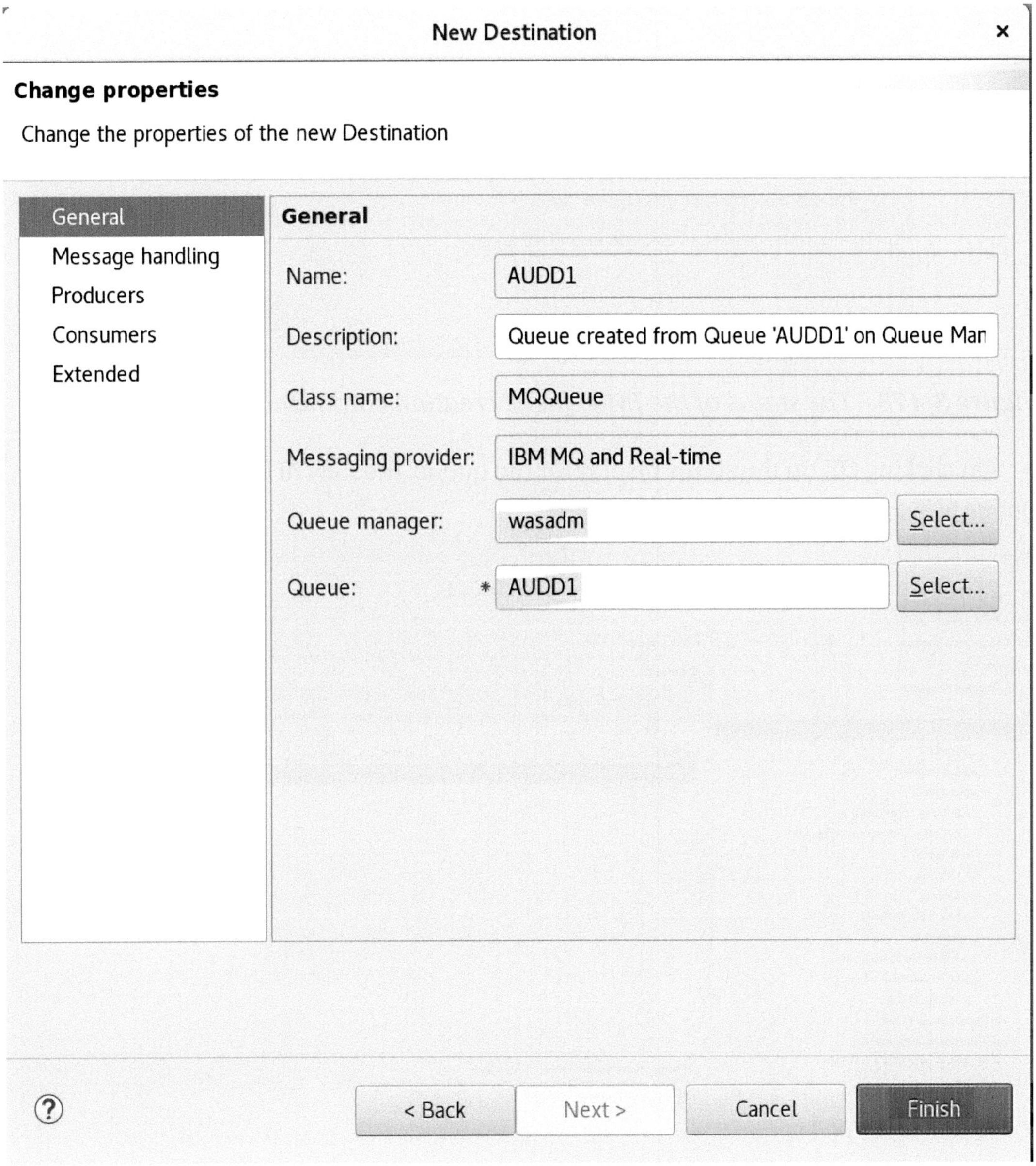

CHAPTER 3 IBM JMS INTERFACE DEVELOPMENT IBM FILENET 5.5.X WORKFLOW

Figure 3-174. The Finish command is clicked to complete the JMS Queue link

CHAPTER 3   IBM JMS INTERFACE DEVELOPMENT IBM FILENET 5.5.X WORKFLOW

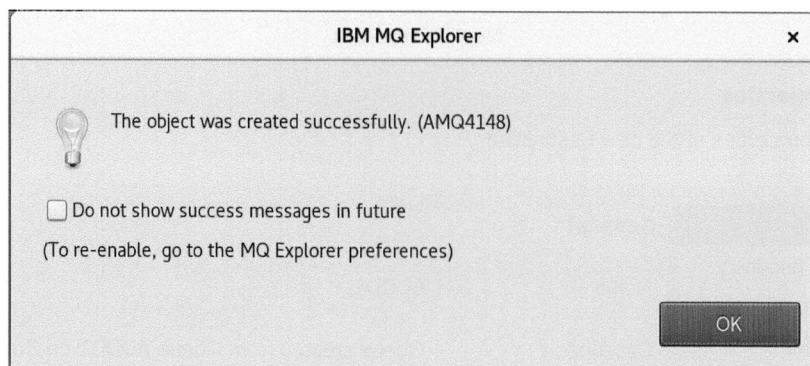

Figure 3-175. *The status of the JMS Queue creation command is displayed*

On clicking OK on the status display for the queue, the new AUDD1 Dead-Letter Queue is displayed.

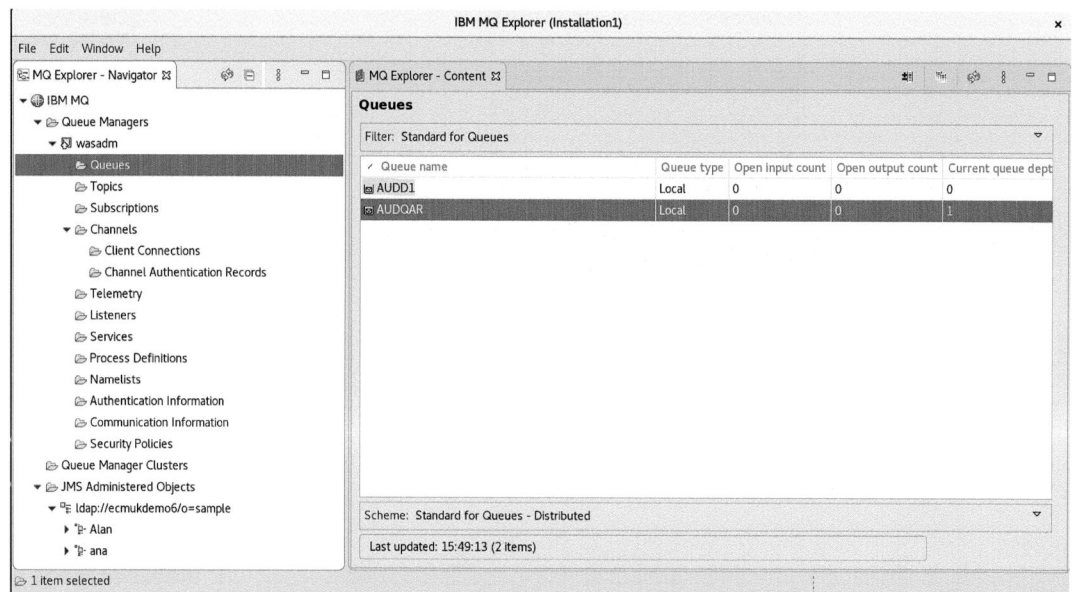

Figure 3-176. *The new AUDD1 Dead-Letter Queue is displayed*

Test the wasadm Queue Manager again.

CHAPTER 3   IBM JMS INTERFACE DEVELOPMENT IBM FILENET 5.5.X WORKFLOW

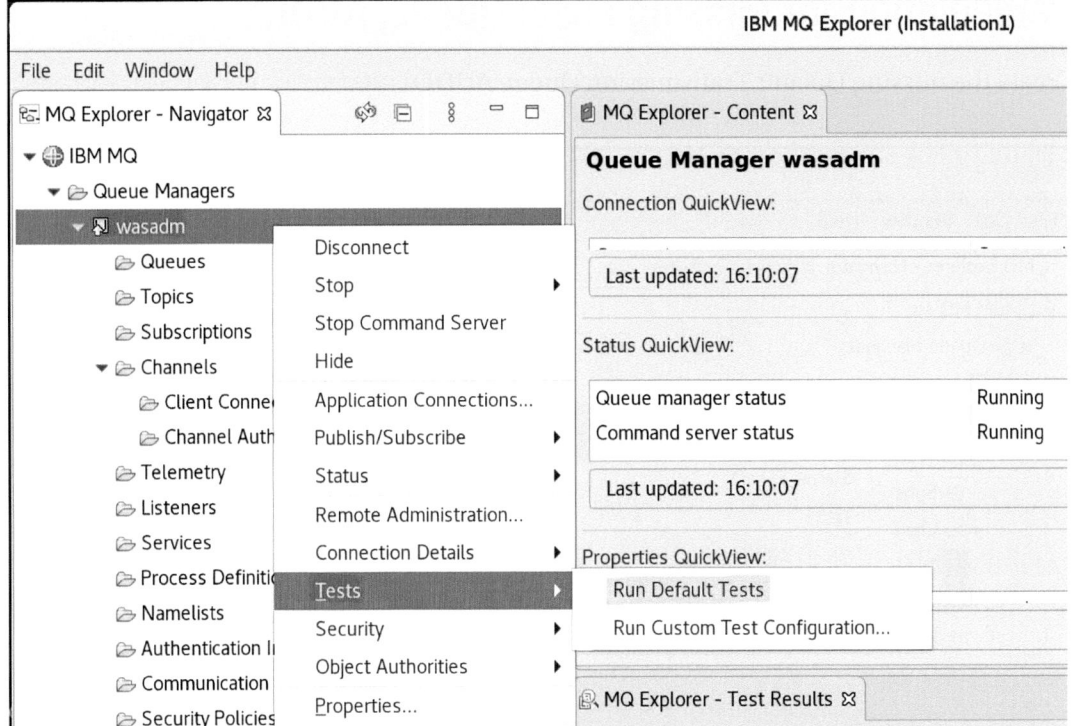

Figure 3-177. *The changes are tested, using the **Tests ➤ Run Default Tests** menu*

The Test error results show we need to create the AUDQ1 Queue.

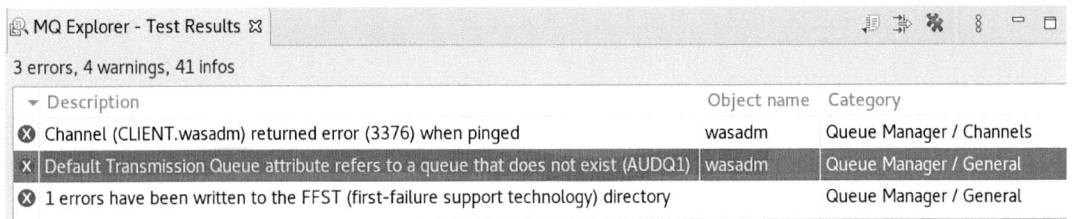

Figure 3-178. *The Run Default Tests are conducted on the wasadm Queue Manager*

Now, we see the error as follows:

```
Error    Default Transmission Queue attribute refers to a queue that does
not exist (AUDQ1)    wasadm    Queue Manager / General
```

Chapter 3 IBM JMS Interface Development IBM FileNet 5.5.x Workflow

Remediation for Missing Local Queue, AUDQ1

Create the missing Default Transmission Queue AUDQ1.

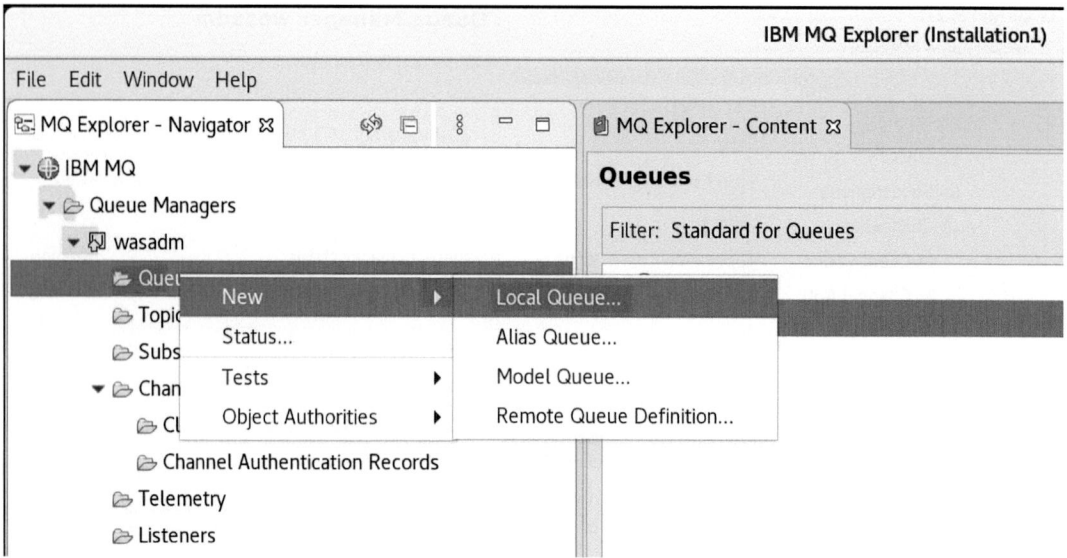

Figure 3-179. *The menu option to create a New Local Queue is selected using the right-mouse button click*

The Queue template Object type is selected as **SYSTEM.DEFAULT.LOCAL.QUEUE** from the Object type list.

CHAPTER 3 IBM JMS INTERFACE DEVELOPMENT IBM FILENET 5.5.X WORKFLOW

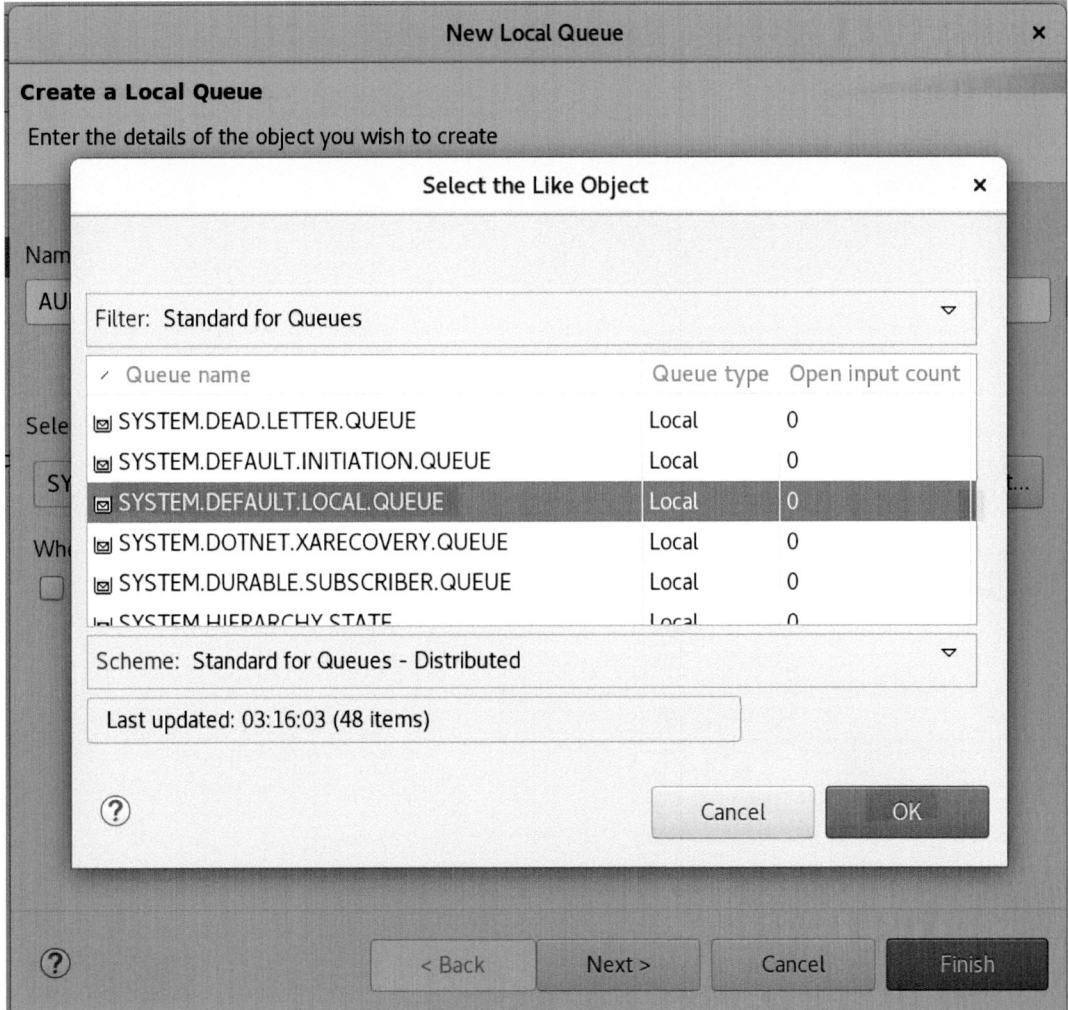

Figure 3-180. The drop-down list is scrolled down to select the required Object type

The AUDQ1 name is entered for the Local Queue to be created.

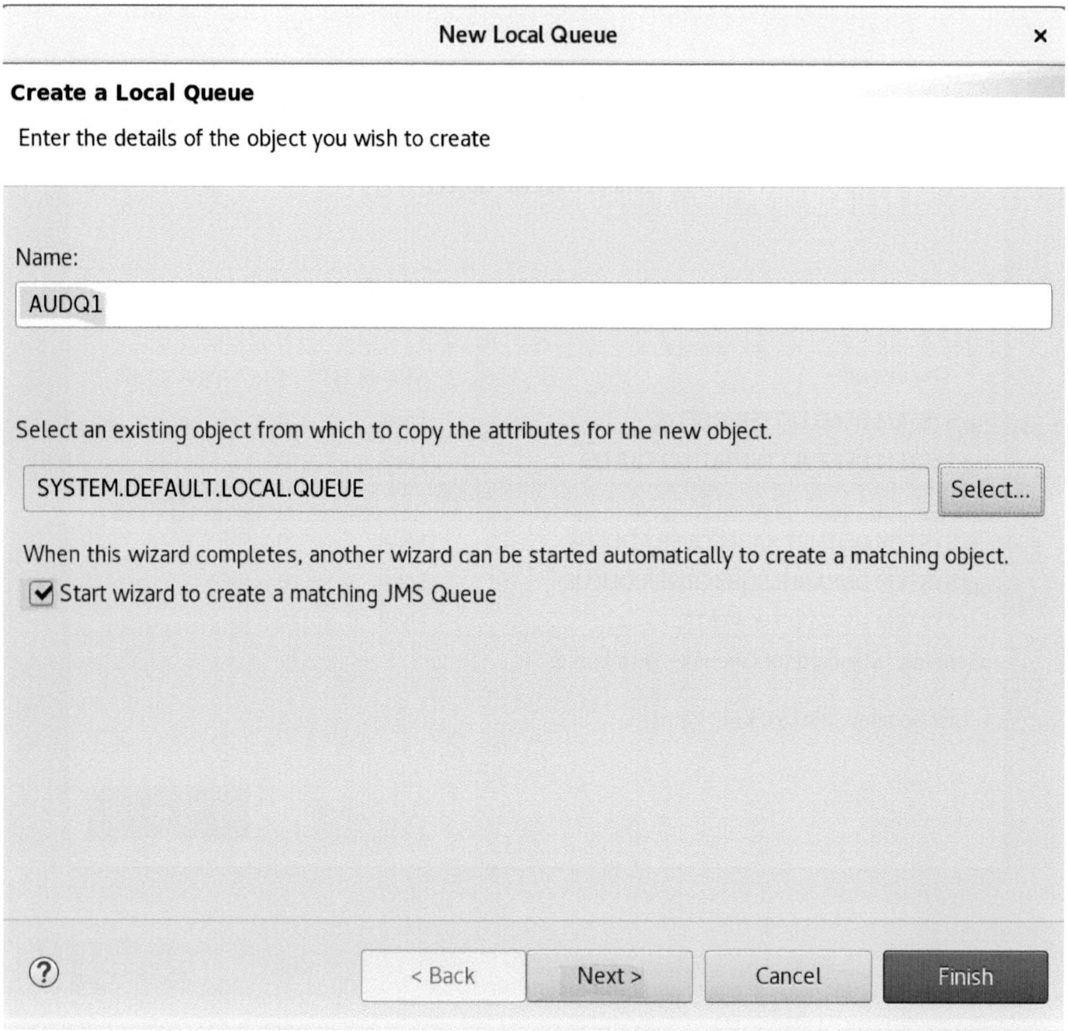

CHAPTER 3 IBM JMS INTERFACE DEVELOPMENT IBM FILENET 5.5.X WORKFLOW

New Local Queue	✕

Create a Local Queue

Enter the details of the object you wish to create

Name:

AUDQ1

Select an existing object from which to copy the attributes for the new object.

SYSTEM.DEFAULT.LOCAL.QUEUE   Select...

When this wizard completes, another wizard can be started automatically to create a matching object.
☑ Start wizard to create a matching JMS Queue

⑦   < Back   Next >   Cancel   Finish

Figure 3-181. *The tick box for JMS Queue creation is selected and Next> clicked*

The **AUDQ1** Queue creation options are set with the tick box to start the wizard for creating the **JMS Queue** as highlighted in Figure 3-181, and the Next> command button is then clicked.

CHAPTER 3   IBM JMS INTERFACE DEVELOPMENT IBM FILENET 5.5.X WORKFLOW

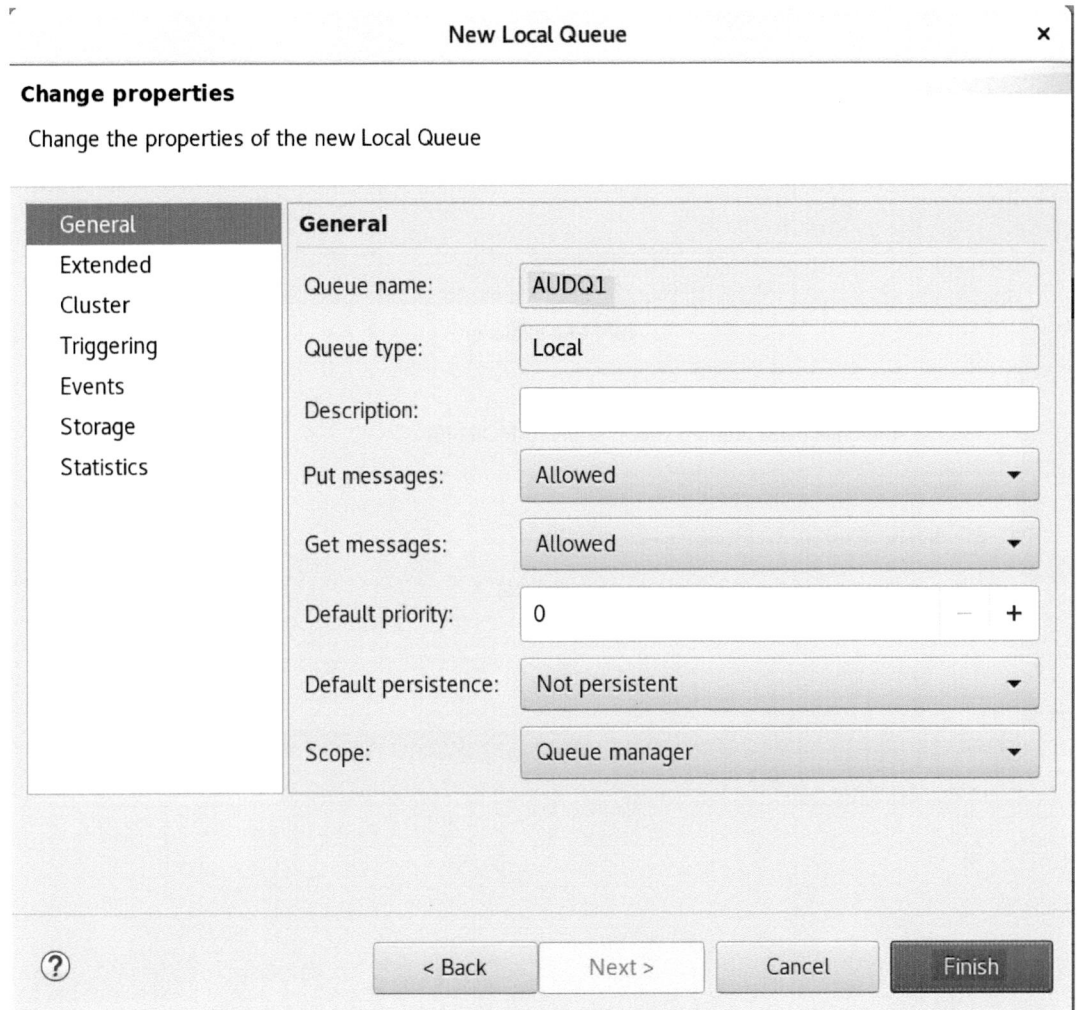

Figure 3-182. The Finish command button is clicked

The status pop-up window appears after clicking the Finish command button.

CHAPTER 3 IBM JMS INTERFACE DEVELOPMENT IBM FILENET 5.5.X WORKFLOW

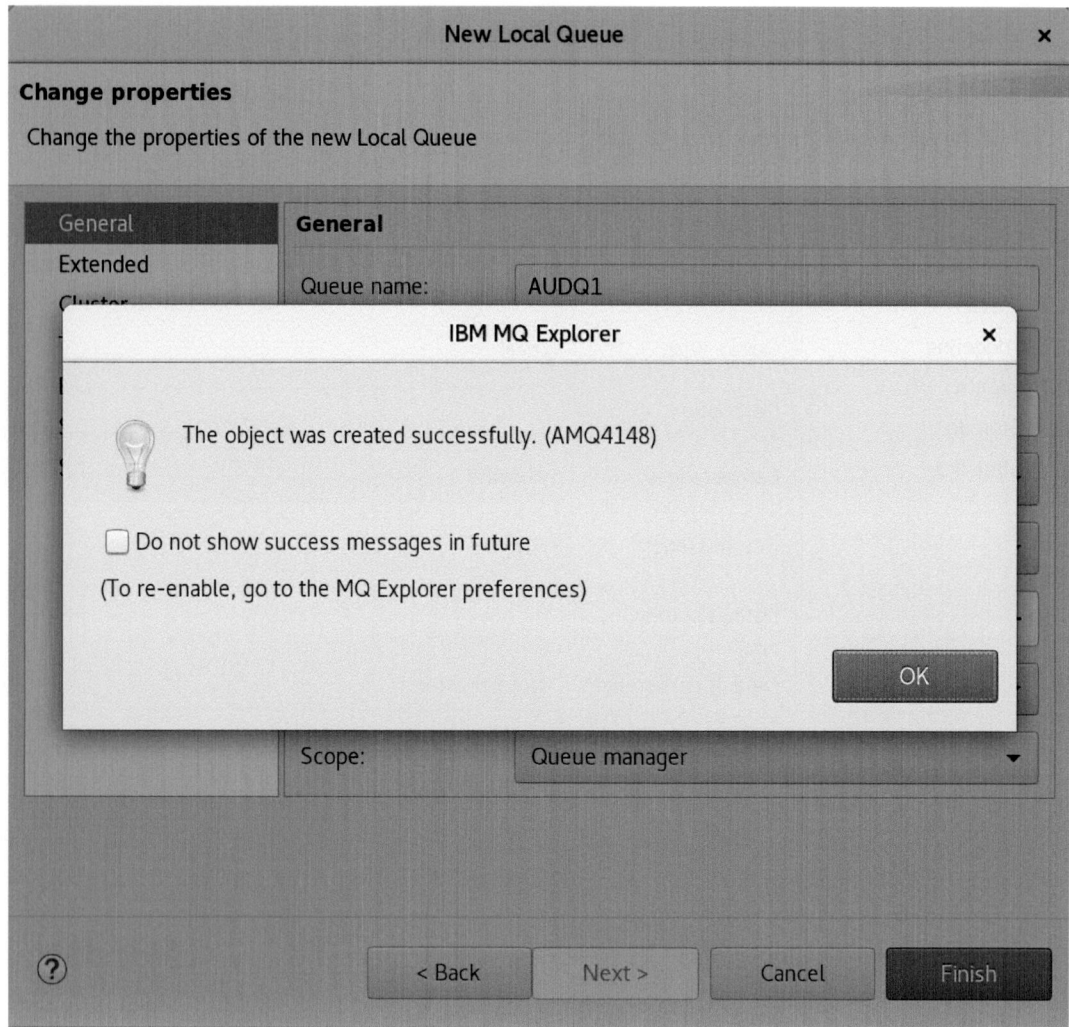

Figure 3-183. *The OK command button is clicked which launches the **New JMS Queue***

The New JMS Queue wizard is launched as shown in Figure 3-184.

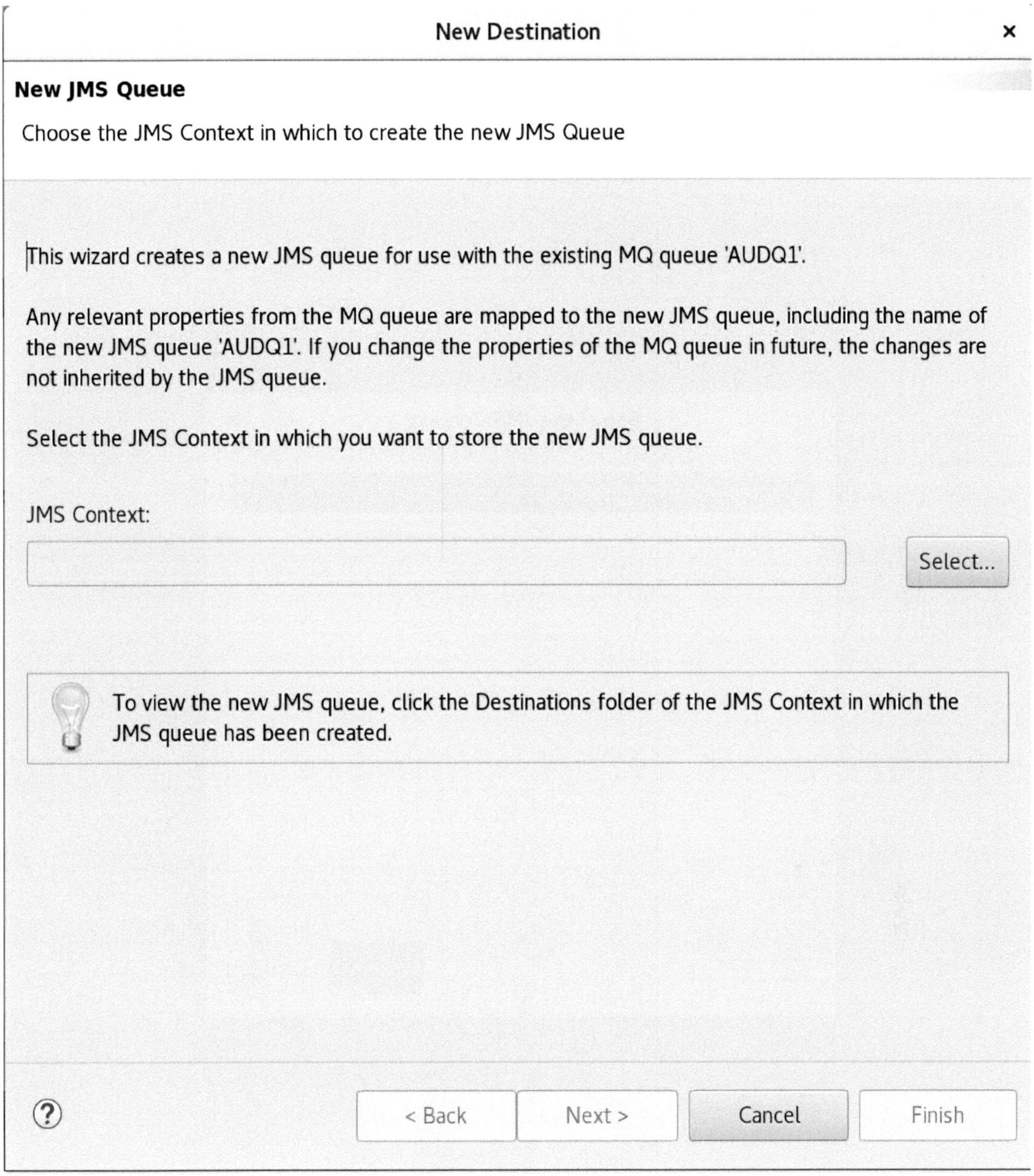

CHAPTER 3 IBM JMS INTERFACE DEVELOPMENT IBM FILENET 5.5.X WORKFLOW

New Destination

New JMS Queue

Choose the JMS Context in which to create the new JMS Queue

This wizard creates a new JMS queue for use with the existing MQ queue 'AUDQ1'.

Any relevant properties from the MQ queue are mapped to the new JMS queue, including the name of the new JMS queue 'AUDQ1'. If you change the properties of the MQ queue in future, the changes are not inherited by the JMS queue.

Select the JMS Context in which you want to store the new JMS queue.

JMS Context:

Select...

To view the new JMS queue, click the Destinations folder of the JMS Context in which the JMS queue has been created.

< Back   Next >   Cancel   Finish

Figure 3-184. *The LDAP server JMS Context created earlier is shown for selection*

CHAPTER 3 IBM JMS INTERFACE DEVELOPMENT IBM FILENET 5.5.X WORKFLOW

The JMS Context is supported by the **ecmukdemo6** LDAP server **o=sample** root for searching.

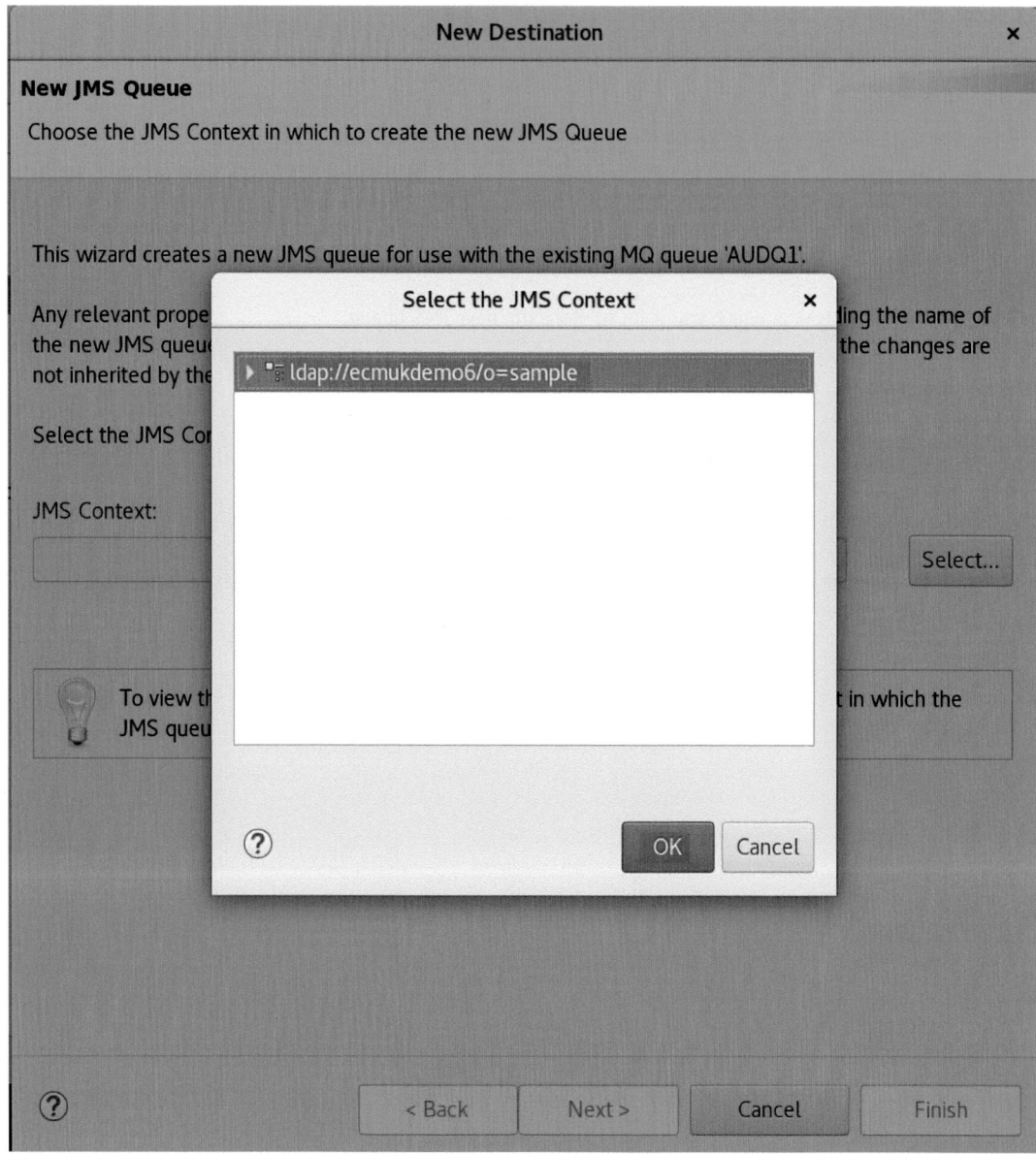

Figure 3-185. *The LDAP server, ecmukdemo6, URL is shown to be selected for JMS*

CHAPTER 3   IBM JMS INTERFACE DEVELOPMENT IBM FILENET 5.5.X WORKFLOW

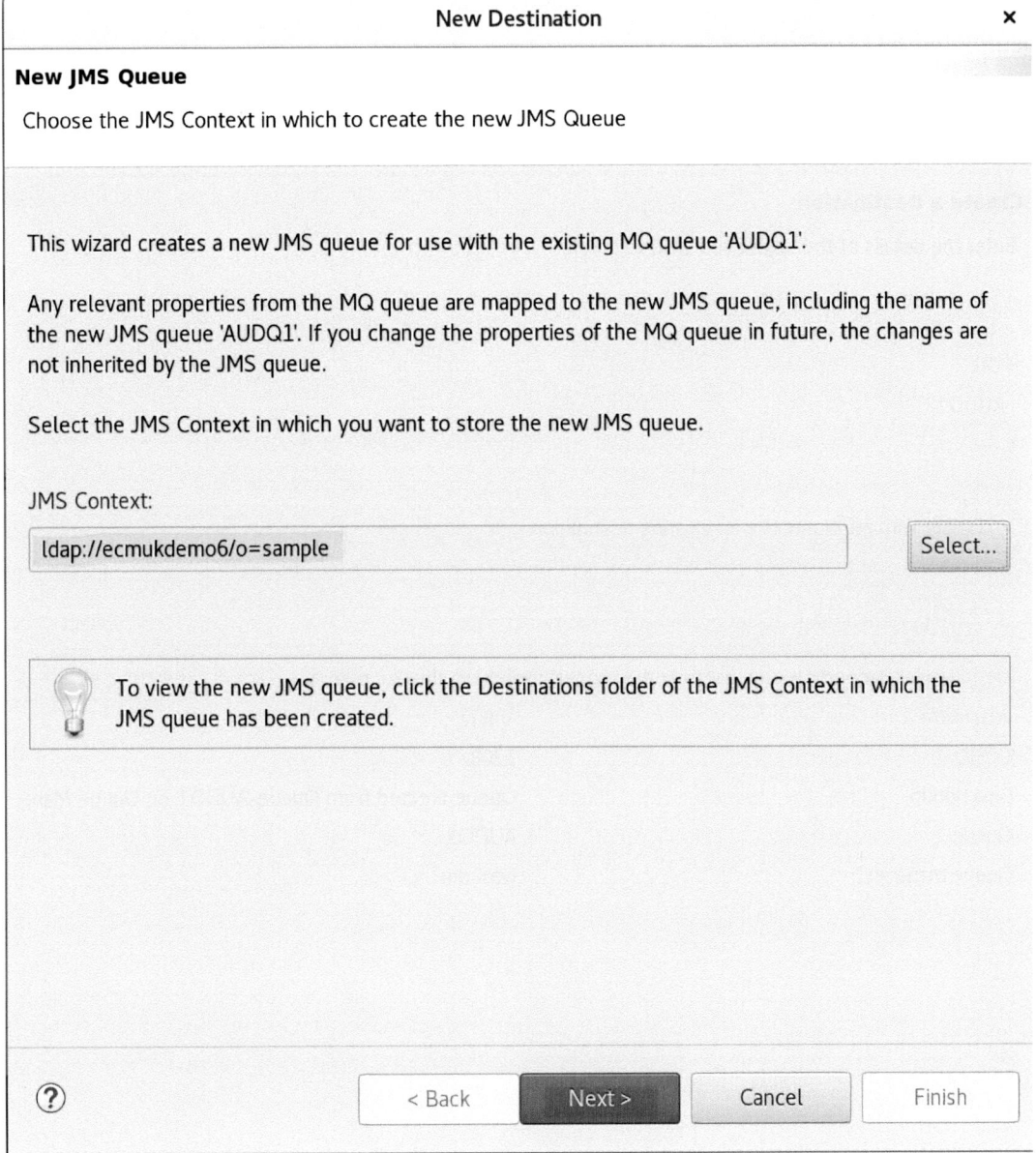

Figure 3-186. *The Next command is used to start the creation of the new JMS queue*

CHAPTER 3   IBM JMS INTERFACE DEVELOPMENT IBM FILENET 5.5.X WORKFLOW

The AUDQ1 Queue now has to be linked to the JMS Context Queue object in the ecmukdemo6 LDAP server.

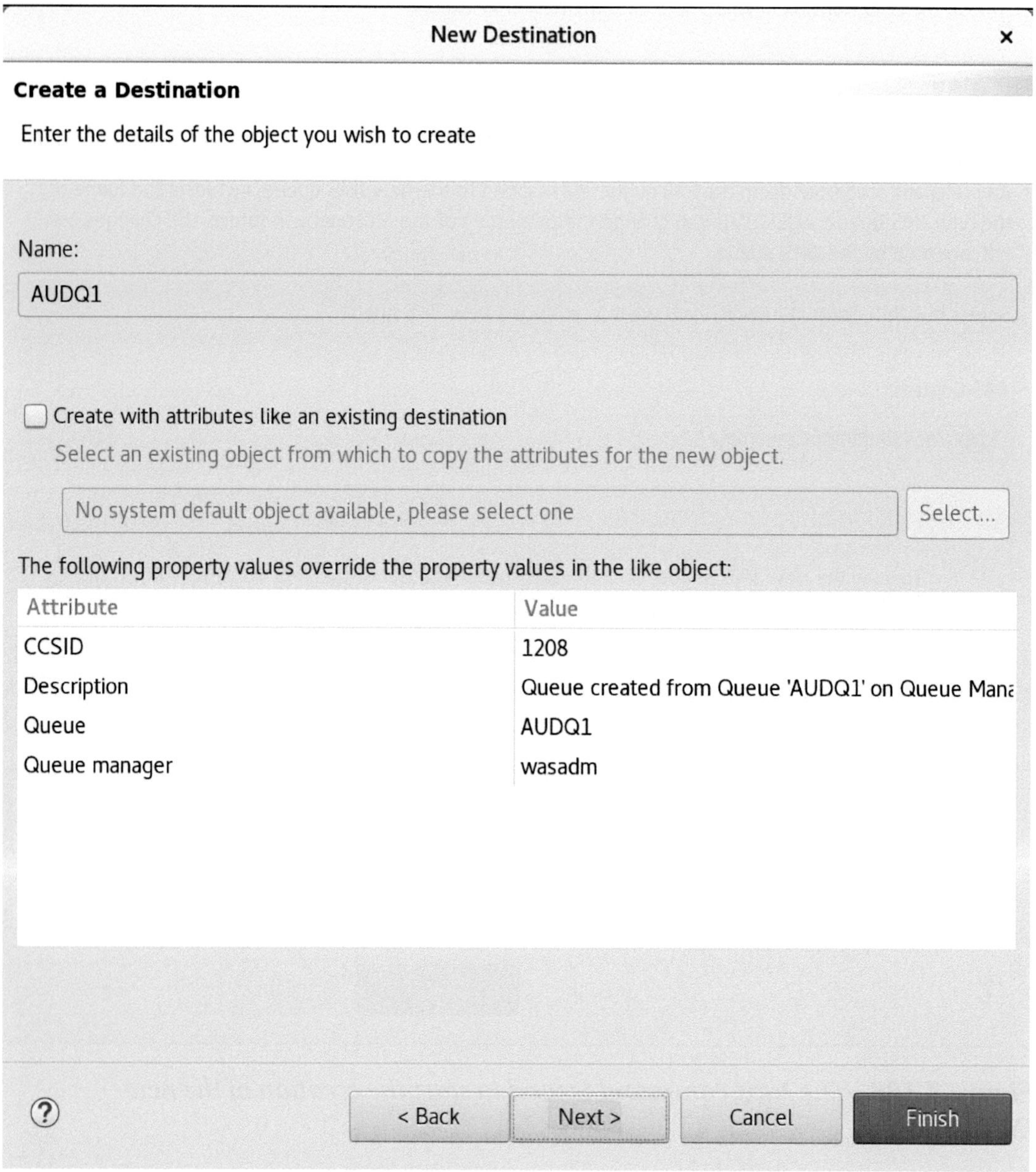

New Destination	✕

Create a Destination
Enter the details of the object you wish to create

Name:
AUDQ1

☐ Create with attributes like an existing destination
Select an existing object from which to copy the attributes for the new object.

No system default object available, please select one	Select...

The following property values override the property values in the like object:

Attribute	Value
CCSID	1208
Description	Queue created from Queue 'AUDQ1' on Queue Mana
Queue	AUDQ1
Queue manager	wasadm

[< Back] [Next >] [Cancel] [Finish]

Figure 3-187. *The details of the New AUDQ1 JMS Context Queue creation are displayed*

CHAPTER 3   IBM JMS INTERFACE DEVELOPMENT IBM FILENET 5.5.X WORKFLOW

The AUDQ1 property values for the JMS Context Queue are displayed after clicking Next.

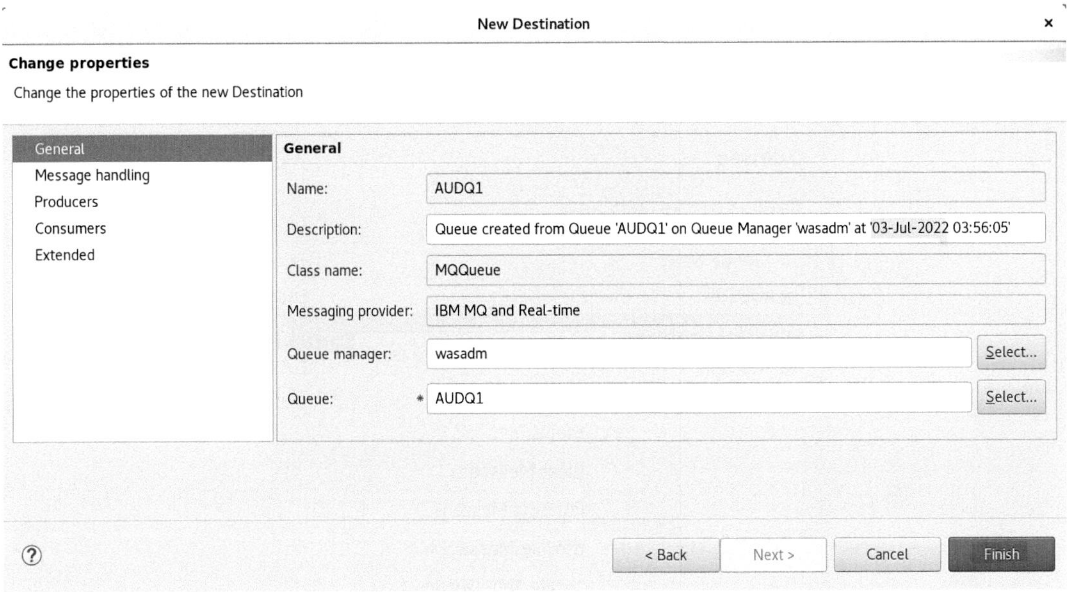

Figure 3-188. The Finish command is clicked

The New AUDQ1 Queue is now created and shown, displayed in the MQ Explorer.

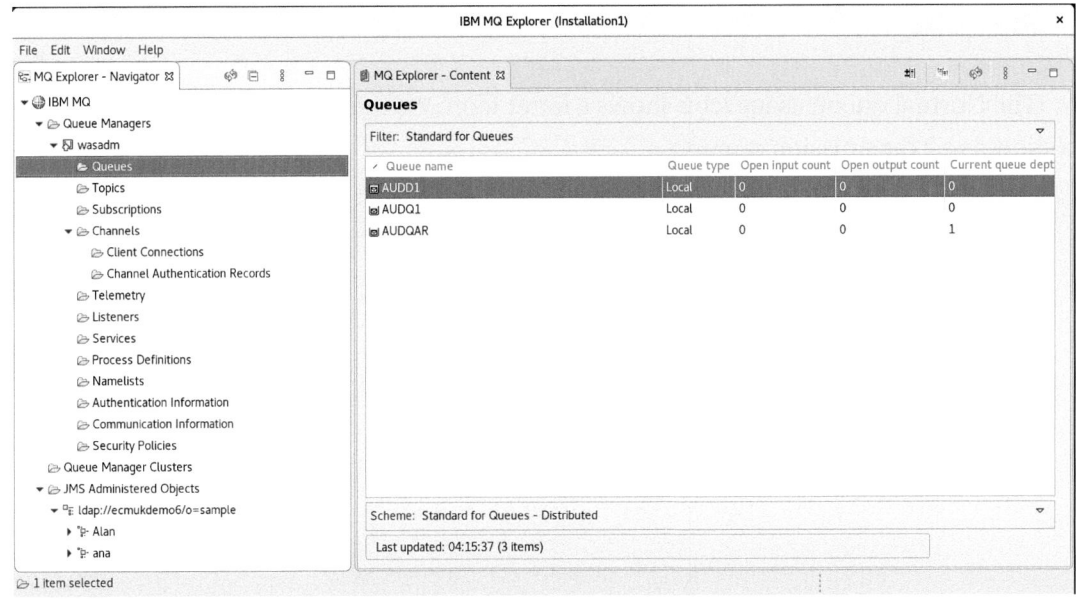

Figure 3-189. The New AUDQ1 Queue which was created is displayed

517

Chapter 3 IBM JMS Interface Development IBM FileNet 5.5.x Workflow

Now we need to change the **AUDQ1** to Transmission usage.

First, we select the **AUDQ1** entry properties menu option using the right-mouse click.

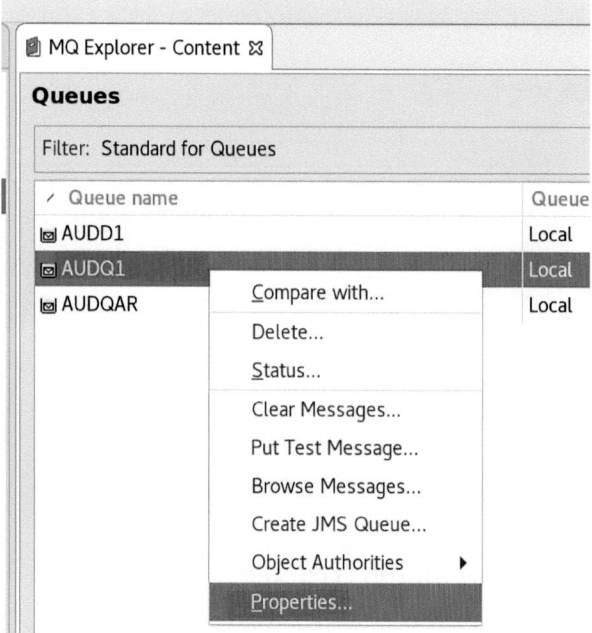

Figure 3-190. *The current **Normal** usage option property is changed to Transmission*

The Normal queue usage icon shows a letter icon. We need to change this property to show as a Transmission queue.

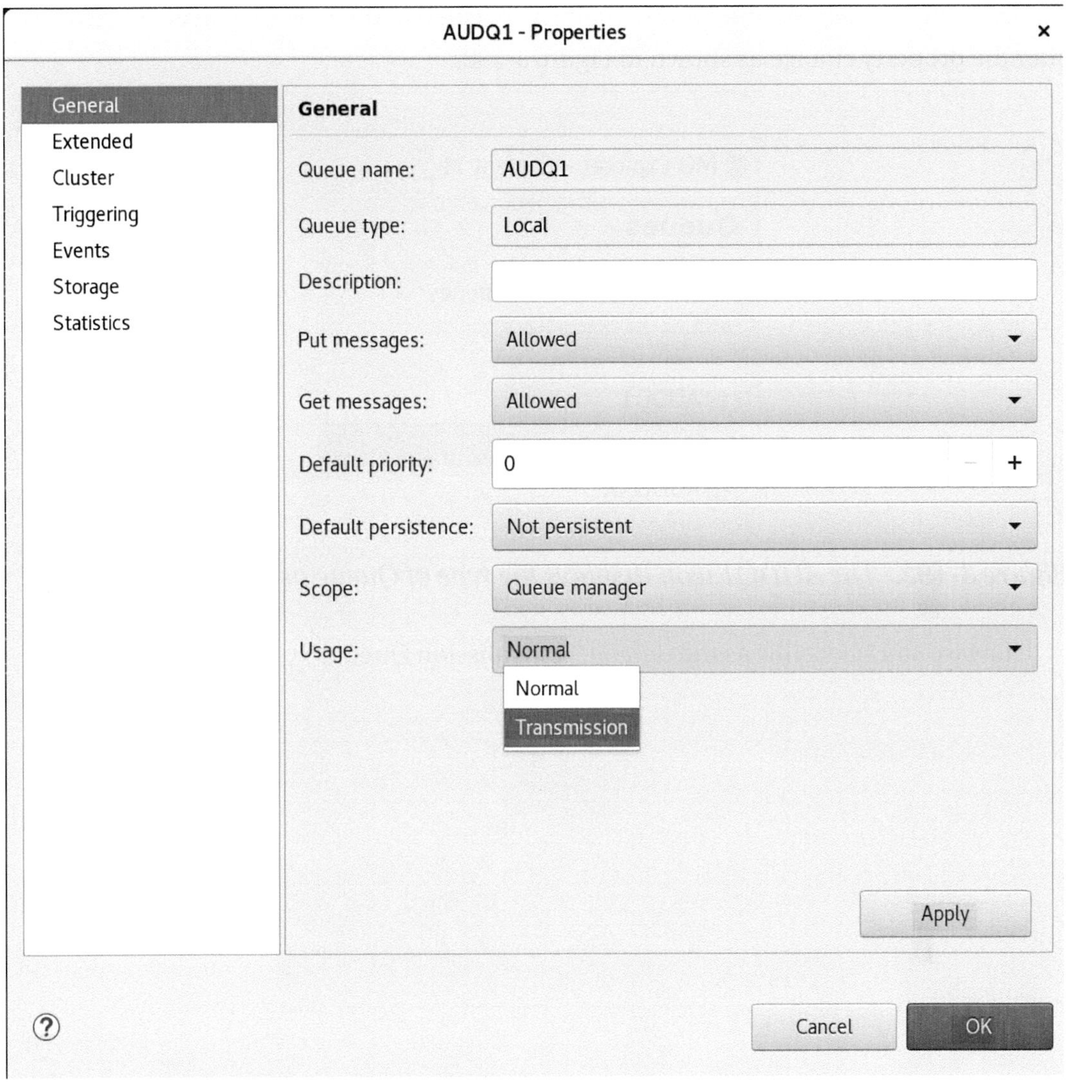

Figure 3-191. The Usage is changed from Normal to Transmission

CHAPTER 3 IBM JMS INTERFACE DEVELOPMENT IBM FILENET 5.5.X WORKFLOW

The Transmission option for the queue is reflected in the icon displayed for AUDQ1 after the property change as shown in Figure 3-192.

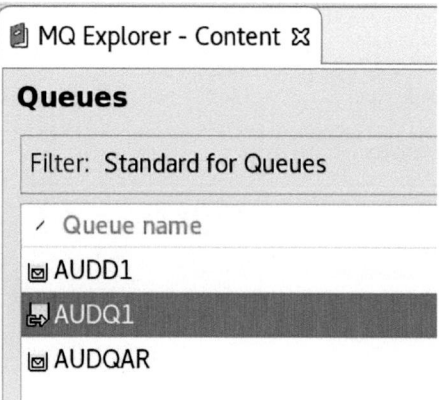

Figure 3-192. *The AUDQ1 icon displays the type of Queue as Transmission*

Now we can select this as the default Transmission Queue.

CHAPTER 3 IBM JMS INTERFACE DEVELOPMENT IBM FILENET 5.5.X WORKFLOW

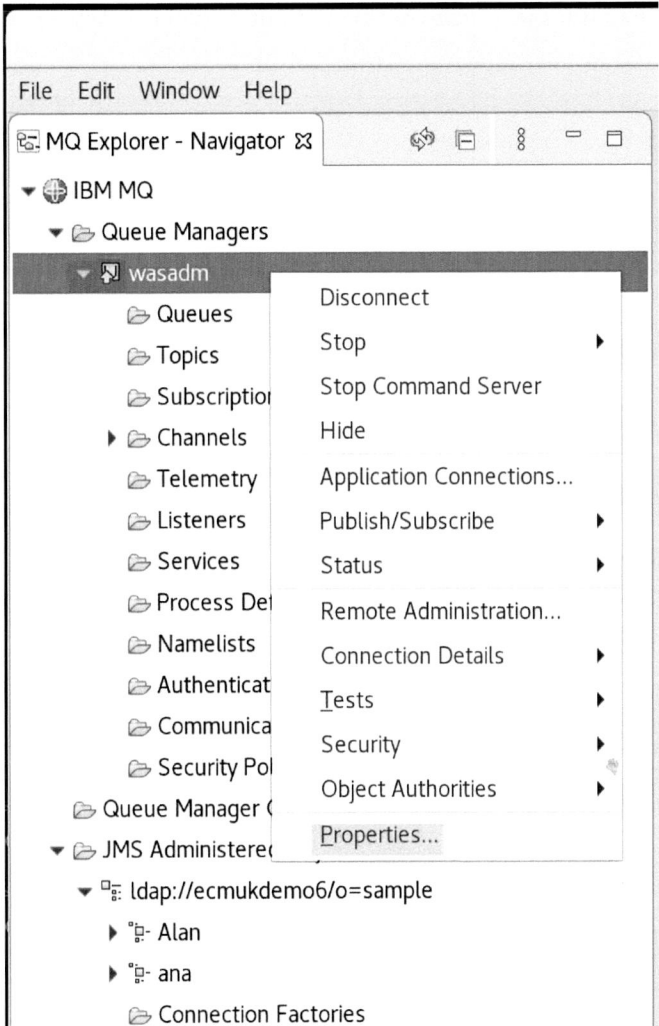

Figure 3-193. *The wasadm Queue Manager properties are selected*

CHAPTER 3 IBM JMS INTERFACE DEVELOPMENT IBM FILENET 5.5.X WORKFLOW

Now we can select the Communications tab and select the AUDQ1 Queue.

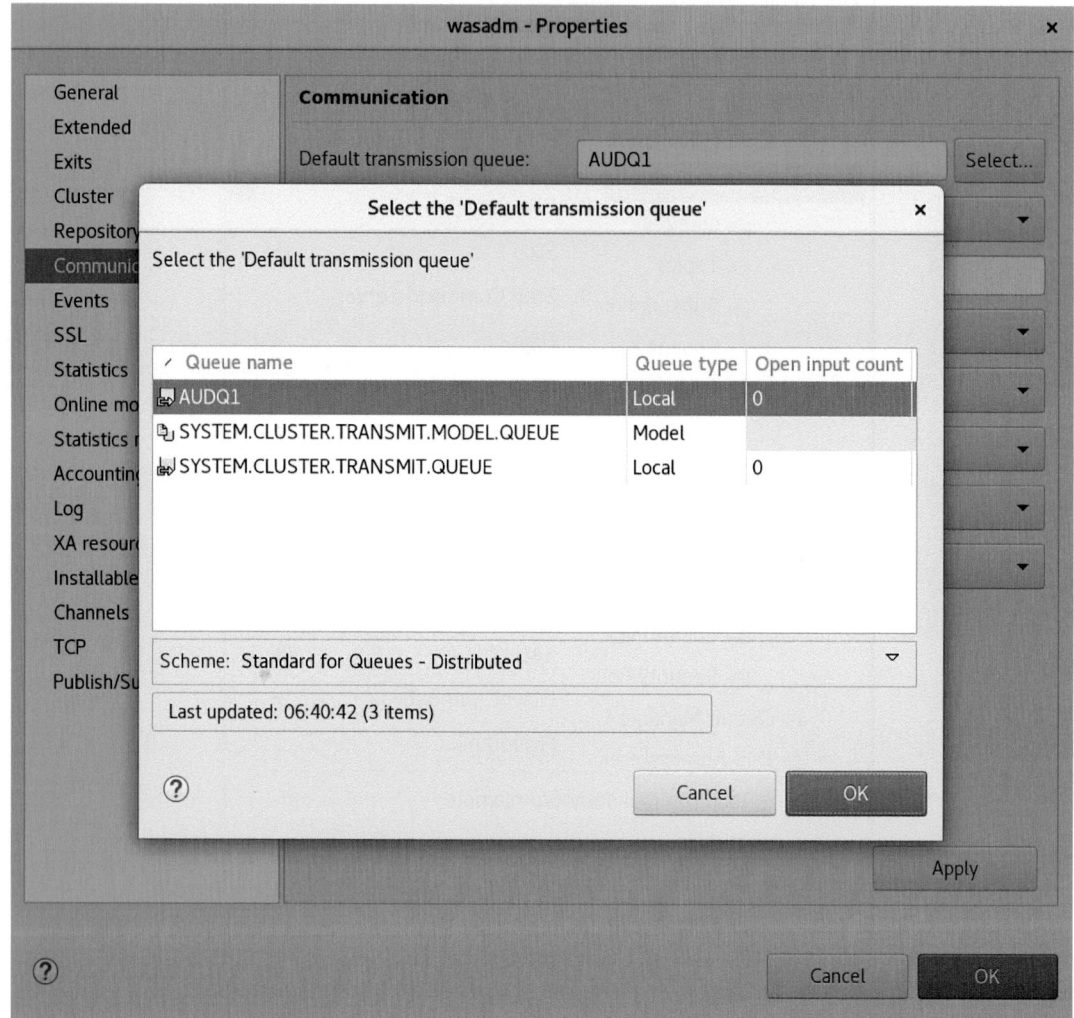

Figure 3-194. *The wasadm Queue Default transmission queue is set as AUDQ1*

Error Log Location

The error logs can be located on the **ecmukdemo6** IBM MQ series 9.2.5 server in the following location.

```
[mqm@ECMUKDEMO6 ~]$ pwd
/var/mqm
[mqm@ECMUKDEMO6 ~]$ ls
config   errors   exits64   mqclient.ini   mqs.ini   service.env   sockets   web
conv     exits    log       mqft           qmgrs     shared        trace
[mqm@ECMUKDEMO6 ~]$ cd errors
[mqm@ECMUKDEMO6 errors]$ ls -lsa
total 812
    0 drwxrwsrwx.  2 mqm mqm     111 Jul  3 03:05 .
    0 drwxrwsr-x. 14 mqm mqm     217 Jun 24 10:59 ..
  528 -rw-r-----.  1 mqm mqm  537369 Jul  3 05:07 AMQ124808.0.FDC
  264 -rw-r-----.  1 mqm mqm  269738 Jun 29 09:46 AMQ20154.0.FDC
   20 -rw-rw-r--.  1 mqm mqm   17608 Jul  3 05:07 AMQERR01.LOG
    0 -rw-rw-r--.  1 mqm mqm       0 Jun 23 15:33 AMQERR02.LOG
    0 -rw-rw-r--.  1 mqm mqm       0 Jun 23 15:33 AMQERR03.LOG
[mqm@ECMUKDEMO6 errors]$
```

Figure 3-195. *The error logs are found located in the /var/mqm/errors subfolder*

com.ibm.mq.jms.MQQueue incompatible with javax.jms.ConnectionFactory
com.ibm.mq.jms.MQQueue incompatible with javax.jms.ConnectionFactory

This was returned from the existing project.

Add a new JNDI Queue Connection Factory, AUDCF, in MQ Explorer.

CHAPTER 3 IBM JMS INTERFACE DEVELOPMENT IBM FILENET 5.5.X WORKFLOW

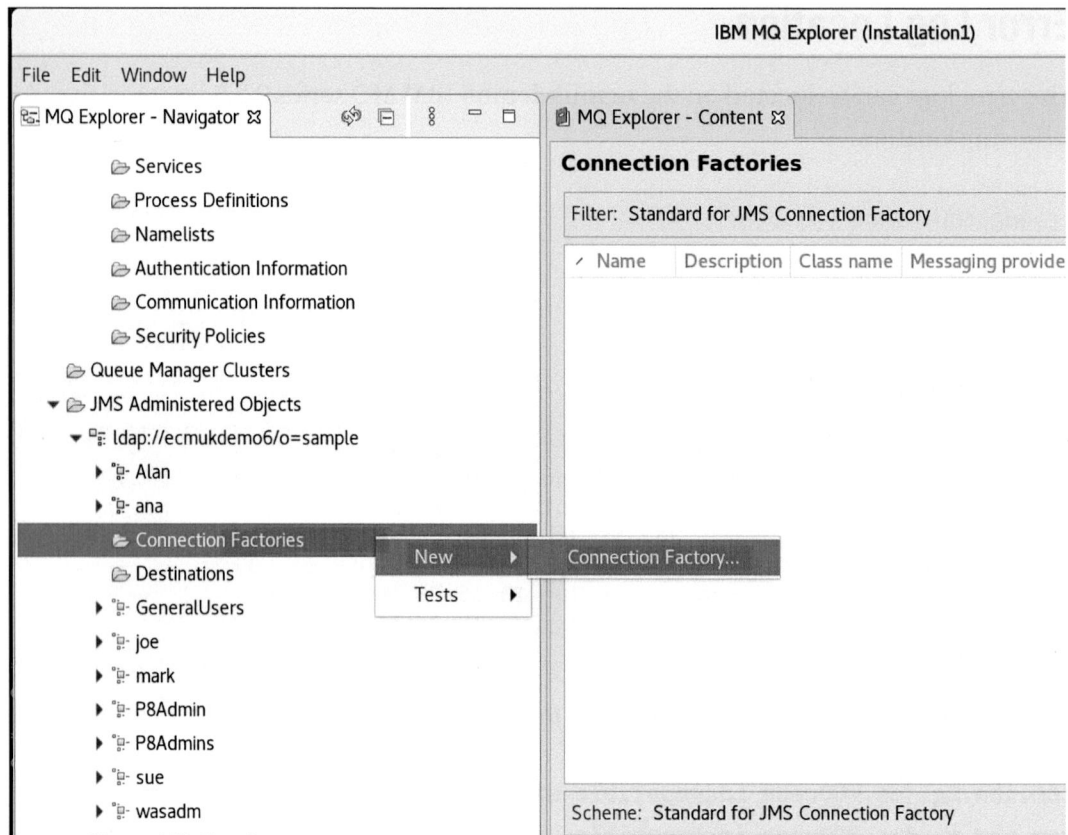

Figure 3-196. *A new Connection Factory, AUDCF, is required*

The Connection Factory wizard parameter window values are entered.

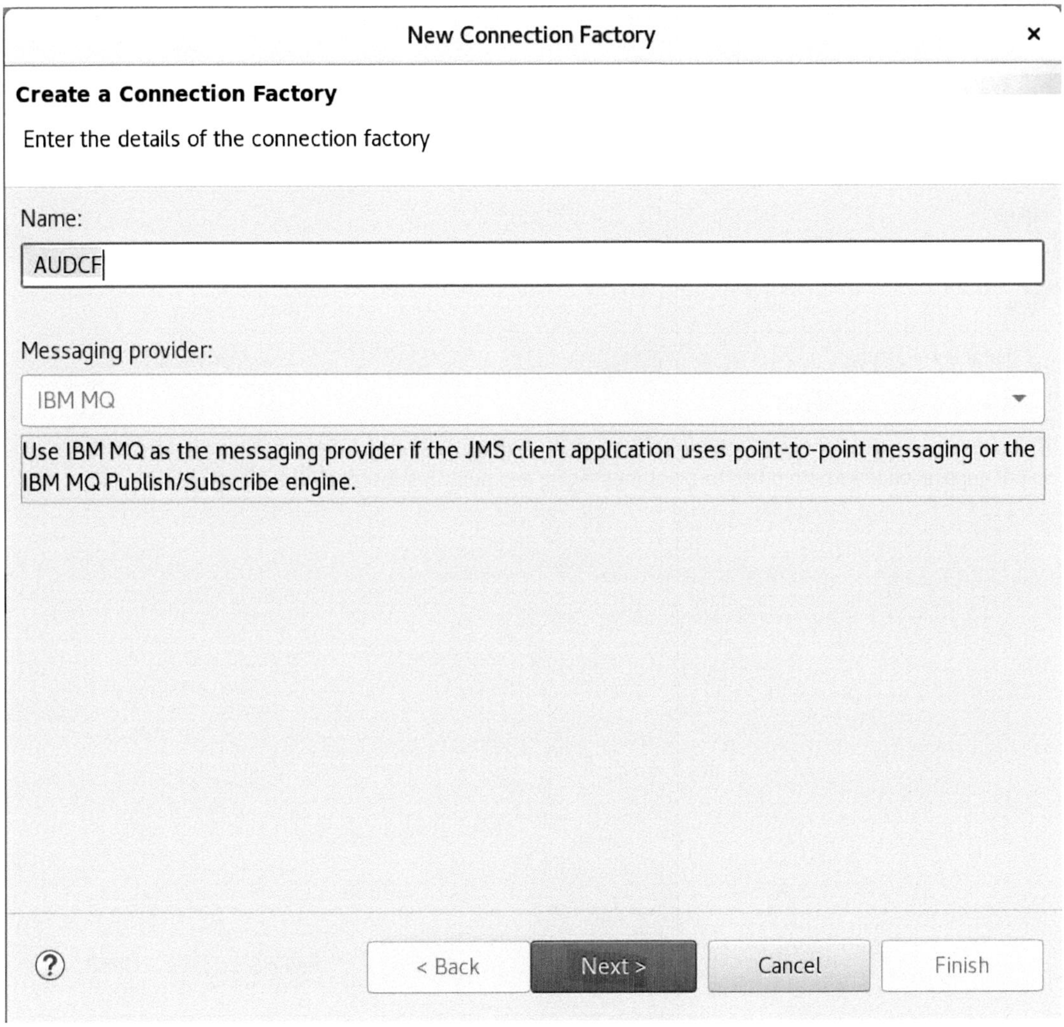

CHAPTER 3 IBM JMS INTERFACE DEVELOPMENT IBM FILENET 5.5.X WORKFLOW

New Connection Factory

Create a Connection Factory

Enter the details of the connection factory

Name:

AUDCF

Messaging provider:

IBM MQ

Use IBM MQ as the messaging provider if the JMS client application uses point-to-point messaging or the IBM MQ Publish/Subscribe engine.

Figure 3-197. A new Connection Factory name, AUDCF, is entered and Next clicked

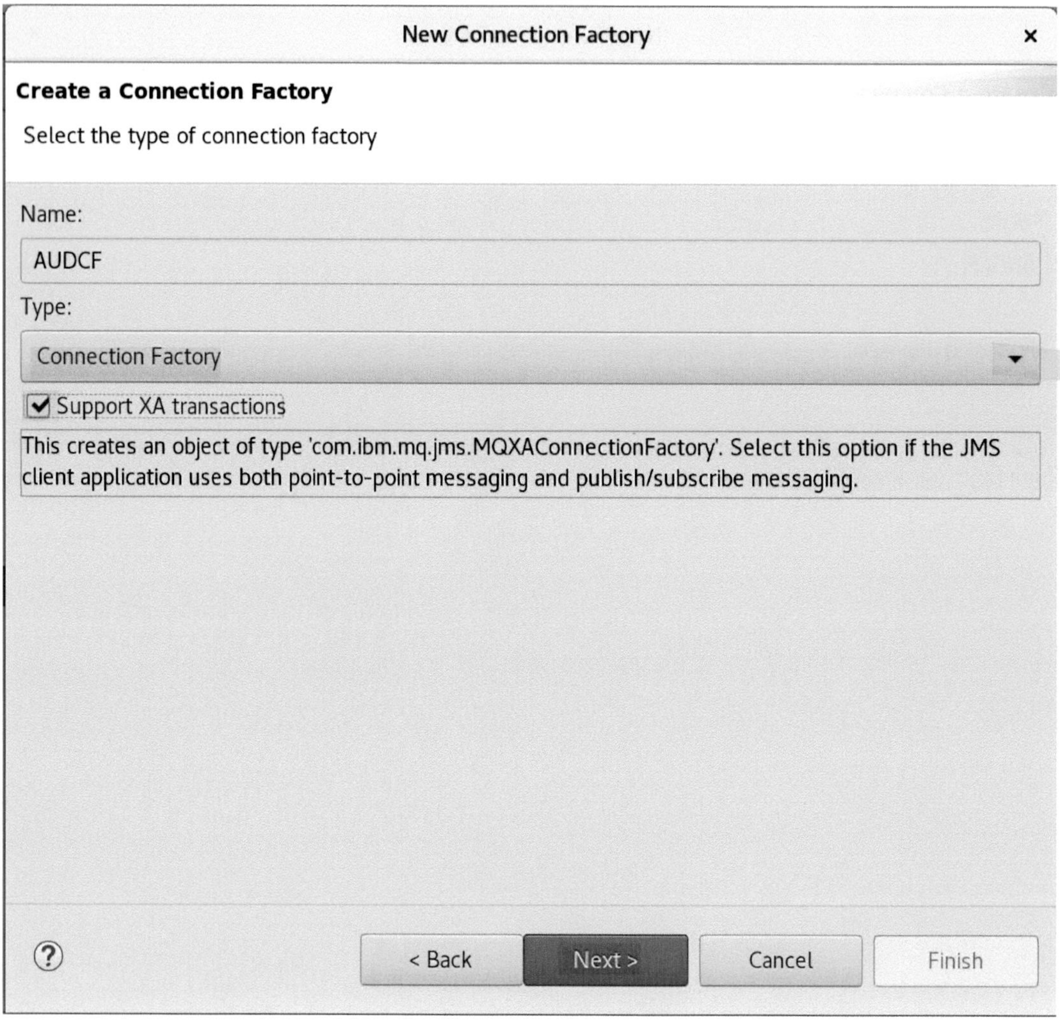

CHAPTER 3　IBM JMS INTERFACE DEVELOPMENT IBM FILENET 5.5.X WORKFLOW

New Connection Factory
Create a Connection Factory
Select the type of connection factory
Name:
AUDCF
Type:
Connection Factory
☑ Support XA transactions
This creates an object of type 'com.ibm.mq.jms.MQXAConnectionFactory'. Select this option if the JMS client application uses both point-to-point messaging and publish/subscribe messaging.
< Back　Next >　Cancel　Finish

Figure 3-198. *The Connection Factory download option is clicked*

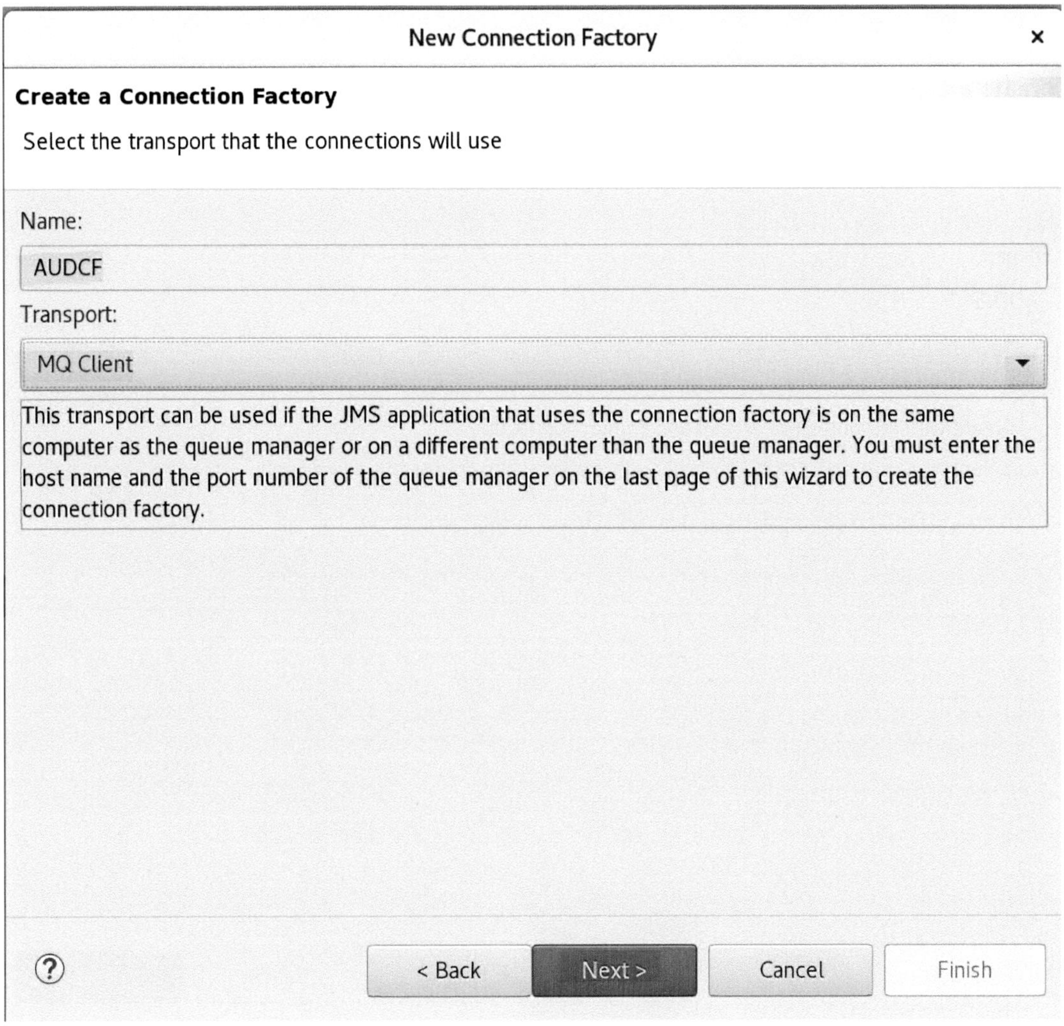

CHAPTER 3 IBM JMS INTERFACE DEVELOPMENT IBM FILENET 5.5.X WORKFLOW

New Connection Factory ✕

Create a Connection Factory

Select the transport that the connections will use

Name:

AUDCF

Transport:

MQ Client ▼

This transport can be used if the JMS application that uses the connection factory is on the same computer as the queue manager or on a different computer than the queue manager. You must enter the host name and the port number of the queue manager on the last page of this wizard to create the connection factory.

ⓘ   < Back   Next >   Cancel   Finish

Figure 3-199. *The Connection Factory option for Transport is selected as MQ Client*

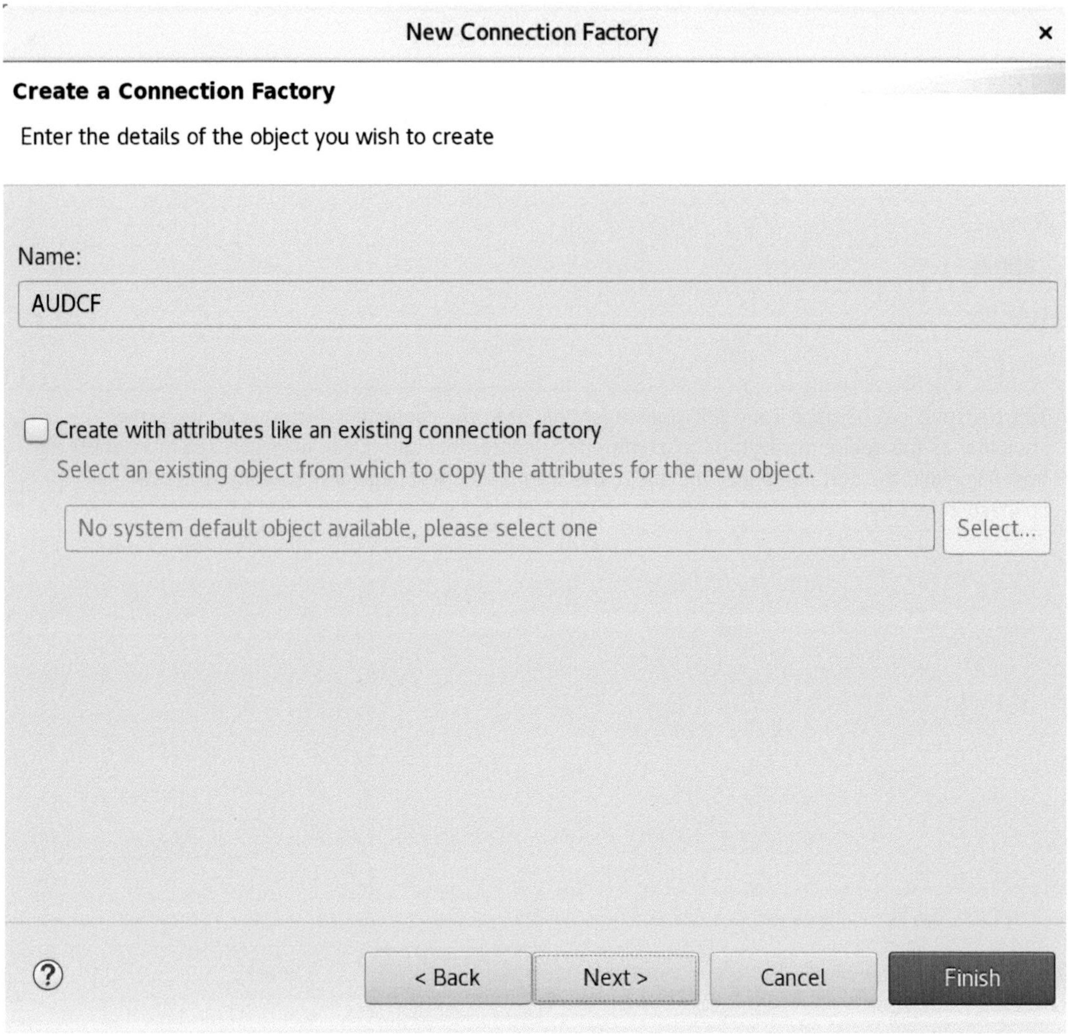

CHAPTER 3   IBM JMS INTERFACE DEVELOPMENT IBM FILENET 5.5.X WORKFLOW

Figure 3-200. The Next command button is selected to leave the defaults

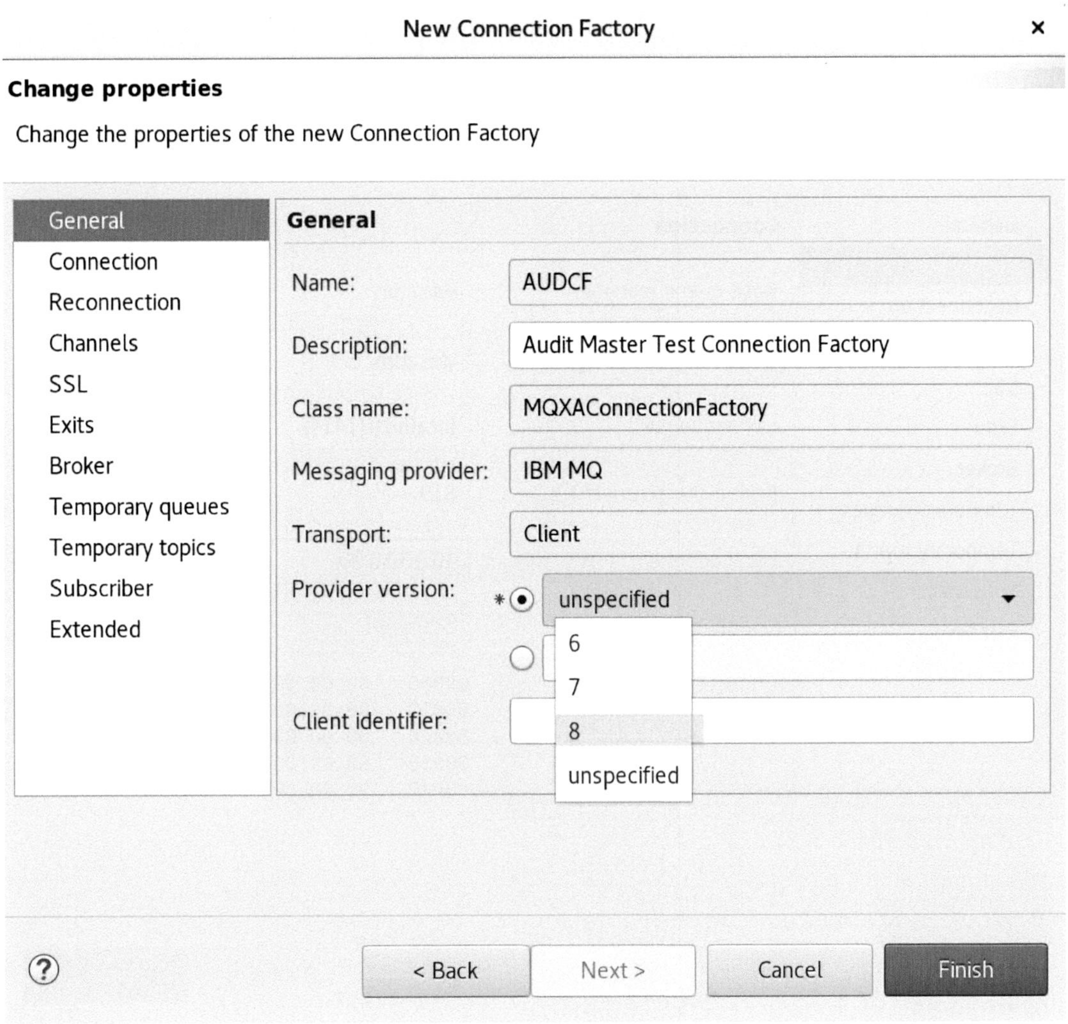

CHAPTER 3   IBM JMS INTERFACE DEVELOPMENT IBM FILENET 5.5.X WORKFLOW

Figure 3-201. The Description is added and a Provider Version selected as highlighted

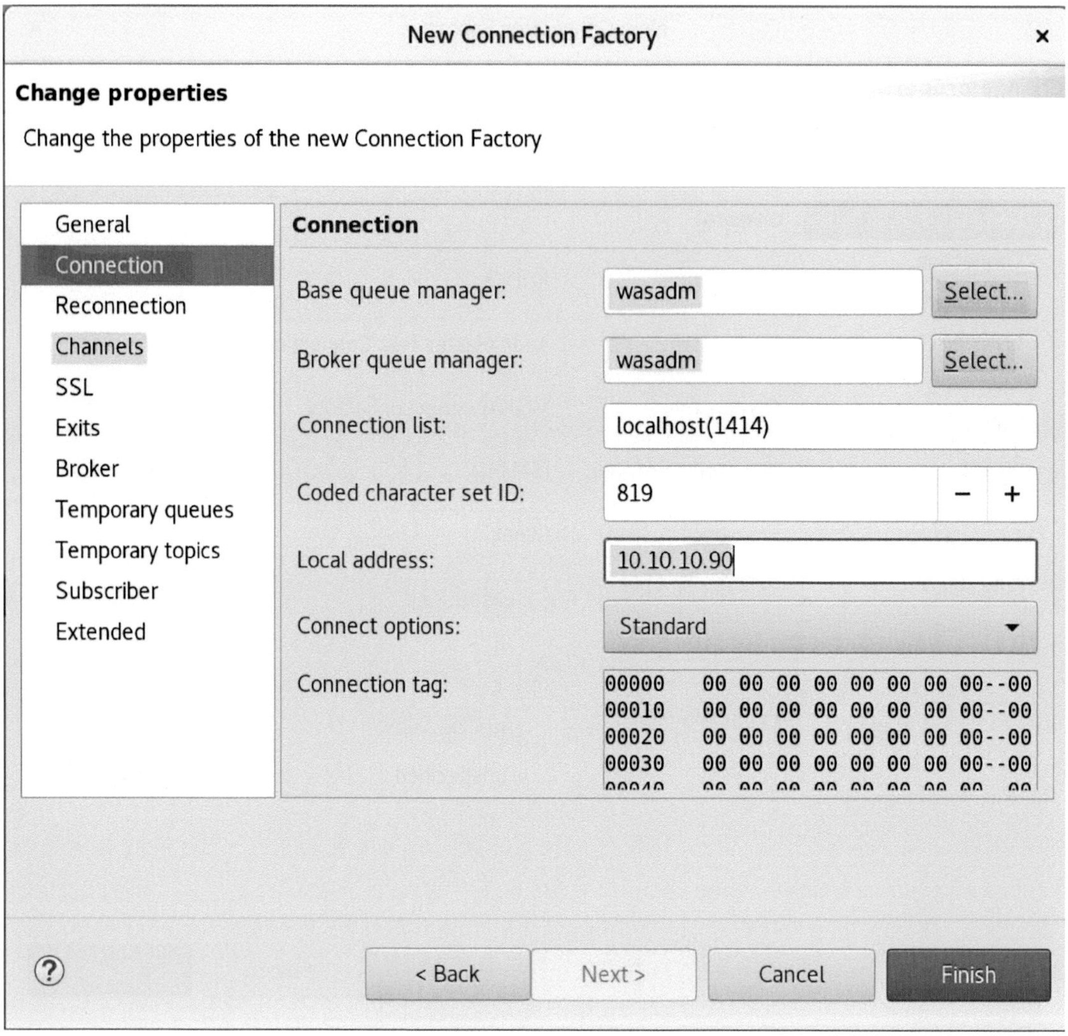

CHAPTER 3   IBM JMS INTERFACE DEVELOPMENT IBM FILENET 5.5.X WORKFLOW

Figure 3-202.   The Connection options are entered

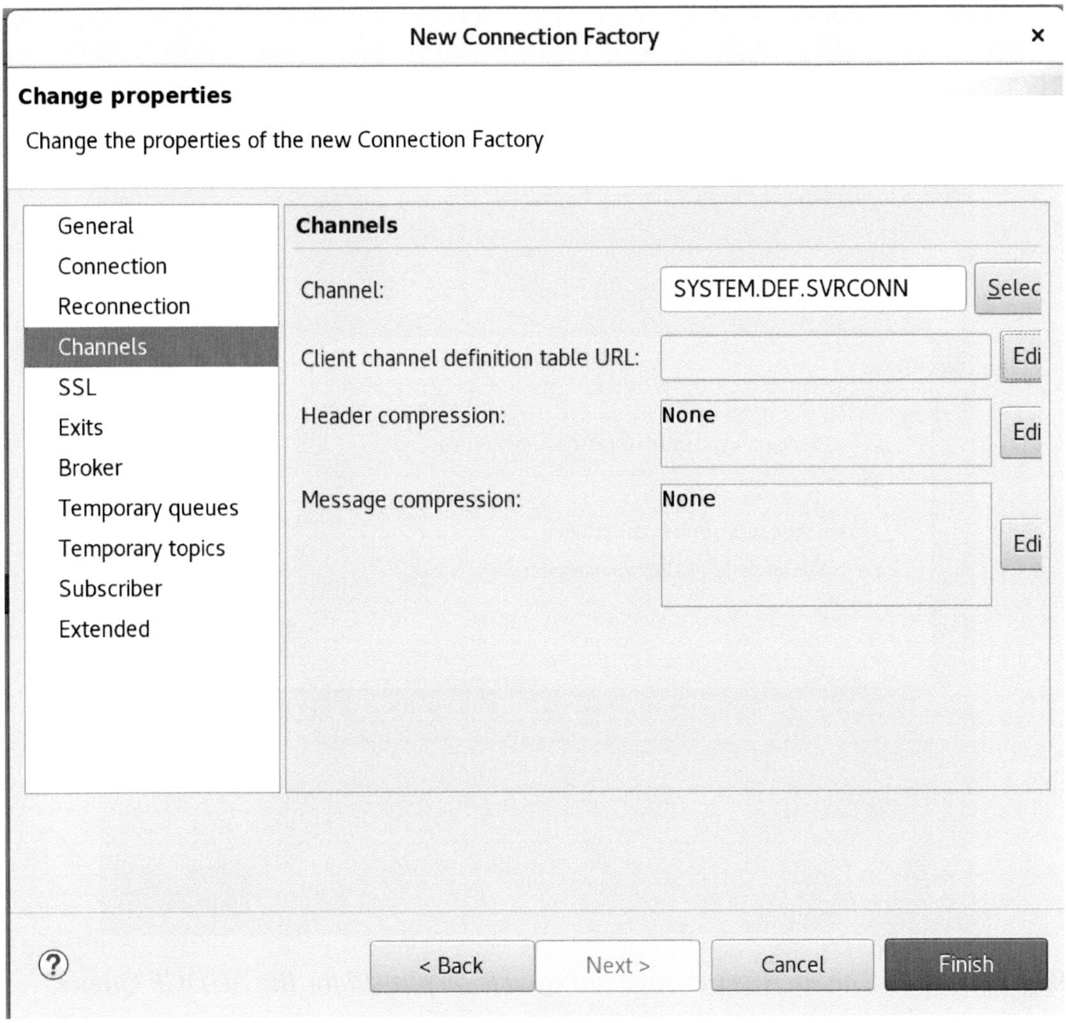
New Connection Factory
Change properties
Change the properties of the new Connection Factory
General
Connection
Reconnection
Channels
SSL
Exits
Broker
Temporary queues
Temporary topics
Subscriber
Extended
Channels
Channel:
SYSTEM.DEF.SVRCONN
Client channel definition table URL:
Header compression:
None
Message compression:
None
< Back
Next >
Cancel
Finish

Figure 3-203. *The Finish command is used to create the AUDCF Queue*

CHAPTER 3　IBM JMS INTERFACE DEVELOPMENT IBM FILENET 5.5.X WORKFLOW

The default Channel is used, and the Finish command is clicked to create the Queue.

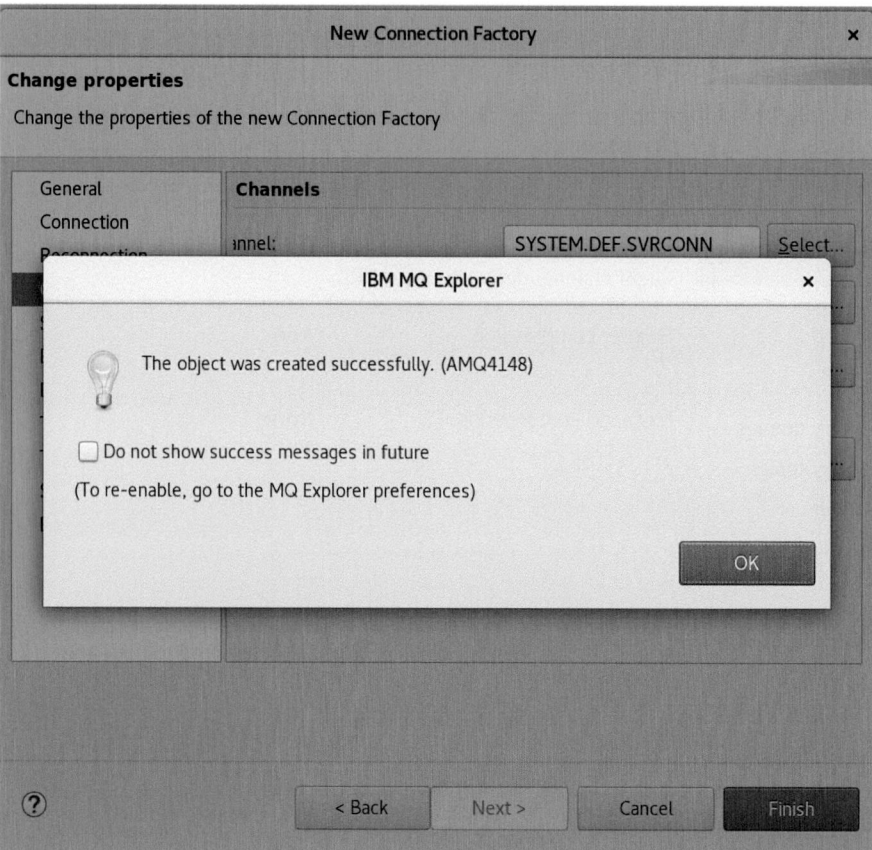

Figure 3-204. *The success message window is displayed for the AUDCF Queue*

CHAPTER 3   IBM JMS INTERFACE DEVELOPMENT IBM FILENET 5.5.X WORKFLOW

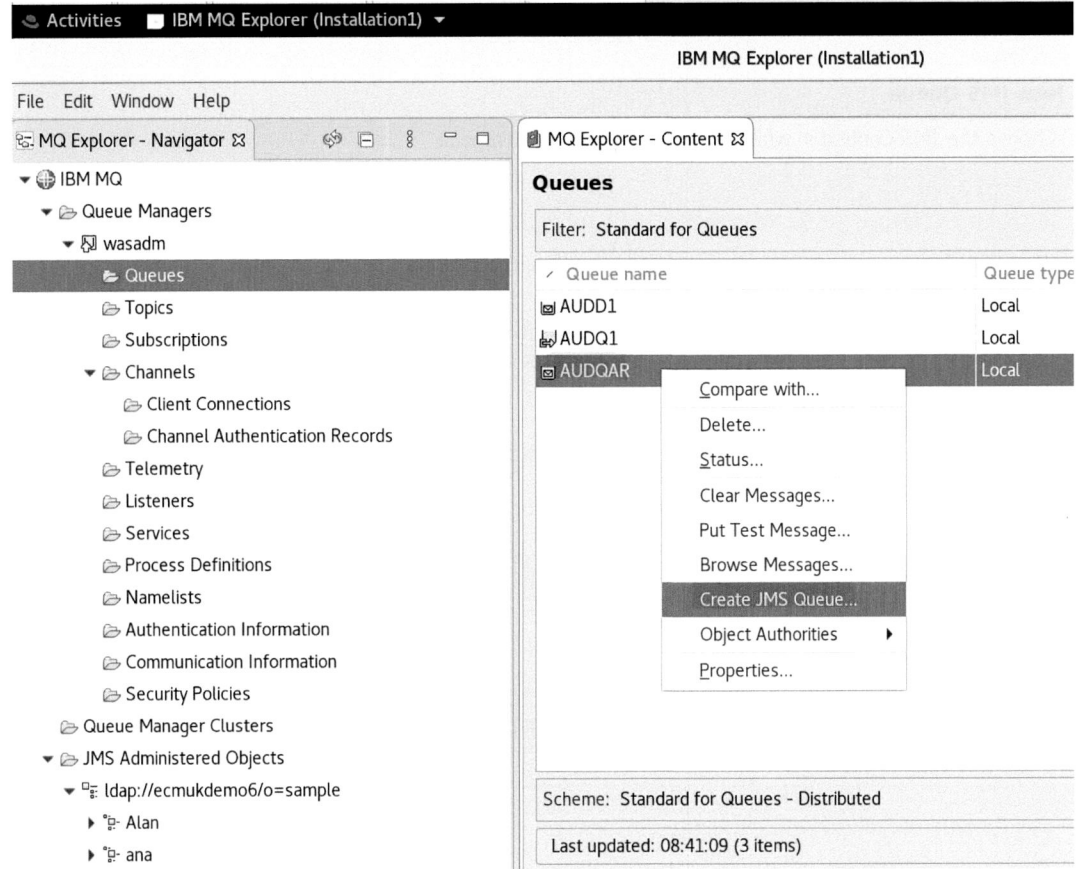

Figure 3-205. *The AUDQAR Queue is selected to create a JMS Queue*

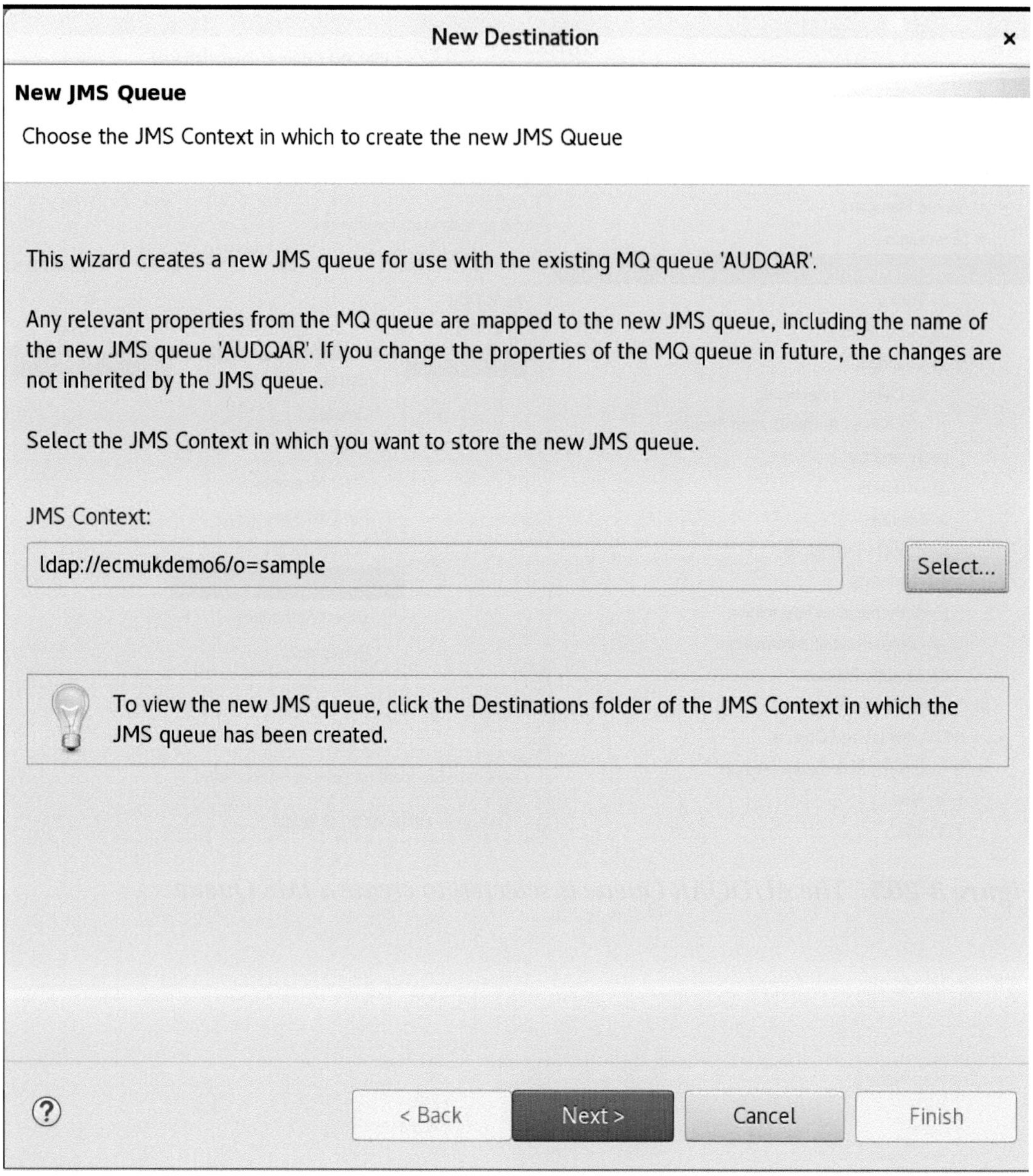

CHAPTER 3 IBM JMS INTERFACE DEVELOPMENT IBM FILENET 5.5.X WORKFLOW

New Destination	✕

New JMS Queue

Choose the JMS Context in which to create the new JMS Queue

This wizard creates a new JMS queue for use with the existing MQ queue 'AUDQAR'.

Any relevant properties from the MQ queue are mapped to the new JMS queue, including the name of the new JMS queue 'AUDQAR'. If you change the properties of the MQ queue in future, the changes are not inherited by the JMS queue.

Select the JMS Context in which you want to store the new JMS queue.

JMS Context:

ldap://ecmukdemo6/o=sample	Select...

💡 To view the new JMS queue, click the Destinations folder of the JMS Context in which the JMS queue has been created.

[?]   [< Back]   [Next >]   [Cancel]   [Finish]

Figure 3-206.* The JMS LDAP Context created earlier is linked to the AUDQAR Queue*

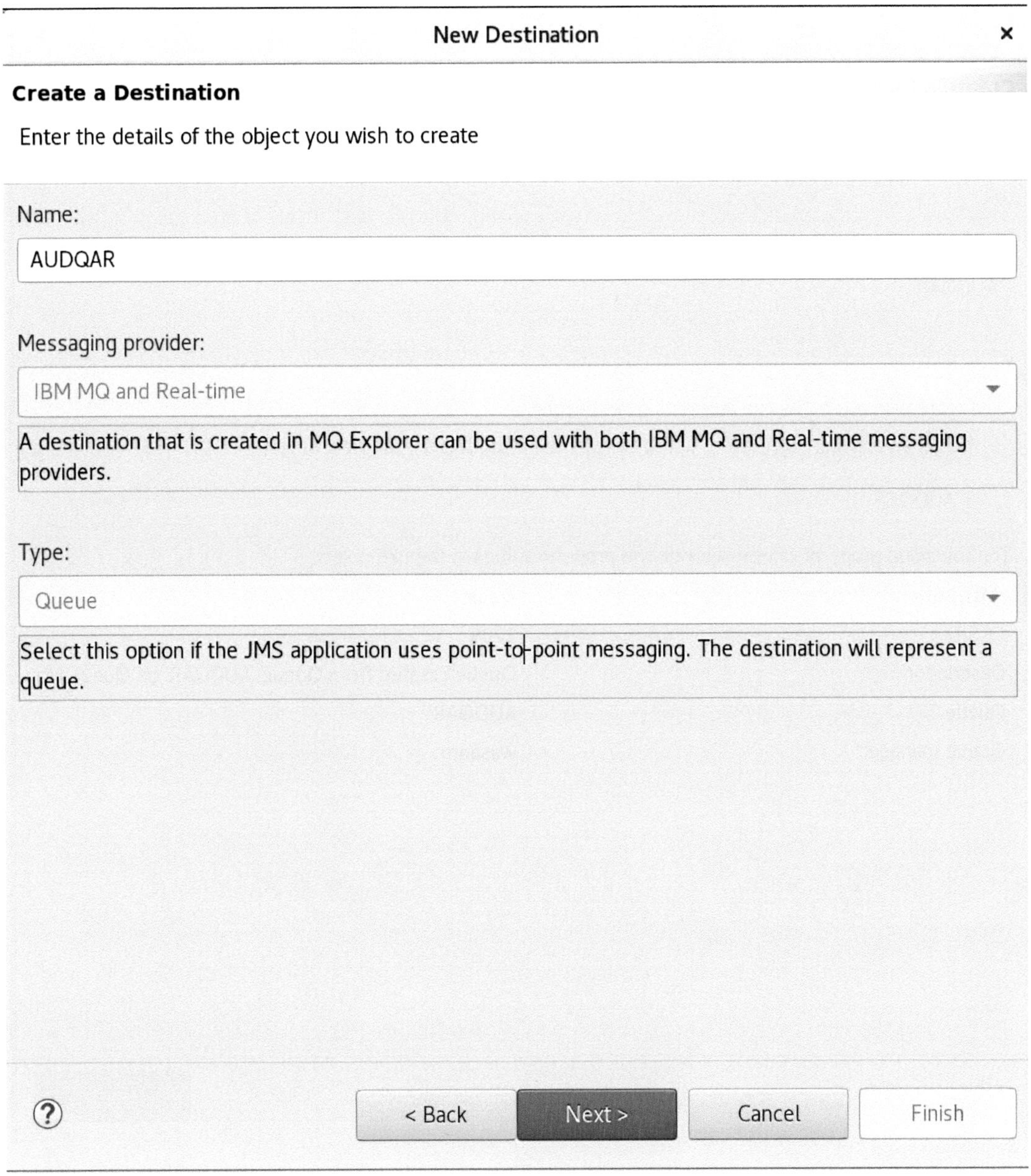

CHAPTER 3   IBM JMS INTERFACE DEVELOPMENT IBM FILENET 5.5.X WORKFLOW

New Destination   ×

Create a Destination

Enter the details of the object you wish to create

Name:

| AUDQAR |

Messaging provider:

| IBM MQ and Real-time ▼ |

A destination that is created in MQ Explorer can be used with both IBM MQ and Real-time messaging providers.

Type:

| Queue ▼ |

Select this option if the JMS application uses point-to-point messaging. The destination will represent a queue.

⑦   < Back   Next >   Cancel   Finish

Figure 3-207. *The JMS Object is set with the name **AUDQAR** and **Next>** clicked*

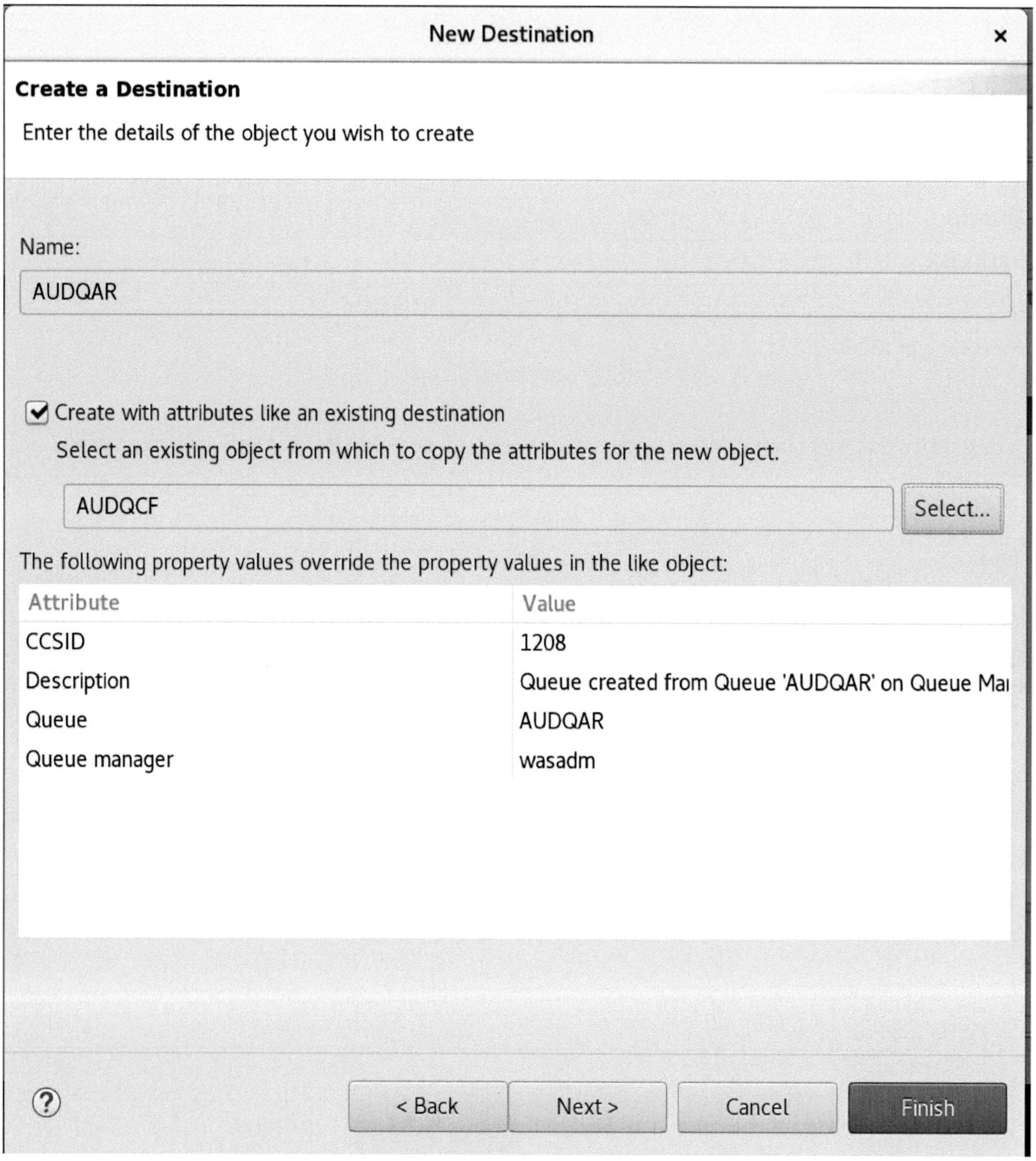

CHAPTER 3   IBM JMS INTERFACE DEVELOPMENT IBM FILENET 5.5.X WORKFLOW

New Destination

Create a Destination
Enter the details of the object you wish to create

Name:
AUDQAR

☑ Create with attributes like an existing destination
   Select an existing object from which to copy the attributes for the new object.
   AUDQCF   Select...

The following property values override the property values in the like object:

Attribute	Value
CCSID	1208
Description	Queue created from Queue 'AUDQAR' on Queue Mai
Queue	AUDQAR
Queue manager	wasadm

[< Back]   [Next >]   [Cancel]   [Finish]

Figure 3-208. *The Attributes to be created for the AUDQAR JMS entry are displayed*

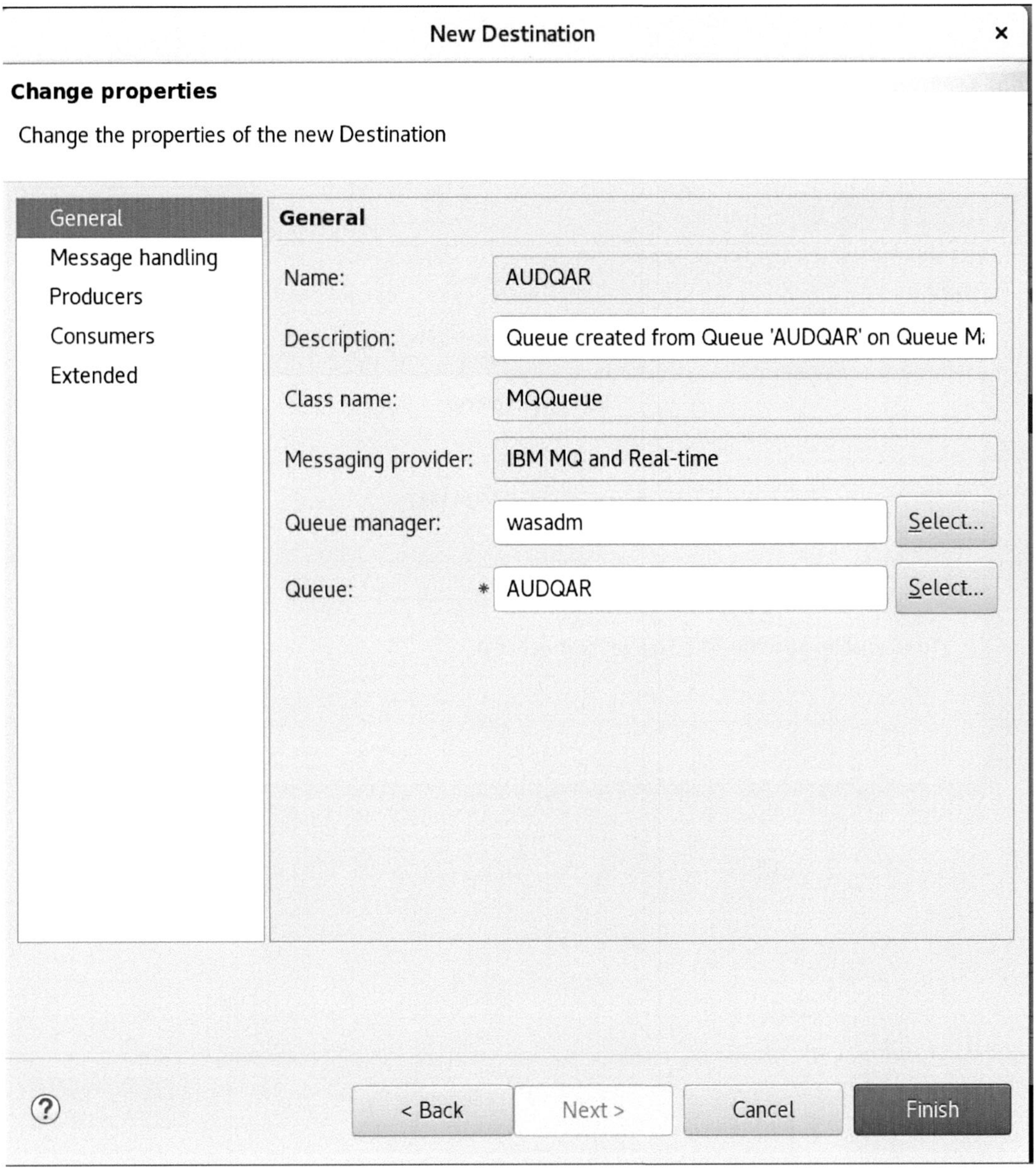

CHAPTER 3 IBM JMS INTERFACE DEVELOPMENT IBM FILENET 5.5.X WORKFLOW

New Destination

Change properties

Change the properties of the new Destination

- General
- Message handling
- Producers
- Consumers
- Extended

General

Name:	AUDQAR
Description:	Queue created from Queue 'AUDQAR' on Queue M:
Class name:	MQQueue
Messaging provider:	IBM MQ and Real-time
Queue manager:	wasadm [Select...]
Queue:	* AUDQAR [Select...]

[< Back] [Next >] [Cancel] [Finish]

Figure 3-209. The Finish command is clicked to create the AUDQAR JMS Queue

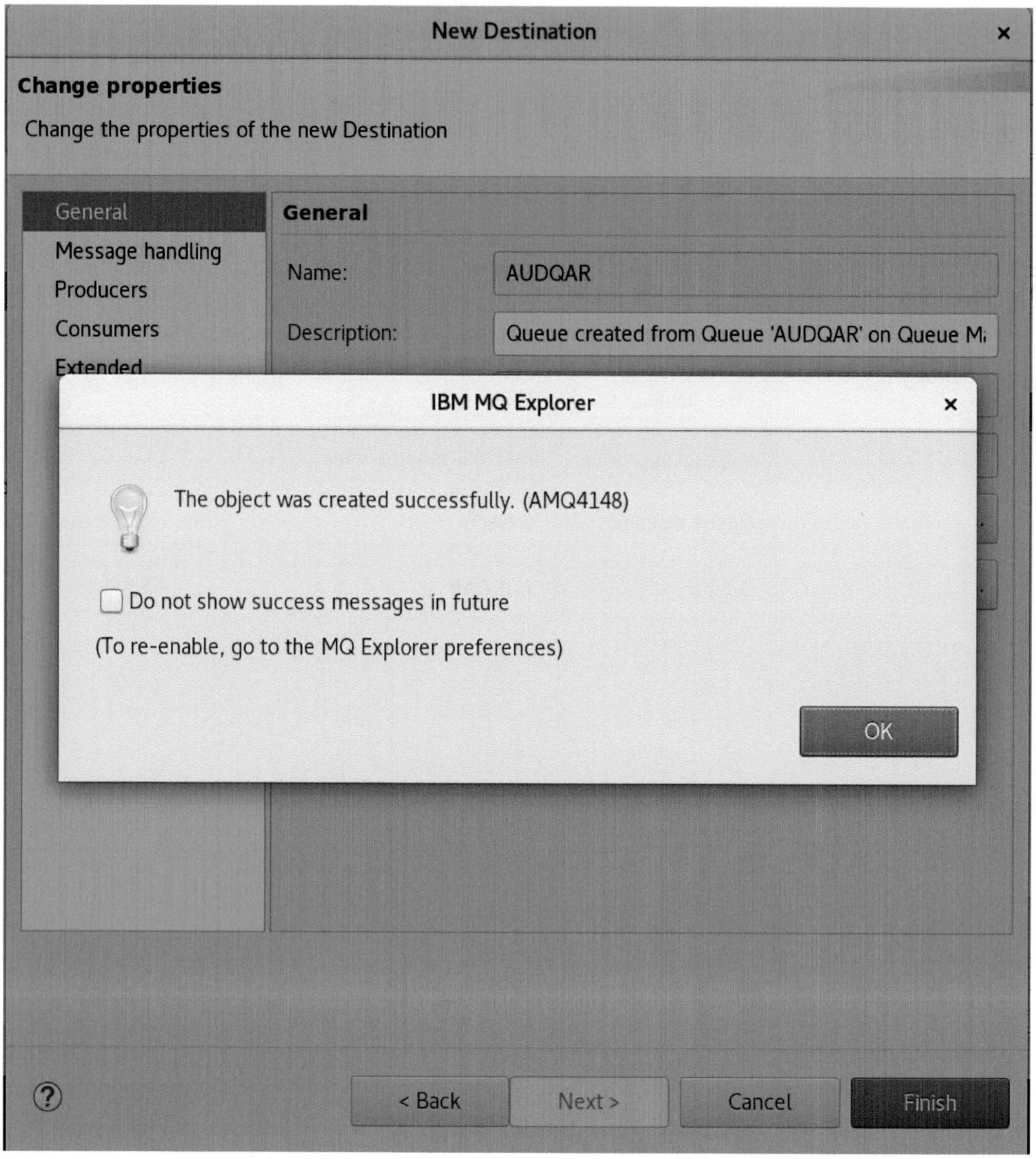

CHAPTER 3   IBM JMS INTERFACE DEVELOPMENT IBM FILENET 5.5.X WORKFLOW

Figure 3-210. The AUDQAR JMS Queue entry is shown as successfully created

CHAPTER 3　IBM JMS INTERFACE DEVELOPMENT IBM FILENET 5.5.X WORKFLOW

Figure 3-211. *The Find Accumulated Authorities are searched for the AUDQCF factory*

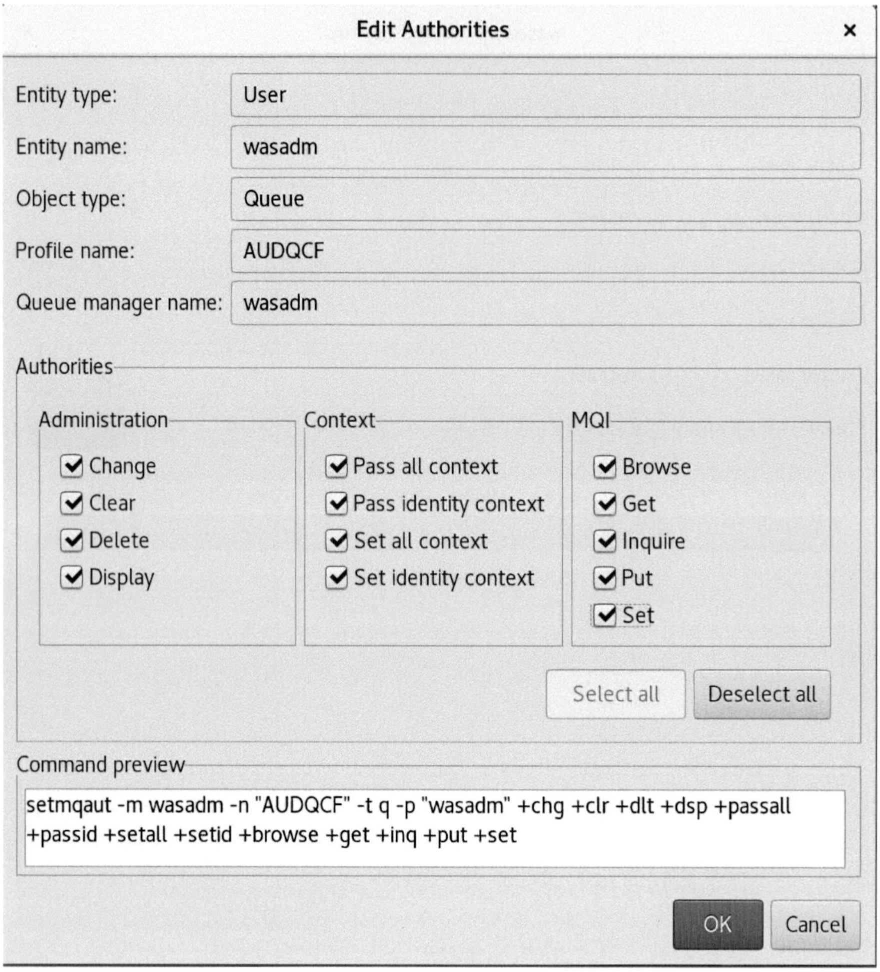

CHAPTER 3   IBM JMS INTERFACE DEVELOPMENT IBM FILENET 5.5.X WORKFLOW

Figure 3-212. The Edit Authorities attributes are all selected using the Select all button

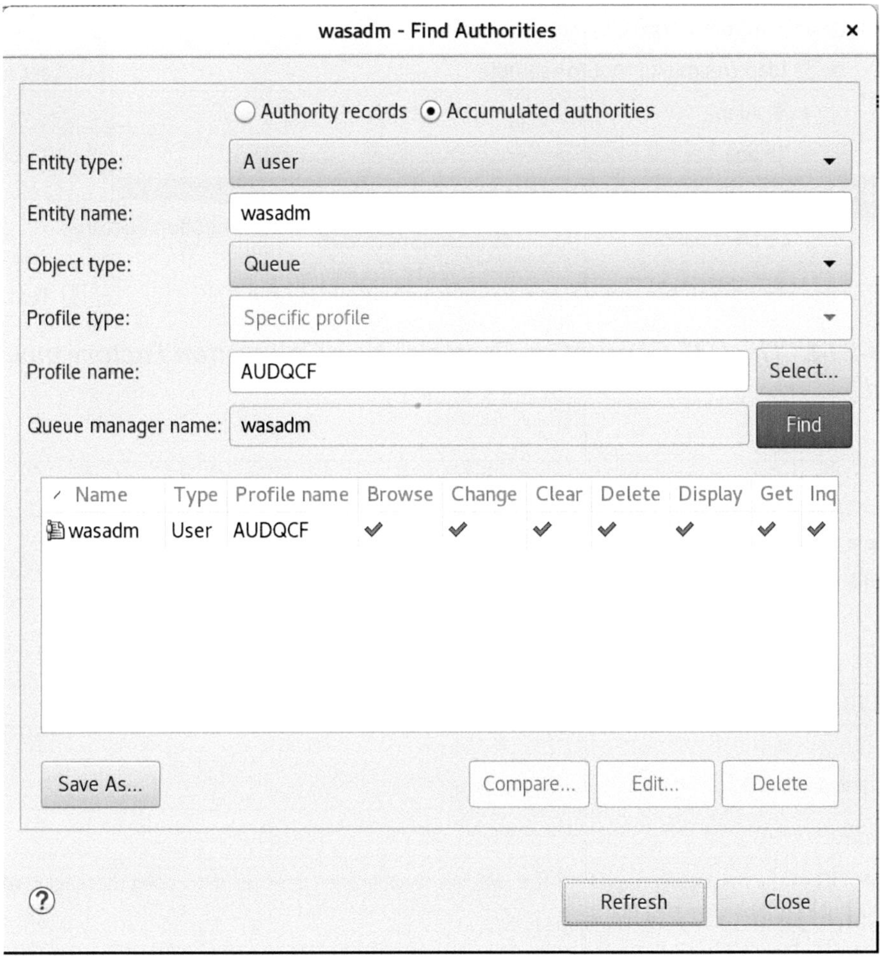

CHAPTER 3   IBM JMS INTERFACE DEVELOPMENT IBM FILENET 5.5.X WORKFLOW

Figure 3-213. *The wasadm user is shown to have the full set of security options*

CHAPTER 3 IBM JMS INTERFACE DEVELOPMENT IBM FILENET 5.5.X WORKFLOW

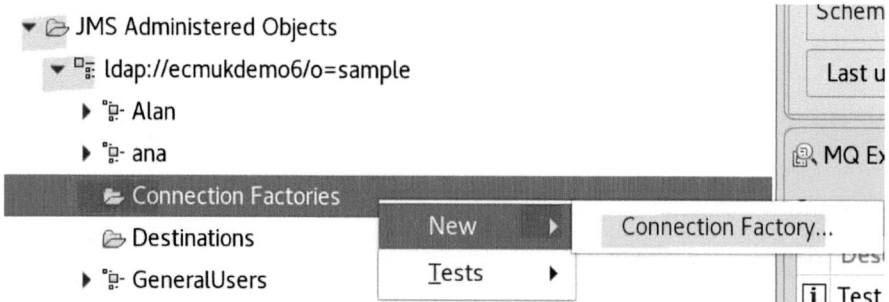

Figure 3-214. *The JMS Connection Factories New Connection Factory menu is clicked*

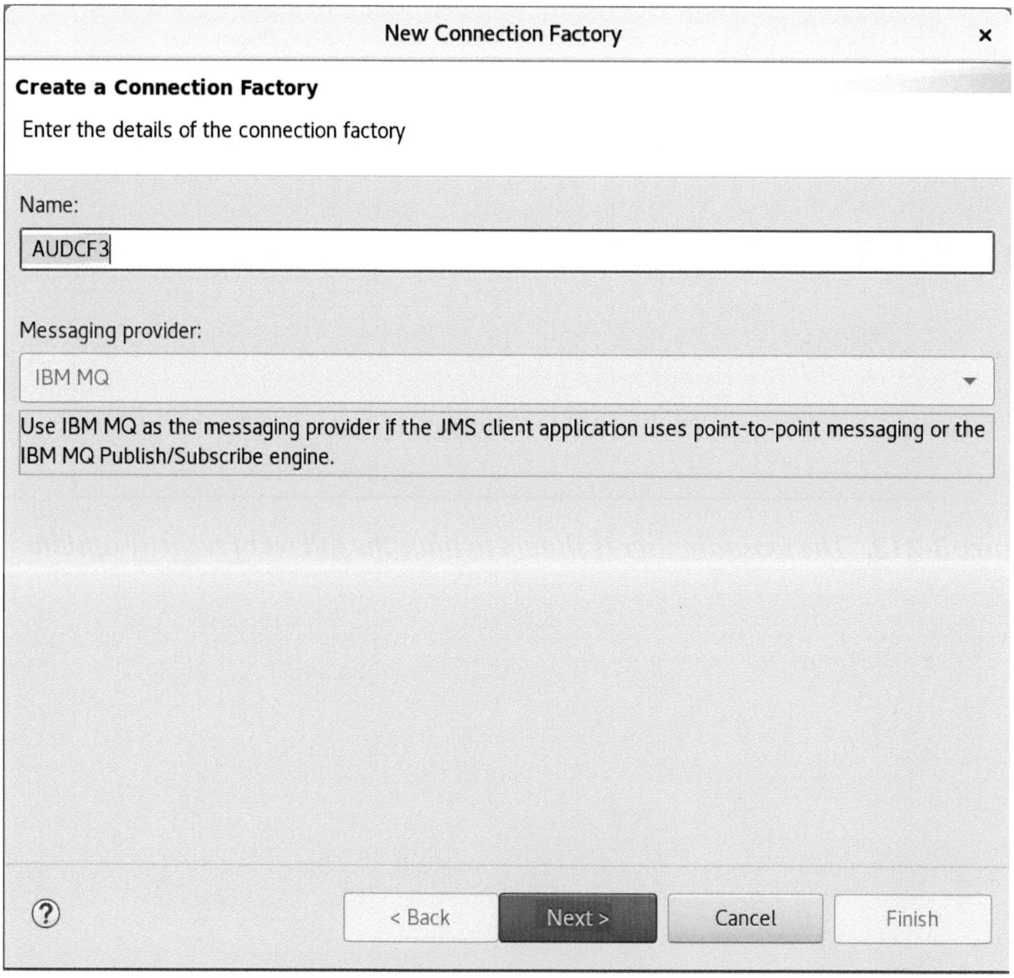

Figure 3-215. *The name of the new Connection Factory is set as AUDCF3*

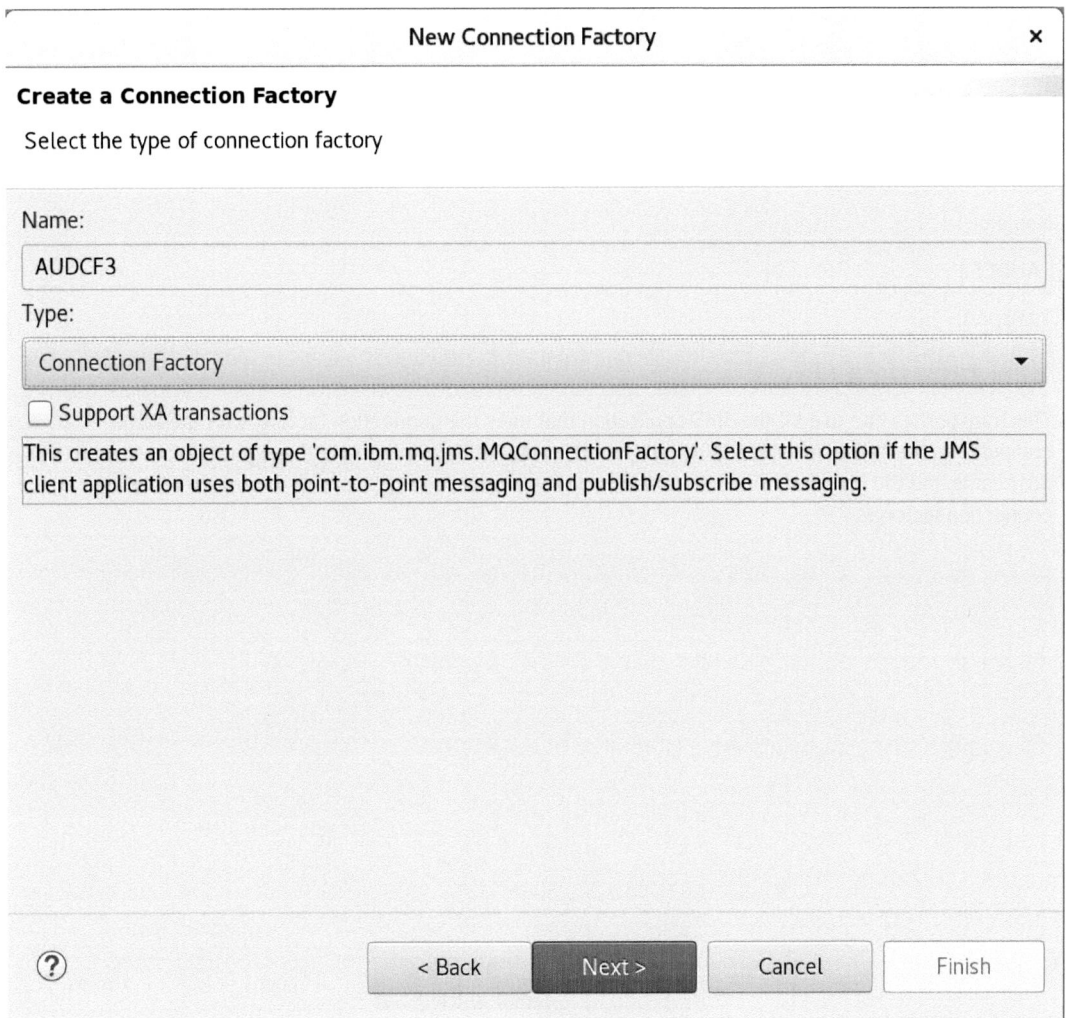

Figure 3-216. *The AUDCF3 Connection Factory is set without the Support XA Transactions option*

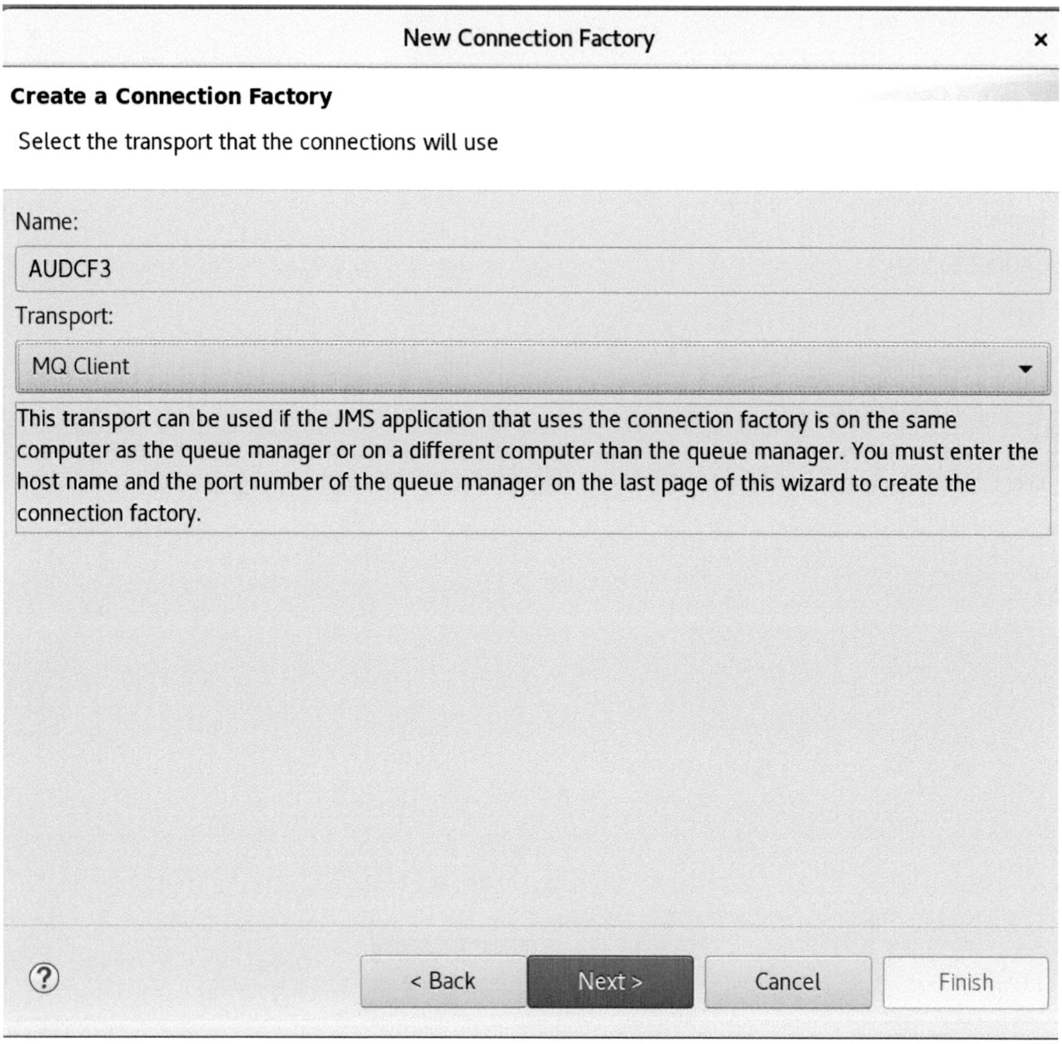

CHAPTER 3 IBM JMS INTERFACE DEVELOPMENT IBM FILENET 5.5.X WORKFLOW

New Connection Factory   ✕

Create a Connection Factory

Select the transport that the connections will use

Name:

AUDCF3

Transport:

MQ Client   ▼

This transport can be used if the JMS application that uses the connection factory is on the same computer as the queue manager or on a different computer than the queue manager. You must enter the host name and the port number of the queue manager on the last page of this wizard to create the connection factory.

⓷   < Back   Next >   Cancel   Finish

Figure 3-217. The AUDCF3 Connection Factory attributes are displayed and Next> clicked

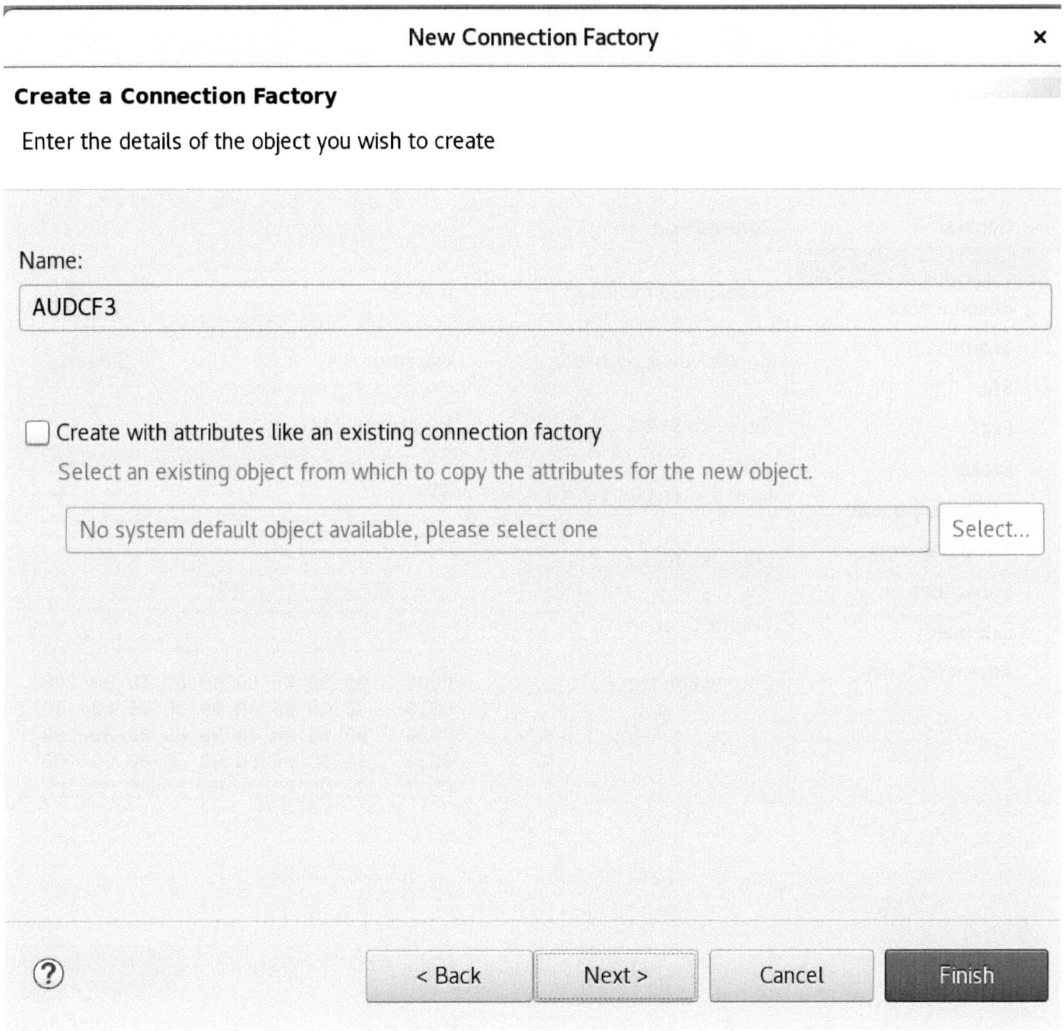

Figure 3-218. The AUDCF3 Connection Factory defaults are used

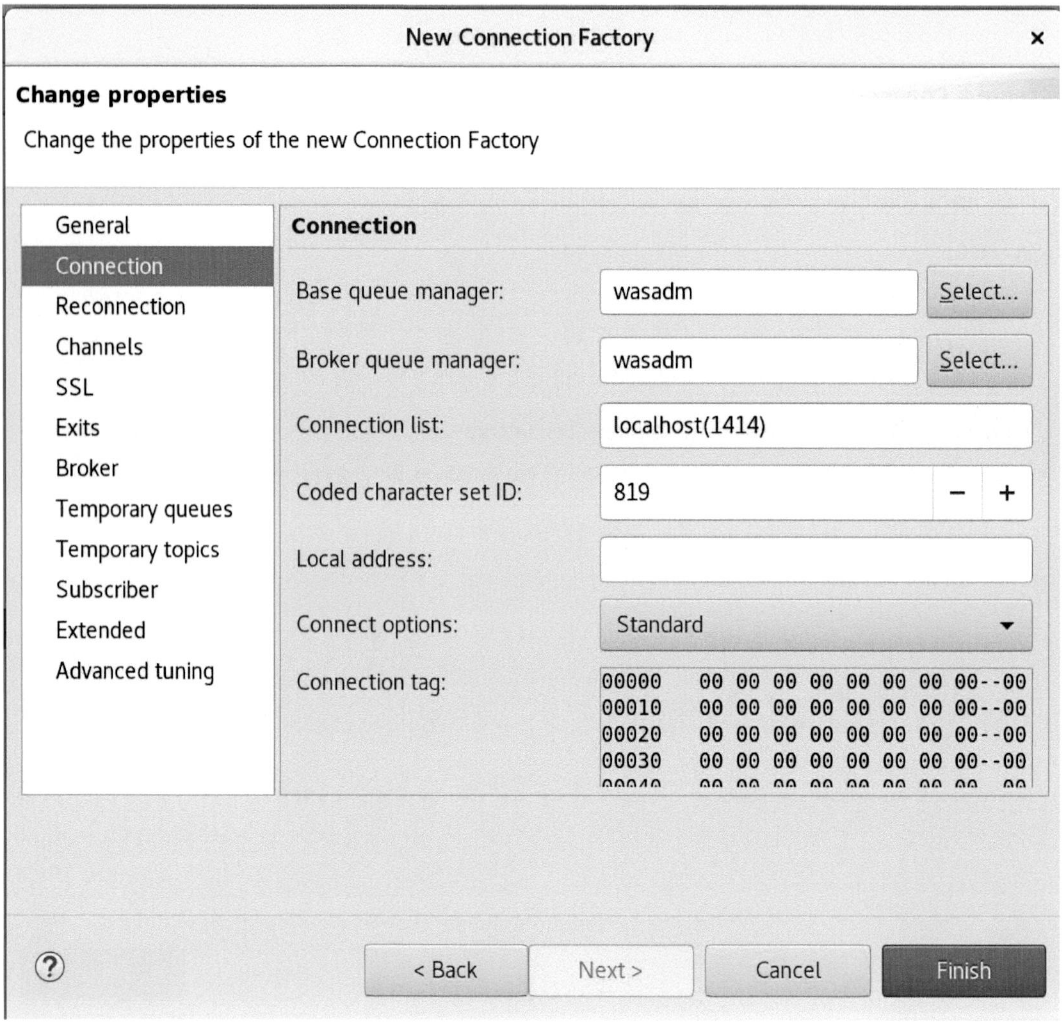

CHAPTER 3   IBM JMS INTERFACE DEVELOPMENT IBM FILENET 5.5.X WORKFLOW

Figure 3-219. The Connection details are set

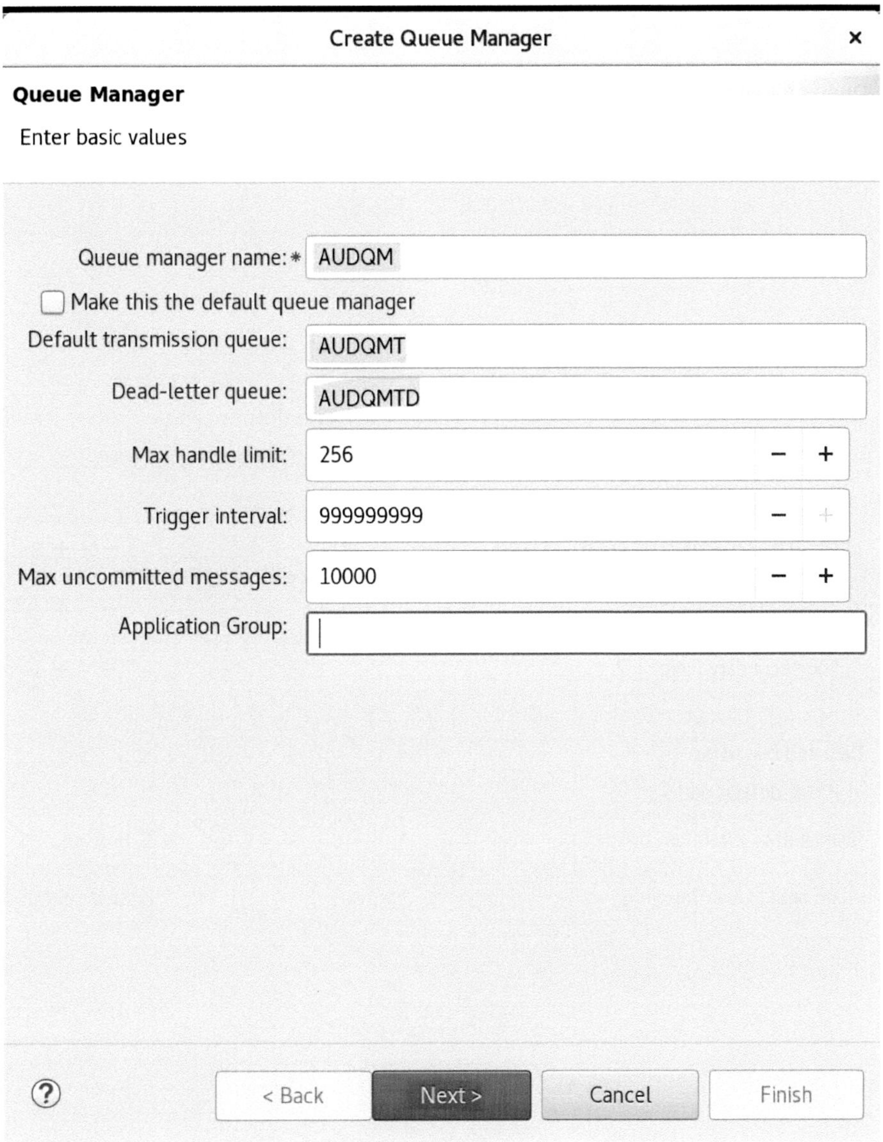

CHAPTER 3　IBM JMS INTERFACE DEVELOPMENT IBM FILENET 5.5.X WORKFLOW

Figure 3-220. *The AUDQM manager and Transmission and Dead-Letter Queue details are entered*

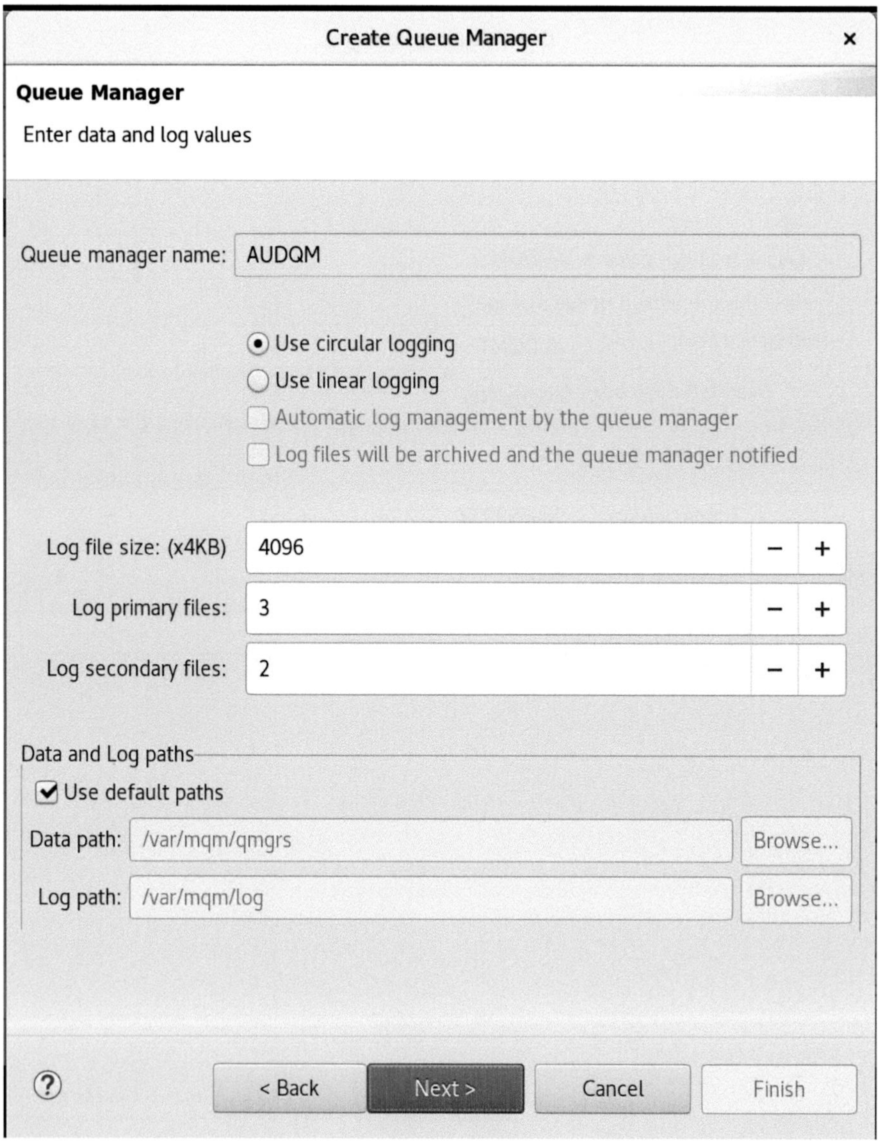

Figure 3-221. The logging details are left as the defaults and Next> clicked

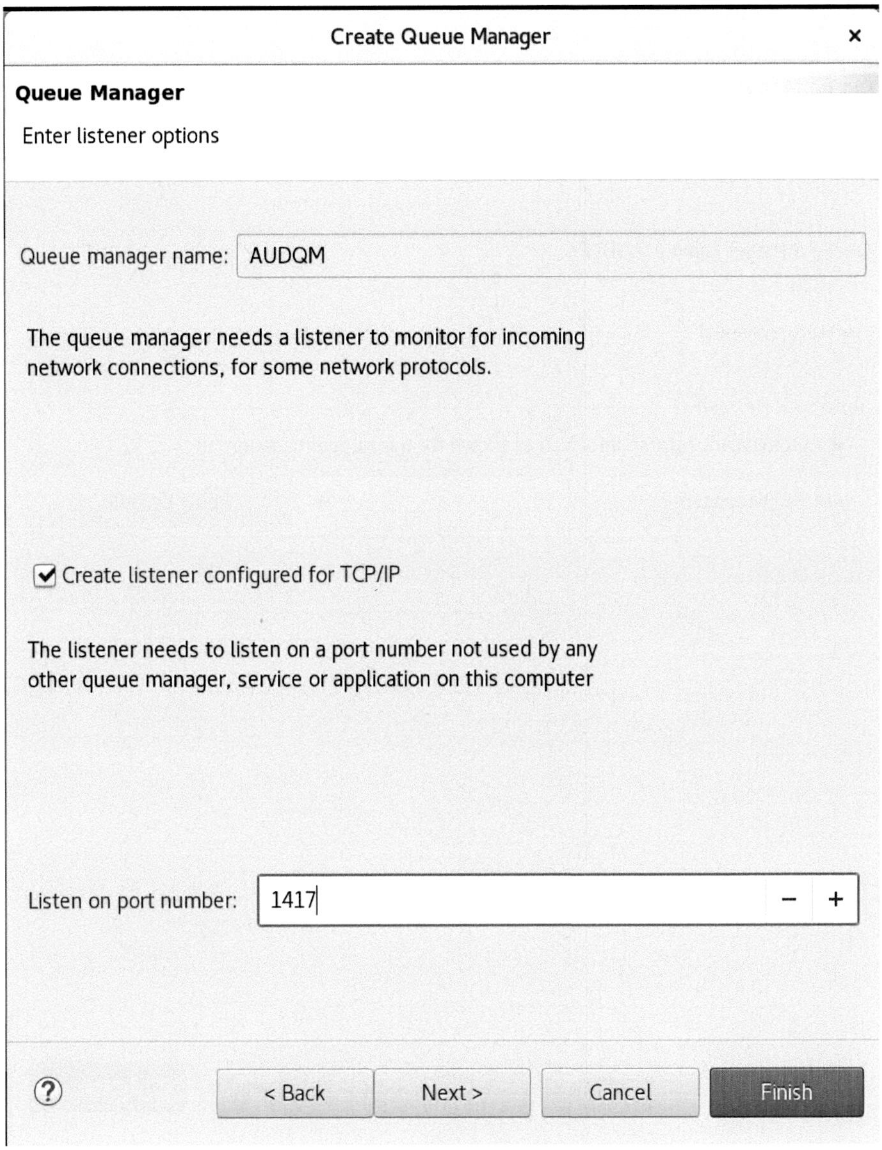

CHAPTER 3 IBM JMS INTERFACE DEVELOPMENT IBM FILENET 5.5.X WORKFLOW

Figure 3-222. *The Listener option is ticked for TCP/IP, and the port number is set to 1417*

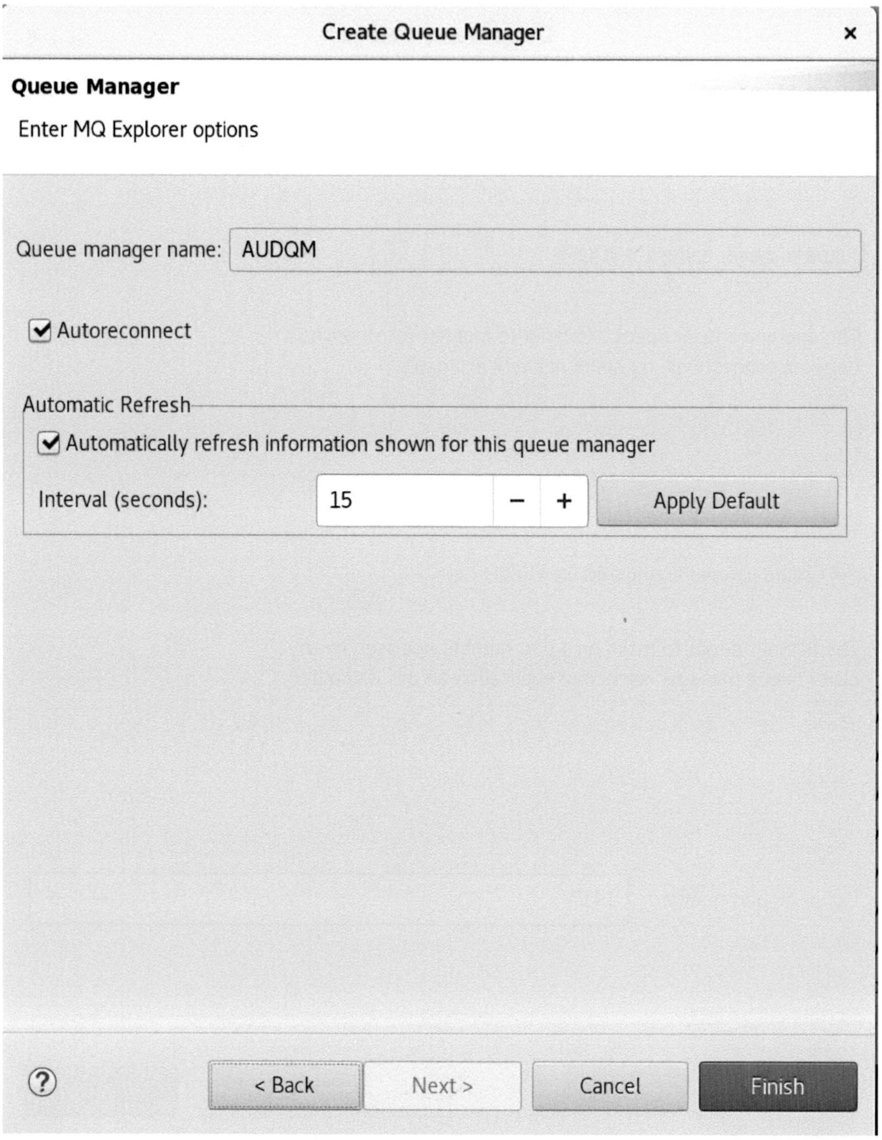

CHAPTER 3 IBM JMS INTERFACE DEVELOPMENT IBM FILENET 5.5.X WORKFLOW

Figure 3-223. The autoreconnect and refresh options are set with 15-second default

CHAPTER 3   IBM JMS INTERFACE DEVELOPMENT IBM FILENET 5.5.X WORKFLOW

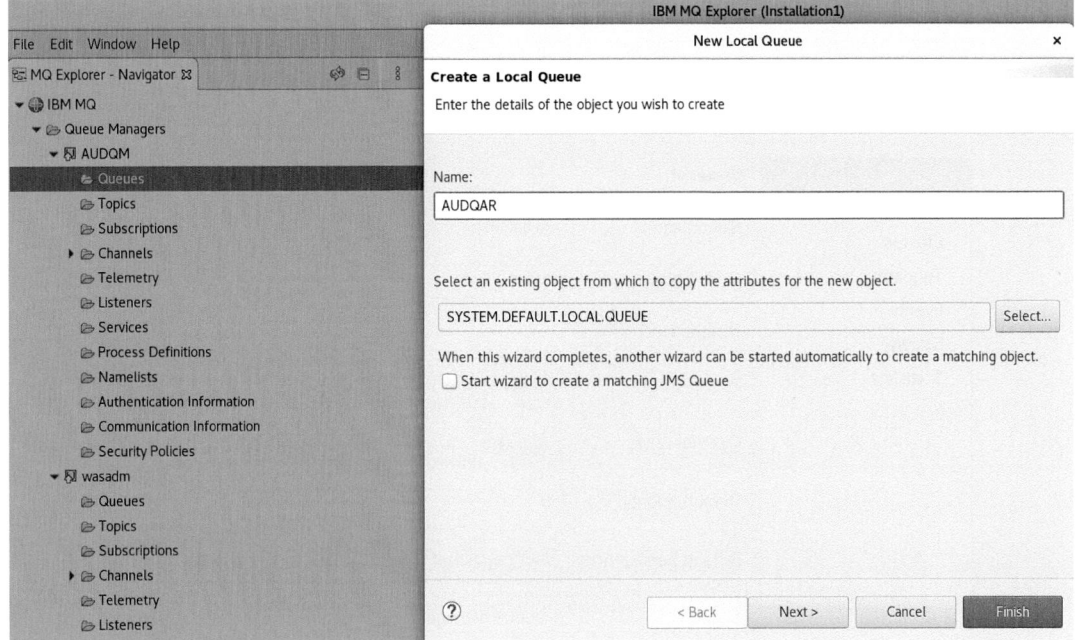

Figure 3-224. *The attributes of the Queue are based on the default Local Queue*

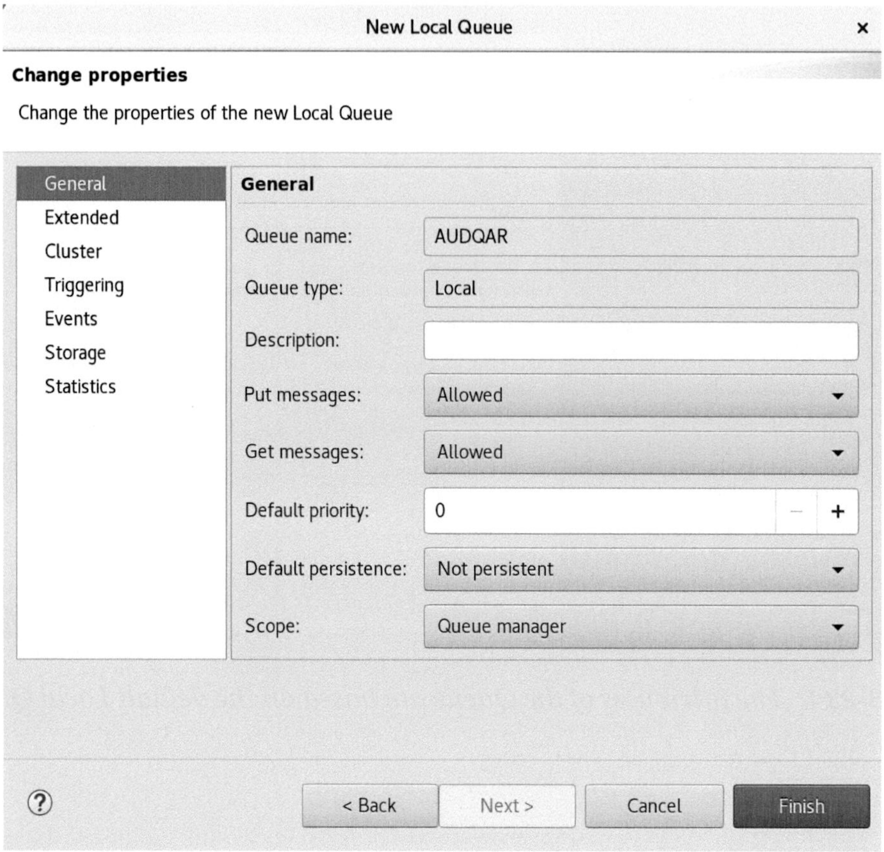

CHAPTER 3   IBM JMS INTERFACE DEVELOPMENT IBM FILENET 5.5.X WORKFLOW

Figure 3-225. *The Finish command is used to create the AUDQAR Local Queue*

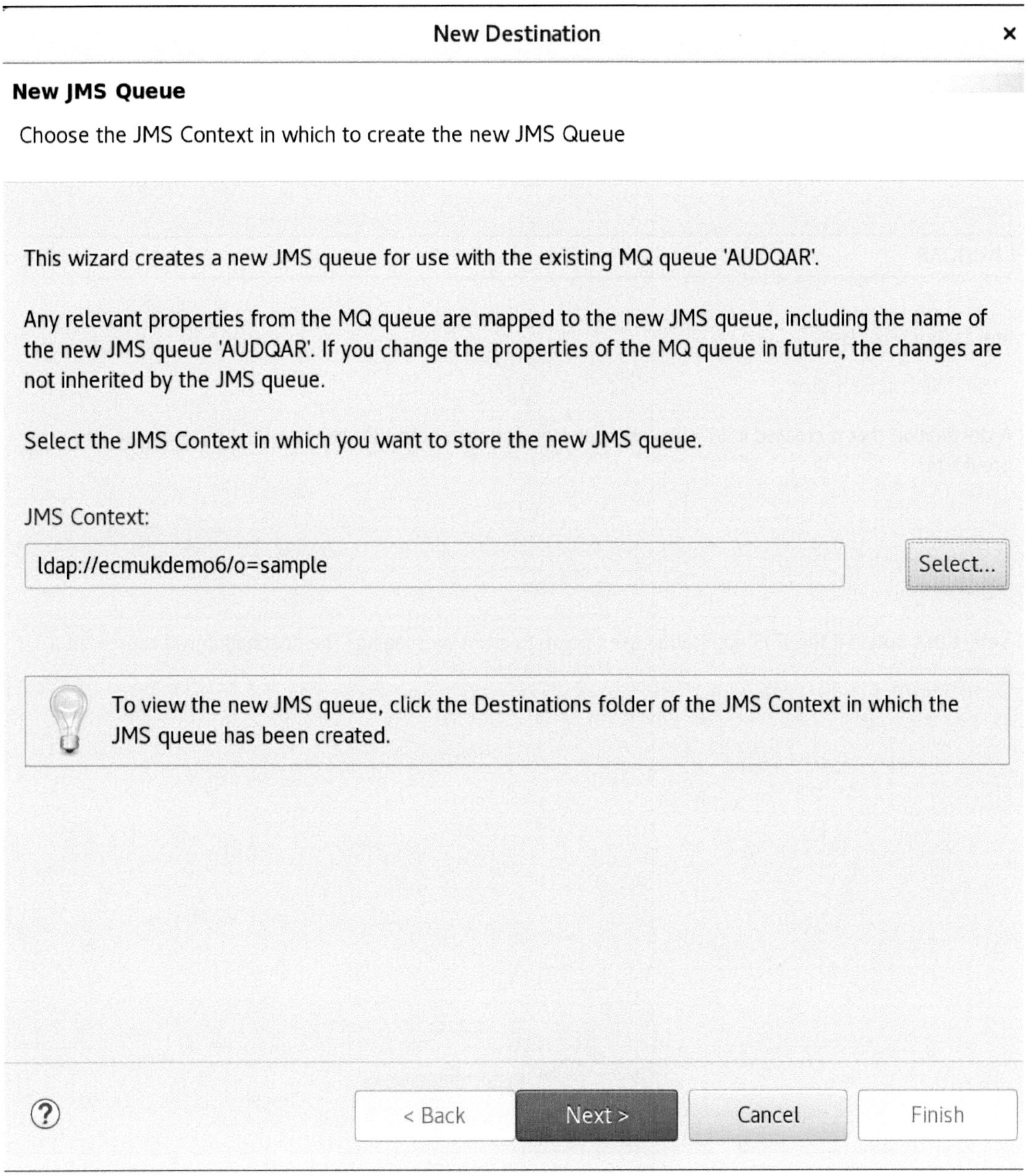

CHAPTER 3 IBM JMS INTERFACE DEVELOPMENT IBM FILENET 5.5.X WORKFLOW

New Destination   ✕

New JMS Queue

Choose the JMS Context in which to create the new JMS Queue

This wizard creates a new JMS queue for use with the existing MQ queue 'AUDQAR'.

Any relevant properties from the MQ queue are mapped to the new JMS queue, including the name of the new JMS queue 'AUDQAR'. If you change the properties of the MQ queue in future, the changes are not inherited by the JMS queue.

Select the JMS Context in which you want to store the new JMS queue.

JMS Context:

| ldap://ecmukdemo6/o=sample | Select... |

💡 To view the new JMS queue, click the Destinations folder of the JMS Context in which the JMS queue has been created.

❓   < Back   Next >   Cancel   Finish

Figure 3-226. *The JMS Queue Creation wizard is launched automatically for AUDQAR*

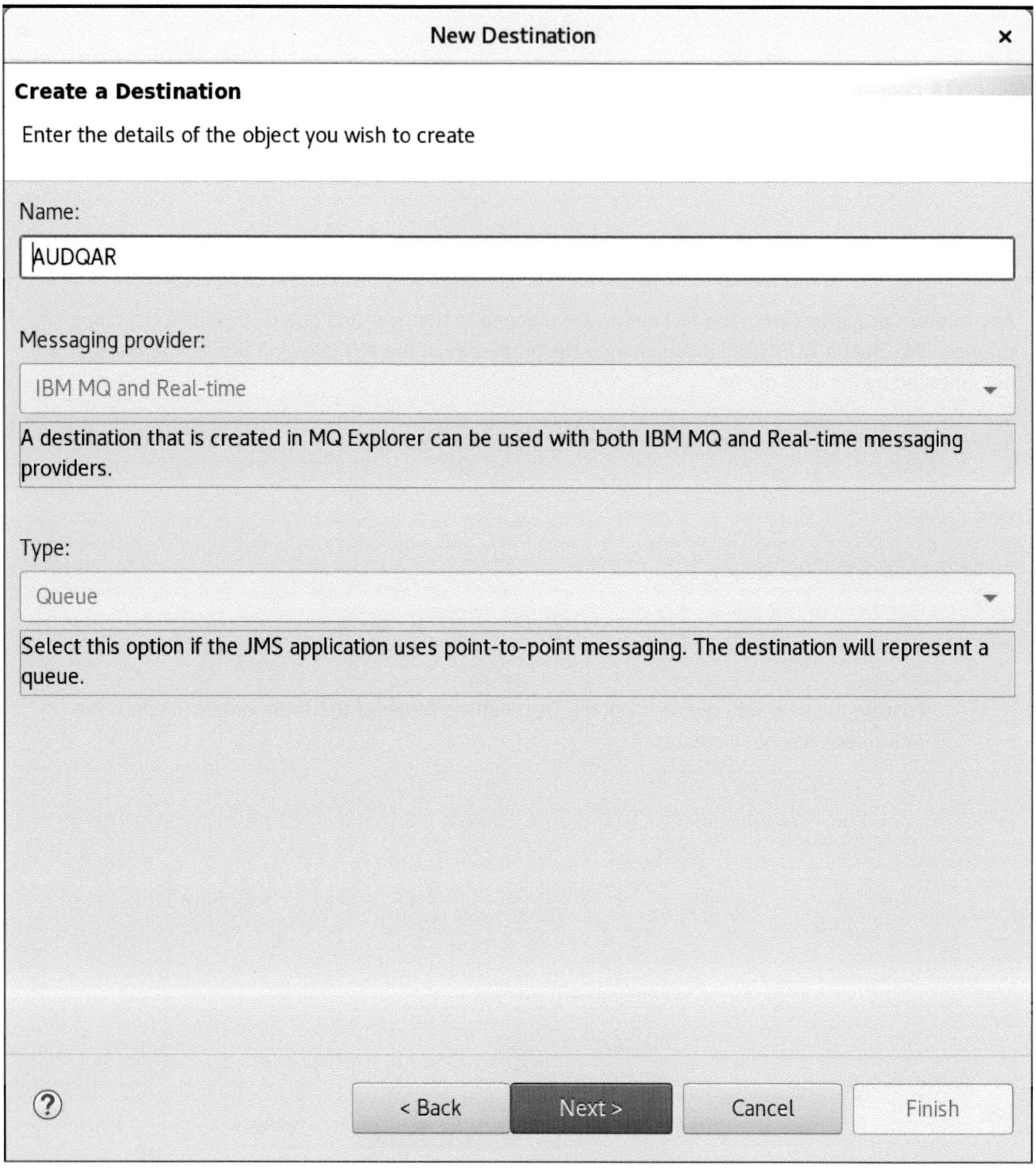

CHAPTER 3 IBM JMS INTERFACE DEVELOPMENT IBM FILENET 5.5.X WORKFLOW

Figure 3-227. The JMS Queue name, AUDQAR, is left as for the Local Queue name

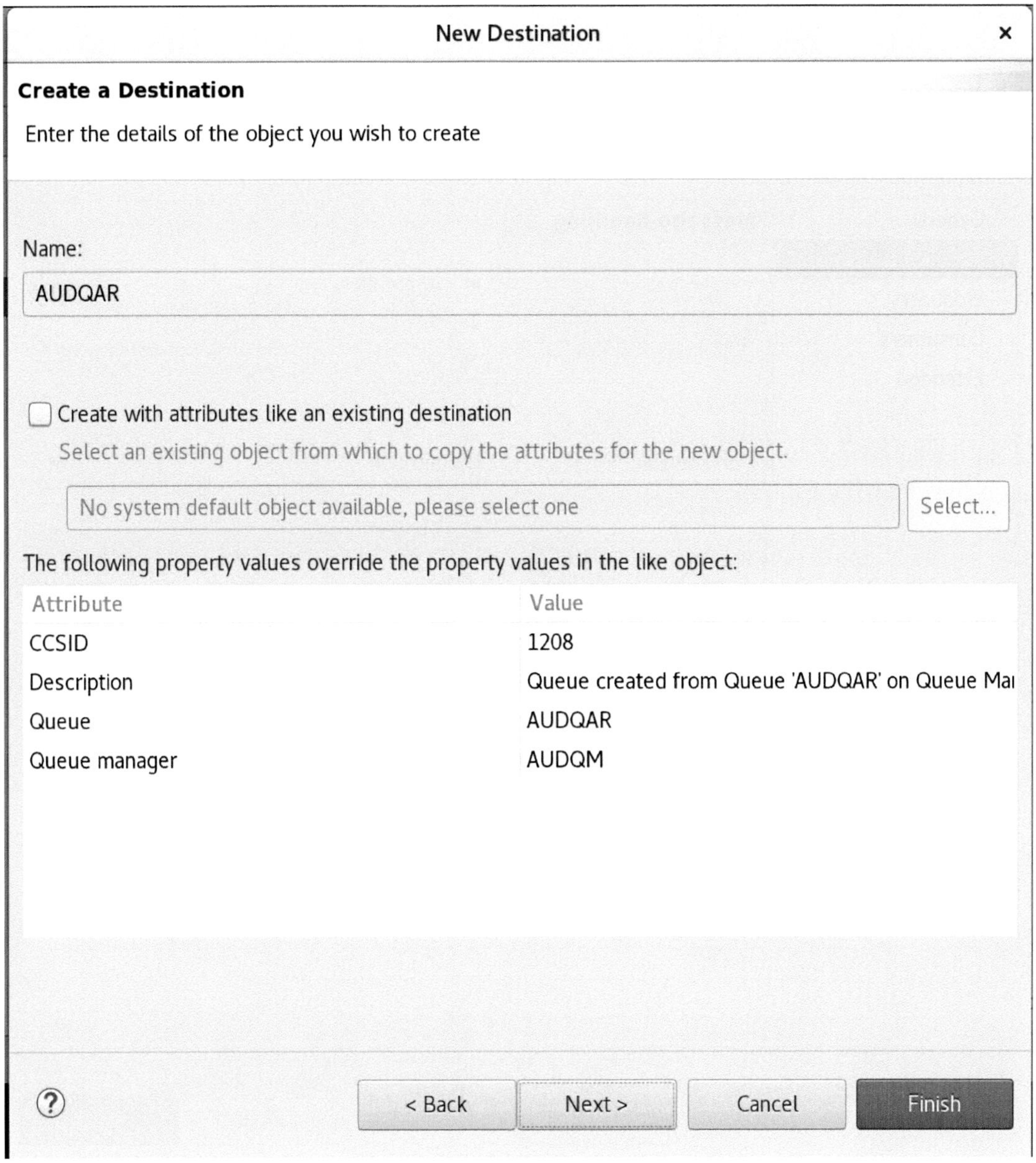

New Destination	✕

Create a Destination

Enter the details of the object you wish to create

Name:

AUDQAR

☐ Create with attributes like an existing destination

Select an existing object from which to copy the attributes for the new object.

| No system default object available, please select one | Select... |

The following property values override the property values in the like object:

Attribute	Value
CCSID	1208
Description	Queue created from Queue 'AUDQAR' on Queue Mar
Queue	AUDQAR
Queue manager	AUDQM

[< Back] [Next >] [Cancel] [Finish]

Figure 3-228. *The AUDQAR JMS Queue details are displayed for review*

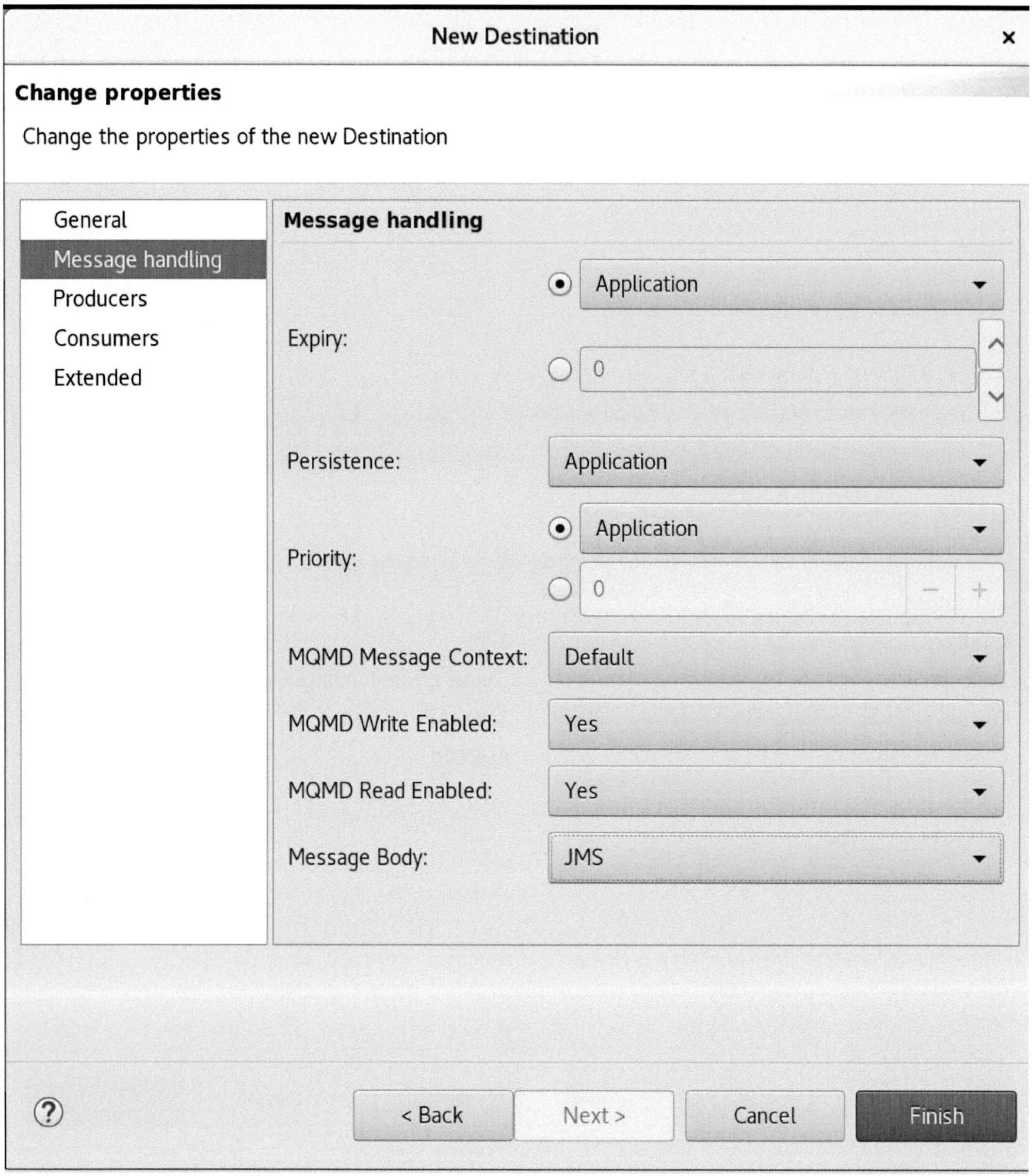

CHAPTER 3 IBM JMS INTERFACE DEVELOPMENT IBM FILENET 5.5.X WORKFLOW

Figure 3-229. *The Message handling properties are set as shown and reviewed*

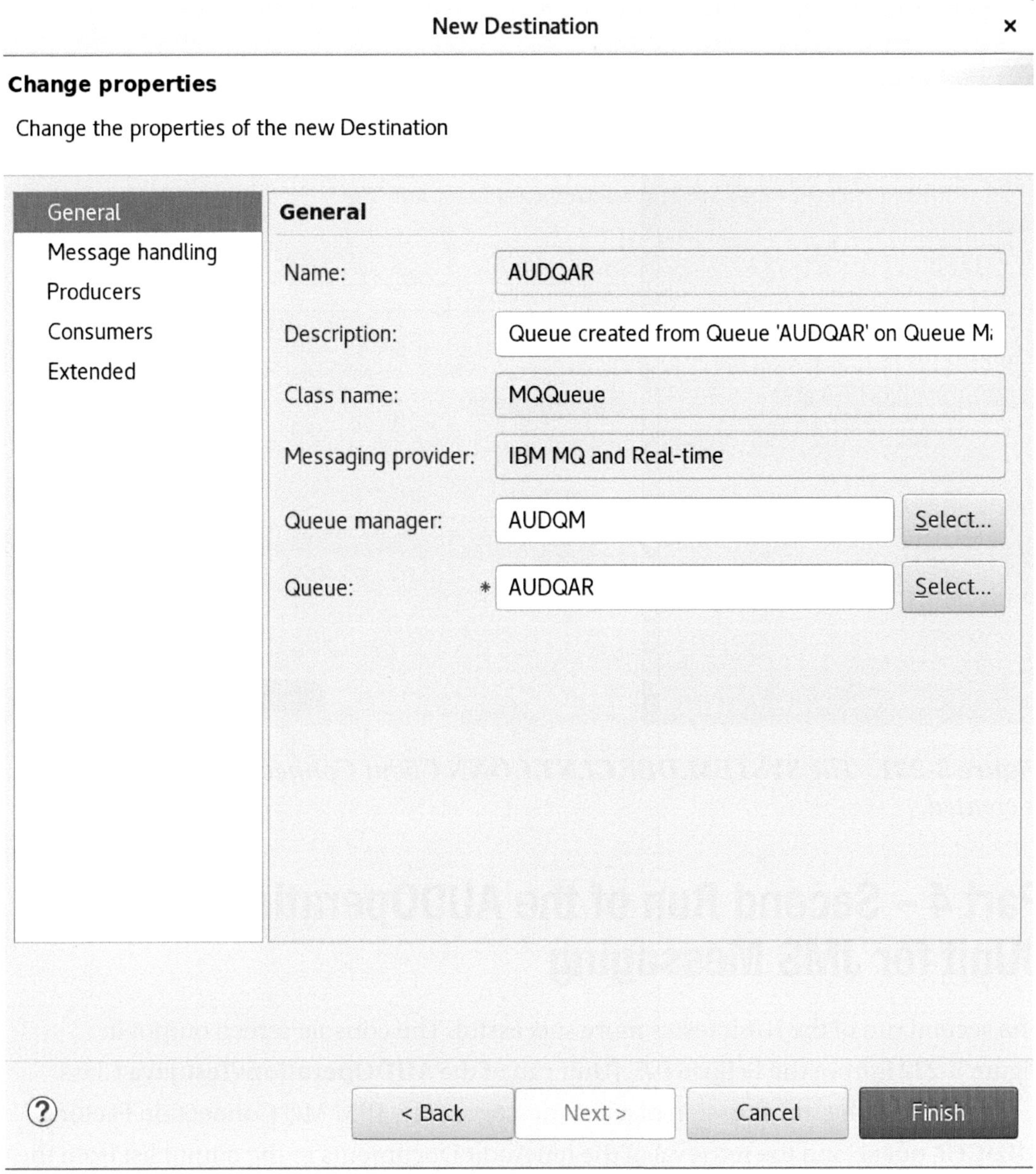

CHAPTER 3   IBM JMS INTERFACE DEVELOPMENT IBM FILENET 5.5.X WORKFLOW

Change properties

Change the properties of the new Destination

General	
Name:	AUDQAR
Description:	Queue created from Queue 'AUDQAR' on Queue Mi
Class name:	MQQueue
Messaging provider:	IBM MQ and Real-time
Queue manager:	AUDQM
Queue:	* AUDQAR

Sidebar: General, Message handling, Producers, Consumers, Extended

Figure 3-230. *The Queue Description is entered and Finish clicked to create AUDQAR*

CHAPTER 3   IBM JMS INTERFACE DEVELOPMENT IBM FILENET 5.5.X WORKFLOW

A new Client Connection Channel is created as shown in Figure 3-231.

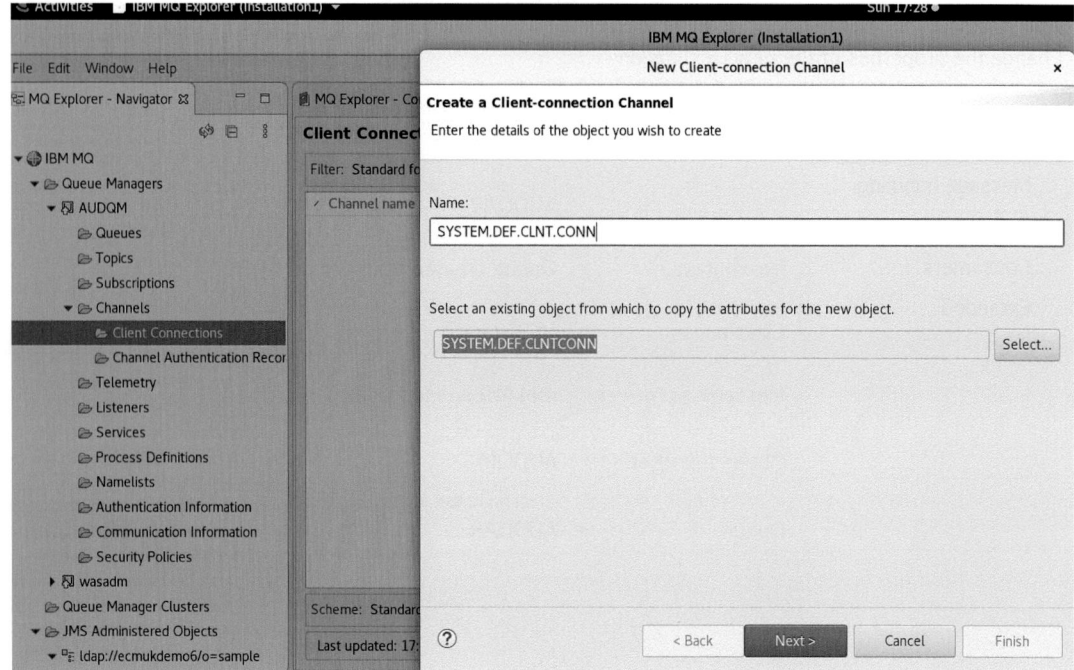

*Figure 3-231. The **SYSTEM.DEF.CLNT.CONN** Client Connection Channel is created*

Part 4 – Second Run of the AUDOperationsTest.java JUnit for JMS Messaging

The second run of the JUnit test is more successful. The console screen output in Figure 3-232 is from the Eclipse IDE JUnit run of the **AUDOperationsTest.java** Class.

This shows the transmission of a test message to the IBM MQ Connection Factory, **AUDCF3**, queue and the retrieval of the ten Audit Documents in the output list from the Audit Master Case Manager solution, Target Object Store, /**AUDIT_TEST** folder.

```
<terminated> AUDOperationsTest [JUnit] /opt/ibm/java-x86_64-80/bin/javaw (Jul 4, 2022, 1:08:32 PM – 1:08:39 PM)
log4j:WARN No appenders could be found for logger (filenet_error.api.com.filenet.apiimpl.util.ConfigValueLookup).
log4j:WARN Please initialize the log4j system properly.
log4j:WARN See http://logging.apache.org/log4j/1.2/faq.html#noconfig for more info.
[Perf Log] No interval found. Auditor disabled.
JSAS1480I: Security is not enabled because the ConfigURL property file is not set.
<?xml version=1.0 encoding= utf-8/?><AUDIT_REPORT><CaseID>00001</CaseID><COMMENTS>AUD Test on 17th February pm</COM
Lookup connection factory cn=AUDCF3
Sending a message...
Lookup connection factory cn=AUDCF3
Create and start the connection
Create the session
10 documents found
AC.pdf||2|3|OS2|{00D14A81-0000-C31E-A7D2-F2B7CF2D4803}
AC.pdf||2|3|OS2|{00D14A81-0000-C31E-A7D2-F2B7CF2D4803}
AC.pdf||2|3|OS2|{00D14A81-0000-C31E-A7D2-F2B7CF2D4803}
AC.pdf||2|3|OS2|{00D14A81-0000-C31E-A7D2-F2B7CF2D4803}
AC.pdf||2|3|OS2|{00D14A81-0000-C31E-A7D2-F2B7CF2D4803}
AC.pdf||2|3|OS2|{00D14A81-0000-C31E-A7D2-F2B7CF2D4803}
AC.pdf||2|3|OS2|{00D14A81-0000-C31E-A7D2-F2B7CF2D4803}
Audit Event Test Version 2||2|3|OS2|{A0DA5481-0000-C216-A96D-41296308BB2E}
Automatic Folder ID added||2|3|OS2|{30085581-0000-C416-9A60-74FB38E99E4D}
Test Audit Events||2|3|OS2|{40D35481-0000-CA14-BF15-512387B70CF4}
```

Figure 3-232. *The Eclipse IDE Console Log output of the second JUnit test run*

Code Now Updated As Follows for AUDOperations.java

The following Java code is now ready to build into the **AUDOperations.jar** Content Integrator method file for testing in the IBM Process Designer Workflow environment.

Listing 3-9. The AUDOperations.java code is updated for rebuild and deployment

```java
package com.ibm.filenet.ps.ciops;

import java.security.AccessController;
import java.util.ArrayList;
import java.util.Iterator;
import java.util.Set;

//JMSDEMO
import javax.jms.*;
import javax.naming.*;
import javax.naming.directory.*;
import java.util.Hashtable;
import javax.jms.ConnectionFactory;
import javax.jms.Destination;
```

CHAPTER 3   IBM JMS INTERFACE DEVELOPMENT IBM FILENET 5.5.X WORKFLOW

```java
import javax.jms.JMSException;
import javax.jms.MessageProducer;
import javax.jms.Queue;
import javax.jms.TextMessage;
import javax.jms.Topic;
import javax.mail.Session;
import javax.mail.Transport;
import javax.mail.internet.InternetAddress;
import javax.mail.internet.MimeMessage;
import javax.naming.Context;
import javax.naming.directory.InitialDirContext;
import javax.naming.ldap.InitialLdapContext;
import javax.security.auth.Subject;

import com.filenet.api.collection.DocumentSet;
import com.filenet.api.constants.PropertyNames;
import com.filenet.api.core.Connection;
import com.filenet.api.core.Document;
import com.filenet.api.core.Domain;
import com.filenet.api.core.EntireNetwork;
import com.filenet.api.core.Factory;
import com.filenet.api.core.Folder;
import com.filenet.api.core.ObjectStore;
import com.filenet.api.property.FilterElement;
import com.filenet.api.property.PropertyFilter;
import com.ibm.filenet.ps.ciops.database.DatabasePrincipal;
import com.ibm.mq.MQC;
import com.ibm.mq.MQEnvironment;
import com.ibm.mq.MQMessage;
import com.ibm.mq.MQPutMessageOptions;
import com.ibm.mq.MQQueueManager;
import com.ibm.mq.constants.CMQC;
import com.ibm.mq.constants.MQConstants;
import com.ibm.mq.jms.MQConnectionFactory;
import com.ibm.mq.jms.MQQueue;
import com.ibm.msg.client.jms.JmsConnectionFactory;
```

```java
import filenet.vw.api.VWAttachment;
import filenet.vw.api.VWAttachmentType;
import filenet.vw.api.VWException;
import filenet.vw.api.VWLibraryType;
import filenet.vw.api.VWSession;
import filenet.vw.base.logging.Logger;
//Import JMS related packages

/**
* Alan S.Bluck   26th June 2022
* AUD JMS Message sender Component Integrator code for AUD_Operations
* Adapted from:
* Custom Java component serving as an example for the article
* "Developing Java components for the Component Integrator".
*
*
*/
public class AUDOperations {
javax.jms.Session JMSsession =  null;
javax.jms.Connection JMSconnection = null;
javax.jms.ConnectionFactory   connFactory  = null;
String messageString = null;
String dLookup      = "cn=AUDQCF"; // Generic JMS Destination
private static Logger logger = Logger.getLogger( AUDOperations.class );
//JMS Variables
static                  Session     session     = null; // JMS Session
static                  Connection connection  = null; // JMS Connection
public AUDOperations() {
//java.sql.Connection databaseConnection = getDatabaseConnection();
//if ( databaseConnection != null ) {
//     logger.debug( databaseConnection.toString() );
//} else {
logger.debug( "No database connection" );
//}
}
```

```java
protected Connection getConnection() {
String uri = System.getProperty("filenet.pe.bootstrap.ceuri");
return Factory.Connection.getConnection(uri);
}

protected VWSession getVWSession() throws VWException {
String connectionPoint = System.getProperty("filenet.pe.cm.connectionPoint");
return new VWSession(connectionPoint);
}

protected java.sql.Connection getDatabaseConnection() {
Subject subject = Subject.getSubject( AccessController.getContext() );
Set<DatabasePrincipal> principals = subject.getPrincipals( DatabasePrincipal.class );
if ( principals != null && ! principals.isEmpty() ) {
DatabasePrincipal principal = principals.iterator().next();
return principal.getConnection();
}
return null;
}

private Folder getFolderFromAttachment(VWAttachment folderAttachment) {
ObjectStore objectStore = getObjectStore( folderAttachment.getLibraryName() );
Folder folder = (Folder) objectStore.getObject("Folder", folderAttachment.getId() );
return folder;
}

private Document getDocumentFromAttachment(VWAttachment documentAttachment) {
ObjectStore objectStore = getObjectStore( documentAttachment.getLibraryName() );
Document folder = (Document) objectStore.getObject("Document", documentAttachment.getId() );
return folder;
}
```

```java
private ObjectStore getObjectStore( String objectStoreName ) {
Connection connection = getConnection();
EntireNetwork entireNetwork = Factory.EntireNetwork.fetchInstance(connecti
on, null);
Domain domain = entireNetwork.get_LocalDomain();
return Factory.ObjectStore.getInstance( domain, objectStoreName );
}

/**
 * Returns the documents filed in the folder.
 *
 * @param folderAttachment the input folder.
 * @return an array of documents filed in the folder.
 * @throws Exception
 */
public VWAttachment[] getFolderDocuments(VWAttachment folderAttachment )
throws Exception {

Folder folder = getFolderFromAttachment(folderAttachment);
DocumentSet containedDocuments = getContainedDocuments(folder);
Iterator<?> iterator = containedDocuments.iterator();
ArrayList<VWAttachment> containedDocumentList = new
ArrayList<VWAttachment>();

while ( iterator.hasNext() ) {
Document document = (Document) iterator.next();
VWAttachment documentAttachment = getAsVWAttachment(document);
containedDocumentList.add( documentAttachment );
}

return containedDocumentList.toArray( new VWAttachment[0] );
}

private DocumentSet getContainedDocuments(Folder folder) {
PropertyFilter propertyFilter = getContainedDocumentsPropertyFilter();
folder.fetchProperties( propertyFilter );
DocumentSet containedDocuments = folder.get_ContainedDocuments();
return containedDocuments;
```

```java
}

private PropertyFilter getContainedDocumentsPropertyFilter() {
    PropertyFilter propertyFilter = new PropertyFilter();
    propertyFilter.addIncludeProperty( new FilterElement( null, null, null,
    PropertyNames.CONTAINED_DOCUMENTS, null ) );
    propertyFilter.addIncludeProperty( new FilterElement( 2, null, null,
    PropertyNames.ID, null ) );
    propertyFilter.addIncludeProperty( new FilterElement( 2, null, null,
    "DocumentTitle", null ) );
    return propertyFilter;
}

private VWAttachment getAsVWAttachment(Document document) throws
VWException {

    VWAttachment documentAttachment = new VWAttachment();

    documentAttachment.setLibraryType( VWLibraryType.LIBRARY_TYPE_CONTENT_ENGINE );
    ObjectStore objectStore = document.getObjectStore();
    objectStore.fetchProperties( new String[] { PropertyNames.NAME } );
    documentAttachment.setLibraryName( objectStore.get_Name() );

    document.fetchProperties( new String[] { PropertyNames.ID,
    PropertyNames.NAME } );
    documentAttachment.setId( document.get_Id().toString() );
    documentAttachment.setAttachmentName( document.get_Name() );
    documentAttachment.setType( VWAttachmentType.ATTACHMENT_TYPE_FOLDER );

    return documentAttachment;
}

/**
 * Sends a JMS message using a JMS Queue as the target of the Message.
 *
 * @param documentAttachment the initiating document from the Workflow
 * @param Message
 *            xml Message to be sent to the Queue.
```

```
* @param url
*          LDAP server URL eg "ldap://ecmukdemo6/o=sample"; where
ecmukdemo6 is the server and o=sample is the root organisation for
searching.
* @param icf
*          the context factory package used for the JNDI object retrieval
usually "com.sun.jndi.ldap.LdapCtxFactory";
* @param qManager
*          the name of the MQ Queue Manager eg wasadm
* @param qName
*          the name of the MQ Queue eg AUDQAR
* @param cfLookup
*          the name of the LDAP lookup Connection Factory to use eg
cn=AUDQCF
* @param JNDITopic
*          the name of the LDAP lookup for the MQ Topic eg cn=AUDTopic
(Not currently used)
* @param JNDIQueue
*          the name of the LDAP lookup for the MQ Queue eg cn=AUDQCF
* @param dLookup
*          the name of the LDAP lookup for the Message Destination
cn=AUDQAR
* @param hostName
*          the MQ Server hostname eg ecmukdemo6
* @param channel
*          the MQ Server listener channel eg SYSTEM.DEF.CLNT.CONN
* @param MQport
*          the MQ Server port eg 1417
* @param MQuser
*          the MQ Server connection user name - this is a Linux/windows
O/S user on the MQ Server with access authority to the Queue eg wasadm
* @param MQpassword
*          the MQ Server connection user name password eg filenet
*
* @throws Exception
```

CHAPTER 3   IBM JMS INTERFACE DEVELOPMENT IBM FILENET 5.5.X WORKFLOW

```java
*/
public void sendJMSMessage(VWAttachment docAttachment, String
Message,    String url,           String icf,        String qManager, String qName,
                                                    String cfLookup, String
JNDITopic, String JNDIQueue, String dLookup,
                                                    String hostName, String
channel, String sMQport, String MQuser, String MQpassword
) throws Exception {
{
messageString = Message;
// TODO Production code will be required to have much finer grained
// exception handling.
try {
//
//Lookup initial context
//TODO Get MQ Parameters from passed string values

//ASB - Using LDAP Context
//create initial context properties
//String url = "ldap://ecmukdemo6/o=sample";
//String icf = "com.sun.jndi.ldap.LdapCtxFactory";
java.util.Hashtable environment = new java.util.Hashtable();
environment.put(Context.PROVIDER_URL, url);
environment.put(Context.INITIAL_CONTEXT_FACTORY, icf);
Context ctx = new InitialDirContext(environment);
Destination        myDest      = null;
//ASB - Initial Content code End
// define the name of the QueueManager
//String qManager = "wasadm";
// and define the name of the Queue
//String qName = "AUDQAR";

// Put the message to the queue
System.out.println("Sending a message...");
// Note that the generic Connection Factory works for both queues & topics
// String      cfLookup     = "cn=AUDQCF";      //ASB Tests
```

```
System.out.println("Lookup connection factory " + cfLookup);
// String      JNDITopic    = "cn=AUDTopic";
// String      JNDIQueue    = "cn=AUDQCF";
// Class variables
javax.jms.Session    jmssession     = null; // JMS Session
javax.jms.Connection jmsconnection  = null; // JMS Connection
//String      dLookup      = "cn=AUDQCF";       // LDAP JMS Destination
String       myMode       = null; // Program mode
String       destType     = null; // Destination type
//JMSDEMO

//MQEnvironment.hostname = "ecmukdemo6";
//MQEnvironment.channel  = "SYSTEM.DEF.CLNT.CONN";
//MQEnvironment.port = 1417;
MQEnvironment.hostname = hostName;
MQEnvironment.channel  = channel;
int MQport = Integer.parseInt(sMQport);
MQEnvironment.port = MQport;
MQEnvironment.properties.put(CMQC.TRANSPORT_PROPERTY, CMQC.TRANSPORT_MQSERIES_CLIENT);

//ASB TODO
//Use two lines of code below for the XML Message
//MQMessage hello_world = new MQMessage();
//hello_world.writeUTF(Message);
// Create a connection to the queue manager

ConnectionFactory connFactory = (ConnectionFactory) ctx.lookup( cfLookup );
ConnectionFactory factory;
Queue queue;
Topic topic;
//Retrieve the Queue object details from the JNDI lookup
queue = (com.ibm.mq.jms.MQQueue)ctx.lookup(JNDIQueue); //ASB Initial code
line Changed from (Queue)
System.out.println("Create and start the connection");
jmsconnection = connFactory.createConnection(MQuser,MQpassword);   //ASB
Note that this a server Unix user for MQ not the
```

```java
                                                          //     LDAP user!
jmsconnection.start();
//
//ConnectionFactory connFactory = (ConnectionFactory) ctx.lookup(
cfLookup );
//JmsConnectionFactory connFactory = (JmsConnectionFactory) ctx.lookup(
cfLookup );   //ASB FIX
//

System.out.println("Create the session");

//Create the session
boolean transacted = true;
javax.jms.Session session= jmsconnection.createSession(transacted, javax.
jms.Session.AUTO_ACKNOWLEDGE);
myDest = (Destination)ctx.lookup(dLookup);

putMsg( myDest, Message,session );

// Clean up session and connection
session.close();
session = null;

jmsconnection.close();
jmsconnection = null;

} catch( JMSException je ) {
System.out.println("caught JMSException: " + je);
Exception le = je.getLinkedException();
if (le != null)  System.out.println("linked exception: "+le);
//TODO Log exception message

} catch( Exception e ) {
//TODO Log exception message
System.out.println("Error : " + e.getMessage());
System.out.println("Error Stack : " + e.toString());
// A finally block is a good place to ensure that we don't forget
// to close the most important JMS objects
} finally {
```

```
if (JMSsession != null) {
//Closing Session
JMSsession.close();
}
if (JMSconnection != null) {
//Closing Connection
JMSconnection.close();
}
}
//if (le.getMessage() != null)
//TODO Add to logger
//Finished
}
}

// A single
private void putMsg( Destination myDest, String outString, javax.jms.
Session jmssession )
throws JMSException, Exception
{
//MessageProducer myProducer = JMSsession.createProducer(myDest);//
REFLECTION ISSUE!
     if (Message.length() > 0) {
     TextMessage outMessage = JMSsession.createTextMessage();
      //myProducer.send(outMessage); //REFLECTION ISSUE!
        JMSsession.commit();
}
myProducer.close();
}
}
```

CHAPTER 3 IBM JMS INTERFACE DEVELOPMENT IBM FILENET 5.5.X WORKFLOW

AUDOperations Rebuild and Deploy .jar: Final Prebuild Test

The following Test output was output after adding the **log4j.xml** and **log4j.dtd** files from the Audit Event Handlers Java project we covered in Chapter 2.

Listing 3-10. The JUnit test output after adding the log4j.xml and log4j.dtd files

```
Configuration property file search order:
    user.dir = /root/eclipse-workspace/AUD_CIOps
    user.home = /root
    java.home = /opt/ibm/java-x86_64-80/jre
Loading configuration resource: FileNetBuild.properties
Configuration for:Config.AutoRefreshEnabled value:null applied
Configuration for:com.filenet.engine.LoggerRefreshInterval
value:null applied
[Perf Log] No interval found. Auditor disabled.
Configuration for:FileNet.EJB.ContextProperties value:null applied
Configuration for:com.filenet.AppServerType value:null applied
Configuration for:com.filenet.AppServer value:null applied
JSAS1480I: Security is not enabled because the ConfigURL property file is
not set.
Configuration for:com.filenet.api.useLocalAuthentication value:null applied
Configuration for:com.filenet.api.crcl.logging.chatty value:null applied
Configuration for:FileNet.crcl.implementation.api.location
value:null applied
Configuration for:FileNet.crcl.implementation.api.locations
value:null applied
Configuration for:FileNet.crcl.implementation.api.urls value:null applied
Configuration for:com.filenet.api.util.Id.SequentialDBType
value:null applied
Configuration for:CheetahBCMode value:false applied
Configuration for:com.filenet.engine.LogRequestWithError value:null applied
Configuration for:com.filenet.engine.init.DefaultLocale value:en-us applied
Configuration for:com.filenet.locale.equivalents value:|,he_*,iw_*,|,nb_
NO,nb_*,no_*,| applied
```

```
Configuration for:FileNet.WSI.HttpCookieMaxAge value:null applied
Configuration for:MaximumRetry value:null applied
Configuration for:ExpireSessionInterval value:null applied
Configuration for:BuildVersion value:dap553.1500 applied
Configuration for:FileNet.WSI.HttpChunkSize value:null applied
Configuration for:FileNet.WSI.custom.credential.class value:null applied
Configuration for:FileNet.WSI.custom.credential.usermethod
value:getUserName applied
Configuration for:FileNet.WSI.custom.credential.passwordmethod
value:getPassword applied
Configuration for:FileNet.WSI.TransportConnectionTimeout value:null applied
Configuration for:FileNet.WSI.stax.XMLOutputFactory value:com.ibm.xml.xlxp.
api.stax.XMLOutputFactoryImpl applied
Configuration for:WSITypeAttributeRequired value:null applied
Configuration for:WSIDefaultNamespacePermitted value:null applied
Configuration for:FileNet.WSI.AutoDetectLTPAToken value:null applied
Configuration for:FileNet.WSI.stax.XMLInputFactory value:com.ibm.xml.xlxp.
api.stax.XMLInputFactoryImpl applied
<?xml version=1.0 encoding= utf-8/?><AUDIT_REPORT><CaseID>00001</
CaseID><COMMENTS>AUD Test on 17th February pm</COMMENTS><AUDIT_
DATE>2022-06-27</AUDIT_DATE><AUDIT_STATUS>Date of Audit Report Document</
AUDIT_STATUS></AUDIT_REPORT>
Lookup connection factory cn=AUDCF3
Sending a message...
Lookup connection factory cn=AUDCF3
Create and start the connection
Create the session
10 documents found
AC.pdf||2|3|OS2|{00D14A81-0000-C31E-A7D2-F2B7CF2D4803}
AC.pdf||2|3|OS2|{00D14A81-0000-C31E-A7D2-F2B7CF2D4803}
AC.pdf||2|3|OS2|{00D14A81-0000-C31E-A7D2-F2B7CF2D4803}
AC.pdf||2|3|OS2|{00D14A81-0000-C31E-A7D2-F2B7CF2D4803}
AC.pdf||2|3|OS2|{00D14A81-0000-C31E-A7D2-F2B7CF2D4803}
AC.pdf||2|3|OS2|{00D14A81-0000-C31E-A7D2-F2B7CF2D4803}
AC.pdf||2|3|OS2|{00D14A81-0000-C31E-A7D2-F2B7CF2D4803}
```

Audit Event Test Version 2||2|3|OS2|{A0DA5481-0000-C216-A96D-41296308BB2E}
Automatic Folder ID added||2|3|OS2|{30085581-0000-C416-9A60-74FB38E99E4D}
Test Audit Events||2|3|OS2|{40D35481-0000-CA14-BF15-512387B70CF4}

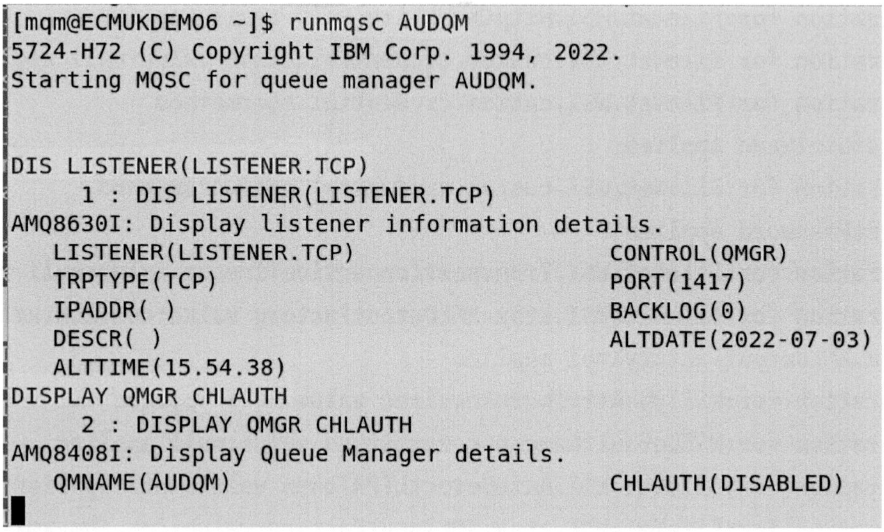

```
[mqm@ECMUKDEMO6 ~]$ runmqsc AUDQM
5724-H72 (C) Copyright IBM Corp. 1994, 2022.
Starting MQSC for queue manager AUDQM.


DIS LISTENER(LISTENER.TCP)
     1 : DIS LISTENER(LISTENER.TCP)
AMQ8630I: Display listener information details.
   LISTENER(LISTENER.TCP)                    CONTROL(QMGR)
   TRPTYPE(TCP)                              PORT(1417)
   IPADDR( )                                 BACKLOG(0)
   DESCR( )                                  ALTDATE(2022-07-03)
   ALTTIME(15.54.38)
DISPLAY QMGR CHLAUTH
     2 : DISPLAY QMGR CHLAUTH
AMQ8408I: Display Queue Manager details.
   QMNAME(AUDQM)                             CHLAUTH(DISABLED)
```

Figure 3-233. *The runmqsc command displays the details of AUDQM Queue Manager*

CHAPTER 3   IBM JMS INTERFACE DEVELOPMENT IBM FILENET 5.5.X WORKFLOW

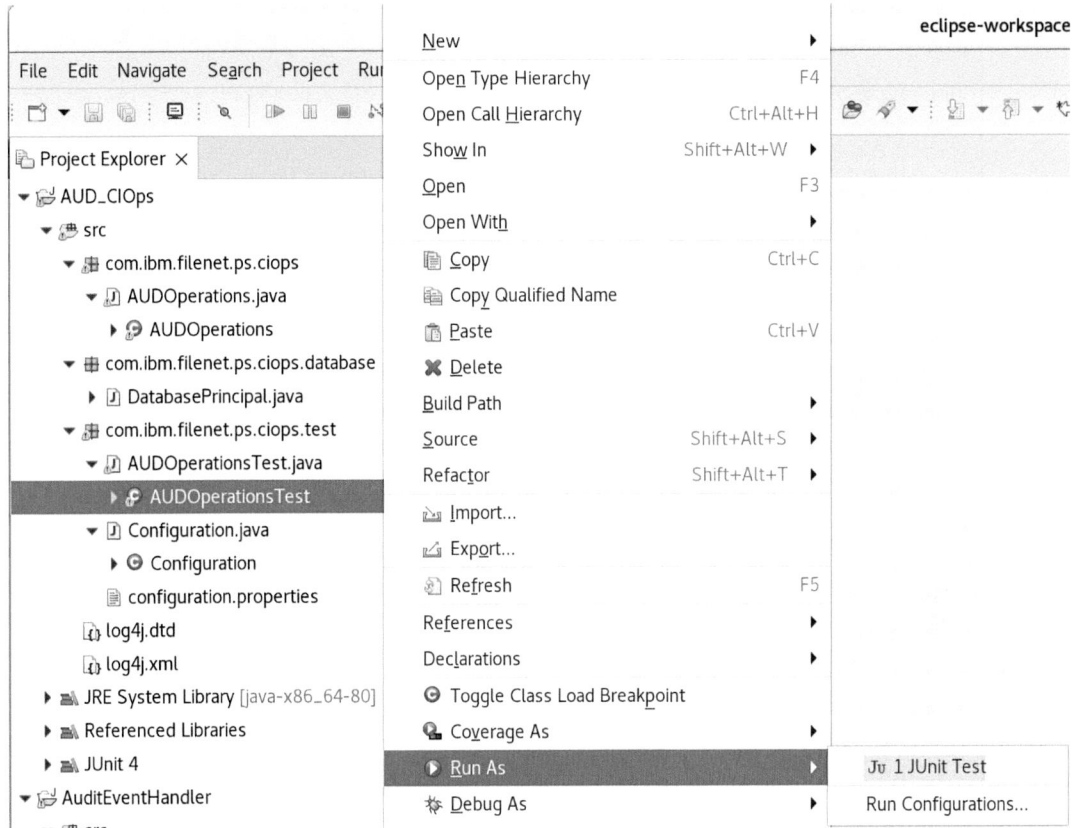

Figure 3-234. *The AUDOperationsTest run for JUnit testing is run*

CHAPTER 3   IBM JMS INTERFACE DEVELOPMENT IBM FILENET 5.5.X WORKFLOW

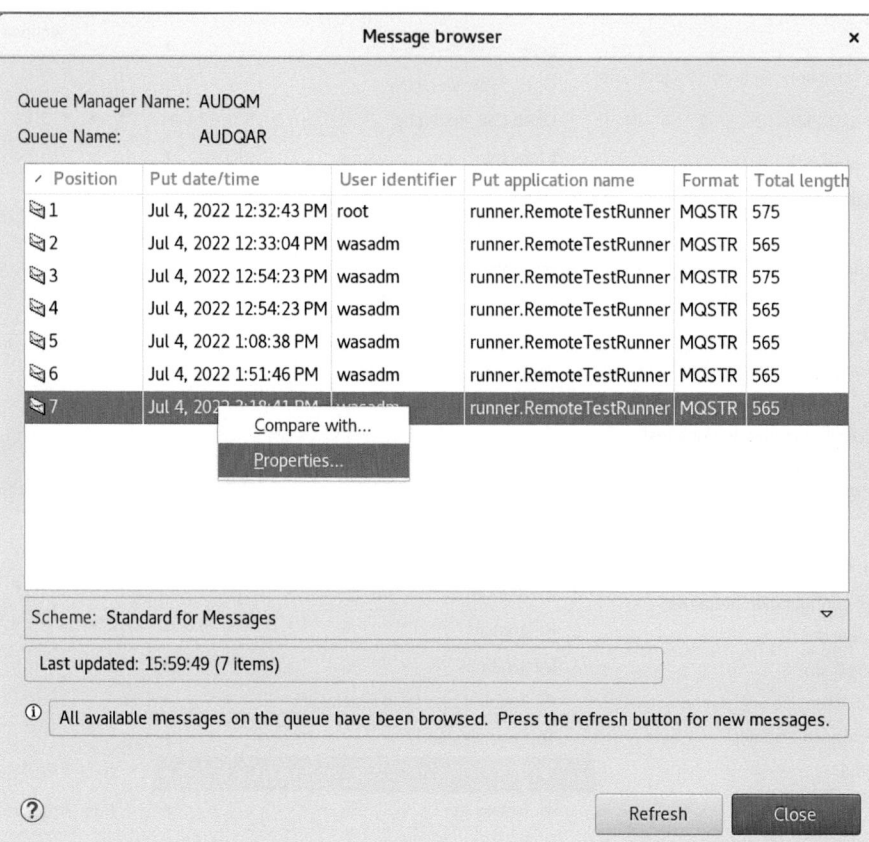

Figure 3-235. *The results of the Test Run are displayed in the Properties of the Queue Entry (dated July 4, 2022)*

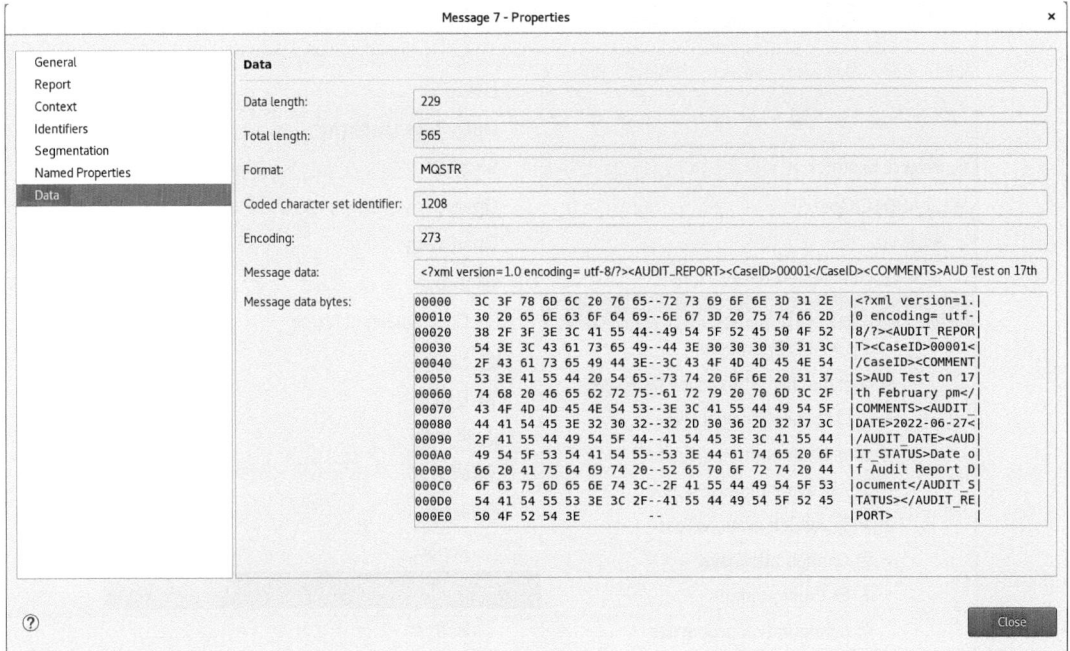

Figure 3-236. *The Data property of the Queue Entry shows the message from the Java*

Part 5 – Building the AUDOperations.jar

Next, we need to export the Java program in the **AUDOperations.jar** file with the exposed methods we wish to call from our IBM Process Designer Workflow.

CHAPTER 3   IBM JMS INTERFACE DEVELOPMENT IBM FILENET 5.5.X WORKFLOW

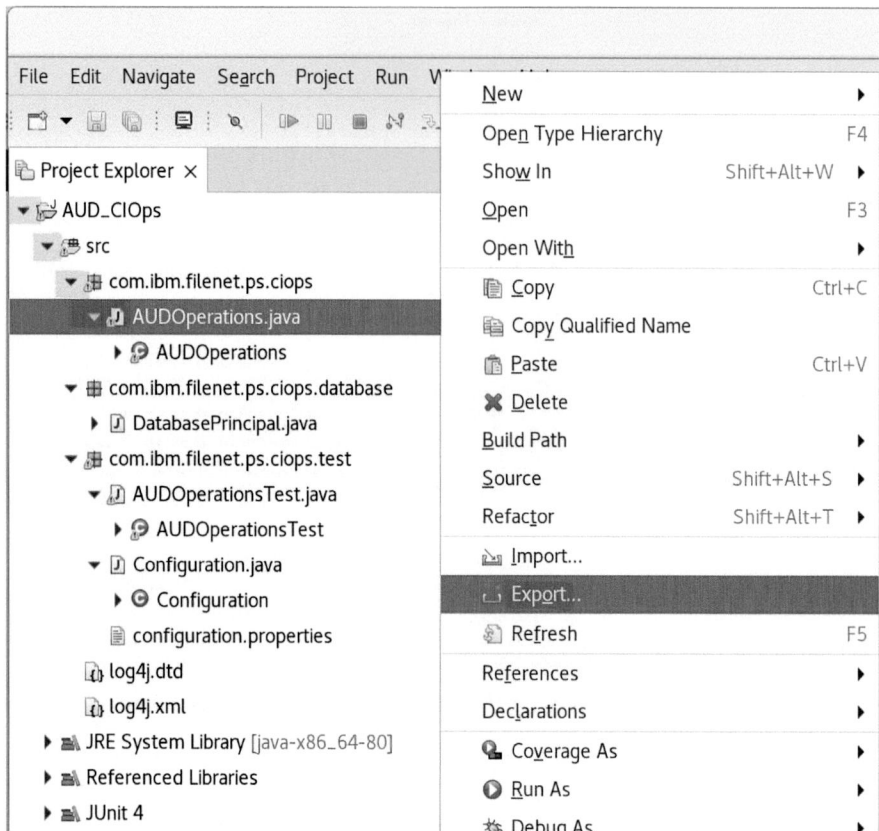

Figure 3-237. *The AUDOperations.java is compiled and exported as a jar file*

CHAPTER 3　IBM JMS INTERFACE DEVELOPMENT IBM FILENET 5.5.X WORKFLOW

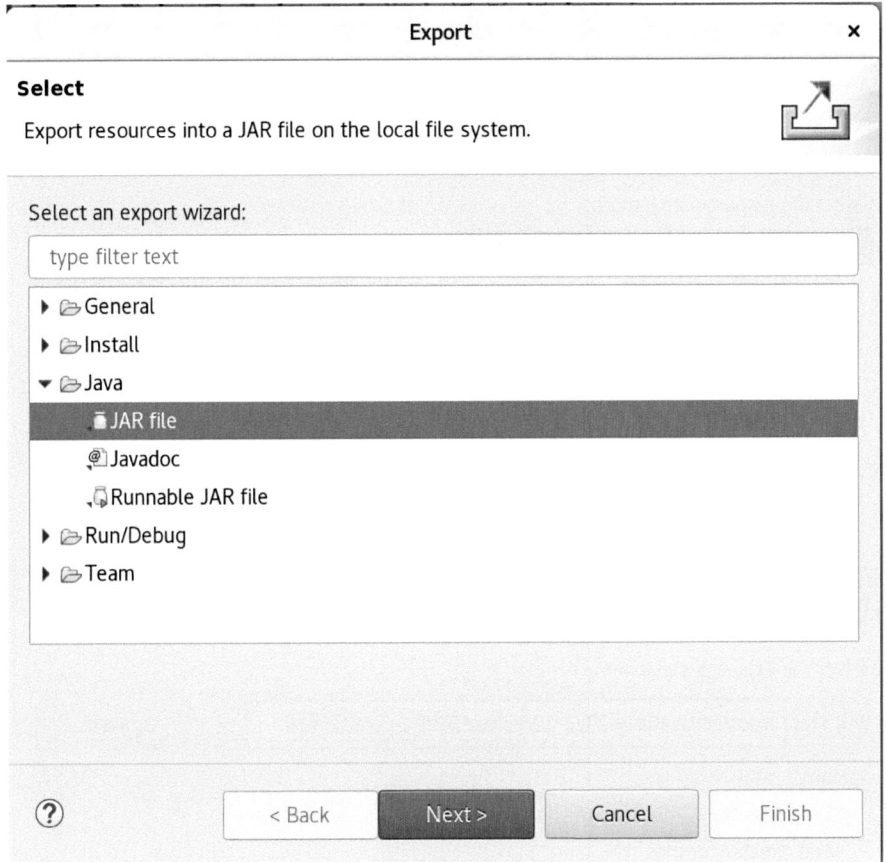

Figure 3-238. *The Eclipse IDE Java ➤ JAR file export wizard is selected*

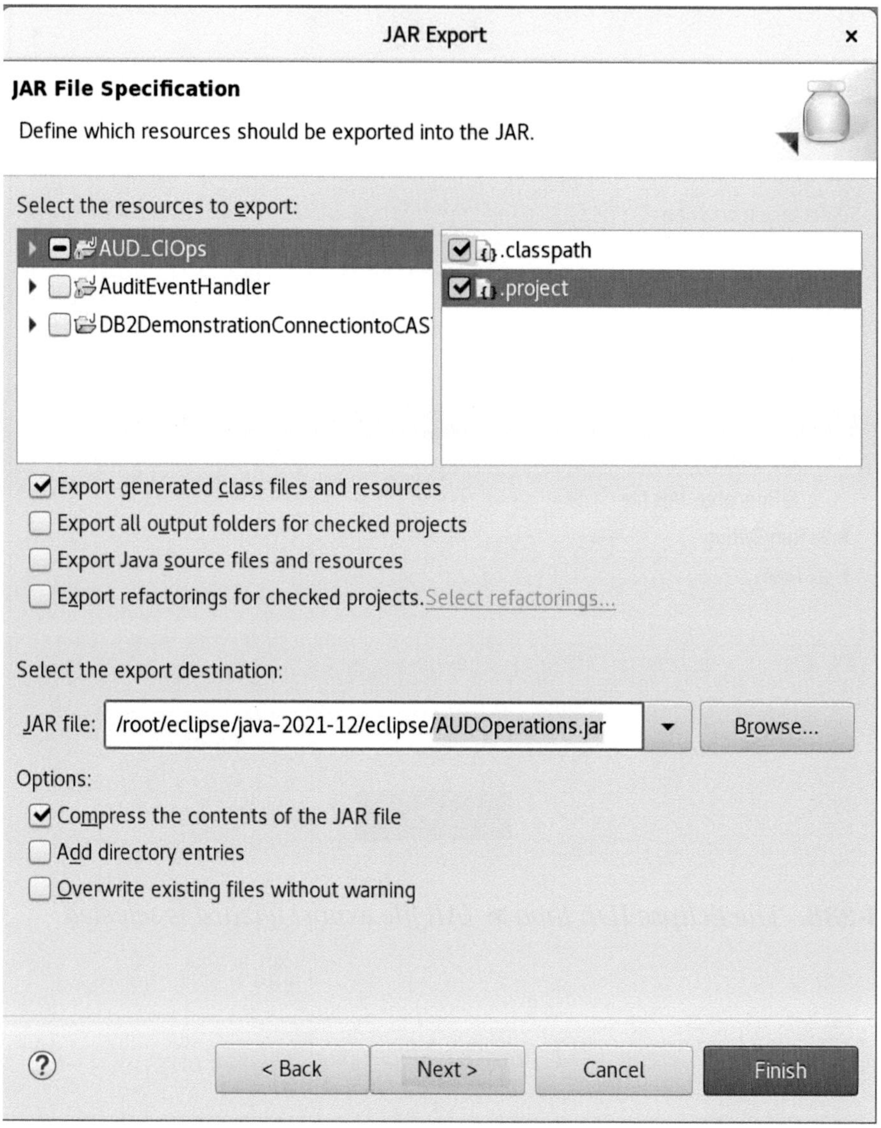

CHAPTER 3   IBM JMS INTERFACE DEVELOPMENT IBM FILENET 5.5.X WORKFLOW

Figure 3-239.   The options for exporting the Java JAR file are selected

CHAPTER 3 IBM JMS INTERFACE DEVELOPMENT IBM FILENET 5.5.X WORKFLOW

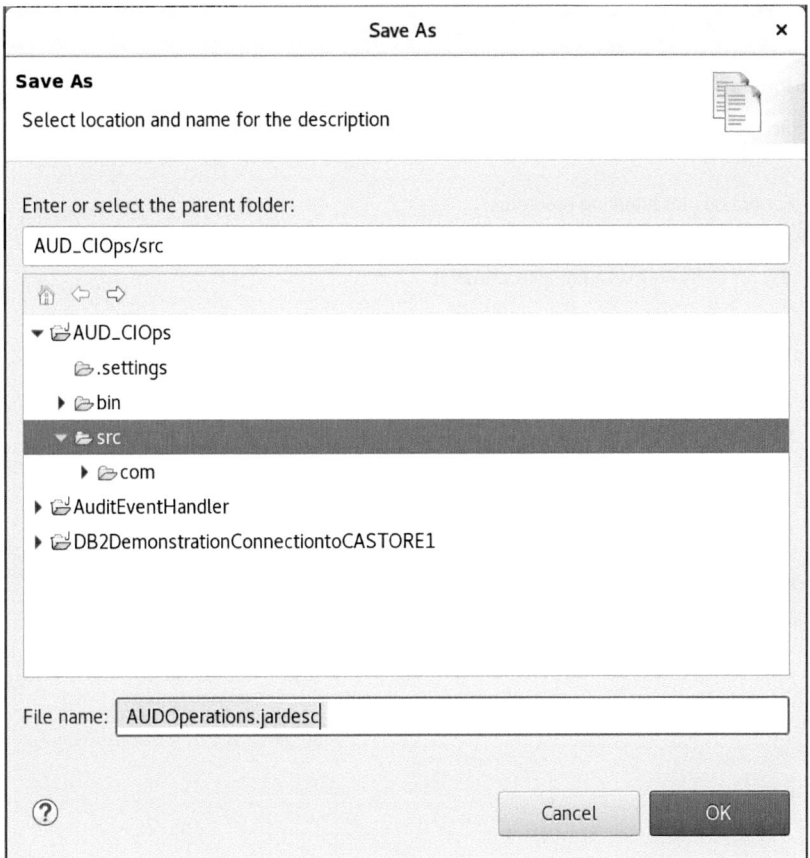

Figure 3-240. *The **AUDOperations.jardesc** file is saved to the Project Source folder*

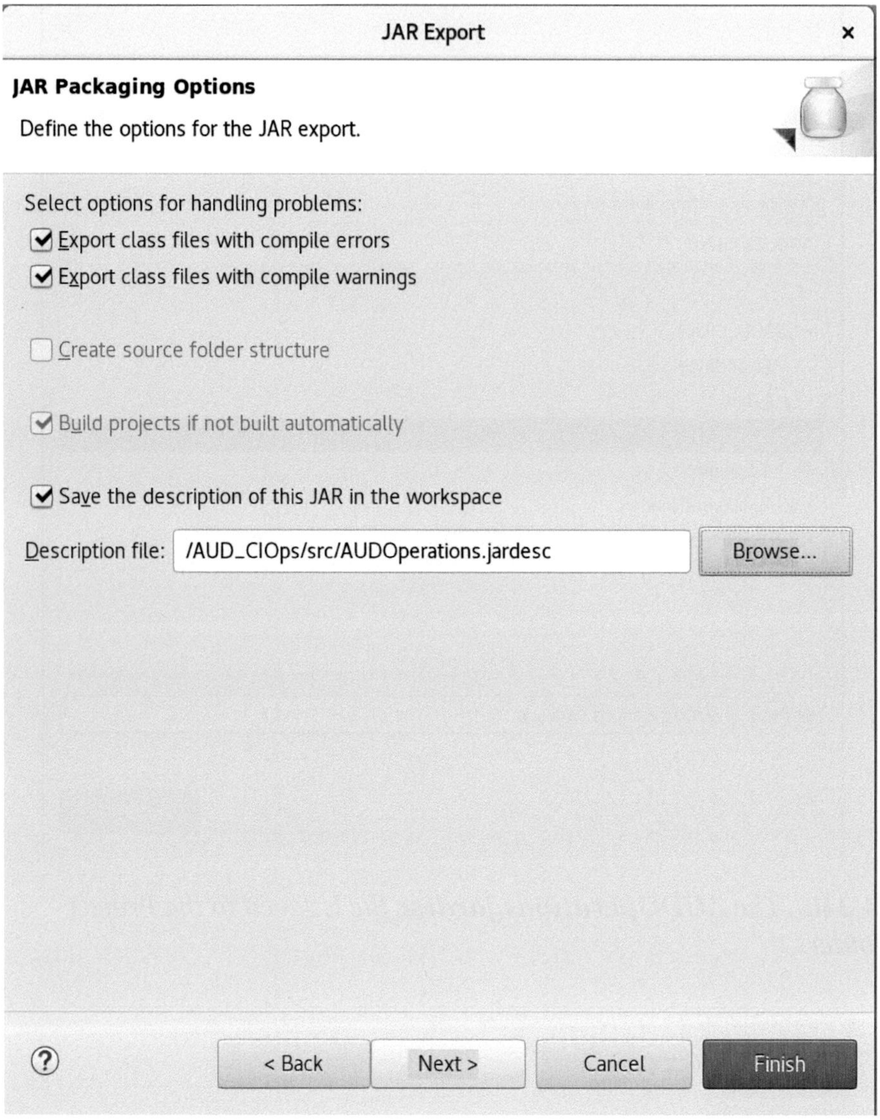

CHAPTER 3   IBM JMS INTERFACE DEVELOPMENT IBM FILENET 5.5.X WORKFLOW

Figure 3-241. The Browse command is used to select the path for the .jardesc file

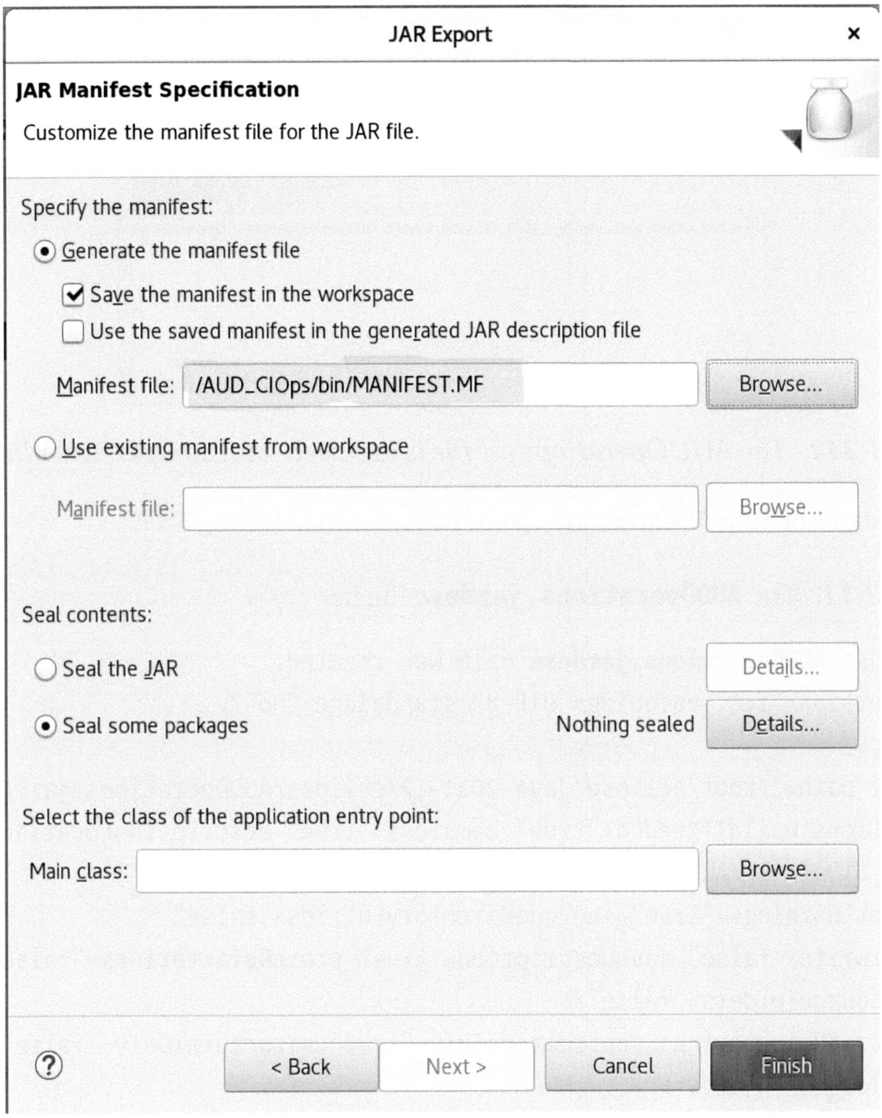

Figure 3-242. The MANIFEST.MF generation option is selected and defined for save

CHAPTER 3 IBM JMS INTERFACE DEVELOPMENT IBM FILENET 5.5.X WORKFLOW

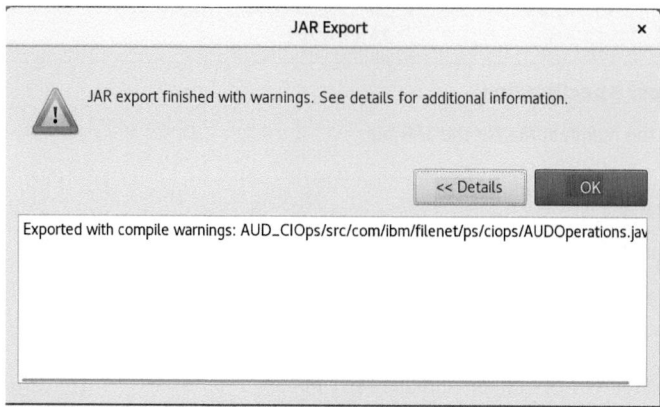

Figure 3-243. *The AUDOperations jar file is exported with compile warnings*

The status of the AUDOperations.jar build is shown in Listing 3-11.

Listing 3-11. The **AUDOperations.jardesc** file list

```
An Eclipse AUDOperations.jardesc file was created.
<?xml version="1.0" encoding="UTF-8" standalone="no"?>
<jardesc>
    <jar path="/root/eclipse/java-2021-12/eclipse/AUDOperations.jar"/>
    <options buildIfNeeded="true" compress="true" descriptionLocation="/
    AUD_CIOps/src/AUDOperations.jardesc" exportErrors="true"
    exportWarnings="true" includeDirectoryEntries="false"
    overwrite="false" saveDescription="true" storeRefactorings="false"
    useSourceFolders="false"/>
    <storedRefactorings deprecationInfo="true" structuralOnly="false"/>
    <selectedProjects/>
    <manifest generateManifest="true" manifestLocation="/AUD_CIOps/
    bin/MANIFEST.MF" manifestVersion="1.0" reuseManifest="false"
    saveManifest="true" usesManifest="true">
        <sealing sealJar="false">
            <packagesToSeal/>
            <packagesToUnSeal/>
        </sealing>
    </manifest>
```

```
    <selectedElements exportClassFiles="true" exportJavaFiles="false"
    exportOutputFolder="false">
        <file path="/AUD_CIOps/.classpath"/>
        <file path="/AUD_CIOps/.project"/>
        <javaElement handleIdentifier="=AUD_CIOps/src&lt;com.ibm.filenet.
          ps.ciops"/>
    </selectedElements>
</jardesc>
```

Rebuilding the AUDOperations.jar File

To rebuild the AUDOperations.jar file, we can make use of the **AUDOperations.jardesc** file we created.

> **Note** Make a backup (if required) of the previous **AUDOperations.jar** first!

Right-click this **AUDOperations.jardesc** file in the Project area **AUD_CIOPS** and select **Create Jar** from the menu, then click **Yes** when prompted to overwrite the existing **AUDOperations.jar** file. (Click **OK**.)

Copy the compressed **AUDOperations.jar** file to the **Installs** folder for import as a **System Component Integrator** step.

The **configuration.properties** file contains the following (used for the JUnit test harness).

Listing 3-12. The latest `configuration.properties` JUnit test file parameters

```
AUDOperationsTest.CEWsiUrl=http://ecmukdemo6:9080/wsi/FNCEWS40MTOM/
AUDOperationsTest.ConnectionPointName=CP1
AUDOperationsTest.EmailFrom=alan.bluck@asbsoftware.co.uk
AUDOperationsTest.EmailTemplatePath=/AUDIT_TEST/AC.pdf
AUDOperationsTest.EmailTo= alan.bluck@asbsoftware.co.uk
AUDOperationsTest.ObjectstoreName=OS2
AUDOperationsTest.Password=filenet
AUDOperationsTest.TestFolderName=/AUDIT_TEST
AUDOperationsTest.Username=Alan
```

CHAPTER 3   IBM JMS INTERFACE DEVELOPMENT IBM FILENET 5.5.X WORKFLOW

AUDOperationsTest.WaspLocation=/opt/IBM/WebSphere/AppServer/profiles/
AppSrv01/installedApps/localhostNode01Cell/FileNetEngine.ear/cews.war/WEB-
INF/classes/com/filenet/engine/wsi
AUDOperationsTest.WsiJaasConfigFile=/opt/IBM/ECMClient/configure/CE_
API/config/jaas.conf.WSI

FileNet Workflow System Component AUDOperations.jar Deployment

Create a new Component Queue, **AUD_Operations**, in the Workflow System:

a) Log in to the FileNet Administration Console for Content Engine (acce) web application configuration tool.

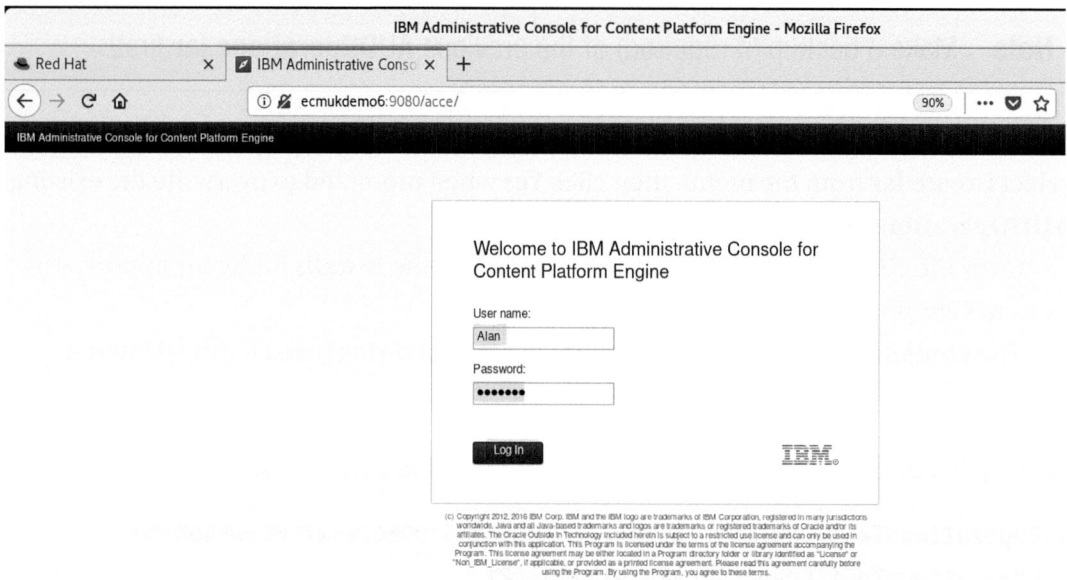

Figure 3-244. *The acce administration web application is logged in to*

b) First, import the AUDOperations.jar file as a new Code Module.

CHAPTER 3   IBM JMS INTERFACE DEVELOPMENT IBM FILENET 5.5.X WORKFLOW

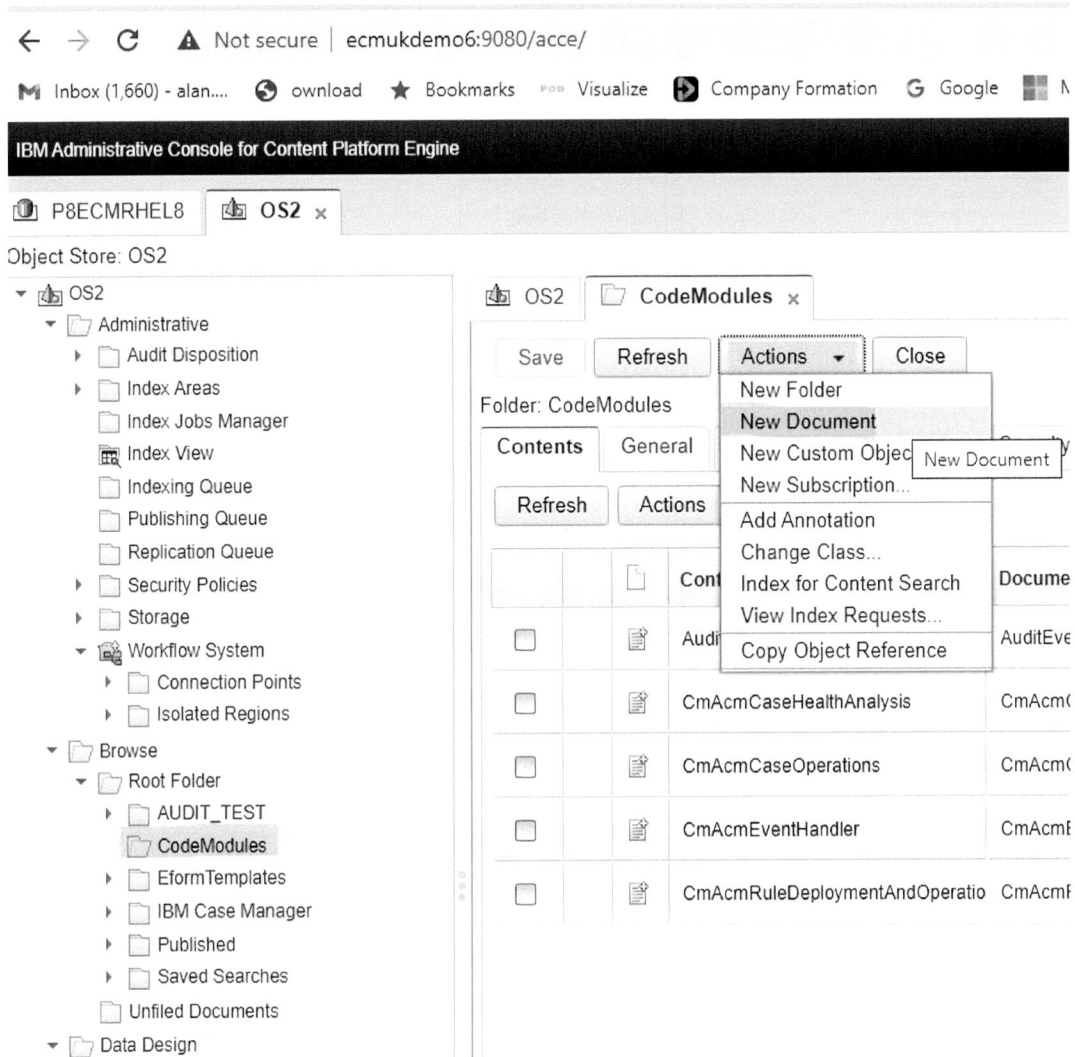

*Figure 3-245.   Select the **Actions ▶ New Document** menu item on the CodeModules tab*

CHAPTER 3 IBM JMS INTERFACE DEVELOPMENT IBM FILENET 5.5.X WORKFLOW

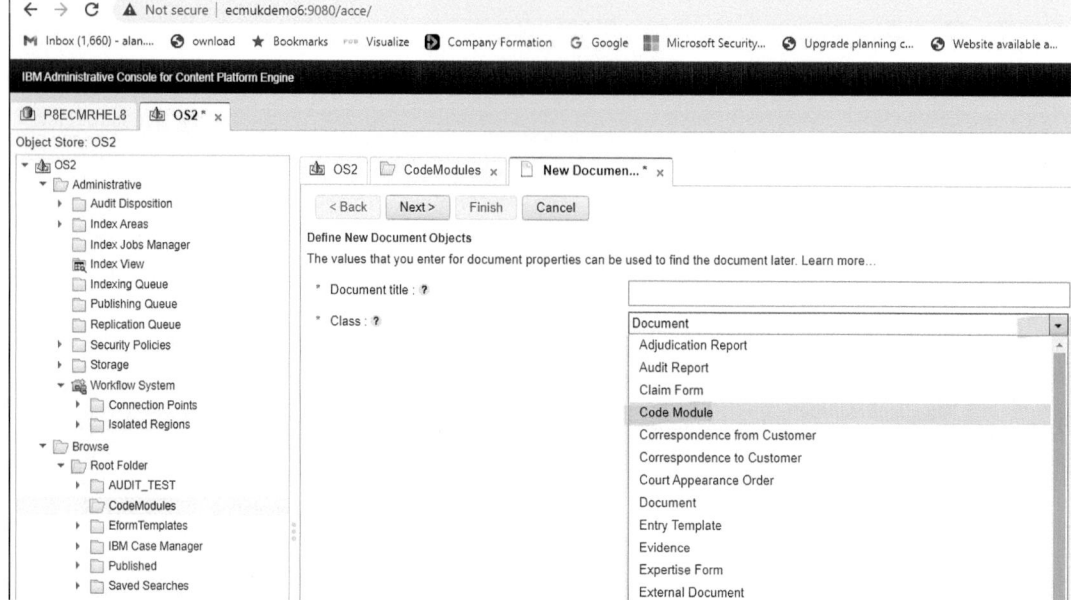

Figure 3-246. *The **Code Module** Document Class is selected from the drop-down*

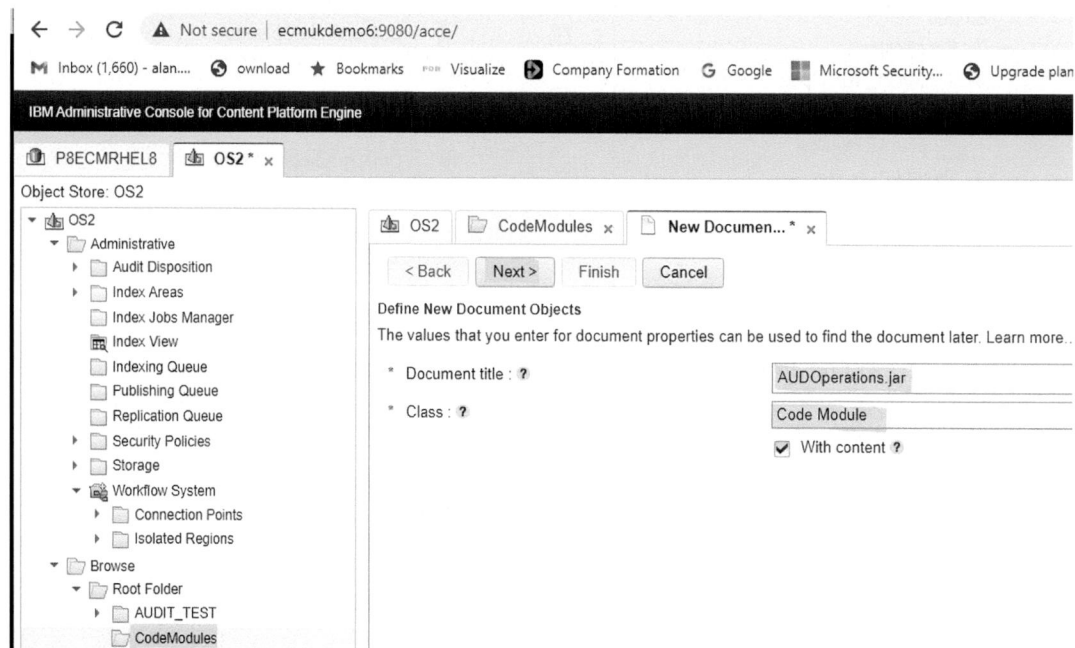

Figure 3-247. *The Document title is entered as **AUDOperations.jar** and **Next>** clicked*

CHAPTER 3 IBM JMS INTERFACE DEVELOPMENT IBM FILENET 5.5.X WORKFLOW

The Document title is entered as **AUDOperations.jar** and **Next>** is clicked with the tick box selected to allow the **AUDOperations.jar** file to be included as the Code Module Document Object content.

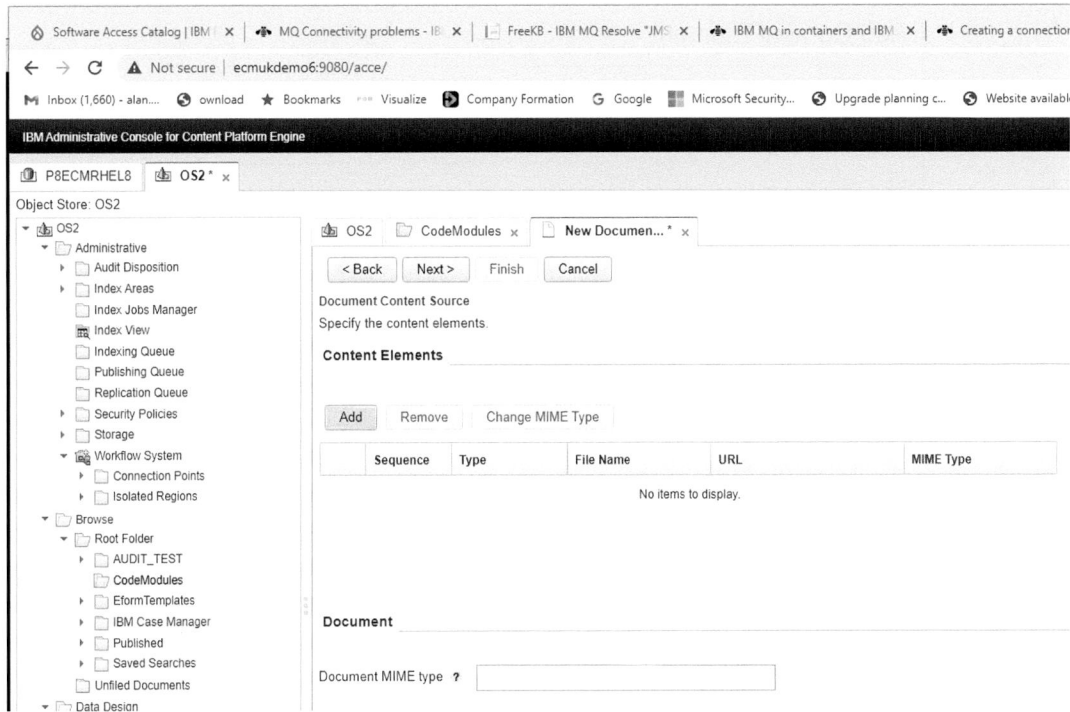

Figure 3-248. *The **Add** button is clicked to add the **AUDOperations.jar** as content*

The **Add** button loads a Web Browser window which can be used from the **acce** web application to locate and select the built **AUDOperations.jar** file.

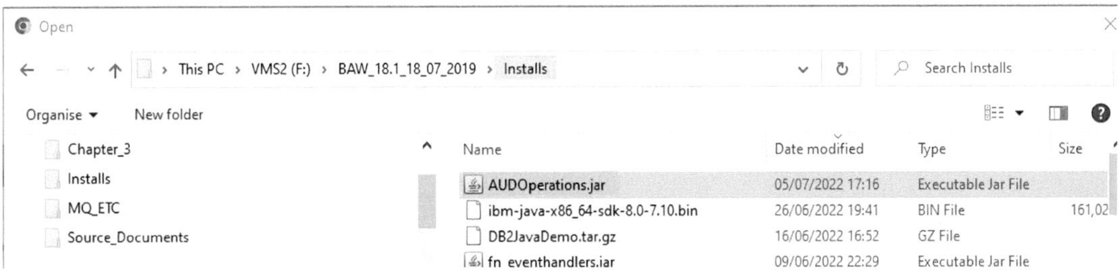

Figure 3-249. *The **AUDOperations.jar** is selected from the Installs subfolder*

CHAPTER 3   IBM JMS INTERFACE DEVELOPMENT IBM FILENET 5.5.X WORKFLOW

The **Add Content Element** pop-up window is shown allowing the browsed **AUDOperations.jar** file to be confirmed and its content added using the **Add Content** command button.

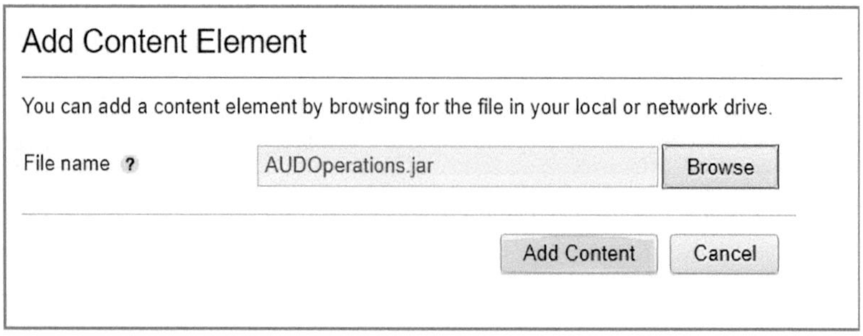

*Figure 3-250.   The **Add Content** button is clicked to complete the Content addition*

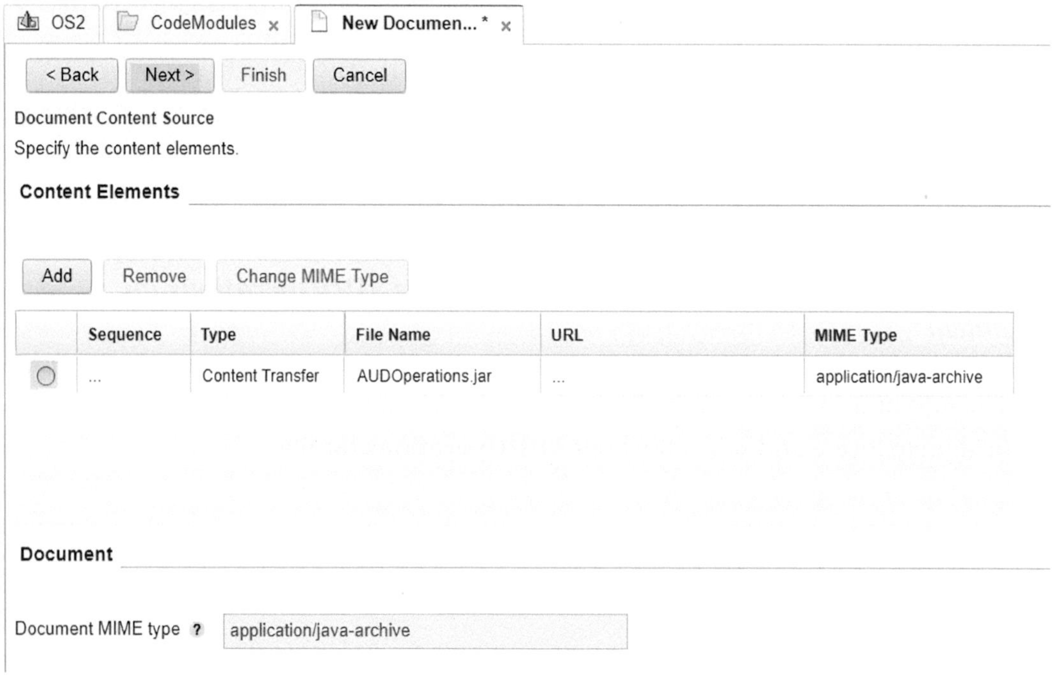

*Figure 3-251.   The **Next>** button is clicked after upload of the **AUDOperations.jar***

CHAPTER 3   IBM JMS INTERFACE DEVELOPMENT IBM FILENET 5.5.X WORKFLOW

The **Next>** command button shows the following Object Property Name values which we leave blank and then click **Next>** again.

Property Name	Property Value		Data Type	Cardinality	Settability
Component Binding Label	<Value not set>		8 <String>	0 <Single>	0 <Read-write>
Publication Source	<Value not set>		7 <Object>	0 <Single>	1 <Settable only before checkin>
Owner Document	<Value not set>		7 <Object>	0 <Single>	1 <Settable only before checkin>
Publication Info	<Value not set>		1 <Binary>	0 <Single>	1 <Settable only before checkin>
Publishing Subsidiary Folder	<Value not set>		7 <Object>	0 <Single>	1 <Settable only before checkin>
Ignore Redirect	<None> (not set)		2 <Boolean>	0 <Single>	0 <Read-write>
Associated Case	<Value not set>		7 <Object>	0 <Single>	0 <Read-write>
Entry Template Object Store Name	<Value not set>		8 <String>	0 <Single>	0 <Read-write>
Entry Template Launched Workflow Number	<Value not set>		8 <String>	0 <Single>	0 <Read-write>
Entry Template Id	<Value not set>		5 <GUID>	0 <Single>	0 <Read-write>
Deployment Version Number	<Value not set>		8 <String>	0 <Single>	0 <Read-write>

***Figure 3-252.**   The **Next>** command button is clicked leaving all the default values*

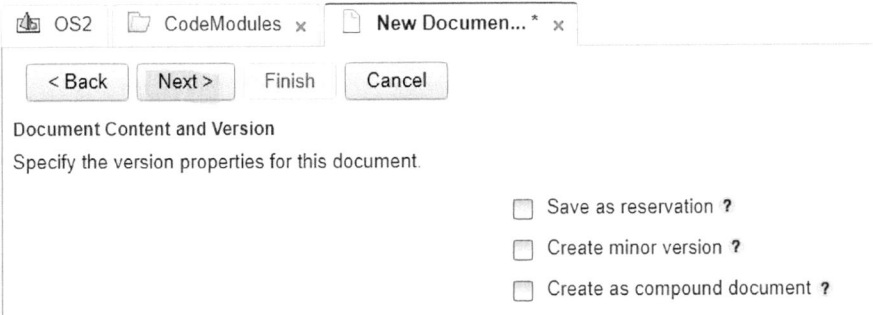

***Figure 3-253.**   The **Next>** command button is clicked leaving all the default values*

580

CHAPTER 3 IBM JMS INTERFACE DEVELOPMENT IBM FILENET 5.5.X WORKFLOW

Figure 3-254. *The Next> command button is clicked leaving a Default retention period*

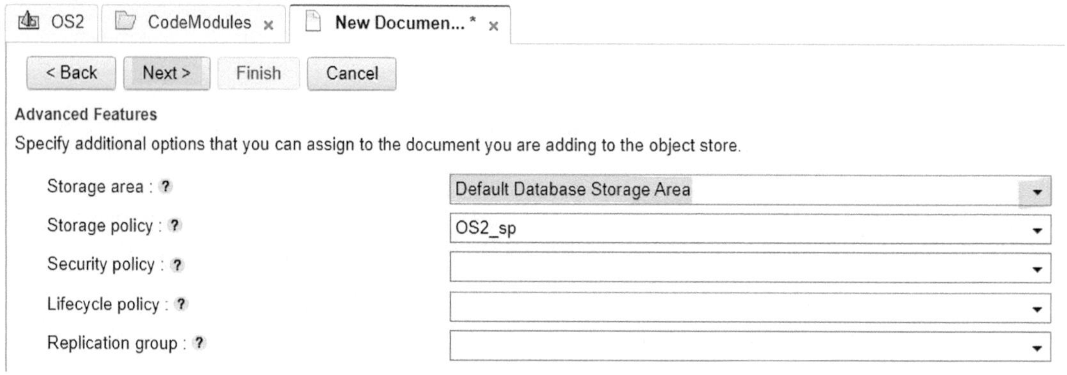

Figure 3-255. *The Default Database Storage Area is selected from the drop-down*

After clicking **Next>** for the **Code Module** Document **Advanced Features**, the **Code Module** Document build parameters are displayed and created using the **Finish** command button.

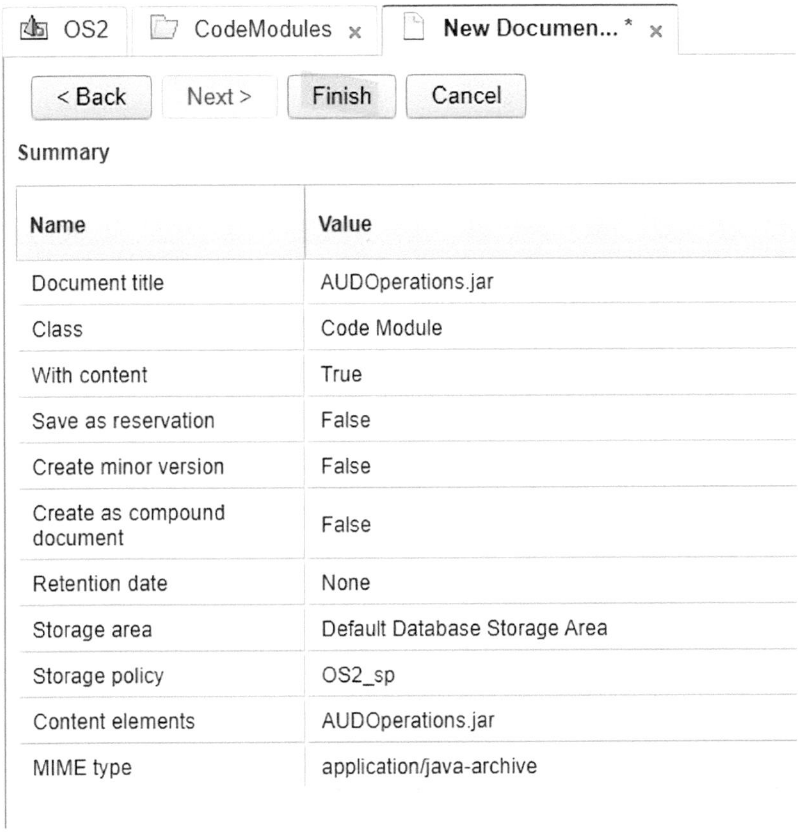

CHAPTER 3   IBM JMS INTERFACE DEVELOPMENT IBM FILENET 5.5.X WORKFLOW

Summary

Name	Value
Document title	AUDOperations.jar
Class	Code Module
With content	True
Save as reservation	False
Create minor version	False
Create as compound document	False
Retention date	None
Storage area	Default Database Storage Area
Storage policy	OS2_sp
Content elements	AUDOperations.jar
MIME type	application/java-archive

Figure 3-256. *The summary of the parameters for creating the Code Module is shown*

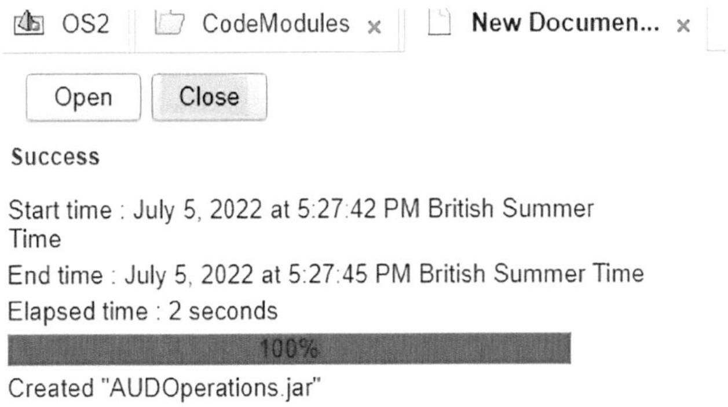

Figure 3-257. *The AUDOperations.jar Code Module is created successfully*

CHAPTER 3   IBM JMS INTERFACE DEVELOPMENT IBM FILENET 5.5.X WORKFLOW

c) Select the required Object Store Workflow System node as shown in Figure 3-258.

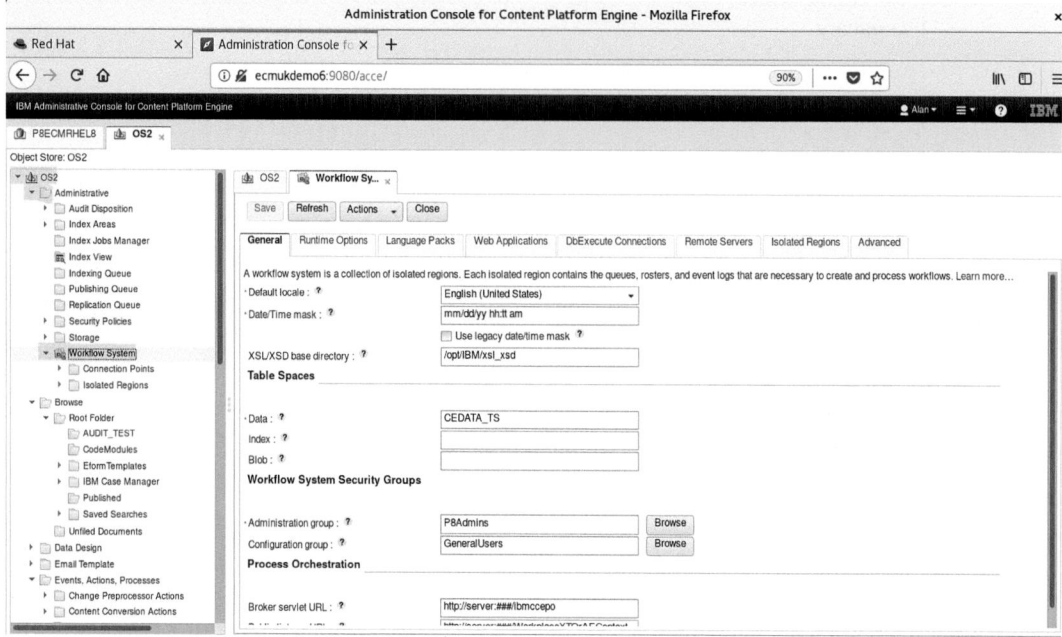

Figure 3-258. *The Workflow System node is selected*

a) In the object store navigation pane, click the **Administrative ➤ Workflow System ➤ Isolated Regions** folder and click the isolated region that you want to modify.

CHAPTER 3   IBM JMS INTERFACE DEVELOPMENT IBM FILENET 5.5.X WORKFLOW

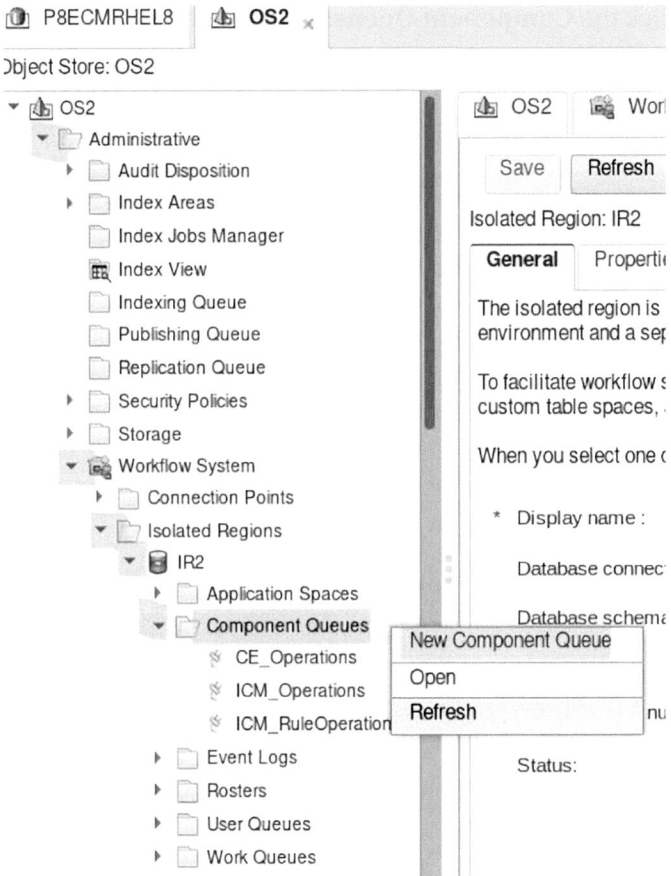

Figure 3-259. *The New Component Queue menu item is selected*

CHAPTER 3   IBM JMS INTERFACE DEVELOPMENT IBM FILENET 5.5.X WORKFLOW

b) Right-click the **Component Queues** folder and click **New Component Queue**.

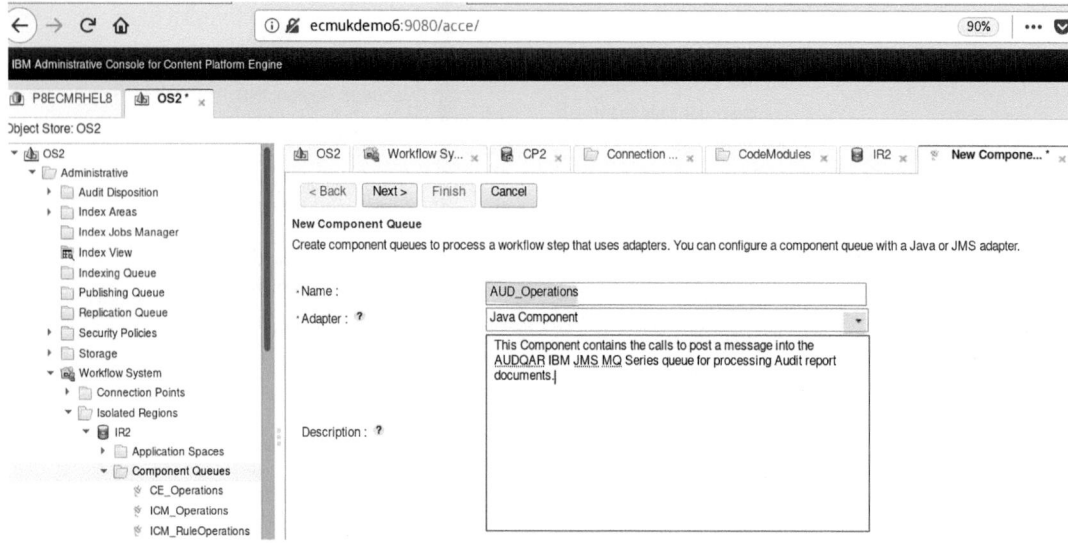

Figure 3-260. The Component Queue is named AUD_Operations

c) The AUDOperations.jar Code Module we created earlier is loaded.

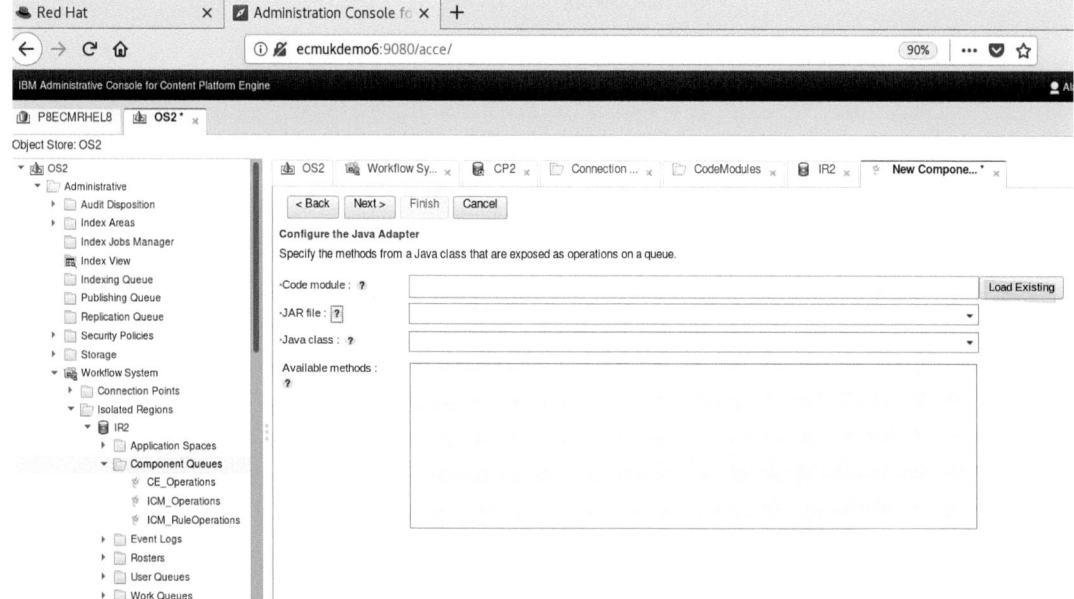

Figure 3-261. The Load Existing command button is selected

594

CHAPTER 3 IBM JMS INTERFACE DEVELOPMENT IBM FILENET 5.5.X WORKFLOW

d) Then, the **AUDOperations.jar** file **Code Module** we added earlier is selected.

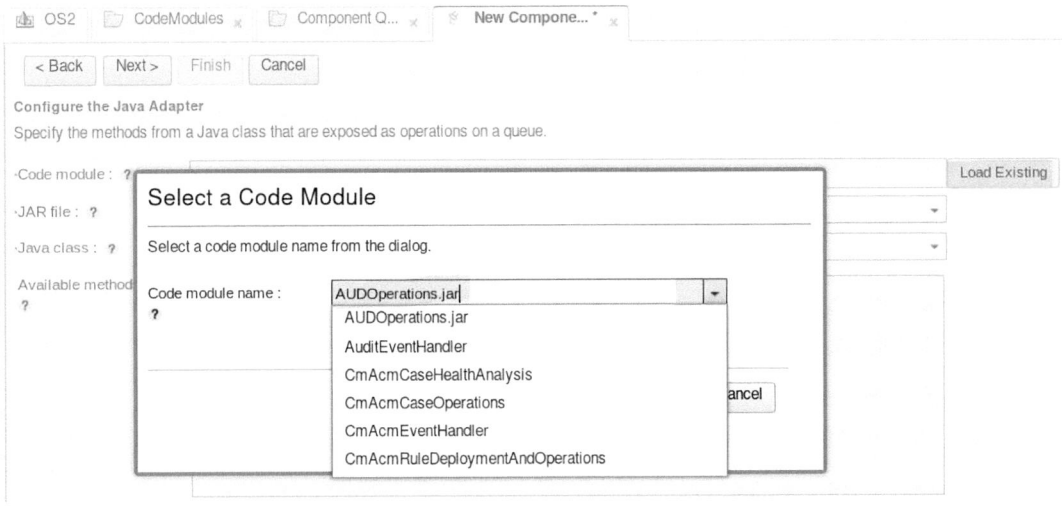

Figure 3-262. *The AUDOperations.jar Code Module is selected*

e) Select the previously loaded AUD_Operations code module.

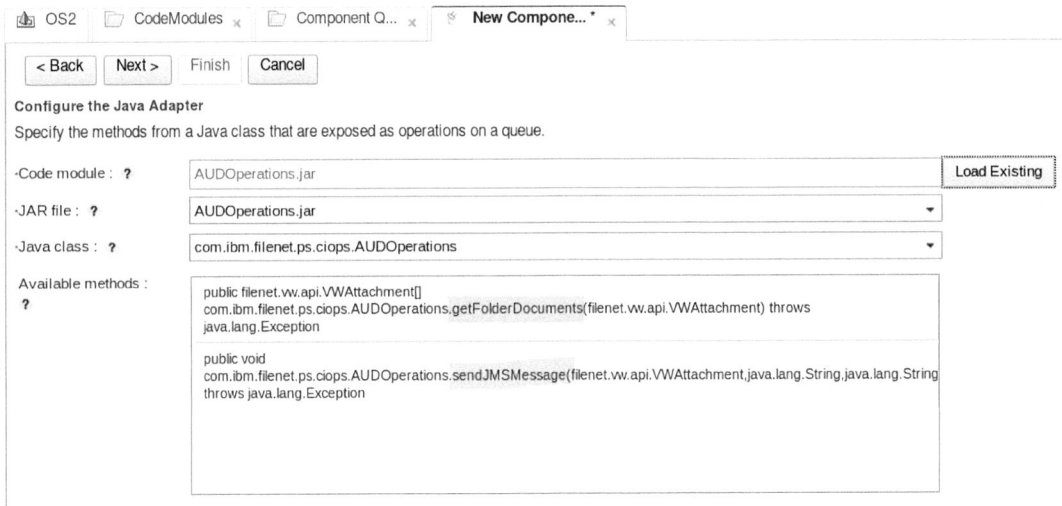

Figure 3-263. *The Available methods list is shown as expected*

595

CHAPTER 3   IBM JMS INTERFACE DEVELOPMENT IBM FILENET 5.5.X WORKFLOW

On loading the AUDOperations.jar file, the system uses Java reflection to identify and display the public methods which can be called in a Component Integrator step in the IBM Case Manager workflow steps, which we will cover later in this chapter.

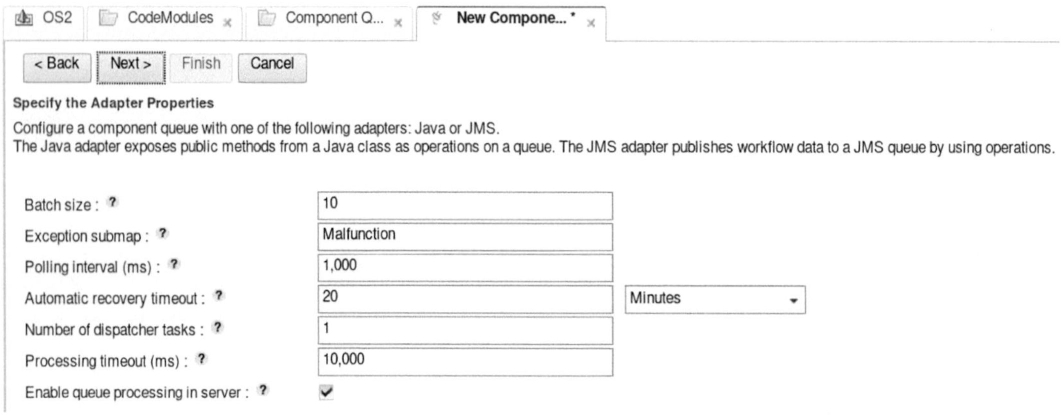

*Figure 3-264. The **Adapter Properties** are shown and the defaults used*

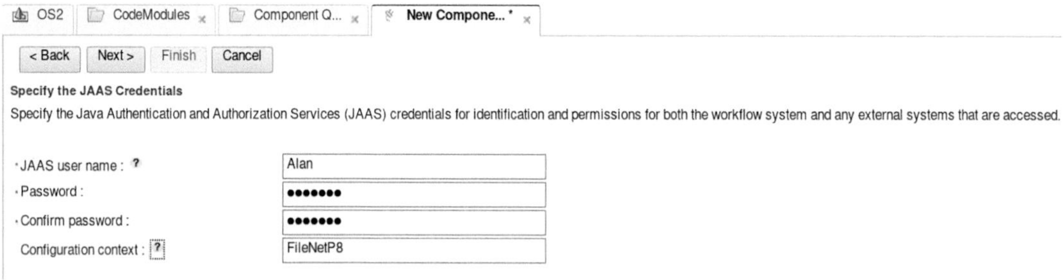

Figure 3-265. The New Component tab is displayed with JAAS authentication prompts

It should be noted that the **Configuration context** was changed later to **CELogin** in Figure 3-265.

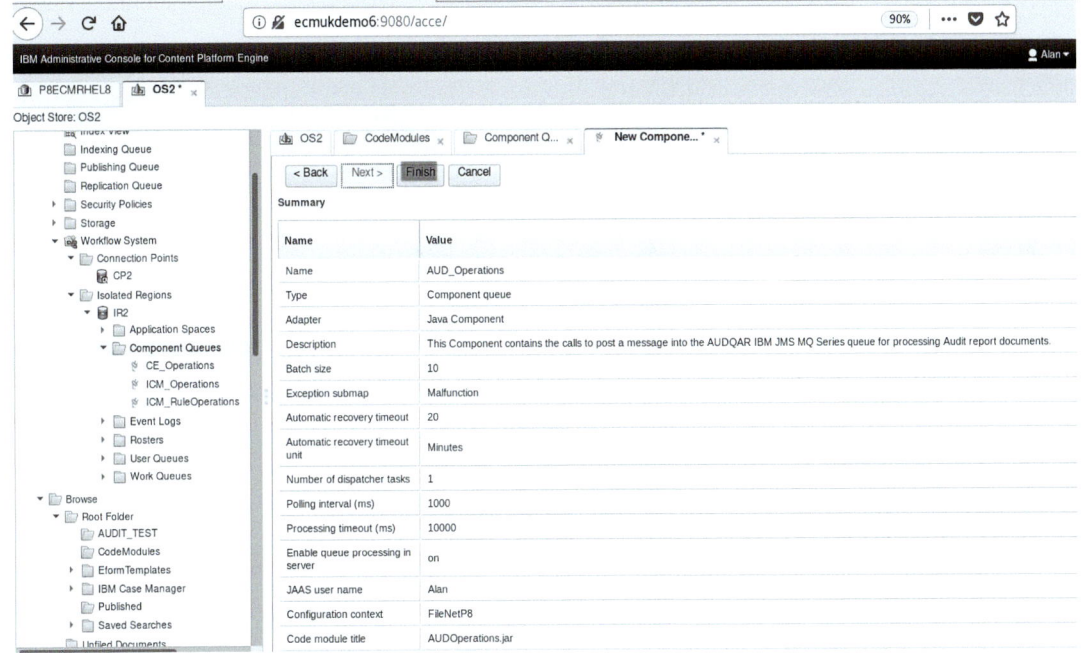

Figure 3-266. *The Finish command is used to create the AUD_Operations Queue*

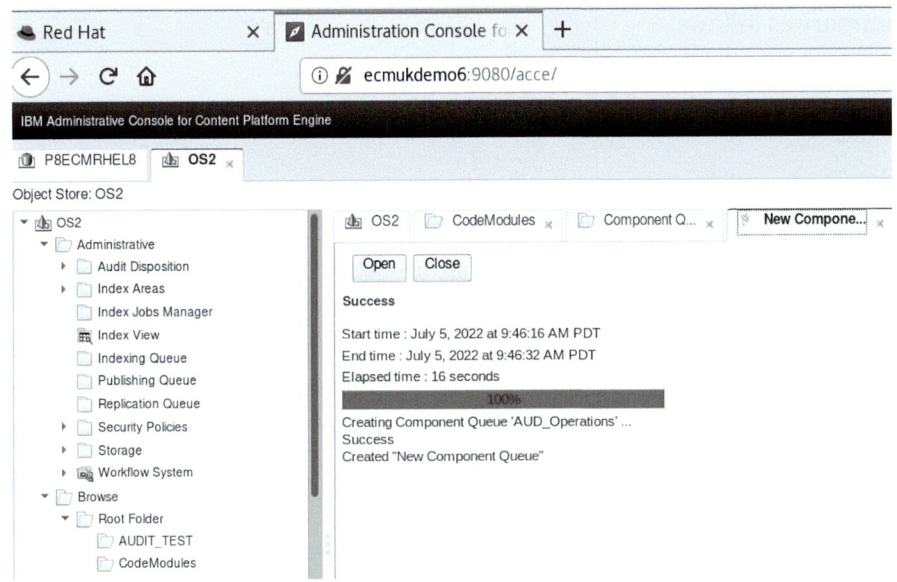

Figure 3-267. *The **AUD_Operations** Queue is shown to be created successfully*

CHAPTER 3   IBM JMS INTERFACE DEVELOPMENT IBM FILENET 5.5.X WORKFLOW

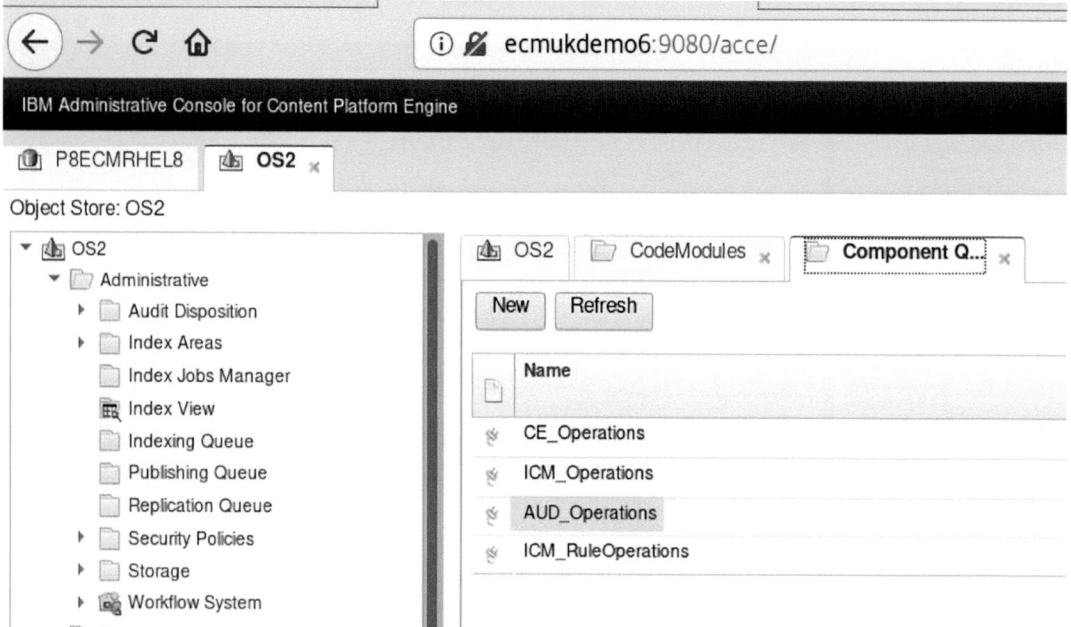

Figure 3-268. *The **AUD_Operations** Component Queue is now available for use*

The preceding steps were repeated for the OS1 Design Object Store.
In summary as follows:

CHAPTER 3   IBM JMS INTERFACE DEVELOPMENT IBM FILENET 5.5.X WORKFLOW

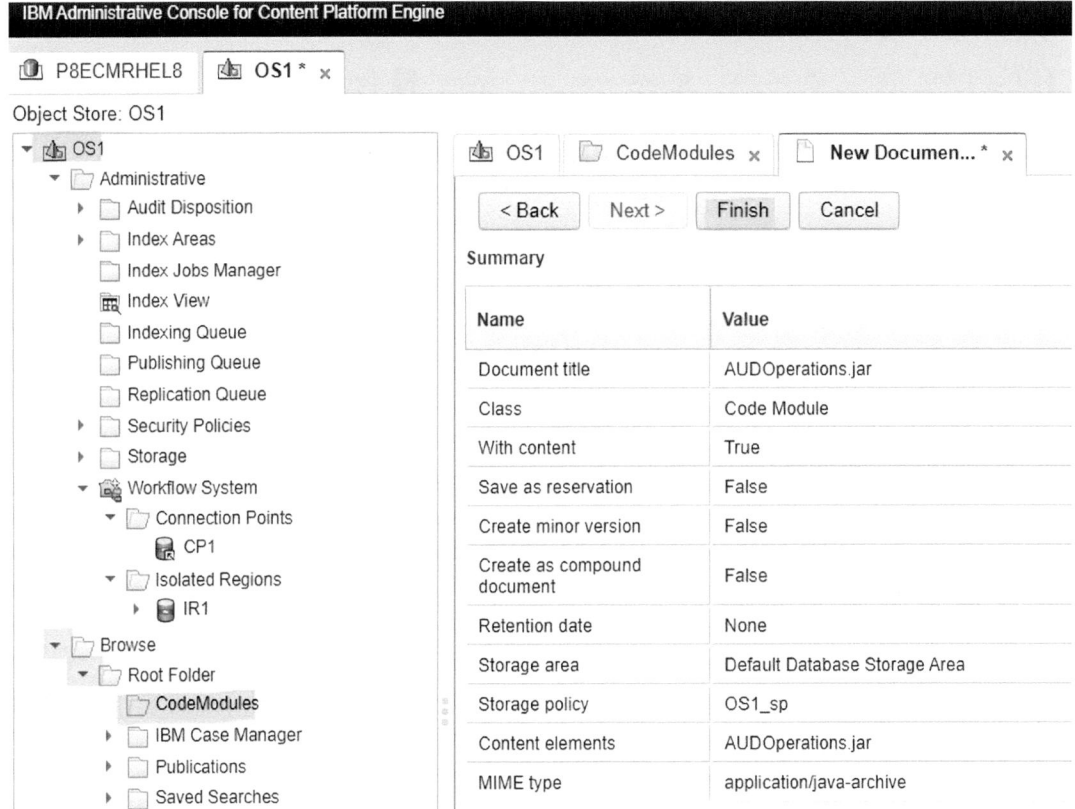

Figure 3-269. *The AUDOperations.jar is created in the OS1, Design Object Store*

CHAPTER 3　IBM JMS INTERFACE DEVELOPMENT IBM FILENET 5.5.X WORKFLOW

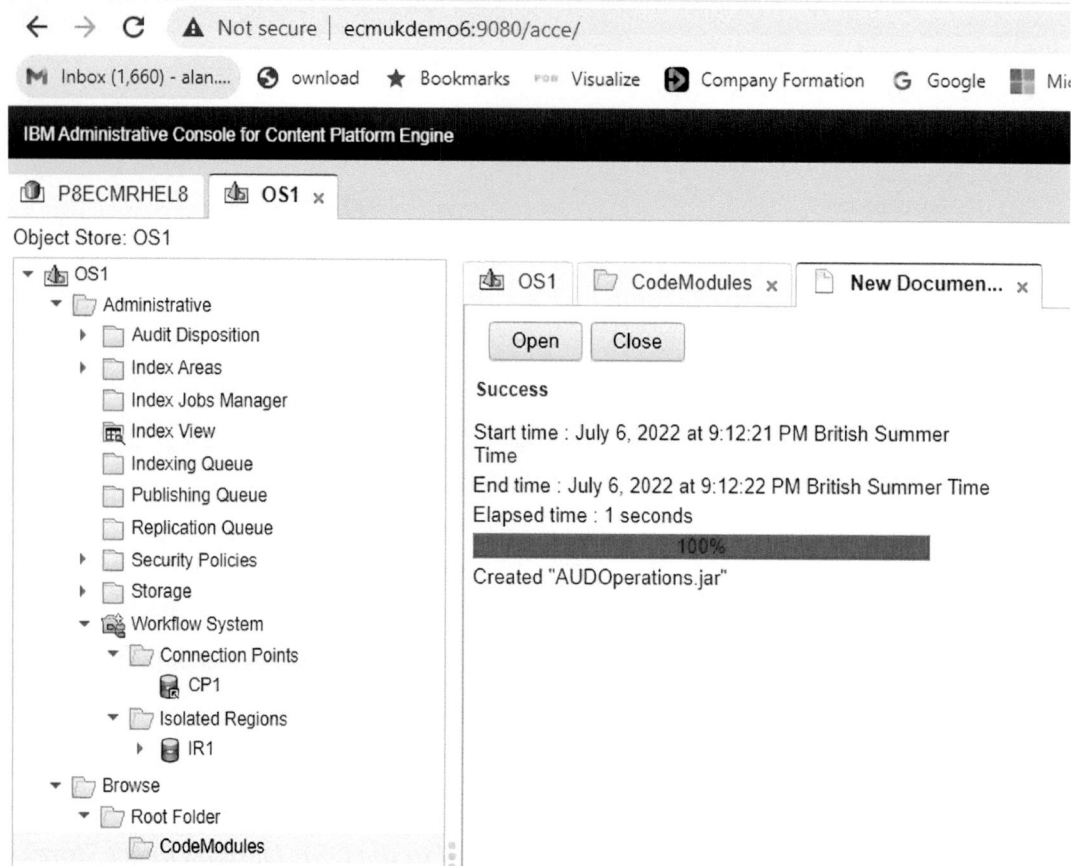

Figure 3-270. *The created AUDOperations.jar in the OS1 Design Object store*

CHAPTER 3 IBM JMS INTERFACE DEVELOPMENT IBM FILENET 5.5.X WORKFLOW

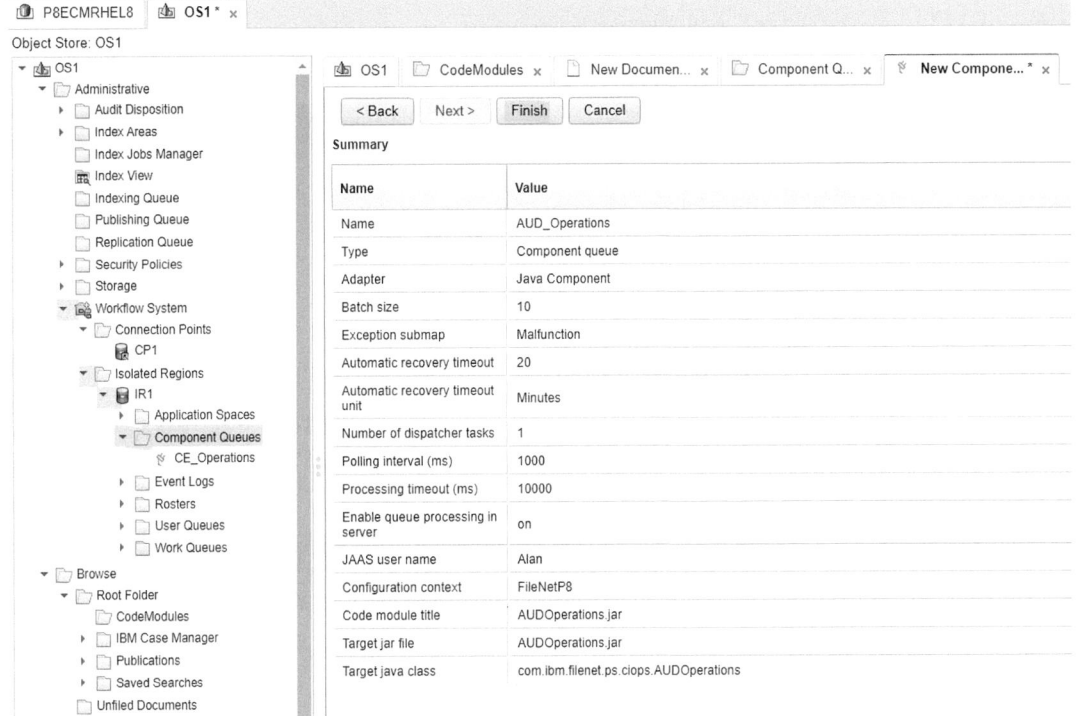

Figure 3-271. The parameters for the AUD_Operations Component Queue

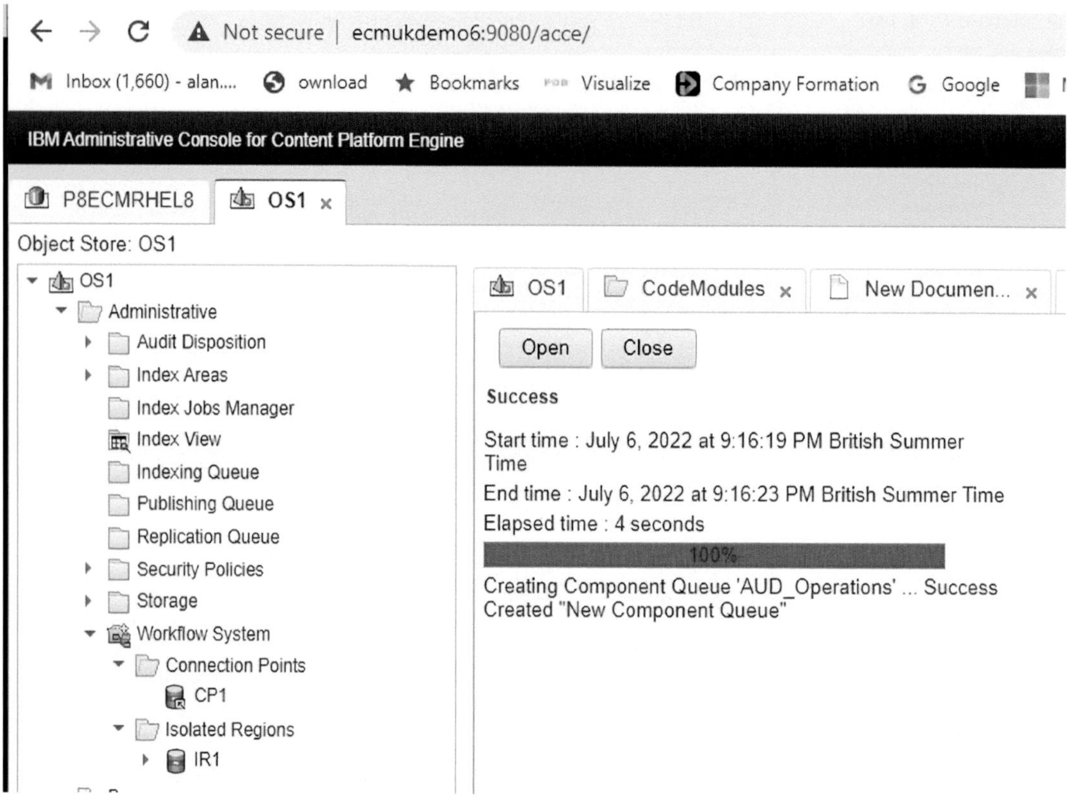

CHAPTER 3   IBM JMS INTERFACE DEVELOPMENT IBM FILENET 5.5.X WORKFLOW

Figure 3-272.   *The AUD_Operations Component Queue is created*

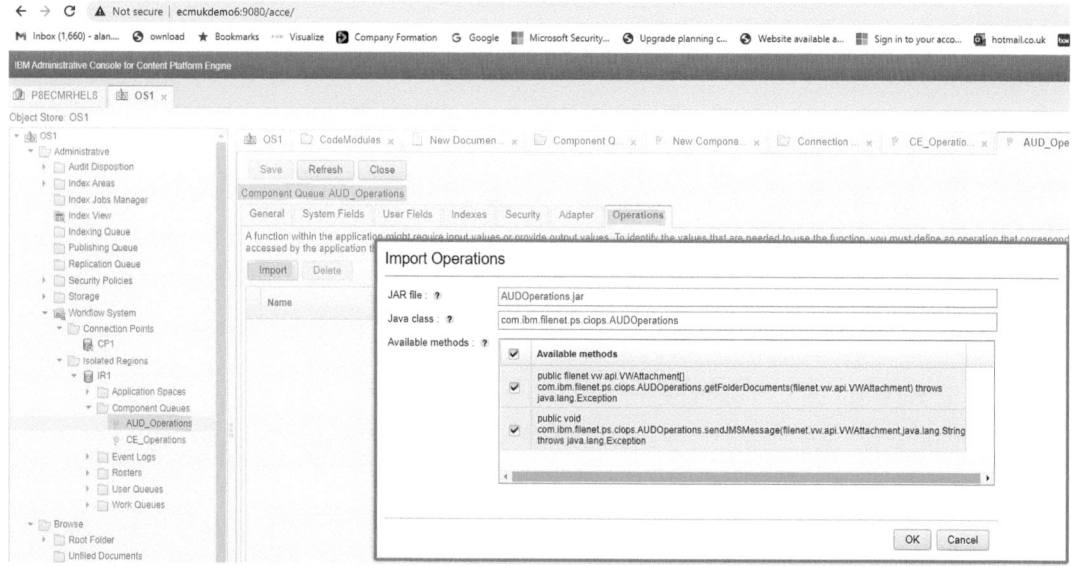

Figure 3-273.   *The methods need to be exposed for use in the Workflow*

CHAPTER 3 IBM JMS INTERFACE DEVELOPMENT IBM FILENET 5.5.X WORKFLOW

The Component Queues **Import Operations** system uses Java reflection to identify the available public methods to be called, from which the option is available to tick a selection box as shown in Figure 3-273. We selected the two methods, **getFolderDocuments** and **sendJMSMessages**, which were tested in the Eclipse JUnit test harness, **AUDOperationsTest.java**.

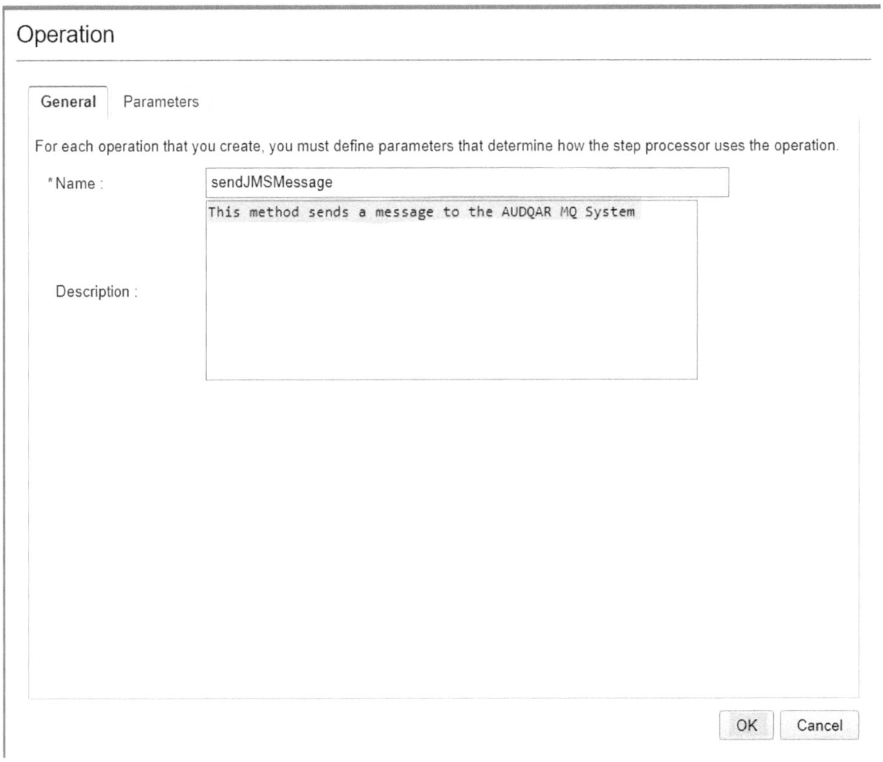

Figure 3-274. The General tab is used to add the Description for the method

The sendJMSMessage method sends a message to the AUDQAR MQ System. First, a description is entered for the General tab.

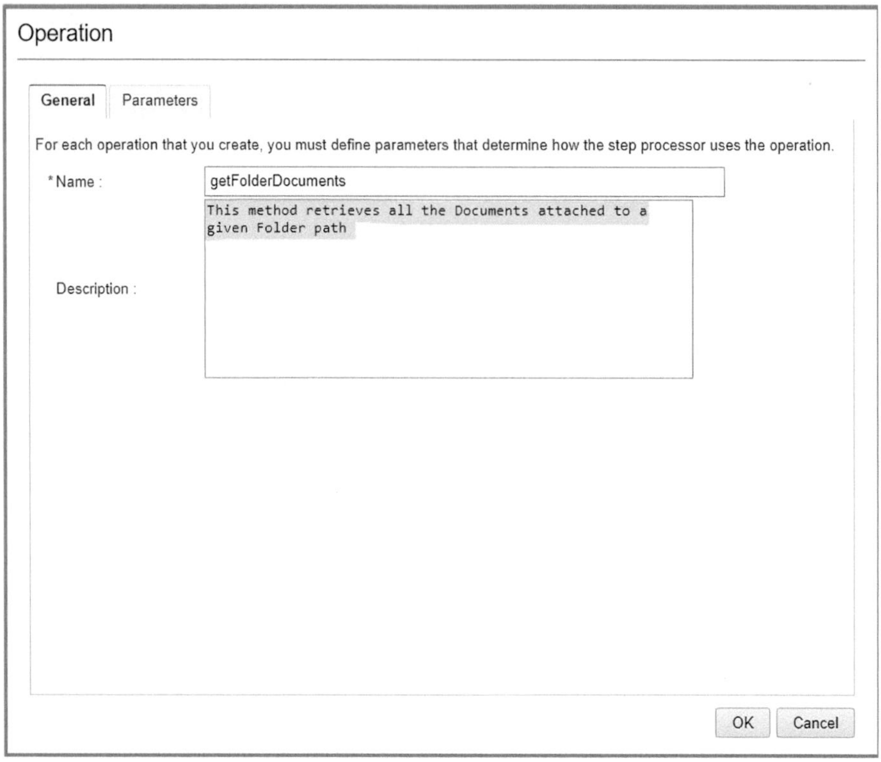

CHAPTER 3   IBM JMS INTERFACE DEVELOPMENT IBM FILENET 5.5.X WORKFLOW

Figure 3-275. *The description for the getFolderDocuments method is entered*

Editing the Parameters

Next, the 15 parameters are edited with their names on the Parameters tab.

Because of the way in which these parameters are sorted, you might find it easier to keep track of the order by deleting all the template parameters and then adding them back one at a time since it is easy for the parameter order to be set mismatching the Java method argument call order.

The parameters for the **sendJMSMessage** method are viewed as templates on the **Parameters** tab. Notice that the param names are not displayed in strict numeric order, and the first attempt to edit with the parameter names led to some being out of order, when displayed as a workflow component step set of arguments. For accuracy of entry, all the automatically created parameters were first deleted and then added back and set up with their names in the correct order for the calling sendJMSMessage method.

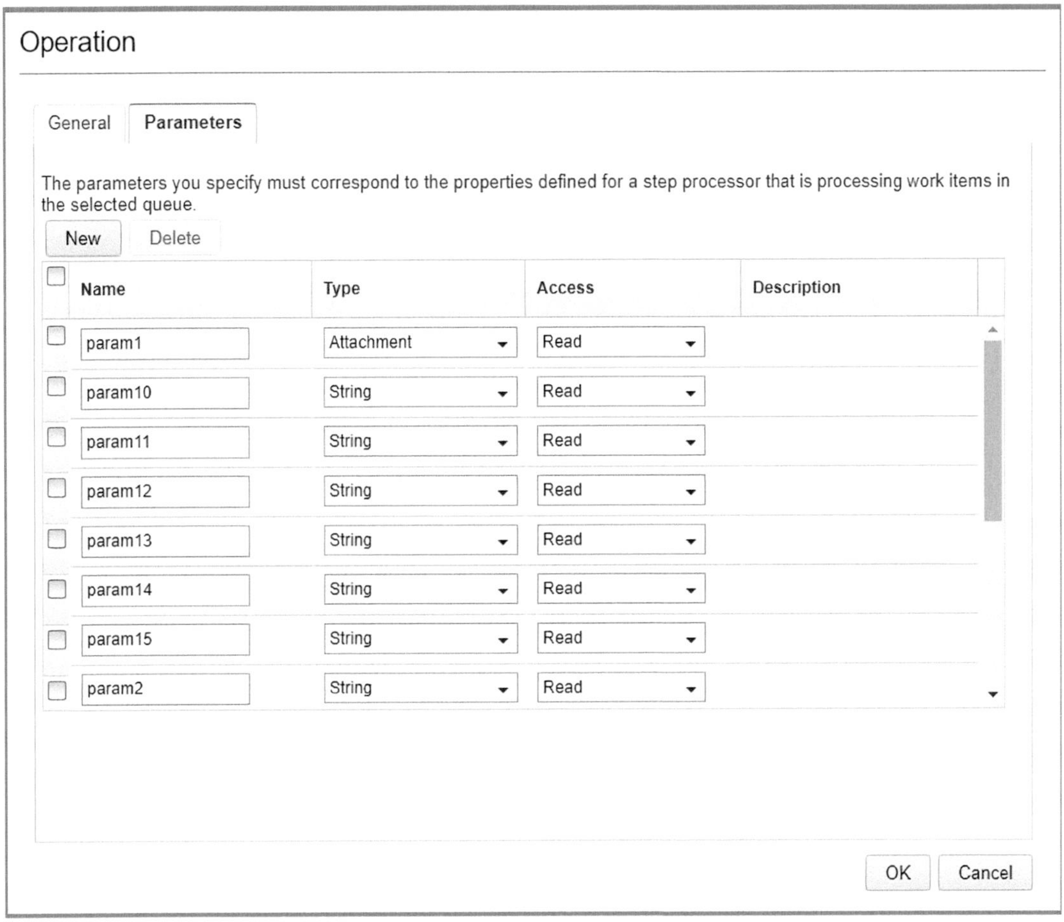

CHAPTER 3   IBM JMS INTERFACE DEVELOPMENT IBM FILENET 5.5.X WORKFLOW

Operation

General | **Parameters**

The parameters you specify must correspond to the properties defined for a step processor that is processing work items in the selected queue.

New   Delete

	Name	Type	Access	Description
☐	param1	Attachment	Read	
☐	param10	String	Read	
☐	param11	String	Read	
☐	param12	String	Read	
☐	param13	String	Read	
☐	param14	String	Read	
☐	param15	String	Read	
☐	param2	String	Read	

OK   Cancel

Figure 3-276.   *The 15 parameters of the sendJMSMessage method*

CHAPTER 3   IBM JMS INTERFACE DEVELOPMENT IBM FILENET 5.5.X WORKFLOW

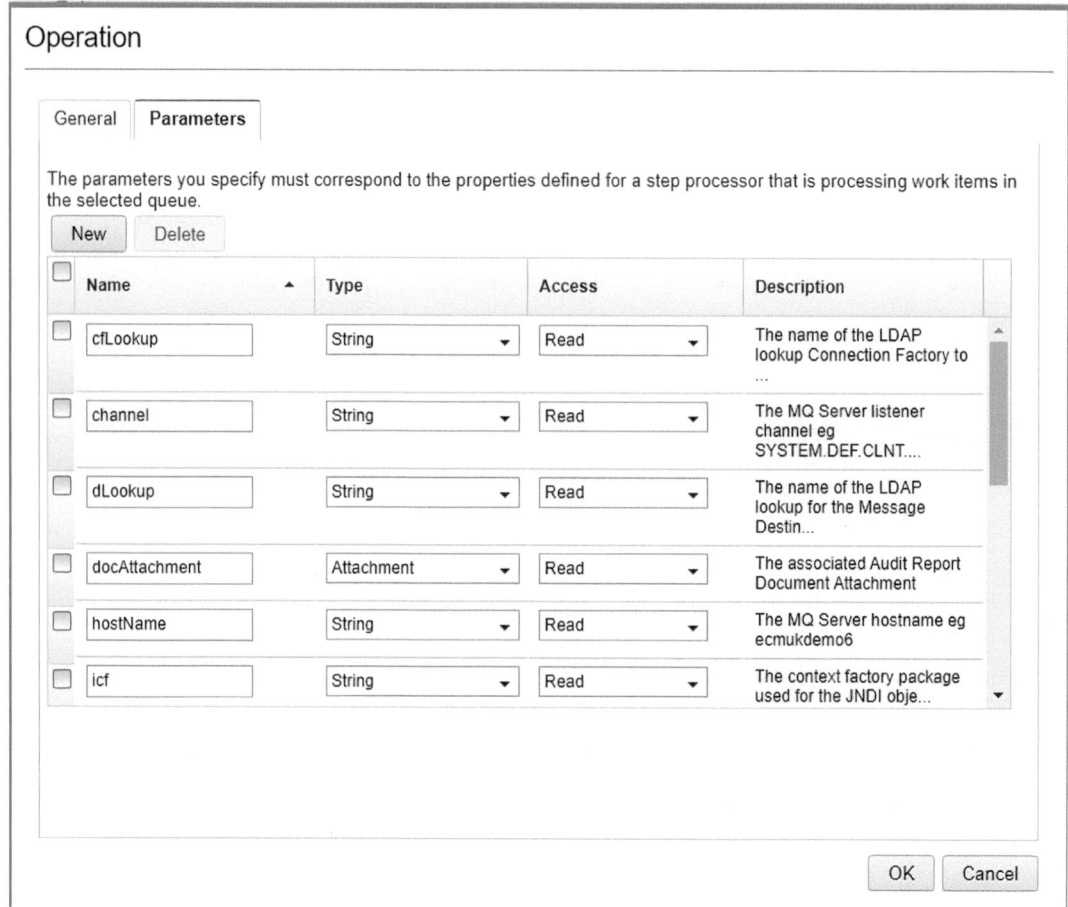

Figure 3-277.   The first attempt at entering the parameters by overwriting the templates

Checking the Deployment in Component Manager and Workflow

We can now launch the Standalone Process Designer Java program using the Windows cmd.exe DOS window (which we launch as the Windows administrator), with the commands as follows:

```
cd c:\PROGRA~2\IBM\FileNet\ContentEngine\tools\PE
pedesigner.bat CP1
```

CHAPTER 3 IBM JMS INTERFACE DEVELOPMENT IBM FILENET 5.5.X WORKFLOW

This loads a pop-up window prompting us for a Process Engine administrator login (see Chapter 2, Figure 2-190).

IBM Process Designer Component Queue Configuration

After the parameters for the **AUD_Operations** have been configured using the IBM FileNet Content Engine **acce** web administration tool, the Workflow Component Queues have to be configured using the IBM Process Designer Java application. The menu item **View ▶ Configuration** allows the Component Queues to be configured, as shown in Figure 3-278.

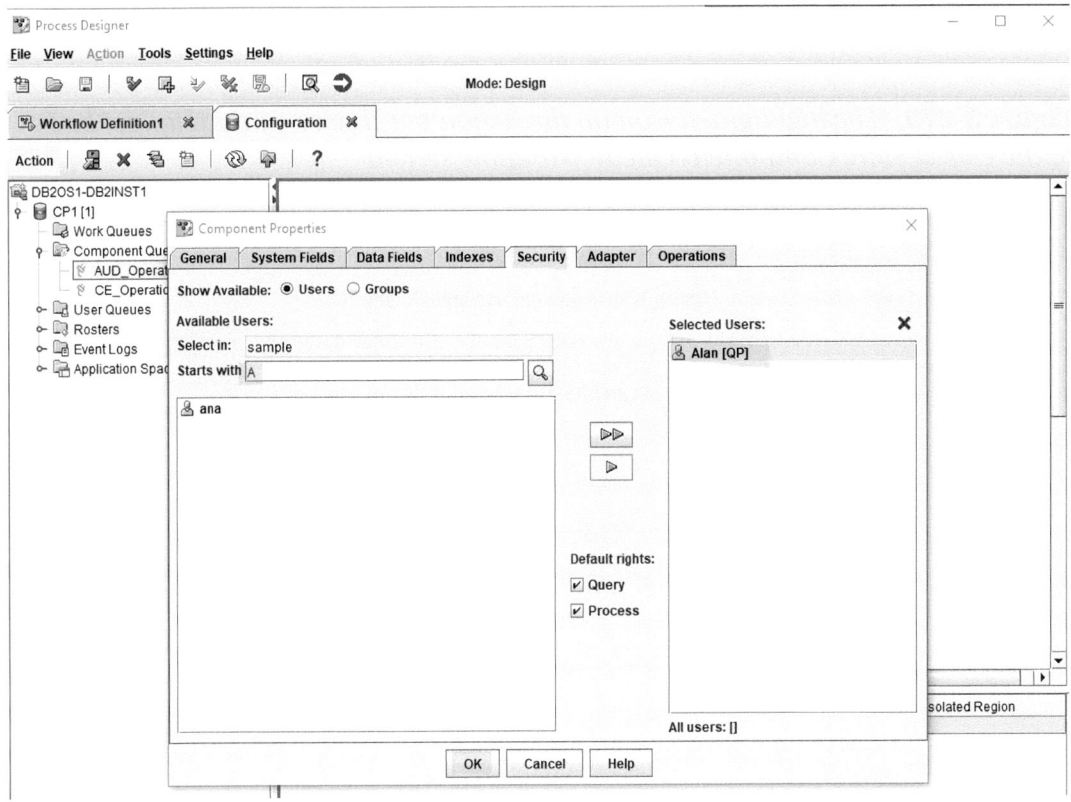

*Figure 3-278. The **View ▶ Configuration** menu launches the Configuration tool*

CHAPTER 3   IBM JMS INTERFACE DEVELOPMENT IBM FILENET 5.5.X WORKFLOW

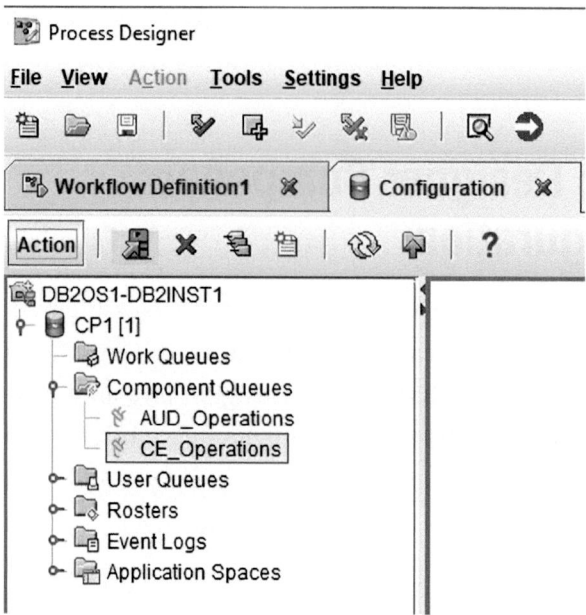

Figure 3-279. *The highlighted icon on the Action bar line commits changes back to the Object Store Component Queue Workflow system*

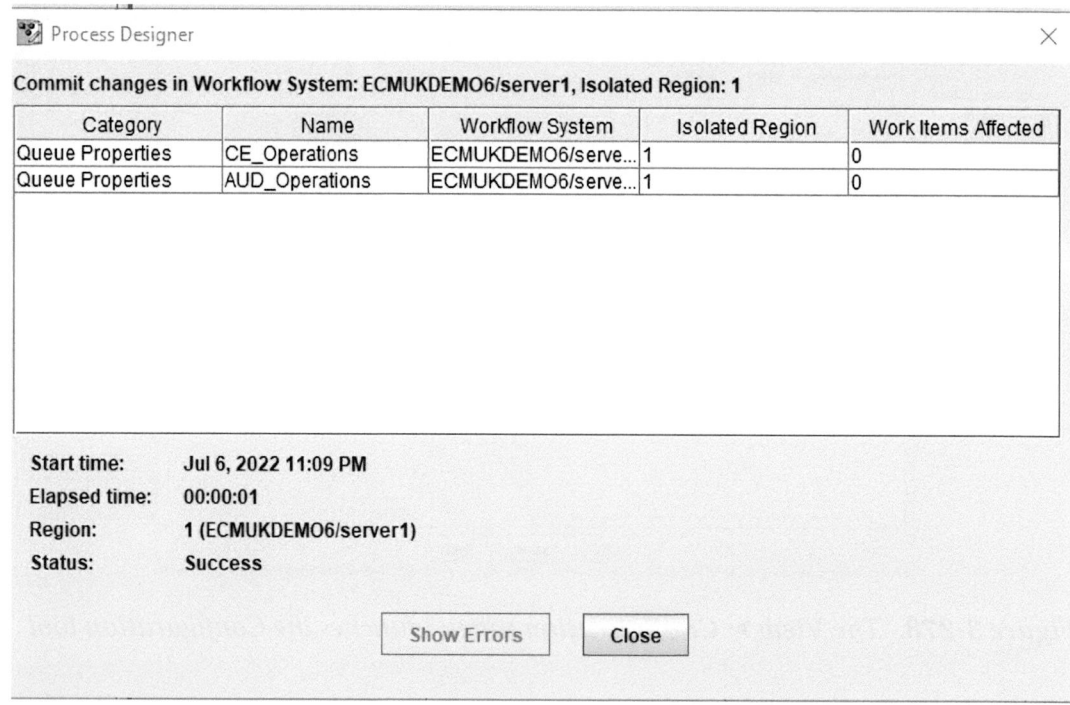

Figure 3-280. *The list of Queue Properties to be updated is displayed*

CHAPTER 3 IBM JMS INTERFACE DEVELOPMENT IBM FILENET 5.5.X WORKFLOW

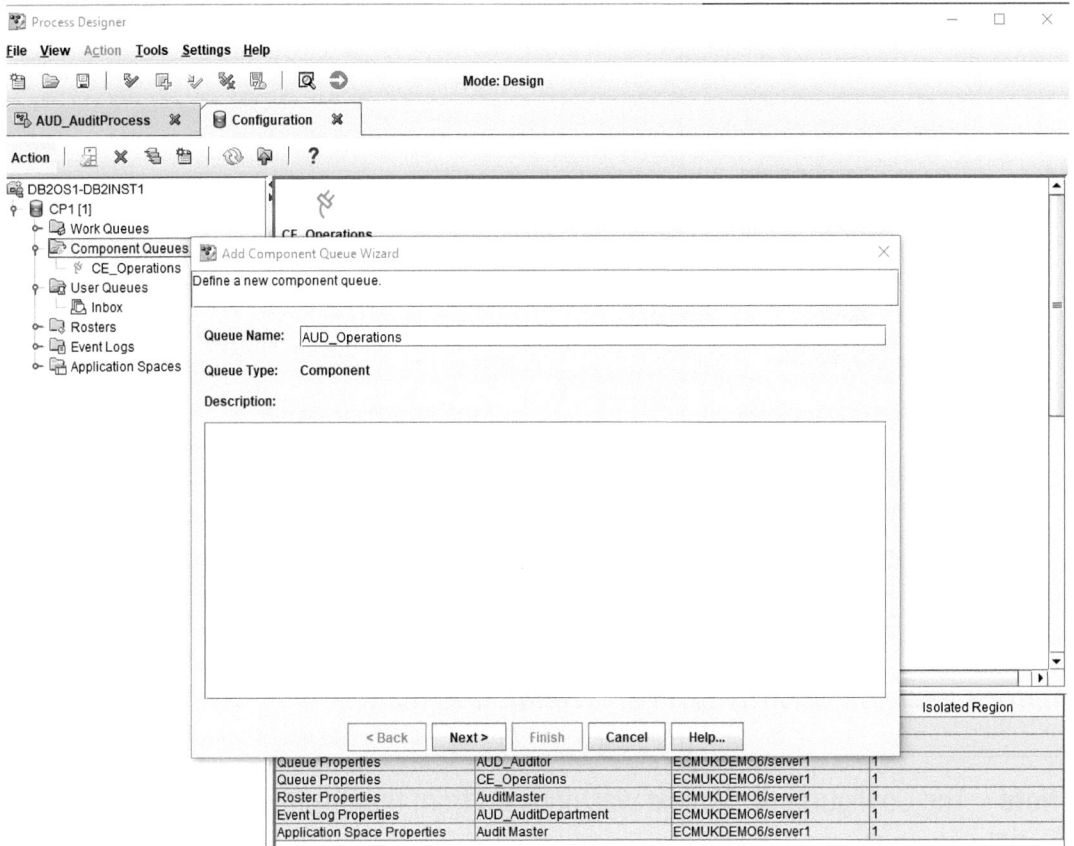

Figure 3-281. *The AUD_Operations queue is created by right-clicking the Component Queues node in the Process Designer Configuration tool view*

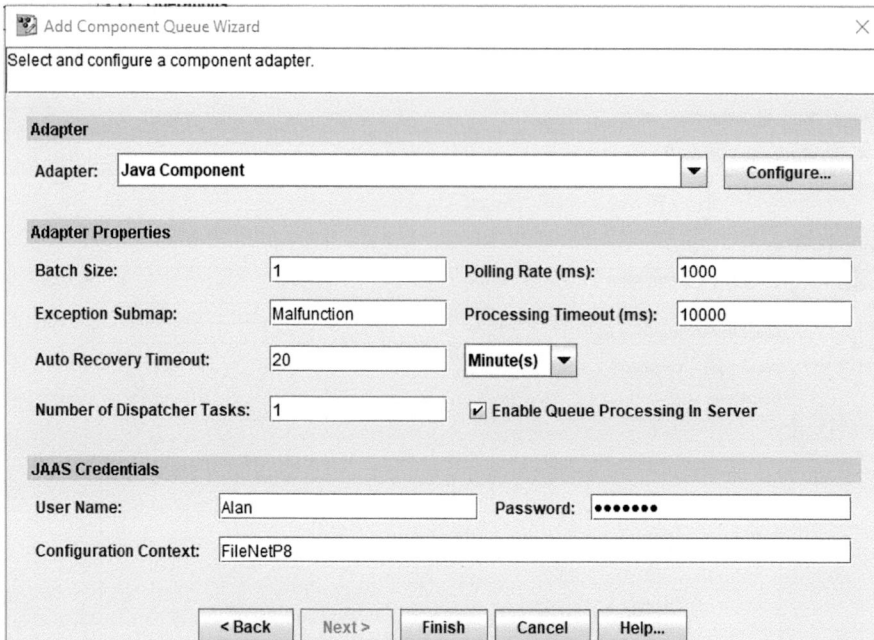

Figure 3-282. *The Security JAAS Credentials are entered*

Note The Configuration Context was updated from the displayed FileNetP8 value to **CELogin** in Figure 3-282.

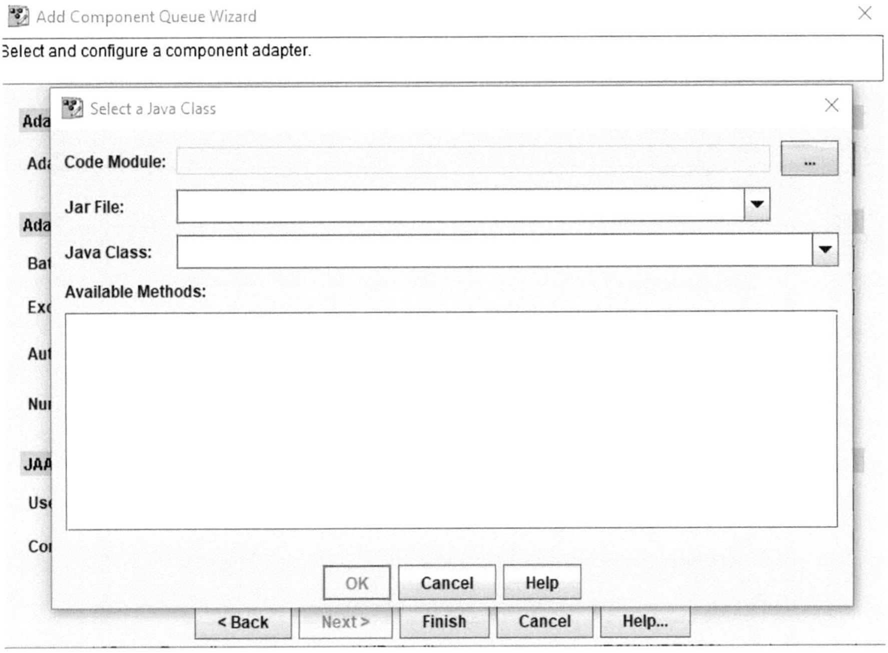

Figure 3-283. The Code Module is selected using the highlighted icon

CHAPTER 3 IBM JMS INTERFACE DEVELOPMENT IBM FILENET 5.5.X WORKFLOW

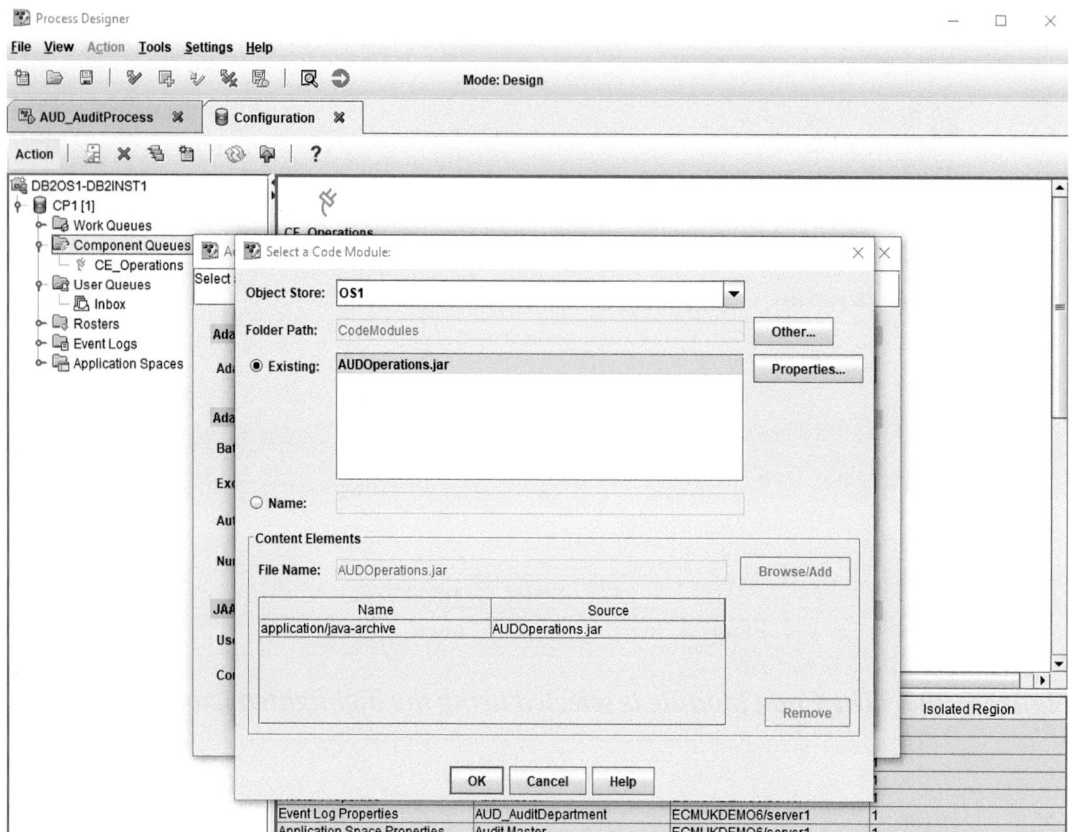

Figure 3-284. *The AUDOperations.jar file is selected from the OS1 Design Object store*

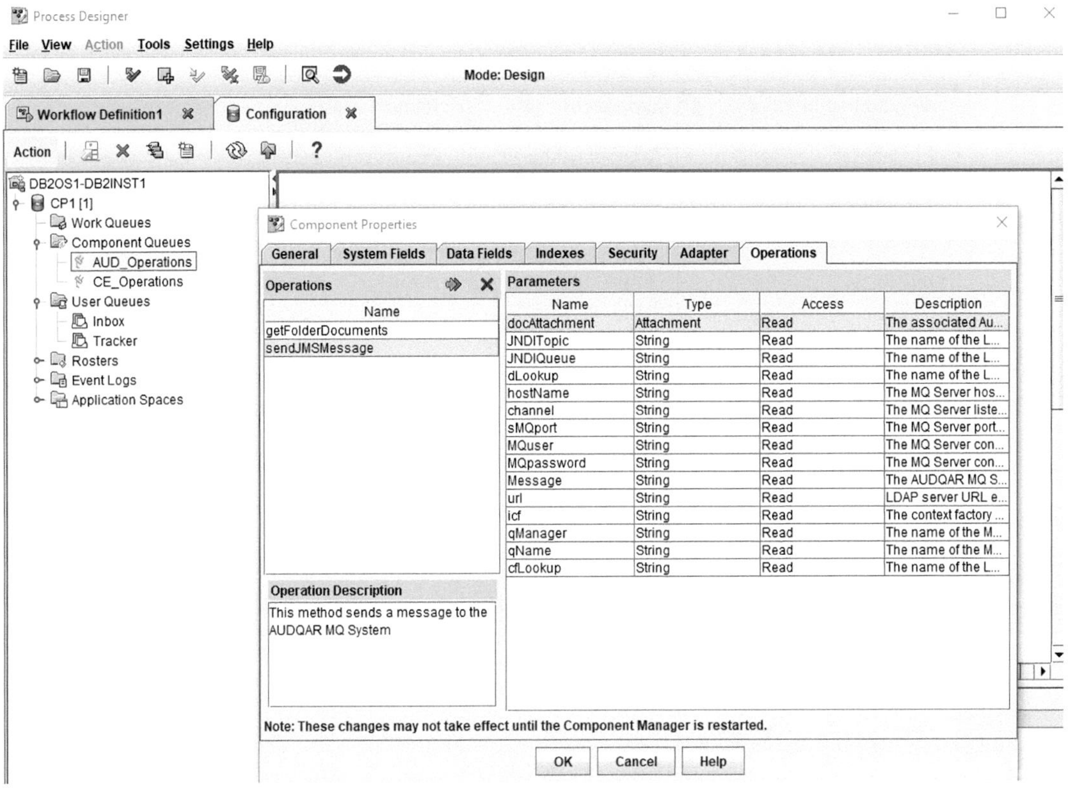

CHAPTER 3   IBM JMS INTERFACE DEVELOPMENT IBM FILENET 5.5.X WORKFLOW

Figure 3-285.   The sendJMSMessage parameters are displayed

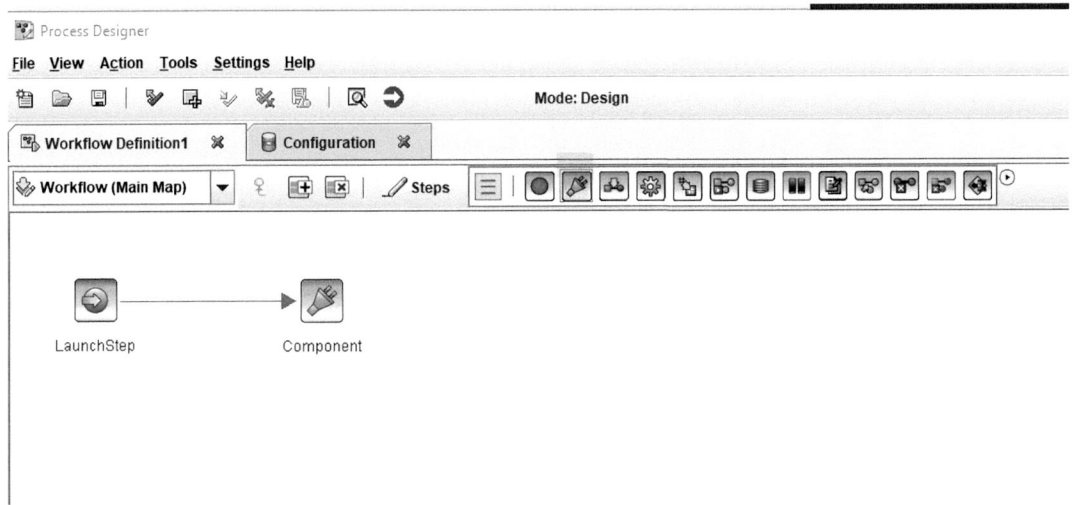

Figure 3-286.   The Component step is dragged and dropped to the new Workflow

CHAPTER 3 IBM JMS INTERFACE DEVELOPMENT IBM FILENET 5.5.X WORKFLOW

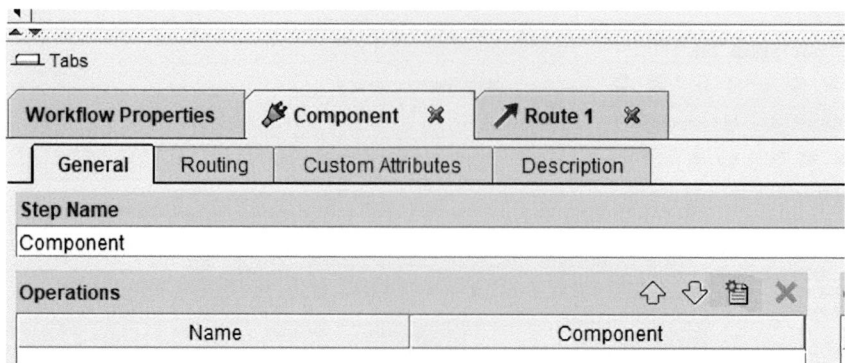

Figure 3-287. *The highlighted icon is selected to allow selection of the required queue*

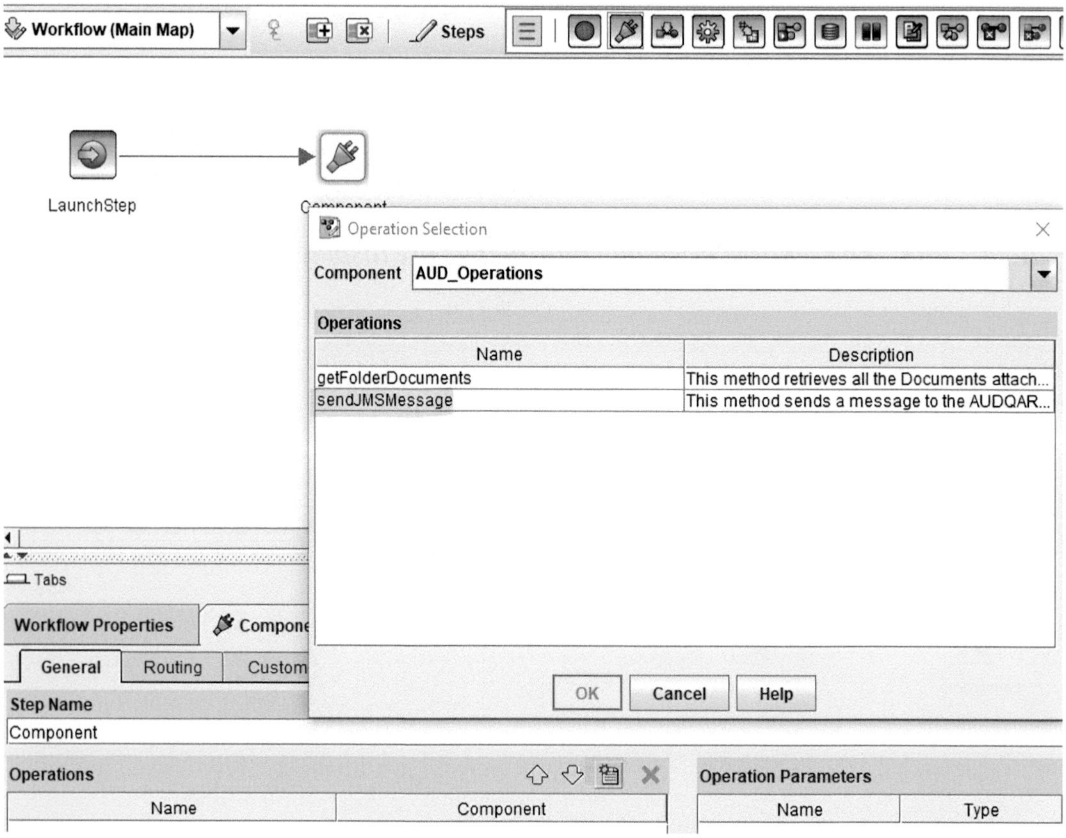

Figure 3-288. *The sendJMSMessage method is clicked to load the call parameters*

CHAPTER 3   IBM JMS INTERFACE DEVELOPMENT IBM FILENET 5.5.X WORKFLOW

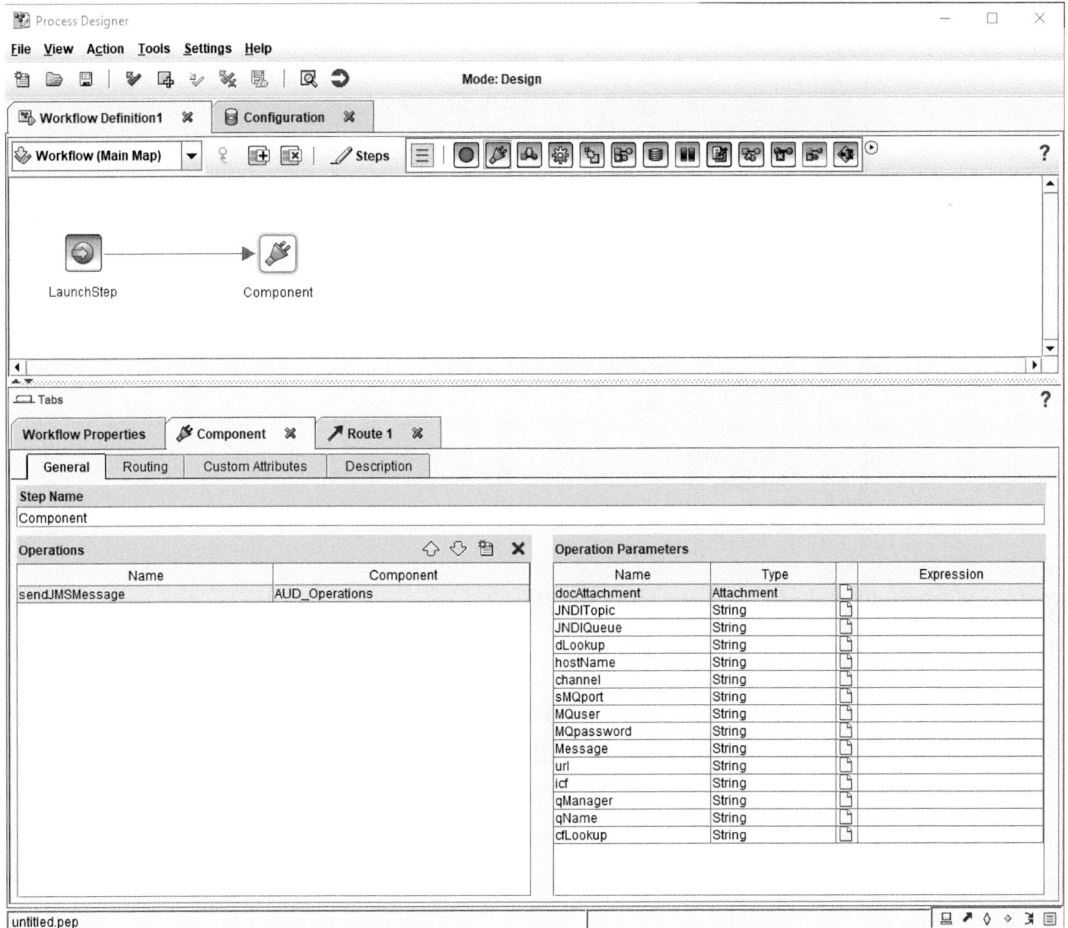

Figure 3-289. *The Expression values can be filled with String values*

CHAPTER 3   IBM JMS INTERFACE DEVELOPMENT IBM FILENET 5.5.X WORKFLOW

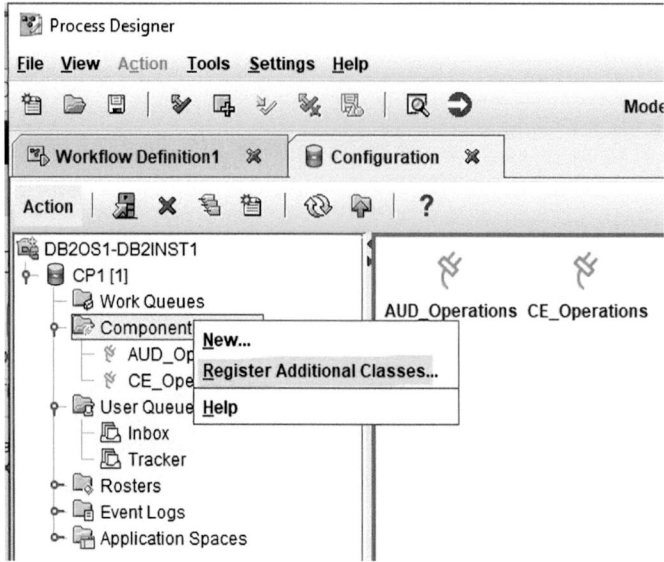

Figure 3-290. *The right-mouse click on the Component node is used to add additional Java classes*

Additional supporting MQ libraries are required in the **AUD_Operations** Classpath to allow the installed methods to run correctly. These are added as follows.

```
(base) [root@ECMUKDEMO6 opt]# cd MQJars/
(base) [root@ECMUKDEMO6 MQJars]# ls
activation.jar                        com.ibm.ws.emf.jar
bootstrap.jar                         com.ibm.ws.runtime.jar
com.ibm.ffdc.jar                      ibmorb.jar
com.ibm.mq.allclient-9.2.4.0.jar      j2ee.jar
com.ibm.mq.jar                        javax.jms-api-2.0.1.jar
com.ibm.mq.jmqi.jar                   json-20211205.jar
com.ibm.msg.client.jms.internal.jar   org.eclipse.emf.common.jar
com.ibm.msg.client.jms.jar            org.eclipse.emf.ecore.jar
com.ibm.msg.client.provider.jar       providerutil.jar
com.ibm.ws.ejb.thinclient_8.5.0.jar
(base) [root@ECMUKDEMO6 MQJars]# cp com.ibm.mq.allclient-9.2.4.0.jar /mnt/hgfs/Installs/
(base) [root@ECMUKDEMO6 MQJars]# cp javax.jms-api-2.0.1.jar /mnt/hgfs/Installs/
(base) [root@ECMUKDEMO6 MQJars]# cp com.ibm.msg.*.jar /mnt/hgfs/Installs/
(base) [root@ECMUKDEMO6 MQJars]# cp com.ibm.mq.jar /mnt/hgfs/Installs/
(base) [root@ECMUKDEMO6 MQJars]# cp com.ibm.mq.jmqi.jar /mnt/hgfs/Installs/
(base) [root@ECMUKDEMO6 MQJars]#
```

Figure 3-291. *The required MQ Series JMS message libraries are copied for installation*

CHAPTER 3   IBM JMS INTERFACE DEVELOPMENT IBM FILENET 5.5.X WORKFLOW

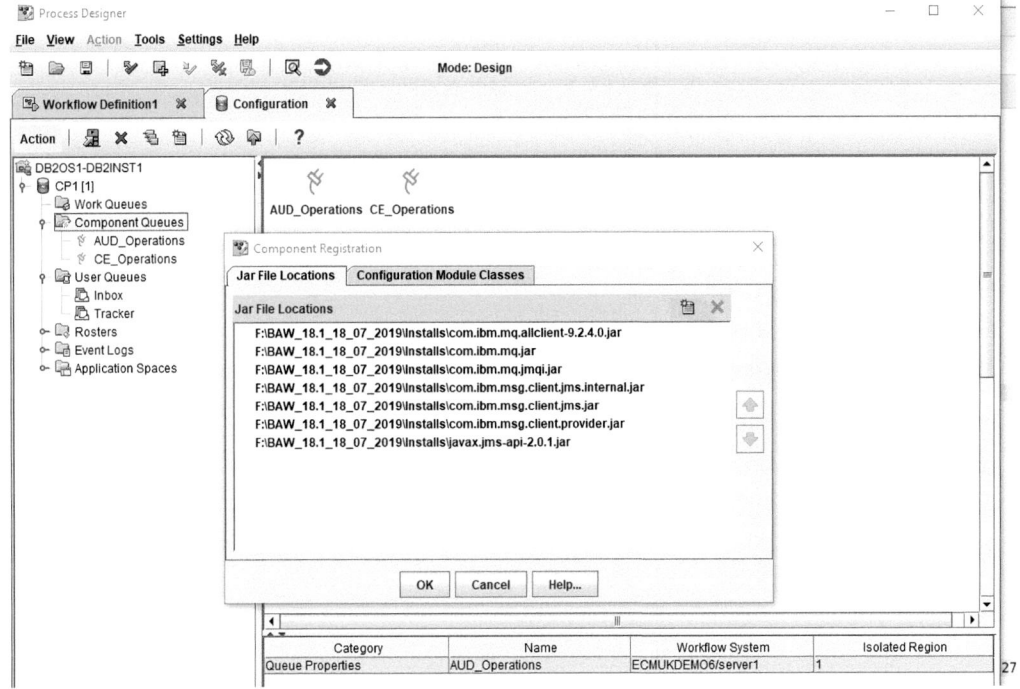

Figure 3-292. *The required .jar libraries are imported from the Installs directory*

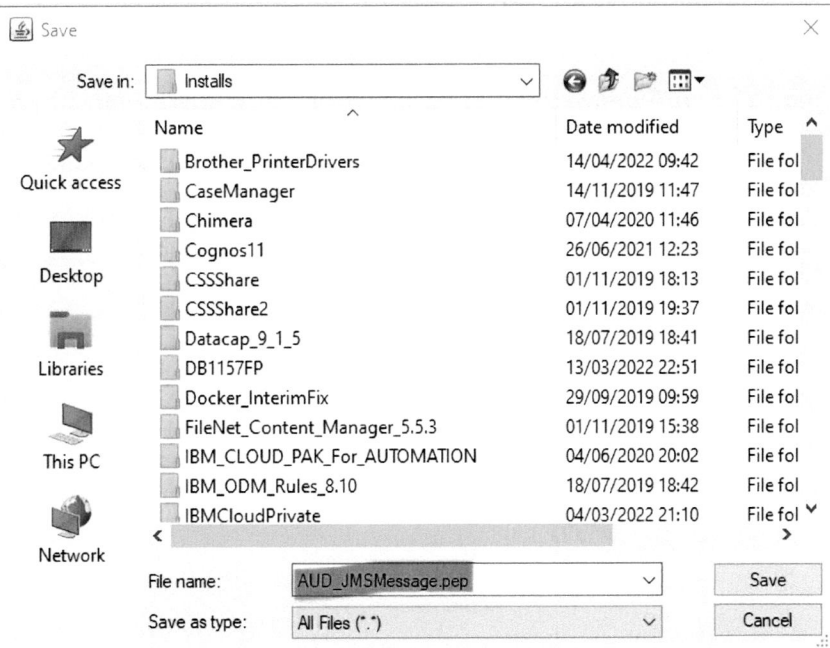

Figure 3-293. *The workflow is saved as **AUD_JMSMessage.pep***

CHAPTER 3 IBM JMS INTERFACE DEVELOPMENT IBM FILENET 5.5.X WORKFLOW

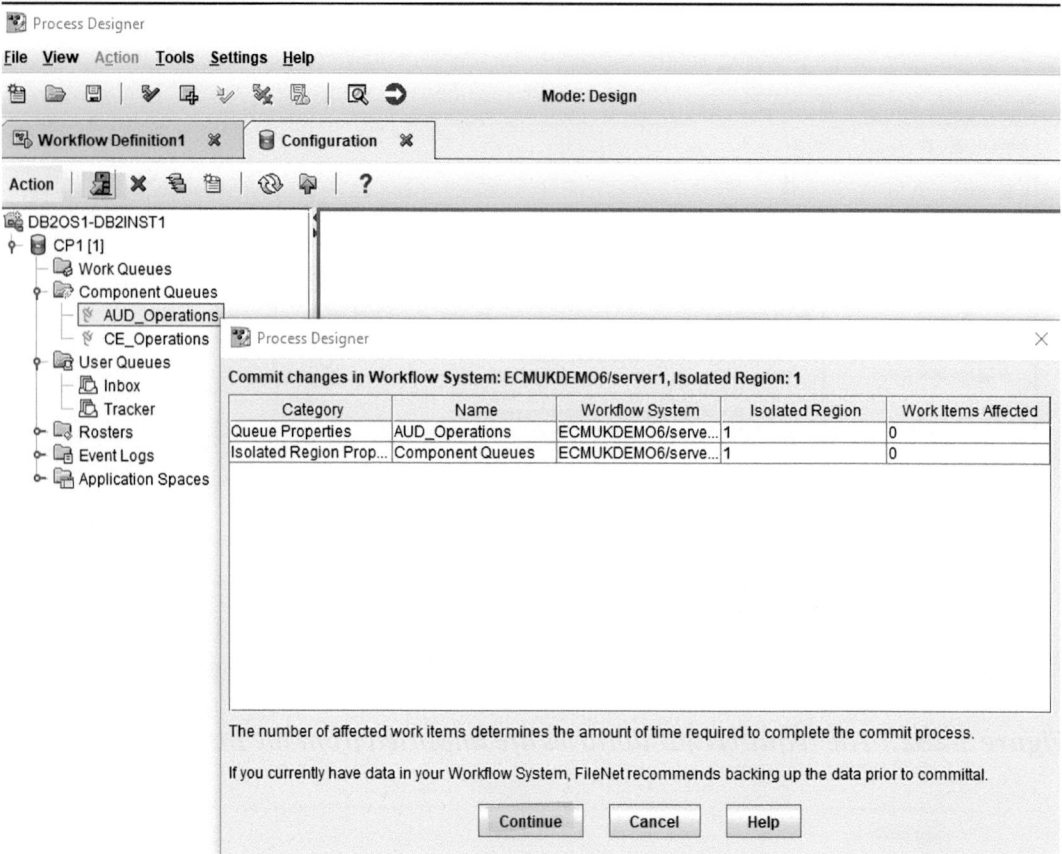

Figure 3-294. *The changes to the AUD_Operations queue are committed*

618

CHAPTER 3 IBM JMS INTERFACE DEVELOPMENT IBM FILENET 5.5.X WORKFLOW

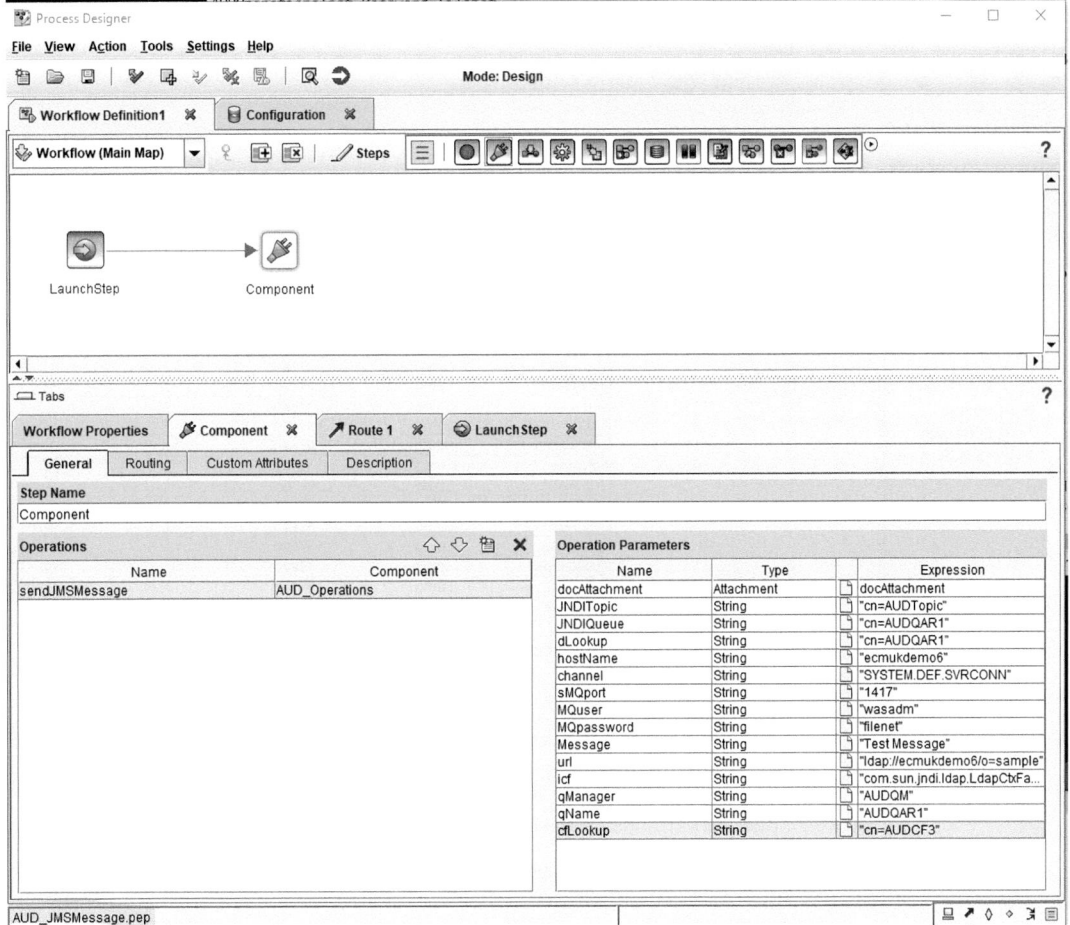

Figure 3-295. *The initial parameter order is shown to be incorrect!*

As shown in Figure 3-295, the initial sendJMSMessage method argument list order was incorrect, because the template parameters are not correctly ordered in the list box! The parameters were then reentered singly.

CHAPTER 3   IBM JMS INTERFACE DEVELOPMENT IBM FILENET 5.5.X WORKFLOW

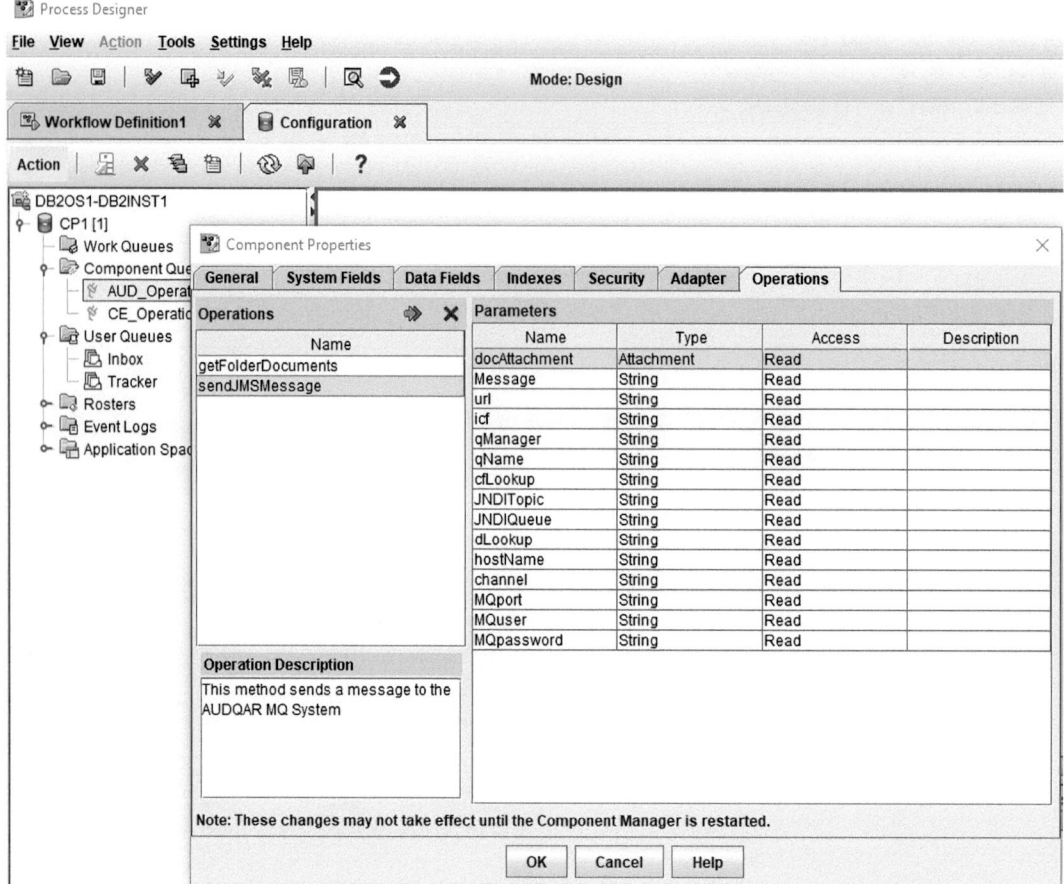

Figure 3-296. *The AUD_Operations parameters are reentered one at a time*

CHAPTER 3   IBM JMS INTERFACE DEVELOPMENT IBM FILENET 5.5.X WORKFLOW

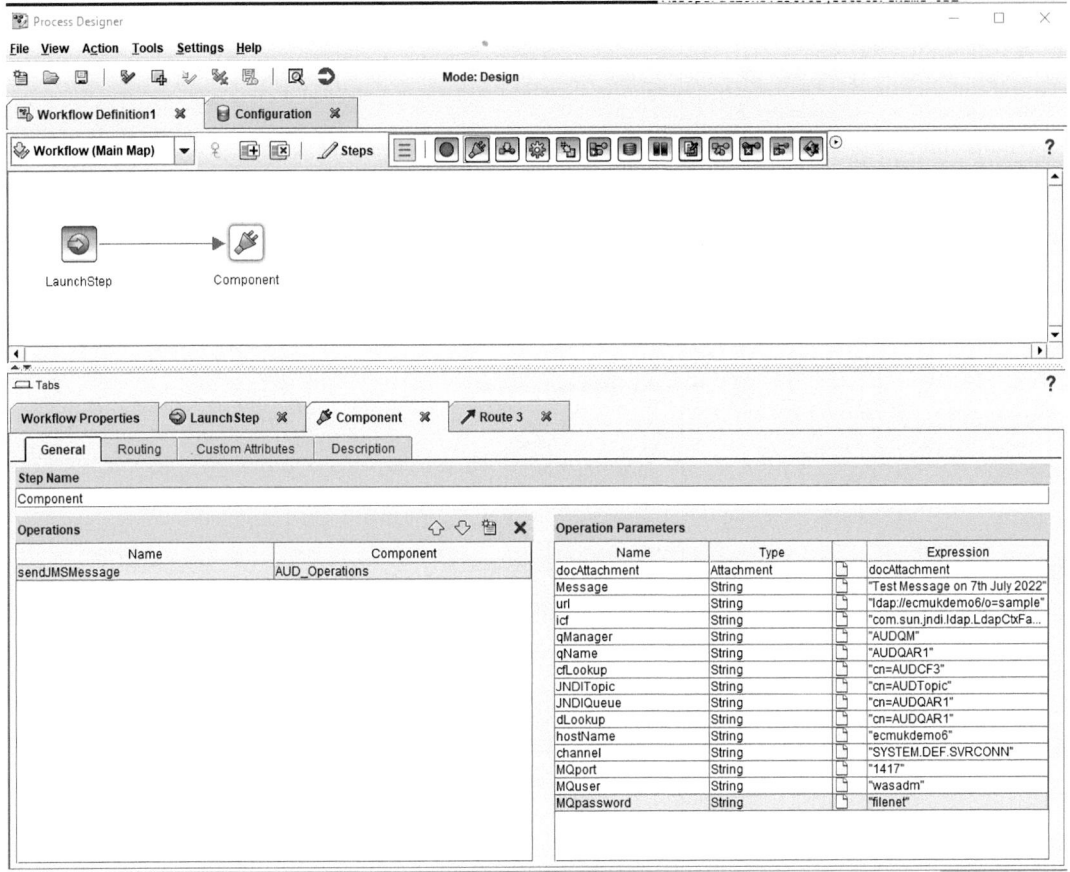

Figure 3-297. *The Component step parameters are now ordered correctly*

CHAPTER 3   IBM JMS INTERFACE DEVELOPMENT IBM FILENET 5.5.X WORKFLOW

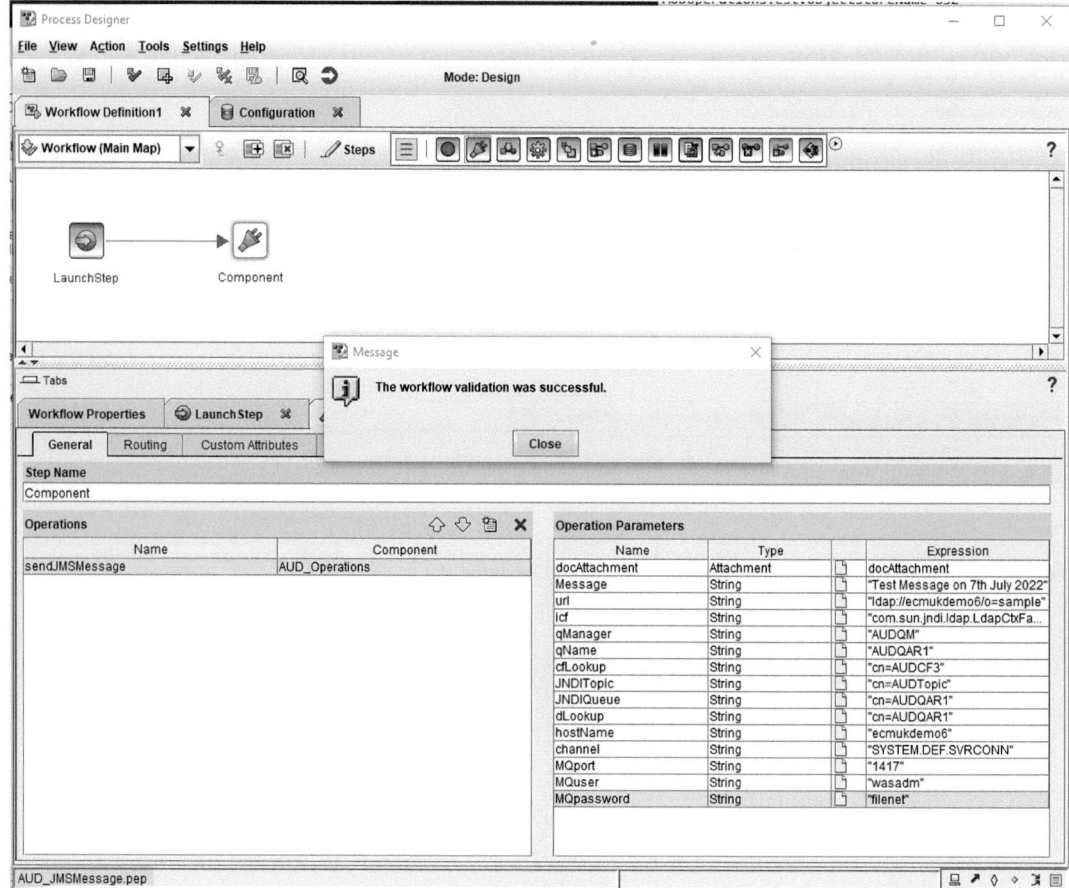

Figure 3-298.   *The workflow is validated and resaved*

Note   It was discovered that the following original section of code (lines in red bold) caused Java reflection to fail in the import tool, so parameters were not exposed!

```
MessageProducer myProducer = JMSsession.createProducer(myDest);
if (Message.length() > 0) {
  TextMessage outMessage = JMSsession.createTextMessage();
  outMessage.setText(Message);
  myProducer.send(outMessage);
    JMSsession.commit();
}
```

622

The new code is as follows (using a different Queue Factory class):

```
private void putMsg( Destination myDest, String outString, javax.jms.
Session jmssession )
throws JMSException, Exception
{
// Use generic MessageProducer instead of Queue/Topic Producer
MessageProducer myProducer = jmssession.createProducer(myDest);

// Get user input and create messages.  Loop until user sends CR.
if (outString.length() > 0) {
TextMessage outMessage = jmssession.createTextMessage();
outMessage.setText(outString);
myProducer.send(outMessage);
jmssession.commit();
}
myProducer.close();
}
```

Part 6 – Transferring Workflow and Setting Up Workflow Subscriptions

Creating a Workflow Subscription on a New Audit Report Document Event

The Target Object store, OS2, with a Workflow Definition document class (**AUD_JMSMessage.pep** file) is used to run the Workflow we created, after a new Audit Report document is created in the Target Object store from an Audit Master solution Case Manager Audit Department Case.

Logging in to the acce web Administration console, we can create the required Workflow Subscription as follows.

CHAPTER 3 IBM JMS INTERFACE DEVELOPMENT IBM FILENET 5.5.X WORKFLOW

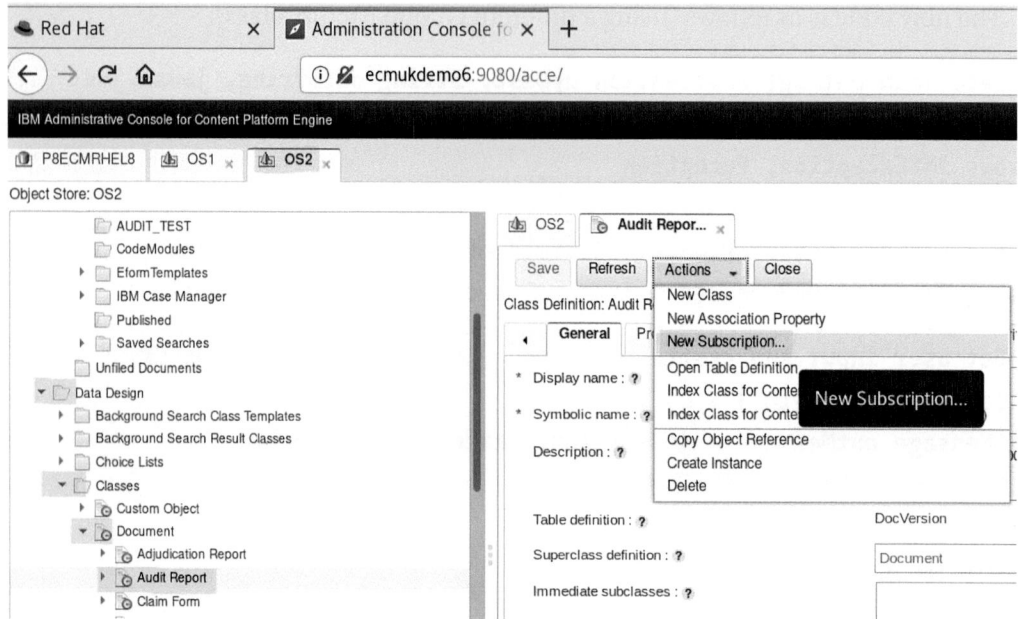

Figure 3-299. *The Audit Report class is used for the basis of a new Workflow subscription*

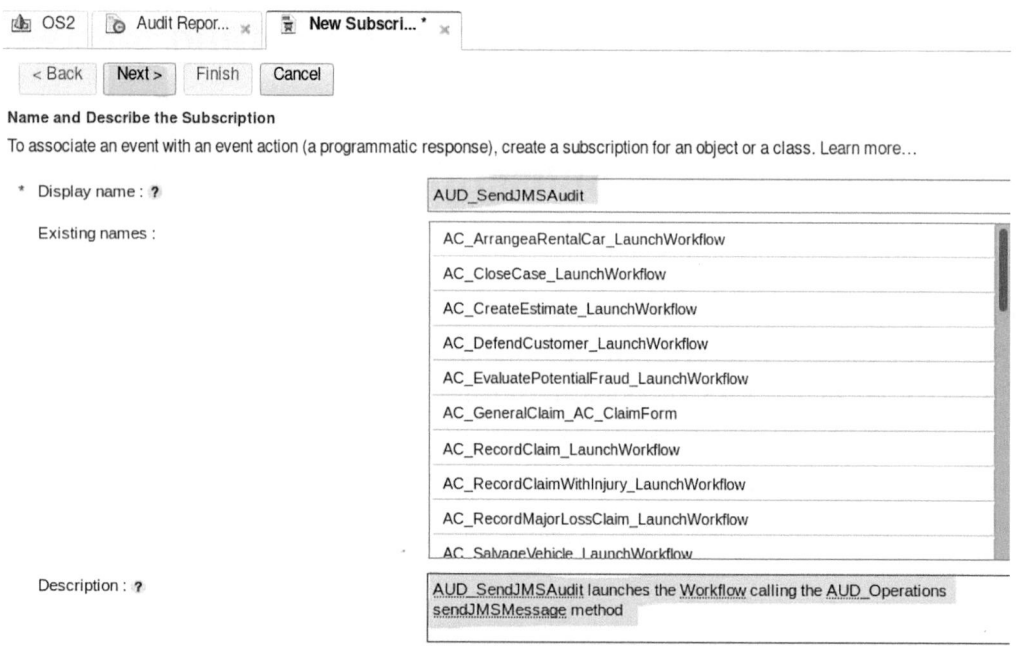

Figure 3-300. *The AUD_SendJMSAudit subscription is created to launch the Workflow*

CHAPTER 3   IBM JMS INTERFACE DEVELOPMENT IBM FILENET 5.5.X WORKFLOW

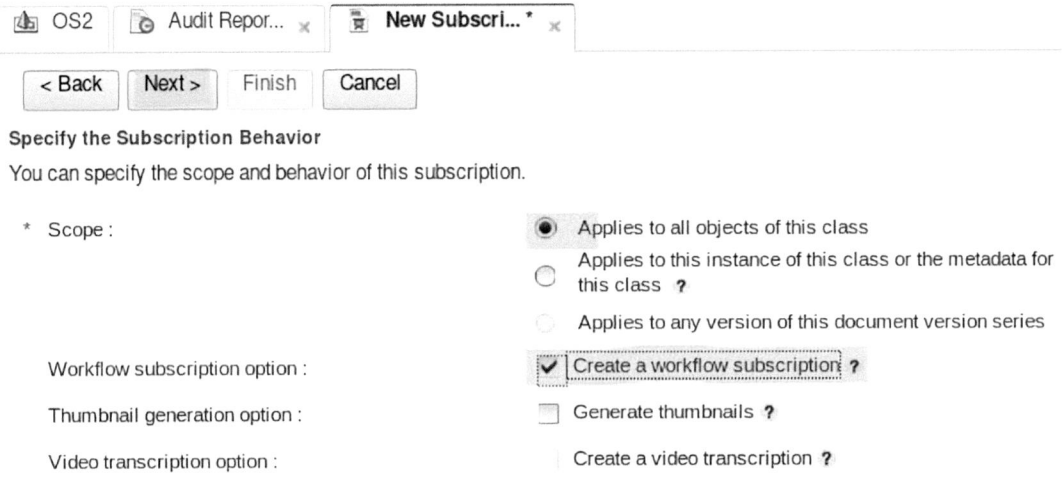

Figure 3-301. The Create a workflow subscription must be ticked

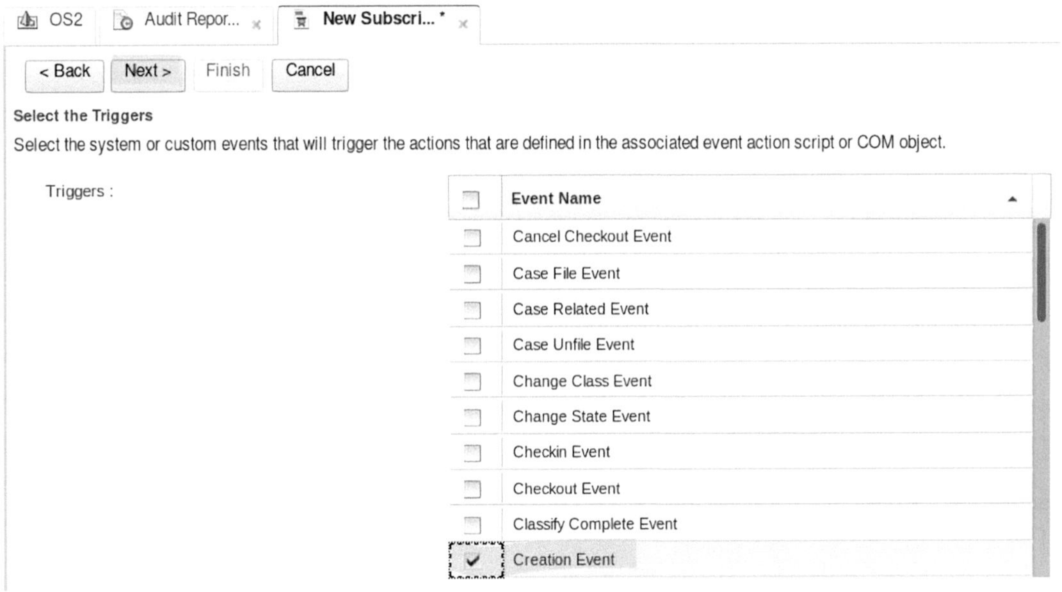

Figure 3-302. The Creation Event is selected to trigger the Workflow

CHAPTER 3   IBM JMS INTERFACE DEVELOPMENT IBM FILENET 5.5.X WORKFLOW

Transferring the JMS Test Workflow to the Target Object Store

The Workflow created in the standalone Process Designer Java applet must be transferred to the Target Object Store to allow it to be triggered by the Workflow Subscription Event we created in the previous section.

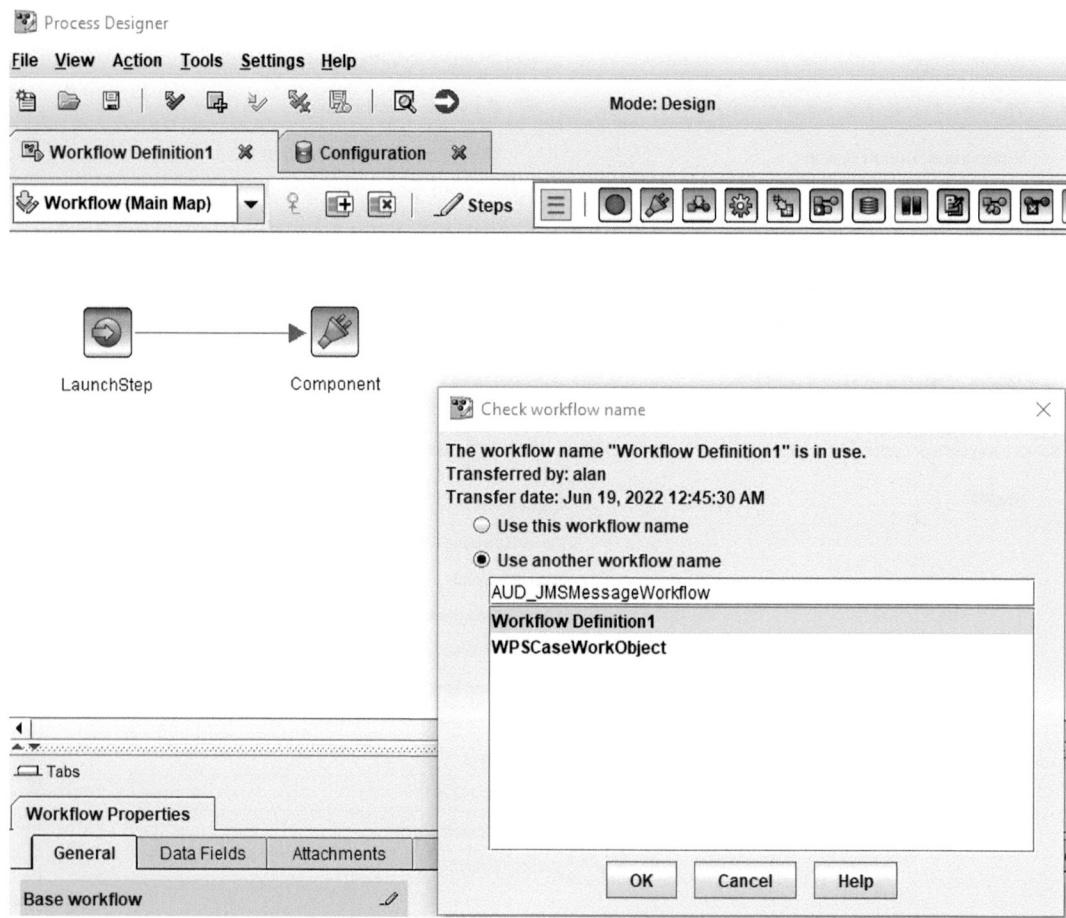

Figure 3-303. *The JMS test workflow is transferred to the Target Object store*

CHAPTER 3 IBM JMS INTERFACE DEVELOPMENT IBM FILENET 5.5.X WORKFLOW

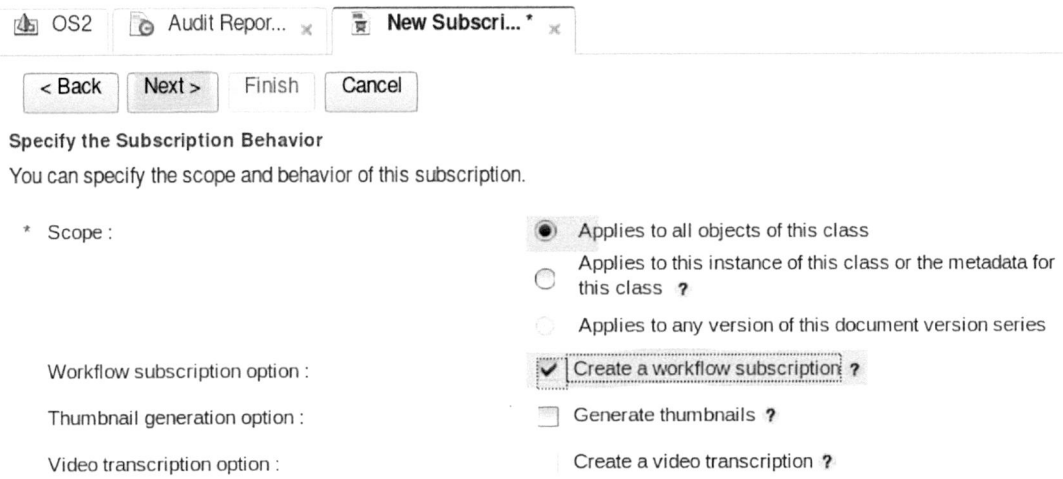

Figure 3-301. *The Create a workflow subscription must be ticked*

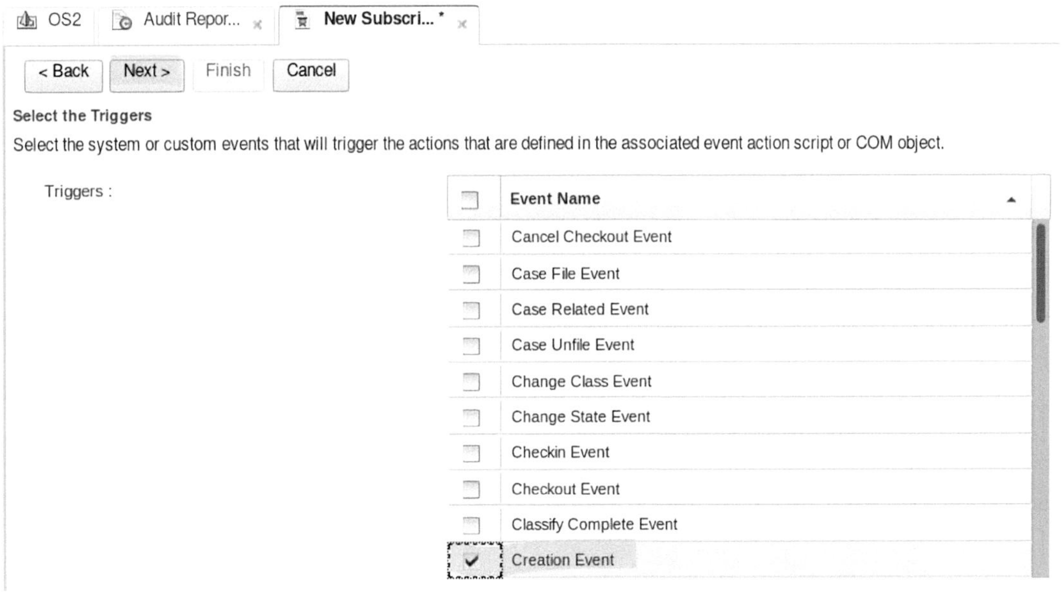

Figure 3-302. *The Creation Event is selected to trigger the Workflow*

625

CHAPTER 3   IBM JMS INTERFACE DEVELOPMENT IBM FILENET 5.5.X WORKFLOW

Transferring the JMS Test Workflow to the Target Object Store

The Workflow created in the standalone Process Designer Java applet must be transferred to the Target Object Store to allow it to be triggered by the Workflow Subscription Event we created in the previous section.

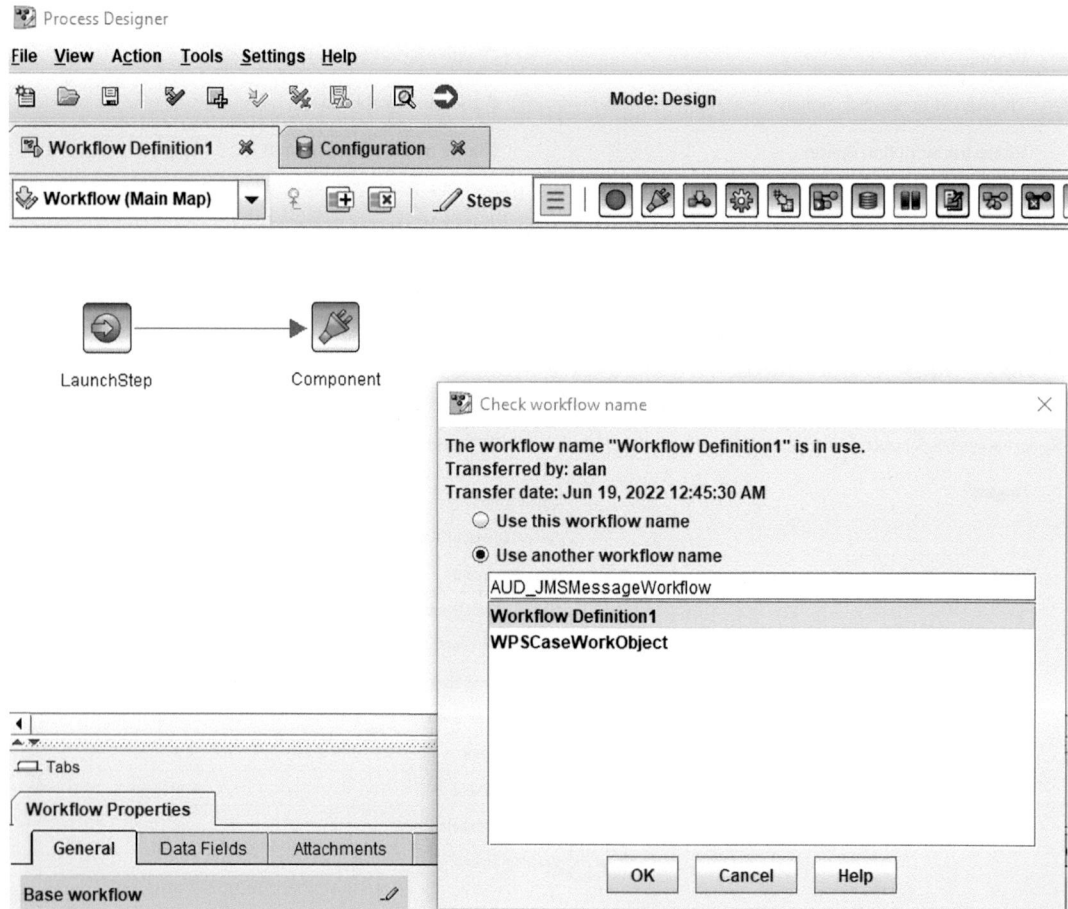

Figure 3-303. *The JMS test workflow is transferred to the Target Object store*

CHAPTER 3 IBM JMS INTERFACE DEVELOPMENT IBM FILENET 5.5.X WORKFLOW

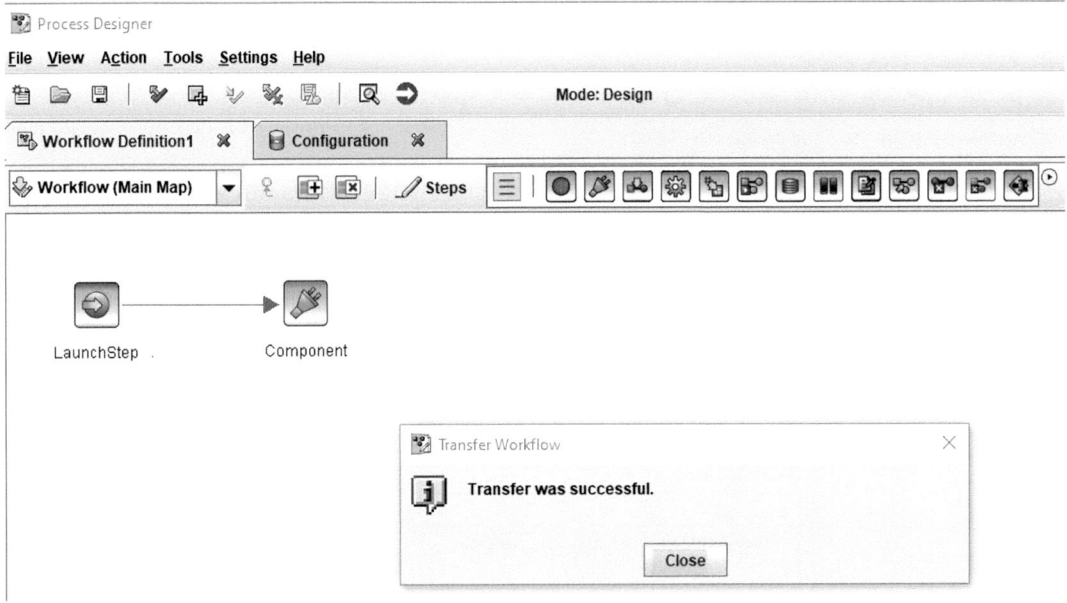

Figure 3-304. *The status of the Workflow Transfer is displayed by the Process Designer*

Figure 3-305. *The OS2:\Published folder is used to save the Workflow Definition*

CHAPTER 3 IBM JMS INTERFACE DEVELOPMENT IBM FILENET 5.5.X WORKFLOW

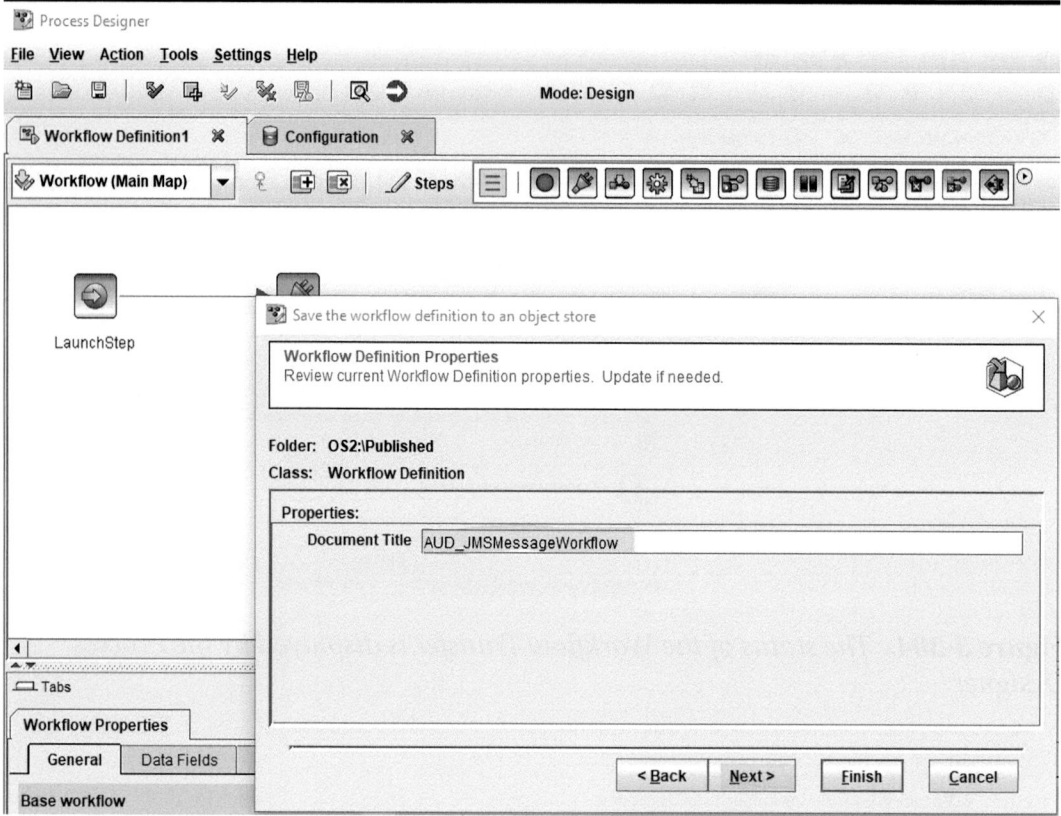

Figure 3-306. *The Workflow Definition is saved as AUD_JMSMessageWorkflow*

CHAPTER 3 IBM JMS INTERFACE DEVELOPMENT IBM FILENET 5.5.X WORKFLOW

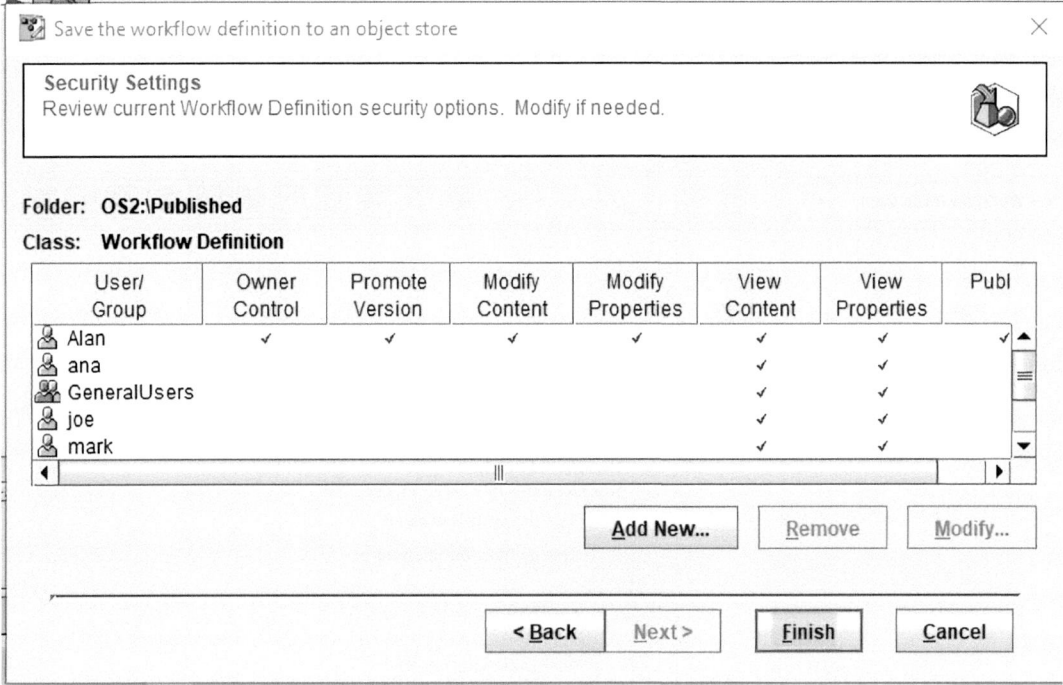

Figure 3-307. *The Security for the Workflow Definition is displayed. Click Finish to create it*

CHAPTER 3   IBM JMS INTERFACE DEVELOPMENT IBM FILENET 5.5.X WORKFLOW

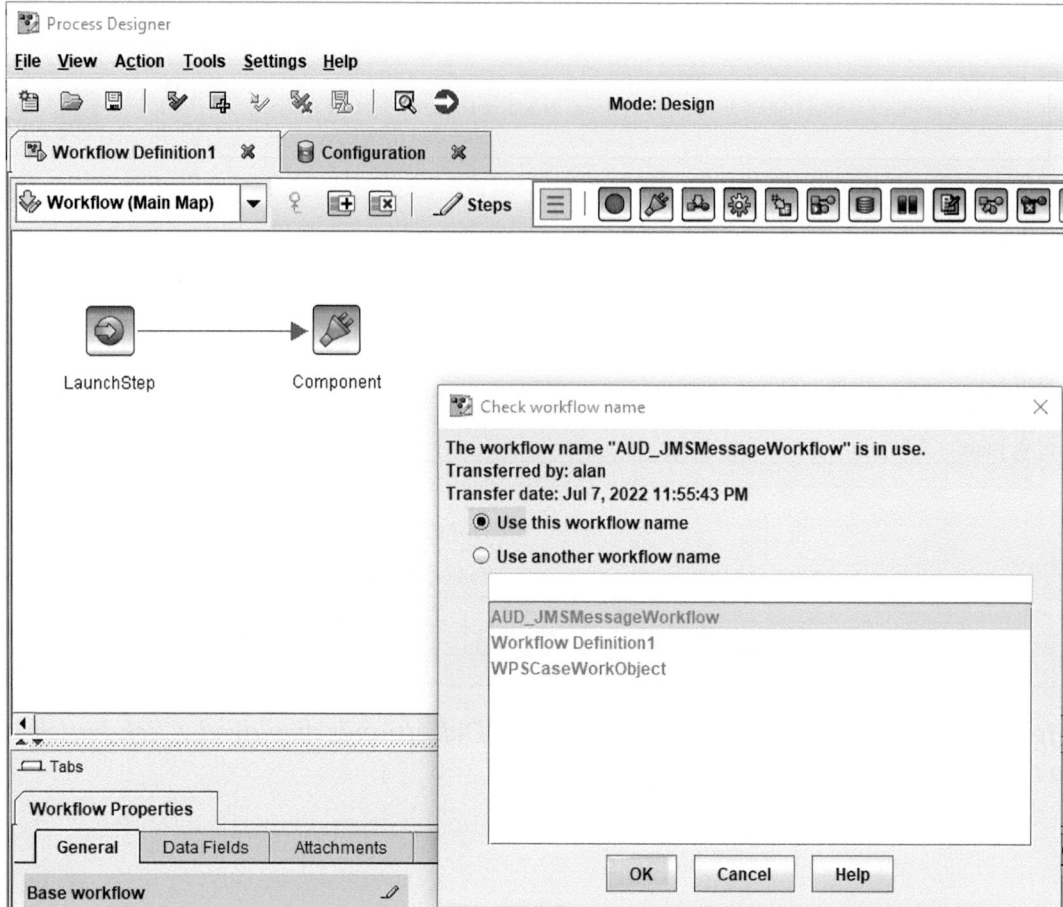

Figure 3-308. *The AUD_JMSMessageWorkflow is transferred to the Target Object Store*

CHAPTER 3   IBM JMS INTERFACE DEVELOPMENT IBM FILENET 5.5.X WORKFLOW

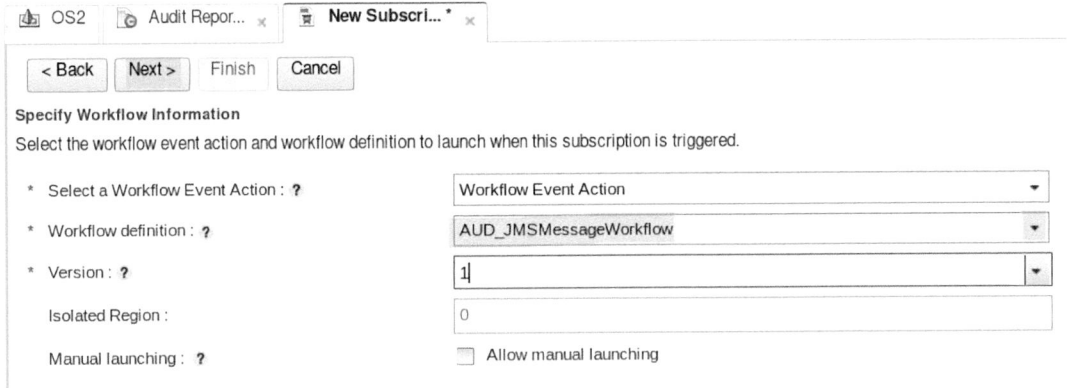

Figure 3-309. *The Workflow Event Action can now be selected with the transferred definition*

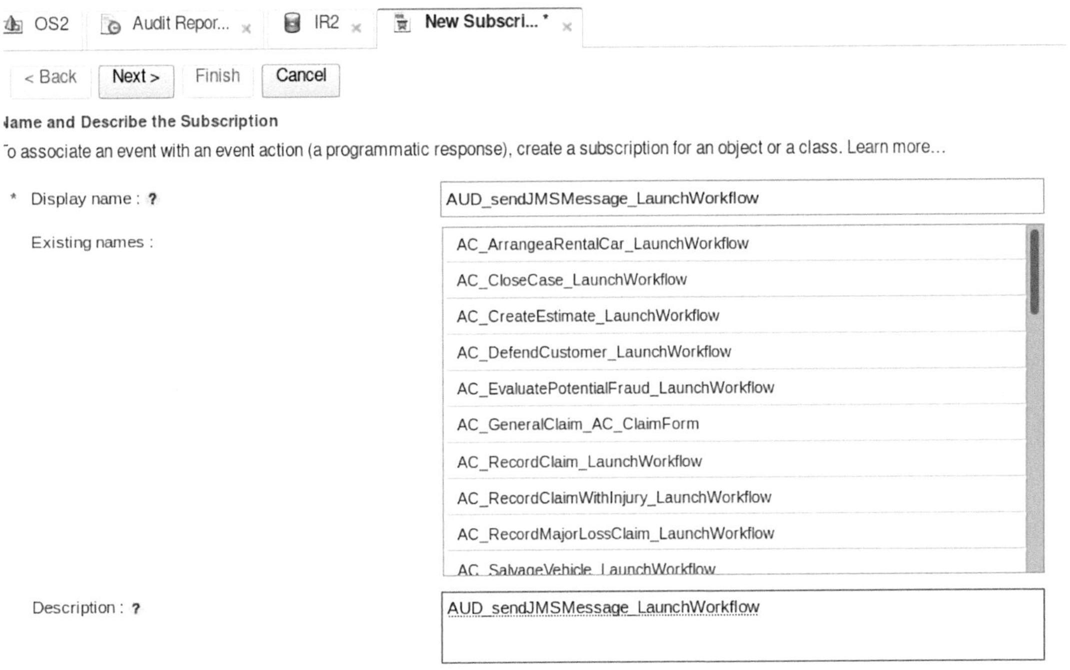

Figure 3-310. *The AUD_sendJMSMessage_LaunchWorkflow subscription name is entered*

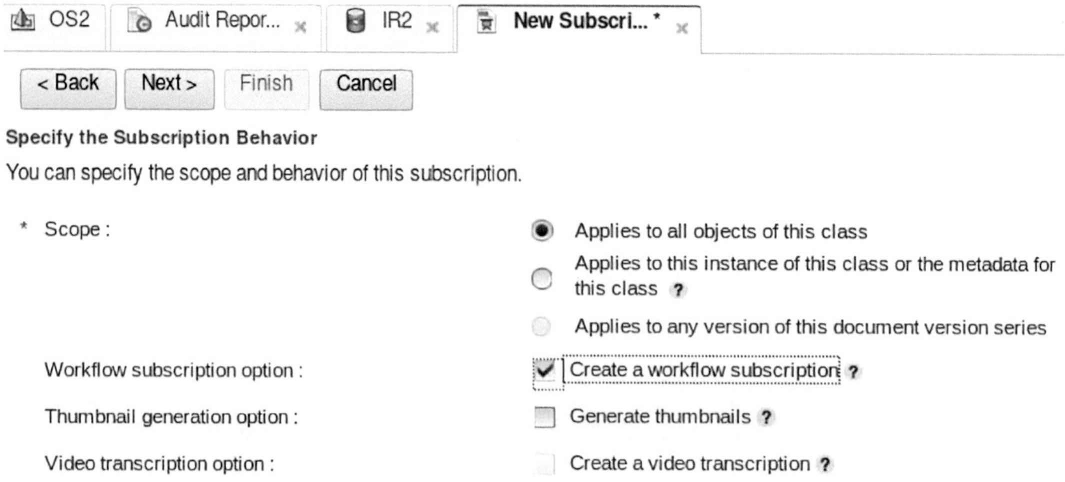

CHAPTER 3   IBM JMS INTERFACE DEVELOPMENT IBM FILENET 5.5.X WORKFLOW

Figure 3-311.   The Create a workflow subscription option is selected

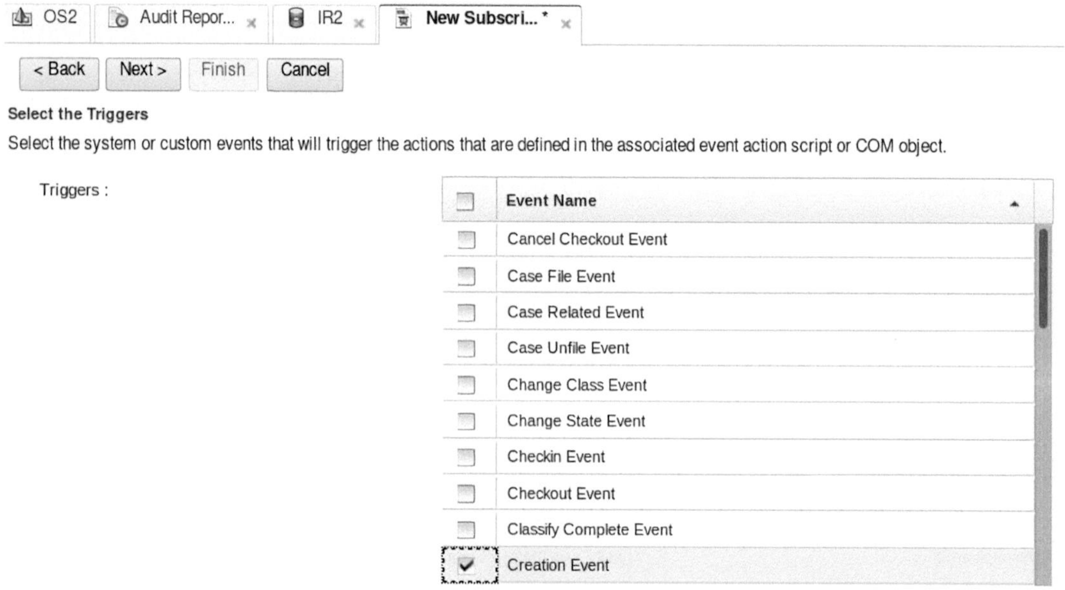

Figure 3-312.   The Creation Event is selected

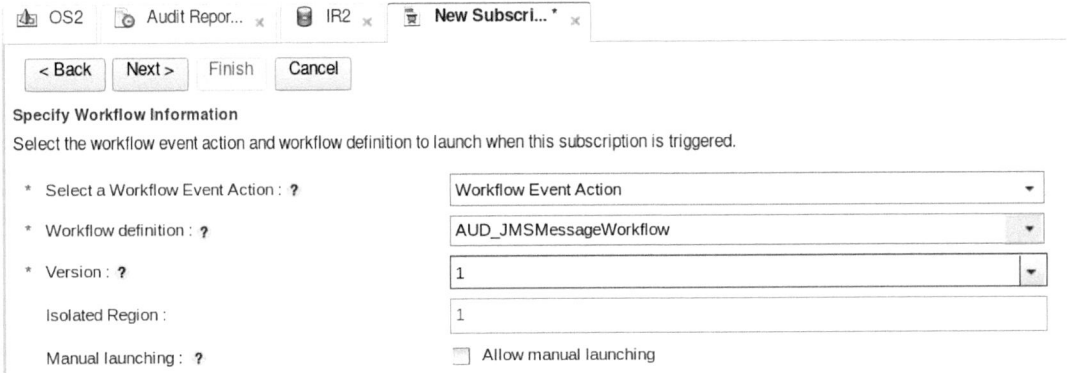

Figure 3-313. *The AUD_JMSMessageWorkflow is selected from the drop-down and Next> clicked*

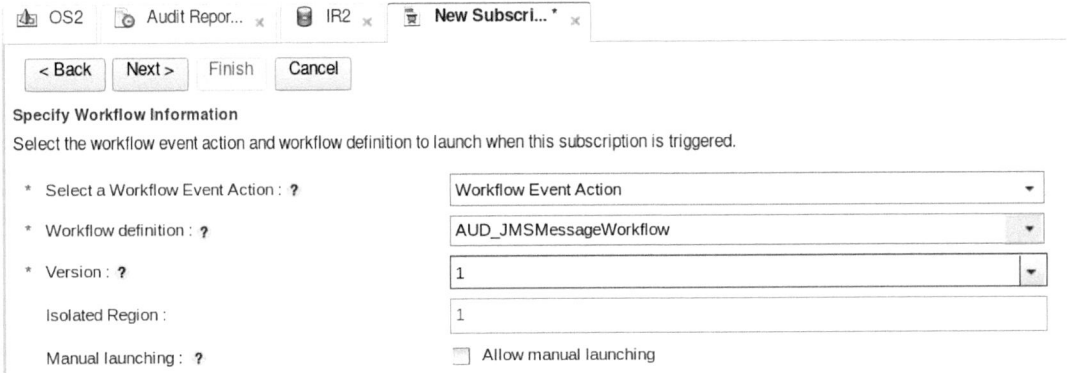

Figure 3-314. *The Workflow Fields are mapped to the Audit Report Document class properties*

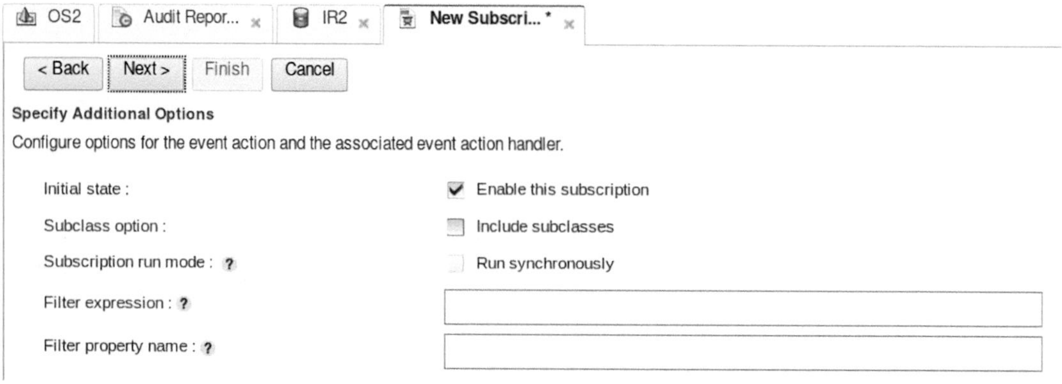

CHAPTER 3   IBM JMS INTERFACE DEVELOPMENT IBM FILENET 5.5.X WORKFLOW

Figure 3-315. The Enable this subscription flag property is ticked and Next> clicked

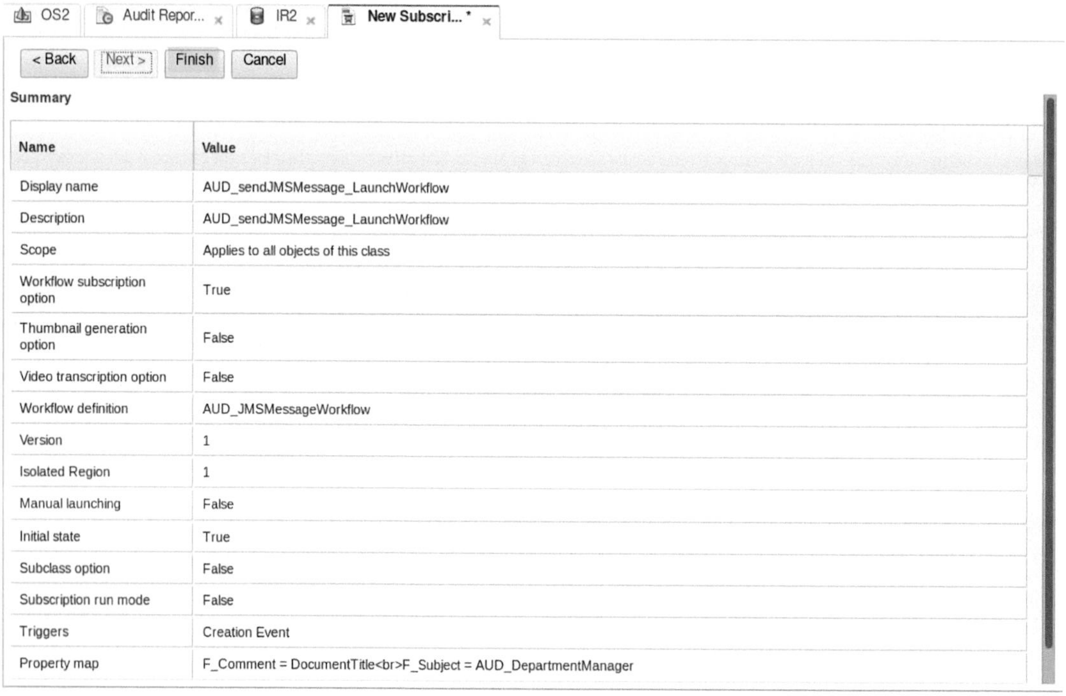

Figure 3-316. The parameters for creation of the Workflow Subscription are displayed

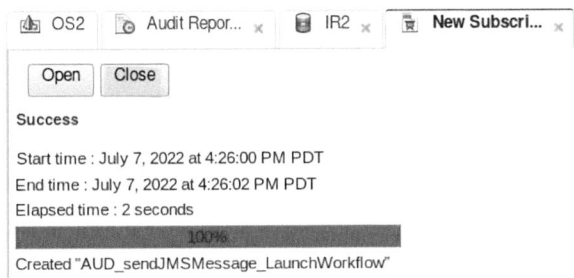

Figure 3-317. The status of the Workflow subscription Object creation is shown

Testing the Updated Audit Master Solution

After transferring the Workflow, we can now test the Audit Report triggering document class creation, by launching the Audit Master solution in the IBM Case Builder test environment.

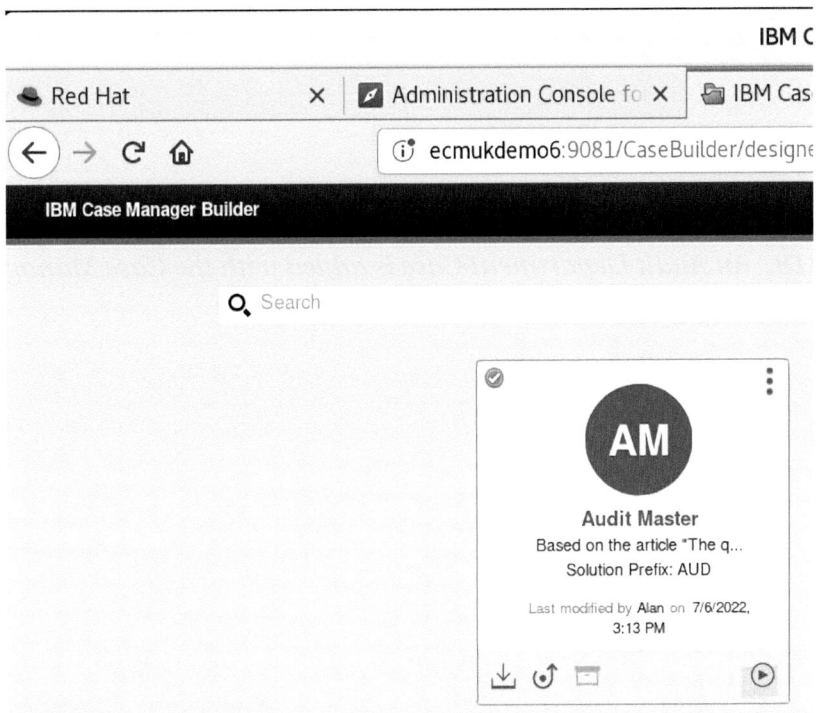

Figure 3-318. The highlighted triangle icon is clicked to launch the IBM Case Manager

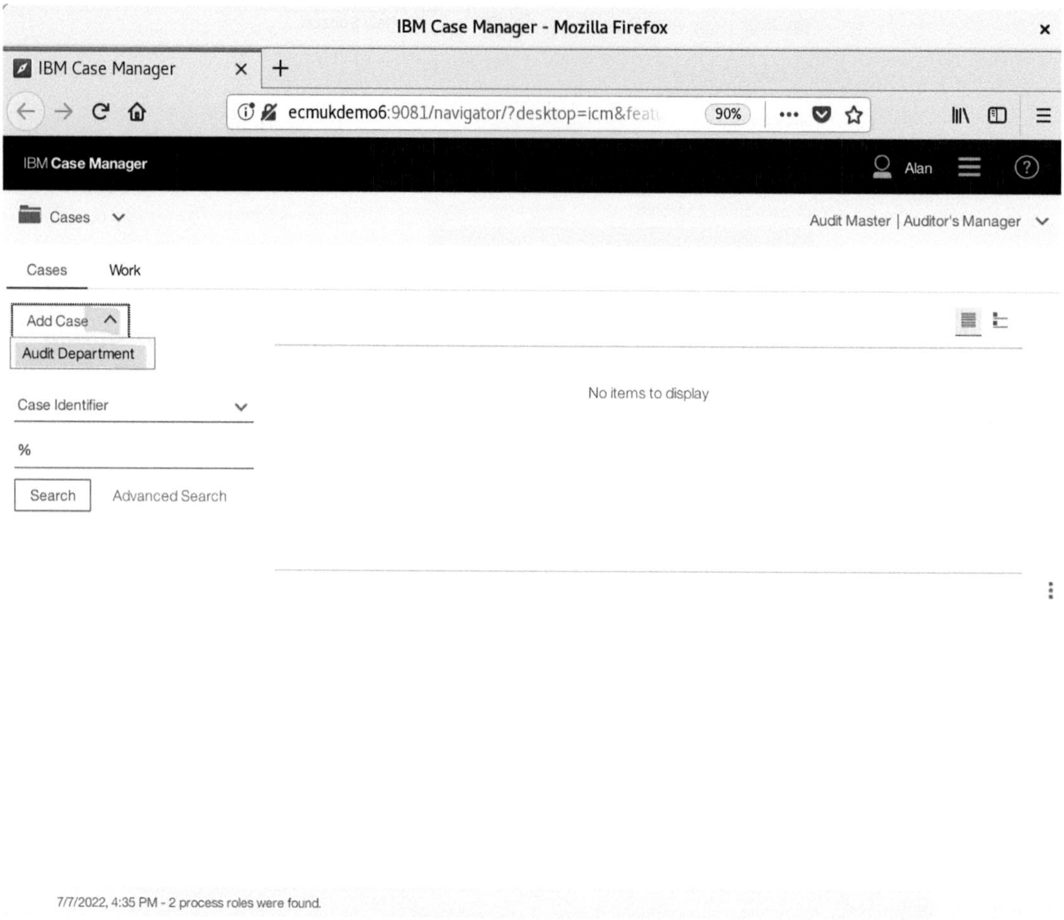

CHAPTER 3   IBM JMS INTERFACE DEVELOPMENT IBM FILENET 5.5.X WORKFLOW

Figure 3-319. *An Audit Department Case is added with the Case Manager Audit Solution*

CHAPTER 3 IBM JMS INTERFACE DEVELOPMENT IBM FILENET 5.5.X WORKFLOW

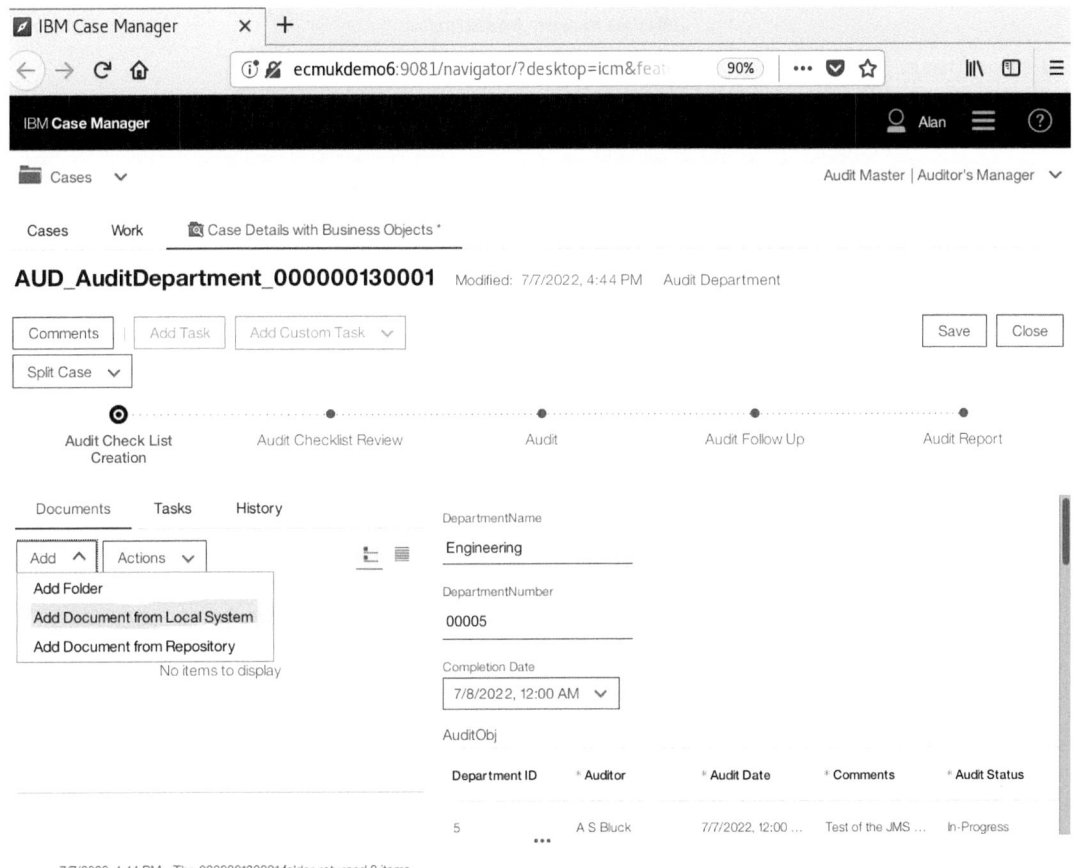

*Figure 3-320. The **Add Document from Local System** Document add option is selected*

CHAPTER 3   IBM JMS INTERFACE DEVELOPMENT IBM FILENET 5.5.X WORKFLOW

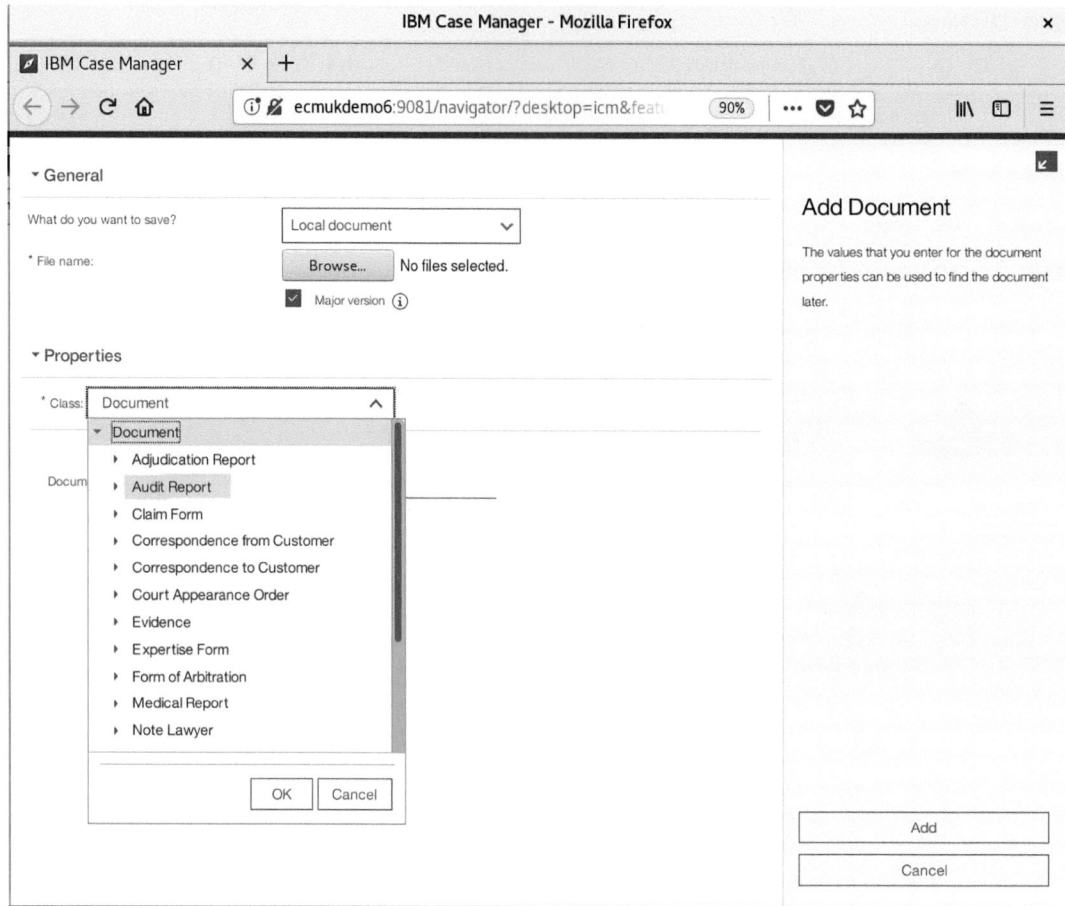

Figure 3-321. *The Audit Report Document Class is selected from the list*

CHAPTER 3 IBM JMS INTERFACE DEVELOPMENT IBM FILENET 5.5.X WORKFLOW

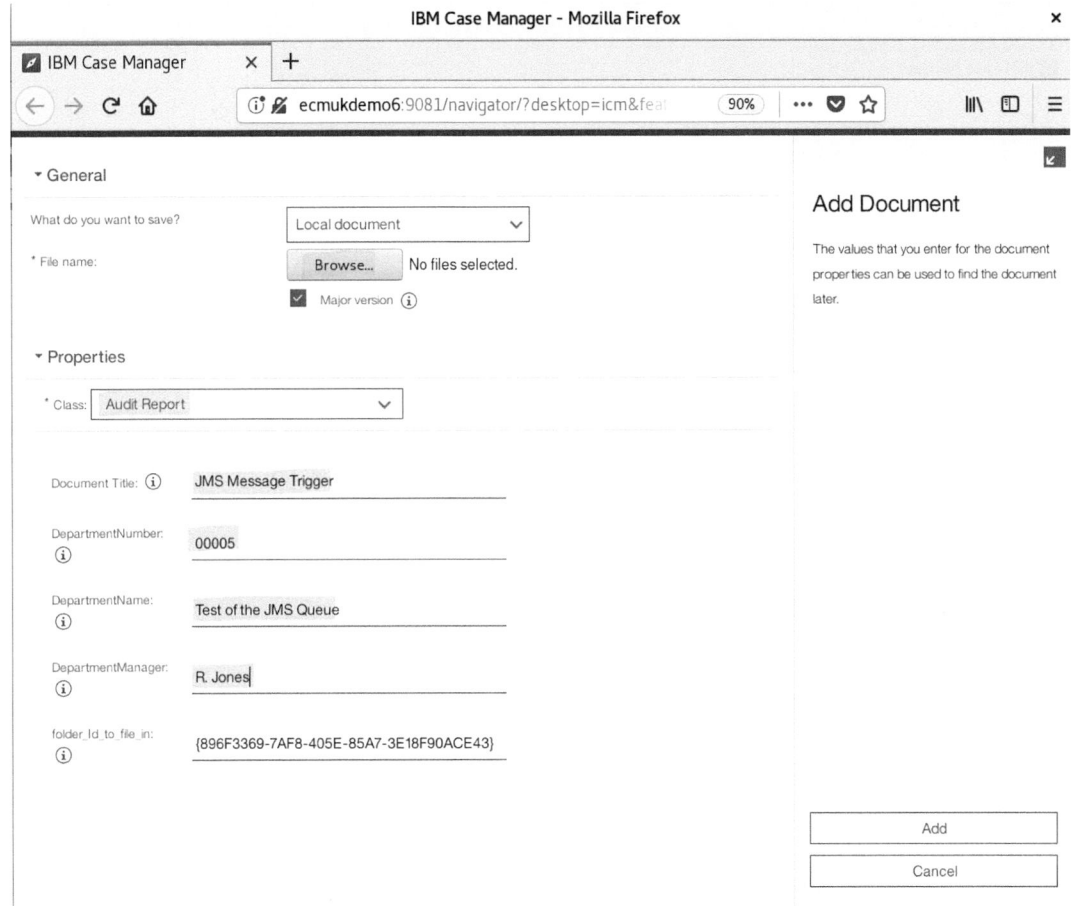

Figure 3-322. *The property details for the Audit Report Document are entered*

CHAPTER 3   IBM JMS INTERFACE DEVELOPMENT IBM FILENET 5.5.X WORKFLOW

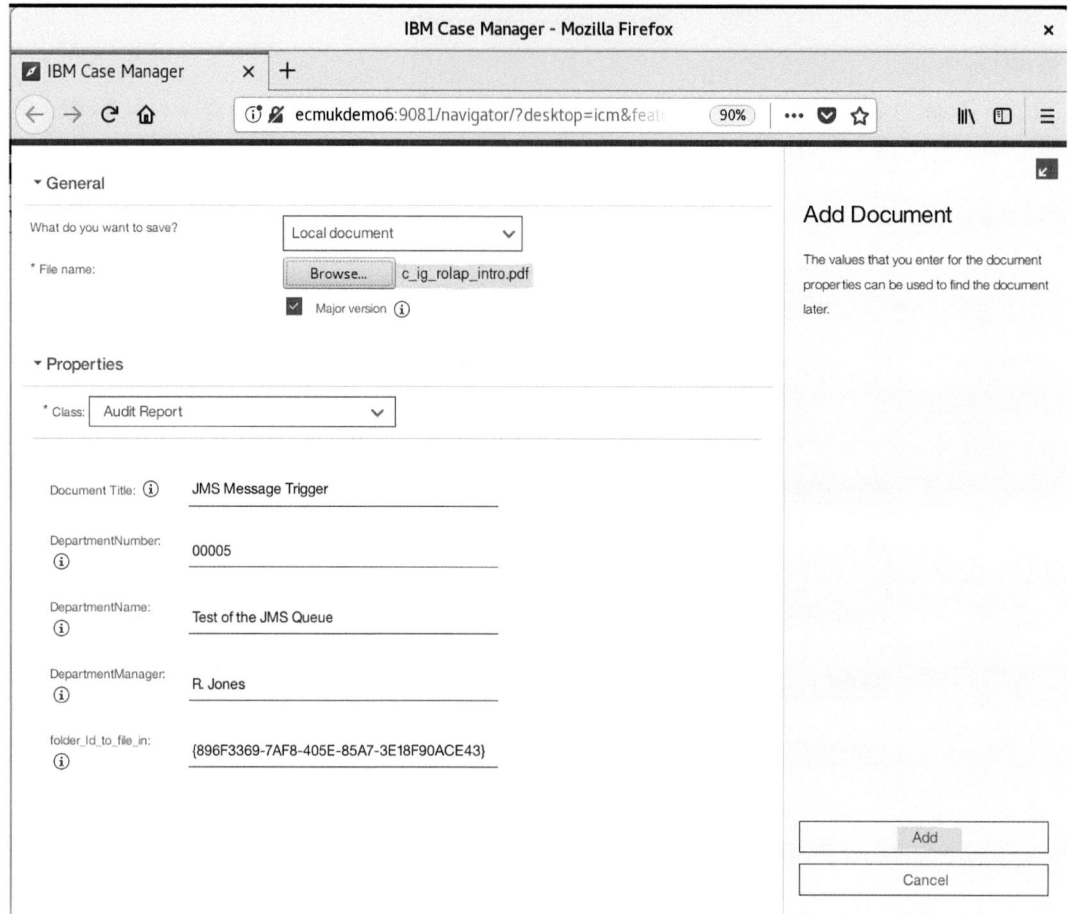

Figure 3-323. *The Add command button is used to create the Audit Report document*

CHAPTER 3   IBM JMS INTERFACE DEVELOPMENT IBM FILENET 5.5.X WORKFLOW

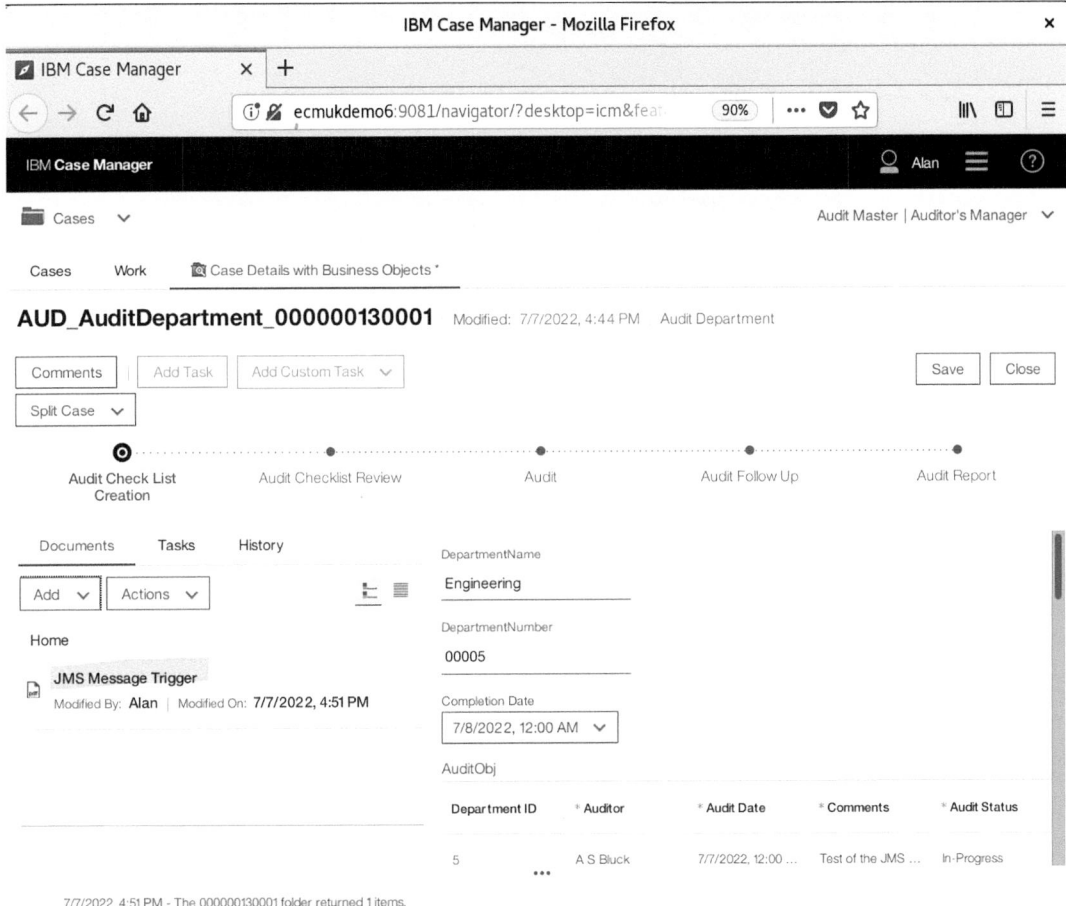

Figure 3-324. *The new Document,* **JMS Message Trigger***, is displayed in the Case*

CHAPTER 3 IBM JMS INTERFACE DEVELOPMENT IBM FILENET 5.5.X WORKFLOW

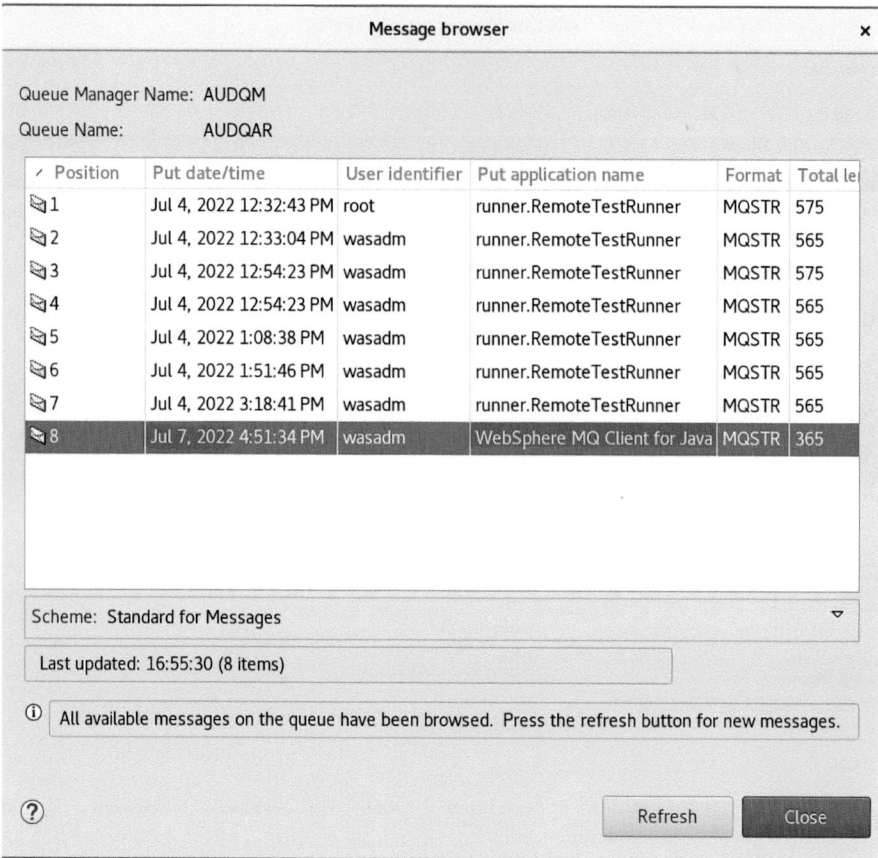

Figure 3-325. *A new JMS message can be seen in the IBM MQ AUDQAR queue*

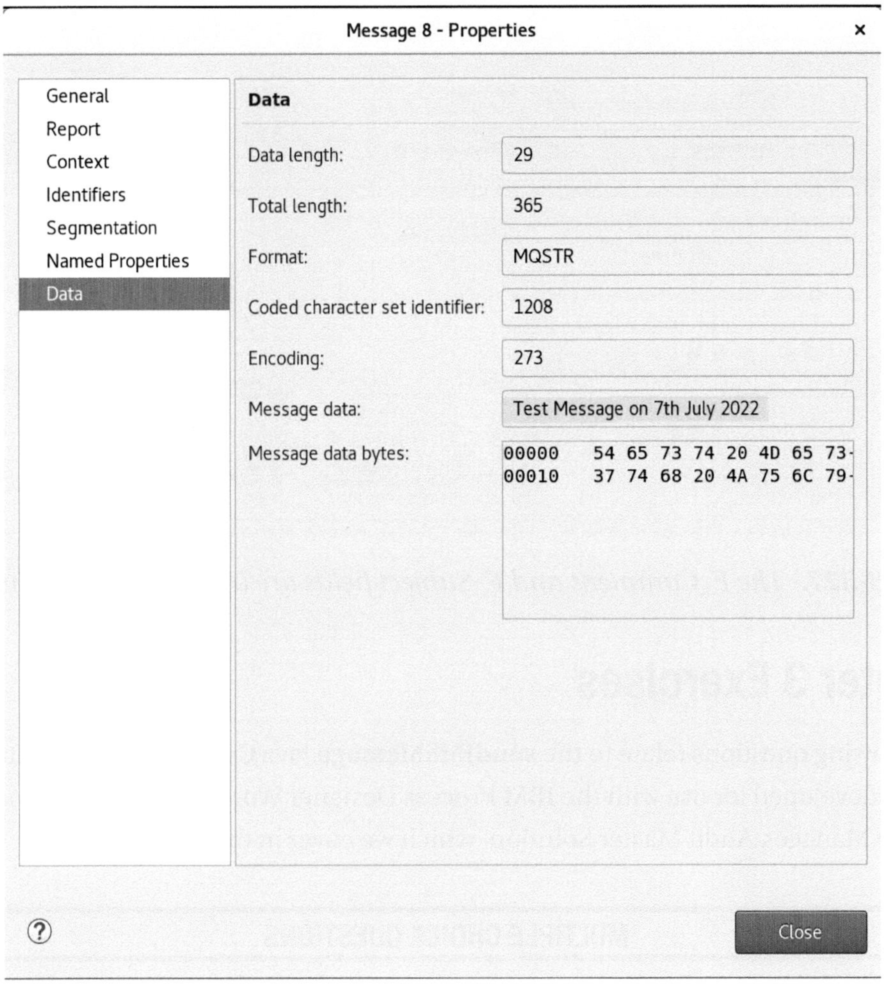

Message 8 - Properties	✕

General
Report
Context
Identifiers
Segmentation
Named Properties
Data

Data

Data length:	29
Total length:	365
Format:	MQSTR
Coded character set identifier:	1208
Encoding:	273
Message data:	Test Message on 7th July 2022
Message data bytes:	00000 54 65 73 74 20 4D 65 73 00010 37 74 68 20 4A 75 6C 79

Figure 3-326. *The Message data from the Workflow JMS component call*

CHAPTER 3 IBM JMS INTERFACE DEVELOPMENT IBM FILENET 5.5.X WORKFLOW

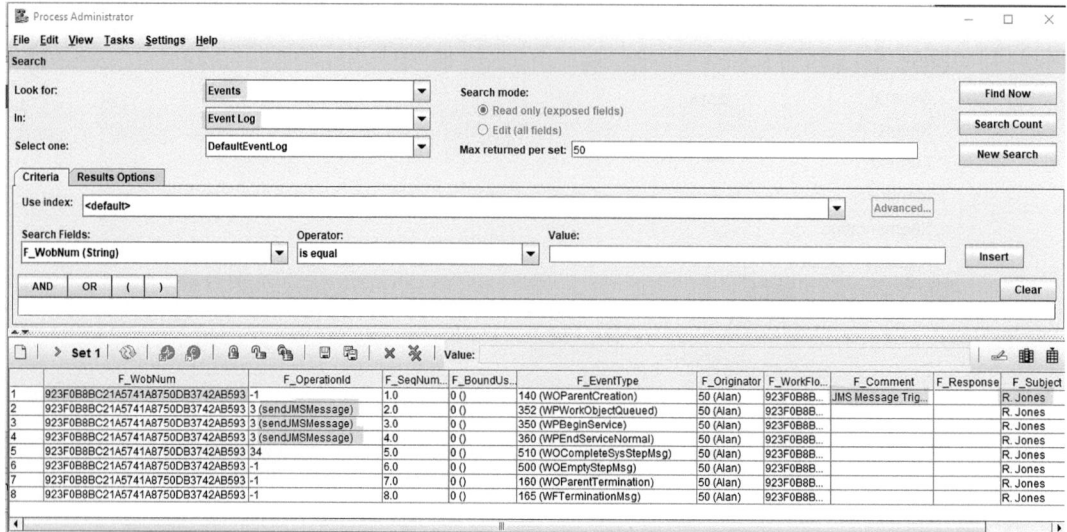

Figure 3-327. The F_Comment and F_Subject fields are in the Workflow Events

Chapter 3 Exercises

The following questions relate to the **sendJMSMessage** Java Component Integrator method developed for use with the IBM Process Designer Workflow subscriptions and the Case Manager, Audit Master Solution, which we cover in this chapter.

MULTIPLE CHOICE QUESTIONS
---

1. What script would you run to accept the IBM MQ 9.2.x license before installation?

    a) pedesigner.bat

    b) cmd.exe

    c) mqlicense.sh

    d) amqsputc

2. What is the solution if you get the following JMS Java message:

   "**the class com.ibm.mq.jms.MQQueue cannot be cast to javax.jms. ConnectionFactory**"?

   a) A queue connection factory must be defined as a WebSphere MQ Connection Factory.

   b) A queue connection factory must be defined as a WebSphere MQ Queue Connection Factory.

   c) Use a javax.jms.QueueConnectionFactory object.

   d) Use the Java code: **TextMessage outMessage = JMSsession. createTextMessage();**

3. What change do you need to make if you get the Java runtime error in Eclipse as follows:

   "**JVMCFRE003 bad major version; class=com/ibm/filenet/ps/ciops/test/ AUDOperationsTest, offset=6**" ?

   a) Add the JMS Message library, javax.jms-api-2.0.1.jar, to the Eclipse IDE classpath.

   b) Add two new lines to the /etc/security/limits.conf file; @root soft nofile 10240 and @root hard nofile 16384.

   c) Change the JVM version in the Eclipse project from 1.7 to 1.8.

   d) Add jars from the /opt/MQJars folder to the Eclipse Classpath.

4. The main purpose of a Workflow Subscription is

   a) To enable the creation of a Folder Object

   b) To enable the creation of a Document Object

   c) To enable a workflow to be launched on the storage of a document

   d) To enable a Case to be created in IBM Case Manager

CHAPTER 3   IBM JMS INTERFACE DEVELOPMENT IBM FILENET 5.5.X WORKFLOW

MULTIPLE CHOICE ANSWERS

1. c) mqlicense.sh

2. b) A queue connection factory must be defined as a WebSphere MQ Queue Connection Factory.

3. c) Change the JVM version in the Eclipse project from 1.7 to 1.8.

4. c) To enable a workflow to be launched on the storage of a document

QUESTIONS

1. Describe what steps you would use to install a new Component Integrator, given that you have been given a working **AUDOperations.jar** file and have been asked to configure a **sendJMSMessage** custom operations method call step in a Workflow.

2. What security group(s) do you need to have to run the IBM MQ Explorer program and what commands would you need to run to use this from a Unix command window?

3. What Java libraries would you install in an Eclipse IDE Classpath to develop a Component Integrator Java class and where would you find them?

In Chapter 4, we will cover the use of the Java language for the IBM FileNet Java API calls and the configuration required to replicate an IBM FileNet Document Management Object Store.

CHAPTER 4

A Replication Java Program for IBM FileNet Object Stores

This chapter covers the development of the Java API calls and the configuration required to replicate an IBM FileNet Document Management Object Store.

See also:

www.ibm.com/docs/en/filenet-p8-platform/5.2.1?topic=guide-replication

Although this link states:

> "Each object in a Content Engine object store has a unique ID that is controlled by the Content Engine. However, a replicated object in an external repository possesses a **different unique ID** from the original object over which the Content Engine has no control. An **ExternalIdentity** object represents the identity of a replicated object in an external repository."

I will demonstrate a Java program process which will maintain an **exact** replication of the original Object Store ID Documents and their containing folder structure. Unlike the standard replication discussed in the preceding link, the Java Replication program covered in this chapter is designed to even copy the unique GUIDs of the original IBM FileNet Content Object Store Objects (including Documents, Folders, and Property objects).

This functionality is useful for larger organizations with geographically separate sites, where the Standard Operating Procedures and Quality Manuals have to be consistently maintained, but usually require a single point of authorship to maintain the same standards of production across all the sites.

CHAPTER 4   A REPLICATION JAVA PROGRAM FOR IBM FILENET OBJECT STORES

Chapter Organization

This chapter contains the following six core Parts, each of which is organized by sections within:

Part 1 – Bill of Materials. This Part lists the prerequisite IBM Software components, including the IBM product code numbers, required to download the latest IBM FileNet Content Manager software components, used to test the Java Replication Program FileNet API calls. It also covers the download of the document describing the installation of the software on a Red Hat Linux server.

Part 2 – The Replication Program Introduction. This part describes the main Replication Program functionality and the development tools and procedure.

Part 3 – Code Developed for the Replication Program. This Part lists the main Replication Program Java Code.

Part 4 – Code Listing – CEReplicateConfig. In this Part, the Java Code for the supporting utility methods is listed.

Part 5 – Code Listing – CEMigrateConfig. In this Part, the supporting Java Code for the **CEMigrateConfig** code is listed. This code reads the parameters from the **config.xml** file which defines the source and target object stores and the required login and security parameters.

Part 6 – Code Listing – Supporting Utility Code. In this Part, the supporting utility Java Code for the Replication Program is listed. This is defined for the compiled fn_utils.jar, fn_connect.jar, and the fn_connection.jar supporting libraries.

Part 1 – Bill of Materials

The installation steps required for the base IBM Case Manager 5.3.3 system we are using are covered step by step in the following free ResearchGate document, downloaded using the DOI URL as follows:

https://doi.org/10.13140/RG.2.2.21708.16001

entitled *"Case Manager 5.3.3 Installation on RHEL 8.0 with Content Navigator 3.0.6"*
Click the **Download file PDF** command button in Figure 4-1.

CHAPTER 4   A REPLICATION JAVA PROGRAM FOR IBM FILENET OBJECT STORES

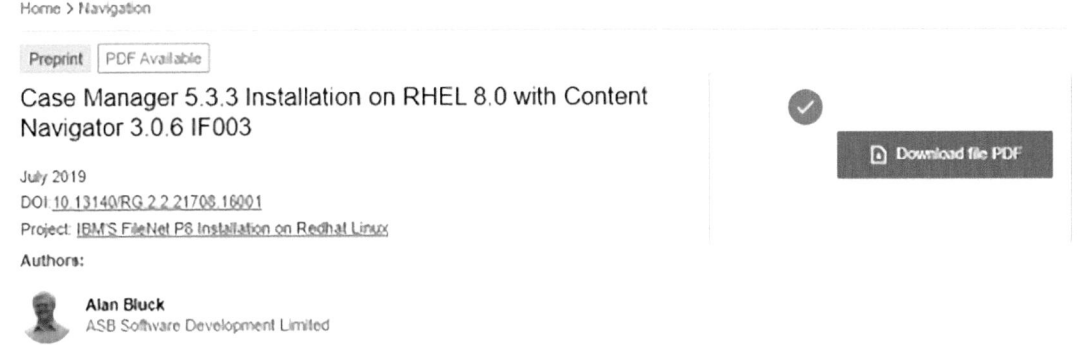

Figure 4-1.   Download the IBM Case Manager Installation Document

This gives the downloaded document entitled *CaseManagerInstallationonRHEL8.0_V3.docx*.

The installed IBM software product codes required are listed as follows:

IBM FileNet Content Manager V5.5.0 for IBM Case Foundation V5.3.0 Multiplatform Multilingual eAssembly (CJ2VNML)

FileNet Content Manager V5.5.0 Quick Start Guide Multiplatform Multilingual (CNP8XML)

IBM FileNet Content Platform Engine V5.5.0 Linux Multilingual (CNP8ZML)

IBM FileNet Content Platform Engine V5.5.0 Windows Multilingual (CNP90ML)

IBM FileNet Content Platform Engine Client V5.5.0 Linux English (CNP93EN)

IBM FileNet Content Platform Engine Client V5.5.0 Windows English (CNP94EN)

IBM DB2 Enterprise Server Edition Restricted Use Quick Start and Activation V11.1 for Linux, UNIX and Windows Multilingual (CNB25ML)

Quick Start Guide for IBM WebSphere Application Server V9.0 (CNA8LML)

IBM WebSphere Application Server V9.0 (CND1AML)

IBM WebSphere Application Server V9.0 Supplements – Application Client (CND1CML)

IBM WebSphere Application Server V9.0 Supplements – IBM HTTP Server (CND1DML)

IBM WebSphere Application Server V9.0 Supplements – Web Server Plugins (CND1EML)

CHAPTER 4   A REPLICATION JAVA PROGRAM FOR IBM FILENET OBJECT STORES

IBM WebSphere Application Server V9.0 Supplements – WebSphere Customization Toolkit (CND1FML)

IBM WebSphere Application Server Liberty (IBM Installation Manager install) (CND1GML)

IBM Installation Manager V1.8.5 for Linux x86_64 (CND0ZML)

IBM SDK, Java (TM) Technology Edition, Version 8 for Windows (CND15ML)

IBM SDK, Java (TM) Technology Edition, Version 8 for Linux (CND18ML)

Note   WebSphere 8.5.5 Fix pack 15 automatically installs and defaults the JDK to 1.8.

Part 2 – The Replication Program Introduction

This Java program consists of 4116 lines of Java code calling methods from the FileNet 5.x Java APIs to replicate documents and folders from a selected Folder location in a source Object Store to a selected Target Object Store on a different server. (See the section code lists in this Chapter.)

The Target documents are copied with **exact GUIDS**, Version Series ID, and Security profile and folder linkage as in the source Object Store, except that the Group and User security are set to Read Only for the Replicated Folder and Document Properties and Content.

One exception Group and User are left with their originally defined security levels.

The Replication program also detects security and property changes in the Source Object store and the removal/move of Documents and Folders and faithfully copies the changes to the target object store.

This program was built to satisfy the following requirements:

a) It can be rerun from any starting date and updates the Target replica with all changes; there is no need to reset the target object store as the program has been designed to check and ignore existing Documents or Folders already replicated.

CHAPTER 4 A REPLICATION JAVA PROGRAM FOR IBM FILENET OBJECT STORES

b) The tests for specifications and functionality of the program are as follows:

1. First replication:
   - Verify the object store and folders.
   - The same object store and folders are found at target.
2. Verify the document at the target side.
   - The number of documents are the same.
3. Verify the permission setup.
   - The permission setups are the same.
4. Check whether the copy at the target side is editable or not.
   - The documents on the target Object Store must be READ-ONLY.
5. Verify the document change – Upload a new document at the source.
   - A new document should be synced to the target, with identical permissions.
6. Upload a new version of the document at the source.
   - A new version of the document should be found at the target.
7. Change the permissions of the document at the source.
   - The permissions must be updated at the target.

   (Within the READONLY specifications!)
8. Remove a document at the source.
   - The same document should be removed at the target.
9. Modify the document attribute.
   - The same attribute change is applied at the target.
10. Move the document from one folder to another folder at the source.
    - The same change is applied at the target.

11. Change the replication frequency.

    - The target copy should be refreshed as per the frequency setting.

12. Verify the folder change – Create a new folder at the source.

    - A new folder should be synced to the target, with identical permissions.

13. Verify the folder change – Amend the name of an existing folder at the source.

    - The Folder Name change should be synced to the target Object Store.

14. Remove a folder at the source Object Store.

    - The same folder should be removed at the target Object Store.

15. Verify the target Object Store copy is still accessible if the source Object Store copy is stopped.

    - The target copy should still be accessible for read.

Nonfunctional Requirements

1) An administrator should be able to adjust the replication frequency.

   (Originally defined as 1 hour, but run at present every 15 minutes)

2) The full Java code source is in Part 6 of this chapter.

3) The Replication script should be compatible for future FileNet 5.x versions. (All API calls are compatible with the base FileNet 5.2 version.)

CHAPTER 4 A REPLICATION JAVA PROGRAM FOR IBM FILENET OBJECT STORES

Development Tools Used

1) VMware Red Hat Enterprise Linux Server RHEL 8.0 with full IBM FileNet P8 5.5.5 system

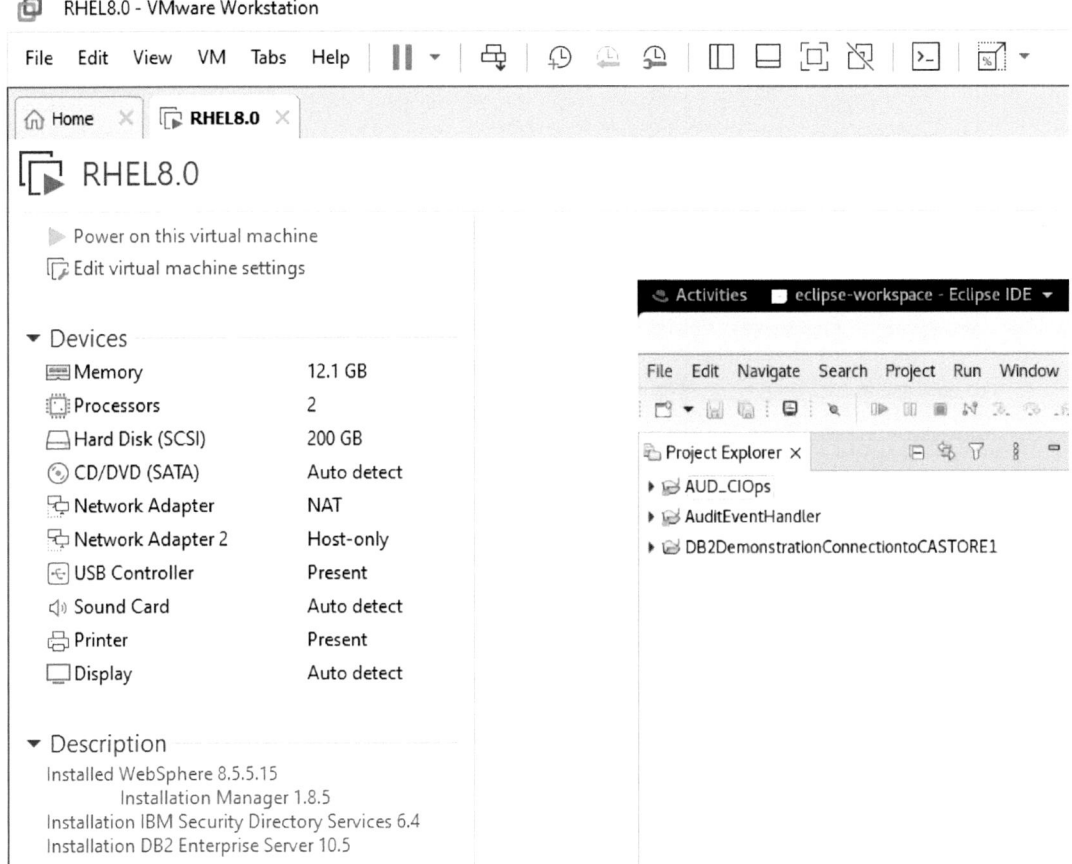

Figure 4-2. The Red Hat Linux server specification used for the development

CHAPTER 4 A REPLICATION JAVA PROGRAM FOR IBM FILENET OBJECT STORES

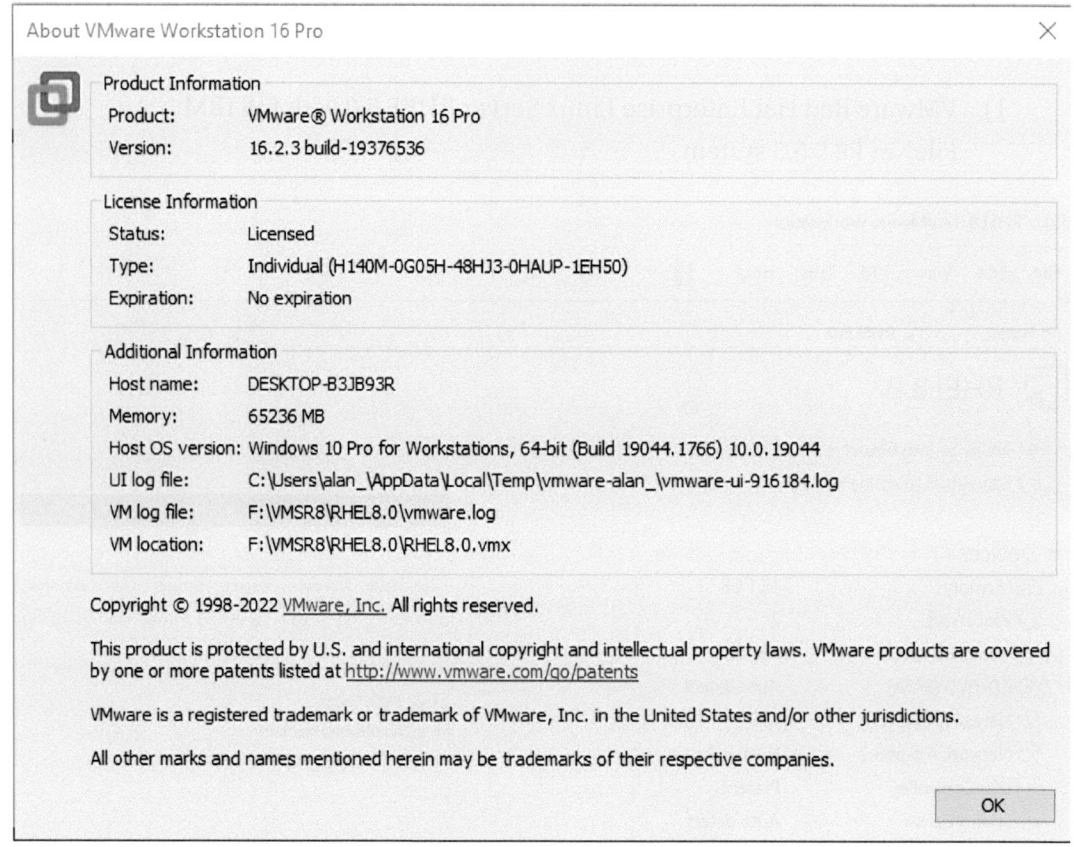

Figure 4-3. *The VMware Host server and version details*

The P8 5.5.5 Environment

Machine Name: ECMUKDEMO6
    Red Hat Linux Enterprise 8.0 Server
    Main User: Alan
    WebSphere User: wasadm
    DB2 User: db2inst1
    Installed WebSphere 8.5.5.15
    Installation Manager 1.8.5
    Installation IBM Security Directory Services 6.4
    Installation DB2 Enterprise Server 11.5

```
Product name:              "IBM Data Server Client"
Product identifier:        "db2client"
```

CHAPTER 4   A REPLICATION JAVA PROGRAM FOR IBM FILENET OBJECT STORES

Version information:	"11.5"
Product name:	"IBM DB2 Developer-C Edition"
License type:	"Community"
Expiry date:	"Permanent"
Product identifier:	"db2dec"
Version information:	"11.5"
Max amount of memory (GB):	"16"
Max number of cores:	"4"
Max amount of table space (GB):	"100"

Login: root
IBM FileNet Case Manager 5.5.3
IBM FileNet Content Engine 5.5.5
Eclipse IDE Version 2021-12 (4.22.0) for Red Hat Linux 8.0

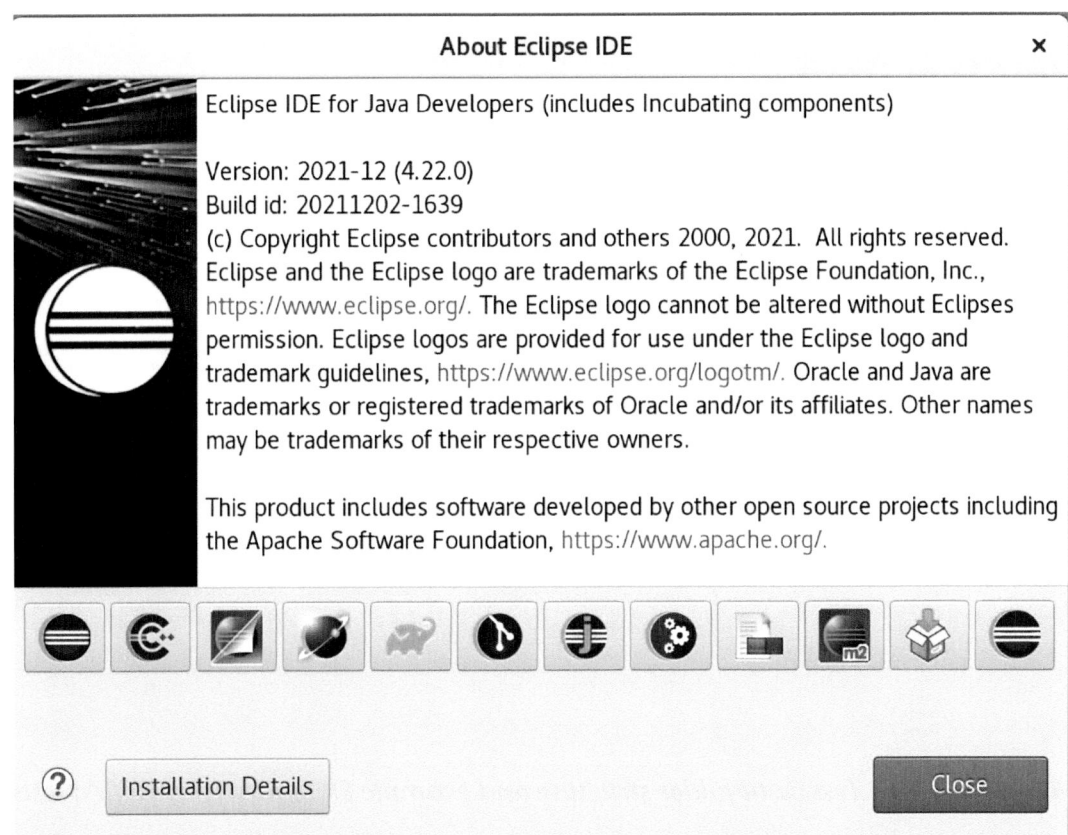

Figure 4-4. *The Eclipse IDE Java version used for the development in this book*

655

CHAPTER 4   A REPLICATION JAVA PROGRAM FOR IBM FILENET OBJECT STORES

www.eclipse.org/downloads/download.php?file=/technology/epp/downloads/release/2022-03/R/eclipse-java-2022-03-R-linux-gtk-x86_64.tar.gz

(Download **eclipse-java-2022-03-R-linux-gtk-x86_64.tar.gz** for unpacking and installation.)

At IBM FileNet Content Engine 5.x, there are Java packages and classes which help with the Java API calls used in the Replication code, especially as follows:

com.ibm.ce.rep.utils.CEReplicate
com.ibm.ce.utils.CEReplicateConfig

New Configuration Files:
(These can all be found in the **/opt/IBM/FileNet/CEClient** folder path.)

WcmApiConfig_ejb.properties
WcmApiConfig_wsi.properties

Unit Test Data

The Root "**/AUDIT_TEST**" folder in the **OS2** "Target" object store was used for initial testing, Test Data, as shown in Figure 4-5.

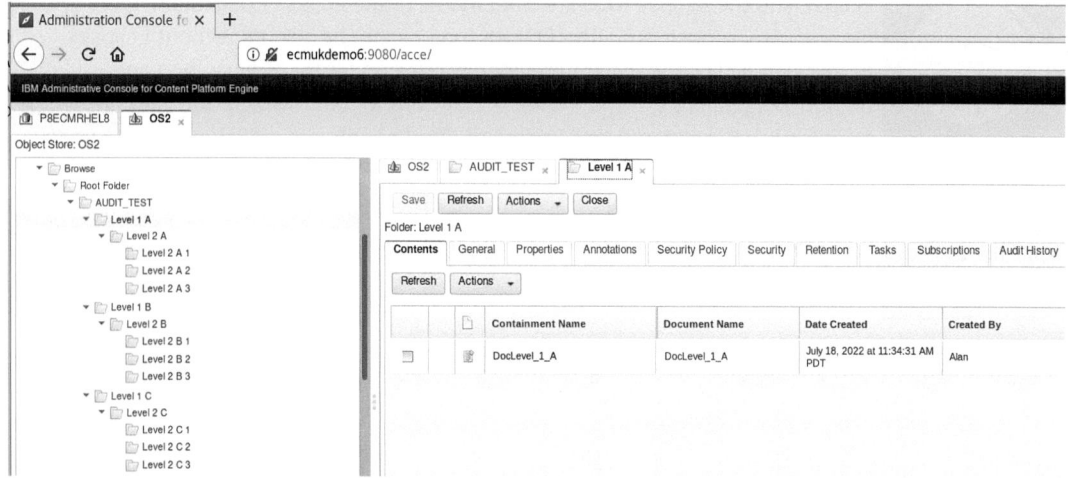

Figure 4-5. *The Test Data folder structure and example Document for the first tests*

CHAPTER 4 A REPLICATION JAVA PROGRAM FOR IBM FILENET OBJECT STORES

Project references required for the Java Replication program

- AUDIT_Common
- AUDIT_Utils
- FileNetConnection
- FN_Connect

Folder Referenced Object properties replicated

a) Permissions (Security ACLs)

b) Containers

c) Containees

d) SubFolders

e) Contained Documents (retrieved separately in the **importDocuments** method)

Folder Referenced Object properties NOT replicated

a) Replication Group

b) External Replica Identities

c) Active Markings

d) Annotations

e) Security Policy

f) Coordinated Tasks

g) Workflow Subscriptions

Note The following is required to write **DateCreated** and **Creator** properties, etc.

Creator Property

See the following links for updating this:
   www.ibm.com/docs/en/filenet-p8-platform/5.5.0?topic=security-working
   This indicates the name of the user assigned as the creator of the object.

www.ibm.com/docs/en/filenet-p8-platform/5.5.x?topic=comfilenetapiconstants-accessright

Settability of this property is read-only for most users. For users who have been granted privileged write access (**AccessRight.PRIVILEGED_WRITE**), this property is settable only on create. After initial object creation, this property is read-only for all users.

Setting Object Store Access Rights

The permissions (access rights) associated with an object store control the degree of access that users have to the objects within the object store. Granting full control means that the specified user is granted permission to connect to the object store, store objects, modify objects, and remove objects. (For additional examples, see Document security levels – IBM Documentation, www.ibm.com/docs/en/filenet-p8-platform/5.5.x?topic=rights-document-security-levels.)

```
static AccessRight        PRIVILEGED_WRITE
                          Specifies that the user or group is granted or denied permission to set certain system-level properties
                          (Creator, DateCreated, LastModifier, DateLastModified).
```

Figure 4-6. *The static **AccessRight** privilege we need to keep the same system properties*

Shell Script Batch Jobs

The following Shell Script jobs were developed to run a scheduled replication service using the Linux cron system to run the Java Replication program which we develop. For cron details, see www.redhat.com/sysadmin/automate-linux-tasks-cron.

For 15-minute repeats of the Replication, we use

*/15 * * * * /opt/replication/Replicate.sh

The following also works:

0,15,30,45 * * * * /opt/replication/Replicate.sh

CHAPTER 4   A REPLICATION JAVA PROGRAM FOR IBM FILENET OBJECT STORES

Shell Script Name: Replicate.sh

```sh
#!/bin/sh
# This command file launches the Replication application.
# Java copied to the path /opt/replication/applibs/sdk/bin
STARTING_FOLDER=/opt/replication
export STARTING_FOLDER
CLASSPATH=${STARTING_FOLDER}/jars/replication.jar:${STARTING_FOLDER}/jars/Jace.jar:${STARTING_FOLDER}/jars/log4j-1.2.17.jar
export CLASSPATH
JAVA_HOME=${STARTING_FOLDER}/applibs/sdk
export JAVA_HOME
JAVA_BIN=${JAVA_HOME}/bin
export JAVA_BIN
source ${STARTING_FOLDER}/envCommonReplicate.sh
source ${STARTING_FOLDER}/env5x.sh
echo CLASSPATH= ${CLASSPATH}
echo Starting Replication job ... Output to console.log in this directory
${JAVA_BIN}/java com.ibm.ce.rep.utils.CEReplicate > console.log
```

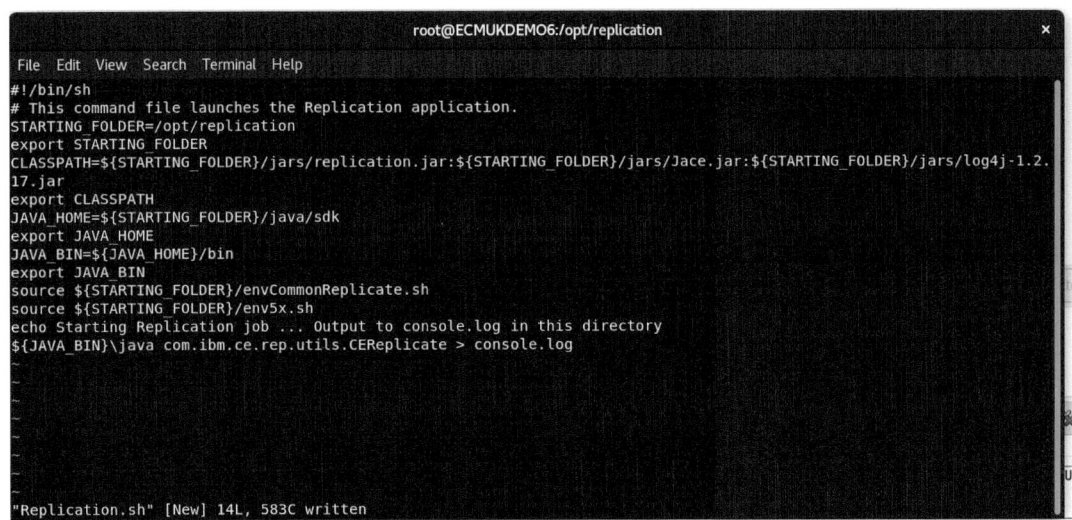

Figure 4-6A. *The main **Replication.sh** shell script for running in the cron scheduled job system*

Shell Script Name: envCommonReplicate.sh

```sh
#!/bin/sh
# This command file sets the CLASSPATH environment for the
# Replication application
STARTING_FOLDER=/opt/replication
export STARTING_FOLDER
export APPSLIBS=${STARTING_FOLDER}/libs/applibs/
export OTHER=${STARTING_FOLDER}/libs/other/
CLASSPATH=${APPSLIBS}replication.jar
export CLASSPATH
CLASSPATH=${CLASSPATH}:${APPSLIBS}fn_common.jar
export CLASSPATH
CLASSPATH=${CLASSPATH}:${APPSLIBS}fn_connection.jar
export CLASSPATH
CLASSPATH=${CLASSPATH}:${APPSLIBS}fn_utils.jar;
export CLASSPATH
CLASSPATH=${CLASSPATH}:${APPSLIBS}fn_connect.jar;
export CLASSPATH
CLASSPATH=${CLASSPATH}:./config
export CLASSPATH
```

CHAPTER 4   A REPLICATION JAVA PROGRAM FOR IBM FILENET OBJECT STORES

```
root@ECMUKDEMO6:/opt/replication
File Edit View Search Terminal Help
# This command file sets the CLASSPATH environment for the
# Replication application
STARTING_FOLDER=/opt/replication
export STARTING_FOLDER
export APPSLIBS=${STARTING_FOLDER}/libs/applibs/
export OTHER=${STARTING_FOLDER}/libs/other/
CLASSPATH=${APPSLIBS}replication.jar
export CLASSPATH
CLASSPATH=${CLASSPATH}:${APPSLIBS}fn_common.jar
export CLASSPATH
CLASSPATH=${CLASSPATH}:${APPSLIBS}fn_connection.jar
export CLASSPATH
CLASSPATH=${CLASSPATH}:${APPSLIBS}fn_utils.jar;
export CLASSPATH
CLASSPATH=${CLASSPATH}:${APPSLIBS}fn_connect.jar;
export CLASSPATH
CLASSPATH=${CLASSPATH}:../config
export CLASSPATH
~
~
~
"envCommonReplicate.sh" [New] 20L, 614C written
```

Figure 4-6B. *The CLASSPATH environment is set using this shell script*

Shell Script Name: env5x.sh

```
#!/bin/sh
# This command file launches the IBM FileNet library CLASSPATH settings
application.
STARTING_FOLDER=/opt/replication
export STARTING_FOLDER
FIVEXLIBS=${STARTING_FOLDER}/libs/5xlibs/
export FIVEXLIBS
CLASSPATH=${CLASSPATH}:${FIVEXLIBS}Jace.jar
export CLASSPATH
CLASSPATH=${CLASSPATH}:${FIVEXLIBS}log4j.jar
export CLASSPATH
CLASSPATH=${CLASSPATH}:${FIVEXLIBS}log4j-1.2.17.jar
export CLASSPATH
CLASSPATH=${CLASSPATH}:${FIVEXLIBS}xerces.jar
export CLASSPATH
```

CHAPTER 4   A REPLICATION JAVA PROGRAM FOR IBM FILENET OBJECT STORES

```
# jars for Encryption 1.6
CLASSPATH=${CLASSPATH}:${FIVEXLIBS}commons-codec-1.15.jar
export CLASSPATH
```

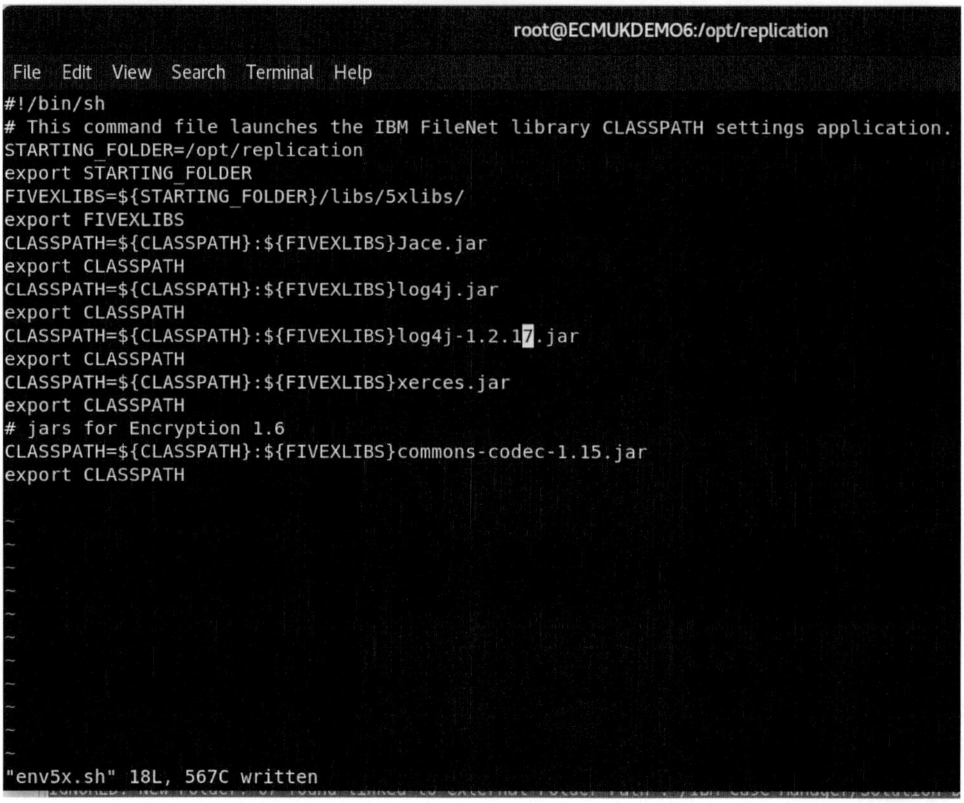

```
root@ECMUKDEMO6:/opt/replication
File  Edit  View  Search  Terminal  Help
#!/bin/sh
# This command file launches the IBM FileNet library CLASSPATH settings application.
STARTING_FOLDER=/opt/replication
export STARTING_FOLDER
FIVEXLIBS=${STARTING_FOLDER}/libs/5xlibs/
export FIVEXLIBS
CLASSPATH=${CLASSPATH}:${FIVEXLIBS}Jace.jar
export CLASSPATH
CLASSPATH=${CLASSPATH}:${FIVEXLIBS}log4j.jar
export CLASSPATH
CLASSPATH=${CLASSPATH}:${FIVEXLIBS}log4j-1.2.17.jar
export CLASSPATH
CLASSPATH=${CLASSPATH}:${FIVEXLIBS}xerces.jar
export CLASSPATH
# jars for Encryption 1.6
CLASSPATH=${CLASSPATH}:${FIVEXLIBS}commons-codec-1.15.jar
export CLASSPATH




"env5x.sh" 18L, 567C written
```

Figure 4-6C.   *The IBM FileNet API jar files added to the **CLASSPATH** with the **env5x.sh** shell script*

The edited shell script files have the Linux security updated as shown in Figure 4-6D.

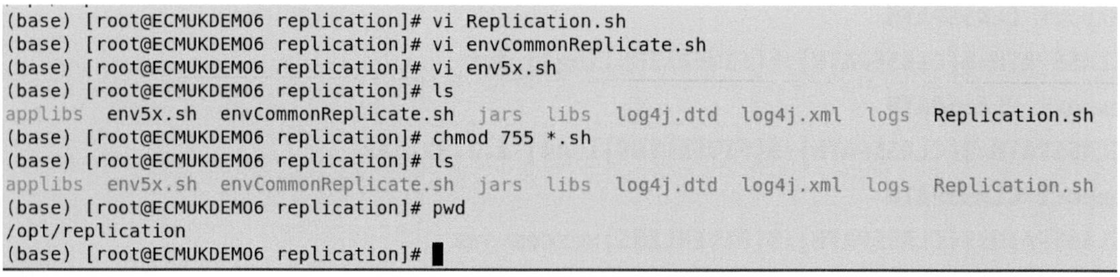

```
(base) [root@ECMUKDEMO6 replication]# vi Replication.sh
(base) [root@ECMUKDEMO6 replication]# vi envCommonReplicate.sh
(base) [root@ECMUKDEMO6 replication]# vi env5x.sh
(base) [root@ECMUKDEMO6 replication]# ls
applibs   env5x.sh  envCommonReplicate.sh   jars   libs   log4j.dtd   log4j.xml   logs   Replication.sh
(base) [root@ECMUKDEMO6 replication]# chmod 755 *.sh
(base) [root@ECMUKDEMO6 replication]# ls
applibs   env5x.sh  envCommonReplicate.sh   jars   libs   log4j.dtd   log4j.xml   logs   Replication.sh
(base) [root@ECMUKDEMO6 replication]# pwd
/opt/replication
(base) [root@ECMUKDEMO6 replication]#
```

Figure 4-6D.   *The security on the shell script files is updated using **chmod 755 *.sh***

CHAPTER 4 A REPLICATION JAVA PROGRAM FOR IBM FILENET OBJECT STORES

Linux Directory Paths Required

The following library paths are created for holding the Replication program Java .jar library files:

/opt/replication
/opt/replication/config

```
(base) [root@ECMUKDEMO6 replication]# mkdir config
(base) [root@ECMUKDEMO6 replication]# ls
applibs  config  env5x.sh  envCommonReplicate.sh  jars  libs  log4j.dtd  log4j.xml  logs  Replication.sh
(base) [root@ECMUKDEMO6 replication]# cd config
(base) [root@ECMUKDEMO6 config]# cp /root/eclipse-workspace/AUDIT_CEReplicate/config/config.xml .
(base) [root@ECMUKDEMO6 config]# ls
config.xml
(base) [root@ECMUKDEMO6 config]# pwd
/opt/replication/config
(base) [root@ECMUKDEMO6 config]#
```

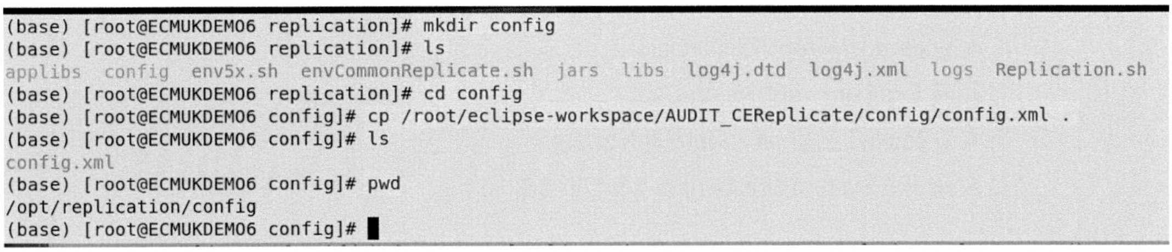

Figure 4-6E. *The /opt/replication/config subdirectory is created and **config.xml** copied from Eclipse*

/opt/replication/libs
/opt/replication/jars
/opt/replication/jars/sdk
/opt/replication/jars/sdk/bin
/opt/replication/libs/5xlibs
/opt/replication/libs/other

The Replication.jar file program uses the following .jars on the **AUDIT_CEReplication** Project Build Path.

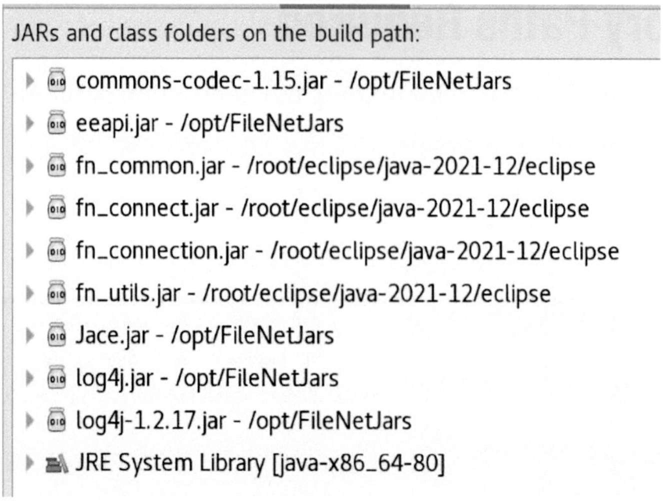

CHAPTER 4   A REPLICATION JAVA PROGRAM FOR IBM FILENET OBJECT STORES

```
JARs and class folders on the build path:
  commons-codec-1.15.jar - /opt/FileNetJars
  eeapi.jar - /opt/FileNetJars
  fn_common.jar - /root/eclipse/java-2021-12/eclipse
  fn_connect.jar - /root/eclipse/java-2021-12/eclipse
  fn_connection.jar - /root/eclipse/java-2021-12/eclipse
  fn_utils.jar - /root/eclipse/java-2021-12/eclipse
  Jace.jar - /opt/FileNetJars
  log4j.jar - /opt/FileNetJars
  log4j-1.2.17.jar - /opt/FileNetJars
  JRE System Library [java-x86_64-80]
```

Figure 4-7. *The supporting .jar files required by the main **Replication.jar** program*

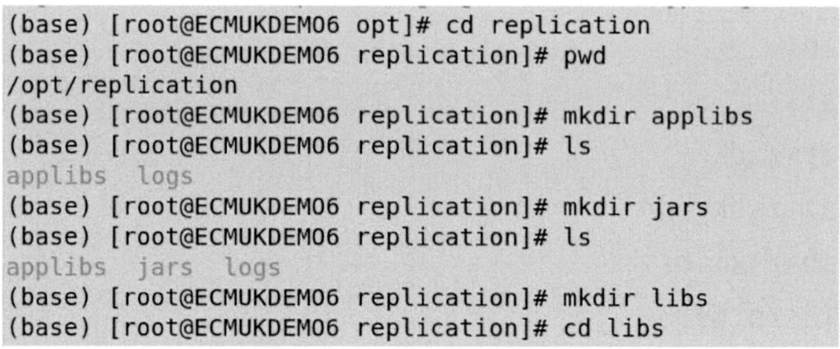

```
(base) [root@ECMUKDEMO6 opt]# cd replication
(base) [root@ECMUKDEMO6 replication]# pwd
/opt/replication
(base) [root@ECMUKDEMO6 replication]# mkdir applibs
(base) [root@ECMUKDEMO6 replication]# ls
applibs  logs
(base) [root@ECMUKDEMO6 replication]# mkdir jars
(base) [root@ECMUKDEMO6 replication]# ls
applibs  jars  logs
(base) [root@ECMUKDEMO6 replication]# mkdir libs
(base) [root@ECMUKDEMO6 replication]# cd libs
```

Figure 4-7A. *The **/opt/replication** jar library folders are created*

(base) [root@ECMUKDEMO6~]# **cd /opt/replication**
(base) [root@ECMUKDEMO6 replication]# **pwd**
/opt/replication
(base) [root@ECMUKDEMO6 replication]# **mkdir applibs**
(base) [root@ECMUKDEMO6 replication]# **ls**
applibs   logs
(base) [root@ECMUKDEMO6 replication]# **mkdir jars**
(base) [root@ECMUKDEMO6 replication]# **ls**
applibs   jars   logs
(base) [root@ECMUKDEMO6 replication]# **mkdir libs**
(base) [root@ECMUKDEMO6 replication]# **cd libs**

CHAPTER 4 A REPLICATION JAVA PROGRAM FOR IBM FILENET OBJECT STORES

```
(base) [root@ECMUKDEMO6 libs]# pwd
/opt/replication/libs
(base) [root@ECMUKDEMO6 libs]# ls
(base) [root@ECMUKDEMO6 libs]# mkdir applibs
(base) [root@ECMUKDEMO6 libs]# mkdir other
(base) [root@ECMUKDEMO6 libs]# ls
applibs  other
(base) [root@ECMUKDEMO6 libs]#
```

Figure 4-7B. *The /**opt/replication/libs** jar library folders are created*

The following files are copied into the folders created earlier:

(base) [root@ECMUKDEMO6 libs]# **pwd**

/opt/replication/libs

(base) [root@ECMUKDEMO6 libs]# **ls**

(base) [root@ECMUKDEMO6 libs]# **mkdir applibs**

(base) [root@ECMUKDEMO6 libs]# **mkdir other**

(base) [root@ECMUKDEMO6 libs]# **ls**

applibs   other

(base) [root@ECMUKDEMO6 libs]#

```
root@ECMUKDEMO6:/opt/replication/applibs
File Edit View Search Terminal Help
(base) [root@ECMUKDEMO6 replication]# ls
applibs  config  env5x.sh  envCommonReplicate.sh  jars  libs  log4j.dtd  log4j.xml  logs  Replication.sh
(base) [root@ECMUKDEMO6 replication]# cd applibs
(base) [root@ECMUKDEMO6 applibs]# ls
fn_common.jar  fn_connection.jar  fn_connect.jar  fn_eventhandlers.jar  fn_utils.jar  jars  java  sdk
(base) [root@ECMUKDEMO6 applibs]# cp /root/eclipse/java-2021-12/eclipse/replication.jar .
(base) [root@ECMUKDEMO6 applibs]# ls -lsa
total 80
 0 drwxr-xr-x.  5 root root    180 Aug  1 14:15 .
 0 drwxr-xr-x.  7 root root    172 Aug  1 14:03 ..
16 -rwxr-xr-x.  1 root root  14684 Aug  1 12:35 fn_common.jar
 8 -rwxr-xr-x.  1 root root   4289 Aug  1 12:35 fn_connection.jar
 4 -rwxr-xr-x.  1 root root   2504 Aug  1 12:35 fn_connect.jar
12 -rwxr-xr-x.  1 root root  11136 Aug  1 12:35 fn_eventhandlers.jar
 4 -rwxr-xr-x.  1 root root   2971 Aug  1 12:35 fn_utils.jar
 0 drwxr-xr-x.  2 root root      6 Aug  1 13:34 jars
 0 drwxr-xr-x.  3 root root     17 Aug  1 13:39 java
36 -rw-r--r--.  1 root root  35840 Aug  1 14:15 replication.jar
 0 drwxr-xr-x. 10 root root    216 Aug  1 13:40 sdk
(base) [root@ECMUKDEMO6 applibs]# chmod 755 replication.jar
(base) [root@ECMUKDEMO6 applibs]# ls
fn_common.jar       fn_connect.jar        fn_utils.jar  java             sdk
fn_connection.jar   fn_eventhandlers.jar  jars          replication.jar
(base) [root@ECMUKDEMO6 applibs]#
```

Figure 4-7C. *The supporting jar files copied to the /**opt/application/applibs** subdirectory*

CHAPTER 4   A REPLICATION JAVA PROGRAM FOR IBM FILENET OBJECT STORES

In summary, the folders are created as follows:

/opt/replication
/opt/replication/config
/opt/replication/libs
/opt/replication/jars
/opt/replication/jars/sdk
/opt/replication/jars/sdk/bin
/opt/replication/libs/5xlibs
/opt/replication/libs/applibs
/opt/replication/libs/other

Jar files and shell script files are located as follows:

/opt/replication/Replicate.sh
/opt/replication/envCommonReplicate.sh
/opt/replication/env5x.sh
/opt/replication/log4j.dtd
/opt/replication/log4j.xml
/opt/replication/config/config.xml
/opt/replication/jars/replication.jar
/opt/replication/jars/Jace.jar
/opt/replication/jars/log4j-1.2.17.jar
/opt/replication/jars/log4j.jar
/opt/replication/jars/eeapi.jar
/opt/replication/jars/
/opt/replication/libs/applibs/replication.jar
/opt/replication/libs/applibs/fn_common.jar
/opt/replication/libs/applibs/fn_connection.jar
/opt/replication/libs/applibs/fn_utils.jar
/opt/replication/libs/applibs/fn_connect.jar
/opt/replication/libs/5xlibs/Jace.jar
/opt/replication/libs/5xlibs/log4.jar
/opt/replication/libs/5xlibs/xerces.jar
/opt/replication/libs/5xlibs/commons-codec-1.15.jar

CHAPTER 4   A REPLICATION JAVA PROGRAM FOR IBM FILENET OBJECT STORES

Jars Required for Content Engine Client

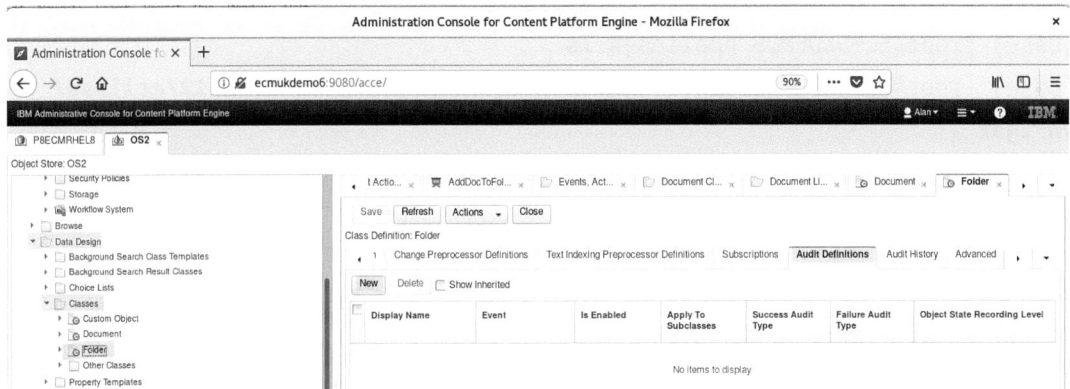

Figure 4-7D. *The list of supporting IBM FileNet P8 Content Engine Version 5.5.x jars*

The files are copied as follows:

(base) [root@ECMUKDEMO6 libs]# **mkdir 5xlibs**
(base) [root@ECMUKDEMO6 libs]# **ls**
5xlibs applibs other
(base) [root@ECMUKDEMO6 libs]# **cd 5xlibs/**
(base) [root@ECMUKDEMO6 5xlibs]# **cp /opt/FileNetJars/Jace.jar .**
(base) [root@ECMUKDEMO6 5xlibs]# **cp /opt/FileNetJars/commons-codec-1.15.jar .**
(base) [root@ECMUKDEMO6 5xlibs]# **cp /opt/FileNetJars/log4j.* .**
(base) [root@ECMUKDEMO6 5xlibs]# **cp /opt/FileNetJars/xerces.jar .**
(base) [root@ECMUKDEMO6 5xlibs]# **cp /opt/FileNetJars/pe*.jar .**
(base) [root@ECMUKDEMO6 5xlibs]# **cp /opt/FileNetJars/eeapi.jar .**
(base) [root@ECMUKDEMO6 5xlibs]# **chmod 755 xerces.jar**
(base) [root@ECMUKDEMO6 5xlibs]# **chmod 755 commons-codec-1.15.jar**
(base) [root@ECMUKDEMO6 5xlibs]# **pwd**
/opt/replication/libs/5xlibs
(base) [root@ECMUKDEMO6 5xlibs]#

CHAPTER 4   A REPLICATION JAVA PROGRAM FOR IBM FILENET OBJECT STORES

For **/opt/replication/libs/applibs:**

```
(base) [root@ECMUKDEMO6 libs]# cd applibs
(base) [root@ECMUKDEMO6 applibs]# ls
(base) [root@ECMUKDEMO6 applibs]# cp /root/eclipse/java-2021-12/eclipse/fn*.jar .
(base) [root@ECMUKDEMO6 applibs]# cp /root/eclipse/java-2021-12/eclipse/replication.jar .
(base) [root@ECMUKDEMO6 applibs]# ls
fn_common.jar  fn_connection.jar  fn_connect.jar  fn_eventhandlers.jar
fn_utils.jar   replication.jar
(base) [root@ECMUKDEMO6 applibs]# pwd
/opt/replication/libs/applibs
(base) [root@ECMUKDEMO6 applibs]#
```

For **/opt/replication/jars:**

```
(base) [root@ECMUKDEMO6 replication]# cd jars
(base) [root@ECMUKDEMO6 jars]# ls
(base) [root@ECMUKDEMO6 jars]# cp /root/eclipse/java-2021-12/eclipse/replication.jar .
(base) [root@ECMUKDEMO6 jars]# cp /opt/FileNetJars/Jace.jar .
(base) [root@ECMUKDEMO6 jars]# cp /opt/FileNetJars/log4j*.jar .
(base) [root@ECMUKDEMO6 jars]# cp /opt/FileNetJars/eeapi.jar .
(base) [root@ECMUKDEMO6 jars]# ls
eeapi.jar  Jace.jar  log4j-1.2.17.jar  log4j.jar  replication.jar
(base) [root@ECMUKDEMO6 jars]# chmod 755 replication.jar
(base) [root@ECMUKDEMO6 jars]#
(base) [root@ECMUKDEMO6 jars]# ls
eeapi.jar  Jace.jar  log4j-1.2.17.jar  log4j.jar  replication.jar
(base) [root@ECMUKDEMO6 jars]# pwd
/opt/replication/jars
(base) [root@ECMUKDEMO6 jars]#
```

The Java sdk used by Case Manager is copied to the **/opt/replication/applibs** subfolder using the command:

cp -R /opt/IBM/CaseManagement/java/ /opt/replication/applibs

Config Area

```
(base) [root@ECMUKDEMO6 replication]# cd config
(base) [root@ECMUKDEMO6 config]# ls -lsa
total 4
0 drwxr-xr-x. 2 root root   24 Aug  1 15:25 .
0 drwxr-xr-x. 7 root root  191 Aug  2 04:53 ..
4 -rwxr-xr-x. 1 root root 2495 Aug  1 15:16 config.xml (NB Date Modified
and Size is updated every run!)
(base) [root@ECMUKDEMO6 config]# pwd
/opt/replication/config
(base) [root@ECMUKDEMO6 config]#
```

Encryption of the Administrator User Password in config.xml by the Program

The processing is as follows:

a. Just for the first run, the **Config.xml** password setting is unencrypted.

b. The first run of the Replication program will load the unencrypted password (remember this is the **exact "clear" unencrypted password** string and so is a vulnerable security risk just for the first run of the program!)

c. It will attempt to decrypt the string found.

d. It will test the decrypted string is wrapped as follows: ***Encrypt3dpasswordvalueP4ssw0rd***

e. If the decrypted string does not contain the start and end strings shown earlier, it will assume the config contains a clear password.

f. In this case, the program will wrap the start and end strings ***Encrypt3d*** and ***P4ssw0rd*** around the *input clear password* string.

g. Encrypt the resulting ***Encrypt3dpasswordvalueP4ssw0rd*** string and write out the result to the **config.xml** file.

h. If the decrypted string is in the form ***Encrypt3dpasswordvalue P4ssw0rd***, **passwordvalue** will be set as the password.

i. Then the read in password will be set in the replication program for use.

Update of the Config.xml Start Date

At the point in time the replication program is started, the current run date and time from the local server system clock will be used as the next start date/time and written back to the **config .xml** file in the following format:

20220719T114828Z

NB Delta set at -1 indicates no overlap.

0 is 1 hour prior to last run date (to the nearest whole hour).

1 is 2 hours prior "" etc.

Config.xml

This was renamed on Linux as **config.xml** (all lowercase).

Note here that the **source** Object Store is defined in the **<ceimport>** tag of the XML!

```xml
<?xml version="1.0" encoding="UTF-8"?><config>
    <ceimport>
        <osname>OS1</osname>
        <osrootfolder>/AUDIT_TEST</osrootfolder>
        <ceuser>Alan</ceuser>
        <cepassword>RW5jcnlwdDNkZmlsZW5ldFAOc3N3MHJk</cepassword>
        <ceurl>http://ecmukdemo6:9080/wsi/FNCEWS40MTOM/</ceurl>
<celoginconfig>/opt/IBM/FileNet/CEClient/config/jaas.conf.WSI</celoginconfig>
        <cestanza>FileNetP8WSI</cestanza>
        <UpdatingBatchSize>250</UpdatingBatchSize>
        <MaxDocCount>50</MaxDocCount>
        <MaxRunTimeMinutes>30</MaxRunTimeMinutes>
        <MaxConsecutiveErrors>1000</MaxConsecutiveErrors>
        <AuditFile>/opt/replication/logs/AuditImportAuditDocs.log</AuditFile>
        <AuditFileFolders>/opt/replication/logs/AuditImportAuditFolders.log</AuditFileFolders>
    </ceimport>
```

```xml
<cereplicate>
    <MaxSearchSize>2000</MaxSearchSize>
<readonlygroup>remoteauditors</readonlygroup>
    <startsearchdate>20220801T151626Z</startsearchdate>
    <DeleteAuditFile>/opt/replication/logs/AuditImportAuditDeleteDocs.log</DeleteAuditFile>
    <deleteMaxDocCount>500</deleteMaxDocCount>
    <DeleteFolderAuditFile>/opt/replication/logs/AuditImportAuditDeleteFolders.log</DeleteFolderAuditFile>
    <DeleteMaxFolderCount>500</DeleteMaxFolderCount>
    <excludedgroup>p8admins</excludedgroup>
    <excludeduser>Alan</excludeduser>
<UpdateDocumentAuditFile>/opt/replication/logs/AuditImportAuditUpdateDocuments.log</UpdateDocumentAuditFile>
    <UpdateMaxDocCount>500</UpdateMaxDocCount>
    <MaxConsecutiveUpdateErrors>500</MaxConsecutiveUpdateErrors>
    <DebugOutputFlag>off</DebugOutputFlag>
    <LDAPSearchFlag>on</LDAPSearchFlag>
<MaxConsecutiveDeleteErrors>500</MaxConsecutiveDeleteErrors>
    <DeltaHours>-1</DeltaHours>
<FolderSubclasses>Folder,FOLDER_ATTRIBUTE</FolderSubclasses>
</cereplicate>
    <ceexport>
        <osname>OS2</osname>
        <ceuser>Alan</ceuser>
        <cepassword>RW5jcnlwdDNkZmlsZW5ldFAOc3N3MHJk</cepassword>
        <MaxDocCount>500</MaxDocCount>
        <MaxRunTimeMinutes>0</MaxRunTimeMinutes>
        <MaxConsecutiveErrors>1000</MaxConsecutiveErrors>
        <AuditFile>/opt/replication/logs/AuditExportAudit.log</AuditFile>
        <SQLBatchSize>250</SQLBatchSize>
        <celoginconfig>/opt/IBM/FileNet/CEClient/config/jaas.conf.WSI</celoginconfig>
        <ceurl>http://ecmukdemo6:9080/wsi/FNCEWS40MTOM</ceurl>
        <cestanza>FileNetP8WSI</cestanza>
    </ceexport>
```

log4j.xml (Updated to Fix Error Issues)

The first run of the Replication program produced the following two errors:

```
log4j:ERROR No appender named [ rfa_threads ] could be found.
log4j:ERROR No appender named [rfa2] could be found.
```

In the following XML, the new appender sections are highlighted:

```xml
<?xml version="1.0" encoding="UTF-8"?>
<!DOCTYPE log4j:configuration SYSTEM "log4j.dtd" >
 <log4j:configuration>
    <appender name="stdout" class="org.apache.log4j.ConsoleAppender">
        <layout class="org.apache.log4j.PatternLayout">
            <param name="ConversionPattern" value="%m%n" />
        </layout>
        <filter class="org.apache.log4j.varia.LevelRangeFilter">
            <param name="levelMin" value="DEBUG" />
            <param name="levelMax" value="INFO" />
        </filter>
    </appender>
    <appender name="rfa" class="org.apache.log4j.RollingFileAppender">
        <param name="file" value="/opt/filenet_app.log"/>
        <param name="MaxFileSize" value="100KB"/>
        <!-- Keep some backup files -->
        <param name="MaxBackupIndex" value="10"/>
        <layout class="org.apache.log4j.PatternLayout">
            <param name="ConversionPattern" value="%p %t %c - %m%n"/>
        </layout>
        <filter class="org.apache.log4j.varia.LevelRangeFilter">
            <param name="levelMin" value="WARN" />
          <param name="levelMax" value="FATAL" />
        </filter>
    </appender>
   <appender name="rfa2" class="org.apache.log4j.RollingFileAppender">
       <param name="file" value="/opt/filenet_app2.log"/>
       <param name="MaxFileSize" value="100KB"/>
```

```xml
    <!-- Keep some backup files -->
    <param name="MaxBackupIndex" value="10"/>
    <layout class="org.apache.log4j.PatternLayout">
         <param name="ConversionPattern" value="%p %t %c - %m%n"/>
    </layout>
    <filter class="org.apache.log4j.varia.LevelRangeFilter">
        <param name="levelMin" value="WARN" />
        <param name="levelMax" value="FATAL" />
    </filter>
  </appender>
<appender name=" rfa_threads " class="org.apache.log4j.RollingFileAppender">
    <param name="file" value="/opt/filenet_appthreads.log"/>
    <param name="MaxFileSize" value="100KB"/>
    <!-- Keep some backup files -->
    <param name="MaxBackupIndex" value="10"/>
    <layout class="org.apache.log4j.PatternLayout">
         <param name="ConversionPattern" value="%p %t %c - %m%n"/>
    </layout>
    <filter class="org.apache.log4j.varia.LevelRangeFilter">
        <param name="levelMin" value="WARN" />
        <param name="levelMax" value="FATAL" />
    </filter>
  </appender>
<logger name="com.ibm.filenet.ce" additivity="false">
    <level value="debug" />
    <appender-ref ref="rfa" />
    <appender-ref ref="stdout" />
</logger>
<logger name="filenet_error.api.com.filenet.apiimpl.util.ConfigValueLookup" additivity="false">
    <level value="debug" />
    <appender-ref ref="rfa" />
    <appender-ref ref="stdout" />
</logger>
```

```xml
<logger name="filenet_tracing.api.detail.moderate.summary.com.filenet.
apiimpl.util.ConfigValueLookup" additivity="false">
    <level value="debug" />
    <appender-ref ref="rfa" />
    <appender-ref ref="stdout" />
</logger>
<logger name=" com.ibm.ce.utils.FileCopier " additivity="false">
    <level value="debug" />
    <appender-ref ref=" rfa_threads " />
</logger>

<logger name="com.ibm.ce.rep.utils.CEReplicate" additivity="false">
    <level value="info" />
    <appender-ref ref="rfa2" />
    <appender-ref ref="stdout" />
</logger>
</log4j:configuration>
```

Original log4j.xml File

```xml
<?xml version="1.0" encoding="UTF-8"?>
<!DOCTYPE log4j:configuration SYSTEM "log4j.dtd" >
<log4j:configuration>
    <appender name="stdout" class="org.apache.log4j.ConsoleAppender">
        <layout class="org.apache.log4j.PatternLayout">
            <param name="ConversionPattern" value="%m%n" />
        </layout>
        <filter class="org.apache.log4j.varia.LevelRangeFilter">
           <param name="levelMin" value="DEBUG" />
          <param name="levelMax" value="INFO" />
      </filter>
     </appender>
    <appender name="rfa" class="org.apache.log4j.RollingFileAppender">
       <param name="file" value="/opt/filenet_app.log"/>
       <param name="MaxFileSize" value="100KB"/>
       <!-- Keep some backup files -->
```

```xml
    <!-- Keep some backup files -->
    <param name="MaxBackupIndex" value="10"/>
    <layout class="org.apache.log4j.PatternLayout">
          <param name="ConversionPattern" value="%p %t %c - %m%n"/>
    </layout>
    <filter class="org.apache.log4j.varia.LevelRangeFilter">
        <param name="levelMin" value="WARN" />
        <param name="levelMax" value="FATAL" />
    </filter>
  </appender>
<appender name=" rfa_threads " class="org.apache.log4j.RollingFileAppender">
    <param name="file" value="/opt/filenet_appthreads.log"/>
    <param name="MaxFileSize" value="100KB"/>
    <!-- Keep some backup files -->
    <param name="MaxBackupIndex" value="10"/>
    <layout class="org.apache.log4j.PatternLayout">
          <param name="ConversionPattern" value="%p %t %c - %m%n"/>
    </layout>
    <filter class="org.apache.log4j.varia.LevelRangeFilter">
        <param name="levelMin" value="WARN" />
        <param name="levelMax" value="FATAL" />
    </filter>
  </appender>
<logger name="com.ibm.filenet.ce" additivity="false">
    <level value="debug" />
    <appender-ref ref="rfa" />
    <appender-ref ref="stdout" />
</logger>
<logger name="filenet_error.api.com.filenet.apiimpl.util.ConfigValueLookup" additivity="false">
    <level value="debug" />
    <appender-ref ref="rfa" />
    <appender-ref ref="stdout" />
</logger>
```

```xml
<logger name="filenet_tracing.api.detail.moderate.summary.com.filenet.
apiimpl.util.ConfigValueLookup" additivity="false">
    <level value="debug" />
    <appender-ref ref="rfa" />
    <appender-ref ref="stdout" />
</logger>
<logger name=" com.ibm.ce.utils.FileCopier " additivity="false">
    <level value="debug" />
    <appender-ref ref=" rfa_threads " />
</logger>

<logger name="com.ibm.ce.rep.utils.CEReplicate" additivity="false">
    <level value="info" />
    <appender-ref ref="rfa2" />
    <appender-ref ref="stdout" />
</logger>
</log4j:configuration>
```

Original log4j.xml File

```xml
<?xml version="1.0" encoding="UTF-8"?>
<!DOCTYPE log4j:configuration SYSTEM "log4j.dtd" >
 <log4j:configuration>
    <appender name="stdout" class="org.apache.log4j.ConsoleAppender">
        <layout class="org.apache.log4j.PatternLayout">
            <param name="ConversionPattern" value="%m%n" />
        </layout>
        <filter class="org.apache.log4j.varia.LevelRangeFilter">
           <param name="levelMin" value="DEBUG" />
           <param name="levelMax" value="INFO" />
      </filter>
    </appender>
    <appender name="rfa" class="org.apache.log4j.RollingFileAppender">
        <param name="file" value="/opt/filenet_app.log"/>
        <param name="MaxFileSize" value="100KB"/>
        <!-- Keep some backup files -->
```

```xml
    <param name="MaxBackupIndex" value="10"/>
    <layout class="org.apache.log4j.PatternLayout">
          <param name="ConversionPattern" value="%p %t %c - %m%n"/>
    </layout>
    <filter class="org.apache.log4j.varia.LevelRangeFilter">
        <param name="levelMin" value="WARN" />
        <param name="levelMax" value="FATAL" />
    </filter>
  </appender>
<logger name="com.ibm.filenet.ce" additivity="false">
     <level value="debug" />
     <appender-ref ref="rfa" />
     <appender-ref ref="stdout" />
</logger>
<logger name="filenet_error.api.com.filenet.apiimpl.util.
ConfigValueLookup" additivity="false">
     <level value="debug" />
     <appender-ref ref="rfa" />
     <appender-ref ref="stdout" />
</logger>
<logger name="filenet_tracing.api.detail.moderate.summary.com.filenet.
apiimpl.util.ConfigValueLookup" additivity="false">
     <level value="debug" />
     <appender-ref ref="rfa" />
     <appender-ref ref="stdout" />
</logger>
<logger name=" com.ibm.ce.utils.FileCopier " additivity="false">
     <level value="debug" />
     <appender-ref ref=" rfa_threads " />
</logger>
```

```
        <logger name="com.ibm.ce.rep.utils.CEReplicate" additivity="false">
            <level value="info" />
            <appender-ref ref="rfa2" />
            <appender-ref ref="stdout" />
        </logger>
</log4j:configuration>
```

log4j.dtd

```
<?xml version="1.0" encoding="UTF-8" ?>
<!--
Licensed to the Apache Software Foundation (ASF) under one or more
contributor license agreements.  See the NOTICE file distributed with
this work for additional information regarding copyright ownership.
The ASF licenses this file to You under the Apache License, Version 2.0
(the "License"); you may not use this file except in compliance with
the License.  You may obtain a copy of the License at

     http://www.apache.org/licenses/LICENSE-2.0

Unless required by applicable law or agreed to in writing, software
distributed under the License is distributed on an "AS IS" BASIS,
WITHOUT WARRANTIES OR CONDITIONS OF ANY KIND, either express or implied.
See the License for the specific language governing permissions and
limitations under the License.
-->

<!-- Authors: Chris Taylor, Ceki Gulcu. -->

<!-- Version: 1.2 -->

<!-- A configuration element consists of optional renderer
elements,appender elements, categories and an optional root
element. -->

<!ELEMENT log4j:configuration (renderer*, appender*,plugin*,
(category|logger)*,root?,
                              (categoryFactory|loggerFactory)?)>
```

```xml
<!-- The "threshold" attribute takes a level value below which -->
<!-- all logging statements are disabled. -->

<!-- Setting the "debug" enable the printing of internal log4j logging   -->
<!-- statements.           -->

<!-- By default, debug attribute is "null", meaning that we not do touch -->
<!-- internal log4j logging settings. The "null" value for the threshold -->
<!-- attribute can be misleading. The threshold field of a repository    -->
<!-- cannot be set to null. The "null" value for the threshold attribute -->
<!-- simply means don't touch the threshold field, the threshold field   -->
<!-- keeps its old value.       -->

<!ATTLIST log4j:configuration
  xmlns:log4j     CDATA #FIXED "http://jakarta.apache.org/log4j/"
  threshold       (all|trace|debug|info|warn|error|fatal|off|null) "null"
  debug           (true|false|null)  "null"
  reset           (true|false) "false"
>

<!-- renderer elements allow the user to customize the conversion of -->
<!-- message objects to String.                                      -->

<!ELEMENT renderer EMPTY>
<!ATTLIST renderer
  renderedClass   CDATA #REQUIRED
  renderingClass  CDATA #REQUIRED
>

<!-- Appenders must have a name and a class. -->
<!-- Appenders may contain an error handler, a layout, optional parameters -->
<!-- and filters. They may also reference (or include) other appenders. -->
```

```
<!ELEMENT appender (errorHandler?, param*,
      rollingPolicy?, triggeringPolicy?, connectionSource?,
      layout?, filter*, appender-ref*)>
<!ATTLIST appender
  name        CDATA      #REQUIRED
  class       CDATA      #REQUIRED
>

<!ELEMENT layout (param*)>
<!ATTLIST layout
  class          CDATA      #REQUIRED
>

<!ELEMENT filter (param*)>
<!ATTLIST filter
  class          CDATA      #REQUIRED
>

<!-- ErrorHandlers can be of any class. They can admit any number of -->
<!-- parameters. -->

<!ELEMENT errorHandler (param*, root-ref?, logger-ref*,  appender-ref?)>
<!ATTLIST errorHandler
   class        CDATA    #REQUIRED
>

<!ELEMENT root-ref EMPTY>

<!ELEMENT logger-ref EMPTY>
<!ATTLIST logger-ref
  ref CDATA #REQUIRED
>

<!ELEMENT param EMPTY>
<!ATTLIST param
  name         CDATA      #REQUIRED
  value        CDATA      #REQUIRED
>
```

```
<!-- The priority class is org.apache.log4j.Level by default -->
<!ELEMENT priority (param*)>
<!ATTLIST priority
  class        CDATA     #IMPLIED
  value        CDATA     #REQUIRED
>

<!-- The level class is org.apache.log4j.Level by default -->
<!ELEMENT level (param*)>
<!ATTLIST level
  class        CDATA     #IMPLIED
  value        CDATA     #REQUIRED
>

<!-- If no level element is specified, then the configurator MUST not -->
<!-- touch the level of the named category. -->
<!ELEMENT category (param*,(priority|level)?,appender-ref*)>
<!ATTLIST category
  class        CDATA     #IMPLIED
  name         CDATA     #REQUIRED
  additivity   (true|false) "true"
>

<!-- If no level element is specified, then the configurator MUST not -->
<!-- touch the level of the named logger. -->
<!ELEMENT logger (level?,appender-ref*)>
<!ATTLIST logger
  name         CDATA     #REQUIRED
  additivity   (true|false) "true"
>

<!ELEMENT categoryFactory (param*)>
<!ATTLIST categoryFactory
   class       CDATA #REQUIRED>

<!ELEMENT loggerFactory (param*)>
<!ATTLIST loggerFactory
   class       CDATA #REQUIRED>
```

```
<!ELEMENT appender-ref EMPTY>
<!ATTLIST appender-ref
  ref CDATA #REQUIRED
>

<!-- plugins must have a name and class and can have optional
parameters -->
<!ELEMENT plugin (param*, connectionSource?)>
<!ATTLIST plugin
  name          CDATA           #REQUIRED
  class     CDATA   #REQUIRED
>

<!ELEMENT connectionSource (dataSource?, param*)>
<!ATTLIST connectionSource
  class       CDATA   #REQUIRED
>

<!ELEMENT dataSource (param*)>
<!ATTLIST dataSource
  class       CDATA   #REQUIRED
>

<!ELEMENT triggeringPolicy ((param|filter)*)>
<!ATTLIST triggeringPolicy
  name      CDATA    #IMPLIED
  class     CDATA    #REQUIRED
>

<!ELEMENT rollingPolicy (param*)>
<!ATTLIST rollingPolicy
  name      CDATA    #IMPLIED
  class     CDATA    #REQUIRED
>

<!-- If no priority element is specified, then the configurator MUST not -->
<!-- touch the priority of root. -->
<!-- The root category always exists and cannot be subclassed. -->
```

```
<!ELEMENT root (param*, (priority|level)?, appender-ref*)>

<!-- ================================================================ -->
<!--                      A logging event                              -->
<!-- ================================================================ -->
<!ELEMENT log4j:eventSet (log4j:event*)>
<!ATTLIST log4j:eventSet
  xmlns:log4j           CDATA #FIXED "http://jakarta.apache.org/log4j/"
  version               (1.1|1.2) "1.2"
  includesLocationInfo  (true|false) "true"
>
<!ELEMENT log4j:event (log4j:message, log4j:NDC?, log4j:throwable?,
                      log4j:locationInfo?, log4j:properties?) >

<!-- The timestamp format is application dependent. -->
<!ATTLIST log4j:event
    logger     CDATA #REQUIRED
    level      CDATA #REQUIRED
    thread     CDATA #REQUIRED
    timestamp  CDATA #REQUIRED
    time       CDATA #IMPLIED
>

<!ELEMENT log4j:message (#PCDATA)>
<!ELEMENT log4j:NDC (#PCDATA)>

<!ELEMENT log4j:throwable (#PCDATA)>

<!ELEMENT log4j:locationInfo EMPTY>
<!ATTLIST log4j:locationInfo
  class   CDATA    #REQUIRED
  method  CDATA    #REQUIRED
  file    CDATA    #REQUIRED
  line    CDATA    #REQUIRED
>
```

```
<!ELEMENT log4j:properties (log4j:data*)>

<!ELEMENT log4j:data EMPTY>
<!ATTLIST log4j:data
  name   CDATA    #REQUIRED
  value  CDATA    #REQUIRED
>
```

Static Constants for Property Types

The DataType property can have one of the values in the following table.

Name	Value	Description
BINARY	1	Specifies a binary data type. Represents binary data by using an array of unsigned 8-bit bytes.
BOOLEAN	2	Specifies a Boolean data type. Represents Boolean data having a value of true or false.
DATE	3	Specifies a DateTime data type. Represents an instance in time as a date and time of day in accordance with ISO 8601.
DOUBLE	4	Specifies a double (Float64) data type. Represents an IEEE-standard 64-bit floating-point number, which has a value ranging from -1.79769313486232e308 to +1.79769313486232e308.
GUID	5	Specifies a GUID (ID) data type. Represents a Globally Unique Identifier (GUID) or DCE Universally Unique Identifier (UUID), which is a unique 128-bit number, as a string of 32 hexadecimal characters enclosed by brackets in the following format: {XXXXXXXX-XXXX-XXXX-XXXX-XXXXXXXXXXXX}. For example, {3F2504E0-4F89-11D3-9A0C-0305E82C3301}.
LONG	6	Specifies an integer data type. Represents a signed 32-bit integer, which has a value ranging from -2,147, 483,648 to +2,147,483,647.
OBJECT	7	Specifies an object data type. Represents an object that is instantiated from a Content Engine class.
STRING	8	Specifies a string data type. Represents text consisting of a sequential collection of 16-bit Unicode characters.

CHAPTER 4 A REPLICATION JAVA PROGRAM FOR IBM FILENET OBJECT STORES

The Cardinality property can have one of the values in the following table.

Name	Value	Description
ENUM	1	Specifies a property with enumeration cardinality. A property with enumeration cardinality is an object-valued property that returns a set collection. A set collection is a read-only collection of unique, unordered, independent objects that must be traversed sequentially. You can iterate through the items of a set collection one page of elements at a time from the server to your client application. However, if the collection changes on the server while you are iterating through it, the number, order, and values of the items in your client copy can change, even if you maintain the same reference to it. A set collection cannot hold any items other than independent objects. By contrast, a list collection can hold items of any data type, with the exception of independent objects.
LIST	2	Specifies a property with list cardinality. A property with list cardinality returns a list collection. A list collection is a collection of ordered items that can either be modifiable (allowing items to be inserted, replaced, or deleted) or read-only. These items need not be unique and can be traversed in any order. When you access a list collection from the server, a complete copy of it is created on your client application, which you can iterate through one element at a time. The items in a list collection must all be of the same data type and must match the data type of the property that returns it. If the property returning a list collection is an object-valued property, all of the objects in the list collection must be dependent objects. A list collection can hold items of any data type (provided each item is of the same data type. However, if a list collection holds objects, they must all be dependent objects; only a set collection can hold independent objects. You cannot create a custom property with list cardinality.
SINGLE	0	Specifies a property with single cardinality. A property with single cardinality returns a single value of the data type that the property can hold.

CHAPTER 4   A REPLICATION JAVA PROGRAM FOR IBM FILENET OBJECT STORES

Event Setup

Events recorded example query:
Events should be set up using the **acce** web application for the following classes.

Folder Class

- File Event
- Update Security Event
- Unfile Event
- Update Event
- Deletion Event

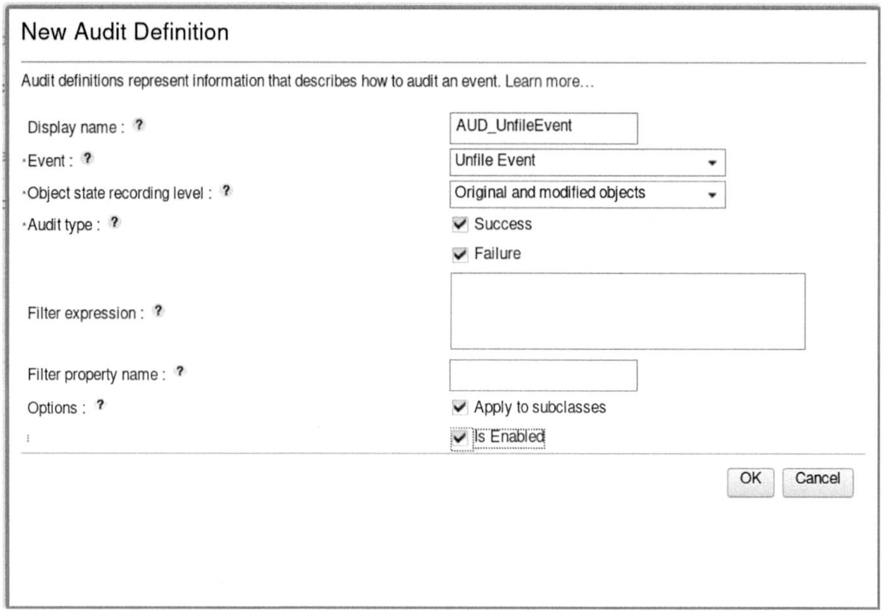

*Figure 4-8.   The **Folder** Class, **Audit Definitions** tab is selected to add events using **New***

The **acce** web application (`http://ecmukdemo6:9080/acce/`) is used to set the Source (OS2) Folder Class Event records which then can be used to identify changes to the Object Store Folder objects, used by the Replication program.

CHAPTER 4　A REPLICATION JAVA PROGRAM FOR IBM FILENET OBJECT STORES

The **New** command button is used to create a new **Event** definition, for example, the Unfile Event to remove a Folder Object from another Folder.

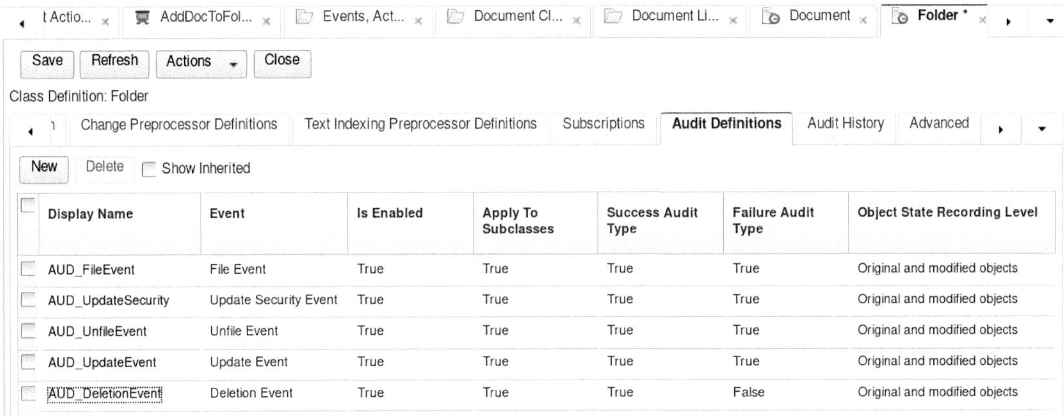

Figure 4-9. The Folder Class Unfile Event

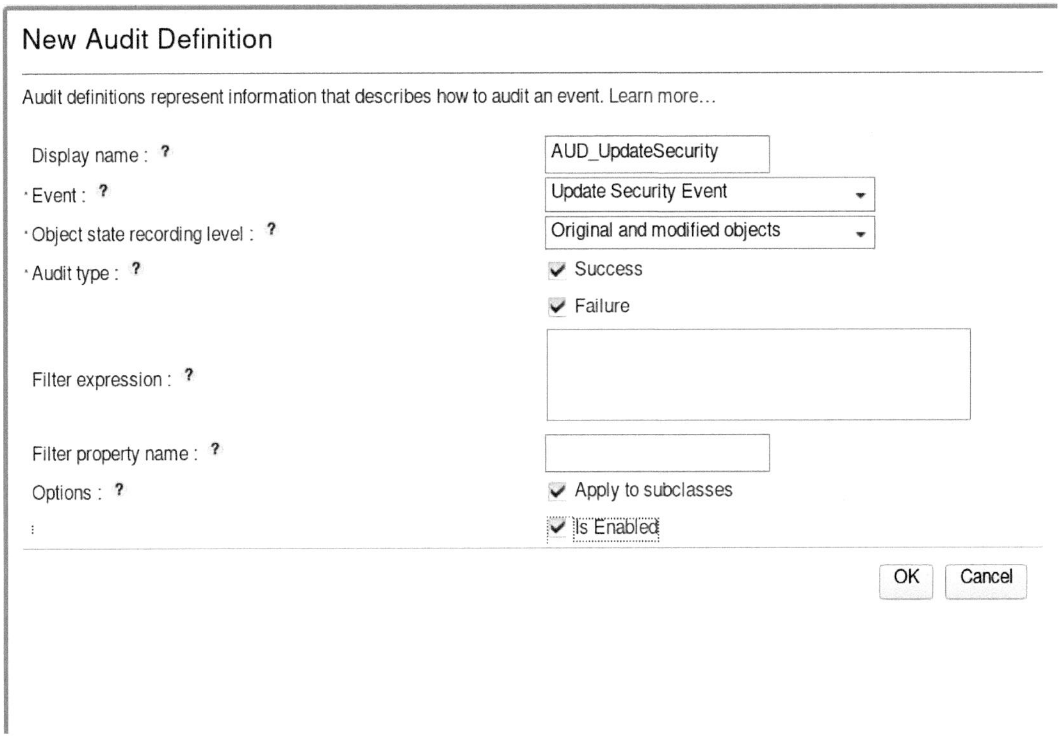

Figure 4-10. The Folder Events set for Audit Definitions

CHAPTER 4 A REPLICATION JAVA PROGRAM FOR IBM FILENET OBJECT STORES

Note Don't forget to click the Save command button or nothing will be preserved!

Document Class

- Update Event
- Checkin Event
- Deletion Event
- Update Security Event

The **acce** web application (http://ecmukdemo6:9080/acce/) is used to set the Source (OS2) Document Class Event records which then can be used to identify changes to the Object Store Document objects, used by the Replication program.

The **New** command button is used to create a new **Event** definition, for example, the **Update Security** Event to record changes to a Document Object Security.

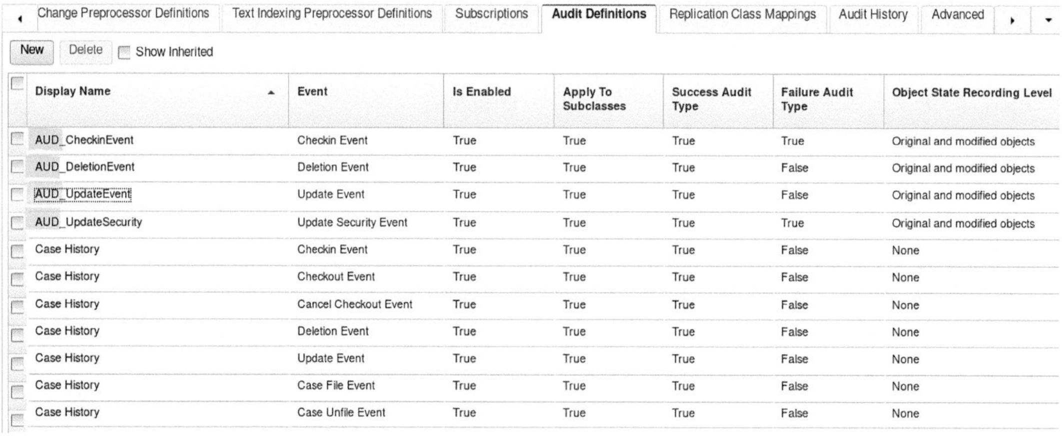

Display Name	Event	Is Enabled	Apply To Subclasses	Success Audit Type	Failure Audit Type	Object State Recording Level
AUD_CheckinEvent	Checkin Event	True	True	True	True	Original and modified objects
AUD_DeletionEvent	Deletion Event	True	True	True	False	Original and modified objects
AUD_UpdateEvent	Update Event	True	True	True	False	Original and modified objects
AUD_UpdateSecurity	Update Security Event	True	True	True	True	Original and modified objects
Case History	Checkin Event	True	True	True	False	None
Case History	Checkout Event	True	True	True	False	None
Case History	Cancel Checkout Event	True	True	True	False	None
Case History	Deletion Event	True	True	True	False	None
Case History	Update Event	True	True	True	False	None
Case History	Case File Event	True	True	True	False	None
Case History	Case Unfile Event	True	True	True	False	None

Figure 4-11. The Audit Definition Event for the Document Update Security Event

CHAPTER 4 A REPLICATION JAVA PROGRAM FOR IBM FILENET OBJECT STORES

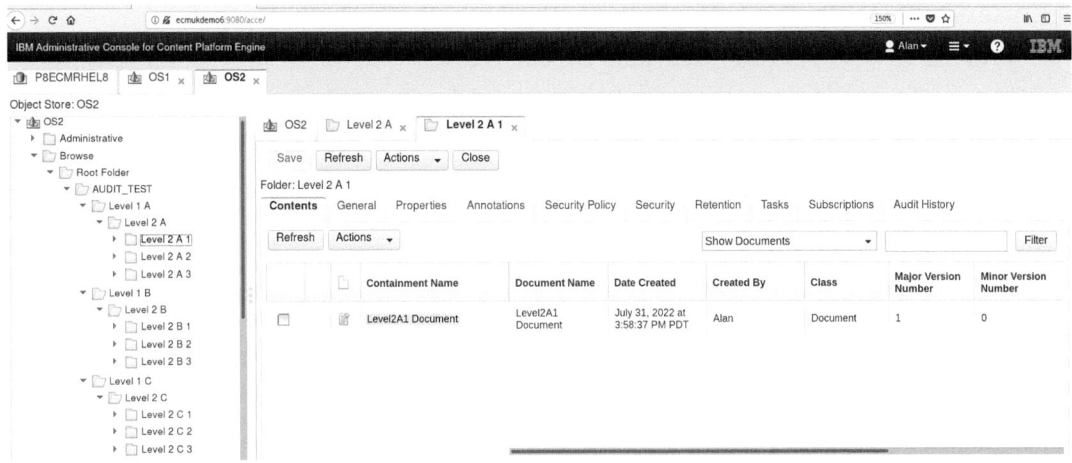

Figure 4-12. The Document Events set for Audit Definitions

Note Don't forget to click the Save command button or nothing will be preserved!

Unit Test Phases

Phase 1

a) Source Object Store Connection is successful.

b) Target Object Store Connection is successful.

c) C) All Test folders under a source Object root Folder, Replicate Test 1, are created in the Target Object Store irrespective of the number of overlapping runs.

Console Log output of the first Phase 1 run:

```
log4j:ERROR No appender named [ rfa_threads ] could be found.
log4j:ERROR No appender named [rfa2] could be found.
CE Replicate Version 3.6
Configuration property file search order:
    user.dir = /root/eclipse-workspace/AUDIT_CEReplicate
```

CHAPTER 4   A REPLICATION JAVA PROGRAM FOR IBM FILENET OBJECT STORES

```
    user.home = /root
    java.home = /opt/ibm/java-x86_64-80/jre
Loading configuration resource: FileNetBuild.properties
Configuration for:Config.AutoRefreshEnabled value:null applied
Configuration for:com.filenet.engine.LoggerRefreshInterval
value:null applied
[Perf Log] No interval found. Auditor disabled.
Configuration for:FileNet.EJB.ContextProperties value:null applied
Configuration for:com.filenet.AppServerType value:null applied
Configuration for:com.filenet.AppServer value:null applied
Configuration for:com.filenet.api.useLocalAuthentication value:null applied
Configuration for:com.filenet.api.crcl.logging.chatty value:null applied
Configuration for:FileNet.crcl.implementation.api.location
value:null applied
Configuration for:FileNet.crcl.implementation.api.locations
value:null applied
Configuration for:FileNet.crcl.implementation.api.urls value:null applied
Configuration for:com.filenet.api.util.Id.SequentialDBType
value:null applied
Configuration for:CheetahBCMode value:false applied
Configuration for:com.filenet.engine.LogRequestWithError value:null applied
Configuration for:com.filenet.engine.init.DefaultLocale value:en-us applied
Configuration for:com.filenet.locale.equivalents value:|,he_*,iw_*,|,
nb_NO,nb_*,no_*,| applied
Configuration for:FileNet.WSI.HttpCookieMaxAge value:null applied
Configuration for:MaximumRetry value:null applied
Configuration for:ExpireSessionInterval value:null applied
Configuration for:BuildVersion value:dap553.1500 applied
Configuration for:FileNet.WSI.HttpChunkSize value:null applied
Configuration for:FileNet.WSI.custom.credential.class value:null applied
Configuration for:FileNet.WSI.custom.credential.usermethod
value:getUserName applied
Configuration for:FileNet.WSI.custom.credential.passwordmethod
value:getPassword applied
Configuration for:FileNet.WSI.TransportConnectionTimeout value:null applied
```

Configuration for:FileNet.WSI.stax.XMLOutputFactory value:com.ibm.xml.xlxp.api.stax.XMLOutputFactoryImpl applied
Configuration for:WSITypeAttributeRequired value:null applied
Configuration for:WSIDefaultNamespacePermitted value:null applied
Configuration for:FileNet.WSI.AutoDetectLTPAToken value:null applied
Configuration for:FileNet.WSI.stax.XMLInputFactory value:com.ibm.xml.xlxp.api.stax.XMLInputFactoryImpl applied
Begin Import Folders ...
Got ObjectStore OS2
Got ObjectStore OS1
SQL: SELECT TOP 500 f.foldername, f.Id, f.Creator, f.DateCreated, f.LastModifier, f.DateLastModified, f.Name, f.Owner, f.LockToken, f.LockTimeout, f.LockOwner, f.PathName, f.IndexationId, f.CmIndexingFailureCode, f.CmRetentionDate, f.ContainerType, f.InheritParentPermissions, f.IsHiddenContainer FROM Folder AS f WITH INCLUDESUBCLASSES WHERE f.DateCreated > 20220530T080541Z ORDER BY f.DateCreated
Configuration for:FileNet.WSI.DeserializationBufferSize value:null applied
Configuration for:FileNet.WSI.SpillCutover value:null applied
IGNORED: New Folder: Audit Master found linked to external Folder Path : /IBM Case Manager/Solution Deployments/Audit Master Should be linked to :/AUDIT_TEST
IGNORED: New Folder: Case Types found linked to external Folder Path : /IBM Case Manager/Solution Deployments/Audit Master/Case Types Should be linked to :/AUDIT_TEST
IGNORED: New Folder: AUD_AuditDepartment found linked to external Folder Path : /IBM Case Manager/Solution Deployments/Audit Master/Case Types/AUD_AuditDepartment Should be linked to :/AUDIT_TEST
IGNORED: New Folder: Cases found linked to external Folder Path : /IBM Case Manager/Solution Deployments/Audit Master/Case Types/AUD_AuditDepartment/Cases Should be linked to :/AUDIT_TEST
IGNORED: New Folder: CaseStructure found linked to external Folder Path : /IBM Case Manager/Solution Deployments/Audit Master/Case Types/AUD_AuditDepartment/CaseStructure Should be linked to :/AUDIT_TEST
Realm Name: o=sample

IGNORED: New Folder: 2022 found linked to external Folder Path : /IBM Case Manager/Solution Deployments/Audit Master/Case Types/AUD_AuditDepartment/Cases/2022 Should be linked to :/AUDIT_TEST
IGNORED: New Folder: 06 found linked to external Folder Path : /IBM Case Manager/Solution Deployments/Audit Master/Case Types/AUD_AuditDepartment/Cases/2022/06 Should be linked to :/AUDIT_TEST
IGNORED: New Folder: 09 found linked to external Folder Path : /IBM Case Manager/Solution Deployments/Audit Master/Case Types/AUD_AuditDepartment/Cases/2022/06/09 Should be linked to :/AUDIT_TEST
IGNORED: New Folder: 0044 found linked to external Folder Path : /IBM Case Manager/Solution Deployments/Audit Master/Case Types/AUD_AuditDepartment/Cases/2022/06/09/0044 Should be linked to :/AUDIT_TEST
IGNORED: New Folder: 000000100001 found linked to external Folder Path : /IBM Case Manager/Solution Deployments/Audit Master/Case Types/AUD_AuditDepartment/Cases/2022/06/09/0044/000000100001 Should be linked to :/AUDIT_TEST
IGNORED: New Folder: 11 found linked to external Folder Path : /IBM Case Manager/Solution Deployments/Audit Master/Case Types/AUD_AuditDepartment/Cases/2022/06/11 Should be linked to :/AUDIT_TEST
IGNORED: New Folder: 0234 found linked to external Folder Path : /IBM Case Manager/Solution Deployments/Audit Master/Case Types/AUD_AuditDepartment/Cases/2022/06/11/0234 Should be linked to :/AUDIT_TEST
IGNORED: New Folder: 000000110001 found linked to external Folder Path : /IBM Case Manager/Solution Deployments/Audit Master/Case Types/AUD_AuditDepartment/Cases/2022/06/11/0234/000000110001 Should be linked to :/AUDIT_TEST
IGNORED: New Folder: 18 found linked to external Folder Path : /IBM Case Manager/Solution Deployments/Audit Master/Case Types/AUD_AuditDepartment/Cases/2022/06/18 Should be linked to :/AUDIT_TEST
IGNORED: New Folder: 0118 found linked to external Folder Path : /IBM Case Manager/Solution Deployments/Audit Master/Case Types/AUD_AuditDepartment/Cases/2022/06/18/0118 Should be linked to :/AUDIT_TEST
IGNORED: New Folder: 000000120001 found linked to external Folder Path : /IBM Case Manager/Solution Deployments/Audit Master/Case Types/AUD_

AuditDepartment/Cases/2022/06/18/0118/000000120001 Should be linked to :/AUDIT_TEST
IGNORED: New Folder: 0101 found linked to external Folder Path : /IBM Case Manager/Solution Deployments/Audit Master/Case Types/AUD_AuditDepartment/Cases/2022/06/18/0101 Should be linked to :/AUDIT_TEST
IGNORED: New Folder: 000000120002 found linked to external Folder Path : /IBM Case Manager/Solution Deployments/Audit Master/Case Types/AUD_AuditDepartment/Cases/2022/06/18/0101/000000120002 Should be linked to :/AUDIT_TEST
IGNORED: New Folder: 0042 found linked to external Folder Path : /IBM Case Manager/Solution Deployments/Audit Master/Case Types/AUD_AuditDepartment/Cases/2022/06/18/0042 Should be linked to :/AUDIT_TEST
IGNORED: New Folder: 000000120003 found linked to external Folder Path : /IBM Case Manager/Solution Deployments/Audit Master/Case Types/AUD_AuditDepartment/Cases/2022/06/18/0042/000000120003 Should be linked to :/AUDIT_TEST
IGNORED: New Folder: 0141 found linked to external Folder Path : /IBM Case Manager/Solution Deployments/Audit Master/Case Types/AUD_AuditDepartment/Cases/2022/06/18/0141 Should be linked to :/AUDIT_TEST
IGNORED: New Folder: 000000120004 found linked to external Folder Path : /IBM Case Manager/Solution Deployments/Audit Master/Case Types/AUD_AuditDepartment/Cases/2022/06/18/0141/000000120004 Should be linked to :/AUDIT_TEST
IGNORED: New Folder: 0250 found linked to external Folder Path : /IBM Case Manager/Solution Deployments/Audit Master/Case Types/AUD_AuditDepartment/Cases/2022/06/18/0250 Should be linked to :/AUDIT_TEST
IGNORED: New Folder: 000000120005 found linked to external Folder Path : /IBM Case Manager/Solution Deployments/Audit Master/Case Types/AUD_AuditDepartment/Cases/2022/06/18/0250/000000120005 Should be linked to :/AUDIT_TEST
IGNORED: New Folder: 0156 found linked to external Folder Path : /IBM Case Manager/Solution Deployments/Audit Master/Case Types/AUD_AuditDepartment/Cases/2022/06/18/0156 Should be linked to :/AUDIT_TEST
IGNORED: New Folder: 000000120006 found linked to external Folder Path : /IBM Case Manager/Solution Deployments/Audit Master/Case Types/AUD_

AuditDepartment/Cases/2022/06/18/0156/000000120006 Should be linked to :/AUDIT_TEST
IGNORED: New Folder: 19 found linked to external Folder Path : /IBM Case Manager/Solution Deployments/Audit Master/Case Types/AUD_AuditDepartment/Cases/2022/06/19 Should be linked to :/AUDIT_TEST
IGNORED: New Folder: 0248 found linked to external Folder Path : /IBM Case Manager/Solution Deployments/Audit Master/Case Types/AUD_AuditDepartment/Cases/2022/06/19/0248 Should be linked to :/AUDIT_TEST
IGNORED: New Folder: 000000120007 found linked to external Folder Path : /IBM Case Manager/Solution Deployments/Audit Master/Case Types/AUD_AuditDepartment/Cases/2022/06/19/0248/000000120007 Should be linked to :/AUDIT_TEST
IGNORED: New Folder: 07 found linked to external Folder Path : /IBM Case Manager/Solution Deployments/Audit Master/Case Types/AUD_AuditDepartment/Cases/2022/07 Should be linked to :/AUDIT_TEST
IGNORED: New Folder: 07 found linked to external Folder Path : /IBM Case Manager/Solution Deployments/Audit Master/Case Types/AUD_AuditDepartment/Cases/2022/07/07 Should be linked to :/AUDIT_TEST
IGNORED: New Folder: 0279 found linked to external Folder Path : /IBM Case Manager/Solution Deployments/Audit Master/Case Types/AUD_AuditDepartment/Cases/2022/07/07/0279 Should be linked to :/AUDIT_TEST
IGNORED: New Folder: 000000130001 found linked to external Folder Path : /IBM Case Manager/Solution Deployments/Audit Master/Case Types/AUD_AuditDepartment/Cases/2022/07/07/0279/000000130001 Should be linked to :/AUDIT_TEST
Realm Name: o=sample
Realm Name: o=sample
Realm Name: o=sample
Realm Name: o=sample
Realm Name: o=sample
Realm Name: o=sample
Realm Name: o=sample
Realm Name: o=sample
Realm Name: o=sample
Realm Name: o=sample

Realm Name: o=sample
Realm Name: o=sample
Realm Name: o=sample
Realm Name: o=sample
Realm Name: o=sample
Import 16 Folders took 6043 milliseconds
Begin Delete Docs ...
SQL: SELECT TOP 500 e.CmAuditSequence,e.ClassDescription,e.Creator,e.DateCreated,e.SourceObjectId,e.DateLastModified, e.EventStatus, e.Id, e.InitiatingUser, e.LastModifier, e.Name, e.Owner, e.SourceClassId FROM Event AS e WITH INCLUDESUBCLASSES INNER JOIN CLASSDEFINITION AS cd WITH INCLUDESUBCLASSES ON e.SourceClassId = cd.Id WHERE (e.DateCreated >= 20220530T080541Z) AND (cd.SymbolicName <>'Folder') AND (cd.SymbolicName <>'FOLDER_ATTRIBUTE') AND (e.EventStatus = 0) ORDER BY e.CmAuditSequence,e.DateCreated
Delete 10 Docs took 6933 milliseconds
Begin Import Docs ...
Start Search Date : 20220530T080541Z
Start Delta Hours : -1
SQL: SELECT TOP 500 d.* FROM Document AS d WITH INCLUDESUBCLASSES WHERE d.DateLastModified > 20220530T080541Z and d.IsCurrentVersion = true and d.IsReserved = false ORDER BY d.DateLastModified,d.MajorVersionNumber, d.MinorVersionNumber
IGNORED: New Document: PageResources.zip found linked to external Folder Path : /IBM Case Manager/Solution Deployments/Audit Master Should be linked to :/AUDIT_TEST
IGNORED: New Document: ViewResources.zip found linked to external Folder Path : /IBM Case Manager/Solution Deployments/Audit Master Should be linked to :/AUDIT_TEST
IGNORED: New Document: AC.pdf found linked to external Folder Path : /IBM Case Manager/Solution Deployments/Audit Master/Case Types/AUD_AuditDepartment/Cases/2022/06/09/0044/000000100001 Should be linked to :/AUDIT_TEST
IGNORED: New Document: AuditEventHandler found linked to external Folder Path : /CodeModules Should be linked to :/AUDIT_TEST

CHAPTER 4   A REPLICATION JAVA PROGRAM FOR IBM FILENET OBJECT STORES

IGNORED: New Document: Test of the Audit Event found linked to external Folder Path : /IBM Case Manager/Solution Deployments/Audit Master/Case Types/AUD_AuditDepartment/Cases/2022/06/09/0044/000000100001 Should be linked to :/AUDIT_TEST
IGNORED: New Document: Test Audit Events found linked to external Folder Path : /Published Should be linked to :/AUDIT_TEST
IGNORED: New Document: Audit Event Test Version 2 found linked to external Folder Path : /IBM Case Manager/Solution Deployments/Audit Master/Case Types/AUD_AuditDepartment/Cases/2022/06/09/0044/000000100001 Should be linked to :/AUDIT_TEST
IGNORED: New Document: Automatic Folder ID added found linked to external Folder Path : /IBM Case Manager/Solution Deployments/Audit Master/Case Types/AUD_AuditDepartment/Cases/2022/06/11/0234/000000110001 Should be linked to :/AUDIT_TEST
IGNORED: New Document: Solution Workflow Collection not linked to any Folder Path !
IGNORED: New Document: Detail Deployment Log found linked to external Folder Path : /IBM Case Manager/Solution Deployments/Audit Master Should be linked to :/AUDIT_TEST
IGNORED: New Document: Error Deployment Log found linked to external Folder Path : /IBM Case Manager/Solution Deployments/Audit Master Should be linked to :/AUDIT_TEST
IGNORED: New Document: AUDOperations.jar found linked to external Folder Path : /CodeModules Should be linked to :/AUDIT_TEST
IGNORED: New Document: Solution Workflow Collection not linked to any Folder Path !
IGNORED: New Document: AUDOperations.jar found linked to external Folder Path : /CodeModules Should be linked to :/AUDIT_TEST
IGNORED: New Document: JMS Message Trigger found linked to external Folder Path : /IBM Case Manager/Solution Deployments/Audit Master/Case Types/AUD_AuditDepartment/Cases/2022/07/07/0279/000000130001 Should be linked to :/AUDIT_TEST
IGNORED: New Document: Second Audit Report Example 08-July-2022 found linked to external Folder Path : /IBM Case Manager/Solution Deployments/Audit Master/Case Types/AUD_AuditDepartment/Cases/2022/07/07/0279/000000130001 Should be linked to :/AUDIT_TEST

IGNORED: New Document: AUD_JMSMessageWorkflow found linked to external Folder Path : /Published Should be linked to :/AUDIT_TEST
IGNORED: New Document: AuditReport00005_Engineering.pdf found linked to external Folder Path : /IBM Case Manager/Solution Deployments/Audit Master/Case Types/AUD_AuditDepartment/Cases/2022/07/07/0279/000000130001 Should be linked to :/AUDIT_TEST
Realm Name: o=sample
IGNORED: New Document: Resources.js not linked to any Folder Path !
IGNORED: New Document: Solution Definition not linked to any Folder Path !
IGNORED: New Document: PE Configuration not linked to any Folder Path !
IGNORED: New Document: General Claim Workflow Definition not linked to any Folder Path !
IGNORED: New Document: Solution Definition not linked to any Folder Path !
IGNORED: New Document: PE Configuration not linked to any Folder Path !
IGNORED: New Document: Audit Department Workflow Definition not linked to any Folder Path !
Realm Name: o=sample
Realm Name: o=sample
Realm Name: o=sample
Import 13 Docs took 13420 milliseconds
Begin Delete Folders ...
SQL: SELECT TOP 500 e.CmAuditSequence,e.ClassDescription,e.Creator,e.DateCreated,e.SourceObjectId,e.SourceObject,e.DateLastModified, e.EventStatus, e.Id, e.InitiatingUser, e.LastModifier, e.Name, e.Owner, e.SourceClassId FROM Event AS e WITH INCLUDESUBCLASSES INNER JOIN CLASSDEFINITION AS cd WITH INCLUDESUBCLASSES ON e.SourceClassId = cd.Id WHERE (e.DateCreated > 20220530T080541Z) AND ((cd.SymbolicName = 'Folder') OR (cd.SymbolicName = 'FOLDER_ATTRIBUTE')) AND (e.EventStatus = 0) ORDER BY e.CmAuditSequence,e.DateCreated
Delete 0 Folders took 838 milliseconds
Begin Update Docs ...
SQL: SELECT TOP 500 e.CmAuditSequence,e.ClassDescription,e.Creator,e.DateCreated,e.SourceObjectId,e.DateLastModified, e.EventStatus, e.Id, e.InitiatingUser, e.LastModifier, e.Name, e.Owner, e.SourceClassId FROM Event AS e WITH INCLUDESUBCLASSES INNER JOIN CLASSDEFINITION AS cd WITH INCLUDESUBCLASSES ON e.SourceClassId = cd.Id WHERE (e.DateCreated

CHAPTER 4   A REPLICATION JAVA PROGRAM FOR IBM FILENET OBJECT STORES

```
> 20220530T080541Z) AND (cd.SymbolicName <>'Folder') AND (cd.
SymbolicName <>'FOLDER_ATTRIBUTE') AND (e.EventStatus = 0) ORDER BY
e.CmAuditSequence,e.DateCreated
Update 242 Docs took 1591 milliseconds~
```

The **OS1** replicate /**AUDIT_TEST** folder Object Store now has the replicated folder structure and linked documents exported from the **OS2** Object Store source.

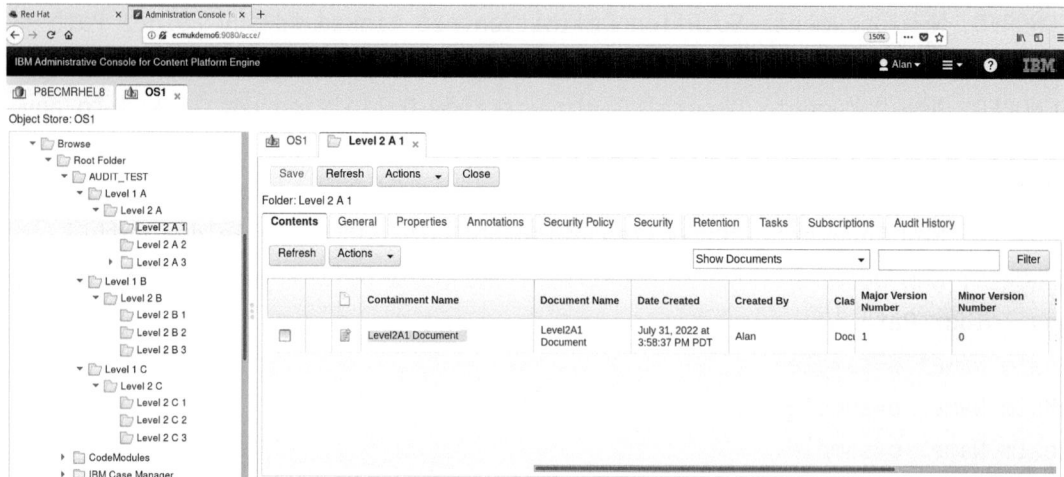

Figure 4-12A. *The original OS2 test data folder and document structure to be replicated*

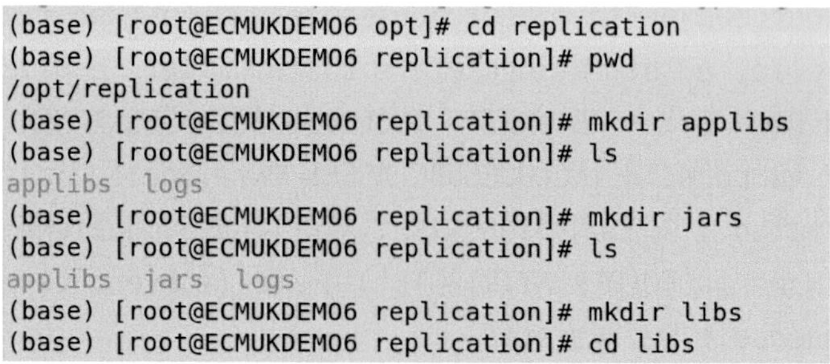

Figure 4-12B. *The OS1 test data folder and document structure from the OS2 original*

Notice that the exported **OS1**, **AUDIT_TEST** folder document, **Date Created** property entry (July 31, 2022 at 3:58:37 PM PDT), is exactly the same as from the original document in the **OS2** Object Store, as is required.

Phase 2

a) Ensure matching property values in the Source Object and with the same security ACLs and GUID values.

b) Ensure major/minor versions of documents under the source Object Folders, Replicate Test 1 and subfolders, are created and linked correctly in the Target Object.

c) Ensure Folders and Documents created in other Folder paths outside the defined path are not copied.

Note This was seen in the preceding output log, for example:

IGNORED: New Folder: Audit Master found linked to external Folder Path : /IBM Case Manager/Solution Deployments/Audit Master Should be linked to :/AUDIT_TEST

---

d) Add new code to test for an existing document in the Target Object Store and delete and recreate it.

Phase 3

a) First replication – Verify the object store and folders.
The same object store and folders are found at the target.
(Checked: See Figures 4-12A and 4-12B.)

b) Remove a document at the source – The same document is removed at the target.

c) Verify the folder change – Amend the name of the existing folder at the source.
The change is synced to the target.

d) Remove a folder at the source – The same folder is removed at the target.

CHAPTER 4 A REPLICATION JAVA PROGRAM FOR IBM FILENET OBJECT STORES

Phase 4

a) Complete Deployment Manager full replication Test.
    Update to encrypt the Alan password – Automatic config.xml updates.
    (This is in the code listing, tested and working!)
    Update to automatically update the next replication start date in the config.xml file.
    (This was checked in the config.xml file and updated as expected.)

b) Release Batch Replication.sh and Java program for testing.

c) Create the documentation (this chapter).

Creating the Java Projects

The Eclipse Java IDE program used in Chapters 2 and 3 was used in this chapter to create a set of projects for the development of the Java Replication program .jar files with the outline settings as shown in Table 4-1.

Table 4-1. The Java Eclipse Projects used for support of the Replication program

Project Name	Library .Jar Name	Source Java File(s)	Brief Description
FileNetConnection	fn_connection.jar	CEConnection.java	Loads the user and password
FN_Connect	fn_connect.jar	CEConnect.java	Uses the FileNet Web Service Interface to connect (using the FileNetP8WSI stanza)
AUDIT_CEReplicate	replication.jar	CEReplicate.java	The main Replication program in the package **com.ibm.ce.rep.utils**
AUDIT_Common	fn_common.jar	CEReplicateConfig.java CEMigrateConfig.java	Reads in the XML configuration parameters for the Replication
AUDIT_Utils	fn_utils.jar	PropsUtil.java	Loads configuration details from a properties file

CHAPTER 4   A REPLICATION JAVA PROGRAM FOR IBM FILENET OBJECT STORES

Part 3 – Code Developed for the Replication Program

CEReplicate – 2957 Lines of Java Code – replication.jar

The supporting compiled jar files and standard encryption .jars are copied into the **/opt/replication/libs/applibs** subfolder.

```
(base) [root@ECMUKDEMO6 replication]# ls
applibs  jars  libs  logs
(base) [root@ECMUKDEMO6 replication]# cp /root/eclipse-workspace/AUDIT_CEReplicate/src/log4j.dtd
(base) [root@ECMUKDEMO6 replication]# cp /root/eclipse-workspace/AUDIT_CEReplicate/src/log4j.xml
(base) [root@ECMUKDEMO6 replication]# ls
applibs  jars  libs  log4j.dtd  log4j.xml  logs
(base) [root@ECMUKDEMO6 replication]# chmod 775 log4j.*
(base) [root@ECMUKDEMO6 replication]# ls
applibs  jars  libs  log4j.dtd  log4j.xml  logs
(base) [root@ECMUKDEMO6 replication]# cp /root/eclipse/java-2021-12/eclipse/fn_*.jar .
(base) [root@ECMUKDEMO6 replication]# ls
applibs         fn_connection.jar  fn_eventhandlers.jar  jars  log4j.dtd  logs
fn_common.jar   fn_connect.jar     fn_utils.jar          libs  log4j.xml
(base) [root@ECMUKDEMO6 replication]# cd applibs
(base) [root@ECMUKDEMO6 applibs]# mv ../fn*.jar .
(base) [root@ECMUKDEMO6 applibs]# ls
fn_common.jar  fn_connection.jar  fn_connect.jar  fn_eventhandlers.jar  fn_utils.jar
```

Figure 4-12C. *The applibs, jars, and libs subdirectories are built under /opt/replication*

```
(base) [root@ECMUKDEMO6 replication]# ./Replication.sh
CLASSPATH= /opt/replication/libs/applibs/replication.jar:/opt/replication/libs/applibs/fn_common.jar:/opt/replication/libs/applibs/fn_connection.jar:/opt/replication/libs/applibs/fn_utils.jar:/opt/replication/libs/applibs/fn_connect.jar:./config:/opt/replication/libs/5xlibs/Jace.jar:/opt/replication/libs/5xlibs/log4j.jar:/opt/replication/libs/5xlibs/log4j-1.2.17.jar:/opt/replication/libs/5xlibs/xerces.jar:/opt/replication/libs/5xlibs/commons-codec-1.15.jar
Starting Replication job ... Output to console.log in this directory
log4j:ERROR No appender named [ rfa_threads ] could be found.
log4j:ERROR No appender named [rfa2] could be found.
(base) [root@ECMUKDEMO6 replication]#
```

Figure 4-12D. *The /opt/replication/libs/applibs subfolder .jar files*

```
(base) [root@ECMUKDEMO6 libs]# ls
applibs   other
(base) [root@ECMUKDEMO6 libs]# pwd
/opt/replication/libs
(base) [root@ECMUKDEMO6 libs]# cd ..
(base) [root@ECMUKDEMO6 replication]# ls
```

```
applibs  jars  libs  logs
(base) [root@ECMUKDEMO6 replication]# cp /root/eclipse-workspace/AUDIT_
CEReplicate/src/log4j.dtd .
(base) [root@ECMUKDEMO6 replication]# cp /root/eclipse-workspace/AUDIT_
CEReplicate/src/log4j.xml .
(base) [root@ECMUKDEMO6 replication]# ls
applibs  jars  libs  log4j.dtd  log4j.xml  logs
(base) [root@ECMUKDEMO6 replication]# chmod 775 log4j.*
(base) [root@ECMUKDEMO6 replication]# ls
applibs  jars  libs  log4j.dtd  log4j.xml  logs
(base) [root@ECMUKDEMO6 replication]# cp /root/eclipse/java-2021-12/
eclipse/fn_*.jar .
(base) [root@ECMUKDEMO6 replication]# ls
applibs         fn_connection.jar   fn_eventhandlers.jar   jars   log4j.
dtd  logs
fn_common.jar  fn_connect.jar   fn_utils.jar             libs   log4j.xml
(base) [root@ECMUKDEMO6 replication]# cd applibs
(base) [root@ECMUKDEMO6 applibs]# mv ../fn*.jar .
(base) [root@ECMUKDEMO6 applibs]# ls
fn_common.jar  fn_connection.jar  fn_connect.jar  fn_eventhandlers.jar  fn_
utils.jar
```

The latest release of the apache codec jar can be downloaded from the following URL:

https://commons.apache.org/proper/commons-codec/download_codec.cgi
A Xerces Java library file can be downloaded from the zip file:
http://archive.apache.org/dist/xml/xerces-j/Xerces-J-bin.1.4.4.zip
(An import used in CEMigrateConfig.java code for XML parsing)

CHAPTER 4 A REPLICATION JAVA PROGRAM FOR IBM FILENET OBJECT STORES

```
root@ECMUKDEMO6:/opt/replication
File Edit View Search Terminal Help
CE Replicate Version 3.6
Configuration property file search order:
    user.dir = /opt/replication
    user.home = /root
    java.home = /opt/replication/applibs/sdk/jre
Loading configuration resource: FileNetBuild.properties
Configuration for:Config.AutoRefreshEnabled value:null applied
Configuration for:com.filenet.engine.LoggerRefreshInterval value:null applied
[Perf Log] No interval found. Auditor disabled.
Configuration for:FileNet.EJB.ContextProperties value:null applied
Configuration for:com.filenet.AppServerType value:null applied
Configuration for:com.filenet.AppServer value:null applied
Configuration for:com.filenet.api.useLocalAuthentication value:null applied
Configuration for:com.filenet.api.crcl.logging.chatty value:null applied
Configuration for:FileNet.crcl.implementation.api.location value:null applied
Configuration for:FileNet.crcl.implementation.api.locations value:null applied
Configuration for:FileNet.crcl.implementation.api.urls value:null applied
Configuration for:com.filenet.api.util.Id.SequentialDBType value:null applied
Configuration for:CheetahBCMode value:false applied
Configuration for:com.filenet.engine.LogRequestWithError value:null applied
Configuration for:com.filenet.engine.init.DefaultLocale value:en-us applied
Configuration for:com.filenet.locale.equivalents value:|,he_*,iw_*,|,nb_NO,nb_*,no_*,| applied
Configuration for:FileNet.WSI.HttpCookieMaxAge value:null applied
Configuration for:MaximumRetry value:null applied
Configuration for:ExpireSessionInterval value:null applied
Configuration for:BuildVersion value:dap553.1500 applied
Configuration for:FileNet.WSI.HttpChunkSize value:null applied
Configuration for:FileNet.WSI.custom.credential.class value:null applied
Configuration for:FileNet.WSI.custom.credential.usermethod value:getUserName applied
"console.log" 135L, 15981C
```

Figure 4-12E. *The first run of the Replication.sh shell script (the **CLASSPATH** is echoed)*

CHAPTER 4   A REPLICATION JAVA PROGRAM FOR IBM FILENET OBJECT STORES

```
root@ECMUKDEMO6:/opt/replication                                          x
File  Edit  View  Search  Terminal  Help
IGNORED: New Document: AuditReport00005_Engineering.pdf found linked to external Folder Path : /IBM Case Manager/Solu
tion Deployments/Audit Master/Case Types/AUD_AuditDepartment/Cases/2022/07/07/0279/000000130001 Should be linked to :
/AUDIT_TEST
Realm Name: o=sample
IGNORED: New Document: Resources.js not linked to any Folder Path !
IGNORED: New Document: Solution Definition not linked to any Folder Path !
IGNORED: New Document: PE Configuration not linked to any Folder Path !
IGNORED: New Document: General Claim Workflow Definition not linked to any Folder Path !
IGNORED: New Document: Solution Definition not linked to any Folder Path !
IGNORED: New Document: PE Configuration not linked to any Folder Path !
IGNORED: New Document: Audit Department Workflow Definition not linked to any Folder Path !
Realm Name: o=sample
Realm Name: o=sample
Realm Name: o=sample
Import 13 Docs took 9745 milliseconds
Begin Delete Folders ...
SQL: SELECT TOP 500 e.CmAuditSequence,e.ClassDescription,e.Creator,e.DateCreated,e.SourceObjectId,e.SourceObject,e.Da
teLastModified, e.EventStatus, e.Id, e.InitiatingUser, e.LastModifier FROM Event AS
 e WITH INCLUDESUBCLASSES INNER JOIN CLASSDEFINITION AS cd WITH INCLUDESUBCLASSES ON e.SourceClassId = cd.Id WHERE (e
.DateCreated > 20220531T080942Z) AND ((cd.SymbolicName = 'Folder') OR (cd.SymbolicName = 'FOLDER_ATTRIBUTE') ) AND (e
.EventStatus = 0) ORDER BY e.CmAuditSequence,e.DateCreated
Delete 0 Folders took 572 milliseconds
Begin Update Docs ...
SQL: SELECT TOP 500 e.CmAuditSequence,e.ClassDescription,e.Creator,e.DateCreated,e.SourceObjectId,e.DateLastModified,
  e.EventStatus, e.Id, e.InitiatingUser, e.LastModifier, e.Name, e.Owner, e.SourceClassId FROM Event AS e WITH INCLUDE
SUBCLASSES INNER JOIN CLASSDEFINITION AS cd WITH INCLUDESUBCLASSES ON e.SourceClassId = cd.Id WHERE (e.DateCreated >
20220531T080942Z) AND (cd.SymbolicName <>'Folder')  AND (cd.SymbolicName <>'FOLDER_ATTRIBUTE')  AND (e.EventStatus =
0) ORDER BY e.CmAuditSequence,e.DateCreated
Update 242 Docs took 1434 milliseconds
```

Figure 4-12F. *The console.log file from the first run of the Replication.sh shell script*

The end of the console.log shows the successful transfer of the folders and documents.

CHAPTER 4   A REPLICATION JAVA PROGRAM FOR IBM FILENET OBJECT STORES

```
<?xml version="1.0" encoding="UTF-8"?><config>
    <ceimport>
        <osname>OS1</osname>
        <osrootfolder>/AUDIT_TEST</osrootfolder>
        <ceuser>Alan</ceuser>
        <cepassword>RW5jcnlwdDNkZmlsZW5ldFA0c3N3MHJk</cepassword>
        <ceurl>http://ecmukdemo6:9080/wsi/FNCEWS40MTOM/</ceurl>
<celoginconfig>/opt/IBM/FileNet/CEClient/config/jaas.conf.WSI</celoginconfig>
        <cestanza>FileNetP8WSI</cestanza>
        <UpdatingBatchSize>250</UpdatingBatchSize>
        <MaxDocCount>50</MaxDocCount>
        <MaxRunTimeMinutes>30</MaxRunTimeMinutes>
        <MaxConsecutiveErrors>1000</MaxConsecutiveErrors>
        <AuditFile>/opt/replication/logs/AuditImportAuditDocs.log</AuditFile>
        <AuditFileFolders>/opt/replication/logs/AuditImportAuditFolders.log</AuditFileFolders>
    </ceimport>
    <cereplicate>
        <MaxSearchSize>2000</MaxSearchSize>
<readonlygroup>remoteauditors</readonlygroup>
        <startsearchdate>20220801T151626Z</startsearchdate>
        <DeleteAuditFile>/opt/replication/logs/AuditImportAuditDeleteDocs.log</DeleteAuditFile>
        <deleteMaxDocCount>500</deleteMaxDocCount>
        <DeleteFolderAuditFile>/opt/replication/logs/AuditImportAuditDeleteFolders.log</DeleteFolderAuditFile>
        <DeleteMaxFolderCount>500</DeleteMaxFolderCount>
        <excludedgroup>p8admins</excludedgroup>
        <excludeduser>Alan</excludeduser>
        <UpdateDocumentAuditFile>/opt/replication/logs/AuditImportAuditUpdateDocuments.log</UpdateDocumentAuditFile>
        <UpdateMaxDocCount>500</UpdateMaxDocCount>
        <MaxConsecutiveUpdateErrors>500</MaxConsecutiveUpdateErrors>
        <DebugOutputFlag>off</DebugOutputFlag>
        <LDAPSearchFlag>on</LDAPSearchFlag>
<MaxConsecutiveDeleteErrors>500</MaxConsecutiveDeleteErrors>
        <DeltaHours>-1</DeltaHours>
<FolderSubclasses>Folder,FOLDER_ATTRIBUTE</FolderSubclasses>
</cereplicate>
    <ceexport>
        <osname>OS2</osname>
        <ceuser>Alan</ceuser>
        <cepassword>RW5jcnlwdDNkZmlsZW5ldFA0c3N3MHJk</cepassword>
        <MaxDocCount>500</MaxDocCount>
        <MaxRunTimeMinutes>0</MaxRunTimeMinutes>
        <MaxConsecutiveErrors>1000</MaxConsecutiveErrors>
        <AuditFile>/opt/replication/logs/AuditExportAudit.log</AuditFile>
        <SQLBatchSize>250</SQLBatchSize>
```

Figure 4-12G. *The Replication program run statistics of the console.log file is shown*

```
(base) [root@ECMUKDEMO6 logs]# cat AuditImportAuditDocs.log
Import Start -
GUID, SOURCEVSID, DESTDOCID, STATE, Millisecs
{20971282-0000-C812-A324-A297D40C4583},{20971282-0000-C040-BACC-E8F1283325E2},{20971282-0000-C812-A324-A297D40C4583},Imported,234
{907B5682-0000-C91D-B246-DA41D0421E00},{907B5682-0000-C446-B8D2-AFCE4A3EC989},{907B5682-0000-C91D-B246-DA41D0421E00},Imported,84
{F07C5682-0000-CC1D-9AD4-475BBAE585EA},{F07C5682-0000-CA3B-ACC9-B2D5BF5F4DF5},{F07C5682-0000-CC1D-9AD4-475BBAE585EA},Imported,100
{A07E5682-0000-CB18-B67F-58A7D3319184},{A07E5682-0000-CD38-9B5F-8A80CB75457C},{A07E5682-0000-CB18-B67F-58A7D3319184},Imported,110
Finished - time taken 9745milliseconds
Finished - documents processed - 13
Finished - documents failed to be processed - 0
Finished - document total 'Add Content Retry Count' - 0
(base) [root@ECMUKDEMO6 logs]#
```

Figure 4-12H. *The Replication program has successfully updated the start search date*

CHAPTER 4 A REPLICATION JAVA PROGRAM FOR IBM FILENET OBJECT STORES

```
(base) [root@ECMUKDEMO6 replication]# ls
applibs  jars  libs  logs
(base) [root@ECMUKDEMO6 replication]# cp /root/eclipse-workspace/AUDIT_CEReplicate/src/log4j.dtd
(base) [root@ECMUKDEMO6 replication]# cp /root/eclipse-workspace/AUDIT_CEReplicate/src/log4j.xml
(base) [root@ECMUKDEMO6 replication]# ls
applibs  jars  libs  log4j.dtd  log4j.xml  logs
(base) [root@ECMUKDEMO6 replication]# chmod 775 log4j.*
(base) [root@ECMUKDEMO6 replication]# ls
applibs  jars  libs  log4j.dtd  log4j.xml  logs
(base) [root@ECMUKDEMO6 replication]# cp /root/eclipse/java-2021-12/eclipse/fn_*.jar .
(base) [root@ECMUKDEMO6 replication]# ls
applibs         fn_connection.jar  fn_eventhandlers.jar  jars  log4j.dtd  logs
fn_common.jar   fn_connect.jar     fn_utils.jar          libs  log4j.xml
(base) [root@ECMUKDEMO6 replication]# cd applibs
(base) [root@ECMUKDEMO6 applibs]# mv ../fn*.jar .
(base) [root@ECMUKDEMO6 applibs]# ls
fn_common.jar  fn_connection.jar  fn_connect.jar  fn_eventhandlers.jar  fn_utils.jar
```

Figure 4-12I. *The AuditImportAuditDocs.log shows the details of imported documents*

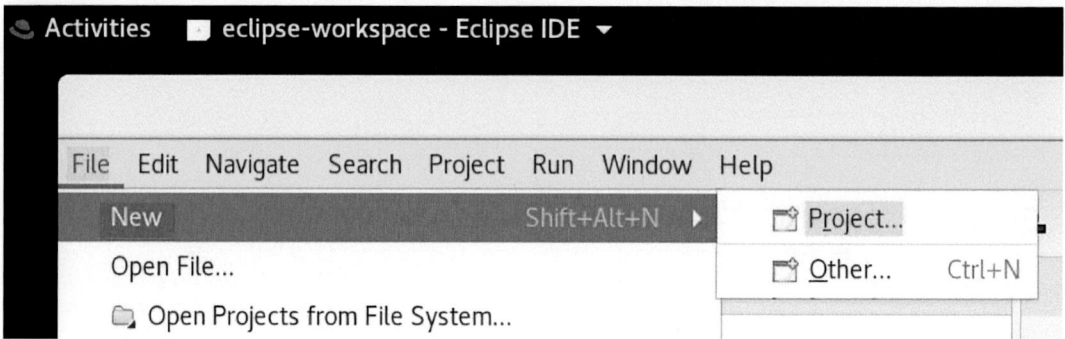

Figure 4-13. *The setup of the supporting file systems for the Replication program*

Project Creation

The Eclipse New project, **AUDIT_CEReplicate**, is created.

CHAPTER 4 A REPLICATION JAVA PROGRAM FOR IBM FILENET OBJECT STORES

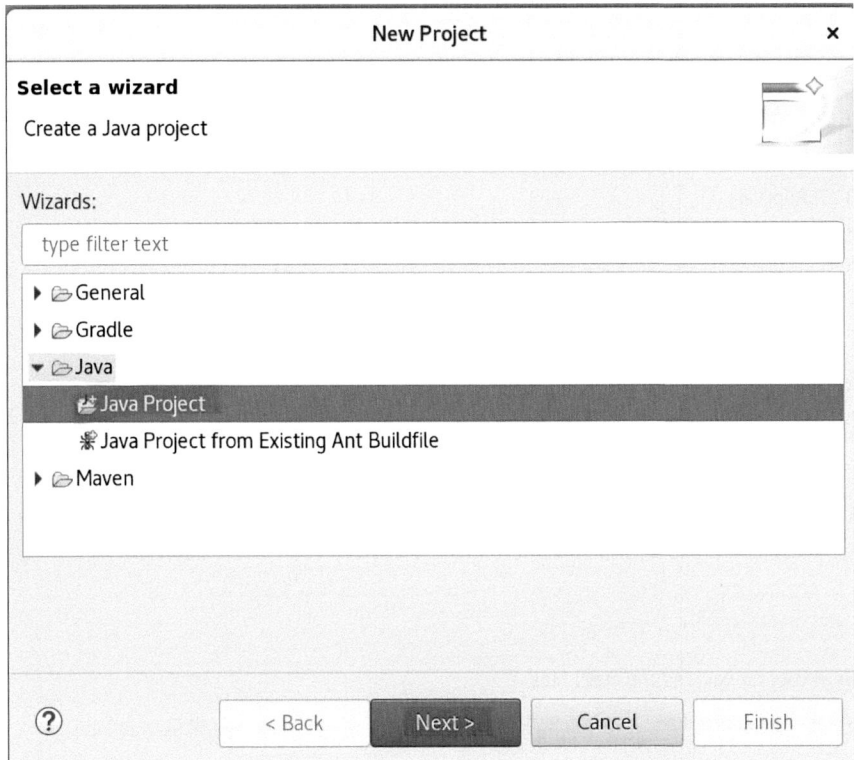

Figure 4-14. *The File* ➤ *New* ➤ *Project menu option is selected for the New Project wizard*

The Java Project type is selected.

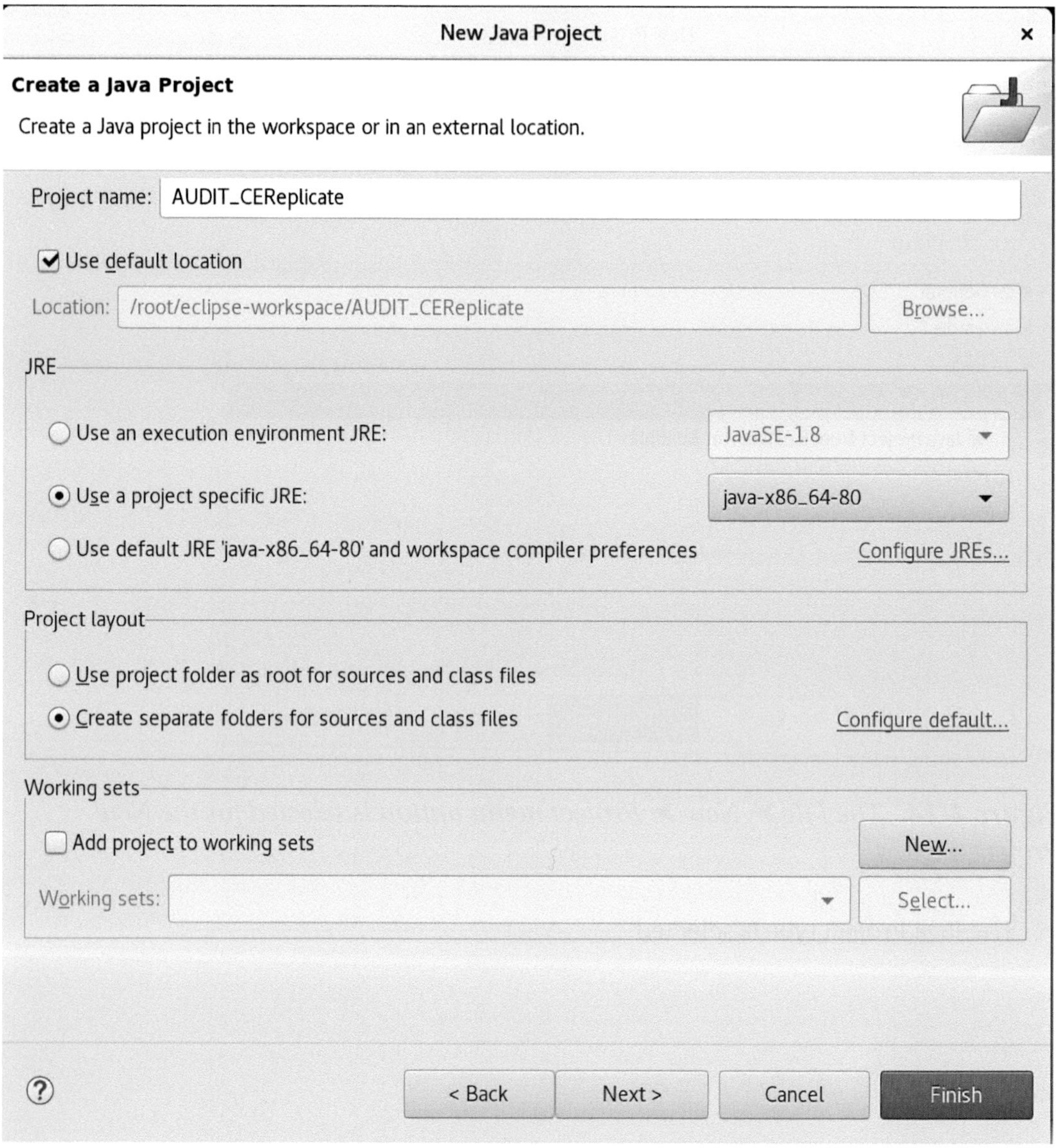

CHAPTER 4   A REPLICATION JAVA PROGRAM FOR IBM FILENET OBJECT STORES

Figure 4-15. The Java Project is selected with the Eclipse Project wizard

A number of separate projects will be created for each custom utility .jar file we will reference.

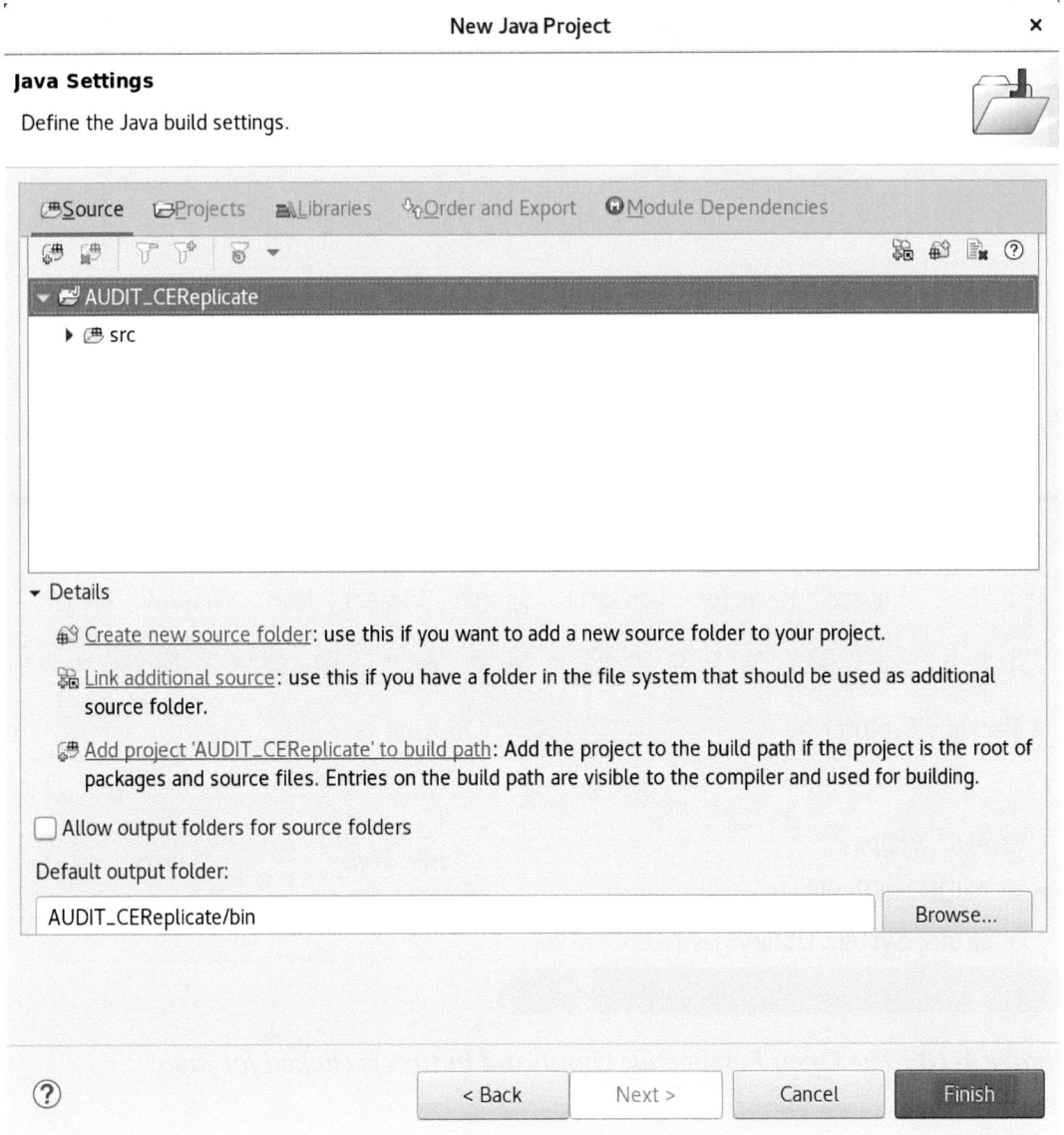

Figure 4-16. *The Project Name, AUDIT_CEReplicate, is entered and the 1.8 JRE selected*

CHAPTER 4   A REPLICATION JAVA PROGRAM FOR IBM FILENET OBJECT STORES

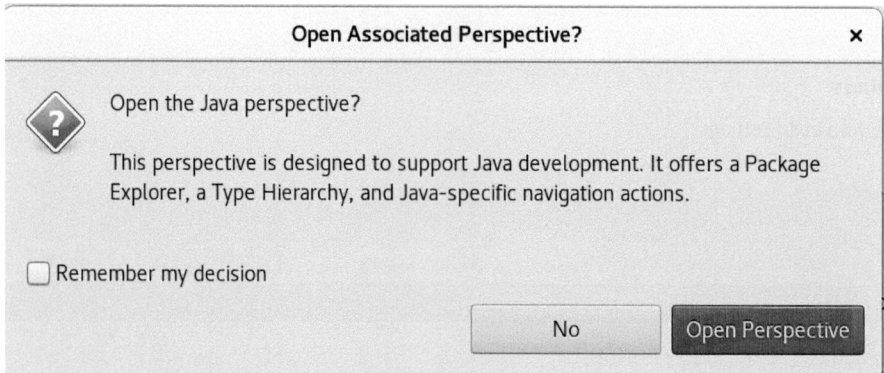

Figure 4-17. *The parameters are set for the Java project, and Finish is clicked*

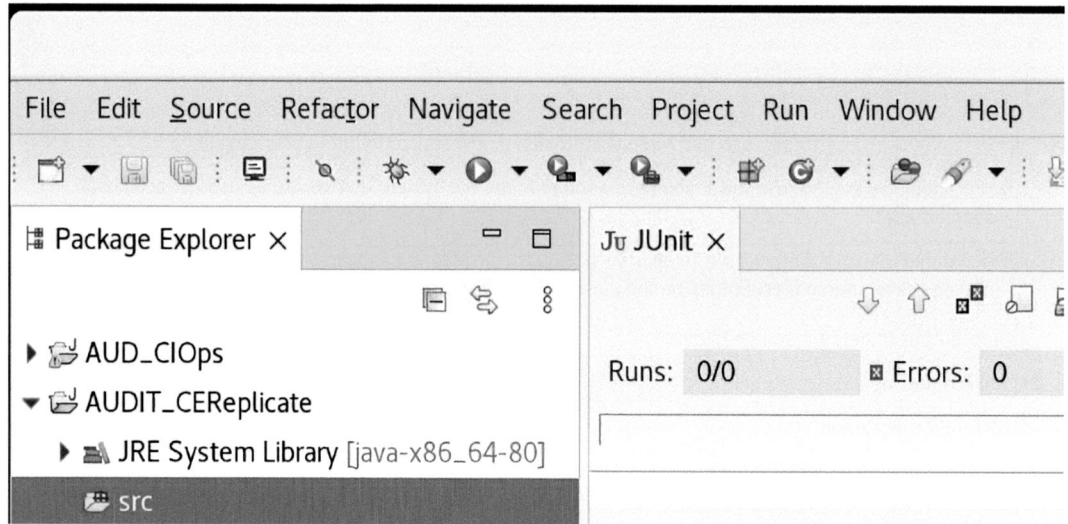

Figure 4-18. *The Open Perspective command button is clicked for Java development*

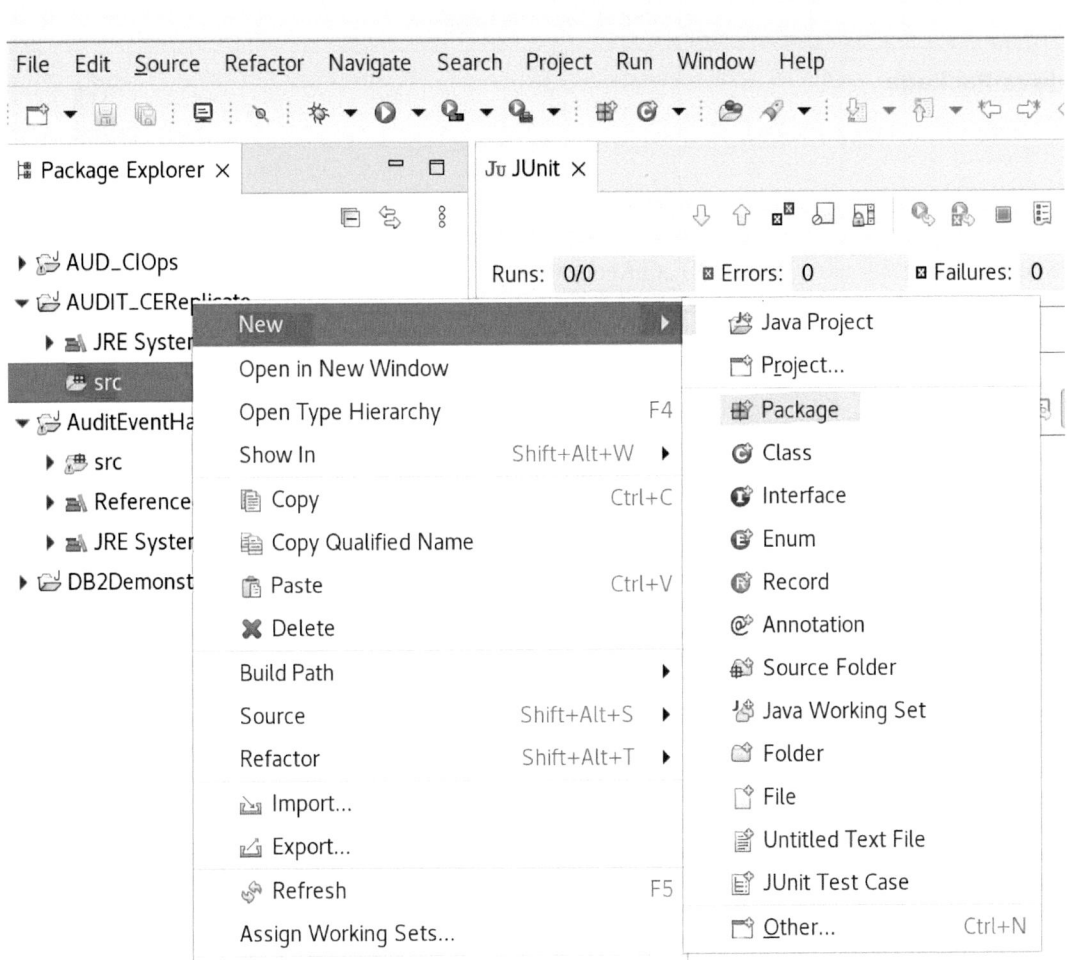

Figure 4-19. *The AUDIT_CEReplicate Project is now available for Java packages and code*

A new Java package is created for the CEReplicate class.

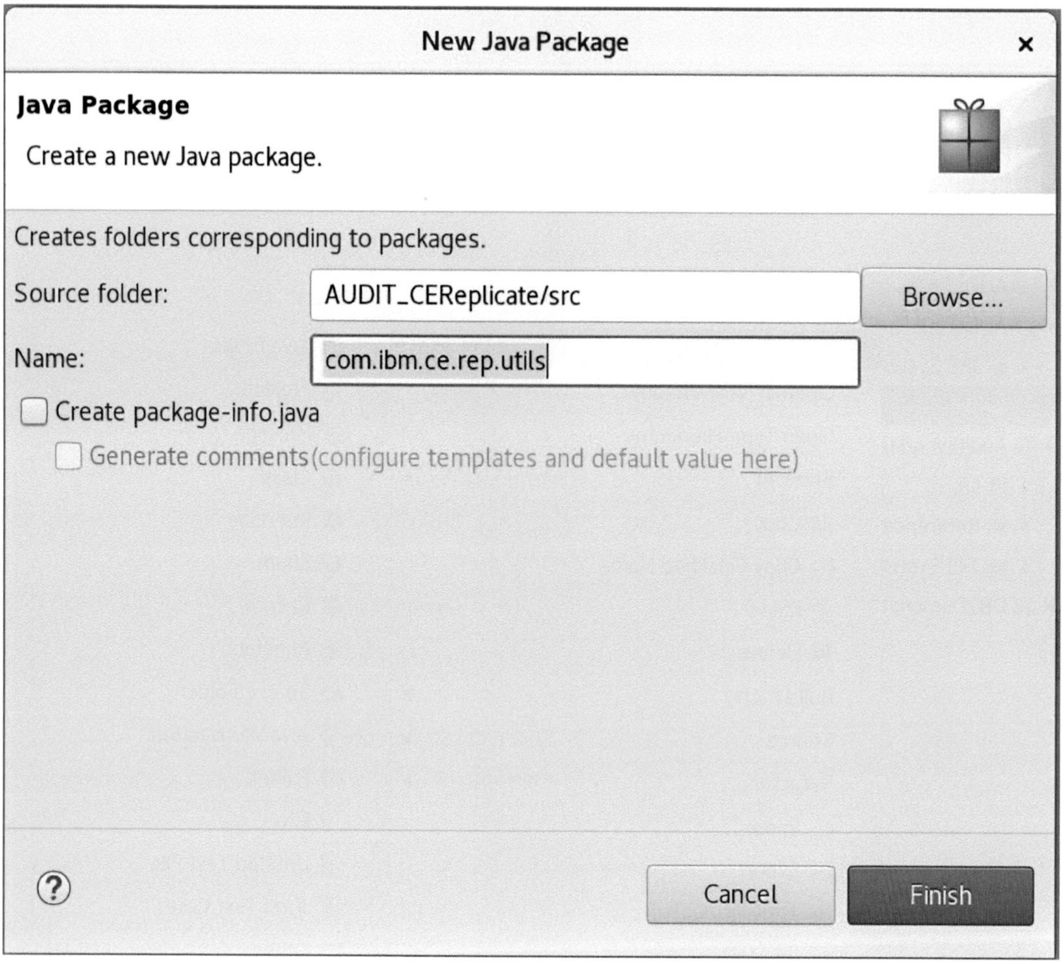

CHAPTER 4   A REPLICATION JAVA PROGRAM FOR IBM FILENET OBJECT STORES

Figure 4-20. The right-mouse click is used to select the New ➤ Package menu option

CHAPTER 4 A REPLICATION JAVA PROGRAM FOR IBM FILENET OBJECT STORES

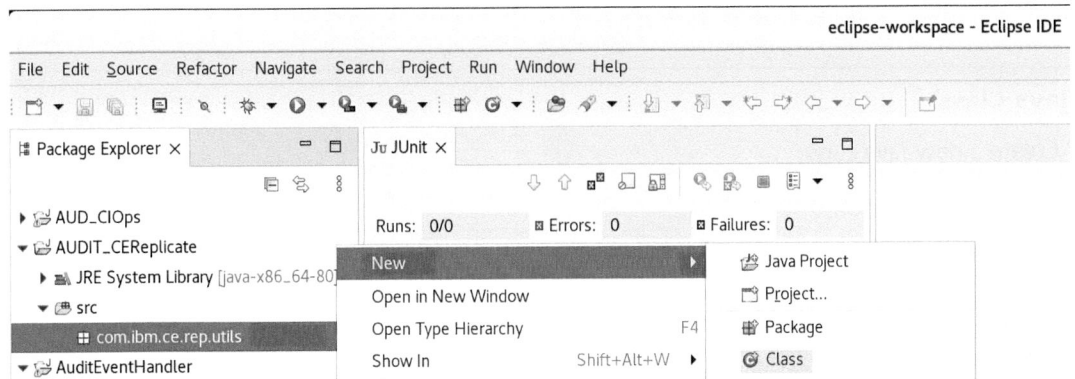

Figure 4-21. *The Java package name,* ***com.ibm.ce.rep.utils****, is created by clicking Finish*

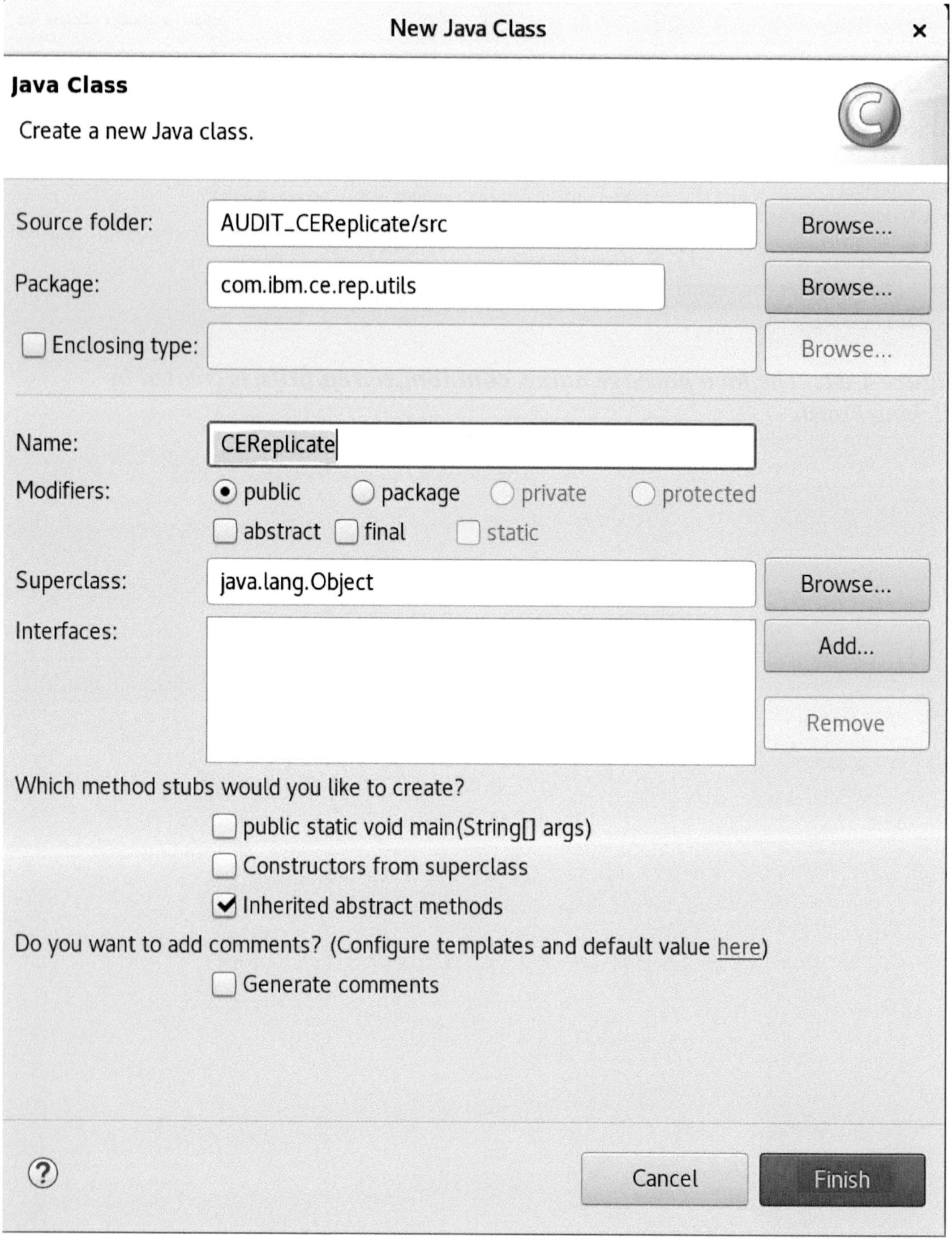

CHAPTER 4 A REPLICATION JAVA PROGRAM FOR IBM FILENET OBJECT STORES

*Figure 4-22. The **CEReplicate** Class is added to the **com.ibm.ce.rep.utils** package*

CHAPTER 4 A REPLICATION JAVA PROGRAM FOR IBM FILENET OBJECT STORES

Figure 4-23. *The CEReplicate class name is entered and the Finish command clicked*

CHAPTER 4   A REPLICATION JAVA PROGRAM FOR IBM FILENET OBJECT STORES

Adding the Base CEReplicate Java Code

The base **CEReplicate.java** code is pasted in from the standard code module as a starting point for the development and comments and new Java code added for changes for our functionality.

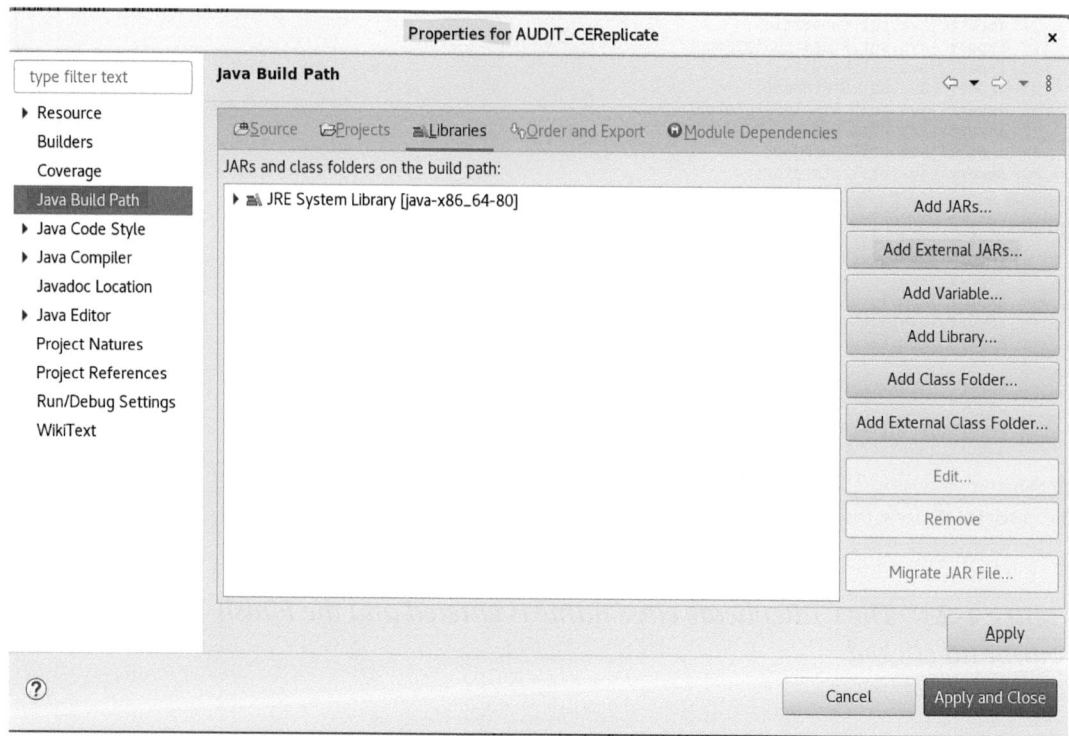

Figure 4-24.   The base Java Code is added for CEReplicate.java

Adding the IBM FileNet Libraries to the AUDIT_CEReplicate Classpath

IBM FileNet P8 5.5.x standard libraries are added to the **AUDIT_CEReplicate** Project Classpath. This should correct the flagged issues on the missing imports from the standard IBM FileNet API jace.jar and its supporting library .jar files.

CHAPTER 4   A REPLICATION JAVA PROGRAM FOR IBM FILENET OBJECT STORES

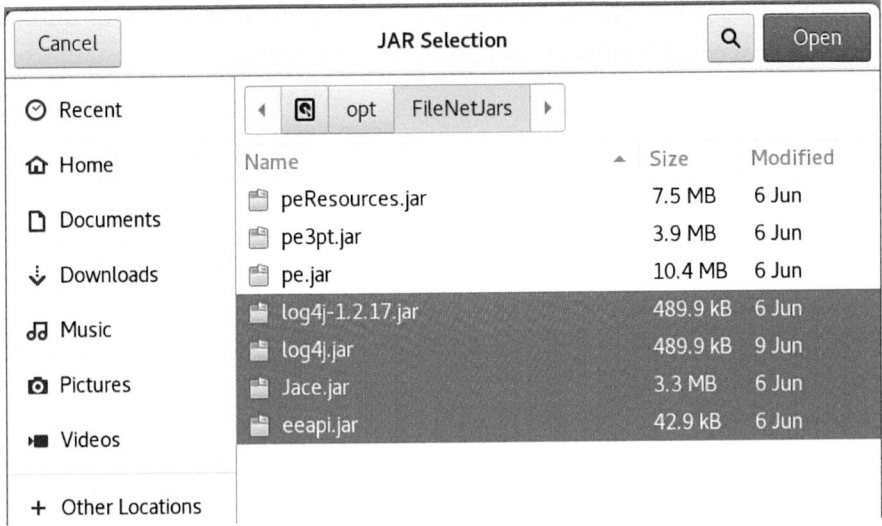

Figure 4-25. IBM FileNet P8 .jar libraries are added to the Classpath

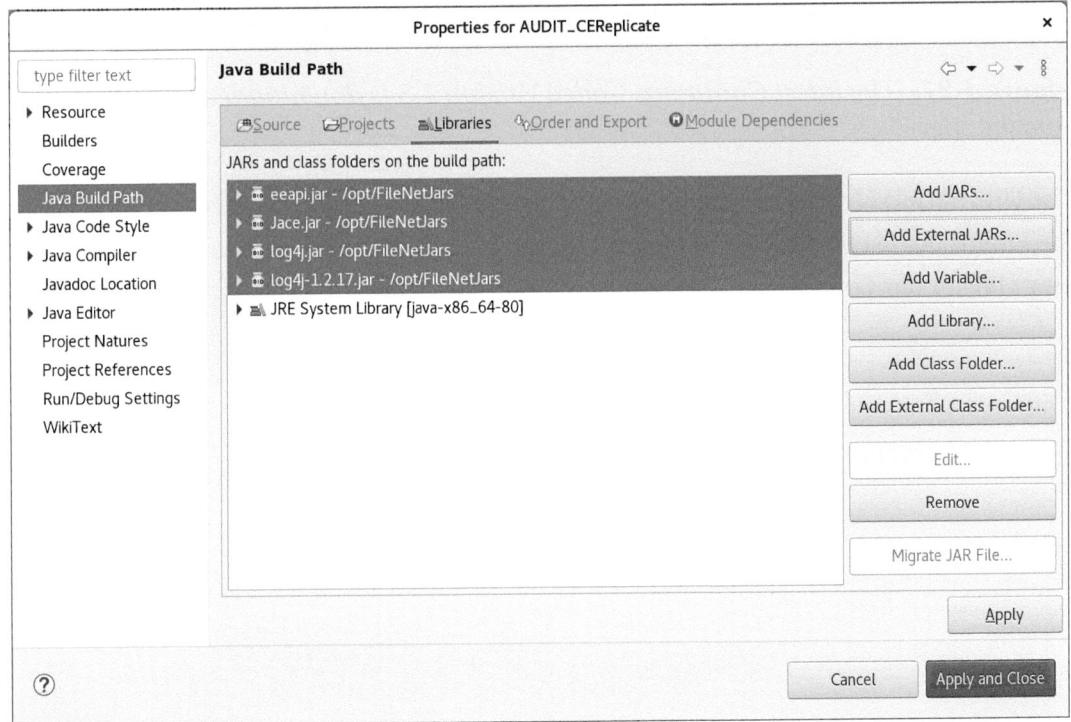

Figure 4-26. The Apply and Close command button is clicked to add the library .jar files

CHAPTER 4   A REPLICATION JAVA PROGRAM FOR IBM FILENET OBJECT STORES

The other projects are similarly created as described in Table 4-1 (with the code copied from the Java code listings in the following subsections).

Adding the Apache Commons Codec 1.5 Library

Figure 4-27. *The latest Commons Codec Version 1.5 is downloaded from the Apache site*

The URL https://commons.apache.org/proper/commons-code.codec.cgi is entered in the Firefox browser, as shown in Figure 4-27, to download the commons-codec-1.15-bin.tar.gz file for our Linux Eclipse project.

CHAPTER 4 A REPLICATION JAVA PROGRAM FOR IBM FILENET OBJECT STORES

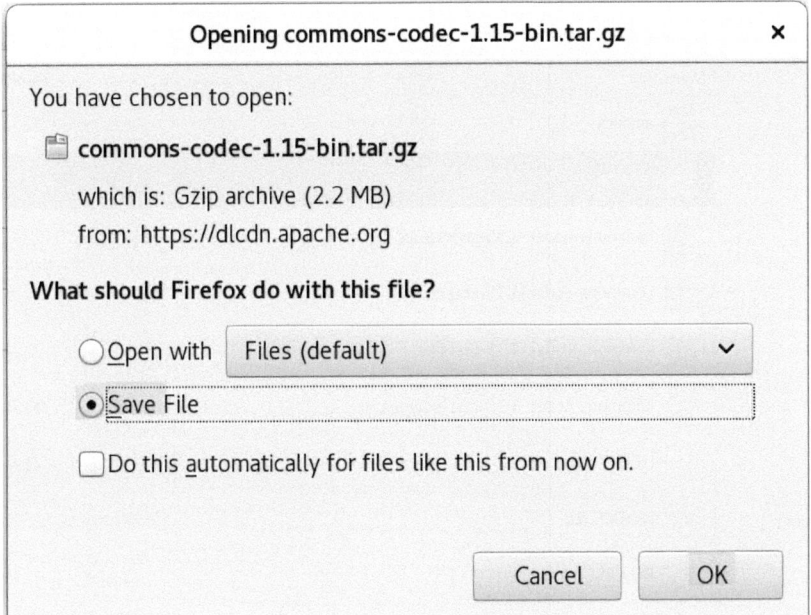

Figure 4-28. *The downloaded commons-codec-1.15-bin.tar.gz file is saved*

The downloaded **commons-codec-1.15-bin.tar.gz** file is saved to the **/root/Downloads** browser folder.

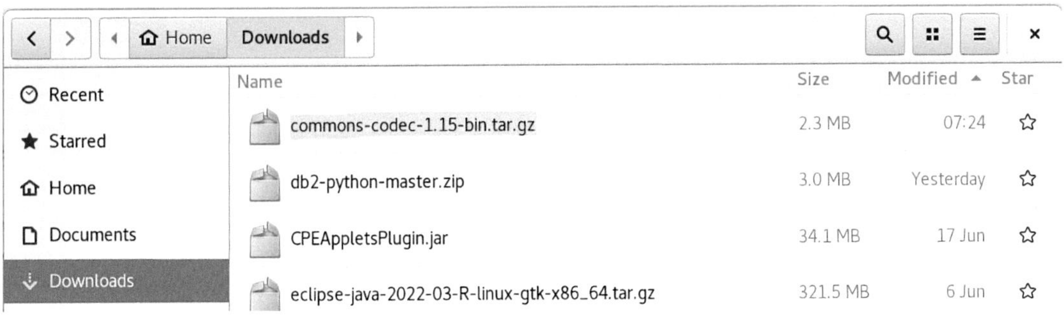

Figure 4-29. *The downloaded commons-codec-1.15-bin.tar.gz file is unpacked*

The downloaded **commons-codec-1.15-bin.tar.gz** file is unpacked using the Linux File Explorer desktop utility by double-clicking the .gz file to view the packed files.

717

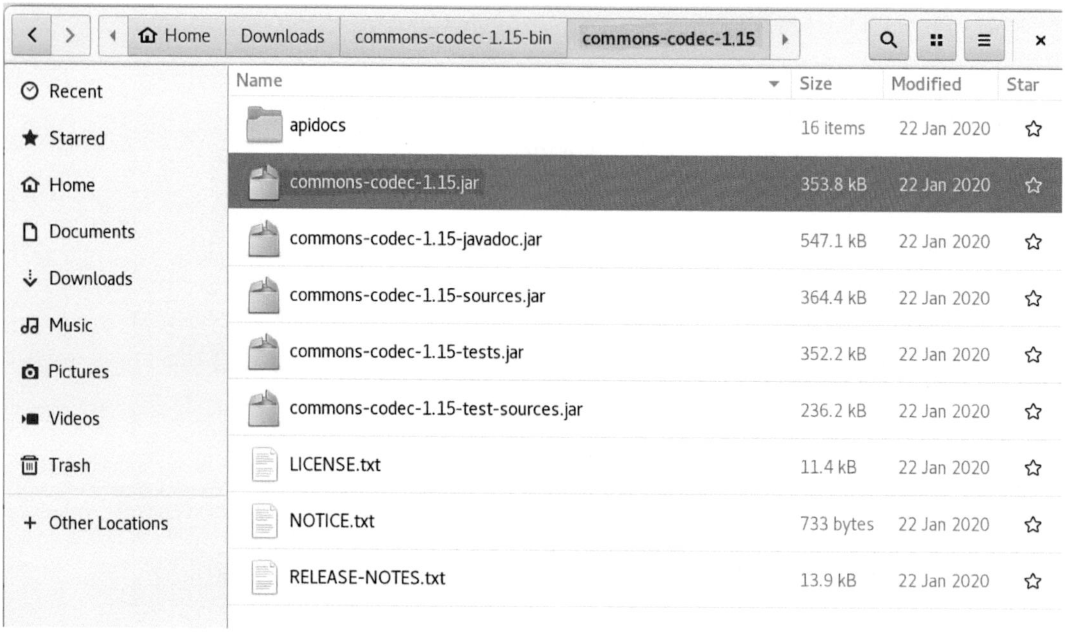

CHAPTER 4 A REPLICATION JAVA PROGRAM FOR IBM FILENET OBJECT STORES

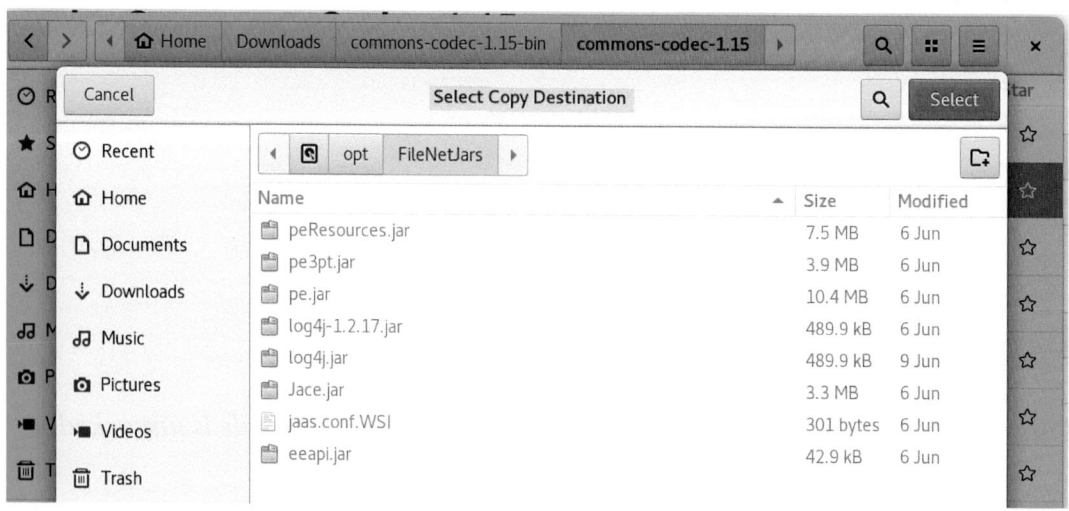

*Figure 4-30. The unpacked .gz file displays the **commons-codec-1.15.jar** file*

The **commons-codec-1.15.jar** file is now copied to the **/opt/FileNetJars** folder.

*Figure 4-31. The /opt/FileNetJars folder is browsed to copy the **commons-codec-1.15.jar***

Adding the Supporting External Library JAR Files

The properties for the **AUDIT_CEReplicate** project show the **Java Build Path**, and we can use the **Add External JARs** command button to browse and select the highlighted **commons-code-1.15.jar** file from the **/opt/FileNetJars** folder path.

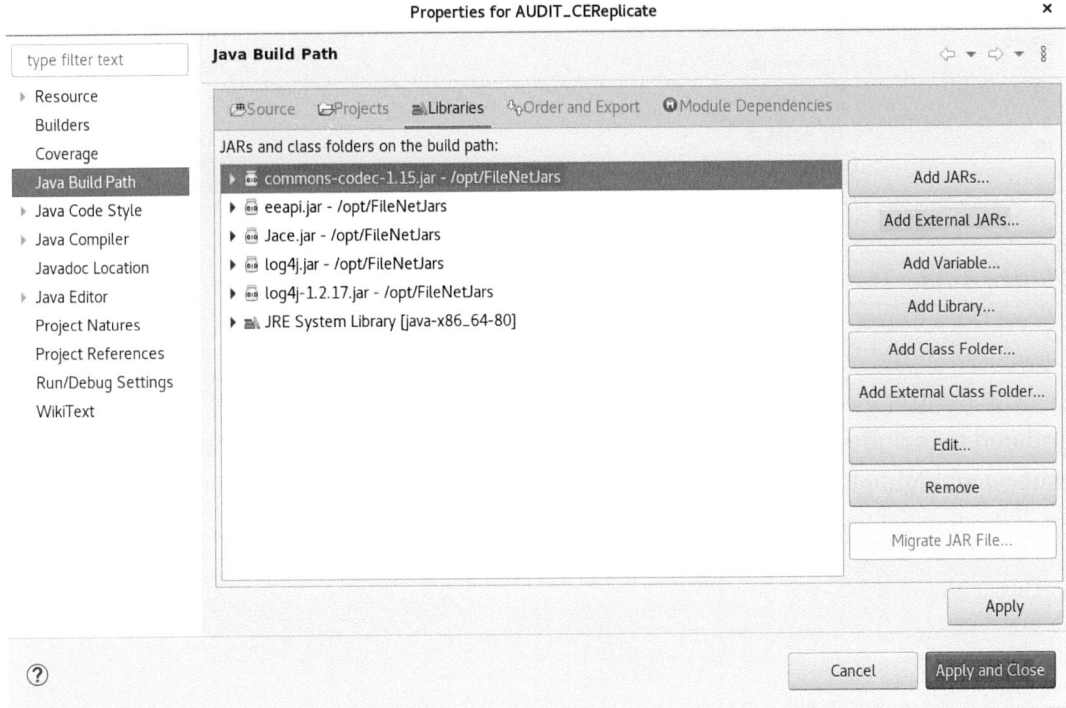

Figure 4-32. *The Add External JARs command button is used to load the JAR files required*

The **commons-code-1.15.jar** file we downloaded earlier is used for the password encryption in the **config.xml** file.

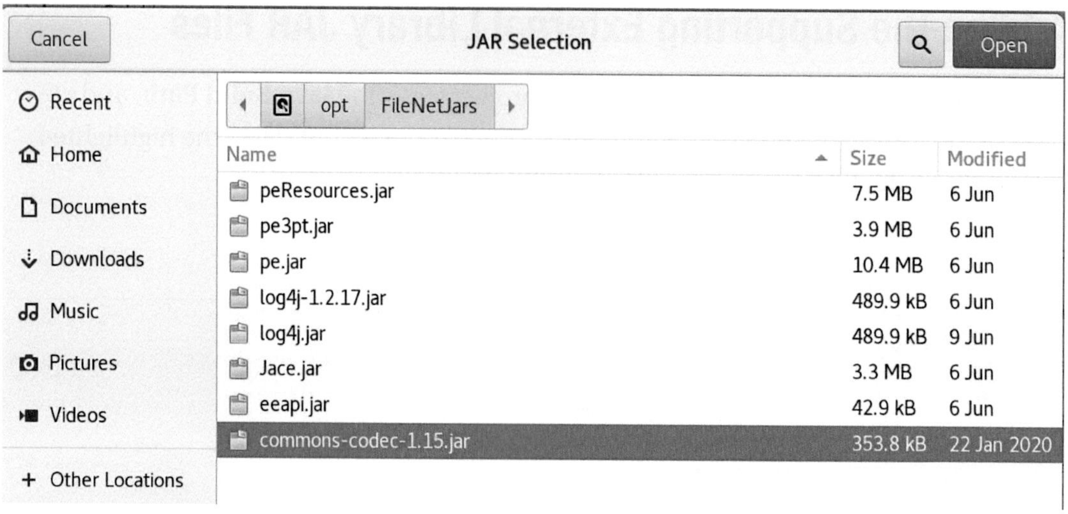

CHAPTER 4   A REPLICATION JAVA PROGRAM FOR IBM FILENET OBJECT STORES

Cancel		JAR Selection			Open
⊙ Recent	◀ ▣ opt FileNetJars ▶				
⌂ Home	Name	▲	Size	Modified	
▯ Documents	peResources.jar		7.5 MB	6 Jun	
	pe3pt.jar		3.9 MB	6 Jun	
⇩ Downloads	pe.jar		10.4 MB	6 Jun	
♫ Music	log4j-1.2.17.jar		489.9 kB	6 Jun	
	log4j.jar		489.9 kB	9 Jun	
◉ Pictures	Jace.jar		3.3 MB	6 Jun	
▶■ Videos	eeapi.jar		42.9 kB	6 Jun	
	commons-codec-1.15.jar		353.8 kB	22 Jan 2020	
+ Other Locations					

Figure 4-33. *The commons-codec-1.15.jar is selected from the /opt/FileNetJars folder*

The **Eclipse** properties **Java Build Path** for the **AUDIT_CEReplicate** project is updated to include the **commons-codec-1.15.jar** file which is missing from the import statement, highlighted in Figure 4-34, it can be seen that the addition of the Jar file has fixed the flagged issue of the missing **Base64** class.

```
CEReplicate.java ×   Arrays_Sqlpl.java
41  import java.security.spec.AlgorithmParameterSpec;
42  import java.security.spec.KeySpec;
43
44  import javax.crypto.Cipher;
45  import javax.crypto.IllegalBlockSizeException;
46  import javax.crypto.SecretKey;
47  import javax.crypto.SecretKeyFactory;
48  import javax.crypto.spec.PBEKeySpec;
49  import javax.crypto.spec.PBEParameterSpec;
50  //ASB ... 4 END
51
52  import com.filenet.api.admin.PropertyDefinition;
53  import com.filenet.api.core.BatchItemHandle;
54
55  import org.apache.commons.codec.binary.Base64;
56  import org.apache.log4j.Logger;
57  import org.w3c.dom.Node;
58  import org.w3c.dom.NodeList;
59
60  import com.ibm.filenet.ce.connection.CEConnection;
61
62  import com.filenet.api.collection.AccessPermissionList;
63  import com.filenet.api.collection.ContentElementList;
```

Figure 4-34. *The import org.apache.commons.codec.binary.Base64 class is now fixed*

CHAPTER 4   A REPLICATION JAVA PROGRAM FOR IBM FILENET OBJECT STORES

Adding the Xerces Java Library for XML File Processing

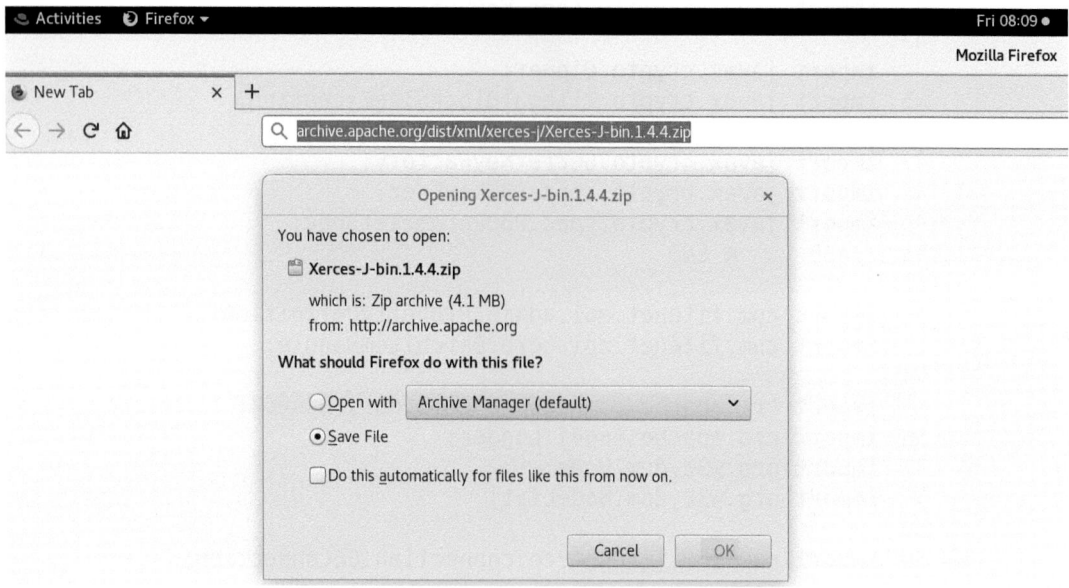

*Figure 4-35. The **Xerces-J-bin.1.4.4.zip** file is downloaded from the Apache website*

The **Xerces-J-bin.1.4.4.zip** file is downloaded from the `https://archive.apache.org/dist/xml/xerces-j/Xerces-J-bin.1.4.4.zip` Apache jar download site.

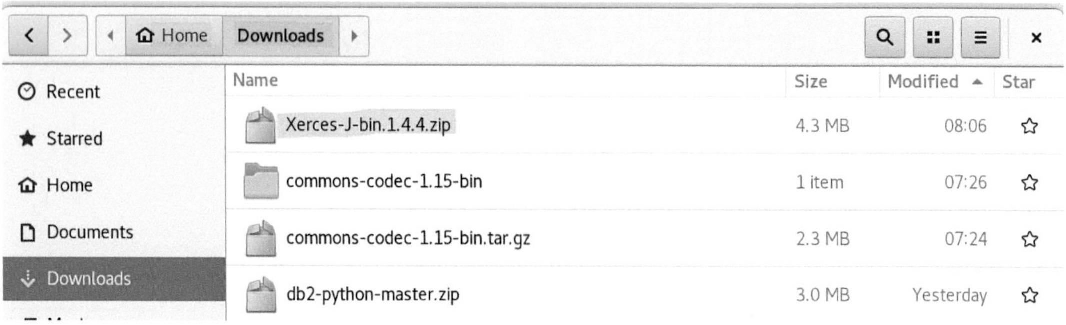

*Figure 4-36. The **Xerces-J-bin.1.4.4.zip** file is shown in the **/root/Downloads** Linux folder*

CHAPTER 4 A REPLICATION JAVA PROGRAM FOR IBM FILENET OBJECT STORES

The Xerces-J-bin.1.4.4.zip file is downloaded and unpacked using the Extract function of the Linux File Explorer desktop utility.

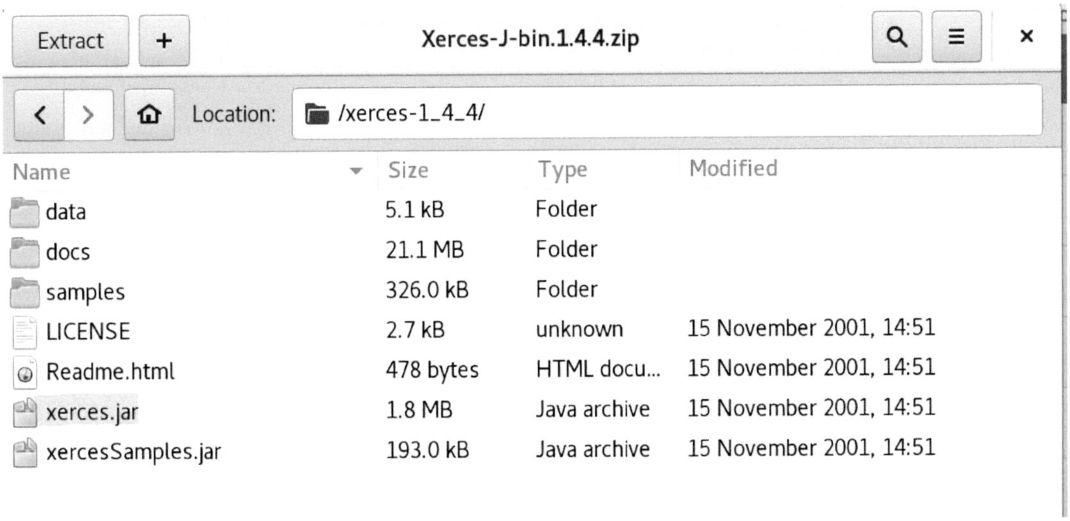

Name	Size	Type	Modified
data	5.1 kB	Folder	
docs	21.1 MB	Folder	
samples	326.0 kB	Folder	
LICENSE	2.7 kB	unknown	15 November 2001, 14:51
Readme.html	478 bytes	HTML docu...	15 November 2001, 14:51
xerces.jar	1.8 MB	Java archive	15 November 2001, 14:51
xercesSamples.jar	193.0 kB	Java archive	15 November 2001, 14:51

Figure 4-37. *The **xerces.jar** file is extracted from the **Xerces-J-bin.1.4.4.zip** file*

The unpacked **xerces.jar** file is extracted to the **/opt/FileNetJars** folder location.

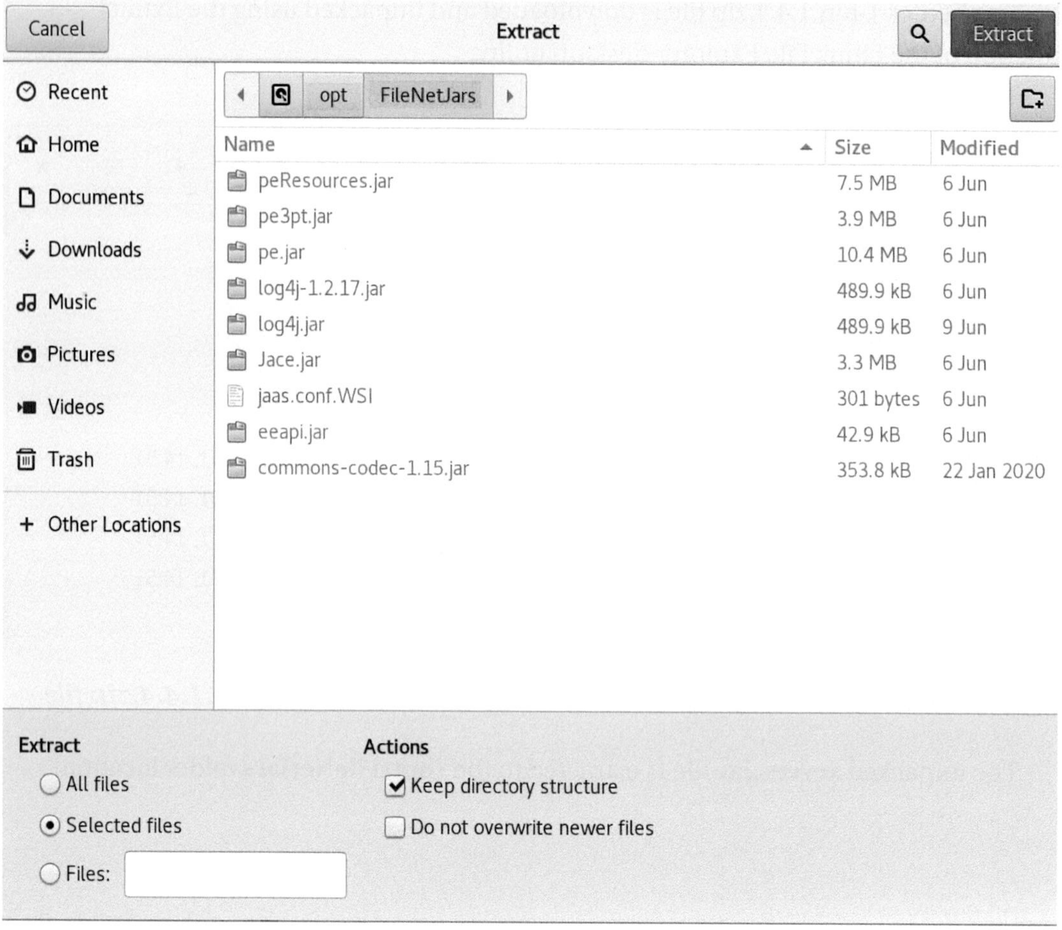

CHAPTER 4   A REPLICATION JAVA PROGRAM FOR IBM FILENET OBJECT STORES

Name	Size	Modified
peResources.jar	7.5 MB	6 Jun
pe3pt.jar	3.9 MB	6 Jun
pe.jar	10.4 MB	6 Jun
log4j-1.2.17.jar	489.9 kB	6 Jun
log4j.jar	489.9 kB	9 Jun
Jace.jar	3.3 MB	6 Jun
jaas.conf.WSI	301 bytes	6 Jun
eeapi.jar	42.9 kB	6 Jun
commons-codec-1.15.jar	353.8 kB	22 Jan 2020

Figure 4-38. *The unpacked **xerces.jar** file is extracted to the /**opt/ FileNetJars** folder*

The **xerces.jar** file is used to satisfy the XML parsing requirements for the Java replication program.

CHAPTER 4   A REPLICATION JAVA PROGRAM FOR IBM FILENET OBJECT STORES

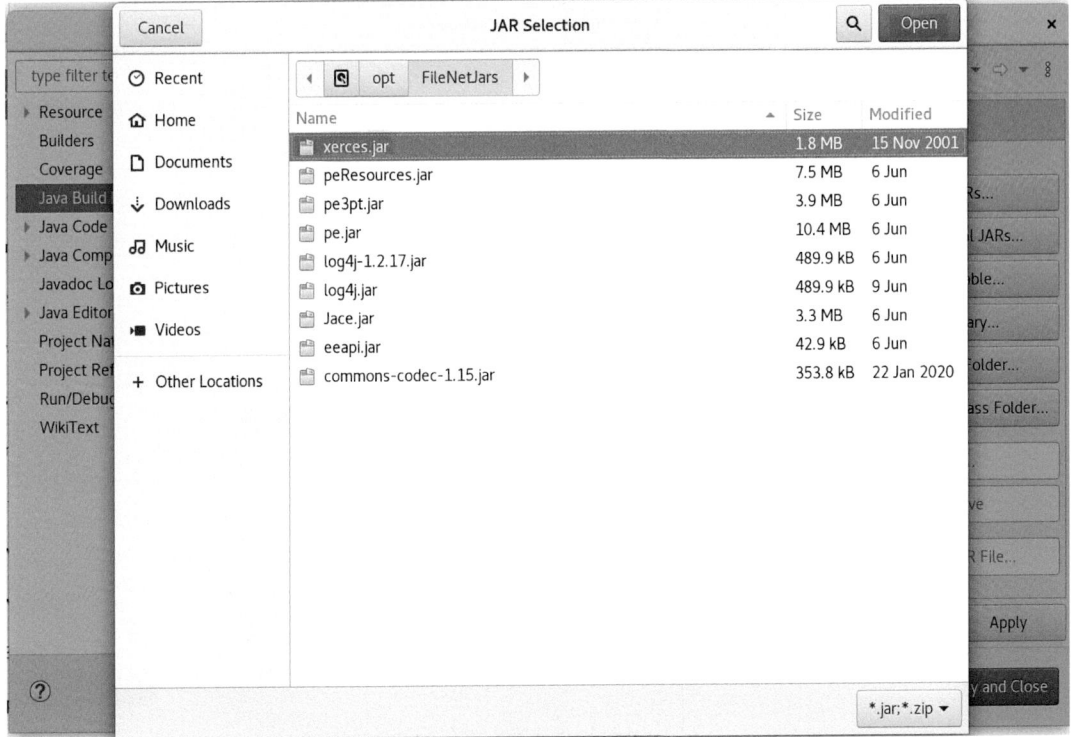

Figure 4-39. *The unpacked **xerces.jar** file is added to the Java Build Path*

The properties of the AUDIT_CEReplicate project are selected to add the xerces.jar file to the Java Build Path property.

CHAPTER 4   A REPLICATION JAVA PROGRAM FOR IBM FILENET OBJECT STORES

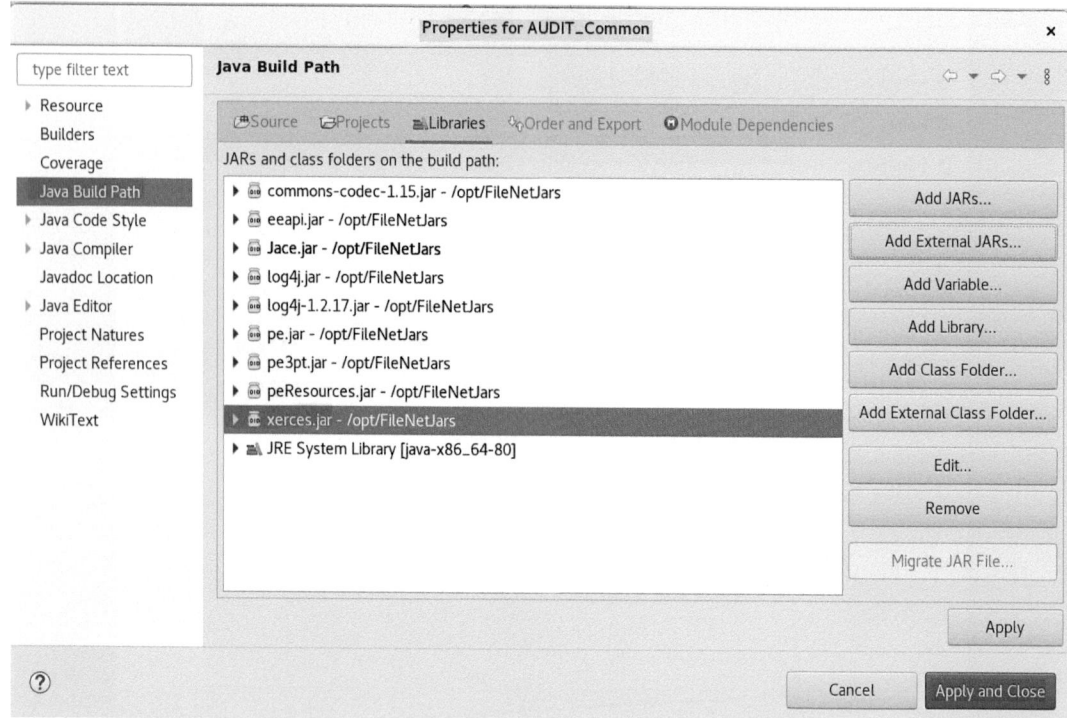

*Figure 4-40.   The **xerces.jar** is now added to the Java Build Path*

The **import** statement needed to be updated to adjust the package path of the **OutputFormat** class as shown in Figure 4-41.

```
// ASB Software Development Limited - import updated to point to the Apache equivalent class 29/07/2022
//import com.sun.org.apache.xml.internal.serialize.OutputFormat;
import org.apache.xml.serialize.OutputFormat;
```

Figure 4-41.   The adjusted import statement for the OutputFormat class

Adding the Supporting Java Projects

The four supporting projects, FileNetConnection, FN_Connect, AUDIT_Common, and AUDIT_Utils, are run as follows to create the supporting .jar files (as defined in Table 4-1).

In the **AUDIT_CEReplicate** project source folder, we first create a **config** subfolder.

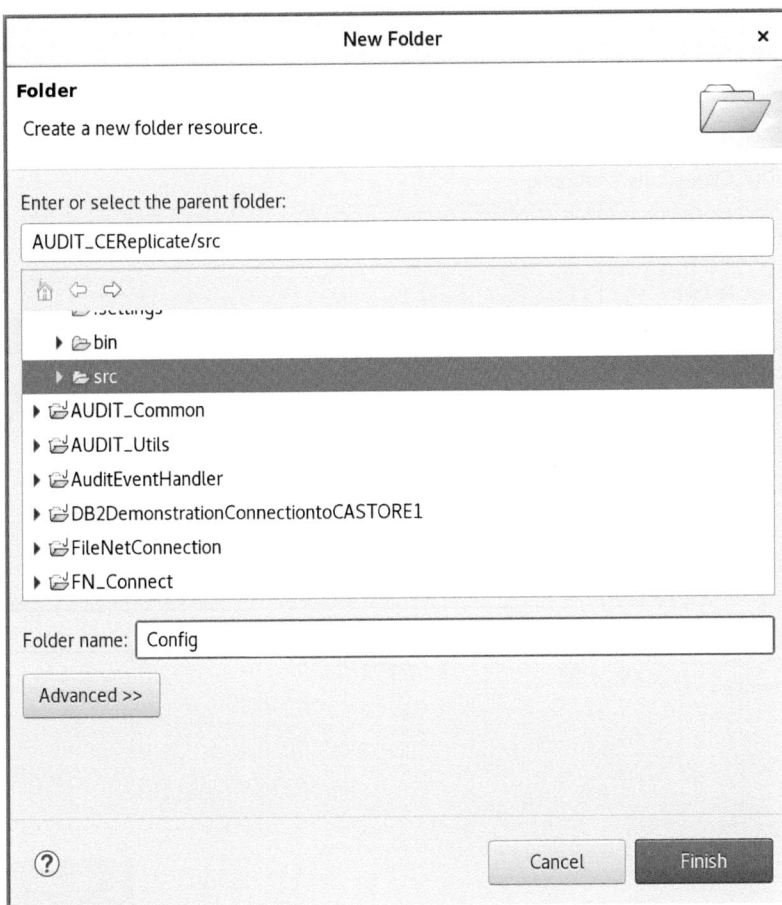

Figure 4-42. *The Config folder is created for the config.xml file*

The config.xml file is created to hold the Replication program parameters for the source and target Object Store names, the administrator name and encrypted password for the IBM FileNet content engine system account, and the included users and group security for access to the replicated folders and documents.

CHAPTER 4   A REPLICATION JAVA PROGRAM FOR IBM FILENET OBJECT STORES

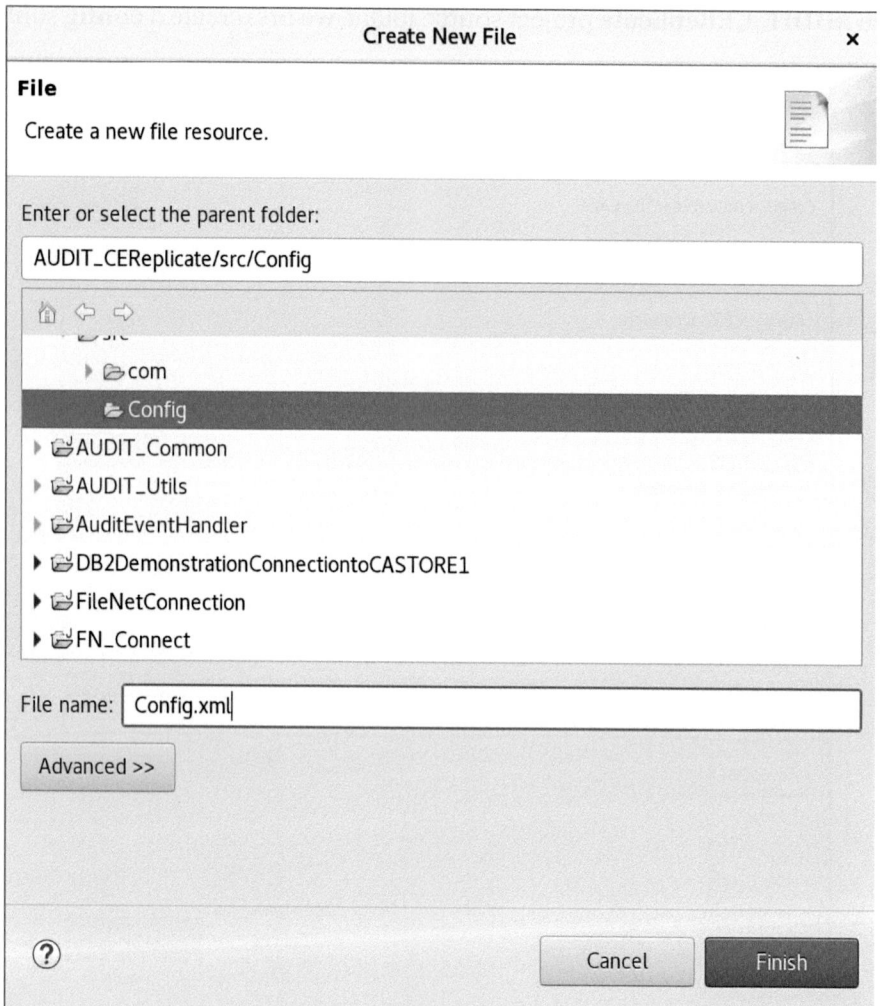

Figure 4-43. *The **Config.xml** file is created – note later changed to **config.xml** (lowercase)*

The supporting project jar files are generated as shown in Table 4-1.

CHAPTER 4   A REPLICATION JAVA PROGRAM FOR IBM FILENET OBJECT STORES

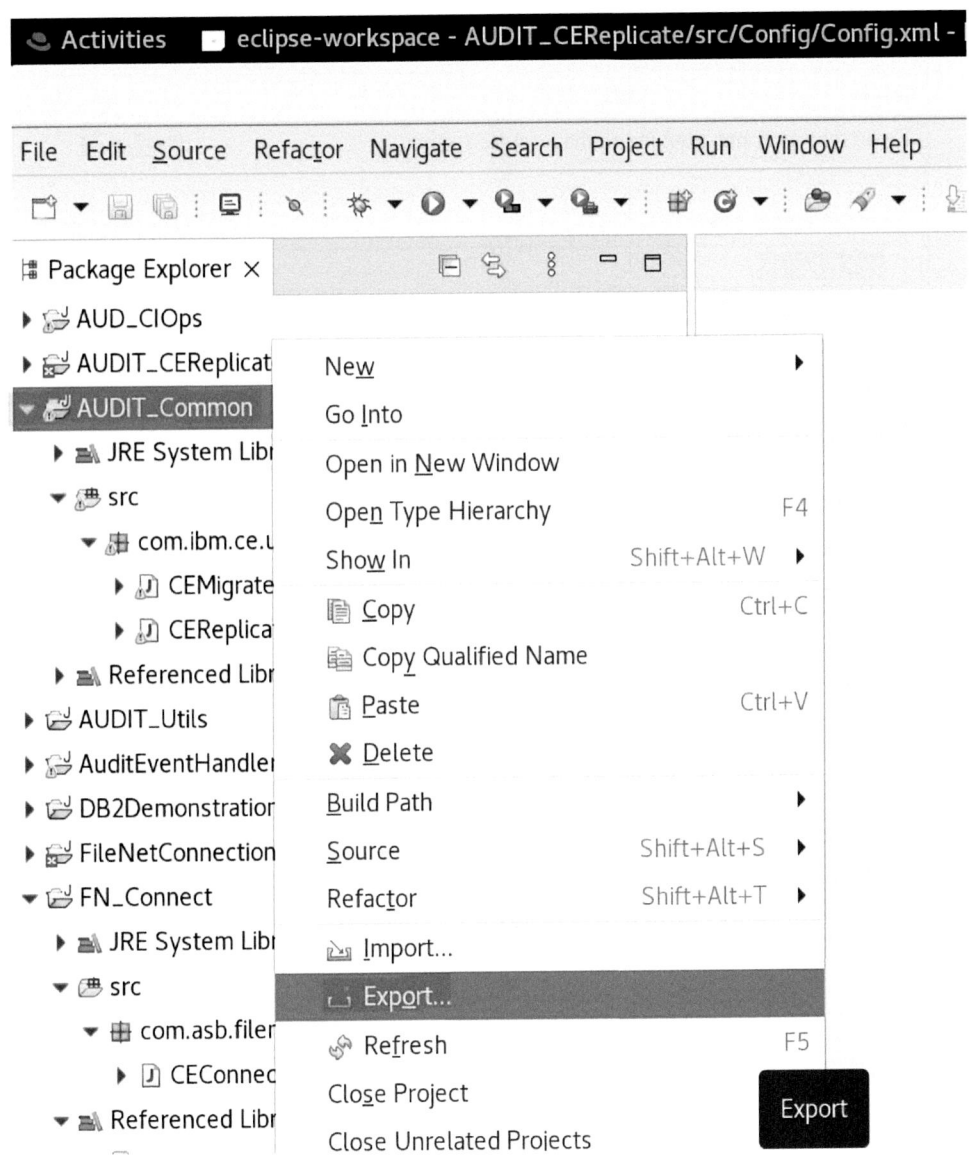

Figure 4-44. *The AUDIT_Common project is selected to export the fn_common.jar file*

CHAPTER 4 A REPLICATION JAVA PROGRAM FOR IBM FILENET OBJECT STORES

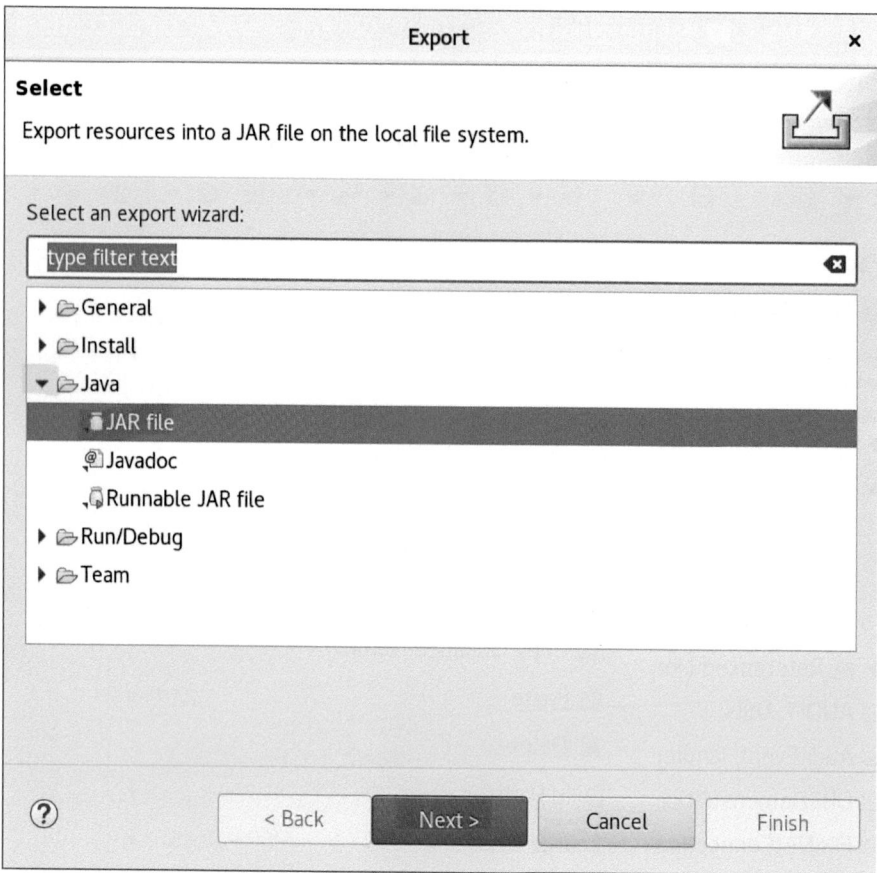

Figure 4-45. *The Java JAR file export option is selected*

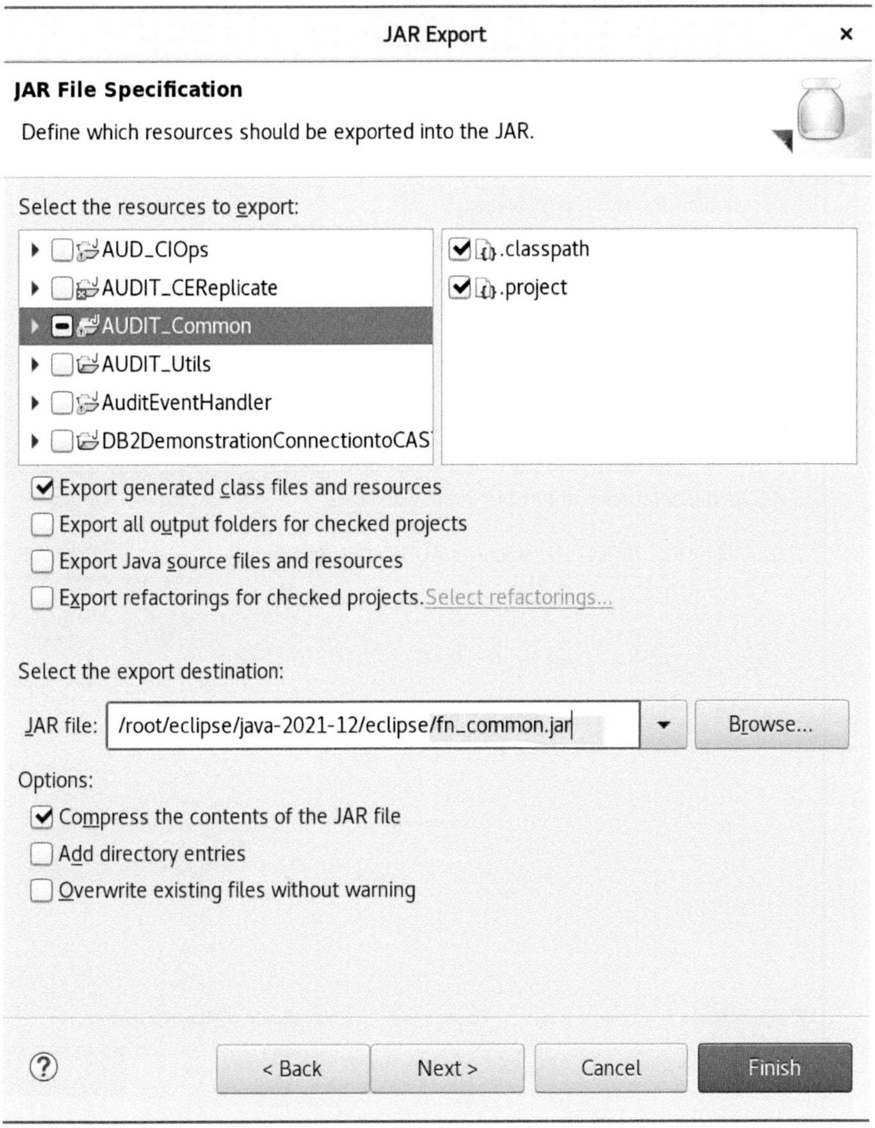

CHAPTER 4   A REPLICATION JAVA PROGRAM FOR IBM FILENET OBJECT STORES

*Figure 4-46. The **fn_common.jar** export path is selected*

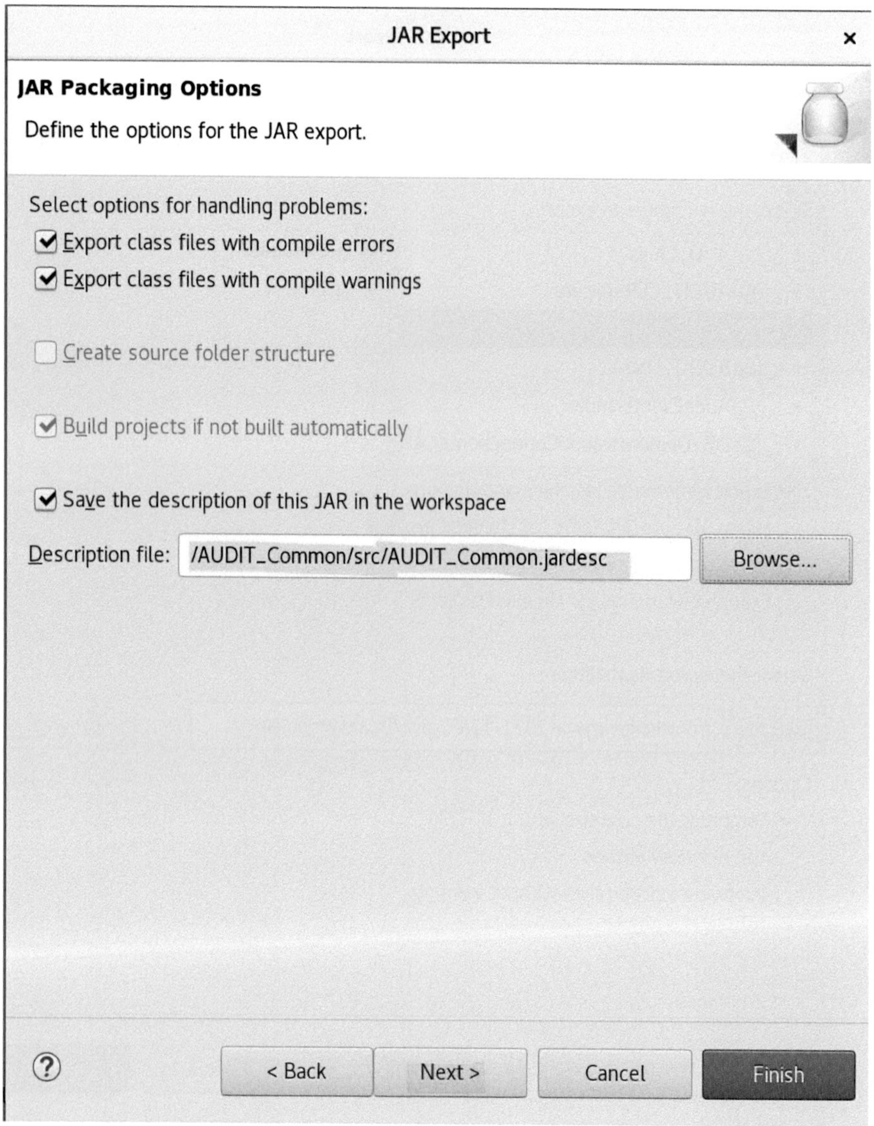

CHAPTER 4 A REPLICATION JAVA PROGRAM FOR IBM FILENET OBJECT STORES

Figure 4-47. The AUDIT_Common.jardesc file is selected for build, to use for faster rebuilds

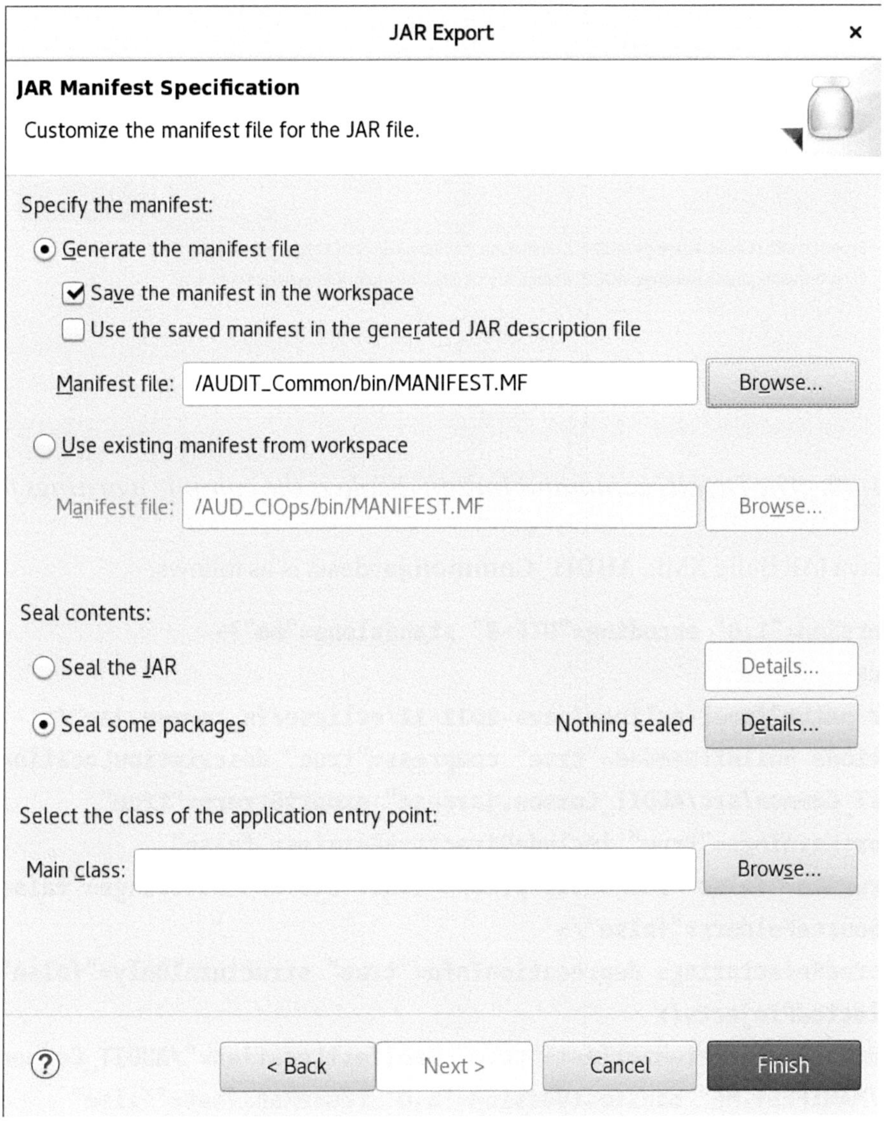

Figure 4-48. The jar MANIFEST.MF file is selected for creation and Finish clicked

CHAPTER 4   A REPLICATION JAVA PROGRAM FOR IBM FILENET OBJECT STORES

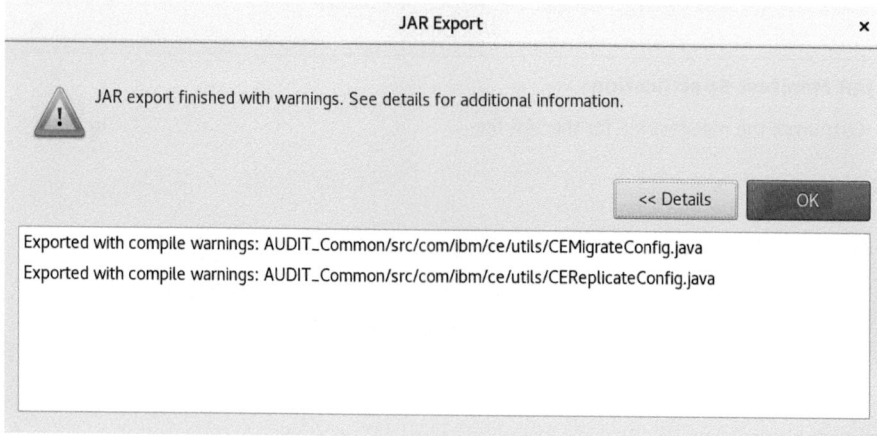

Figure 4-49. *The Details command button displays the compile warnings found*

The Java JAR Build XML, **AUDIT_Common.jardesc**, is as follows:

```xml
<?xml version="1.0" encoding="UTF-8" standalone="no"?>
<jardesc>
    <jar path="/root/eclipse/java-2021-12/eclipse/fn_common.jar"/>
    <options buildIfNeeded="true" compress="true" descriptionLocation="/AUDIT_Common/src/AUDIT_Common.jardesc" exportErrors="true" exportWarnings="true" includeDirectoryEntries="false" overwrite="false" saveDescription="true" storeRefactorings="false" useSourceFolders="false"/>
    <storedRefactorings deprecationInfo="true" structuralOnly="false"/>
    <selectedProjects/>
    <manifest generateManifest="true" manifestLocation="/AUDIT_Common/bin/MANIFEST.MF" manifestVersion="1.0" reuseManifest="false" saveManifest="true" usesManifest="true">
        <sealing sealJar="false">
            <packagesToSeal/>
            <packagesToUnSeal/>
        </sealing>
    </manifest>
    <selectedElements exportClassFiles="true" exportJavaFiles="false" exportOutputFolder="false">
        <file path="/AUDIT_Common/.classpath"/>
```

CHAPTER 4   A REPLICATION JAVA PROGRAM FOR IBM FILENET OBJECT STORES

```
        <file path="/AUDIT_Common/.project"/>
        <javaElement handleIdentifier="=AUDIT_Common/src"/>
    </selectedElements>
</jardesc>
```

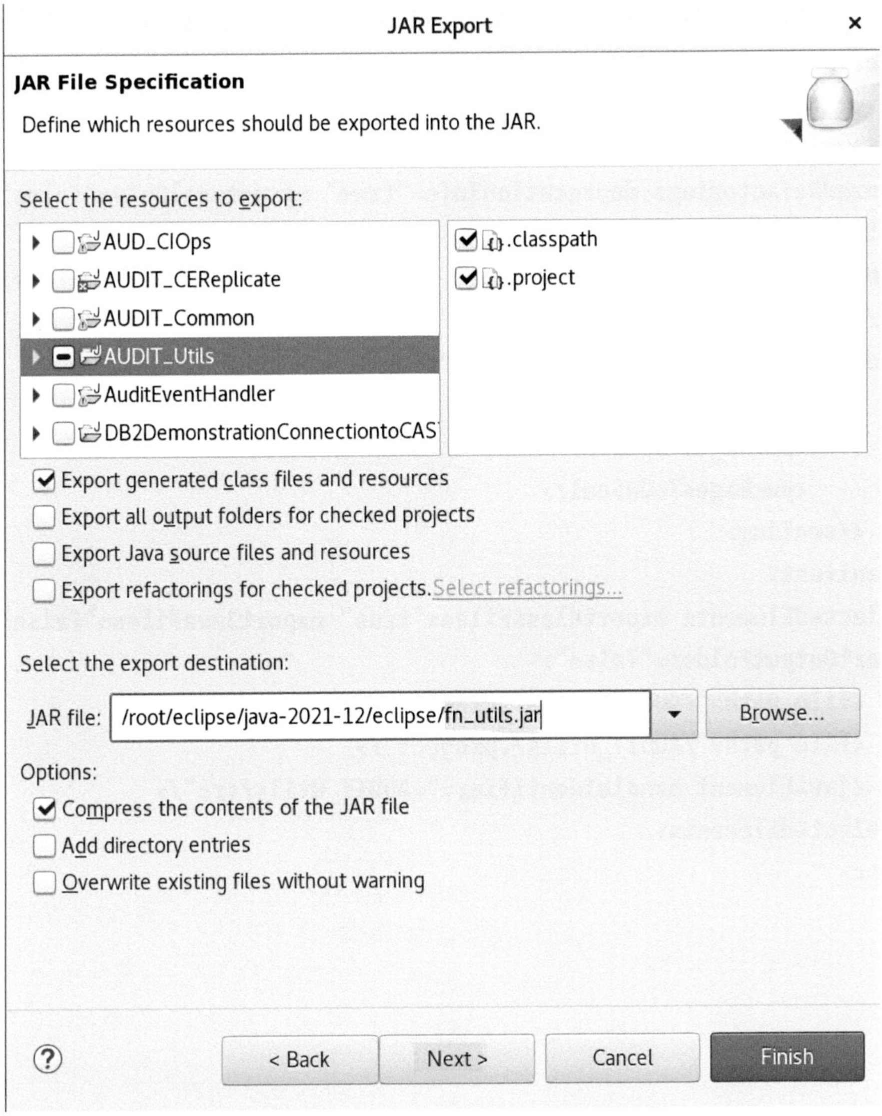

*Figure 4-50.   The **fn_utils.jar** export path is selected*

CHAPTER 4　A REPLICATION JAVA PROGRAM FOR IBM FILENET OBJECT STORES

The Java JAR Build XML, **AUDIT_Utils.jardesc**, for the **fn_utils.jar** is as follows:

```xml
<?xml version="1.0" encoding="UTF-8" standalone="no"?>
<jardesc>
    <jar path="/root/eclipse/java-2021-12/eclipse/fn_utils.jar"/>
    <options buildIfNeeded="true" compress="true" descriptionLocation="/AUDIT_Utils/src/AUDIT_Utils.jardesc" exportErrors="true" exportWarnings="true" includeDirectoryEntries="false" overwrite="false" saveDescription="true" storeRefactorings="false" useSourceFolders="false"/>
    <storedRefactorings deprecationInfo="true" structuralOnly="false"/>
    <selectedProjects/>
    <manifest generateManifest="true" manifestLocation="/AUDIT_Utils/bin/MANIFEST.MF" manifestVersion="1.0" reuseManifest="false" saveManifest="true" usesManifest="true">
        <sealing sealJar="false">
            <packagesToSeal/>
            <packagesToUnSeal/>
        </sealing>
    </manifest>
    <selectedElements exportClassFiles="true" exportJavaFiles="false" exportOutputFolder="false">
        <file path="/AUDIT_Utils/.classpath"/>
        <file path="/AUDIT_Utils/.project"/>
        <javaElement handleIdentifier="=AUDIT_Utils/src"/>
    </selectedElements>
</jardesc>
```

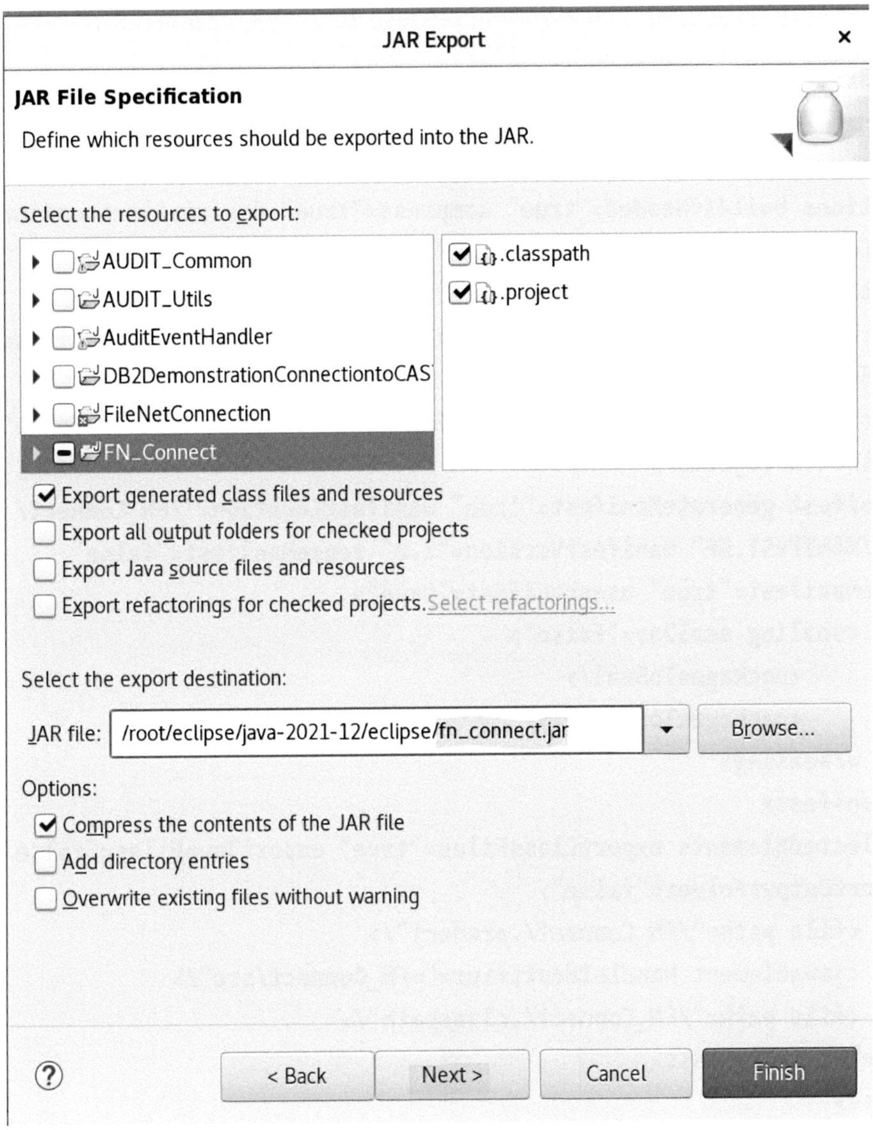

CHAPTER 4 A REPLICATION JAVA PROGRAM FOR IBM FILENET OBJECT STORES

Figure 4-51. The fn_connect.jar export path is selected

CHAPTER 4 A REPLICATION JAVA PROGRAM FOR IBM FILENET OBJECT STORES

The Java JAR Build XML, **FN_Connect.jardesc**, for the **fn_connect.jar** is as follows:

```xml
<?xml version="1.0" encoding="UTF-8" standalone="no"?>
<jardesc>
    <jar path="/root/eclipse/java-2021-12/eclipse/fn_connect.jar"/>
    <options buildIfNeeded="true" compress="true" descriptionLocation="/FN_Connect/src/FN_Connect.jardesc" exportErrors="true" exportWarnings="true" includeDirectoryEntries="false" overwrite="false" saveDescription="true" storeRefactorings="false" useSourceFolders="false"/>
    <storedRefactorings deprecationInfo="true" structuralOnly="false"/>
    <selectedProjects/>
    <manifest generateManifest="true" manifestLocation="/FN_Connect/bin/MANIFEST.MF" manifestVersion="1.0" reuseManifest="false" saveManifest="true" usesManifest="true">
        <sealing sealJar="false">
            <packagesToSeal/>
            <packagesToUnSeal/>
        </sealing>
    </manifest>
    <selectedElements exportClassFiles="true" exportJavaFiles="false" exportOutputFolder="false">
        <file path="/FN_Connect/.project"/>
        <javaElement handleIdentifier="=FN_Connect/src"/>
        <file path="/FN_Connect/.classpath"/>
    </selectedElements>
</jardesc>
```

CHAPTER 4   A REPLICATION JAVA PROGRAM FOR IBM FILENET OBJECT STORES

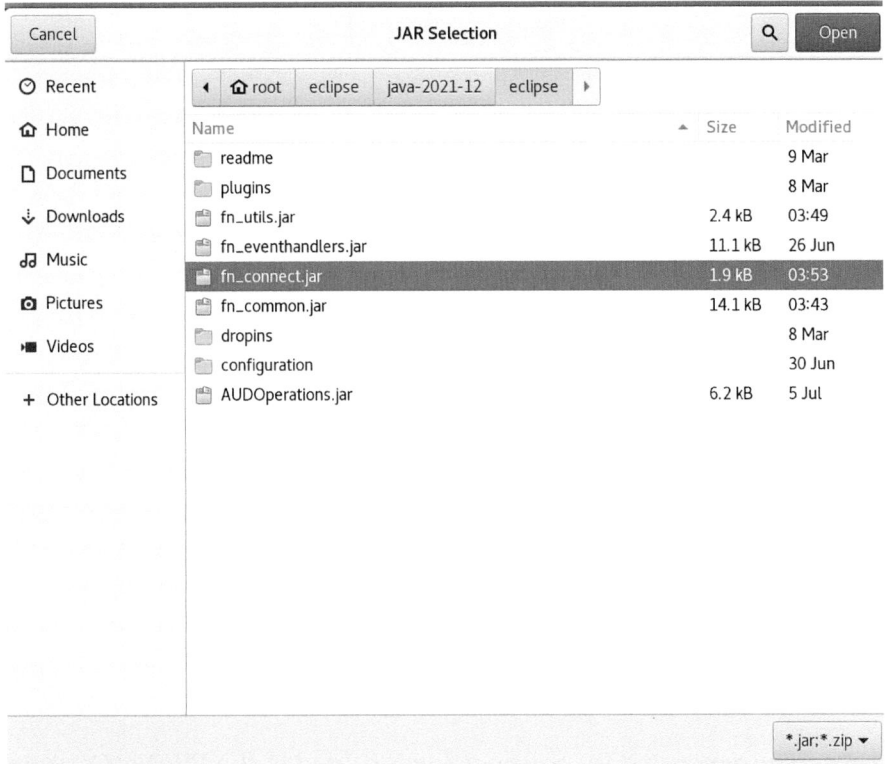

Figure 4-52. *The generated **fn_connect.jar** supporting jar file for the Replication program*

The supporting jar file **fn_connect.jar** is added to the Java Build Path of the FileNet **fn_connection.jar** creation project **FileNetConnection**.

CHAPTER 4   A REPLICATION JAVA PROGRAM FOR IBM FILENET OBJECT STORES

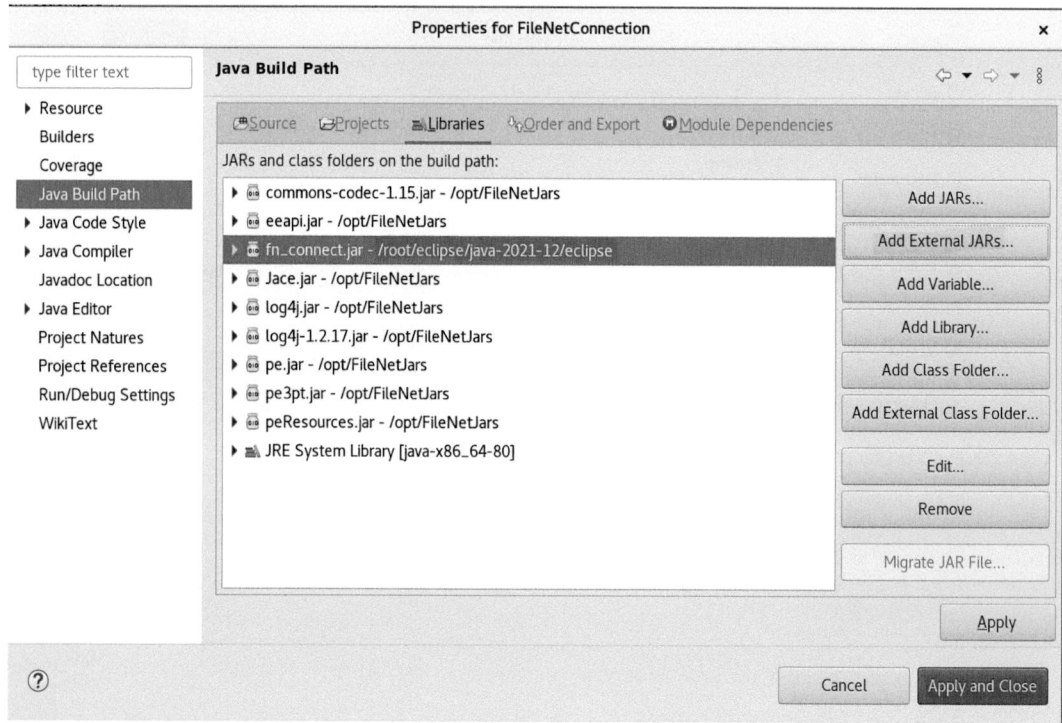

Figure 4-53.   *The generated **fn_connect.jar** is added to the **FileNetConnection** project*

CHAPTER 4   A REPLICATION JAVA PROGRAM FOR IBM FILENET OBJECT STORES

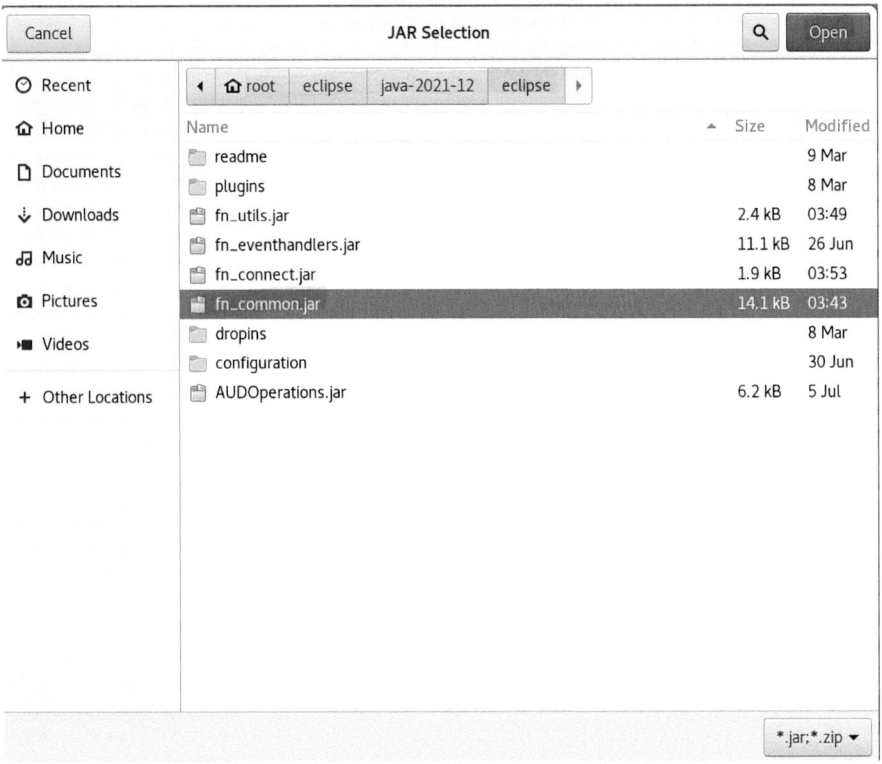

Figure 4-54. *The generated **fn_common.jar** file is selected from the eclipse root path*

CHAPTER 4   A REPLICATION JAVA PROGRAM FOR IBM FILENET OBJECT STORES

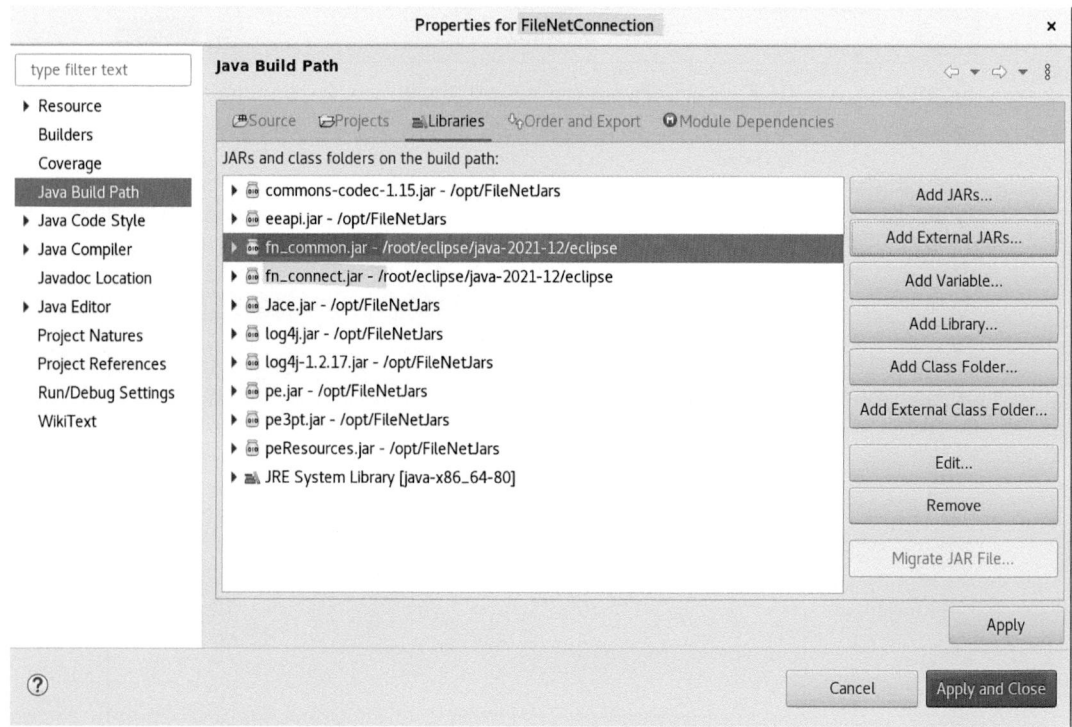

Figure 4-55. *The generated **fn_common.jar** is added to the **FileNetConnection** project*

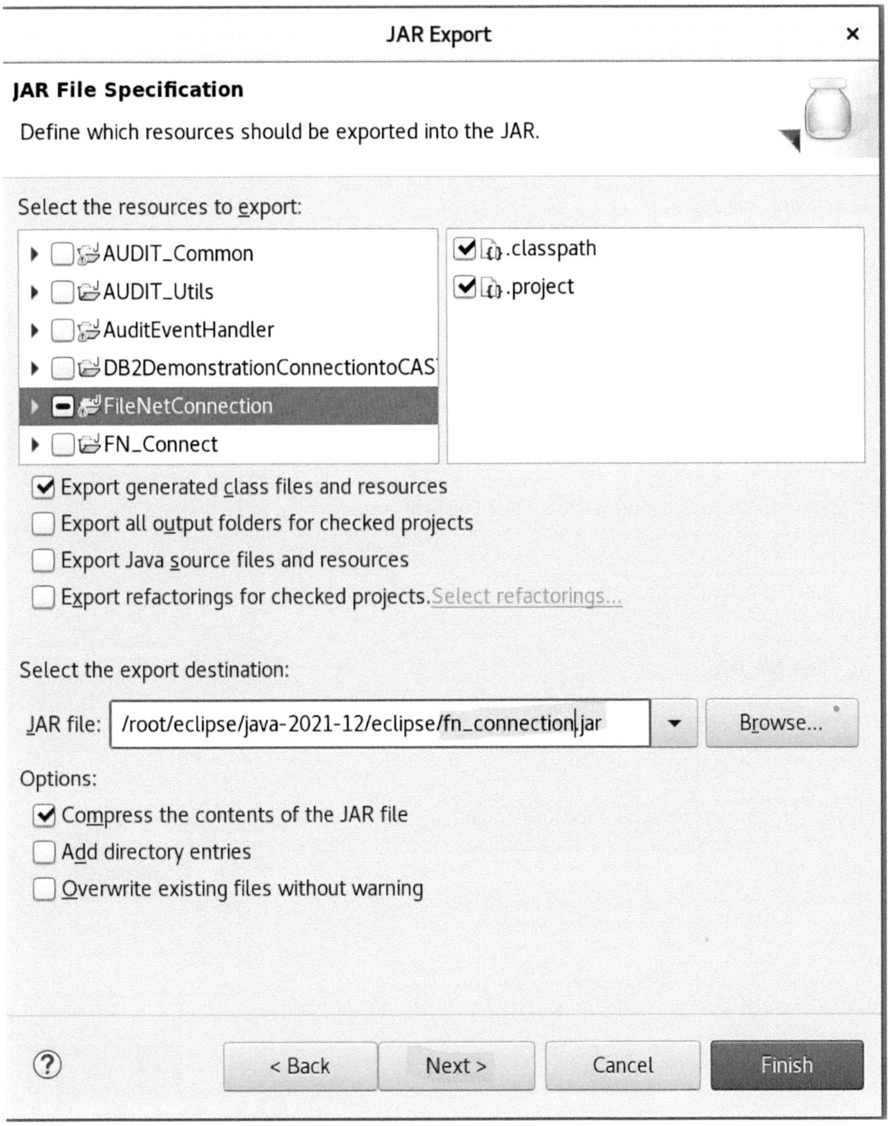

CHAPTER 4　A REPLICATION JAVA PROGRAM FOR IBM FILENET OBJECT STORES

*Figure 4-56. The **fn_connection.jar** export path is selected*

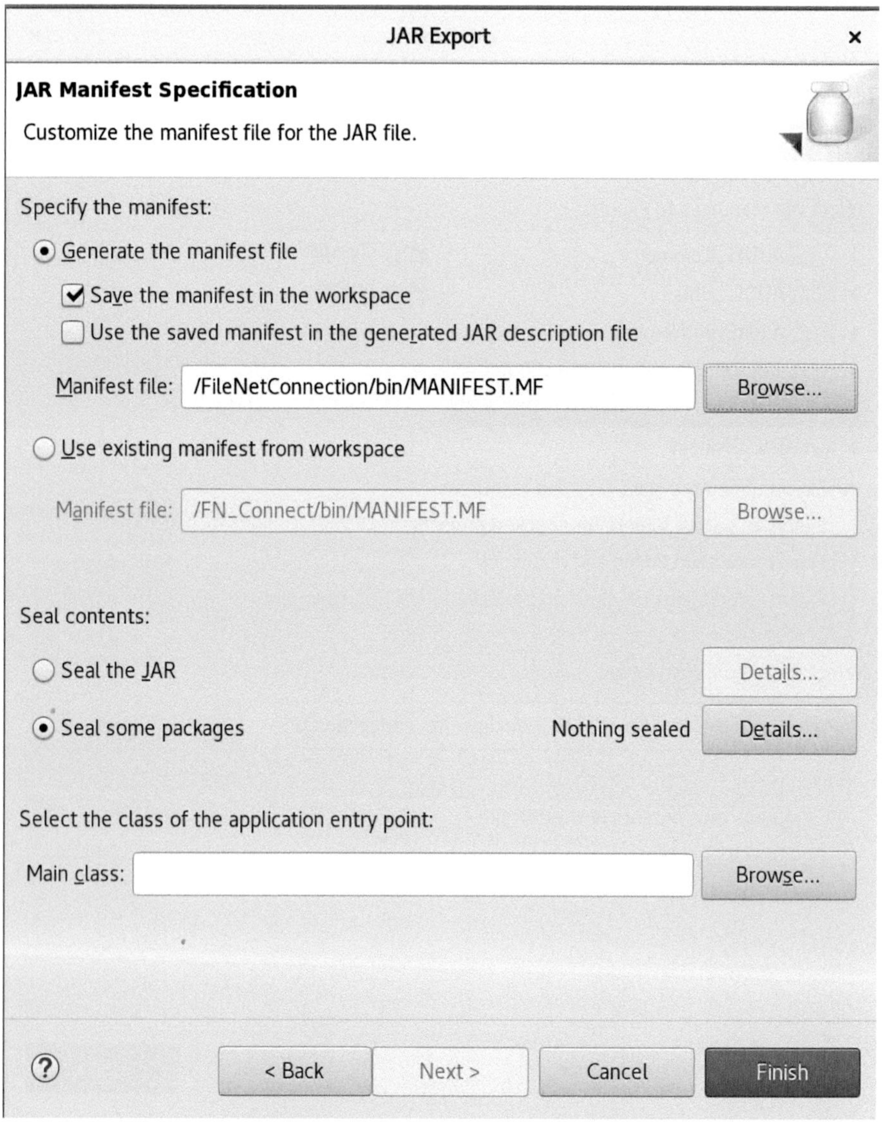

CHAPTER 4   A REPLICATION JAVA PROGRAM FOR IBM FILENET OBJECT STORES

*Figure 4-57. The **fn_connection.jar** export MANIFEST.MF file path is selected*

The Java JAR Build XML, **FileNetConnection.jardesc**, is as follows:

```
<?xml version="1.0" encoding="UTF-8" standalone="no"?>
<jardesc>
    <jar path="/root/eclipse/java-2021-12/eclipse/fn_connection.jar"/>
    <options buildIfNeeded="true" compress="true" descriptionLocation="/
    FileNetConnection/src/FileNetConnection.jardesc" exportErrors="true"
```

```xml
        exportWarnings="true" includeDirectoryEntries="false"
        overwrite="false" saveDescription="true" storeRefactorings="false"
        useSourceFolders="false"/>
    <storedRefactorings deprecationInfo="true" structuralOnly="false"/>
    <selectedProjects/>
    <manifest generateManifest="true" manifestLocation="/FileNetConnection/bin/MANIFEST.MF" manifestVersion="1.0" reuseManifest="false" saveManifest="true" usesManifest="true">
        <sealing sealJar="false">
            <packagesToSeal/>
            <packagesToUnSeal/>
        </sealing>
    </manifest>
    <selectedElements exportClassFiles="true" exportJavaFiles="false" exportOutputFolder="false">
        <file path="/FileNetConnection/.project"/>
        <file path="/FileNetConnection/.classpath"/>
        <javaElement handleIdentifier="=FileNetConnection/src"/>
    </selectedElements>
</jardesc>
```

CHAPTER 4   A REPLICATION JAVA PROGRAM FOR IBM FILENET OBJECT STORES

Updating the Java Build Path with the Required Projects

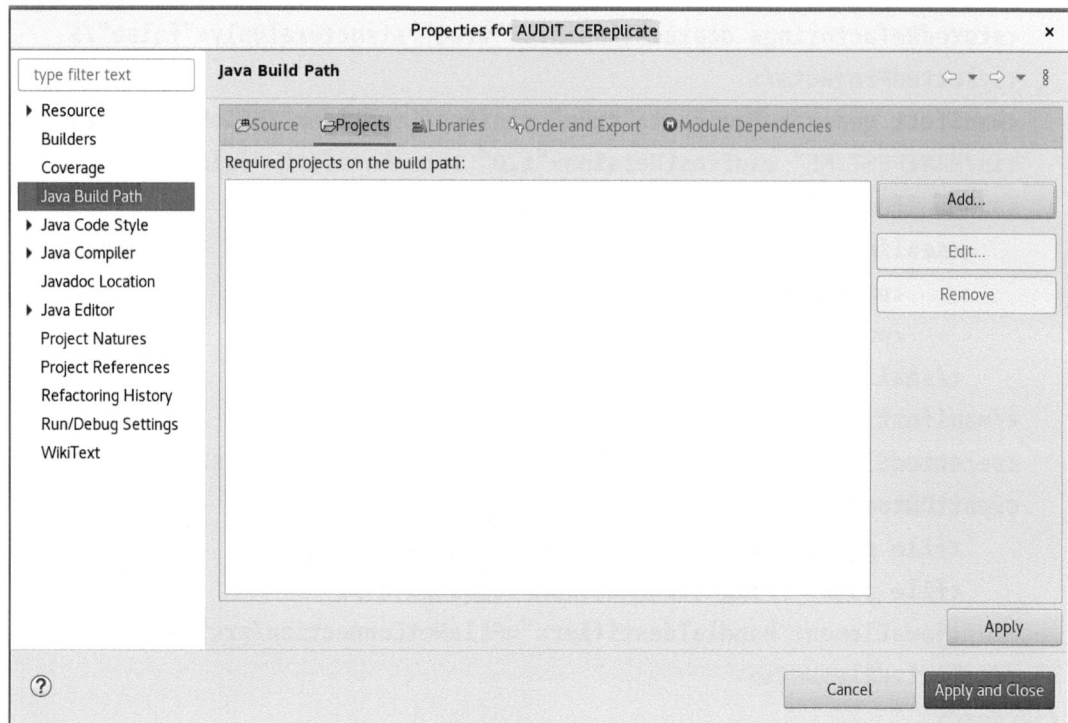

*Figure 4-58. The **Java Build Path** Projects tab is selected and the **Add** command clicked*

In Figure 4-58, we select the **AUDIT_CEReplicate, Replication.jar** build project properties and click the **Java Build Path** menu item and then select the **Projects** tab to identify the Required projects on the build path.

CHAPTER 4 A REPLICATION JAVA PROGRAM FOR IBM FILENET OBJECT STORES

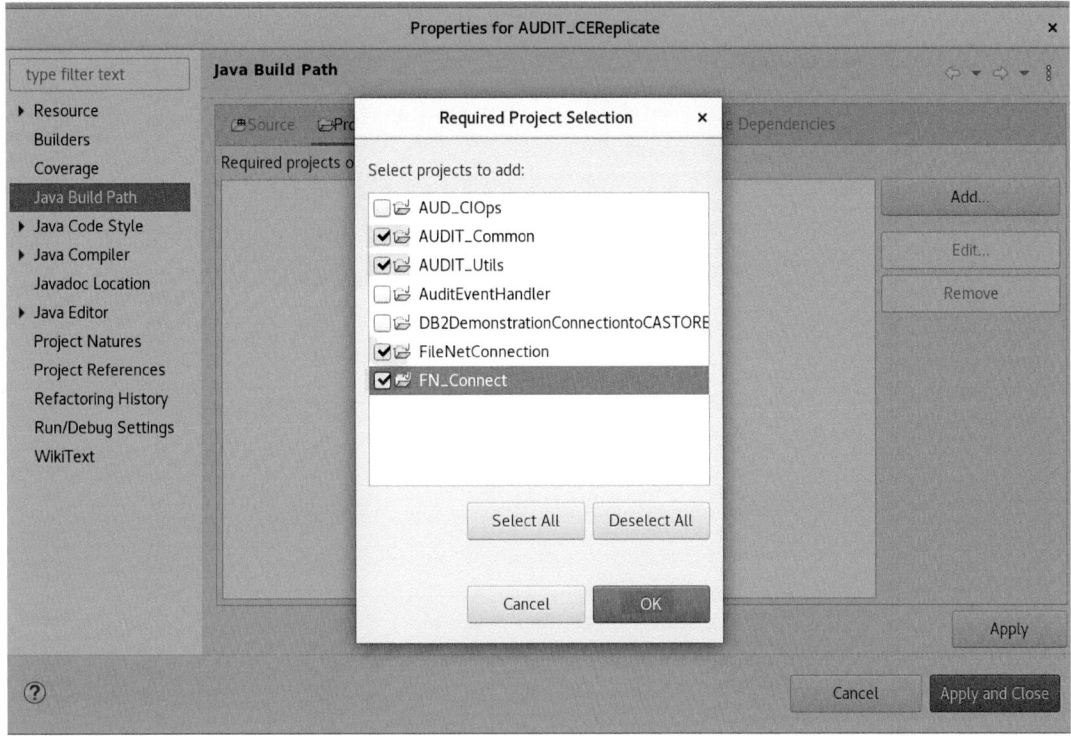

Figure 4-59. *The supporting Projects selected for addition to the Eclipse Java Build Path*

CHAPTER 4   A REPLICATION JAVA PROGRAM FOR IBM FILENET OBJECT STORES

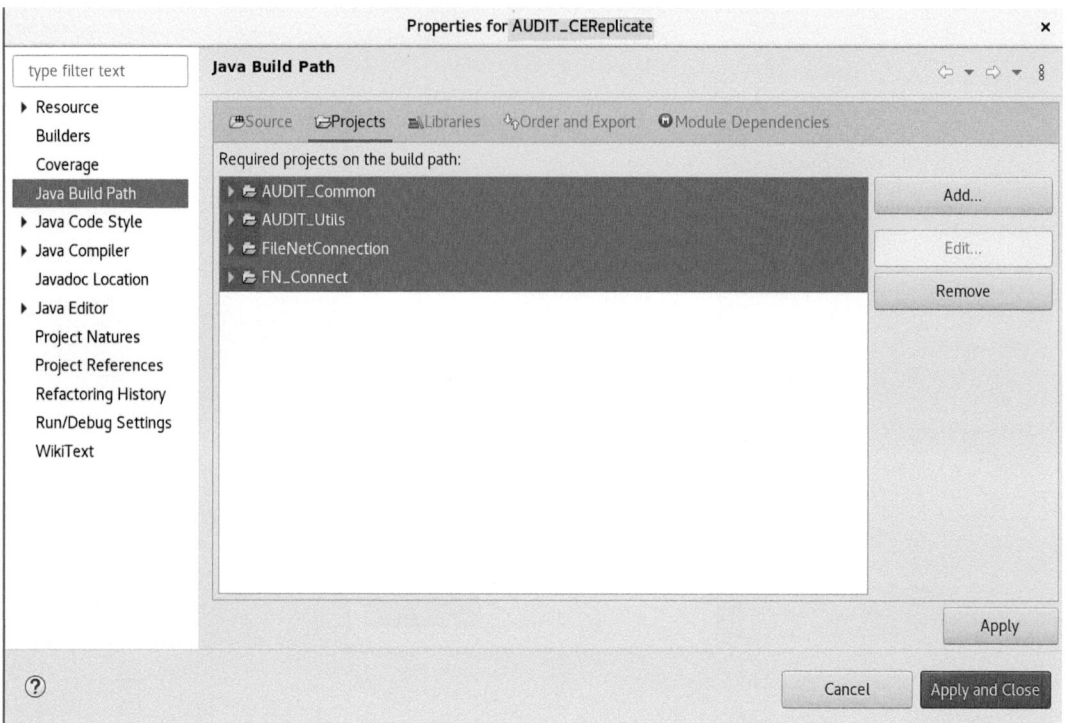

Figure 4-60. *The supporting Projects are added to the **AUDIT_CEReplicate** project*

CHAPTER 4 A REPLICATION JAVA PROGRAM FOR IBM FILENET OBJECT STORES

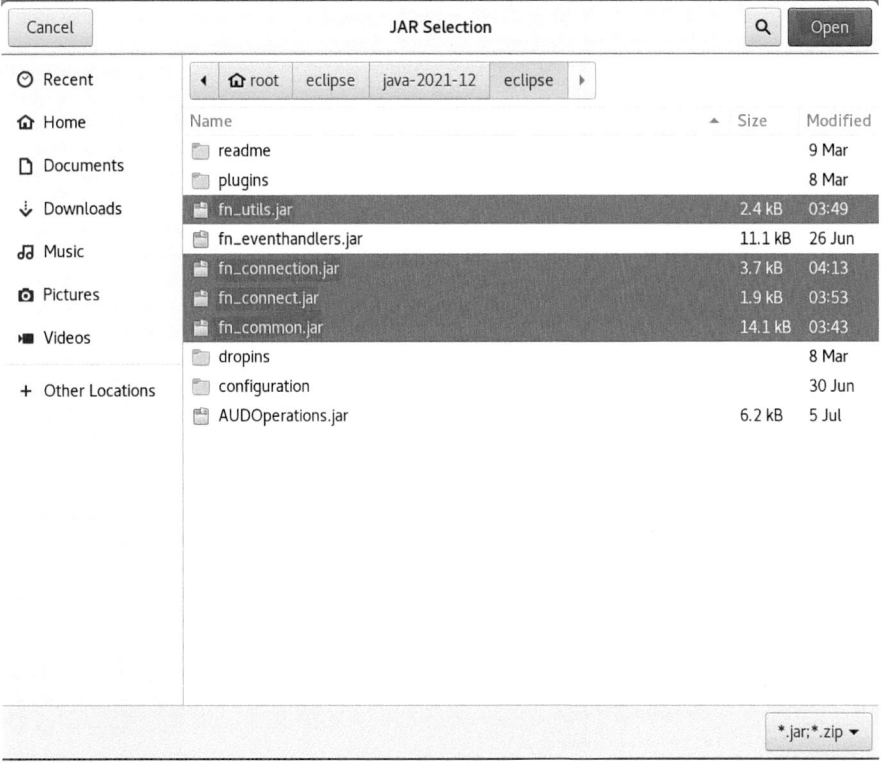

Figure 4-61. *The four created supporting jar files in the Eclipse root folder on Linux*

The **Add External JARs** command is selected, and the supporting jars displayed in Figure 4-61 are added to the **AUDIT_CEReplicate project** in the **Java Build Path** Libraries tab list.

CHAPTER 4   A REPLICATION JAVA PROGRAM FOR IBM FILENET OBJECT STORES

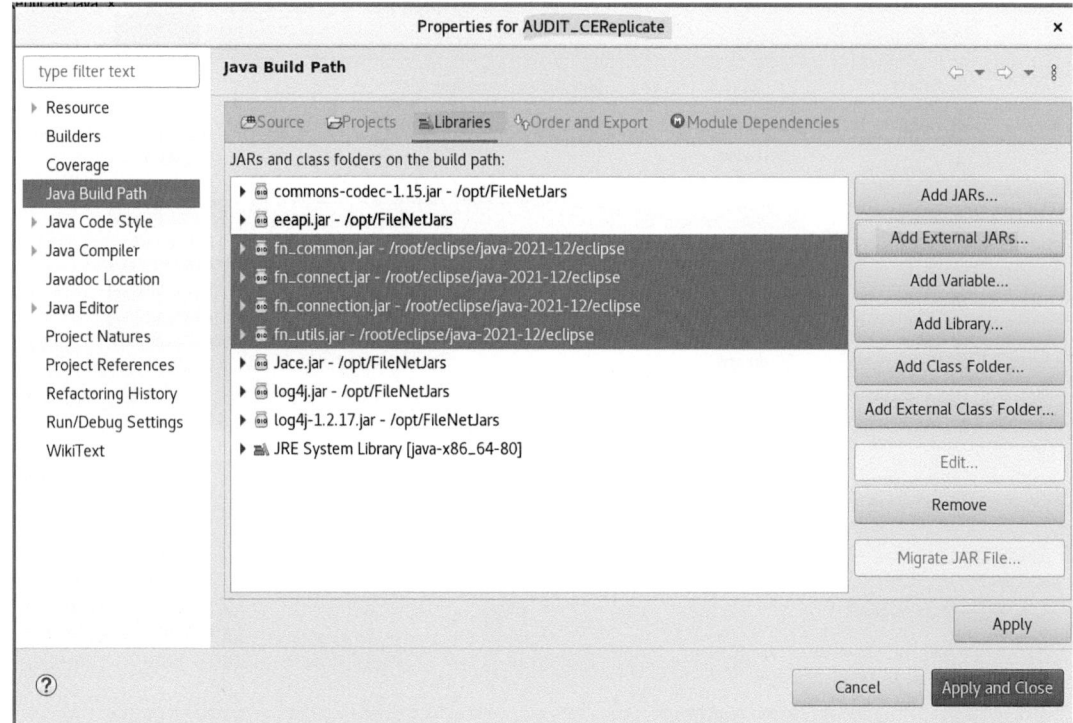

Figure 4-62. *The four supporting .jar libraries are added to the main Project*

Updating the Supporting Jar Library File Security

```
(base) [root@ECMUKDEMO6 eclipse]# pwd
/root/eclipse/java-2021-12/eclipse
(base) [root@ECMUKDEMO6 eclipse]# ls
AUDOperations.jar  dropins   eclipse.ini    fn_connection.jar  fn_eventhandlers.jar  icon.xpm  readme
configuration      eclipse   fn_common.jar  fn_connect.jar     fn_utils.jar          plugins
(base) [root@ECMUKDEMO6 eclipse]# chmod 755 *.jar
(base) [root@ECMUKDEMO6 eclipse]# ls
AUDOperations.jar  dropins   eclipse.ini    fn_connection.jar  fn_eventhandlers.jar  icon.xpm  readme
configuration      eclipse   fn_common.jar  fn_connect.jar     fn_utils.jar          plugins
(base) [root@ECMUKDEMO6 eclipse]#
```

Figure 4-63. *The security on the four generated supporting jar files is changed to allow access for the project build*

750

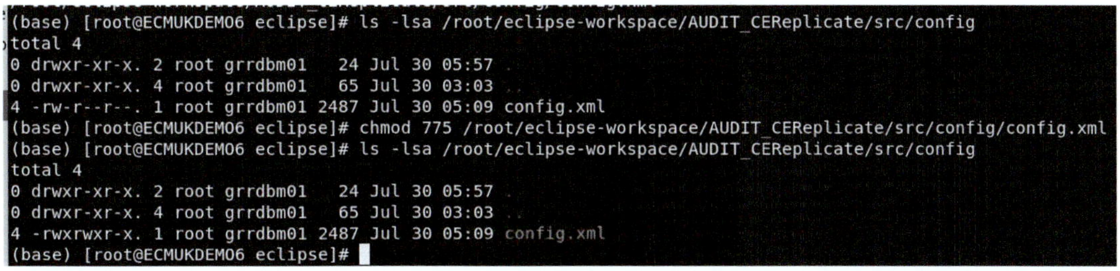
(base) [root@ECMUKDEMO6 eclipse]# ls -lsa /root/eclipse-workspace/AUDIT_CEReplicate/src/config
total 4
0 drwxr-xr-x. 2 root grrdbm01   24 Jul 30 05:57 .
0 drwxr-xr-x. 4 root grrdbm01   65 Jul 30 03:03 ..
4 -rw-r--r--. 1 root grrdbm01 2487 Jul 30 05:09 config.xml
(base) [root@ECMUKDEMO6 eclipse]# chmod 775 /root/eclipse-workspace/AUDIT_CEReplicate/src/config/config.xml
(base) [root@ECMUKDEMO6 eclipse]# ls -lsa /root/eclipse-workspace/AUDIT_CEReplicate/src/config
total 4
0 drwxr-xr-x. 2 root grrdbm01   24 Jul 30 05:57 .
0 drwxr-xr-x. 4 root grrdbm01   65 Jul 30 03:03 ..
4 -rwxrwxr-x. 1 root grrdbm01 2487 Jul 30 05:09 config.xml
(base) [root@ECMUKDEMO6 eclipse]#

Figure 4-64. *The security on the* ***config.xml*** *file is updated to support the search date updates which set the next search start date for source object store changes*

Updating the IBM FileNet Object Store Security

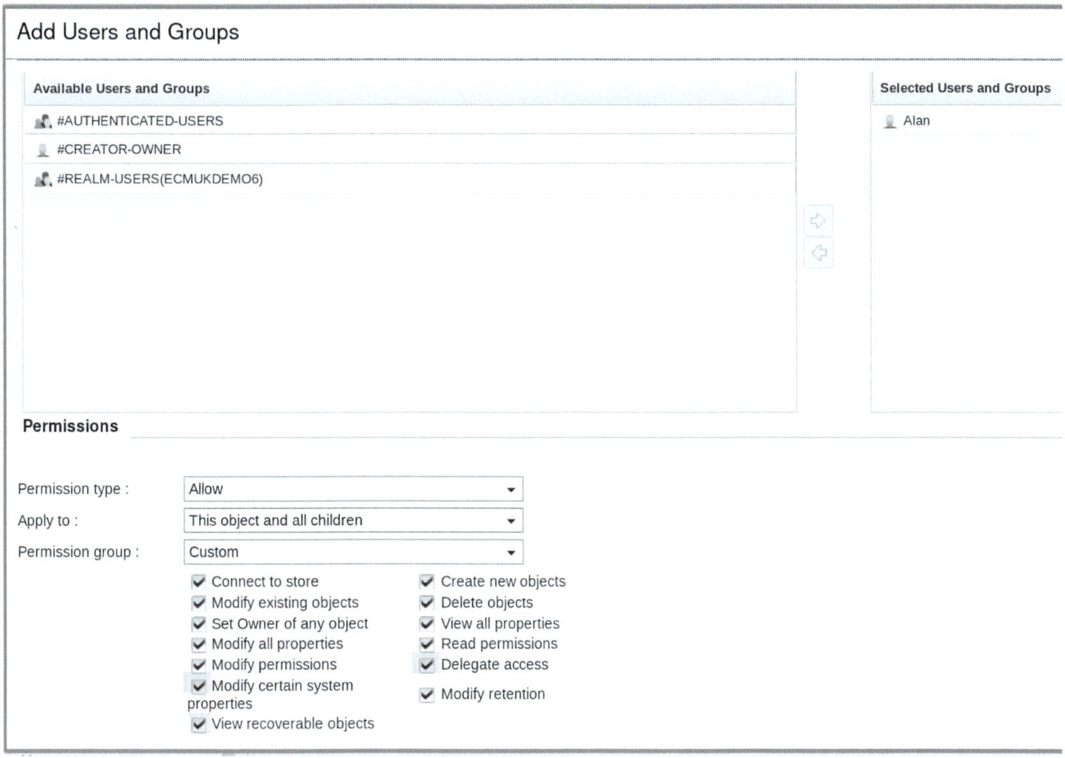
Add Users and Groups
Available Users and Groups
#AUTHENTICATED-USERS
#CREATOR-OWNER
#REALM-USERS(ECMUKDEMO6)
Selected Users and Groups
Alan
Permissions
Permission type :
Allow
Apply to :
This object and all children
Permission group :
Custom
Connect to store
Modify existing objects
Set Owner of any object
Modify all properties
Modify permissions
Modify certain system properties
View recoverable objects
Create new objects
Delete objects
View all properties
Read permissions
Delegate access
Modify retention

Figure 4-65. *The security on the two Object Stores OS1 and OS2 is updated for the connected Content Engine administrator user used by the Replication program*

CHAPTER 4   A REPLICATION JAVA PROGRAM FOR IBM FILENET OBJECT STORES

The security displayed in Figure 4-65 is used to ensure the user has the ACL security options for **Modify certain system properties** (also programmatically added using the API).

```
root@ECMUKDEMO6:/opt/replication/logs
File  Edit  View  Search  Terminal  Help
(base) [root@ECMUKDEMO6 opt]# mkdir replication
(base) [root@ECMUKDEMO6 opt]# cd replication
(base) [root@ECMUKDEMO6 replication]# mkdir logs
(base) [root@ECMUKDEMO6 replication]# cd logs
(base) [root@ECMUKDEMO6 logs]# touch AuditImportAuditFolders.log
(base) [root@ECMUKDEMO6 logs]# ls -lsa
total 0
0 drwxr-xr-x. 2 root root 41 Jul 30 13:15 .
0 drwxr-xr-x. 3 root root 18 Jul 30 13:14 ..
0 -rw-r--r--. 1 root root  0 Jul 30 13:15 AuditImportAuditFolders.log
(base) [root@ECMUKDEMO6 logs]# chmod 775 AuditImportAuditFolders.log
(base) [root@ECMUKDEMO6 logs]# ls
AuditImportAuditFolders.log
(base) [root@ECMUKDEMO6 logs]# touch AuditImportAuditDocs.log
(base) [root@ECMUKDEMO6 logs]# chmod 775 AuditImportAuditDocs.log
(base) [root@ECMUKDEMO6 logs]# touch AuditImportAuditDeleteDocs.log
(base) [root@ECMUKDEMO6 logs]# chmod 775 AuditImportAuditDeleteDocs.log
(base) [root@ECMUKDEMO6 logs]# touch AuditImportAuditDeleteFolders.log
(base) [root@ECMUKDEMO6 logs]# chmod 775 AuditImportAuditDeleteFolders.log
(base) [root@ECMUKDEMO6 logs]# touch AuditImportAuditUpdateDocuments.log
(base) [root@ECMUKDEMO6 logs]# chmod 775 AuditImportAuditUpdateDocuments.log
(base) [root@ECMUKDEMO6 logs]# touch AuditExportAudit.log
(base) [root@ECMUKDEMO6 logs]# chmod 775 AuditExportAudit.log
(base) [root@ECMUKDEMO6 logs]# ls -lsa
total 0
0 drwxr-xr-x. 2 root root 223 Jul 30 13:20 .
0 drwxr-xr-x. 3 root root  18 Jul 30 13:14 ..
0 -rwxrwxr-x. 1 root root   0 Jul 30 13:20 AuditExportAudit.log
0 -rwxrwxr-x. 1 root root   0 Jul 30 13:18 AuditImportAuditDeleteDocs.log
0 -rwxrwxr-x. 1 root root   0 Jul 30 13:19 AuditImportAuditDeleteFolders.log
0 -rwxrwxr-x. 1 root root   0 Jul 30 13:16 AuditImportAuditDocs.log
0 -rwxrwxr-x. 1 root root   0 Jul 30 13:15 AuditImportAuditFolders.log
0 -rwxrwxr-x. 1 root root   0 Jul 30 13:20 AuditImportAuditUpdateDocuments.log
(base) [root@ECMUKDEMO6 logs]# pwd
/opt/replication/logs
(base) [root@ECMUKDEMO6 logs]#
```

Figure 4-66. *The Audit log file security is changed to support the program writes*

CHAPTER 4   A REPLICATION JAVA PROGRAM FOR IBM FILENET OBJECT STORES

Running the Replication Program Tests

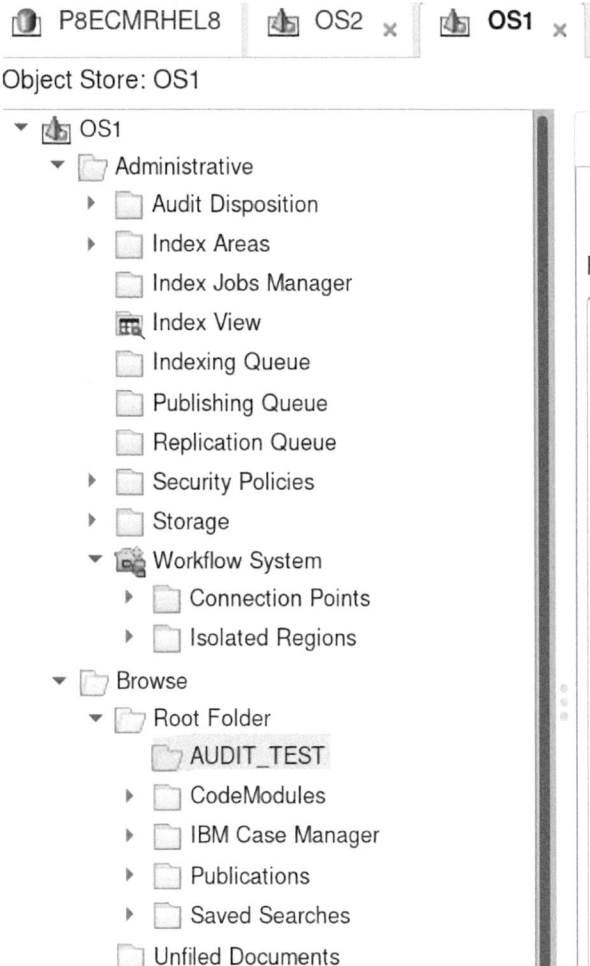

Figure 4-67. *The root folder /AUDIT_TEST is created in the target Object Store, OS1*

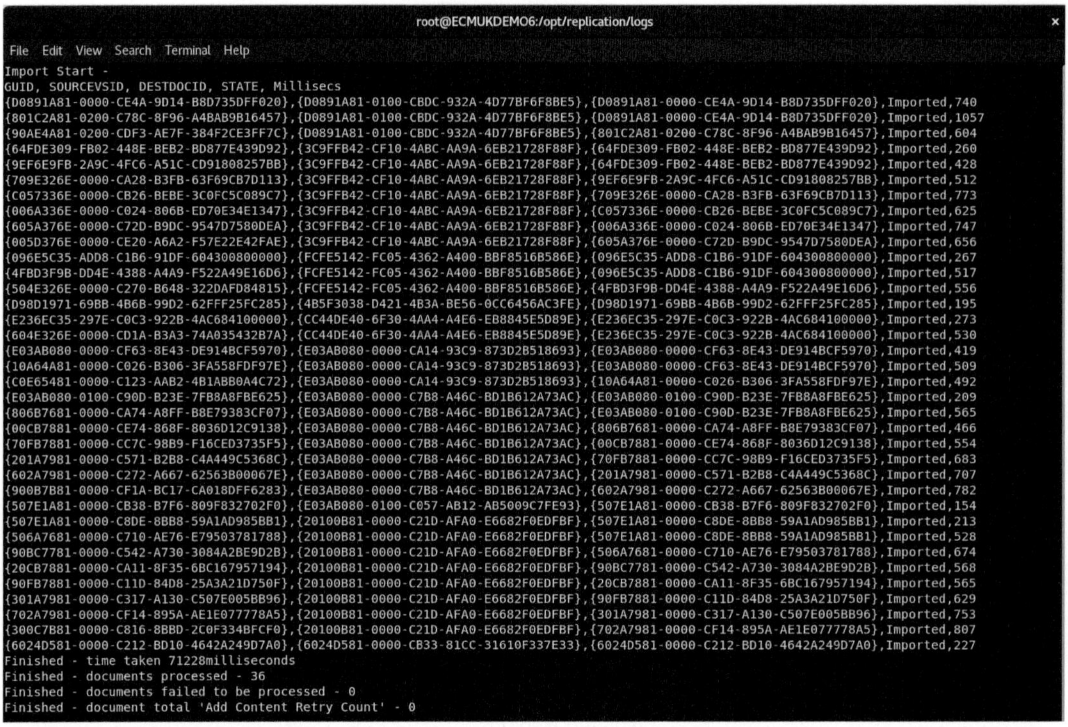

CHAPTER 4   A REPLICATION JAVA PROGRAM FOR IBM FILENET OBJECT STORES

Figure 4-68.   The 36 Documents processed displayed in the AuditImportAuditDocs.log

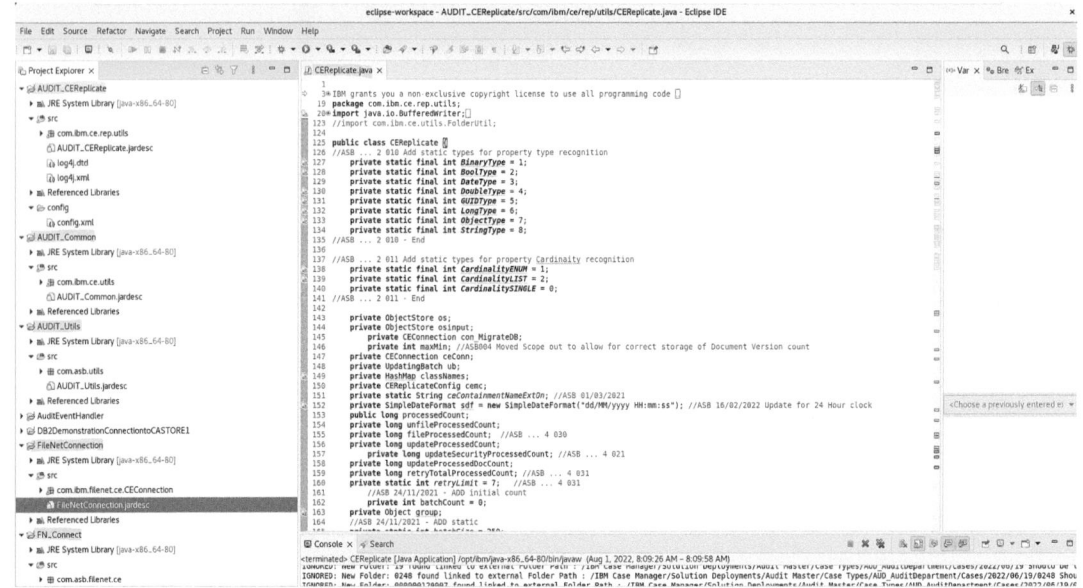

Figure 4-69.   The completed project set for the Replication.jar file build

754

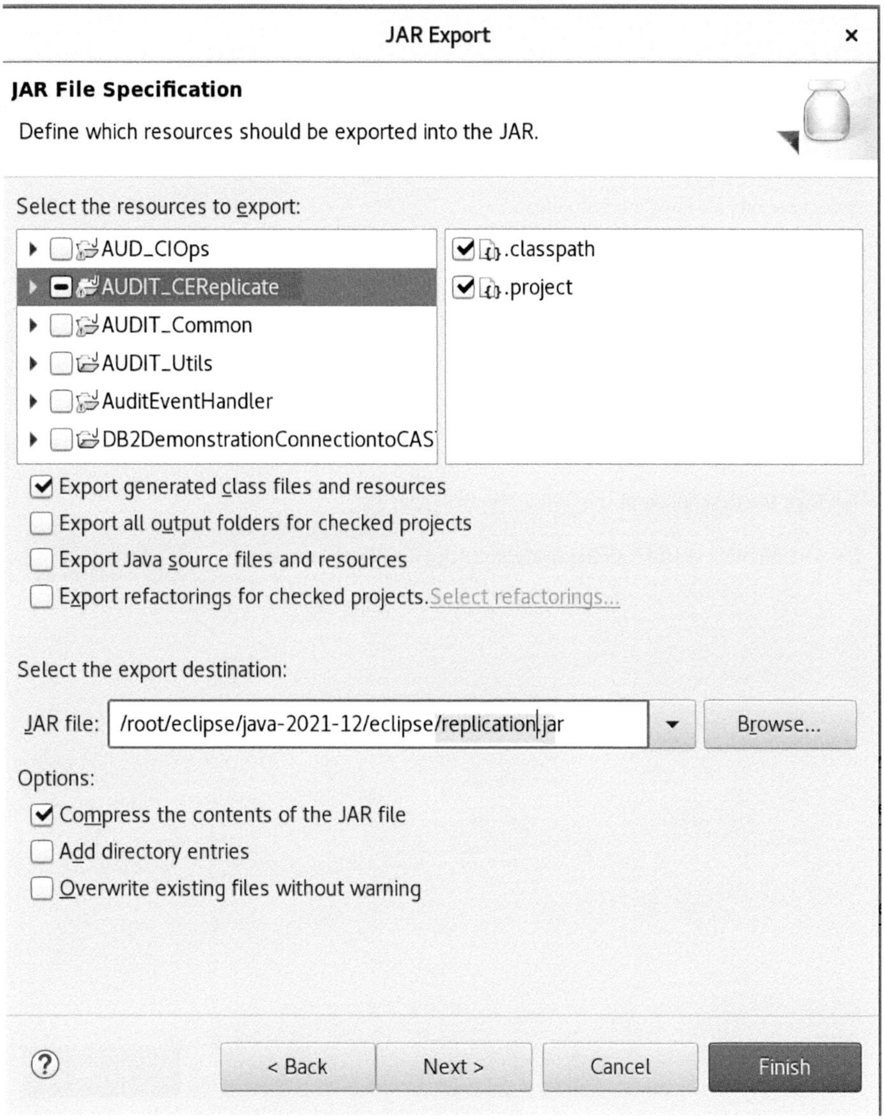

CHAPTER 4   A REPLICATION JAVA PROGRAM FOR IBM FILENET OBJECT STORES

Figure 4-70.*   The main **replication.jar** file is exported for use*

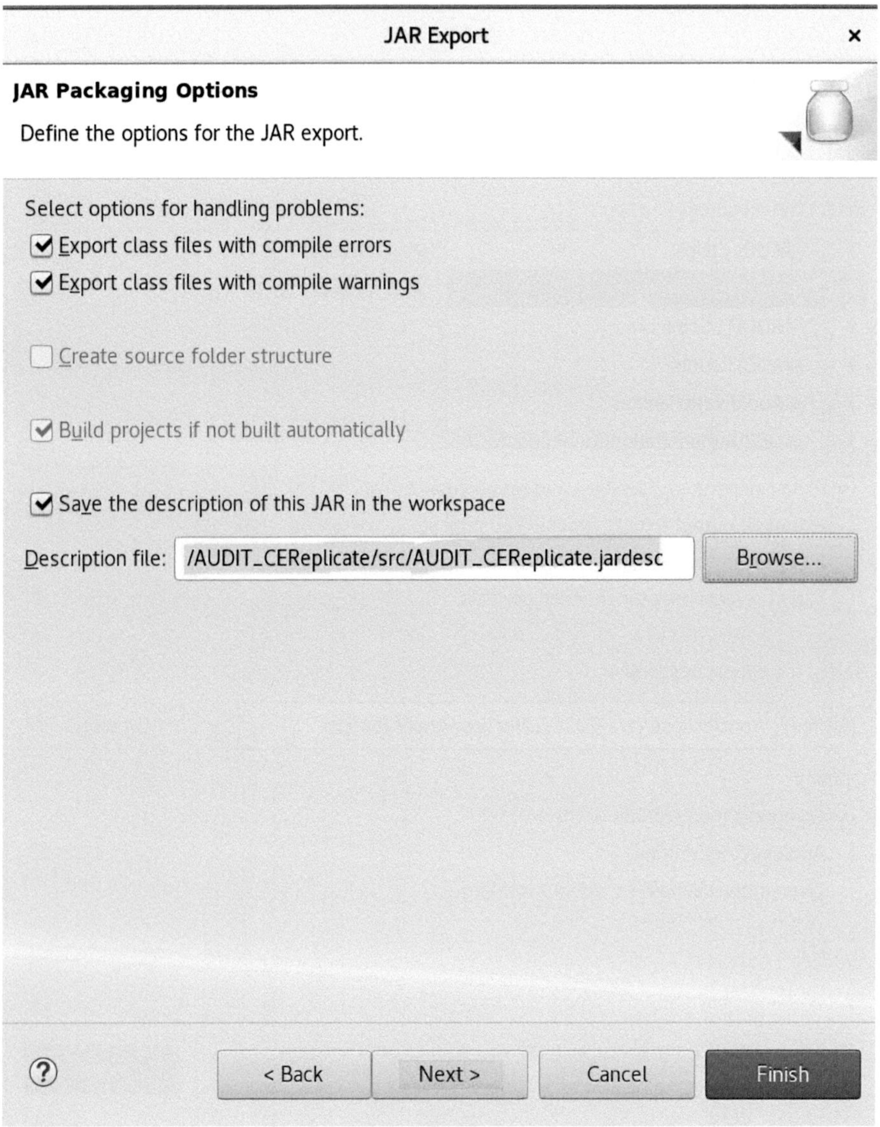

CHAPTER 4 A REPLICATION JAVA PROGRAM FOR IBM FILENET OBJECT STORES

*Figure 4-71. The **AUDIT_CEReplicate.jardesc** file is created for easy rebuilds*

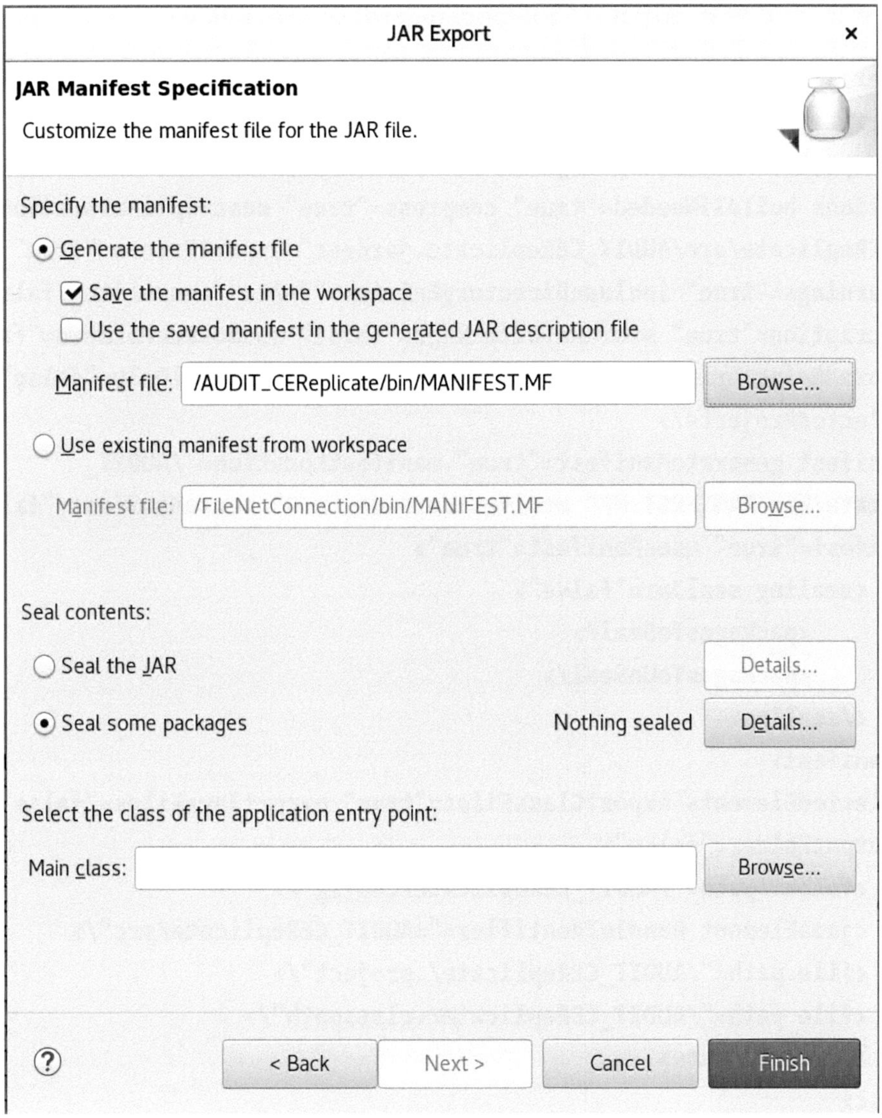

Figure 4-72. The MANIFEST.MF file is selected for addition to the bin folder

The Java JAR Build XML, **AUDIT_CEReplicate.jardesc**, is as follows:

```xml
<?xml version="1.0" encoding="UTF-8" standalone="no"?>
<jardesc>
    <jar path="/root/eclipse/java-2021-12/eclipse/replication.jar"/>
    <options buildIfNeeded="true" compress="true" descriptionLocation="/AUDIT_CEReplicate/src/AUDIT_CEReplicate.jardesc" exportErrors="true" exportWarnings="true" includeDirectoryEntries="false" overwrite="false" saveDescription="true" storeRefactorings="false" useSourceFolders="false"/>
    <storedRefactorings deprecationInfo="true" structuralOnly="false"/>
    <selectedProjects/>
    <manifest generateManifest="true" manifestLocation="/AUDIT_CEReplicate/bin/MANIFEST.MF" manifestVersion="1.0" reuseManifest="false" saveManifest="true" usesManifest="true">
        <sealing sealJar="false">
            <packagesToSeal/>
            <packagesToUnSeal/>
        </sealing>
    </manifest>
    <selectedElements exportClassFiles="true" exportJavaFiles="false" exportOutputFolder="false">
        <folder path="/AUDIT_CEReplicate/config"/>
        <javaElement handleIdentifier="=AUDIT_CEReplicate/src"/>
        <file path="/AUDIT_CEReplicate/.project"/>
        <file path="/AUDIT_CEReplicate/.classpath"/>
    </selectedElements>
</jardesc>
```

Result with the Exact GUID Copy of a Replicated Document

The following two screenshots demonstrate the unique Replication we have achieved where the source and target documents have exactly the same core system properties, including the Creation Date and GUID (ID) values.

The OS2 source of the replicated Document Level2A1 Document is shown in Figure 4-73.

CHAPTER 4   A REPLICATION JAVA PROGRAM FOR IBM FILENET OBJECT STORES

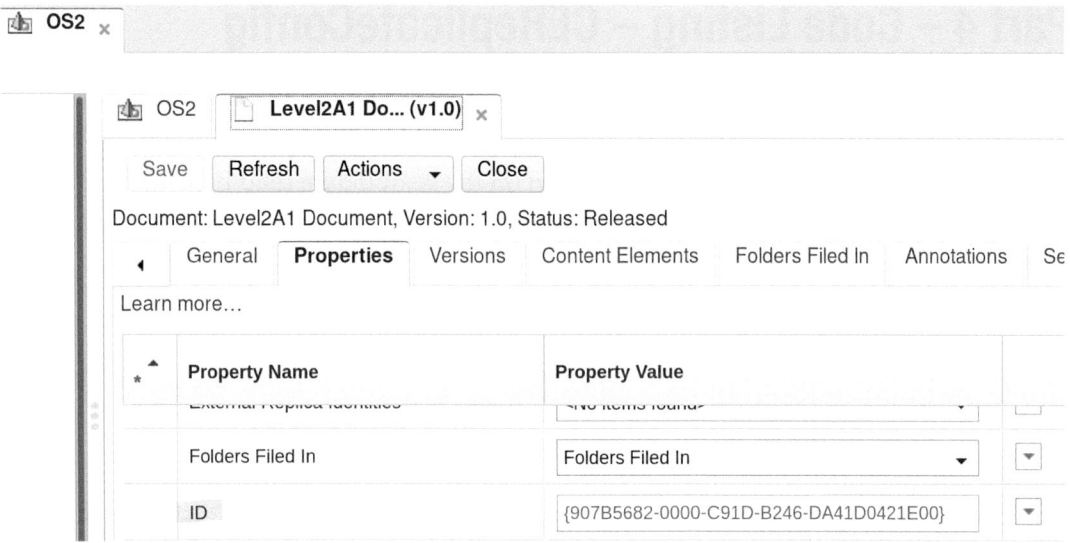

Figure 4-73. *The Document in OS2 GUID {907B5682-0000-C91D-B246-DA41D0421E00}*

The OS1 target of the replicated Document Level2A1 Document is shown in Figure 4-74.

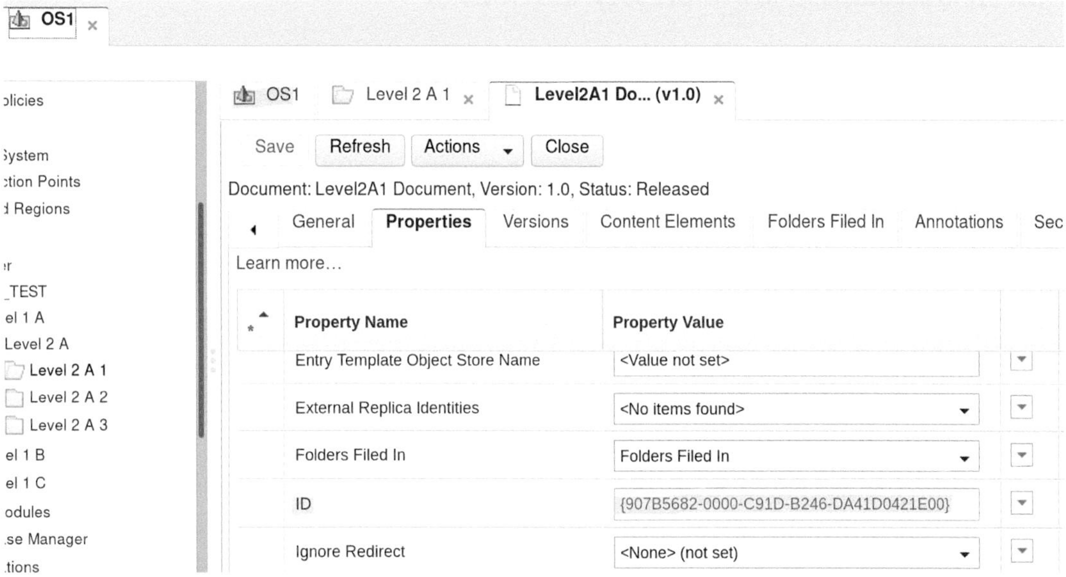

Figure 4-74. *The Document in OS1 GUID {907B5682-0000-C91D-B246-DA41D0421E00}*

759

CHAPTER 4   A REPLICATION JAVA PROGRAM FOR IBM FILENET OBJECT STORES

Part 4 – Code Listing – CEReplicateConfig

The CEReplicateConfig Java code contains the Main program calling the Java methods to load the config.xml file parameters and query the source Target Object store for the changed Document and Folder objects based on the search date read in from the file.

The program then builds an exact replica of all the folders and documents under the designated "root" folder (AUDIT_TEST in the example) of the Target Object Store.

```
/**
IBM grants you a non-exclusive copyright license to use all programming
code examples from which you can generate similar function tailored to your
own specific needs.

All sample code is provided by IBM for illustrative purposes only.
These examples have not been thoroughly tested under all conditions.  IBM,
therefore cannot guarantee or imply reliability, serviceability, or
function of these programs.

All Programs or code component contained herein are provided to you "AS IS"
without any warranties of any kind.
The implied warranties of non-infringement, merchantability and fitness for
a particular purpose are expressly disclaimed.

© Copyright IBM Corporation 2013, ALL RIGHTS RESERVED.
*/
package com.ibm.ce.rep.utils;
import java.io.BufferedWriter;
import java.io.File;
import java.io.FileInputStream;
import java.io.FileWriter;
import java.io.InputStream;
import java.io.UnsupportedEncodingException;
import java.lang.reflect.TypeVariable;
import java.sql.Connection;
import java.sql.PreparedStatement;
import java.sql.ResultSet;
```

```java
import java.sql.Statement;
import java.text.SimpleDateFormat;
import java.util.Date;
import java.util.HashMap;
import java.util.Iterator;

//ASB ... 4  001 Encryption / Decryption Java Classes
import javax.crypto.KeyGenerator;
import javax.crypto.SecretKey;
import java.io.UnsupportedEncodingException;
import java.security.spec.AlgorithmParameterSpec;
import java.security.spec.KeySpec;

import javax.crypto.Cipher;
import javax.crypto.IllegalBlockSizeException;
import javax.crypto.SecretKey;
import javax.crypto.SecretKeyFactory;
import javax.crypto.spec.PBEKeySpec;
import javax.crypto.spec.PBEParameterSpec;
//ASB ... 4 END

import com.filenet.api.admin.PropertyDefinition;
import com.filenet.api.core.BatchItemHandle;

import org.apache.commons.codec.binary.Base64;
import org.apache.log4j.Logger;
import org.w3c.dom.Node;
import org.w3c.dom.NodeList;

//import com.ibm.filenet.ce.CEConnection;

import com.filenet.api.collection.AccessPermissionList;
import com.filenet.api.collection.ContentElementList;
import com.filenet.api.collection.GroupSet;
import com.filenet.api.collection.IndependentObjectSet;
import com.filenet.api.collection.PropertyDefinitionList;
import com.filenet.api.collection.StringList;
```

CHAPTER 4   A REPLICATION JAVA PROGRAM FOR IBM FILENET OBJECT STORES

```java
import com.filenet.api.collection.UserSet;
import com.filenet.api.collection.VersionableSet;
import com.filenet.api.constants.AccessRight;
import com.filenet.api.constants.AccessType;
import com.filenet.api.constants.AutoClassify;
import com.filenet.api.constants.AutoUniqueName;
import com.filenet.api.constants.CheckinType;
import com.filenet.api.constants.ClassNames;
import com.filenet.api.constants.CompoundDocumentState;
import com.filenet.api.constants.DefineSecurityParentage;
import com.filenet.api.constants.FilteredPropertyType;
import com.filenet.api.constants.JoinComparison;
import com.filenet.api.constants.JoinOperator;
import com.filenet.api.constants.PrincipalSearchAttribute;
import com.filenet.api.constants.PrincipalSearchType;
import com.filenet.api.constants.PropertyNames;
import com.filenet.api.constants.RefreshMode;
import com.filenet.api.constants.ReservationType;
import com.filenet.api.constants.VersionStatus;
import com.filenet.api.core.ContentTransfer;
import com.filenet.api.core.Document;
import com.filenet.api.core.EngineObject;
import com.filenet.api.core.Factory;
import com.filenet.api.core.Factory.AccessPermission;
import com.filenet.api.core.Factory.ClassDefinition;
import com.filenet.api.core.Factory.ClassDescription;
import com.filenet.api.core.Factory.Domain;
import com.filenet.api.core.Factory.DynamicReferentialContainmentRelationship;
//import com.filenet.api.core.Factory.PropertyDefinitionDateTime;
import com.filenet.api.core.Factory.Realm;
import com.filenet.api.core.Factory.User;
import com.filenet.api.core.Folder;
import com.filenet.api.core.IndependentObject;
import com.filenet.api.core.IndependentlyPersistableObject;
```

```java
import java.sql.Statement;
import java.text.SimpleDateFormat;
import java.util.Date;
import java.util.HashMap;
import java.util.Iterator;

//ASB ... 4  001 Encryption / Decryption Java Classes
import javax.crypto.KeyGenerator;
import javax.crypto.SecretKey;
import java.io.UnsupportedEncodingException;
import java.security.spec.AlgorithmParameterSpec;
import java.security.spec.KeySpec;

import javax.crypto.Cipher;
import javax.crypto.IllegalBlockSizeException;
import javax.crypto.SecretKey;
import javax.crypto.SecretKeyFactory;
import javax.crypto.spec.PBEKeySpec;
import javax.crypto.spec.PBEParameterSpec;
//ASB ... 4 END

import com.filenet.api.admin.PropertyDefinition;
import com.filenet.api.core.BatchItemHandle;

import org.apache.commons.codec.binary.Base64;
import org.apache.log4j.Logger;
import org.w3c.dom.Node;
import org.w3c.dom.NodeList;

//import com.ibm.filenet.ce.CEConnection;

import com.filenet.api.collection.AccessPermissionList;
import com.filenet.api.collection.ContentElementList;
import com.filenet.api.collection.GroupSet;
import com.filenet.api.collection.IndependentObjectSet;
import com.filenet.api.collection.PropertyDefinitionList;
import com.filenet.api.collection.StringList;
```

```java
import com.filenet.api.collection.UserSet;
import com.filenet.api.collection.VersionableSet;
import com.filenet.api.constants.AccessRight;
import com.filenet.api.constants.AccessType;
import com.filenet.api.constants.AutoClassify;
import com.filenet.api.constants.AutoUniqueName;
import com.filenet.api.constants.CheckinType;
import com.filenet.api.constants.ClassNames;
import com.filenet.api.constants.CompoundDocumentState;
import com.filenet.api.constants.DefineSecurityParentage;
import com.filenet.api.constants.FilteredPropertyType;
import com.filenet.api.constants.JoinComparison;
import com.filenet.api.constants.JoinOperator;
import com.filenet.api.constants.PrincipalSearchAttribute;
import com.filenet.api.constants.PrincipalSearchType;
import com.filenet.api.constants.PropertyNames;
import com.filenet.api.constants.RefreshMode;
import com.filenet.api.constants.ReservationType;
import com.filenet.api.constants.VersionStatus;
import com.filenet.api.core.ContentTransfer;
import com.filenet.api.core.Document;
import com.filenet.api.core.EngineObject;
import com.filenet.api.core.Factory;
import com.filenet.api.core.Factory.AccessPermission;
import com.filenet.api.core.Factory.ClassDefinition;
import com.filenet.api.core.Factory.ClassDescription;
import com.filenet.api.core.Factory.Domain;
import com.filenet.api.core.Factory.DynamicReferentialContainmentRelationship;
//import com.filenet.api.core.Factory.PropertyDefinitionDateTime;
import com.filenet.api.core.Factory.Realm;
import com.filenet.api.core.Factory.User;
import com.filenet.api.core.Folder;
import com.filenet.api.core.IndependentObject;
import com.filenet.api.core.IndependentlyPersistableObject;
```

```java
import com.filenet.api.core.ObjectStore;
import com.filenet.api.core.ReferentialContainmentRelationship;
import com.filenet.api.core.UpdatingBatch;
import com.filenet.api.collection.FolderSet;
import com.filenet.api.collection.ReferentialContainmentRelationshipSet;

import com.filenet.api.exception.EngineRuntimeException;
import com.filenet.api.exception.ExceptionCode;
import com.filenet.api.property.FilterElement;
import com.filenet.api.property.Properties;
import com.filenet.api.property.Property;
import com.filenet.api.property.PropertyFilter;
import com.filenet.api.query.SearchSQL;
import com.filenet.api.query.SearchScope;
import com.filenet.api.security.Group;
import com.filenet.api.security.Permission;
import com.filenet.api.util.Id;
import com.filenet.api.core.VersionSeries;
import com.ibm.ce.rep.utils.CEReplicate;
import com.ibm.ce.utils.CEReplicateConfig;
import com.ibm.filenet.ce.CEConnection.CEConnection;
//import com.ibm.ce.utils.FolderUtil;

public class CEReplicate {
//ASB ... 2 010 Add static types for property type recognition
    private static final int BinaryType = 1;
    private static final int BoolType = 2;
    private static final int DateType = 3;
    private static final int DoubleType = 4;
    private static final int GUIDType =    5;
    private static final int LongType = 6;
    private static final int ObjectType = 7;
    private static final int StringType = 8;
//ASB ... 2 010 - End

//ASB ... 2 011 Add static types for property Cardinaity recognition
    private static final int CardinalityENUM = 1;
```

```java
    private static final int CardinalityLIST = 2;
    private static final int CardinalitySINGLE = 0;
//ASB ... 2 011 - End

    private ObjectStore os;
    private ObjectStore osinput;
    private CEConnection con_MigrateDB;
    private int maxMin; //ASB004 Moved Scope out to allow for correct storage of Document Version count
    private CEConnection ceConn;
    private UpdatingBatch ub;
    private HashMap classNames;
    private CEReplicateConfig cemc;
    private static String ceContainmentNameExtOn; //ASB 01/03/2022
    private SimpleDateFormat sdf = new SimpleDateFormat("dd/MM/yyyy HH:mm:ss"); //ASB 16/02/2022 Update for 24 Hour clock
    public long processedCount;
    private long unfileProcessedCount;
    private long fileProcessedCount;   //ASB ... 4 030
    private long updateProcessedCount;
    private long updateSecurityProcessedCount; //ASB ... 4 021
    private long updateProcessedDocCount;
    private long retryTotalProcessedCount; //ASB ... 4 031
    private static int retryLimit = 7;    //ASB ... 4 031
    //ASB 24/11/2021 - ADD initial count
    private int batchCount = 0;
    private Object group;
    //ASB 24/11/2021 - ADD static
    private static int batchSize = 250;
    //ASB ... 2 021 Additional Replication Parameters
    private String readOnlyGroup;
    private Date startSearchDate;
    private int maxSearchRecords;
    //ASB ... 2 021 End
    //ASB ... 4 022 Folder Symbolic Name List
    private String folderSymbolList = "";
```

```java
    private String folderExcludeSymbolList = "";
    private static Logger logger =                    Logger.getLogger(CEReplicate.class.getName());

/**
 * @param args
 */
public static void main(String[] args) throws Exception {
    logger.info("CE Replicate Version 3.7");
    //ASB set up the Containment Name
    CEReplicate ceReplicate = new CEReplicate();
    ceReplicate.init();
        long msecs = System.currentTimeMillis();
        //ASB Comment Out Arguments - ALL in the same run now
        /*
         * if (args[0].compareToIgnoreCase("importfolders") == 0){
         */
            logger.info("Begin Import Folders ...");

            ceReplicate.importFolders();
            msecs = System.currentTimeMillis() - msecs;
            logger.info("Import " + ceReplicate.getProcessedCount() + "
            Folders took " + msecs + " milliseconds");
        //}
            msecs = System.currentTimeMillis();
            //ASB ... 3 003 Add deletion processing - Documents
            //ASB ... 4 011 Moved to delete documents before
            importing them!
            /*
             * if (args[0].compareToIgnoreCase("deletedocs") == 0){
             */
            logger.info("Begin Delete Docs ...");
                ceReplicate.deleteDocuments();
            msecs = System.currentTimeMillis() - msecs;
            logger.info("Delete " + ceReplicate.getProcessedCount() + "
            Docs took " + msecs + " milliseconds");
            //}
```

```java
            /*
             * if (args[0].compareToIgnoreCase("importdocs") == 0){
             */
                msecs = System.currentTimeMillis();
                logger.info("Begin Import Docs ...");
            ceReplicate.importDocuments();
                msecs = System.currentTimeMillis() - msecs;
                logger.info("Import " + ceReplicate.getProcessedCount() + " 
                Docs took " + msecs + " milliseconds");
                //}
                /*
                 * if (args[0].compareToIgnoreCase("deletefolders") == 0){
                 */
                //ASB ... 3 003 Add deletion processing - Folders
                msecs = System.currentTimeMillis();
                logger.info("Begin Delete Folders ...");
            ceReplicate.deleteFolders();
                msecs = System.currentTimeMillis() - msecs;
                logger.info("Delete " + ceReplicate.getProcessedCount() + " 
                Folders took " + msecs + " milliseconds");
                //}
                //ASB ... 4 012 New Update documents process
                /*
                 * if (args[0].compareToIgnoreCase("updatedocs") == 0){
                 */
                msecs = System.currentTimeMillis();
                logger.info("Begin Update Docs ...");
                    ceReplicate.updateDocuments();
                msecs = System.currentTimeMillis() - msecs;
                logger.info("Update " + ceReplicate.
                getUpdateProcessedDocCount() + " Docs took " + msecs + " 
                milliseconds");
                //}

            //}
```

```java
        //}
}
private String getProcessedCount(){
    return String.valueOf(processedCount);
}
private String getUnfileProcessedCount(){
    return String.valueOf(unfileProcessedCount);
}
//ASB ... 4 030
private String getFileProcessedCount(){
    return String.valueOf(fileProcessedCount);
}
private String getUpdateProcessedCount(){
    return String.valueOf(updateProcessedCount);
}
private String getUpdateProcessedDocCount(){
    return String.valueOf(updateProcessedDocCount);
}
//Deletes all the versions associated with the document
//Used when a document is being re-imported
private boolean deleteDoc(String docId) throws Exception
{
    try
    {
        //ASB ... 4 019 Retrieve input doc to check the state
         //             Need to ignore checked out docs
          Document docSource = Factory.Document.fetchInstance(osinput,
          docId, null);
          if(docSource.get_VersionSeries().get_IsReserved()){
              return false;
          }
          Document doc = null;
       // S T A R T Deletion list
          //Get Version Series ID
          VersionableSet verSet  = docSource.get_Versions();
```

CHAPTER 4 A REPLICATION JAVA PROGRAM FOR IBM FILENET OBJECT STORES

```java
        int versionCount=0;
        Iterator versIt = verSet.iterator();
        int versCount = 0;
        Id delId = null;
        Document docDel = null; //Input Document versions
        Id delDocFound = null;
        VersionSeries vs = null; //ASB ... 4 23  16 11 2021
        while (versIt.hasNext()){
            versCount ++;
          docDel = (Document)versIt.next();
        //Get id to delete
        delId = docDel.get_Id();
        try{
                doc = Factory.Document.fetchInstance(os, delId, null);
                delDocFound = doc.get_Id();
                vs = doc.get_VersionSeries();
                break;
        }catch(Exception delEx){
            //Here because this version does not exist
        }

    }
    // E N D
      // Return in the following cases
      // a) IsCurrentVersion = false
      // b) Version series Current version is reserved
      //ASB ... 4 23 VersionSeries vs = doc.get_VersionSeries();
      //ASB ... 4 019 Check if the target doc is reserved and
      return if so!
      // Most unlikely as it is supposed to be read-only!
      //ASB ... 4 23if (vs.get_CurrentVersion().get_IsReserved()){
      //ASB ... 4 23     return false;
      //ASB ... 4 23 }
      vs.delete();
      vs.save(RefreshMode.REFRESH);
```

```java
        }
        catch(Exception e)
        {
            logger.error("Error deleting document - " + docId);
            logger.error(e.getMessage(), e);
            throw e;
        }
        return true;
}
//Deletes all the versions associated with the document
//Used when a document is being re-imported
private boolean deleteDocTargetOnly(String docId) throws Exception
{
    try
    {
        //ASB ... 4 019 Retrieve input doc to check the state
        //              Need to ignore checked out docs
        Document doc = Factory.Document.fetchInstance(os, docId, null);
        // S T A R T Deletion list
        //Get Version Series ID
        VersionableSet verSet  = doc.get_Versions();
        int versionCount=0;
        Iterator versIt = verSet.iterator();
        int versCount = 0;
        Id delId = null;
        Document docDel = null; //Input Document versions
        Id delDocFound = null;
        VersionSeries vs =  null; //ASB ... 4 23  16 11 2021
        while (versIt.hasNext()){
            versCount ++;
            docDel = (Document)versIt.next();
            //Get id to delete
            delId = docDel.get_Id();
            try{
```

CHAPTER 4   A REPLICATION JAVA PROGRAM FOR IBM FILENET OBJECT STORES

```
                    docDel = Factory.Document.fetchInstance(os,
                    delId, null);
                    delDocFound = doc.get_Id();
                    vs = doc.get_VersionSeries();
                    break;
                }catch(Exception delEx){
                    //Here because this version does not exist
                }

            }
        // E N D
            // Return in the following cases
            // a) IsCurrentVersion = false
            // b) Version series Current version is reserved
           //ASB ... 4 23 VersionSeries vs = doc.get_VersionSeries();
           //ASB ... 4 019 Check if the target doc is reserved and
           return if so!
           // Most unlikely as it is supposed to be read-only!
           //ASB ... 4 23if (vs.get_CurrentVersion().get_IsReserved()){
           //ASB ... 4 23     return false;
           //ASB ... 4 23 }
           vs.delete();
           vs.save(RefreshMode.REFRESH);

    }
    catch(Exception e)
    {
            logger.error("Error deleting document - " + docId);
            logger.error(e.getMessage(), e);
          throw e;
    }
      return true;
}
//ASB ... 3 003
//Deletes the Folder whose GUID is passed in
//Used when a folder has been deleted from the source Object Store
```

```java
//And there is a deletion Event record for the Folder Class
private void deleteFolder(String folderId) throws Exception
{
    try
    {
        Folder folder = Factory.Folder.fetchInstance(os, folderId, null);
        folder.delete();
        folder.save(RefreshMode.REFRESH);

    }
    catch(Exception e)
    {
        logger.error("Error deleting folder - " + folderId);
        logger.error(e.getMessage(), e);
        throw e;
    }
}

//Main Process loop for the Import Documents process.
//Outer loop executes the retrieval of Documents from the source Object
Store until no matches are returned.
//Inner loop processes each Document returned from the retrieval.
//
private void importDocuments() throws  Exception {
    BufferedWriter auditFileWriter;
    Long importStartTime;
    Long docImportStartTime;
    Long timeToImport;
        long maxRunTimeMillis;
        //Long maxDocCount = new Long(0);
    boolean moreDocsToImport=true;
    Integer consecutiveErrs=0;
    Integer errorCount=0;

    boolean completed=false;

    this.processedCount = 0;
```

```java
importStartTime = System.currentTimeMillis();
maxRunTimeMillis = cemc.getImportMaxRunTimeMinutes()*60*1000;
Long maxDocCount = new Long(cemc.getImportMaxDocCount());

auditFileWriter = new BufferedWriter(new FileWriter(cemc.
getImportAuditFile()));
auditFileWriter.write("Import Start - ");
auditFileWriter.newLine();
auditFileWriter.write("GUID, SOURCEVSID, DESTDOCID, STATE,
Millisecs");
auditFileWriter.newLine();
while (moreDocsToImport && completed==false)
{
    moreDocsToImport=false;
    SearchSQL sqlObject = new SearchSQL();
    //sqlObject.setSelectList("d.DocumentTitle, d.Id, d.Creator,
d.DateCreated, d.LastModifier, d.DateLastModified, d.Name,
d.Owner, d.IsReserved, d.IsCurrentVersion, d.IsFrozenVersion,
d.IsVersioningEnabled, d.MajorVersionNumber, d.MinorVersionNumber,d.
VersionStatus, d.IsInExceptionState, d.LockToken, d.LockTimeout,
d.LockOwner, d.ReservationType,d.DateCheckedIn, d.StorageLocation,
d.ContentElementsPresent, d.ContentSize, d.MimeType,
d.DateContentLastAccessed, d.ContentRetentionDate, d.CurrentState,
d.ClassificationStatus, d.IndexationId, d.CmIndexingFailureCode,
d.CompoundDocumentState, d.CmRetentionDate, d.ComponentBindingLabel,d.
EntryTemplateObjectStoreName, d.EntryTemplateLaunchedWorkflowNumber,
d.EntryTemplateId, d.PublicationInfo, d.PublishingSubsidiaryFolder,
d.IgnoreRedirect, d.FoldersFiledIn");
    //ASB ... 4 006 Need to allow for additional properties required
    sqlObject.setSelectList("d.*");
    //              sqlObject.setSelectList("d.DocumentTitle,
d.Id, d.Creator, d.DateCreated, d.LastModifier, d.DateLastModified,
d.Name, d.Owner, d.IsReserved, d.IsCurrentVersion, d.IsFrozenVersion,
d.IsVersioningEnabled, d.MajorVersionNumber, d.MinorVersionNumber,d.
VersionStatus, d.IsInExceptionState, d.LockToken, d.LockTimeout,
d.LockOwner, d.ReservationType,d.DateCheckedIn, d.StorageLocation,
```

```java
d.ContentElementsPresent, d.ContentSize, d.MimeType,
d.DateContentLastAccessed, d.ContentRetentionDate, d.CurrentState,
d.ClassificationStatus, d.IndexationId, d.CmIndexingFailureCode,
d.CompoundDocumentState, d.CmRetentionDate, d.ComponentBindingLabel,d.
EntryTemplateObjectStoreName, d.EntryTemplateLaunchedWorkflowNumber,
d.EntryTemplateId, d.PublicationInfo, d.PublishingSubsidiaryFolder,
d.IgnoreRedirect, d.FoldersFiledIn");
        //ASB ... 2   022
        Integer maxRecords = cemc.getExportMaxDocCount();
        sqlObject.setMaxRecords(maxRecords); //was hard coded to 20 !
        sqlObject.setFromClauseInitialValue("Document", "d", true); //set
        true to include subclasses
        //ASB ... 022 - Retrieve Date start from config.xml
        String sDeltaHours = cemc.getDeltaHours();
        String sSearchDate = cemc.getStartSearchDate();
        //ASB ... 4 002 Print Out the Current Date used for searching
        //System.out.print("\r\tStart Search Date : " + sSearchDate);
        logger.info("Start Search Date : " + sSearchDate);
        logger.info("Start Delta Hours : " + sDeltaHours);
         //ASB ... 4 002 Now dynamically update the next Replication Date
         cemc.updateProcessDate(new Date());

        //Date searchDate = sdf.parse(sSearchDate);
        //sqlObject.setWhereClause("d.DateCreated > " + "20210923T125628Z
        and d.IsCurrentVersion = true" );
        //ASB ... 022 - enter date to search
        //ASB ... 4 019 Exclude reservation objects
          //sqlObject.setWhereClause("d.DateCreated > " +  sSearchDate.
          trim() + " and d.IsCurrentVersion = true and d.IsReserved =
          false"   ); //ASB ... 4 003
        //ASB ... 4 023 Set documents to be imported based on Date Last
        Modified
        //           Rather than DateCreated to ensure documents left
          checked out passed the delta time are included when checked in !!
```

```java
sqlObject.setWhereClause("d.DateLastModified > " + sSearchDate.
trim() + " and d.IsCurrentVersion = true and d.IsReserved =
false"   ); //ASB ... 4 003
sqlObject.setOrderByClause("d.DateLastModified,d.
MajorVersionNumber, d.MinorVersionNumber");
// Check the SQL statement.
logger.info("SQL: " + sqlObject.toString());

// Create a SearchScope instance. (Assumes you have the
object store
// object.)
SearchScope search = new SearchScope(osinput);

// Set the page size (Long) to use for a page of query result
data. This value is passed
// in the pageSize parameter. If null, this defaults to the
value of
// ServerCacheConfiguration.QueryPageDefaultSize.
Integer myPageSize = new Integer(1000);

// Specify a property filter to use for the filter parameter,
if needed.
// This can be null if you are not filtering properties.
PropertyFilter myFilter = new PropertyFilter();
int myFilterLevel = 1;
myFilter.setMaxRecursion(myFilterLevel);
//ASB ... 4 020
if(cemc.getDebugOutputFlag().equalsIgnoreCase("off")){
    myFilter.addIncludeType(new FilterElement(null, null, null,
    FilteredPropertyType.ANY_LIST, null));
    myFilter.addIncludeType(new FilterElement(null, null, null,
    FilteredPropertyType.ANY_SINGLETON, null));
}else{
    myFilter.addIncludeType(new FilterElement(null, null, null,
    FilteredPropertyType.ANY, null));
}
```

```java
// Set the (Boolean) value for the continuable parameter. This indicates
// whether to iterate requests for subsequent pages of result data when the end of the
// first page of results is reached. If null or false, only a single page of results is
// returned.
Boolean continuable = new Boolean(true);

// Execute the fetchObjects method using the specified parameters.
IndependentObjectSet myObjects = search.fetchObjects(sqlObject, myPageSize, myFilter, continuable);

// You can then iterate through the collection of rows to access the properties.
int rowCount = 0;
Iterator iter = myObjects.iterator();
Id  GUID = null;
//ASB ... 4 019 Need to set completed to true in the createversions method
while (iter.hasNext() && completed==false){
    //moreDocsToImport=true; Assume All brought in in one batch At the moment ASB ...2 .001
    docImportStartTime = System.currentTimeMillis();
    Boolean importFailure = false;
    Boolean filedDoc = false;
    Document d = null;
    //String fileName = rs.getString("EXPORTFILE").trim();
    String documentTitle = "";
  //ASB ... 3 005 - Retrieve Folder list for checks
  //               - (Only replicate documents in config Path)
    String [] sDocFolderPath = new String [1000]; //TODO set max document Folders Filed In from config.xml
  int folderCount = -1;
    //String guid = rs.getString("GUID").trim(); //ASB 11/03/2021 - Added Trimmed string
```

```
                //ASB ... 1 Add to pass to new version of the
                createDocument Method
                IndependentObject object = (IndependentObject) iter.next();
                //ASB Software Development Limited example found to illustrate
                full property retrieval 01-08-2022
                // Id id = document.get_Id(); //get the doc ID back on this
                shallow document copy.
                // *retrieve a complete document copy back with complete
                property set using doc ID*/
                // Document docCopy = (Document)Factory.Document.
                fetchInstance(objectStore, id, null);
                // FolderSet pFolders = doc.get_FoldersFiledIn();//Get the
                full "FolderFiledIn" property on the new copy
                //
                // ASB Properties props = object.getProperties();
                //ASB Iterator iterProps = props.iterator();
                GUID = ((Document) object).get_Id();
                //ASB Software Development Limited - get full document
                property set which we need! 01-08-2022
                Document docCopy = (Document)Factory.
                Document.fetchInstance(osinput, GUID, null);
                String guid = GUID.toString().trim();
                Boolean isCurrent = ((Document) docCopy).get_
                IsCurrentVersion();
                Boolean isReserved = ((Document) docCopy).get_IsReserved();
                Properties props = docCopy.getProperties();
                Iterator iterProps = props.iterator();
                //Fetch object
                String lastGoodPropName = "";
                String propName = "";
                while (iterProps.hasNext() )
                {
                // ASB ... 4 0018 Additional try/catch block to
                //                 Trap property error here!
                 try{
```

```java
com.filenet.api.property.Property prop = (com.filenet.
api.property.Property)iterProps.next();
if(cemc.getDebugOutputFlag().equalsIgnoreCase("off")){
}else{
    logger.info("Property: " + prop.getPropertyName() );
}
propName = prop.getPropertyName(); //ASB ... 4 018
For error
if ( ((com.filenet.api.property.Property) prop).
getObjectValue() != null )
    if(cemc.getDebugOutputFlag().
    equalsIgnoreCase("off")){
    }else{
        logger.info("  Value: " + prop.getObjectValue().
        toString() );
    }
if (prop.getPropertyName().equalsIgnoreCase("Docume
ntTitle"))
{
    documentTitle = prop.getStringValue();
}
//GET FOLDERS FILED IN FOR THIS DOC
if (prop.getPropertyName().equalsIgnoreCase("Folders
FiledIn"))
{
    if ( prop.getObjectValue() != null )
    {
        FolderSet fs = (FolderSet)prop.
        getIndependentObjectSetValue();
        Iterator iterFs = fs.iterator();
        //ASB ... 3 005 - Retrieve Folder list for checks
        //           - (Only replicate documents in
                        config Path)
        while (iterFs.hasNext())
        {
```

```java
                            folderCount ++;
                        Folder folder = (Folder)iterFs.next();
                            if(cemc.getDebugOutputFlag().
                            equalsIgnoreCase("off")){
                            }else{
                                logger.info("Folder Name: " + folder.
                                get_FolderName() +
                        "      Folder Path: " + folder.get_
                            PathName());
                            }
                        sDocFolderPath[folderCount] = folder.get_
                            PathName();
                        filedDoc = true;
                    }
                }
            }
        }catch(Exception errprop){
            logger.info("Document GUID :" + guid + " : Property   " +
            propName + " Caused Error : " + errprop.getMessage() + "
            : Last Good property :" + lastGoodPropName);
        }
        lastGoodPropName = propName;
    }
    //E N D -- DEBUG OFF/ON SECTION

    //ASB ... 3 005 - Retrieve Folder list for checks
    //              - (Only replicate documents in config Path)
    int checkFolderCount = -1;
    Boolean filterDoc = false;
    while(checkFolderCount < folderCount){
        checkFolderCount ++;
        if ((sDocFolderPath[checkFolderCount] != null) &&
        !(sDocFolderPath[checkFolderCount].startsWith( cemc.
        getImportOSRootFoler())))){
            //Document selected is new but not in our import
            Folder Path!
```

CHAPTER 4 A REPLICATION JAVA PROGRAM FOR IBM FILENET OBJECT STORES

```java
            logger.info("IGNORED: New Document: " +
            documentTitle + " found linked to external
            Folder Path : " + sDocFolderPath[checkFolde
            rCount] + " Should be linked to :" + cemc.
            getImportOSRootFoler());
            filterDoc = true;
            break;
        }

    }
    try {
        //isCurrent=false; //ASB ... 1
        if(!filedDoc){
        //Document selected is new but not in our import
        Folder Path!
            logger.info("IGNORED: New Document: " +
            documentTitle + " not linked to any Folder
            Path ! ");
            this.processedCount = this.processedCount + 1;
        }

        if (!importFailure && !filterDoc && filedDoc)
        {
            //  Add Version series ID and ID to retain
            the GUIDs
            //     NB already set-up in GUID and VSID
            docImportStartTime = System.currentTimeMillis();
            //ASB Use docCopy now
            //ASB d = createDocument(documentTitle,
            GUID, object, props, iterProps, isCurrent,
            isReserved,auditFileWriter); // ASB 01-08-2022
            d = createDocument(documentTitle, GUID,
            docCopy, props, iterProps, isCurrent,
            isReserved,auditFileWriter);
        //ASB ... 4 019 d is null at this point if all the
        versions have been created
```

779

```java
                Document doc = Factory.Document.fetchInstance(os,
                GUID, null);

                Long runTimeMillis = System.currentTimeMillis() -
                importStartTime;
                if (this.processedCount == maxDocCount  &&
                maxDocCount>0)
                {
                    completed = true;
                    logger.info("Processing Completed Max Document
                    Count Reached. ");
                }
                else if ((runTimeMillis > maxRunTimeMillis)&&
                maxRunTimeMillis>0 )
                {
                    completed = true;
                    logger.info("Processing Completed Max
                    Processing Time Reached.");
                }
                consecutiveErrs=0;   // success so initialise the
                consecutive errs count
            }
        }
        catch (Exception ceme){
         //ASB ... 4 Need to check if d is null, if so then not
         created!
            try {      // getting occasional unexpected errors on the
            server, try to make sure doc is deleted
                if(d != null){ // ASB ... 4 008 Then valid to
                delete d otherwise it did not get creaetd
                    String docId = d.get_Id().toString();
                    deleteDoc(docId);
                    //ASB ... 2 013 - Set Import failure to
                    true here
                    importFailure = true;
                    consecutiveErrs ++;
```

```java
                    errorCount++;
                    logger.warn(ceme.getMessage());
                    if (consecutiveErrs > cemc.getImportMaxErrorCount())
                    {
                            auditFileWriter.write("Maximum number of consecutive errors reached - Exiting.");
                        auditFileWriter.newLine();
                         logger.info("Maximum number of consecutive errors reached - Exiting.");
                        throw ceme;

                    }
                }
                else{
                    //Assume we need to delete the Current Document GUID
                    //String docId = GUID.toString();
                    //deleteDoc(docId);
                }
            }catch (Exception e){
                //                              }
            //ASB ... 4 019 If d is null There are no errors ... moved message inside the try block above
        }
    }
}
auditFileWriter.write("Finished - time taken " + (System.currentTimeMillis() - importStartTime) + "milliseconds");
auditFileWriter.newLine();
auditFileWriter.write("Finished - documents processed - " + String.valueOf(processedCount));
auditFileWriter.newLine();
auditFileWriter.write("Finished - documents failed to be processed - " + errorCount.toString());
```

CHAPTER 4   A REPLICATION JAVA PROGRAM FOR IBM FILENET OBJECT STORES

```
        auditFileWriter.newLine();
      //ASB ... 4 031 Add retry count for Content
        auditFileWriter.write("Finished - document total 'Add Content Retry
        Count' - " + String.valueOf(retryTotalProcessedCount));
        auditFileWriter.newLine();
        auditFileWriter.close();
        }
}

//Creates a org.w3c.dom.Document object
//Loops through each version calling createVersion to create the document /
add the version
//to it.
private Document createDocument(String documentTitle, Id GUID,
IndependentObject object, Properties props, Iterator iterProps,Boolean
isCurrent, Boolean isReserved, BufferedWriter auditFileWriter) throws Excep
tion,EngineRuntimeException {
    try
    {
        Document currentDocument = (Document) object;
        String documentclass = currentDocument.getClassName();
        FolderSet nFolders = currentDocument.get_FoldersFiledIn();
      String sCreator = currentDocument.get_Creator();
      Date dateCreated = currentDocument.get_DateCreated();
      Date dateLastModified = currentDocument.get_DateLastModified();
      Date dateCheckedIn = currentDocument.get_DateCheckedIn();
      String sModifier = currentDocument.get_LastModifier();
      String sOwner = currentDocument.get_Owner();
      Boolean bCurrent = currentDocument.isCurrent();
       Long docImportStartTime;
       Long timeToImport;
       docImportStartTime = System.currentTimeMillis();
      //ASB ... 2 014 Get the Version Numbers of the Document
      Integer docMajorVersion = currentDocument.get_MajorVersionNumber();
      Integer docMinorVersion = currentDocument.get_MinorVersionNumber();
      //ASB ... 2 015 Check if this document has previous versions
```

```java
isCurrent = bCurrent; //ASB ... 1
//ASB ... 2 007 Fetch the Source Object Document to be re-created
 PropertyFilter myFilter = new PropertyFilter();
 int myFilterLevel = 2; //Changed from 4 to 2 -- TODO Set as a
 parameter
 myFilter.setMaxRecursion(myFilterLevel);
 myFilter.addIncludeType(new FilterElement(null, null, null,
 FilteredPropertyType.ANY, null));
Document dInput = Factory.Document.fetchInstance(osinput, GUID,
myFilter);
 VersionSeries versions =dInput.get_VersionSeries();
//ASB ... 2 Get the version Series ID and number of Document
Versions and set it in the copy
 Id vGUID = versions.get_Id();
//this.minMax = 0; // Set to return maximum value of the
minimum version
 Integer docVersionCount = countVersions(versions);
 //ASB ... 2 016 Need to create an array of document versions from
 oldest to
 //           newest since the iterator returns the newest
 document first!
 Document[][] docVersions = new Document[docVersionCount +1]
 [maxMin+1];
 docVersions = getDocumentVersionsArray (versions,docVersions,
 maxMin);
 Document d = null;
 if (!docVersionCount.equals(1)){
//Create versions here
     createVersions(documentTitle, GUID, object, props, iterProps,
     isCurrent,  isReserved, docVersions,docVersionCount,maxMin,au
     ditFileWriter);
 } else {
     //ASB ... 3 006 New Version of the first document create to
     store Version series ID
```

```java
//ASB ... 3 006 Many thanks to David Greenhouse for supplying
the call :-)
//ASB ... 4 008
try {
    docImportStartTime = System.currentTimeMillis();
    d = Factory.Document.createInstance(os, documentclass,
    GUID, vGUID, com.filenet.api.constants.
    ReservationType.EXCLUSIVE);
}catch(Exception e)
  {
  //If the Exception is this exists then delete and retry
  add again
    logger.error("Document already exists: " +
    documentTitle + "");
    boolean deletedDoc = false;
     //docImportStartTime = System.currentTimeMillis();
    deletedDoc = this.deleteDoc(GUID.toString());
    //ASB ... 4 008 And retry
    d = Factory.Document.createInstance(os, documentclass,
    GUID, vGUID, com.filenet.api.constants.
    ReservationType.EXCLUSIVE);
        //ASB ... 4 023
        //String docId = GUID.toString();
        //String sourceVSID = vGUID.toString();
        //Write to audit log
        //timeToImport = System.currentTimeMillis() -
        docImportStartTime;
        //auditFileWriter.write(docId + "," + sourceVSID
        + "," + docId + ",Imported," + timeToImport.
        toString());
        //auditFileWriter.newLine();
         //this.processedCount = this.processedCount +1;
}

    //Create Document with the same GUID as the Source Object
    //ASB ... 2 002 :  Phase 2 - Add properties from Iterator
```

```java
            writeDocProps(d, props, currentDocument);
             d.set_DateCreated(dateCreated); //ASB 17/02/2022
             d.set_Owner(sOwner);
             d.save(RefreshMode.REFRESH); //ASB XXX
                this.processedCount = this.processedCount +1;
              String guid = GUID.toString().trim();
                //Write to audit log
               timeToImport = System.currentTimeMillis() -
               docImportStartTime;
               auditFileWriter.write(guid + "," + vGUID.toString()
               + "," + d.get_Id().toString() + ",Imported," +
               timeToImport.toString());
               auditFileWriter.newLine();
            //ADD CONTENT
            //ASB ... 4 024      -JUST ADDED
//ASB ... 4 031 Add retry 7 times to fix API bug!!
    int retryCount = 0;
    Document currDoc = null;
    while (retryCount < retryLimit){
        try {
            currDoc = (Document)osinput.fetchObject(documentclass, GUID,
            myFilter);
            //Document currDoc = (Document)osinput.
            fetchObject("Document", GUID, myFilter);
            ContentTransfer ct = Factory.ContentTransfer.createI
            nstance();
            InputStream str = currDoc.accessContentStream(0);
            ct.setCaptureSource(str);
            // Add Document Title
            ct.set_RetrievalName(documentTitle);
            // Add Content Mime Type
            ct.set_ContentType(currDoc.get_MimeType());
            ContentElementList cel = Factory.
            ContentElement.createList();
            cel.add(ct);
```

```java
                d.set_ContentElements(cel);
                break;
            }
            catch (Exception e){
                String testMessage = e.getMessage();
                String testCode = e.toString();
                if( testMessage.contains("A uniqueness requirement has
                been violated")){
                    retryCount = retryLimit;
                    break;
                }
                //ASB ... 4 Check if the Content could not be added
                because the document already exists, if so
                //         Remove document and retry (next run)
                //  ASB ... 4 025 Log Exception here
                logger.error(e.getMessage(), e);
                retryCount ++;
                retryTotalProcessedCount ++;
                logger.info("Retry Count - createDocument method :
                " + retryCount + " of " + retryLimit + " On : " +
                documentTitle + " : with GUID : " + GUID);
            }
    } //ASB ... 4 031 Above loops Test retryCount less than the
    retryLimit of 7
            //Add security from the Source Document
            //ASB ... 2 020 Update Document ACL from source Document
            security
            AccessPermissionList  aclIn = currDoc.get_Permissions();
        //ASB ... 2 021 Add in Read Only Group

            // Add the permission to the list for the Object Store.
            //Get The group and user from the config.xml file
            //
            String sExcludeGroup = cemc.getExcludedGroup();
            String sExcludeUser =  cemc.getExcludedUser();
```

```java
    //ASB ... 4 027 For Performance check if a full LDAP search
    is required
    if(cemc.getLDAPSearchFlag().equalsIgnoreCase("on")){
        aclIn = addPermissions(aclIn, sExcludeGroup,
        sExcludeUser);
    }else {
        aclIn = addPermissionsNoSearch(aclIn, sExcludeGroup,
        sExcludeUser);
    }
    d.set_Permissions(aclIn);
    Boolean isMajorVersion = false;
    if(currDoc.get_MinorVersionNumber() == 0)
    isMajorVersion = true;
//ASB ... 4 028 Change to do not Classify from AutoClassify.
AUTO_CLASSIFY, to
    if (isMajorVersion){
        d.checkin(AutoClassify.DO_NOT_AUTO_
        CLASSIFY,CheckinType.MAJOR_VERSION);
    }
    else {
        d.checkin(AutoClassify.DO_NOT_AUTO_
        CLASSIFY,CheckinType.MINOR_VERSION);
    }
    //d.set_DateCreated(dateCreated); //ASB ... 4 029
    d.set_DateLastModified(dateLastModified);
    d.set_LastModifier(sModifier);
    d.set_DateCheckedIn(dateCheckedIn); //ASB ... 2 005
    d.save(RefreshMode.REFRESH);   //ASB XXX
FolderSet fs = currentDocument.get_FoldersFiledIn();
Iterator iterFs = fs.iterator();
while (iterFs.hasNext())
{
    Folder folder = (Folder)iterFs.next();
    if(cemc.getDebugOutputFlag().equalsIgnoreCase("off")){
    }else{
```

```java
            logger.info("\r\tFolder Name: " + folder.get_FolderName() +
                "    Folder Path: " + folder.get_PathName());
            }
                folderDoc(d,folder.get_PathName(),documentTitle);
    }
        } //ASB ... 2 014 Count of Versions of Documents
        return d;
}catch(EngineRuntimeException ere)
    {
        //ASB ... 2 005
        // Create failed.  See if it's because the Document exists.
        ExceptionCode code = ere.getExceptionCode();
        if (code.getErrorId() != ExceptionCode.DB_NOT_UNIQUE.
        getErrorId() )
        {
            logger.error("Unexpected Error : " + documentTitle + "
            Error stack: " + ere.getStackTrace());

            throw ere;
        }
        logger.error("Document already exists: " +
        documentTitle + "");
            //ASB ... 2 006 Delete document in the Target Object
            Store and Recreate
            this.deleteDoc(GUID.toString());
                //ASB ... 2 007 Fetch the Source Object Document to
                be re-created
            PropertyFilter myFilter = new PropertyFilter();
            int myFilterLevel = 2; //Changed from 4 to 2 -- TODO Set
            as a parameter
            myFilter.setMaxRecursion(myFilterLevel);
            myFilter.addIncludeType(new FilterElement(null, null,
            null, FilteredPropertyType.ANY, null));
            Document currentDocument = (Document) object;
            String documentclass = currentDocument.getClassName();
```

CHAPTER 4   A REPLICATION JAVA PROGRAM FOR IBM FILENET OBJECT STORES

```java
  Document dInput = Factory.Document.fetchInstance(osinput,
  GUID, myFilter);
//Create Version Series for the document in the Target
Object Store
  VersionSeries versions = dInput.get_VersionSeries();
  //Get the Document Versions to be added from the Source
  Document
  VersionableSet verSet = versions.get_Versions();
  Iterator versIt = verSet.iterator();
  int versCount = 0;
  int majorVn = 0;
  int minorVn = 0;
  Document doc = null; //Input Document versions
  Document d = null;   //Output document versions
  isReserved = false;
  Id createdGuid = null;
  while (versIt.hasNext()){
       versCount ++;
      doc = (Document)versIt.next();
      majorVn = doc.get_MajorVersionNumber();
      minorVn = doc.get_MinorVersionNumber();
      if (!isReserved)isReserved = doc.get_IsReserved();
      Id nextGUID = doc.get_Id();
      //Check for error here (sometimes!)
       d = createVersion(versCount, d, doc, isReserved,
        majorVn,minorVn, documentTitle, GUID, createdGuid,
        nextGUID,auditFileWriter);
      createdGuid = d.get_Id();
           if (versCount == 1){
             FolderSet fs = doc.get_FoldersFiledIn();
             Iterator iterFs = fs.iterator();
             while (iterFs.hasNext())
             {
                 Folder folder = (Folder)iterFs.next();
```

```java
                        //Create/Check and Make link in Target
                        for the Document to this Folder
                           if(cemc.getDebugOutputFlag().
                           equalsIgnoreCase("off")){
                           }else{
                               logger.info("\r\tFolder Name: "
                               + folder.get_FolderName() +
                           "    Folder Path: " + folder.get_
                           PathName());
                           }
                           folderDoc(d,folder.get_
                           PathName(),documentTitle);
                    }
                    }

                    if (isReserved){
                        break;
                    }
                }
                if (isReserved)
                     d = doReservationProperties(versCount,
                     doc, d, isReserved, majorVn,minorVn,docum
                     entTitle);
                //ASB ... 2 020 Update Document ACL from source
                Document security
                AccessPermissionList  aclIn = dInput.get_
                Permissions();
                // Add the permission to the list for the
                Object Store.
                //Get This group from the config.xml file
                String sExcludeGroup = cemc.getExcludedGroup();
                String sExcludeUser =  cemc.getExcludedUser();
                //ASB ... 4 027 For Performance check if a full
                LDAP search is required
                if(cemc.getLDAPSearchFlag().
                equalsIgnoreCase("on")){
```

```
                        aclIn = addPermissions(aclIn, sExcludeGroup,
                        sExcludeUser);
                    }else {
                        aclIn = addPermissionsNoSearch(aclIn,
                        sExcludeGroup, sExcludeUser);
                    }
                    //ASB ... 4 027 aclIn = addPermissions(aclIn,
                    sExcludeGroup, sExcludeUser);
                    d.set_Permissions(aclIn);
            return d;
        }
}
//This method returns the count of the number of Document versions
//(Unfortunately java Iterator class does not return size!)
private int countVersions(VersionSeries versions){
    int versionCount=0;
    VersionableSet verSet = versions.get_Versions();
    Iterator versIt = verSet.iterator();
    int versCount = 0;
    maxMin = 0;
    int min = 0;
    Document doc = null; //Input Document versions
    while (versIt.hasNext()){
        versCount ++;
        doc = (Document)versIt.next();
        //Record largest minor version value for loop
        min = doc.get_MinorVersionNumber();
        if (min > maxMin){
            this.maxMin = min;
        }
    }
    versionCount = versCount;
    return versionCount;
}
//Main Process loop for the Delete Documents process.
```

CHAPTER 4   A REPLICATION JAVA PROGRAM FOR IBM FILENET OBJECT STORES

```
//Outer loop executes the retrieval of Documents from the source Object
Store Events until no matches are returned.
//Inner loop processes each Document Deletion Event returned from the
retrieval.
//ASB ... 3 003 deleteDocuments Method
private void deleteDocuments() throws  Exception {
    Id GUID = null;
    BufferedWriter auditFileWriter;
    Long deleteStartTime;
    Long docDeleteStartTime;
    Long timeToDelete;
       long maxRunTimeMillis;
    boolean moreDocsToDelete=true;
    Integer consecutiveErrs=0;
    Integer errorCount=0;

    boolean completed=false;

    processedCount = 0;
    deleteStartTime = System.currentTimeMillis();
    maxRunTimeMillis = cemc.getImportMaxRunTimeMinutes()*60*1000;
    Long maxDocCount = new Long(cemc.getDeleteMaxDocCount());

    auditFileWriter = new BufferedWriter(new FileWriter(cemc.
    getDeleteDocumentAuditFile()));
    auditFileWriter.write("Delete Start - ");
    auditFileWriter.newLine();
    auditFileWriter.write("GUID, INITIATING USER, EVENT TYPE, AUDIT
    SEQUENCE");
    auditFileWriter.newLine();
    while (moreDocsToDelete && completed==false)
    {
    /*
    Query to use is as follows:
    SELECT [EVENT].[This], [EVENT].[CmAuditSequence],
    [EVENT].[ClassDescription], [EVENT].[Creator], [EVENT].
    [DateCreated],   [EVENT].[SourceObjectId],
```

```
[EVENT].[DateLastModified], [EVENT].[EventStatus], [EVENT].[Id], [EVENT].
[InitiatingUser], [EVENT].[LastModifier],
[EVENT].[Name], [EVENT].[Owner]
FROM [EVENT] INNER JOIN [CLASSDEFINITION]
ON [EVENT].SourceClassId = [CLASSDEFINITION].[Id]
WHERE ([EVENT].[DateCreated] >= 20210920T210827Z) AND ([ClassDefinition].
[SymbolicName] <>'Folder')
ORDER BY [EVENT].[DateCreated]
    */
         moreDocsToDelete=false;
        SearchSQL sqlObject = new SearchSQL();
        sqlObject.setSelectList("e.CmAuditSequence,e.ClassDescription,
        e.Creator,e.DateCreated,e.SourceObjectId,e.DateLastModified,
        e.EventStatus, e.Id, e.InitiatingUser, e.LastModifier, e.Name,
        e.Owner, e.SourceClassId");
        //ASB ... 2  022
        Integer maxRecords = cemc.getDeleteMaxDocCount();
        sqlObject.setMaxRecords(maxRecords); //was hard coded to 20 !
        sqlObject.setFromClauseInitialValue("Event", "e", true);
        sqlObject.setFromClauseAdditionalJoin(JoinOperator.INNER,
        "CLASSDEFINITION", "cd", "e.SourceClassId", JoinComparison.EQUAL,
        "cd.Id", true);
        //ASB ... 022 - Retrieve Date start from config.xml
        String sSearchDate = cemc.getStartSearchDate();
       //ASB ... 022 - enter date to search
       //ASB ... 4 013 Changed to cd.SymbolicName <> 'Folder'
       // sqlObject.setWhereClause("(e.DateCreated > " +  sSearchDate.
       trim() + ") and (cd.SymbolicName = 'Document')" );
       //ASB ... 4 017 Changed to filter for just Success Audit Events
       sqlObject.setWhereClause("(e.DateCreated >= " +  sSearchDate.
       trim() + ")" + folderExcludeSymbolList + " AND (e.EventStatus
       = 0)" );
       sqlObject.setOrderByClause("e.CmAuditSequence,e.DateCreated");
       // Check the SQL statement.
       logger.info("SQL: " + sqlObject.toString());
```

```java
// Create a SearchScope instance. (Assumes you have the object store
// object.)
SearchScope search = new SearchScope(osinput);

// Set the page size (Long) to use for a page of query result data. This value is passed
// in the pageSize parameter. If null, this defaults to the value of
// ServerCacheConfiguration.QueryPageDefaultSize.
Integer myPageSize = new Integer(1000);

// Specify a property filter to use for the filter parameter, if needed.
// This can be null if you are not filtering properties.
PropertyFilter myFilter = new PropertyFilter();
int myFilterLevel = 1;
myFilter.setMaxRecursion(myFilterLevel);
//ASB ... 4 020
if(cemc.getDebugOutputFlag().equalsIgnoreCase("off")){
    myFilter.addIncludeType(new FilterElement(null, null, null,
        FilteredPropertyType.ANY_LIST, null));
    myFilter.addIncludeType(new FilterElement(null, null, null,
        FilteredPropertyType.ANY_SINGLETON, null));
}else{
    myFilter.addIncludeType(new FilterElement(null, null, null,
        FilteredPropertyType.ANY, null));
}
// Set the (Boolean) value for the continuable parameter. This indicates
// whether to iterate requests for subsequent pages of result data when the end of the
// first page of results is reached. If null or false, only a single page of results is
// returned.
Boolean continuable = new Boolean(true);
```

```java
// Execute the fetchObjects method using the specified parameters.
IndependentObjectSet myObjects = search.fetchObjects(sqlObject,
myPageSize, myFilter, continuable);

// You can then iterate through the collection of rows to access
the properties.
int rowCount = 0;
Iterator iter = myObjects.iterator();

while (iter.hasNext() && completed==false){
    //moreDocsToImport=true; Assume All brought in in one batch
    At the moment ASB ...2 .001
    docDeleteStartTime = System.currentTimeMillis();
    Boolean deleteFailure = false;
    //Document d = null;
    String documentTitle = "";
    //String guid = rs.getString("GUID").trim(); //ASB
    11/03/2022 - Added Trimmed string
    //ASB ... 1 Add to pass to new version of the
    createDocument Method
    IndependentObject object = (IndependentObject) iter.next();
    //ClassDescription object.get_ClassDescription();
    Properties props = object.getProperties(); //TODO ASB ... 3
    String eventDesc = "unknown";
    Iterator prop = props.iterator();
    String initiatingUser = "";
    String eventSequence = "";
    while (prop.hasNext()){
        Property eventProp = (Property) prop.next();
        //ASB ... 3 004 Get Event Type
        if (eventProp.getPropertyName().equalsIgnoreCase("ClassDe
        scription")){
            eventDesc = eventProp.getEngineObjectValue().
            getProperties().get("SymbolicName").getStringValue();
        }
```

```java
                //ASB ... 3 003 Check for the Document ID property
                to delete
                if (eventProp.getPropertyName().equalsIgnoreCase("SourceO
                bjectId")){
                    GUID = eventProp.getIdValue();
                }
                if (eventProp.getPropertyName().equalsIgnoreCase("Initiat
                ingUser")){
                    initiatingUser = eventProp.getStringValue();
                }
                if (eventProp.getPropertyName().equalsIgnoreCase("CmAudit
                Sequence")){
                    eventSequence = eventProp.getFloat64Value().
                    toString();
                }

        } //prop.hasNet()
        String guid = GUID.toString().trim();
        boolean docDeleted = false;
        try {
            if (eventDesc.equalsIgnoreCase("CheckinEvent") ){
                            //ASB ... 3 003 Delete the Target
                            document with the Delete event/Checkin
                            Event in the Source
                docDeleted = deleteDoc(GUID.toString()); //ASB ... 4
                019 - Test if reservation Object
                    if (docDeleted){
                        auditFileWriter.write(GUID.toString() + ", "
                        + initiatingUser+ ", " + eventDesc + ", " +
                        eventSequence);
                        auditFileWriter.newLine();
                        processedCount++;
                    }
            }
            //ASB ... 4 023 Split to deal with DeletionEvent separately
                if (eventDesc.equalsIgnoreCase("DeletionEvent")  ){
```

```
                //ASB ... 3 003 Delete the Target document with the
                Delete event/Checkin Event in the Source
                    docDeleted = deleteDocTargetOnly(GUID.
                    toString()); //ASB ... 4 019 - Test if
                    reservation Object
                    if (docDeleted){
                        auditFileWriter.write(GUID.toString() +
                        ", " + initiatingUser+ ", " + eventDesc +
                        ", " + eventSequence);
                        auditFileWriter.newLine();
                        processedCount++;
                    }
            }
        }
    //catch (CEMigrateException ceme){
    catch (Exception ceme){
        try {      // getting occasional unexpected errors on the
        server, try to make sure doc is deleted
            String docId = GUID.toString();
            //deleteDoc(docId);
          //ASB ... 2 013 - Set Delete failure to true here
            deleteFailure = true;
        }catch (Exception e){
            //
        }
        consecutiveErrs ++;
        errorCount++;
        logger.warn(ceme.getMessage());
        if (consecutiveErrs > cemc.getDeleteMaxErrorCount())
        {
                auditFileWriter.write("Maximum number of
                consecutive delete errors reached - Exiting.");
            auditFileWriter.newLine();
            logger.info("Maximum number of consecutive errors
            reached - Exiting.");
            throw ceme;
```

CHAPTER 4 A REPLICATION JAVA PROGRAM FOR IBM FILENET OBJECT STORES

```
                }
            }
        }
    auditFileWriter.write("Finished - time taken " +
    (System.currentTimeMillis() - deleteStartTime) + "milliseconds");
    auditFileWriter.newLine();
    auditFileWriter.write("Finished - documents deleted - " + String.value
    Of(processedCount));
    auditFileWriter.newLine();
    auditFileWriter.write("Finished - documents updated - " + String.value
    Of(updateProcessedDocCount));
    auditFileWriter.newLine();
    auditFileWriter.write("Finished - documents failed to be deleted - " +
    errorCount.toString());
    auditFileWriter.newLine();
    auditFileWriter.close();
  }
}
//Main Process loop for the Update Documents process.
//Outer loop executes the retrieval of Documents from the source Object
Store Events until no matches are returned.
//Inner loop processes each Document Update Event returned from the
retrieval.
//ASB ... 4 012 updateDocuments Method
private void updateDocuments() throws  Exception {
    Id GUID = null;
    BufferedWriter auditFileWriter;
    Long updateStartTime;
    Long docUpdateStartTime;
    Long timeToUpdate;
        long maxRunTimeMillis;
    boolean moreDocsToUpdate=true;
    Integer consecutiveErrs=0;
    Integer errorCount=0;
```

```java
    boolean completed=false;

    processedCount = 0;
    updateStartTime = System.currentTimeMillis();
    maxRunTimeMillis = cemc.getImportMaxRunTimeMinutes()*60*1000;
    Long maxDocCount = new Long(cemc.getUpdateMaxDocCount());

    auditFileWriter = new BufferedWriter(new FileWriter(cemc.
    getUpdateDocumentAuditFile()));
    auditFileWriter.write("Update Start - ");
    auditFileWriter.newLine();
    auditFileWriter.write("GUID, INITIATING USER, EVENT TYPE, AUDIT
    SEQUENCE");
    auditFileWriter.newLine();
    while (moreDocsToUpdate && completed==false)
    {
    /*
    Query to use is as follows:
    SELECT [EVENT].[This], [EVENT].[CmAuditSequence],
    [EVENT].[ClassDescription], [EVENT].[Creator], [EVENT].
    [DateCreated],  [EVENT].[SourceObjectId],
[EVENT].[DateLastModified], [EVENT].[EventStatus], [EVENT].[Id], [EVENT].
[InitiatingUser], [EVENT].[LastModifier],
[EVENT].[Name], [EVENT].[Owner]
FROM [EVENT] INNER JOIN [CLASSDEFINITION]
ON [EVENT].SourceClassId = [CLASSDEFINITION].[Id]
WHERE ([EVENT].[DateCreated] >= 20210920T210827Z) AND [ClassDefinition].
[SymbolicName]='Document'
ORDER BY [EVENT].[DateCreated
    */
         moreDocsToUpdate=false;
         SearchSQL sqlObject = new SearchSQL();
         sqlObject.setSelectList("e.CmAuditSequence,e.ClassDescription,e.
         Creator,e.DateCreated,e.SourceObjectId,e.DateLastModified,
         e.EventStatus, e.Id, e.InitiatingUser, e.LastModifier, e.Name,
         e.Owner, e.SourceClassId");
```

```
//ASB ... 2  022
Integer maxRecords = cemc.getUpdateMaxDocCount();
sqlObject.setMaxRecords(maxRecords); //was hard coded to 20 !
sqlObject.setFromClauseInitialValue("Event", "e", true);
sqlObject.setFromClauseAdditionalJoin(JoinOperator.INNER,
"CLASSDEFINITION", "cd", "e.SourceClassId", JoinComparison.EQUAL,
"cd.Id", true);
//ASB ... 022 - Retrieve Date start from config.xml
String sSearchDate = cemc.getStartSearchDate();
//ASB ... 022 - enter date to search
//sqlObject.setWhereClause("(e.DateCreated > " +  sSearchDate.
trim() + ") and (cd.SymbolicName = 'Document')" );
//ASB ... 4 017 Search limit to Success Audit Events
//ASB ... 4 022 Changed " and (cd.SymbolicName <> 'Folder')" to
allow multiple sub-classes of Folder
sqlObject.setWhereClause("(e.DateCreated > " +  sSearchDate.trim()
+ ")" + folderExcludeSymbolList + " AND (e.EventStatus = 0)" );
sqlObject.setOrderByClause("e.CmAuditSequence,e.DateCreated");
// Check the SQL statement.
logger.info("SQL: " + sqlObject.toString());

// Create a SearchScope instance. (Assumes you have the
object store
// object.)
SearchScope search = new SearchScope(osinput);

// Set the page size (Long) to use for a page of query result
data. This value is passed
// in the pageSize parameter. If null, this defaults to the
value of
// ServerCacheConfiguration.QueryPageDefaultSize.
Integer myPageSize = new Integer(1000);

// Specify a property filter to use for the filter parameter,
if needed.
// This can be null if you are not filtering properties.
PropertyFilter myFilter = new PropertyFilter();
```

```java
int myFilterLevel = 1;
myFilter.setMaxRecursion(myFilterLevel);
//ASB ... 4 O2O
if(cemc.getDebugOutputFlag().equalsIgnoreCase("off")){
    myFilter.addIncludeType(new FilterElement(null, null, null,
    FilteredPropertyType.ANY_SINGLETON, null));
}else{
    myFilter.addIncludeType(new FilterElement(null, null, null,
    FilteredPropertyType.ANY, null));
}
// Set the (Boolean) value for the continuable parameter. This indicates
// whether to iterate requests for subsequent pages of result data when the end of the
// first page of results is reached. If null or false, only a single page of results is
// returned.
Boolean continuable = new Boolean(true);

// Execute the fetchObjects method using the specified parameters.
IndependentObjectSet myObjects = search.fetchObjects(sqlObject, myPageSize, myFilter, continuable);

// You can then iterate through the collection of rows to access the properties.
int rowCount = 0;
Iterator iter = myObjects.iterator();

while (iter.hasNext() && completed==false){
    //moreDocsToImport=true; Assume All brought in in one batch
    At the moment ASB ...2 .001
    docUpdateStartTime = System.currentTimeMillis();
    Boolean updateFailure = false;
    //Document d = null;
    String documentTitle = "";
    //String guid = rs.getString("GUID").trim(); //ASB
    11/03/2022 - Added Trimmed string
```

```java
        //ASB ... 1 Add to pass to new version of the
        createDocument Method
        IndependentObject object = (IndependentObject) iter.next();
        //ClassDescription object.get_ClassDescription();
        Properties props = object.getProperties(); //TODO ASB ... 3
        String eventDesc = "unknown";
        Iterator prop = props.iterator();
        String initiatingUser = "";
        String eventSequence = "";
        while (prop.hasNext()){
            Property eventProp = (Property) prop.next();
            //ASB ... 3 004 Get Event Type
            if (eventProp.getPropertyName().equalsIgnoreCase("ClassDe
            scription")){
                eventDesc = eventProp.getEngineObjectValue().
                getProperties().get("SymbolicName").getStringValue();
              }
            //ASB ... 3 003 Check for the Document ID property
            to update
            if (eventProp.getPropertyName().equalsIgnoreCase("SourceO
            bjectId")){
                GUID = eventProp.getIdValue();
            }
            if (eventProp.getPropertyName().equalsIgnoreCase("Initiat
            ingUser")){
                initiatingUser = eventProp.getStringValue();
            }
            if (eventProp.getPropertyName().equalsIgnoreCase("CmAudit
            Sequence")){
                eventSequence = eventProp.getFloat64Value().
                toString();
            }
        } //prop.hasNet()
        String guid = GUID.toString().trim();
         try {
```

```java
    //ASB ... 4 011 Check The Event Type is a UpdateEvent
    for the Document
    //ASB ... 4 021 Update Security on Document Objects
    if (eventDesc.equalsIgnoreCase("UpdateEvent")||eventDe
    sc.equalsIgnoreCase("UpdateSecurityEvent")){
        //Update of the Folder Detected - copy the new
        Source Folder properties
        updateDocument(GUID);
        auditFileWriter.write(GUID.toString() + ", "
        + initiatingUser+ ", " + eventDesc + ", " +
        eventSequence);
        auditFileWriter.newLine();
         updateProcessedDocCount++;
    }
}
//catch (CEMigrateException ceme){
catch (Exception ceme){
    try {     // getting occasional unexpected errors on the
    server, try to make sure doc is updated
        String docId = GUID.toString();
      //ASB ... 2 013 - Set update failure to true here
        updateFailure = true;
    }catch (Exception e){
        //swallow, don't care
    }
    consecutiveErrs ++;
    errorCount++;
    logger.warn(ceme.getMessage());
    if (consecutiveErrs > cemc.getUpdateMaxErrorCount())
    {
            auditFileWriter.write("Maximum number of
            consecutive update errors reached - Exiting.");
        auditFileWriter.newLine();
        logger.info("Maximum number of consecutive errors
        reached - Exiting.");
        throw ceme;
```

```java
            }
          }
        }
        auditFileWriter.write("Finished - time taken " +
        (System.currentTimeMillis() - updateStartTime) + "milliseconds");
        auditFileWriter.newLine();
        //auditFileWriter.write("Finished - documents updated - " + String.
        valueOf(processedCount));
        //auditFileWriter.newLine();
        auditFileWriter.write("Finished - documents updated - " + String.value
        Of(updateProcessedDocCount));
        auditFileWriter.newLine();
        auditFileWriter.write("Finished - documents failed to be updated - " +
        errorCount.toString());
        auditFileWriter.newLine();
        auditFileWriter.close();
    }
}

//Main Process loop for the Delete Folders process.
//Outer loop executes the retrieval of Folders from the source Object Store
Events until no matches are returned.
//Inner loop processes each Folder Deletion Event returned from the
retrieval.
// It also detects and processes Update Folder and Unfile Document Events
//ASB ... 3 003 deleteFolders Method
private void deleteFolders() throws Exception {
    Id GUID = null;
    BufferedWriter auditFileWriter;
    Long deleteStartTime;
    Long folderDeleteStartTime;
    Long timeToDelete;
        long maxRunTimeMillis;
    boolean moreFoldersToDelete=true;
    Integer consecutiveErrs=0;
```

```java
Integer errorCount=0;

boolean completed=false;

processedCount = 0;
updateSecurityProcessedCount = 0;
deleteStartTime = System.currentTimeMillis();
maxRunTimeMillis = cemc.getImportMaxRunTimeMinutes()*60*1000;
Long maxFolderCount = new Long(cemc.getDeleteMaxFolderCount());

auditFileWriter = new BufferedWriter(new FileWriter(cemc.
getDeleteFolderAuditFile()));
auditFileWriter.write("Delete Start - ");
auditFileWriter.newLine();
auditFileWriter.write("GUID, SOURCEVSID, DESTDOCID, STATE,
Millisecs");
auditFileWriter.newLine();
//ASB...1 Should be already connected here
while (moreFoldersToDelete && completed==false)
{
/*
Query to use is as follows:

*/
    moreFoldersToDelete=false;
    SearchSQL sqlObject = new SearchSQL();
    sqlObject.setSelectList("e.CmAuditSequence,e.ClassDescription,
    e.Creator,e.DateCreated,e.SourceObjectId,e.SourceObject,
    e.DateLastModified, e.EventStatus, e.Id, e.InitiatingUser,
    e.LastModifier, e.Name, e.Owner, e.SourceClassId");
    //ASB ... 2  022
    Integer maxRecords = cemc.getDeleteMaxFolderCount();
    sqlObject.setMaxRecords(maxRecords); //was hard coded to 20 !
    sqlObject.setFromClauseInitialValue("Event", "e", true);
```

```
sqlObject.setFromClauseAdditionalJoin(JoinOperator.INNER,
"CLASSDEFINITION", "cd", "e.SourceClassId", JoinComparison.EQUAL,
"cd.Id", true);
//ASB ... 022 - Retrieve Date start from config.xml
String sSearchDate = cemc.getStartSearchDate();
//ASB ... 022 - enter date to search
//ASB ... 4 017 Search for just the Success Audit Events
sqlObject.setWhereClause("(e.DateCreated > " + sSearchDate.trim()
+ ") AND (" + folderSymbolList + ") AND (e.EventStatus = 0)"  );
sqlObject.setOrderByClause("e.CmAuditSequence,e.DateCreated");
// Check the SQL statement.
logger.info("SQL: " + sqlObject.toString());

// Create a SearchScope instance. (Assumes you have the object store
// object.)
SearchScope search = new SearchScope(osinput);

// Set the page size (Long) to use for a page of query result data. This value is passed
// in the pageSize parameter. If null, this defaults to the value of
// ServerCacheConfiguration.QueryPageDefaultSize.
Integer myPageSize = new Integer(1000);

// Specify a property filter to use for the filter parameter, if needed.
// This can be null if you are not filtering properties.
PropertyFilter myFilter = new PropertyFilter();
int myFilterLevel = 1;
myFilter.setMaxRecursion(myFilterLevel);
//ASB ... 4 020
myFilter.addIncludeType(new FilterElement(null, null, null,
FilteredPropertyType.ANY, null));

// Set the (Boolean) value for the continuable parameter. This indicates
```

```java
// whether to iterate requests for subsequent pages of result data when the end of the
// first page of results is reached. If null or false, only a single page of results is
// returned.
Boolean continuable = new Boolean(true);

// Execute the fetchObjects method using the specified parameters.
IndependentObjectSet myObjects = search.fetchObjects(sqlObject, myPageSize, myFilter, continuable);

// You can then iterate through the collection of rows to access the properties.
int rowCount = 0;
Iterator iter = myObjects.iterator();
EngineObject unFileDoc = null;
Long lastFileEvent = null;
int lastEventSequence = 0;
while (iter.hasNext() && completed==false){
    folderDeleteStartTime = System.currentTimeMillis();
    Boolean deleteFailure = false;
    IndependentObject object = (IndependentObject) iter.next();
    Properties props = object.getProperties();
    String eventDesc = "unknown";  //ASB ... 3 004 Set to Event Type default
    Iterator prop = props.iterator();
    int eventSequence = 0;
    while (prop.hasNext()){
        Property eventProp = (Property) prop.next();
        //ASB ... 3 004 Get Event Type
        if (eventProp.getPropertyName().equalsIgnoreCase("ClassDescription")){
            eventDesc = eventProp.getEngineObjectValue().
            getProperties().get("SymbolicName").getStringValue();
        }
```

```java
            //ASB ... 3 003 Check for the Folder ID property 
            to delete
            if (eventProp.getPropertyName().equalsIgnoreCase("SourceO
            bjectId")){
                GUID = eventProp.getIdValue();
            }
            //ASB ... 3 004 Get The Source Object (for unfile)
            if (eventProp.getPropertyName().equalsIgnoreCase("Source 
            Object")){
                unFileDoc = eventProp.getEngineObjectValue();
            }
            if (eventProp.getPropertyName().equalsIgnoreCase("CmAudit
            Sequence")){
                eventSequence = eventProp.getFloat64Value().
                intValue();
            }
        } //prop.hasNet()
    String guid = GUID.toString().trim();
     try {
            //ASB ... 3 004 Check The Event Type is a DeletionEvent
            if (eventDesc.equalsIgnoreCase("DeletionEvent")){
            //ASB ... 3 003 Delete the Target Folder with the Delete 
            event in the Source
               deleteFolder(GUID.toString());
            processedCount++;
            }
            //ASB ... 3 004 Check The Event Type is a UpdateEvent
            if (eventDesc.equalsIgnoreCase("UpdateEvent")){
                //Update of the Folder Detected - copy the new 
                Source Folder properties
                updateFolder(GUID);
                updateProcessedCount++;
            }
            //ASB ... 4 021 Check Folder Security Update Event
            if (eventDesc.equalsIgnoreCase("UpdateSecurityEvent")){
```

```java
        //Update of the Folder Detected - copy the new
        Source Folder properties
        updateFolder(GUID);
        updateSecurityProcessedCount++;
}
//ASB ... 3 004 Check The Event Type is a UnFile
if (eventDesc.equalsIgnoreCase("UnfileEvent")){
    myFilter.addIncludeType(new FilterElement(null,
    null, null, FilteredPropertyType.ANY, null)); //ASB
    ... 4 020 added here
    Folder f = Factory.Folder.fetchInstance(os, GUID,
    myFilter);
    //Fetch Document Object DRCR Object
    // Dynamic Referential Containment Relationship
    //ASB ... 4 032 Detect  a File/Unfile event and
    ignore unfile here!!
    if (unFileDoc != null) {
        EngineObject sourceObject =  unFileDoc;
        String objClass = sourceObject.getClassName();
        if(objClass.equalsIgnoreCase("DynamicReferential
        ContainmentRelationship")){
            com.filenet.api.core.
            DynamicReferentialContainmentRelationship
            DRCR = (com.filenet.api.core.
            DynamicReferentialContainmentRelationship)
            sourceObject;
            Document doc = (Document) DRCR.get_Head();
            Id removeGUID = doc.get_Id();//Get
            Source Doc
        Document drem = Factory.
        Document.fetchInstance(os, removeGUID,
        myFilter);
            com.filenet.api.core.
            ReferentialContainmentRelationship RCR_
            returned = f.unfile(drem);
```

```java
                        RCR_returned.save(RefreshMode.REFRESH);
                        unfileProcessedCount++;
                    }
                }
            }
            //ASB ... 4 030 Check The Event Type is a File Event
            if (eventDesc.equalsIgnoreCase("FileEvent")){
                lastEventSequence = eventSequence;
               myFilter.addIncludeType(new FilterElement(null,
               null, null, FilteredPropertyType.ANY, null)); //ASB
               ... 4 020 added here
               Folder f = Factory.Folder.fetchInstance(os, GUID,
               myFilter);
               //Fetch Document Object DRCR Object
               // Dynamic Referential Containment Relationship
               if (unFileDoc != null){
                   EngineObject sourceObject =  unFileDoc;
                   String objClass = sourceObject.getClassName();
                   if(objClass.equalsIgnoreCase("DynamicReferential
                   ContainmentRelationship")){
                       com.filenet.api.core.
                       DynamicReferentialContainmentRelationship
                       DRCRsource = (com.filenet.api.core.
                       DynamicReferentialContainmentRelationship)
                       sourceObject;
                       Document DOCsource = (Document) DRCRsource.
                       get_Head();
                  Id DRCR_id = DRCRsource.get_Id();
                       Id addGUID = DOCsource.get_Id();//Get
                       Source Doc;
                       //Check if document is not already linked
                       Document drem = Factory.
                       Document.fetchInstance(os, addGUID,
                       myFilter);
                       //Unfile Document First!!
```

```java
            try{
                com.filenet.api.core.
                ReferentialContainmentRelationship RCR_
                returned = f.unfile(drem);
                RCR_returned.save(RefreshMode.REFRESH);
            }catch(Exception eUnfile){

            }
                com.filenet.api.core.DynamicReferential
                ContainmentRelationship DRCR =
                Factory.DynamicReferentialContain
                mentRelationship.createInstance(
                os, ClassNames.DYNAMIC_REFERENTIAL_
                CONTAINMENT_RELATIONSHIP, DRCR_id);
                DRCR.set_Head(drem);
                DRCR.set_Tail(f);
                DRCR.save(RefreshMode.NO_REFRESH);
                fileProcessedCount++;
             }
           }
         }
      }
//catch (CEMigrateException ceme){
catch (Exception ceme){
    try {       // getting occasional unexpected errors on the
    server, try to make sure doc is deleted
        if (eventDesc.equalsIgnoreCase("DeletionEvent")){
        //Protect from other exceptions which might Occur!!
            String folderId = GUID.toString();
            deleteFolder(folderId);
            //ASB ... 2 013 - Set Import failure to
            true here
            deleteFailure = true;
        }
    }catch (Exception e){
        //Nothing to throw here
```

```
                }
                consecutiveErrs ++;
                errorCount++;
                logger.warn(ceme.getMessage());
                if (consecutiveErrs > cemc.getDeleteMaxErrorCount())
                {
                        auditFileWriter.write("Maximum number of
                        consecutive delete errors reached - Exiting.");
                    auditFileWriter.newLine();
                     logger.info("Maximum number of consecutive errors
                     reached - Exiting.");
                    throw ceme;
                }
            
            }
        }
//} Moved While
auditFileWriter.write("Finished - time taken " +
(System.currentTimeMillis() - deleteStartTime) + "milliseconds");
auditFileWriter.newLine();
auditFileWriter.write("Finished - folders deleted - " + String.valueOf
(processedCount));
auditFileWriter.newLine();
auditFileWriter.write("Finished - folders updated - " + String.valueOf
(updateProcessedCount));
auditFileWriter.newLine();
auditFileWriter.write("Finished - folder security updated - " + String
.valueOf(updateSecurityProcessedCount));
auditFileWriter.newLine();
auditFileWriter.write("Finished - Documents unfiled - " + String.value
Of(unfileProcessedCount));
auditFileWriter.newLine();
auditFileWriter.write("Finished - Documents filed - " + String.valueOf
(fileProcessedCount)); //ASB ... 4 030
auditFileWriter.newLine();
```

```java
        auditFileWriter.write("Finished - folders failed to be deleted - " +
        errorCount.toString());
        auditFileWriter.newLine();
        auditFileWriter.close();
    }
}
//ASB ... 3 004 Update the Target Folder ACL and properties from the
updated source Folder
private void updateFolder(Id GUID) throws Exception {
    Folder parent   = os.get_RootFolder(); //SET AS DEFAULT
    int myFilterLevel = 1; //Changed from 4 to 1 for performance!
    PropertyFilter myFilter = new PropertyFilter();
    myFilter.setMaxRecursion(myFilterLevel);
    //ASB ... 4 020
    if(cemc.getDebugOutputFlag().equalsIgnoreCase("off")){
        myFilter.addIncludeType(new FilterElement(null, null, null,
        FilteredPropertyType.ANY_LIST, null));
        myFilter.addIncludeType(new FilterElement(null, null, null,
        FilteredPropertyType.ANY_SINGLETON, null));
    }else{
        myFilter.addIncludeType(new FilterElement(null, null, null,
        FilteredPropertyType.ANY, null));
    }
    //myFilter.addIncludeType(new FilterElement(null, null, null,
     FilteredPropertyType.ANY, null));
    Folder f = Factory.Folder.fetchInstance(os, GUID, myFilter);

            try {
                    //Get the Current Source Folder Object referenced in
                    the Event table

        Folder currentFolder = Factory.Folder.fetchInstance(osinput, GUID,
        myFilter);
        //Fetch The Target Folder requiring modification
```

```java
        Date dateLastModified = currentFolder.get_DateLastModified();
        String sModifier = currentFolder.get_LastModifier();
        String sOwner = currentFolder.get_Owner();
      String path = currentFolder.get_PathName();
   //Get Target Object Store Folder's Parent Folder
      String pathParts[] = path.split("/");
      if (pathParts.length > 2){
         String parentname = path.substring(0, path.
         lastIndexOf("/"));
         parent = (Folder)os.getObject("Folder", parentname);
      }

        String folderclass = currentFolder.getClassName();
       String name = currentFolder.get_Name();

   //Update the properties and security from the Source Object Folder
   properties and ACL
     f.set_Parent(parent);
     f.set_FolderName(name);
     f.set_DateLastModified(dateLastModified);
     f.set_LastModifier(sModifier);
     f.set_Owner(sOwner);

        AccessPermissionList  aclIn = currentFolder.get_Permissions();
         // Add the permission to the list for the Object Store.
         //Get This group from the config.xml file
         String sExcludeGroup = cemc.getExcludedGroup();
         String sExcludeUser =  cemc.getExcludedUser();
         //ASB ... 4 027 For Performance check if a full LDAP search
         is required
         if(cemc.getLDAPSearchFlag().equalsIgnoreCase("on")){
             aclIn = addPermissions(aclIn, sExcludeGroup,
             sExcludeUser);
         }else {
             aclIn = addPermissionsNoSearch(aclIn, sExcludeGroup,
             sExcludeUser);
         }
```

```java
            //aclIn = addPermissions(aclIn, sExcludeGroup, sExcludeUser);
            f.set_Permissions(aclIn);

            f.save(RefreshMode.REFRESH);
        }
        catch (Exception e){
            logger.error(e.getMessage(), e);
        }
    }
    //ASB ... 4 011 Update the Target Document ACL and properties from the updated source Document
    private void updateDocument(Id GUID) {
        //ASB ... 2 007 Fetch the Source Object Document to be Updated
        try {
            PropertyFilter myFilter = new PropertyFilter();
            //ASB ... 4 Changed from 2 to 1 was 12 seconds for 3 updates
            //                       now 8.5 Seconds for  3 updtes
            int myFilterLevel = 1; //Changed from 4 to 2 -- TODO Set as a parameter
            myFilter.setMaxRecursion(myFilterLevel);
            myFilter.addIncludeType(new FilterElement(null, null, null,
            FilteredPropertyType.ANY, null));
            Document d = Factory.Document.fetchInstance(os, GUID, myFilter);

            //Get the Current Source Document Object referenced in the Event table
            Document currentDocument = Factory.Document.fetchInstance(osinput,
            GUID, myFilter);
            //ASB ... 4 020 Check if this is a reserved Document Object, if so Don't process!!
            if(currentDocument.get_IsReserved()){
                return;
            }
            //Fetch The Target document requiring modification
            Properties props = currentDocument.getProperties(); //TODO
            ASB ... 3
```

```java
            updateDocProps(d, props, currentDocument);

    //Update the properties and security from the Source Object
    Document properties and ACL

            AccessPermissionList  aclIn = currentDocument.get_
            Permissions();
             // Add the permission to the list for the Object Store.
             //Get This group from the config.xml file
             String sExcludeGroup = cemc.getExcludedGroup();
             String sExcludeUser =  cemc.getExcludedUser();
             //ASB ... 4 027 For Performance check if a full LDAP search
             is required
             if(cemc.getLDAPSearchFlag().equalsIgnoreCase("on")){
                 aclIn = addPermissions(aclIn, sExcludeGroup,
                 sExcludeUser);
             }else {
                 aclIn = addPermissionsNoSearch(aclIn, sExcludeGroup,
                 sExcludeUser);
             }
             //aclIn = addPermissions(aclIn, sExcludeGroup, sExcludeUser);
             d.set_Permissions(aclIn);

         d.save(RefreshMode.REFRESH);
    }
    catch (Exception e){
        logger.error(e.getMessage(), e);
    }
}
//This method returns the  Document versions in ascending order
//(Unfortunately FileNet API iterator returns latest version first! )
private Document[][] getDocumentVersionsArray(VersionSeries versions,
Document [][] docVersions,int maxMin){
    int versionCount=0;
    VersionableSet verSet = versions.get_Versions();
    Iterator versIt = verSet.iterator();
    int versCount = 0;
```

```java
        Document doc = null; //Input Document versions
        maxMin = 0;
        int min=0;
        int maj=0;
        while (versIt.hasNext()){
            versCount ++;
            doc = (Document)versIt.next();
            maj = doc.get_MajorVersionNumber();
            min = doc.get_MinorVersionNumber();
            //Record largest minor version value for loop
            if (min > maxMin){
                maxMin = min;
            }
          docVersions[maj][min] = doc;
        }
        versionCount = versCount;
        return docVersions;
}
//This method updates the security ACLs from a source Object
//to a target Object
private  Boolean  upDateSecurityACLSet(IndependentlyPersistableObject ipoInput, IndependentlyPersistableObject ipoOutput){
Boolean changedACLs = false;
Properties propsIn = ipoInput.getProperties();
Properties propsOut = ipoOutput.getProperties();
//Check to ensure we have ACL properties to update
if (!propsIn.isPropertyPresent(PropertyNames.PERMISSIONS))
{
                // No ACLs so return false
    return changedACLs;
}
if (!propsOut.isPropertyPresent(PropertyNames.PERMISSIONS))
{
                // No ACLs so return false
```

```java
    return changedACLs;
}
        //Get source ACL list
        AccessPermissionList aclIn =(AccessPermissionList)propsIn.get
        DependentObjectListValue(PropertyNames.PERMISSIONS);
        AccessPermissionList aclOut =(AccessPermissionList)propsOut.get
        DependentObjectListValue(PropertyNames.PERMISSIONS);
//Replace ACL list in target object
        //Remove existing ACLs
        Iterator iterOut = aclOut.iterator();
        while (iterOut.hasNext())
        {
            AccessPermission apIn = (AccessPermission)iterOut.next();
            aclOut.remove(apIn);
        }
        //Add Source ACLs
        Iterator iterIn = aclIn.iterator();
        while (iterIn.hasNext())
        {
            AccessPermission apIn = (AccessPermission)iterIn.next();
            aclOut.add(apIn);
        }
//Set in target Object
        changedACLs = true;
            return changedACLs;

}
//Changes all groups and user ACLs
//To 'read-only' except the passed User and Group for a Document or Folder
access permission list.
//User access is limited  by altering the ACL permission

private AccessPermissionList addPermissions(AccessPermissionList apl,
String excludeGroupName, String excludeUserName)
{

//ASB ... 4 007 Changed to exclude a user and a group
```

```java
//ASB ... 2 021 - Property Security Filter
PropertyFilter PropF = new PropertyFilter();
PropF.addIncludeProperty(new FilterElement(null, null, null,
PropertyNames.ID, null));
//Save the original APL
AccessPermissionList aplSaved = apl;

AccessPermissionList aplWorking = Factory.AccessPermission.createList();
Boolean excludeGranteeFound = false; //Set true if at least one APL is reserved
//Create a new access permission object.
//com.filenet.api.security.AccessPermission ap = Factory.AccessPermission.createInstance();

int PropFLevel = 1; // TODO Set as a parameter
PropF.setMaxRecursion(PropFLevel);
PropF.addIncludeType(new FilterElement(null, null, null,
FilteredPropertyType.ANY, null));

//Get Security Realm
com.filenet.api.security.Realm realm = Factory.Realm.fetchCurrent(os.getConnection(), PropF);

GroupSet g = realm.findGroups(excludeGroupName,
    PrincipalSearchType.EXACT,              //search Type Needed
    PrincipalSearchAttribute.DISPLAY_NAME,  //searchAttribute Needed
    null,                                   //Sort Type
    100,                                    //Page Size
    PropF);
Iterator iterG = g.iterator();

UserSet  u = realm.findUsers(excludeUserName,
    PrincipalSearchType.EXACT,              //search Type Needed
    PrincipalSearchAttribute.DISPLAY_NAME,  //searchAttribute Needed
    null,                                   //Sort Type
    100,                                    //Page Size
    PropF);
Iterator iterU = u.iterator();
```

CHAPTER 4   A REPLICATION JAVA PROGRAM FOR IBM FILENET OBJECT STORES

```java
logger.info("Realm Name: " + realm.get_Name()); //print the name of the
realm for the current user

//Set Read Only Access Bits
final int ACCESS_READONLY = AccessRight.READ.getValue() |
AccessRight.VIEW_CONTENT.getValue() |
AccessRight.CONNECT.getValue();

Iterator iterAp = apl.iterator();
//Go through each ACE in turn setting the required permissions
while (iterAp.hasNext())
{
com.filenet.api.security.AccessPermission   apWorking = Factory.AccessPermis
sion.createInstance();
com.filenet.api.security.AccessPermission   ap = (com.filenet.api.security.
AccessPermission) iterAp.next();
//Set the returned permission
// Get Grantee Name
   Boolean groupFound = false;
   Boolean userFound = false;
String sGrantee =    ap.get_GranteeName();
iterG = g.iterator();
while (iterG != null && iterG.hasNext() == true)
{
   Group group = (Group) iterG.next();
   String sGroupDisplay = group.get_DisplayName();
   String sGroupShortName = group.get_ShortName();
   String sGroupName = group.get_Name();
   String sGroupDistinguishedName = group.get_DistinguishedName();
   // Set permissions based on whether Group is in the Read Only Group.
   if (group.get_DisplayName().equalsIgnoreCase(sGrantee)
          ||group.get_Name().equalsIgnoreCase(sGrantee)
          ||group.get_ShortName().equalsIgnoreCase(sGrantee)
          ||group.get_DistinguishedName().equalsIgnoreCase(sGrantee) )
   {
        groupFound = true;
```

```java
            excludeGranteeFound = true;
    }   //Group Search in apl
}
iterU = u.iterator();
while (iterU != null && iterU.hasNext() == true)
{
     com.filenet.api.security.User user = (com.filenet.api.security.User) iterU.next();
    String sUserDisplay = user.get_DisplayName();
    String sUserShortName = user.get_ShortName();
    String sUserName = user.get_Name();
    String sUserDistinguishedName = user.get_DistinguishedName();
   // Set permissions based on whether Group is in the Read Only Group.
    if (sUserDisplay.equalsIgnoreCase(sGrantee)
         ||sUserName.equalsIgnoreCase(sGrantee)
         ||sUserShortName.equalsIgnoreCase(sGrantee)
         ||sUserDistinguishedName.equalsIgnoreCase(sGrantee))
         {
            userFound = true;
           excludeGranteeFound = true;
    }   //UserSearch in apl
}

// Set Read Only access permissions if the user and group are not excluded.
if(!(groupFound || userFound) ){
     apWorking.set_GranteeName(sGrantee);
     apWorking.set_AccessType(AccessType.ALLOW);
     apWorking.set_AccessMask(new Integer(ACCESS_READONLY));
     aplWorking.add(apWorking);
} else{
     aplWorking.add(ap);
}

}

//Set and save the new permissions.
```

```java
// Check if we have rescued at least one ACE as excluded
if (!excludeGranteeFound){
    apl = aplSaved;
} else{
    apl = aplWorking;
}
return apl;
}
//Changes all groups and user ACLs
//To 'read-only' except the passed User and Group for a Document or Folder access permission list.
//User access is limited  by altering the ACL permission

private AccessPermissionList addPermissionsNoSearch(AccessPermissionList apl, String excludeGroupName, String excludeUserName)
{

//ASB ... 4 007 Changed to exclude a user and a group
//ASB ... 2 021 - Property Security Filter
PropertyFilter PropF = new PropertyFilter();
PropF.addIncludeProperty(new FilterElement(null, null, null, PropertyNames.ID, null));
//Save the original APL
AccessPermissionList aplSaved = apl;

AccessPermissionList aplWorking = Factory.AccessPermission.createList();
Boolean excludeGranteeFound = false; //Set true if at least one APL is reserved
//Create a new access permission object.
//com.filenet.api.security.AccessPermission ap = Factory.AccessPermission.createInstance();

int PropFLevel = 1; // TODO Set as a parameter
PropF.setMaxRecursion(PropFLevel);
PropF.addIncludeType(new FilterElement(null, null, null, FilteredPropertyType.ANY, null));

//Set Read Only Access Bits
```

```java
            excludeGranteeFound = true;
    }   //Group Search in apl
}
iterU = u.iterator();
while (iterU != null && iterU.hasNext() == true)
{
       com.filenet.api.security.User user = (com.filenet.api.security.User) iterU.next();
    String sUserDisplay = user.get_DisplayName();
    String sUserShortName = user.get_ShortName();
    String sUserName = user.get_Name();
    String sUserDistinguishedName = user.get_DistinguishedName();
    // Set permissions based on whether Group is in the Read Only Group.
    if (sUserDisplay.equalsIgnoreCase(sGrantee)
        ||sUserName.equalsIgnoreCase(sGrantee)
        ||sUserShortName.equalsIgnoreCase(sGrantee)
        ||sUserDistinguishedName.equalsIgnoreCase(sGrantee))
        {
          userFound = true;
          excludeGranteeFound = true;
    }   //UserSearch in apl
}

// Set Read Only access permissions if the user and group are not excluded.
if(!(groupFound || userFound) ){
    apWorking.set_GranteeName(sGrantee);
    apWorking.set_AccessType(AccessType.ALLOW);
    apWorking.set_AccessMask(new Integer(ACCESS_READONLY));
    aplWorking.add(apWorking);
} else{
    aplWorking.add(ap);
}

}

//Set and save the new permissions.
```

CHAPTER 4   A REPLICATION JAVA PROGRAM FOR IBM FILENET OBJECT STORES

```java
// Check if we have rescued at least one ACE as excluded
if (!excludeGranteeFound){
    apl = aplSaved;
} else{
    apl = aplWorking;
}
return apl;
}
//Changes all groups and user ACLs
//To 'read-only' except the passed User and Group for a Document or Folder access permission list.
//User access is limited  by altering the ACL permission

private AccessPermissionList addPermissionsNoSearch(AccessPermissionList apl, String excludeGroupName, String excludeUserName)
{

//ASB ... 4 007 Changed to exclude a user and a group
//ASB ... 2 021 - Property Security Filter
PropertyFilter PropF = new PropertyFilter();
PropF.addIncludeProperty(new FilterElement(null, null, null, PropertyNames.ID, null));
//Save the original APL
AccessPermissionList aplSaved = apl;

AccessPermissionList aplWorking = Factory.AccessPermission.createList();
Boolean excludeGranteeFound = false; //Set true if at least one APL is reserved
//Create a new access permission object.
//com.filenet.api.security.AccessPermission ap = Factory.AccessPermission.createInstance();

int PropFLevel = 1; // TODO Set as a parameter
PropF.setMaxRecursion(PropFLevel);
PropF.addIncludeType(new FilterElement(null, null, null, FilteredPropertyType.ANY, null));

//Set Read Only Access Bits
```

```java
final int ACCESS_READONLY = AccessRight.READ.getValue() |
AccessRight.VIEW_CONTENT.getValue() |
AccessRight.CONNECT.getValue();

Iterator iterAp = apl.iterator();
//Go through each ACE in turn setting the required permissions
while (iterAp.hasNext())
{
com.filenet.api.security.AccessPermission  apWorking = Factory.AccessPermission.createInstance();
com.filenet.api.security.AccessPermission  ap = (com.filenet.api.security.AccessPermission) iterAp.next();
//Set the returned permission
// Get Grantee Name
   Boolean groupFound = false;
   Boolean userFound = false;
   excludeGranteeFound = false;
String sGrantee =   ap.get_GranteeName();
if (excludeGroupName.equalsIgnoreCase(sGrantee)){
       groupFound = true;
       excludeGranteeFound = true;
        if(cemc.getDebugOutputFlag().equalsIgnoreCase("on")){
            logger.info("Group Excluded : " + sGrantee); //print the name
            of the Group for the current user
       }
 }
if (excludeUserName.equalsIgnoreCase(sGrantee)){
       userFound = true;
       excludeGranteeFound = true;
        if(cemc.getDebugOutputFlag().equalsIgnoreCase("on")){
            logger.info("User Excluded : " + sGrantee); //print the name
            of the User for the current user
       }
}

// Set Read Only access permissions if the user and group are not excluded.
```

```java
if(!(groupFound || userFound) ){
    apWorking.set_GranteeName(sGrantee);
    apWorking.set_AccessType(AccessType.ALLOW);
    apWorking.set_AccessMask(new Integer(ACCESS_READONLY));
    aplWorking.add(apWorking);
   if(cemc.getDebugOutputFlag().equalsIgnoreCase("on")){
       logger.info("Group/User set with Read Only : " + sGrantee); //
       print the name of the Group for the current user
    }
  } else{
     if(cemc.getDebugOutputFlag().equalsIgnoreCase("on")){
        logger.info("User or Group Excluded from Read Only : " +
        sGrantee); //print the name of the User for the current user
     }
         aplWorking.add(ap);
}

}

//Set and save the new permissions.

// Check if we have rescued at least one ACE as excluded
if (!excludeGranteeFound){
    apl = aplSaved;
} else{
    apl = aplWorking;
}
return apl;
}
private Document createVersions(String documentTitle, Id GUID,
IndependentObject object, Properties props, Iterator iterProps,Boolean
isCurrent, Boolean isReserved,Document[][] docVersions,Integer majCount,int
minCount, BufferedWriter auditFileWriter)throws Exception{
PropertyFilter myFilter = new PropertyFilter();
int myFilterLevel = 2; //Changed from 4 to 2 -- TODO Set as a parameter
myFilter.setMaxRecursion(myFilterLevel);
```

```java
myFilter.addIncludeType(new FilterElement(null, null, null,
FilteredPropertyType.ANY, null));
Document currentDocument = (Document) object;
String documentclass = currentDocument.getClassName();
Document dInput = Factory.Document.fetchInstance(osinput, GUID, myFilter);
//Create Version Series for the document in the Target Object Store
VersionSeries versions = dInput.get_VersionSeries();
//Get the Document Versions to be added from the Source Document
VersionableSet verSet = versions.get_Versions();
Iterator versIt = verSet.iterator();
int versCount = 0;
int majorVn = 0;
int minorVn = 0;
Document doc = null; //Input Document versions
Document d = null;   //Output document versions
isReserved = false;
Id createdGuid = null;
Id nextGUID = null;
int foldCount = 0;
//while (versIt.hasNext()){ ASB ... 2
int   minVersCount = -1;
while (versCount < majCount){
      versCount ++;
    while (minVersCount < minCount){
            minVersCount ++;
        doc = docVersions[versCount][minVersCount];
        if(doc == null){
           break;
        }
        foldCount ++;
        majorVn = doc.get_MajorVersionNumber();
        minorVn = doc.get_MinorVersionNumber();
        nextGUID = doc.get_Id();
        //ASB ... 4 004 Get the Release Version
        VersionStatus dVersionStatus = doc.get_VersionStatus();
```

```
    if (!isReserved)isReserved = doc.get_IsReserved();
//Need to update for each version in turn
    d = createVersion(versCount, d, doc, isReserved, majorVn,minorVn,
    documentTitle, nextGUID, createdGuid, nextGUID,auditFileWriter);
    createdGuid = d.get_Id(); //ASB ... 4 025 Try/Catch Here especially
    as reservation has no folders!!
    try {
        if (foldCount == majCount){
            FolderSet fs = doc.get_FoldersFiledIn();
            Iterator iterFs = fs.iterator();
            while (iterFs.hasNext())
            {
                Folder folder = (Folder)iterFs.next();
                //Create/Check and Make link in Target for the Document
                to this Folder
                    if(cemc.getDebugOutputFlag().
                    equalsIgnoreCase("off")){
                    }else{
                        logger.info("\r\tFolder Name: " + folder.get_
                        FolderName() +
                "   Folder Path: " + folder.get_PathName());
                    }
                    folderDoc(d,folder.get_PathName(),documentTitle);
            }
        }
    }catch(Exception FileErr){
            logger.error(FileErr.getMessage(), FileErr);
    }
        if (isReserved){
        break;
    }
}
```

```java
    //ASB ... 4 019 Check at this point if we have reached the completed
    creation
        minVersCount = -1;
} // Next Major Version
    if (isReserved)
        d = doReservationProperties(versCount, doc, d, isReserved,
        majorVn,minorVn,documentTitle);
return d;

}   //This method adds the content specified in the XML Nodes passed in
to the specified document object.
//Creates a Content Transfer object for each node.  Sets RetrievalName to
the original filename
//to ensure that when the version is downloaded it as the correct file
name    Sets mimetype.
private void addContent(Document dSource, Document d,  String
documentTitle, PropertyFilter myFilterId, Id GUID, String DocClass) throws
Exception {
    //ADD CONTENT   NEW CODE HERE
    //ASB ... 4 024
    int retryCount = 0;
    Document currDoc = null;
    while (retryCount < retryLimit){
        try {
            currDoc = (Document)osinput.fetchObject(DocClass, GUID,
            myFilterId);
            //ASB ... 4 031 Add retry 7 times to fix API bug!!
            ContentTransfer ct = Factory.ContentTransfer.create
            Instance();
            InputStream str = currDoc.accessContentStream(0);
            ct.setCaptureSource(str);
            // Add Document Title
            ct.set_RetrievalName(documentTitle);
            // Add Content Mime Type
            ct.set_ContentType(currDoc.get_MimeType());
```

```java
                ContentElementList cel = Factory.
                ContentElement.createList();
                cel.add(ct);
                d.set_ContentElements(cel);
                break;
                }
                catch (Exception e){
                    String testMessage = e.getMessage();
                    String testCode = e.toString();
                    if( testMessage.contains("A uniqueness requirement has
                    been violated")){
                        retryCount = retryLimit;
                        break;
                    }
                    //ASB ... 4 Check if the Content could not be added
                    because the document already exists, if so
                    //          Remove document and retry (next run)
                    //   ASB ... 4 025 Log Exception here
                    logger.error(e.getMessage(), e);
                    retryCount ++;
                    retryTotalProcessedCount ++;
                    logger.info("Retry Count - createVersion method :
                    " + retryCount + " of " + retryLimit + " On : " +
                    documentTitle + " : with GUID : " + GUID);
                }
} //ASB ... 4 031 Above loops Test retryCount less than the retryLimit of 7
// Add Security ACL

    //ASB ... 2 020 Update Document ACL from source Document security

    AccessPermissionList   aclIn = currDoc.get_Permissions();
    // Add the permission to the list for the Object Store.
    //Get This group from the config.xml file
    String sExcludeGroup = cemc.getExcludedGroup();
    String sExcludeUser =  cemc.getExcludedUser();
```

```java
    //ASB ... 4 027 For Performance check if a full LDAP search is
    required
if(cemc.getLDAPSearchFlag().equalsIgnoreCase("on")){
        aclIn = addPermissions(aclIn, sExcludeGroup, sExcludeUser);
    }else {
        aclIn = addPermissionsNoSearch(aclIn, sExcludeGroup,
        sExcludeUser);
    }
    //ASB ... 4 027 aclIn = addPermissions(aclIn, sExcludeGroup,
    sExcludeUser);
    d.set_Permissions(aclIn);
    Boolean isMajorVersion = false;
    if(currDoc.get_MinorVersionNumber() == 0) isMajorVersion = true;
    //ASB ... 4 028 Change to do not Classify from AutoClassify.AUTO_
    CLASSIFY, to
     if (isMajorVersion){
         d.checkin(AutoClassify.DO_NOT_AUTO_CLASSIFY,CheckinType.MAJOR_
         VERSION);
     }
     else {
         d.checkin(AutoClassify.DO_NOT_AUTO_CLASSIFY,CheckinType.MINOR_
         VERSION);
     }
     //d.checkin(AutoClassify.AUTO_CLASSIFY, CheckinType.MAJOR_VERSION);
    //ASB ... 4 25 Try/Catch here ?
    Date dateLastModified = dSource.get_DateLastModified();
    //Date dateCreated = dSource.get_DateCreated(); //ASB ... 4 029
    String sModifier = dSource.get_LastModifier();
    Date dateCheckedIn = dSource.get_DateCheckedIn();
    // Update Checkin Date etc from Source here
    //d.set_DateCreated(dateCreated);   //ASB ... 4 029 READONLY AT
THIS POINT!!
    d.set_DateLastModified(dateLastModified);
    d.set_LastModifier(sModifier);
    d.set_DateCheckedIn(dateCheckedIn); //ASB ... 2 005
```

CHAPTER 4   A REPLICATION JAVA PROGRAM FOR IBM FILENET OBJECT STORES

```
    d.save(RefreshMode.REFRESH); //ASB XXX
}
//Sets the properties on the reservation object for docs that were checked
out on the source system.
//Sets properties but no content.  No need to set mime type here - causes
problems.
//private Document doReservationProperties(Node verNode, Document d) throws
Exception {
private Document doReservationProperties(int versCount, Document
dInput, Document d, Boolean isReserved, int majorVn,int minorVn, String
documentTitle     ) throws Exception{
    Properties props = Factory.Document.createInstance(os, null).
    getProperties();
    String sCreator = dInput.get_Creator();
    Date dateCreated = dInput.get_DateCreated();
    String sLastModifier = dInput.get_LastModifier();
    Date dateLastModified = dInput.get_DateLastModified();
    // This collection is used when we do the checkout
    props.putValue("Creator", sCreator);
    props.putValue("DateCreated", dateCreated);
    props.putValue("LastModifier", sLastModifier);
    props.putValue("DateLastModified", dateLastModified);
    Date dateCheckedIn = dInput.get_DateCheckedIn();
    //Update Checkin Date etc from Source here

    d.set_LastModifier(sCreator);
    d.set_DateLastModified(dateCreated);
    d.set_DateCreated(dateCreated); //ASB ... 4 029
    Id reservationId_null = null;
    String reservationClass_null = null; // to remind of the
    parameter types!
    try
    {
       // Get Document properties
          Properties docProps = d.getProperties();
          Iterator iterProps = props.iterator();
```

```java
        Boolean justCreated = false;
        writeSpecialDocProps(d, iterProps, justCreated,documentTitle,
        dInput);
    }
    catch (Exception e){
        throw new Exception(e.getMessage(), e);
    }

    d.set_LastModifier(sLastModifier);

    // ASB Compare : d.checkout(ReservationType.EXCLUSIVE, GUID,
    docClass, null);
    d.checkout(ReservationType.OBJECT_STORE_DEFAULT, reservationId_null,
    reservationClass_null, props);
    d.set_DateLastModified(dateLastModified);
    d.set_DateCreated(dateCreated); //ASB ... 4 029
    d.set_LastModifier(sLastModifier);
    d.save(RefreshMode.REFRESH);
    return d;
}

//Called by CreateDocument,  creates a version from the information in the XML node specified.
//Creates a new reservation object, specifying properties from XML.
//private Document createVersion(Document d) throws CEMigrateException, Exception {
    private Document createVersion(int     versCount, Document d,
    Document dInput, Boolean isReserved, int majorVn, int minorVn,String
    documentTitle, Id GUID, Id createdGUID, Id nextGUID,BufferedWriter
    auditFileWriter)     throws Exception {
        Document res = null;
    Long docImportStartTime = System.currentTimeMillis();
        PropertyFilter myFilter = new PropertyFilter();
       int myFilterLevel = 2; //Changed from 4 to 2 -- For performance!
        myFilter.setMaxRecursion(myFilterLevel);
        myFilter.addIncludeType(new FilterElement(null, null, null,
        FilteredPropertyType.ANY, null));
```

CHAPTER 4　A REPLICATION JAVA PROGRAM FOR IBM FILENET OBJECT STORES

```java
Document currDoc = null;
 try
    {
 int thisMajorVersionNumber = majorVn;
 int thisMinorVersionNumber = minorVn; //ASB ... 3 004
 String docClass = dInput.getClassName();
 //ASB ... 3 006 Now get the current source Doc Version Series ID
 Id vId = dInput.get_VersionSeries().get_Id();
 if (!isClassExist(docClass))
     throw new Exception("Document class " + docClass + " Does
     not exist");
 Id reservationId_null = null; String reservationClass_null =
 null; // to remind of the parameter types!
 boolean isMajorVersion = false;
 boolean justCreated = true;

 if (d == null){
     if (thisMajorVersionNumber > 0)
         isMajorVersion = true;
     if (thisMinorVersionNumber > 0)
         isMajorVersion = false;
         //ASB ... 2 015 Added GUID to ensure we get the
         same later
         //ASB ... 3 006 New Version of the first document create
         to store Version series ID
         //ASB ... 3 006 Many thanks to David Greenhouse for
         supplying the call :-)
         res = Factory.Document.createInstance(os,
         docClass, GUID, vId, com.filenet.api.constants.
         ReservationType.EXCLUSIVE);

           //res = (Document)os.createObject(docClass,GUID); //ASB
           ... 3 006 Removed Old Creation

           //Added this to ensure that these values were correctly
           set on the initial version
```

```
            String mimetype = dInput.get_MimeType();
            String filename = dInput.get_Name();
        Date dateCreated = dInput.get_DateCreated();
            res.set_MimeType(mimetype);
            res.set_DateCreated(dateCreated); //ASB ... 4 029
            //ASB 04/04/2022 - Update Creator and last Modifier
            and dates
            String sCreator = dInput.get_Creator();
            res.set_Creator(sCreator);                          //
            ASB 04/04/2022
            String sLastModifier = dInput.get_LastModifier();
            res.set_LastModifier(sLastModifier);        //ASB
            04/04/2022
            Date dateLastModified = dInput.get_DateLastModified();
            res.set_DateLastModified(dateLastModified); //ASB
            04/04/2022
        currDoc = dInput;
            Properties docprops = Factory.
            Document.createInstance(os, null).
            getProperties(); //?????
         Properties props = res.getProperties();
         Iterator iterProps = props.iterator();

            writeSpecialDocProps(res, iterProps, justCreated,
            documentTitle, currDoc);
        //ASB ... 4 023 Add Count for first version
            this.processedCount = this.processedCount +1;
          String guid = GUID.toString().trim();
            //Write to audit log
            Long timeToImport = System.currentTimeMillis() -
            docImportStartTime;
            auditFileWriter.write(guid + "," + vId.toString() + ","
            + guid + ",Imported," + timeToImport.toString());
            auditFileWriter.newLine();
    }
    else {
```

```java
                    justCreated = false;
                //GUID here should be from next iteration set above
                    //ASB ... 2 016 Changed to osinput from os! try nextGUID
                    //currDoc = (Document)os.fetchObject("Document",
                    createdGUID, myFilter);
                //ASB ... 4 024 Update to support other Document Types
                    currDoc = (Document)os.fetchObject(docClass, createdGUID,
                    myFilter);
                    Properties props = Factory.Document.createInstance(os,
                    null).getProperties();
                    String sCreator = currDoc.get_Creator();
                    Date dateCreated = currDoc.get_DateCreated();
                    String sLastModifier = currDoc.get_LastModifier();
                    Date dateLastModified = currDoc.get_DateLastModified();
                    // This collection is used when we do the checkout
                    props.putValue("Creator", sCreator);
                    props.putValue("DateCreated", dateCreated);
                    props.putValue("LastModifier", sLastModifier);
                    props.putValue("DateLastModified", dateLastModified);
                    //ASB ... 4 024
                    //d = (Document)os.fetchObject(docClass, createdGUID,
                    myFilter);
                    d = currDoc; //No need to fetch again!
                    //d = (Document)os.fetchObject("Document", createdGUID,
                    myFilter);
                    //ASB ... 4 014 New Version needs to use the GUID
                    passed in !!
                    String sNewGuid = GUID.toString();
                    // ASB Compare : d.checkout(ReservationType.EXCLUSIVE, GUID,
                    docClass, null);
                    d.checkout(ReservationType.EXCLUSIVE, GUID, docClass,
                    props); //ASB ... 4 029 Changed from null to props
                    //d.set_DateCreated(dateCreated); //ASB004 029 READONLY AT
                    THIS POINT!!
                    d.save(RefreshMode.REFRESH);
```

```java
            this.processedCount = this.processedCount +1;
     String guid = GUID.toString().trim();
        //Write to audit log
     Long timeToImport = System.currentTimeMillis() -
     docImportStartTime;
     auditFileWriter.write(guid + "," + d.get_VersionSeries().get_
     Id().toString() + "," + d.get_Id().toString() + ",Imported,"
     + timeToImport.toString());
     auditFileWriter.newLine();
      // Get the reservation object
      res = (Document)d.get_Reservation();
      res.set_LastModifier(sLastModifier);          //ASB00 4
      res.set_DateLastModified(dateLastModified); //ASB00 4 Uodate
      from dateCreated ??!!
      //res.set_DateCreated(dateCreated); //ASB004 029 READ ONLY
      AT THIS POINT!!!

      String mimetype = currDoc.get_MimeType();
      props.putValue("MimeType", mimetype);

      int mvn = majorVn;
      int minvn = minorVn;
      if (thisMajorVersionNumber > 0) isMajorVersion = true;
      if (thisMinorVersionNumber > 0) isMajorVersion = false;
   dInput = currDoc;
   GUID = nextGUID; //Get Current Content from version
 }
             try {
             addContent(dInput, res, documentTitle,  myFilter,
             GUID,docClass);
             }
             catch (Exception e){
                 //ASB ... 4 Check if the Content could not be
                 added because the document already exists, if so
                 //          Remove document and retry (next run)
                 //   ASB ... 4 025 Log Exception here
```

```java
                    logger.error(e.getMessage(), e);
                }

            if(isReserved){ //ASB ... 2 012 Required For Checkin the Document
            must be Status of reserved!!
                if (isMajorVersion){
                    //ASB ... 4 028 Changed from NULL to AutoClassify.
                    DO_NOT_AUTO_CLASSIFY
                    res.checkin(AutoClassify.DO_NOT_AUTO_
                    CLASSIFY,CheckinType.MAJOR_VERSION);
                    res.save(RefreshMode.NO_REFRESH);
                }
                else {
                    res.checkin(AutoClassify.DO_NOT_AUTO_
                    CLASSIFY,CheckinType.MINOR_VERSION);
                    res.save(RefreshMode.NO_REFRESH);
                }
            }
        }
        catch (Exception e)
        {
            e.printStackTrace();
            //TODO If caused by VSID issue attempt Delete Here
            throw e;
        }
        return res;
}
//Adds properties such as LastModifier to the document properties.
//Called by doReservationProperties, createVersion
private void writeSpecialDocProps(Document d,Iterator iterProps, Boolean
justCreated, String documentTitle, Document dInput) throws Exception {
    //ASB ... 2 008 Updated for new property definition objects
    //Retrieve the current property list to retrieve based on the Document
    Class definitions from the Object Store
```

```java
com.filenet.api.admin.ClassDefinition classDefs = Factory.ClassDefinit
ion.fetchInstance(osinput,d.getClassName(), null);
PropertyDefinitionList propList = classDefs.get_PropertyDefinitions();
StringBuffer tempBuf = new StringBuffer();
//Retrieve property names
Iterator iter = propList.iterator();
while (iter.hasNext())
{
    //ASB ... 4 025 Catch any property exceptions here
    try{
        PropertyDefinition propDef = (PropertyDefinition)
        iter.next();
            String propName = propDef.get_SymbolicName();
            //ASB ... 2 011 Need to skip Date Last Accessed
            etc - ReadOnly error here
            if (!(propName.equalsIgnoreCase("DateContentLast
            Accessed")
                    ||propName.equalsIgnoreCase("LockOwner")
                    ||propName.equalsIgnoreCase("Name")
                    ||propName.equalsIgnoreCase("Storage
                    Location")
                    ||propName.equalsIgnoreCase("ContentElem
                    entsPresent")
                    ||propName.equalsIgnoreCase("DateC
                    heckedIn")
                    ||propName.equalsIgnoreCase("ContentRete
                    ntionDate")
                    ||propName.equalsIgnoreCase("CurrentS
                    tate") )){
                //ASB ... 2 011 Set to check cardinality
                int CardinalityVal = propDef.get_
                Cardinality().getValue();
                switch (propDef.get_DataType().getValue()){
                case StringType:
```

```java
            if (CardinalityVal ==
CardinalitySINGLE){
                String val = dInput.
                getProperties().
                getStringValue(propName);
                d.getProperties().
                putValue(propName, val);
                if (propName.equalsIgnoreCase("last
                modifier")){
                    d.set_LastModifier(val);
                }
                if (!propName.
                equalsIgnoreCase("creator") &&
                !propName.equalsIgnoreCase("lastmo
                difier")){
                    d.getProperties().
                    putValue(propName, val);
                }
            }else{
                //List
                StringList valList =
                dInput.getProperties().
                getStringListValue(propName);
                d.getProperties().
                putValue(propName, valList);
                
            }
        break;
    case DateType:
        if (CardinalityVal ==
CardinalitySINGLE){
            Date dateVal = dInput.
            getProperties().
            getDateTimeValue(propName);
            if (propName.equalsIgnoreCase("date
            lastmodified")){
```

```java
                                if (!justCreated){
                                        dInput.getProperties().re
                                        moveFromCache("DateLastM
                                        odified");
                                }
                                d.set_
                                DateLastModified(dateVal);
                            }
                            if (!propName.
                            equalsIgnoreCase("datecreated") &&
                            !propName.equalsIgnoreCase("datelas
                            tmodified")){
                                    d.getProperties().
                                    putValue(propName, dateVal);
                            }
                        }
                        //TODO Process Multi-value Dates (if
                        exists)?
                        break;
                    default:

                    }
                }

            }catch(Exception writeProp){
                logger.error(writeProp.getMessage(), writeProp);
        }

    }
//ASB ... 2 020 Update Document ACL from source Document security
AccessPermissionList   aclIn = dInput.get_Permissions();
// Add the permission to the list for the Object Store.
//Get This group from the config.xml file
String sExcludeGroup = cemc.getExcludedGroup();
String sExcludeUser  =  cemc.getExcludedUser();
```

```
    //ASB ... 4 027 For Performance check if a full LDAP search is
    required
    if(cemc.getLDAPSearchFlag().equalsIgnoreCase("on")){
        aclIn = addPermissions(aclIn, sExcludeGroup, sExcludeUser);
    }else {
        aclIn = addPermissionsNoSearch(aclIn, sExcludeGroup,
        sExcludeUser);
    }
    //ASB ... 4 027 aclIn = addPermissions(aclIn, sExcludeGroup,
    sExcludeUser);
    d.set_Permissions(aclIn);
}

//Called by doReservationProperties, createVersion.  Adds document
properties.
private void writeDocProps(Document d,Properties props,Document currentDoc)
throws Exception {
    //Retrieve the current property list to retrieve based on the Document
    Class definitions from the Object Store
    com.filenet.api.admin.ClassDefinition classDefs = Factory.ClassDefinit
    ion.fetchInstance(osinput,d.getClassName(), null);
    PropertyDefinitionList propList = classDefs.get_PropertyDefinitions();
    StringBuffer tempBuf = new StringBuffer();
    //Retrieve property names
    Iterator iter = propList.iterator();
    while (iter.hasNext())
    {
       try {
            PropertyDefinition propDef = (PropertyDefinition) iter.next();
            //PropertyDefinition propDef = (PropertyDefinition)
            iter.next();
            String propName = propDef.get_SymbolicName();
            //ASB ... 2 011 Need to skip Date Last Accessed etc - ReadOnly
            error here
            if (!(propName.equalsIgnoreCase("DateContentLastAccessed")
                    ||propName.equalsIgnoreCase("LockOwner")
```

```java
        ||propName.equalsIgnoreCase("Name")
        ||propName.equalsIgnoreCase("StorageLocation")
        ||propName.equalsIgnoreCase("ContentRetentionDate")
        ||propName.equalsIgnoreCase("ContentElement
    sPresent")
        ||propName.equalsIgnoreCase("DateCheckedIn")
        ||propName.equalsIgnoreCase("CurrentState") )){

//ASB ... 2 011 Set to check cardinality
int CardinalityVal = propDef.get_Cardinality().getValue();
switch (propDef.get_DataType().getValue()){
case StringType:

    if (CardinalityVal == CardinalitySINGLE){
        String val = currentDoc.getProperties().
        getStringValue(propName);
        d.getProperties().putValue(propName, val);
        if (propName.equalsIgnoreCase("lastmodifier")){
            d.set_LastModifier(val);
        }
        if (!propName.equalsIgnoreCase("creator") &&
        !propName.equalsIgnoreCase("lastmodifier")){
            d.getProperties().putValue(propName, val);
        }
    }else{
        //List
        StringList valList = currentDoc.getProperties().
        getStringListValue(propName);
        d.getProperties().putValue(propName, valList);
    }

    // Process Multi-value Strings (if required)?
    break;
case DateType:
    if (CardinalityVal == CardinalitySINGLE){
        Date dateVal = currentDoc.getProperties().
        getDateTimeValue(propName);
```

```java
                            if (propName.equalsIgnoreCase("datelastmo
                        dified")){
                                currentDoc.getProperties().removeFromCache
                                ("DateLastModified");
                                d.set_DateLastModified(dateVal);
                            }
                            //ASB ... 4 029 Start
                            if (propName.equalsIgnoreCase("datecreated")){
                                currentDoc.getProperties().removeFromCache
                                ("DateCreated");
                                d.set_DateLastModified(dateVal);
                            }
                            //ASB ... 4 029 End
                        if (!propName.equalsIgnoreCase("datecreated") &&
                        !propName.equalsIgnoreCase("datelastmodified")){
                                d.getProperties().putValue(propName, dateVal);
                            }
                        }
                    //TODO Process Multi-value Dates (if exists)?
                    break;
                    // }
                     default:
                    }
                }
            }catch(Exception writeProp){
                logger.error(writeProp.getMessage(), writeProp);
            }

    } //for (int iProp = 0;iProp < xmlprops.getLength();iProp ++){
}

private void updateDocProps(Document d,Properties props,Document
currentDoc) {
    //Retrieve the current property list to retrieve based on the Document
      Class definitions from the Object Store
try {
```

```java
    PropertyFilter myFilter = new PropertyFilter();
//ASB ... 4 017 Changed from 2 to 1 for performance
//              Changed from 8.5 seconds to 5.3 Seconds
 int myFilterLevel = 1; //Changed from 4 to 1 -- TODO Set as a
 parameter
myFilter.setMaxRecursion(myFilterLevel);
//ASB ... 4 020
if(cemc.getDebugOutputFlag().equalsIgnoreCase("off")){
    myFilter.addIncludeType(new FilterElement(null, null, null,
    FilteredPropertyType.ANY_LIST, null));
    myFilter.addIncludeType(new FilterElement(null, null, null,
    FilteredPropertyType.ANY_SINGLETON, null));
}else{
    myFilter.addIncludeType(new FilterElement(null, null, null,
    FilteredPropertyType.ANY, null));
}
//myFilter.addIncludeType(new FilterElement(null, null, null,
FilteredPropertyType.ANY, null));
String sDocClassName = d.getClassName();
 com.filenet.api.admin.ClassDefinition classDefs = Factory.ClassDefinit
 ion.fetchInstance(osinput,sDocClassName, myFilter);
 PropertyDefinitionList propList = classDefs.get_PropertyDefinitions();
 StringBuffer tempBuf = new StringBuffer();
 // ASB ... 4 Following has to be commented out because ct is readonly!
 // Set Content Retrieval name - Can't because it is Read-Only
 //ContentElementList ctElementList = d.get_ContentElements();
 //Iterator iterCTList = ctElementList.iterator();
 //ContentTransfer ct = null;
 // Add Document Title
 //Retrieve property names
 Iterator iter = propList.iterator();
 while (iter.hasNext())
 {
     try {
```

```
PropertyDefinition propDef = (PropertyDefinition)
iter.next();
//PropertyDefinition propDef = (PropertyDefinition)
iter.next();
String propName = propDef.get_SymbolicName();
//ASB ... 2 011 Need to skip Date Last Accessed etc -
ReadOnly error here
if (!(propName.equalsIgnoreCase("DateContentLastAccessed")
        ||propName.equalsIgnoreCase("LockOwner")
        ||propName.equalsIgnoreCase("Name")
        ||propName.equalsIgnoreCase("Creator")
        ||propName.equalsIgnoreCase("StorageLocation")
        ||propName.equalsIgnoreCase("ContentReten
      tionDate")
        ||propName.equalsIgnoreCase("ContentElement
      sPresent")
        ||propName.equalsIgnoreCase("DateCheckedIn")
        ||propName.equalsIgnoreCase("MimeType")
        ||propName.equalsIgnoreCase("CurrentState") )){

    //ASB ... 2 011 Set to check cardinality
    int CardinalityVal = propDef.get_Cardinality().
    getValue();
    switch (propDef.get_DataType().getValue()){
    case StringType:
        if (CardinalityVal == CardinalitySINGLE){
            String val = currentDoc.getProperties().
            getStringValue(propName);
            d.getProperties().putValue(propName, val);
            if (propName.equalsIgnoreCase("lastmo
            difier")){
                d.set_LastModifier(val);
            }
            if (!propName.equalsIgnoreCase("creator") &&
            !propName.equalsIgnoreCase("lastmodifier")){
```

```java
                d.getProperties().
                putValue(propName, val);
            }
            if (propName.equalsIgnoreCase("Documen
            tTitle")){
    /*
                while (iterCTList.hasNext())
                {
                    ct = (ContentTransfer)
                    iterCTList.next();
                    ct.set_RetrievalName(val);
                }
    */
            }
        }else{
            //List
            StringList valList = currentDoc.
            getProperties().getStringListValue(propName);
            d.getProperties().putValue(propName,
            valList);
        }
        //TODO Process Multi-value Strings (if required)?
        break;
    case DateType:
        if (CardinalityVal == CardinalitySINGLE){
            Date dateVal = currentDoc.getProperties().
            getDateTimeValue(propName);
            if (propName.equalsIgnoreCase("datelastmo
            dified")){
                currentDoc.getProperties().removeFrom
                Cache("DateLastModified");
                d.set_DateLastModified(dateVal);
            }
            //ASB ... 4 029 Start
```

```
                                if (propName.
                                equalsIgnoreCase("datecreated")){
                                        currentDoc.getProperties().removeFromCac
                                        he("DateCreated");
                                        d.set_DateLastModified(dateVal);
                                }
                                //ASB ... 4 029 End
                                if (!propName.equalsIgnoreCase("datecreated")
                                && !propName.equalsIgnoreCase("datelastmo
                                dified")){
                                        d.getProperties().putValue(propName,
                                        dateVal);
                                }
                            }
                            //TODO Process Multi-value Dates (if exists)?
                            break;
                            // }
                        default:
                        }
                    }
                }catch(Exception writeProp){
                    logger.error(writeProp.getMessage(), writeProp);
                }
        } //for (int iProp = 0;iProp < xmlprops.getLength();iProp ++){
} catch (Exception e){
        logger.error(e.getMessage(), e);
}
}

//OLD Comment Called from main and imports Folders from Folders table.
//Called from main and imports Folders from Source Folders Event table.
private void importFolders() throws Exception {

    BufferedWriter auditFileWriter;
    long importStartTime;
    long folderImportStartTime;
```

```java
        Long timeToImport;
//      long maxRunTimeMillis;
        long consecutiveErrs=0;
        Long errorCount= new Long(0);

        importStartTime = System.currentTimeMillis();

        auditFileWriter = new BufferedWriter(new FileWriter(cemc.
        getImportAuditFileFolders()));
        auditFileWriter.write("Import Start - ");
        auditFileWriter.newLine();
        auditFileWriter.write("GUID, PATH, STATE, EXPORTFILE, Millisecs");
        auditFileWriter.newLine();
        String OSName = cemc.getExportOSName().trim();
        osinput = con_MigrateDB.getObjectStore(OSName);
        os = ceConn.getObjectStore(cemc.getImportOSName());
        ub = UpdatingBatch.createUpdatingBatchInstance(ceConn.getDomain(),
        RefreshMode.NO_REFRESH);
        String exportFile = ""; //ASB Fix Imported Issue
        String guid ="";
        batchCount = 0;    //ASB 24/02/2021 ADD
        batchSize = cemc.getUpdatingBatchSize(); //ASB 24/02/2021 ADD
        SearchSQL sqlObject = new SearchSQL();
        sqlObject.setSelectList("f.foldername, f.Id, f.Creator, 
f.DateCreated, f.LastModifier, f.DateLastModified, f.Name, f.Owner, 
f.LockToken, f.LockTimeout, f.LockOwner, f.PathName, f.IndexationId, 
f.CmIndexingFailureCode, f.CmRetentionDate, f.ContainerType, 
f.InheritParentPermissions, f.IsHiddenContainer");
        Integer maxRecords = cemc.getExportMaxDocCount();
        sqlObject.setMaxRecords(maxRecords); //was hard coded to 20 !
        sqlObject.setFromClauseInitialValue("Folder", "f", true); //Update to 
        look for other Folder types
        //Retrieve Date start from config.xml
        String sSearchDate = cemc.getStartSearchDate();
        //sqlObject.setWhereClause("f.DateCreated > " + "20200921T124517Z" );
        sqlObject.setWhereClause("f.DateCreated > " + sSearchDate );
```

```java
sqlObject.setOrderByClause("f.DateCreated");
// Check the SQL statement.
logger.info("SQL: " + sqlObject.toString());

// Create a SearchScope instance. (Assumes you have the object store
// object.)
SearchScope search = new SearchScope(osinput);

// Set the page size (Long) to use for a page of query result data. This value is passed
// in the pageSize parameter. If null, this defaults to the value of
// ServerCacheConfiguration.QueryPageDefaultSize.
Integer myPageSize = new Integer(1000);

// Specify a property filter to use for the filter parameter, if needed.
// This can be null if you are not filtering properties.
PropertyFilter myFilter = new PropertyFilter();
int myFilterLevel = 2; //Changed from 1 to 2 -- For performance !
myFilter.setMaxRecursion(myFilterLevel);
myFilter.addIncludeType(new FilterElement(null, null, null,
FilteredPropertyType.ANY, null));

// Set the (Boolean) value for the continuable parameter. This indicates
// whether to iterate requests for subsequent pages of result data when the end of the
// first page of results is reached. If null or false, only a single page of results is
// returned.
Boolean continuable = new Boolean(true); //SET To allow more results to be fetched

// Execute the fetchObjects method using the specified parameters.
IndependentObjectSet myObjects = search.fetchObjects(sqlObject,
myPageSize, myFilter, continuable);
```

```java
// You can then iterate through the collection of rows to access the properties.
int rowCount = 0;
Iterator iter = myObjects.iterator();
 while (iter.hasNext()){
     try
     {
        folderImportStartTime = System.currentTimeMillis();
        Boolean moreFoldersToImport=true;
        Boolean importFailure = false;
        Folder f = null;
        IndependentObject object = (IndependentObject) iter.next();
        //ASB Software Development Limited Fetch the full folder Object
        //ASB Software Development Limited example found to illustrate full property retrieval 01-08-2022
        // Id id = document.get_Id(); //get the doc ID back on this shallow document copy.
        // *retrieve a complete document copy back with complete property set using doc ID*/
        // Document docCopy = (Document)Factory.Document.fetchInstance(objectStore, id, null);
        // FolderSet pFolders = doc.get_FoldersFiledIn();//Get the full "FolderFiledIn" property on the new copy
        //
        // ASB Properties props = object.getProperties();
        //ASB Iterator iterProps = props.iterator();
        Id  GUID = ((Folder) object).get_Id();
        //ASB Software Development Limited - get full document property set which we need! 01-08-2022
        Folder folderCopy = (Folder)Factory.Folder.fetchInstance(osinput, GUID, null);
        // ASB Properties props = object.getProperties();
        Properties props = folderCopy.getProperties();
        Iterator iterProps = props.iterator();
```

CHAPTER 4   A REPLICATION JAVA PROGRAM FOR IBM FILENET OBJECT STORES

```java
        String folderName = "";
        String folderPath = "";
     guid = GUID.toString().trim();
     while (iterProps.hasNext() )
     {
         com.filenet.api.property.Property prop = (com.filenet.
         api.property.Property)iterProps.next();
         if(cemc.getDebugOutputFlag().equalsIgnoreCase("off")){
         }else{
             logger.info("\nProperty: " + prop.
             getPropertyName() );
             if ( ((com.filenet.api.property.Property) prop).
             getObjectValue() != null )
                 logger.info("  Value: " + prop.
                 getObjectValue().toString() );
         }
         if (prop.getPropertyName().equalsIgnoreCase("fol
         dername"))
         {
             folderName = prop.getStringValue();
         }
          if (prop.getPropertyName().equalsIgnoreCase("PathName"))
         {
             folderPath = prop.getStringValue();
         }
     }
     //ASB ... 3 005 Check Folder is in the path required
     if (folderPath.startsWith( cemc.getImportOSRootFoler())){
           //ASB f = createFolder(folderName,GUID,folderPat
           h,object);
           f = createFolder(folderName,GUID,folderPath,fo
           lderCopy);
     }else{
         //Folder selected is new but not in our import Folder Path!
```

```java
                logger.info("IGNORED: New Folder: " + folderName + "
                found linked to external Folder Path : " + folderPath + "
                Should be linked to :" +  cemc.getImportOSRootFoler());
        }
        if (f != null){
            //ASB 24/02/2022 New Code here
            if (batchSize > 0)
                ub.add(f,null);
            doUpdateBatch(false);

            processedCount++;

            timeToImport = System.currentTimeMillis() -
            folderImportStartTime;
            auditFileWriter.write(guid + "," + folderPath +
            ",Imported," + folderName + "," + timeToImport.
            toString());
            auditFileWriter.newLine();
        }

    }
    catch(Exception e)
    {
        consecutiveErrs ++;
        errorCount++;
        logger.warn(e.getMessage());
        if (consecutiveErrs > cemc.getImportMaxErrorCount())
        {
                auditFileWriter.write("Maximum number of conecutive
                errors reached - Exiting.");
            auditFileWriter.newLine();
             logger.info("Maximum number of consecutive errors
             reached - Exiting.");
            throw e;

        }
    }
```

```java
        }
        doUpdateBatch(true);
        auditFileWriter.write("Finished - time taken " +
        (System.currentTimeMillis() - importStartTime) + "milliseconds");
        auditFileWriter.newLine();
        auditFileWriter.write("Finished - folder processed - " +
        getProcessedCount());
        auditFileWriter.newLine();
        auditFileWriter.write("Finished - folder failed to be processed - " +
        errorCount.toString());
        auditFileWriter.newLine();
        auditFileWriter.close();

}
private void doUpdateBatch(boolean last){
    if (batchSize > 0){
        batchCount ++;
        if (last || (batchCount >= batchSize)){
            logger.debug("Commit batch");
            try {
                ub.updateBatch();
            }
            catch (Exception e){
                Iterator iterator1 = ub.getBatchItemHandles(null).
                iterator();
                while (iterator1.hasNext()) {
                    BatchItemHandle obj = (BatchItemHandle)
                    iterator1.next();
                    if (obj.hasException())
                    {
                        // Displays the exception, for the purpose of
                        brevity here.
                        EngineRuntimeException thrown = obj.
                        getException();
                        logger.error("Exception: " + thrown.
                        getMessage());
```

```
                        }
                    }
                }
                logger.debug("Commit batch OK");
                batchCount = 0;
            }
        }
    }
    //Creates the folder defined by the XML passed in.
    private Folder createFolder(String folderName, Id GUID,String folderPath,
    IndependentObject object) throws Exception {
        Folder parent  = os.get_RootFolder(); //SET AS DEFAULT
        //Read from folder path from input Folder object (passed in)
        String path = folderPath;
        logger.debug("Importing Folder " + path); //ASB 24/02/2022
        //ASB ... 2 020 Need to change to fetch for security ACL settings
        Folder currentFolder = (Folder) object;
        String folderclass = currentFolder.getClassName();
        String name = folderName;
        String sCreator = currentFolder.get_Creator();
        Date dateCreated = currentFolder.get_DateCreated();
        Date dateLastModified = currentFolder.get_DateLastModified();
        String sModifier = currentFolder.get_LastModifier();
        String sOwner = currentFolder.get_Owner();

        String pathParts[] = path.split("/");
        if (pathParts.length > 2){
            // the parent should already exist, so just get it
            String parentname = path.substring(0, path.lastIndexOf("/"));
            parent = (Folder)os.getObject("Folder", parentname);
        }
        try {
            //Create Folder with the same GUID as the Source Object
            Folder f = Factory.Folder.createInstance(os, folderclass,GUID);
            f.set_Parent(parent);
            f.set_FolderName(name);
```

```
            f.set_Creator(sCreator);               //ASB 17/02/2022
            f.set_DateCreated(dateCreated); //ASB 17/02/2022
            f.set_DateLastModified(dateLastModified);
            f.set_LastModifier(sModifier);
            f.set_Owner(sOwner);
            try {
            //ASB ... 2 020 Update Folder ACL from source Folder security
                int myFilterLevel = 1; //Changed from 4 to 1 -- For
                Performance
                PropertyFilter myFilter = new PropertyFilter();
                myFilter.setMaxRecursion(myFilterLevel);
                //ASB ... 4 020
                if(cemc.getDebugOutputFlag().equalsIgnoreCase("off")){
                    myFilter.addIncludeType(new FilterElement(null, null,
                    null, FilteredPropertyType.ANY_LIST, null));
                    myFilter.addIncludeType(new FilterElement(null, null,
                    null, FilteredPropertyType.ANY_SINGLETON, null));
                }else{
                    myFilter.addIncludeType(new FilterElement(null, null,
                    null, FilteredPropertyType.ANY, null));
                }
                Folder currentFolderACL = Factory.
                Folder.fetchInstance(osinput, GUID, myFilter);
                AccessPermissionList  aclIn = currentFolderACL.get_
                Permissions();
                // Add the permission to the list for the Object Store.
                //Get This group from the config.xml file
                String sExcludeGroup = cemc.getExcludedGroup();
                String sExcludeUser =  cemc.getExcludedUser();
                //ASB ... 4 027 For Performance check if a full LDAP search
                is required
                if(cemc.getLDAPSearchFlag().equalsIgnoreCase("on")){
                    aclIn = addPermissions(aclIn, sExcludeGroup,
                    sExcludeUser);
                }else {
```

```
                    }
                }
            }
            logger.debug("Commit batch OK");
            batchCount = 0;
        }
    }
}
//Creates the folder defined by the XML passed in.
private Folder createFolder(String folderName, Id GUID,String folderPath,
IndependentObject object) throws Exception {
    Folder parent  = os.get_RootFolder(); //SET AS DEFAULT
    //Read from folder path from input Folder object (passed in)
    String path = folderPath;
    logger.debug("Importing Folder " + path); //ASB 24/02/2022
    //ASB ... 2 020 Need to change to fetch for security ACL settings
    Folder currentFolder = (Folder) object;
    String folderclass = currentFolder.getClassName();
    String name = folderName;
    String sCreator = currentFolder.get_Creator();
    Date dateCreated = currentFolder.get_DateCreated();
    Date dateLastModified = currentFolder.get_DateLastModified();
    String sModifier = currentFolder.get_LastModifier();
    String sOwner = currentFolder.get_Owner();

    String pathParts[] = path.split("/");
    if (pathParts.length > 2){
        // the parent should already exist, so just get it
        String parentname = path.substring(0, path.lastIndexOf("/"));
        parent = (Folder)os.getObject("Folder", parentname);
    }
    try {
        //Create Folder with the same GUID as the Source Object
        Folder f = Factory.Folder.createInstance(os, folderclass,GUID);
        f.set_Parent(parent);
        f.set_FolderName(name);
```

```java
        f.set_Creator(sCreator);              //ASB 17/02/2022
        f.set_DateCreated(dateCreated); //ASB 17/02/2022
        f.set_DateLastModified(dateLastModified);
        f.set_LastModifier(sModifier);
        f.set_Owner(sOwner);
        try {
        //ASB ... 2 020 Update Folder ACL from source Folder security
            int myFilterLevel = 1; //Changed from 4 to 1 -- For
            Performance
            PropertyFilter myFilter = new PropertyFilter();
            myFilter.setMaxRecursion(myFilterLevel);
            //ASB ... 4 020
            if(cemc.getDebugOutputFlag().equalsIgnoreCase("off")){
                myFilter.addIncludeType(new FilterElement(null, null,
                null, FilteredPropertyType.ANY_LIST, null));
                myFilter.addIncludeType(new FilterElement(null, null,
                null, FilteredPropertyType.ANY_SINGLETON, null));
            }else{
                myFilter.addIncludeType(new FilterElement(null, null,
                null, FilteredPropertyType.ANY, null));
            }
            Folder currentFolderACL = Factory.
            Folder.fetchInstance(osinput, GUID, myFilter);
            AccessPermissionList  aclIn = currentFolderACL.get_
            Permissions();
             // Add the permission to the list for the Object Store.
             //Get This group from the config.xml file
             String sExcludeGroup = cemc.getExcludedGroup();
             String sExcludeUser =  cemc.getExcludedUser();
             //ASB ... 4 027 For Performance check if a full LDAP search
             is required
             if(cemc.getLDAPSearchFlag().equalsIgnoreCase("on")){
                 aclIn = addPermissions(aclIn, sExcludeGroup,
                 sExcludeUser);
             }else {
```

```java
                    aclIn = addPermissionsNoSearch(aclIn, sExcludeGroup,
                    sExcludeUser);
                }
                //ASB ... 4 027 aclIn = addPermissions(aclIn, sExcludeGroup,
                sExcludeUser);
                f.set_Permissions(aclIn);
            }catch(Exception aclErr){
                //Just warning here if we can't set for some obscure reason!
                logger.warn("Couldn't Add ACL Security for Folder " + aclErr.
                getMessage());
            }

            f.save(RefreshMode.REFRESH);
            return f;
        }
        catch (Exception e){
            logger.error(e.getMessage(), e);
        }
        return null;
    }
    private void init() throws Exception {
        cemc = new CEReplicateConfig();
        classNames = new HashMap();
        logger.debug("Getting Destination (5x) CE Connection ...");
        ceConn = new CEConnection("Target");
        logger.debug("Dest (5x) CE Connection OK");

        logger.debug("Getting Source (5x) CE Connection ...");
        con_MigrateDB = new CEConnection("Source");
        logger.debug("Migration Database Connection OK");
        //ASB ... 4 022 Cater for Folder subclasses
        String folderSubclasses = cemc.getFolderSubclasses();
        String[] folderItems = folderSubclasses.split(",");
        for (String item : folderItems)
        {
            //Add repeating units of
```

CHAPTER 4   A REPLICATION JAVA PROGRAM FOR IBM FILENET OBJECT STORES

```
            //" and (cd.SymbolicName = 'Folder') " etc
            //(cd.SymbolicName = 'Folder')  OR (cd.SymbolicName = 'FOLDER_
            ATTRIBUTE') etc
            String nextFolder = item;
            folderSymbolList = folderSymbolList + " (cd.SymbolicName = '" +
            nextFolder + "') ";
            folderExcludeSymbolList = folderExcludeSymbolList + " AND (cd.
            SymbolicName <>'" + nextFolder + "') ";
            folderSymbolList = folderSymbolList + "OR";
        }
        //Remove last OR
        folderSymbolList = folderSymbolList.substring(1,folderSymbolList.
        length()-2 );
    }
    private void sampleCode() {
        final String createUser = "CreateUser";     // we cant this on
                                                            version 1
        final String modUser = "ModUser";      // we cant this on version 2

        // Get the ObjectStore

        os = ceConn.getObjectStore("ECM");

        // Create a new properties collection
        Properties props = Factory.Document.createInstance(os, null).
        getProperties();
        // This collection is used when we do the checkout
        props.putValue("Creator", createUser);
        //props.putValue("DateCreated", theDate);
        props.putValue("LastModifier", modUser);
        //props.putValue("DateLastModified", theDate);

        // Folder here for easy find in Workplace
        Folder f = (Folder)os.getObject("Folder", "/asb");

        // Create the new document
        Document d = (Document)os.createObject("Document");
        d.getProperties().putValue("DocumentTitle", "Doc1234567890");
```

```java
    d.set_Creator(createUser);
    d.set_LastModifier(createUser);
    d.checkin(null, CheckinType.MAJOR_VERSION);
    d.save(RefreshMode.NO_REFRESH); // If you refresh, you cant reset the
    last modifier
    // Folder it (not needed, just makes it easier to find to see results)
    if (ceContainmentNameExtOn.equalsIgnoreCase("yes")){
        //ASB 01/03/2022 - Add Containment Name from the Title with no
        stripped extension
          folderDocWithExtension(d, f);
    }else if (ceContainmentNameExtOn.equalsIgnoreCase("no"))
    {
        //ASB 01/03/2022 - Add Containment Name from the Title with
        stripped extension
          folderDoc(d,f);
    }else{
//          ASB 01/03/2022 - Add Containment Name using API Default
          folderDocDefault(d,f);
    }

    // This makes sure the last modifier stays intact
    d.set_LastModifier(createUser);

    Id reservationId_null = null; String reservationClass_null = null; //
    to remind of the parameter types!
    // ASB Compare : d.checkout(ReservationType.EXCLUSIVE, GUID,
    docClass, null);
    d.checkout(ReservationType.EXCLUSIVE, reservationId_null,
    reservationClass_null, props);
    // Have to do a save or the doc doesnt get checked out, needs to be
    refresh or we cant get the reservation
    // object to check it back in
    d.save(RefreshMode.REFRESH);

    Document res = (Document)d.get_Reservation();
    res.getProperties().removeFromCache("LastModifier");
    res.set_LastModifier(modUser);
```

```java
            res.checkin(null,CheckinType.MAJOR_VERSION);
            res.save(RefreshMode.NO_REFRESH);
            //Encryption
            try {
                // Generate a temporary key. In practice, you would save this key.
                // See also Encrypting with DES Using a Pass Phrase.
                SecretKey key = KeyGenerator.getInstance("DES").generateKey();
                // Encrypt
                String encrypted = this.encrypt("Don't tell anybody!");

                // Decrypt
                String decrypted = this.decrypt(encrypted);
            } catch (Exception e) {
            }

    }
    private void folderDoc(Document doc, Folder folder){
            //ASB 01/03/2022 - Add Containment Name from the Title with stripped
            extension
        if (folder != null){
                String sContainmentName = "";        //FOR RCR:   //ASB 17/02/2022
                Add Containment name
                sContainmentName = doc.getProperties().get("DocumentTitle").
                getStringValue();       //FOR RCR: //ASB 17/02/2022 Add
                Containment name
                ReferentialContainmentRelationship rcr = folder.
                file(doc,AutoUniqueName.AUTO_UNIQUE,null,DefineSecurityParentage.
                DO_NOT_DEFINE_SECURITY_PARENTAGE);
                if(sContainmentName.lastIndexOf(".") > 0){        //FOR RCR:      //
                ASB 17/02/2022 Add Containment name
                        sContainmentName = sContainmentName.
                        substring(0,sContainmentName.lastIndexOf(".")); //FOR
                        RCR: //ASB 17/02/2022 extract filename
                } //FOR RCR:   //ASB 17/02/2022 Add Containment name
                 if(!sContainmentName.equalsIgnoreCase("")){  //FOR RCR:   //ASB
                17/02/2022 Add Containment name
```

```java
            rcr.set_ContainmentName(sContainmentName); //FOR RCR: //ASB
            17/02/2022 Add Containment name
        }   //FOR RCR: //ASB 17/02/2022 Add Containment name
    rcr.save(RefreshMode.NO_REFRESH);
    }

}

private void folderDoc(Document doc, String folderPathName, String
documentTitle){
    String sContainmentName = documentTitle;        //FOR RCR: //ASB
    17/02/2022 Add Containment name
    if (folderPathName != null && folderPathName.length() > 0){
        Folder folder = null;
        try {
            folder = Factory.Folder.getInstance(os, null,
            folderPathName);
            //folder = Factory.Folder.fetchInstance(os,
            folderName, null);
            com.filenet.api.core.
            DynamicReferentialContainmentRelationship rcr =
                    Factory.DynamicReferentialContainmentRelationship.c
                    reateInstance(os, null,
                            AutoUniqueName.AUTO_UNIQUE,
                            DefineSecurityParentage.DO_NOT_
                            DEFINE_SECURITY_PARENTAGE);
            rcr.set_Tail(folder);
            rcr.set_Head(doc);
            rcr.set_ContainmentName(sContainmentName);
          rcr.save(RefreshMode.NO_REFRESH);
        }
        catch (Exception e){

        }
    }
}
```

```java
private void folderDocWithExtension(Document doc, Folder folder){
    //ASB 01/03/2022 - Add Containment Name from the Title with no stripped extension
    if (folder != null){
        String sContainmentName = "";       //FOR RCR:  //ASB 17/02/2022 Add Containment name
        sContainmentName = doc.getProperties().get("DocumentTitle").getStringValue();       //FOR RCR: //ASB 17/02/2022 Add Containment name
        ReferentialContainmentRelationship rcr = folder.file(doc,AutoUniqueName.AUTO_UNIQUE,null,DefineSecurityParentage.DO_NOT_DEFINE_SECURITY_PARENTAGE);
         if(!sContainmentName.equalsIgnoreCase("")){   //FOR RCR:  //ASB 17/02/2022 Add Containment name
                rcr.set_ContainmentName(sContainmentName); //FOR RCR: // ASB 17/02/2022 Add Containment name
            }   //FOR RCR: //ASB 17/02/2022 Add Containment name
        rcr.save(RefreshMode.NO_REFRESH);
    }

}
private void folderDocDefault(Document doc, Folder folder){
    //ASB 01/03/2022 - Add Containment Name using the API Default
    if (folder != null){
        ReferentialContainmentRelationship rcr = folder.file(doc,AutoUniqueName.NOT_AUTO_UNIQUE,null,DefineSecurityParentage.DO_NOT_DEFINE_SECURITY_PARENTAGE);
        rcr.save(RefreshMode.NO_REFRESH);
    }

}
private boolean isClassExist(String className) throws Exception {
    if (classNames.containsKey(className))
        return true;
    try {
```

```java
            com.filenet.api.admin.ClassDefinition test = Factory.ClassDefinit
            ion.fetchInstance(os, className, null);
            classNames.put(test.get_Name(), test.get_Name());
            return true;
        }
        catch (EngineRuntimeException e){
            if (e.getExceptionCode() == ExceptionCode.E_BAD_CLASSID)
                return false;
            throw e;
        }
    }

    public static String encrypt(String str) {
        try {
            //encoding  byte array into base 64
            byte[] encoded = Base64.encodeBase64(str.getBytes());
            String encString = new String(encoded);
            return encString;
        } catch (Exception e) {
        }
        return null;
    }

    public static String decrypt(String str) {
        try {
            //decoding byte array into base64
            byte[] decoded = Base64.decodeBase64(str);
            String sDecoded =  new String(decoded);
            return sDecoded;
        } catch (Exception e) {
        }
        return null;
    }

}
```

CEReplicateConfig – 502 Lines of Java Code –fn_common.jar

```java
package com.ibm.ce.utils;
import java.io.BufferedWriter;
import java.io.File;
import java.io.FileInputStream;
import java.io.FileNotFoundException;
import java.io.FileWriter;
import java.io.IOException;
import java.io.InputStream;
import java.io.StringWriter;
import java.text.DateFormat;
import java.text.ParseException;
import java.text.SimpleDateFormat;
import java.util.Date;
import java.util.Locale;
import java.util.Vector;

import javax.xml.parsers.DocumentBuilder;
import javax.xml.parsers.DocumentBuilderFactory;
import javax.xml.parsers.ParserConfigurationException;
import javax.xml.transform.OutputKeys;
import javax.xml.transform.Transformer;
import javax.xml.transform.TransformerFactory;
import javax.xml.transform.dom.DOMSource;
import javax.xml.transform.stream.StreamResult;
import javax.xml.xpath.XPath;
import javax.xml.xpath.XPathConstants;
import javax.xml.xpath.XPathFactory;

import org.apache.commons.codec.binary.Base64;
import org.apache.log4j.Logger;
import org.w3c.dom.Document;
import org.w3c.dom.Node;
import org.w3c.dom.NodeList;
```

```java
import org.xml.sax.SAXException;

public class CEReplicateConfig {
    Logger logger = Logger.getLogger(CEReplicateConfig.class.getName());
    private Document dom;

    // Export Parameters
    private String exportOSName;
    private String exportCEUser;
    private String exportCEPassword;
//     private String exportSql;
    private Integer exportMaxDocCount;
    private Integer deleteMaxFolderCount;
    private Integer exportMaxRunTimeMinutes;
    private Integer exportMaxErrorCount;
    private Integer deleteMaxErrorCount; //ASB ... 3 003 Max Count of
    deleted Document errors
//ASB ... 4 004 006 add Document propertyList values
    private String exportAuditFile;
    private Integer exportFeedBatchSize;
    private Integer updateMaxErrorCount; //ASB ... 4 012 Max Count of
    updated Document errors
    // Import parameters
    private String importOSName;
    private String importOSRootFolder;
    private String importCEUrl;
    private String importCELoginConfig; //ASB
    private String importCEStanza;
    private String importCEUser;
    private String importCEPassword;
    private Integer importMaxErrorCount;
    private String importAuditFile;
    //ASB ... 3 003 Set up Delete Document Audit Log file
    private String deleteDocumentAuditFile;
    //ASB ... 3 003 Set up Delete Folder Audit Log file
    private String deleteFolderAuditFile;
```

```java
//ASB ... 4 012 Set up Update Document Audit Log file
private String updateDocumentAuditFile;
private Integer importMaxDocCount;
private Integer deleteMaxDocCount;
private Integer updateMaxDocCount; //ASB ... 4 012 Separate update document process
private Integer importMaxRunTimeMinutes;
private String importAuditFileFolder;
private Integer updatingBatchSize;

// ASB ... 2 022 Replicate Parameters
//private String   readOnlyGroup;
private String   excludedGroup;
private String   excludedUser;
private String   startSearchDate;
private String   deltaSearchDate; //ASB ... 4 009
private String   deltaHours = "01"; //ASB ... 4 020
private Integer maxSearchSize;
//ASB ... 4 020 Debug Flag for performance improvement
private String debugFlag;
//ASB ... 4 027 Get LDAP Search Flag for performance improvement
private String LDAPSearchFlag;
//ASB ... 4 022 Get Folder Subclass list
private String folderSubclasses;
private XPath xPath;
public static void main(String args[]) throws Exception {
    CEReplicateConfig cemc = new CEReplicateConfig();
}

public CEReplicateConfig() throws Exception {
    XPathFactory   factory= XPathFactory.newInstance();
    xPath=factory.newXPath();
    xPath.reset();

    getDocument();
    readConfig();
}
```

```java
private void readConfig() throws Exception {

    //ASB ... 2 022 Replication Configuration Parameters
    //readOnlyGroup = getXMLVal("/config/cereplicate/readonlygroup/
    text()");
    excludedGroup = getXMLVal("/config/cereplicate/excludedgroup/
    text()");
    excludedUser = getXMLVal("/config/cereplicate/excludeduser/
    text()");
    startSearchDate = getXMLVal("/config/cereplicate/startsearchdate/
    text()");
    maxSearchSize = Integer.parseInt(getXMLVal("/config/cereplicate/
    MaxSearchSize/text()"));
    //ASB ... 4 020 Retrieve Debug Output flag (off/on) value
    debugFlag = getXMLVal("/config/cereplicate/DebugOutputFlag/
    text()");
    //ASB ... 4 27 Retrieve LDAP Search flag (off/on) value
    LDAPSearchFlag = getXMLVal("/config/cereplicate/LDAPSearchFlag/
    text()");
    deltaHours = getXMLVal("/config/cereplicate/DeltaHours/text()");
    //ASB ... 4 022 Add FolderSubclasses retrieval
    folderSubclasses = getXMLVal("/config/cereplicate/
    FolderSubclasses/text()");
    // Export Parameters
    exportOSName = getXMLVal("/config/ceexport/osname/text()");
    exportCEUser = getXMLVal("/config/ceexport/ceuser/text()");
    exportCEPassword = getXMLVal("/config/ceexport/cepassword/
    text()");
    //ASB ... 4 001 Not in the form Encrypt3dpasswordvalueP4sswOrd
    String decryptedExportCEPassword = decrypt(exportCEPassword);
    if (decryptedExportCEPassword.startsWith("Encrypt3d")&&
    decryptedExportCEPassword.endsWith("P4sswOrd")  ){
        //ASB ... 4 001 Extract the actual password
            //
            exportCEPassword = decryptedExportCEPassword.substring(9,
            decryptedExportCEPassword.length()-8);
```

```
}else{
    //ASB ... 4 001 update the clear password
    String encryptedExportCEPassword = "Encrypt3d" +
    exportCEPassword + "P4ssw0rd";
    encryptedExportCEPassword = encrypt(encryptedExportCEP
    assword);
    System.out.print("\r\tENCRYPTED CE Export PASSWORD = " +
    encryptedExportCEPassword);
    System.out.print("\r\t");
    // ASB ... 4 001 Write encrypted password back to the
    config.xml file
    // ASB Note Forward slashes for a Linux path!!
      String file = System.getProperty("user.dir") + "/config/
      config.xml";                    //String file = "config.xml";
      // SET    file Name, root element, tag element, old value,
      new value
     updateXML(file,"ceexport","cepassword", exportCEPassword,
     encryptedExportCEPassword);
}

exportMaxDocCount = Integer.parseInt(getXMLVal("/config/ceexport/
MaxDocCount/text()"));
exportMaxRunTimeMinutes = Integer.parseInt(getXMLVal("/config/
ceexport/MaxRunTimeMinutes/text()"));
exportMaxErrorCount = Integer.parseInt(getXMLVal("/config/
ceexport/MaxConsecutiveErrors/text()"));
exportAuditFile = getXMLVal("/config/ceexport/AuditFile/text()");

String tmp = getXMLVal("/config/ceimport/UpdatingBatchSize/
text()");
if (tmp.length() > 0){
    try {
        updatingBatchSize = Integer.parseInt(tmp);
    }
```

```java
        catch (Exception e){
            updatingBatchSize = 250;
        }
    }
    else {
        updatingBatchSize = 250;
    }
    // import parameters
    importOSName = getXMLVal("/config/ceimport/osname/text()");
    importOSRootFolder = getXMLVal("/config/ceimport/osrootfolder/text()");
    importCEUrl = getXMLVal("/config/ceimport/ceurl/text()");
    importCEStanza = getXMLVal("/config/ceimport/cestanza/text()");
    importCELoginConfig = getXMLVal("/config/ceimport/celoginconfig/text()");
    importCEUser = getXMLVal("/config/ceimport/ceuser/text()");
    importCEPassword = getXMLVal("/config/ceimport/cepassword/text()");
    //ASB ... 4 001 Check if this is still clear ie
    //ASB ... 4 001 Not in the form Encrypt3dpasswordvalueP4ssw0rd
    String decryptedImportCEPassword = decrypt(importCEPassword);
    if (decryptedImportCEPassword.startsWith("Encrypt3d")&&
    decryptedImportCEPassword.endsWith("P4ssw0rd")   ){
        //ASB ... 4 001 Extract the actual password
            //
            importCEPassword = decryptedImportCEPassword.substring(9,
            decryptedImportCEPassword.length()-8);
    }else{
        //ASB ... 4 001 update the clear password
        String encryptedImportCEPassword = "Encrypt3d" +
        importCEPassword + "P4ssw0rd";
        encryptedImportCEPassword = encrypt(encryptedImportCEPassword);
        System.out.print("\r\tENCRYPTED CE IMPORT PASSWORD = " +
        encryptedImportCEPassword);
```

```java
                System.out.print("\r\t");
                // ASB ... 4 001 Write encrypted password back to the
                config.xml file
                  String file = System.getProperty("user.dir") + "/config/
                  config.xml";
                  // SET    file Name, root element, tag element, old value,
                  new value
                  updateXML(file,"ceimport","cepassword", importCEPassword,
                  encryptedImportCEPassword);
            }
            importMaxErrorCount = Integer.parseInt(getXMLVal("/config/
            ceimport/MaxConsecutiveErrors/text()"));
            importMaxDocCount = Integer.parseInt(getXMLVal("/config/ceimport/
            MaxDocCount/text()"));
            importMaxRunTimeMinutes = Integer.parseInt(getXMLVal("/config/
            ceimport/MaxRunTimeMinutes/text()"));
            importAuditFile = getXMLVal("/config/ceimport/AuditFile/text()");
            importAuditFileFolder= getXMLVal("/config/ceimport/
            AuditFileFolders/text()");
//ASB ... 3 003 Get Delete Document and Delete Folder parameters
            deleteMaxDocCount = Integer.parseInt(getXMLVal("/config/
            cereplicate/deleteMaxDocCount/text()"));
            deleteDocumentAuditFile = getXMLVal("/config/cereplicate/
            DeleteAuditFile/text()");
            deleteMaxErrorCount = Integer.parseInt(getXMLVal("/config/
            cereplicate/MaxConsecutiveDeleteErrors/text()"));
            deleteMaxFolderCount = Integer.parseInt(getXMLVal("/config/
            cereplicate/DeleteMaxFolderCount/text()"));
            deleteFolderAuditFile = getXMLVal("/config/cereplicate/
            DeleteFolderAuditFile/text()");
            //ASB ... 4 012 Get Update Document parameters
            updateMaxDocCount = Integer.parseInt(getXMLVal("/config/
            cereplicate/UpdateMaxDocCount/text()"));
            updateDocumentAuditFile = getXMLVal("/config/cereplicate/
            UpdateDocumentAuditFile/text()");
```

```java
        updateMaxErrorCount = Integer.parseInt(getXMLVal("/config/
        cereplicate/MaxConsecutiveUpdateErrors/text()"));
    }
    public void updateProcessDate(Date processDate){
        // ASB ... 4 001 Write Date String back to the config.xml file
        // Need date in the format 20130923T125628Z
          String file = System.getProperty("user.dir") + "/config/
          config.xml";
         SimpleDateFormat sdf = new SimpleDateFormat("yyyyMMdd");
         //Need 24 Hour clock 00 -> 23 !! above is 12 hour
         SimpleDateFormat sdfTime = new SimpleDateFormat("HHmmss");
         String newStartSearchDate = sdf.format(processDate)+ "T" +
         sdfTime.format(processDate)+ "Z";
            updateXML(file,"cereplicate","startsearchdate",
            startSearchDate, newStartSearchDate);
    }
     private void getDocument() throws Exception {
         DocumentBuilderFactory dbf = DocumentBuilderFactory.newI
         nstance();
         DocumentBuilder db = null;
         try {
             db = dbf.newDocumentBuilder();
         } catch (ParserConfigurationException e) {
             logger.error(e);
             throw e;
         }
         //ASB ... 4 001 Normalise path to the config we want to use
         InputStream in = new FileInputStream(System.getProperty("user.
         dir") + "/config/config.xml");
         dom = db.parse(in);
         in.close(); //ASB ... 4 001 Now close the config.xml stream
     }
     public String getXMLVal(String expression) throws Exception {
         Node n = (Node) xPath.evaluate(expression, dom,
         XPathConstants.NODE);
```

CHAPTER 4   A REPLICATION JAVA PROGRAM FOR IBM FILENET OBJECT STORES

```java
            if (n != null)
                return n.getNodeValue();
            return "";
    }
    //ASB ... 4 001 Put new value into the dom node
    private void updateXML(String file, String mainElement, String sTag,
    String strOldValue, String strNewValue) {
        //DocumentBuilderFactory docFactory = DocumentBuilderFactory.
        newInstance();
         Document doc = null;
        try {
            DocumentBuilderFactory docFactory = DocumentBuilderFactory.ne
            wInstance();
            DocumentBuilder docBuilder = docFactory.newDocumentBuilder();
            doc = docBuilder.parse(file);

            // Get the root element
            Node config = doc.getFirstChild();
            // Get the main element by tag name directly
            Node ceMain = doc.getElementsByTagName(mainElement).item(0);
            // loop the configuration main section child node
            NodeList list = ceMain.getChildNodes();

            for (int i = 0; i < list.getLength(); i++) {

                    Node node = list.item(i);

                // get the passaword element, and update the value
                if (sTag.equals(node.getNodeName())) {
                  node.setTextContent(strNewValue);
                }

            }
        } catch (ParserConfigurationException pce) {
          pce.printStackTrace();
        } catch (IOException ioe) {
        ioe.printStackTrace();
```

```java
        } catch (SAXException sae) {
          sae.printStackTrace();
        }
        catch(Exception ex) {
            String exError = ex.getMessage();
        }
        File filer = new File(file);

        saveToXML(filer,doc);

}       // Save the updated DOM into the XML back
 // Save the updated DOM into the XML back
 private void saveToXML(File file, Document doc) {
    try {
          TransformerFactory factory = TransformerFactory.newInstance();
          Transformer transformer = factory.newTransformer();
          transformer.setOutputProperty(OutputKeys.INDENT, "yes");

          StringWriter writer = new StringWriter();
          StreamResult result = new StreamResult(writer);
          DOMSource source = new DOMSource(doc);

          transformer.transform(source, result);

          String strTemp = writer.toString();

          FileWriter fileWriter = new FileWriter(file);
          BufferedWriter bufferedWriter = new
        BufferedWriter(fileWriter);

          bufferedWriter.write(strTemp);
          bufferedWriter.flush();
          bufferedWriter.close();
    }
    catch(Exception ex) {
    }
}
```

```java
public String[] getXMLVals(String expression) throws Exception {
    Vector vtmp = null;
    NodeList nl = null;
    vtmp = new Vector();
    nl = (NodeList) xPath.evaluate(expression, dom,
    XPathConstants.NODESET);
    for (int x = 0; x < nl.getLength(); x ++){
        vtmp.add(nl.item(x).getNodeValue());
    }
    return (String[])vtmp.toArray(new String[vtmp.size()]);
}
public String getExportOSName(){
    return exportOSName;
}
public String getExportCEUser(){
    return exportCEUser;
}
public String getExportCEPassword(){
    return exportCEPassword;
}

public Integer getExportMaxDocCount(){
    return exportMaxDocCount;
}
//ASB ... 3 003 Add Getter for exportMaxFolderCount the maximum number
of Folders to be deleted
public Integer getDeleteMaxFolderCount(){
    return deleteMaxFolderCount;
}
public Integer getExportMaxRunTimeMinutes(){
    return exportMaxRunTimeMinutes;
}
public Integer getMaxSearchSize(){
    return maxSearchSize;   //ASB ... 2 022
}
```

```java
public String getExcludedGroup(){
    return excludedGroup;   //ASB ... 4 007
}
public String getExcludedUser(){
    return excludedUser;   //ASB ... 4 007
}
public String getDebugOutputFlag(){
    return debugFlag;   //ASB ... 4 020
}
public String getLDAPSearchFlag(){
    return LDAPSearchFlag;   //ASB ... 4 027
}
public String getDeltaHours(){
    return deltaHours;   //ASB ... 4 020
}
public String getFolderSubclasses(){
    return folderSubclasses;   //ASB ... 4 007
}
public String getStartSearchDate(){
     Date processDate = new Date();
    SimpleDateFormat sdf = new SimpleDateFormat("yyyyMMdd",Locale.ENGLISH); //ASB ... 4 009 Added Locale
    Date javaStartDate = null;
    //Need 24 Hour clock 00 -> 23 !! hh is 12 hour
    SimpleDateFormat sdfTime = new SimpleDateFormat("HHmmss",Locale.ENGLISH); //ASB ... 4 009 Added Locale
   //deltaSearchDate = sdf.format(processDate)+ "T" + sdfTime.format(processDate).substring(0,2)+"0000Z";
   //ASB ... 4 013 Temporary change to midnight delta
      //deltaSearchDate = sdf.format(processDate)+ "T" + sdfTime.format(processDate)+ "Z";
    //deltaSearchDate = sdf.format(processDate)+ "T000000Z";
   //ASB ... 4 009 Check if the current date was manually set - ie is greater
   //               than one hour difference from the current time.
```

```
//                 If date was set manually then use this date
                   instead as the search date.
DateFormat diffDateFormat = DateFormat.getDateInstance();
String sDiffDate = startSearchDate.substring(0, 8);
String sDiffTime = startSearchDate.substring(9,15);
Date diffDate = null;
Date diffTime = null;
Date dDeltaTime = null;
Long dateSeconds = null;
try {
        //ASB ... 4 010 Need to subtract off the java start date
        javaStartDate = sdf.parse("19700101"); //ASB ... 4 010
        diffDate = sdf.parse(sDiffDate) ;
        diffTime = sdfTime.parse(sDiffTime) ;
        //Test for zero
        Long searchhours = (long) 0;
        if((diffTime.getTime() - javaStartDate.getTime() - Long.parse
        Long(deltaHours)*3600000)  > 0)
            searchhours =(diffTime.getTime() - javaStartDate.
            getTime() - Long.parseLong(deltaHours)*3600000)/3600000;
        String sSearchHours = searchhours.toString();
        if (sSearchHours.length() == 1)sSearchHours = "0"+
        sSearchHours;
        dDeltaTime =  sdfTime.parse(sSearchHours + "0000");

           deltaSearchDate = sdf.format(processDate) + "T" +
           sSearchHours + "0000Z";
        //ASB ... 4 010 Added subtraction of the Java Start Date
        milliseconds
        //dateSeconds = diffDate.getTime() + diffTime.getTime() -
        javaStartDate.getTime();
        dateSeconds = (diffDate.getTime() + (diffTime.getTime()) -
        (dDeltaTime.getTime()));
} catch (ParseException e) {
        // Should be a controlled format - set to default if we
        get here
```

```java
public String getExcludedGroup(){
    return excludedGroup;   //ASB ... 4 007
}
public String getExcludedUser(){
    return excludedUser;   //ASB ... 4 007
}
public String getDebugOutputFlag(){
    return debugFlag;   //ASB ... 4 020
}
public String getLDAPSearchFlag(){
    return LDAPSearchFlag;   //ASB ... 4 027
}
public String getDeltaHours(){
    return deltaHours;   //ASB ... 4 020
}
public String getFolderSubclasses(){
    return folderSubclasses;   //ASB ... 4 007
}
public String getStartSearchDate(){
     Date processDate = new Date();
    SimpleDateFormat sdf = new SimpleDateFormat("yyyyMMdd",Locale.ENGLISH); //ASB ... 4 009 Added Locale
    Date javaStartDate = null;
    //Need 24 Hour clock 00 -> 23 !! hh is 12 hour
    SimpleDateFormat sdfTime = new SimpleDateFormat("HHmmss",Locale.ENGLISH); //ASB ... 4 009 Added Locale
 //deltaSearchDate = sdf.format(processDate)+ "T" + sdfTime.format(processDate).substring(0,2)+"0000Z";
 //ASB ... 4 013 Temporary change to midnight delta
   //deltaSearchDate = sdf.format(processDate)+ "T" + sdfTime.format(processDate)+ "Z";
    //deltaSearchDate = sdf.format(processDate)+ "T000000Z";
 //ASB ... 4 009 Check if the current date was manually set - ie is greater
 //             than one hour difference from the current time.
```

CHAPTER 4   A REPLICATION JAVA PROGRAM FOR IBM FILENET OBJECT STORES

```java
            //              If date was set manually then use this date
                        instead as the search date.
            DateFormat diffDateFormat = DateFormat.getDateInstance();
            String sDiffDate = startSearchDate.substring(0, 8);
            String sDiffTime = startSearchDate.substring(9,15);
            Date diffDate = null;
            Date diffTime = null;
            Date dDeltaTime = null;
            Long dateSeconds = null;
            try {
                //ASB ... 4 010 Need to subtract off the java start date
                javaStartDate = sdf.parse("19700101"); //ASB ... 4 010
                diffDate = sdf.parse(sDiffDate) ;
                diffTime = sdfTime.parse(sDiffTime) ;
                //Test for zero
                Long searchhours = (long) 0;
                if((diffTime.getTime() - javaStartDate.getTime() - Long.parse
                Long(deltaHours)*3600000)  > 0)
                    searchhours =(diffTime.getTime() - javaStartDate.
                    getTime() - Long.parseLong(deltaHours)*3600000)/3600000;
                String sSearchHours = searchhours.toString();
                if (sSearchHours.length() == 1)sSearchHours = "0"+
                sSearchHours;
                dDeltaTime =  sdfTime.parse(sSearchHours + "0000");

                    deltaSearchDate = sdf.format(processDate) + "T" +
                    sSearchHours + "0000Z";
                //ASB ... 4 010 Added subtraction of the Java Start Date
                milliseconds
                //dateSeconds = diffDate.getTime() + diffTime.getTime() -
                javaStartDate.getTime();
                dateSeconds = (diffDate.getTime() + (diffTime.getTime()) -
                (dDeltaTime.getTime()));
            } catch (ParseException e) {
                // Should be a controlled format - set to default if we
                get here
```

```java
            e.printStackTrace();
     }
    //ASB ... 4 009 Get Process Date in milliseconds
    Long milliSecondsLastDate  = processDate.getTime();
    //ASB ... 4 013 Temporary change to midnight delta //WAS 3600000
    //if ((milliSecondsLastDate - dateSeconds) < 86400000 ){
    //ASB ... 4 020 Allow 30 Minute delta on top
    if ((milliSecondsLastDate - dateSeconds) < (diffTime.getTime()-
    javaStartDate.getTime() )){
        startSearchDate = deltaSearchDate; //Allow 60 Minutes to fix
        missing documents
    }
     return startSearchDate; //ASB ... 2 022
}
public Integer getExportMaxErrorCount(){
    return exportMaxErrorCount;
}
//ASB ... 3 003 Retrieve the maximum allowed Delete Document errors
public Integer getDeleteMaxErrorCount(){
    return deleteMaxErrorCount;
}
//ASB ... 4 012 Retrieve the maximum allowed Update Document errors
public Integer getUpdateMaxErrorCount(){
    return updateMaxErrorCount;
}
public String getExportAuditFile(){
    return exportAuditFile;
}
public String getImportOSName(){
    return importOSName;
}
public String getImportOSRootFoler(){
    return importOSRootFolder;
}
public String getImportCEUrl(){
```

```java
        return importCEUrl;
    }
    public String getImportCEStanza(){
        return importCEStanza;
    }
    public String getImportCELoginConfig(){
        return importCELoginConfig;
    }
    public String getImportCEUser(){
        return importCEUser;
    }
    public String getImportCEPassword(){
        return importCEPassword;
    }
    public Integer getImportMaxErrorCount(){
        return importMaxErrorCount;
    }
    public Integer getImportMaxDocCount(){
        return importMaxDocCount;
    }
    //ASB ... 3 003 Getter Method for Delete Max Document Count
    public Integer getDeleteMaxDocCount(){
        return deleteMaxDocCount;
    }
    //ASB ... 4 012 Getter Method for Update Max Document Count
    public Integer getUpdateMaxDocCount(){
        return updateMaxDocCount;
    }
    public Integer getImportMaxRunTimeMinutes(){
        return importMaxRunTimeMinutes;
    }

    public String getImportAuditFile(){
        return importAuditFile;
    }
//ASB ... 3 003 Set Delete Document Audit File
```

```java
public String getDeleteDocumentAuditFile(){
    return deleteDocumentAuditFile;
}
//ASB ... 3 003 Set Delete Folder Audit File
public String getDeleteFolderAuditFile(){
    return deleteFolderAuditFile;
}
//ASB ... 4 012 Set Update Document Audit File
    public String getUpdateDocumentAuditFile(){
        return updateDocumentAuditFile;
    }
public String getImportAuditFileFolders(){
    return importAuditFileFolder;
}
public int getUpdatingBatchSize(){
    return updatingBatchSize;
}
 private static String encrypt(String str) {
        try {
            //encoding  byte array into base 64
            byte[] encoded = Base64.encodeBase64(str.getBytes());
            String encString = new String(encoded);
            return encString;
        } catch (Exception e) {
        }
        return null;
    }

    private static String decrypt(String str) {
        try {
            //decoding byte array into base64
            byte[] decoded = Base64.decodeBase64(str);
            String sDecoded =  new String(decoded);
    return sDecoded;
        } catch (Exception e) {
        }
```

```
            return null;
        }
}
```

Part 5 – Code Listing – CEMigrateConfig

The **CEMigrateConfig.java** Java code file supports the Main Java Replication program by retrieving the Object Store administration username and password credentials. On a first call to access the config.xml file, it checks if the password requires encryption and updates the config.xml to store an encrypted version of the password for additional security.

It also reads in the replication XML parameters in the **config.xml** file for the search date and source and target object store names and the root folder name to be replicated.

CEMigrateConfig – 425 Lines of Java Code – fn_common.jar

```java
package com.ibm.ce.utils;

import java.io.File;
import java.io.FileInputStream;
import java.io.FileNotFoundException;
import java.io.FileOutputStream;
import java.io.IOException;
import java.io.InputStream;
import java.io.OutputStream;
import java.util.Enumeration;
import java.util.Vector;

import javax.xml.parsers.DocumentBuilder;
import javax.xml.parsers.DocumentBuilderFactory;
import javax.xml.parsers.ParserConfigurationException;
import javax.xml.xpath.XPath;
import javax.xml.xpath.XPathConstants;
import javax.xml.xpath.XPathFactory;
```

//ASB ... 4 001 Encryption / Decryption Java Classes
```java
import javax.crypto.KeyGenerator;
import javax.crypto.SecretKey;
import java.io.UnsupportedEncodingException;
import java.security.spec.AlgorithmParameterSpec;
import java.security.spec.KeySpec;

import javax.crypto.Cipher;
import javax.crypto.IllegalBlockSizeException;
import javax.crypto.SecretKey;
import javax.crypto.SecretKeyFactory;
import javax.crypto.spec.PBEKeySpec;
import javax.crypto.spec.PBEParameterSpec;
import javax.lang.model.element.Element;

import java.io.BufferedWriter;
import java.io.File;
import java.io.FileWriter;
import java.io.StringWriter;
import java.net.URL;

import javax.xml.parsers.DocumentBuilder;
import javax.xml.parsers.DocumentBuilderFactory;
import javax.xml.transform.OutputKeys;
import javax.xml.transform.Transformer;
import javax.xml.transform.TransformerException;
import javax.xml.transform.TransformerFactory;
import javax.xml.transform.dom.DOMSource;
import javax.xml.transform.stream.StreamResult;
```
//ASB ... 4 END

```java
import org.apache.commons.codec.binary.Base64;
import org.apache.log4j.Logger;
import org.apache.xml.serialize.XMLSerializer;
import org.w3c.dom.Document;
import org.w3c.dom.Node;
```

CHAPTER 4   A REPLICATION JAVA PROGRAM FOR IBM FILENET OBJECT STORES

```java
import org.w3c.dom.NodeList;
import org.xml.sax.SAXException;

// ASB Software Development Limited - import updated to point to the Apache
equivalent class 29/07/2022
//import com.sun.org.apache.xml.internal.serialize.OutputFormat;
import org.apache.xml.serialize.OutputFormat;

public class CEMigrateConfig {
    Logger logger = Logger.getLogger(CEMigrateConfig.class.getName());
    private Document dom;

    private String sourceCEDBDriverClasses[];
    private String sourceCEDBUserName;
    private String sourceCEDBPassword;
    private String sourceCEDBURL;

    //ASB ... 3 001 - if needed!
    private String targetCEDBURL;
    private String targetCEDBUserName;
    private String targetCEDBPassword;
    private String targetStanza;

    private String CEMigrateDBDriverClasses[];
    private String CEMigrateDBUserName;
    private String CEMigrateDBPassword;
    private String CEMigrateDBURL;

    // Export Parameters
    private boolean isOracle;
    private String exportPath;
    private String exportOSName;
    private String exportCEUser;
    private String exportCEPassword;
    private String exportSql;
    private Integer exportMaxDocCount;
    private Integer exportMaxRunTimeMinutes;
    private Integer exportMaxErrorCount;
```

```java
private String exportSelectFromDocsForExportSQL;
private String exportAuditFile;
private Integer exportFeedBatchSize;
private int bytesize;
private int threadCount;
private int threadPoolMaxQueueSize;
private int SQLBatchSize;

//ASB ... 3 001 Export parameters
private String exportCEUrl;
private String exportCELoginConfig; //ASB
private String exportCEWaspLocation;//ASB
private String exportContainmentNameExtension;   //ASB 01/03/2011
Switch = Yes then set Containment Name Extension
private String exportCEStanza;

// Import parameters
private String importPath;
private String importOSName;
private String importOSRootFolder;
private String importCEUrl;
private String importCELoginConfig; //ASB
private String importCEWaspLocation;//ASB
private String importContainmentNameExtension;   //ASB 01/03/2011
Switch = Yes then set Containment Name Extension
private String importCEStanza;
private String importCEUser;
private String importCEPassword;
private Integer importMaxErrorCount;
private String importAuditFile;
private String importSelectFromDocsForImportSQL;
private Integer importMaxDocCount;
private Integer importMaxRunTimeMinutes;
private String importAuditFileFolder;
private Integer updatingBatchSize;

private XPath xPath;
```

```java
public static void main(String args[]) throws Exception {
    CEMigrateConfig cemc = new CEMigrateConfig();
}

public CEMigrateConfig() throws Exception {
    XPathFactory  factory= XPathFactory.newInstance();
    xPath=factory.newXPath();
    xPath.reset();

    getDocument();
    readConfig();
}

private void readConfig() throws Exception {
    // Export Parameters
    //exportPath = getXMLVal("/config/ceexport/exportpath/text()");
    exportOSName = getXMLVal("/config/ceexport/osname/text()");
    exportCEUser = getXMLVal("/config/ceexport/ceuser/text()");
  //ASB ... 4 001 Password is returned encrypted now
    exportCEPassword = getXMLVal("/config/ceexport/cepassword/text()");
    //ASB ... 4 001 Not in the form Encrypt3dpasswordvalueP4ssw0rd
    String decryptedExportCEPassword = decrypt(exportCEPassword);
    if (decryptedExportCEPassword.startsWith("Encrypt3d")&&
    decryptedExportCEPassword.endsWith("P4ssw0rd")   ){
       //ASB ... 4 001 Extract the actual password
          //
          exportCEPassword = decryptedExportCEPassword.substring(9,
          decryptedExportCEPassword.length()-8);
    }else{
         //ASB ... 4 001 update the clear password
         String encryptedExportCEPassword = "Encrypt3d" +
       exportCEPassword + "P4ssw0rd";
         encryptedExportCEPassword = encrypt(encryptedExportCEP
         assword);
         System.out.print("\r\tENCRYPTED CE Export PASSWORD = " +
         encryptedExportCEPassword);
```

```java
    System.out.print("\r\t");
    // ASB ... 4 001 Write encrypted password back to the
    config.xml file
      String file = System.getProperty("user.dir") + "/config/
      config.xml";
      //String file = "config.xml";
      // SET    file Name, root element, tag element, old value,
      new value
    updateXML(file,"ceexport","cepassword", exportCEPassword,
    encryptedExportCEPassword);
}
exportMaxDocCount = Integer.parseInt(getXMLVal("/config/ceexport/
MaxDocCount/text()"));

exportMaxRunTimeMinutes = Integer.parseInt(getXMLVal("/config/
ceexport/MaxRunTimeMinutes/text()"));
exportMaxErrorCount = Integer.parseInt(getXMLVal("/config/
ceexport/MaxConsecutiveErrors/text()"));
exportAuditFile = getXMLVal("/config/ceexport/AuditFile/text()");
String tmp = getXMLVal("/config/ceimport/UpdatingBatchSize/
text()");
if (tmp.length() > 0){
    try {
        updatingBatchSize = Integer.parseInt(tmp);
    }
    catch (Exception e){
        updatingBatchSize = 250;
    }
}
else {
    updatingBatchSize = 250;
}
// import parameters
importOSName = getXMLVal("/config/ceimport/osname/text()");
importOSRootFolder = getXMLVal("/config/ceimport/osrootfolder/
text()");
```

CHAPTER 4   A REPLICATION JAVA PROGRAM FOR IBM FILENET OBJECT STORES

```java
        importCEUrl = getXMLVal("/config/ceimport/ceurl/text()");
        importCEStanza = getXMLVal("/config/ceimport/cestanza/text()");
        importCELoginConfig = getXMLVal("/config/ceimport/celoginconfig/
        text()");
        importCEUser = getXMLVal("/config/ceimport/ceuser/text()");
    //ASB ... 4 001 Password is returned encrypted now
        importCEPassword = getXMLVal("/config/ceimport/cepassword/
        text()");
        //ASB ... 4 001 Check if this is still clear ie
        //ASB ... 4 001 Not in the form Encrypt3dpasswordvalueP4sswOrd
        String decryptedImportCEPassword = decrypt(importCEPassword);
        if (decryptedImportCEPassword.startsWith("Encrypt3d")&&
        decryptedImportCEPassword.endsWith("P4sswOrd")   ){
          //ASB ... 4 001 Extract the actual password
            //
            importCEPassword = decryptedImportCEPassword.substring(9,
            decryptedImportCEPassword.length()-8);
    }else{
            //ASB ... 4 001 update the clear password
            String encryptedImportCEPassword = "Encrypt3d" +
            importCEPassword + "P4sswOrd";
            encryptedImportCEPassword = encrypt(encryptedImportCEP
            assword);
            System.out.print("\r\tENCRYPTED CE IMPORT PASSWORD = " +
            encryptedImportCEPassword);
            System.out.print("\r\t");
            // ASB ... 4 001 Write encrypted password back to the
            config.xml file
              String file = System.getProperty("user.dir") + "/config/
              config.xml";
              // SET    file Name, root element, tag element, old value,
              new value
             updateXML(file,"ceimport","cepassword", importCEPassword,
             encryptedImportCEPassword);
        }
```

```java
        importMaxErrorCount = Integer.parseInt(getXMLVal("/config/
        ceimport/MaxConsecutiveErrors/text()"));
        importMaxDocCount = Integer.parseInt(getXMLVal("/config/ceimport/
        MaxDocCount/text()"));
        importMaxRunTimeMinutes = Integer.parseInt(getXMLVal("/config/
        ceimport/MaxRunTimeMinutes/text()"));
        importAuditFile = getXMLVal("/config/ceimport/AuditFile/text()");
        importAuditFileFolder= getXMLVal("/config/ceimport/
        AuditFileFolders/text()");

        //ASB ... 3 001 export parameters for Session Object
        exportCEUrl = getXMLVal("/config/ceexport/ceurl/text()");
        exportCEStanza = getXMLVal("/config/ceexport/cestanza/
        text()"); //null
        exportCELoginConfig = getXMLVal("/config/ceexport/celoginconfig/
        text()");
    }
    private void getDocument() throws Exception {
        DocumentBuilderFactory dbf = DocumentBuilderFactory.newI
        nstance();
        DocumentBuilder db = null;
        try {
            db = dbf.newDocumentBuilder();
        } catch (ParserConfigurationException e) {
            logger.error(e);
            throw e;
        }
        //ASB ... 4 001 Normalise path to the config we want to use
        InputStream in = new FileInputStream(System.getProperty("user.
        dir") + "/config/config.xml");
        dom = db.parse(in);
        in.close(); //ASB ... 4 001 Now close the confog.xml file
    }
    //ASB ... 4 001 Create new Write method
// Save the updated DOM into the XML back
```

```java
    public String getXMLVal(String expression) throws Exception {
        Node n = (Node) xPath.evaluate(expression, dom,
        XPathConstants.NODE);
        if (n != null)
            return n.getNodeValue();
        return "";
    }
    //ASB ... 4 001 Put new value into the dom node
    private void updateXML(String file, String mainElement, String sTag,
    String strOldValue, String strNewValue) {
        Document doc = null;
        try {
            DocumentBuilderFactory docFactory = DocumentBuilderFactory.ne
            wInstance();
            DocumentBuilder docBuilder = docFactory.newDocumentBuilder();
            doc = docBuilder.parse(file);

            // Get the root element
            Node config = doc.getFirstChild();
            // Get the main element by tag name directly
            Node ceMain = doc.getElementsByTagName(mainElement).item(0);
            // loop the configuration main section child node
            NodeList list = ceMain.getChildNodes();

            for (int i = 0; i < list.getLength(); i++) {

                Node node = list.item(i);

                // get the passaword element, and update the value
                if (sTag.equals(node.getNodeName())) {
                    node.setTextContent(strNewValue);
                }

            }
        } catch (ParserConfigurationException pce) {
            pce.printStackTrace();
        } catch (IOException ioe) {
        ioe.printStackTrace();
```

```java
    } catch (SAXException sae) {
      sae.printStackTrace();
    }
    catch(Exception ex) {
        String exError = ex.getMessage();
    }
    File filer = new File(file);

    saveToXML(filer,doc);

}    // Save the updated DOM into the XML back
// Save the updated DOM into the XML back
private void saveToXML(File file, Document doc) {
    try {
        TransformerFactory factory = TransformerFactory.newInstance();
        Transformer transformer = factory.newTransformer();
        transformer.setOutputProperty(OutputKeys.INDENT, "yes");

        StringWriter writer = new StringWriter();
        StreamResult result = new StreamResult(writer);
        DOMSource source = new DOMSource(doc);

        transformer.transform(source, result);

        String strTemp = writer.toString();

        FileWriter fileWriter = new FileWriter(file);
        BufferedWriter bufferedWriter = new
        BufferedWriter(fileWriter);

        bufferedWriter.write(strTemp);
        bufferedWriter.flush();
        bufferedWriter.close();
    }
    catch(Exception ex) {
    }
}
```

```java
    public String[] getXMLVals(String expression) throws Exception {
        Vector vtmp = null;
        NodeList nl = null;
        vtmp = new Vector();
        nl = (NodeList) xPath.evaluate(expression, dom,
        XPathConstants.NODESET);
        for (int x = 0; x < nl.getLength(); x ++){
            vtmp.add(nl.item(x).getNodeValue());
        }
        return (String[])vtmp.toArray(new String[vtmp.size()]);
    }
    public String getExportOSName(){
        return exportOSName;
    }
    public String getExportCEUser(){
        return exportCEUser;
    }
    public String getExportCEPassword(){
        return exportCEPassword;
    }
    public Integer getExportMaxDocCount(){
        return exportMaxDocCount;
    }
    public Integer getExportMaxRunTimeMinutes(){
        return exportMaxRunTimeMinutes;
    }
    public Integer getExportMaxErrorCount(){
        return exportMaxErrorCount;
    }
    public String getExportAuditFile(){
        return exportAuditFile;
    }
    public String getImportOSName(){
        return importOSName;
    }
```

```java
public String getImportOSRootFoler(){
    return importOSRootFolder;
}

//ASB ... 3 001 add getter Methods for export Session Object
public String getExportCEUrl(){
    return exportCEUrl;
}
public String getExportCEStanza(){
    return exportCEStanza;
}
public String getExportCELoginConfig(){
    return exportCELoginConfig;
}
//ASB ... 3 001 add END

public String getImportCEUrl(){
    return importCEUrl;
}
public String getImportCEStanza(){
    return importCEStanza;
}
public String getImportCELoginConfig(){
    return importCELoginConfig;
}
public String getImportCEUser(){
    return importCEUser;
}
public String getImportCEPassword(){
    return importCEPassword;
}
public Integer getImportMaxErrorCount(){
    return importMaxErrorCount;
}
public Integer getImportMaxDocCount(){
    return importMaxDocCount;
}
```

```java
    public Integer getImportMaxRunTimeMinutes(){
        return importMaxRunTimeMinutes;
    }
    public String getImportAuditFile(){
        return importAuditFile;
    }
    public String getImportAuditFileFolders(){
        return importAuditFileFolder;
    }
    public int getUpdatingBatchSize(){
        return updatingBatchSize;
    }
     private static String encrypt(String str) {
            try {
                //encoding  byte array into base 64
                byte[] encoded = Base64.encodeBase64(str.getBytes());
                String encString = new String(encoded);
                return encString;
            } catch (Exception e) {
            }
            return null;
        }

        private static String decrypt(String str) {
            try {
                //decoding byte array into base64
                byte[] decoded = Base64.decodeBase64(str);
                String sDecoded =  new String(decoded);
            return sDecoded;
            } catch (Exception e) {
            }
            return null;
        }

}
```

Part 6 – Code Listing – Supporting Utility Code

This PropsUtil Java code is one of a number of general supporting utility methods described in this part. The code is used to retrieve properties from a cebase.properties file containing external parameters to support access to Object Stores. (It is not used in the example replication program described in this chapter, but is bundled as a utility method.)

PropsUtil – 68 Lines of Java Code – fn_utils.jar

```java
package com.asb.utils;

import org.apache.log4j.Logger;

import java.util.Properties;
import java.io.IOException;
import java.io.InputStream;
import java.io.FileInputStream;

/**
 *
 */
public class PropsUtil {
    public PropsUtil() {}

    public String getProperty(Properties theProperties, String strProperty,
    String strDefault){
        if (theProperties.getProperty(strProperty) != null){
            return theProperties.getProperty(strProperty);
        }
        else {
            theLog.warn("Did not find " + strProperty + " property in
            cebase.properties");
            theProperties.setProperty(strProperty, strDefault);
            return strDefault;
        }
    }
}
```

```java
    public Properties getProps(String propFile) throws Exception {
        return pgetProps(propFile,false);
    }
    public Properties getProps(Properties theProps, String propFile) throws Exception {

        return pgetProps(theProps,propFile,false);
    }
    public Properties getProps(String propFile, boolean fullPath) throws Exception {
        return pgetProps(propFile,fullPath);
    }
    private Properties pgetProps(String propFile, boolean fullPath) throws Exception {
        Properties theProps = new Properties();
        return pgetProps(theProps, propFile, fullPath);
    }
    private Properties pgetProps(Properties theProps, String propFile, boolean fullPath) throws Exception {
        if (theProps == null)
            theProps = new Properties();

        InputStream in = null;
        if (fullPath == false){
            in = PropsUtil.class.getClassLoader().
            getResourceAsStream(propFile);
        }
        else {
            in = new FileInputStream(propFile);
        }

        try {
            theProps.load(in);

        }
        catch (IOException e) {
            theLog.debug(e.getMessage());
        }
```

```java
        catch (NullPointerException e){
            Exception theE = new Exception("Properties file " + propFile +
            " not found");
            throw theE;
        }
        return theProps;
    }
    static Logger theLog = Logger.getLogger(PropsUtil.class.getName());
}
```

The CEConnect Java code is one of a number of general supporting utility methods described in this part. The method is a helper used to facilitate connection to the IBM FileNet Object stores through the Web Service Interface SOAP calls.

CEConnect – 67 Lines of Java Code – fn_connect.jar

```java
package com.asb.filenet.ce;

import javax.security.auth.Subject;

//import org.apache.log4j.Logger;

import com.filenet.api.core.Connection;
import com.filenet.api.core.Factory;
import com.filenet.api.util.UserContext;

public class CEConnect {
    private String stanza = null;
    private Connection conn = null;
    private UserContext uc = null;
    private String ceUri = null;
    private Subject subject = null;

    //private static Logger logger = Logger.getLogger(CEConnect.class.getName());
```

CHAPTER 4   A REPLICATION JAVA PROGRAM FOR IBM FILENET OBJECT STORES

```java
/**
 *
 * @param ceUrl
 */
/**
 * @param ceUrl
 */
public CEConnect(String ceUrl){
    this.stanza = "FileNetP8WSI";
    this.ceUri = ceUrl;
    connect();
}
/**
 * @param ceUrl
 * @param stanza
 */
public CEConnect(String ceUrl, String stanza){
    this.stanza = stanza;
    this.ceUri = ceUrl;
    connect();
}
private void connect(){
    conn = Factory.Connection.getConnection(this.ceUri);
    uc = UserContext.get();
}
/**
 * @param userName
 * @param password
 */
public void logon(String userName, String password){
    subject = UserContext.createSubject(conn, userName, password,
    this.stanza);
    uc.pushSubject(subject);
}
```

```java
/**
 * @return
 */
public String getCEUri(){
    return this.ceUri;
}
public Connection getConnection(){
    return conn;
}
public void popSubject(){
    uc.popSubject();
}
public void pushSubject(){
    uc.pushSubject(subject);
}
}
```

CEConnection – 119 Lines of Java Code – fn_connection.jar

The CEConnection Java code is one of a number of general supporting utility methods described in this part. It is the main method called by the Replication program to sign in the IBM FileNet administration user to the source and target Object Stores. It also sets the necessary user security access to give special access levels to enable the Document and Folder object system properties for Date Created and Date Modified and the unique GUID identifier to be overwritten to create an exact replica of the source objects.

```java
package com.ibm.filenet.ce.CEConnection;
import org.apache.log4j.Level;
import org.apache.log4j.Logger;

import com.asb.filenet.ce.CEConnect;

import com.filenet.api.collection.AccessPermissionList;
import com.filenet.api.constants.AccessRight;
```

```java
import com.filenet.api.constants.AccessType;
import com.filenet.api.constants.RefreshMode;
import com.filenet.api.core.Domain;
import com.filenet.api.core.Factory;
import com.filenet.api.core.Factory.AccessPermission;
import com.filenet.api.core.ObjectStore;
import com.ibm.ce.utils.CEMigrateConfig;
//import com.ibm.ce.utils.Config;
//ASB Software Development Limited   30th July 2022 - Added as below
import com.ibm.ce.utils.CEReplicateConfig;

public class CEConnection {

    private static String uri;
    private static String username;
    private static String password;
    private static String stanza;
    private static String targetURI;
    private static String targetUserName;
    private static String targetPassword;
    private static String targetStanza;

    private CEConnect ceConnect = null;
    private Domain d = null;

    private static Logger logger = Logger.getLogger(CEConnection.class.getName());

    static {
        try {
        CEMigrateConfig cemc = new CEMigrateConfig();
        uri = cemc.getExportCEUrl();
        username = cemc.getExportCEUser();//Administrator
        password = cemc.getExportCEPassword(); //filenet
        stanza = cemc.getExportCEStanza(); //null
        targetURI = cemc.getImportCEUrl();
        targetUserName = cemc.getImportCEUser();
        targetPassword = cemc.getImportCEPassword(); //filenet
```

```java
            targetStanza = cemc.getImportCEStanza();
            //ASB Need these for session
             System.setProperty("java.security.auth.login.config",cemc.
             getImportCELoginConfig());

        }
        catch (Exception e){
            logger.error(e.getMessage(),e);
        }
}
public CEConnection(String sourceType){
    //ASB ... 3 Check if this is Source or Target Object Store
if(sourceType.equalsIgnoreCase("Source")){
            ceConnect = new CEConnect(uri, stanza);
            ceConnect.logon(username, password);
            d = Factory.Domain.fetchInstance(ceConnect.
            getConnection(),"",null);

        }else{ //Target Object Store - get Target parameters
            ceConnect = new CEConnect(targetURI, targetStanza);
            ceConnect.logon(targetUserName, targetPassword);
            d = Factory.Domain.fetchInstance(ceConnect.
            getConnection(),"",null);
        }

}
public void popSubject(){
        ceConnect.popSubject();
}
public void pushSubect(){
    ceConnect.pushSubject();
}
public ObjectStore getObjectStore(String osName){

    ObjectStore os = Factory.ObjectStore.fetchInstance(d,osName, null);
    //ObjectStore os = (ObjectStore)oStores.get(osName);
```

```
            logger.log(Level.DEBUG, "Got ObjectStore " + os.get_Name());
            // Add special rights to the Object Store
            setAccessRights(
                    os.get_Domain(),
                    username,       // Example: "CEMPAdmin"
                    osName);        // Example: "ObjectStore1"

        return os;
    }
    public Domain getDomain(){
        return d;
    }
    @SuppressWarnings("static-access")
    public static void setAccessRights(
            Domain domain,
            String granteeName,     // Example: "CEMPAdmin"
            String objStoreName)    // Example: "ObjectStore1"
        {
            final int ACCESS_REQUIRED = AccessRight.WRITE_ANY_OWNER.
            getValue() |
            AccessRight.REMOVE_OBJECTS.getValue() | AccessRight.STORE_
            OBJECTS.getValue() |
            AccessRight.CONNECT.getValue() | AccessRight.WRITE_ACL.
            getValue() |
            AccessRight.READ_ACL.getValue() | AccessRight.MODIFY_OBJECTS.
            getValue() | AccessRight.PRIVILEGED_WRITE_AS_INT;

            // Create a new access permission object.
            com.filenet.api.security.AccessPermission ap = Factory.Access
            Permission.createInstance();

            // Set access permissions.
            ap.set_GranteeName(granteeName);
            ap.set_AccessType(AccessType.ALLOW);
            ap.set_AccessMask(new Integer(ACCESS_REQUIRED));

            // Set and save the new permissions.
```

```
        ObjectStore objStore = Factory.ObjectStore.fetchInstance(doma
        in, objStoreName, null);
        AccessPermissionList apl = objStore.get_Permissions();
        apl.add(ap);
        objStore.set_Permissions(apl);
        objStore.save(RefreshMode.REFRESH);
    }
}
```

Chapter 4 Exercises

The following questions relate to the **Replication Java program**, CEReplicate class, and its supporting Java library components we covered in this chapter.

MULTIPLE CHOICE QUESTIONS

1. What shell script would you run to copy the documents and folders from a source Object Store to a Target replicated Object Store?

   a) envCommonReplicate.sh

   b) Replicate.sh

   c) http://ecmukdemo6:9080/acce/

   d) env5x.sh

2. Which file is used to provide the Java for XML parsing requirements for the Java replication program?

   a) config.xml

   b) Jace.jar

   c) xerces.jar

   d) fn_utils.jar

CHAPTER 4   A REPLICATION JAVA PROGRAM FOR IBM FILENET OBJECT STORES

3. In which folder can the Replication.sh file be found?

   a) /opt/replication/config

   b) /opt/replication

   c) /opt/replication/libs

   d) /opt/replication/jars

4. Which of the following Events are **not** required for replicating the Folder Objects?

   a) Update Security Event

   b) Unfile Event

   c) Checkin Event

   d) Deletion Event

MULTIPLE CHOICE ANSWERS

1. b) Replicate.sh
2. c) xerces.jar
3. b) /opt/replication
4. c) Checkin Event

QUESTIONS

1. Outline what steps you would use to validate that the Replication program was functioning correctly. Describe the tests you would perform to prove the correct functionality for repeat runs of the Replication program.

2. What security does the connecting program user need to have to run the Replication program?

3. What Java libraries would you install in an Eclipse IDE Classpath to support the Replication program and where would you find them?

In Chapter 5, we will cover the use of the IBM Cognos BI Analytics, Real-Time Monitor to create a new Custom View for a Multilevel Time Dimension.

Note: Although Cognos BI 10.2.2 is at EOS as of December 31, 2022 (`www.ibm.com/support/pages/end-continuing-support-cognos-business-intelligence-102x-effective-december-31-2022`), the following IBM link announcement indicates that it is included in continuing support:

`www.ibm.com/support/pages/cognos-analytics-continuing-support`

CHAPTER 5

IBM Cognos Analytics Custom Development

In this chapter, we cover the use of **IBM Cognos Analytics** to create a new **Custom View** for a **Multilevel Time Dimension**.

We also cover the step-by-step procedure to use **IBM Cognos Analytics** to analyze **Event** records with an **IBM Case Analytics CASTORE OLAP** database from **Events** collected from the **IBM Case Manager Audit System**.

A short section describing the use of **Microsoft SQL Server Analysis Manager** is also covered for integration with the **Process Analyzer CASTORE Event** tables and the extension for **Multilevel Time Dimensions**.

Bill of Materials

For **IBM Cognos Analytics** installation, there is a **ResearchGate** free download of a **MS Word** document, named as follows:

"IBMCognosAnalytics11.1.4InstallationandConfigurationOnWindows10andRHEL8.0.docx"

Note This is a Word document although its download is initiated using a Download PDF command button.

See https://doi.org/10.13140/RG.2.2.11774.13129 entitled "IBM Cognos Analytics 11.1.4 Installation and Configuration On Windows 10 and RHEL 8.0."

There is also a related document entitled "IBM Case Analyzer 5.5 Installation and Configuration On RHEL 8.0 for Case Monitor Dashboard 5.3.3 plugin":

https://doi.org/10.13140/RG.2.2.11682.38089

CHAPTER 5 IBM COGNOS ANALYTICS CUSTOM DEVELOPMENT

For the latest Microsoft SQL Server Analysis Manager support tools, the following URL was used:

https://docs.microsoft.com/en-us/sql/ssdt/download-sql-server-data-tools-ssdt

In Google Search, the linked URL was

https://docs.microsoft.com/en-us/sql/ssdt/download-sql-server-data-tools-ssdt?view=sql-server-ver16

This opens a Download page; this also gives the link to the Microsoft Visual Studio 2022 edition with the Free download option:

https://visualstudio.microsoft.com/downloads/

The Installer has a link to the following:

https://docs.microsoft.com/en-us/visualstudio/install/workload-and-component-ids?view=vs-2022

There is also a Blog Link from the Microsoft Visual Studio 2022 Installer:

https://devblogs.microsoft.com/visualstudio/analysis-services-and-reporting-services-extensions-for-visual-studio-2022-are-here/

There is also a link from this:

https://marketplace.visualstudio.com/items?itemName=ProBITools.MicrosoftAnalysisServicesModelingProjects2022

Microsoft makes available early releases from their documentation:

https://marketplace.visualstudio.com/items?itemName=ProBITools.MicrosoftAnalysisServicesModelingProjects2022

"Preview Release Candidate Builds If you'd like to have early access to release candidate builds before they are published to the VS Gallery, we're making available to give customers an opportunity to provide feedback prior to finalizing releases. To access the release candidate of the extensions in Visual Studio, you may do so by creating a "private gallery" and entering the following settings.

- Navigate to Tools ➤ Options and then select Extensions and Updates under General.

- Under "Additional Extension Galleries," enter the following details:

- Name: Microsoft BI VSIX Preview

- URL: http://aka.ms/VSIX2022

After entering these settings, your "Extensions and Updates" dialog will show updates for the extension when there is a new release candidate that you can install to provide feedback for a day or two before the VS Gallery VSIX is updated."

There is also a free SQL Server Developer edition available:

www.microsoft.com/en-gb/sql-server/sql-server-downloads

The SQL Server Management Studio can be installed next:

https://docs.microsoft.com/en-us/sql/ssms/download-sql-server-management-studio-ssms?redirectedfrom=MSDN&view=sql-server-ver16

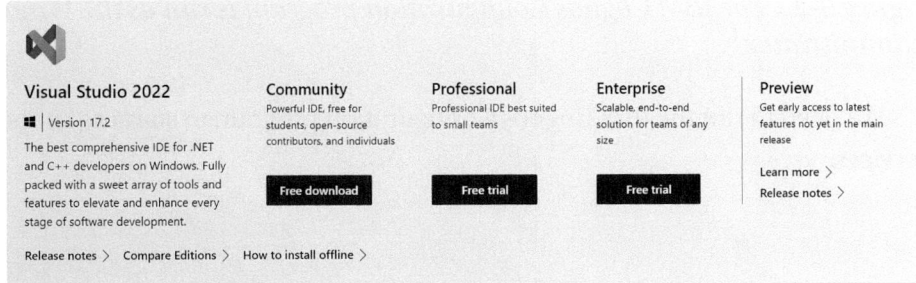

Figure 5-1. The latest downloads for Visual Studio

Development Introduction

Ideally, the approach to the provision of calculating days in a queue would be to use an integer workflow field and then assigning a value to this field in a workflow step to store the time the work was held in a queue step. This field could then be exposed as a measure in Process Analyzer, and this would provide the maximum flexibility for pivot table reporting.

When this approach cannot be used for retrospective reporting requirements or where the calculations require the system date for a current "view" of the status of the queues, then the following mechanism can be used to create a "dynamic" shared dimension based on a table view.

CHAPTER 5 IBM COGNOS ANALYTICS CUSTOM DEVELOPMENT

Time Dimension SLA Calculations on Cognos

First, we need to start the IBM Cognos Analytics services. We will use the host Windows system with the network access to the CASTORE1 Database which is installed on the ecmukdemo6 Linux RHEL 8.0 VMware server.

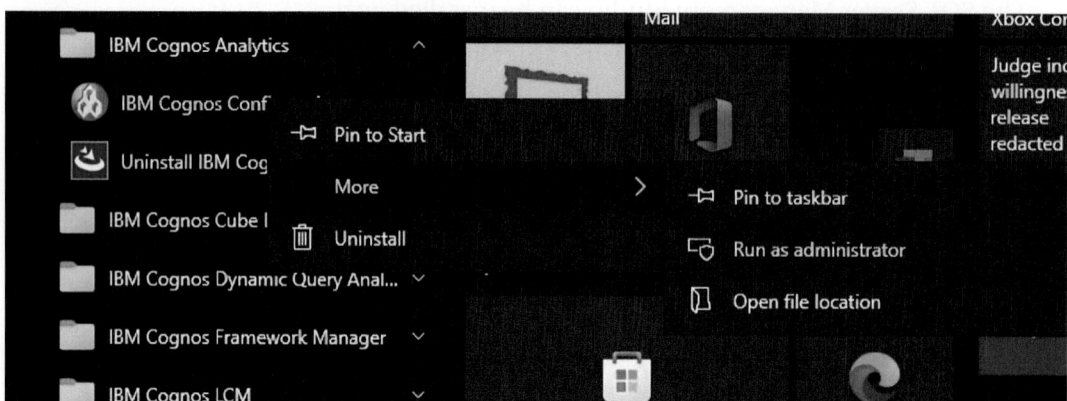

Figure 5-2. *The IBM Cognos Configuration program is run as the Windows Administrator*

We need to run the IBM Cognos Configuration program to start the supporting services.

CHAPTER 5   IBM COGNOS ANALYTICS CUSTOM DEVELOPMENT

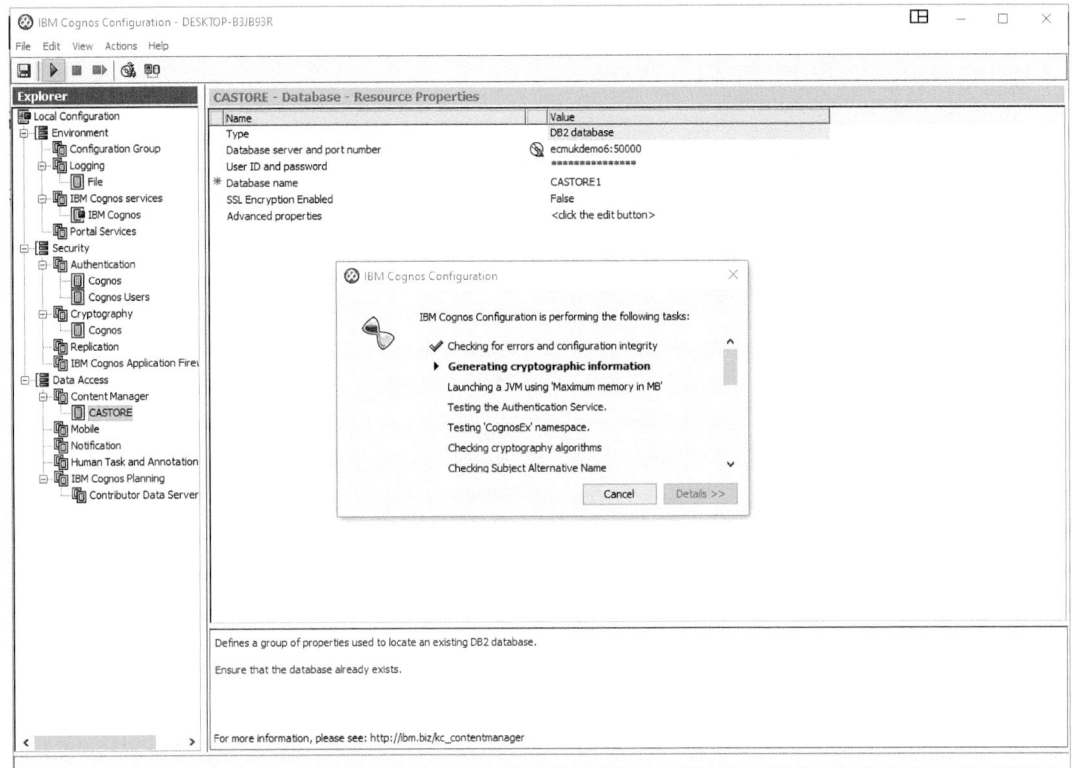

Figure 5-3. *The highlighted green arrow icon is clicked to start the supporting Services*

The services take around 6 minutes to start on a Windows server with an 8-core, 16-thread processor, running at base frequency of 1.8 GHz.

a) In IBM Data Studio 4.1.x, on the Linux server, a new table view can be created in the Db2 CASTORE database, CASTORE1, to utilize an existing date dimension as the basis of the calculation.

This query calculation was tested with the SQL as follows:

```
SELECT
days(current date) - days(TIMEINTERVAL) as DaysIn,
DMCASE_KEY
FROM CAUSER.F_DMCASELOAD
ORDER BY DaysIn
```

CHAPTER 5 IBM COGNOS ANALYTICS CUSTOM DEVELOPMENT

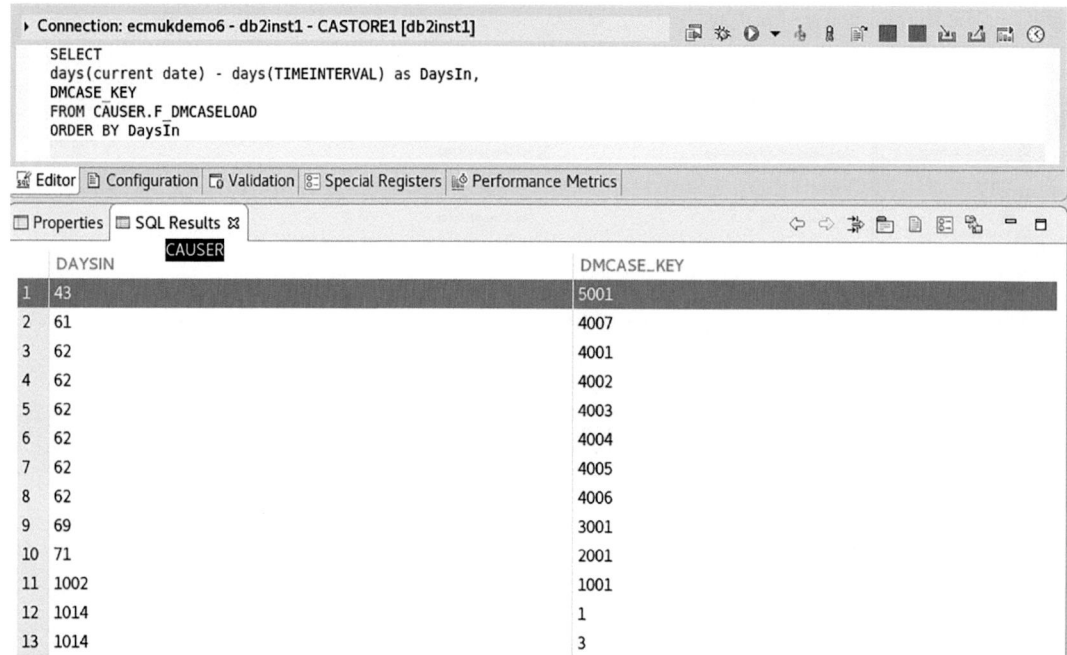

***Figure 5-4.** The query is tested in the SQL query window of IBM Data Studio 4.1.3*

The Create View statement used to create the View, **CAUSER.DMDaysIn_Case**:

CREATE VIEW CAUSER.DMDaysIn_Case AS
SELECT
days(current date) - days(TIMEINTERVAL) as DaysIn,
DMCASE_KEY
FROM CAUSER.F_DMCASELOAD;

CHAPTER 5 IBM COGNOS ANALYTICS CUSTOM DEVELOPMENT

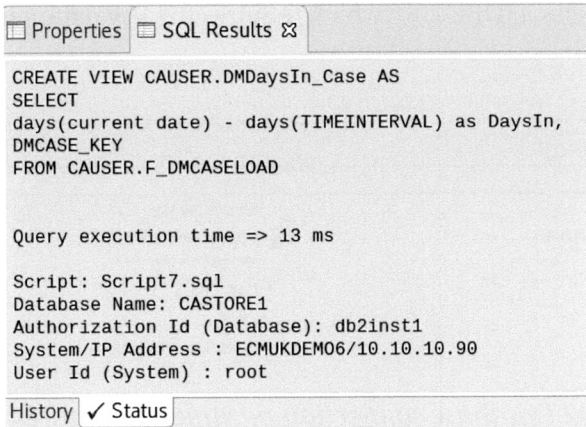

Figure 5-5. *The CREATE VIEW SQL status is shown in IBM Data Studio 4.1.3*

The access to the ecmukdemo6 Linux Server, Db2 V10.6, CASTORE1 database is configured in the IBM Cognos Configuration program on the Windows server. This configuration is in the Data Access section as shown in Figure 5-6.

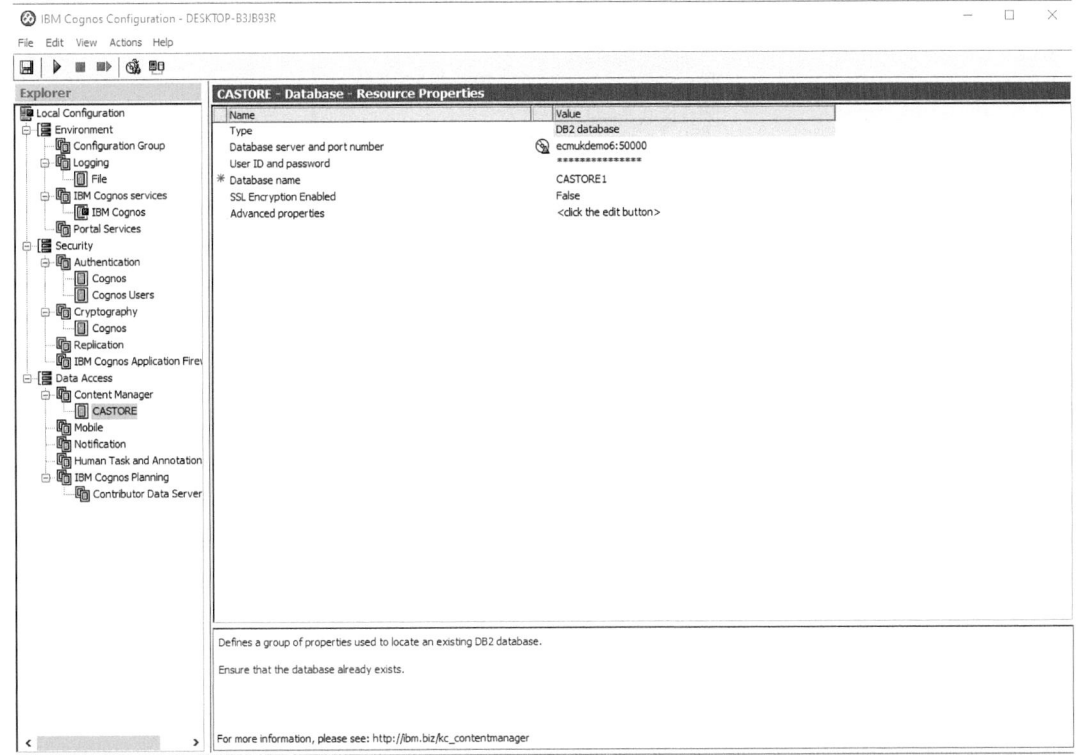

Figure 5-6. *The Linux server CASTORE1 Database connection configuration*

CHAPTER 5 IBM COGNOS ANALYTICS CUSTOM DEVELOPMENT

The connection uses a JDBC URL which requires the key parameters as shown in Figure 5-7.

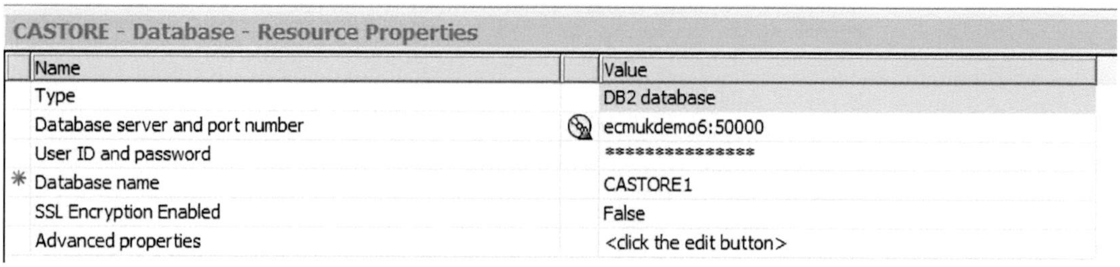

Name	Value
Type	DB2 database
Database server and port number	ecmukdemo6:50000
User ID and password	***************
* Database name	CASTORE1
SSL Encryption Enabled	False
Advanced properties	<click the edit button>

CASTORE - Database - Resource Properties

Figure 5-7. *The DB2 Database connection parameters required to connect to CASTORE1*

The connection to the Linux CASTORE1 is checked using the Test connection menu displayed by using the right-mouse click on the CASTORE entry as shown in Figure 5-8.

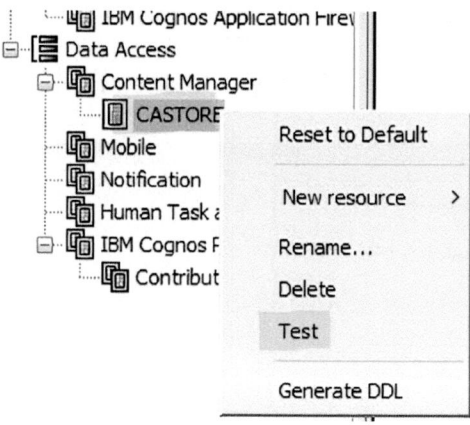

Figure 5-8. *The CASTORE database connection is tested using the right-mouse click*

The **CAUSER.DMDaysIn_Case** View is checked in the Linux IBM Data Studio 4.1.3 program to ensure the expected rows are returned.

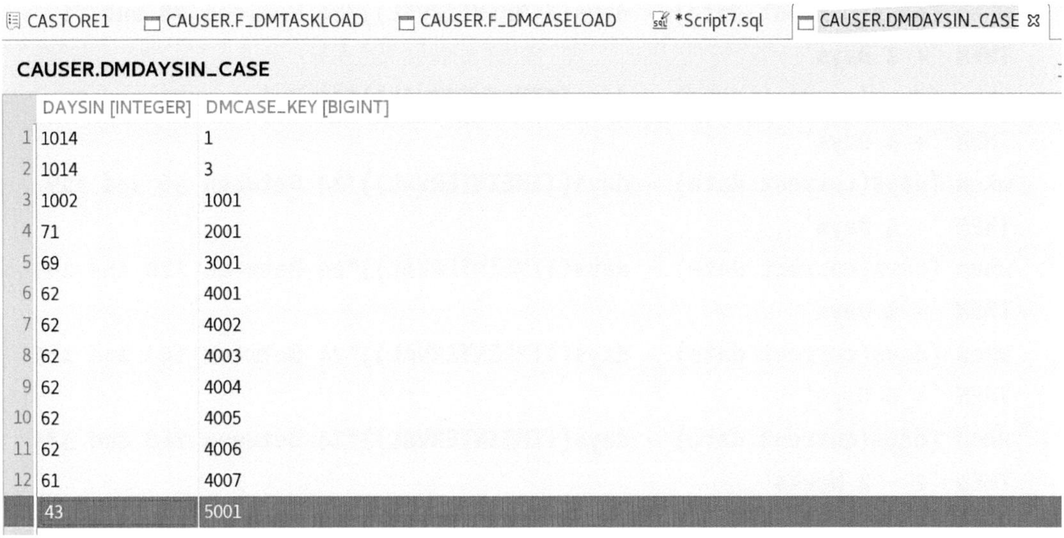

CHAPTER 5 IBM COGNOS ANALYTICS CUSTOM DEVELOPMENT

	DAYSIN [INTEGER]	DMCASE_KEY [BIGINT]
1	1014	1
2	1014	3
3	1002	1001
4	71	2001
5	69	3001
6	62	4001
7	62	4002
8	62	4003
9	62	4004
10	62	4005
11	62	4006
12	61	4007
	43	5001

Figure 5-9. *The CAUSER.DMDaysIn_Case View is checked to show the expected rows*

Note It is important to note here that the custom view we created earlier may need to be reapplied if the IBM Case Analyzer CASTORE1 database system is updated.

Dynamic Dimensions for Aging Bands

The procedure used to create the simple calculated Dimension earlier can be extended to create a "range" Dimension to allow the display of counts for ranges of dates, for example, in bands, for example, Today, '+ 1 Day', '+ 2 Days', '+ 3 Days', '+ 4 Days', '+ 5 Days', '+ 6 Days', '1 - 2 Weeks', '2 - 3 Weeks', '3 - 4 Weeks', '4 - 5 Weeks', '5 - 120 Weeks', '120 - 240 Weeks', '>= 240 Weeks'.

This query calculation was tested with

```
select
  case
    when (days(current date) - days(TIMEINTERVAL))*24 < 24 THEN 'Today'
    when (days(current date) - days(TIMEINTERVAL))*24 Between 24 and 47
    THEN '+ 1 Day'
```

```
when (days(current date) - days(TIMEINTERVAL))*24 Between 48 and 71
THEN '+ 2 Days'
when (days(current date) - days(TIMEINTERVAL))*24 Between 72 and 95
THEN '+ 3 Days'
when (days(current date) - days(TIMEINTERVAL))*24 Between 96 and 119
THEN '+ 4 Days'
when (days(current date) - days(TIMEINTERVAL))*24 Between 120 and 143
THEN '+ 5 Days'
when (days(current date) - days(TIMEINTERVAL))*24 Between 144 and 167
THEN '+ 6 Days'
when (days(current date) - days(TIMEINTERVAL))*24 Between 168 and 335
THEN '1 - 2 Weeks'
when (days(current date) - days(TIMEINTERVAL))*24 Between 336 and 503
THEN '2 - 3 Weeks'
when (days(current date) - days(TIMEINTERVAL))*24 Between 672 and 839
THEN '4 - 5 Weeks'
when (days(current date) - days(TIMEINTERVAL))*24 Between 840 and 20160
THEN '5 - 120 Weeks'
when (days(current date) - days(TIMEINTERVAL))*24 Between 20161 and
40319 THEN '120 - 240 Weeks'
when (days(current date) - days(TIMEINTERVAL))*24 > 40319 THEN '>=
240 Weeks'
else 'Not In tested Range'
end as DaysIn,
DMCASE_KEY
FROM CAUSER.F_DMCASELOAD;
```

CHAPTER 5   IBM COGNOS ANALYTICS CUSTOM DEVELOPMENT

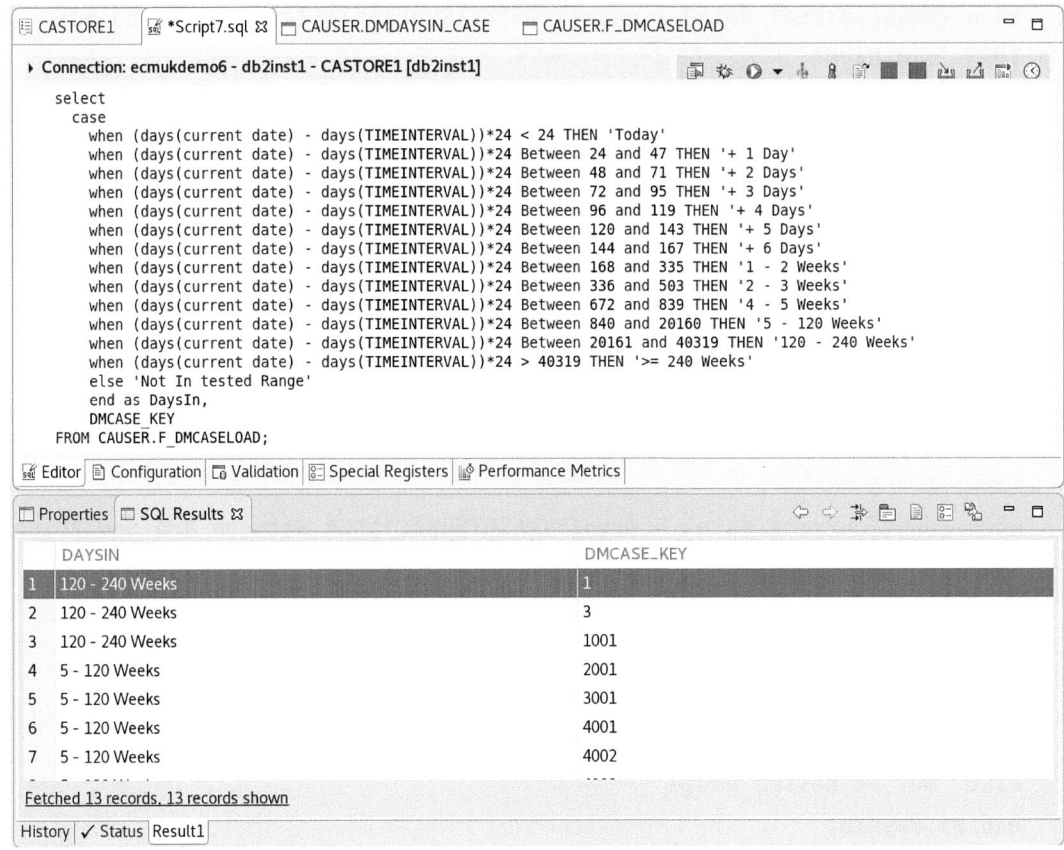

Figure 5-10. The SQL for the Case Aging ranges is tested in Linux IBM Data Studio 4.1.3

And the Create View statement used to create the View, **CAUSER.DMAgeing_Case**:

```
CREATE VIEW CAUSER.DMAgeing_Case AS
select
  case
    when (days(current date) - days(TIMEINTERVAL))*24 < 24 THEN 'Today'
    when (days(current date) - days(TIMEINTERVAL))*24 Between 24 and 47
    THEN '+ 1 Day'
    when (days(current date) - days(TIMEINTERVAL))*24 Between 48 and 71
    THEN '+ 2 Days'
    when (days(current date) - days(TIMEINTERVAL))*24 Between 72 and 95
    THEN '+ 3 Days'
```

```
when (days(current date) - days(TIMEINTERVAL))*24 Between 96 and 119
THEN '+ 4 Days'
when (days(current date) - days(TIMEINTERVAL))*24 Between 120 and 143
THEN '+ 5 Days'
when (days(current date) - days(TIMEINTERVAL))*24 Between 144 and 167
THEN '+ 6 Days'
when (days(current date) - days(TIMEINTERVAL))*24 Between 168 and 335
THEN '1 - 2 Weeks'
when (days(current date) - days(TIMEINTERVAL))*24 Between 336 and 503
THEN '2 - 3 Weeks'
when (days(current date) - days(TIMEINTERVAL))*24 Between 672 and 839
THEN '4 - 5 Weeks'
when (days(current date) - days(TIMEINTERVAL))*24 Between 840 and 20160
THEN '5 - 120 Weeks'
when (days(current date) - days(TIMEINTERVAL))*24 Between 20161 and
40319 THEN '120 - 240 Weeks'
when (days(current date) - days(TIMEINTERVAL))*24 > 40319 THEN '>=
240 Weeks'
else 'Not In tested Range'
end as DaysIn,
DMCASE_KEY
FROM CAUSER.F_DMCASELOAD;
```

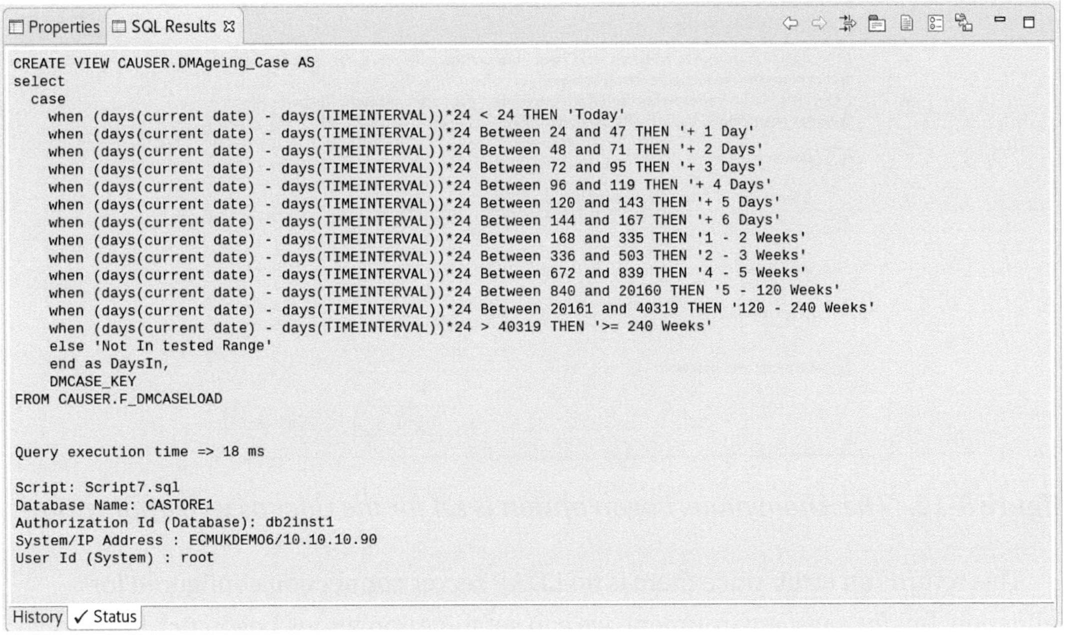

CHAPTER 5 IBM COGNOS ANALYTICS CUSTOM DEVELOPMENT

```
CREATE VIEW CAUSER.DMAgeing_Case AS
select
    case
        when (days(current date) - days(TIMEINTERVAL))*24 < 24 THEN 'Today'
        when (days(current date) - days(TIMEINTERVAL))*24 Between 24 and 47 THEN '+ 1 Day'
        when (days(current date) - days(TIMEINTERVAL))*24 Between 48 and 71 THEN '+ 2 Days'
        when (days(current date) - days(TIMEINTERVAL))*24 Between 72 and 95 THEN '+ 3 Days'
        when (days(current date) - days(TIMEINTERVAL))*24 Between 96 and 119 THEN '+ 4 Days'
        when (days(current date) - days(TIMEINTERVAL))*24 Between 120 and 143 THEN '+ 5 Days'
        when (days(current date) - days(TIMEINTERVAL))*24 Between 144 and 167 THEN '+ 6 Days'
        when (days(current date) - days(TIMEINTERVAL))*24 Between 168 and 335 THEN '1 - 2 Weeks'
        when (days(current date) - days(TIMEINTERVAL))*24 Between 336 and 503 THEN '2 - 3 Weeks'
        when (days(current date) - days(TIMEINTERVAL))*24 Between 672 and 839 THEN '4 - 5 Weeks'
        when (days(current date) - days(TIMEINTERVAL))*24 Between 840 and 20160 THEN '5 - 120 Weeks'
        when (days(current date) - days(TIMEINTERVAL))*24 Between 20161 and 40319 THEN '120 - 240 Weeks'
        when (days(current date) - days(TIMEINTERVAL))*24 > 40319 THEN '>= 240 Weeks'
        else 'Not In tested Range'
    end as DaysIn,
    DMCASE_KEY
FROM CAUSER.F_DMCASELOAD

Query execution time => 18 ms

Script: Script7.sql
Database Name: CASTORE1
Authorization Id (Database): db2inst1
System/IP Address : ECMUKDEMO6/10.10.10.90
User Id (System) : root
```

***Figure 5-11.** The Create View statement is run for the CAUSER.DMAgeing_Case view*

The next step after creating the preceding views is to load the Windows 10, IBM Cognos Analytics application, to set up a Datasource from the Linux CASTORE1 database and then create the Dashboards with the graphics used to display the OLAP database tables and our newly created views.

To access the IBM Cognos Analytics web application, the URL link is used:

http://DESKTOP-B3JB93R:9300/bi/v1/disp (http://*servername:9300/bi/v1/disp*)

This can be set up with an anonymous login by launching the **IBM Cognos Dynamic Query Analyzer** program, after which a pop-up window is displayed as follows.

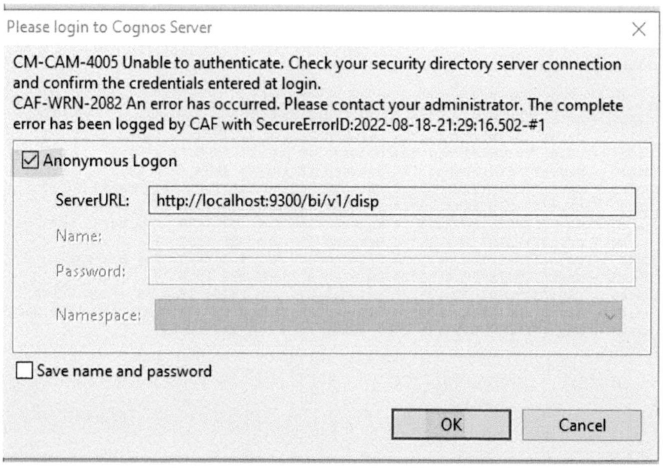

CHAPTER 5 IBM COGNOS ANALYTICS CUSTOM DEVELOPMENT

Figure 5-12. *The Anonymous Logon option is set for the Cognos web application*

This returns an error, since there is no LDAP server connection configured for validation, but for a test environment, we can set the Anonymous Logon tick box shown in Figure 5-12.

The URL (http://DESKTOP-B3JB93R:9300/bi/v1/disp) then launches the first page of the IBM Cognos Analytics 11.2 web application for processing Dashboard reports as shown in Figure 5-13.

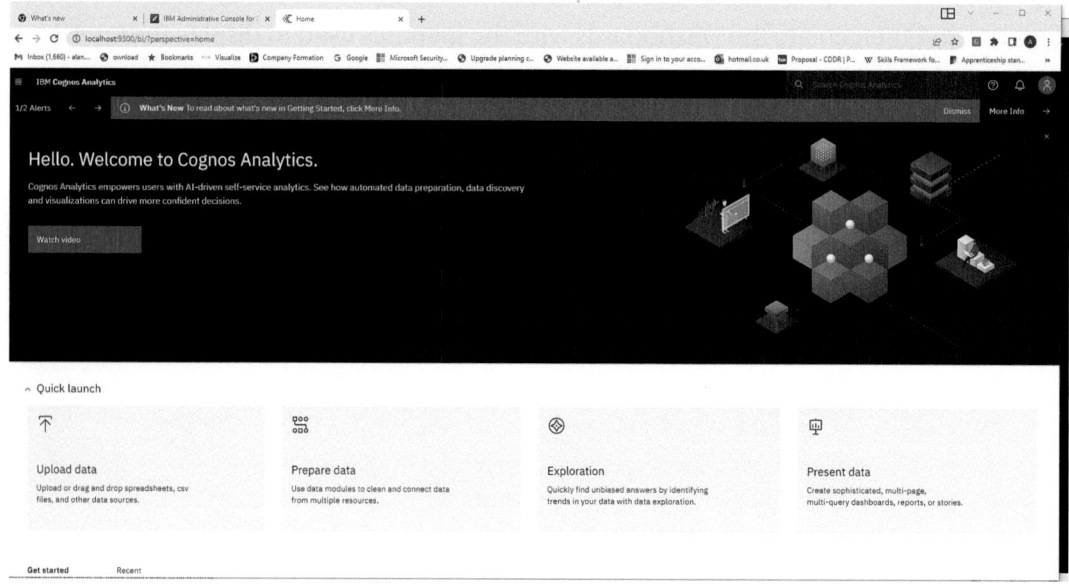

Figure 5-13. *The IBM Cognos Analytics 11.2 web application main screen is displayed*

CHAPTER 5   IBM COGNOS ANALYTICS CUSTOM DEVELOPMENT

The top-left "hamburger" icon is clicked, and the Manage menu option is selected as shown in Figure 5-14.

Initial Datasource Setup Before Adding the Custom Views

The following screens show the setup initially made **before** adding the Custom Views to the CASTORE1 database.

This section is used to highlight the procedure which must then be followed to include the additional custom Views, since a local Cognos datasource copy is made for the loaded tables, and on making any additional changes, the Datasource must be refreshed.

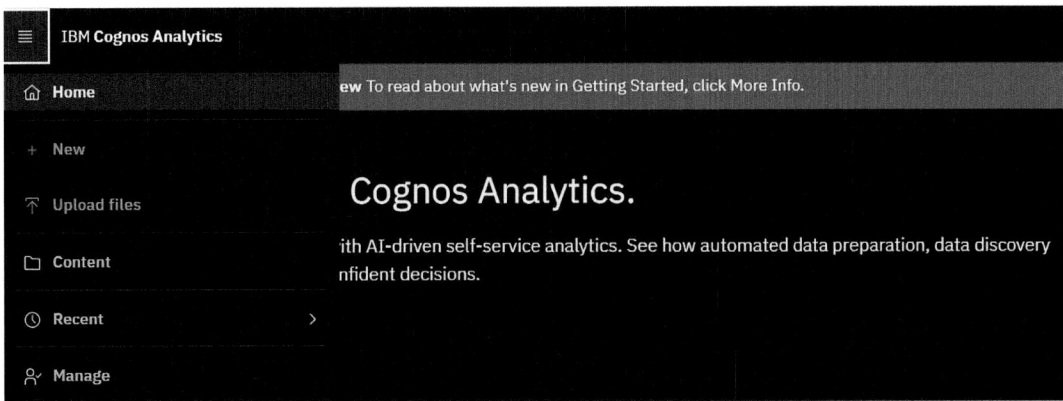

Figure 5-14. *The Manage menu option is selected from the drop-down menu list*

From the **Manage** menu, a long list of options is displayed as shown in Figure 5-15, from which we select the "**Data server connections**" menu.

917

CHAPTER 5   IBM COGNOS ANALYTICS CUSTOM DEVELOPMENT

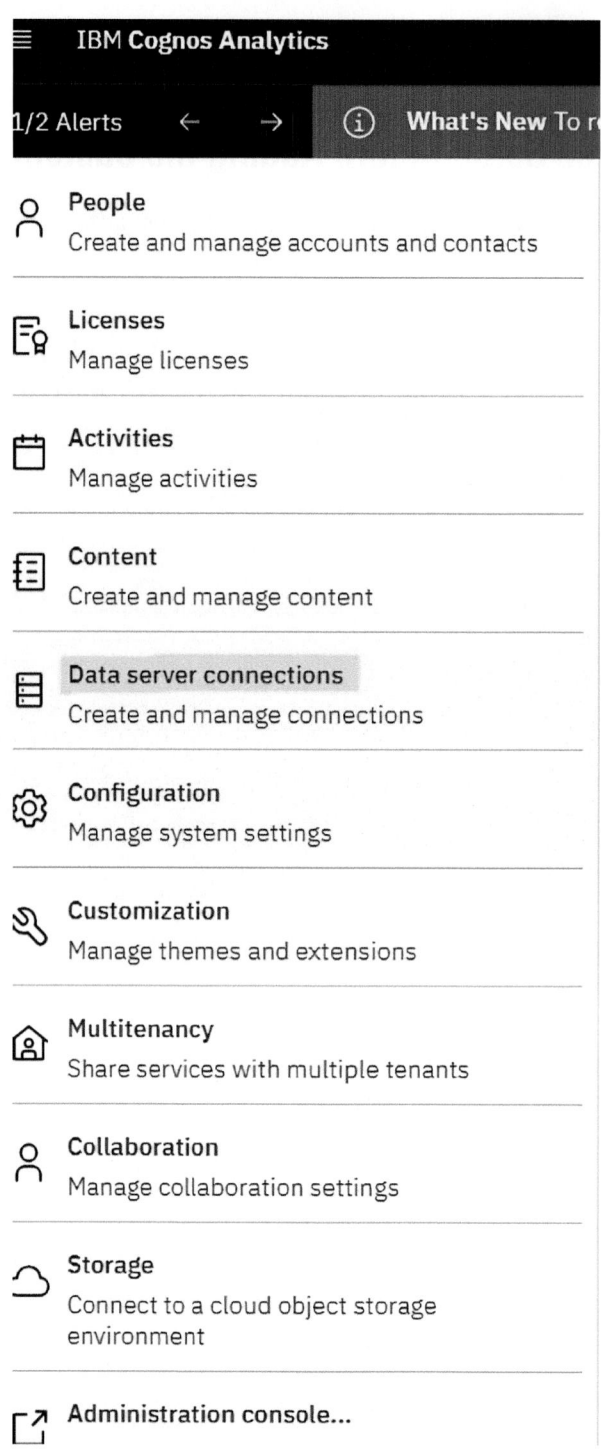

Figure 5-15. *The Data server connections submenu is selected*

CHAPTER 5 IBM COGNOS ANALYTICS CUSTOM DEVELOPMENT

In the next slide-out frame, we can then enter the Data connection name, **AUDIT_DB_CASTORE1**, and then click the + Add data server icon.

Figure 5-16. *The Data server connection name, AUDIT_DB_CASTORE1, is entered*

The Data server connection name, **AUDIT_DB_CASTORE1**, is entered, and then the Add data server + icon is clicked to give a list of Data server connection types.

919

CHAPTER 5 IBM COGNOS ANALYTICS CUSTOM DEVELOPMENT

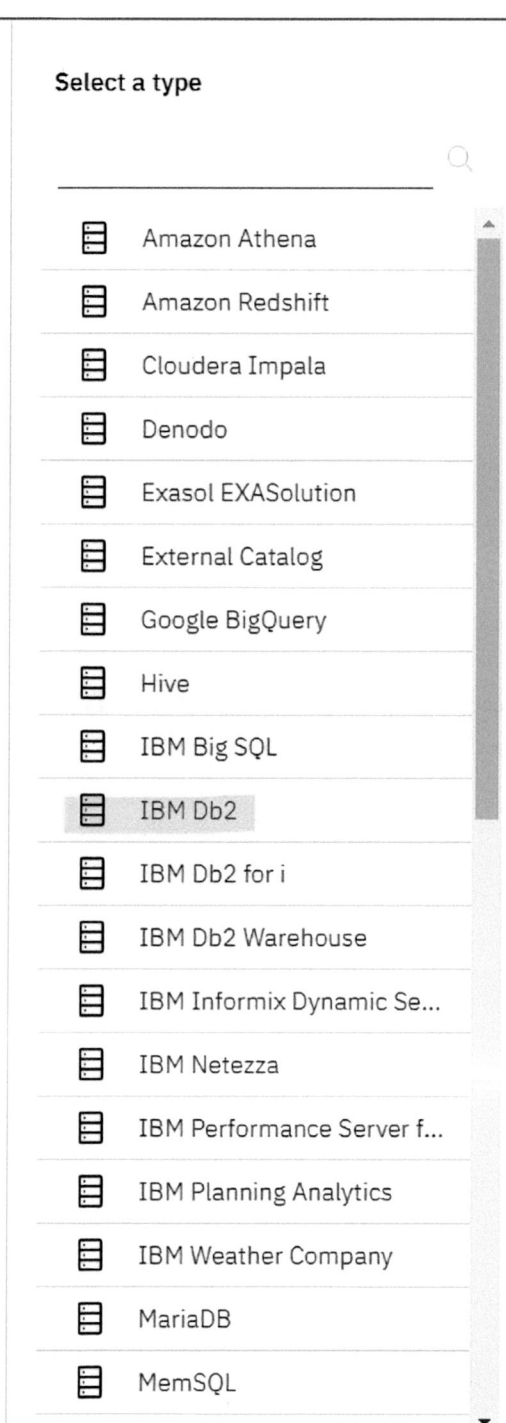

Figure 5-17. *The slide-out panel now gives a list of datasource types, and IBM Db2 is selected*

CHAPTER 5   IBM COGNOS ANALYTICS CUSTOM DEVELOPMENT

There is a long scrollable list of Database and Datasource types (as shown in Figure 5-17) from which we select the **IBM Db2** database type.

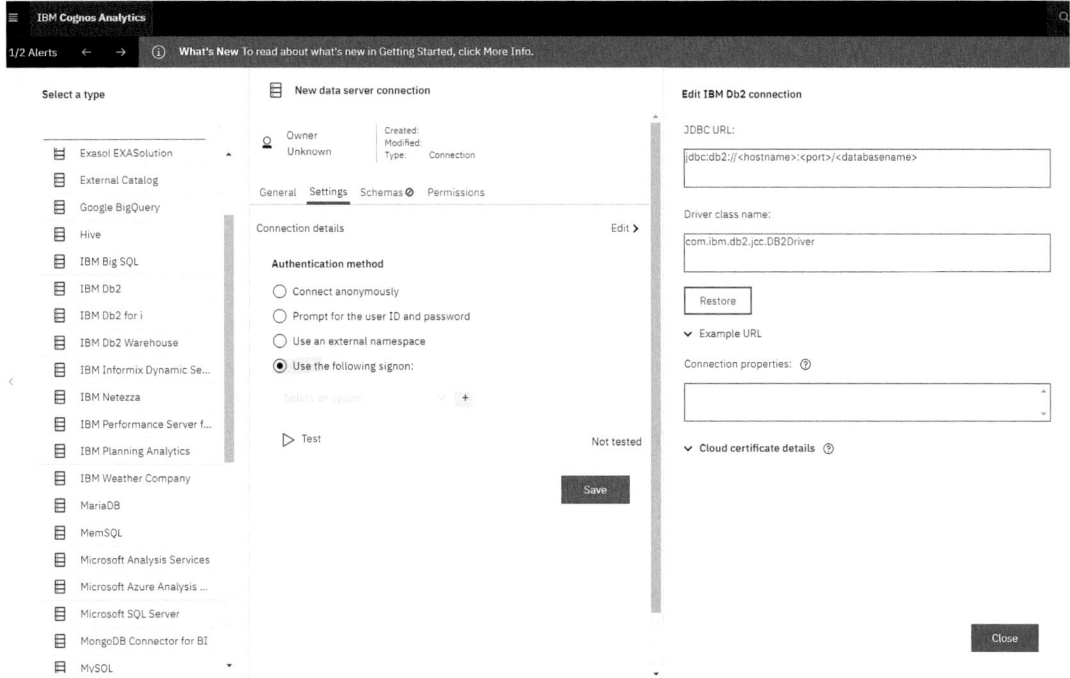

Figure 5-18. *The option to use a signon for the IBM Db2 connection is selected*

First, we need to enter a value (**AUDIT_DB_CASTORE1**) to replace the **New data server connection** default value shown at the top of the second panel.

Next, as shown in Figure 5-18, the option to use a signon is selected, and the + icon causes the third panel to slide open with a JDBC URL template for entry of the **<hostname>**, **<port>**, and **<databasename>** values replacing these placeholders.

921

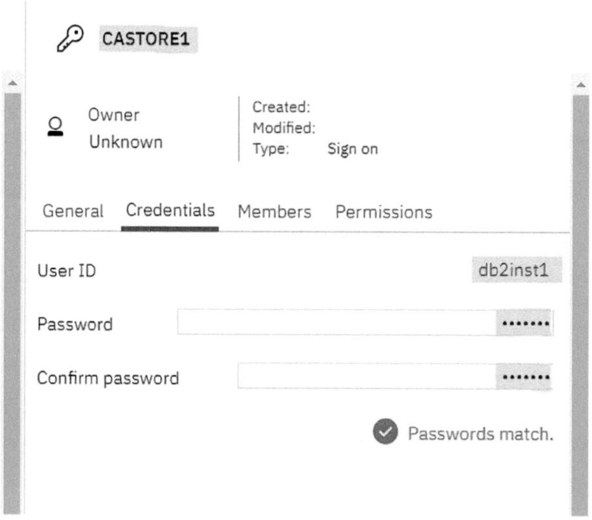

CHAPTER 5   IBM COGNOS ANALYTICS CUSTOM DEVELOPMENT

Figure 5-19. *The CASTORE1 db2inst1 user signon credentials are entered*

The entered credentials are tested.

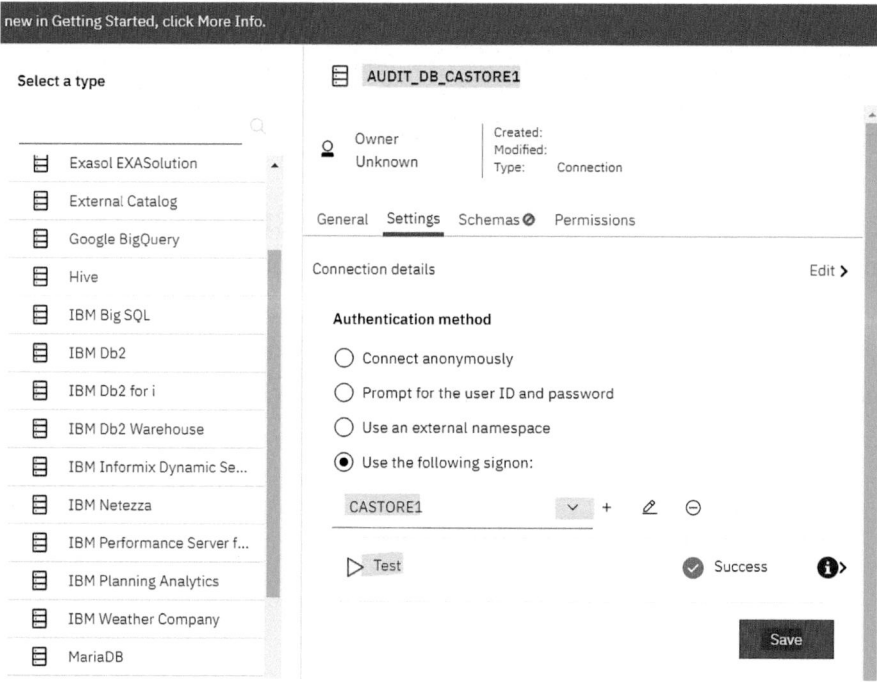

Figure 5-20. *The AUDIT_DB_CASTORE1 Datasource is tested with the entered credentials*

CHAPTER 5   IBM COGNOS ANALYTICS CUSTOM DEVELOPMENT

The Manage menu is selected again.

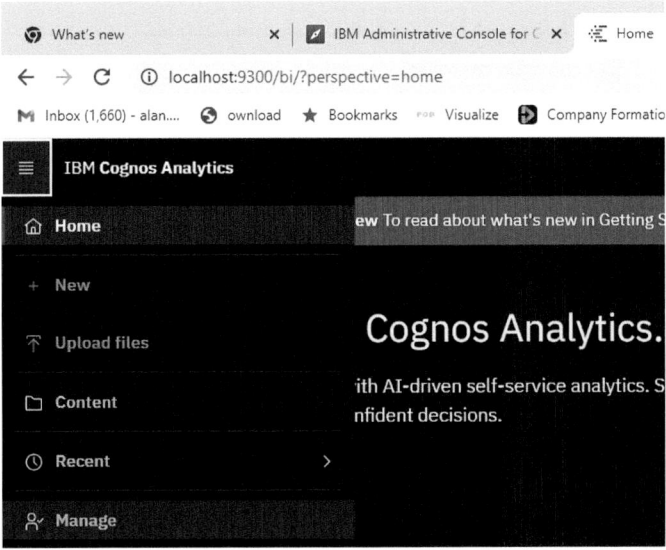

Figure 5-21. The Manage menu option is selected

The Manage option is selected, which then displays the **Data server connection** we created.

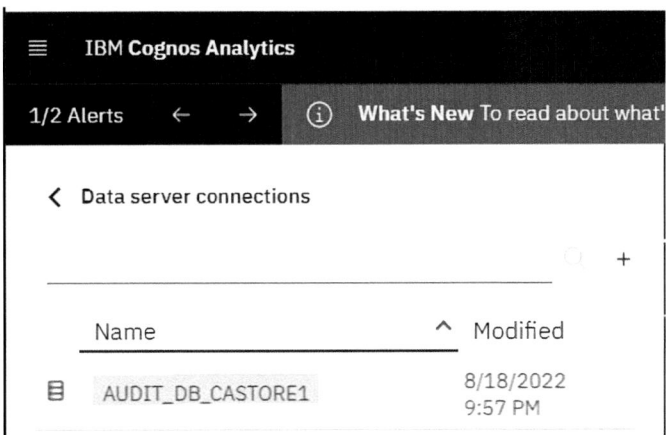

*Figure 5-22. The Data server connections showing the **AUDIT_DB_CASTORE1** connection*

CHAPTER 5   IBM COGNOS ANALYTICS CUSTOM DEVELOPMENT

The Cognos user permissions can now be updated using the screen shown in Figure 5-23 after clicking the **AUDIT_DB_CASTORE1** Data server connection.

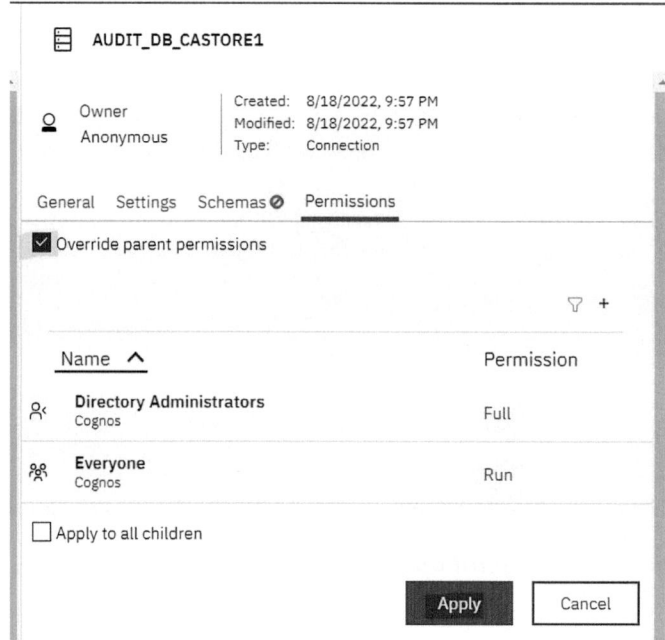

Figure 5-23.** The permissions are set for the Cognos users for **AUDIT_DB_CASTORE1

CHAPTER 5 IBM COGNOS ANALYTICS CUSTOM DEVELOPMENT

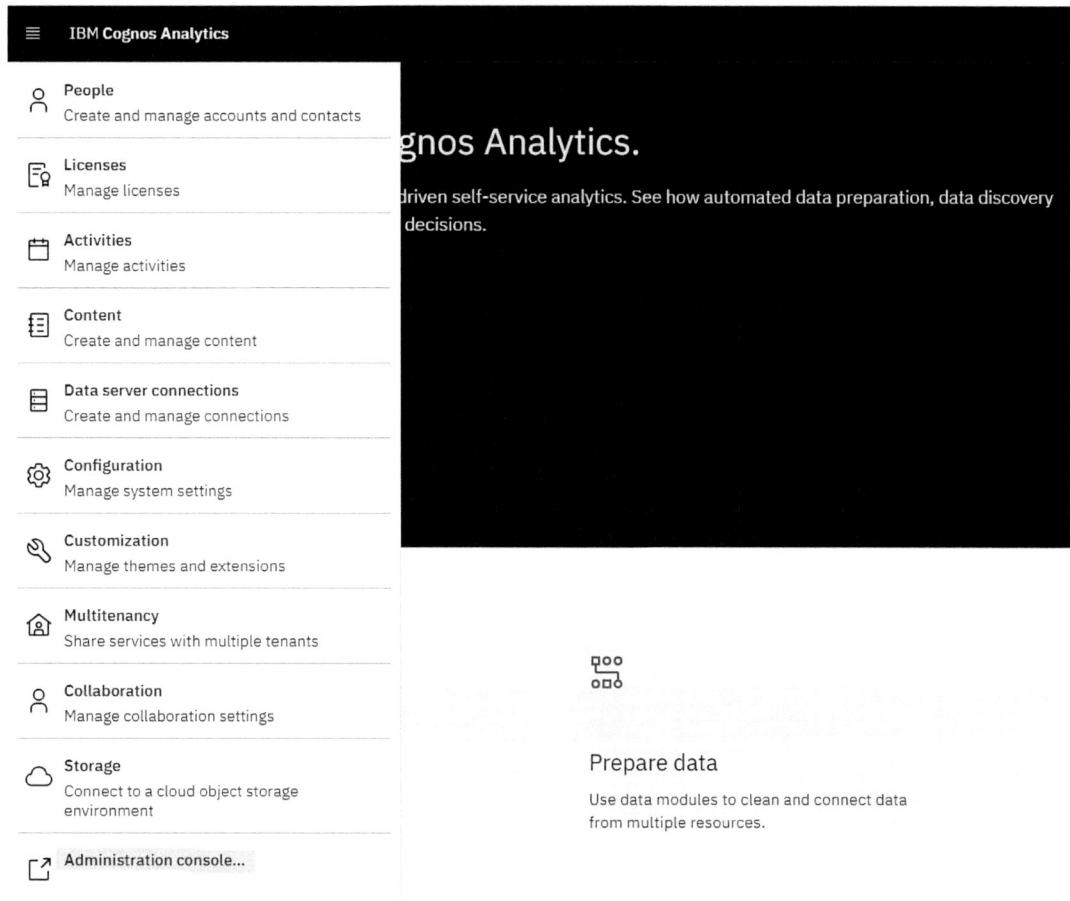

***Figure 5-24.** The **Administration console...** is selected next*

From the top-left "**Hamburger**" icon in Figure 5-24, we now select the **Administration console...** menu item, and then we select the **Configuration** tab shown in Figure 5-25.

CHAPTER 5   IBM COGNOS ANALYTICS CUSTOM DEVELOPMENT

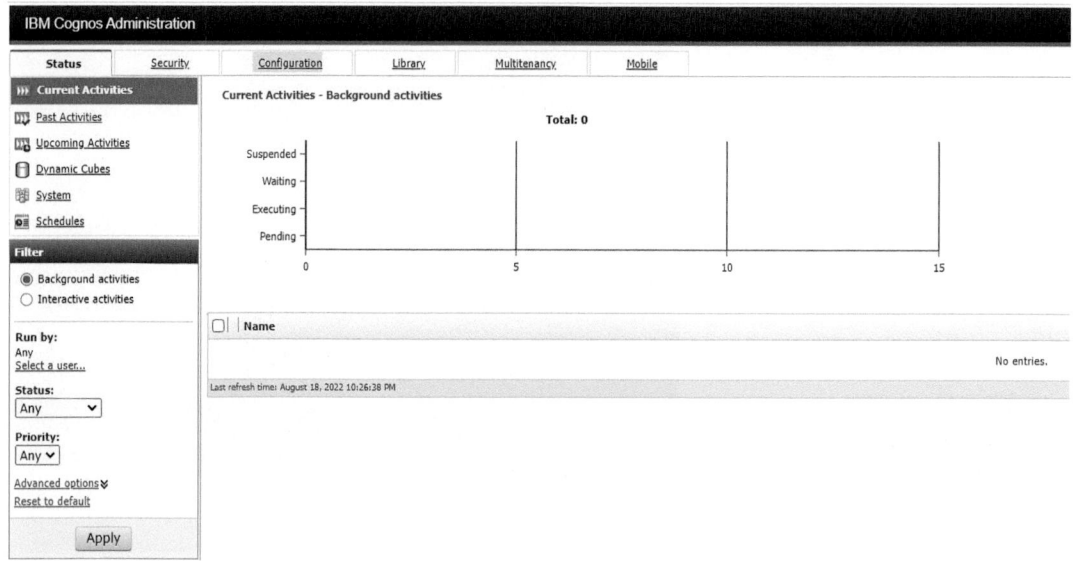

Figure 5-25. *The **Configuration** tab, as highlighted, is selected*

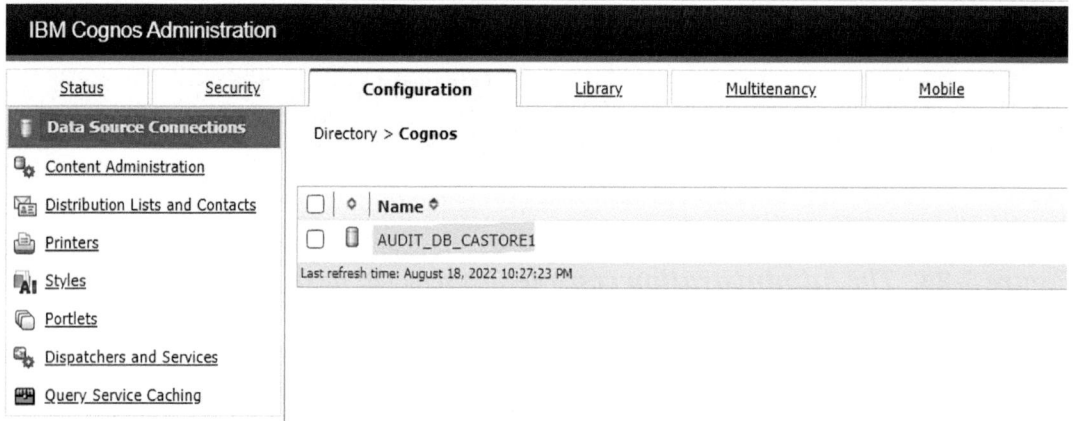

Figure 5-26. *The **AUDIT_DB_CASTORE1** datasource connection is selected*

On clicking the **AUDIT_DB_CASTORE1** datasource connection link, we can select the **Schemas** tab shown in Figure 5-27, which allows us to select a **CAUSER** Schema to download from the Linux server **CASTORE1** database.

CHAPTER 5 IBM COGNOS ANALYTICS CUSTOM DEVELOPMENT

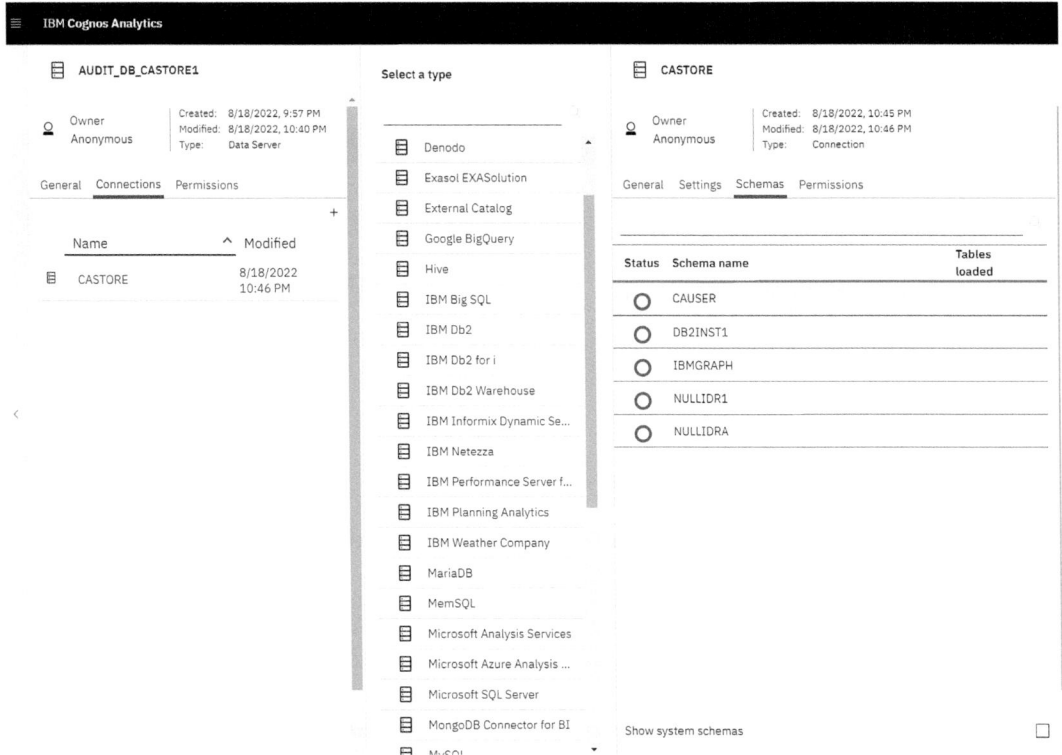

Figure 5-27. *The **CAUSER** Schema is clicked to use for the table data load for analysis*

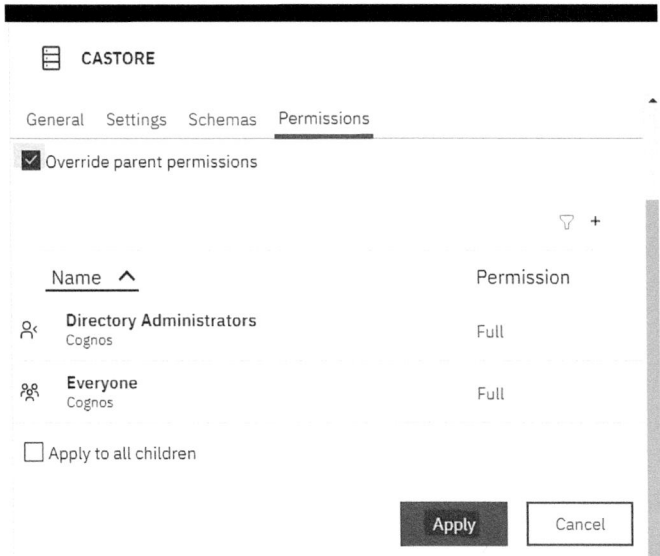

Figure 5-28. *The **Permissions** tab can be used to provide the required security levels*

927

CHAPTER 5   IBM COGNOS ANALYTICS CUSTOM DEVELOPMENT

The right-mouse click on the **CAUSER** schema allows the selection of the **Load metadata** menu, highlighted in Figure 5-29.

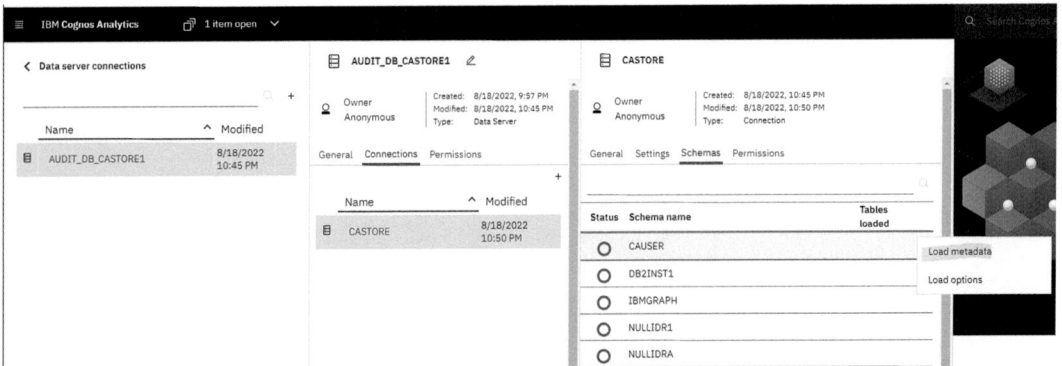

Figure 5-29. *The **Load metadata** menu item is selected from the **CAUSER** Schema entry*

The percentage progress of the "table load" is shown as highlighted in Figure 5-30.

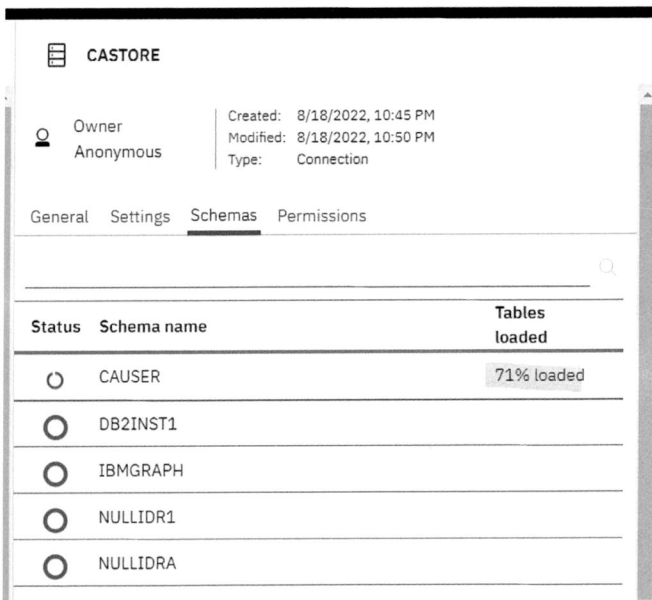

Figure 5-30. *The status of the table load for the CAUSER Schema tables is shown*

CHAPTER 5　IBM COGNOS ANALYTICS CUSTOM DEVELOPMENT

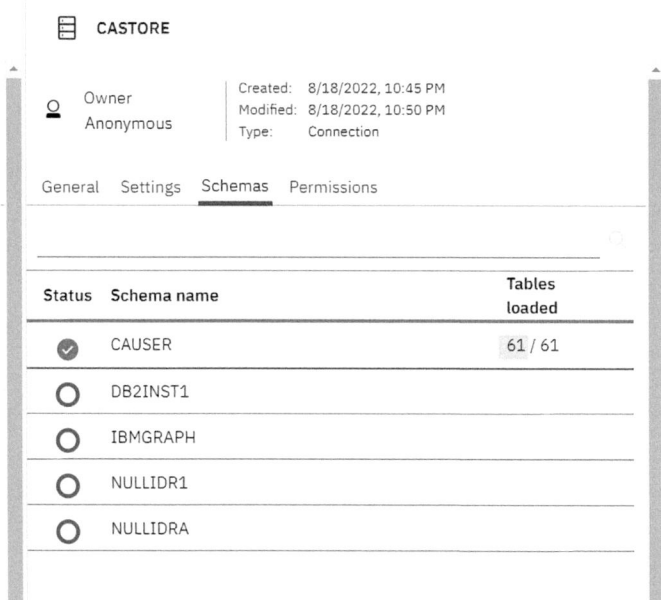

Figure 5-31.　*The number of tables for the Schema and the number loaded are shown*

The loaded Schema is now saved as **CaseAgeing** under a new folder, **CaseAnalytics**, in the **My content** tab.

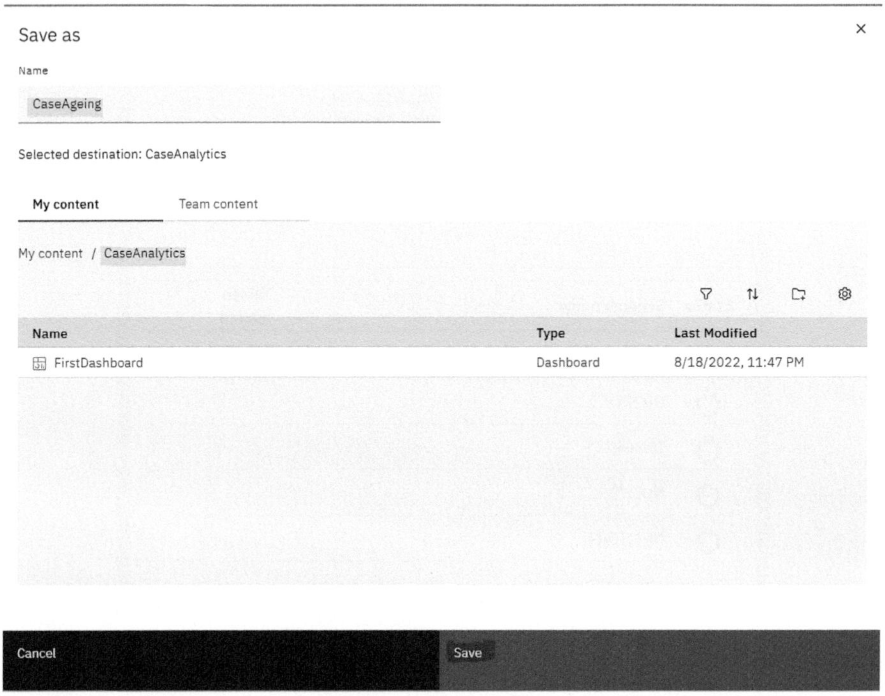

CHAPTER 5   IBM COGNOS ANALYTICS CUSTOM DEVELOPMENT

Figure 5-32. *The File icon is clicked and the Save as menu option selected*

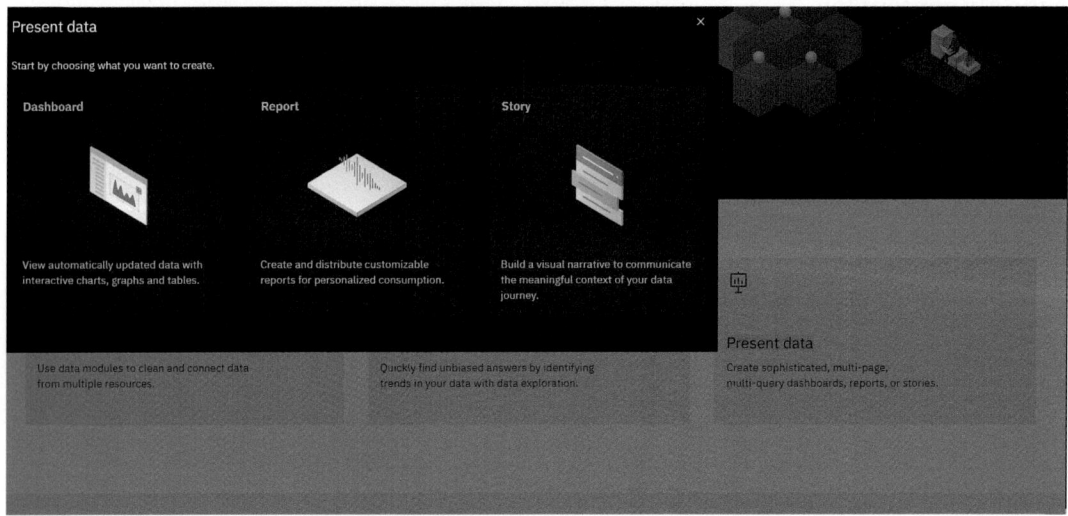

Figure 5-33. *The Present data ➤ Dashboard options are selected*

CHAPTER 5 IBM COGNOS ANALYTICS CUSTOM DEVELOPMENT

After selecting the **Present data ➤ Dashboard** panels on the main page of the Cognos web application, the **Create a dashboard** page is loaded as shown in Figure 5-34.

We can now select a Framework layout as highlighted.

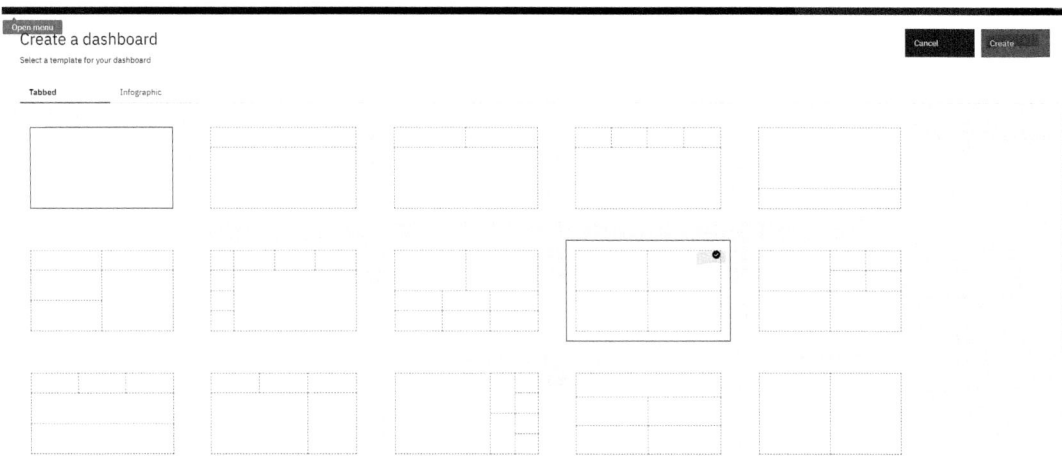

Figure 5-34. *The Framework layout for the Dashboard is selected as highlighted*

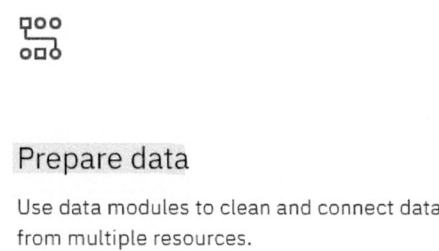

Figure 5-35. *The Prepare data panel is used to create a Datasource for the Dashboard*

We can select a Datasource from the **Prepare data** panel shown in Figure 5-35 to be used with the Dashboard.

931

CHAPTER 5   IBM COGNOS ANALYTICS CUSTOM DEVELOPMENT

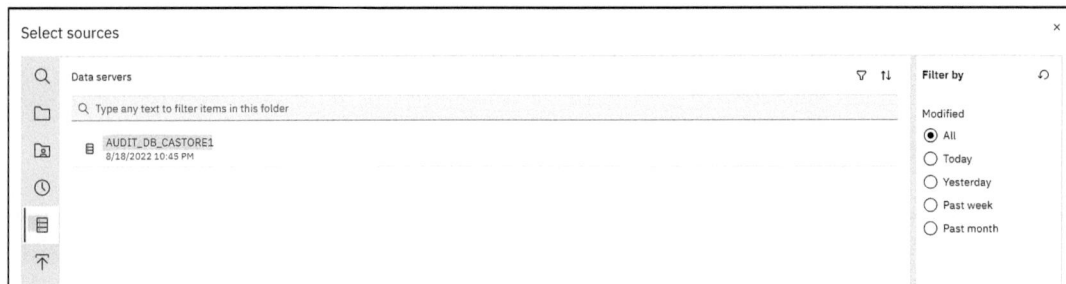

Figure 5-36. *The AUDIT_DB_CASTORE1 database datasource is selected*

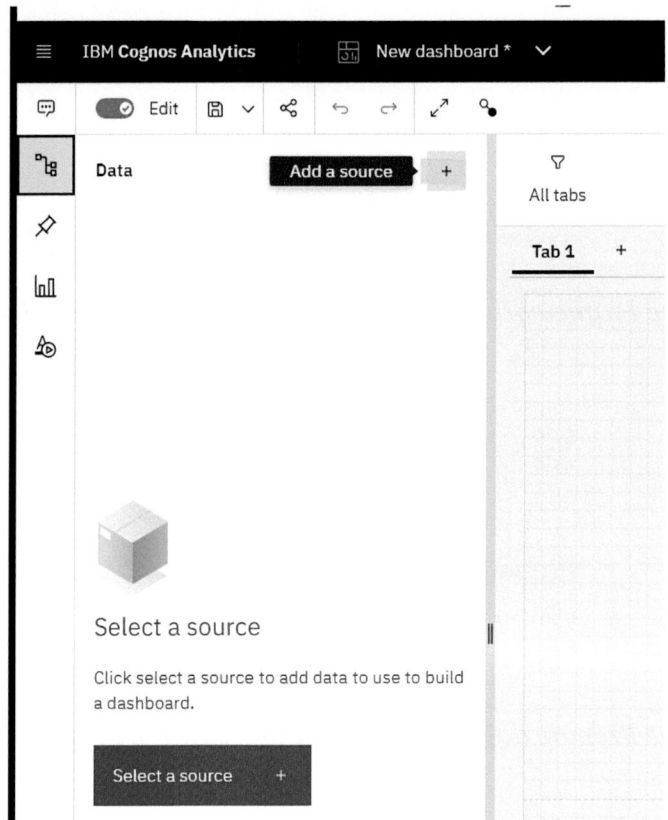

Figure 5-37. *The Add a source + icon is selected as highlighted*

A new source can be selected from the saved **CAUSER** schema of the AUDIT_DB_CASTORE server connection as shown in Figure 5-38.

CHAPTER 5   IBM COGNOS ANALYTICS CUSTOM DEVELOPMENT

Figure 5-38. *The CAUSER schema we loaded earlier is selected*

The **OK** command button is then clicked to confirm the use of the **CAUSER** schema we selected in Figure 5-38.

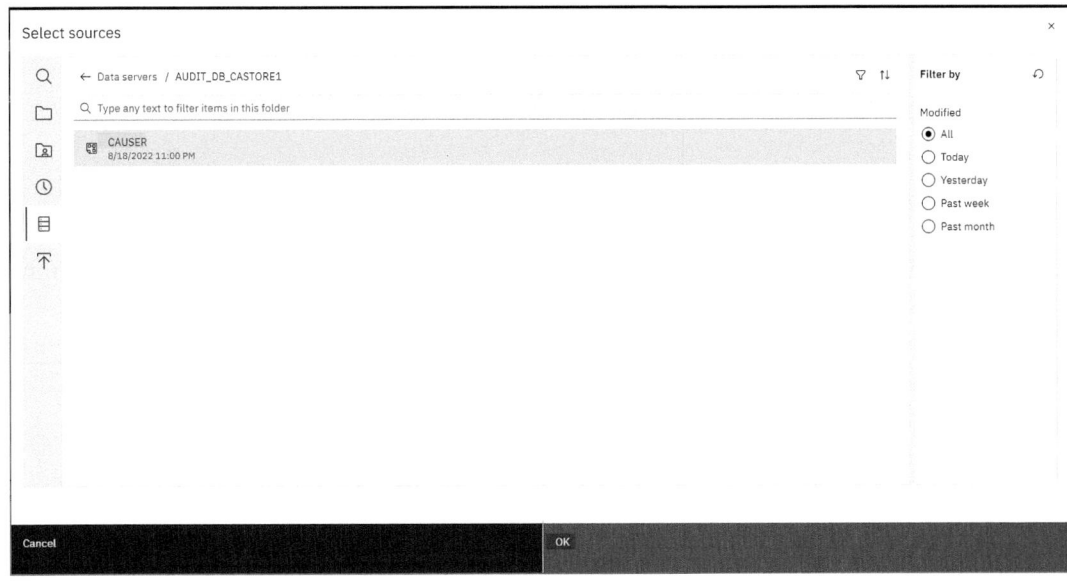

Figure 5-39. *The **OK** button is clicked to use the **CAUSER** schema we selected*

933

CHAPTER 5   IBM COGNOS ANALYTICS CUSTOM DEVELOPMENT

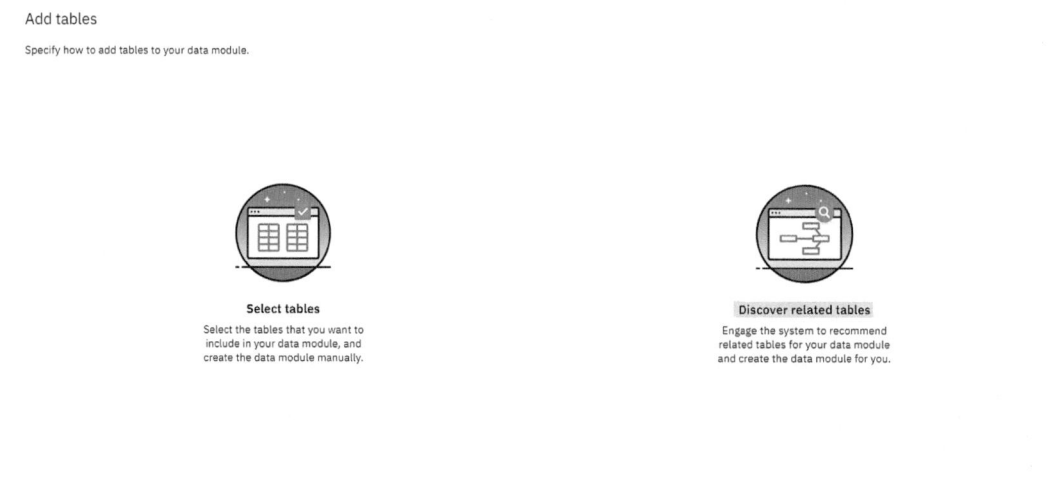

Figure 5-40. *The Discover related tables option is chosen first to use the Cognos system*

Next, we can add tables from the **CAUSER** schema by either manually selecting the tables (using the left icon, **Select tables** option) or by allowing the Cognos system to process the table rows (using the right icon, **Discover related tables** option) which then automatically displays the most significant words to be used as shown in Figure 5-41.

Figure 5-41. *The Cognos system displays the status word which we have selected*

CHAPTER 5 IBM COGNOS ANALYTICS CUSTOM DEVELOPMENT

We can click the word we wish to use as the keyword using the mouse button (I selected **status** in Figure 5-41). Then click the **Next** command button on the bottom right of the screen.

Figure 5-42. The automatically selected CAUSER schema table is displayed

The system selected the **CAUSER.F_Ceevents** Fact table from the **CASTORE1 OLAP** database. We can now add additional tables as required using the **Add more tables** menu option as shown in Figure 5-43.

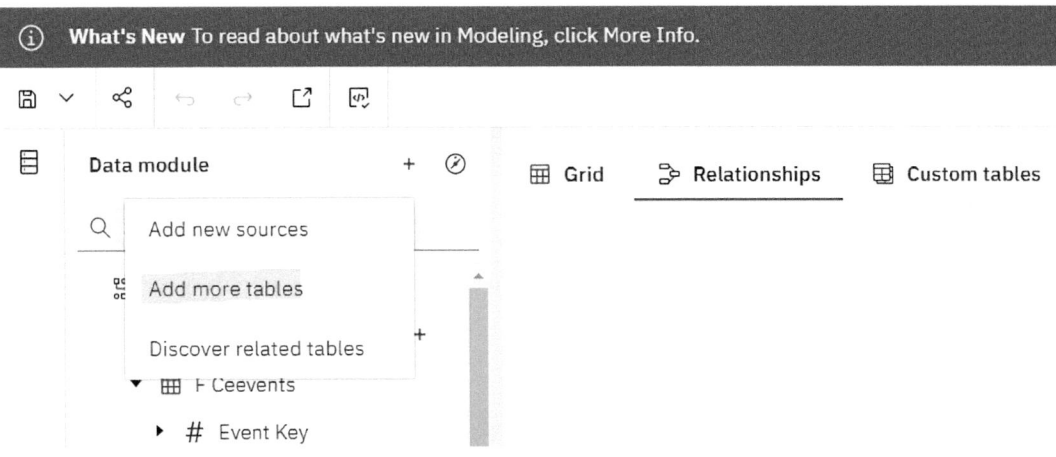

Figure 5-43. The Add more tables option is selected

935

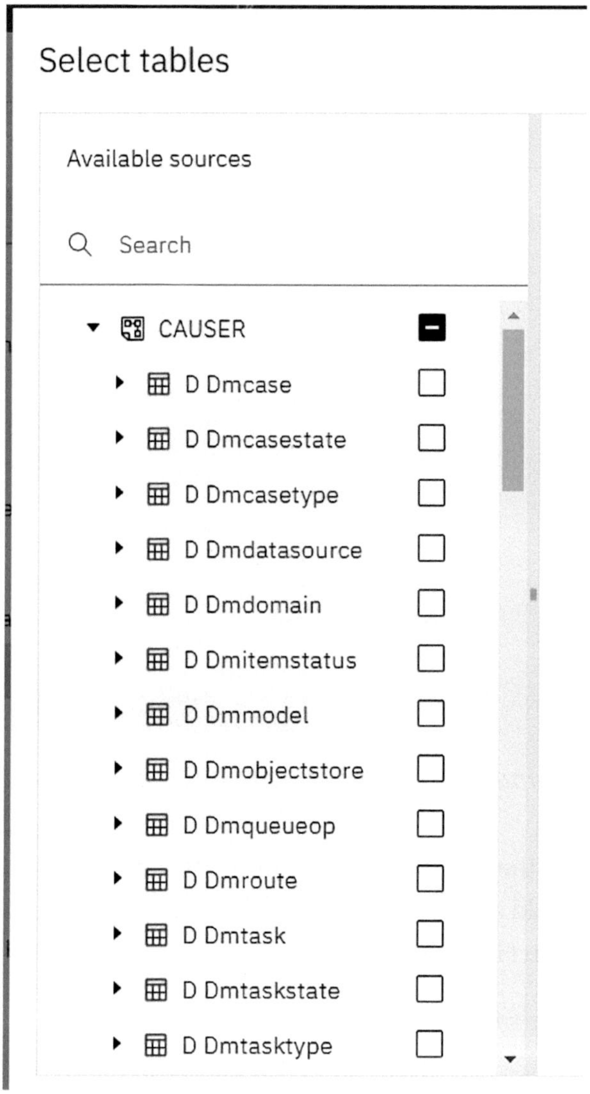

CHAPTER 5   IBM COGNOS ANALYTICS CUSTOM DEVELOPMENT

Figure 5-44.   The scrollable list of CASTORE tables for the CAUSER schema is shown

The required tables can be selected/deselected by clicking the tick box shown to the right of each Table name in the list; this list also shows any Views which may have been created.

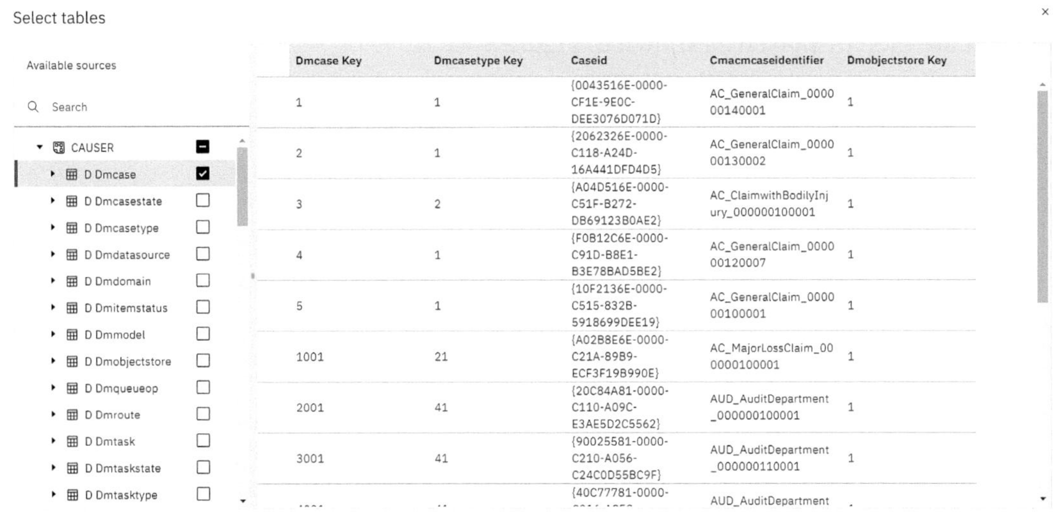

CHAPTER 5 IBM COGNOS ANALYTICS CUSTOM DEVELOPMENT

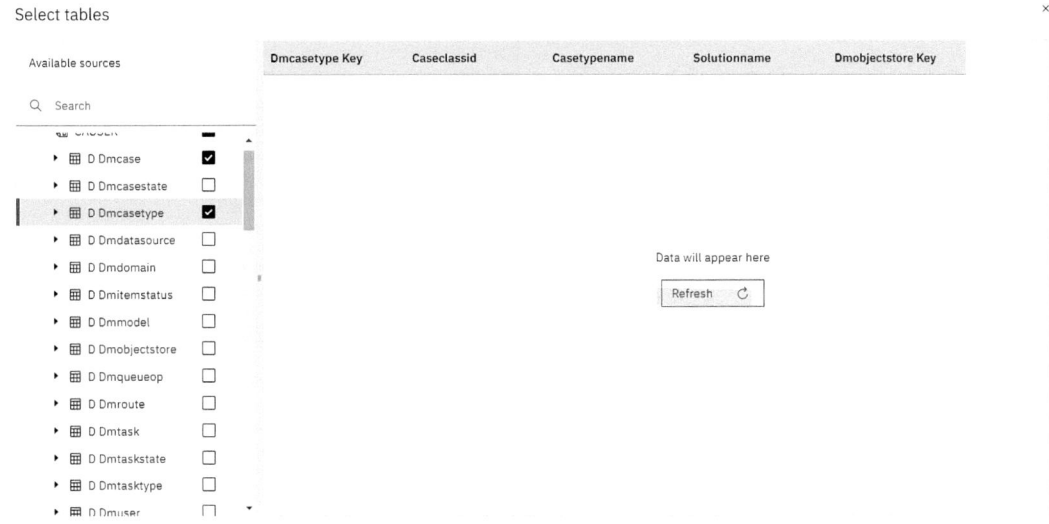

*Figure 5-45. The required Dimension table, **CAUSER.D_Dmcase**, is selected*

On first selecting a table, as shown in Figure 5-46, the **Refresh** command button in the central panel is clicked in order to display the table row data.

*Figure 5-46. The **CAUSER.D_Dmcasetype** table data is displayed using **Refresh***

CHAPTER 5   IBM COGNOS ANALYTICS CUSTOM DEVELOPMENT

The rows loaded from the Linux CASTORE Events database for the IBM Case Manager are then displayed as shown in Figure 5-47.

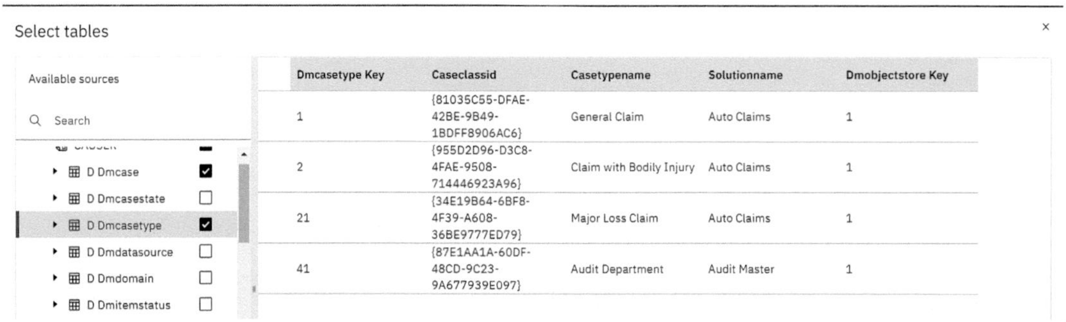

Figure 5-47. The CAUSER.D_Dmcasetype Dimension table rows are displayed

The **CAUSER.D_Dmtasktype** Dimension table is also selected as shown in Figure 5-48.

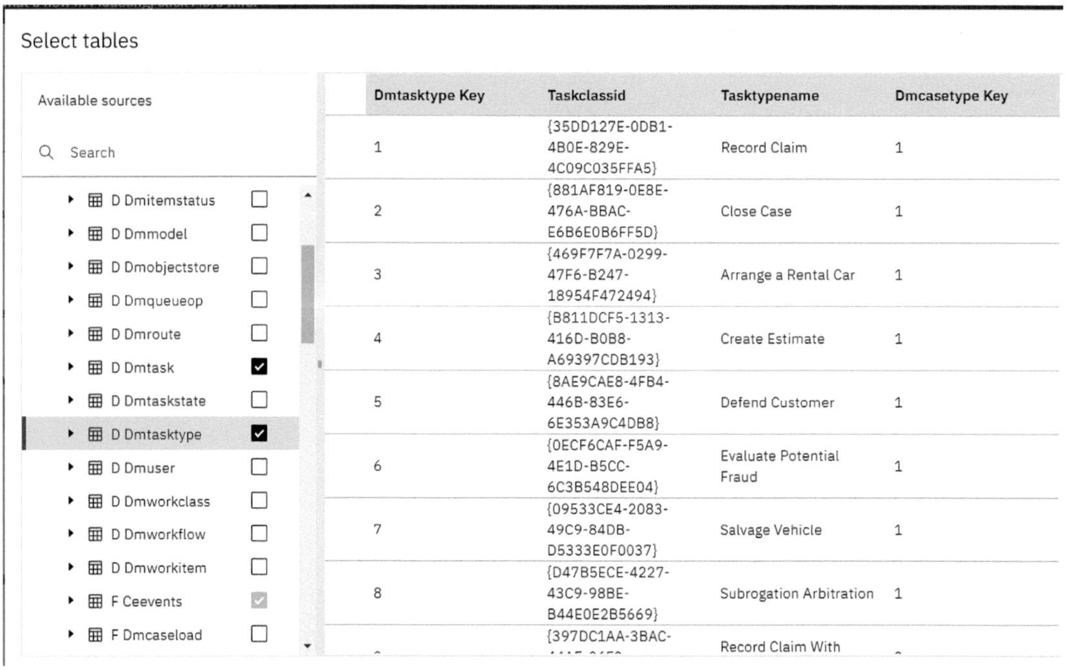

Figure 5-48. The CAUSER.D_Dmtasktype Dimension table rows are displayed

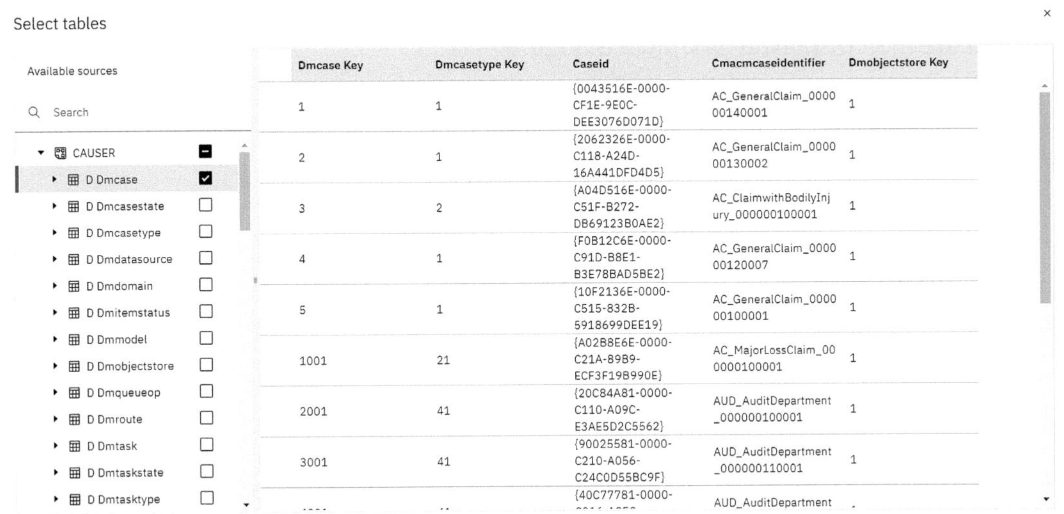

CHAPTER 5 IBM COGNOS ANALYTICS CUSTOM DEVELOPMENT

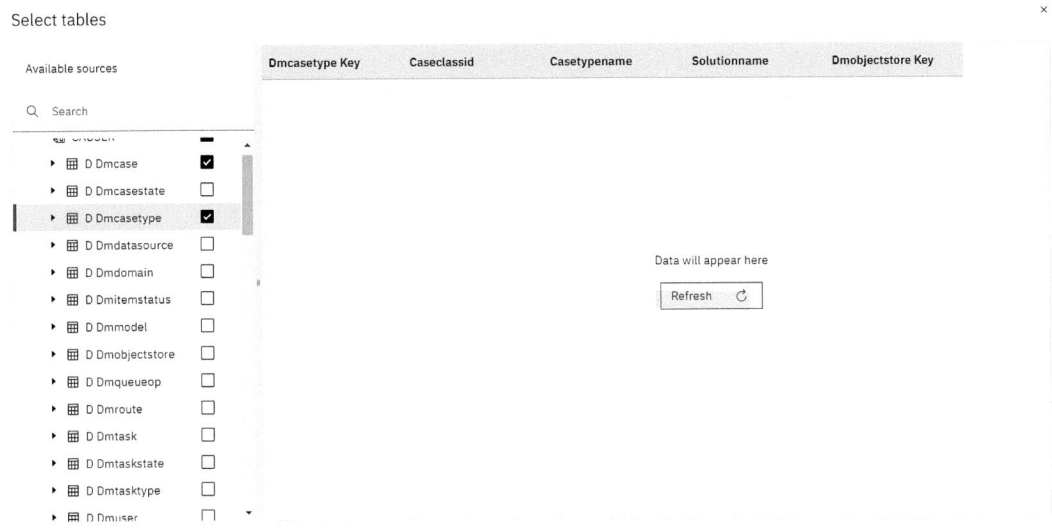

*Figure 5-45. The required Dimension table, **CAUSER.D_Dmcase**, is selected*

On first selecting a table, as shown in Figure 5-46, the **Refresh** command button in the central panel is clicked in order to display the table row data.

*Figure 5-46. The **CAUSER.D_Dmcasetype** table data is displayed using **Refresh***

CHAPTER 5   IBM COGNOS ANALYTICS CUSTOM DEVELOPMENT

The rows loaded from the Linux CASTORE Events database for the IBM Case Manager are then displayed as shown in Figure 5-47.

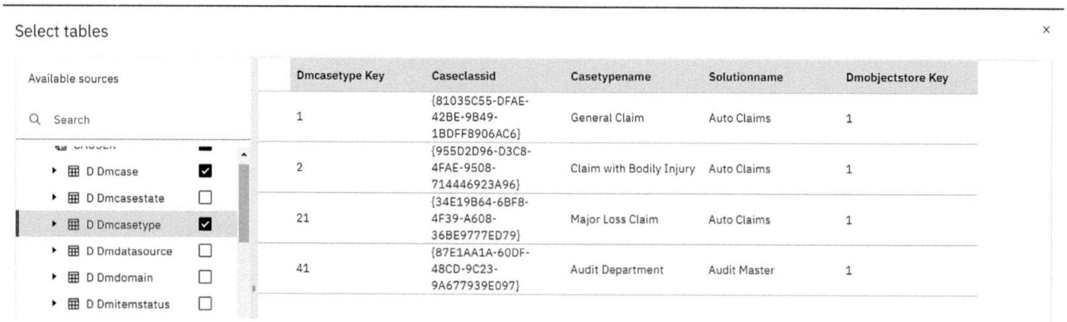

*Figure 5-47. The **CAUSER.D_Dmcasetype** Dimension table rows are displayed*

The **CAUSER.D_Dmtasktype** Dimension table is also selected as shown in Figure 5-48.

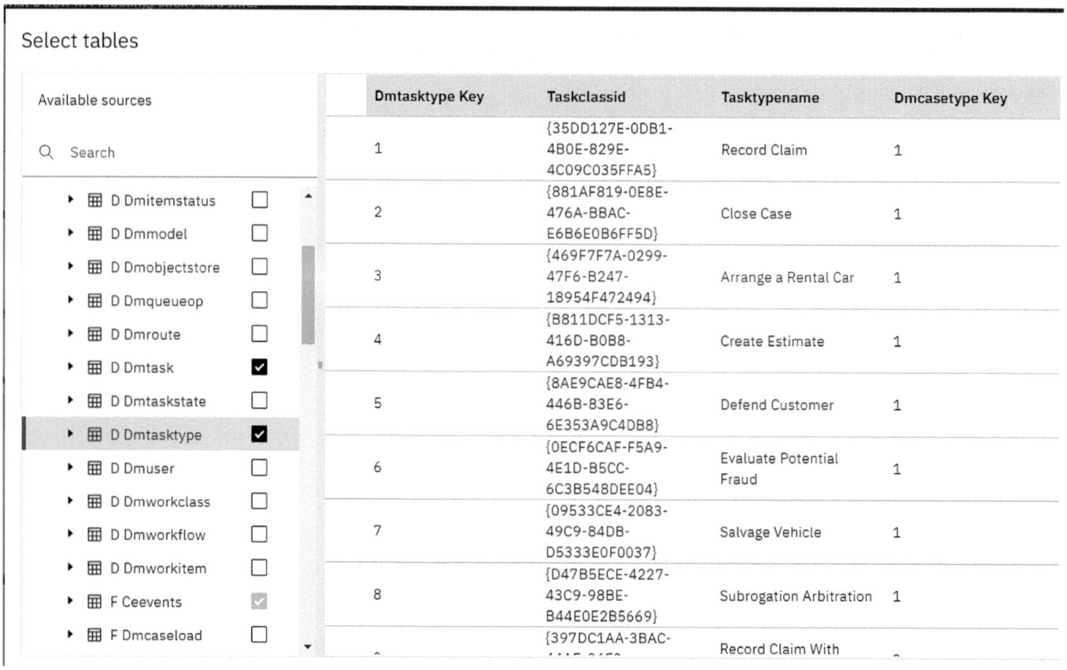

*Figure 5-48. The **CAUSER.D_Dmtasktype** Dimension table rows are displayed*

The validation of the Data module is selected as Automatic which monitors changes as they are edited to provide fast feedback of any errors to the user.

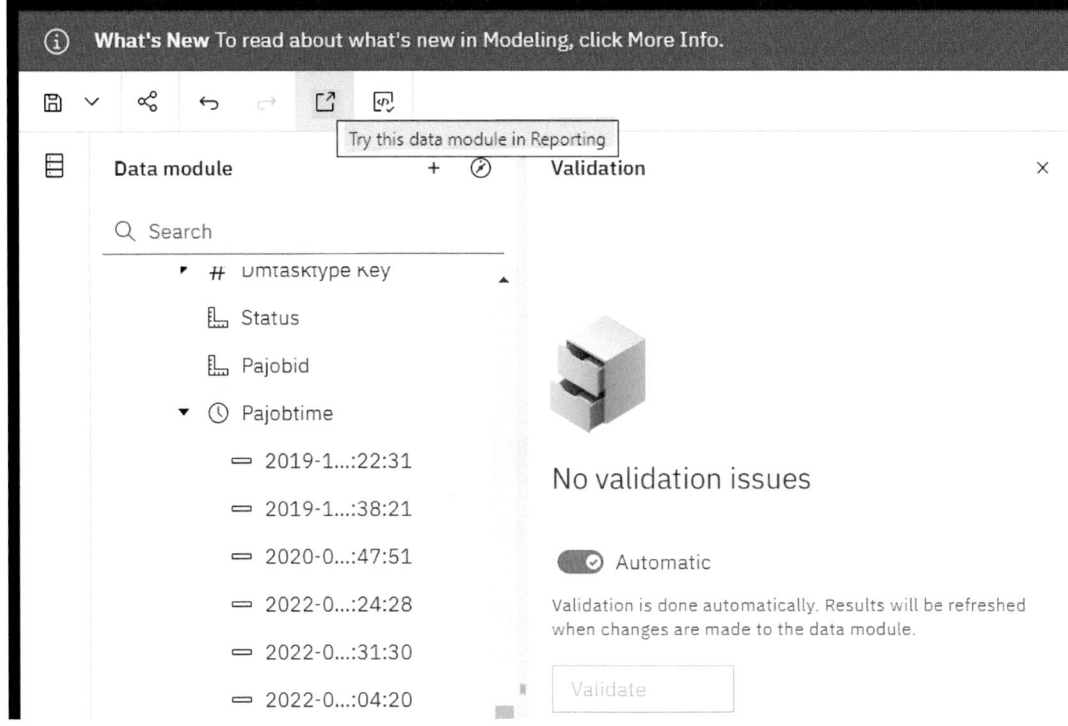

Figure 5-49. *The validation for the selected tables shows no issues*

The final selection of tables is highlighted in the left panel in Figure 5-50, and the **Data module** we have created, with these tables selected, can now be saved as **CaseLoad** in the **My content** tab root location. You can see that we have also created a Folder called **CaseAnalytics** under the **My content** root.

CHAPTER 5   IBM COGNOS ANALYTICS CUSTOM DEVELOPMENT

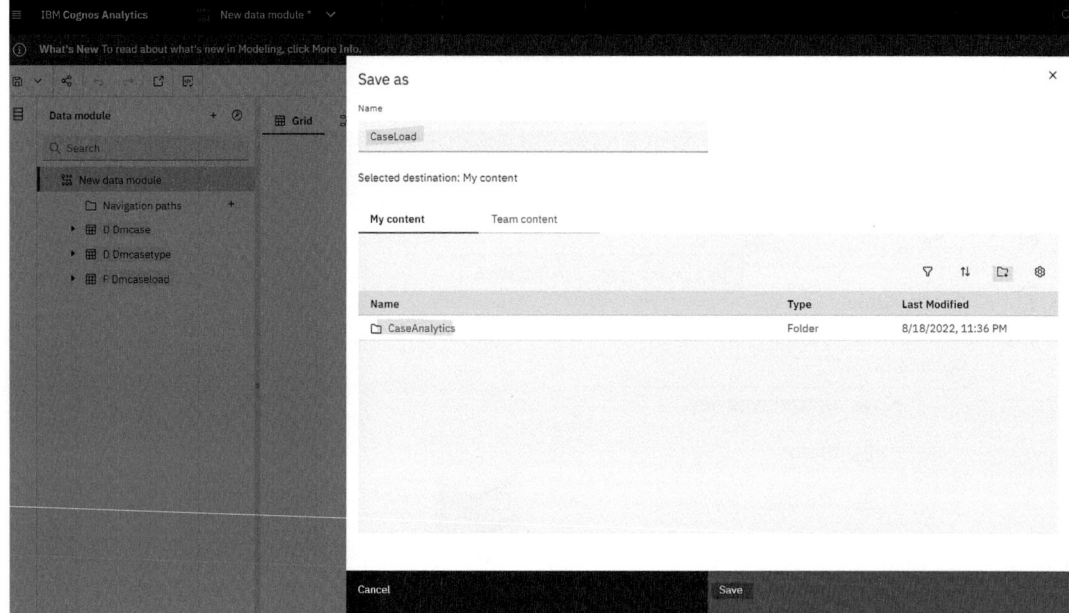

Figure 5-50. *The selected tables are saved as the **CaseLoad** Data module*

After the **Save** command button is clicked, the **Data module**, **CaseLoad**, is now shown and is ready to be used in charts in a new Cognos **Dashboard**.

CHAPTER 5   IBM COGNOS ANALYTICS CUSTOM DEVELOPMENT

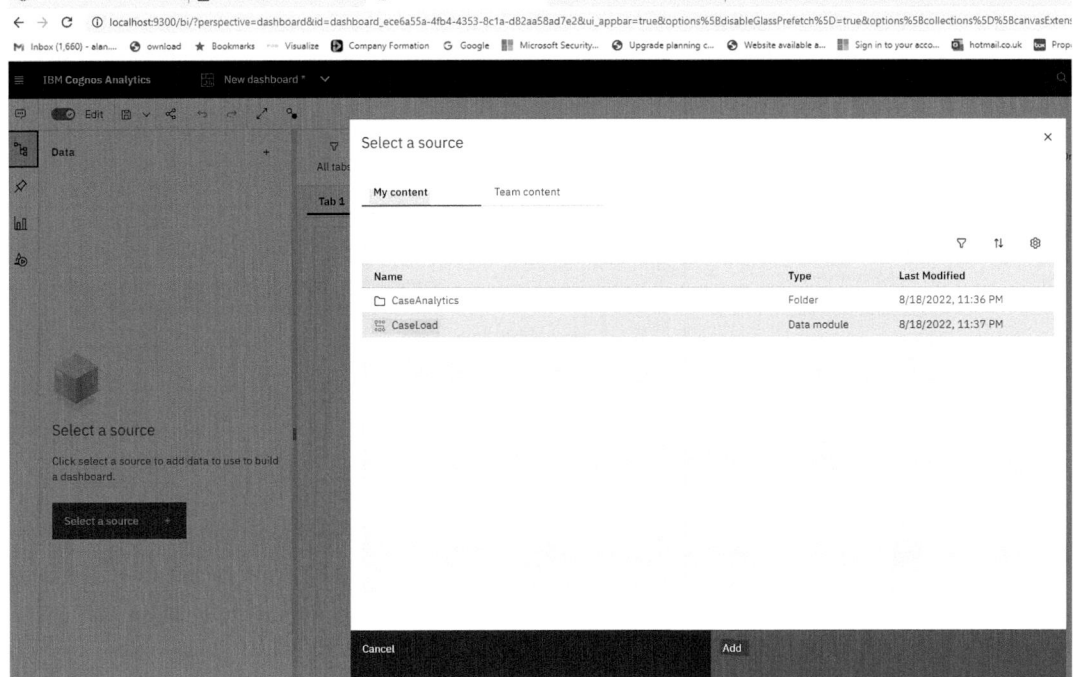

Figure 5-51. *The CaseLoad Data module is displayed under the My content root*

In the **New dashboard** Cognos frame, we can load the **Datasource** module, **CaseLoad**, highlighted in Figure 5-51, and drag and drop the required measure fields for analysis and display in the Cognos frame.

CHAPTER 5 IBM COGNOS ANALYTICS CUSTOM DEVELOPMENT

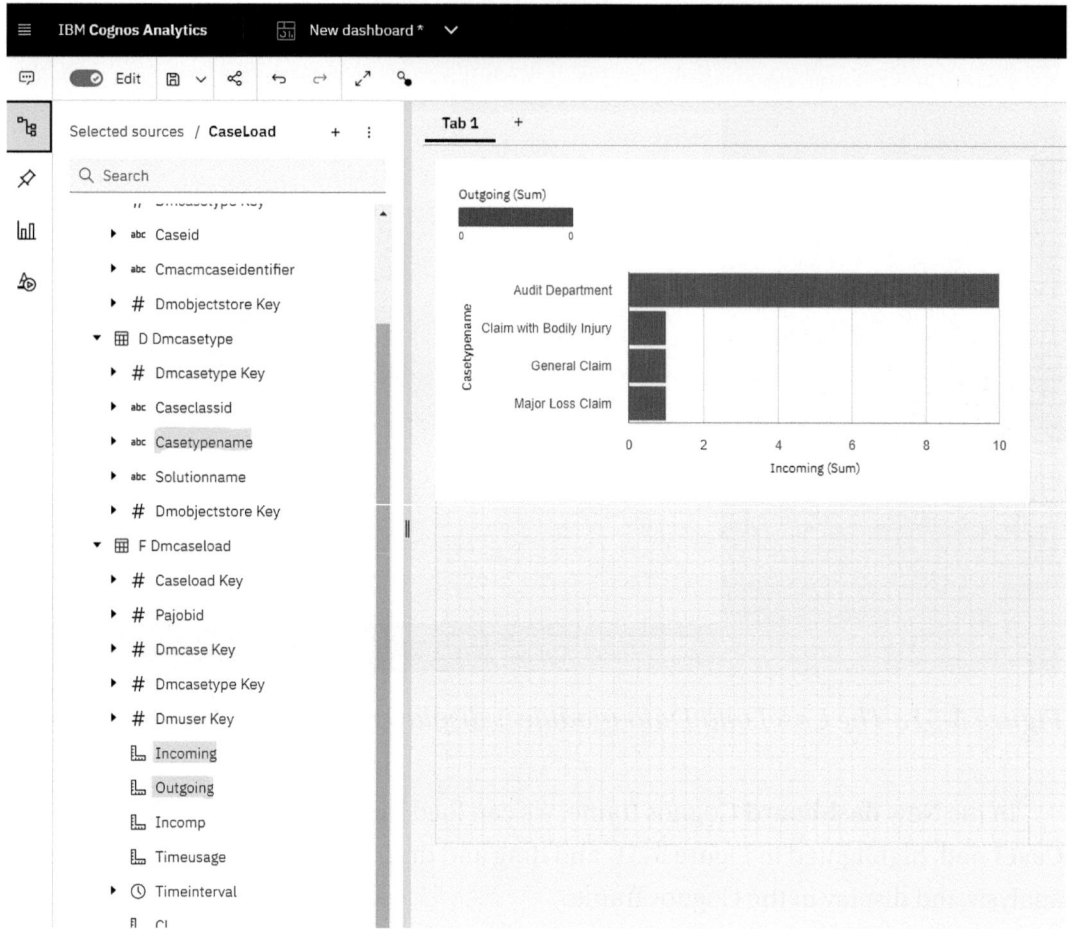

Figure 5-52. *The Incoming case count for each Case Type from Case Manager is displayed*

CHAPTER 5   IBM COGNOS ANALYTICS CUSTOM DEVELOPMENT

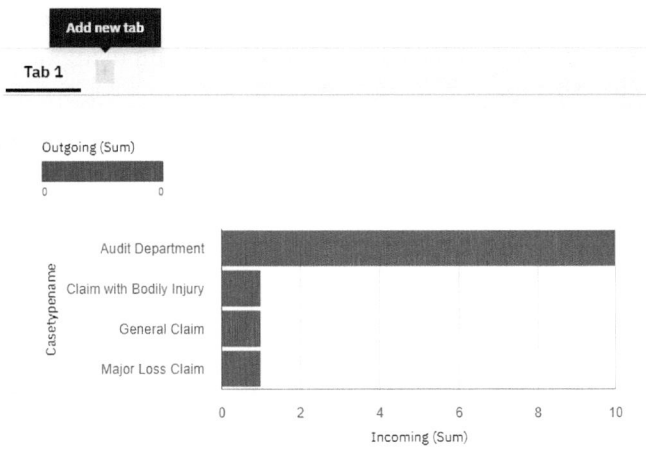

*Figure 5-53. The **Add new tab** + icon can now be selected to add additional graphs*

Updated Datasource Setup After Adding the Custom Views

In this section, we will now refresh the initial **CAUSER** schema from the IBM Case Manager Case Analyzer **CASTORE1** Linux **Db2** database Events server, **ecmukdemo6**, to include the two custom Views we created earlier (**CAUSER.DMDaysIn_Case** and **CAUSER.DMAgeing_Case**).

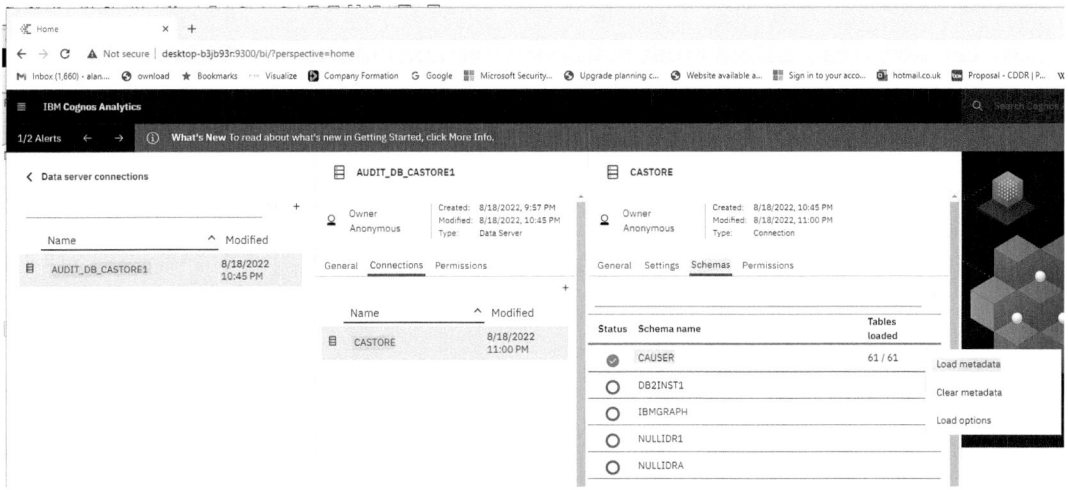

Figure 5-54. The CAUSER schema of the AUDIT_DB_CASTORE1 connection is refreshed

943

CHAPTER 5   IBM COGNOS ANALYTICS CUSTOM DEVELOPMENT

We right-click the selected **CAUSER** schema (showing the initial 61 tables transferred), as shown in Figure 5-54, to reload the metadata using the highlighted **Load metadata** command menu item.

After the reload, it can be seen that the count of tables is now increased by two to 63, reflecting the discovery of the two additional views we created.

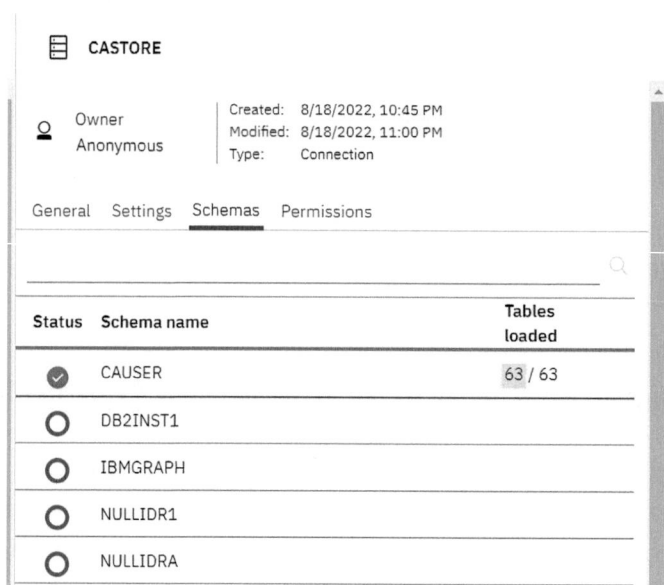

Figure 5-55. *The two new views are now included in the **CASTORE** datasource*

We can now create a **Data module** where we can include the new **Views** in the selected **CAUSER schema** tables and save this new **Data module** for analysis.

CHAPTER 5 IBM COGNOS ANALYTICS CUSTOM DEVELOPMENT

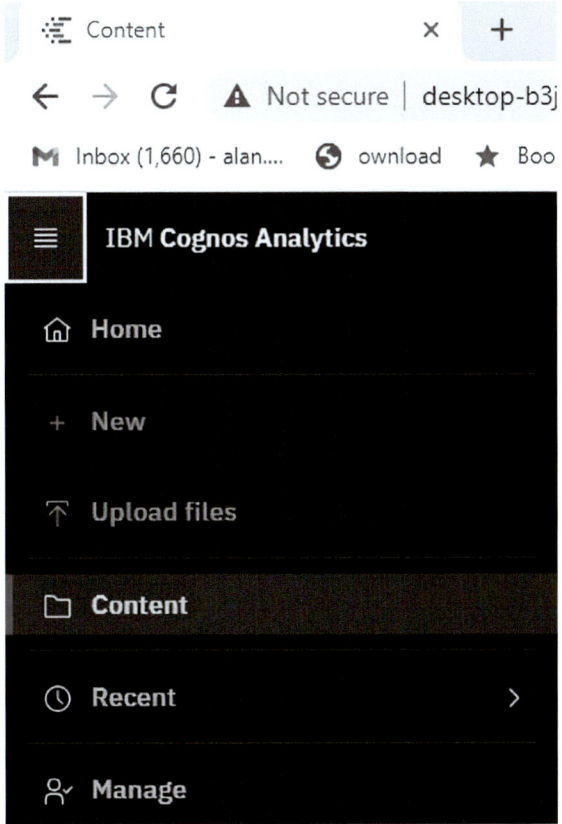

Figure 5-56. *The Cognos **Content** folder option is selected*

The **New +** drop-down menu list command (highlighted in Figure 5-57) is selected.

CHAPTER 5   IBM COGNOS ANALYTICS CUSTOM DEVELOPMENT

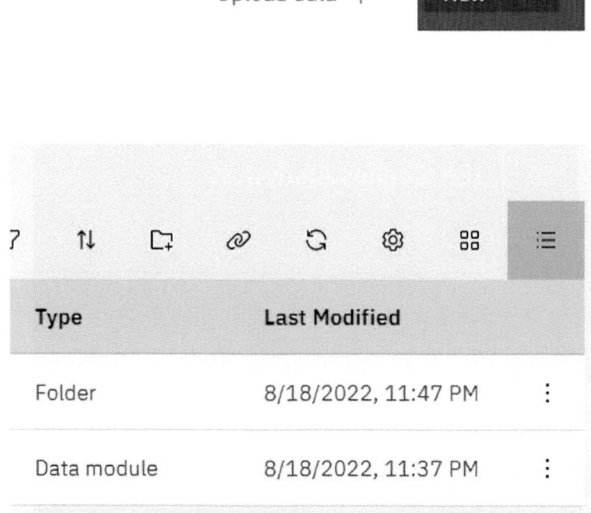

*Figure 5-57.   The **New +** command button is selected*

We can then select the **Data module** menu item from the drop-down menu list.

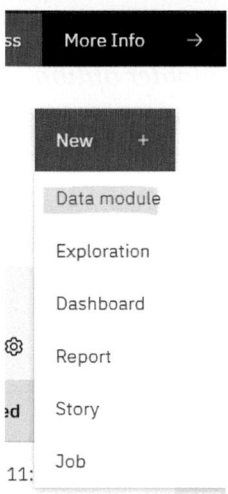

*Figure 5-58.   The **Data module** menu item is selected from the **New +** drop-down list*

CHAPTER 5   IBM COGNOS ANALYTICS CUSTOM DEVELOPMENT

The **Select sources** pop-up window is shown to allow us to select the **CAUSER** source we just updated from the **AUDIT_DB_CASTORE1** data server connection.

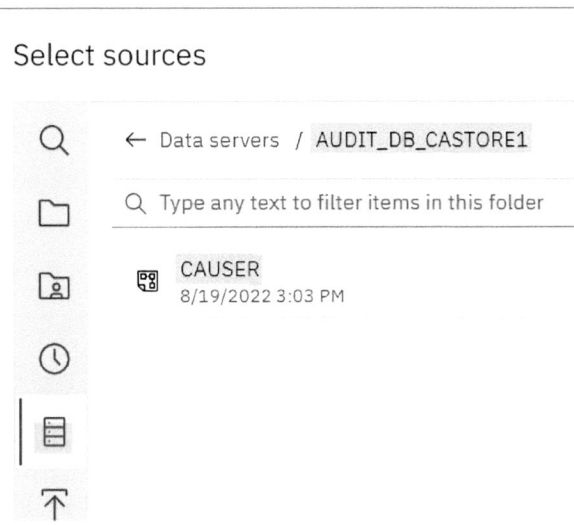

Figure 5-59.   *The updated **CAUSER** Datasource is selected for analysis*

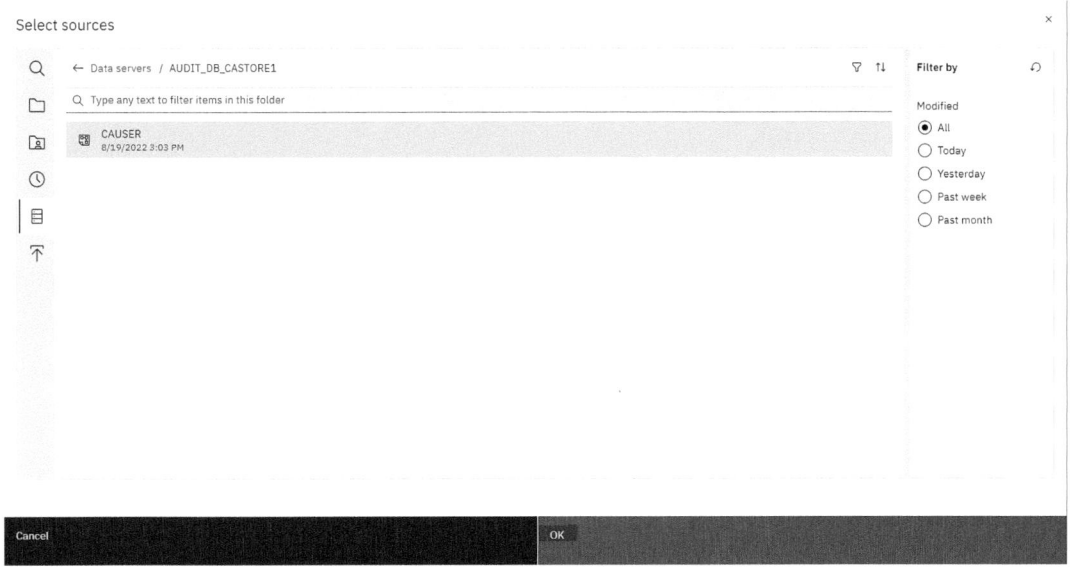

Figure 5-60.   *The **OK** command button is used to confirm the **CAUSER** selection we made*

CHAPTER 5   IBM COGNOS ANALYTICS CUSTOM DEVELOPMENT

The **Select tables** option is now selected to select the required tables for the analysis.

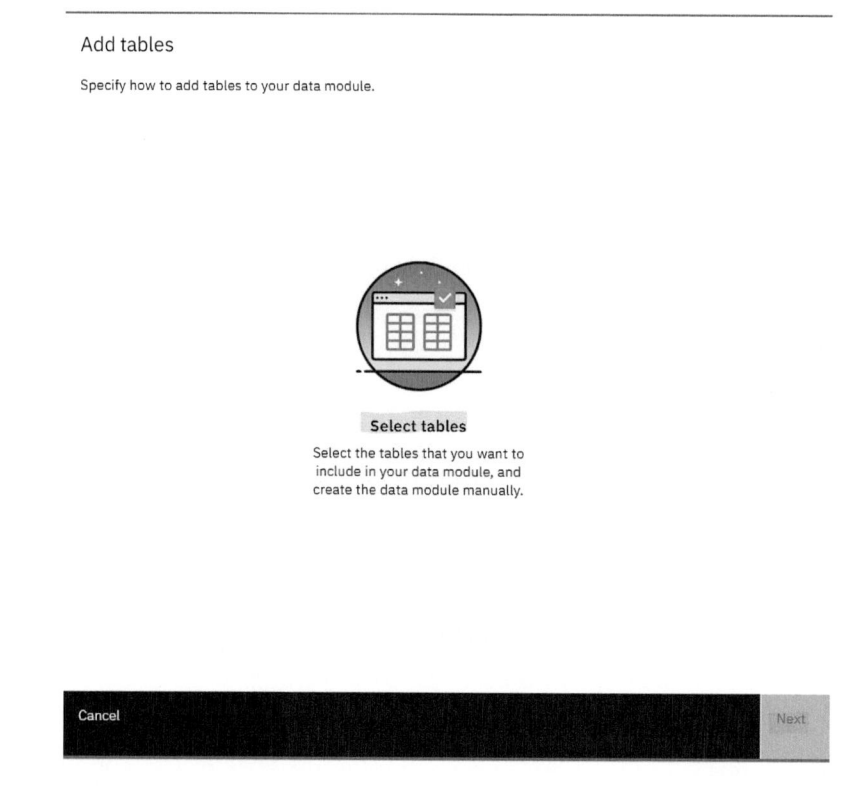

*Figure 5-61. The **Select tables** option is selected for manual table selection*

The alternative access to creating a **Data module** is shown in the following screens.

CHAPTER 5 IBM COGNOS ANALYTICS CUSTOM DEVELOPMENT

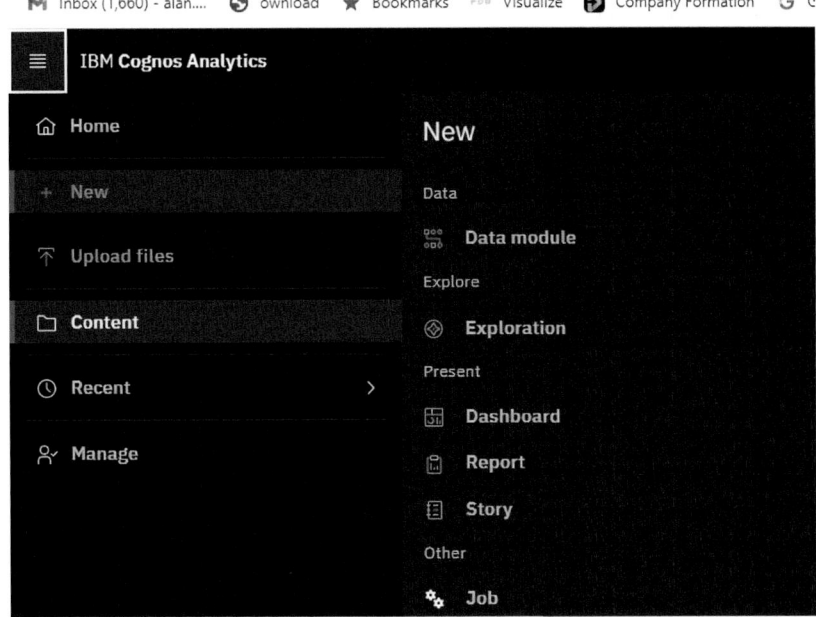

Figure 5-62. *The* **+ New** *option allows the Data module menu option to be selected*

We can then see the same **Select sources** pop-up window we used earlier.

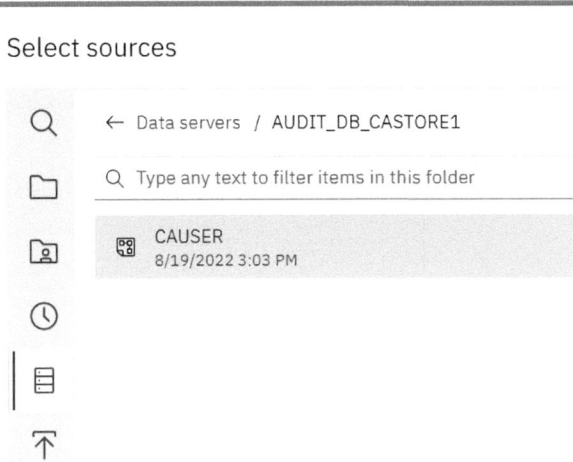

Figure 5-63. *The updated* **CAUSER** *Datasource is available for selection*

949

CHAPTER 5   IBM COGNOS ANALYTICS CUSTOM DEVELOPMENT

On opening the CAUSER datasource, we can select from the required tables as before with the previous example.

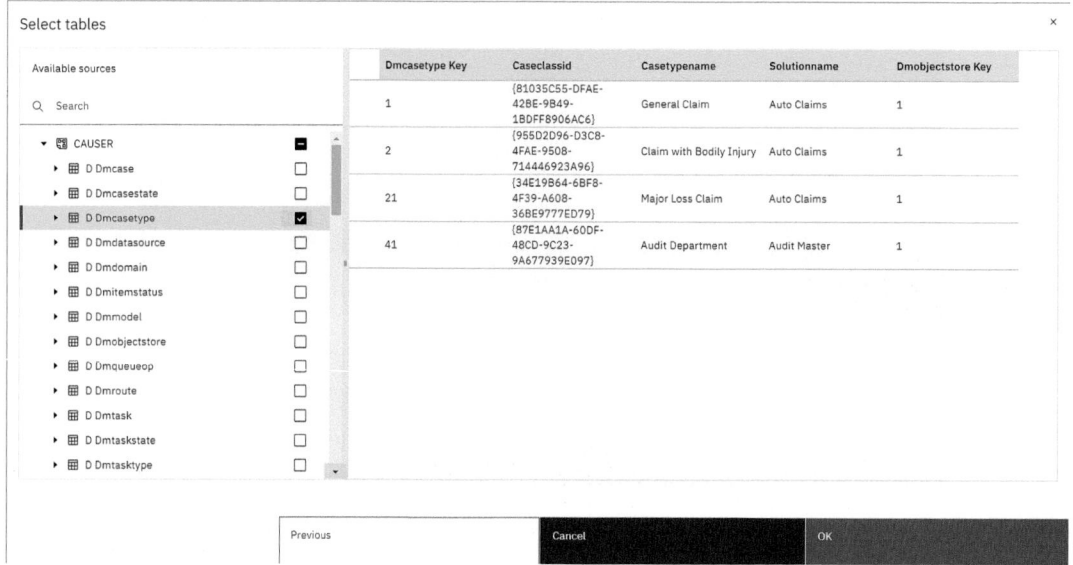

*Figure 5-64. The **CAUSER.D_Dmcasetype** Dimension table is selected*

The Fact table **CAUSER.F_Dmcaseload** is selected next, as shown in Figure 5-65.

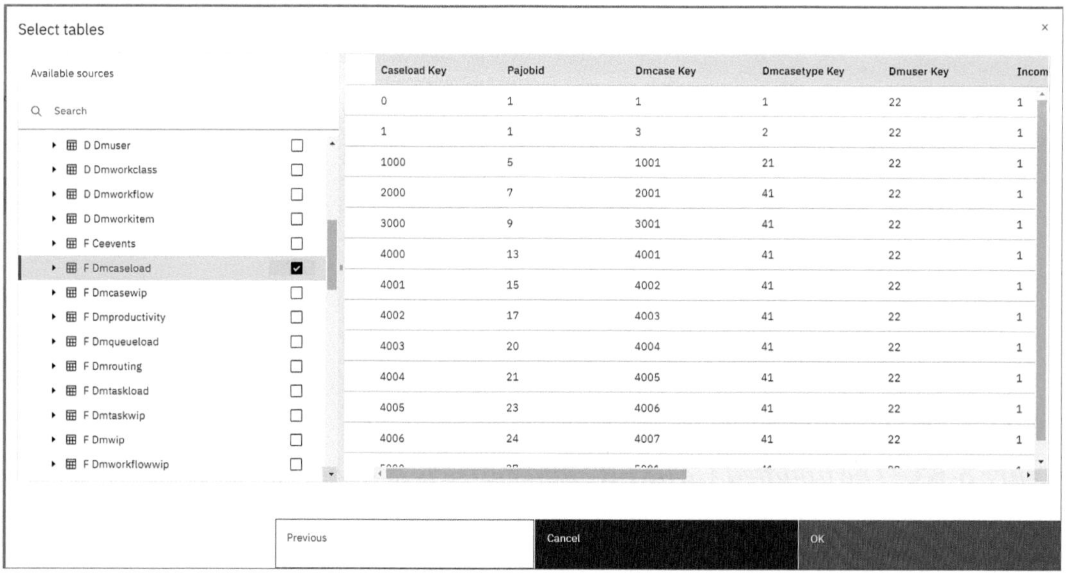

*Figure 5-65. The **CAUSER.F_Dmcaseload** Fact table is selected*

CHAPTER 5 IBM COGNOS ANALYTICS CUSTOM DEVELOPMENT

On scrolling further down the list of tables, we can now see we have access to the two custom views we created earlier, **CAUSER.DmdaysIn_Case** and **CAUSER.Dmageing_Case**.

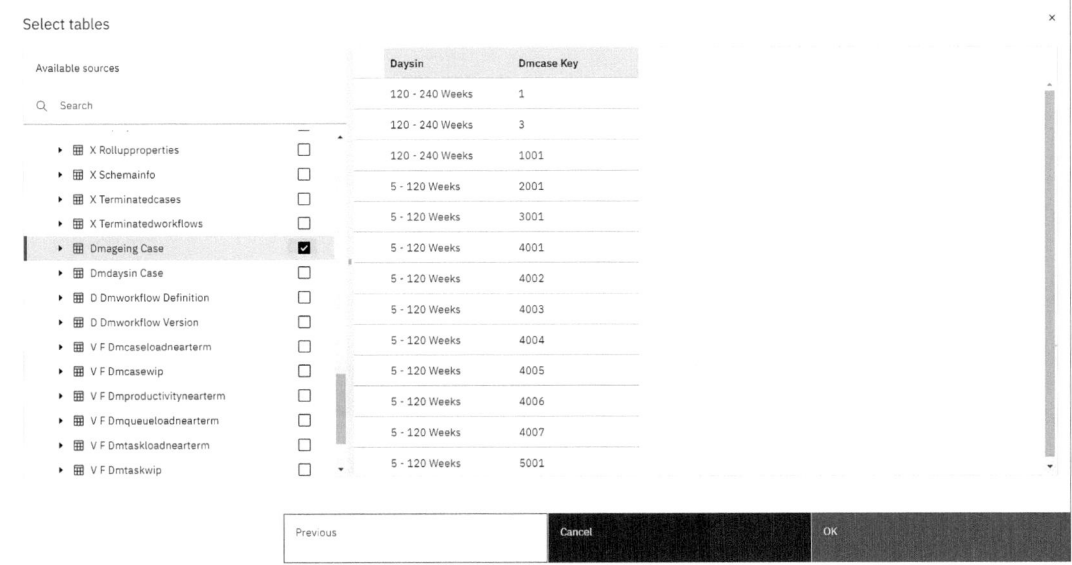

Figure 5-66. *The custom view,* ***CAUSER.Dmageing_Case****, is selected*

On selecting the Refresh command button, the expected rows for the View are returned in the data pane as shown in Figure 5-66.

951

CHAPTER 5   IBM COGNOS ANALYTICS CUSTOM DEVELOPMENT

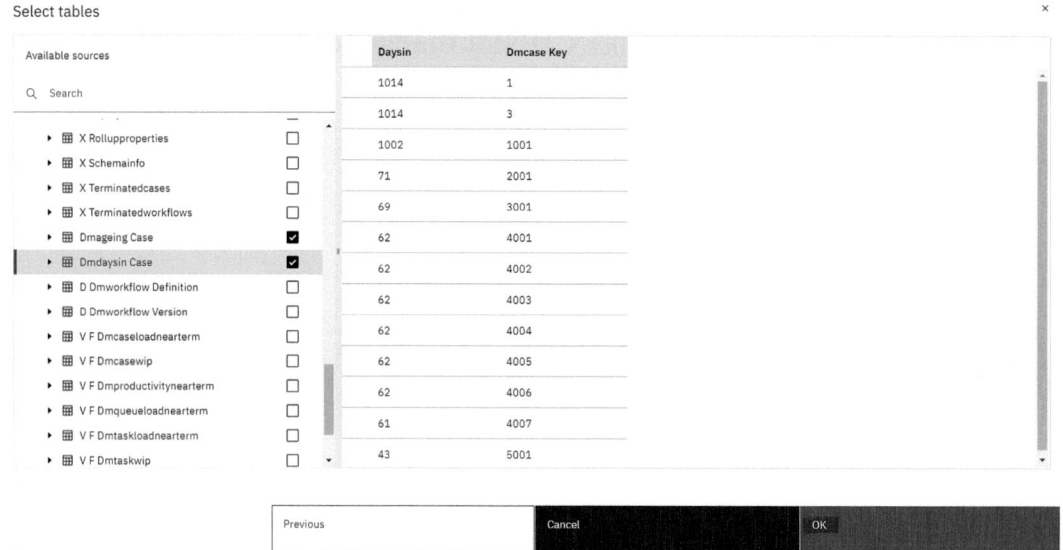

*Figure 5-67. The custom view, **CAUSER.Dmdaysin_Case**, is selected*

The **New data module** is now ready to be edited to supply the necessary relationships we will need to associate the custom **Daysin** values with the Case types.

CHAPTER 5 IBM COGNOS ANALYTICS CUSTOM DEVELOPMENT

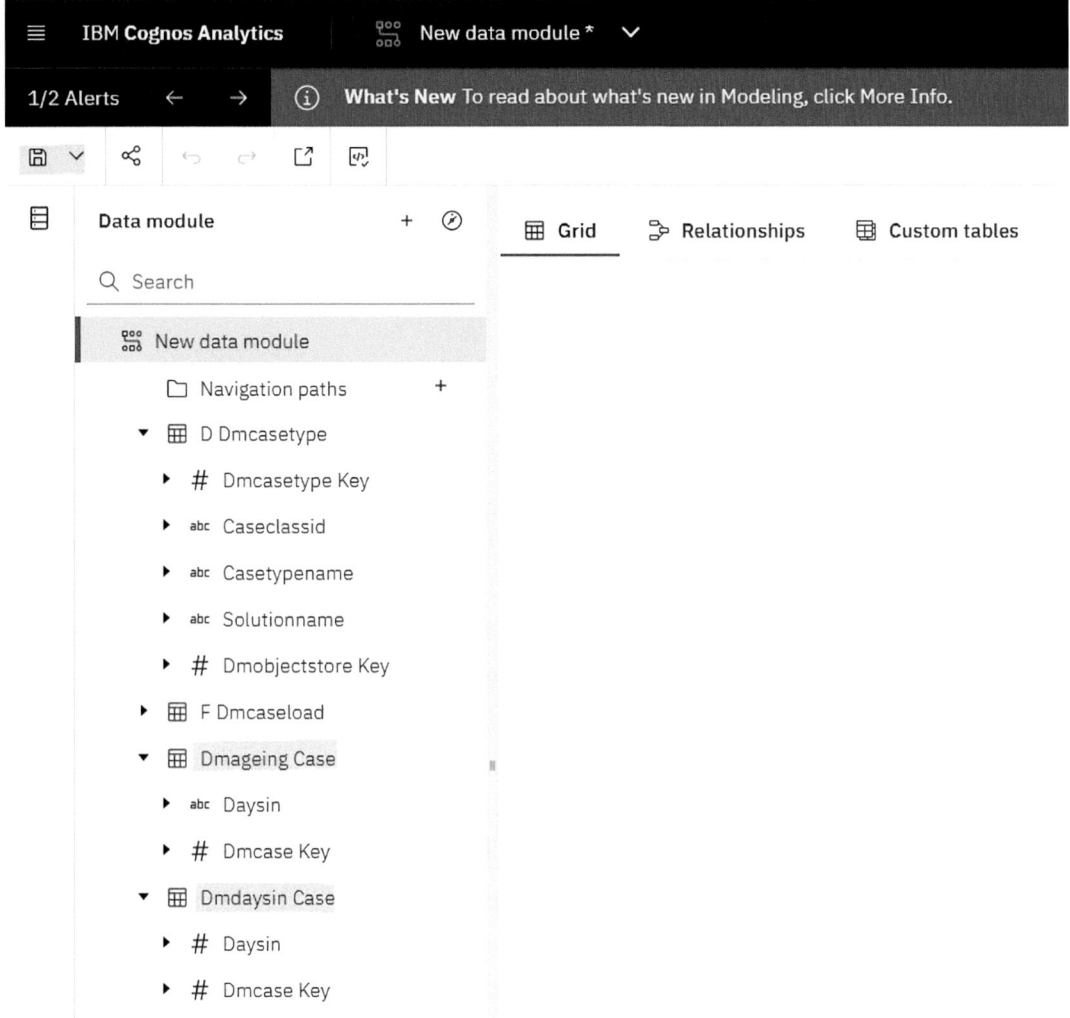

Figure 5-68. *The new custom views are highlighted in the **New data module***

We can now drag and drop the required tables into the **Relationships** tab pane, as shown in Figure 5-68, and then we select the **F_Dmcaseload** Fact table and then right-click this table's box and select the **Relationship** menu option from the menu list.

953

CHAPTER 5   IBM COGNOS ANALYTICS CUSTOM DEVELOPMENT

Figure 5-69. *The highlighted Relationship menu item is selected for*
F_Dmcaseload

The Relationship window opens, and we can then select our Custom view, **Dmageing_Case**, to enable selection to the matching Key field, **Dmcase_Key**, to the Key field, **Dmcase_Key**, we select from the **F_Dmcaseload** table (shown in Figure 5-70).

CHAPTER 5   IBM COGNOS ANALYTICS CUSTOM DEVELOPMENT

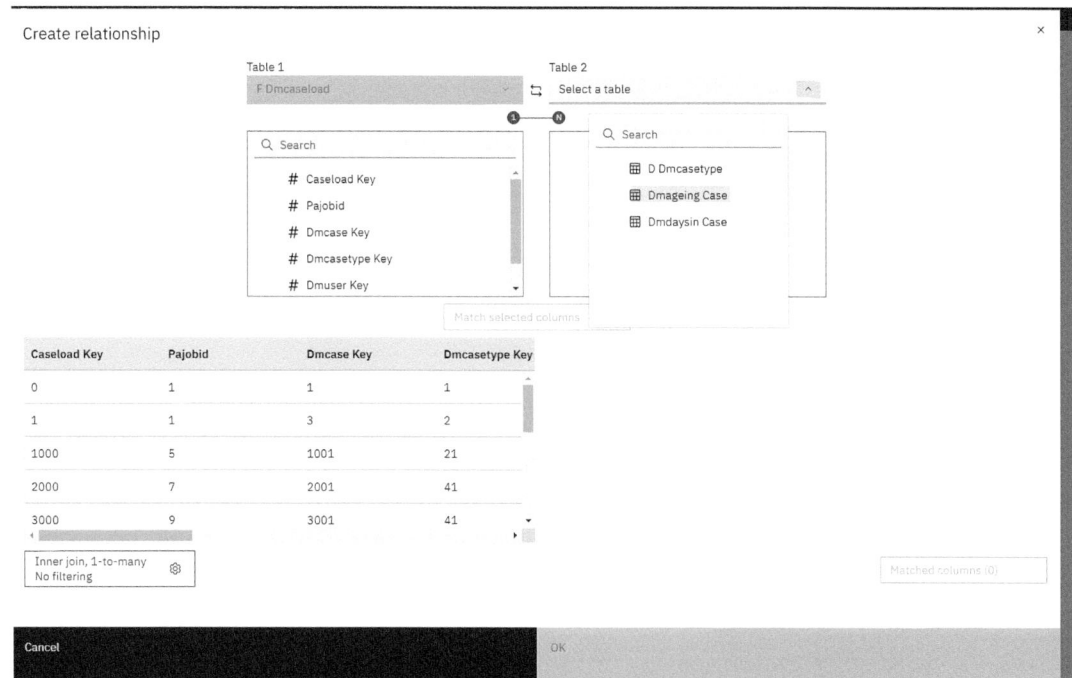

Figure 5-70.* The custom view, Dmageing_Case, is selected from the drop-down list*

On selection of the two fields, we believe are the relationship fields (both named **Dmcase_Key**), we can click the Match selected columns command button.

CHAPTER 5 IBM COGNOS ANALYTICS CUSTOM DEVELOPMENT

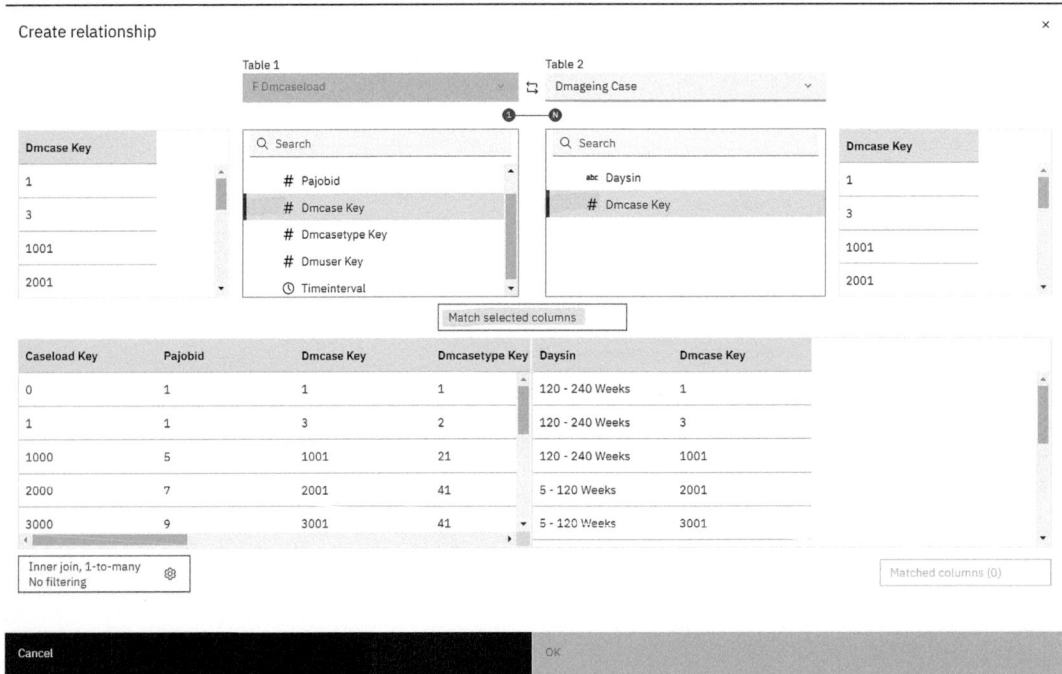

Figure 5-71. *The two relationship fields are selected, then* **Match selected columns** *clicked*

Next, we are presented with a **Refresh** command button which will show the resulting columns from the joined Key fields.

CHAPTER 5 IBM COGNOS ANALYTICS CUSTOM DEVELOPMENT

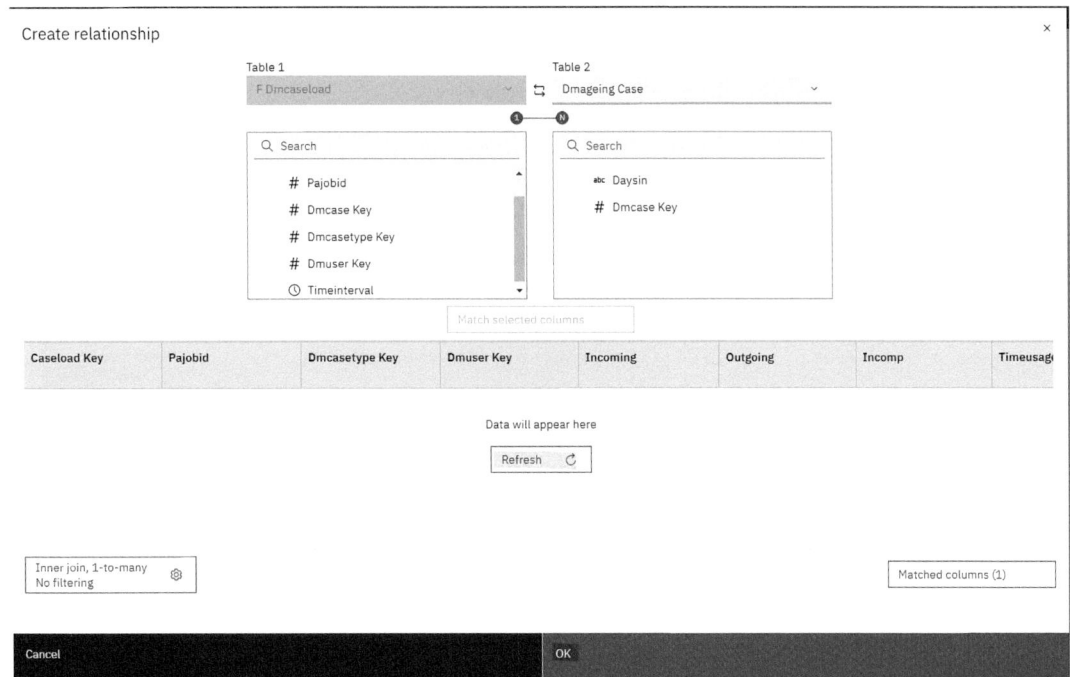

Figure 5-72. *The **Refresh** command button is clicked to display the join rows*

In Figure 5-73, you can see the rows resulting from joining the custom view with the Fact table.

CHAPTER 5   IBM COGNOS ANALYTICS CUSTOM DEVELOPMENT

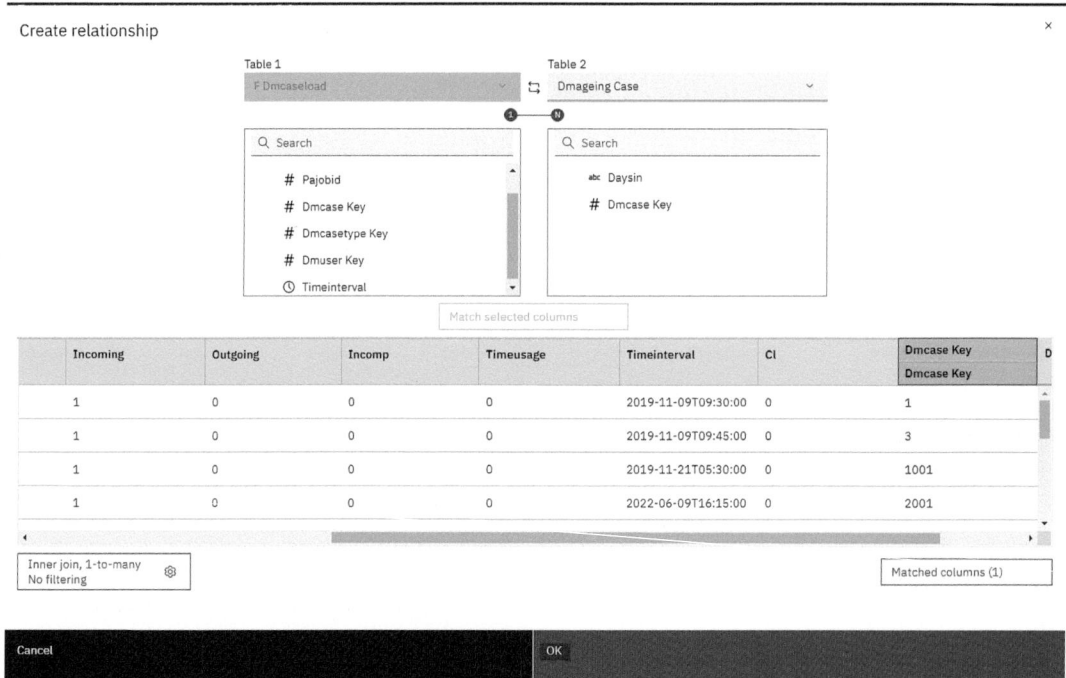

Figure 5-73. *The Inner join displays the resulting field values (scrolling right shows Daysin)*

On clicking the **OK** command button, we can now see that a new **Relationship** diagram is created with a new link to our Custom view, **Dmageing_Case**.

We can now repeat the steps displayed in Figure 5-69 to Figure 5-73 for the second custom view **Dmdaysin_Case**.

CHAPTER 5 IBM COGNOS ANALYTICS CUSTOM DEVELOPMENT

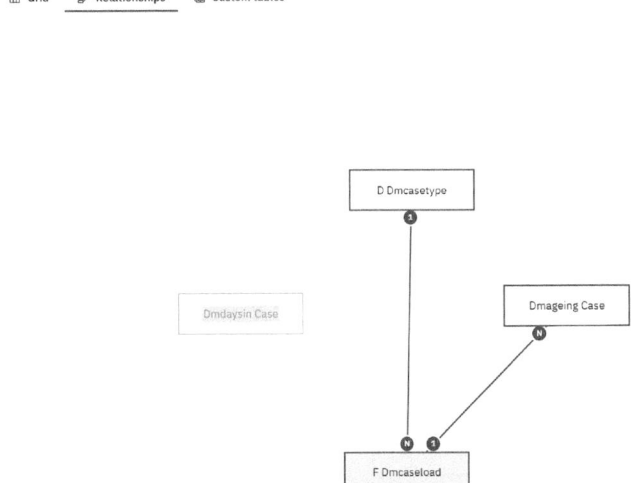

*Figure 5-74. The new **Relationship** link is drawn to the **Dmageing_Case** view*

We have now repeated the Relationship link procedure for the custom Dmdaysin_Case view we created.

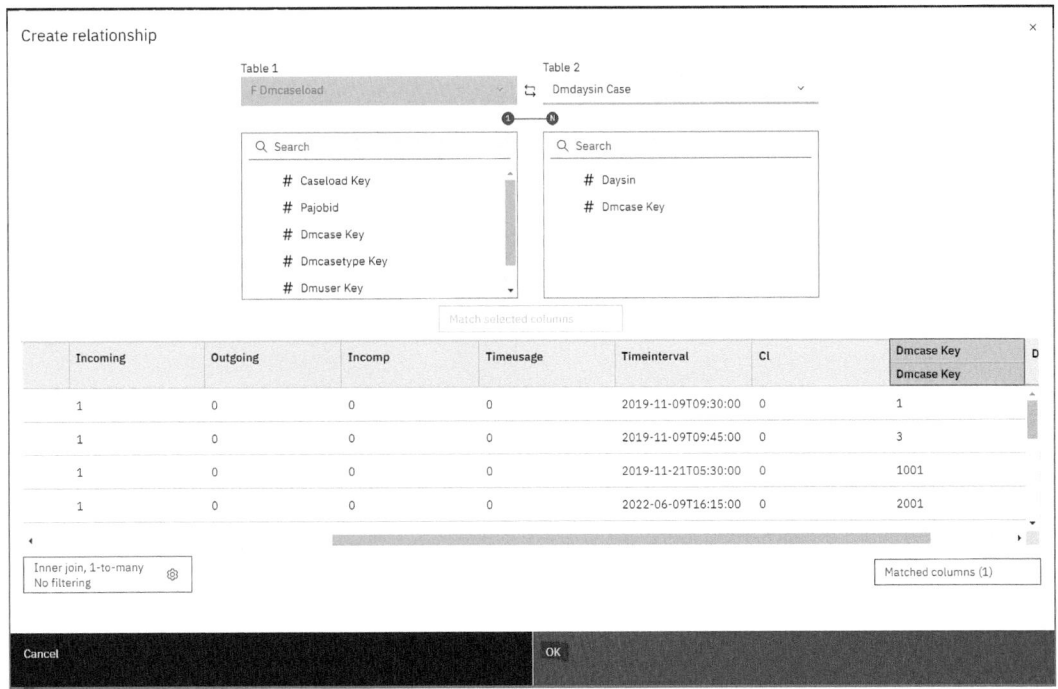

*Figure 5-75. The new **Relationship** link is now made for the **Dmduysln_Case** view*

CHAPTER 5   IBM COGNOS ANALYTICS CUSTOM DEVELOPMENT

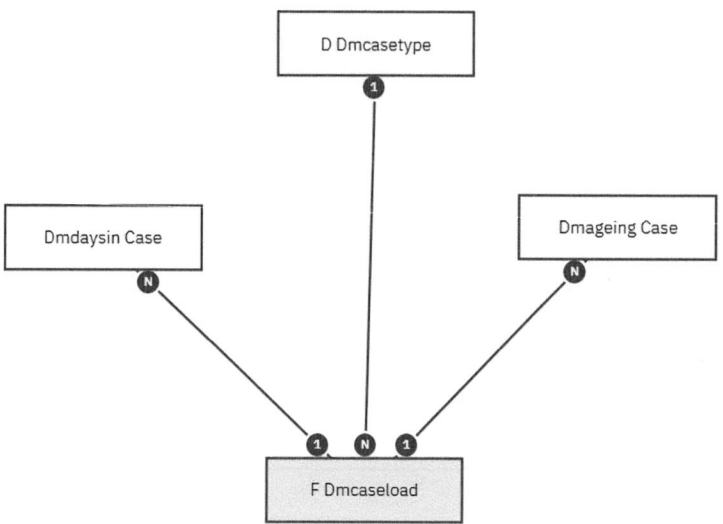

Figure 5-76. *The completed Relationship diagram for the new Data module*

We can now save the **Data module** and select the **Present data** main panel to create a new Dashboard.

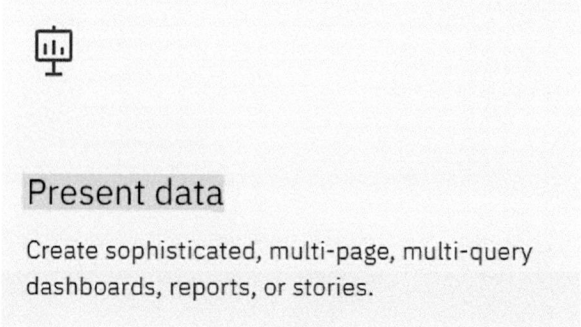

Figure 5-77. *The Present data panel is clicked*

CHAPTER 5   IBM COGNOS ANALYTICS CUSTOM DEVELOPMENT

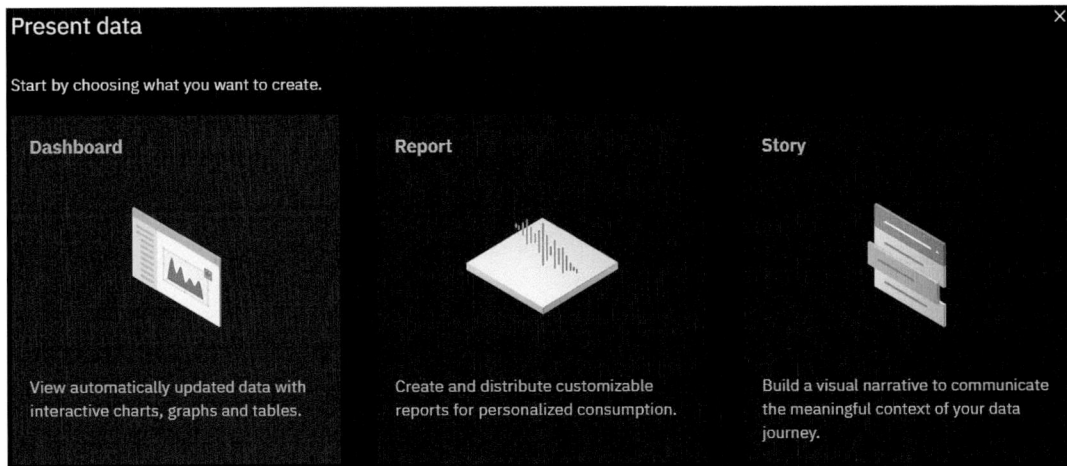

Figure 5-78.   *The Dashboard option is selected for creating a new Dashboard*

After clicking the Dashboard panel, we are shown a list of Frame templates to use which assists with the new Dashboard layout.

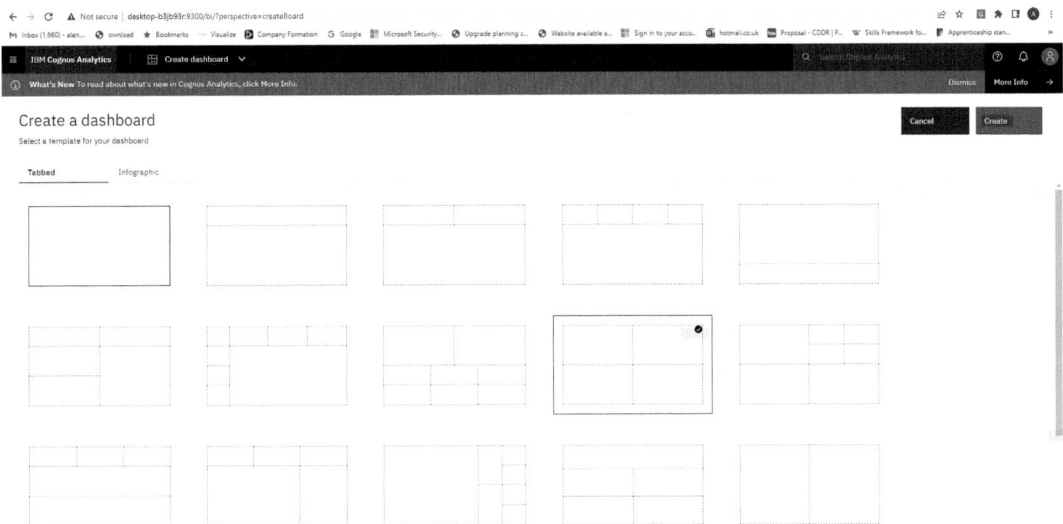

Figure 5-79.   *The highlighted template is selected for our Dashboard example*

After selecting the template (highlighted in Figure 5-79), we can click the **Create** command button highlighted top right in Figure 5-79 to create the new **Dashboard**.

961

CHAPTER 5　IBM COGNOS ANALYTICS CUSTOM DEVELOPMENT

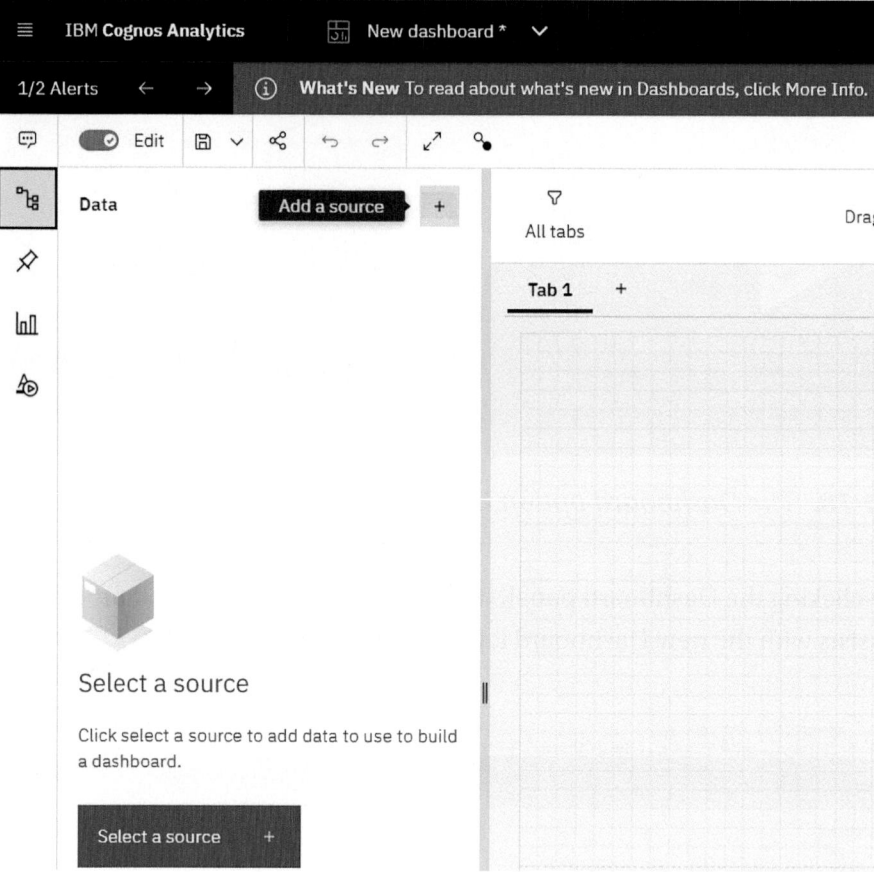

Figure 5-80. *The Add a source + icon is clicked to select our new datasource*

We have to navigate to the **CaseAnalytics** subfolder we created earlier as this is where we saved the Data module with our custom views.

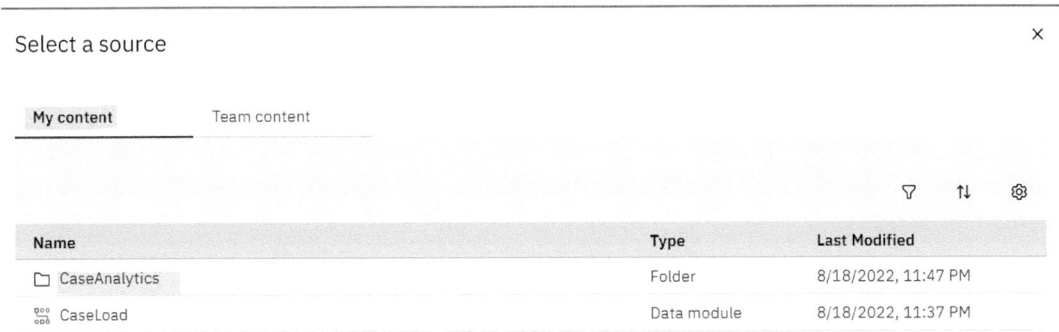

Figure 5-81. The **CaseAnalytics** folder is double-clicked to display the saved Data module

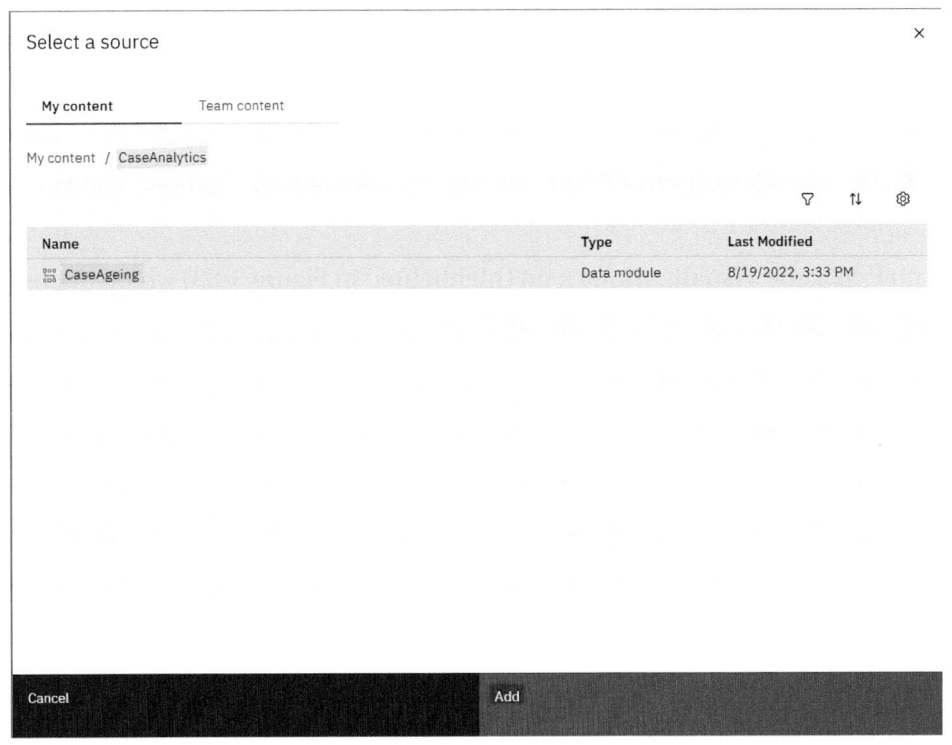

Figure 5-82. The **Add** command is used to load the saved **CaseAgeing** Data module

CHAPTER 5 IBM COGNOS ANALYTICS CUSTOM DEVELOPMENT

Now, the loaded **CaseAgeing** Data module shows the two tables and two custom views we are going to use for a Graph Visualization.

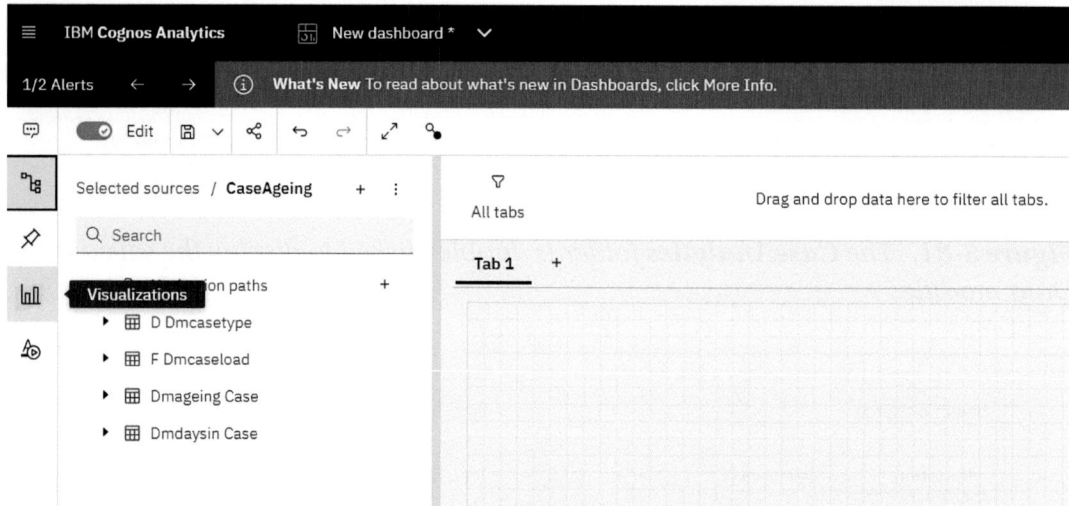

Figure 5-83. *The **CaseAgeing** Data module shows the four "Tables" we can use*

We now click the **Visualizations** icon (highlighted in Figure 5-83) which opens a large scrollable list of Cognos Analytics graph types.

CHAPTER 5 IBM COGNOS ANALYTICS CUSTOM DEVELOPMENT

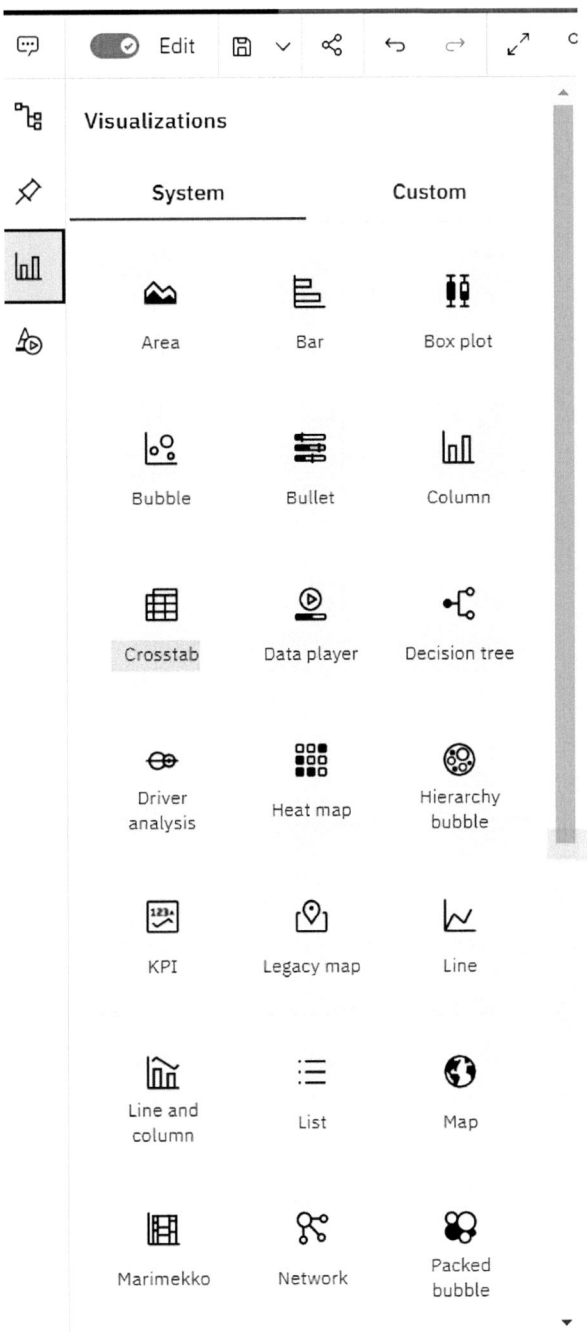

Figure 5-84. *The graph types are displayed, from which we can select a type*

CHAPTER 5 IBM COGNOS ANALYTICS CUSTOM DEVELOPMENT

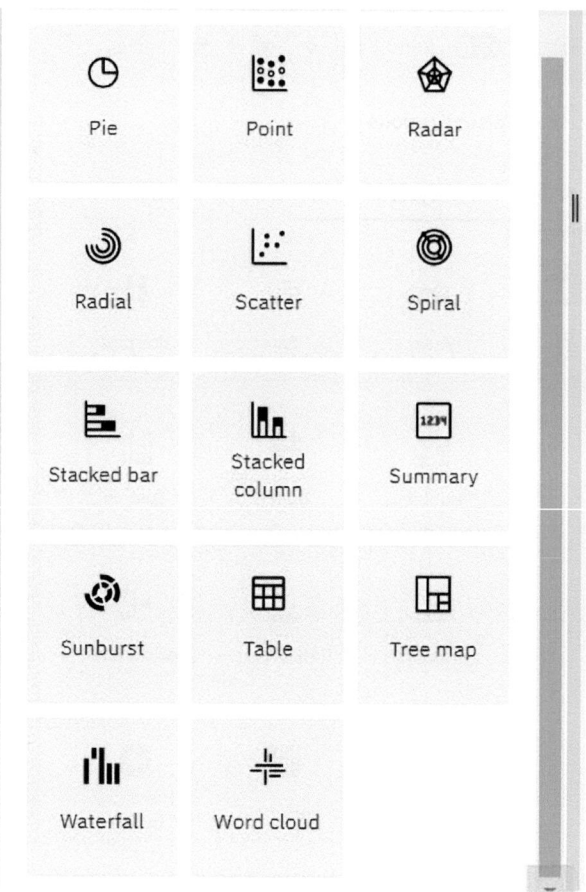

Figure 5-85. *The **Stacked bar** graph type is selected on scrolling down*

Eventually, we selected the **Stacked bar** graph type, which displays a template for the graph which we can fill by using drag and drop to select the appropriate fields from our table list. Each table has an opener (filled > icon) which opens a sublist of the table's fields.

CHAPTER 5 IBM COGNOS ANALYTICS CUSTOM DEVELOPMENT

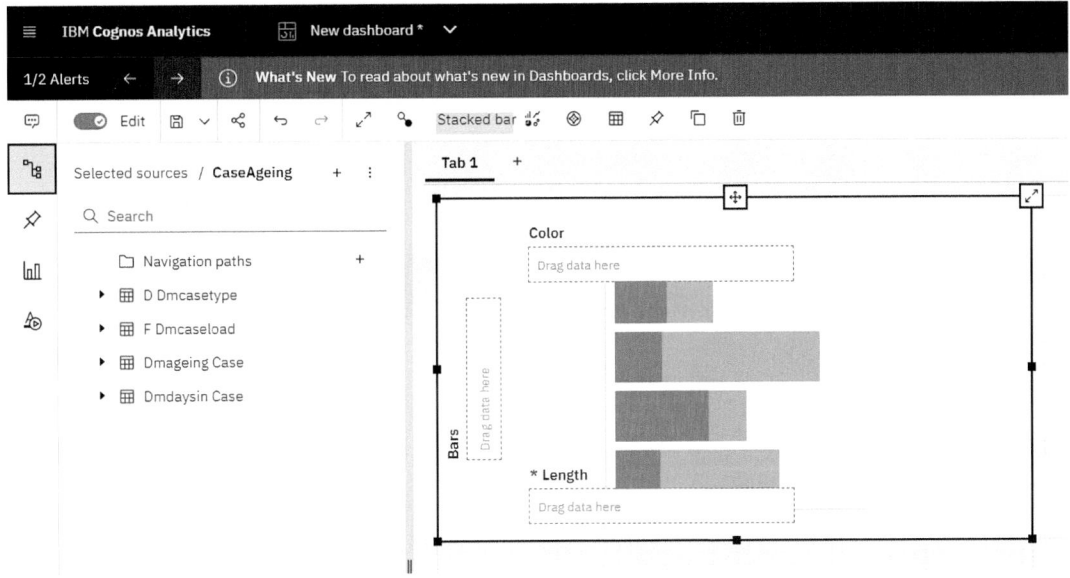

Figure 5-86. The Stacked bar graph type is selected for our data

We drag and drop the **Incoming** Case Events and the **Casetypename** fields for the y axis with the custom **Daysin** field values used for the count on the x axis.

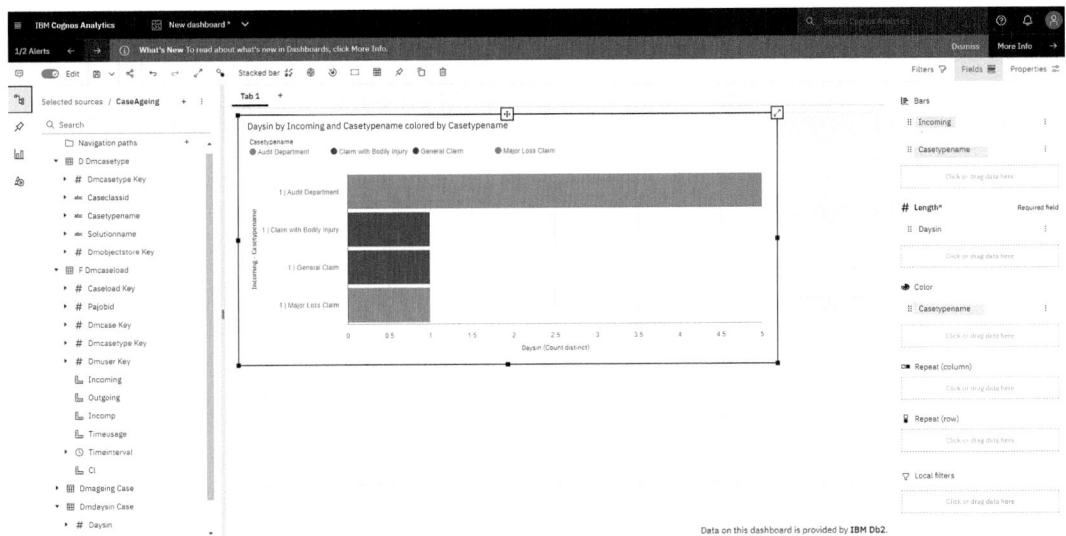

*Figure 5-87. The graph created from our custom **Dmdaysin_Case** view*

967

CHAPTER 5   IBM COGNOS ANALYTICS CUSTOM DEVELOPMENT

The final graph displayed is as follows.

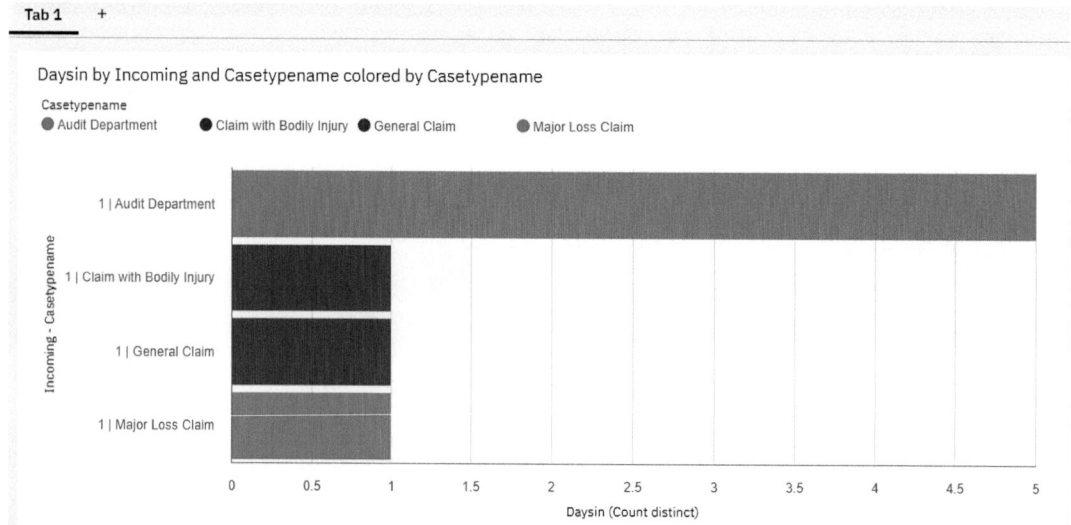

Figure 5-88. *The Daysin count by Incoming and Casetypename graph*

Now we can select a new tab on the Dashboard and create a second (similar) graph using the **Dmageing_Case** view.

CHAPTER 5 IBM COGNOS ANALYTICS CUSTOM DEVELOPMENT

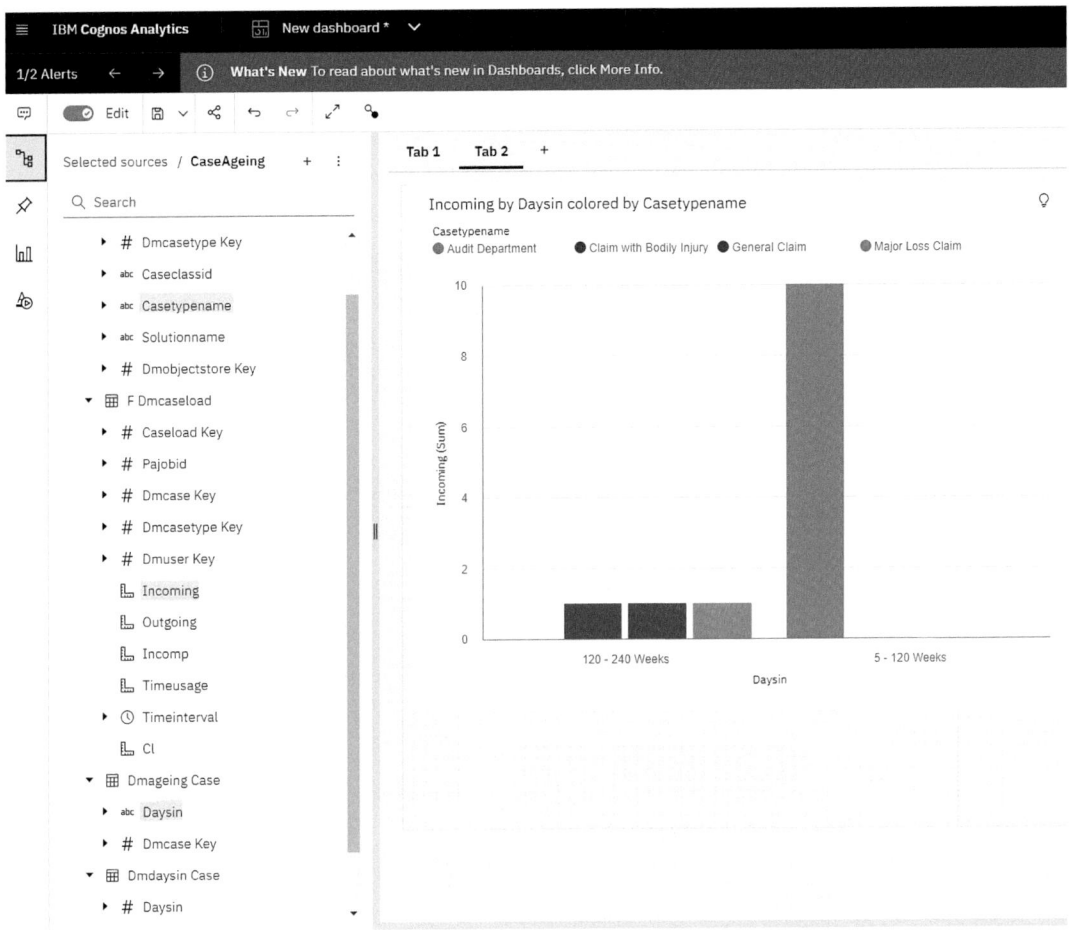

Figure 5-89. *The Incoming by Daysin (using time ranges) colored by Casetypename graph*

CHAPTER 5   IBM COGNOS ANALYTICS CUSTOM DEVELOPMENT

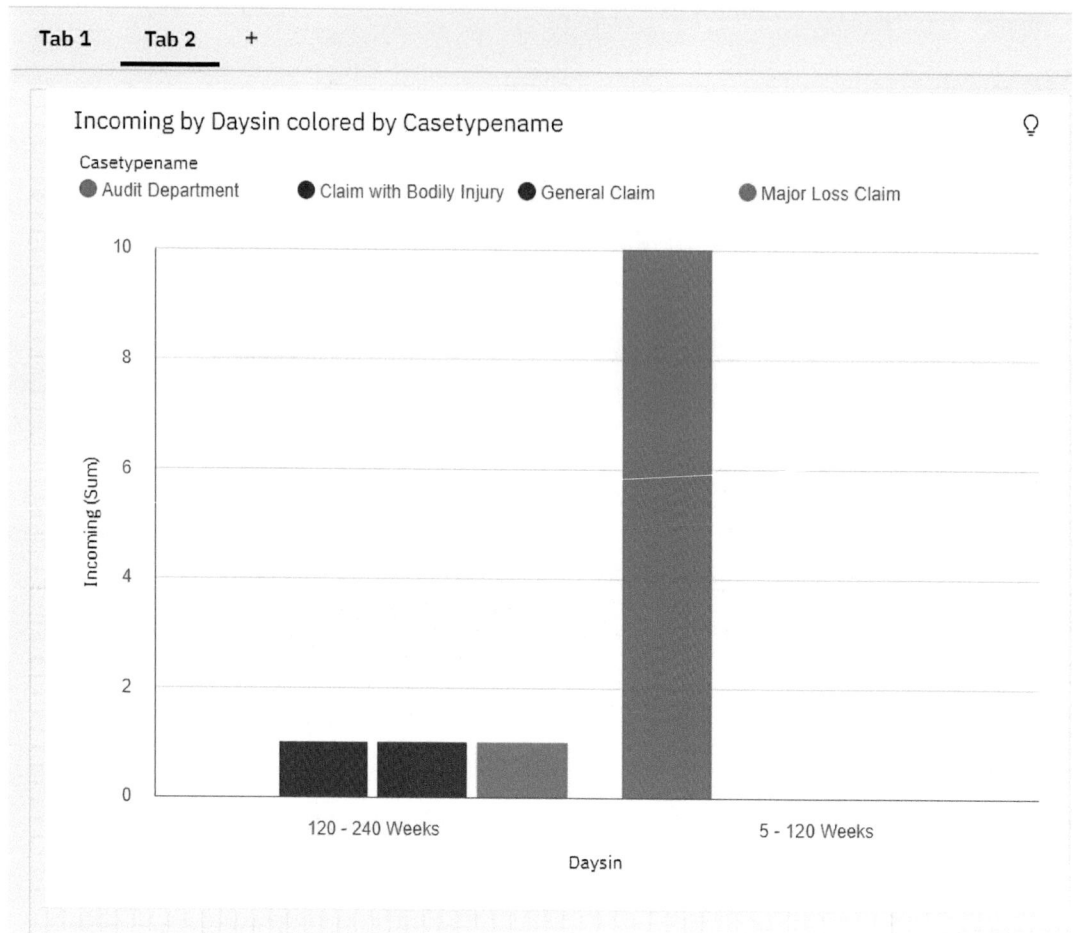

Figure 5-90. *The Daysin (week ranges) by Incoming (count) and Casetypename graph*

The IBM Cognos Analytics main page shows the items we have open for editing.

CHAPTER 5 IBM COGNOS ANALYTICS CUSTOM DEVELOPMENT

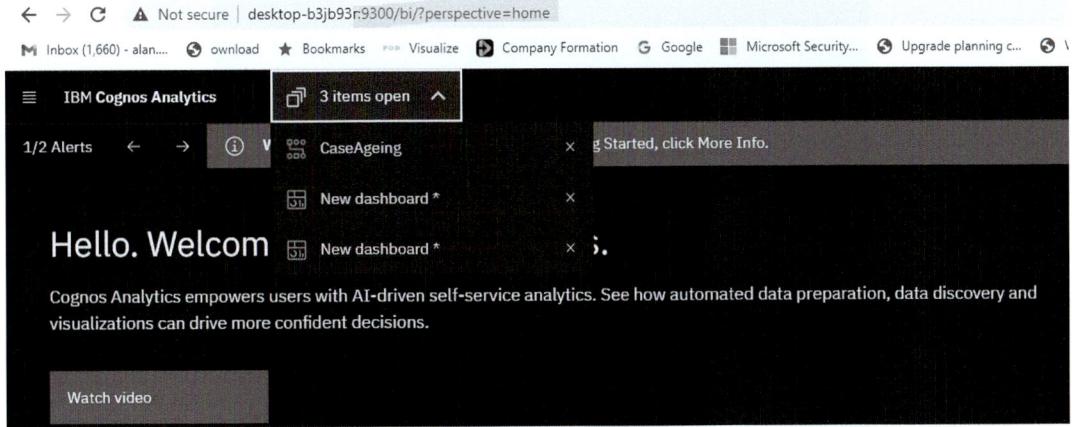

Figure 5-91. *The list of pages we have open which we used for the preceding custom views*

CHAPTER 5 IBM COGNOS ANALYTICS CUSTOM DEVELOPMENT

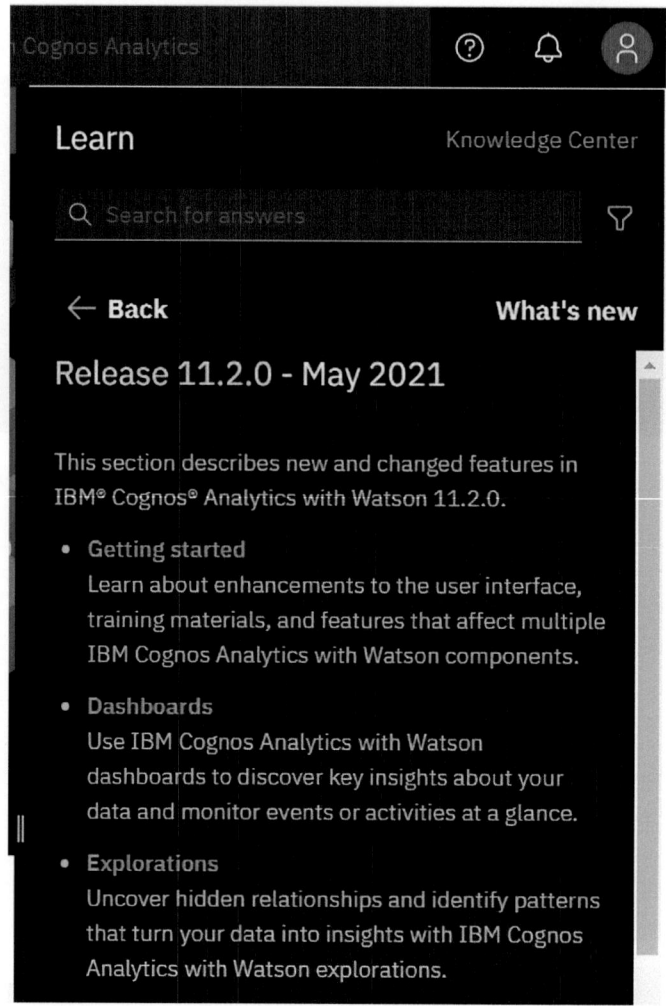

*Figure 5-92. The ? icon shows the **What's new** for Cognos Analytics version 11.2*

CHAPTER 5   IBM COGNOS ANALYTICS CUSTOM DEVELOPMENT

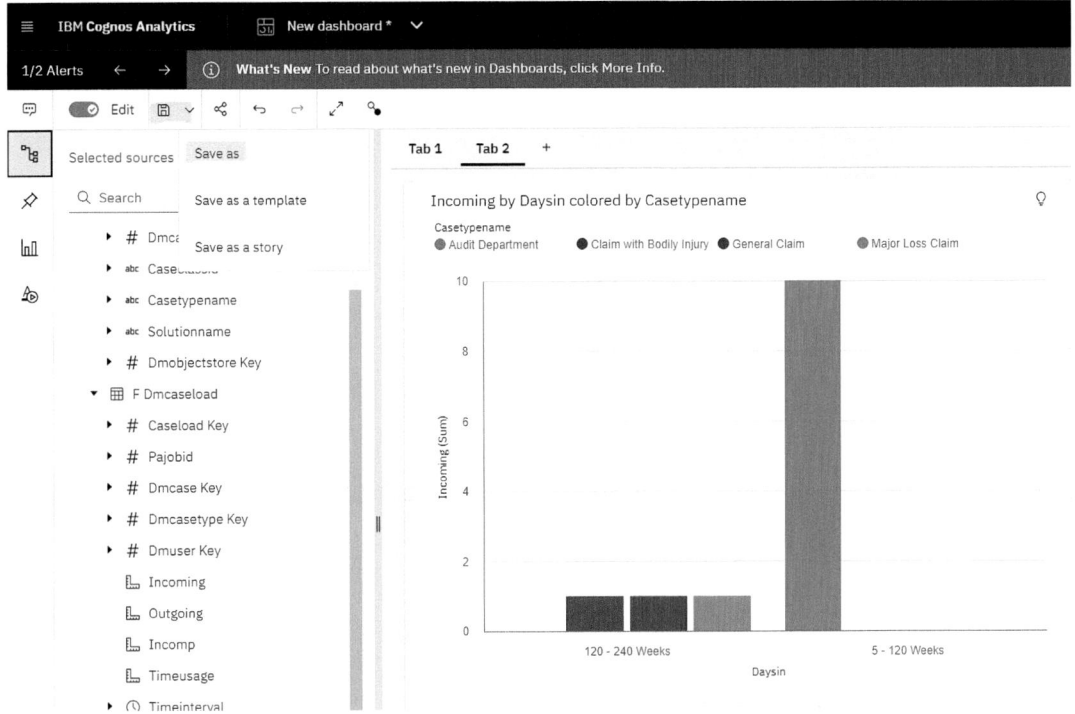

*Figure 5-93.   The Dashboards can be saved using the highlighted **Save as** menu item*

We can save the Dashboard graph as "**Audit and Insurance Solutions Case Ageing Dashboard**" to load later under the **CaseAnalytics** subfolder we created.

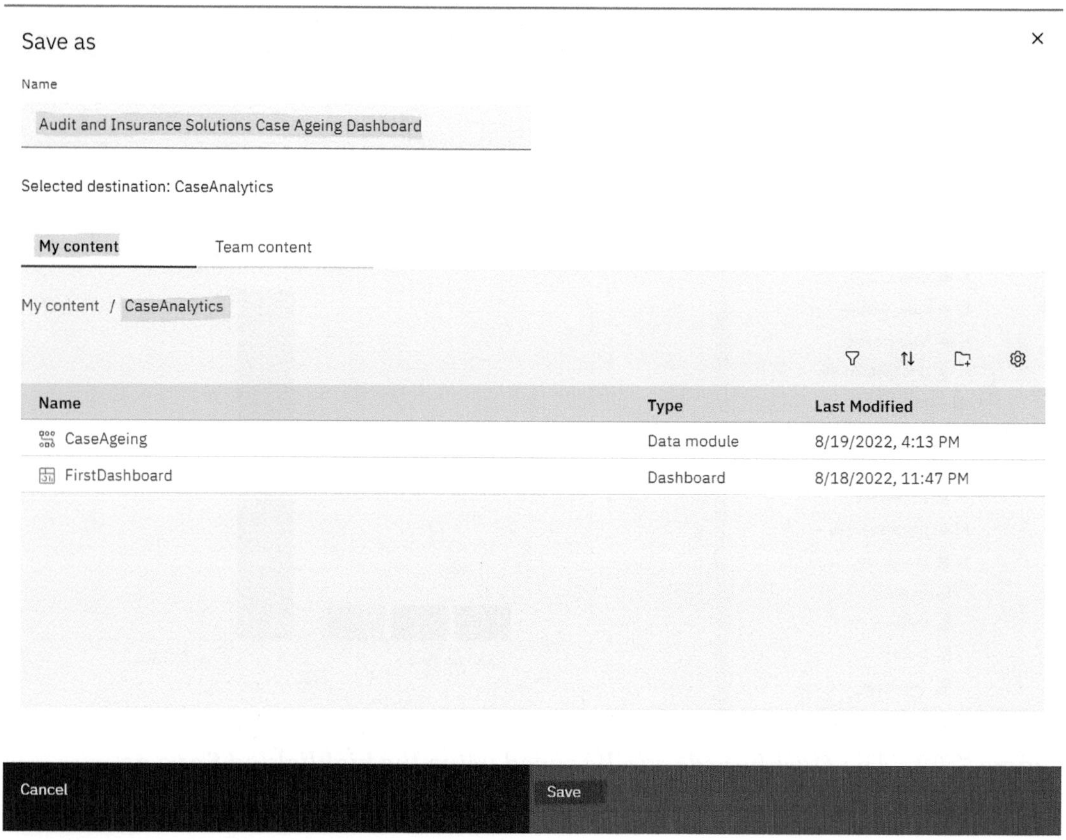

CHAPTER 5   IBM COGNOS ANALYTICS CUSTOM DEVELOPMENT

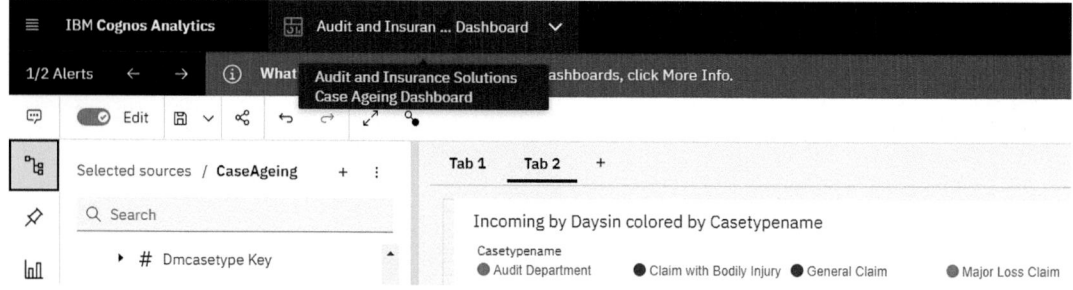

Figure 5-94. *The Dashboard graph is saved as* ***"Audit and Insurance Solutions Case Ageing Dashboard"***

Figure 5-95. *The saved Dashboard is selectable from the highlighted drop-down*

Some More Complex Examples

To illustrate how the preceding example can be extended, the following screenshots show the additional tables added with new relationships and the resulting graphs.

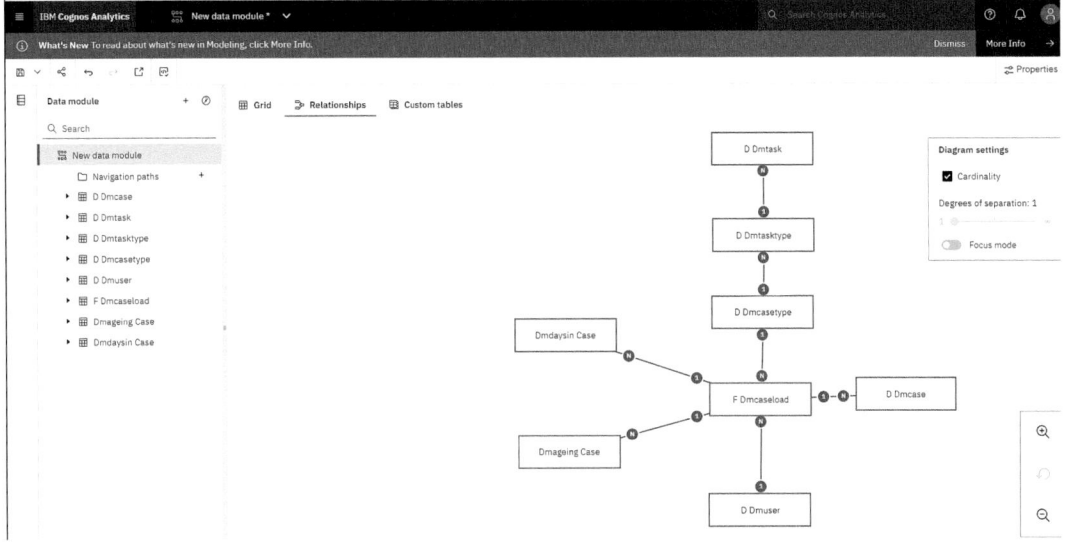

Figure 5-96. *The Relationship diagram created from an extended table selection*

CHAPTER 5 IBM COGNOS ANALYTICS CUSTOM DEVELOPMENT

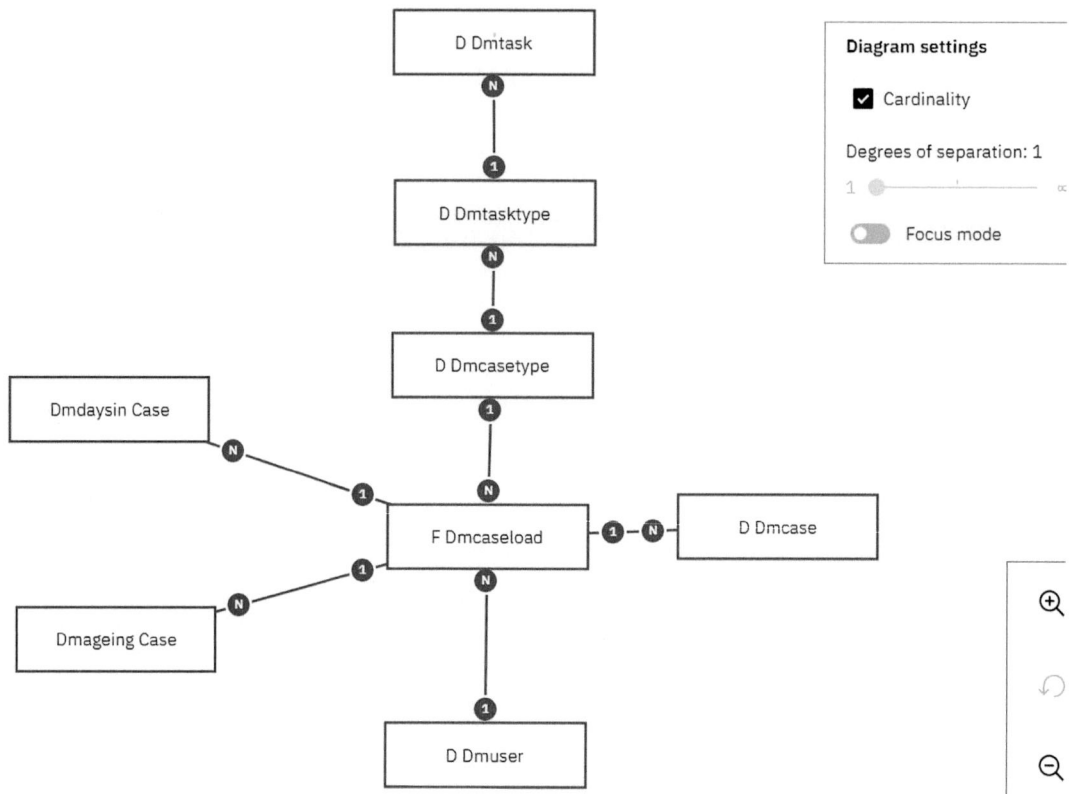

Figure 5-97. *The Relationship diagram zoomed in from the Figure 5-96 screenshot*

The resulting graph from the drag and drop of fields as shown in Figure 5-98:

CHAPTER 5 IBM COGNOS ANALYTICS CUSTOM DEVELOPMENT

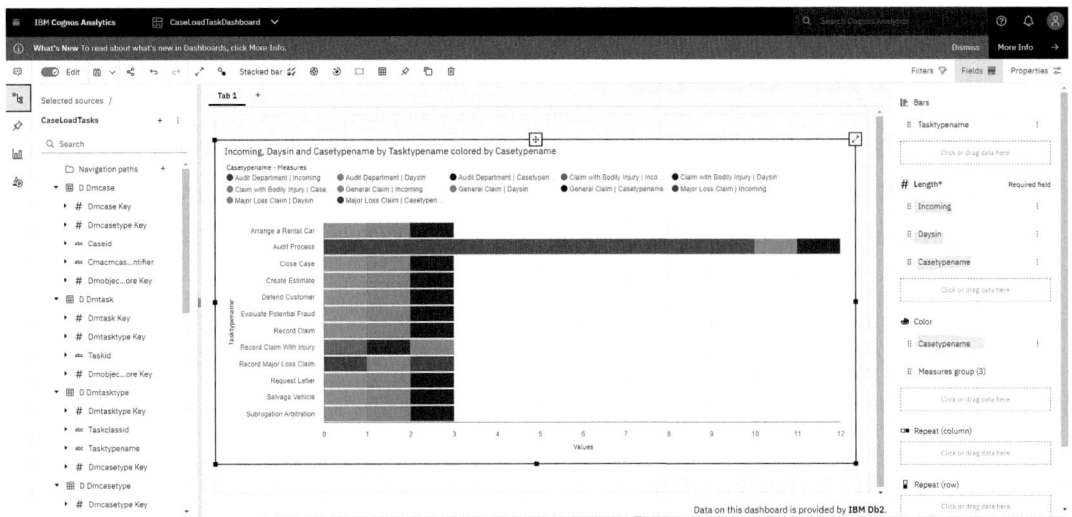

***Figure 5-98.** The Incoming Daysin and Casetypename by Tasktypename graph*

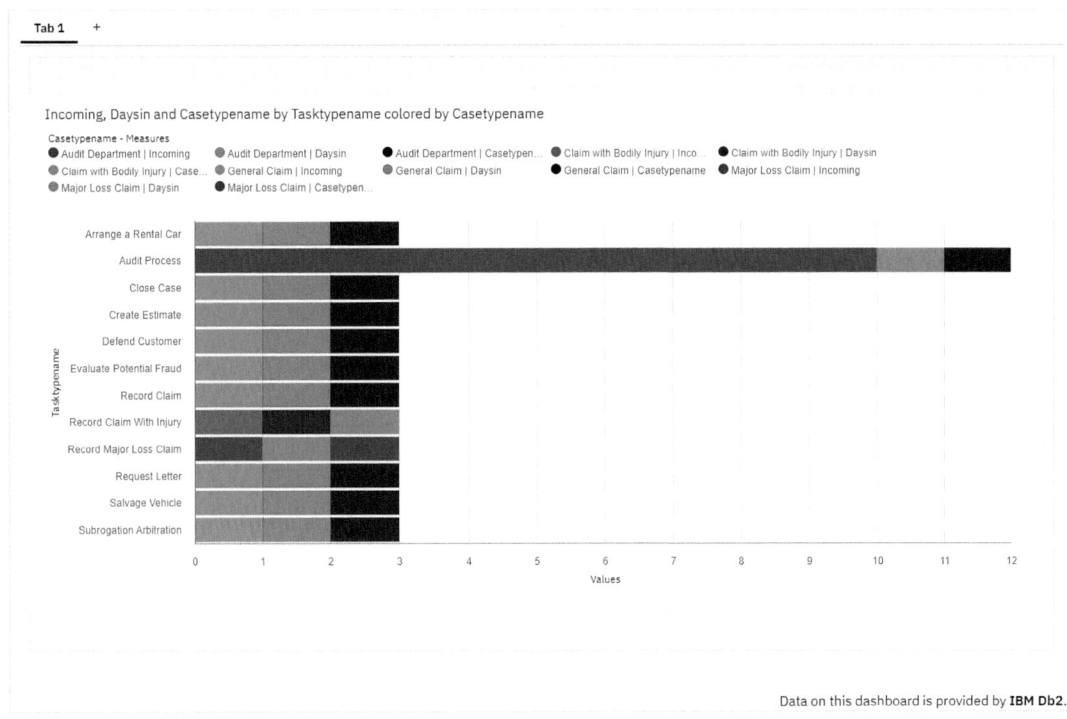

***Figure 5-99.** The zoomed in graph from Figure 5-98*

CHAPTER 5   IBM COGNOS ANALYTICS CUSTOM DEVELOPMENT

Time Dimension SLA Calculations on Microsoft SQL Server Analysis Services

The following **URL** link describes the Architecture which **IBM Case Analyzer** provides for support of **OLAP** cubes defined in **Microsoft SQL Server Analysis Services**:

www.ibm.com/docs/en/filenet-p8-platform/5.5.x?topic=analyzer-as-analysis-tool

IBM Case Analyzer services can use **OLAP** (online analytical processing). This is configured as an optional **OLAP** integration; the **IBM Case Analyzer CASTORE** must then be hosted on a Microsoft **SQL Server** database.

Enabling OLAP Integration

If **OLAP** integration is enabled, the **IBM FileNet Content Platform Engine** calls the **IBM Case Analyzer SSAS Connector** to perform cube processing. The **IBM Case Analyzer SSAS Connector** communicates with the Microsoft **SQL Server Analysis Services** **(SSAS)** cube processing system and then builds the **OLAP** cubes, which are stored in the **Case Analyzer OLAP** database and used for **Cognos Business Intelligence** reports or Microsoft **Excel** charts.

The following Microsoft **URLs** describe the installation of Microsoft **SQL Server Analysis Services**:

https://docs.microsoft.com/en-us/analysis-services/instances/install-windows/install-analysis-services?view=asallproducts-allversions

and

https://docs.microsoft.com/en-us/sql/ssdt/download-sql-server-data-tools-ssdt?view=sql-server-ver16

For Analysis Services or Reporting Services projects, you can install Microsoft SQL Server Analysis Services using MS Visual Studio with **Extensions ➤ Manage Extensions**.

In **SqlServer Query Analyzer**, a new table view can be created in the cube database, **VMAEDM**, to utilize an existing date dimension as the basis of the calculation:

```
CREATE VIEW dbo.DMDaysIn_Workflow AS
SELECT
TOP 100 PERCENT DATEDIFF([day], ReceivedDate, { fn NOW() }) AS DaysIn,
ReceivedDate_key
FROM   dbo.D_DMDataField_ReceivedDate
```

```
WHERE     (ReceivedDate <> '')
ORDER BY DaysIn
```

This uses the **DateDiff** function to calculate the number of days from the current date (from the **{ fn NOW() }** function) that the case data has been received (from the **ReceivedDate** Dimension field).

Use the cube editor and select to add an existing dimension. The link to the **Queue load** fact table must be made manually by dragging from the fact table key to the Dimension table key.

This can then be used to display the counts of cases in each of the calculated day "slots."

Dynamic Dimensions for Aging Bands

The procedure used to create the simple calculated Dimension earlier can be extended to create a "range" Dimension to allow the display of counts for ranges of dates, for example, in bands, for example, Today, '+ 1 Day', '+ 2 Days', '+ 3 Days', '+ 4 Days', '+ 5 Days', '+ 6 Days', '1 - 2 Weeks', '2 - 3 Weeks', '3 - 4 Weeks', '4 - 5 Weeks', '>= 5 Weeks'.

```
CREATE VIEW dbo.DMAgeing_Workflow AS
SELECT [AgeBand] =  CASE
WHEN  DATEDIFF([hour], ReceivedDate, { fn NOW() }) < 24 THEN 'Today'
WHEN  DATEDIFF([hour], ReceivedDate, { fn NOW() }) Between 24 and 47 THEN
'+ 1 Day'
WHEN  DATEDIFF([hour], ReceivedDate, { fn NOW() }) Between 48 and 71 THEN
'+ 2 Days'
WHEN  DATEDIFF([hour], ReceivedDate, { fn NOW() }) Between 72 and 95 THEN
'+ 3 Days'
WHEN  DATEDIFF([hour], ReceivedDate, { fn NOW() }) Between 96 and 119 THEN
'+ 4 Days'
WHEN  DATEDIFF([hour], ReceivedDate, { fn NOW() }) Between 120 and 143 THEN
'+ 5 Days'
WHEN  DATEDIFF([hour], ReceivedDate, { fn NOW() }) Between 144 and 167 THEN
'+ 6 Days'
WHEN  DATEDIFF([hour], ReceivedDate, { fn NOW() }) Between 168 and 335 THEN
'1 - 2 Weeks'
```

```
WHEN   DATEDIFF([hour], ReceivedDate, { fn NOW() }) Between 336 and 503 THEN
'2 - 3 Weeks'
WHEN   DATEDIFF([hour], ReceivedDate, { fn NOW() }) Between 504 and 671 THEN
'3 - 4 Weeks'
WHEN   DATEDIFF([hour], ReceivedDate, { fn NOW() }) Between 672 and 839 THEN
'4 - 5 Weeks'
ELSE  '>= 5 Weeks'
END
, ReceivedDate_key
FROM   dbo.D_DMDataField_ReceivedDate
WHERE      (ReceivedDate <> '')
```

On adding the new shared Dimension to the Queue Load cube, the link to the Dimension table **ReceivedDate_key** field must be manually made by dragging this field to the **VMAE_ReceivedDate_key** field of the **dbo.F_DMQueueLoad** table.

This can then be used to display the counts of cases in each of the calculated aging bands.

Procedure for Creating a Multilevel Dimension

The following procedure was used to test the creation of a hierarchical Dimension in a custom cube based on the **Queue Load** fact table. This was set up to demonstrate the feasibility of using a *department-employee* table structure allowing the possibility of an employee moving from one team to another, for example.

1) A copy of **VMAE** database was made to preserve the standard **PA** architecture as **VMAETEST** in SQL Enterprise Manager.

2) Create a new cube **QueueLoad1** using **SQL Analysis Manager**.

3) Add **Department-Employee** Dimension as follows:

4) In **SQL Analysis Manager**:

   a) Select **Snowflake Schema** from the **Add New Dimension** wizard.

   b) Select the three linked tables for the **Department-Employee**.

(See the section **"SQL Statements for Table Creation and Load to Test the DepartmentName Dimension"** with the **SQL** script used to build and populate the tables.)

dbo.D_Department
dbo.D_Team
dbo.D_User

Click next; this should automatically make the correct joins.
Click **Next**.

c) Select the levels for the **Dimension** and click **Next**:

DepartmentName
Team Name
User Name

d) Select the member key columns for each level as

DepartmentName	dbo.D_DepartmentName.Department_key
Team Name	dbo.D_Team.Team_key
User Name	dbo.D_User.User_key

and click **Next**.

e) In the **Select Advanced** options, select all three options:

Changing Dimensions option
Ordering and Uniqueness of members
Storage Mode and member groups
and click **Next**.

f) In the Set changing property option, click **No, the Dimension is not changing** radio button.

Note This MUST be set since, although it requires a full reprocess for changes to the cubes, it is the only option which allows users to exist in more than one team.

(Only click Yes, new dimension is a changing dimension radio button if the members are unique – not the case for users who move teams.)

Click Next.

g) For the **Sort option**, ensure that ordering at each level is by **Name**. Leave the defaults as **Unique among members** at the top level and among siblings at the next levels down and click **Next**.

h) Select create **MOLAP** (should be the default) and leave other fields as default.

i) Click **Next** and then enter the **Dimension Name** as **DepartmentName_Multi** (leave the share this dimension with other Hierarchies box ticked). Click the **Finish** command button.

j) Link the new multilevel **Dimension DepartmentName_Multi** from the **User name** key, dbo.D_User.User_key, to the exposed fact table key field, **DMUser_key**, field name on the **Queue Load** Fact table.

In the test **VMware** system used, the field is set as an **int** to allow linkage to the standard **Queue Load DMUser_key** field.

Reuse of Existing Exposed Dimensions to Create a Hierarchy

The preceding procedure requires the maintenance of the source tables for the **Department**, **Team**, and **User**. It is possible, as an alternative, to create a Dynamic Hierarchy from existing exposed user **Dimension** tables, using table views, to allow the relinking of related user **Dimensions**. This has the advantage of being a self-maintaining hierarchy.

In the following example, **Company Name**, **Account Number**, and the **User** dimensions are linked to create a Hierarchy as follows.

SQL is used to create new views containing the unique values for each Dimension, allowing the necessary link fields to be created to support the required Hierarchy:

Company Name

```
CREATE VIEW dbo.DMCompany_List AS
SELECT DISTINCT dbo.D_DMDataField_CompanyName.CompanyName_key,
CASE dbo.D_DMDataField_CompanyName.CompanyName_key
WHEN 0 THEN 'No Company Defined'
ELSE dbo.D_DMDataField_CompanyName.CompanyName
```

CHAPTER 5 IBM COGNOS ANALYTICS CUSTOM DEVELOPMENT

```
END AS CompanyName
FROM
    dbo.D_DMDataField_CompanyName INNER JOIN
                    dbo.F_DMQueueLoad ON dbo.D_DMDataField_CompanyName.
                    CompanyName_key = dbo.F_DMQueueLoad.VMAE_
                    CompanyName_key
WHERE
(dbo.D_DMDataField_CompanyName.CompanyName <> '')
```

Account Number

```
CREATE VIEW dbo.DMAccountNumber_List AS
SELECT DISTINCT dbo.D_DMDataField_AccountNumber. AccountNumber_key,
CASE dbo.D_DMDataField_AccountNumber. AccountNumber_key
WHEN 0 THEN 'No Account Defined'
ELSE dbo.D_DMDataField_AccountNumber. AccountNumber
END AS AccountNumber,
     dbo.F_DMQueueLoad.VMAE_CompanyName_key
FROM
    dbo.F_DMQueueLoad INNER JOIN
                    dbo.D_DMDataField_AccountNumber ON dbo.F_DMQueueLoad.
                    VMAE_AccountNumber_key = dbo.D_DMDataField_
                    AccountNumber.AccountNumber_key
WHERE
(dbo.D_DMDataField_AccountNumber.AccountNumber <> '')
```

Note In the preceding SQL, the use of the associated Fact table **CompanyName** Key field to provide the hierarchical link to the top-level **Company name** dimension.

Users

```
CREATE VIEW dbo.DMUser_List AS
SELECT DISTINCT dbo.D_DMUser.DMUser_key,
CASE dbo.D_DMUser.DMUser_key
WHEN 0 THEN 'No User Defined'
ELSE dbo.D_DMUser.UserName
```

```
END AS UserName, dbo.F_DMQueueLoad.VMAE_AccountNumber_key
FROM           dbo.D_DMUser INNER JOIN
dbo.F_DMQueueLoad ON dbo.D_DMUser.DMUser_key = dbo.F_DMQueueLoad.DMUser_key
WHERE
(dbo.D_DMUser.UserName <> '')
```

Note In the preceding SQL, the use of the associated Fact table **AccountNumber** Key field to provide the hierarchical link to the middle-level **Account Number** dimension.

The three new views can now be linked using the Dimension wizard.

Select the **Snowflake Schema**.

Select the three new views created for the hierarchical "**Company Accounts**" **Dimension**.

Link the associated key fields to give the required hierarchical association.

Select the **Name Fields** for each level in the hierarchical Dimension.

Select the **key fields** to use at each dimension level.

Enter the **Hierarchical Dimension** name.

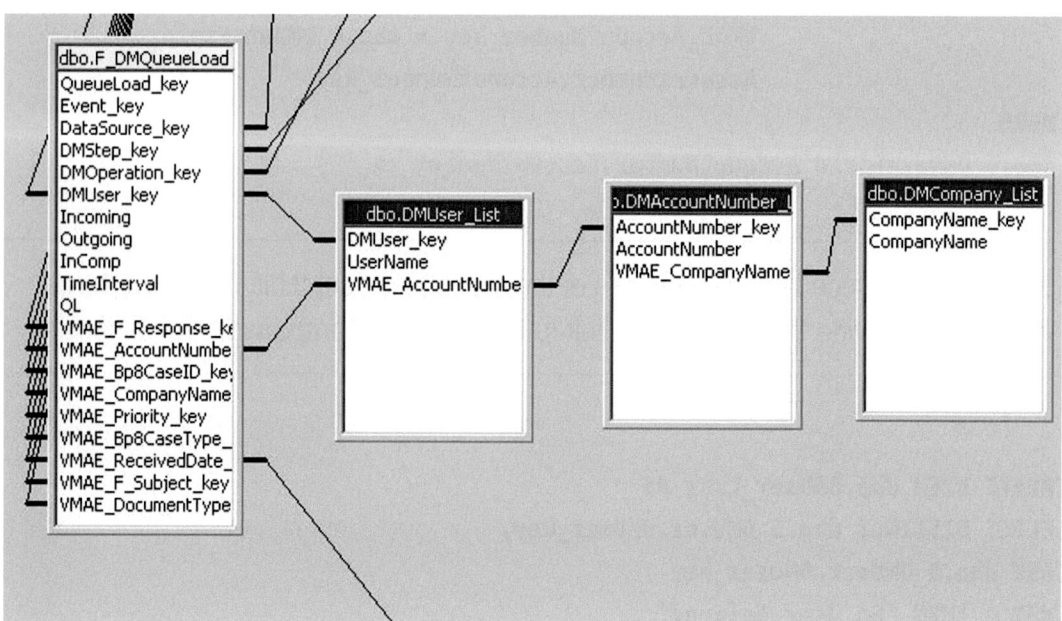

Figure 5-100. The multilevel "Company Accounts" Dimension

Add the new ("existing") shared **Hierarchical Dimension** to the **Queue Load** fact table using the cube editor.

After adding the "**CompanyAccounts**" **Dimension** and reprocessing, the results can be displayed.

SQL Statements for Table Creation and Load to Test the DepartmentName Dimension

SQL for a unique multilevel list

```
CREATE TABLE [D_DepartmentName] (
    [DepartmentName_key] [int] NOT NULL ,
    [DepartmentName] [varchar] (50) COLLATE Latin1_General_CI_AS NOT NULL ,
    CONSTRAINT [PK_D_DepartmentName] PRIMARY KEY  CLUSTERED
    (
        [DepartmentName_key]
    ) ON [PRIMARY]
) ON [PRIMARY]
GO

CREATE TABLE [D_Team] (
    [Team_key] [char] (5) COLLATE Latin1_General_CI_AS NOT NULL ,
    [TeamName] [varchar] (50) COLLATE Latin1_General_CI_AS NOT NULL ,
    [DepartmentName_key] [int] NOT NULL ,
    CONSTRAINT [PK_D_Team] PRIMARY KEY  CLUSTERED
    (
        [Team_key]
    ) ON [PRIMARY] ,
    CONSTRAINT [FK_D_Team_D_DepartmentName] FOREIGN KEY
    (
        [DepartmentName_key]
    ) REFERENCES [D_DepartmentName] (
        [DepartmentName_key]
    )
) ON [PRIMARY]
GO
```

CHAPTER 5 IBM COGNOS ANALYTICS CUSTOM DEVELOPMENT

```sql
CREATE TABLE [D_User] (
    [User_key] [varchar] (20) COLLATE Latin1_General_CI_AS NOT NULL ,
    [UserName] [varchar] (30) COLLATE Latin1_General_CI_AS NOT NULL ,
    [Team_key] [char] (5) COLLATE Latin1_General_CI_AS NOT NULL ,
    CONSTRAINT [PK_D_User] PRIMARY KEY  CLUSTERED
    (
        [User_key]
    )  ON [PRIMARY] ,
    CONSTRAINT [FK_D_User_D_Team] FOREIGN KEY
    (
        [Team_key]
    ) REFERENCES [D_Team] (
        [Team_key]
    )
) ON [PRIMARY]
GO
```

A New Test D_User table version was developed for multiple users with a change on the field to allow a link to the Queue Load fact table from the D_User table. This has been altered to have a compound Primary Key of User_Key + Team_Key which allows users who swap teams to be added to the new Team:

```sql
if exists (select * from dbo.sysobjects where id = object_id(N'[D_User]')
and OBJECTPROPERTY(id, N'IsUserTable') = 1)
drop table [D_User]
GO

CREATE TABLE [D_User] (
    [User_key] [varchar] (20) COLLATE Latin1_General_CI_AS NOT NULL ,
    [UserName] [varchar] (30) COLLATE Latin1_General_CI_AS NOT NULL ,
    [Team_key] [char] (5) COLLATE Latin1_General_CI_AS NOT NULL ,
    CONSTRAINT [PK_D_User] PRIMARY KEY  CLUSTERED
    (
        [User_key],
        [Team_key]
    )  ON [PRIMARY] ,
    CONSTRAINT [FK_D_User_D_Team] FOREIGN KEY
```

```
        (
            [Team_key]
        ) REFERENCES [D_Team] (
            [Team_key]
        )
) ON [PRIMARY]
GO
```

TEST SYSTEM WITH STANDARD Queue Load Fact table using int DMuser_key

```
if exists (select * from dbo.sysobjects where id = object_id(N'[D_User]')
and OBJECTPROPERTY(id, N'IsUserTable') = 1)
drop table [D_User]
GO

CREATE TABLE [D_User] (
    [User_key] [int] NOT NULL ,
    [UserName] [varchar] (30) COLLATE Latin1_General_CI_AS NOT NULL ,
    [Team_key] [char] (5) COLLATE Latin1_General_CI_AS NOT NULL ,
    CONSTRAINT [PK_D_User] PRIMARY KEY  CLUSTERED
    (
        [User_key],
        [Team_key]
    )  ON [PRIMARY] ,
    CONSTRAINT [FK_D_User_D_Team] FOREIGN KEY
    (
        [Team_key]
    ) REFERENCES [D_Team] (
        [Team_key]
    )
) ON [PRIMARY]
GO
```

Insert test users for initial Pivot Table with no user movement

```
INSERT INTO VMAEDMTEST.dbo.D_User(User_key, UserName, Team_key) VALUES(50,
'Administrator','18_S1')
GO
```

```sql
INSERT INTO VMAEDMTEST.dbo.D_User(User_key, UserName, Team_key) VALUES(52,
'suser','09_I1')
GO
INSERT INTO VMAEDMTEST.dbo.D_User(User_key, UserName, Team_key) VALUES(57,
'Tester','09_I1')
GO
INSERT INTO VMAEDMTEST.dbo.D_User(User_key, UserName, Team_key) VALUES(55,
'User 55','32_S1')
GO
INSERT INTO VMAEDMTEST.dbo.D_User(User_key, UserName, Team_key) VALUES(56,
'User 56','32_S1')
GO
INSERT INTO VMAEDMTEST.dbo.D_User(User_key, UserName, Team_key) VALUES(61,
'joe','18_S1')
GO
INSERT INTO VMAEDMTEST.dbo.D_User(User_key, UserName, Team_key) VALUES(58,
'sue','18_S1')
GO
```

Additional Users Added to show affect of Users added to multiple Teams

```sql
INSERT INTO VMAEDMTEST.dbo.D_User(User_key, UserName, Team_key) VALUES(50,
'Administrator','09_I1')
GO
INSERT INTO VMAEDMTEST.dbo.D_User(User_key, UserName, Team_key) VALUES(56,
'User 56','18_S1')
GO
INSERT INTO VMAEDMTEST.dbo.D_User(User_key, UserName, Team_key) VALUES(50,
'Administrator','32_S1')
GO
INSERT INTO VMAEDMTEST.dbo.D_User(User_key, UserName, Team_key) VALUES(50,
'Administrator','09_N1')
GO
INSERT INTO VMAEDMTEST.dbo.D_User(User_key, UserName, Team_key) VALUES(15,
'joe,'58_I2')
GO
```

```
INSERT INTO VMAEDMTEST.dbo.D_User(User_key, UserName, Team_key) VALUES(15,
'sue,'58_N1')
GO
```

Note Although Cognos BI 10.2.2 is at EOS as of December 31, 2022 (www.ibm.com/support/pages/end-continuing-support-cognos-business-intelligence-102x-effective-december-31-2022), the following IBM link announcement indicates that it is included in continuing support:

www.ibm.com/support/pages/cognos-analytics-continuing-support

Chapter 5 Exercises

The following questions relate to the **IBM Cognos Analytics** custom **Views** and the Graph Dashboards we covered in this chapter.

MULTIPLE CHOICE QUESTIONS

1. What Cognos Analytics web application main menu item would you select to show the **Data server connections** submenu option?

    a) Content menu

    b) Manage menu

    c) Home menu

    d) New menu

2. To create the simple calculated Dimension for aging, we used a custom Db2 database

    a) Stored procedure

    b) Table

    c) Function

    d) View

3. To start the IBM Cognos Windows Services, we used

   a) The IBM Cognos Dynamic Query Analyzer program

   b) The IBM Cognos Configuration program

   c) The IBM Data Studio 4.1.3 program

   d) The Microsoft SQL Server Analysis Services program

4. From the list of data server connection types, we used

   a) IBM Big SQL

   b) IBM Db2 for i

   c) IBM Db2

   d) IBM Db2 Warehouse

MULTIPLE CHOICE ANSWERS

1. b) Manage menu
2. d) View
3. b) The IBM Cognos Configuration program
4. c) IBM Db2

QUESTIONS

1. Outline what steps you would use to add a Custom View for a user's age. The age should be the age in years and months as of today's date.

2. Describe the process you would use to set the permissions for a Cognos Analytics Datasource.

3. Give an example and the steps you would use to create a Dynamic Hierarchy from an existing exposed user **Dimension** table for use with Microsoft SQL Server Analysis Manager.

In Chapter 6, we will describe a complete Java program to generate PDF documents using the Java iText library.

```
INSERT INTO VMAEDMTEST.dbo.D_User(User_key, UserName, Team_key) VALUES(15,
'sue,'58_N1')
GO
```

Note Although Cognos BI 10.2.2 is at EOS as of December 31, 2022 (www.ibm.com/support/pages/end-continuing-support-cognos-business-intelligence-102x-effective-december-31-2022), the following IBM link announcement indicates that it is included in continuing support:

www.ibm.com/support/pages/cognos-analytics-continuing-support

Chapter 5 Exercises

The following questions relate to the **IBM Cognos Analytics** custom **Views** and the Graph Dashboards we covered in this chapter.

MULTIPLE CHOICE QUESTIONS

1. What Cognos Analytics web application main menu item would you select to show the **Data server connections** submenu option?

    a) Content menu

    b) Manage menu

    c) Home menu

    d) New menu

2. To create the simple calculated Dimension for aging, we used a custom Db2 database

    a) Stored procedure

    b) Table

    c) Function

    d) View

3. To start the IBM Cognos Windows Services, we used

   a) The IBM Cognos Dynamic Query Analyzer program

   b) The IBM Cognos Configuration program

   c) The IBM Data Studio 4.1.3 program

   d) The Microsoft SQL Server Analysis Services program

4. From the list of data server connection types, we used

   a) IBM Big SQL

   b) IBM Db2 for i

   c) IBM Db2

   d) IBM Db2 Warehouse

MULTIPLE CHOICE ANSWERS

1. b) Manage menu
2. d) View
3. b) The IBM Cognos Configuration program
4. c) IBM Db2

QUESTIONS

1. Outline what steps you would use to add a Custom View for a user's age. The age should be the age in years and months as of today's date.

2. Describe the process you would use to set the permissions for a Cognos Analytics Datasource.

3. Give an example and the steps you would use to create a Dynamic Hierarchy from an existing exposed user **Dimension** table for use with Microsoft SQL Server Analysis Manager.

In Chapter 6, we will describe a complete Java program to generate PDF documents using the Java iText library.

CHAPTER 6

PDF Document Creation Using Java

This chapter covers the step-by-step procedure to develop Java programs to generate PDF documents using the iText library.

This is shown first with simple code steps to

- Add an Image to a PDF
- Create a PdfWriter object
- Create a PdfDocument object
- Create the Document object
- Create an Image object
- Add the image to the Document
- Close the Document
- Add a Table to a PDF

Chapter Organization

This chapter contains the following four Parts:

Part 1 – Bill of Materials. This Part describes the iText import packages used in the example pdf generation calls used in this chapter. It also shows some simple example Java code used to demonstrate the calls which can be made.

Part 2 – Example 3 – An Audit Report from the Audit Master. This Part lists the main Audit Report program Java Code.

© Alan S. Bluck 2023
A. S. Bluck, *IBM Software Systems Integration*, https://doi.org/10.1007/978-1-4842-8861-0_6

Part 3 – Supporting Java Classes for the Main Audit Report Program. In this Part, the Java Code is listed for the supporting methods used by the Audit Report program.

Part 4 – Example 4 – Create an Auditor Calendar Table in a PDF Document. This Part demonstrates the calls required to call the iText methods to create a pdf Table structure.

See the following URL link for a tutorial provided by IBM on the use of Java for program development for IBM FileNet systems:

www.ibm.com/docs/en/filenet-p8-platform/5.5.x?topic=transport-adding-connection-code ("Adding Connection Code – IBM Documentation")

An overview of the API concepts is also covered in this link:

www.ibm.com/docs/en/filenet-p8-platform/5.5.x?topic=guide-getting-started#gs_concepts__gs_requirements

IBM C# .NET Web Service Code Examples can be found at the following URL link:

www.ibm.com/docs/en/filenet-p8-platform/5.5.x?topic=guide-code-examples

The following IBM Redbook link is available:

www.redbooks.ibm.com/abstracts/sg247743.html?Open

Note This last URL link is Archive documentation, but some sections are (surprisingly) still useful as the IBM FileNet API is relatively stable, and many of the Content Engine Architecture concepts are still relevant.

The following site is run by Ricardo Belfor:

https://ecmdeveloper.com/plugin/intro/

It is described as follows:

"ECM Developer is an **open-source Eclipse plug-in** aimed at supporting the development of applications using the IBM FileNet P8 Content Engine."

Note "The old version also supported CMIS repositories, but due to limited time and resources this is no longer the case. The focus will be on IBM FileNet P8 Content Engine repositories."

This can be downloaded from the URL page: https://ecmdeveloper.com/plugin/getting-started/

"This first step is to download the software for the plug-in. The plug-in can be downloaded at https://ecmdeveloper.com/eclipse-plugin/com.ecmdeveloper.plugin.repository-2.3.0.zip *or by using the Eclipse update site* https://ecmdeveloper.com/eclipse-plugin."

I have a ResearchGate publication as follows for IBM FileNet Java Code development:

"IBM FileNet P8 Java Development on ECM Cloud Private Container P8 Examples" https://doi.org/10.13140/RG.2.2.20160.69129

Part 1 – Bill of Materials

The Java build used in this chapter is largely based on the Eclipse IDE and support jar files we used in Chapter 2.

Imports Used with the Test Audit Report Stub from the iText jar Library

The following code package import statements are used for the iText pdf generation examples:

```
import com.lowagie.text.Document;
import com.lowagie.text.Paragraph;
import com.lowagie.text.Image;
import com.lowagie.text.PageSize;
import com.lowagie.text.Rectangle;
import com.lowagie.text.Anchor;
import com.lowagie.text.BadElementException;
import com.lowagie.text.Chapter;
import com.lowagie.text.DocumentException;
import com.lowagie.text.Element;
import com.lowagie.text.Font;
import com.lowagie.text.List;
import com.lowagie.text.ListItem;
import com.lowagie.text.Phrase;
```

```java
import com.lowagie.text.Section;
import com.lowagie.text.pdf.PdfContentByte;
import com.lowagie.text.pdf.PdfPCell;
import com.lowagie.text.pdf.PdfPTable;
import com.lowagie.text.pdf.PdfWriter;
import com.lowagie.text.pdf.RandomAccessFileOrArray;
import com.lowagie.text.pdf.codec.TiffImage;
```

Example 1 – A Simple Audit PDF Document

The following example code uses the iText paragraph object and sets the pdf Header metadata settings.

Listing 6-1. The AuditTest.java test pdf Text creation code

```java
package com.ibm.filenet.ps.ciops.test;

//Java I/O library imports
import java.io.File;
import java.io.FileOutputStream;
import java.io.IOException;
import java.io.OutputStream;
import java.sql.Date;
//iText jar Library imports
import com.lowagie.text.Document;
import com.lowagie.text.Paragraph;
import com.lowagie.text.pdf.PdfWriter;
import java.time.LocalDateTime; // import the LocalDateTime class
import java.time.LocalDate; // import the LocalDate class
import java.time.LocalTime; // import the LocalTime class
import java.time.format.DateTimeFormatter; // import the Java Date formatter class
/**
 * Java Program to create a Simple Audit PDF document using the iText library.
 *
 * @author Alan S. Bluck
```

```java
*/
public class AuditTest {
public static void main(String args[]) {
//Set up the Date / Time system variables
LocalDate AuditReportDate = LocalDate.now();
LocalDateTime AuditReportDateTime = LocalDateTime.now();
LocalTime AuditReportTime = LocalTime.now();
DateTimeFormatter AuditDateFormat = DateTimeFormatter.ofPattern("dd-MM-yyyy HH:mm:ss");

String sAuditDate = AuditDateFormat.format(AuditReportDateTime);
OutputStream AuditPDFfile = null;
try {
    AuditPDFfile = new FileOutputStream(new File("/opt/AuditReport/
    AuditTest.pdf"));
    //Create a new Audit Document object
    Document audit_document = new Document();
    //You need the iText PdfWriter class for a PDF document
    PdfWriter.getInstance(audit_document, AuditPDFfile);
    //Open the Audit document for writing a PDF
    audit_document.open();
    //Write the test content
    audit_document.add(new Paragraph("Audit Report: Document Test PDF
    Paragraph"));
    audit_document.add(new Paragraph("Auditor Name: R. Jones"));
    audit_document.add(new Paragraph("Audit Date: " + sAuditDate));
    //Add Header meta-data information to the PDF file
    audit_document.addCreationDate();
    audit_document.addAuthor("ASB Software Development Limited");
    audit_document.addTitle("Audit Report Test PDF Generation");
    audit_document.addCreator("Program AudiTest");
    //close the document
    audit_document.close();
    System.out.println("Audit Report Created:" + sAuditDate);
} catch (Exception e) {
        e.printStackTrace();
```

```
        } finally {
//close the FileOutputStream
        try {
            if (AuditPDFfile != null) {
                AuditPDFfile.close();
            }
        } catch (IOException io) {
            //Failed to close
            System.out.println("The Audit PDF file failed to close!");
        }
    }
  }
}
```

Expected Output from the AuditTest.java Code

As shown in the preceding code, the generated PDF can be set with a metadata Header. This is a standard PDF format which can be used to set attributes such as the pdf author name, a title, a file description, etc. This is a useful additional feature for use with an Auditing system, where traceability is important (especially since the Audit system itself is audited by the Auditing standards body!).

Adding an Image to a PDF Document

An empty PDF Document is created using the **Document** class. A **PdfDocument** object is then passed as an argument to the **Document** class constructor. To add an image to the PDF, we can create an object of the image that is required to be added and add this image using the **add()** method of the **Document** class.

The following are the steps to add a TIFF image to the PDF document.

> **Note** I am using iText version 2.1.7 for this chapter; there is a new iText version 7 available which has more features (support for jpeg images, for example), but the 2.1.7 version is supported by an Apache 2.0 license (embedded in the **com.lowagie.text-2.1.7.jar** file) and is fine for the application we are using, and the whole system is held in this one .jar file.
>
> Later versions are split into multiple .jar files, and some of the Java classes are coded differently.

Creating a PdfWriter Object

The **PdfWriter** class is a Java **DocWriter** class for a PDF. This class is found in the Java package **com.lowagie.text.pdf**. The constructor of this class accepts a string containing the path of the file where the PDF is to be created.

The **PdfWriter** class is created by first passing a string value file path (defining the folder path of the PDF file) as shown in Listing 6-2.

Listing 6-2. Code snippet for creating the iText Document Class

```
// Create a PdfWriter
    String sTargetPDFFile = "/opt/AuditReport/AuditTest/AuditImage.pdf";
    File pdfFile = new File(sTargetPDFFile);
    com.lowagie.text.Document document = new com.lowagie.text.
    Document(PageSize.A3.rotate(), 50, 50, 100, 100);
```

When an object of this type is passed to an iText **Document** Java class, every element added to the document object will be written to the file we defined earlier.

Creating an iText Document Object Class

The **Document** Java class object holds an image of the PDF Document used in iText. This Java class is defined in the package **com.lowagie.text.Document**. To instantiate this class, a **PdfWriter** class object is passed to the **Document** Java class. This relationship can be demonstrated using the Java code in Listing 6-3.

CHAPTER 6 PDF DOCUMENT CREATION USING JAVA

Listing 6-3. Code to create an iText PdfWriter class

```
PdfWriter AuditPDFwriter = PdfWriter.getInstance(document, new
FileOutputStream(pdfFile));
// Create a PdfDocument
    document.open();
```

After a PDF Document class object is created, you can add the elements like page, font, file attachment, and event handler using iText methods provided by the class as shown in Listing 6-4 for adding sections of Auditor questions.

Listing 6-4. Code to add paragraphs, sections, and New pages to a pdf document

```
//AUDITOR START OUTPUT TO PDF
for(int iSection=0 ; iSection < noAUDITORSections ; iSection++) {
// We add one empty line
addEmptyLine(sections, 1);
// Lets write a big header
//TODO Pick all text up from the config.xml file
Paragraph paragraph = new Paragraph(TabSECTIONS_AUDITOR[iSection],
catFont);
    paragraph.setAlignment(Element.ALIGN_CENTER);
    sections.add(paragraph);
    addEmptyLine(sections, 1);
for(int iProp=0 ; iProp < noAUDITORprops[iSection] ; iProp++) {
    paragraph = new Paragraph(AUDITOR_propNames[iSection][iProp] + " : " +
    AUDITOR_propValues[iSection][iProp], catFont);
paragraph.setAlignment(Element.ALIGN_CENTER);
    sections.add(paragraph);
    addEmptyLine(sections, 1);
    }
    }
    document.add(sections);
    // Start a new page
    document.newPage();
```

Creating the Document Object

The **Document** class defined in the Java package **com.lowagie.text.Document** is the root element of the PDF image object.

You can instantiate the **Document** class by passing an object of the class **PdfWriter** (in package **com.lowagie.text.pdf**) created in the previous steps, as shown in Listing 6-5, which can be set to a specific PDF version (we selected version 1.3).

Listing 6-5. Code to set the version of PDF generated

```
// Creating an Audit Report Document as class Document
    AuditPDFwriter.setPdfVersion(PdfWriter.VERSION_1_3);
```

Creating an Image Object

To create the **TiffImage** class object, we can use the package **com.lowagie.text.pdf.codec** and then use the iText **RandomAccessFileOrArray** class object class in the constructor method. As an argument of the **RandomAccessFileOrArray** constructor method, we pass a string argument defining the path of the image, as shown in Listing 6-6.

Listing 6-6. Code to create an iText Image object RandomAccessFileOrArray

```
// Creating an Image object
    String imageLogoFile = "/opt/AuditReport/images/AuditMasterLogo.tif";
    RandomAccessFileOrArray ra = new RandomAccessFileOrArray(image
    LogoFile);
    int pages = TiffImage.getNumberOfPages(ra);
```

Now we can instantiate the **Image** class of the iText library package, **com.lowagie.text.Image**. In the constructor, we can pass the preceding **RandomAccessFileOrArray** class object as an argument, as shown in Listing 6-7.

Listing 6-7. Code to create the iText Image class object

```
// Creating an Image object
Image image;
image = TiffImage.getTiffImage(ra, 1);
```

Adding Images to the Document

We can now add the image object created in the previous step using the **add()** method of the **Document** class, as shown in Listing 6-8.

Listing 6-8. Code to set the page size and add the iText Image object to a document object

```
// Adding the Audit Master Logo image to the Audit report document
    Rectangle pageSize = new Rectangle(image.getWidth(),
    image.getHeight());
    document.setPageSize(pageSize);
    document.add(image);
```

Closing the Document

Finally, we must close the document using the **close()** method of the **Document** class, as shown in Listing 6-9.

Listing 6-9. Code to close the pdf Document object

```
// Save to the Linux folder path /opt/AuditMaster/AuditTest
// Closing the Audit report document
document.close();
        System.out.println("Audit Master Logo Image added");
    }
}
```

Example 2 – Test Code to Add the Audit Master Logo

The following Test Java code was used to add the Audit Master Logo TIFF image to a PDF document using the iText library. It creates a PDF Audit Report document with the name **AuditImage.pdf**, adds the Audit Master TIFF image to it, and saves it in the Linux folder path **/opt/AuditReport/AuditTest/**.

We have saved this code in a file with name **AuditMasterLogoTest.java**.

Listing 6-10. Complete example Java code to add a TIFF image to a pdf Document

```java
package com.ibm.filenet.ps.ciops.test;
import java.io.File;
import java.io.FileOutputStream;
// iText jar Library imports
import com.lowagie.text.Document;
import com.lowagie.text.Image;
import com.lowagie.text.PageSize;
import com.lowagie.text.Paragraph;
import com.lowagie.text.Rectangle;
import com.lowagie.text.pdf.PdfDocument;
import com.lowagie.text.pdf.PdfWriter;
import com.lowagie.text.pdf.RandomAccessFileOrArray;
import com.lowagie.text.pdf.codec.TiffImage;
public class AuditMasterLogoTest {
   public static void main(String args[]) throws Exception {
// Create a PdfWriter
      String sTargetPDFFile = "/opt/AuditReport/AuditTest/AuditImage.pdf";
      File pdfFile = new File(sTargetPDFFile);
      com.lowagie.text.Document document = new com.lowagie.text.
      Document(PageSize.A3.rotate(), 50, 50, 100, 100);
      PdfWriter AuditPDFwriter = PdfWriter.getInstance(document, new
      FileOutputStream(pdfFile));
// Create a PdfDocument
       document.open();
// Creating an Audit Report Document as class Document
      AuditPDFwriter.setPdfVersion(PdfWriter.VERSION_1_3);
// Creating an ImageData object
      String imageLogoFile = "/opt/AuditReport/images/AuditMasterLogo.tif";
      RandomAccessFileOrArray ra = new RandomAccessFileOrArray(image
      LogoFile);
      int pages = TiffImage.getNumberOfPages(ra);
```

```java
// Creating an Image object
    Image image;
    image = TiffImage.getTiffImage(ra, 1);
// Adding the Audit Master Logo image to the Audit report document
    Rectangle pageSize = new Rectangle(image.getWidth(),
    image.getHeight());
    document.setPageSize(pageSize);
    document.add(image);
// Save to the Linux folder path /opt/AuditMaster/AuditTest
// Closing the Audit report document
document.close();
        System.out.println("Audit Master Logo Image added");
    }
}
```

Part 2 – Example 3 – An Audit Report from the Audit Master

A complete program to provide a pdf Audit report from the Audit questions from the Audit Master using Cases from the solution created in Chapter 1 is as follows.

We added Audit questions as follows:

Check management responsibility

1. Check quality system.

2. Latest/relevant issue of documentation available.

3. Authorization of documentation.

4. No unauthorized additions/alterations to documentation.

5. Documentation marked with issue status.

6. Procedures manual exists and is followed.

7. No uncontrolled documents present.

8. Transaction records are used for identification.

9. Inventory records are kept.

CHAPTER 6 PDF DOCUMENT CREATION USING JAVA

10. System for determining the current status of parts in production.

11. Correct paperwork accompanying parts during the production process.

12. Evidence of the use of procedures and work instructions.

13. Measuring equipment is calibrated.

14. Procedures for in-process inspection.

15. Identification of jigs and tools.

Creating the Audit_Report Java Project

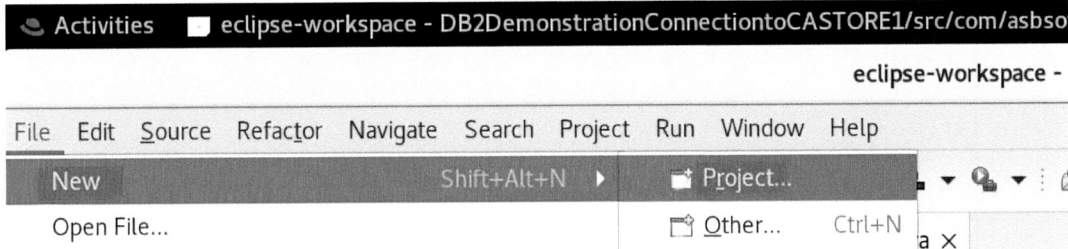

Figure 6-1. A new Java Project is selected in the Eclipse IDE

CHAPTER 6 PDF DOCUMENT CREATION USING JAVA

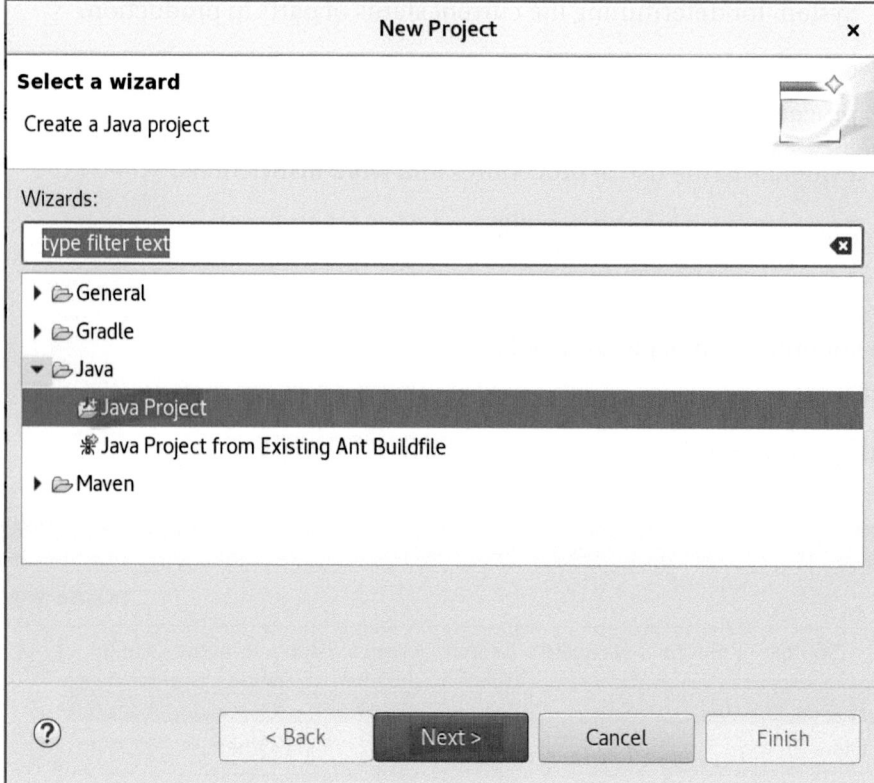

Figure 6-2. *The Java Project type is selected*

For compatibility with the FileNet jar files, we select the same JRE Library we used for the Java in Chapter 2.

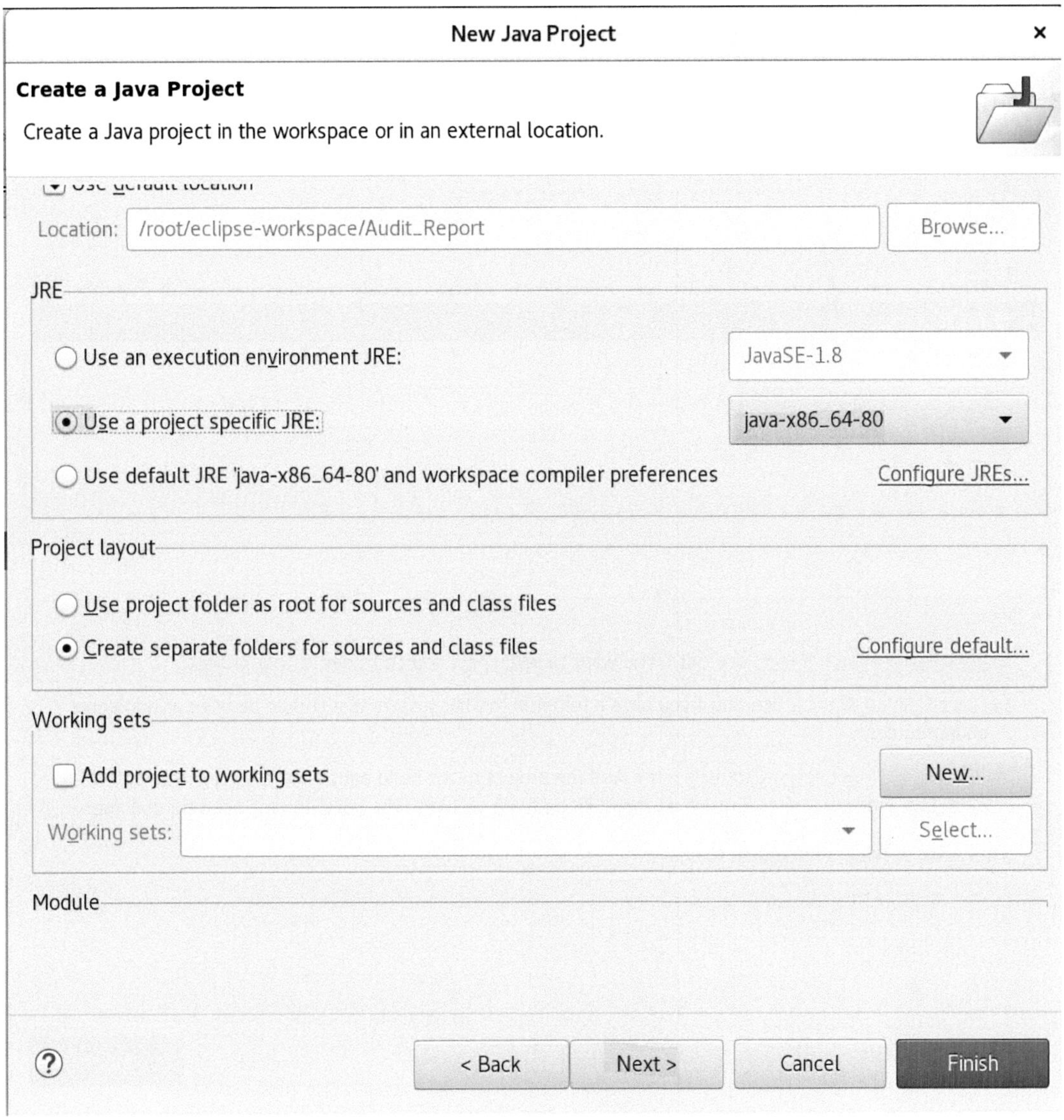

Figure 6-3. The JRE is selected for compatibility with the FileNet API jar files

CHAPTER 6 PDF DOCUMENT CREATION USING JAVA

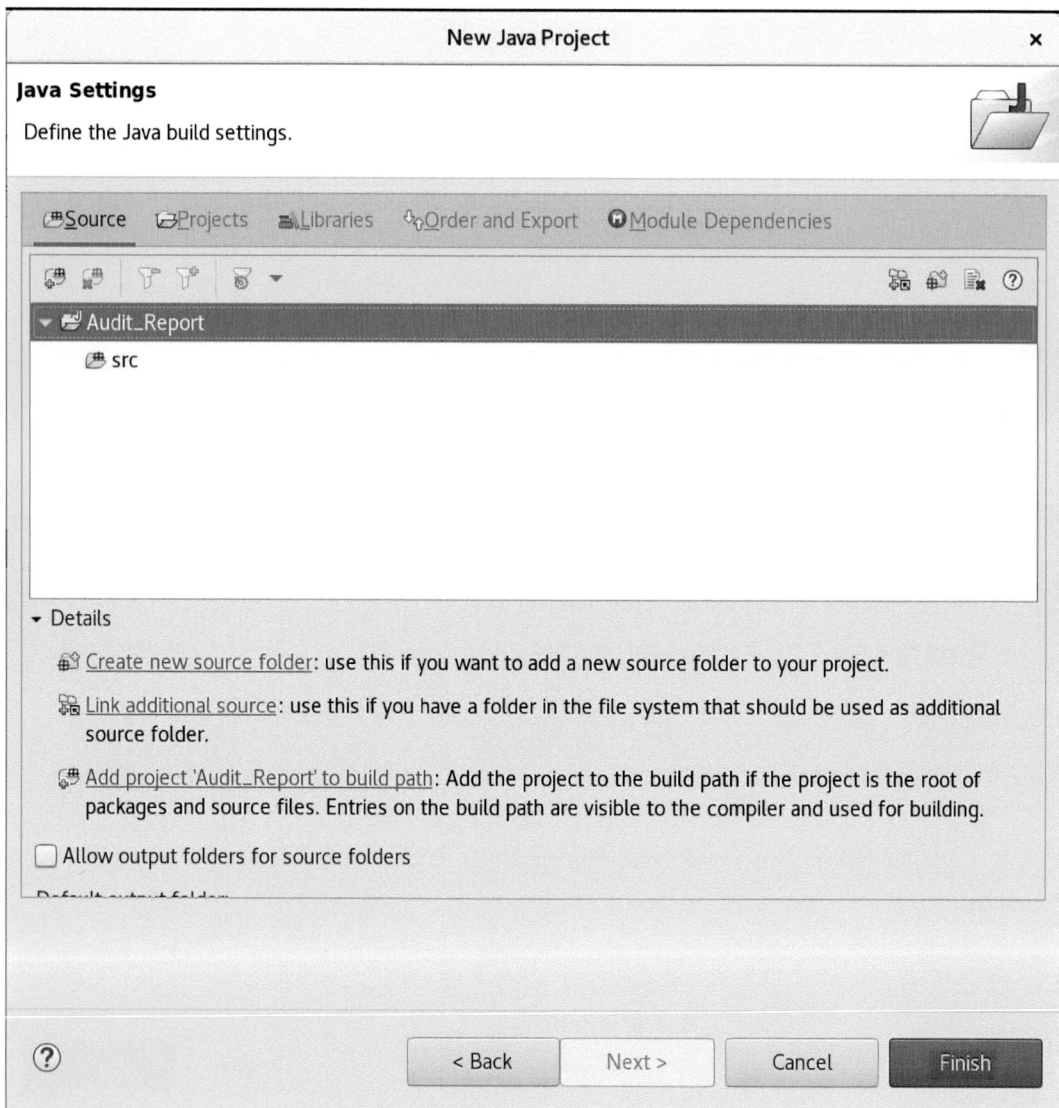

Figure 6-4. *The name is set as Audit_Report*

A pop-up window asks if we want to use the Eclipse IDE Java perspective, which automatically sets a number of window pane views and a tree browser pane on the left of the window to allow the program classes to be split into a logical package structure for the Java project.

CHAPTER 6   PDF DOCUMENT CREATION USING JAVA

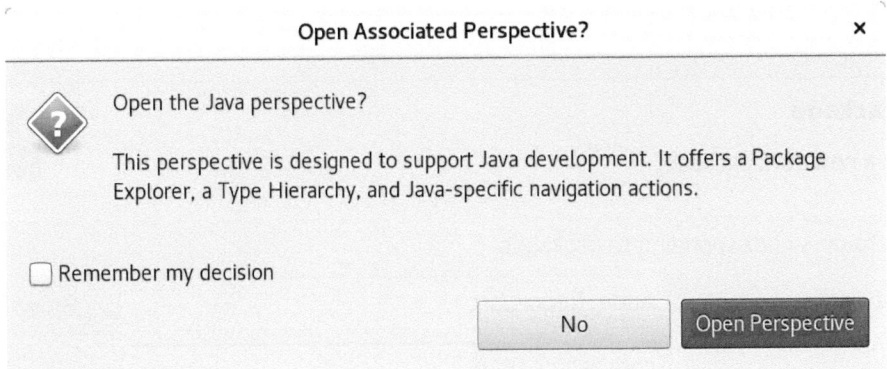

Figure 6-5. *The Java perspective is selected*

Creating the com.asb.ce.utils Java Package

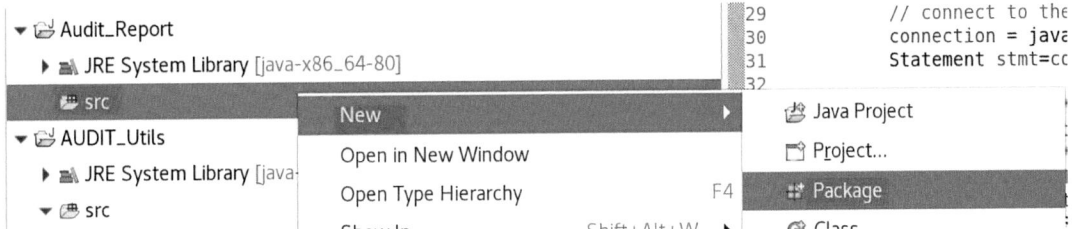

Figure 6-6. *The New Package option is selected for the Audit Report program*

In Figure 6-6, we select a new class Package to add supporting Java code for the main Audit Report program.

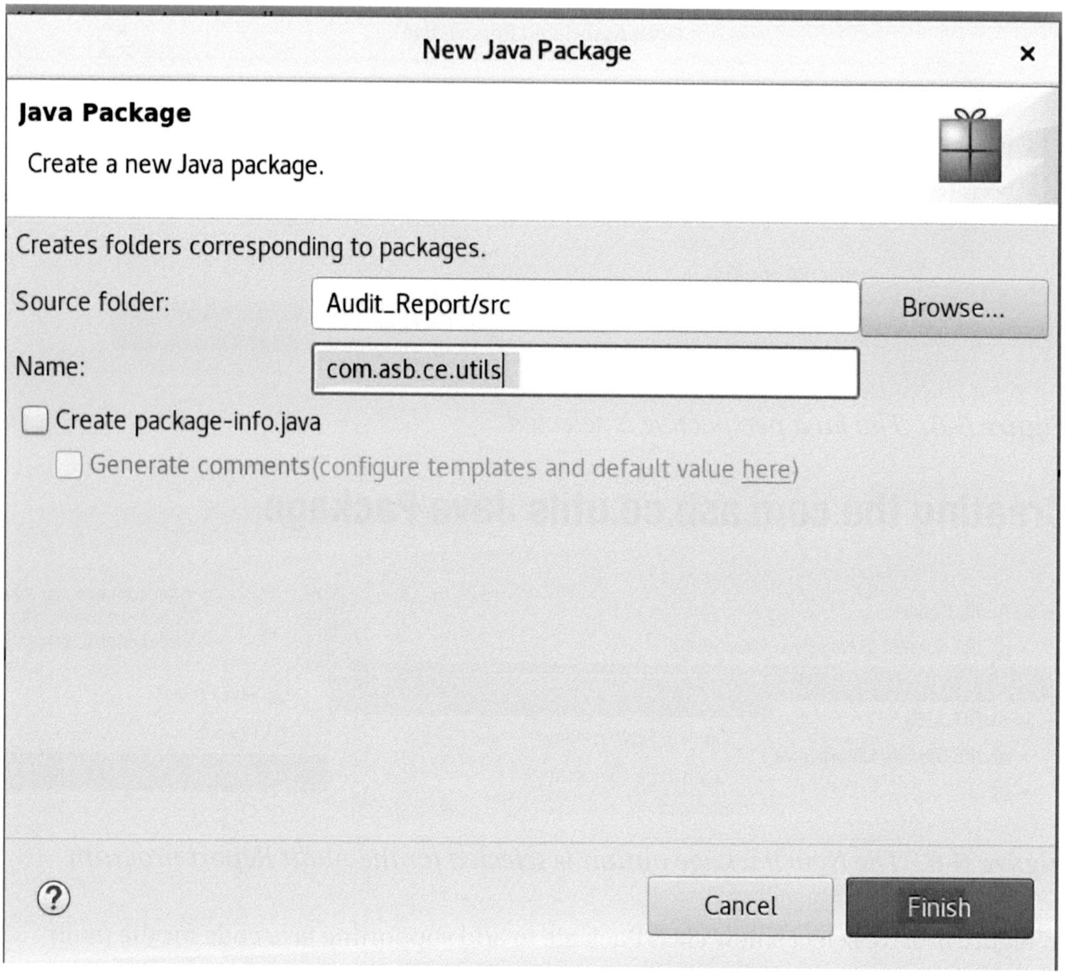

CHAPTER 6   PDF DOCUMENT CREATION USING JAVA

Figure 6-7. *The Package name is entered as com.asb.ce.utils*

Underneath the com.asb.ce.utils package (which is actually an Eclipse nested folder structure), we now add a Java class, which will hold the Java code we need for some utility class methods.

CHAPTER 6 PDF DOCUMENT CREATION USING JAVA

Creating the AuditReportConfig Java Class

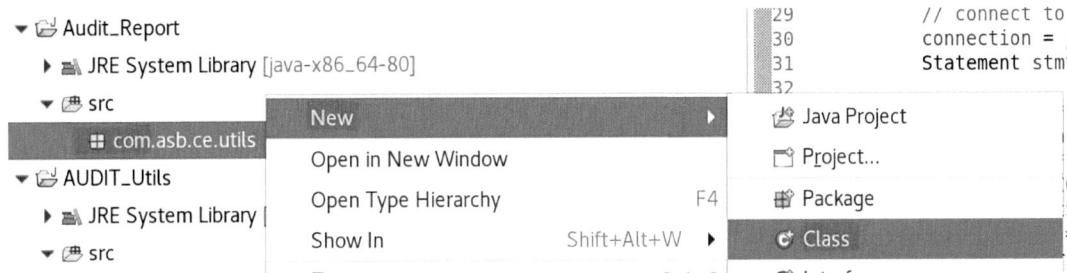

Figure 6-8. *The right-clicked package gives the New ➤ Class menu*

We create an **AUDITReportConfig** class for Java code which will read in the parameters we require to drive the Audit Report program. These are held in a **config.xml** file which we structure to make entry of the variables easier and to allow the right values to be read for each of the program variables (and to avoid hard-coding for maximum flexibility).

The **config.xml** file path is defined for access using a platform-independent section of Java code which detects if this is a Windows-based system or a Linux/Unix-based system directory path structure, as shown in the following Java code snippet.

Listing 6-11. The section of code to read the **config.xml** with platform independence

```
//ASB ... 11-08-2022 - Make platform independent
String OS = System.getProperty("os.name").toLowerCase();
String file;
if (OS.indexOf("win") >= 0) {
    file = System.getProperty("user.dir") + "\\config\\config.xml";
} else { //Set to forward slash
    file = System.getProperty("user.dir") + "/config/config.xml";
}
```

1009

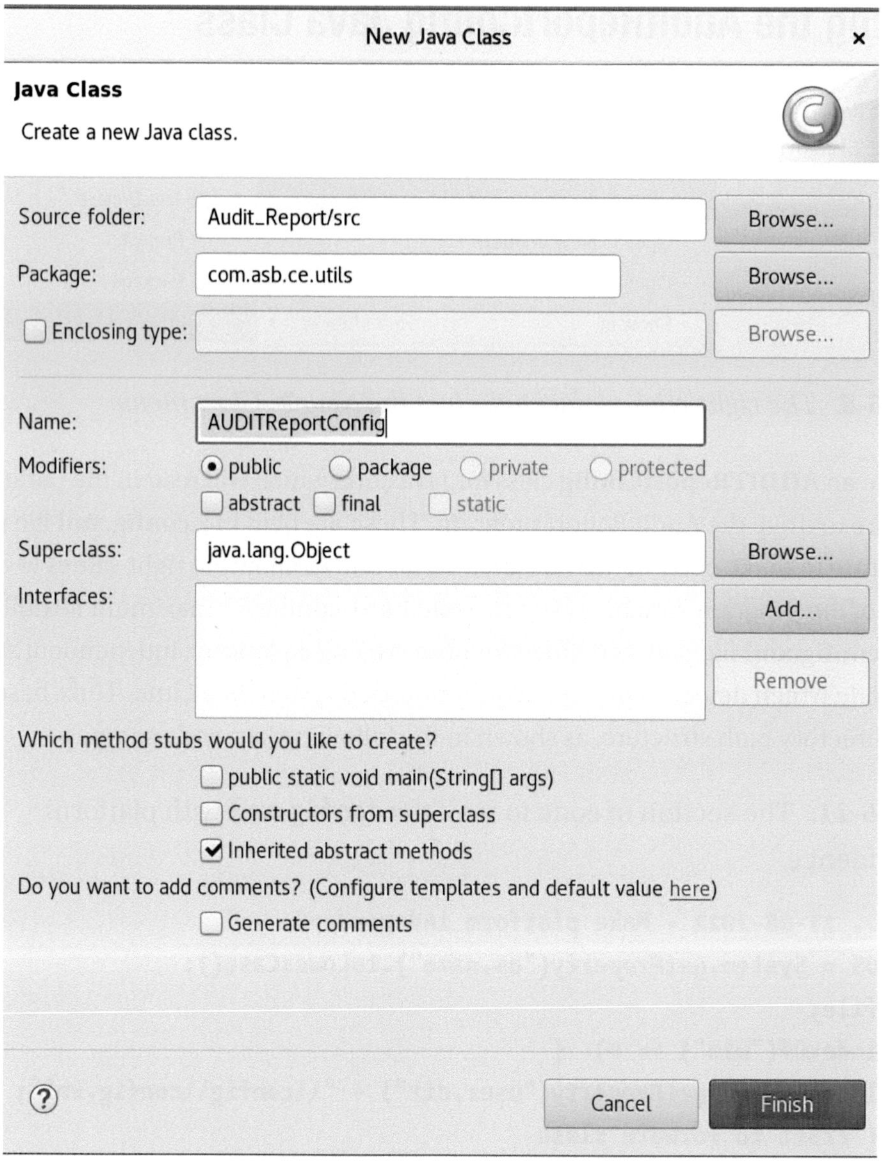

CHAPTER 6 PDF DOCUMENT CREATION USING JAVA

*Figure 6-9. The **AUDITReportConfig** class is defined for entry of the code*

The code in Listing 6-12 is entered next for the **AuditReportConfig.java**.

Listing 6-12. The **AUDITReportConfig**.java code for reading the **config.xml** file

```java
package com.asb.ce.utils;
/**
IBM grants you a non-exclusive copyright license to use all programming
code examples from which you can generate similar function tailored to your
own specific needs.

All sample code is provided by IBM for illustrative purposes only.
These examples have not been thoroughly tested under all conditions. IBM,
therefore cannot guarantee or imply reliability, serviceability, or
function of these programs.

All Programs or code component contained herein are provided to you "AS IS"
without any warranties of any kind.
The implied warranties of non-infringement, merchantability and fitness for
a particular purpose are expressly disclaimed.

© Copyright IBM Corporation 2013, ALL RIGHTS RESERVED.
*/
    /**
     *
     */
    /**
     * @author Alan S. Bluck, ASB Software Development Limited
     *
     */
    import java.io.BufferedWriter;
    import java.io.File;
    import java.io.FileInputStream;
    import java.io.FileWriter;
    import java.io.IOException;
    import java.io.InputStream;
    import java.io.StringWriter;
    import java.text.DateFormat;
    import java.text.ParseException;
    import java.text.SimpleDateFormat;
```

```java
import java.util.Calendar;
import java.util.Date;
import java.util.Locale;
import java.util.Vector;

import javax.xml.parsers.DocumentBuilder;
import javax.xml.parsers.DocumentBuilderFactory;
import javax.xml.parsers.ParserConfigurationException;
import javax.xml.transform.OutputKeys;
import javax.xml.transform.Transformer;
import javax.xml.transform.TransformerFactory;
import javax.xml.transform.dom.DOMSource;
import javax.xml.transform.stream.StreamResult;
import javax.xml.xpath.XPath;
import javax.xml.xpath.XPathConstants;
import javax.xml.xpath.XPathFactory;

import org.apache.commons.codec.binary.Base64;
import org.apache.log4j.Logger;
import org.w3c.dom.Document;
import org.w3c.dom.Node;
import org.w3c.dom.NodeList;
import org.xml.sax.SAXException;

import com.asb.config.AUDITConfig;

    public class AUDITReportConfig {
        Logger logger = Logger.getLogger(AUDITReportConfig.class.
        getName());
        private Document dom;

        // Export Parameters
        private String SolutionPrefix;
        private String AUDITORMainTask;
        private String exportOSName;
        private String exportCEUser;
        private String exportCEPassword;
        private String exportCEUrl;
```

```java
    //ASB Added for Workflow System calls 10-08-2022
    private String exportConnectionPoint;

    private String exportCEStanza;
    private String exportWaspLocation;
    private String MimeType;

    private String QSelectList;
    private String QWhereClause;
    private String QOrderByClause;
    //ASB 17-05-2022 add Folder Clause strings for Case Folder queries
    private String QCaseFolderOrderByClause;
    private String QCaseFolderWhereClause;
    private String ReportDirectoryName;
    private String TestFolderName;
//ASB ... add Document propertyList values
    private String exportAuditFile;
    private Integer maximumDocs;
    private Integer noQQuestions;
    private Integer noQColumns;
    //Loader
    private String[] propNames;
    private String[] propFlags;
    private String[] propTypes;
    private String[] propCardinality;
    // Import parameters
    private String importOSName;
    private String importOSRootFolder;
    private String docPathName;
    private String CEFolderPath;
    private String DocClassName;
    private String ReportPath;
    private String ReportExtension; //ASB 11-08-2022 Added for
                                    report files
    private String importCEUrl;
    private String importCELoginConfig; //ASB
```

```java
            private String importCEStanza;
            private String importCEUser;
            private String importCEPassword;
            private String importAuditFile;

            //ASB Define the new Email getter/setter values 11-08-2022
            private String sourceEmailAddress;
            private String sourceEmailPassword;
            private String SMTPTLSRequired;
            private String SMTPSSLRequired;
            private String SMTPUser;
            private String TLSPort;
            private Integer emailTLSPort;
            private String SSLPort;
            private Integer emailSSLPort;
            private String SMTPType;
            private String SMTPHost;
            private String SMTPEmailTemplates;
            private Integer importMaxErrorCount;
            private Integer importMaxRunTimeMinutes;
            private String exportCELoginConfig;
            private Integer importMaxDocCount;
            private String importAuditFileFolder;
            private Integer updatingBatchSize;
            private String caseStatus; //ASB   - Search status for GUID
                                            deletion
            private String caseStatusField; //ASB   - Search status Field Name
                                            for GUID deletion
            private String[] PropCardinality;

            private String EmailListName;
            private String EmailFlagListName;

            private String  STOPState;
            private String  ChoiceListFlag;
            private String  startSearchDate;
```

```java
        private String   QstartSearchDate;   //ASB 11-08-2022
        private String   QSearchDays;        //ASB 11-08-2022
        private String   SearchSubClasses;   //ASB Flag set to YES will
                                             search the subclasses of the
                                             Document class used as the
                                             root class
        private String   deltaSearchDate;    //ASB         Calculated End
                                             Date/Time
        private String   deltaHours = "01"; //ASB Delta for "   "   "
        private Integer  SleepTime = 2;      //ASB Delta for "   "   "
        private String   deltaMinutes = "01"; //ASB Delta for "   "   "
        private String   searchProperty = "Id"; //ASB Symbolic property
                                                   name for Delete search
        private String   searchPropertyDocs = "Id"; //ASB Symbolic
                                                    property name for
                                                    Delete search
        private String   searchPropertyFolders = "CmAcmCaseFolder";
        //ASB Symbolic property name for Delete search
        private Integer maxSearchSize;
        //ASB ...   Debug Flag for performance improvement
        private String debugFlag;
        //ASB ...   Get LDAP Search Flag for performance improvement
        private String LDAPSearchFlag;
        //ASB ...   Get Case Folder Class list
        private String folderCaseClasses;
        //ASB ...   Get JDBC Database info
        private String JDBC;
        private String JDBCDriverClass;
        private String JDBCUser;
        private String JDBCPassword;
        private XPath xPath;
        public static void main(String args[]) throws Exception {
            AUDITReportConfig cemc = new AUDITReportConfig();
        }
```

CHAPTER 6 PDF DOCUMENT CREATION USING JAVA

```java
public AUDITReportConfig() throws Exception {
    XPathFactory  factory= XPathFactory.newInstance();
    xPath=factory.newXPath();
    xPath.reset();

    getDocument();
    readConfig();
}

private void readConfig() throws Exception {
//ASB ...  Deletion Configuration Parameters
//readOnlyGroup = getXMLVal("/config/AUDITProcessor/readonlygroup/text()");
SolutionPrefix = getXMLVal("/config/AUDITProcessor/SolutionPrefix/text()"); //CHECK
//excludedGroup = getXMLVal("/config/AUDITProcessor/excludedgroup/text()");
//excludedUser = getXMLVal("/config/AUDITProcessor/excludeduser/text()");
startSearchDate = getXMLVal("/config/AUDITProcessor/startsearchdate/text()");
//ASB 18-05-2022 For Calculation of QstartSearchDate
QSearchDays = getXMLVal("/config/AUDITProcessor/searchdays/text()");
maxSearchSize = Integer.parseInt(getXMLVal("/config/AUDITProcessor/MaxSearchSize/text()"));
noQQuestions = Integer.parseInt(getXMLVal("/config/NumberOfFullQQuestions/text()"));
noQColumns = Integer.parseInt(getXMLVal("/config/NumberOfFullQColumns/text()"));
SleepTime = Integer.parseInt(getXMLVal("/config/SleepTime/text()"));
//ASB ...  Retrieve Debug Output flag (off/on) value
debugFlag = getXMLVal("/config/AUDITProcessor/DebugOutputFlag/text()");
//ASB ...  Retrieve LDAP Search flag (off/on) value
//LDAPSearchFlag = getXMLVal("/config/AUDITProcessor/LDAPSearchFlag/text()");
```

```java
deltaHours = getXMLVal("/config/AUDITProcessor/DeltaHours/text()");
deltaMinutes = getXMLVal("/config/AUDITProcessor/DeltaMinutes/text()");
searchProperty = getXMLVal("/config/AUDITProcessor/searchProperty/text()");
//ASB 18-01-2022 Added for proc processing
searchPropertyDocs = getXMLVal("/config/AUDITProcessor/searchPropertyDocuments/text()");
searchPropertyFolders = getXMLVal("/config/AUDITProcessor/searchPropertyFolders/text()");
//ASB ...  Add FolderSubclasses retrieval
folderCaseClasses = getXMLVal("/config/AUDITProcessor/folderCaseClasses/text()"); //CHANGE
//JDBC Parameters for the Database
JDBC = getXMLVal("/config/AUDITProcessor/JDBC/text()");
JDBCDriverClass = getXMLVal("/config/AUDITProcessor/JDBCDriverClass/text()");
JDBCUser = getXMLVal("/config/AUDITProcessor/JDBCUser/text()");
JDBCPassword = getXMLVal("/config/AUDITProcessor/JDBCPassword/text()");
//Email
EmailListName = getXMLVal("/config/AUDITProcessor/EmailListName/text()");
EmailFlagListName = getXMLVal("/config/AUDITProcessor/EmailFlagListName/text()");
//ASB 13-08-2022 Added parameters for the pdf file import processing
importMaxRunTimeMinutes = Integer.parseInt(getXMLVal("/config/ceimport/MaxRunTimeMinutes/text()"));
importAuditFile = getXMLVal("/config/ceimport/AuditFile/text()");
importAuditFileFolder= getXMLVal("/config/ceimport/AuditFileFolders/text()");
//ASB ...  Check if this is still clear ie
//ASB ...   Not in the form Encrypt3dpasswordvalueP4ssw0rd
String decryptedJDBCPassword = decrypt(JDBCPassword);
if (decryptedJDBCPassword.startsWith("Encrypt3d")&&
decryptedJDBCPassword.endsWith("P4ssw0rd")   ){
```

```java
    //ASB ...  Extract the actual password
    //
        JDBCPassword = decryptedJDBCPassword.substring(9,
        decryptedJDBCPassword.length()-8);
    }else{
    //ASB ...  update the clear password
    String encryptedJDBCPassword = "Encrypt3d" + JDBCPassword + "P4ssw0rd";
    encryptedJDBCPassword = encrypt(encryptedJDBCPassword);
    System.out.print("\r\tENCRYPTED JDBC CDDR PASSWORD = " +
    encryptedJDBCPassword);
    System.out.print("\r\t");
    // ASB ...  Write encrypted password back to the config.xml file
    String dirPath = System.getProperty("user.dir");
    String OS = System.getProperty("os.name").toLowerCase();
    String file = null;
    // ASB
    if (OS.indexOf("win") >= 0)  {
       file = dirPath + "\\config\\config.xml";
    } else { //Set to forward slash
       file = dirPath + "/config/config.xml";
    }
     // SET    file Name, root element, tag element, old value, new value
     updateXML(file,"AUDITProcessor","JDBCPassword", JDBCPassword,
     encryptedJDBCPassword);
     }
// Export Parameters
    exportOSName = getXMLVal("/config/ceexport/osname/text()");
    exportConnectionPoint = getXMLVal("/config/ceexport/
    ConnectionPointName/text()");
    exportCEUser = getXMLVal("/config/ceexport/ceuser/text()");
    exportCEPassword = getXMLVal("/config/ceexport/cepassword/text()");
    importMaxDocCount = Integer.parseInt(getXMLVal("/config/ceimport/
    MaxDocCount/text()"));
    docPathName = getXMLVal("/config/ceexport/DocPathName/text()");
    DocClassName =  getXMLVal("/config/ceexport/DocClassName/text()");
```

```java
caseStatus = getXMLVal("/config/ceexport/CaseStatus/text()");
caseStatusField = getXMLVal("/config/ceexport/CaseStatusField/text()");
SearchSubClasses = getXMLVal("/config/database/SearchSubClasses/text()"); //ASB 11-08-2022
MimeType = getXMLVal("/config/ceexport/MimeType/text()");
AUDITORMainTask = getXMLVal("/config/AUDITOR_Task/text()");
QSelectList = getXMLVal("/config/AUD_Select_List/text()");
QWhereClause = getXMLVal("/config/AUD_where_clause/text()");
QOrderByClause = getXMLVal("/config/AUD_orderby_clause/text()");
//ASB Add code to retrieve new Order and Where query clauses for Case Folder search 11-08-2022
QCaseFolderWhereClause = getXMLVal("/config/QCase_Folder_where_clause/text()");
QCaseFolderOrderByClause = getXMLVal("/config/QCase_Folder_orderby_clause/text()");
ReportDirectoryName = getXMLVal("/config/ReportDirectoryName/text()");
STOPState = getXMLVal("/config/SET_TO_STOP/text()");
ChoiceListFlag = getXMLVal("/config/ChoiceListProcessor/SET_TO_GENERATE_CHOICELISTS/text()");//ChoiceListFlag
TestFolderName = getXMLVal("/config/TestFolderName/text()");
// ASB New Email Parameters 11-08-2022
sourceEmailAddress = getXMLVal("/config/SMTPreport/sourceEmailAddress/text()");
sourceEmailPassword = getXMLVal("/config/SMTPreport/sourceEmailPassword/text()");
SMTPTLSRequired = getXMLVal("/config/SMTPreport/SMTPTLSRequired/text()");
SMTPSSLRequired = getXMLVal("/config/SMTPreport/SMTPSSLRequired/text()");
SMTPHost = getXMLVal("/config/SMTPreport/SMTPHost/text()");
SMTPUser = getXMLVal("/config/SMTPreport/SMTPUser/text()");
SMTPHost = getXMLVal("/config/SMTPreport/SMTPHost/text()");
SMTPEmailTemplates = getXMLVal("/config/SMTPreport/EmailTemplatePath/text()");
TLSPort = getXMLVal("/config/SMTPreport/TLSPort/text()");
emailTLSPort = Integer.parseInt(TLSPort);
SSLPort = getXMLVal("/config/SMTPreport/SSLPort/text()");
```

CHAPTER 6   PDF DOCUMENT CREATION USING JAVA

```java
emailSSLPort = Integer.parseInt(SSLPort);
SMTPType = getXMLVal("/config/SMTPreport/SMTPType/text()");
//ASB ...  Not in the form Encrypt3dpasswordvalueP4ssw0rd
String decryptedEmailPassword = decrypt(sourceEmailPassword);
if (decryptedEmailPassword.startsWith("Encrypt3d")&&
decryptedEmailPassword.endsWith("P4ssw0rd")   ){
//ASB ...  Extract the actual password
//
sourceEmailPassword = decryptedEmailPassword.substring(9,
decryptedEmailPassword.length()-8);
}else{
//ASB ...  update the clear password
String encryptedEmailPassword = "Encrypt3d" + sourceEmailPassword +
"P4ssw0rd";
encryptedEmailPassword = encrypt(encryptedEmailPassword);
System.out.print("\r\tENCRYPTED eMail PASSWORD = " +
encryptedEmailPassword);
System.out.print("\r\t");
// ASB ...  Write encrypted password back to the config.xml file
String dirPath = System.getProperty("user.dir");
String OS = System.getProperty("os.name").toLowerCase();
String file = null;
if (OS.indexOf("win") >= 0)  {
   file = dirPath + "\\config\\config.xml";
} else { //Set to forward slash
   file = dirPath + "/config/config.xml";
}
  // SET    file Name, root element, tag element, old value, new value
     updateXML(file,"SMTPreport","sourceEmailPassword",
     sourceEmailPassword, encryptedEmailPassword);
}

//ASB ...  Not in the form Encrypt3dpasswordvalueP4ssw0rd
String decryptedExportCEPassword = decrypt(exportCEPassword);
```

```java
if (decryptedExportCEPassword.startsWith("Encrypt3d")&&
decryptedExportCEPassword.endsWith("P4ssw0rd")  ){
    //ASB ...  Extract the actual password
    //
exportCEPassword = decryptedExportCEPassword.substring(9,
decryptedExportCEPassword.length()-8);
}else{
//ASB ...  update the clear password
String encryptedExportCEPassword = "Encrypt3d" + exportCEPassword +
"P4ssw0rd";
encryptedExportCEPassword = encrypt(encryptedExportCEPassword);
System.out.print("\r\tENCRYPTED CE Export PASSWORD = " +
encryptedExportCEPassword);
System.out.print("\r\t");
// ASB ...  Write encrypted password back to the config.xml file
String dirPath = System.getProperty("user.dir");
String OS = System.getProperty("os.name").toLowerCase();
String file = null;
if (OS.indexOf("win") >= 0)  {
  file = dirPath + "\\config\\config.xml";
} else { //Set to forward slash
  file = dirPath + "/config/config.xml";
}
  // SET    file Name, root element, tag element, old value, new value
      updateXML(file,"ceexport","cepassword", exportCEPassword,
      encryptedExportCEPassword);
}

exportAuditFile = getXMLVal("/config/ceexport/AuditFile/text()");

String tmp = getXMLVal("/config/ceimport/UpdatingBatchSize/text()");
if (tmp.length() > 0){
    try {
      updatingBatchSize = Integer.parseInt(tmp);
    }
```

```java
        catch (Exception e){
            updatingBatchSize = 250;
        }
    }
    else {
        updatingBatchSize = 250;
    }
    // import parameters
    importOSName = getXMLVal("/config/ceimport/osname/text()");
    //importOSRootFolder = getXMLVal("/config/ceimport/osrootfolder/text()");
    importCEUrl = getXMLVal("/config/ceimport/ceurl/text()");
    importCEStanza = getXMLVal("/config/ceimport/cestanza/text()");
    importCELoginConfig = getXMLVal("/config/ceimport/celoginconfig/text()");
    importCEUser = getXMLVal("/config/ceimport/ceuser/text()");
    importCEPassword = getXMLVal("/config/ceimport/cepassword/text()");
    //ASB ...  Check if this is still clear ie
    //ASB ...  Not in the form Encrypt3dpasswordvalueP4ssw0rd
    String decryptedImportCEPassword = decrypt(importCEPassword);
    if (decryptedImportCEPassword.startsWith("Encrypt3d")&&
    decryptedImportCEPassword.endsWith("P4ssw0rd")   ){
    //ASB ...  Extract the actual password
    //
      importCEPassword = decryptedImportCEPassword.substring(9,
      decryptedImportCEPassword.length()-8);
    }else{
        //ASB ...  update the clear password
    String encryptedImportCEPassword = "Encrypt3d" + importCEPassword + "P4ssw0rd";
    encryptedImportCEPassword = encrypt(encryptedImportCEPassword);
    System.out.print("\r\tENCRYPTED CE IMPORT PASSWORD = " + encryptedImportCEPassword);
    System.out.print("\r\t");
```

```java
   // ASB ...   Write encrypted password back to the config.xml file
   String dirPath = System.getProperty("user.dir");
   String OS = System.getProperty("os.name").toLowerCase();
        String file = null;
      if (OS.indexOf("win") >= 0)  {
        file = dirPath + "\\config\\config.xml";
      } else { //Set to forward slash
        file = dirPath + "/config/config.xml";
      }
// SET     file Name, root element, tag element, old value, new value
   updateXML(file,"ceimport","cepassword", importCEPassword,
   encryptedImportCEPassword);
   }
   //ASB ...  export parameters for Session Object
   exportCEUrl = getXMLVal("/config/ceexport/ceurl/text()");
   exportCEStanza = getXMLVal("/config/ceexport/cestanza/text()"); //null
   exportWaspLocation = getXMLVal("/config/ceexport/WaspLocation/
   text()"); //null
   exportCELoginConfig = getXMLVal("/config/ceexport/celoginconfig/
   text()");
   //ASB ...  Get Delete Document and Delete Folder parameters
   propCardinality =  getXMLVals("/config/AUDITProcessor/
   PropCardinalityList/PropCardinality/text()");
   propNames = getXMLVals("/config/AUDITProcessor/DocumentPropertyNames/
   PropertyName/text()");
   //propFlags ASB
   propFlags = getXMLVals("/config/AUDITProcessor/DocumentWhiteListFlags/
   PropFlag/text()");
   propTypes = getXMLVals("/config/AUDITProcessor/DocumentPropertyTypes/
   PropertyType/text()");
   }
      public void updateProcessDate(Date processDate){
      if (startSearchDate.length() == 0) {
          return;
      }
```

```
// ASB ...  Write Date String back to the config.xml file
    // Need date in the format 20220823T125628Z
String dirPath = System.getProperty("user.dir");
String OS = System.getProperty("os.name").toLowerCase();
String file = null;
if (OS.indexOf("win") >= 0)  {
  file = dirPath + "\\config\\config.xml";
} else { //Set to forward slash
  file = dirPath + "/config/config.xml";
}
    SimpleDateFormat sdf = new SimpleDateFormat("yyyyMMdd");
    //Need 24 Hour clock 00 -> 23 !! above is 12 hour
    SimpleDateFormat sdfTime = new SimpleDateFormat("HHmmss");
    String newStartSearchDate = sdf.format(processDate)+ "T" +
    sdfTime.format(processDate)+ "Z";
  updateXML(file,"AUDITProcessor","startsearchdate", startSearchDate,
  newStartSearchDate);
    }
    public void updateChoiceListFlag(String choiceListFlag){
        if (choiceListFlag.length() == 0) {
            return;
        }
// ASB ...  Write ChoiceListFlag String back to the config.xml file
// Need to set the System path for the config.xml file dependent on
the platform
String dirPath = System.getProperty("user.dir");
String OS = System.getProperty("os.name").toLowerCase();
    String file = null;
    if (OS.indexOf("win") >= 0)  {
      file = dirPath + "\\config\\config.xml";
    } else { //Set to forward slash
      file = dirPath + "/config/config.xml";
    }
```

```java
        String newchoiceListFlag = "No";
            updateXML(file,"ChoiceListProcessor","SET_TO_GENERATE_
            CHOICELISTS", choiceListFlag, newchoiceListFlag);
    }
    private void getDocument() throws Exception {
      DocumentBuilderFactory dbf = DocumentBuilderFactory.newI
      nstance();
      DocumentBuilder db = null;
      try {
         db = dbf.newDocumentBuilder();
      } catch (ParserConfigurationException e) {
      logger.error(e);
      throw e;
}
//ASB ...  Normalise path to the config we want to use
String dirPath = System.getProperty("user.dir");
//ASB We can determine the Operatiing System (Windows v Linux) to set
correct folder delimiters
String OS = System.getProperty("os.name").toLowerCase();
InputStream in = null;
if (OS.indexOf("win") >= 0)  {
 in = new FileInputStream(dirPath + "\\config\\config.xml");
} else { //Set to forward slash
 in = new FileInputStream(dirPath + "/config/config.xml");
}
dom = db.parse(in);
in.close(); //ASB ...  Now close the config.xml stream
}
public String getXMLVal(String expression) throws Exception {
Node n = (Node) xPath.evaluate(expression, dom, XPathConstants.NODE);
if (n != null)
    return n.getNodeValue();
    return "";
}
```

```java
//ASB ...   Put new value into the dom node
private void updateXML(String file, String mainElement, String
sTag, String strOldValue, String strNewValue) {
Document doc = null;
try {
    DocumentBuilderFactory docFactory = DocumentBuilderFactory.newInstance();
    DocumentBuilder docBuilder = docFactory.newDocumentBuilder();
    doc = docBuilder.parse(file);

    // Get the root element
    Node config = doc.getFirstChild();
    // Get the main element by tag name directly
    Node ceMain = doc.getElementsByTagName(mainElement).item(0);
    // loop the configuration main section child node
    NodeList list = ceMain.getChildNodes();

    for (int i = 0; i < list.getLength(); i++) {

    Node node = list.item(i);

    // get the password element, and update the value
        if (sTag.equals(node.getNodeName())) {
          node.setTextContent(strNewValue);
          break;
        }

    }
} catch (ParserConfigurationException pce) {
  pce.printStackTrace();
  } catch (IOException ioe) {
  ioe.printStackTrace();
} catch (SAXException sae) {
  sae.printStackTrace();
  }
  catch(Exception ex) {
      String exError = ex.getMessage();
  }
```

CHAPTER 6 PDF DOCUMENT CREATION USING JAVA

```java
        File filer = new File(file);

        saveToXML(filer,doc);

    }   // Save the updated DOM into the XML back
    // Save the updated DOM into the XML back
    private void saveToXML(File file, Document doc) {
        try {
            TransformerFactory factory = TransformerFactory.newI
            nstance();
            Transformer transformer = factory.newTransformer();
            transformer.setOutputProperty(OutputKeys.INDENT, "yes");

            StringWriter writer = new StringWriter();
            StreamResult result = new StreamResult(writer);
            DOMSource source = new DOMSource(doc);

            transformer.transform(source, result);

            String strTemp = writer.toString();

            FileWriter fileWriter = new FileWriter(file);
            BufferedWriter bufferedWriter = new
            BufferedWriter(fileWriter);

            bufferedWriter.write(strTemp);
            bufferedWriter.flush();
            bufferedWriter.close();
        }
        catch(Exception ex) {
        }
    }
public String[] getXMLVals(String expression) throws Exception {
    Vector vtmp = null;
    NodeList nl = null;
    vtmp = new Vector();
    nl = (NodeList) xPath.evaluate(expression, dom,
    XPathConstants.NODESET);
```

```java
        for (int x = 0; x < nl.getLength(); x ++){
            vtmp.add(nl.item(x).getNodeValue());
        }
         return (String[])vtmp.toArray(new String[vtmp.size()]);
    }
    public String getImportAuditFile(){
        return importAuditFile;
    }
    public String getImportAuditFileFolders(){
        return importAuditFileFolder;
    }
    public String getExportOSName(){
        return exportOSName;
    }
    public String getExportConnectionPointName(){
        return exportConnectionPoint;
    }
    public String getExportCEUser(){
        return exportCEUser;
    }
    public String getExportCEPassword(){
        return exportCEPassword;
    }
//ASB ... add getter Methods for export Session Object
    public String getExportCEUrl(){
        return exportCEUrl;
    }
    public String getExportCEStanza(){
        return exportCEStanza;
    }
    public String getExportWaspLocation(){
        return exportWaspLocation;
    }
    public String getExportCELoginConfig(){
        return exportCELoginConfig;
    }
```

CHAPTER 6   PDF DOCUMENT CREATION USING JAVA

```java
    public Integer getmaximumDocs(){
        return maximumDocs;
    }
    public Integer getNoQQuestions(){
        return noQQuestions;
    }
    public Integer getNoQColumns(){
        return noQColumns;
    }
    public Integer getMaxSearchSize(){
        return maxSearchSize;   //ASB ...
    }
//New Getters for the eMail configuration
    public String getSourceEmailAddress(){
        return sourceEmailAddress;   //ASB ...
    }
    public String getsourceEmailPassword(){
        return sourceEmailPassword;   //ASB ...
    }
    public String getSMTPTLSRequired(){
        return SMTPTLSRequired;   //ASB ...
    }
    public String getSMTPHost(){
        return SMTPHost;   //ASB ...
    }
    public String getSMTPEmailTemplatePath(){
        return SMTPEmailTemplates;   //ASB ...
    }
    public String getSMTPUser(){
        return SMTPUser;   //ASB ...
    }
    public String getSMTPSSLRequired(){
        return SMTPSSLRequired;   //ASB ...
    }
```

```java
    public Integer getEmailTLSPort(){
        return emailTLSPort;   //ASB ...
    }
    public Integer getEmailSSLPort(){
        return emailSSLPort;   //ASB ...
    }
    public String getStringTLSPort(){
        return TLSPort;   //ASB ...
    }
    public String getStringSSLPort(){
        return SSLPort;   //ASB ...
    }
    public String getDebugOutputFlag(){
        return debugFlag;   //ASB ... 4 020
    }
    public String getLDAPSearchFlag(){
        return LDAPSearchFlag;   //ASB ... 4 027
    }
    public String getDeltaHours(){
        return deltaHours;   //ASB ...
    }
    public String getSTOPState(){ //
        return STOPState;   //ASB ... ChoiceListFlag
    }
    public String getChoiceListFlag(){ //
        return ChoiceListFlag;   //ASB ...
    }
    public Integer getSleepTime(){
        return SleepTime;   //ASB ...
    }
    public String getDeltaMinutes(){
        return deltaMinutes;   //ASB ...
    }
    public String getsearchProperty(){
        return searchProperty;   //ASB ...
    }
```

```java
public String getsearchPropertyDocs(){
    return searchPropertyDocs;   //ASB ...
}
public String getsearchPropertyFolders(){
    return searchPropertyFolders;   //ASB ...
}
public String getFolderSubclasses(){
    return folderCaseClasses;   //ASB ...
}
public String getJDBC(){
    return JDBC;                //ASB ...
}
public String getJDBCDriverClass(){
    return JDBCDriverClass;     //ASB ...
}
public String getJDBCUser(){
    return JDBCUser;            //ASB ...
}
public String getJDBCPassword(){
    return this.JDBCPassword;   //ASB ...
}
public String getEmailListName(){
    return EmailListName;       //ASB ...
}
public String getEmailFlagListName(){
    return EmailFlagListName;   //ASB ...
}
public String getStartDateraw(){
  if (startSearchDate.length() == 0){
     Date processDate = new Date();
     Date dateBefore = new Date(processDate.getTime() -
     processDate.getHours() * 3600*1000
```

```java
            - processDate.getMinutes()*60*1000 - processDate.getSeconds()*
    1000 -60000); //Subtract n days
            SimpleDateFormat sdf = new SimpleDateFormat("yyyyMMdd");
            //Need 24 Hour clock 00 -> 23 !! above is 12 hour
            SimpleDateFormat sdfTime = new SimpleDateFormat("HHmmss");
                String newStartSearchDate = sdf.format(dateBefore)+
                "T" + sdfTime.format(dateBefore)+ "Z";
    return newStartSearchDate; //ASB ...
    } else{
        return startSearchDate;
    }
    }
    public String getStartSearchDate(){
    //ASB allow search date to be set from the current system
    processing date
    if (startSearchDate.length() == 0){
        Date processDate = new Date();
        Date dateBefore = new Date(processDate.getTime() -
        processDate.getHours() * 3600*1000
        - processDate.getMinutes()*60*1000 - processDate.getSeconds()*
        1000 -60000); //Subtract n days
          SimpleDateFormat sdf = new SimpleDateFormat("yyyyMMdd");
          //Need 24 Hour clock 00 -> 23 !! above is 12 hour
          SimpleDateFormat sdfTime = new SimpleDateFormat("HHmmss");
          String newStartSearchDate = sdf.format(dateBefore)+ "T" +
          sdfTime.format(dateBefore)+ "Z";
            return newStartSearchDate; //ASB ...
    }else{
      Date processDate = new Date();
            SimpleDateFormat sdf = new SimpleDateFormat("yyyyMMdd",
            Locale.ENGLISH); //ASB ...   Added Locale
            Date javaStartDate = null;
            //Need 24 Hour clock 00 -> 23 !! hh is 12 hour
            SimpleDateFormat sdfTime = new SimpleDateFormat("HHmmss",
            Locale.ENGLISH); //ASB ...   Added Locale
```

//ASB ..Check if the current date was manually set - _ie_ is greater
//   than one hour difference from the current time.
//   If date was set manually then use this date instead as the search date.

```
            DateFormat diffDateFormat = DateFormat.getDateInstance();
            String sDiffDate = startSearchDate.substring(0, 8);
            String sDiffTime = startSearchDate.substring(9,15);
            Date diffDate = null;
            Date diffTime = null;
            Date dDeltaTime = null;
            Long dateSeconds = null;
            try {
                //ASB ... Need to subtract off the java start date
                javaStartDate = sdf.parse("19700101"); //ASB ...
                diffDate = sdf.parse(sDiffDate) ;
                diffTime = sdfTime.parse(sDiffTime) ;
                //Test for zero
                Long searchhours = (long) 0;
                if((diffTime.getTime() - javaStartDate.getTime() - Long.
                parseLong(deltaHours)*3600000)  > 0)
               searchhours =(diffTime.getTime() - javaStartDate.
               getTime() - Long.parseLong(deltaHours)*3600000)/3600000;
                String sSearchHours = searchhours.toString();
                if (sSearchHours.length() == 1)sSearchHours = "0"+
                sSearchHours;
                dDeltaTime =  sdfTime.parse(sSearchHours + "0000");
          deltaSearchDate = sdf.format(processDate) + "T" + sSearchHours
          + "0000Z";
          //ASB ...  Added subtraction of the Java Start Date
          milliseconds
           dateSeconds = (diffDate.getTime() + (diffTime.getTime()) -
           (dDeltaTime.getTime()));
            } catch (ParseException e) {
```

```java
        // Should be a controlled format - set to default if we get here
        e.printStackTrace();
    }
        //ASB ...   Get Process Date in milliseconds
        Long milliSecondsLastDate  = processDate.getTime();
        //ASB ...   Allow 30 Minute delta on top
        if ((milliSecondsLastDate - dateSeconds) < (diffTime.
        getTime()- javaStartDate.getTime() )){
            startSearchDate = deltaSearchDate; //Allow 60 Minutes to
            fix missing documents
        }
         return startSearchDate; //ASB ...
         }
}
public String getQStartSearchDate(){
     //ASB ...   Now dynamically update the search date from the
     System Date
      //ASB ... 18-05-2022
     Date newDateSearch = new Date();
      SimpleDateFormat sdf = new SimpleDateFormat("yyyyMMdd",
      Locale.ENGLISH);
      //Need 24 Hour clock 00 -> 23 !! above is 12 hour
     SimpleDateFormat sdfTime = new SimpleDateFormat("HHmmss",
     Locale.ENGLISH);
     // Need date in the format 20221805T152528Z
      int DeltaHours = Integer.parseInt(QSearchDays) * 24;
      String sDiffDate = sdf.format(newDateSearch);
      String sDiffTime = sdfTime.format(newDateSearch);
      String sSearchDate = sDiffDate + "T" + sDiffTime + "Z" ;

     SimpleDateFormat formatter = new SimpleDateFormat("yyyyMM
     ddHHmmss");

     try {
            String sDate = sSearchDate.substring(0, 8);
            String sTime = sSearchDate.substring(9,15);
```

//ASB ..Check if the current date was manually set - ie is greater
//   than one hour difference from the current time.
//   If date was set manually then use this date instead as the search date.
```
            DateFormat diffDateFormat = DateFormat.getDateInstance();
            String sDiffDate = startSearchDate.substring(0, 8);
            String sDiffTime = startSearchDate.substring(9,15);
            Date diffDate = null;
            Date diffTime = null;
            Date dDeltaTime = null;
            Long dateSeconds = null;
            try {
                //ASB ...  Need to subtract off the java start date
                javaStartDate = sdf.parse("19700101"); //ASB ...
                diffDate = sdf.parse(sDiffDate) ;
                diffTime = sdfTime.parse(sDiffTime) ;
                //Test for zero
                Long searchhours = (long) 0;
                if((diffTime.getTime() - javaStartDate.getTime() - Long.
                parseLong(deltaHours)*3600000)  > 0)
               searchhours =(diffTime.getTime() - javaStartDate.
               getTime() - Long.parseLong(deltaHours)*3600000)/3600000;
                String sSearchHours = searchhours.toString();
                if (sSearchHours.length() == 1)sSearchHours = "0"+
                sSearchHours;
                dDeltaTime =  sdfTime.parse(sSearchHours + "0000");
           deltaSearchDate = sdf.format(processDate) + "T" + sSearchHours
           + "0000Z";
         //ASB ...  Added subtraction of the Java Start Date
         milliseconds
            dateSeconds = (diffDate.getTime() + (diffTime.getTime()) -
            (dDeltaTime.getTime()));
             } catch (ParseException e) {
```

```java
        // Should be a controlled format - set to default if we get here
        e.printStackTrace();
    }
        //ASB ...  Get Process Date in milliseconds
        Long milliSecondsLastDate  = processDate.getTime();
        //ASB ...  Allow 30 Minute delta on top
        if ((milliSecondsLastDate - dateSeconds) < (diffTime.
        getTime()- javaStartDate.getTime() )){
            startSearchDate = deltaSearchDate; //Allow 60 Minutes to
            fix missing documents
        }
         return startSearchDate; //ASB ...
        }
}
public String getQStartSearchDate(){
    //ASB ...  Now dynamically update the search date from the
    System Date
     //ASB ... 18-05-2022
    Date newDateSearch = new Date();
     SimpleDateFormat sdf = new SimpleDateFormat("yyyyMMdd",
     Locale.ENGLISH);
     //Need 24 Hour clock 00 -> 23 !! above is 12 hour
    SimpleDateFormat sdfTime = new SimpleDateFormat("HHmmss",
    Locale.ENGLISH);
    // Need date in the format 20221805T152528Z
     int DeltaHours = Integer.parseInt(QSearchDays) * 24;
     String sDiffDate = sdf.format(newDateSearch);
     String sDiffTime = sdfTime.format(newDateSearch);
     String sSearchDate = sDiffDate + "T" + sDiffTime + "Z" ;

    SimpleDateFormat formatter = new SimpleDateFormat("yyyyMM
    ddHHmmss");

    try {
            String sDate = sSearchDate.substring(0, 8);
            String sTime = sSearchDate.substring(9,15);
```

```java
                newDateSearch = formatter.parse(sDate + sTime);
                Calendar cal = Calendar.getInstance();
                cal.setTime(newDateSearch);
                cal.add(Calendar.HOUR, -DeltaHours);
                newDateSearch = cal.getTime();
        } catch (ParseException e) {
                e.printStackTrace();
        }
// Need date in the format 20220823T125628Z
        QstartSearchDate = sdf.format(newDateSearch)+ "T" + sdfTime.
        format(newDateSearch)+ "Z";
         return QstartSearchDate; //ASB ... 2 022
    }
    public String getExportedAuditFile(){
        return exportAuditFile;
    }
    public String getImportOSName(){
        return importOSName;
    }
    public String getImportOSRootFoler(){
        return importOSRootFolder;
    }
    public String getImportCEUrl(){
        return importCEUrl;
    }
    public String getImportCEStanza(){
        return importCEStanza;
    }
    public String getImportCELoginConfig(){
        return importCELoginConfig;
    }
    public String getImportCEUser(){
        return importCEUser;
    }
```

```java
    public String getImportCEPassword(){
        return importCEPassword;
    }
    public Integer getImportMaxErrorCount(){
        return importMaxErrorCount;
    }
    public Integer getImportMaxDocCount(){
        return importMaxDocCount;
    }
    public String getQSearchDays(){ //QSearchDays
        return QSearchDays;
    }
    public String getQSelectList(){ //QSearchDays
        return QSelectList;   //
    }
    public String getQWhereClause(){
        return QWhereClause;   //
    }
    public String getAUDITORMainTask(){
        return AUDITORMainTask;   //
    }
    public String getQCaseFolderWhereClause(){
        return QCaseFolderWhereClause;   // ASB 11-08-2022
    }
    public String getQOrderByClause(){
        return QOrderByClause;   //
    }
    public String getQCaseFolderOrderByClause(){
        return QCaseFolderOrderByClause;  // ASB 11-08-2022
    }
    public String getReportDirectoryName(){
        return ReportDirectoryName;
    }
    public String getTestFolderName(){
        return TestFolderName;
    }
```

```java
    public String getDateInOracleFormat(String ISO8601Date){
//ASB Return as ddmmyyyyhhmmss from     ISO 8601 format
//ASB eg Change from   <upper_bound_date>20220723T041232Z</upper_bound_date>
        //                      to                       yyyymmddhhmmss
String OracleFormat = ISO8601Date.substring(6, 8) + ISO8601Date.substring(4, 6) + ISO8601Date.substring(0, 4) + ISO8601Date.substring(9, 15);
    return OracleFormat;
}
    public void updateUpperBoundDate(Date newUpperBoundDate){
        // ASB ... 11-08-2022 Write Date String back to the config.xml file
            // Need date in the format 20220823T125628Z
            //ASB ... 11-08-2022 - Make platform independent
            String OS = System.getProperty("os.name").toLowerCase();
            String file;
            if (OS.indexOf("win") >= 0)  {
                file = System.getProperty("user.dir") + "\\config\\config.xml";
            } else { //Set to forward slash
                file = System.getProperty("user.dir") + "/config/config.xml";
            }
               SimpleDateFormat sdf = new SimpleDateFormat("yyyyMMdd");
               //Need 24 Hour clock 00 -> 23 !! above is 12 hour
               SimpleDateFormat sdfTime = new SimpleDateFormat("HHmmss");
               String sNewUpperBoundDate = sdf.format(newUpperBoundDate)+
               "T" + sdfTime.format(newUpperBoundDate)+ "Z";
    updateXML(file,"config","upper_bound_date", startSearchDate, sNewUpperBoundDate);
   }
     public int getUpdatingBatchSize(){
         return updatingBatchSize;
     }
```

```java
    public Integer getImportMaxRunTimeMinutes(){
        return importMaxRunTimeMinutes;
    }          public String getPropName(int i){
        return propNames[i];
    }
    public int getPropType(int i){
        return Integer.parseInt(propTypes[i]);
    }
    public int getPropCardinality(int i){
        return Integer.parseInt(propCardinality[i]);
    }
    //ASB IBM -  Get Boolean Property Flag name or "NONE"
    public String getPropFlags(int i){
        return propFlags[i];
    }
    public String get_DocPathName(){
        return docPathName;
    }
    public String get_CEPathName(){
        return CEFolderPath;
    }
    //getDocClassName()
    public String getDocClassName(){
        return DocClassName;
    }
    //ASB Flag set to YES will search the subclasses 11-08-2022
    public String getSearchSubClasses(){
        return SearchSubClasses;
    }
    public String getReportPath(){
        return ReportPath;
    }

    public String getReportExtension(){
        return ReportExtension;
    }
```

```java
//getCaseStatus
public String getCaseStatus(){
    return caseStatus;
}
//getCaseStatusField
public String getCaseStatusField(){
    return caseStatusField;
}
//getMimeType
public String getMimeType(){
    return MimeType;
}
//getSolutionPrefix
public String getSolutionPrefix(){
    return SolutionPrefix;
}
 private static String encrypt(String str) {
        try {
            //encoding  byte array into base 64
            byte[] encoded = Base64.encodeBase64(str.getBytes());
            String encString = new String(encoded);
            return encString;
        } catch (Exception e) {
        }
        return null;
    }

    private static String decrypt(String str) {
        try {
            //decoding byte array into base64
            byte[] decoded = Base64.decodeBase64(str);
            String sDecoded =  new String(decoded);
    return sDecoded;
```

```
            } catch (Exception e) {
            }
            return null;
        }
}
```

Creating the com.asb.config Package

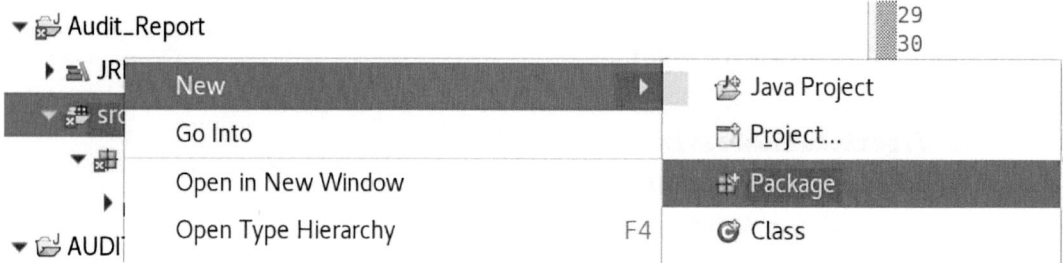

Figure 6-10. *The second Java package is added*

A config package, named com.asb.config, is created as follows.

CHAPTER 6   PDF DOCUMENT CREATION USING JAVA

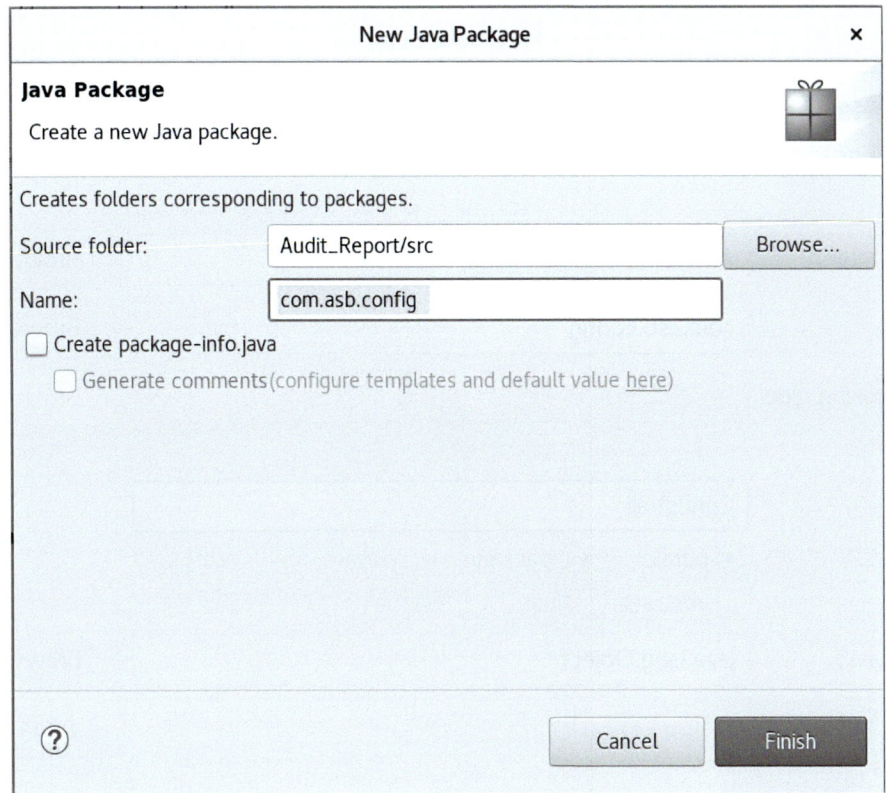

Figure 6-11. *The **com.asb.config** package is added to **Audit_Report/src***

Creating the ConfigVal Java Class

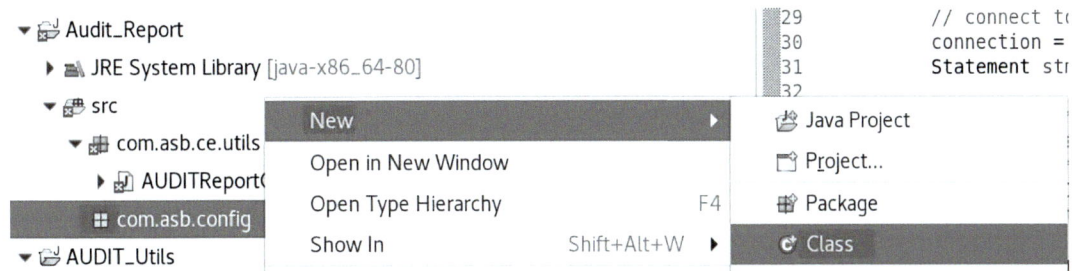

Figure 6-12. *A New Java Class is added to the **com.asb.config** package*

The **ConfigVal** Class is added to the **com.asb.config** package.

1041

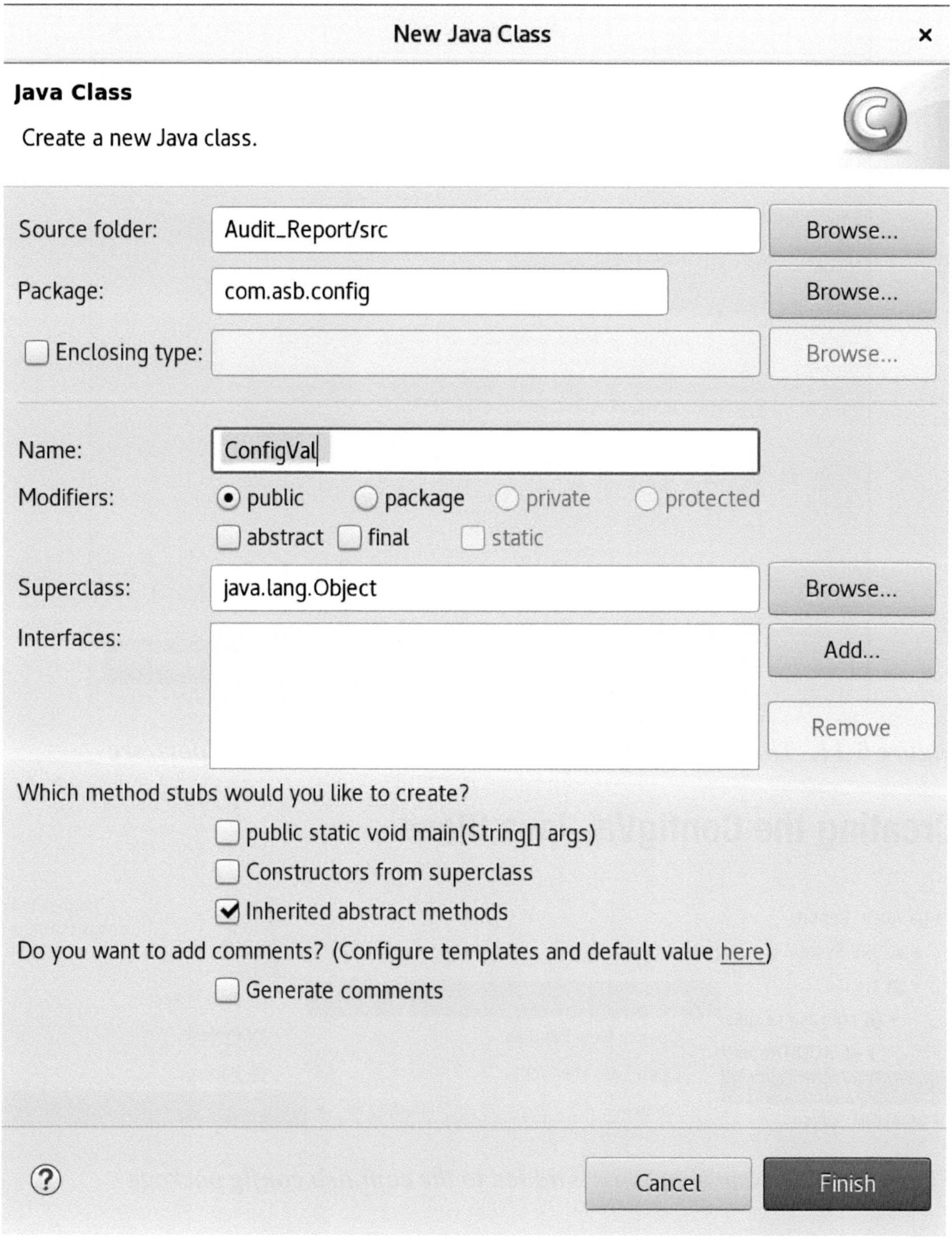

*Figure 6-13. The **ConfigVal** class is created under **com.asb.config***

The Java code for the **ConfigVal.java** source is as follows.

Listing 6-13. The Java code for the ConfigVal.java file

```java
package com.asb.config;
public class ConfigVal {
    public String setting;
    public String default_val;
    public ConfigVal(String setting, String val){
        this.setting = setting;
        this.default_val = val;
    }
}
```

This Java code has a **ConfigVal** public method, used as an initial test code for getter and setter methods. (It is not currently used in the **Audit_Report** program.)

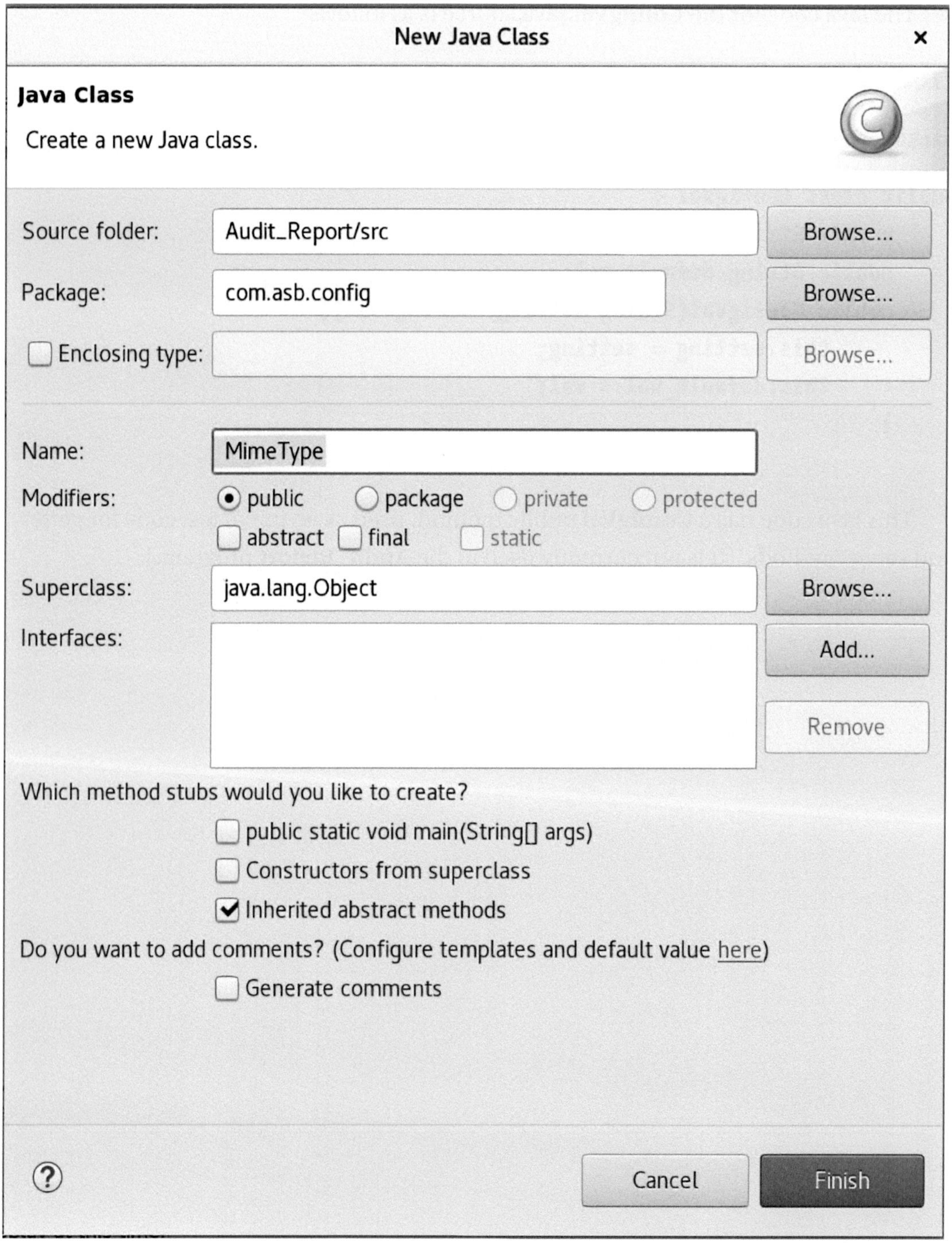

CHAPTER 6   PDF DOCUMENT CREATION USING JAVA

Figure 6-14. *The **MimeType** Java class is created under **com.asb.config***

Adding the MimeType Java Code

The Java code for the **MimeType.java** source is as follows.

Listing 6-14. The MimeType.java code file

```java
package com.asb.config;

public class MimeType {
    private String mimeString;
    private String extension;
    public void setMimeString(String mimeString) {
        this.mimeString = mimeString;
    }
    public String getMimeString() {
        return mimeString;
    }
    public void setExtension(String extension) {
        this.extension = extension;
    }
    public String getExtension() {
        return extension;
    }
}
```

This Java code has a **MimeType** public method, used as initial test code for getter and setter methods. (It is not currently used in the **Audit_Report** program.)

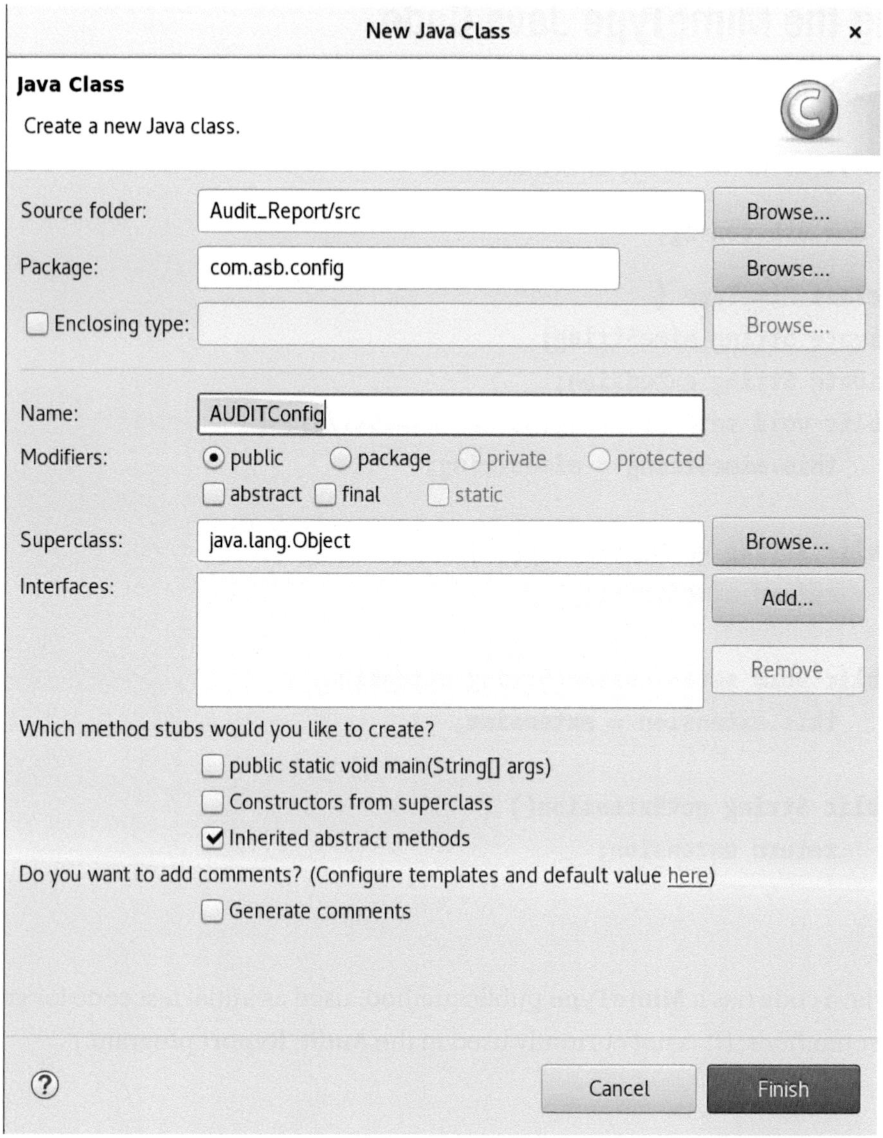

CHAPTER 6   PDF DOCUMENT CREATION USING JAVA

*Figure 6-15.   The **AUDITConfig** class is created under **com.asb.config***

Adding the AUDITConfig Java Code

The Java code for the **AUDITConfig.java** source is as follows.

Listing 6-15. The AuditConfig.java code file

```java
/**
IBM grants you a non-exclusive copyright license to use all programming
code examples from which you can generate similar function tailored to your
own specific needs.

All sample code is provided by IBM for illustrative purposes only.
These examples have not been thoroughly tested under all conditions. IBM,
therefore cannot guarantee or imply reliability, serviceability, or
function of these programs.

All Programs or code component contained herein are provided to you "AS IS
"without any warranties of any kind.
The implied warranties of non-infringement, merchantability and fitness for
a particular purpose are expressly disclaimed.

© Copyright IBM Corporation 2013, ALL RIGHTS RESERVED.
*/
/**
 *
 */
/**
 * @author Alan S. Bluck, ASB Software Development Limited
 *
 */
package com.asb.config;

import java.io.File;
import java.util.HashMap;
import java.util.LinkedHashMap;
import java.util.Vector;
```

```java
import javax.xml.xpath.XPathConstants;
import javax.xml.xpath.XPathExpressionException;
import javax.xml.xpath.XPathFactory;

import org.apache.log4j.Logger;
import org.w3c.dom.Node;
import org.w3c.dom.NodeList;

public class AUDITConfig extends XMLConfigReader {
    static Logger logger = Logger.getLogger(AUDITConfig.class.getName());

    private static AUDITConfig theInstance;
    private static HashMap mimeTypes;

    static {
        if (theInstance == null)
            try {
                theInstance = new AUDITConfig();
            } catch (Exception e) {
                // TODO Auto-generated catch block
                e.printStackTrace();
            }
    }
    public static String getStrVal(ConfigVal cv){
        return AUDITConfig.getConfigStringValue(cv.setting,
        cv.default_val);
    }
    public static int getIntVal(ConfigVal cv){
        return AUDITConfig.getConfigIntValue(cv.setting, cv.default_val);
    }
    public static int getConfigIntValue(String expression, String
    sdefault) {
        try {
            return Integer.parseInt(theInstance.get_
            ConfigValue(expression, sdefault));
        } catch (Exception e) {
            logger.warn(e.getMessage(), e);
```

```java
            return Integer.parseInt(sdefault);
    }
}
public static String getConfigStringValue(String expression, String sdefault) {
    try {
        return theInstance.get_ConfigValue(expression, sdefault);
    } catch (Exception e) {
        logger.warn(e.getMessage(), e);
        return sdefault;
    }
}
public AUDITConfig() throws Exception {
    XPathFactory  factory= XPathFactory.newInstance();
    String fullPath = System.getProperty("user.dir") + File.separator
    + "config" + File.separator + "config.xml";
    logger.debug("AUDITConfig reading " + fullPath);
    setConfigXml(fullPath, true);

    NodeList nlMimeTypes = getNodeList("//config/mimetypes/*");
    theInstance.mimeTypes = new HashMap();
    for (int x = 0; x < nlMimeTypes.getLength(); x++){
        MimeType mt = new MimeType();
        Node mtn = nlMimeTypes.item(x);

        String mimeString = getConfigValue(mtn,"mimestring");
        mt.setMimeString(mimeString);
        String ext = getConfigValue(mtn,"extension");
        if (!ext.startsWith("."))
            ext = "." + ext;
        mt.setExtension(ext);
        theInstance.mimeTypes.put(mimeString, mt);
    }
}
```

```java
    public static Node getConfigNode(String expression) throws Exception {
        return theInstance.getNode(expression);
    }
    public static String getNodeStringValue(String expression,Node n)
    throws XPathExpressionException{
        return theInstance.get_NodeStringValue(expression, n);
    }
    public static HashMap getMimeTypes() throws Exception {
        return theInstance.mimeTypes;
    }
    public static String[] getXMLVals(String expression, Node n) throws
    Exception {
        return theInstance.get_XMLVals(expression, n);
    }
}
```

The Java class, **AUDITConfig**, is an example of the object-oriented features of the Java language, as this class extends the abstract **XMLConfigReader** class (see the following Java code). This class is used to read the XML tag elements of the **config.xml** file we use for the Audit Report Java program.

> **Note** An abstract class, like **XMLConfigReader**, allows the method functionality to be created that the subclass can implement or override; however, the subclass (**AUDITConfig** in our implementation) can extend only one abstract class.

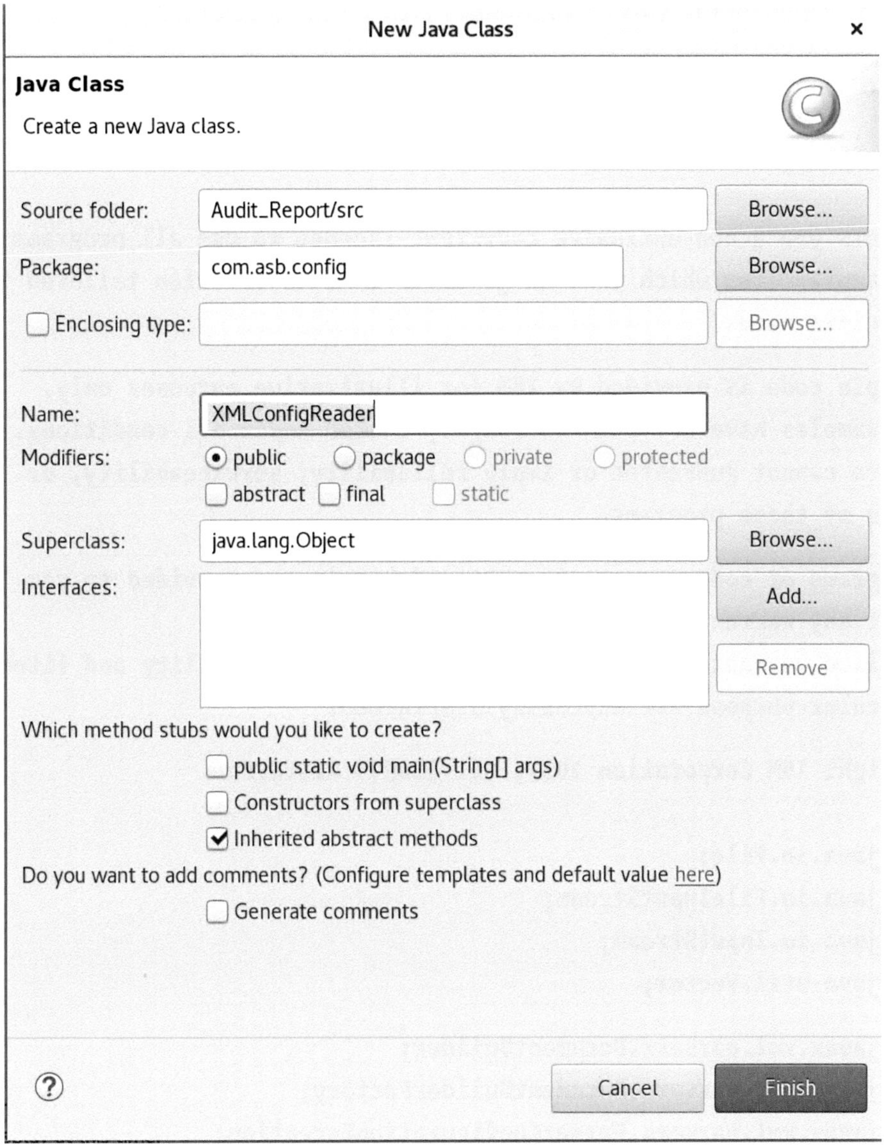

Figure 6-16. The abstract class, XMLConfigReader, is created

The code for this **Java** class is used by the **AUDITConfig** class defined in the following line of code:

```
public class AUDITConfig extends XMLConfigReader {
```

The Java code for the **XMLConfigReader.java** source is as follows.

Listing 6-16. The `XMLConfigReader.java` source code file

```
package com.asb.config;
/**
IBM grants you a non-exclusive copyright license to use all programming
code examples from which you can generate similar function tailored to your
own specific needs.

All sample code is provided by IBM for illustrative purposes only.
These examples have not been thoroughly tested under all conditions. IBM,
therefore cannot guarantee or imply reliability, serviceability, or
function of these programs.

All Programs or code component contained herein are provided to you "AS IS
"without any warranties of any kind.
The implied warranties of non-infringement, merchantability and fitness for
a particular purpose are expressly disclaimed.

© Copyright IBM Corporation 2013, ALL RIGHTS RESERVED.
*/
import java.io.File;
import java.io.FileInputStream;
import java.io.InputStream;
import java.util.Vector;

import javax.xml.parsers.DocumentBuilder;
import javax.xml.parsers.DocumentBuilderFactory;
import javax.xml.parsers.ParserConfigurationException;
import javax.xml.transform.Result;
import javax.xml.transform.Source;
import javax.xml.transform.Transformer;
import javax.xml.transform.TransformerFactory;
import javax.xml.transform.dom.DOMSource;
import javax.xml.transform.stream.StreamResult;
import javax.xml.xpath.XPath;
import javax.xml.xpath.XPathConstants;
```

```java
import javax.xml.xpath.XPathExpressionException;
import javax.xml.xpath.XPathFactory;

import org.apache.log4j.Logger;
import org.w3c.dom.Document;
import org.w3c.dom.Node;
import org.w3c.dom.NodeList;

public abstract class XMLConfigReader {

    private static Logger logger = Logger.getLogger(XMLConfigReader.class.getName());
    private XPath xPath;
    private String xmlConfig;
    private Document dom;
    private String fullPath;

    protected String get_NodeStringValue(String expression,Node n) throws XPathExpressionException{
        return (String)xPath.evaluate(expression, n, XPathConstants.STRING);
    }
    protected Node getNode(String expression) throws Exception {
        return (Node)xPath.evaluate(expression, dom, XPathConstants.NODE);
    }
    protected NodeList getNodeList(String expression) throws XPathExpressionException{
        return (NodeList) xPath.evaluate(expression, dom, XPathConstants.NODESET);
    }
    protected void setConfigXml(String fileName, boolean fullPath){

        XPathFactory  factory= XPathFactory.newInstance();
        xPath=factory.newXPath();
        xPath.reset();
```

```java
            if (fullPath)
                this.fullPath = fileName;
        xmlConfig = fileName;
        try {
            getDocument(fullPath);
        } catch (Exception e) {
            // TODO Auto-generated catch block
            e.printStackTrace();
        }
    }
    public String getConfigValue(String expression) throws Exception {
        return get_ConfigValue(expression,null);
    }
    public String getConfigValue(Node n,String expression) throws Exception {
        return getConfigValue(n,expression,null);
    }
    public void setConfigValue(String expression,String val){
        expression = expression.replace(".", "/");
        Node n1 = null;
        try {
            n1 = (Node) xPath.evaluate(expression, dom,
                XPathConstants.NODE);
        } catch (XPathExpressionException e) {
            // TODO Auto-generated catch block
            e.printStackTrace();
        }
        if (n1 != null)
            n1.setTextContent(val);
    }
    public String get_ConfigValue(String expression,String sdefault)
    throws Exception {
        expression = expression.replace(".", "/");
        expression = "//" + expression + "/text()";
        Node n1 = (Node) xPath.evaluate(expression, dom,
            XPathConstants.NODE);
```

```java
        if (n1 == null)
            return sdefault;
        return n1.getNodeValue();
    }
    public String getConfigValue(Node n, String expression, String sdefault) throws Exception {
        expression = expression.replace(".", "/");
        expression = expression + "/text()";
        Node n1 = (Node) xPath.evaluate(expression, n,
        XPathConstants.NODE);
        if (n1 == null)
            return sdefault;
        return n1.getNodeValue();
    }
    private void getDocument(boolean fullPath) throws Exception {
        DocumentBuilderFactory dbf = DocumentBuilderFactory.newInstance();
        DocumentBuilder db = null;
        try {
            db = dbf.newDocumentBuilder();
        } catch (ParserConfigurationException e) {
            logger.error(e);
            throw e;
        }
        InputStream in;
        if (fullPath){
            in = new FileInputStream(xmlConfig);
        }
        else {
            in = XMLConfigReader.class.getClassLoader().getResourceAsStream(xmlConfig);

        }

        dom = db.parse(in);

    }
```

```java
protected String getXMLStringVal(String expression) throws Exception {
    Node n = (Node) xPath.evaluate(expression, dom,
    XPathConstants.NODE);
    if (n != null)
        return n.getNodeValue();
    return "";
}
protected void setXMLStringVal(String expression, String val) throws
Exception {
    Node n = (Node) xPath.evaluate(expression, dom,
    XPathConstants.NODE);
    if (n != null)
        n.setNodeValue(val);

}

protected int getXMLIntVal(String expression) throws Exception {
    Node n = (Node) xPath.evaluate(expression, dom,
    XPathConstants.NODE);
    if (n != null){
        String test = n.getNodeValue();
        return Integer.parseInt(n.getNodeValue());
    }
    return -1;
}
public String[] get_XMLVals(String expression, Node n) throws
Exception {
    Vector vtmp = null;
    NodeList nl = null;
    vtmp = new Vector();
    nl = (NodeList) xPath.evaluate(expression, n,
    XPathConstants.NODESET);
    for (int x = 0; x < nl.getLength(); x ++){
        vtmp.add(nl.item(x).getNodeValue());
    }
```

```java
            return (String[])vtmp.toArray(new String[vtmp.size()]);
    }
    public String[] get_XMLVals(String expression) throws Exception {
        Vector vtmp = null;
        NodeList nl = null;
        vtmp = new Vector();
        nl = (NodeList) xPath.evaluate(expression, dom,
        XPathConstants.NODESET);
        for (int x = 0; x < nl.getLength(); x ++){
            vtmp.add(nl.item(x).getNodeValue());
        }
        return (String[])vtmp.toArray(new String[vtmp.size()]);
    }
    public void save() throws Exception {
        if (fullPath != null){
            logger.debug("XMLConfigReader saving " + fullPath);
          Source source = new DOMSource(dom);

            // Prepare the output file
            File file = new File(fullPath);
            Result result = new StreamResult(file);

            // Write the DOM document to the file
            Transformer xformer = TransformerFactory.newInstance().
            newTransformer();
            xformer.transform(source, result);
        }
    }
}
```

CHAPTER 6 PDF DOCUMENT CREATION USING JAVA

Adding the com.ibm.filenet.ps.ciops Package

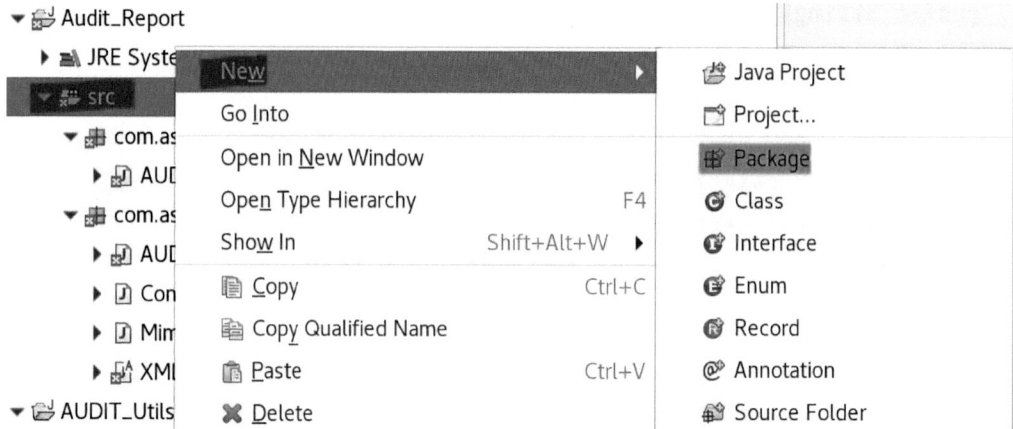

*Figure 6-17. A new package option for **com.ibm.filenet.ps.ciops** is selected*

This is a base code package for a FileNet Component Integration Java code.

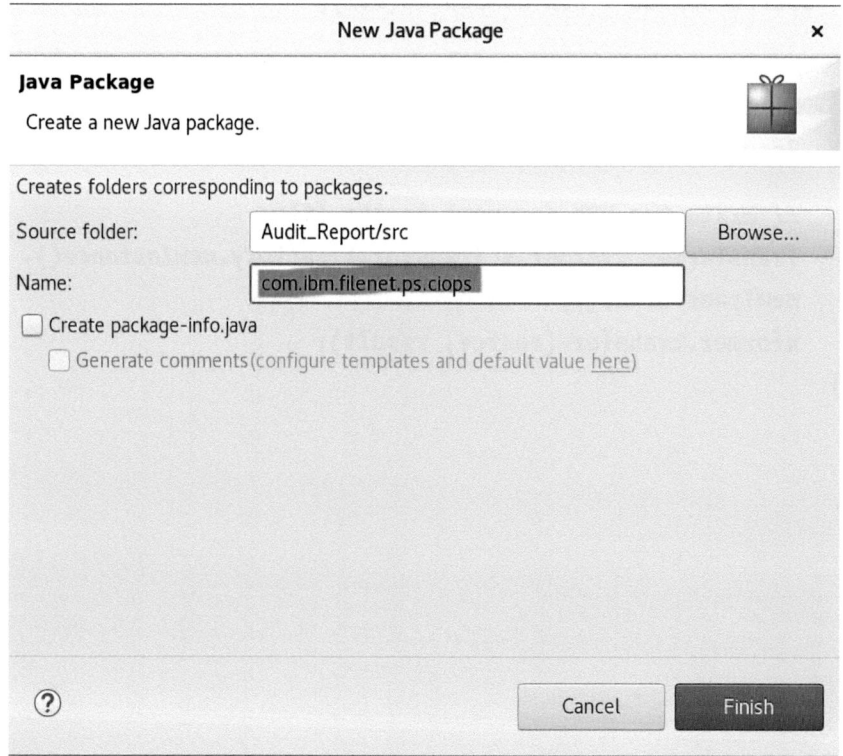

*Figure 6-18. The package name **com.ibm.filenet.ps.ciops** is entered*

CHAPTER 6   PDF DOCUMENT CREATION USING JAVA

Adding the ContentElementDataSource Java Class

The **ContentElementDataSource** Java class is created under the new **com.ibm.filenet.ps.ciops** Java package as follows.

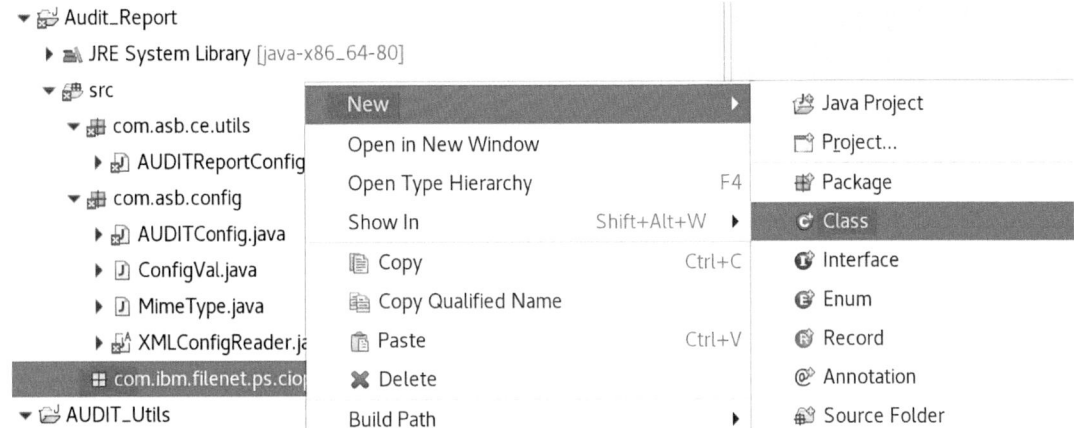

Figure 6-19. *The New **ContentElementDataSource** class is created*

1059

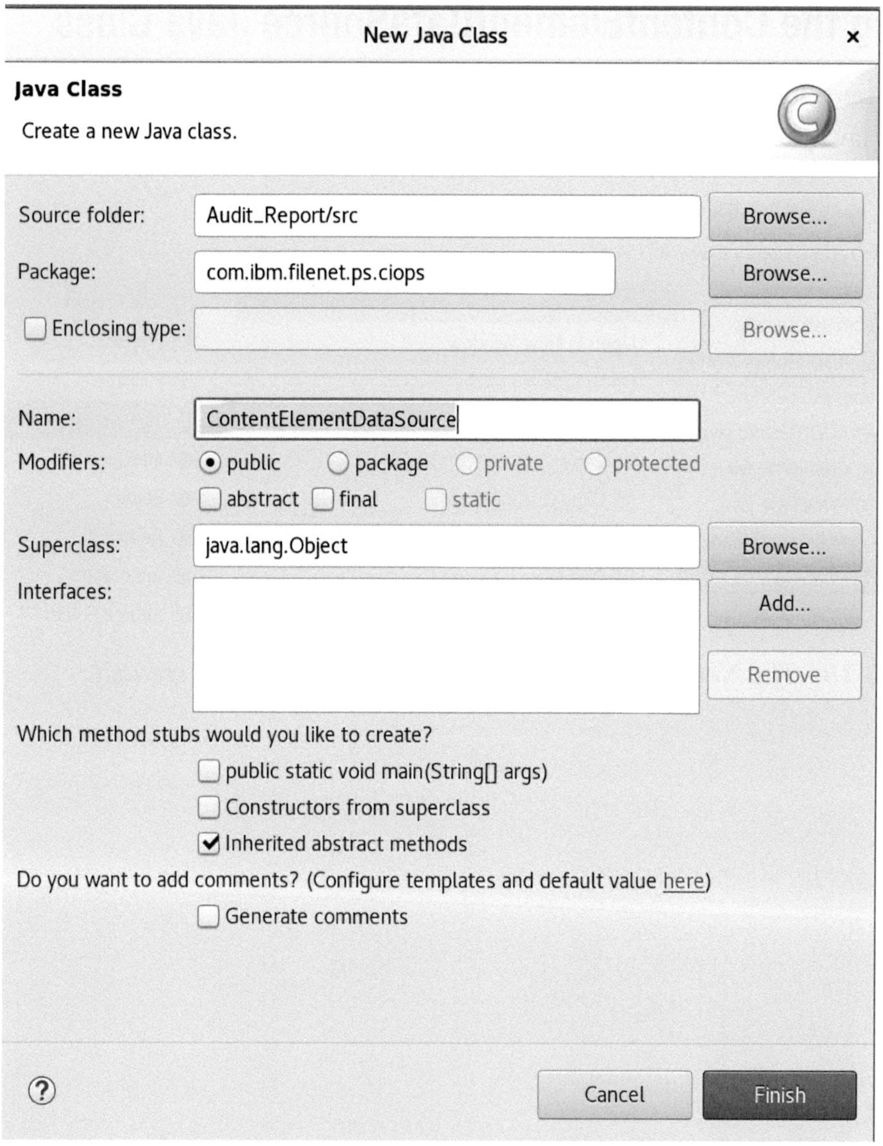

CHAPTER 6   PDF DOCUMENT CREATION USING JAVA

*Figure 6-20. The **ContentElementDataSource** Class name is entered*

The Java code for the **ContentElementDataSource.java** source is as follows.

Listing 6-17. The **ContentElementDataSource.java** source code file

```java
package com.ibm.filenet.ps.ciops;
/**
IBM grants you a non-exclusive copyright license to use all programming
code examples from which you can generate similar function tailored to your
own specific needs.

All sample code is provided by IBM for illustrative purposes only.
These examples have not been thoroughly tested under all conditions.  IBM,
therefore cannot guarantee or imply reliability, serviceability, or
function of these programs.

All Programs or code component contained herein are provided to you "AS IS"
without any warranties of any kind.
The implied warranties of non-infringement, merchantability and fitness for
a particular purpose are expressly disclaimed.

© Copyright IBM Corporation 2013, ALL RIGHTS RESERVED.
*/

import java.io.IOException;
import java.io.InputStream;
import java.io.OutputStream;

import javax.activation.DataSource;

import com.filenet.api.core.ContentTransfer;

/**
 * This class wraps a ContentTransfer object in a DataSource object. This way
 * the content can be used as a part of an e-mail message.
 *
 * @author Ricardo Belfor
 *
 */
```

```java
public class ContentElementDataSource implements DataSource {

    private ContentTransfer contentTransfer;

    public ContentElementDataSource(ContentTransfer contentTransfer) {
        this.contentTransfer = contentTransfer;
    }

    public String getContentType() {
        return contentTransfer.get_ContentType();
    }

    public InputStream getInputStream() throws IOException {
        return contentTransfer.accessContentStream();
    }

    public String getName() {
        return contentTransfer.get_RetrievalName();
    }

    public OutputStream getOutputStream() throws IOException {
        return null;
    }
}
```

The preceding code supports the email sendmessage Java code which is used as an example IBM FileNet Component Integration Workflow step call.

Adding the QOperations Java Class

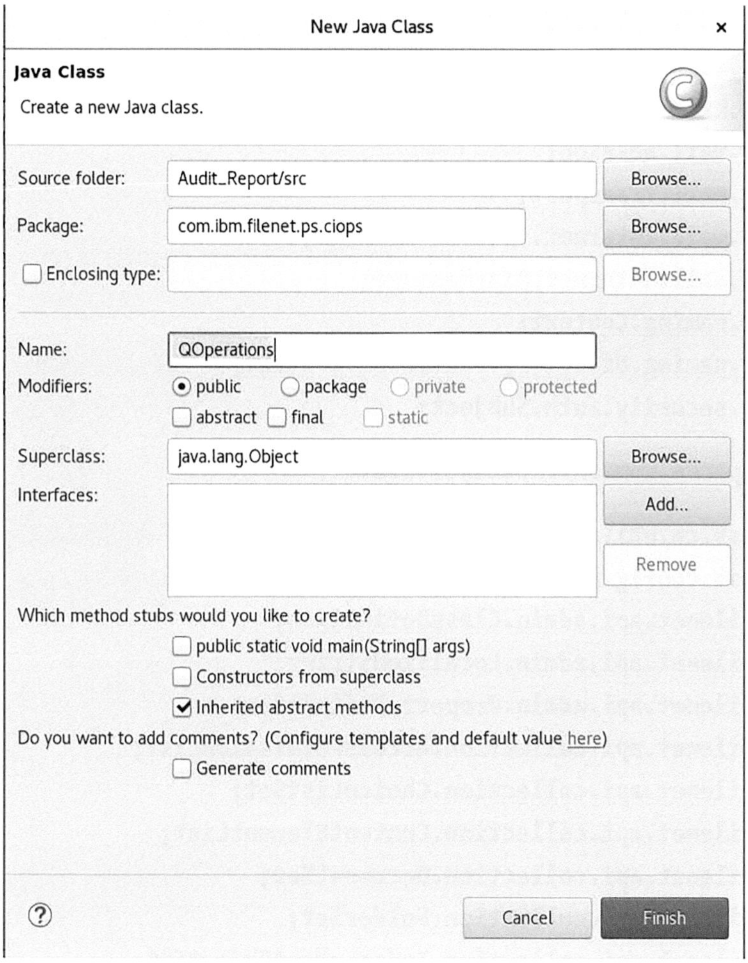

*Figure 6-21. The **QOperations** class code is added to the package*

The Java code for the **QOperations.java** source is as follows.

Listing 6-18. The **QOperations.java** source code file

```
package com.ibm.filenet.ps.ciops;

import java.security.AccessController;
import java.util.ArrayList;
import java.util.Iterator;
```

```java
import java.util.Properties;
import java.util.Set;

import javax.mail.Message;
import javax.mail.MessagingException;
import javax.mail.PasswordAuthentication;
import javax.mail.Session;
import javax.mail.Transport;
import javax.mail.internet.InternetAddress;
import javax.mail.internet.MimeMessage;
import javax.naming.Context;
import javax.naming.directory.InitialDirContext;
import javax.security.auth.Subject;

import org.apache.commons.io.LineIterator;

import com.asb.ce.utils.AUDITReportConfig;
import com.asb.config.AUDITConfig;
import com.filenet.api.admin.ClassDefinition;
import com.filenet.api.admin.LocalizedString;
import com.filenet.api.admin.PropertyDefinition;
import com.filenet.api.collection.AccessPermissionList;
import com.filenet.api.collection.ChoiceListSet;
import com.filenet.api.collection.ContentElementList;
import com.filenet.api.collection.DocumentSet;
import com.filenet.api.collection.FolderSet;
import com.filenet.api.collection.IndependentObjectSet;
import com.filenet.api.collection.PropertyDefinitionList;
import com.filenet.api.collection.StringList;
import com.filenet.api.collection.VersionableSet;
import com.filenet.api.constants.AutoClassify;
import com.filenet.api.constants.AutoUniqueName;
import com.filenet.api.constants.CheckinType;
import com.filenet.api.constants.ChoiceType;
import com.filenet.api.constants.DefineSecurityParentage;
import com.filenet.api.constants.FilteredPropertyType;
import com.filenet.api.constants.PropertyNames;
```

```java
import com.filenet.api.constants.RefreshMode;
import com.filenet.api.constants.ReservationType;
import com.filenet.api.constants.TypeID;
import com.filenet.api.constants.VersionStatus;
import com.filenet.api.core.Connection;
import com.filenet.api.core.Document;
import com.filenet.api.core.Domain;
import com.filenet.api.core.EntireNetwork;
import com.filenet.api.core.Factory;
import com.filenet.api.core.Factory.Choice;
import com.filenet.api.core.Factory.ChoiceList;
import com.filenet.api.core.Factory.ContentTransfer;
import com.filenet.api.core.Factory.ReferentialContainmentRelationship;
import com.filenet.api.core.Factory.VersionSeries;
import com.filenet.api.core.Folder;
import com.filenet.api.core.CmTask; //ASB added for Task property retrieval
import com.filenet.api.core.IndependentObject;
import com.filenet.api.core.ObjectStore;
import com.filenet.api.core.UpdatingBatch;
import com.filenet.api.exception.EngineRuntimeException;
import com.filenet.api.exception.ExceptionCode;
import com.filenet.api.property.FilterElement;
import com.filenet.api.property.PropertyFilter;
import com.filenet.api.query.SearchSQL;
import com.filenet.api.query.SearchScope;
import com.filenet.api.util.Id;
import com.ibm.filenet.ps.ciops.database.DatabasePrincipal;

import filenet.vw.api.VWAttachment;
import filenet.vw.api.VWAttachmentType;
import filenet.vw.api.VWException;
import filenet.vw.api.VWLibraryType;
import filenet.vw.api.VWSession;
import filenet.vw.base.logging.Logger;

import java.io.File;
```

//Q PDF Creation Operations and supporting methods 29th July 2022
import java.io.FileOutputStream;
import java.util.Date;
import java.util.HashMap;

import com.lowagie.text.Anchor;
import com.lowagie.text.BadElementException;
//import com.lowagie.text.BaseColor;
import com.lowagie.text.Chapter;
import com.lowagie.text.DocumentException;
import com.lowagie.text.Element;
import com.lowagie.text.Font;
import com.lowagie.text.Image;
import com.lowagie.text.List;
import com.lowagie.text.ListItem;
import com.lowagie.text.PageSize;
import com.lowagie.text.Paragraph;
import com.lowagie.text.Phrase;
import com.lowagie.text.Rectangle;
import com.lowagie.text.Section;
import com.lowagie.text.pdf.PdfContentByte;
import com.lowagie.text.pdf.PdfPCell;
import com.lowagie.text.pdf.PdfPTable;
import com.lowagie.text.pdf.PdfWriter;
import com.lowagie.text.pdf.RandomAccessFileOrArray;
import com.lowagie.text.pdf.codec.TiffImage;
//
import java.io.BufferedWriter;
import java.io.FileInputStream;
import java.io.FileNotFoundException;
import java.io.FileWriter;
import java.io.IOException;
import java.io.InputStream;
import java.sql.DriverManager;
import java.sql.ResultSet;
import java.sql.SQLException;

```java
import java.sql.Statement;
import java.sql.Timestamp;
import java.text.SimpleDateFormat;
import java.util.Vector;

import org.apache.commons.io.FileUtils;
//import org.apache.commons.lang.math.RandomUtils; //ASB removed this
                                                    import 26-08-2022
//
/**
IBM grants you a non-exclusive copyright license to use all programming
code examples from which you can generate similar function tailored to your
own specific needs.

All sample code is provided by IBM for illustrative purposes only.
These examples have not been thoroughly tested under all conditions. IBM,
therefore cannot guarantee or imply reliability, serviceability, or
function of these programs.

All Programs or code component contained herein are provided to you "AS IS"
without any warranties of any kind.
The implied warranties of non-infringement, merchantability and fitness for
a particular purpose are expressly disclaimed.

© Copyright IBM Corporation 2013, ALL RIGHTS RESERVED.
*/
/**
 * Alan S.Bluck   29th July 2022
 * Code for QOperations
 * Adapted from:
 * Custom Java component serving as an example for the article
 * "Developing Java components for the Component Integrator".
 *
 * @author Ricardo Belfor
 *
 */
```

```java
public class QOperations {
    private static Logger logger = Logger.getLogger( QOperations.class );
    private static final Properties QQidMap = new Properties();
    private static final int CardinalityENUM = 1;
    private static final int CardinalityLIST = 2;
    private static final int CardinalitySINGLE = 0;
    //ASB ... Add static types for property type recognition
        private static final int BinaryType = 1;
        private static final int BoolType = 2;
        private static final int DateType = 3;
        private static final int DoubleType = 4;
        private static final int GUIDType =     5;
        private static final int LongType = 6;
        private static final int ObjectType = 7;
        private static final int StringType = 8;
        private static AUDITReportConfig cemc;
        private static Connection connection;
        private static int processedCount;
        private ObjectStore os;
      private int maxMin;
        private UpdatingBatch ub;
        private HashMap classNames;
        //PDF Column and Table Section properties
        public static String [] ColumnNameT4 = new String[3];
        //ASB Array of the Q Table 4 Columns
        public static String [] ColumnWidthT4 = new String[3];
        //ASB Array of the Q Table 4 Columns
        public static String [] TableParams = new String[10];
        //ASB Array of the Case task type parameters
      //AUDITOR Tab driving arrays
        public static String [] TableParams_AUDITOR = new String[10];
        //ASB Array of the Case task type parameters
        public static String [] TabSECTIONS_AUDITOR = new String[10];
        public static String [][] AUDITOR_props = new String[10][50];
```

```java
//Audit Tab driving arrays
    public static String [] TableParams_Audit = new String[10]; //ASB
    Array of the Case task type parameters
    public static String [] TabSECTIONS_Audit = new String[10];
    public static String [][] Audit_props = new String[10][50];
//Department Tab driving arrays
    public static String [] TableParams_Department = new String[10];
    //ASB Array of the Case task type parameters
    public static String [] TabSECTIONS_Department = new String[10];
    public static String [][] Department_props = new String[10][50];
    //PDF Fonts required
    private static Font catFont = new Font(Font.TIMES_ROMAN, 18,
            Font.BOLD);
    private static Font redFont = new Font(Font.TIMES_ROMAN, 12,
            Font.NORMAL);
    private static Font subFont = new Font(Font.TIMES_ROMAN, 16,
            Font.BOLD);
        private static Font smallBold = new Font(Font.TIMES_
        ROMAN, 12,
            Font.BOLD);
    private static String sTab = "\u0009";

public QOperations() {
    try {
        cemc = new AUDITReportConfig();
    } catch (Exception e1) {
        // TODO Auto-generated catch block
        e1.printStackTrace();
    } //ASB add our config.xml file methods here!

    String fullPath = System.getProperty("user.dir") + File.separator
            + "config" + File.separator + "QQID.properties";
    try {
        QQidMap.load(new FileInputStream(new File(fullPath)));
        return;
```

```java
        } catch (FileNotFoundException e) {
            // TODO Auto-generated catch block
            e.printStackTrace();
        } catch (IOException e) {
            // TODO Auto-generated catch block
            e.printStackTrace();
        }
        throw new RuntimeException(fullPath + " cannot load QQid map");

        //java.sql.Connection databaseConnection = getDatabaseConnection();
        //if ( databaseConnection != null ) {
        //    logger.debug( databaseConnection.toString() );
        //} else {
        //    logger.debug( "No database connection" );
        //}
    }

    protected Connection getConnection() {
        String uri = System.getProperty("filenet.pe.bootstrap.ceuri");
        return Factory.Connection.getConnection(uri);
    }

    protected VWSession getVWSession() throws VWException {
        String connectionPoint = System.getProperty("filenet.pe.cm.connectionPoint");
        return new VWSession(connectionPoint);
    }

    protected java.sql.Connection getDatabaseConnection() {
        Subject subject = Subject.getSubject( AccessController.getContext() );
        Set<DatabasePrincipal> principals = subject.getPrincipals( DatabasePrincipal.class );
        if ( principals != null && ! principals.isEmpty() ) {
            DatabasePrincipal principal = principals.iterator().next();
            return principal.getConnection();
```

```java
        }
        return null;
}

private Folder getFolderFromAttachment(VWAttachment
folderAttachment) {
    ObjectStore objectStore = getObjectStore( folderAttachment.
    getLibraryName() );
    Folder folder = (Folder) objectStore.getObject("Folder",
    folderAttachment.getId() );
    return folder;
}

private Document getDocumentFromAttachment(VWAttachment
documentAttachment) {
    ObjectStore objectStore = getObjectStore( documentAttachment.
    getLibraryName() );
    Document folder = (Document) objectStore.getObject("Document",
    documentAttachment.getId() );
    return folder;
}

public ObjectStore getObjectStore( String objectStoreName ) {
    Connection connection = getConnection();
    EntireNetwork entireNetwork = Factory.EntireNetwork.fetchInstance
    (connection, null);
    Domain domain = entireNetwork.get_LocalDomain();
  String OSName = cemc.getExportOSName().trim();
    //os = Factory.ObjectStore.getInstance( domain, objectStoreName
    ); //Initial version
    os = Factory.ObjectStore.fetchInstance( domain, objectStoreName,
    null );    //ASB 8th August 2022
    return os;
}
```

```java
/**
 * Returns the documents filed in the folder.
 *
 * @param folderAttachment the input folder.
 * @return an array of documents filed in the folder.
 * @throws Exception
 */
public VWAttachment[] getFolderDocuments(VWAttachment folderAttachment
) throws Exception {

    Folder folder = getFolderFromAttachment(folderAttachment);
    DocumentSet containedDocuments = getContainedDocuments(folder);
    Iterator<?> iterator = containedDocuments.iterator();
    ArrayList<VWAttachment> containedDocumentList = new
    ArrayList<VWAttachment>();

    while ( iterator.hasNext() ) {
        com.filenet.api.core.Document document = (com.filenet.api.
        core.Document) iterator.next();
        VWAttachment documentAttachment =
        getAsVWAttachment(document);
        containedDocumentList.add( documentAttachment );
    }

    return containedDocumentList.toArray( new VWAttachment[0] );
}

private DocumentSet getContainedDocuments(Folder folder) {
    PropertyFilter propertyFilter =
    getContainedDocumentsPropertyFilter();
    folder.fetchProperties( propertyFilter );
    DocumentSet containedDocuments = folder.get_ContainedDocuments();
    return containedDocuments;
}
```

```java
private PropertyFilter getContainedDocumentsPropertyFilter() {
    PropertyFilter propertyFilter = new PropertyFilter();
    propertyFilter.addIncludeProperty( new FilterElement( null, null,
    null, PropertyNames.CONTAINED_DOCUMENTS, null ) );
    propertyFilter.addIncludeProperty( new FilterElement( 2, null,
    null, PropertyNames.ID, null ) );
    propertyFilter.addIncludeProperty( new FilterElement( 2, null,
    null, "DocumentTitle", null ) );
    return propertyFilter;
}

private VWAttachment getAsVWAttachment(Document document) throws
VWException {

    VWAttachment documentAttachment = new VWAttachment();

    documentAttachment.setLibraryType( VWLibraryType.LIBRARY_TYPE_
    CONTENT_ENGINE );
    ObjectStore objectStore = document.getObjectStore();
    objectStore.fetchProperties( new String[] {
    PropertyNames.NAME } );
    documentAttachment.setLibraryName( objectStore.get_Name() );

    document.fetchProperties( new String[] { PropertyNames.ID,
    PropertyNames.NAME } );
    documentAttachment.setId( document.get_Id().toString() );
    documentAttachment.setAttachmentName( document.get_Name() );
    documentAttachment.setType( VWAttachmentType.ATTACHMENT_TYPE_
    FOLDER );

    return documentAttachment;
}
/**
 * Sends an e-mail message using the mail session of the
 process engine
 * session. It uses a document stored in the Content Engine as a
 template.
```

```
* The first content element is considered the main part of the
template,
* containing place holders for template values. Template value place
* holders are prefixed in the template with a $-sign.
* <p>
* It may also contain references to images contained in the
other content
* elements of the document. The references have the following form:
* <code>cid:contentX</code>, where <code>X</code> stands for the
index of
* the content element. So a reference to the first content element
after the
* main template will look like <code>cid:content1</code>. This
reference
* can be used in an HTML image tag as follows:<br>
* <code>&lt;img src='cid:content1'&gt;</code><br>
* The samples folder contains a example template with a
background image
* and an embedded logo.
*
* @param templateAttachment the template document
* @param from
*           from address.
* @param to
*           to address.
* @param subject
*           the subject of the message.
* @param templateNames
*           the name of the template variables.
* @param templateValues
*           the values of the template variables.
*
* @throws Exception
*/
```

```java
public void sendMailMessage(VWAttachment templateAttachment, String
from, String to, String subject, String[] templateNames, String
templateValues[] ) throws Exception {
    Document template = getDocumentFromAttachment(templateAt
    tachment);
    EmailTemplate emailTemplate = new EmailTemplate(template);
    Session session = getVWSession().createMailSession();
    MimeMessage message = emailTemplate.getMessage(session, new
    InternetAddress(from), InternetAddress.parse( to ), subject,
    templateNames, templateValues );
    Transport.send(message);
}
public void writePDFToCaseFolder(String CaseGUID, String TaskGUID,
String TaskName, String pdfFileName, String WorkingDirectory, String
sQuestions,String sQuestionColumns, String StartQNo, String EndQNo) {
    //ASB takes the generated pdf File and imports it to a given
    Case Folder
        try {
            // Get Contained Documents in the Case Folder
            Folder caseFolder =fetchFolderById( os, CaseGUID);

            DocumentSet containedDocuments = getContainedDocuments(ca
            seFolder);
            Iterator<?> iterator = containedDocuments.iterator();
            String [] docTitles = new String[100]; //TODO set this as a
            config.xml parameter
            com.filenet.api.core.Document [] docList = new com.filenet.
            api.core.Document[100];
            int nDocs = 0;
            while ( iterator.hasNext() ) {
                com.filenet.api.core.Document document = (com.filenet.
                api.core.Document) iterator.next();
                docTitles[nDocs] = document.get_Name();
                docList[nDocs] = document;
                nDocs ++;
            }
```

```
            importDocuments(CaseGUID,TaskGUID,TaskName,pdfFileName,
            WorkingDirectory,docTitles,docList,nDocs);
    } catch (Exception e) {
        // TODO Auto-generated catch block
        e.printStackTrace();
    }
}
public String generatePDF(String CaseGUID, String TaskGUID, String
TaskName, String pdfFileName, String WorkingDirectory, String
sQuestions,String sQuestionColumns, String StartQNo, String EndQNo)
throws IOException {

    // Call the pdf Builder method
    int nQuestions = Integer.parseInt(sQuestions);
    int questionColumns = Integer.parseInt(sQuestionColumns);
    int nStartQNO = Integer.parseInt(StartQNo);
    int nEndQNO = Integer.parseInt(EndQNo);
    String FileCreated = createPDFFile(CaseGUID, TaskGUID, TaskName,
    pdfFileName, WorkingDirectory, nQuestions+1,questionColumns,
    nStartQNO,nEndQNO);
    return FileCreated;
}
private String  createPDFFile(String CaseGUID, String TaskGUID,
String TaskName, String pdfFileName, String WorkingDirectory,int
nQuestions,int nQuestionColumns, int nStartQNO,int nEndQNO) throws
IOException{
    //TODO Detect Linux or Windows environment and set paths
    accordingly
        String MainAUDITORTask =cemc.getAUDITORMainTask();
        File fTextFile = new File( pdfFileName);
        LineIterator lineIterator = null;
        String contentDirectoryName = WorkingDirectory;
        //File.separator
        String directoryName = contentDirectoryName +  File.separator +
        TaskName +  File.separator;
        String     txtFile = directoryName +  pdfFileName;
```

CHAPTER 6 PDF DOCUMENT CREATION USING JAVA

```java
    fTextFile = new File(txtFile);
    //Create pdf File from text fileName
    String  pdfFile = fTextFile.getName().substring(0, fTextFile.
    getName().lastIndexOf('.') + 1) + "pdf";

//Add the TaskID to give a more unique file name
    pdfFile = directoryName + TaskGUID + pdfFile;

    // TODO Auto-generated method stub
    //Document document = new Document();
    com.lowagie.text.Document document = new com.lowagie.text.
    Document(PageSize.A3.rotate(), 50, 50, 100, 100);
    String status = "pdf_create_started";
    String sTab = "\u0009";
    String [][] AUDITOR_propValues;
    String [][] AUDITOR_propNames;
    String [][] Department_propValues;
    String [][] Department_propNames;
    String [][] Audit_propValues;
    String [][] Audit_propNames;

    try {
        //Set the PDF parameters required
        PdfWriter writer = PdfWriter.getInstance(document, new
        FileOutputStream(pdfFile));
        writer.setPdfVersion(PdfWriter.VERSION_1_3);
        //PdfWriter.getInstance(document, new
        FileOutputStream(pdfFile));
         document.open();
        //document.add(new Chunk(""));
         //TODO Get values from config.xml
        document.setMargins(50, 50, 100, 100);
        document.addTitle(pdfFileName);
        if(TaskName.equalsIgnoreCase(MainAUDITORTask)) {
        //ASB 22-08-2022 Read from config.xml
            document.addSubject("AUDITOR Document");
```

```java
}else {
    document.addSubject("Q Document");
}
document.addKeywords("AUDITOR, Q, Audit, Department");
document.addAuthor("ASB Software Development Limited");
document.addCreator("createPDFFile");
//Retrieve Tabs 1 to 5 Q Case Fields (Audit)
if(TaskName.equalsIgnoreCase(MainAUDITORTask)) {
    addAUDITORTitlePage(document);
}else {
    addQTitlePage(document);
}
//============P D Q =============================
Integer paramCount = 0;
Integer fieldCount = 0; //set for loading Fields
String fieldVal="";
int nProps = 0;
Integer [] noAUDITORprops = new Integer[10];
//ASB  Retrieve the Tab Name parameters from the config.xml
for (String field:TabSECTIONS_AUDITOR){
    //ASB Returns NotUsed string for any elements not in
    the config.xml file
    fieldVal = AUDITConfig.getConfigStringValue("config.
    TabSECTIONS_AUDITOR.T" + String.valueOf(fieldCount),
    "NotUsed");
     if (!fieldVal.equalsIgnoreCase("NotUsed")){
        TableParams_AUDITOR = fieldVal.split(",");
        TabSECTIONS_AUDITOR[paramCount] = TableParams_
        AUDITOR[0];
        nProps = Integer.parseInt(TableParams_
        AUDITOR[1]);
        noAUDITORprops[paramCount] = nProps;
        for (int i = 0; i < nProps; i++) {
            AUDITOR_props[paramCount][i] = TableParams_
            AUDITOR[i+2];
        }
```

```java
    fTextFile = new File(txtFile);
    //Create pdf File from text fileName
    String  pdfFile = fTextFile.getName().substring(0, fTextFile.
    getName().lastIndexOf('.') + 1) + "pdf";

//Add the TaskID to give a more unique file name
    pdfFile = directoryName + TaskGUID + pdfFile;

    // TODO Auto-generated method stub
    //Document document = new Document();
    com.lowagie.text.Document document = new com.lowagie.text.
    Document(PageSize.A3.rotate(), 50, 50, 100, 100);
    String status = "pdf_create_started";
    String sTab = "\u0009";
    String [][] AUDITOR_propValues;
    String [][] AUDITOR_propNames;
    String [][] Department_propValues;
    String [][] Department_propNames;
    String [][] Audit_propValues;
    String [][] Audit_propNames;

    try {
        //Set the PDF parameters required
        PdfWriter writer = PdfWriter.getInstance(document, new
        FileOutputStream(pdfFile));
        writer.setPdfVersion(PdfWriter.VERSION_1_3);
        //PdfWriter.getInstance(document, new
        FileOutputStream(pdfFile));
         document.open();
        //document.add(new Chunk(""));
         //TODO Get values from config.xml
        document.setMargins(50, 50, 100, 100);
        document.addTitle(pdfFileName);
        if(TaskName.equalsIgnoreCase(MainAUDITORTask)) {
        //ASB 22-08-2022 Read from config.xml
            document.addSubject("AUDITOR Document");
```

```java
}else {
    document.addSubject("Q Document");
}
document.addKeywords("AUDITOR, Q, Audit, Department");
document.addAuthor("ASB Software Development Limited");
document.addCreator("createPDFFile");
//Retrieve Tabs 1 to 5 Q Case Fields (Audit)
if(TaskName.equalsIgnoreCase(MainAUDITORTask)) {
    addAUDITORTitlePage(document);
}else {
    addQTitlePage(document);
}
//===========P D Q ============================
Integer paramCount = 0;
Integer fieldCount = 0; //set for loading Fields
String fieldVal="";
int nProps = 0;
Integer [] noAUDITORprops = new Integer[10];
//ASB  Retrieve the Tab Name parameters from the config.xml
for (String field:TabSECTIONS_AUDITOR){
    //ASB Returns NotUsed string for any elements not in
    the config.xml file
    fieldVal = AUDITConfig.getConfigStringValue("config.
    TabSECTIONS_AUDITOR.T" + String.valueOf(fieldCount),
    "NotUsed");
     if (!fieldVal.equalsIgnoreCase("NotUsed")){
         TableParams_AUDITOR = fieldVal.split(",");
         TabSECTIONS_AUDITOR[paramCount] = TableParams_
         AUDITOR[0];
         nProps = Integer.parseInt(TableParams_
         AUDITOR[1]);
         noAUDITORprops[paramCount] = nProps;
         for (int i = 0; i < nProps; i++) {
             AUDITOR_props[paramCount][i] = TableParams_
             AUDITOR[i+2];
         }
```

```java
            paramCount++;
      }
      fieldCount++;
}
Integer noAUDITORSections = paramCount;
fieldCount = 0;
//=============================================
paramCount = 0;
fieldCount = 0; //set for loading Fields
fieldVal="";
nProps = 0;
Integer [] noAuditprops = new Integer[10];
//ASB  Retrieve the Tab Name parameters from the config.xml
for (String field:TabSECTIONS_Audit){
      //ASB Returns NotUsed string for any elements not in
      the config.xml file
      fieldVal = AUDITConfig.getConfigStringValue("config.
      TabSECTIONS_Audit.T" + String.valueOf(fieldCount),
      "NotUsed");
       if (!fieldVal.equalsIgnoreCase("NotUsed")){
            TableParams_Audit = fieldVal.split(",");
            TabSECTIONS_Audit[paramCount] = TableParams_
            Audit[0];
            nProps = Integer.parseInt(TableParams_Audit[1]);
            noAuditprops[paramCount] = nProps;
            for (int i = 0; i < nProps; i++) {
                  Audit_props[paramCount][i] = TableParams_
                  Audit[i+2];
            }
            paramCount++;
      }
      fieldCount++;
}
Integer noAuditSections = paramCount;
fieldCount = 0;
```

```java
//Retrieve Tabs 1 to 5 Q Case Fields (Department)
paramCount = 0;
fieldCount = 0; //set for loading Fields
fieldVal="";
Integer [] noDepartmentprops = new Integer[10];
//ASB  Retrieve the Tab Name parameters from the config.xml
for (String field:TabSECTIONS_Department){
    //ASB Returns NotUsed string for any elements not in
    the config.xml file
    fieldVal = AUDITConfig.getConfigStringValue("config.
    TabSECTIONS_Department.T" + String.valueOf(fieldCount),
    "NotUsed");
     if (!fieldVal.equalsIgnoreCase("NotUsed")){
        TableParams_Department = fieldVal.split(",");
        TabSECTIONS_Department[paramCount] = TableParams_
        Department[0];
        nProps = Integer.parseInt(TableParams_
        Department[1]);
        noDepartmentprops[paramCount] = nProps;
        for (int i = 0; i < nProps; i++)  {
            Department_props[paramCount][i] =
            TableParams_Department[i+2];
        }
        paramCount++;
    }
    fieldCount++;
}
Integer noDepartmentSections = paramCount;
//ASB 15-08-2022 Create tab headers and properties for each
section in the pdf
fieldCount = 0;
//Retrieve Tab Q Table parameters and Case Fields
 paramCount = 0;
 fieldCount = 0; //set for loading Fields
 fieldVal="";
```

```java
float[] columnWidths = {5, 1, 3.2f}; //Defaults for 3 columns
//ASB  Retrieve the Column Name parameters from the config.xml
for (String field:ColumnNameT4){
    //ASB Returns NotUsed string for any elements not in the config.xml file
    fieldVal = AUDITConfig.getConfigStringValue("config.PDFTable4.C" + String.valueOf(fieldCount), "NotUsed");
     if (!fieldVal.equalsIgnoreCase("NotUsed")){
        TableParams = fieldVal.split(",");
        ColumnNameT4[paramCount] = TableParams[0];
        ColumnWidthT4[paramCount] = TableParams[1];
        columnWidths[paramCount] = Float.parseFloat(ColumnWidthT4[paramCount]);
        paramCount++;
     }
    fieldCount++;
}
fieldCount = 0;
PdfPTable table = new PdfPTable(columnWidths);
try
{   // Get data from the passed  Task ID Where CmAcmTaskName = AUD_ProduceFullQ
    // From CmAcmCaseTask
    SearchSQL sqlObject = new SearchSQL();
    sqlObject.setSelectList("f.*");
    Integer maxRecords = 10 ;//Should always be just one returned Object anyway!
    sqlObject.setFromClauseInitialValue(TaskName, "f", true); //set true to include subclasses
    sqlObject.setWhereClause("[Id] = " + TaskGUID );
    //ASB 11-08-2022
     ObjectStore objectStore = getObjectStore("OS2");
     //ASB TODO Change to load from properties
```

```java
SearchScope search = new SearchScope(objectStore);
PropertyFilter myFilter = new PropertyFilter();
int myFilterLevel = 2; //Changed from 1 to 2 -- For performance !
myFilter.setMaxRecursion(myFilterLevel);
myFilter.addIncludeType(new FilterElement(null, null, null, FilteredPropertyType.ANY, null));
// Set the (Boolean) value for the continuable parameter. This indicates
// whether to iterate requests for subsequent pages of result data when the end of the
// first page of results is reached. If null or false, only a single page of results is
// returned.
Boolean continuable = new Boolean(true); //SET To allow more results to be fetched
// Set the page size (Long) to use for a page of query result data. This value is passed
// in the pageSize parameter. If null, this defaults to the value of
// ServerCacheConfiguration.QueryPageDefaultSize.
Integer myPageSize = new Integer(1000);

// Execute the fetchObjects method using the specified parameters.
//ASB added set myFilter to null REF Best Practice API on FNRCA0024E: API_PROPERTY_NOT_IN_CACHE

IndependentObjectSet myObjects = search.fetchObjects(sqlObject, myPageSize, null, continuable);
//ASB myFilter  changed to null
//===============================================================S T A R T
// You can then iterate through the collection of rows to access the properties.
```

```java
        String [][] sTableCells = new String[nQuestions +1 ]
        [nQuestionColumns +1]; //Could be max of nQuestion rows
        in the table
        String [] publishYes = new String[nQuestions +1];
        String [] questionValue = new String[nQuestions +1];
        String [] questionDetails =  new String[nQuestions +1];
          int rowCount = 0;
          Iterator iter = myObjects.iterator();
        Paragraph sections = new Paragraph();
           while (iter.hasNext()){   //Should be just one
           Task ID  !
                try
                {
                    //Folder f = null;
                    IndependentObject object = (IndependentObject)
                    iter.next();
                    com.filenet.api.property.Properties props =
                    object.getProperties();
                    Iterator iterProps = props.iterator();
                  for (int index = 0; index < nQuestions  ;
                  index++)
                  {
                      //Initialise to empty
                      publishYes[index] = "No";
                      questionDetails[index] ="Not Used";
                      questionValue[index]  ="Not Used";
                      sTableCells[index][1] = questionValue[index];
                      sTableCells[index][2] = publishYes[index];
                      sTableCells[index][3] =
                      questionDetails[index];
                  }
//============================ S T A R T ================================
                // Call Routine to retrieve the Question arrays
                    CmTask currentTask = (CmTask) object;
                    String parentTaskClass = "CmTask";
                    //ASB 11-08-2022
```

```java
String taskClass = currentTask.getClassName();

//ASB 06-08-2022 Get the Current Case Folder Task Property definitions
 com.filenet.api.admin.ClassDefinition tClassDefs = Factory.ClassDefinition.fetch Instance(objectStore,taskClass, null);
//update myFilter to null!!! ASB 06/08/2022
PropertyDefinitionList tPropList = tClassDefs.get_PropertyDefinitions();
//ASB Get the Current Task Object properties
//AUDITOR================================
AUDITOR_propValues = new String[noAUDITORSections ][100];
//Could be max of property values in the section
AUDITOR_propNames = new String[noAUDITORSections ][100];
//Could be max of property names in the section
getTaskAUDITORPropsSections(currentTask, taskClass,tClassDefs,tPropList,props,iterProps, AUDITOR_props,TabSECTIONS_AUDITOR,AUDITOR_ propValues,AUDITOR_propNames,noAUDITORSections, noAUDITORprops);
//AUDITOR================================
Department_propValues = new String[noDepartmentSections ][20];
//Could be max of property values in the section
Department_propNames = new String[noDepartmentSections ][20];
//Could be max of property names in the section
getTaskDepartmentPropsSections(currentTask, taskClass,tClassDefs,tPropList,props, iterProps,Department_props,TabSECTIONS_ Department,Department_propValues,Department_ propNames,noDepartmentSections,noDepartm entprops);
```

```
            Audit_propValues = new String[noAuditSections]
            [20]; //Could be max of property values in
            the section
            Audit_propNames = new String[noAuditSections
            ][20]; //Could be max of property names in
            the section
            getTaskAuditPropsSections(currentTask, taskClass,
            tClassDefs,tPropList,props,iterProps,Audit_
            props,TabSECTIONS_Audit,Audit_propValues,Audit_
            propNames,noAuditSections,noAuditprops);
            // 4. Populate the Case level Task Folder
            properties.
            getTaskProps(currentTask, taskClass,tC
            lassDefs,tPropList,props,iterProps,sT
            ableCells, questionValue, publishYes,
            questionDetails,nQuestions +1,nQuestionColumns);
//===================================== E N D ==============================
            //Set Table Column widths (Relative)
            //float[] columnWidths = {1, 1, 1, 3, 4};
            //float[] columnWidths = {1, 1, 1, 2, 5};
            //AUDITOR START OUTPUT TO PDF
            for(int iSection=0 ; iSection <
            noAUDITORSections ; iSection++) {
                // We add one empty line
                addEmptyLine(sections, 1);
                // Lets write a big header
                //TODO Pick all text up from the config.
                xml file
                Paragraph paragraph = new
                Paragraph(TabSECTIONS_AUDITOR[iSection],
                catFont);
                paragraph.setAlignment(Element.ALIGN_
                CENTER);
                sections.add(paragraph);
                    addEmptyLine(sections, 1);
```

```java
                    for(int iProp=0 ; iProp <
                    noAUDITORprops[iSection] ; iProp++) {
                        paragraph = new Paragraph(AUDITOR_
                        propNames[iSection][iProp] + " : " +
                        AUDITOR_propValues[iSection][iProp],
                        catFont);
                        paragraph.setAlignment(Element.ALIGN_
                        CENTER);
                        sections.add(paragraph);
                            addEmptyLine(sections, 1);
                    }

                }
                document.add(sections);
                // Start a new page
                document.newPage();

                    //END


                //AUDITOR END OUTPUT TO PDF
                //SECTION 0 TO 3 TABS
                for(int iSection=0 ; iSection < 4 ;
                iSection++) {
                    // We add one empty line
                    addEmptyLine(sections, 1);
                    // Lets write a big header
                    //TODO Pick all text up from the config.
                    xml file
                    Paragraph paragraph = new
                    Paragraph(TabSECTIONS_Department[iSection],
                    catFont);
                    paragraph.setAlignment(Element.ALIGN_
                    CENTER);
                    sections.add(paragraph);
                        addEmptyLine(sections, 1);
```

```java
            for(int iProp=0 ; iProp <
            noDepartmentprops[iSection] ; iProp++) {
                paragraph = new Paragraph(Department_
                propNames[iSection][iProp] + " : " +
                Department_propValues[iSection][iProp],
                catFont);
                paragraph.setAlignment(Element.ALIGN_
                CENTER);
                sections.add(paragraph);
                    addEmptyLine(sections, 1);
            }

    }
    document.add(sections);
    sections = new Paragraph();
    for(int iSection=4 ; iSection < noAuditSections ;
    iSection++) {
        // We add one empty line
        addEmptyLine(sections, 1);
        // Lets write a big header
        //TODO Pick all text up from the config.
        xml file
        Paragraph paragraph = new
        Paragraph(TabSECTIONS_Department[iSection],
        catFont);
        paragraph.setAlignment(Element.ALIGN_
        CENTER);
        sections.add(paragraph);
            addEmptyLine(sections, 1);
         for(int iProp=0 ; iProp <
         noDepartmentprops[iSection] ; iProp++) {
                paragraph = new Paragraph(Department_
                propNames[iSection][iProp] + " : " +
                Department_propValues[iSection][iProp],
                catFont);
```

```java
                    paragraph.setAlignment(Element.ALIGN_
                    CENTER);
                    sections.add(paragraph);
                        addEmptyLine(sections, 1);
            }

        }
        document.add(sections);
        // Start a new page
        document.newPage();
        //SECTION 4 TAB
          table.setWidthPercentage(95);
          table.getDefaultCell().setUseAscender(true);

          PdfPCell c1 = new PdfPCell(new
          Phrase(ColumnNameT4[0]));
          c1.setHorizontalAlignment(Element.ALIGN_
          CENTER);
          table.addCell(c1);

          PdfPCell c2 = new PdfPCell(new
          Phrase(ColumnNameT4[1]));
          c2.setHorizontalAlignment(Element.ALIGN_
          CENTER);
          table.addCell(c2);

          PdfPCell c3 = new PdfPCell(new
          Phrase(ColumnNameT4[2]));
          c3.setHorizontalAlignment(Element.ALIGN_
          CENTER);
          table.addCell(c3);

          //PdfPCell c4 = new PdfPCell(new
          Phrase("Description"));
          //c4.setHorizontalAlignment(Element.
          ALIGN_LEFT);
          //table.addCell(c4);
```

```java
//PdfPCell c5 = new PdfPCell(new Phrase("Description"));
//c5.setHorizontalAlignment(Element.ALIGN_LEFT);
//table.addCell(c5);
table.setHeaderRows(1);
    //We have all the Questions loaded now
int iCell = 0;
int iRow = 0;
 String titleVal = null;
      for  (String qRow:publishYes) {
           //Limit to the passed question
           number range //ASB 09-08-2022 14:00
         if (qRow != null ) {
                String [] cellRow = new String[nQuestionColumns];
                cellRow[0] = questionValue[iRow+1];
                cellRow[1] = publishYes[iRow+1];
                cellRow[2] = questionDetails[iRow+1];
                if((iRow+1 >= nStartQNO ) && ( iRow+1 <= nEndQNO)) {
                   for (String sCell:cellRow){
                      titleVal =sCell;
                      table.addCell(sCell);
                      iCell ++;
                   }
                }
          }
              iCell = 0;
            iRow ++;
   }
      //START
```

```java
                        //Audit SECTION 0 TO 9 TABS
                        for(int iSection=0 ; iSection <
                        noAuditSections ; iSection++) {
                            // We add one empty line
                            addEmptyLine(sections, 1);
                            // Lets write a big header
                            //TODO Pick all text up from the
                        config.xml file
                            Paragraph paragraph = new
                            Paragraph(TabSECTIONS_Audit[iSection],
                            catFont);
                            paragraph.setAlignment(Element.ALIGN_
                            CENTER);
                            sections.add(paragraph);
                                addEmptyLine(sections, 1);
                            for(int iProp=0 ; iProp <
                            noAuditprops[iSection] ; iProp++) {
                                paragraph = new Paragraph(Audit_
                                propNames[iSection][iProp] + " :
                                " + Audit_propValues[iSection]
                                [iProp], catFont);
                                paragraph.
                                setAlignment(Element.ALIGN_
                                CENTER);
                                sections.add(paragraph);
                                    addEmptyLine(sections, 1);
                            }
                        }
                        document.add(sections);
                        // Start a new page
                        document.newPage();

                            //END
                }
            catch(Exception errprop){
                }
```

```java
            //                        lastGoodPropName = propName;
        }
                //Add the created table to the PDF Document
                document.add(table);
            document.close();
            //Now delete the Text File created earlier
        if(fTextFile.exists()){
                fTextFile.delete();
        }
      } catch (DocumentException e) {
            // TODO Auto-generated catch block
            e.printStackTrace();
            status = e.getMessage();
        }finally {
            status = "processed";
        }
        }catch (FileNotFoundException e) {
                // TODO Auto-generated catch block
                e.printStackTrace();
                status = e.getMessage();
        } catch (DocumentException  e1) {
            // TODO Auto-generaed catch block
            e1.printStackTrace();
        }
        return pdfFile;
 }
public String SendAlertEmailSSL(String to ,String from, String fromPassword,String host,String Subject, String Body,String SSLPort,String TLSPort )
{
    // Uses the SMTP Gateway to send an email notification to the
    passed user(s)
      // Get system properties
      Properties properties = System.getProperties();
      final String sFrom = from;
      final String sFromPassword = fromPassword;
```

```java
// Setup mail server
   properties.setProperty("mail.smtp.host", host);
   properties.setProperty("mail.smtp.port", SSLPort); //SSLPort
   (gmail is 465)
//May need TLS support
      properties.put("mail.smtp.starttls.enable","true");
//Need SSL support - mail.smtp.ssl.enable
      properties.put("mail.smtp.ssl.enable","true");
      properties.put("mail.smtp.auth", "true");  //Authentication
      is required (At the moment!!)
// Need the following for SSL
      properties.put("mail.smtp.socketFactory.port", SSLPort);
      //SSL Port
      properties.put("mail.smtp.socketFactory.class", "javax.
      net.ssl.SSLSocketFactory");
      properties.put("mail.smtp.socketFactory.fallback",
      "false");
   // Get the default Session object.
   Session session = Session.getDefaultInstance(properties,new
   javax.mail.Authenticator() {
      protected PasswordAuthentication
      getPasswordAuthentication() {
         return new PasswordAuthentication(sFrom, sFromPassword);
      }
   });

   try{

      // Create a default MimeMessage object.
      MimeMessage message = new MimeMessage(session);

      // Set From: header field of the header.
      message.setFrom(new InternetAddress(sFrom));

      // Set To: header field of the header.
      message.addRecipient(Message.RecipientType.TO,
                    new InternetAddress(to));
```

```java
            // Set Subject: header field
            message.setSubject(Subject);

            // Now set the actual message
            message.setText(Body);
            Transport t = session.getTransport("smtp");

            try {
                t.connect(host, sFrom, sFromPassword);
                t.sendMessage(message, message.getAllRecipients());
              } finally {
                t.close();
              }

            return "Sent message successfully....";
        }catch (MessagingException mex) {
            return mex.getMessage();
        }
   }
   private void  getTaskProps(CmTask object,String taskClass,com.
   filenet.api.admin.ClassDefinition tClassDefs, PropertyDefinitionList
   tPropList,com.filenet.api.property.Properties props, Iterator
   iterProps, String [][] sTableCells, String [] questionValue,String []
   publishYes,String [] questionDetails, int nQuestions,int
   nQuestionColumns)throws Exception {
//       private void getTaskProps(Folder currentFolder,Folder f,com.
   filenet.api.admin.ClassDefinition fClassDefs,Properties props
   ,Iterator iterProps) throws Exception {

   // called as :
    //                  getTaskProps(currentTask, taskClass,tClassDefs,
     tPropList,props,iterProps,questionValue,publishYes,questionDetails);
      //    String [][] sTableCells = new String[nQuestions +1 ]
     [nQuestionColumns+1]; //Could be max of nQuestion rows in the table
      //    String [] publishYes = new String[nQuestions+1];
      //    String [] questionValue = new String[nQuestions+1];
      //    String [] questionDetails =  new String[nQuestions+1];
```

```java
//ASB
//Retrieve the current property list to retrieve based on the
Task Class definitions from the Object Store
    String tClassName = object.getClassName(); //ASB 03-08-2022
    com.filenet.api.admin.ClassDefinition classDefs =
    tClassDefs;
    PropertyDefinitionList propList = tPropList;
 StringBuffer tempBuf = new StringBuffer();
    //Retrieve property names code From writeDocProps to compare
    Iterator iter =propList.iterator();
    while (iter.hasNext())
    {
       try {
            PropertyDefinition propDef = (PropertyDefinition)
            iter.next();
            String propName = propDef.get_SymbolicName();
            //ASB ... 2 011 Need to skip Date Last Accessed
            etc - ReadOnly error here
//          ||propName.equalsIgnoreCase("Permissions")
    //ASB removed 25/08/2022
                  if (!(propName.equalsIgnoreCase("Name")
                        ||propName.equalsIgnoreCase("DateC
                        heckedIn")
                      ||propName.equalsIgnoreCase("Curr
                      entState")
                      ||propName.equalsIgnoreCase("PathName")
                      ||propName.equalsIgnoreCase("LockOwner")
                      ||propName.equalsIgnoreCase("Owner")
                      //ASB 28-08-2022
                      ||propName.startsWith("CmAcm") )){
              //ASB ... Set to check property cardinality
              int CardinalityVal = propDef.get_Cardinality().
              getValue();
```

```java
switch (propDef.get_DataType().getValue()){
case StringType:
    if (CardinalityVal == CardinalitySINGLE){
        String val = object.getProperties().
        getStringValue(propName);
        for (int index = 1; index < nQuestions;
        index++)
        {
         // Value for Required Question
            String QValue = val ;
            if (propName.
            equalsIgnoreCase("AUD_Report4_
            QYN_" + index))
            {
                //Checks if this is set to a
                required response line
                    publishYes[index]
                    = QValue;
                    sTableCells[index][2] =
                    publishYes[index];
            }
        // Question Value for Required
        Question Details
            if (propName.
            equalsIgnoreCase("AUD_Report4_
            QDetails_" + index))
            {
                    questionDetails[index]
                    = QValue;
                    sTableCells[index][3]
                    = questionDetails
                    [index];
            }
        // Question Value for Required
        Question
```

```
                                if (propName.
                                equalsIgnoreCase("AUD_Report4_
                                AUD_" + index))
                                {
                                        questionValue[index]
                                        = QValue;
                                        sTableCells[index][1]
                                        = questionValue
                                        [index];
                                }
                        }
                        //ASB 11-08-2022
                        //f.getProperties().
                        putValue(propName, val);
                        if (propName.equalsIgnoreCase("lastmo
                        difier")){
                            //f.set_LastModifier(val);
                        }
                        if (!propName.
                        equalsIgnoreCase("creator") &&
                        !propName.equalsIgnoreCase("lastmo
                        difier")){
                            //f.getProperties().
                            putValue(propName, val);
                        }
                }else{
                    //StringList valList =
                    currentFolder.getProperties().
                    getStringListValue(propName);
                    //f.getProperties().putValue(propName,
                    valList);
                }
                    // Process Multi-value Strings (if
                    required)?
                    break;
```

```java
case DateType:
    if (CardinalityVal == CardinalitySINGLE){
        //Date dateVal = currentFolder.
        getProperties().
        getDateTimeValue(propName);
        //if (propName.equalsIgnoreCase("date
        lastmodified")){
        //     currentFolder.getProperties().
        removeFromCache("DateLastModified");
        //     f.set_
        DateLastModified(dateVal);
    }

        //ASB ... 4 029 Start
        if (propName.
        equalsIgnoreCase("datecreated")){
            //currentFolder.getProperties().
            removeFromCache("DateCreated");
            //f.set_DateCreated(dateVal); //
            ASB 19-08-2022 Bug fix on Folder
            Date from dateLast Modified!
            //if (lastFolderDateCreated.
            compareTo(dateVal) < 0) {
            //     lastFolderDateCreated =
            dateVal;
            //}
        }
        //ASB ... 4 029 End
    if (!propName.equalsIgnoreCase("datecreated")
    && !propName.equalsIgnoreCase("datelastmo
    dified")){
        //f.getProperties().putValue(propName,
        dateVal);
    }
  break;
```

CHAPTER 6 PDF DOCUMENT CREATION USING JAVA

```
                            default:
                          }
                      }
                        }catch (Exception writeProp){
                            logger.error(writeProp.getMessage(),
                            writeProp);
                        }

                } //for (int iProp = 0;iProp < xmlprops.
                getLength();iProp ++){
            }
//========S T A R T    C O D E    F O R    P D F   I M P O R T ========
//Main Process loop for the Import Documents process.
//
private void importDocuments(String CaseGUID,String TaskGUID,String
TaskName,String pdfFileName,String WorkingDirectory, String[]
DocTitles,com.filenet.api.core.Document [] docList, int nDocs) throws
Exception {
    BufferedWriter auditFileWriter;
    Long importStartTime;
    Long docImportStartTime;
    Long timeToImport;
        long maxRunTimeMillis;
//      boolean moreDocsToImport=true;
    Integer consecutiveErrs=0;
    Integer errorCount=0;

    boolean completed=false;

    this.processedCount = 0;
    String LineSave="";

    importStartTime = System.currentTimeMillis();
    maxRunTimeMillis = cemc.getImportMaxRunTimeMinutes()*60*1000;
    Long maxDocCount = new Long(cemc.getImportMaxDocCount());

    auditFileWriter = new BufferedWriter(new FileWriter(cemc.
    getImportAuditFile()));
```

```java
auditFileWriter.write("Import Start - ");
auditFileWriter.newLine();
auditFileWriter.write("GUID, SOURCEVSID, DESTDOCID, STATE, Millisecs");
auditFileWriter.newLine();
//ub = UpdatingBatch.createUpdatingBatchInstance(os.get_Domain(), RefreshMode.NO_REFRESH);
 File pdfFile =  new File(pdfFileName);
 //get all the files from a directory, based on the config File Filter (default *.pdf)
//File directory = new File(WorkingDirectory); //Set in config.xml
//     ArrayList<File> files  = new ArrayList<File>();
//     this.listf(WorkingDirectory, files,pdfFileName);
//     File[] fList = files.toArray(new File[]{});
 String  documentFile ="";
 String documentFileName ="";
 String documentFilePath = "";
 //for (int i = 0; i < fList.length; i++) {
     documentFile = pdfFile.toString();
     documentFileName = pdfFile.getName();
     for(int idoc=0;idoc < nDocs;idoc++){
       if(documentFileName.equalsIgnoreCase(DocTitles[idoc])) {
            return; //ASB 14/10/1066
            // Change to just ever allow one pdf production as
            the task is only converted after it is complete
            // ie is already immutable!!!
            //String mimeType = cemc.getMimeType();
            //updateNewDocument(docList[idoc],documentFileName,
            mimeType, documentFile);
              //========END DOCUMENT VERSION PROCESSING==========
            //return;
       }
     }
     documentFilePath = pdfFile.getPath();
```

```java
              //ASB ... 022 - Retrieve Date start from config.xml
              String sDeltaHours = cemc.getDeltaHours();
              String sSearchDate = cemc.getStartSearchDate();

              logger.info("Start Search Date : " + sSearchDate);
              logger.info("Start Delta Hours : " + sDeltaHours);
               //ASB ... 4 002 Now dynamically update the next Loader Date
               cemc.updateProcessDate(new Date());

              logger.info("File Imported: " + documentFile);
//                Integer myPageSize = new Integer(1000);

              // Specify a property filter to use for the filter parameter,
              if needed.
              // This can be null if you are not filtering properties.
              PropertyFilter myFilter = new PropertyFilter();
              int myFilterLevel = 1;
              myFilter.setMaxRecursion(myFilterLevel);
              //ASB ... 4 020
              if(cemc.getDebugOutputFlag().equalsIgnoreCase("off")){
                   myFilter.addIncludeType(new FilterElement(null, null,
                   null, FilteredPropertyType.ANY_LIST, null));
                   myFilter.addIncludeType(new FilterElement(null, null,
                   null, FilteredPropertyType.ANY_SINGLETON, null));
              }else{
                   myFilter.addIncludeType(new FilterElement(null, null,
                   null, FilteredPropertyType.ANY, null));
              }
                        //U P D A T E - Iterate through the Directory of
                        input Documents with pdf properties
                        Id   GUID = null;
                        docImportStartTime = System.currentTimeMillis();
                        Boolean importFailure = false;
                        Boolean filedDoc = false;
                        Document d = null;
                        String documentTitle = "";
```

```
//ASB ... 3 005 - Retrieve Folder list for checks
//                - (Only Load documents in
config Path)
String [] sDocFolderPath = new String [1000];
//TODO set max document Folders Filed In from
config.xml
int folderCount = -1;
String lastGoodPropName = "";
String propName = "";
String propValue = " ";
//Extract the  comma separated lines
//String docProps[] = line.split(",");
//String docPropNames[] = new String[docProps.
length];
//Integer docPropTypes[] = new Integer[docProps.
length];
//Integer docPropCardinality[] = new
Integer[docProps.length];
//String docPropFlags[] = new String[docProps.
length]; //ASB IBM - Added 2nd October

//for (int j = 0; j < docProps.length; j++) {
    //Strip out the double quotes (if present)
//      if(docProps[j] != null && docProps[j].
length() > 1 && docProps[j].substring(1,2).
equalsIgnoreCase("\"")){
//          docProps[j] = docProps[j].
substring(1,docProps[j].length() -1);
//      }
//      propName = SolutionPrefix + cemc.
getPropName(j);
//      if(docProps[j]==null || docProps[j].
length() == 0){
//          docProps[j]=" ";
//      }
```

CHAPTER 6 PDF DOCUMENT CREATION USING JAVA

```
//      propValue = docProps[j].replace("'","");//
Remove any single quote values!!
//      docPropNames[j]= propName;
//      docPropTypes[j]=cemc.getPropType(j);
//      docPropCardinality[j]=cemc.
getPropCardinality(j);
//      docPropFlags[j]=cemc.getPropFlags(j); //ASB
IBM - Added 2nd October
//ASB IBM - Added 2nd August - Check if this is a
placeholder or a Boolean list property
//      if (!docPropFlags[j].
equalsIgnoreCase("NONE")){
//          docPropFlags[j] = SolutionPrefix +
docPropFlags[j];
//      }
//      try{
//          if(cemc.getDebugOutputFlag().
equalsIgnoreCase("off")){
//          }else{
//              logger.info("Property: " +
                  propName );
//              logger.info("Value    : " +
                  propValue );
//          }
//          if (propName.
              equalsIgnoreCase(SolutionPrefix +
              "FileName"))
//          {
//              documentTitle = propValue;
//              documentFile = propValue;
//          }
//          sDocFolderPath[1] = cemc.get_
              DocPathName();
//      }catch(Exception errprop){
```

```java
//              logger.info("Document Title :" +
                documentTitle + " : Property  " +
                propName + " Caused Error : " +
                errprop.getMessage() + " : Last Good
                property :" + lastGoodPropName);
//          }
//          lastGoodPropName = propName;
//}
//E N D -- DEBUG OFF/ON SECTION
        documentTitle = documentFileName;
        if (documentTitle.equalsIgnoreCase("")) return; //
        Changed from break

        try {

            if (!importFailure )
            {
                docImportStartTime = System.currentTimeMillis();
                d = createDocument(documentTitle, auditFileWriter,GUID,documentFile);
                Long runTimeMillis =
                System.currentTimeMillis() -
                importStartTime;
                //Folder the document in the Case Folder
                //Get the Case Folder GUID ,String
                TaskGUID,String TaskName,
                  String sFolderId = CaseGUID;
                Folder caseFolder =fetchFolderById( os,
                sFolderId);
                //Folder caseFolder = getFolderFromID(folderGUID,objectStoreName,CMFolderClass);
                    if (caseFolder != null){
                        com.filenet.api.core.
                        ReferentialContainmentRelationship
                        rcr = caseFolder.file(d,
                        AutoUniqueName.NOT_AUTO_UNIQUE,
```

```java
                            null,DefineSecurityParentage.DO_NOT_
                            DEFINE_SECURITY_PARENTAGE);
                         rcr.save(RefreshMode.NO_REFRESH);
                        }
                        processedCount ++;
                        //importProcessedDocCount ++;
                        if (this.processedCount ==
                        maxDocCount  && maxDocCount>0)
                            {
                            completed = true;
                            logger.info("Processing Completed
                            Max Document Count Reached. ");
                            }
                        else if ((runTimeMillis >
                        maxRunTimeMillis)&& maxRunTimeMillis>0 )
                            {
                            completed = true;
                            logger.info("Processing Completed
                            Max Processing Time Reached.");
                            }
                        consecutiveErrs=0;   // success so
                        initialise the consecutive errs count
                    }
                }
                catch (Exception ceme){
                
                }
        //  }
//}
            //ASB 01/12/2020 Need to close Buffered Reader to allow
            file rename to work
            //br.close(); //ASB Close File I/O
        // {
         //Rename file
```

CHAPTER 6   PDF DOCUMENT CREATION USING JAVA

```java
            //fList[i].renameTo("doc_fact.cvs_" + System.
            currentTimeMillis());
                //File newName = new File(fList[i].getAbsolutePath() + "_"
                + System.currentTimeMillis() );

                //if(fList[i].renameTo(newName)) {
                //    System.out.println("Renamed " + "doc_fact.csv" + " to
                " + newName.getName() );
                //} else {
                //    System.out.println("Error");
                // }

        //}
        auditFileWriter.write("Finished - time taken " +
        (System.currentTimeMillis() - importStartTime) + "milliseconds");
        auditFileWriter.newLine();
        auditFileWriter.write("Finished - documents processed - " + Strin
        g.valueOf(processedCount));
        auditFileWriter.newLine();
        auditFileWriter.write("Finished - documents failed to be
        processed - " + errorCount.toString());
        auditFileWriter.newLine();
    //ASB ... 4 031 Add retry count for Content
        //auditFileWriter.write("Finished - document total 'Add Content
        Retry Count' - " + String.valueOf(retryTotalProcessedCount));
        auditFileWriter.newLine();
        auditFileWriter.close();
        //}
 }
//ASB 19th November 2020
public void listf(String directoryName, ArrayList<File> files,String
filenamePattern) {
    File directory = new File(directoryName);
     // create new filename

    // get all the files from a directory
    File[] fList = directory.listFiles();
```

CHAPTER 6   PDF DOCUMENT CREATION USING JAVA

```java
        for (File file : fList) {
            if (file.isFile()) {
                if(file.toString().endsWith(".pdf")){
                    files.add(file);
                }
            } else if (file.isDirectory()) {
                listf(file.getAbsolutePath(), files,filenamePattern);
            }
        }
    }
    //Creates a com.filenet.api.core.Document object
    private Document createDocument(String documentTitle, BufferedWriter
    auditFileWriter,Id GUID, String documentFile) throws  Exception,
    EngineRuntimeException {
        Document d = null;
         try
         {
            Boolean isReserved = false;
            String documentclass = cemc.getDocClassName(); //Read the
              required Class from cemc
          Long docImportStartTime;
          Long timeToImport;
          docImportStartTime = System.currentTimeMillis();
         //ASB ... 2 014 Get the Version Numbers of the Document
         Integer docMajorVersion = 1;
         Integer docMinorVersion = 0;
            PropertyFilter myFilter = new PropertyFilter();
            int myFilterLevel = 2; //Changed from 4 to 2 -- Set as a
            parameter
            myFilter.setMaxRecursion(myFilterLevel);
            myFilter.addIncludeType(new FilterElement(null, null, null,
            FilteredPropertyType.ANY, null));
                try {
                    docImportStartTime = System.currentTimeMillis();
```

```java
            d = Factory.Document.createInstance(os,
            documentclass, null);
            //Write to audit log
             timeToImport = System.currentTimeMillis() -
             docImportStartTime;
             //auditFileWriter.write(d.get_Id().toString() +
             "," + d.get_VersionSeries().get_Id().toString() +
             " ,Created," + timeToImport.toString());
             //auditFileWriter.newLine();
        }catch(Exception e)
           {
           //If the Exception is this exists then delete and retry
           add again
              logger.error("Error creating the Document: " +
              documentTitle + "");
           }
        d.getProperties().putValue("DocumentTitle",
        documentTitle);
          //writeDocProps(d, docPropNames, docPropFlags,
          docProps, docPropTypes, docPropCardinality,docume
          ntTitle );
               this.processedCount = this.processedCount +1;
               //Write to audit log
             timeToImport = System.currentTimeMillis() -
             docImportStartTime;
             //auditFileWriter.write("GUID :"  + d.get_Id().
             toString() + ",Properties Updated," + timeToImport.
             toString());
             //auditFileWriter.newLine();
          //ADD CONTENT
int retryCount = 0;
FileInputStream fis = null;
int retryLimit = 5; //TODO Maybe set this as a configuration
parameter
while (retryCount < retryLimit){
```

```java
try {
    com.filenet.api.core.ContentTransfer ct = Factory.ContentTransfer.createInstance();
    File file = new File(documentFile );
    fis = new FileInputStream(file);
    InputStream str = fis;
    ct.setCaptureSource(str);
    // Add Document Title
    ct.set_RetrievalName(documentTitle);
    // Add Content Mime Type
    ct.set_ContentType(cemc.getMimeType());
    ContentElementList cel = Factory.ContentElement.createList();
    cel.add(ct);
    d.set_ContentElements(cel);
    break;
}
catch (Exception e){
    String testMessage = e.getMessage();
    String testCode = e.toString();
    if( testMessage.contains("A uniqueness requirement has been violated")){
        retryCount = retryLimit;
        break;
    }
    //  ASB ... 4 025 Log Exception here
    logger.error(e.getMessage(), e);
    retryCount ++;
    //retryTotalProcessedCount ++;
    logger.info("Retry Count - createDocument method : " + retryCount + " of " + retryLimit + " On : " + documentTitle + " : with GUID : " + GUID);
}finally {
}
```

```java
} //ASB ... 4 031 Above loops Test retryCount less than the retryLimit of 7
    //ASB ... 4 028 Change to do not Classify from
    AutoClassify.AUTO_CLASSIFY, to
        d.checkin(AutoClassify.DO_NOT_AUTO_CLASSIFY,
        CheckinType.MAJOR_VERSION);
        d.save(RefreshMode.REFRESH);
         return d;
    }catch(EngineRuntimeException ere)
     {
        //ASB ... 2 005
        // Create failed.  See if it's because the
        Document exists.
        ExceptionCode code = ere.getExceptionCode();
        if (code.getErrorId() != ExceptionCode.DB_NOT_UNIQUE.
        getErrorId() )
        {
            logger.error("Unexpected Error : " + documentTitle
            + " Error stack: " + ere.getStackTrace());

            throw ere;
        }
        logger.error("Document already exists: " +
        documentTitle + "");
            //ASB ... 2 006 Delete document in the Target Object
            Store and Recreate
            //this.deleteDoc(GUID.toString());
            //ASB ... 2 007 Fetch the Source Object Document to
            be re-created
            PropertyFilter myFilter = new PropertyFilter();
            int myFilterLevel = 2; //Changed from 4 to 2 -- Set
            as a parameter
            myFilter.setMaxRecursion(myFilterLevel);
            myFilter.addIncludeType(new FilterElement(null,
            null, null, FilteredPropertyType.ANY, null));
             String documentclass = cemc.getDocClassName();
```

CHAPTER 6 PDF DOCUMENT CREATION USING JAVA

```java
Document dInput = Factory.Document.fetchInstance(os,
GUID, myFilter);
//Create Version Series for the document in the Target
Object Store
com.filenet.api.core.VersionSeries versions =
dInput.get_VersionSeries();
//Get the Document Versions to be added from the
Source Document
VersionableSet verSet = versions.get_Versions();
Iterator versIt = verSet.iterator();
int versCount = 0;
int majorVn = 0;
int minorVn = 0;
Document doc = null; //Input Document versions
Id createdGuid = null;
while (versIt.hasNext()){
    versCount ++;
    doc = (Document)versIt.next();
    majorVn = doc.get_MajorVersionNumber();
    minorVn = doc.get_MinorVersionNumber();
    //if (!isReserved)isReserved = doc.get_
    IsReserved();
    Id nextGUID = doc.get_Id();
    createdGuid = d.get_Id();
        if (versCount == 1){
          FolderSet fs = doc.get_
          FoldersFiledIn();
          Iterator iterFs = fs.iterator();
          while (iterFs.hasNext())
          {
              Folder folder = (Folder)
              iterFs.next();
              //Create/Check and Make link
              in Target for the Document to
              this Folder
```

```java
                        if(cemc.getDebugOutputFlag().
                        equalsIgnoreCase("off")){
                        }else{
                            logger.info("\r\tFolder
                            Name: " + folder.get_
                            FolderName() +
                      "    Folder Path: " + folder.
                        get_PathName());
                        }
                         folderDoc(d,folder.get_
                         PathName(),documentTitle);
                    }
                 }

             }
            //ASB ... 2 020 Update Document ACL from
            source Document security
            AccessPermissionList  aclIn = dInput.get_
            Permissions();
            // Add the permission to the list for the
            Object Store.
            //Get This group from the config.xml file
//              String sExcludeGroup = cemc.
                getExcludedGroup();
//              String sExcludeUser =  cemc.
                getExcludedUser();
            //ASB ... 4 027 For Performance check if a
            full LDAP search is required
//              if(cemc.getLDAPSearchFlag().
                equalsIgnoreCase("on")){
//                  aclIn = addPermissions(aclIn,
                    sExcludeGroup, sExcludeUser);
//              }else {
//                  aclIn = addPermissionsNoSearch(aclIn,
                    sExcludeGroup, sExcludeUser);
//              }
```

```java
                        //ASB ... 4 027 aclIn = addPermissions(aclIn,
                        sExcludeGroup, sExcludeUser);
                        d.set_Permissions(aclIn);
            }
                return d;
    }
    private void folderDoc(Document doc, String folderPathName, String documentTitle){
        String sContainmentName = documentTitle;       //FOR RCR: //ASB 17/02/2011 Add Containment name
        if (folderPathName != null && folderPathName.length() > 0){
            Folder folder = null;
            try {
                folder = Factory.Folder.getInstance(os, null, folderPathName);
                //folder = Factory.Folder.fetchInstance(os, folderName, null);
                com.filenet.api.core.DynamicReferentialContainmentRelationship rcr =
                        Factory.DynamicReferentialContainmentRelationship.createInstance(os, null,
                                    AutoUniqueName.AUTO_UNIQUE,
                                    DefineSecurityParentage.DO_NOT_DEFINE_SECURITY_PARENTAGE);
                    rcr.set_Tail(folder);
                    rcr.set_Head(doc);
                    rcr.set_ContainmentName(sContainmentName);
                rcr.save(RefreshMode.NO_REFRESH);
            }
            catch (Exception e){

            }
        }
    }
```

```java
    private static Folder fetchFolderById(ObjectStore os, String id)
{
    Id id1 = new Id(id);
    Folder folderCase = Factory.Folder.fetchInstance(os, id1, null);
    return folderCase;
}
//This method returns the count of the number of Document versions
//(Unfortunately java Iterator class does not return size!)
private int countVersions(com.filenet.api.core.VersionSeries versions){
    int versionCount=0;
    VersionableSet verSet = versions.get_Versions();
    Iterator versIt = verSet.iterator();
    int versCount = 0;
     maxMin = 0;
    int min = 0;
    Document doc = null; //Input Document versions
    while (versIt.hasNext()){
        versCount ++;
        doc = (Document)versIt.next();
       //Record largest minor version value for loop
       min = doc.get_MinorVersionNumber();
       if (min > maxMin){
           maxMin = min;
       }
    }
    versionCount = versCount;
    return versionCount;
}
//This method returns the  Document versions in ascending order
//(Unfortunately FileNet API iterator returns latest version first! )
private Document[][] getDocumentVersionsArray(com.filenet.api.core.VersionSeries versions, Document [][] docVersions,int maxMin){
    int versionCount=0;
    VersionableSet verSet = versions.get_Versions();
```

CHAPTER 6   PDF DOCUMENT CREATION USING JAVA

```java
        Iterator versIt = verSet.iterator();
        int versCount = 0;
        Document doc = null; //Input Document versions
        maxMin = 0;
        int min=0;
        int maj=0;
        while (versIt.hasNext()){
            versCount ++;
            doc = (Document)versIt.next();
            maj = doc.get_MajorVersionNumber();
            min = doc.get_MinorVersionNumber();
            //Record largest minor version value for loop
            if (min > maxMin){
                maxMin = min;
            }
          docVersions[maj][min] = doc;
        }
        versionCount = versCount;
        return docVersions;
    }
    private Document createVersions(String documentTitle, Id GUID,
    Document object,Boolean isReserved,Document[][] docVersions,Integer
    majCount,int minCount, BufferedWriter auditFileWriter,com.filenet.api.
    core.VersionSeries versions)throws Exception{
        PropertyFilter myFilter = new PropertyFilter();
        int myFilterLevel = 2; //Changed from 4 to 2 -- TODO Set as a
        parameter
        myFilter.setMaxRecursion(myFilterLevel);
        myFilter.addIncludeType(new FilterElement(null, null, null,
        FilteredPropertyType.ANY, null));
        Document currentDocument = (Document) object;
        String documentclass = currentDocument.getClassName();
        Document dInput = Factory.Document.fetchInstance(os, GUID,
        myFilter);
```

```java
//Create Version Series for the document in the Target
Object Store
//VersionSeries versions = dInput.get_VersionSeries();
//Get the Document Versions to be added from the Source Document
VersionableSet verSet = versions.get_Versions();
Iterator versIt = verSet.iterator();
int versCount = 0;
int majorVn = 0;
int minorVn = 0;
Document doc = null; //Input Document versions
Document d = null;   //Output document versions
isReserved = false;
Id createdGuid = null;
Id nextGUID = null;
int foldCount = 0;
//while (versIt.hasNext()){ ASB ... 2
int   minVersCount = -1;
while (versCount < majCount){
      versCount ++;
    while (minVersCount < minCount){
            minVersCount ++;
      doc = docVersions[versCount][minVersCount];
      if(doc == null){
         break;
      }
      foldCount ++;
      majorVn = doc.get_MajorVersionNumber();
      minorVn = doc.get_MinorVersionNumber();
      nextGUID = doc.get_Id();
      //ASB ... 4 004 Get the Release Version
      VersionStatus dVersionStatus = doc.get_VersionStatus();
      if (!isReserved)isReserved = doc.get_IsReserved();
   //Need to update for each version in turn
```

```java
            d = createVersion(versCount, d, doc, isReserved,
            majorVn,minorVn, documentTitle, nextGUID, createdGuid,
            nextGUID,auditFileWriter);
            // try{
                    //createdGuid = nextGUID;
            createdGuid = d.get_Id(); //ASB ... 4 025 Try/Catch Here
            especially as reservation has no folders!!
            // }catch(Exception createdEx){
                //    logger.info("Version number "+ majorVn + "." +
                minorVn + " of " + documentTitle + " caused error : " +
                createdEx.getMessage());
            // }
            try {
                if (foldCount == majCount){
                    FolderSet fs = doc.get_FoldersFiledIn();
                    Iterator iterFs = fs.iterator();
                    while (iterFs.hasNext())
                    {
                        Folder folder = (Folder)iterFs.next();
                        //Create/Check and Make link in Target for the
                        Document to this Folder
                            if(cemc.getDebugOutputFlag().
                            equalsIgnoreCase("off")){
                            }else{
                                logger.info("\r\tFolder Name: " +
                                folder.get_FolderName() +
                        "    Folder Path: " + folder.get_
                            PathName());
                        }
                        folderDoc(d,folder.get_
                        PathName(),documentTitle);
                    }
                }
```

```
            }catch(Exception FileErr){
                    logger.error(FileErr.getMessage(), FileErr);
            }
                if (isReserved){
                break;
            }
        }
          //ASB ... 4 019 Check at this point if we have reached the
          completed creation
          //                If
            minVersCount = -1;
    } // Next Major Version
        if (isReserved)
            d = doReservationProperties(versCount, doc, d,
            isReserved, majorVn,minorVn,documentTitle);
    return d;

    }
//Called by CreateDocument, creates a version from the information in
the XML node specified.
//Creates a new reservation object, specifying properties from XML.
//private Document createVersion(Document d) throws
CEMigrateException, Exception {
        private Document createVersion(int     versCount, Document
        d, Document dInput, Boolean isReserved, int majorVn, int
        minorVn,String documentTitle, Id GUID, Id createdGUID, Id
        nextGUID,BufferedWriter auditFileWriter)    throws Exception {
            String guid = "";
            Long timeToImport = 0l;
            Id vGUID =null;
            Document res = null;
        Long docImportStartTime = System.currentTimeMillis();
            PropertyFilter myFilter = new PropertyFilter();
            int myFilterLevel = 2; //Changed from 4 to 2 -- For
            performance!
```

CHAPTER 6 PDF DOCUMENT CREATION USING JAVA

```java
myFilter.setMaxRecursion(myFilterLevel);
myFilter.addIncludeType(new FilterElement(null, null, null,
FilteredPropertyType.ANY, null));
Document currDoc = null;
 try
    {
 int thisMajorVersionNumber = majorVn;
 int thisMinorVersionNumber = minorVn; //ASB ... 3 004
 String docClass = dInput.getClassName();
 //ASB ... 3 006 Now get the current source Doc Version
 Series ID
 Id vId = dInput.get_VersionSeries().get_Id();
 if (!isClassExist(docClass))
      throw new Exception("Document class " + docClass + "
      Does not exist");
 Id reservationId_null = null; String reservationClass_null =
 null; // to remind of the parameter types!
 boolean isMajorVersion = false;
 boolean justCreated = true;

 if (d == null){
     if (thisMajorVersionNumber > 0)
         isMajorVersion = true;
     if (thisMinorVersionNumber > 0)
         isMajorVersion = false;
         //ASB ... 2 015 Added GUID to ensure we get the
         same later
         //ASB ... 3 006 New Version of the first document
         create to store Version series ID
         //ASB ... 3 006 Many thanks to David Greenhouse for
         supplying the call :-)
         res = Factory.Document.createInstance(os, docClass,
         GUID, vId, com.filenet.api.constants.ReservationTyp
         e.EXCLUSIVE);
```

CHAPTER 6 PDF DOCUMENT CREATION USING JAVA

```
//res = (Document)os.createObject(docClass,GUID);
//ASB ... 3 006 Removed Old Creation

//Added this to ensure that these values were
correctly set on the initial version
String mimetype = dInput.get_MimeType();
String filename = dInput.get_Name();
Date dateCreated = dInput.get_DateCreated();
res.set_MimeType(mimetype);
res.set_DateCreated(dateCreated); //ASB ... 4 029
//ASB 04/04/2011 - Update Creator and last
Modifier and dates
String sCreator = dInput.get_Creator();
res.set_Creator(sCreator);     //ASB 04/04/2011
String sLastModifier = dInput.get_LastModifier();
res.set_LastModifier(sLastModifier);
//ASB 04/04/2011
Date dateLastModified = dInput.get_
DateLastModified();
res.set_DateLastModified(dateLastModified); //ASB
04/04/2011
currDoc = dInput;
com.filenet.api.property.Properties docprops
= Factory.Document.createInstance(os, null).
getProperties(); //????
com.filenet.api.property.Properties props = res.
getProperties();
 Iterator iterProps = props.iterator();

   writeSpecialDocProps(res, iterProps, justCreated,
   documentTitle, currDoc);
//ASB ... 4 023 Add Count for first version
   this.processedCount = this.processedCount +1;
  guid = GUID.toString().trim();
    //Write to audit log
```

```java
            timeToImport = System.currentTimeMillis() -
            docImportStartTime;
            auditFileWriter.write(guid + "," + vId.toString()
            + "," + guid + ",Imported," + timeToImport.
            toString());
            auditFileWriter.newLine();
        }
        else {
            justCreated = false;
        //GUID here should be from next iteration set above
            //ASB ... 2 016 Changed to osinput from os! try
            nextGUID
            //currDoc = (Document)os.fetchObject("Document",
            createdGUID, myFilter);
        //ASB ... 4 024 Update to support other Document Types
            currDoc = (Document)os.fetchObject(docClass,
            createdGUID, myFilter);
            com.filenet.api.property.Properties props = Factory.
            Document.createInstance(os, null).getProperties();
            String sCreator = currDoc.get_Creator();
            Date dateCreated = currDoc.get_DateCreated();
            String sLastModifier = currDoc.get_LastModifier();
            Date dateLastModified = currDoc.get_DateLastModified();
            // This collection is used when we do the checkout
            props.putValue("Creator", sCreator);
            props.putValue("DateCreated", dateCreated);
            props.putValue("LastModifier", sLastModifier);
            props.putValue("DateLastModified", dateLastModified);
            //ASB ... 4 024
            //d = (Document)os.fetchObject(docClass, createdGUID,
            myFilter);
            d = currDoc; //No need to fetch again!
            //d = (Document)os.fetchObject("Document", createdGUID,
            myFilter);
```

CHAPTER 6   PDF DOCUMENT CREATION USING JAVA

```java
    //ASB ... 4 014 New Version needs to use the GUID
    passed in !!
    String sNewGuid = GUID.toString();
    // ASB Compare : d.checkout(ReservationType.EXCLUSIVE,
    GUID, docClass, null);
    d.checkout(ReservationType.EXCLUSIVE, GUID, docClass,
    props); //ASB ... 4 029 Changed from null to props
    //d.set_DateCreated(dateCreated); //ASB004 029 READONLY
    AT THIS POINT!!
    d.save(RefreshMode.REFRESH);
    this.processedCount = this.processedCount +1;
  guid = GUID.toString().trim();
    //Write to audit log
    timeToImport = System.currentTimeMillis() -
    docImportStartTime;
    auditFileWriter.write(guid + "," + d.get_
    VersionSeries().get_Id().toString() + "," + d.get_Id().
    toString() + ",Imported," + timeToImport.toString());
    auditFileWriter.newLine();
    // Get the reservation object
    res = (Document)d.get_Reservation();
    res.set_LastModifier(sLastModifier);          //ASB00 4
    res.set_DateLastModified(dateLastModified); //ASB00 4
    Uodate from dateCreated ??!!
    //res.set_DateCreated(dateCreated); //ASB004 029 READ
    ONLY AT THIS POINT!!!

    String mimetype = currDoc.get_MimeType();
    props.putValue("MimeType", mimetype);

    int mvn = majorVn;
    int minvn = minorVn;
    if (thisMajorVersionNumber > 0) isMajorVersion = true;
    if (thisMinorVersionNumber > 0) isMajorVersion = false;
  dInput = currDoc;
  GUID = nextGUID; //Get Current Content from version
}
```

```java
                    try {
                        addContent(dInput, res, documentTitle,  myFilter,
                        GUID,docClass);
                    }
                    catch (Exception e){
                        //ASB ... 4 Check if the Content could
                        not be added because the document already
                        exists, if so
                        //             Remove document and retry
                        (next run)
                        //   ASB ... 4 025 Log Exception here
                        logger.error(e.getMessage(), e);
                    }

            if(isReserved){ //ASB ... 2 012 Required For Checkin the
            Document must be Status of reserved!!
                if (isMajorVersion){
                    //ASB ... 4 028 Changed from NULL to AutoClassify.
                    DO_NOT_AUTO_CLASSIFY
                    res.checkin(AutoClassify.DO_NOT_AUTO_
                    CLASSIFY,CheckinType.MAJOR_VERSION);
                    res.save(RefreshMode.NO_REFRESH);
                }
                else {
                    res.checkin(AutoClassify.DO_NOT_AUTO_
                    CLASSIFY,CheckinType.MINOR_VERSION);
                    res.save(RefreshMode.NO_REFRESH);
                }
            }
        }
        catch (Exception e)
        {
            e.printStackTrace();
            //TODO If caused by VSID issue attempt Delete Here
            logger.error(e.getMessage(), e);
```

```java
            throw e;
    }
    return res;
}

    private void addContent(Document dSource, Document d, String 
    documentTitle, PropertyFilter myFilterId, Id GUID, String 
    DocClass) throws Exception {
        //ADD CONTENT   NEW CODE HERE
        //ASB ... 4 024
        int retryCount = 0;
        Document currDoc = null;
        //while (retryCount < retryLimit){
            try {
                currDoc = (Document)os.fetchObject(DocClass, GUID, 
                myFilterId);
                //Add substituteDoc above to get Content Stream
                //Document currDoc = (Document)osinput.
                fetchObject("Document", GUID, myFilterId);
                com.filenet.api.core.ContentTransfer ct = Factory.
                ContentTransfer.createInstance();
                String dClassName = currDoc.getClassName();
                String dMimeType = currDoc.get_MimeType();
                String dContentSize = currDoc.get_ContentSize().
                toString();
                String dGUID = currDoc.get_Id().toString();
                InputStream str = currDoc.accessContentStream(0);
                ct.setCaptureSource(str);
                // Add Document Title
                ct.set_RetrievalName(documentTitle);
                // Add Content Mime Type
                ct.set_ContentType(currDoc.get_MimeType());
                ContentElementList cel = Factory.ContentElement.cr
                eateList();
                cel.add(ct);
                d.set_ContentElements(cel);
```

```java
            //break;
        }
        catch (Exception e){
            String testMessage = e.getMessage();
            String testCode = e.toString();
            if( testMessage.contains("A uniqueness
            requirement has been violated")){
                //retryCount = retryLimit;
                return;
            }
            //ASB ... 4 Check if the Content could
            not be added because the document already
            exists, if so
            //           Remove document and retry
            (next run)
            //   ASB ... 4 025 Log Exception here
            logger.error(e.getMessage(), e);
            retryCount ++;
            //retryTotalProcessedCount ++;
            //logger.info("Retry Count - createVersion
            method : " + retryCount + " of " + retryLimit
            + " On : " + documentTitle + " : with GUID :
            " + GUID);
        }
//} //ASB ... 4 031 Above loops Test retryCount less than
the retryLimit of 7

// Add Security ACL

//ASB ... 2 020 Update Document ACL from source Document
security
AccessPermissionList   aclIn = null;
    aclIn = currDoc.get_Permissions();
    d.set_Permissions(aclIn);
Boolean isMajorVersion = false;
```

```java
        if(currDoc.get_MinorVersionNumber() == 0)
        isMajorVersion = true;
        //ASB ... 4 028 Change to do not Classify from AutoClassify.
        AUTO_CLASSIFY, to
        if (isMajorVersion){
            d.checkin(AutoClassify.DO_NOT_AUTO_
            CLASSIFY,CheckinType.MAJOR_VERSION);
        }
        else {
            d.checkin(AutoClassify.DO_NOT_AUTO_
            CLASSIFY,CheckinType.MINOR_VERSION);
        }
        //d.checkin(AutoClassify.AUTO_CLASSIFY, CheckinType.MAJOR_
        VERSION);
        //ASB ... 4 25 Try/Catch here ?
        Date dateLastModified = dSource.get_DateLastModified();
        //Date dateCreated = dSource.get_DateCreated(); //ASB
        ... 4 029
        String sModifier = dSource.get_LastModifier();
        Date dateCheckedIn = dSource.get_DateCheckedIn();
        // Update Checkin Date etc from Source here
        //d.set_DateCreated(dateCreated);   //ASB ... 4 029 READONLY
        AT THIS POINT!!
        d.set_DateLastModified(dateLastModified);
        d.set_LastModifier(sModifier);
        d.set_DateCheckedIn(dateCheckedIn); //ASB ... 2 005
        d.save(RefreshMode.REFRESH); //ASB Only save if Not in Bulk
        mode 21-08-2022
}
//Adds properties such as LastModifier to the document
properties.
//Called by doReservationProperties, createVersion
private void writeSpecialDocProps(Document d,Iterator iterProps,
Boolean justCreated, String documentTitle, Document dInput)
throws Exception {
```

```java
//ASB ... 2 008 Updated for new property definition objects
//Retrieve the current property list to retrieve based on
the Document Class definitions from the Object Store
String dClassName     = d.getClassName();
com.filenet.api.admin.ClassDefinition classDefs = Factory.Cl
assDefinition.fetchInstance(os,dClassName, null);
PropertyDefinitionList propList = classDefs.get_
PropertyDefinitions();
StringBuffer tempBuf = new StringBuffer();
//Retrieve property names
Iterator iter = propList.iterator();
while (iter.hasNext())
{
    //ASB ... 4 025 Catch any property exceptions here
    try{
        PropertyDefinition propDef = (PropertyDefinition)
        iter.next();
            String propName = propDef.get_
            SymbolicName();
            //ASB ... 2 011 Need to skip Date Last
            Accessed etc - ReadOnly error here
            if (!(propName.equalsIgnoreCase("DateCon
            tentLastAccessed")
                ||propName.
                equalsIgnoreCase("LockOwner")
                ||propName.
                equalsIgnoreCase("Name")
                ||propName.equalsIgnoreCase("S
                torageLocation")
                ||propName.equalsIgnoreCase("C
                ontentElementsPresent")
                ||propName.equalsIgnoreCase("D
                ateCheckedIn")
                ||propName.equalsIgnoreCase("C
                ontentRetentionDate")
```

||propName.equalsIgnoreCase("CurrentState")
||propName.equalsIgnoreCase("CmThumbnails"))){ //ASB 21-08-2022

//ASB ... 2 011 Set to check cardinality
```
int CardinalityVal = propDef.get_Cardinality().getValue();
switch (propDef.get_DataType().getValue()){
case StringType:
    if (CardinalityVal == CardinalitySINGLE){
        String val = dInput.getProperties().getStringValue(propName);
        d.getProperties().putValue(propName, val);
        if (propName.equalsIgnoreCase("lastmodifier")){
            d.set_LastModifier(val);
        }
        if (!propName.equalsIgnoreCase("creator") && !propName.equalsIgnoreCase("lastmodifier")){
            d.getProperties().putValue(propName, val);
        }
    }else{
        //List
```

CHAPTER 6   PDF DOCUMENT CREATION USING JAVA

```
                                        StringList valList =
                                        dInput.get
                                        Properties().get
                                        StringListValue
                                        (propName);
                                        d.getProperties().
                                        putValue(propName,
                                        valList);
                                    }
                                break;
                            case DateType:
                                if (CardinalityVal ==
                                CardinalitySINGLE){
                                    Date dateVal = dInput.
                                    getProperties().get
                                    DateTimeValue(propName);
                                    if (propName.equalsIg
                                    noreCase("datelastmo
                                    dified")){
                                        if (!justCreated){
                                            dInput.
getProperties().removeFromCache("DateLastModified");
                                        }
                                        d.set_
DateLastModified(dateVal);
                                    }
                                    if (!propName.
equalsIgnoreCase("datecreated") && !propName.equalsIgnoreCase("datelast
modified")){
                                        d.getProperties().
                                        putValue(propName,
                                        dateVal);
                                    }
                                }
```

```java
                            //TODO Process Multi-value
                            Dates (if exists)?
                            break;
                    default:

                    }
                }
            }catch(Exception writeProp){
                logger.error(writeProp.getMessage(),
                writeProp);
            }
        }
    }
    //ASB ... 2 020 Update Document ACL from source Document
    security
    AccessPermissionList  aclIn = null;
        aclIn = dInput.get_Permissions();
        d.set_Permissions(aclIn);
}
//Sets the properties on the reservation object for docs that
were checked out on the source system.
//Sets properties but no content.  No need to set mime type
here - causes problems.
//private Document doReservationProperties(Node verNode, Document
d) throws Exception {
private Document doReservationProperties(int versCount, Document
dInput, Document d, Boolean isReserved, int majorVn,int minorVn,
String documentTitle    ) throws Exception{
    com.filenet.api.property.Properties props = Factory.
    Document.createInstance(os, null).getProperties();
    String sCreator = dInput.get_Creator();
    Date dateCreated = dInput.get_DateCreated();
    String sLastModifier = dInput.get_LastModifier();
    Date dateLastModified = dInput.get_DateLastModified();
    // This collection is used when we do the checkout
    props.putValue("Creator", sCreator);
```

```java
props.putValue("DateCreated", dateCreated);
props.putValue("LastModifier", sLastModifier);
props.putValue("DateLastModified", dateLastModified);
Date dateCheckedIn = dInput.get_DateCheckedIn();
//Update Checkin Date etc from Source here

d.set_LastModifier(sCreator);
d.set_DateLastModified(dateCreated);
d.set_DateCreated(dateCreated); //ASB ... 4 029
Id reservationId_null = null;
String reservationClass_null = null; // to remind of the
parameter types!
try
{
   // Get Document properties
      com.filenet.api.property.Properties docProps =
      d.getProperties();
    Iterator iterProps = props.iterator();
   Boolean justCreated = false;
      writeSpecialDocProps(d, iterProps,
      justCreated,documentTitle, dInput);
}
catch (Exception e){
    throw new Exception(e.getMessage(), e);
}

d.set_LastModifier(sLastModifier);

// ASB Compare : d.checkout(ReservationType.EXCLUSIVE, GUID,
docClass, null);
d.checkout(ReservationType.OBJECT_STORE_DEFAULT,
reservationId_null, reservationClass_null, props);
d.set_DateLastModified(dateLastModified);
d.set_DateCreated(dateCreated); //ASB ... 4 029
d.set_LastModifier(sLastModifier);
d.save(RefreshMode.REFRESH);
```

```
            return d;
    }
    private boolean isClassExist(String className) throws Exception {
        if (classNames.containsKey(className))
            return true;
        try {
            com.filenet.api.admin.ClassDefinition test = Factory.Cl
            assDefinition.fetchInstance(os, className, null);
            classNames.put(test.get_Name(), test.get_Name());
            return true;
        }
        catch (EngineRuntimeException e){
            if (e.getExceptionCode() == ExceptionCode.E_BAD_
            CLASSID)
                return false;
            throw e;
        }
    }
    private void updateNewDocument(Document doc,String docTitle, String mimeType, String FileName) {
        // REF
        // Page 92-93 of http://www.redbooks.ibm.com/redbooks/pdfs/sg247743.pdf
        //
        // Get the document (saving a round-trip that a fetch would require)
        //Document doc = Factory.Document.getInstance(os,"Document","/Doc1");
        // Checkout the document and save
        doc.checkout(ReservationType.EXCLUSIVE, null, null, null);
        doc.save(RefreshMode.REFRESH);
        // Get the reservation object
        Document res = (Document)doc.get_Reservation();
        // Update the properties
        res.getProperties().putValue("DocumentTitle",docTitle);
```

```java
        // Prepare the content for attaching
        // Create the element list
        ContentElementList list = Factory.ContentElement.createList();
        // Create a content transfer element by attaching a simple
        text file
        com.filenet.api.core.ContentTransfer element = Factory.Content
        Transfer.createInstance();
        // Set the MIME type
        element.set_ContentType(mimeType);
        // Set the retrieval name
        element.set_RetrievalName(FileName);
        // Set the content source
        try {
                element.setCaptureSource(new FileInputStream(new
                File(FileName)));
            } catch (FileNotFoundException e) {
                // TODO Auto-generated catch block
                e.printStackTrace();
            }
        // Add the item to the list
        list.add(element);
        // Add the content transfer list to the document
        res.set_ContentElements(list);
        // Set the PendingAction to be check-in as a major
        version without
        // automatic content classification
        res.checkin(AutoClassify.DO_NOT_AUTO_CLASSIFY,
        CheckinType.MAJOR_VERSION);
        // Save the document to the repository
        res.save(RefreshMode.NO_REFRESH);
    }
    private static void addQTitlePage( com.lowagie.text.Document
    document)
                throws DocumentException, IOException {
                // Add the Audit Master Logo
```

```java
// Creating an ImageData object
String imageLogoFile = "/opt/AuditReport/images/
AuditMasterLogo.tif"; //TODO add to config.xml file
RandomAccessFileOrArray ra = new RandomAccessFileOrArray
(imageLogoFile);
int pages = TiffImage.getNumberOfPages(ra);
// Creating an Image object
Image image;
image = TiffImage.getTiffImage(ra, 1);
// Adding the Audit Master Logo image to the Audit
report document
Rectangle pageSize = new Rectangle(image.getWidth(),
          image.getHeight());
document.setPageSize(pageSize);
document.add(image);
// Start a new page
document.newPage();
Paragraph preface = new Paragraph();
// We add one empty line
addEmptyLine(preface, 1);
// Lets write a big header
//TODO Pick all text up from the config.xml file
Paragraph paragraph = new Paragraph("Audit Master - Audit
Report", catFont);
paragraph.setAlignment(Element.ALIGN_CENTER);
preface.add(paragraph);
   addEmptyLine(preface, 1);
// Lets write another big header
//TODO Pick all text up from the config.xml file
   paragraph = new Paragraph("ASB Software Development
   Limited", catFont);
   paragraph.setAlignment(Element.ALIGN_CENTER);
preface.add(paragraph);
   addEmptyLine(preface, 1);
// Lets write another big header
```

```java
//TODO Pick all text up from the config.xml file
    paragraph = new Paragraph(sTab+ "Department Audit
    Process Task", catFont);
    paragraph.setAlignment(Element.ALIGN_CENTER);
preface.add(paragraph);
// Lets write another big header
//TODO Pick all text up from the config.xml file
    paragraph = new Paragraph(sTab+ "ISO9000/BS5750",
    catFont);
    paragraph.setAlignment(Element.ALIGN_CENTER);
preface.add(paragraph);
addEmptyLine(preface, 1);
// Will create: Report generated by: _name, _date
preface.add(new Paragraph(sTab+ "Report generated by: "
+ System.getProperty("user.name") + ", " + new Date(),
//$NON-NLS-1$ //$NON-NLS-2$ //$NON-NLS-3$
    smallBold));
addEmptyLine(preface, 3);
paragraph = new Paragraph(sTab+ "Note: This is an
automatically generated Audit Report, the full Audit
results and conclusions are produced following the run of
the review tasks. ",
        smallBold);
paragraph.setAlignment(Element.ALIGN_LEFT);
paragraph.setIndentationLeft(50);

preface.add(paragraph);

addEmptyLine(preface, 8);
addEmptyLine(preface, 8);
paragraph = new Paragraph(sTab+ "Once you have completed
the initial Department Audit, you should have:\r\n" +
        "\r\n" + sTab +
        "•      Arranged a non-conformance review and a
        follow-up audit and questions. \r\n" +
        "\r\n" + sTab +
```

```java
                    "•       Allowed the Department sufficient time to
                    correct the non-conformances and   \r\n" +
                    "\r\n" + sTab+
                    "•       Generated the evidence for the required
                    changes to their procedures.  \r\n" +
                    "",
                redFont);
        paragraph.setAlignment(Element.ALIGN_LEFT);
        paragraph.setIndentationLeft(50);

         preface.add(paragraph);

         document.add(preface);
         // Start a new page
         document.newPage();
      }
private static void addEmptyLine(Paragraph paragraph, int number) {
    for (int i = 0; i < number; i++) {
      paragraph.add(new Paragraph(" "));
    }
  }
private static void  getTaskDepartmentPropsSections(CmTask
object,String taskClass,com.filenet.api.admin.ClassDefinition
tClassDefs, PropertyDefinitionList tPropList,com.filenet.api.
property.Properties props, Iterator iterProps, String [][]
Departmentprops, String [] TabSECTIONS_Department, String [][]
Department_propValues,String [][] Department_propNames,Integer
noDepartmentSections, Integer [] noDepartmentprops)throws
Exception {
    // called as :
    //                    getTaskDepartmentPropsSections(currentTa
    sk, taskClass,tClassDefs,tPropList,props,iterProps,Departme
    nt_props,TabSECTIONS_Department,Department_propValues,noDepart
    mentSections,noDepartmentprops);
```

```
//      String [][] Department_props paramCount is the section
        number, 0-6 i is the noDepartmentprops[paramCount]
        contains the symbolic property name
//      TabSECTIONS_Department[]      // has noDepartmentSections
//      Integer noDepartmentSections  The number of sections to
        process 0-3
        //ASB
        //Retrieve the current property list to retrieve based
        on the Task Class definitions from the Object Store
String tClassName = object.getClassName(); //ASB 15-08-2022
com.filenet.api.admin.ClassDefinition classDefs = tClassDefs;
PropertyDefinitionList propList = tPropList;
StringBuffer tempBuf = new StringBuffer();
//Retrieve property names code From writeDocProps to compare
Iterator iter =propList.iterator();
while (iter.hasNext()){
    try {
        PropertyDefinition propDef = (PropertyDefinition)
        iter.next();
        String propName = propDef.get_SymbolicName();
        String propDisplayName = propDef.get_DisplayName();
        //ASB ... 2 011 Need to skip Date Last Accessed etc -
        ReadOnly error here
        if (!(propName.equalsIgnoreCase("Name")
            ||propName.equalsIgnoreCase("DateCheckedIn")
            ||propName.equalsIgnoreCase("CurrentState")
            ||propName.equalsIgnoreCase("PathName")
            ||propName.equalsIgnoreCase("LockOwner")
            ||propName.equalsIgnoreCase("Owner") //ASB
            15-08-2022
            ||propName.startsWith("CmAcm") )){

            //ASB ... Set to check property cardinality
            int CardinalityVal = propDef.get_
            Cardinality().getValue();
            int propIndex;
```

```java
                //===============START
                    switch (propDef.get_DataType().
                    getValue()){
                    case StringType:
                        if (CardinalityVal ==
                        CardinalitySINGLE){
                            String val = object.
getProperties().getStringValue(propName);
                                for (int sectionIndex = 0;
sectionIndex < noDepartmentSections; sectionIndex++) {
                                    for (propIndex = 0;
propIndex < noDepartmentprops[sectionIndex]; propIndex++) { //Department_
                                                    propValues
                                        // Value for Required response
                                        String propValue = val;
                                        String propTest =
Departmentprops[sectionIndex][propIndex];
                                        if (propName.
equalsIgnoreCase(propTest)){
                                            Department_
propValues[sectionIndex][propIndex] =propValue;
                                            Department_
propNames[sectionIndex][propIndex] = propDisplayName;
                                        }
                                    }
                                }
                                //ASB 15-08-2022
                        }else{
                            StringList valList = object.
getProperties().getStringListValue(propName);
                            //Iterator itList = valList.
iterator();
                            //String valLists = " ";
                            //while (itList.hasNext()){
```

```java
                                        //   valLists = valLists +"\r\n"
                                        + sTab + itList.toString();
                                  //}
                                  for (int sectionIndex = 0; sectionIndex < noDepartmentSections; sectionIndex++){
                                          for (propIndex = 0; propIndex < noDepartmentprops[sectionIndex]; propIndex++) { //Department_propValues
                                              // Value for Required response
                                              if (propName.equalsIgnoreCase(Department_props[sectionIndex][propIndex])){
                                                  // Iterate through the list of values
                                                  StringBuffer sb = new StringBuffer();
                                                  for( Iterator i = valList.iterator(); i.hasNext(); )
                                                  {
                                                  // Get the value and cast to an Id property type
                                                      String propertyVal = (String)i.next();
                                                      sb.append( propertyVal + "\r\n");
                                                  }
                                                      Department_propValues[sectionIndex][propIndex] =sb.toString();
                                                      Department_propNames[sectionIndex][propIndex] = propDisplayName;
                                              }
                                          }
                                      }
                                  }
```

```
//===============START
    switch (propDef.get_DataType().
    getValue()){
    case StringType:
        if (CardinalityVal ==
        CardinalitySINGLE){
            String val = object.
getProperties().getStringValue(propName);
            for (int sectionIndex = 0;
sectionIndex < noDepartmentSections; sectionIndex++) {
                for (propIndex = 0;
propIndex < noDepartmentprops[sectionIndex]; propIndex++) { //Department_
                                                              propValues
                    // Value for Required response
                    String propValue = val;
                    String propTest =
Departmentprops[sectionIndex][propIndex];
                    if (propName.
equalsIgnoreCase(propTest)){
                        Department_
propValues[sectionIndex][propIndex] =propValue;
                        Department_
propNames[sectionIndex][propIndex] = propDisplayName;
                    }
                }
            }
            //ASB 15-08-2022
        }else{
            StringList valList = object.
getProperties().getStringListValue(propName);
            //Iterator itList = valList.
iterator();
            //String valLists = " ";
            //while (itList.hasNext()){
```

```java
                                            //   valLists = valLists +"\r\n"
                                            + sTab + itList.toString();
                                        //}
                                        for (int sectionIndex = 0;
sectionIndex < noDepartmentSections; sectionIndex++){
                                                for (propIndex = 0;
propIndex < noDepartmentprops[sectionIndex]; propIndex++) { //Department_
propValues
                                                    // Value for Required
                                                    response
                                                    if (propName.
equalsIgnoreCase(Department_props[sectionIndex][propIndex])){
                                                        // Iterate through
                                                        the list of values
                                                        StringBuffer
                                                        sb = new
                                                        StringBuffer();
                                                        for( Iterator i =
valList.iterator(); i.hasNext(); )
                                                        {
                                                        // Get the value
and cast to an Id property type
                                                        String propertyVal
= (String)i.next();
                                                        sb.append(
propertyVal + "\r\n");
                                                        }
                                                            Department_
propValues[sectionIndex][propIndex] =sb.toString();
                                                            Department_
propNames[sectionIndex][propIndex] = propDisplayName;
                                                        }
                                                    }
                                                }
                                            }
```

```java
            // Process Multi-value Strings (if
            required)?
            break;
        case DateType:
            if (CardinalityVal ==
            CardinalitySINGLE){
                Date dateVal = object.
                getProperties().
                getDateTimeValue(propName);
                for (int sectionIndex = 0;
sectionIndex < noDepartmentSections; sectionIndex++){
                    for (propIndex =
0; propIndex < noDepartmentprops[sectionIndex]; propIndex++){ //Department_
propValues
                        // Value for Required
                        response
                        if (propName.
equalsIgnoreCase(Departmentprops[sectionIndex][propIndex])){
                            if(dateVal
                            != null) {
                Department_propValues[sectionIndex]
                [propIndex] =dateVal.toString();
                Department_propNames[sectionIndex]
                [propIndex] = propDisplayName;
                            }
                        }
                    }
                }
            }else{
                //
            }
            break;
        default:
    }
        //==============END
}
```

```java
            }catch (Exception writeProp){
                logger.error(writeProp.getMessage(), writeProp);
            }
        }
    }
}
private static void  getTaskAuditPropsSections(CmTask object,String taskClass,com.filenet.api.admin.ClassDefinition tClassDefs, PropertyDefinitionList tPropList,com.filenet.api.property. Properties props, Iterator iterProps, String [][] Auditprops, String [] TabSECTIONS_Audit, String [][] Audit_propValues,String [][] Audit_propNames, Integer noAuditSections, Integer [] noAuditprops)throws Exception {
    // called as :
    //              getTaskAuditPropsSections(currentTask, taskClass,tClassDefs,tPropList,props,iterProps,Audit_props,TabSECTIONS_Audit,Audit_propValues,noAuditSections,noAuditprops);
    //    String [][] Audit_props paramCount is the section number, 0-6 i is the noAuditprops[paramCount] contains the symbolic property name
    //    TabSECTIONS_Audit[]     // has noAuditSections
    //    Integer noAuditSections  The number of sections to process 0-3
        //ASB
        //Retrieve the current property list to retrieve based on the Task Class definitions from the Object Store
    String tClassName = object.getClassName(); //ASB 15-08-2022
    com.filenet.api.admin.ClassDefinition classDefs = tClassDefs;
    PropertyDefinitionList propList = tPropList;
    StringBuffer tempBuf = new StringBuffer();
    //Retrieve property names code From writeDocProps to compare
    Iterator iter =propList.iterator();
    while (iter.hasNext()){
        try {
```

```java
                        PropertyDefinition propDef = (PropertyDefinition)
                        iter.next();
                        String propName = propDef.get_SymbolicName();
                        String propDisplayName = propDef.get_DisplayName();
                        //ASB ... 2 011 Need to skip Date Last Accessed etc -
                        ReadOnly error here
                        if (!(propName.equalsIgnoreCase("Name")
                                ||propName.equalsIgnoreCase("DateCheckedIn")
                                ||propName.equalsIgnoreCase("CurrentState")
                                ||propName.equalsIgnoreCase("PathName")
                                ||propName.equalsIgnoreCase("LockOwner")
                                ||propName.equalsIgnoreCase("Owner") //ASB
                        15-08-2022
                                ||propName.startsWith("CmAcm") )){

                            //ASB ... Set to check property cardinality
                            int CardinalityVal = propDef.get_
                            Cardinality().getValue();
                            int propIndex;
                            //===============START
                                switch (propDef.get_DataType().
                                getValue()){
                              case StringType:
                                    if (CardinalityVal ==
                                    CardinalitySINGLE){
                                        String val = object.
getProperties().getStringValue(propName);
                                        for (int sectionIndex = 0;
sectionIndex < noAuditSections; sectionIndex++) {
                                            for (propIndex = 0;
propIndex < noAuditprops[sectionIndex]; propIndex++) { //Audit_propValues
                                                // Value for Required response
                                                    String propValue = val;
                                                    String propTest =
Auditprops[sectionIndex][propIndex];
```

```
                                            if (propName.equalsIgnoreCase(propTest)){
                                                    Audit_propValues[sectionIndex][propIndex] =propValue;
                                                    Audit_propNames[sectionIndex][propIndex] = propDisplayName;
                                            }
                                        }
                                    }
                                    //ASB 15-08-2022
                                }else{
                                    StringList valList = object.getProperties().getStringListValue(propName);
                                    // Get a multi-value String property
                                    //Iterator itList = valList.iterator();
                                    //String valLists = " ";
                                    //while (itList.hasNext()){
                                    //    valLists = valLists +"\r\n" + sTab + itList.toString();
                                    // }
                                    for (int sectionIndex = 0; sectionIndex < noAuditSections; sectionIndex++){
                                        for (propIndex = 0; propIndex < noAuditprops[sectionIndex]; propIndex++) { //Audit_propValues
                                            // Value for Required response
                                            if (propName.equalsIgnoreCase(Audit_props[sectionIndex][propIndex])){
                                                // Iterate through the list of values
                                                StringBuffer sb = new StringBuffer();
                                                for( Iterator i = valList.iterator(); i.hasNext(); )
```

```java
                        and cast to an Id property type
                        propertyVal = (String)i.next();
                        sb.append(propertyVal + "\r\n");
                    }
                    propValues[sectionIndex][propIndex] =sb.toString();
                    propNames[sectionIndex][propIndex] = propDisplayName;
                {
                    // Get the value
                    String sb.
                }
                    Audit_Audit_
                            }
                        }
                    }
                }
                // Process Multi-value Strings (if required)?
                break;
            case DateType:
                if (CardinalityVal == CardinalitySINGLE){
                    Date dateVal = object.getProperties().getDateTimeValue(propName);
                    for (int sectionIndex = 0; sectionIndex < noAuditSections; sectionIndex++){
                        for (propIndex = 0; propIndex < noAuditprops[sectionIndex]; propIndex++){ //Audit_propValues
                            // Value for Required response
                            if (propName.equalsIgnoreCase(Auditprops[sectionIndex][propIndex])){
                                if(dateVal != null) {
```

```java
                                                                Audit_
propValues[sectionIndex][propIndex] =dateVal.toString();
                                                                Audit_
propNames[sectionIndex][propIndex] = propDisplayName;
                                                    }
                                                }
                                            }
                                        }
                                    }else{
                                        //
                                    }
                                break;
                                default:
                            }
                        //===============END
                        }
                    }catch (Exception writeProp){
                        logger.error(writeProp.getMessage(), writeProp);
                    }
                }
            }
        }
        private static void  getTaskAUDITORPropsSections(CmTask
        object,String taskClass,com.filenet.api.admin.ClassDefinition
        tClassDefs, PropertyDefinitionList tPropList,com.filenet.api.
        property.Properties props, Iterator iterProps, String [][]
        AUDITORprops, String [] TabSECTIONS_AUDITOR, String [][] AUDITOR_
        propValues,String [][] AUDITOR_propNames,Integer noAUDITORSections,
        Integer [] noAUDITORprops)throws Exception {
            // called as :
            //             getTaskAUDITORPropsSections(currentTa
            sk, taskClass,tClassDefs,tPropList,props,iterProps,AUDITOR_
            props,TabSECTIONS_AUDITOR,AUDITOR_propValues,noAUDITORSections
            ,noAUDITORprops);
```

```
//      String [][] AUDITOR_props paramCount is the section
number, 0-6 i is the noAUDITORprops[paramCount] contains the
symbolic property name
//      TabSECTIONS_AUDITOR[]      // has noAUDITORSections
//      Integer noAUDITORSections  The number of sections to
process 0-3
        //ASB
        //Retrieve the current property list to retrieve based
        on the Task Class definitions from the Object Store
String tClassName = object.getClassName(); //ASB 15-08-2022
com.filenet.api.admin.ClassDefinition classDefs = tClassDefs;
PropertyDefinitionList propList = tPropList;
StringBuffer tempBuf = new StringBuffer();
//Retrieve property names code From writeDocProps to compare
Iterator iter =propList.iterator();
while (iter.hasNext()){
    try {
        PropertyDefinition propDef = (PropertyDefinition)
        iter.next();
        String propName = propDef.get_SymbolicName();
        String propDisplayName = propDef.get_DisplayName();
        //ASB ... 2 011 Need to skip Date Last Accessed etc -
        ReadOnly error here
        if (!(propName.equalsIgnoreCase("Name")
            ||propName.equalsIgnoreCase("DateCheckedIn")
            ||propName.equalsIgnoreCase("CurrentState")
            ||propName.equalsIgnoreCase("PathName")
            ||propName.equalsIgnoreCase("LockOwner")
            ||propName.equalsIgnoreCase("Owner") //ASB
            15-08-2022
            ||propName.startsWith("CmAcm") )){

            //ASB ... Set to check property cardinality
            int CardinalityVal = propDef.get_
            Cardinality().getValue();
            int propIndex;
```

```java
                            //===============START
                            switch (propDef.get_DataType().getValue()){
                                case StringType:
                                    if (CardinalityVal == CardinalitySINGLE){
                                        String val = object.getProperties().getStringValue(propName);
                                        for (int sectionIndex = 0; sectionIndex < noAUDITORSections; sectionIndex++) {
                                            for (propIndex = 0; propIndex < noAUDITORprops[sectionIndex]; propIndex++) { //AUDITOR_propValues
                                                // Value for Required response
                                                String propValue = val;
                                                String propTest = AUDITORprops[sectionIndex][propIndex];
                                                if (propName.equalsIgnoreCase(propTest)){
                                                    AUDITOR_propValues[sectionIndex][propIndex] =propValue;
                                                    AUDITOR_propNames[sectionIndex][propIndex] = propDisplayName;
                                                }
                                            }
                                        }
                                        //ASB 15-08-2022
                                    }else{
                                        StringList valList = object.getProperties().getStringListValue(propName);
                                        //Iterator itList = valList.iterator();
                                        //String valLists = " ";
                                        //while (itList.hasNext()){
```

```
                                // valLists = valLists +"\r\n" + sTab + itList.toString();
                                //}
                                for (int sectionIndex = 0; sectionIndex < noAUDITORSections; sectionIndex++){
                                    for (propIndex = 0; propIndex < noAUDITORprops[sectionIndex]; propIndex++) { //AUDITOR_propValues
                                        // Value for Required response
                                        if (propName.equalsIgnoreCase(AUDITOR_props[sectionIndex][propIndex])){
                                            // Iterate through the list of values
                                            StringBuffer sb = new StringBuffer();
                                            for( Iterator i = valList.iterator(); i.hasNext(); )
                                            {
                                                // Get the value and cast to an Id property type
                                                String propertyVal = (String)i.next();
                                                sb.append(propertyVal + "\r\n");
                                            }
                                            AUDITOR_propValues[sectionIndex][propIndex] =sb.toString();
                                            AUDITOR_propNames[sectionIndex][propIndex] = propDisplayName;
                                        }
                                    }
                                }
                            }
```

```java
                                    // Process Multi-value Strings (if required)?
                                    break;
                            case DateType:
                                if (CardinalityVal == CardinalitySINGLE){
                                    Date dateVal = object.getProperties().getDateTimeValue(propName);
                                    for (int sectionIndex = 0; sectionIndex < noAUDITORSections; sectionIndex++){
                                        for (propIndex = 0; propIndex < noAUDITORprops[sectionIndex]; propIndex++){ //AUDITOR_propValues
                                            // Value for Required response
                                            if (propName.equalsIgnoreCase(AUDITORprops[sectionIndex][propIndex])){
                                                if(dateVal != null) {
                                                    AUDITOR_propValues[sectionIndex][propIndex] =dateVal.toString();
                                                    AUDITOR_propNames[sectionIndex][propIndex] = propDisplayName;
                                                }
                                            }
                                        }
                                    }
                                }else{
                                    //
                                }
                                break;
                            //        /*
                            case DoubleType:
                                if (CardinalityVal == CardinalitySINGLE){
```

CHAPTER 6 PDF DOCUMENT CREATION USING JAVA

```
                                        Double doubleVal = object.getProperties().getFloat64Value(propName);
                                        for (int sectionIndex = 0; sectionIndex < noAUDITORSections; sectionIndex++){
                                            for (propIndex = 0; propIndex < noAUDITORprops[sectionIndex]; propIndex++){ //AUDITOR_propValues
                                                // Value for Required response
                                                if (propName.equalsIgnoreCase(AUDITORprops[sectionIndex][propIndex])){
                                                    if(doubleVal != null) {
                                                        AUDITOR_propValues[sectionIndex][propIndex] =doubleVal.toString();
                                                        AUDITOR_propNames[sectionIndex][propIndex] = propDisplayName;
                                                    }
                                                }
                                            }
                                        }
                                    }else{
                                        //
                                    }
                                    // */
                                default:
                            }
                            //===============END
                        }
                    }catch (Exception writeProp){
                        logger.error(writeProp.getMessage(), writeProp);
                    }
                }
            }
        }
```

CHAPTER 6 PDF DOCUMENT CREATION USING JAVA

```java
private static void addAUDITORTitlePage( com.lowagie.text.Document document)
        throws DocumentException, IOException {
    // Add the Audit Master Logo
    // Creating an ImageData object
    String imageLogoFile = "/opt/AuditReport/images/AuditMasterLogo.tif"; //TODO add to config.xml file
    RandomAccessFileOrArray ra = new RandomAccessFileOrArray(imageLogoFile);
    int pages = TiffImage.getNumberOfPages(ra);
    // Creating an Image object
    Image image;
    image = TiffImage.getTiffImage(ra, 1);
    // Adding the Audit Master Logo image to the Audit report document
    Rectangle pageSize = new Rectangle(image.getWidth(), image.getHeight());
    document.setPageSize(pageSize);
    document.add(image);
    // Start a new page
    document.newPage();
    Paragraph preface = new Paragraph();
    // We add one empty line
    addEmptyLine(preface, 1);
    // Lets write a big header
    //TODO Pick all text up from the config.xml file
    Paragraph paragraph = new Paragraph("Audit Master - Audit Report", catFont);
    paragraph.setAlignment(Element.ALIGN_CENTER);
    preface.add(paragraph);
        addEmptyLine(preface, 1);
    // Lets write another big header
    //TODO Pick all text up from the config.xml file
        paragraph = new Paragraph("ASB Software Development Limited", catFont);
```

```java
    paragraph.setAlignment(Element.ALIGN_CENTER);
     preface.add(paragraph);
    addEmptyLine(preface, 1);
    paragraph = new Paragraph("Department Audit Process
    Task", catFont);
    paragraph.setAlignment(Element.ALIGN_CENTER);
    preface.add(paragraph);
        addEmptyLine(preface, 1);
// Lets write another big header
//TODO Pick all text up from the config.xml file
    paragraph = new Paragraph(sTab+ "ISO9000/BS5750",
    catFont);
    paragraph.setAlignment(Element.ALIGN_CENTER);
preface.add(paragraph);

addEmptyLine(preface, 1);
// Will create: Report generated by: _name, _date
preface.add(new Paragraph(sTab+ "Report generated by: "
+ System.getProperty("user.name") + ", " + new Date(),
//$NON-NLS-1$ //$NON-NLS-2$ //$NON-NLS-3$
    smallBold));
addEmptyLine(preface, 3);
paragraph = new Paragraph(sTab+ "Note: This is an
automatically generated Audit Report, the full Audit
results and conclusions are produced following the run of
the review tasks. ",
        smallBold);
paragraph.setAlignment(Element.ALIGN_LEFT);
paragraph.setIndentationLeft(50);

 preface.add(paragraph);
```

```java
            addEmptyLine(preface, 8);
            paragraph = new Paragraph(sTab+ "Once you have completed
            the initial Department Audit, you should have:\r\n" +
                    "\r\n" + sTab +
                    "•       Arranged a non-conformance review and a
                    follow-up audit and questions. \r\n" +
                    "\r\n" + sTab +
                    "•       Allowed the Department sufficient time to
                    correct the non-conformances and  \r\n" +
                    "\r\n" + sTab+
                    "•       Generated the evidence for the required
                    changes to their procedures.  \r\n" +
                    "",
                redFont);
            paragraph.setAlignment(Element.ALIGN_LEFT);
            paragraph.setIndentationLeft(50);

            preface.add(paragraph);

            document.add(preface);
            // Start a new page
            document.newPage();
        }
//=========E N D    C O D E     F O R    P D F   I M P O R T ==========
}
```

Part 3 – Supporting Java Classes for the Main Audit Report Program

The supporting Java methods called by the main Audit Report program are described in this part.

CHAPTER 6 PDF DOCUMENT CREATION USING JAVA

Adding the EmailTemplate Utility Java Class

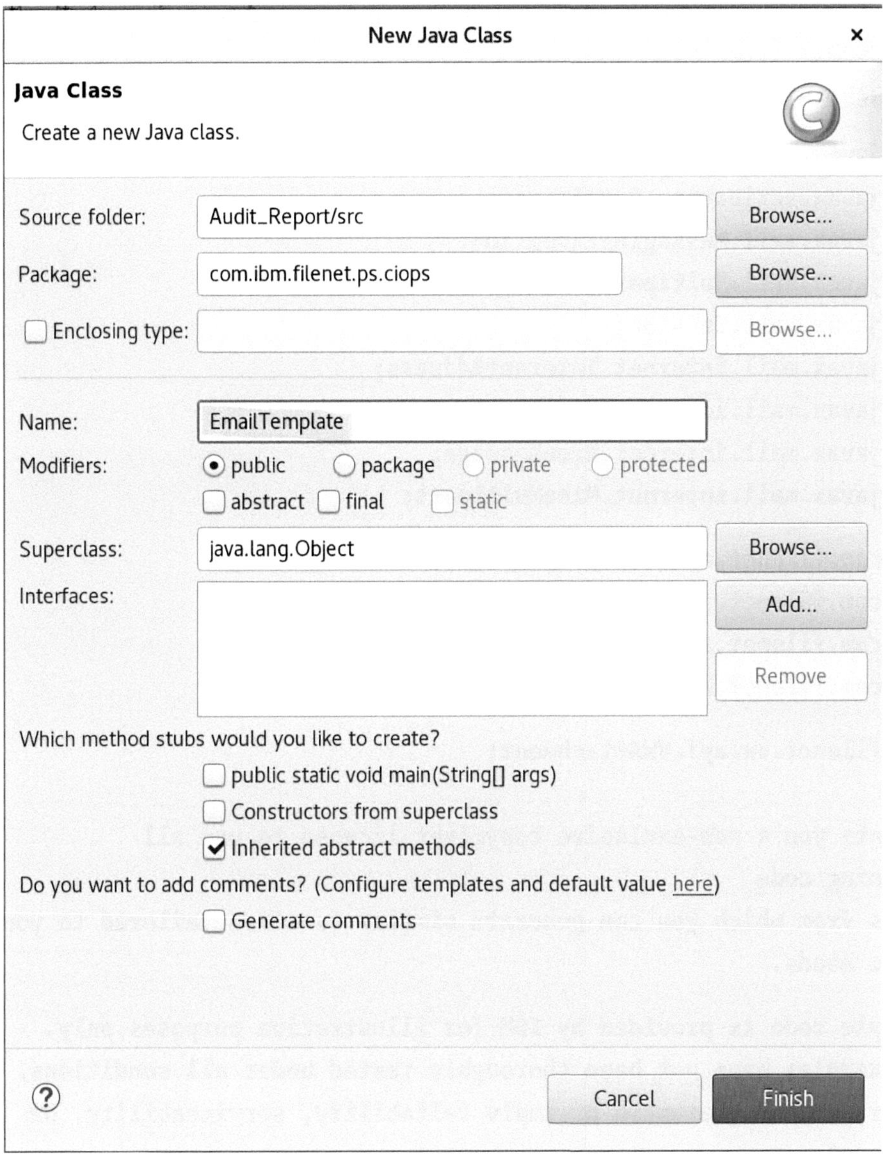

*Figure 6-22. The **EmailTemplate** class code is added to the package*

The Java code for the ***EmailTemplate*.java** source is as follows.

Listing 6-19. The **EmailTemplate.java** source code file

```java
package com.ibm.filenet.ps.ciops;

import java.util.Date;
import java.util.Iterator;

import javax.activation.DataHandler;
import javax.activation.DataSource;
import javax.mail.MessagingException;
import javax.mail.Multipart;
import javax.mail.Session;
import javax.mail.internet.InternetAddress;
import javax.mail.internet.MimeBodyPart;
import javax.mail.internet.MimeMessage;
import javax.mail.internet.MimeMultipart;

import com.filenet.api.collection.ContentElementList;
import com.filenet.api.constants.PropertyNames;
import com.filenet.api.core.ContentTransfer;
import com.filenet.api.core.Document;

import filenet.vw.api.VWAttachment;
/**
IBM grants you a non-exclusive copyright license to use all
programming code
examples from which you can generate similar function tailored to your own
specific needs.

All sample code is provided by IBM for illustrative purposes only.
These examples have not been thoroughly tested under all conditions.  IBM,
therefore cannot guarantee or imply reliability, serviceability, or
function of
these programs.

All Programs or code component contained herein are provided to you
"AS IS "
without any warranties of any kind.
```

The implied warranties of non-infringement, merchantability and fitness for a
particular purpose are expressly disclaimed.

© Copyright IBM Corporation 2013, ALL RIGHTS RESERVED.
*/
/**
 * Utility class for the creation of e-mail messages. See the method
 * {@link mypackage.MyOperations#sendMailMessage(VWAttachment, String, String, String, String[], String[]) sendMailMessage()} for
 * a description of the features.
 *
 * @author Ricardo Belfor
 *
 */
public class EmailTemplate {

    private Document document;

    public EmailTemplate(Document document) {
        this.document = document;
        this.document.fetchProperties(new String[] {
        PropertyNames.*CONTENT_ELEMENTS*, PropertyNames.*CONTENT_TYPE*,
        PropertyNames.*RETRIEVAL_NAME* });
    }

    /**
     * Creates an e-mail message based using the document used to create this class
     * as a template.
     *
     * @param mailSession the mail session
     * @param from the sender of the message
     * @param recipients the recipients of the message
     * @param subject the subject of the message
     * @param templateNames the names of the template parameters.
     * @param templateValues the values of the template parameters.

```java
 *
 * @return the message
 *
 * @throws Exception
 */
public MimeMessage getMessage(Session mailSession, InternetAddress from, InternetAddress[] recipients,
        String subject, String[] templateNames, String
        templateValues[]) throws Exception {

    MimeMessage mimeMessage = createMimeMessage(mailSession, from, recipients, subject);
    Multipart multiPart = new MimeMultipart();
    addMimeBodyParts(multiPart, templateNames, templateValues );
    mimeMessage.setContent( multiPart );

    return mimeMessage;
}

private void addMimeBodyParts(Multipart multiPart, String[] templateNames, String templateValues[] ) throws MessagingException {

    ContentElementList contentElements = document.get_
    ContentElements();
    Iterator<?> iterator = contentElements.iterator();
    if ( !iterator.hasNext() ) {
        return;
    }

    ContentTransfer contentTransfer = (ContentTransfer) iterator.next();
    addMessagePart(multiPart, templateNames, templateValues, contentTransfer);

    addImageParts(multiPart, iterator);
}
```

```java
private void addMessagePart(Multipart multiPart, String[] 
templateNames, String[] templateValues,
        ContentTransfer contentTransfer) throws MessagingException {
    MimeBodyPart mimeBodyPart = new MimeBodyPart();
    DataSource dataSource = new TemplateContentElementDataSource(cont
    entTransfer, templateNames, templateValues);
    mimeBodyPart.setDataHandler(new DataHandler(dataSource ) );
    multiPart.addBodyPart( mimeBodyPart );
}

private void addImageParts(Multipart multiPart, Iterator<?> iterator) 
throws MessagingException {
    int index = 1;
    while (iterator.hasNext() ) {
        ContentTransfer contentTransfer = (ContentTransfer)
        iterator.next();
        addImagePart(multiPart, contentTransfer, index);
        ++index;
    }
}

private void addImagePart(Multipart multiPart, ContentTransfer 
contentTransfer, int index)
        throws MessagingException {

    MimeBodyPart imagePart = new MimeBodyPart();
    DataSource dataSource = new ContentElementDataSource(
    contentTransfer );
    imagePart.setDataHandler(new DataHandler( dataSource) );
    imagePart.setHeader("Content-ID","<content" + index + ">");
    multiPart.addBodyPart( imagePart );
}
```

```java
    private MimeMessage createMimeMessage(Session mailSession,
    InternetAddress from, InternetAddress[] recipients,
            String subject)
            throws MessagingException {
       MimeMessage message = new MimeMessage(mailSession);
      message.setFrom(from);
      message.setRecipients(javax.mail.Message.RecipientType.TO,
      recipients );
       message.setSubject(subject);
      message.setSentDate(new Date());
       return message;
    }
}
```

CHAPTER 6   PDF DOCUMENT CREATION USING JAVA

Adding the TemplateContentElementDataSource Java Class

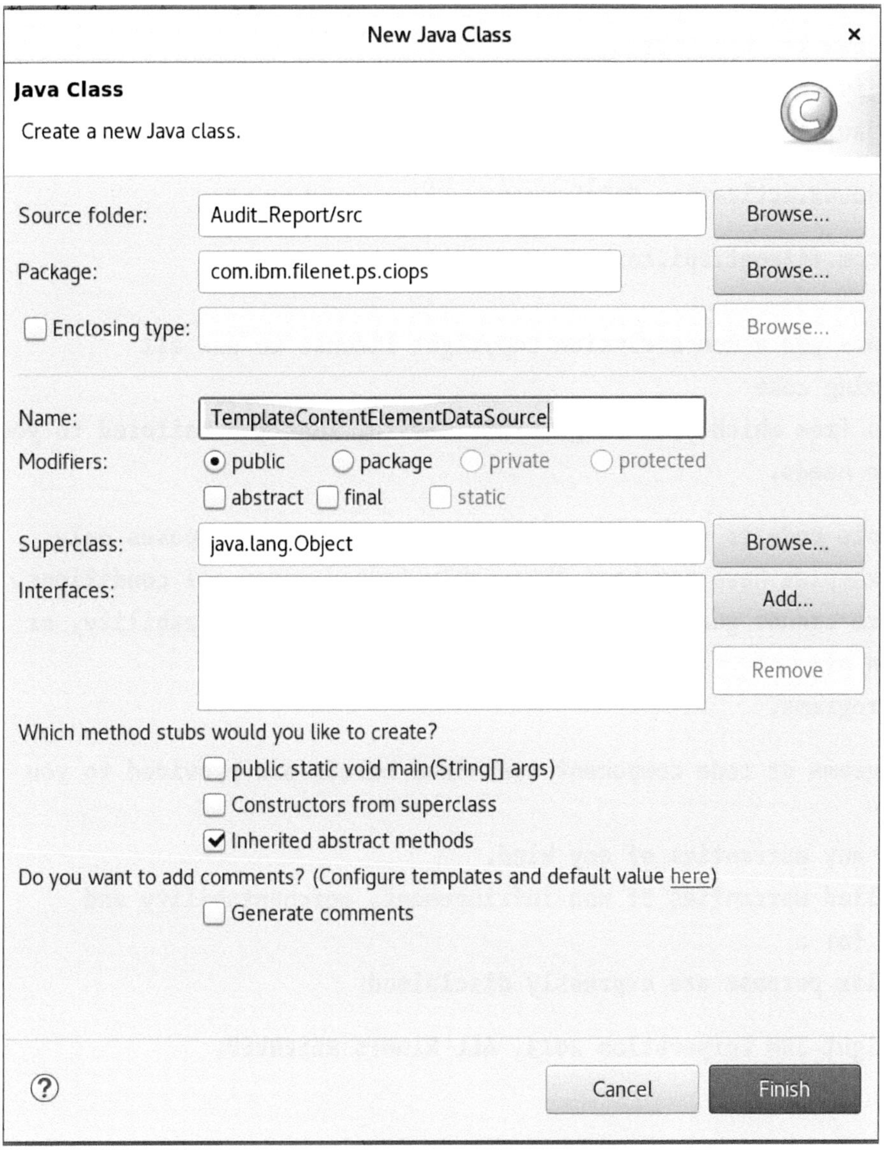

*Figure 6-23. The **TemplateContentElementDataSource** class code is added to the package*

The Java code for the **TemplateContentElementDataSource.java** source is as follows.

Listing 6-20. The TemplateContentElementDataSource.java source code

```
package com.ibm.filenet.ps.ciops;

import java.io.IOException;
import java.io.InputStream;
import java.io.OutputStream;
import java.io.ByteArrayInputStream;

import javax.activation.DataSource;

import com.filenet.api.core.ContentTransfer;
/**
IBM grants you a non-exclusive copyright license to use all
programming code
examples from which you can generate similar function tailored to your own
specific needs.

All sample code is provided by IBM for illustrative purposes only.
These examples have not been thoroughly tested under all conditions.  IBM,
therefore cannot guarantee or imply reliability, serviceability, or
function of
these programs.

All Programs or code component contained herein are provided to you
"AS IS "
without any warranties of any kind.
The implied warranties of non-infringement, merchantability and
fitness for a
particular purpose are expressly disclaimed.

© Copyright IBM Corporation 2013, ALL RIGHTS RESERVED.
*/

/**
 * This class wraps a ContentTransfer object in a DataSource object and
 * substitutes the template place holders with template values.
This way the
 * content is uses a part of an e-mail message.
```

```java
 *
 * @author Ricardo Belfor
 *
 */
public class TemplateContentElementDataSource implements DataSource {

    private ContentTransfer contentTransfer;
    private String[] templateNames;
    private String[] templateValues;

    public TemplateContentElementDataSource(ContentTransfer
    contentTransfer, String[] templateNames, String templateValues[]) {
        this.contentTransfer = contentTransfer;
        this.templateNames = templateNames;
        this.templateValues = templateValues;
    }

    public String getContentType() {
        return contentTransfer.get_ContentType();
    }

    public InputStream getInputStream() throws IOException {
        return new ByteArrayInputStream( getInputString().
        getBytes("UTF-8") );
    }

    private String getInputString() throws IOException {

        String inputString = getContentAsString();

        for (int i = 0; i < templateNames.length; i++) {
            String value = getTemplateValue(i);
            inputString = inputString.replaceAll( "\\$" +
            templateNames[i], value );
        }

        return inputString;
    }
```

```java
    public String getName() {
        return contentTransfer.get_RetrievalName();
    }

    public OutputStream getOutputStream() throws IOException {
        return null;
    }

    private String getTemplateValue(int index) {
        String value = "";
        if (index < templateValues.length && templateValues[index]
        != null) {
            value = templateValues[index];
        }
        return value;
    }

    private String getContentAsString() throws IOException {
        StringBuffer stringContent = new StringBuffer();
        byte buffer[] = new byte[128];
        InputStream inputStream = contentTransfer.accessContentStream();
        int readCount = 0;

        while ((readCount = inputStream.read(buffer)) > 0) {
            stringContent.append(new String(buffer, 0, readCount));

        }
        return stringContent.toString();
    }
}
```

CHAPTER 6 PDF DOCUMENT CREATION USING JAVA

Adding the com.ibm.filenet.ps.ciops.database Package

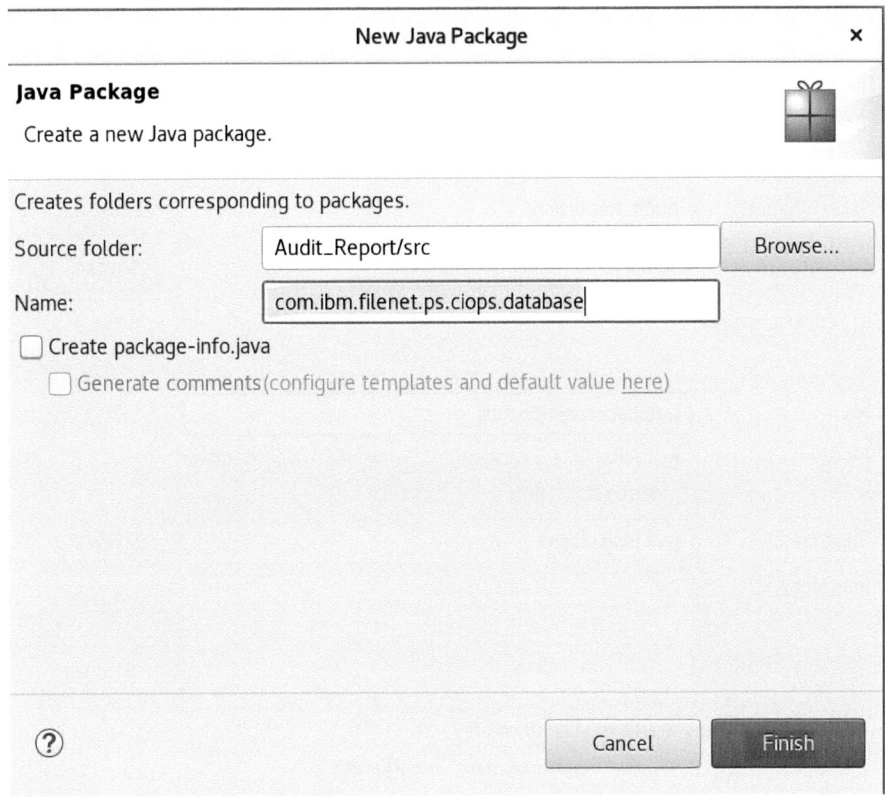

Figure 6-24. *The **com.ibm.filenet.ps.ciops.database** package is added*

The **com.ibm.filenet.ps.ciops.database** package is part of the example IBM FileNet Component Integration Workflow Operations code.

Adding the DatabaseLoginModule Java Class

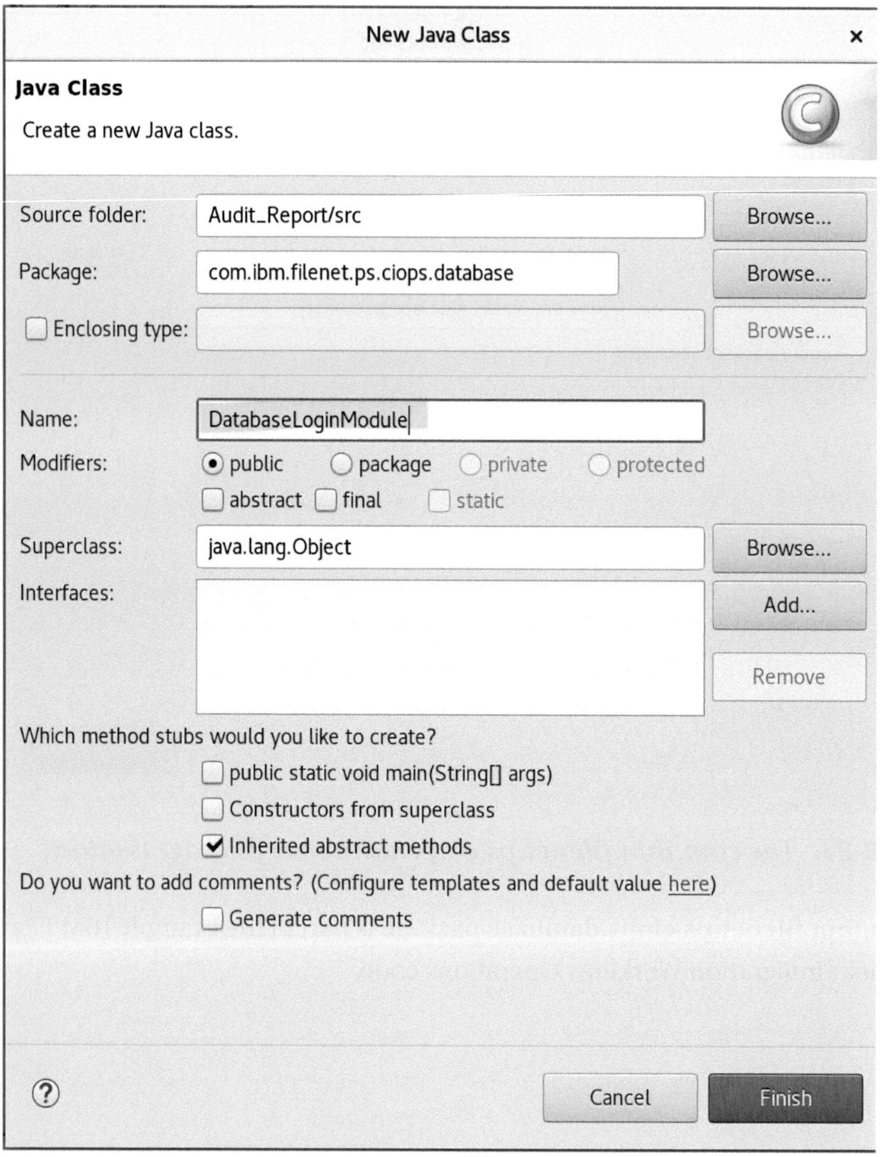

*Figure 6-25. The **DatabaseLoginModule** class code is added to the **com.ibm.filenet.ps.ciops.database** package*

The Java code for the **DatabaseLoginModule.java** source is as follows.

Listing 6-21. The **DatabaseLoginModule.java** source code file

```java
package com.ibm.filenet.ps.ciops.database;

import java.sql.Connection;
import java.sql.DriverManager;
import java.sql.SQLException;
import java.util.Map;

import javax.security.auth.Subject;
import javax.security.auth.callback.CallbackHandler;
import javax.security.auth.login.LoginException;
import javax.security.auth.spi.LoginModule;

import org.apache.log4j.Logger;
/**
IBM grants you a non-exclusive copyright license to use all
programming code
examples from which you can generate similar function tailored to your own
specific needs.

All sample code is provided by IBM for illustrative purposes only.
These examples have not been thoroughly tested under all conditions.  IBM,
therefore cannot guarantee or imply reliability, serviceability, or
function of
these programs.

All Programs or code component contained herein are provided to you
"AS IS "
without any warranties of any kind.
The implied warranties of non-infringement, merchantability and
fitness for a
particular purpose are expressly disclaimed.

© Copyright IBM Corporation 2013, ALL RIGHTS RESERVED.
*/
/**
 *
 * @author Ricardo Belfor
```

```java
 *
 */
public class DatabaseLoginModule implements LoginModule
{
    private static Logger logger = Logger.getLogger( DatabaseLoginModule.
    class );
    @SuppressWarnings("unchecked")
    private Map sharedState;
    private Subject subject;
   private boolean succeeded;
    private String username;
    private String password;
    private String driverClass;
    private String connectionUrl;
    private Connection connection;
    private DatabasePrincipal principal;
    private boolean commitSucceeded = false;
    private boolean debug = true;

    public void initialize(Subject subject, CallbackHandler
    callbackHandler, Map<String, ?> sharedState,
            Map<String, ?> options)
    {
        this.subject = subject;
        this.sharedState = sharedState;

        driverClass = (String) options.get( "driverClass");
        connectionUrl = (String) options.get( "connectionUrl" );
        debug = "true".equals( options.get("debug") );

        debug( "driverClass: " + driverClass );
        debug( "connectionUrl: " + connectionUrl );
    }

    public boolean login() throws LoginException {
        debug( "[enter] login()" );
        succeeded = false;
```

```
        getCredentials();
        createDatabaseConnection();
    succeeded = true;
        debug( "[exit]  login()" );
        return succeeded;
}

private void getCredentials() throws LoginException
{
    username = (String) sharedState.get("javax.security.auth.
    login.name");
    password = (String) sharedState.get("javax.security.auth.login.
    password");

    if ( username == null || password == null ) {
        throw new LoginException( "Username or password not set" );
    }

    debug( "username: " + username );
        debug( "password: " + password );
}

private Connection createDatabaseConnection() throws LoginException
{
    try {
            Class.forName( driverClass );
            connection = DriverManager.getConnection(connectionUrl,
            username, password);
            return connection;
    } catch (Exception exception ) {
            debug( exception.toString(), exception );
            throw new LoginException( exception.getLocalizedMessage() );
        }
}
```

```java
public boolean commit() throws LoginException
{
    debug( "[enter] commit()" );
    if ( succeeded ) {
        checkSubject();
        registerPrincipal();
        clearCredentials();
        commitSucceeded = true;
        debug( "[exit] commit() (true)" );
        return true;
    }
    debug( "[exit] commit() (false)" );
    return false;
}

private void registerPrincipal() throws LoginException {
    debug( "[enter] registerPrincipal" );
    principal = new DatabasePrincipal( username, connection );
    if( !subject.getPrincipals().contains(principal) ) {
        subject.getPrincipals().add(principal);
        debug( principal.toString() +  " added" );
    }
    debug( "[exit]  registerPrincipal" );
}

private void checkSubject() throws LoginException
{
    if ( subject == null ) {
        throw new LoginException("Subject is null");
    }

    if (subject.isReadOnly()) {
        throw new LoginException ("Subject is Readonly");
    }
}

public boolean abort() throws LoginException
```

```java
{
    debug( "[enter] abort()" );
   if( !succeeded ) return false;

    if( !commitSucceeded ) {
       succeeded = false;
       clearCredentials();
       closeDatabaseConnection();
   } else {
       logout();
   }
    debug( "[exit]  abort()" );
   return true;
}

private void closeDatabaseConnection()
{
    debug( "[enter] closeDatabaseConnection()()" );
    try {
         connection.close();
    } catch (SQLException e) {
         debug( "close failed.", e );
    }
    connection = null;
    debug( "[exit]  closeDatabaseConnection()()" );
}

public boolean logout() throws LoginException
{
    debug( "[enter] logout()" );
   unregisterPrincipal();
   succeeded = commitSucceeded;
   clearCredentials();
   closeDatabaseConnection();
    debug( "[exit]  logout()" );
    return true;
}
```

```java
    private void clearCredentials()
    {
        username = null;
        password = null;
    }

    private void unregisterPrincipal()
    {
        if( commitSucceeded ) {
            subject.getPrincipals().remove(principal);
        }
        principal = null;
    }

    private void debug(String message, Throwable throwable) {
        if ( ! debug ) return;

        if ( throwable == null ) {
            logger.debug( message );
        } else {
            logger.error( message, throwable );
        }
    }

    private void debug(String message) {
        debug(message, null );
    }
}
```

Adding the DatabasePrincipal Java Class

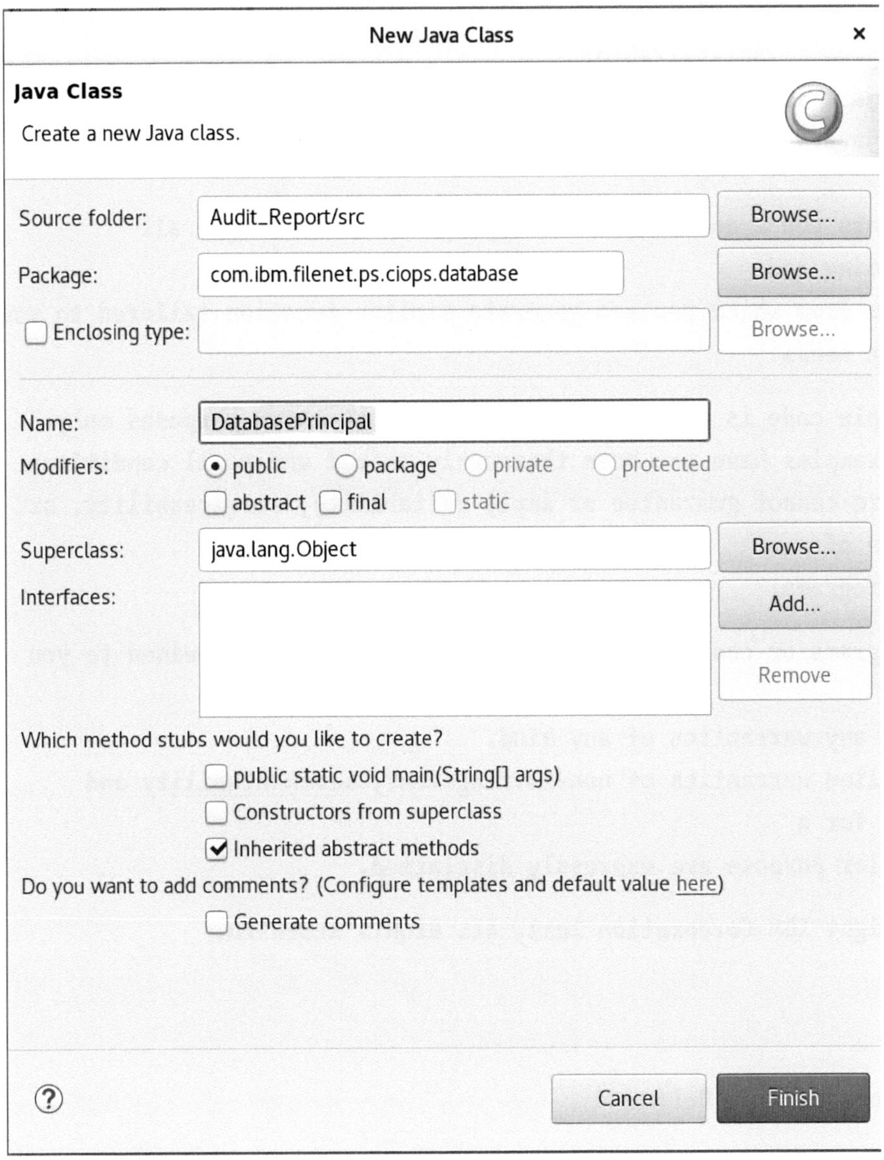

*Figure 6-26. The **DatabasePrincipal** class code is added to the **com.ibm.filenet.ps.ciops.database** package*

The Java code for the **DatabasePrincipal.java** source is as follows.

CHAPTER 6 PDF DOCUMENT CREATION USING JAVA

Listing 6-22. The **DatabasePrincipal.java** source code file

```
package com.ibm.filenet.ps.ciops.database;

import java.io.Serializable;
import java.security.Principal;
import java.sql.Connection;
/**
IBM grants you a non-exclusive copyright license to use all
programming code
examples from which you can generate similar function tailored to your own
specific needs.

All sample code is provided by IBM for illustrative purposes only.
These examples have not been thoroughly tested under all conditions.  IBM,
therefore cannot guarantee or imply reliability, serviceability, or
function of
these programs.

All Programs or code component contained herein are provided to you
"AS IS "
without any warranties of any kind.
The implied warranties of non-infringement, merchantability and
fitness for a
particular purpose are expressly disclaimed.

© Copyright IBM Corporation 2013, ALL RIGHTS RESERVED.
*/
/**
 *
 * @author Ricardo Belfor
 *
 */
public class DatabasePrincipal implements Principal, Serializable
{
    private static final long serialVersionUID = -6417436232744816471L;
    private String name;
    private Connection connection;
```

```java
public DatabasePrincipal(String username, Connection connection2)
{
    this.name = username;
    this.connection = connection2;
}
public String getName()
{
    return name;
}

public Connection getConnection()
{
    return connection;
}

@Override
public boolean equals(Object object)
{
   if (object != null )
   {
       if (this == object) return true;

       if (object instanceof DatabasePrincipal)
       {
           return (((DatabasePrincipal) object).getName().
           equals(name));
       }
   }
   return false;
}

@Override
public int hashCode()
{
    return name.hashCode();
}
```

```
@Override
public String toString()
{
    return "Database connection with " + name;
}
}
```

The **DatabaseLoginModule.java** and **DatabasePrincipal.java** source code is part of the example IBM FileNet Component Integration Workflow Operations code.

Adding the com.ibm.filenet.ps.ciops.test Package

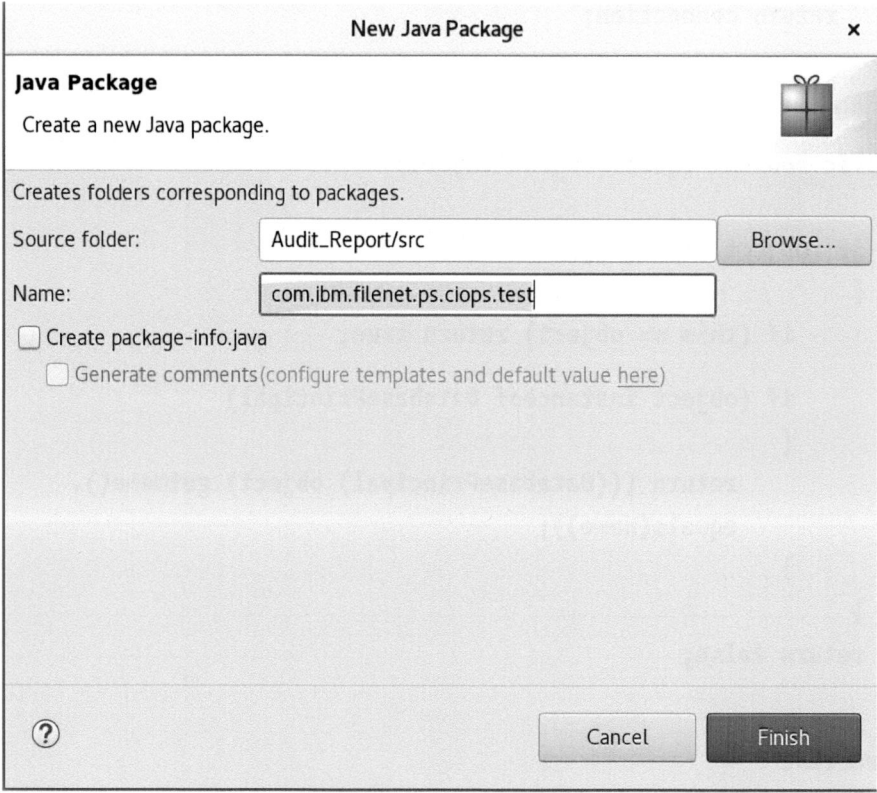

*Figure 6-27.   The **com.ibm.filenet.ps.ciops.test** package is added*

The **com.ibm.filenet.ps.ciops.test** package is part of the example IBM FileNet Component Integration Workflow Operations code.

Adding the Configuration Java Class

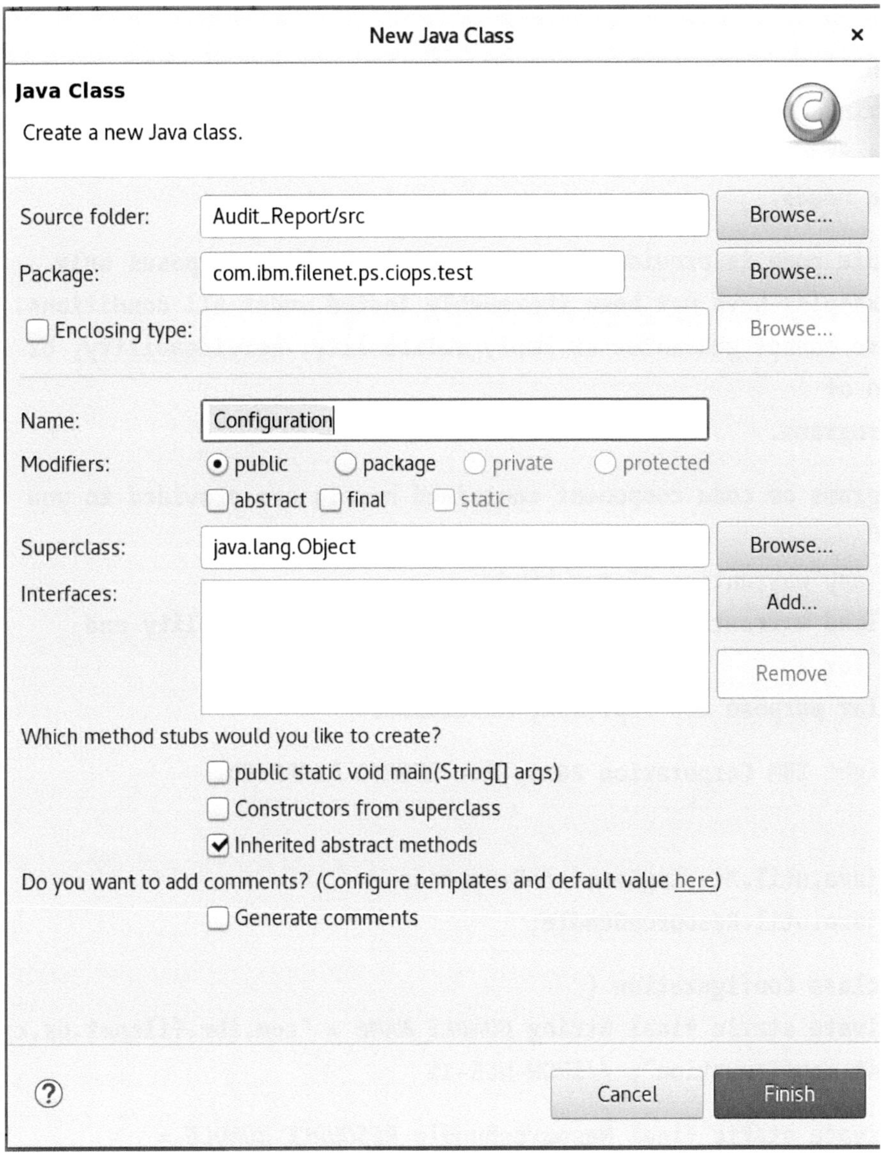

*Figure 6-28. The **Configuration** class code is added to the **com.ibm.filenet. ps.ciops.test** package*

The Java code for the **Configuration.java** source is as follows.

Listing 6-23. The **Configuration.java** source code file

```
package com.ibm.filenet.ps.ciops.test;
/**
IBM grants you a non-exclusive copyright license to use all
programming code
examples from which you can generate similar function tailored to your own
specific needs.

All sample code is provided by IBM for illustrative purposes only.
These examples have not been thoroughly tested under all conditions.  IBM,
therefore cannot guarantee or imply reliability, serviceability, or
function of
these programs.

All Programs or code component contained herein are provided to you
"AS IS "
without any warranties of any kind.
The implied warranties of non-infringement, merchantability and
fitness for a
particular purpose are expressly disclaimed.

© Copyright IBM Corporation 2013, ALL RIGHTS RESERVED.
*/

import java.util.MissingResourceException;
import java.util.ResourceBundle;

public class Configuration {
    private static final String BUNDLE_NAME = "com.ibm.filenet.ps.ciops.
    test.configuration"; //$NON-NLS-1$

    private static final ResourceBundle RESOURCE_BUNDLE =
    ResourceBundle.getBundle(BUNDLE_NAME);

    private Configuration() {
    }
```

```java
    public static String getParameter(String key) {
        try {
            return RESOURCE_BUNDLE.getString(key);
        } catch (MissingResourceException e) {
            return '!' + key + '!';
        }
    }
}
```

Note The preceding code uses a Java utility **ResourceBundle**. Resource bundles contain locale-specific objects. When a program needs a locale-specific resource, such as a String, the program can load it from the resource bundle that is for the current user's locale.

Adding the QOperationsTest Java Class

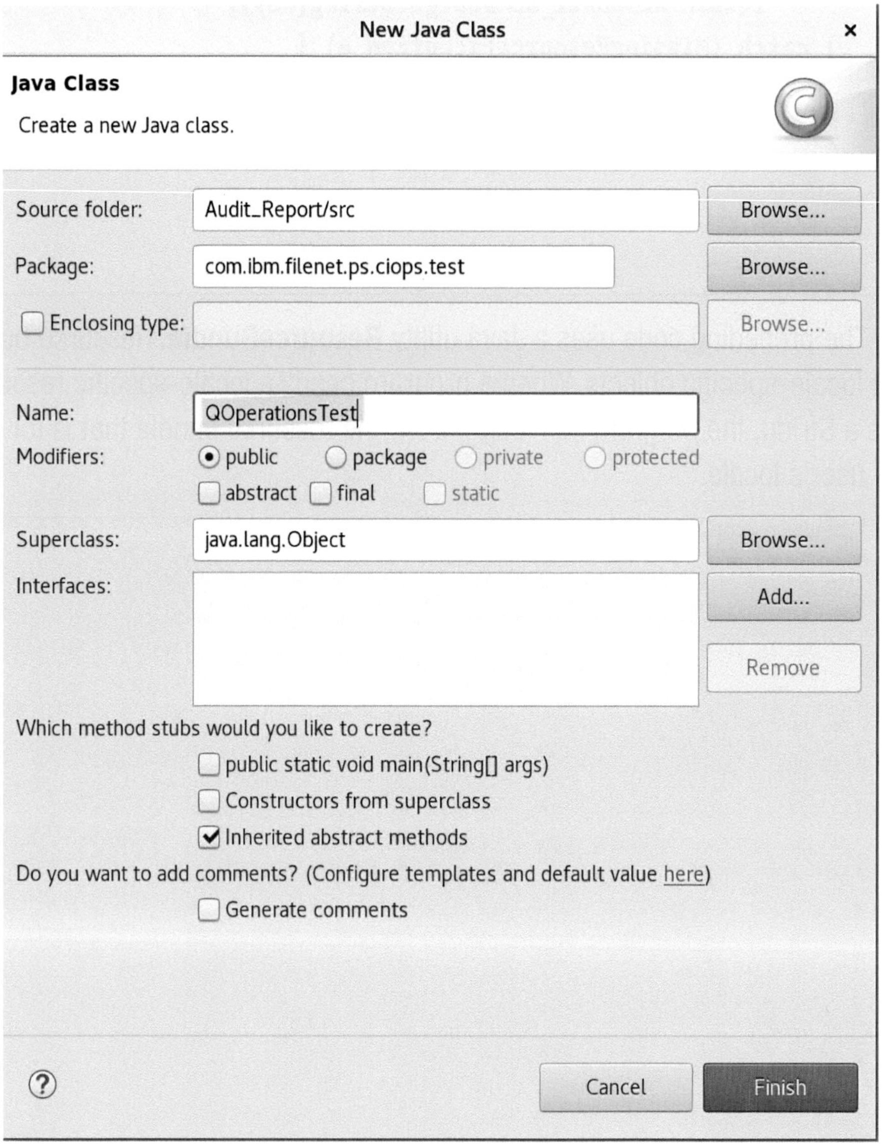

*Figure 6-29. The **QOperationsTest** class code is added to the **com.ibm.filenet.ps.ciops.test** package*

The Java code for the **QOperationsTest.java** source is as follows.

Listing 6-24. The **QOperationsTest.java** source code file

```
package com.ibm.filenet.ps.ciops.test;
/**
IBM grants you a non-exclusive copyright license to use all
programming code
examples from which you can generate similar function tailored to your own
specific needs.

All sample code is provided by IBM for illustrative purposes only.
These examples have not been thoroughly tested under all conditions.  IBM,
therefore cannot guarantee or imply reliability, serviceability, or
function of
these programs.

All Programs or code component contained herein are provided to you
"AS IS "
without any warranties of any kind.
The implied warranties of non-infringement, merchantability and
fitness for a
particular purpose are expressly disclaimed.

© Copyright IBM Corporation 2013, ALL RIGHTS RESERVED.
*/
import static org.junit.Assert.*;

import java.text.SimpleDateFormat;
import java.util.Date;

import javax.security.auth.Subject;

import com.ibm.filenet.ps.ciops.QOperations;

import org.junit.BeforeClass;
import org.junit.Test;
import com.filenet.api.constants.PropertyNames;
import com.filenet.api.core.Connection;
import com.filenet.api.core.Document;
import com.filenet.api.core.Domain;
```

```java
import com.filenet.api.core.EntireNetwork;
import com.filenet.api.core.Factory;
import com.filenet.api.core.Folder;
import com.filenet.api.core.ObjectStore;
import com.filenet.api.util.UserContext;

import filenet.vw.api.VWAttachment;
import filenet.vw.api.VWAttachmentType;
import filenet.vw.api.VWException;
import filenet.vw.api.VWLibraryType;
import filenet.vw.api.VWSession;

/**
 * This class contains code for off line testing of the custom component {@
 link com.ibm.filenet.ps.ciops.QOperations QOperations}. To
 * run this class the different configuration parameters must be set in the
 configuration.properties file.
 *
 * @author Ricardo Belfor
 *
 */
public class QOperationsTest
{
    private static final String EMAIL_TO = Configuration.getParameter("QOp
    erations.EmailTo"); //$NON-NLS-1$
    private static final String EMAIL_FROM = Configuration.getParameter("Q
    Operations.EmailFrom"); //$NON-NLS-1$
    private static final String EMAIL_TEMPLATES_PATH = Configuration.getPa
    rameter("QOperations.EmailTemplatePath"); //$NON-NLS-1$
    private static final String OBJECT_STORE_NAME = Configuration.getParam
    eter("QOperations.ObjectstoreName"); //$NON-NLS-1$
    private static final String TEST_FOLDER_NAME = Configuration.getParame
    ter("QOperations.TestFolderName"); //$NON-NLS-1$
    private static final String USERNAME = Configuration.getParameter("QOp
    erations.Username"); //$NON-NLS-1$
```

```java
private static final String PASSWORD = Configuration.getParameter("QOp
erations.Password"); //$NON-NLS-1$
private static final String CONNECTION_POINT_NAME = Configuration.getP
arameter("QOperations.ConnectionPointName"); //$NON-NLS-1$
private static final String CE_WSI_URL = Configuration.getParameter("Q
Operations.CEWsiUrl"); //$NON-NLS-1$
private static final String WASP_LOCATION = Configuration.getParameter
("QOperations.WaspLocation"); //$NON-NLS-1$
private static final String WSI_JAAS_CONFIG_FILE = Configuration.getPa
rameter("QOperations.WsiJaasConfigFile"); //$NON-NLS-1$

private static Connection connection;
private static VWSession vwSession;

/**
 * This method is run before testing is started. It creates
   connections to the Content Engine and
 * the Process Engine.
 *
 * @throws Exception
 */
@BeforeClass
public static void setUpBeforeClass() throws Exception {
    try {
        System.setProperty("java.security.auth.login.config", WSI_
        JAAS_CONFIG_FILE); //$NON-NLS-1$
        System.setProperty("wasp.location",WASP_LOCATION );
        //$NON-NLS-1$

        String url = CE_WSI_URL;
        String connectionPointName = CONNECTION_POINT_NAME;
        String password = PASSWORD;
        String username = USERNAME;

        createCEConnection(username, password, url);
        createVWSession(username, password, url,
        connectionPointName);
```

```java
        } catch (Exception e) {
            e.printStackTrace();
            throw e;
        }
    }

    private static void createCEConnection(String username, String
    password, String url) {
        connection = Factory.Connection.getConnection(url);

        Subject subject = UserContext.createSubject(connection, username,
        password, "FileNetP8"); //$NON-NLS-1$
        UserContext uc = UserContext.get();
        uc.pushSubject(subject);
    }

    private static void createVWSession(String username, String password,
    String url, String connectionPointName)
            throws VWException {
        vwSession = new VWSession();
        vwSession.setBootstrapCEURI(url);
        vwSession.logon( username, password, connectionPointName );
    }

    private QOperations getQOperations() {
        return new QOperations() {
            @Override
            protected Connection getConnection() {
                return connection;
            }

            @Override
            protected VWSession getVWSession() throws VWException {
                return vwSession;
            }
        };
    }
```

CHAPTER 6 PDF DOCUMENT CREATION USING JAVA

```java
/**
 * Test method for the {@link com.ibm.filenet.ps.ciops.QOperations.#ge
 tFolderDocuments(VWAttachment)} method. It
 * uses a test folder in the object store as input.
 *
 * @throws Exception
 */
@Test
public void testGetFolderDocuments() throws Exception {
    VWAttachment folderAttachment = getTestFolder();
    QOperations  QOperations = getQOperations();
    VWAttachment[] folderDocuments = QOperations.getFolderDocuments(f
    olderAttachment);
    showResults(folderDocuments);
}
/**
 * Test method for the {@link com.ibm.filenet.ps.ciops.QOperatio
 ns.#sendJMSMessage(VWAttachment docAttachment, String Message )
 sendJMSMessage()} method.
 *
 * @throws Exception
 */
@Test
private void showResults(VWAttachment[] folderDocuments) {
    System.out.println( folderDocuments.length +  " documents found"
    ); //$NON-NLS-1$
    for (VWAttachment attachment : folderDocuments) {
        System.out.println( attachment.toString() );
    }
}

private VWAttachment getTestFolder() throws VWException {
    ObjectStore objectStore = getTestObjectStore();
    Folder folder = (Folder) objectStore.getObject( "Folder", TEST_
    FOLDER_NAME ); //$NON-NLS-1$
```

CHAPTER 6   PDF DOCUMENT CREATION USING JAVA

```java
        VWAttachment folderAttachment = getFolderAsVWAttachment(folder);
        return folderAttachment;
    }

    private ObjectStore getTestObjectStore()
    {
        EntireNetwork entireNetwork = Factory.EntireNetwork.fetchInstance
        (connection, null);
        Domain domain = entireNetwork.get_LocalDomain();
        ObjectStore objectStore = Factory.ObjectStore.getInstance(
        domain, OBJECT_STORE_NAME );
        return objectStore;
    }

    /**
     * Test method for the {@link com.ibm.filenet.ps.ciops.QOperations
     * .#sendMailMessage(VWAttachment, String, String, String, String[],
     * String[]) sendMailMessage()} method.
     *
     * @throws Exception
     */
    @Test
    private VWAttachment getDocumentAsVWAttachment(Document document)
    throws VWException {
        VWAttachment folderAttachment = getCEAttachment(document.
        getObjectStore() );
        document.fetchProperties( new String[] { PropertyNames.ID,
        PropertyNames.NAME } );
        folderAttachment.setId( document.get_Id().toString() );
        folderAttachment.setAttachmentName( document.get_Name() );
        folderAttachment.setType( VWAttachmentType.ATTACHMENT_TYPE_
        FOLDER );
        return folderAttachment;
    }
```

```java
private VWAttachment getFolderAsVWAttachment(Folder folder) throws
VWException {
    VWAttachment folderAttachment = getCEAttachment(folder.
    getObjectStore() );
    folder.fetchProperties( new String[] { PropertyNames.ID,
    PropertyNames.NAME } );
    folderAttachment.setId( folder.get_Id().toString() );
    folderAttachment.setAttachmentName( folder.get_Name() );
    folderAttachment.setType( VWAttachmentType.ATTACHMENT_TYPE_
    FOLDER );
    return folderAttachment;
}

private VWAttachment getCEAttachment(ObjectStore objectStore) throws
VWException {
    VWAttachment ceAttachment = new VWAttachment();
    ceAttachment.setLibraryType( VWLibraryType.LIBRARY_TYPE_CONTENT_
    ENGINE );
    objectStore.fetchProperties( new String[] {
    PropertyNames.NAME } );
    ceAttachment.setLibraryName( objectStore.get_Name() );
    return ceAttachment;
}

/**
 * Utility function for time stamping the different things produced by
 the test code. This
 * way the result of different tests can be kept apart.
 *
 * @return a timestamp string.
 */
private static String getTimestamp() {
    SimpleDateFormat timestampFormatter = new
    SimpleDateFormat("yyyyMMdd HHmmSSS"); //$NON-NLS-1$
    return timestampFormatter.format( new Date() );
}
}
```

CHAPTER 6   PDF DOCUMENT CREATION USING JAVA

Adding the AUDITReportMain Java Class

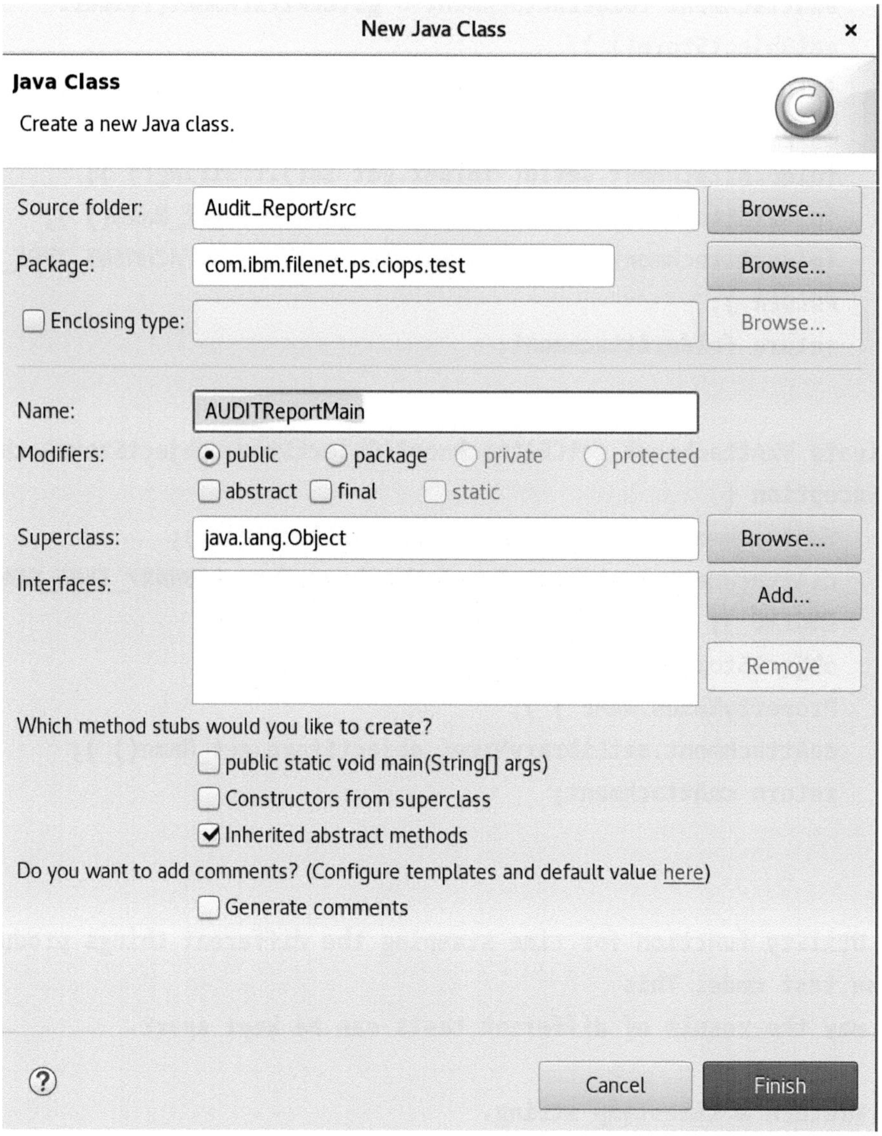

*Figure 6-30.   The **AUDITReportMain** class code is added to the **com.ibm.filenet. ps.ciops.test** package*

The Java code for the **AUDITReportMain.java** source is as follows.

This is the main program class which is called for the production of the Audit Report pdf files.

Listing 6-25. The **AUDITReportMain.java** source code file

```java
package com.ibm.filenet.ps.ciops.test;
import java.security.AccessController;
import java.util.ArrayList;
import java.util.Calendar;
import java.util.Iterator;
import java.util.Locale;
import java.util.Properties;
import java.util.Set;

import javax.mail.Message;
import javax.mail.MessagingException;
import javax.mail.PasswordAuthentication;
import javax.mail.Session;
import javax.mail.Transport;
import javax.mail.internet.InternetAddress;
import javax.mail.internet.MimeMessage;
import javax.naming.Context;
import javax.naming.directory.InitialDirContext;
import javax.security.auth.Subject;

import org.apache.commons.io.LineIterator;

import com.asb.ce.utils.AUDITReportConfig;
import com.asb.config.AUDITConfig;
import com.filenet.api.collection.DocumentSet;
import com.filenet.api.collection.IndependentObjectSet;
import com.filenet.api.constants.FilteredPropertyType;
import com.filenet.api.constants.PropertyNames;
import com.filenet.api.constants.RefreshMode;
import com.filenet.api.core.CmTask;
import com.filenet.api.core.Connection;
import com.filenet.api.core.Document;
import com.filenet.api.core.Domain;
```

CHAPTER 6 PDF DOCUMENT CREATION USING JAVA

```java
import com.filenet.api.core.EntireNetwork;
import com.filenet.api.core.Factory;
import com.filenet.api.core.Folder;
import com.filenet.api.core.IndependentObject;
import com.filenet.api.core.ObjectStore;
import com.filenet.api.property.FilterElement;
import com.filenet.api.property.PropertyFilter;
import com.filenet.api.query.SearchSQL;
import com.filenet.api.query.SearchScope;
import com.filenet.api.util.UserContext;
import com.ibm.filenet.ps.ciops.QOperations;
import com.ibm.filenet.ps.ciops.database.DatabasePrincipal;

import filenet.vw.api.VWAttachment;
import filenet.vw.api.VWAttachmentType;
import filenet.vw.api.VWException;
import filenet.vw.api.VWLibraryType;
import filenet.vw.api.VWSession;
import filenet.vw.base.logging.Logger;

import java.io.File;
//Q PDF Creation Operations and supporting methods 29th July 2022
import java.io.FileOutputStream;
import java.util.Date;

import com.lowagie.text.Anchor;
import com.lowagie.text.BadElementException;
//import com.lowagie.text.BaseColor;
import com.lowagie.text.Chapter;
import com.lowagie.text.DocumentException;
import com.lowagie.text.Element;
import com.lowagie.text.Font;
import com.lowagie.text.Image;
import com.lowagie.text.List;
import com.lowagie.text.ListItem;
import com.lowagie.text.PageSize;
import com.lowagie.text.Paragraph;
```

```java
import com.lowagie.text.Phrase;
import com.lowagie.text.Rectangle;
import com.lowagie.text.Section;
import com.lowagie.text.pdf.PdfContentByte;
import com.lowagie.text.pdf.PdfPCell;
import com.lowagie.text.pdf.PdfPTable;
import com.lowagie.text.pdf.PdfWriter;
import com.lowagie.text.pdf.RandomAccessFileOrArray;
import com.lowagie.text.pdf.codec.TiffImage;
//
import java.io.BufferedWriter;
import java.io.FileInputStream;
import java.io.FileNotFoundException;
import java.io.FileWriter;
import java.io.IOException;
import java.sql.ResultSet;
import java.sql.Statement;
import java.sql.Timestamp;
import java.text.ParseException;
import java.text.SimpleDateFormat;
import java.util.Vector;
import java.util.concurrent.Executors;
import java.util.concurrent.ScheduledExecutorService;
import java.util.concurrent.TimeUnit;

import org.apache.commons.io.FileUtils;
//import org.apache.commons.lang.math.RandomUtils;   //ASB removed this
import 26-08-2022
import com.asb.ce.utils.*;
//
/**
IBM grants you a non-exclusive copyright license to use all
programming code
examples from which you can generate similar function tailored to your own
specific needs.
```

CHAPTER 6 PDF DOCUMENT CREATION USING JAVA

```
All sample code is provided by IBM for illustrative purposes only.
These examples have not been thoroughly tested under all conditions.  IBM,
therefore cannot guarantee or imply reliability, serviceability, or function of
these programs.

All Programs or code component contained herein are provided to you
"AS IS "
without any warranties of any kind.
The implied warranties of non-infringement, merchantability and fitness for a
particular purpose are expressly disclaimed.

© Copyright IBM Corporation 2013, ALL RIGHTS RESERVED.
*/
/**
 * Alan S.Bluck   11th August 2022
 * Code for Testing and Developing AUD_Operations
 * Adapted from:
 * Custom Java component serving as an example for the article
 * "Developing Java components for the Component Integrator".
 *
 * @author Ricardo Belfor
 *
 */
public class AUDITReportMain {
    public static String [] TaskParams = new String[100]; //ASB Array of
    the Case task type parameters
    public static String [] TaskName = new String[100]; //ASB Array of the
    Case task types
    public static String [] QStart = new String[100];    //ASB Array of the
    start question numbers
    public static String [] QEnd = new String[100];      //ASB Array of the
    end question numbers
    public static final Logger logger = Logger.getLogger(AUDITReportMain.
    class.getName());
```

CHAPTER 6　PDF DOCUMENT CREATION USING JAVA

```java
//private static final String EMAIL_TO = Configuration.
getParameter("QOperations.EmailTo"); //$NON-NLS-1$
private static String EMAIL_TO ; //ASB May need to change this to a list of users in the config.xml
private static String EMAIL_FROM; //$NON-NLS-1$
private static String EMAIL_TEMPLATES_PATH ;
private static  String OBJECT_STORE_NAME;
private static String TEST_FOLDER_NAME;
private static String USERNAME;
private static  String PASSWORD;
private static String CONNECTION_POINT_NAME;
private static String CE_WSI_URL;
private static String WASP_LOCATION;
private static String WSI_JAAS_CONFIG_FILE;
// Properties file load
private static final Properties QQidMap = new Properties();

private static Connection connection;
private static VWSession vwSession;
private static QOperations  QOperations;
private static AUDITReportConfig cemc;

private static void init() throws Exception{
    String fullPath = System.getProperty("user.dir") + File.separator
            + "config" + File.separator + "QQID.properties";
    cemc = new AUDITReportConfig(); //ASB add our config.xml file methods here!
    try {
        EMAIL_TEMPLATES_PATH = cemc.getSMTPEmailTemplatePath();
        EMAIL_TO = cemc.getSMTPUser();
        EMAIL_FROM = cemc.getSourceEmailAddress();
        OBJECT_STORE_NAME = cemc.getExportOSName();
        TEST_FOLDER_NAME = cemc.getTestFolderName();
        USERNAME = cemc.getExportCEUser();
        PASSWORD = cemc.getExportCEPassword();
        CONNECTION_POINT_NAME = cemc.getExportConnectionPointName();
```

```java
            CE_WSI_URL = cemc.getExportCEUrl();
            WASP_LOCATION = cemc.getExportWaspLocation();
            WSI_JAAS_CONFIG_FILE = cemc.getExportCELoginConfig();
        String url = cemc.getExportCEUrl();
            //String url = CE_WSI_URL;
        String connectionPointName = cemc.
        getExportConnectionPointName();
            //String connectionPointName = ;
            String password = cemc.getExportCEPassword(); //ASB Let's
            get it encrypted for security from config.xml
            //String password = PASSWORD;
            String username = cemc.getExportCEUser();
            //String username = USERNAME;

            createCEConnection(username, password, url);
            createVWSession(username, password, url,
            connectionPointName); //CHECK
    } catch (Exception e) {
        e.printStackTrace();
        throw e;
    }
    //
    try {
        QQidMap.load(new FileInputStream(new File(fullPath)));
        System.setProperty("java.security.auth.login.config", WSI_
        JAAS_CONFIG_FILE); //$NON-NLS-1$
        System.setProperty("wasp.location",WASP_LOCATION );
        //$NON-NLS-1$
        return;
    } catch (FileNotFoundException e) {
        // TODO Auto-generated catch block
        e.printStackTrace();
    } catch (IOException e) {
        // TODO Auto-generated catch block
        e.printStackTrace();
    }
```

CHAPTER 6   PDF DOCUMENT CREATION USING JAVA

```java
        throw new RuntimeException(fullPath + " cannot load QQid map");

        //java.sql.Connection databaseConnection = getDatabaseConnection();
        //if ( databaseConnection != null ) {
        //     logger.debug( databaseConnection.toString() );
        //} else {
        //     logger.debug( "No database connection" );
        //}
    }

    public static void main(String[] args) throws Exception {
        //We Could use the method below if required!
        //   final ScheduledExecutorService executorService = Executors.newSingleThreadScheduledExecutor();
        init();
          while(cemc.getSTOPState().equalsIgnoreCase("No")) {
           // TODO Auto-generated method stub
          //Test the operations class
       // Set up the Operations Class for testing
          QOperations   = getQOperations();
          String[] TaskId = new String[1000];
          String[] TaskNames = new String[1000];
          String[] CaseId = new String[1000];
          int numberOfTasks=0;
//getCompletedTaskInfo(TaskId,TaskNames,CaseId,numberOfTasks);
//ASB     22-08-2022
              //ASB ...  Dynamically update the next Loader Start
              // Need date in the format 20130923T125628Z
          String sDeltaMinutes = cemc.getDeltaMinutes();
          String sDeltaHours = cemc.getDeltaHours();
          String sSearchDate = cemc.getQStartSearchDate();
          Integer iDeltaHours = Integer.valueOf(sDeltaHours);
          Integer iDeltaMinutes = Integer.valueOf(sDeltaHours);

          SimpleDateFormat formatter = new SimpleDateFormat("yyyyMMddHHmmss");
```

```java
    SimpleDateFormat sdf = new SimpleDateFormat("yyyyMMdd",Locale
    .ENGLISH);
SimpleDateFormat sdfTime = new SimpleDateFormat("HHmmss",Locale
.ENGLISH);
    Date newDateSearch = new Date();
String QstartSearchDate ="";
logger.info("Audit Report Version 3.0 Build 14-08-2022:12-41PM" );

    try {
            //String sDate = sSearchDate.substring(0, 8);
            //String sTime = sSearchDate.substring(9,15);
            //newDateSearch = formatter.parse(sDate + sTime);
            Calendar cal = Calendar.getInstance();
            cal.setTime(newDateSearch);
            cal.add(Calendar.MINUTE, iDeltaMinutes);
            cal.add(Calendar.HOUR, iDeltaHours);
            newDateSearch = cal.getTime();
         // Need date in the format 20130923T125628Z
        QstartSearchDate = sdf.format(newDateSearch)+ "T" + sdfTime.
        format(newDateSearch)+ "Z";
          } catch (Exception e) {
              e.printStackTrace();
          }
    getCompletedTaskInfo(TaskId,TaskNames,CaseId,numberOfTasks,Qstar
    tSearchDate);
     //
    cemc.updateProcessDate(newDateSearch);
    logger.info("Next Batch Date Start from Document :" +
    newDateSearch);
    Integer sleepTime = cemc.getSleepTime();
    TimeUnit.MINUTES.sleep(sleepTime);
     init();
    } //End Batch Loop
}
```

```java
    private static QOperations getQOperations() {
        return new QOperations() {
            @Override
            protected Connection getConnection() {
                return connection;
            }

            @Override
            protected VWSession getVWSession() throws VWException {
                return vwSession;
            }
        };
    }
    private static void getCompletedTaskInfo(String[] TaskId,String[] TaskNames,String[] CaseId, int NumberOfTasks,String sSearchDate) {
        try
        {   // Get data from  all completed Tasks to generate the related pdf files for attachment to the cases
            // From CmAcmCaseTask
            //String sSearchDate = cemc.getQStartSearchDate();
            logger.info("Start Search Date : " + sSearchDate);
            SearchSQL sqlObject = new SearchSQL();
            String selectList = cemc.getQSelectList();
            sqlObject.setSelectList(selectList); //Coordinator is the main Case Folder Object
            Integer maxRecords = 1000 ;//Should only be small numbers at any one time
            sqlObject.setFromClauseInitialValue("CmAcmCaseTask", "f", true); //set true to include subclasses
          String QWhereClause = cemc.getQWhereClause() + " "
          +  sSearchDate.trim();
            sqlObject.setWhereClause(QWhereClause); //ASB 11-08-2022
            //NB 5 is the Complete status
          String osName = cemc.getExportOSName();
```

```
ObjectStore objectStore = QOperations.
getObjectStore(osName);    //ASB TODO Change to load from
properties
SearchScope search = new SearchScope(objectStore);
PropertyFilter myFilter = new PropertyFilter();
int myFilterLevel = 2; //Changed from 1 to 2 -- For
performance !
myFilter.setMaxRecursion(myFilterLevel);
myFilter.addIncludeType(new FilterElement(null, null, null,
FilteredPropertyType.ANY, null));
// Set the (Boolean) value for the continuable parameter.
This indicates
// whether to iterate requests for subsequent pages of result
data when the end of the
// first page of results is reached. If null or false, only a
single page of results is
// returned.
Boolean continuable = new Boolean(true); //SET To allow more
results to be fetched
// Set the page size (Long) to use for a page of query result
data. This value is passed
// in the pageSize parameter. If null, this defaults to the
value of
// ServerCacheConfiguration.QueryPageDefaultSize.
Integer myPageSize = new Integer(1000);
String SQL = sqlObject.toString();
// Execute the fetchObjects method using the specified
parameters.
//ASB added set myFilter to null REF Best Practice API on
FNRCA0024E: API_PROPERTY_NOT_IN_CACHE

IndependentObjectSet myObjects = search.
fetchObjects(sqlObject, myPageSize, null, continuable); //ASB
myFilter   changed to null
```

```
//===============================S T A R ===============================T
    // You can then iterate through the collection of rows to
    access the properties.
    String [][] sTaskInfo = new String[1000][3]; // Index 0=f.Id ,
    1=f.CmAcmTaskName , 2= f.Coordinator Case Folder Id
     int rowCount = 1;
    Iterator iter = myObjects.iterator();
     while (iter.hasNext()){   //Get all Completed Tasks
          NumberOfTasks = rowCount;
          try
          {
             //Folder f = null;
             IndependentObject object = (IndependentObject)
             iter.next();
             //com.filenet.api.property.Properties props =
             object.getProperties();
             //Iterator iterProps = props.iterator();

//=========================== END =======================================
          // Call Routine to retrieve the Question arrays
            CmTask currentTask = (CmTask) object;
            //String parentTaskClass = "CmTask"; //ASB 11-09-2017
            //Update the Task Completed,AUD_FullQDateCompleted
            // Get the properties collection
            //com.filenet.api.property.Properties props
            =  currentTask.getProperties();
            // Get the start date
            //Date start = new Date(System.currentTimeMillis());
            // Set the properties
            //props.putValue("AUD_FullQDateCompleted", start);
            //currentTask.save(RefreshMode.NO_REFRESH);
            TaskNames[rowCount-1] = currentTask.getClassName();
          TaskId[rowCount-1] =  currentTask.get_Id().toString();
          CaseId[rowCount-1] = currentTask.get_Coordinator().
          get_Id().toString();
              //sTaskInfo[rowCount] = ;
```

```
                        //sTaskInfo[rowCount,2] = " ";
                    //ASB 06-08-2022 Get the Current Case Folder Task
                    Property definitions
                     //com.filenet.api.admin.ClassDefinition tClassDefs
                     = Factory.ClassDefinition.fetchInstance(objectStore
                     ,taskClass, null); //update myFilter to null!!! ASB
                     06/08/2022
                     //PropertyDefinitionList tPropList = tClassDefs.
                     get_PropertyDefinitions();
                    //ASB Get the Current Task Object properties
                    // 4. Populate the Case level Task Folder properties.
                    //getTaskProps(currentTask, taskClass,tClass
                    Defs,tPropList,props,iterProps,sTableCells,
                    questionValue, publishYes, questionDetails,nQuestions
                    +1,nQuestionColumns);
//================================= E N D =================================
                }catch(Exception errprop){

                }
                    rowCount ++;
            }
//==========================================================================
            }catch(Exception e) {
                    System.out.println(e.getMessage()) ;
            }
            //ASB Check if we have no tasks to process currently!
            if(NumberOfTasks == 0) {
                return;
            }
        //Generate the pdf files
        //
        String sTaskName ="";
        for (int index = 0; index < NumberOfTasks  ; index++) //ASB
        15/08/2022 Changed from <= to < to fix bug!
```

CHAPTER 6 PDF DOCUMENT CREATION USING JAVA

```
{
  String CaseGUID  = CaseId[index]; //Actually the Task Id whose
  properties are to be loaded
String TaskGUID  = TaskId[index]; //{657A203A-9312-4D95-
AC12-03292DB708F8}
sTaskName = TaskNames[index]; // [AUD_ProduceFullQ]
  // pdfFileName =
  String pdfFileName =sTaskName + ".txt";
  // WorkingDirectory =
  String WorkingDirectory = cemc.getReportDirectoryName();
  //String CaseGUID, String TaskGUID, String TaskName, String
  pdfFileName, String WorkingDirectory
  String nQuestions = cemc.getNoQQuestions().toString();
  String nQuestionColumns =cemc.getNoQColumns().toString();
  //AUD_ProduceFullQ
  Integer paramCount = 0;
  Integer fieldCount = 0; //set for loading Fields
  String fieldVal="";
  //ASB  Retrieve the Task Name parameters from the config.xml
  for (String field:TaskName){
      //ASB Returns NotUsed string for any elements not in the
      config.xml file
      fieldVal = AUDITConfig.getConfigStringValue("config.
      QSECTIONS.S" + String.valueOf(fieldCount), "NotUsed");
       if (!fieldVal.equalsIgnoreCase("NotUsed")){
          TaskParams = fieldVal.split(",");
          TaskName[paramCount] = TaskParams[0];
         QStart[paramCount] = TaskParams[1];
         QEnd[paramCount] = TaskParams[2];
         paramCount++;
      }
      fieldCount++;
  }
  fieldCount = 0;
// Check the property we have
```

```java
try {
    for (String field:TaskName){
        //ASB Returns NotUsed string for any elements
        // not in the config.xml file
        fieldVal = TaskName[fieldCount];
        if(!(TaskName[fieldCount]==null)){
            if(!fieldVal.equalsIgnoreCase("NotUsed")){
              if(!(sTaskName==null)){ //ASB  Check for Null Task Name here
                if(sTaskName.equalsIgnoreCase(TaskName[fieldCount])) {
                 // Process the Section
                 String StartQNo = QStart[fieldCount];
                 String EndQNo = QEnd[fieldCount];
                 String FileCreated = QOperations.generatePDF(CaseGUID, TaskGUID, sTaskName, pdfFileName, WorkingDirectory,nQuestions, nQuestionColumns,StartQNo, EndQNo);
                 QOperations.writePDFToCaseFolder(CaseGUID, TaskGUID, sTaskName,FileCreated, WorkingDirectory,nQuestions, nQuestionColumns,StartQNo, EndQNo);
                 //ASB Now Delete the pdf from the local working folder 15-08-2022
                    File fPDFFile = new File(FileCreated);
                      if(fPDFFile.exists()){
                          fPDFFile.delete();
                      }
                }
            }
           }else {
                logger.error("Warning: No Task listed in config for the following Task retrieved :" + TaskName[fieldCount]);
            }
        }
    }
```

```java
            fieldCount++;
           }
        } catch (Exception e) {
            logger.error(e.getMessage(), e);
        }
      }

        //return status;
    }
    private static void createCEConnection(String username, String
    password, String url) {
        connection = Factory.Connection.getConnection(url);

        Subject subject = UserContext.createSubject(connection, username,
        password, "FileNetP8"); //$NON-NLS-1$
        UserContext uc = UserContext.get();
        uc.pushSubject(subject);
    }

    private static void createVWSession(String username, String password,
    String url, String connectionPointName)
            throws VWException {
        vwSession = new VWSession();
        vwSession.setBootstrapCEURI(url);
        vwSession.logon( username, password, connectionPointName );
    }
}
```

CHAPTER 6   PDF DOCUMENT CREATION USING JAVA

Adding the Audit_Report Project Classpath Properties

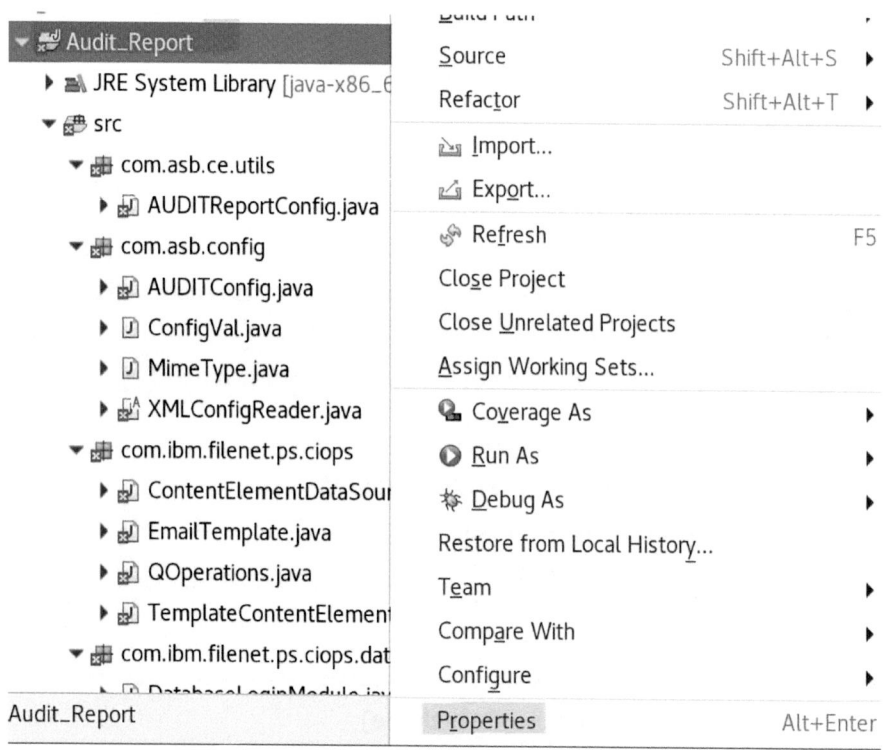

Figure 6-31. *The **Audit_Report** project properties classpath edits*

As can be seen in Figure 6-31, there are flagged issues (with a cross on a red background icon) which we need to resolve using edits to the **Audit_Report** project properties, to add reference jar files for missing Java .jar class imports.

CHAPTER 6 PDF DOCUMENT CREATION USING JAVA

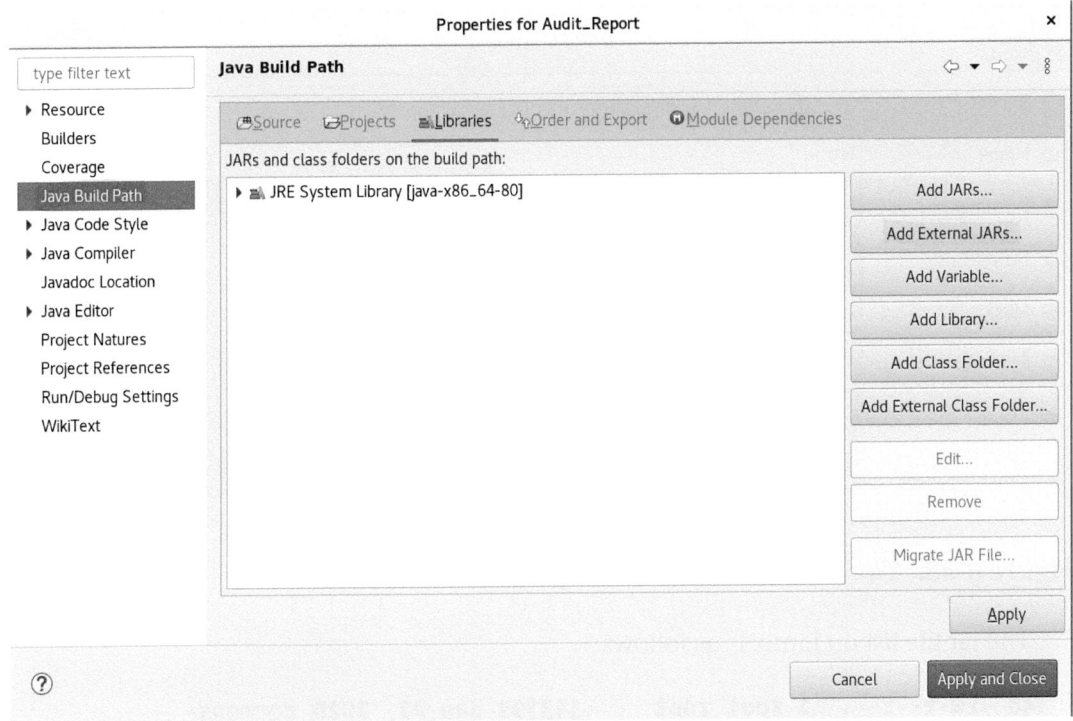

Figure 6-32. *The Add External JARs command button is selected*

We can add the required **IBM FileNet** and **iText** .jar files to the **Eclipse** classpath using the **Add External JARs…** command button in the **Libraries** tab for the **Java Build Path** menu item.

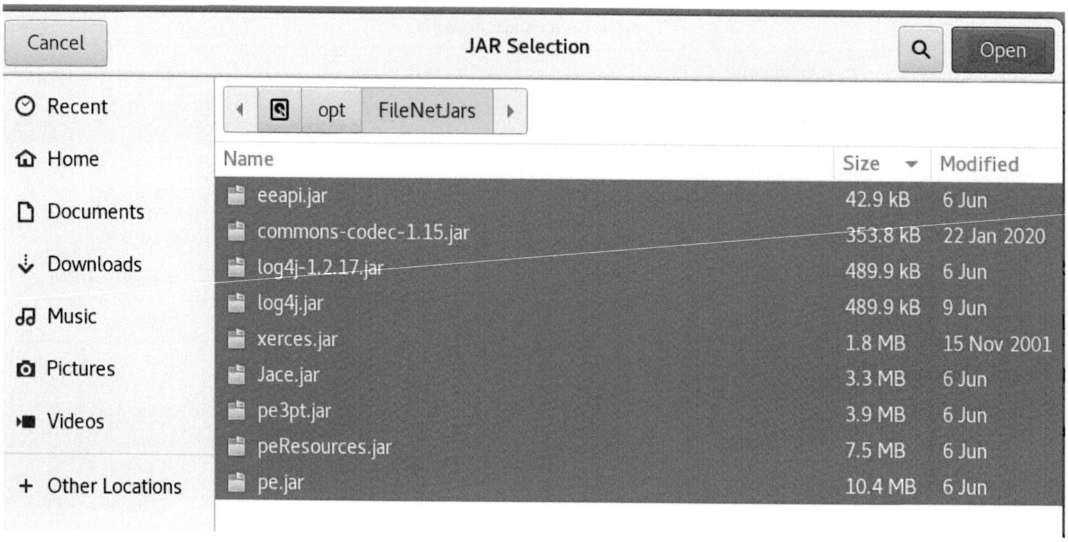

CHAPTER 6   PDF DOCUMENT CREATION USING JAVA

Cancel		JAR Selection			Open
Recent	◀	opt	FileNetJars ▶		
Home	Name			Size ▼	Modified
Documents	eeapi.jar			42.9 kB	6 Jun
	commons-codec-1.15.jar			353.8 kB	22 Jan 2020
Downloads	log4j-1.2.17.jar			489.9 kB	6 Jun
Music	log4j.jar			489.9 kB	9 Jun
Pictures	xerces.jar			1.8 MB	15 Nov 2001
	Jace.jar			3.3 MB	6 Jun
Videos	pe3pt.jar			3.9 MB	6 Jun
	peResources.jar			7.5 MB	6 Jun
Other Locations	pe.jar			10.4 MB	6 Jun

Figure 6-33. *The /opt/FileNetJars API .jar files are selected*

The jar file list on Linux is as follows:

```
   348 -rw-r--r--. 1 root root    353793 Jan 22  2020 commons-codec-1.15.jar
    44 -rwxrwxrwx. 1 root root     42944 Jun  6 16:24 eeapi.jar
  3244 -rwxrwxrwx. 1 root root   3317773 Jun  6 16:24 Jace.jar
   480 -rwxrwxrwx. 1 root root    489883 Jun  6 16:25 log4j-1.2.17.jar
   480 -rwxrwxrwx. 1 root root    489883 Jun  9 08:55 log4j.jar
  3764 -rwxrwxrwx. 1 root root   3850486 Jun  6 16:27 pe3pt.jar
 10192 -rwxrwxrwx. 1 root root  10433010 Jun  6 16:27 pe.jar
  7344 -rwxrwxrwx. 1 root root   7517605 Jun  6 16:28 peResources.jar
  1772 -rw-r--r--. 1 root root   1812019 Nov 15  2001 xerces.jar
(base) [root@ECMUKDEMO6 FileNetJars]#
```

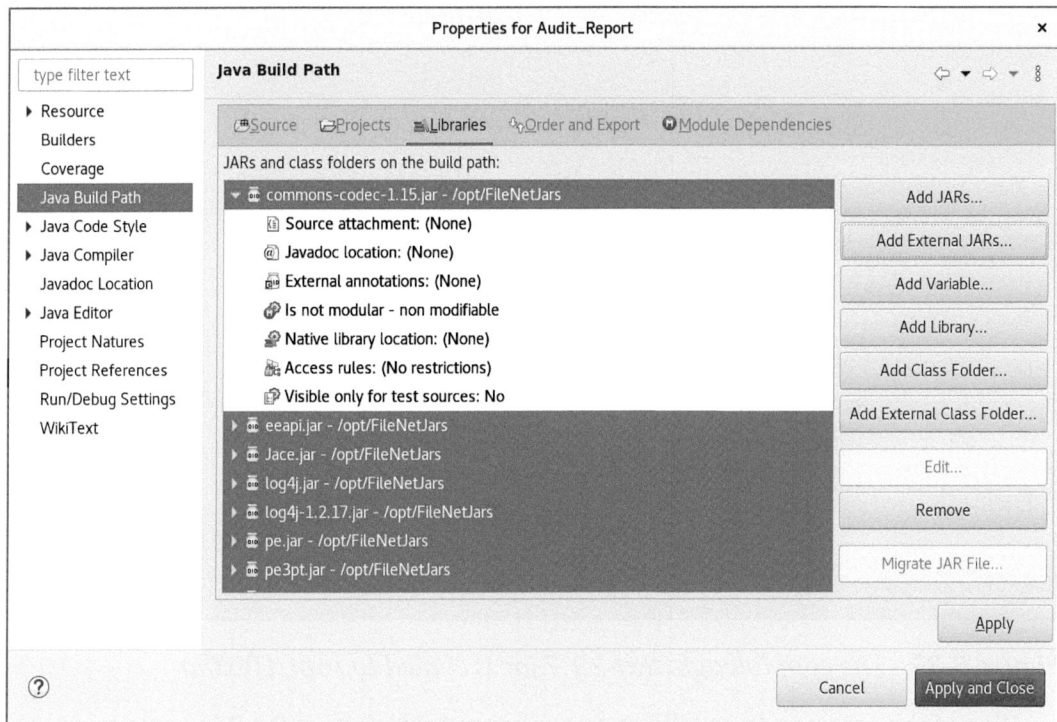

*Figure 6-34. The **Apply and Close** command button is clicked*

Adding the iText 2.1.7 Jar File for the PDF Creation Calls

Next, we copy the iText version 2.1.7 jar file, modified from the project source code (see later) to correct some issues.

CHAPTER 6 PDF DOCUMENT CREATION USING JAVA

```
root@ECMUKDEMO6:/opt/iTextJar
File Edit View Search Terminal Help
(base) [root@ECMUKDEMO6 ~]# pwd
/root
(base) [root@ECMUKDEMO6 ~]# cd /opt
(base) [root@ECMUKDEMO6 opt]# ls
AuditMaster       db2jars                      google             MQJars
box_sample.cpp    Eclipse                      ibm                mqm
CExtension        emboss6-6                    IBM                oneWEX
cloudcli          filenet_app2.log             isai               replication
cognos11          filenet_app.log              jars               WASLiberty19.0.0.11
com               filenet_appthreads.log       jpwidget           watson
containerd        FileNetJars                  jre_1.8_Install    Watson12.0.3.0
CSS               GO                           mono2micro         WatsonFix
(base) [root@ECMUKDEMO6 opt]# mkdir iTextJar
(base) [root@ECMUKDEMO6 opt]# cd iTextJar
(base) [root@ECMUKDEMO6 iTextJar]# ls /mnt/hgfs
Installs   LINUX
(base) [root@ECMUKDEMO6 iTextJar]# cp /mnt/hgfs/Installs/com.lowagie.text-2.1.7.jar .
(base) [root@ECMUKDEMO6 iTextJar]# ls
com.lowagie.text-2.1.7.jar
(base) [root@ECMUKDEMO6 iTextJar]#
```

Figure 6-35. The com.lowagie.text-2.1.7.jar is copied to /opt/iTextJar

We run the commands, as follows, to copy **com.lowagie.text-2.1.7.jar** onto our Linux RHEL 8.0 VMware system:

```
cd /opt
mkdir iTextJar
cd iTextJar
ls /mnt/hgfs
cp /mnt/hgfs/Installs/com.lowagie.text-2.1.7.jar .
```

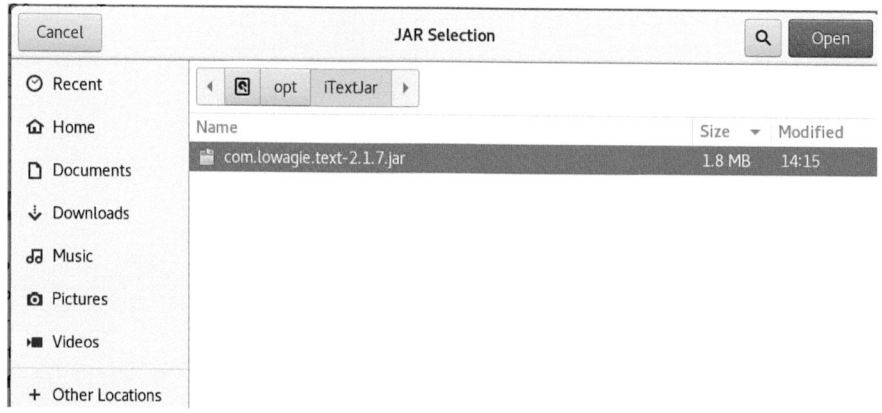

Figure 6-36. The com.lowagie.text-2.1.7.jar is selected

1206

CHAPTER 6   PDF DOCUMENT CREATION USING JAVA

The **com.lowagie.text-2.1.7.jar** is added to the **Audit_Report** Eclipse project classpath.

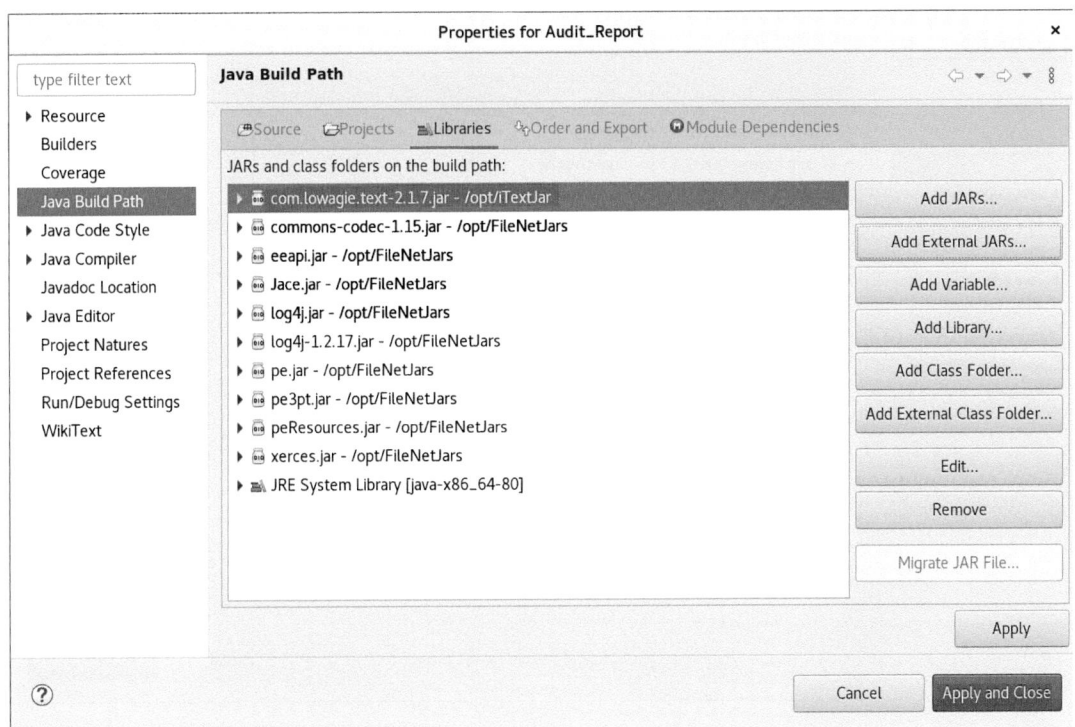

Figure 6-37. *The **com.lowagie.text-2.1.7.jar** is added to the classpath*

CHAPTER 6   PDF DOCUMENT CREATION USING JAVA

Adding the JUnit Java Test Library Jar File

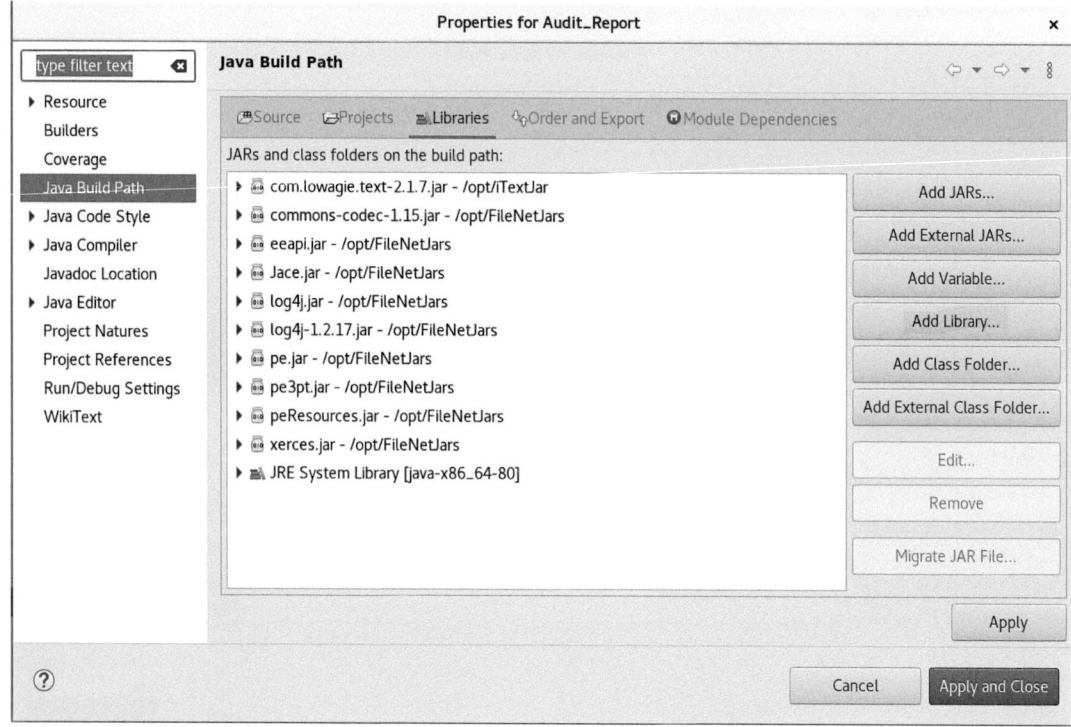

Figure 6-38. *The Add Library command button is selected*

In Figure 6-38, we click the **Add Library...** command to add the **JUnit** library.

CHAPTER 6 PDF DOCUMENT CREATION USING JAVA

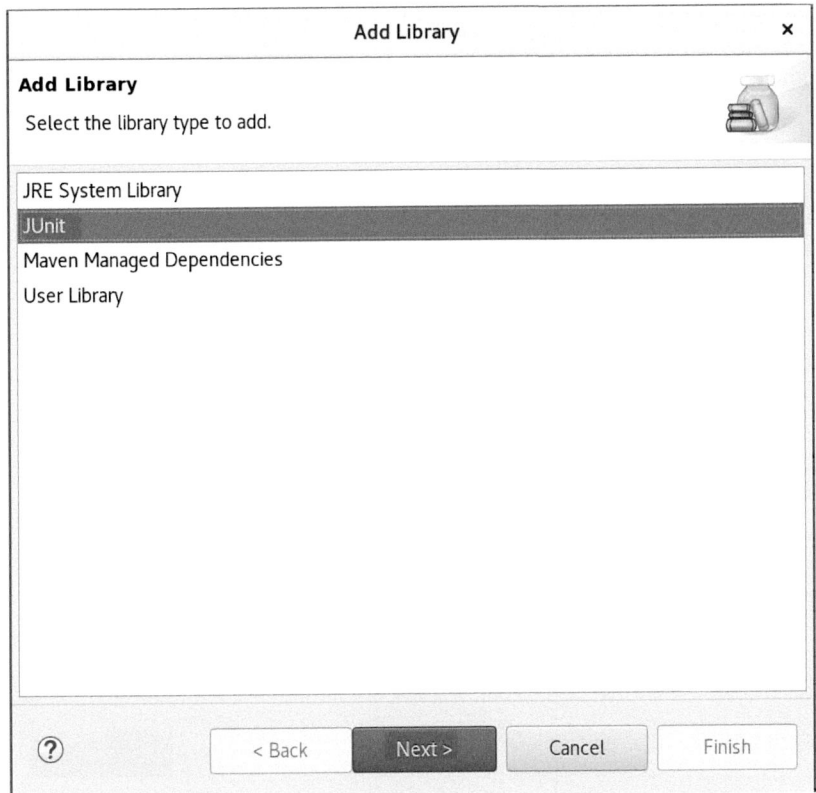

Figure 6-39. *The **JUnit** library is added to the classpath*

We must select the correct JUnit version from the drop-down options as shown in Figure 6-40.

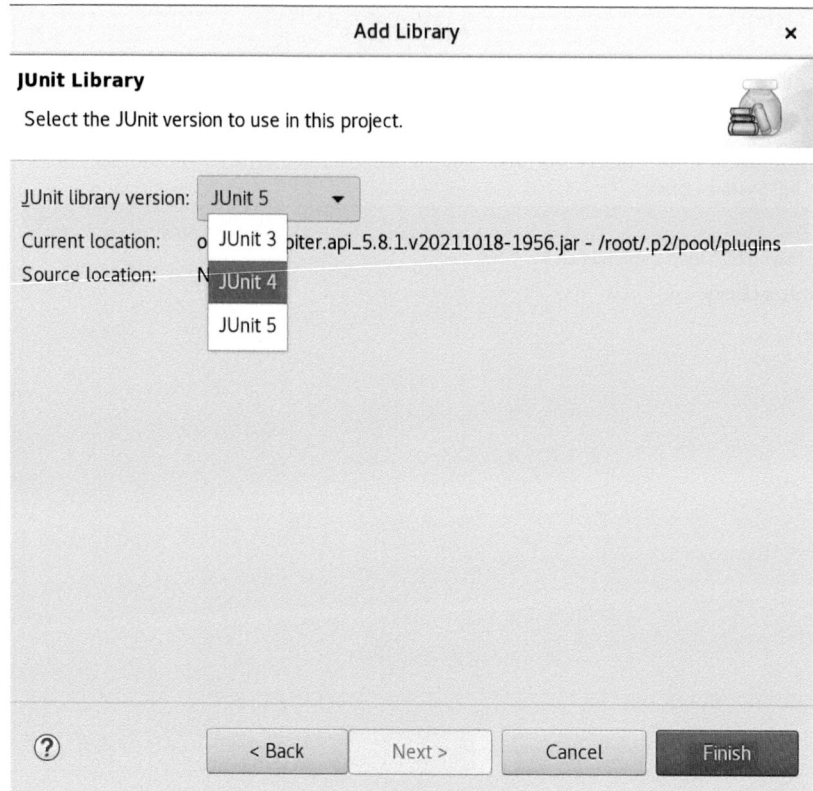

Figure 6-40. *The JUnit 4 version is selected from the drop-down*

We show the JUnit 4 library (used for Java assertion testing in our project).

CHAPTER 6   PDF DOCUMENT CREATION USING JAVA

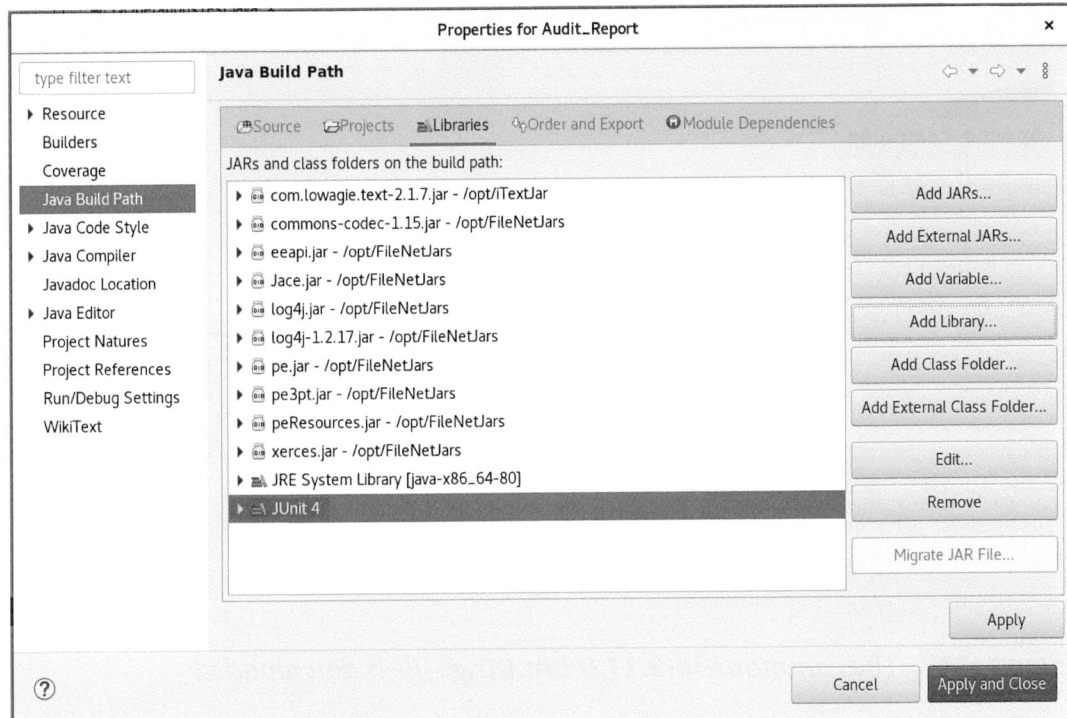

Figure 6-41. *The JUnit 4 library is added and saved using Apply and Close*

Adding the Apache Commons IO Jar File

Next, we need to download the latest Apache Commons IO jar file version 2.11.0 for the supporting I/O file commands we are using with Java 8.

1211

CHAPTER 6   PDF DOCUMENT CREATION USING JAVA

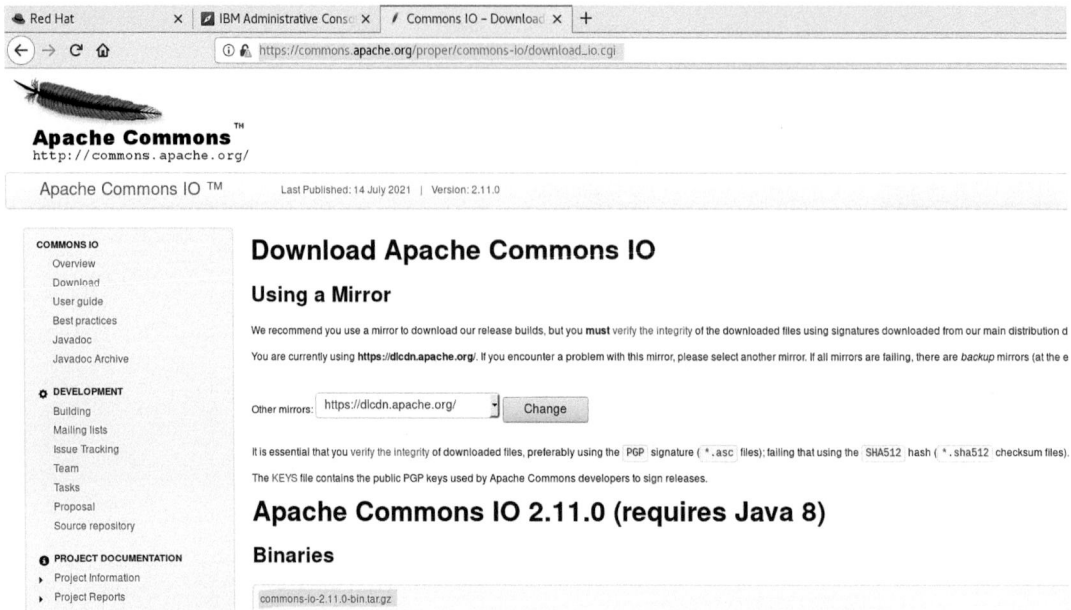

*Figure 6-42.   The **commons-io-2.11.0-bin.tar.gz** file is downloaded*

We can now save the downloaded **commons-io-2.11.0-bin.tar.gz** file to the **/root/Downloads** folder path on Linux.

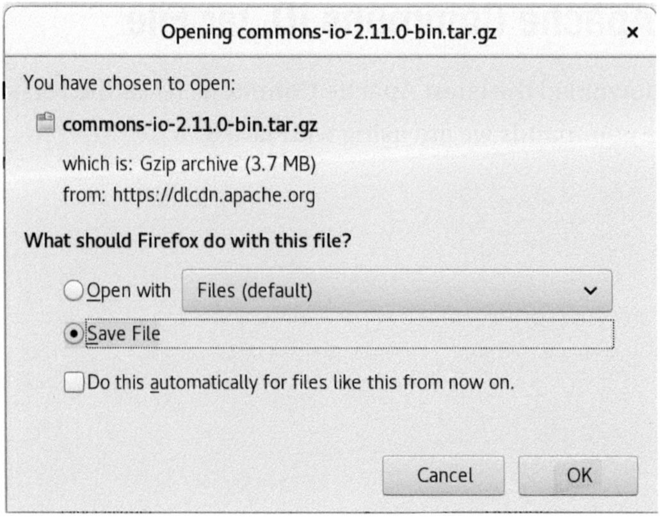

*Figure 6-43.   The **commons-io-2.11.0-bin.tar.gz** file download is saved*

CHAPTER 6   PDF DOCUMENT CREATION USING JAVA

The downloaded **commons-io-2.11.0-bin.tar.gz** is copied to the **/opt/apache_commons** jar file folder and unpacked.

```
(base) [root@ECMUKDEMO6 opt]# mkdir apache_commons
(base) [root@ECMUKDEMO6 opt]# cd apache_commons
(base) [root@ECMUKDEMO6 apache_commons]# cp /root/Downloads/commons-io-2.11.0-bin.tar.gz .
(base) [root@ECMUKDEMO6 apache_commons]# ls
commons-io-2.11.0-bin.tar.gz
(base) [root@ECMUKDEMO6 apache_commons]# tar -zxvf commons-io-2.11.0-bin.tar.gz
```

*Figure 6-44.   The commands to unpack **commons-io-2.11.0-bin.tar.gz***

The following Linux commands were used to unpack the **commons-io-2.11.0-bin.tar.gz** file:

```
cd /opt
mkdir apache_commons
cd apache_commons
cp /root/Downloads/commons-io-2.11.0-bin.tar.gz .
tar -zxvf commons-io-2.11.0-bin.tar.gz
```

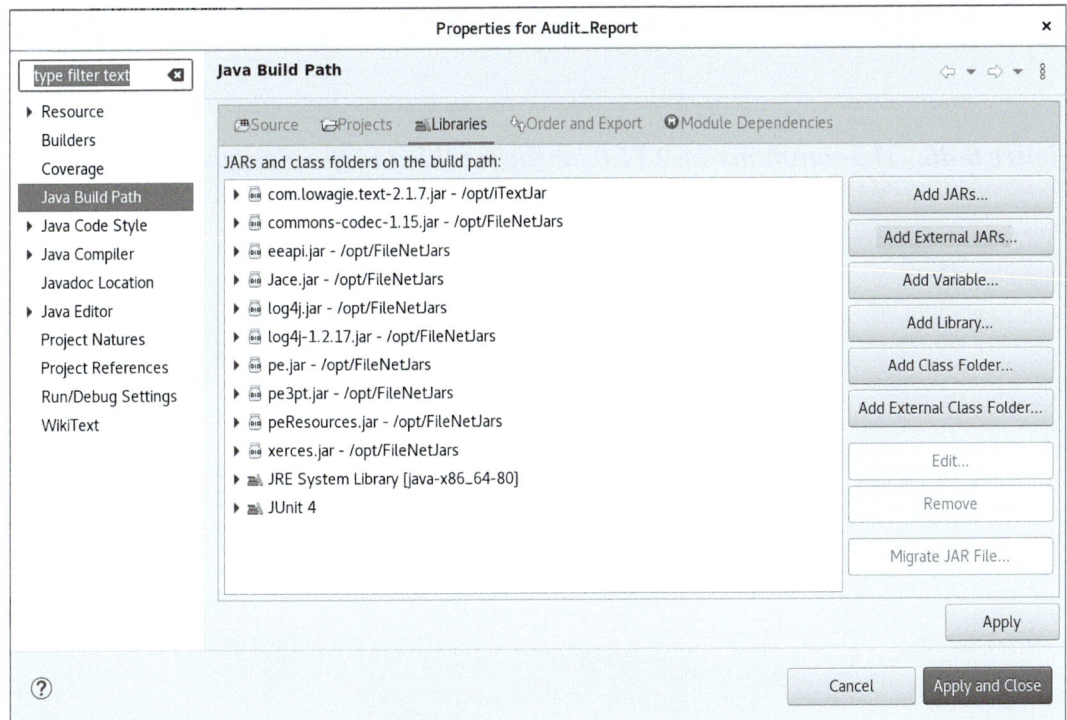

Figure 6-45.   The Add External JARs… command is selected

1213

CHAPTER 6   PDF DOCUMENT CREATION USING JAVA

Now we can add the missing Apache Commons 2.11.0 jar file.

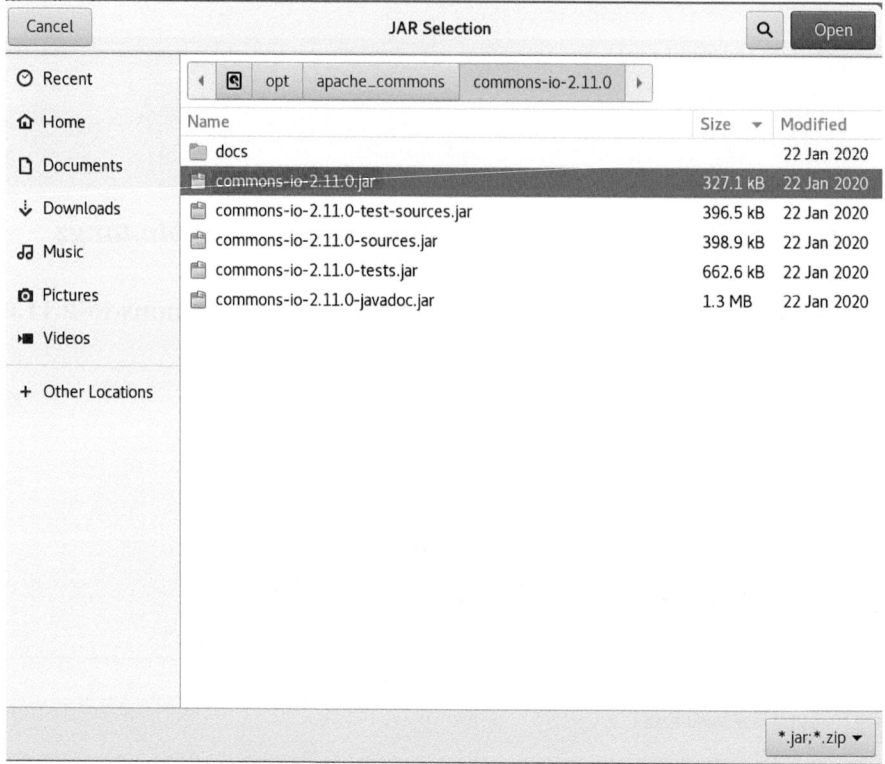

Figure 6-46. *The commons-io-2.11.0.jar file is added to the classpath*

CHAPTER 6 PDF DOCUMENT CREATION USING JAVA

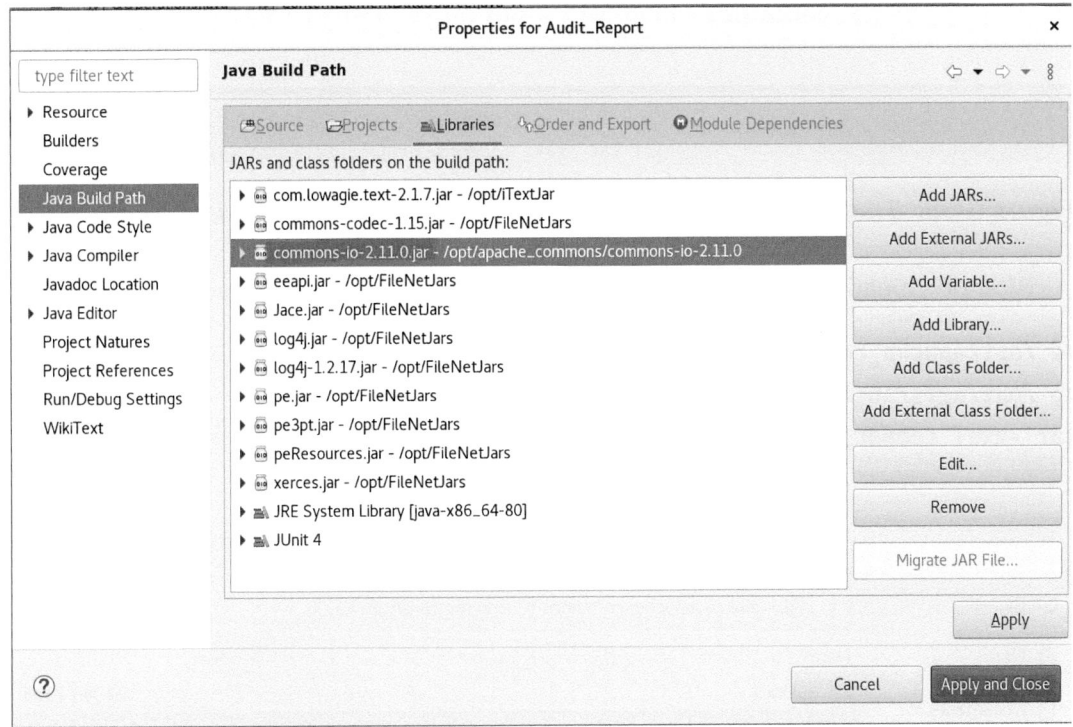

Figure 6-47. *The commons-io-2.11.0.jar file is saved in the classpath*

The **commons-io-2.11.0.jar** file, selected from the **/opt/apache_commons/ commons-i0-2.11.0** folder, is saved in the classpath using the **Apply and Close** command button shown in Figure 6-47.

Adding the Apache Commons Lang Java 8 Language Jar File

We also need the Apache Commons Lang jar file.

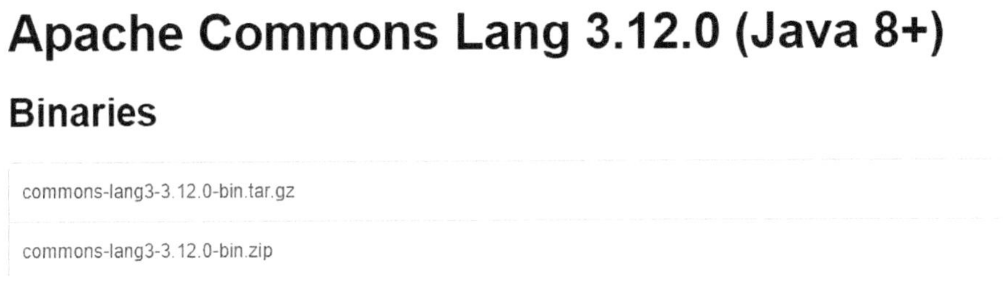

Apache Commons Lang 3.12.0 (Java 8+)

Binaries

commons-lang3-3.12.0-bin.tar.gz

commons-lang3-3.12.0-bin.zip

Figure 6-48. *The Apache Commons Lang 312.0 jar file is downloaded*

1215

CHAPTER 6 PDF DOCUMENT CREATION USING JAVA

We click the link **commons-lang3-3.12.0-bin.tar.gz** for the Linux version.

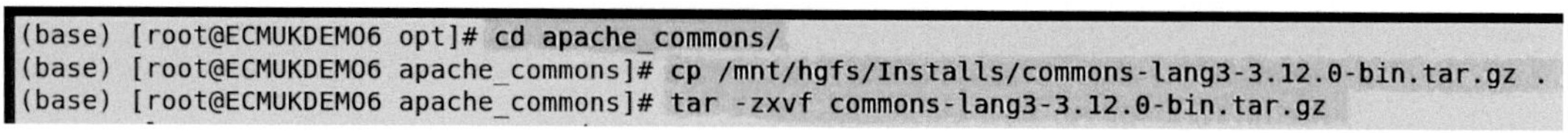

```
(base) [root@ECMUKDEM06 opt]# cd apache_commons/
(base) [root@ECMUKDEM06 apache_commons]# cp /mnt/hgfs/Installs/commons-lang3-3.12.0-bin.tar.gz .
(base) [root@ECMUKDEM06 apache_commons]# tar -zxvf commons-lang3-3.12.0-bin.tar.gz
```

Figure 6-49. *The **commons-lang3-3.12.0-bin.tar.gz** file is unpacked*

After downloading the **commons-lang3-3.12.0-bin.tar.gz**, we unpack it as follows:

```
cd /opt/apache_commons/
cp /mnt/hgfs/Installs/commons-lang3-3.12.0-bin.tar.gz .
tar -zxvf commons-lang3-3.12.0-bin.tar.gz
```

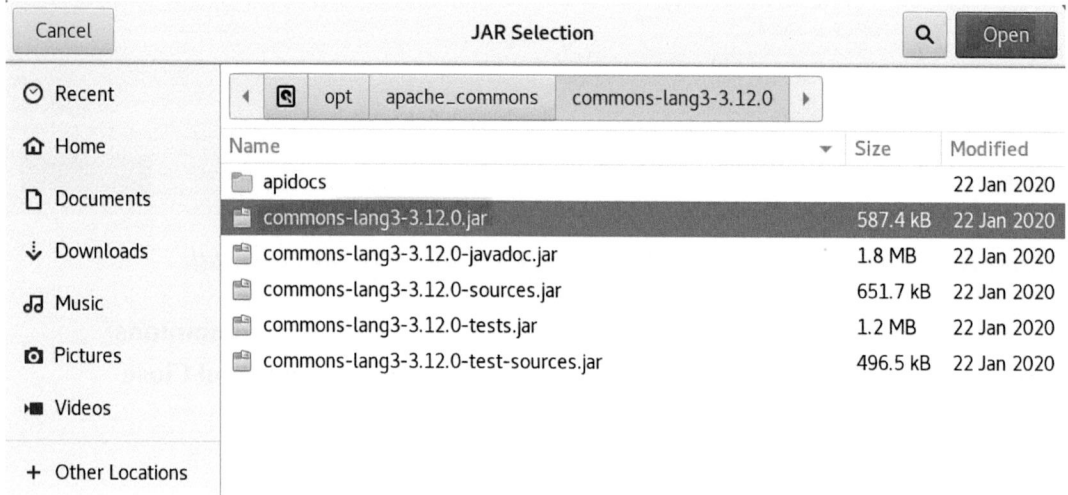

Figure 6-50. *The **commons-lang3-3.12.0.jar** file, selected for the classpath*

We can now add the **commons-lang3-3.12.0.jar** file to the Eclipse **Audit_Report** project classpath.

CHAPTER 6 PDF DOCUMENT CREATION USING JAVA

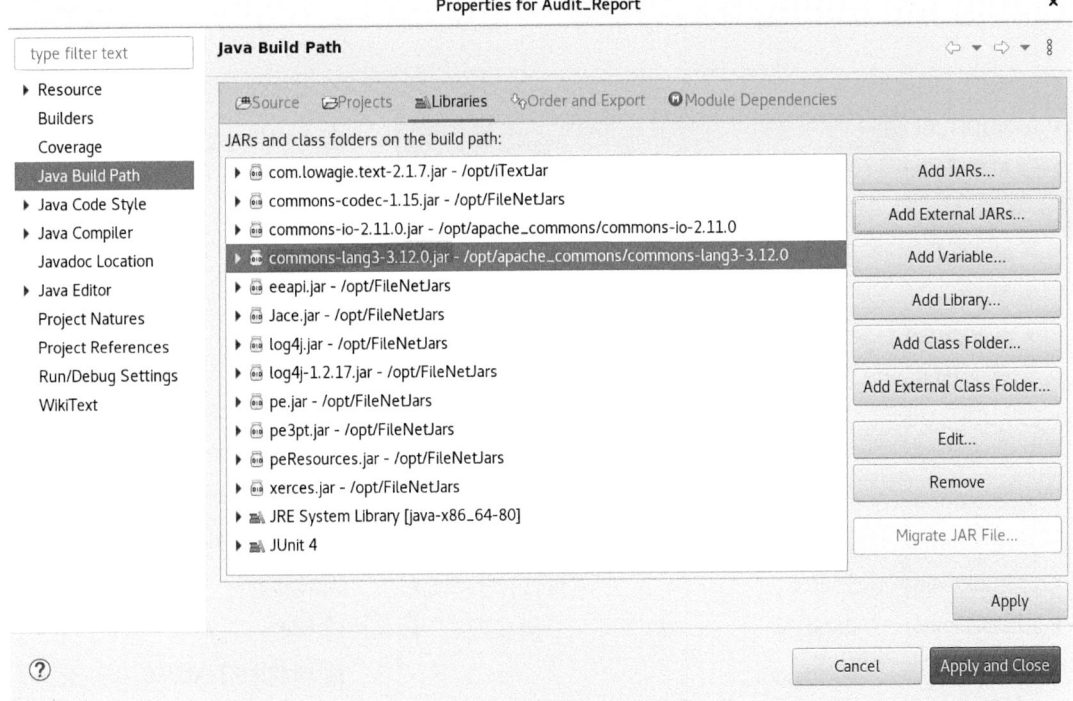

Figure 6-51. *The **commons-lang3-3.12.0.jar** file is saved to the classpath*

We use the **Apply and Close** command button to save the selected **commons-lang3-3.12.0.jar** file to the **Audit_Report** project classpath.

CHAPTER 6　PDF DOCUMENT CREATION USING JAVA

Adding the Config Subfolder to the Audit_Reports Project

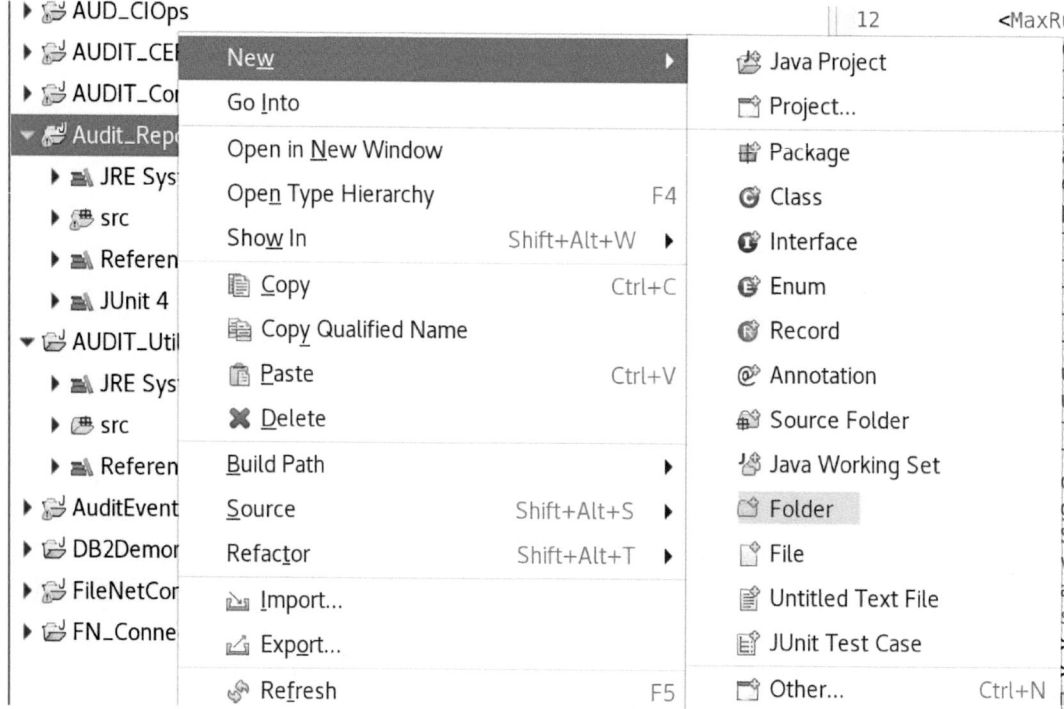

Figure 6-52. *The New Folder menu is selected*

We need to create a **config** subfolder under the **Audit_Report** Eclipse project to hold the **config.xml** and **log4j.xml**.

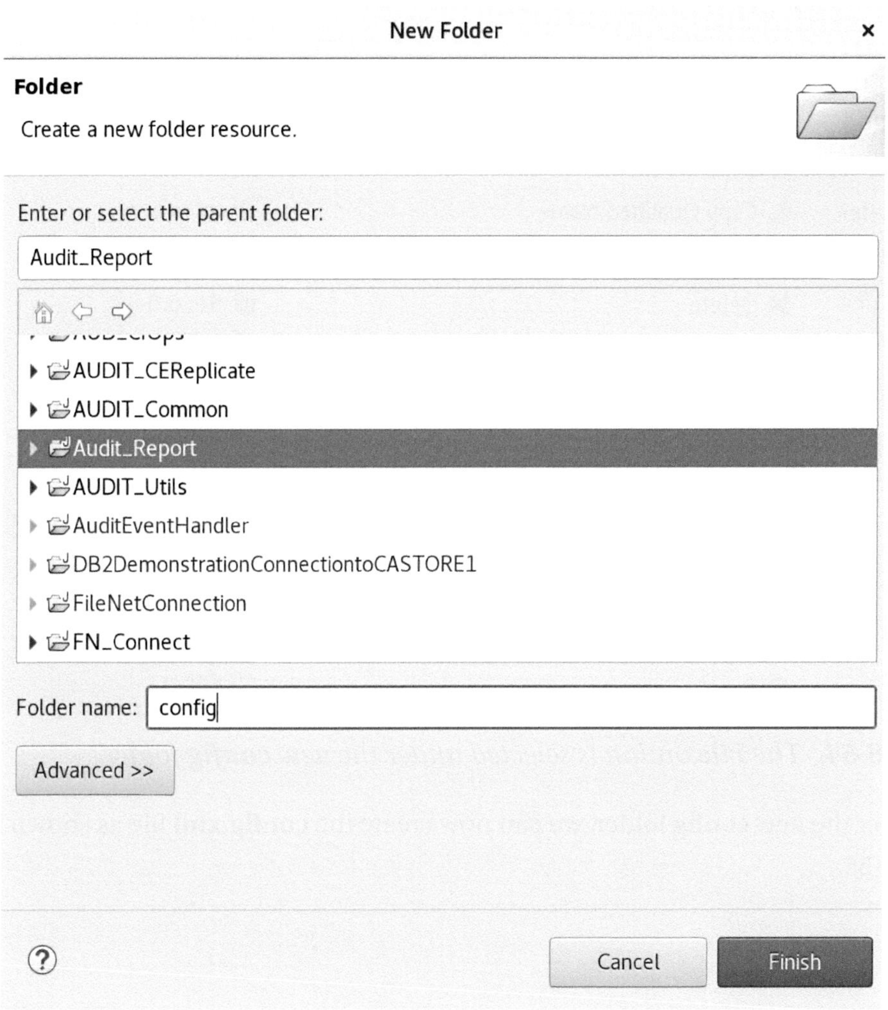

*Figure 6-53. The **config** folder is created under the **Audit_Report** folder*

CHAPTER 6 PDF DOCUMENT CREATION USING JAVA

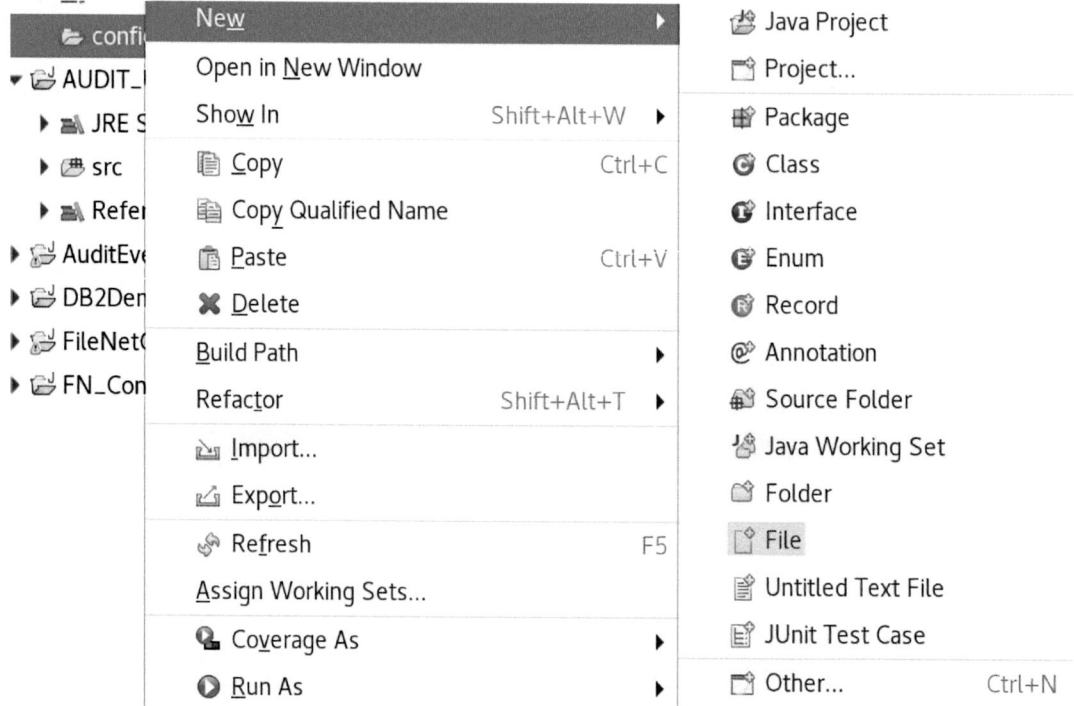

Figure 6-54. *The **File** option is selected under the new **config** folder*

Under the new **config** folder, we can now create the **config.xml** file as shown in Figure 6-55.

CHAPTER 6 PDF DOCUMENT CREATION USING JAVA

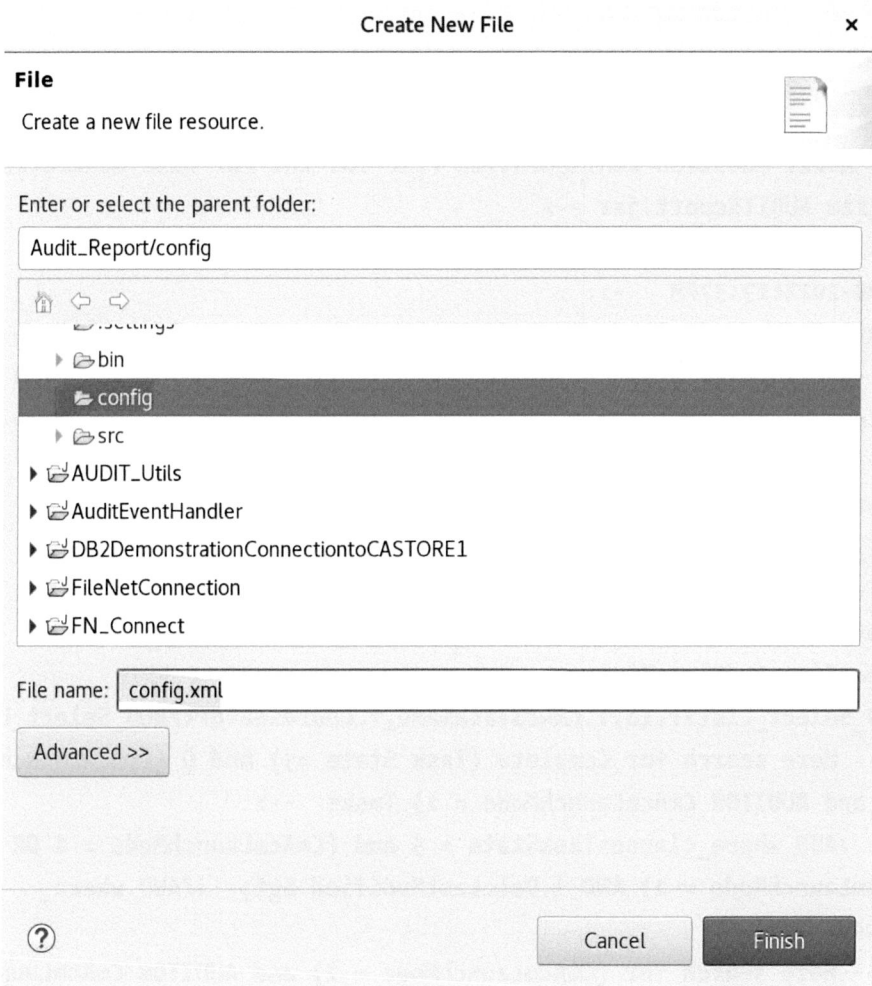

Figure 6-55. *The **config.xml** file is created under the **config** folder*

The new **config.xml** file is used to define the variables we use for driving the **Audit Report** program search for Audit tasks. This is as follows for the single Task we created in the IBM Case Manager, Audit Master solution.

Listing 6-26. The **config.xml** configuration file list for the Audit Report parameters

```xml
<?xml version="1.0" encoding="UTF-8"?><config>
    <!-- AUDIT Question Configuration File for the PDF File Generation
    Program AUDITReport.jar -->
    <!-- Config for AUDIT Question Version 3.0 Build
    14-08-2022:23:57PM  -->
    <maximum_docs>10000</maximum_docs>
    <!-- Setting the value below to Yes will stop the pdf generation
    process on the next run -->
    <!-- Otherwise it will loop ad-infinitum at SleepTime minute
    intervals -->
    <SET_TO_STOP>Yes</SET_TO_STOP>
       <!-- If SET_TO_STOP is No, then loop ad-infinitum at SleepTime minute
       intervals -->
    <SleepTime>2</SleepTime>
    <AUD_Select_List>f.Id,f.CmAcmTaskName,f.Coordinator</AUD_Select_List>
     <!-- Here search for Complete (Task State =5) and Q (CmAcmLaunchMode =
     2) and AUDITOR CmAcmLaunchMode = 1) Tasks  -->
    <!-- <AUD_where_clause>TaskState = 5 and (CmAcmLaunchMode = 1 OR
    CmAcmLaunchMode = 2) AND f.DateLastModified &gt;  </AUD_where_
    clause> -->
       <!-- Here search for (CmAcmLaunchMode = 2) and AUDITOR CmAcmLaunchMode
       = 1 ) Tand CmAcmLaunchMode = 0 Tasks  -->
    <AUD_where_clause>AUD_Status='PDFCreation' and (CmAcmLaunchMode = 0
    OR CmAcmLaunchMode = 1 OR CmAcmLaunchMode = 2) AND f.DateLastModified
    &gt;  </AUD_where_clause>
    <AUD_orderby_clause>d.DateLastModified,d.MajorVersionNumber,
    d.MinorVersionNumber</AUD_orderby_clause>
    <QCase_Folder_where_clause>CmAcmCaseState = 3 AND f.DateLastModified</
    QCase_Folder_where_clause>
    <QCase_Folder_orderby_clause>f.DateLastModified </QCase_Folder_
    orderby_clause>
    <AUDITOR_where_clause>d.DateLastModified</AUDITOR_where_clause>
    <AUDITOR_orderby_clause/>
```

```
<ReportDirectoryName>/opt/AuditReport</ReportDirectoryName>
<NumberOfFullQQuestions>16</NumberOfFullQQuestions>
<NumberOfFullQColumns>3</NumberOfFullQColumns>
<!-- Add Table Column Name, Width for each of 3 columns in the Section
Tables - above defines the number of columns to use -->
<PDFTable4>
        <C0>Change,5.0f</C0>
        <C1>Y/N,1.0f</C1>
        <C2>Details,3.2f</C2>
</PDFTable4>
    <!-- Enter the symbolic name of the main AUDITOR Workflow Task -->
<AUDITOR_Task>AUD_AuditProcess</AUDITOR_Task>
    <!-- Add Tab names as the Header, Number of Properties and
    Symbolic Case Property List additional sections can be added -->
      <!-- as T1 T2 T3 etc -->
<TabSECTIONS_AUDITOR>
        <T0>1. General,2,AUD_DepartmentName,AUD_
        DepartmentNumber</T0>
        <T1>2. Question Section 1,6,AUD_General_1_Q1,AUD_General_1_
        Q2,AUD_General_1_Q3,AUD_General_1_Q4,AUD_General_1_Q5,AUD_
        General_1_Q6</T1>
        <T2>3. Question Section 2,10,AUD_General_1_Q7,AUD_
        General_1_Q8,AUD_General_1_Q9,AUD_General_1_Q10,AUD_
        General_1_Q11,AUD_General_1_Q12,AUD_General_1_Q13,AUD_
        General_1_Q14,AUD_General_1_Q15,AUD_General_1_Q16</T2>
</TabSECTIONS_AUDITOR>
 /<TabSECTIONS_Audit>
 </TabSECTIONS_Audit>
<TabSECTIONS_Department>
</TabSECTIONS_Department>
<!-- Section Question Tables- Add section names as the full Task name ,
start question number, end question number for the section -->
 <!-- Additional sections can be added as S2, S3 etc if more Audit
 properties or Tasks are added to the solution   -->
 <QSECTIONS>
```

CHAPTER 6   PDF DOCUMENT CREATION USING JAVA

```xml
            <S0>AUD_AuditProcess,1,8</S0>
            <S1>AUD_AuditProcess,8,16</S1>
    </QSECTIONS>
    <TestFolderName>/Templates</TestFolderName>
    <database>
        <SearchSubClasses>yes</SearchSubClasses>
        <!-- Note that this section can contain Document Classes or Case
        Folder Classes   -->
        <!-- Note that this section can contain ONLY Document Classes For
        the case manager documents   -->
         <docclass_mapping>
             <d0>AUD_AUDITOR</d0>
             <d1>AUD_AUDITOR</d1>
         <!-- Uncomment below to search additional Document classes as
         required
             <d2>Document</d2>
             <d10>NotUsed</d10>      -->
         </docclass_mapping>
         <!-- Note that this section can contain ONLY Folder Classes For the
         Case Management Folders   -->
          <folderclass_mapping>
             <d0>AUD_AuditReport</d0>
      <!-- Uncomment below to search additional Folder classes as required
             <d10>NotUsed</d10>       -->
         </folderclass_mapping>
    </database>
    <ceimport>
         <osname>OS1</osname>
         <osrootfolder>/AUDIT_TEST</osrootfolder>
         <ceuser>Alan</ceuser>
         <cepassword>RW5jcnlwdDNkZmlsXxXxXxXxXxXxXxXx</cepassword>
         <ceurl>http://ecmukdemo6:9080/wsi/FNCEWS40MTOM/</ceurl>
            <celoginconfig>/opt/IBM/FileNet/CEClient/config/jaas.conf.WSI</
            celoginconfig>
         <cestanza>FileNetP8WSI</cestanza>
```

```xml
        <UpdatingBatchSize>250</UpdatingBatchSize>
        <MaxDocCount>50</MaxDocCount>
        <MaxRunTimeMinutes>30</MaxRunTimeMinutes>
        <MaxConsecutiveErrors>1000</MaxConsecutiveErrors>
        <AuditFile>/opt/replication/logs/AuditImportAuditDocs.log</AuditFile>
        <AuditFileFolders>/opt/replication/logs/AuditImportAuditFolders.log</AuditFileFolders>
</ceimport>
<AUDITProcessor>
        <MaxSearchSize>5000</MaxSearchSize>
        <startsearchdate>20220824T133429Z</startsearchdate>
        <!--  search Date is calculated as System Date - Search Days  -->
        <searchdays>0</searchdays>
        <PropCardinalityList>
                <PropCardinality>2</PropCardinality>
                <PropCardinality>2</PropCardinality>
                <PropCardinality>0</PropCardinality>
        </PropCardinalityList>
        <DocumentPropertyNames>
                <PropertyName>ToEmailAddress</PropertyName>
                <PropertyName>FromEmailAddress</PropertyName>
                <PropertyName>FileName</PropertyName>
         </DocumentPropertyNames>
        <DocumentPropertyTypes>
                <PropertyType>8</PropertyType>
                <PropertyType>8</PropertyType>
                <PropertyType>8</PropertyType>
        </DocumentPropertyTypes>
        <DocumentWhiteListFlags>
                <PropFlag>NONE</PropFlag>
                <PropFlag>NONE</PropFlag>
                <PropFlag>NONE</PropFlag>
        </DocumentWhiteListFlags>
        <DebugOutputFlag>on</DebugOutputFlag>
```

```xml
            <searchProperty>Id</searchProperty>
            <searchPropertyDocuments>Id</searchPropertyDocuments>
            <searchPropertyFolders>Id</searchPropertyFolders>
             <DeltaHours>-96</DeltaHours>
             <DeltaMinutes>0</DeltaMinutes>
             <folderCaseClasses>AuditDepartment</folderCaseClasses>
             <JDBC>jdbc:oracle:thin:@ECMUKDEMO1:1521:orcl</JDBC>
             <JDBCDriverClass>oracle.jdbc.driver.OracleDriver</JDBCDriverClass>
             <JDBCUser>P8DB_ECMUK_04</JDBCUser>
             <JDBCPassword>RW5jcnlwdDNkZmlsXxXxXxXxXxXxXxXx</JDBCPassword>
             <EmailListName>FromEmailAddresses</EmailListName>
             <EmailFlagListName>EmailFlag</EmailFlagListName>
 </AUDITProcessor>
 <ceexport>
      <osname>OS2</osname>
      <ceuser>Alan</ceuser>
      <cepassword>RW5jcnlwdDNkZmlsXxXxXxXxXxXxXxXx</cepassword>
      <MaxDocCount>500</MaxDocCount>
      <MaxRunTimeMinutes>0</MaxRunTimeMinutes>
      <MaxConsecutiveErrors>1000</MaxConsecutiveErrors>
      <AuditFile>/opt/replication/logs/AuditExportAudit.log</AuditFile>
      <SQLBatchSize>250</SQLBatchSize>
      <celoginconfig>/opt/IBM/FileNet/CEClient/config/jaas.conf.WSI </celoginconfig>
      <ceurl>http://ecmukdemo6:9080/wsi/FNCEWS40MTOM</ceurl>
      <cestanza>FileNetP8WSI</cestanza>
      <MimeType>application/pdf</MimeType>
      <DocClassName>AUD_AuditReport</DocClassName>
      <CaseStatus>Initial</CaseStatus>
      <CaseStatusField>CaseStatus</CaseStatusField>
      <ConnectionPointName>CP2</ConnectionPointName>
      <WaspLocation/>
 </ceexport>
 <SMTPreport>
```

```xml
            <sourceEmailAddress>alan.bluck@asbsoftware.co.uk</sourceEmailAddress>
            <sourceEmailPassword>RW5jcnlwdDNkR2xhXxXxXxXxXxXxXxXx</sourceEmailPassword>
            <SMTPTLSRequired>true</SMTPTLSRequired>
            <SMTPSSLRequired>false</SMTPSSLRequired>
            <SMTPHost>smtp-relay.gmail.com</SMTPHost>
            <SMTPUser>alan.bluck@asbsoftware.co.uk</SMTPUser>
            <TLSPort>587</TLSPort>
            <SSLPort>465</SSLPort>
            <SMTPType>TLS</SMTPType>
            <EmailTemplatePath> </EmailTemplatePath>
      </SMTPreport>
</config>
```

CHAPTER 6   PDF DOCUMENT CREATION USING JAVA

Adding the AuditTest Java Class

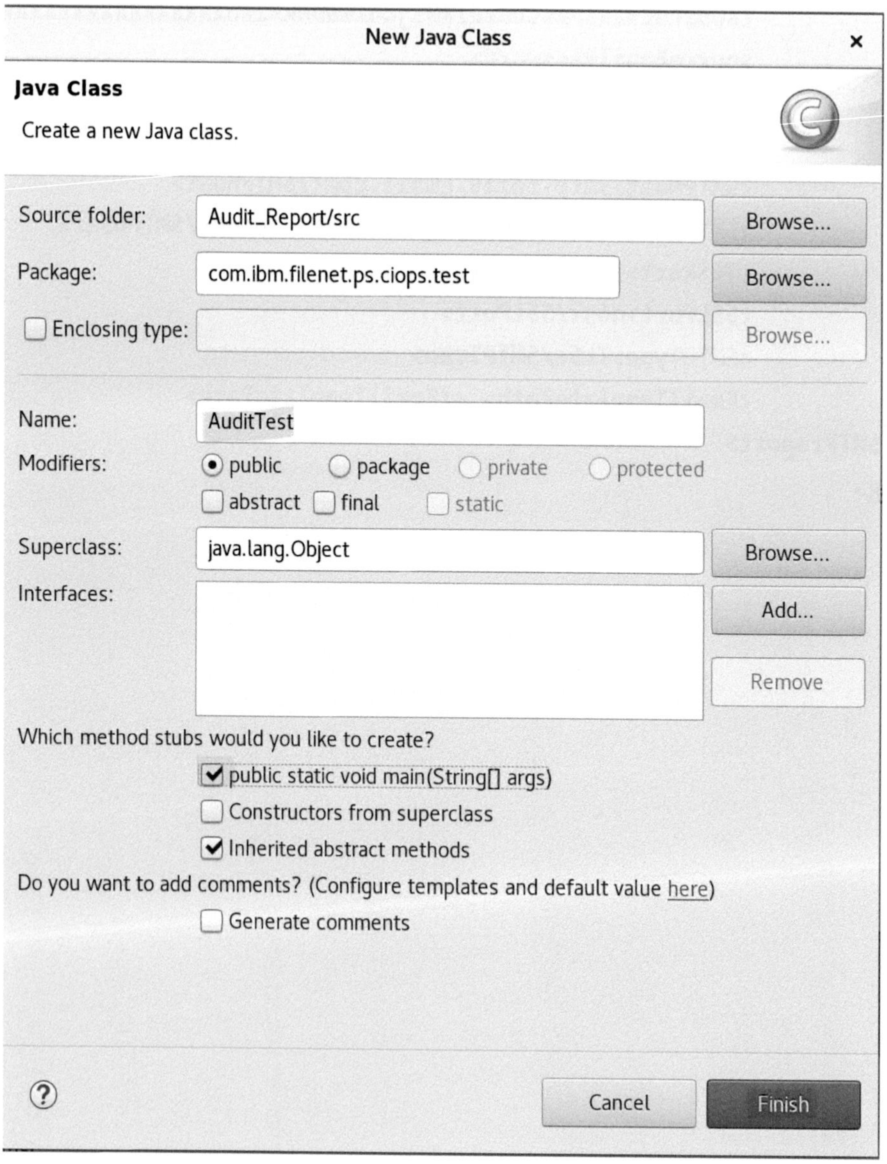

Figure 6-56. *The AuditTest class is created under the Audit Report project*

The full code is shown under the "Part 1 – Bill of Materials" section of the chapter.
   ("A Java Program to create a Simple Audit PDF document using the **iText** library.")

CHAPTER 6   PDF DOCUMENT CREATION USING JAVA

```java
package com.ibm.filenet.ps.ciops.test;

public class AuditTest {

    public static void main(String[] args) {
        // TODO Auto-generated method stub

    }

}
```

Figure 6-57. *The template code is created*

```java
        AuditPDFfile = new FileOutputStream(new File("/opt/AuditReport/AuditTest.pdf"));
        //Create a new Audit Document object
        Document audit_document = new Document();
        //You need the iText PdfWriter class for a PDF document
        PdfWriter.getInstance(audit_document, AuditPDFfile);
        //Open the Audit document for writing a PDF
        audit_document.open();
        //Write the test content
        audit_document.add(new Paragraph("Audit Report: Document Test PDF Paragraph"));
        audit_document.add(new Paragraph("Auditor Name: R. Jones"));
        audit_document.add(new Paragraph("Audit Date: " + sAuditDate));
        //Add Header meta-data information to the PDF file
        audit_document.addCreationDate();
        audit_document.addAuthor("ASB Software Development Limited");
        audit_document.addTitle("Audit Report Test PDF Generation");
        audit_document.addCreator("Program AuditTest");
        //close the document
        audit_document.close();
        System.out.println("Audit Report Created:" + sAuditDate);
    } catch (Exception e) {
        e.printStackTrace();
    } finally {
//close the FileOutputStream
        try {
            if (AuditPDFfile != null) {
                AuditPDFfile.close();
            }
        } catch (IOException io) {
            //Failed to close
            System.out.println("The Audit PDF file failed to close!");
        }
    }
  }
}
```

Console output:
```
AuditTest [Java Application] /opt/ibm/java-x86_64-80/bin/javaw (Aug 27, 2022, 8:32:37 AM)
Audit Report Created:27-08-2022 08:32:52
```

Figure 6-58. *The AuditTest Java code is entered and run*

The output from this test code is shown in the **AuditTest.pdf** file, which was created as shown in Figure 6-59.

CHAPTER 6   PDF DOCUMENT CREATION USING JAVA

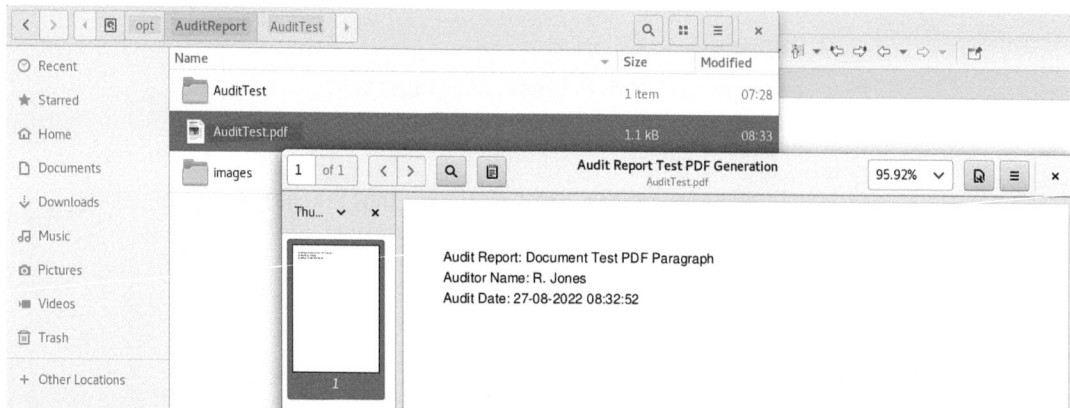

Figure 6-59.   *The **AuditTest.pdf** output from the code in Figure 6-58*

Adding the AuditMasterLogoTest Java Class

Next, we set up the **AuditMasterLogoTest.java** test code to embed a TIF image into the Audit Report pdf document. This code was listed under the "Example 2 – Test Code to Add the Audit Master Logo" section of the chapter. (A Java program for adding the **Audit Master** Logo image to the Audit report document.)

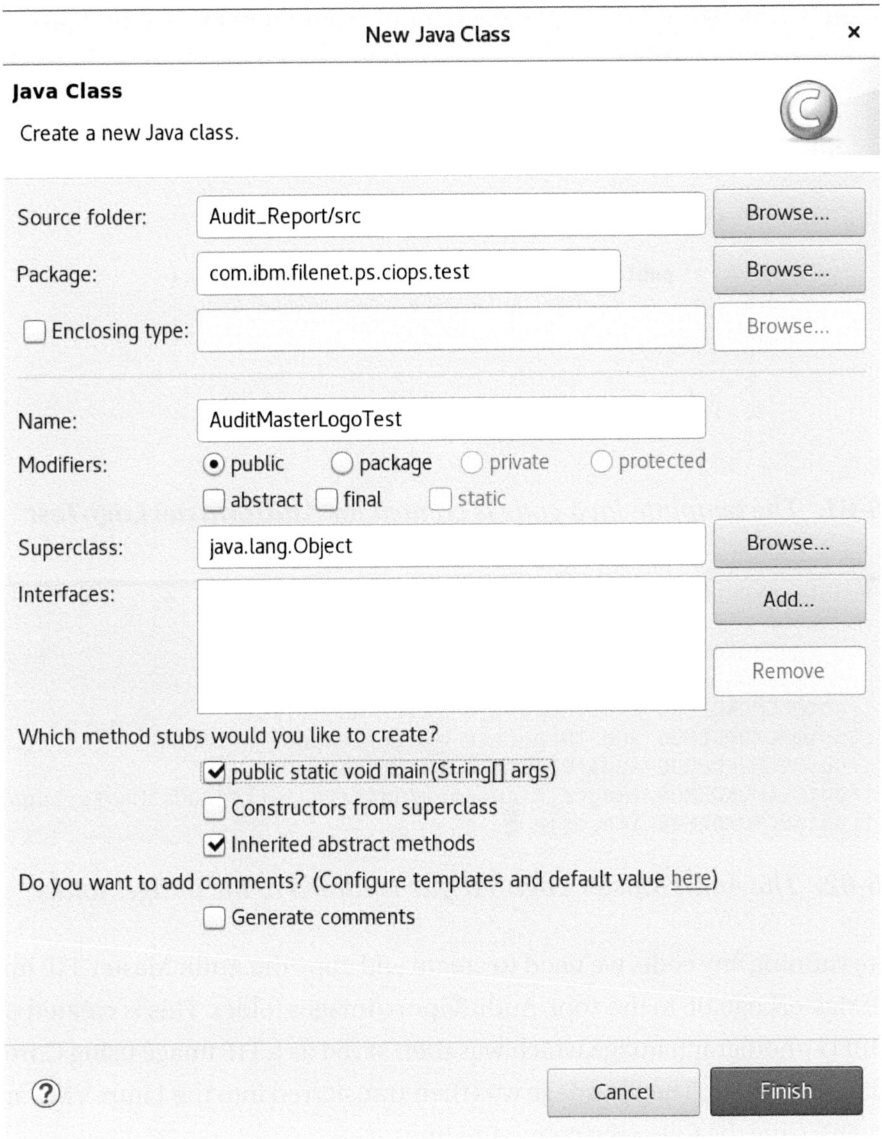

CHAPTER 6 PDF DOCUMENT CREATION USING JAVA

	New Java Class	×

Java Class
Create a new Java class.

Source folder:	Audit_Report/src	Browse...
Package:	com.ibm.filenet.ps.ciops.test	Browse...
☐ Enclosing type:		Browse...

Name:	AuditMasterLogoTest	
Modifiers:	● public ○ package ○ private ○ protected	
	☐ abstract ☐ final ☐ static	
Superclass:	java.lang.Object	Browse...
Interfaces:		Add...
		Remove

Which method stubs would you like to create?
    ☑ public static void main(String[] args)
    ☐ Constructors from superclass
    ☑ Inherited abstract methods

Do you want to add comments? (Configure templates and default value here)
    ☐ Generate comments

[Cancel] [**Finish**]

Figure 6-60. *The **AuditMasterLogoTest** Java Class is created*

CHAPTER 6   PDF DOCUMENT CREATION USING JAVA

The template in Figure 6-61 is generated for the **AuditMasterLogoTest.java**.

```
QOperations.java    AuditMasterLogoTest.java ×
 1  package com.ibm.filenet.ps.ciops.test;
 2
 3  public class AuditMasterLogoTest {
 4
 5      public static void main(String[] args) {
 6          // TODO Auto-generated method stub
 7
 8      }
 9
10  }
11
```

Figure 6-61. *The template Java code is created for* ***AuditMasterLogoTest***

```
(base) [root@ECMUKDEMO6 opt]# pwd
/opt
(base) [root@ECMUKDEMO6 opt]# mkdir AuditReport
(base) [root@ECMUKDEMO6 opt]# cd AuditReport/
(base) [root@ECMUKDEMO6 AuditReport]# mkdir AuditTest
(base) [root@ECMUKDEMO6 AuditReport]# mkdir images
(base) [root@ECMUKDEMO6 AuditReport]# cd images
(base) [root@ECMUKDEMO6 images]# cp /mnt/hgfs/Installs/AuditMasterLogo.tif .
(base) [root@ECMUKDEMO6 images]#
```

Figure 6-62. *The Audit Master Logo TIF file is copied to the images folder*

Before running any code, we need to create and copy the Audit Master TIF image file, AuditMasterLogo.tif, to the /opt/AuditReport/images folder. This is created using an original JPEG photograph image which was then saved as a TIF image using CorelDRAW photo editing software. The file image was then transferred into the Linux VMware environment using the following command lines:

cd /opt
mkdir AuditReport
cd AuditReport
mkdir AuditTest
mkdir images
cd images
cp /mnt/hgfs/Installs/AuditMasterLogo.tif .

1232

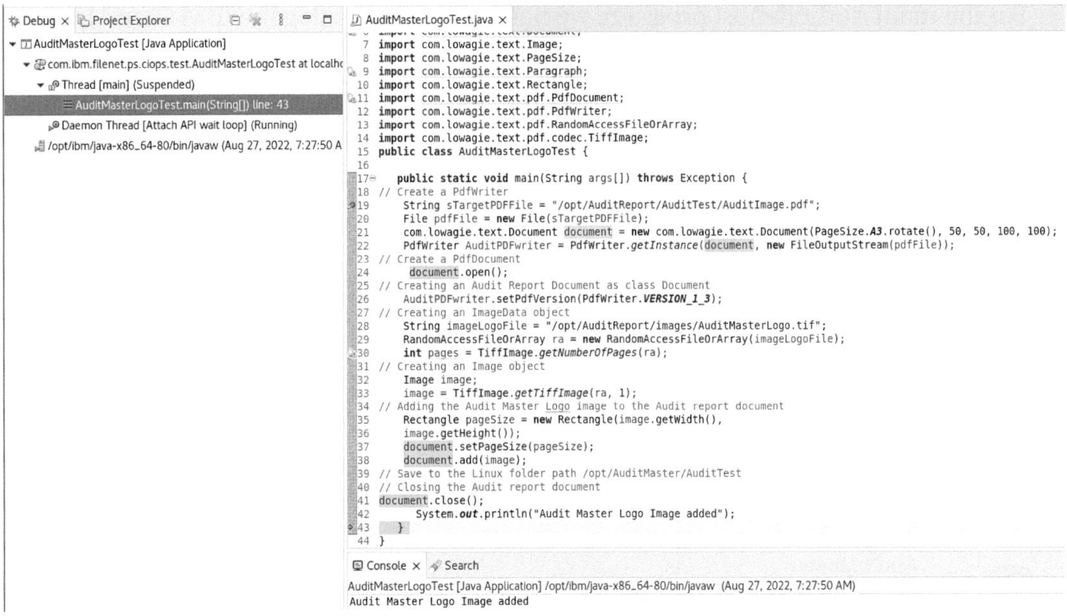

CHAPTER 6   PDF DOCUMENT CREATION USING JAVA

Figure 6-63.   *The AuditMasterLogoTest.java program is successfully run*

After running the Java code shown in Figure 6-63, we can check the contents of the **AuditImage.pdf** output file.

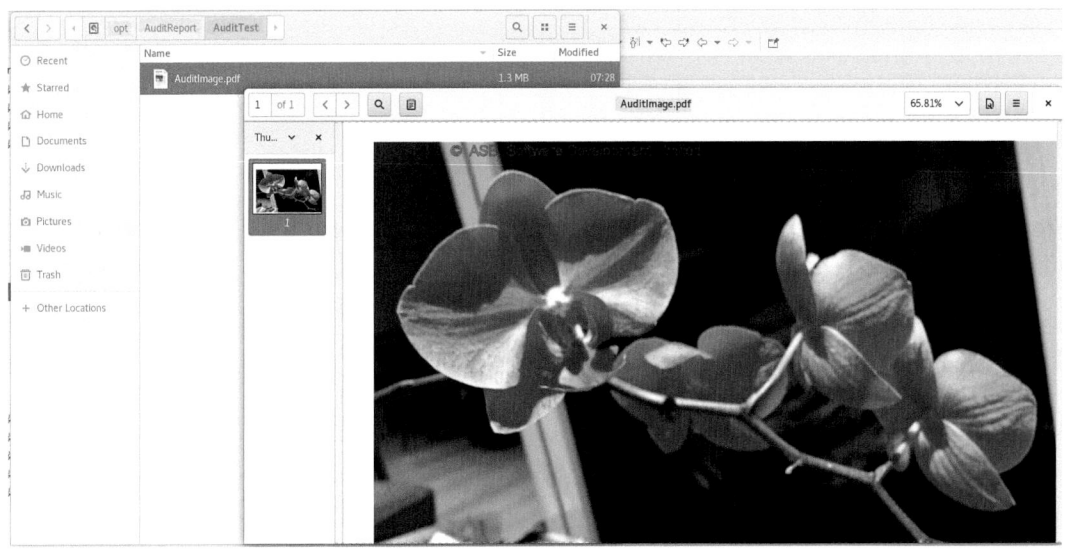

Figure 6-64.   *The AuditImage.pdf output file with the embedded TIF image*

1233

CHAPTER 6   PDF DOCUMENT CREATION USING JAVA

For the main Audit Report program, we need to be able to flag the IBM Case Manager Case Tasks (we have just one, the **AUD_AuditProcess** task).

We can create a new property to identify if we wish to create an Audit Report pdf file for a Case by setting the property with a default String value of "**PDFCreation**" which we can then include in a Content Object Store search string in the **config.xml** file.

Adding the IBM Case Manager Audit Report Questions

We can now update the Audit Master solution using the IBM Case Manager Builder application to add the additional Case properties required by the Audit Report.

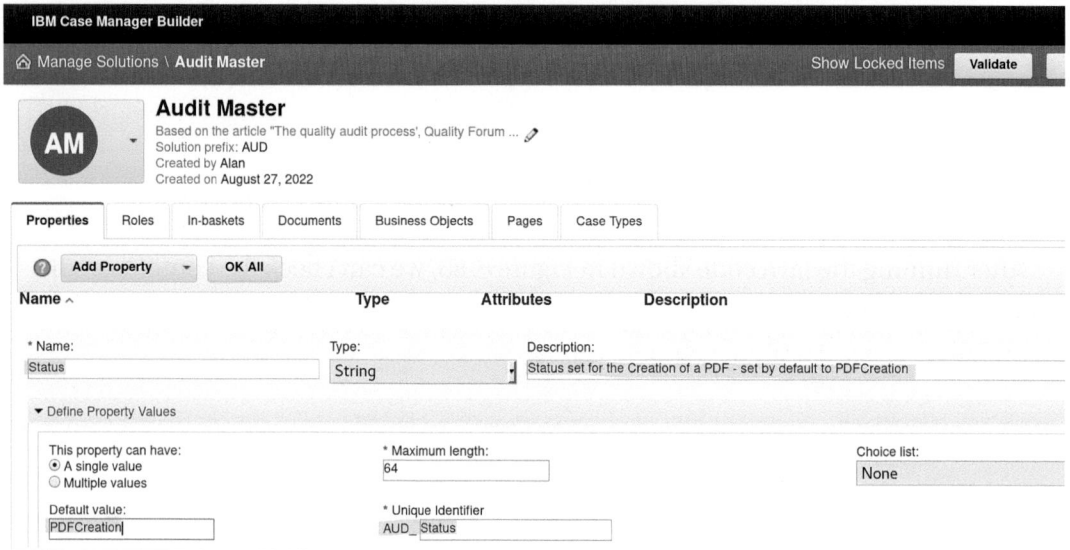

Figure 6-65. *The **Status** for the solution is set by default to "**PDFCreation**"*

Additionally, we add 16 standard Audit Question properties to the **AUD_AuditProcess** task as listed at the beginning of this section (**Example 2**).

The question property names are structured so that they can be referenced in the **config.xml** as a generic stem with the question number defined in a specific place to assist with the Java code loop structure. This scheme provides greater flexibility and allows very efficient code to be produced.

CHAPTER 6   PDF DOCUMENT CREATION USING JAVA

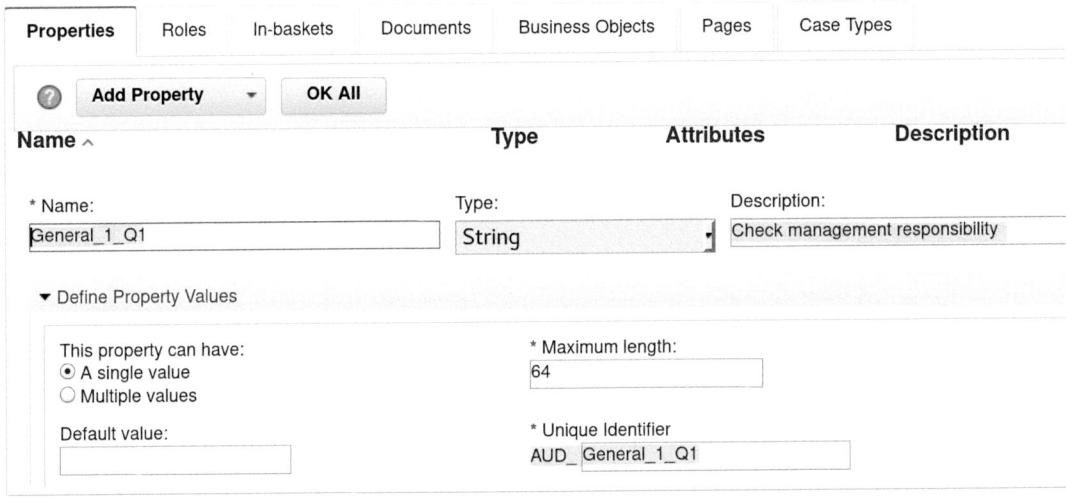

Figure 6-66. The generic questions are added for the Audit Report output

The questions are designed to be output with header section names as shown in the **config.xml** file.

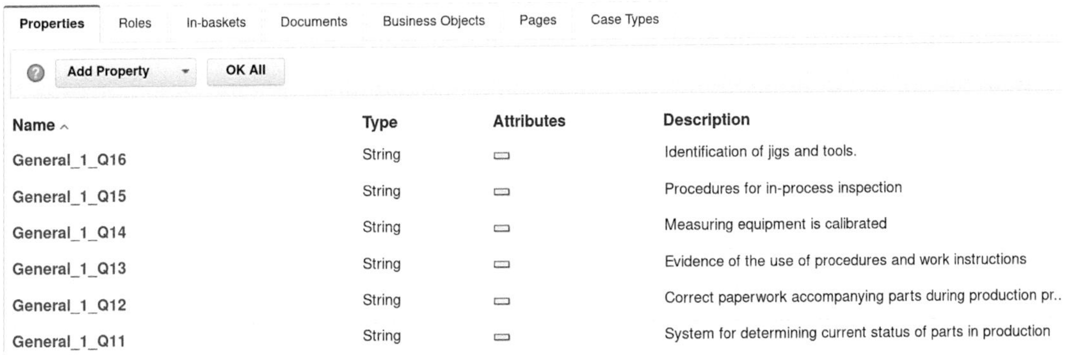

Name	Type	Attributes	Description
General_1_Q16	String		Identification of jigs and tools.
General_1_Q15	String		Procedures for in-process inspection
General_1_Q14	String		Measuring equipment is calibrated
General_1_Q13	String		Evidence of the use of procedures and work instructions
General_1_Q12	String		Correct paperwork accompanying parts during production pr..
General_1_Q11	String		System for determining current status of parts in production

Figure 6-67. The 16 generic questions saved in the Audit Master solution

The 16 generic questions are added to the **AUD_AuditProcess** task properties section, as shown in Figure 6-68.

1235

CHAPTER 6   PDF DOCUMENT CREATION USING JAVA

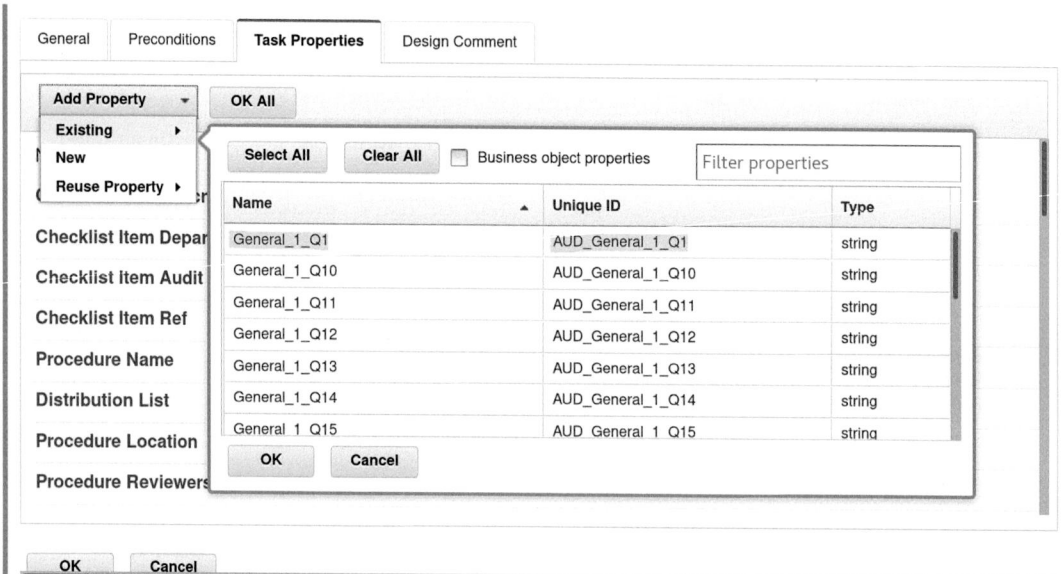

Figure 6-68. *The Task Properties tab of the Audit Process task is shown*

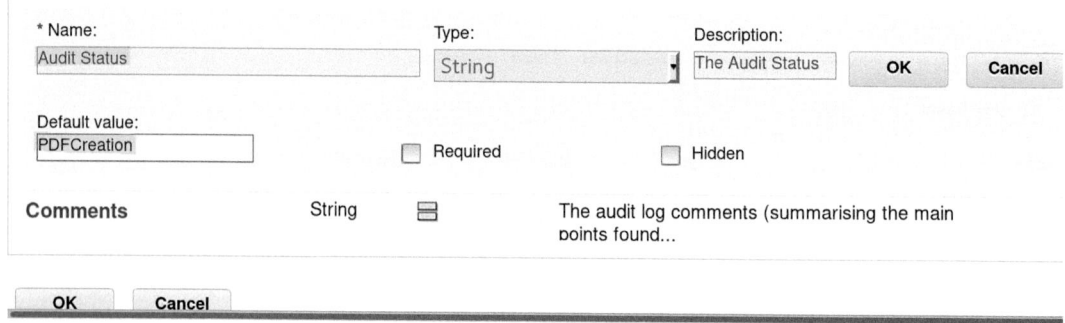

Figure 6-69. *The Audit Status property is created with a default PDFCreation*

The **Audit Master** solution changes are committed and the Solution redeployed. We then need to create new **Audit Department** cases to use for the **Audit Report** tests.

1236

CHAPTER 6   PDF DOCUMENT CREATION USING JAVA

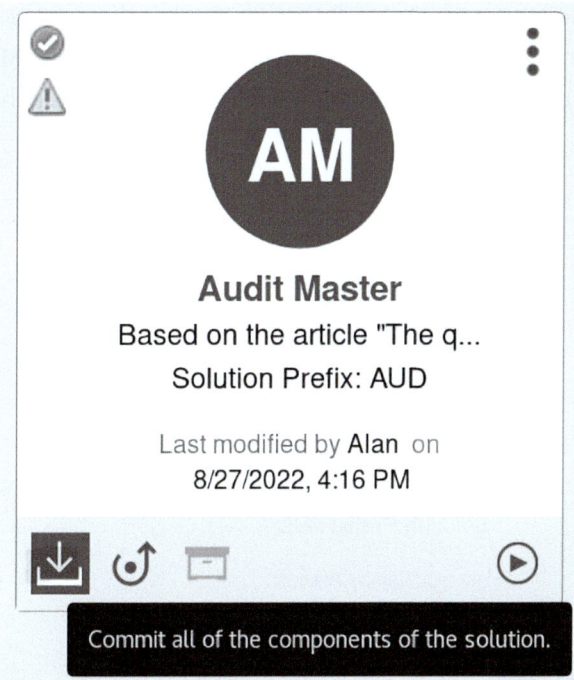

Figure 6-70. *The Commit icon is clicked for the Audit Master solution*

An IBM Case Manager Builder application pop-up window lists the changes to be saved.

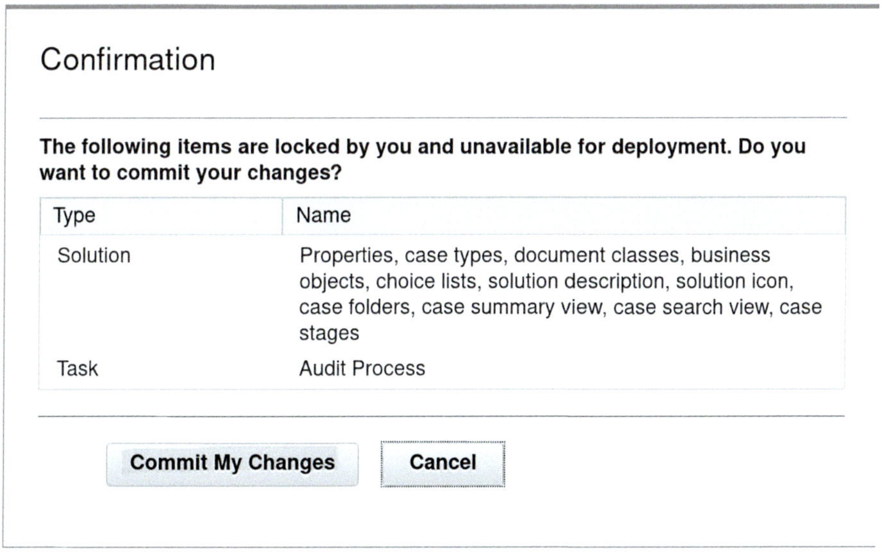

Figure 6-71. *The **Commit My Changes** command button is selected*

The yellow triangle icon with the exclamation mark indicates undeployed edits, so, next, we select the anti-clockwise arrow icon to redeploy the **Audit Master** solution to include the additional Task properties which will be used to create the **Audit Report** pdf file.

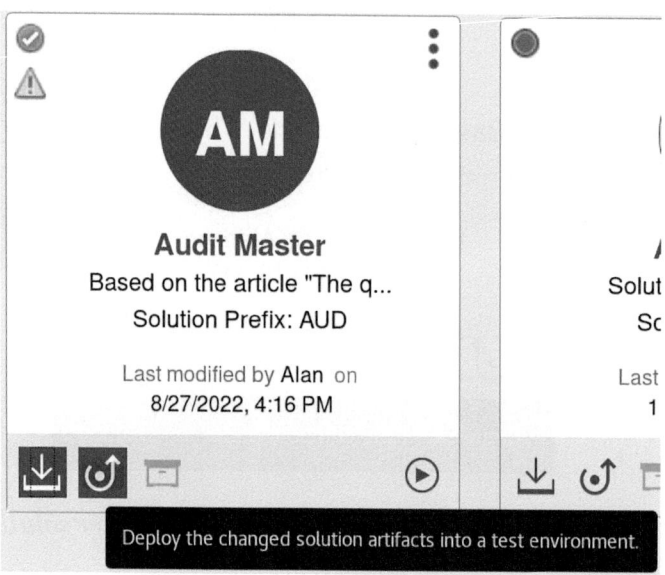

Figure 6-72. *The Audit Master solution is redeployed*

The successful Audit Master Solution deployment is indicated by a white tick icon with a green background (a failed deployment would be indicated by a white cross on a red background).

CHAPTER 6   PDF DOCUMENT CREATION USING JAVA

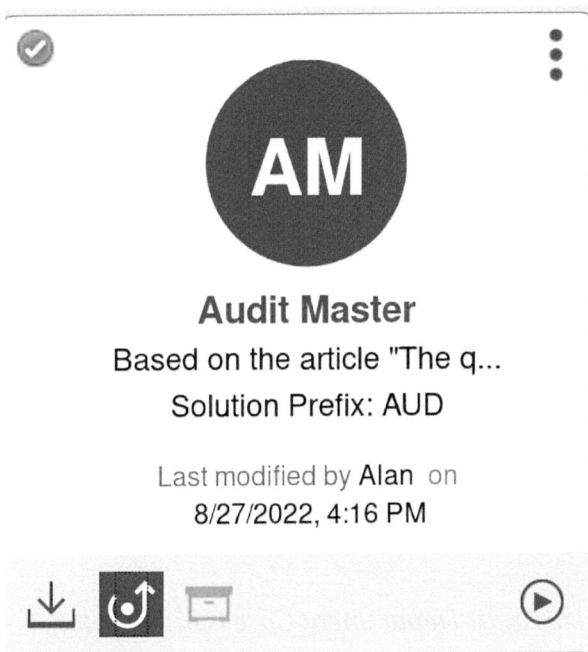

Figure 6-73. *The Audit Master Solution status is indicated with a tick mark*

Testing the Audit Report Program

A new Audit Department Case is created with the additional questions.

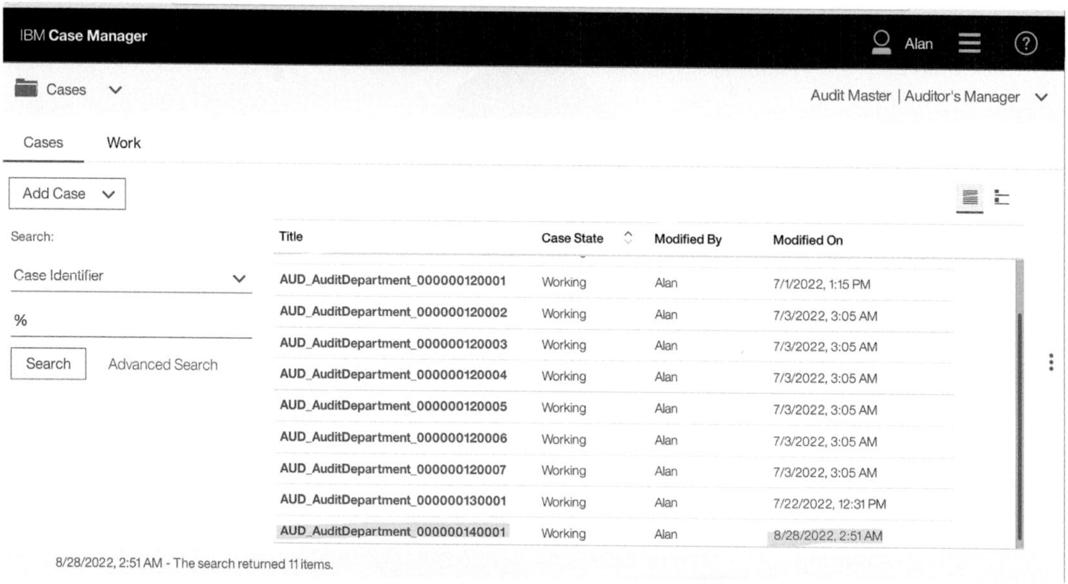

CHAPTER 6   PDF DOCUMENT CREATION USING JAVA

Figure 6-74.   *The new Audit Department Case is highlighted*

The **Audit Report** program is driven by **IBM Case Manager Tasks**. The **config.xml** file is designed to support a Main Case Task, followed by zero or as many additional **Case Manager Task** types as required.

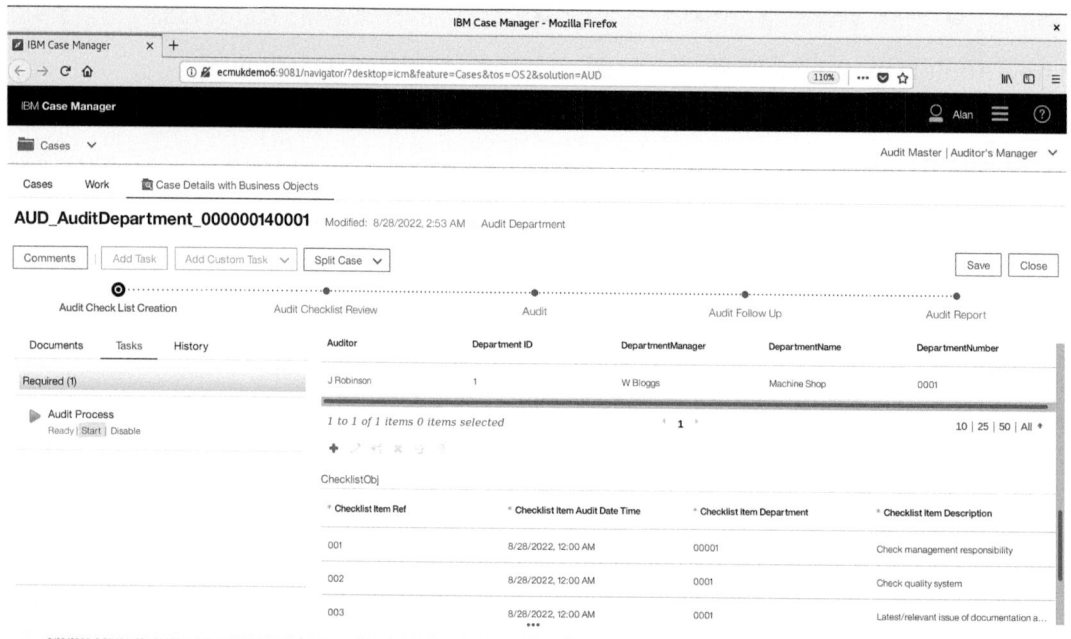

Figure 6-75.   *The optional **Audit Process** Task **Start** command is selected*

1240

The **Task** start is confirmed by selecting the **Yes** command from a pop-up confirmation window.

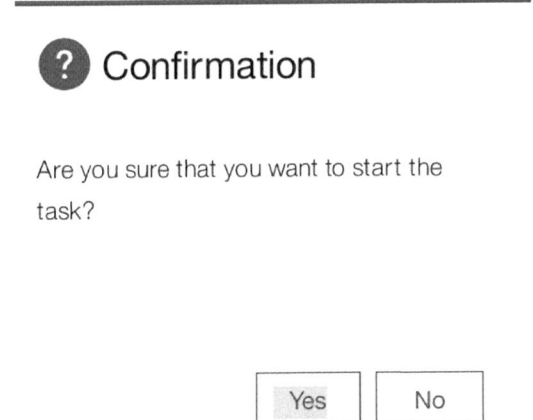

Figure 6-76. *The **Audit Process** Task **Start** command is confirmed*

CHAPTER 6   PDF DOCUMENT CREATION USING JAVA

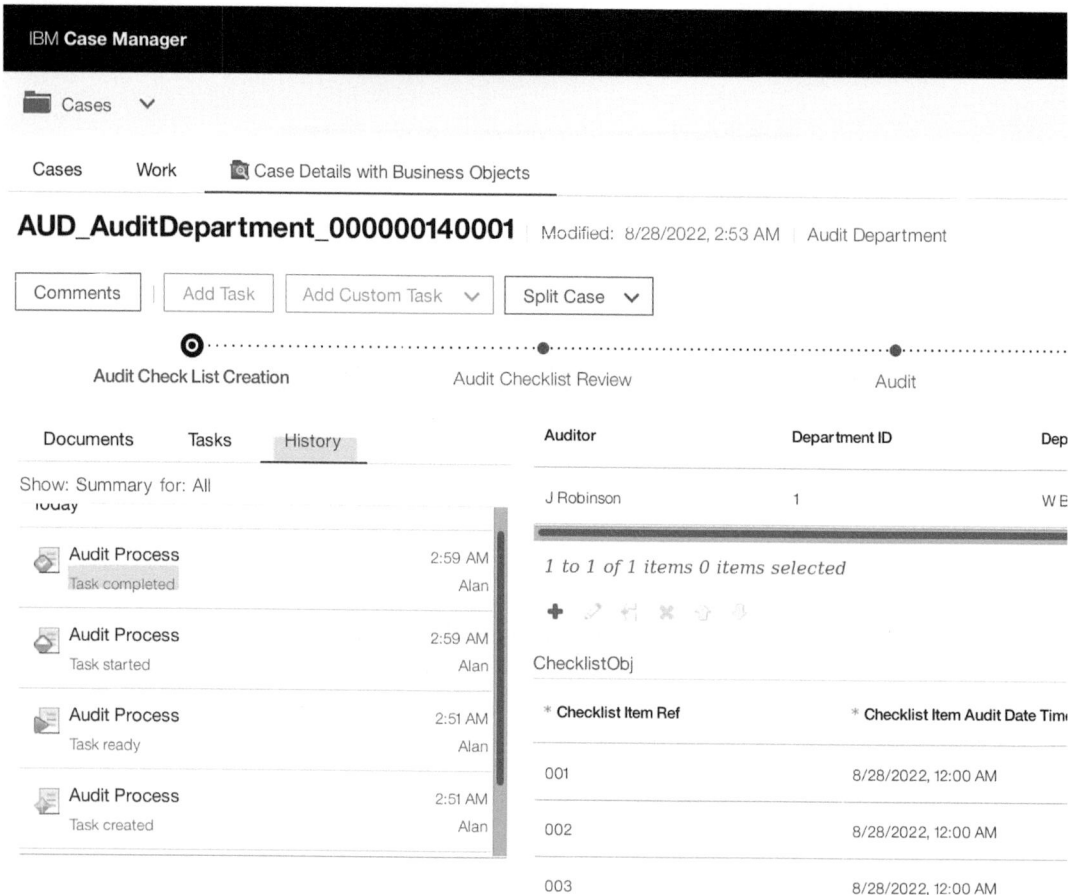

Figure 6-77. *The History tab of the opened Audit Department Case*

The **Audit Department** Case shows that the **Audit Process** task has successfully completed.

Before we run the **Audit Report** program, we have to create a subfolder under the **/opt/AuditReport** directory with the name of the Case Task, since each IBM Case Manager Task is designed to generate a separate Audit Report pdf file. Also, the individual case reports are identified and separated from each other because the program uses the unique **Task Id** of the stored Case as the first part of the Audit report filename.

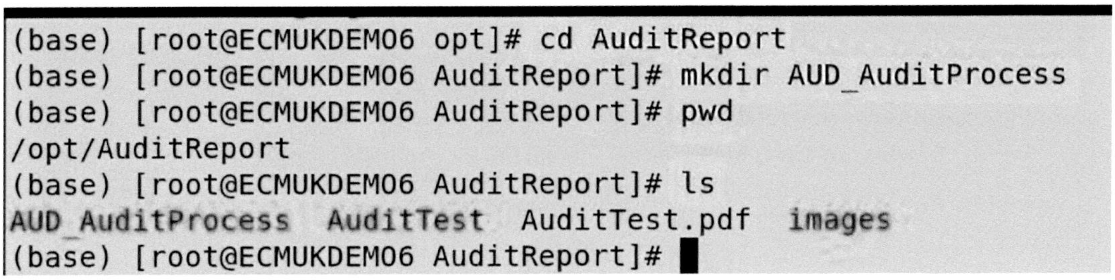

CHAPTER 6   PDF DOCUMENT CREATION USING JAVA

```
(base) [root@ECMUKDEMO6 opt]# cd AuditReport
(base) [root@ECMUKDEMO6 AuditReport]# mkdir AUD_AuditProcess
(base) [root@ECMUKDEMO6 AuditReport]# pwd
/opt/AuditReport
(base) [root@ECMUKDEMO6 AuditReport]# ls
AUD_AuditProcess  AuditTest  AuditTest.pdf  images
(base) [root@ECMUKDEMO6 AuditReport]#
```

Figure 6-78. *The AuditReport task subfolder AUD_AuditProcess is created*

The following command-line code is used to create the **Audit Report** task subfolder:

cd /opt/AuditReport
mkdir AUD_AuditProcess

After the **Audit Report** pdf is created, the file is uploaded back to the Audit Department Case Folder and then deleted from the Linux folder (which is used just as a temporary storage area).

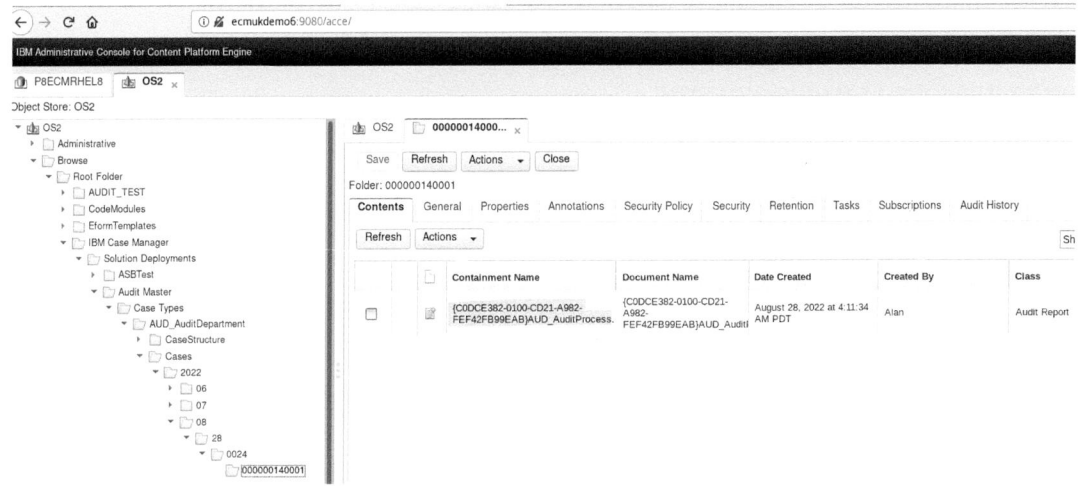

Figure 6-79. *The highlighted Audit Process task pdf is stored with the Case*

This pdf file is stored in the OS2 IBM Case Manager target Object Store as an Audit Report class document.

1243

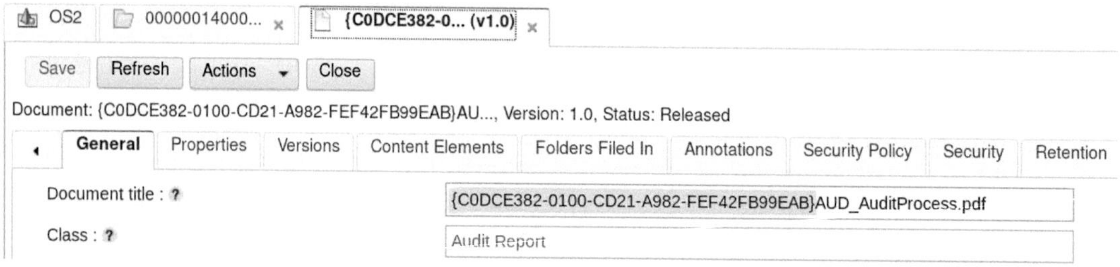

CHAPTER 6   PDF DOCUMENT CREATION USING JAVA

Figure 6-80. *The Audit Process pdf file name shows the Task Id (highlighted)*

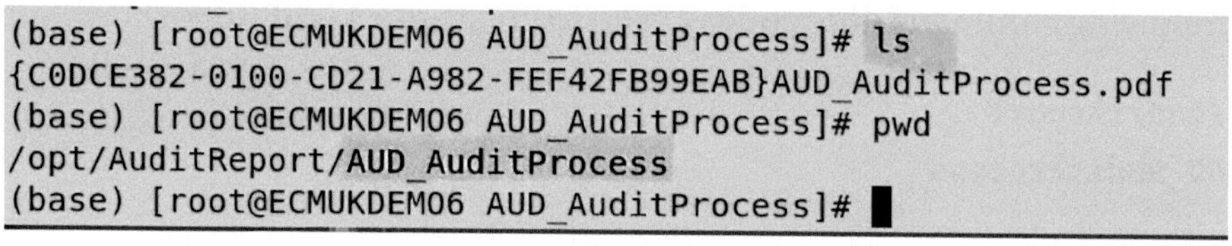

Figure 6-81. *The (temporary) Audit Report file is displayed in the directory*

We have paused the program in debug mode just before running the delete statement, as shown in the code screenshot in Figure 6-82.

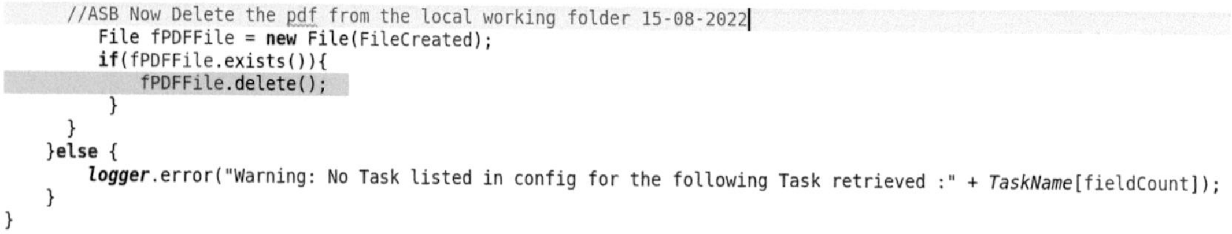

Figure 6-82. *The Java file delete statement will remove the file in Figure 6-81*

Creating the log4j.xml for the Audit Report Logs

The **Log4J** logs subdirectory is created using the following code:

```
cd /opt/AuditReport
mkdir logs
```

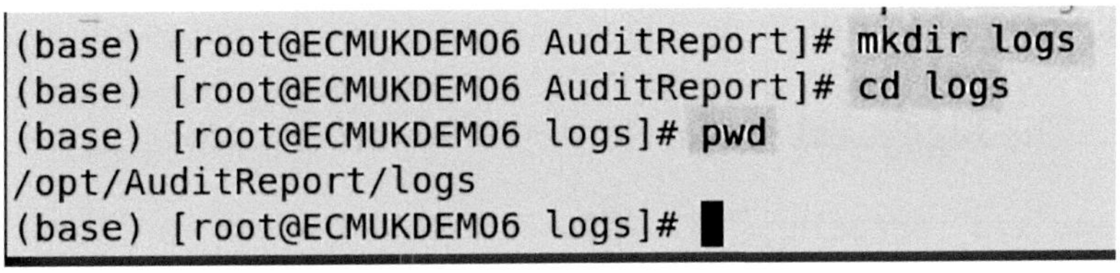

CHAPTER 6   PDF DOCUMENT CREATION USING JAVA

```
(base) [root@ECMUKDEMO6 AuditReport]# mkdir logs
(base) [root@ECMUKDEMO6 AuditReport]# cd logs
(base) [root@ECMUKDEMO6 logs]# pwd
/opt/AuditReport/logs
(base) [root@ECMUKDEMO6 logs]#
```

Figure 6-83. *The Log4J logs subdirectory is created*

The **log4j.xml** file is created under the **Audit_Report/src** project folder.

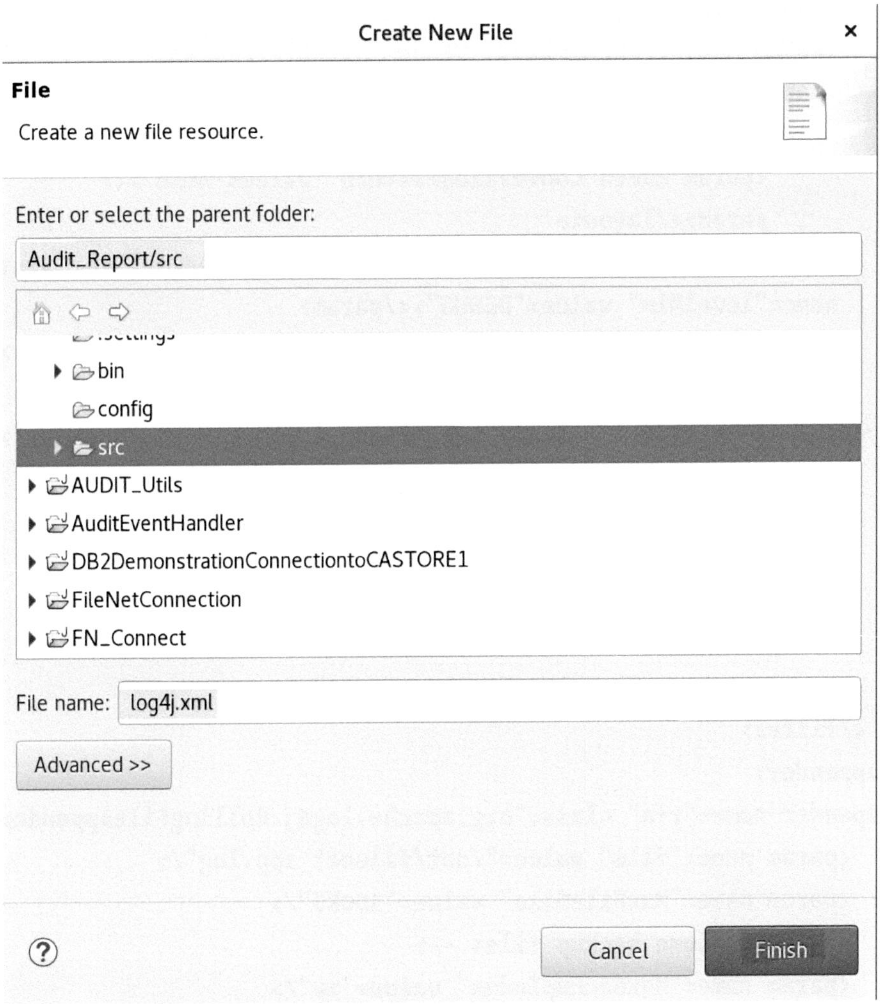

Figure 6-84. *A **log4j.xml** file is created under an **Audit_Report/src** folder*

1245

The **log4j.xml** for the **AuditReport** Java program is configured as follows.

Listing 6-27. The **log4j.xml** list for the Audit Report error logging

```xml
<?xml version="1.0" encoding="UTF-8"?>
<!DOCTYPE log4j:configuration SYSTEM "log4j.dtd" >
<log4j:configuration>
    <appender name="rfa2" class="org.apache.log4j.RollingFileAppender">
        <param name="file" value="/opt/AuditReport/logs/AuditProcessor.
        log"></param>
        <param name="MaxFileSize" value="500KB"></param>
        <param name="MaxBackupIndex" value="100"></param>
        <layout class="org.apache.log4j.PatternLayout">
            <param name="ConversionPattern" value="%m%n"></
            param></layout>
        <filter class="org.apache.log4j.varia.LevelRangeFilter"><param
        name="levelMin" value="DEBUG"></param>
            <param name="levelMax" value="FATAL"></param></filter>
    </appender>
    <appender name="stdout" class="org.apache.log4j.ConsoleAppender">
        <layout class="org.apache.log4j.PatternLayout">
            <param name="ConversionPattern" value="%m%n" />
        </layout>
        <filter class="org.apache.log4j.varia.LevelRangeFilter">
            <param name="levelMin" value="DEBUG" />
            <param name="levelMax" value="INFO" />
        </filter>
    </appender>
    <appender name="rfa" class="org.apache.log4j.RollingFileAppender">
        <param name="file" value="/opt/filenet_app.log"/>
        <param name="MaxFileSize" value="100KB"/>
        <!-- Keep some backup files -->
        <param name="MaxBackupIndex" value="10"/>
        <layout class="org.apache.log4j.PatternLayout">
            <param name="ConversionPattern" value="%p %t %c - %m%n"/>
        </layout>
```

```xml
    <filter class="org.apache.log4j.varia.LevelRangeFilter">
        <param name="levelMin" value="DEBUG" />
        <param name="levelMax" value="FATAL" />
    </filter>
 </appender>
 <appender name="rfa_threads" class="org.apache.log4j.RollingFileAppender">
    <param name="file" value="/opt/AuditReport/logs/filenet_JVM.log"/>
    <param name="MaxFileSize" value="500KB"/>
    <!-- Keep one backup file -->
    <param name="MaxBackupIndex" value="100"/>
    <layout class="org.apache.log4j.PatternLayout">
        <param name="ConversionPattern" value="%m%n"/>
        <!--<param name="ConversionPattern" value="%p %t %c - %m%n"/>-->
    </layout>
    <filter class="org.apache.log4j.varia.LevelRangeFilter">
        <param name="levelMin" value="INFO" />
        <param name="levelMax" value="INFO" />
    </filter>
 </appender>
<logger name="com.ibm.filenet.ce" additivity="false">
    <level value="debug" />
    <appender-ref ref="rfa" />
    <appender-ref ref="stdout" />
</logger>
<logger name="filenet_error.api.com.filenet.apiimpl.util.ConfigValueLookup" additivity="false">
    <level value="debug" />
    <appender-ref ref="rfa" />
    <appender-ref ref="stdout" />
</logger>
<logger name="filenet_tracing.api.detail.moderate.summary.com.filenet.apiimpl.util.ConfigValueLookup" additivity="false">
    <level value="debug" />
```

```
            <appender-ref ref="rfa" />
            <appender-ref ref="stdout" />
    </logger>
    <logger name="com.asb.ce.utils.AUDITProcessorConfig"
    additivity="false">
        <level value="debug" />
            <appender-ref ref="rfa_threads" />
    </logger>
    <logger name="com.ibm.filenet.ps.ciops.test.AUDITReportMain"
    additivity="false">
            <level value="info"></level>
            <appender-ref ref="rfa2"></appender-ref>
            <appender-ref ref="stdout"></appender-ref>
    </logger>
     <root>
            <priority value="error" />
            <appender-ref ref="stdout" />
            <appender-ref ref="rfa" />
            <appender-ref ref="rfa2"></appender-ref></root>
    </log4j:configuration>
```

The **log4j.xml** file also requires a **log4j.dtd** file in the same folder path, which defines the XML file tags used. This **log4j.dtd** contains the following lines.

Listing 6-28. The **log4j.dtd** file for the log4j.xml definitions

```
<?xml version="1.0" encoding="UTF-8" ?>
<!--
Licensed to the Apache Software Foundation (ASF) under one or more
contributor license agreements.  See the NOTICE file distributed with
this work for additional information regarding copyright ownership.
The ASF licenses this file to You under the Apache License, Version 2.0
(the "License"); you may not use this file except in compliance with
the License.  You may obtain a copy of the License at

      http://www.apache.org/licenses/LICENSE-2.0
```

```
Unless required by applicable law or agreed to in writing, software
distributed under the License is distributed on an "AS IS" BASIS,
WITHOUT WARRANTIES OR CONDITIONS OF ANY KIND, either express or implied.
See the License for the specific language governing permissions and
limitations under the License.
-->

<!-- Authors: Chris Taylor, Ceki Gulcu. -->

<!-- Version: 1.2 -->

<!-- A configuration element consists of optional renderer
elements,appender elements, categories and an optional root
element. -->

<!ELEMENT log4j:configuration (renderer*, appender*,plugin*,
(category|logger)*,root?,
                            (categoryFactory|loggerFactory)?)>

<!-- The "threshold" attribute takes a level value below which -->
<!-- all logging statements are disabled. -->

<!-- Setting the "debug" enable the printing of internal log4j
logging    -->
<!--
statements.                                                    -->
<!-- By default, debug attribute is "null", meaning that we not do
touch -->
<!-- internal log4j logging settings. The "null" value for the
threshold -->
<!-- attribute can be misleading. The threshold field of a
repository    -->
<!-- cannot be set to null. The "null" value for the threshold
attribute -->
<!-- simply means don't touch the threshold field, the threshold
field    -->
<!-- keeps its old
value.                                                 -->
```

```
<!ATTLIST log4j:configuration
  xmlns:log4j         CDATA #FIXED "http://jakarta.apache.org/log4j/"
  threshold           (all|trace|debug|info|warn|error|fatal|off|null) "null"
  debug               (true|false|null)  "null"
  reset               (true|false) "false"
>

<!-- renderer elements allow the user to customize the conversion of  -->
<!-- message objects to String.                                       -->

<!ELEMENT renderer EMPTY>
<!ATTLIST renderer
  renderedClass  CDATA #REQUIRED
  renderingClass CDATA #REQUIRED
>

<!-- Appenders must have a name and a class. -->
<!-- Appenders may contain an error handler, a layout, optional parameters -->
<!-- and filters. They may also reference (or include) other appenders. -->
<!ELEMENT appender (errorHandler?, param*,
      rollingPolicy?, triggeringPolicy?, connectionSource?,
      layout?, filter*, appender-ref*)>
<!ATTLIST appender
  name         CDATA      #REQUIRED
  class        CDATA      #REQUIRED
>

<!ELEMENT layout (param*)>
<!ATTLIST layout
  class          CDATA    #REQUIRED
>

<!ELEMENT filter (param*)>
<!ATTLIST filter
  class          CDATA    #REQUIRED
>
```

```
<!-- ErrorHandlers can be of any class. They can admit any number of -->
<!-- parameters. -->

<!ELEMENT errorHandler (param*, root-ref?, logger-ref*,  appender-ref?)>
<!ATTLIST errorHandler
   class         CDATA    #REQUIRED
>

<!ELEMENT root-ref EMPTY>

<!ELEMENT logger-ref EMPTY>
<!ATTLIST logger-ref
  ref CDATA #REQUIRED
>

<!ELEMENT param EMPTY>
<!ATTLIST param
  name           CDATA    #REQUIRED
  value          CDATA    #REQUIRED
>

<!-- The priority class is org.apache.log4j.Level by default -->
<!ELEMENT priority (param*)>
<!ATTLIST priority
  class    CDATA     #IMPLIED
  value         CDATA #REQUIRED
>

<!-- The level class is org.apache.log4j.Level by default -->
<!ELEMENT level (param*)>
<!ATTLIST level
  class    CDATA     #IMPLIED
  value         CDATA #REQUIRED
>

<!-- If no level element is specified, then the configurator MUST not -->
<!-- touch the level of the named category. -->
<!ELEMENT category (param*,(priority|level)?,appender-ref*)>
```

```
<!ATTLIST category
  class         CDATA     #IMPLIED
  name          CDATA     #REQUIRED
  additivity    (true|false) "true"
>

<!-- If no level element is specified, then the configurator MUST not -->
<!-- touch the level of the named logger. -->
<!ELEMENT logger (level?,appender-ref*)>
<!ATTLIST logger
  name          CDATA     #REQUIRED
  additivity    (true|false) "true"
>

<!ELEMENT categoryFactory (param*)>
<!ATTLIST categoryFactory
   class        CDATA #REQUIRED>

<!ELEMENT loggerFactory (param*)>
<!ATTLIST loggerFactory
   class        CDATA #REQUIRED>

<!ELEMENT appender-ref EMPTY>
<!ATTLIST appender-ref
  ref CDATA #REQUIRED
>

<!-- plugins must have a name and class and can have optional parameters -->
<!ELEMENT plugin (param*, connectionSource?)>
<!ATTLIST plugin
  name          CDATA           #REQUIRED
  class      CDATA #REQUIRED
>

<!ELEMENT connectionSource (dataSource?, param*)>
<!ATTLIST connectionSource
   class        CDATA #REQUIRED
>
```

```
<!ELEMENT dataSource (param*)>
<!ATTLIST dataSource
  class       CDATA   #REQUIRED
>

<!ELEMENT triggeringPolicy ((param|filter)*)>
<!ATTLIST triggeringPolicy
  name        CDATA   #IMPLIED
  class       CDATA   #REQUIRED
>

<!ELEMENT rollingPolicy (param*)>
<!ATTLIST rollingPolicy
  name        CDATA   #IMPLIED
  class       CDATA   #REQUIRED
>

<!-- If no priority element is specified, then the configurator MUST not -->
<!-- touch the priority of root. -->
<!-- The root category always exists and cannot be subclassed. -->
<!ELEMENT root (param*, (priority|level)?, appender-ref*)>

<!-- ================================================================ -->
<!--                    A logging event                             -->
<!-- ================================================================ -->
<!ELEMENT log4j:eventSet (log4j:event*)>
<!ATTLIST log4j:eventSet
  xmlns:log4j              CDATA #FIXED "http://jakarta.apache.org/log4j/"
  version                  (1.1|1.2) "1.2"
  includesLocationInfo     (true|false) "true"
>
```

```
<!ELEMENT log4j:event (log4j:message, log4j:NDC?, log4j:throwable?,
                      log4j:locationInfo?, log4j:properties?) >

<!-- The timestamp format is application dependent. -->
<!ATTLIST log4j:event
    logger     CDATA #REQUIRED
    level      CDATA #REQUIRED
    thread     CDATA #REQUIRED
    timestamp  CDATA #REQUIRED
    time       CDATA #IMPLIED
>

<!ELEMENT log4j:message (#PCDATA)>
<!ELEMENT log4j:NDC (#PCDATA)>

<!ELEMENT log4j:throwable (#PCDATA)>

<!ELEMENT log4j:locationInfo EMPTY>
<!ATTLIST log4j:locationInfo
  class   CDATA    #REQUIRED
  method  CDATA    #REQUIRED
  file    CDATA    #REQUIRED
  line    CDATA    #REQUIRED
>

<!ELEMENT log4j:properties (log4j:data*)>

<!ELEMENT log4j:data EMPTY>
<!ATTLIST log4j:data
  name   CDATA    #REQUIRED
  value  CDATA    #REQUIRED
>
```

Verifying the Audit Report pdf File Structure

The output Audit Report pdf file, **AUD_AuditProcess.pdf**, can be viewed in the IBM FileNet Content Engine Administration Console (**acce**) web application Document Viewer.

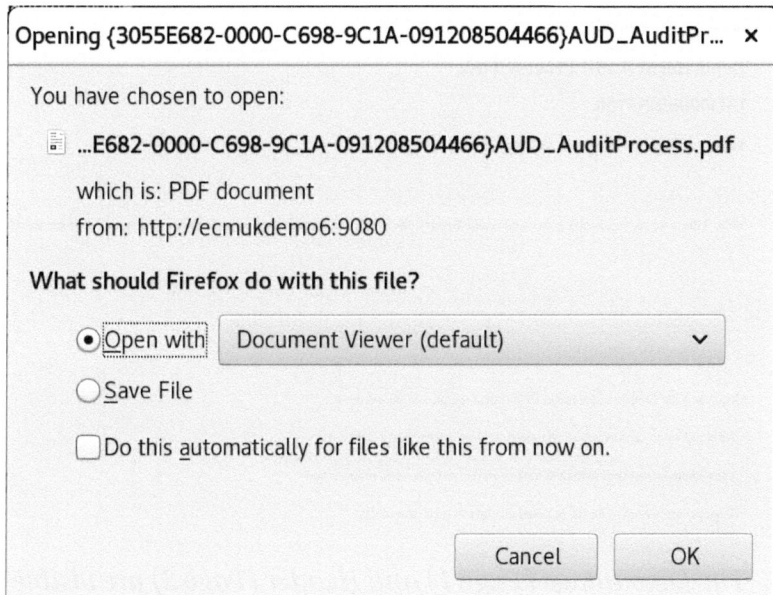

Figure 6-85.   The Audit Report pdf, {3055E682-0000-C698-9C1A-091208504466} AUD_AuditProcess.pdf, is viewed

The IBM FileNet Document Viewer shows the Audit Report pdf thumbnails and the Header page.

CHAPTER 6   PDF DOCUMENT CREATION USING JAVA

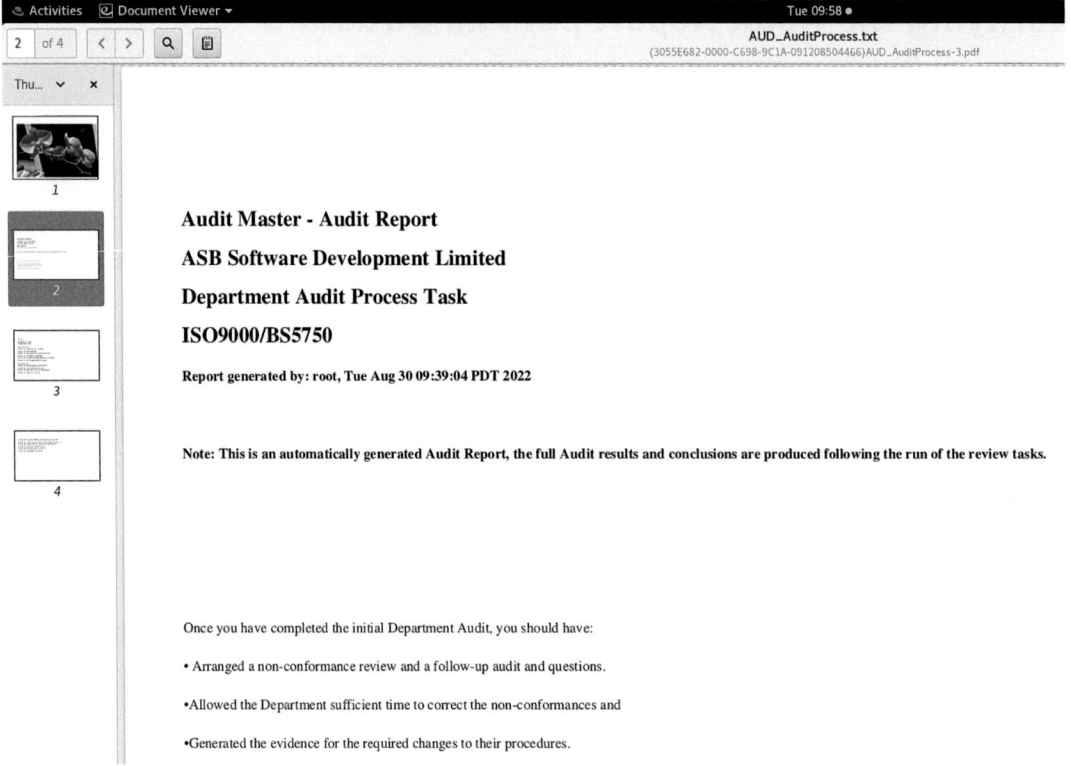

Figure 6-86. *The Logo Image (Page 1) and Header (Page 2) are visible*

The **Audit Report** pdf document properties can be seen which, after clicking the **Properties** menu item, show the Header attributes we set in the Audit Report Java program.

CHAPTER 6　PDF DOCUMENT CREATION USING JAVA

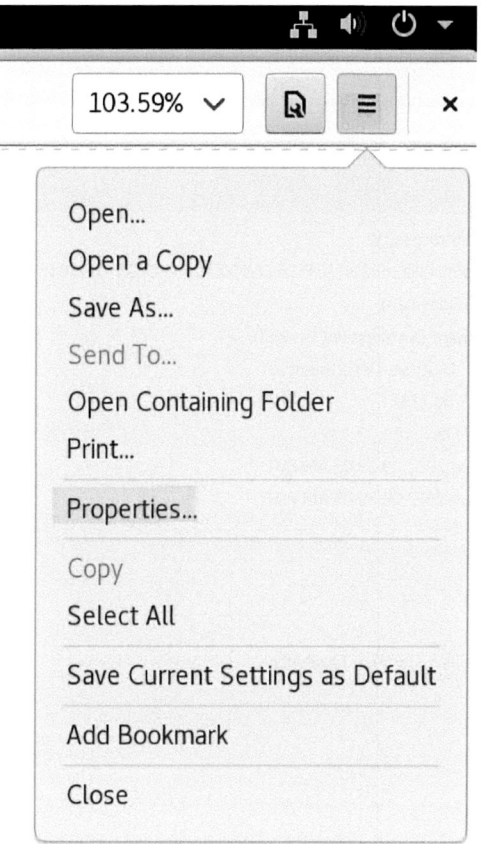

Figure 6-87. *The **Properties...** menu item is selected from the drop-down*

CHAPTER 6   PDF DOCUMENT CREATION USING JAVA

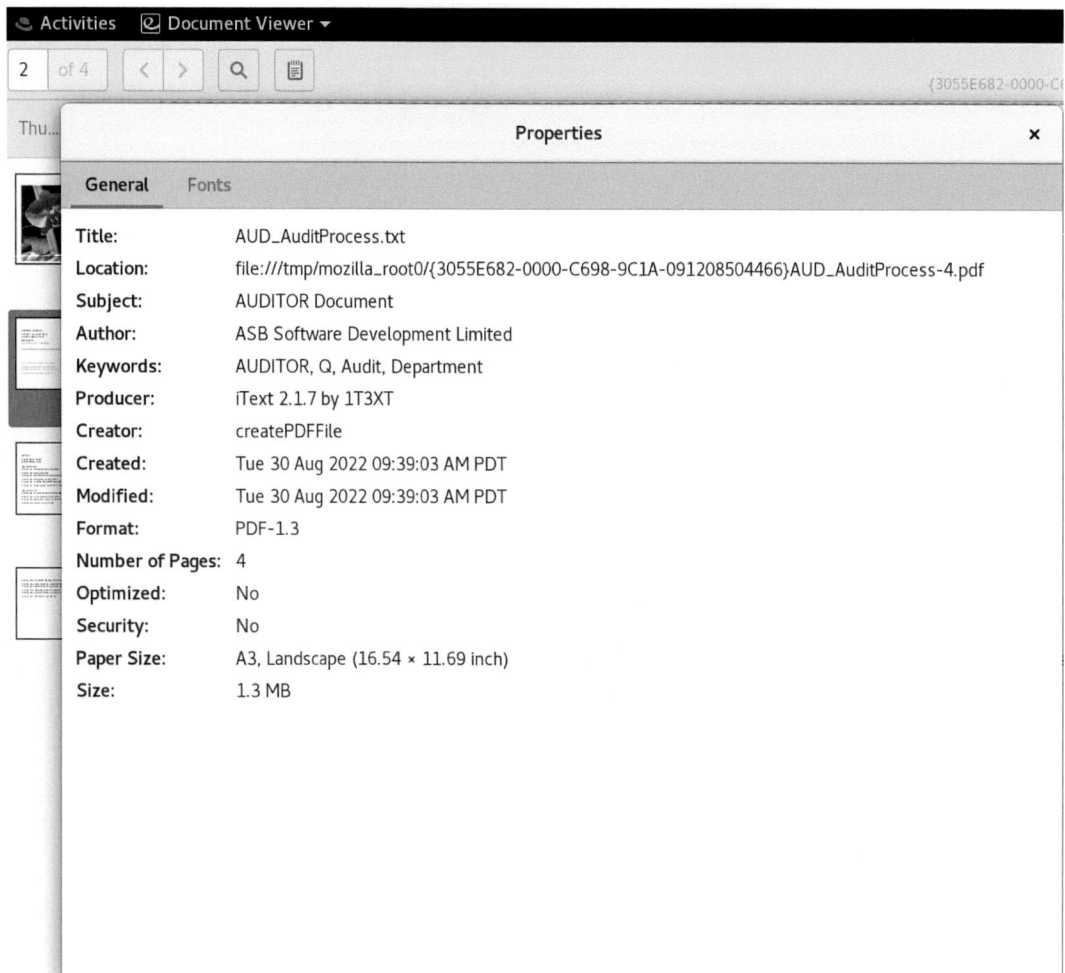

Figure 6-88. *The generated Audit Report pdf Header attributes are shown*

CHAPTER 6 PDF DOCUMENT CREATION USING JAVA

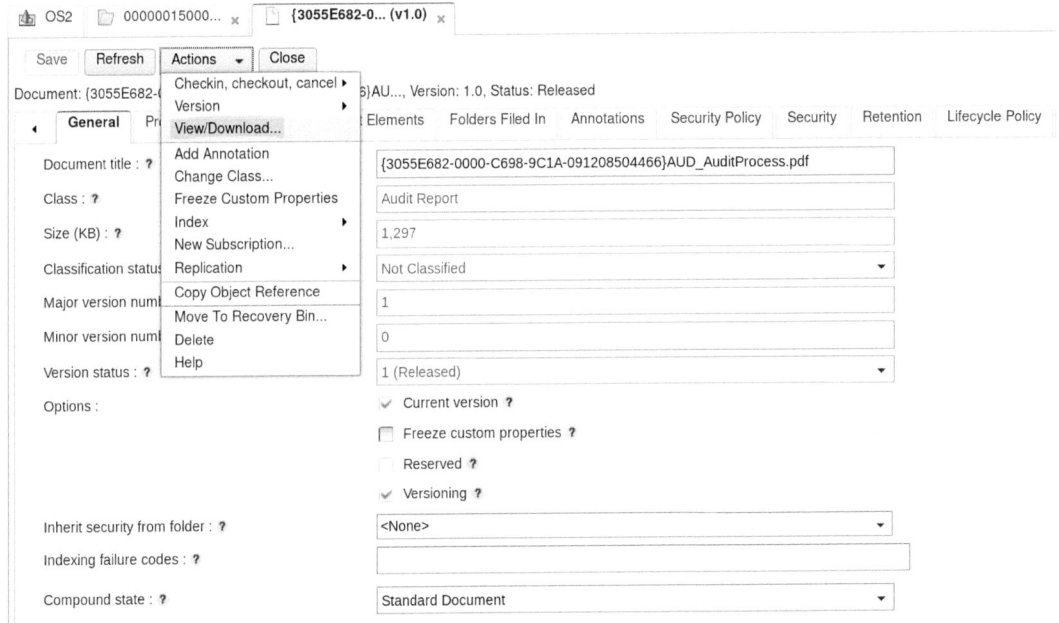

Figure 6-89. The View/Download menu option we have used for the pdf

Creating the Audit Master Solution Workflow Audit Review Step

To provide the necessary review of the Audit Report Questions, a review step is created in the Workflow Designer of the IBM Case Manager for the Auditor role (see Chapter 1 for the full Audit Master solution build).

CHAPTER 6 PDF DOCUMENT CREATION USING JAVA

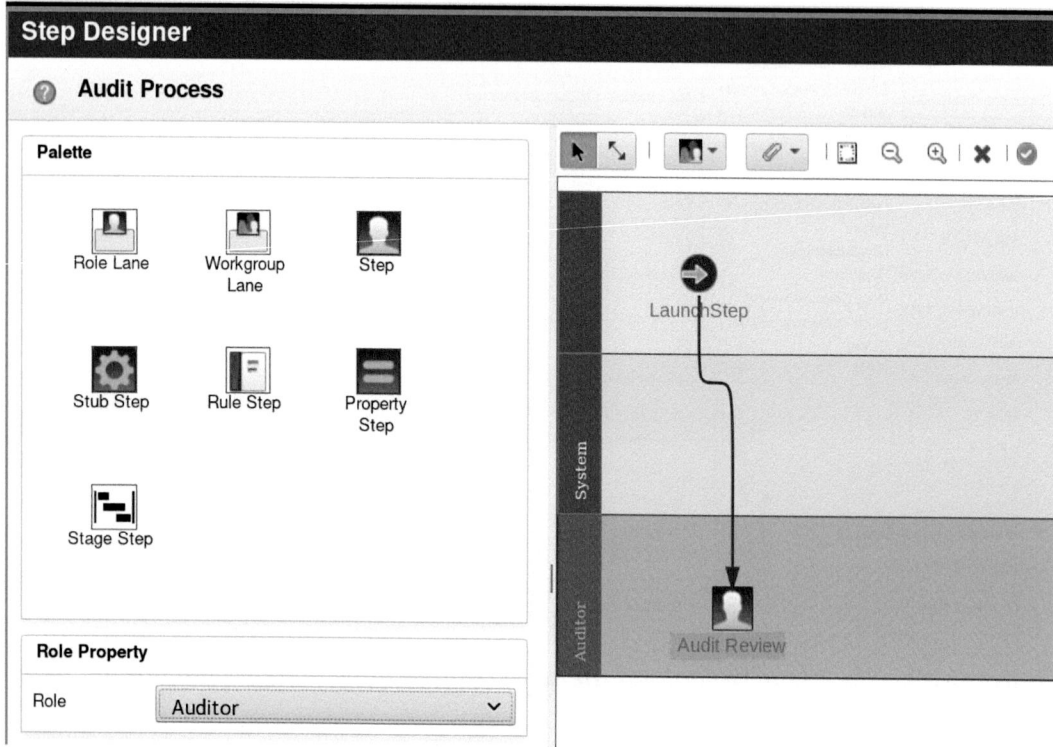

Figure 6-90.* The Audit Review step is created for the Auditor role*

CHAPTER 6　PDF DOCUMENT CREATION USING JAVA

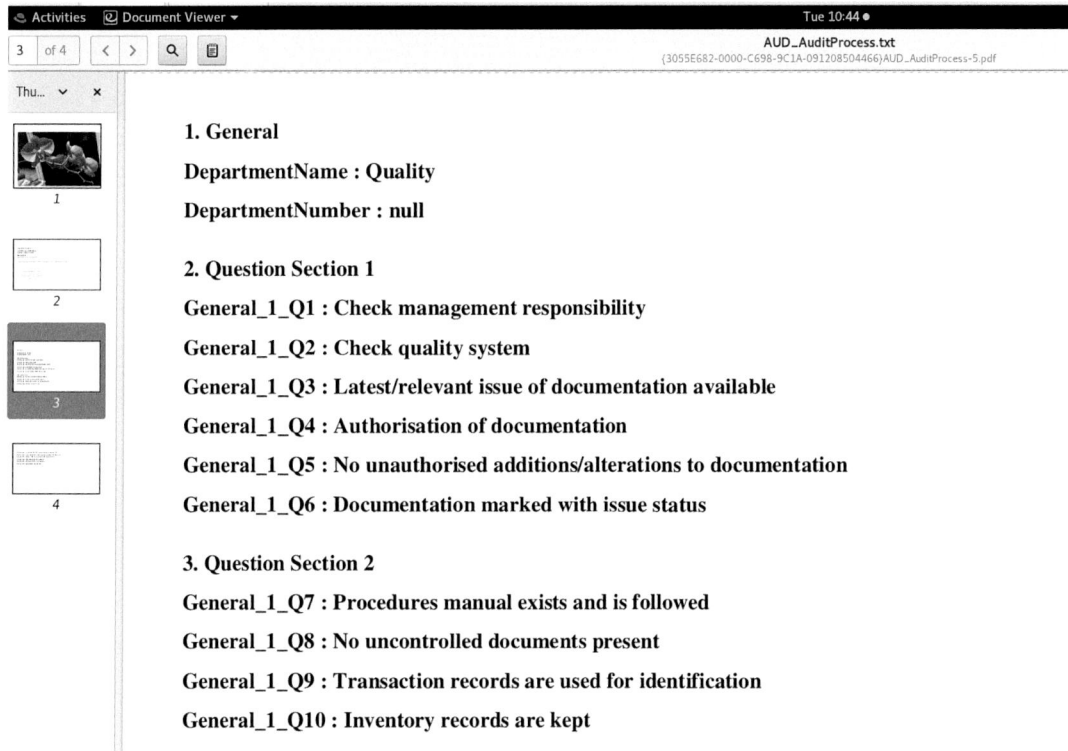

Figure 6-91. *Page 3 of the Audit Report shows the Questions 1 to 10*

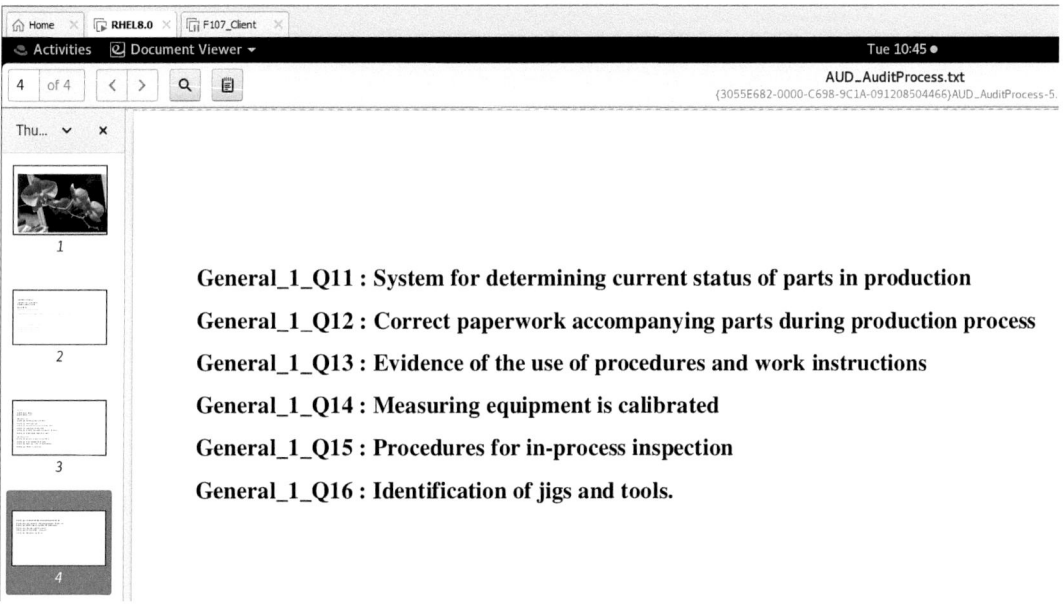

Figure 6-92. *Page 4 of the Audit Report shows the Questions 11 to 16*

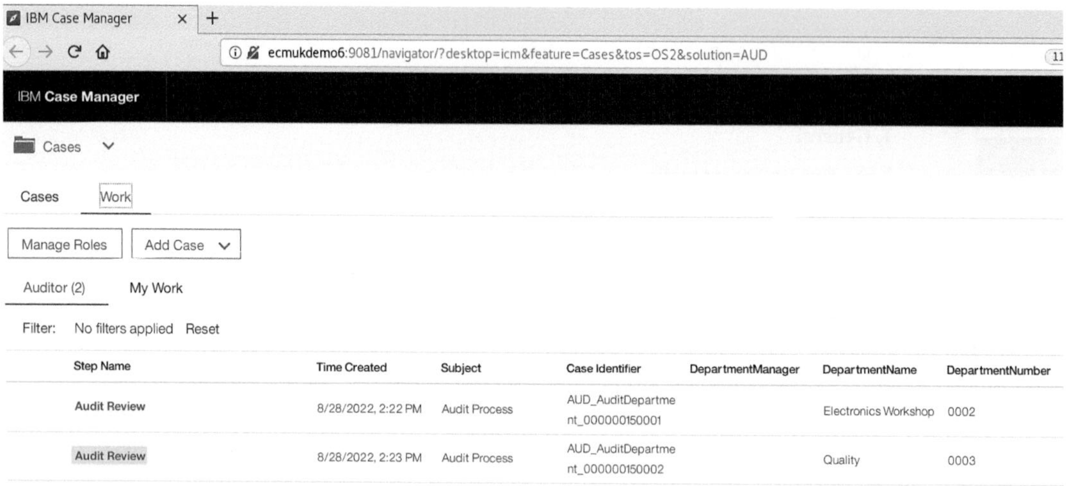

CHAPTER 6   PDF DOCUMENT CREATION USING JAVA

*Figure 6-93. The **Audit Review** Auditor Work Step we created in Figure 6-90*

On clicking the **Work** tab, we can now see that the **Auditor** role has two cases to review. On clicking the highlighted **Audit Review** for the **Quality** department, we can see the Work Details page which also shows us the Audit Report pdf we created, which the Audit Report program linked to the Folder for the Case Documents for the Case.

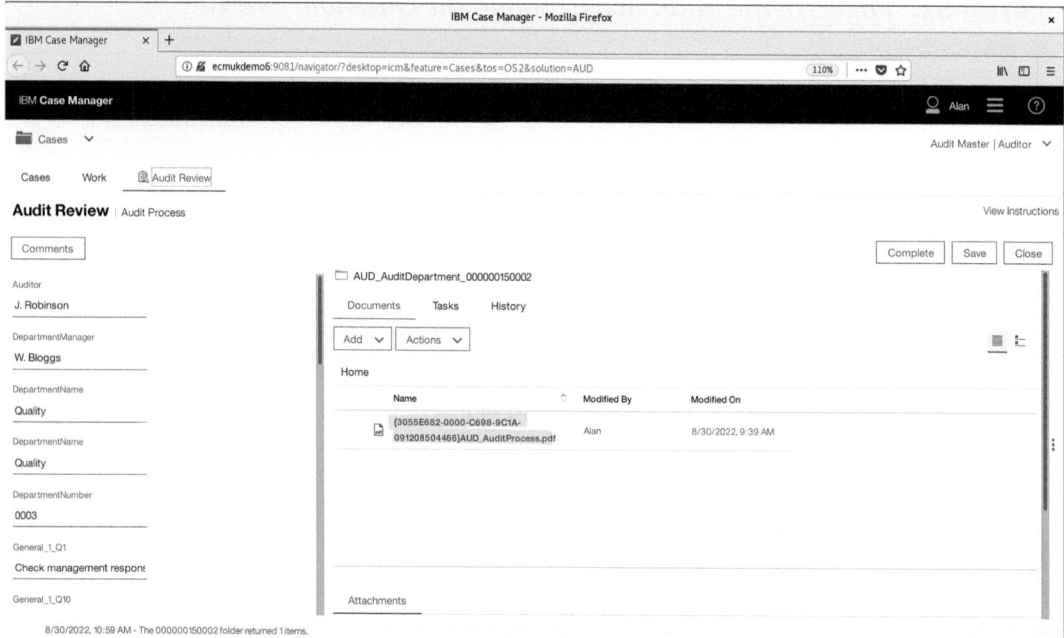

Figure 6-94. The Audit Report pdf for the Audit Review is clicked to open

CHAPTER 6   PDF DOCUMENT CREATION USING JAVA

The Audit Report pdf we created is highlighted in Figure 6-94, and on clicking this link, it launches the IBM Case Manager **ViewOne** Document Viewer which can be seen to be scrollable, so we can see the Logo and the Header page in the Viewer Frame.

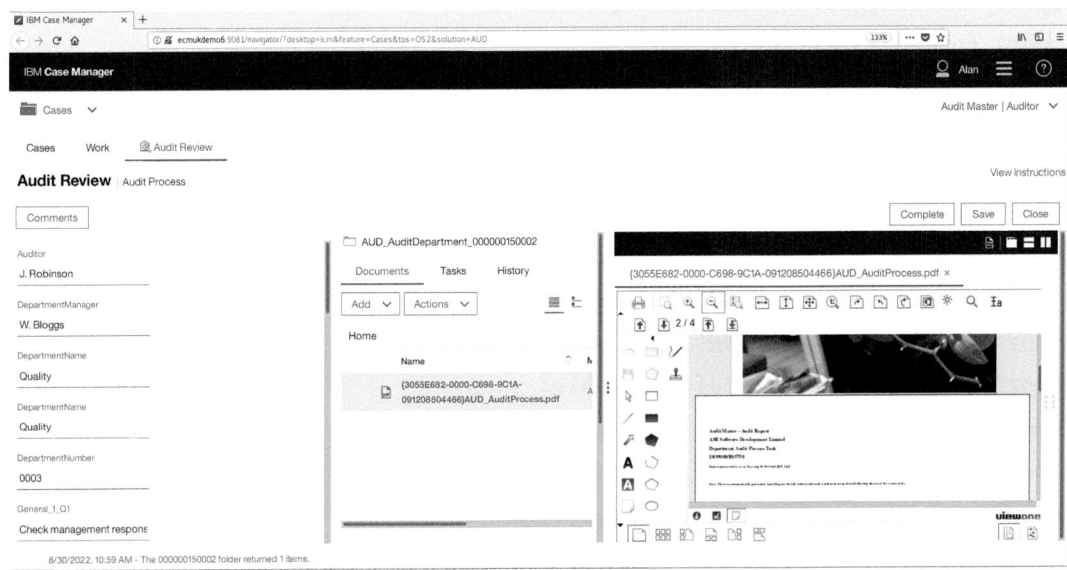

*Figure 6-95.   The **ViewOne** Document Viewer is launched to display the pdf*

Part 4 – Example 4 – Create an Auditor Calendar Table in a PDF Document

The following **AuditTabTest.java** test code provides an example of code to create a Table in a pdf document.

Listing 6-29.   The `AuditTabTest.java` example source code for Table generation in a pdf file

```
package com.ibm.filenet.ps.ciops.test;
import java.io.File;
import java.io.FileOutputStream;
import java.io.IOException;
import java.io.OutputStream;
import java.time.LocalDate;
import java.time.LocalDateTime;
```

1263

```java
import java.time.LocalTime;
import java.time.format.DateTimeFormatter;

// iText jar Library imports
import com.lowagie.text.Document;
import com.lowagie.text.Element;
import com.lowagie.text.Image;
import com.lowagie.text.PageSize;
import com.lowagie.text.Paragraph;
import com.lowagie.text.Phrase;
import com.lowagie.text.Rectangle;
import com.lowagie.text.pdf.PdfDocument;
import com.lowagie.text.pdf.PdfPCell;
import com.lowagie.text.pdf.PdfPTable;
import com.lowagie.text.pdf.PdfWriter;
import com.lowagie.text.pdf.RandomAccessFileOrArray;
import com.lowagie.text.pdf.codec.TiffImage;
public class AuditTabTest {
    public static void main(String args[]) {
        //Set up the Date / Time system variables
        LocalDate AuditReportDate = LocalDate.now();
        LocalDateTime AuditReportDateTime = LocalDateTime.now();
        LocalTime AuditReportTime = LocalTime.now();
        DateTimeFormatter AuditDateFormat =
        DateTimeFormatter.ofPattern("dd-MM-yyyy HH:mm:ss");
        String sAuditDate = AuditDateFormat.format(AuditReportDateTime);
        OutputStream AuditPDFfile = null;
        try {
            AuditPDFfile = new FileOutputStream(new File("/opt/
            AuditReport/AuditTestTable.pdf"));
            //Create a new Audit Document object
            Document audit_document = new Document();
            //You need the iText PdfWriter class for a PDF document
            PdfWriter.getInstance(audit_document, AuditPDFfile);
            //Open the Audit document for writing a PDF
            audit_document.open();
```

CHAPTER 6   PDF DOCUMENT CREATION USING JAVA

```java
            //Write the test content
            audit_document.add(new Paragraph("      Audit Report: 
            Document Test - PDF Table Generation"));
      //Set Table Column widths (Relative)
      float[] columnWidths = {2, 2, 2, 2, 2};
      PdfPTable table = new PdfPTable(columnWidths);
//Set the Table as 95% of the page width
 table.setWidthPercentage(95);
 table.getDefaultCell().setUseAscender(true);
 // Set up the Table data for each column of the Table
 String [] departmentNumber = {"0001","0002","0003","0004","0005"};
 String [] publishYes = {"Y","Y","Y","Y","Y"};
 String [] Auditor = {"N Smith","W Bloggs", "C Black", "F Jones" , "N 
 Robinson"};
 String [] dateDetails1 = {"15-Aug-2022","18-Aug-2022", "20-Aug-2022", 
 "25-Aug-2022" , "30-Aug-2022"};
 String [] departmentDetails2 = {"Quality","Electronics", "Warehouse" , 
 "Engineering", "Engineering"};
 String [] managerDetails3 = {"W Bloggs","C Black", "F Jones" , "N 
 Robinson", "N Smith"};
int nStartQNO = 0;
int nEndQNO   = 4;
 PdfPCell c1 = new PdfPCell(new Phrase("Department No."));
 c1.setHorizontalAlignment(Element.ALIGN_CENTER);
 table.addCell(c1);

 PdfPCell c2 = new PdfPCell(new Phrase("Auditor"));
 c2.setHorizontalAlignment(Element.ALIGN_CENTER);
 table.addCell(c2);

 PdfPCell c3 = new PdfPCell(new Phrase("Date"));
 c3.setHorizontalAlignment(Element.ALIGN_CENTER);
 table.addCell(c3);

 PdfPCell c4 = new PdfPCell(new Phrase("Department"));
 c4.setHorizontalAlignment(Element.ALIGN_LEFT);
 table.addCell(c4);
```

```java
        PdfPCell c5 = new PdfPCell(new Phrase("Manager"));
        c5.setHorizontalAlignment(Element.ALIGN_LEFT);
    table.addCell(c5);
    table.setHeaderRows(1);
        //We have all the Questions loaded now
    int iCell = 0;
    int iRow = 0;
    String titleVal = null;
    int nQuestionColumns = 5;
        for   (String qRow:publishYes) {
            //Add the Table rows for each Element of the
            publishYes array
         if (iRow < 5 ) {
                String [] cellRow = new String[nQuestionColumns];
                cellRow[0] =   departmentNumber[iRow];
                cellRow[1] =   Auditor[iRow];
                cellRow[2] =   dateDetails1[iRow];
                cellRow[3] =   departmentDetails2[iRow];
                cellRow[4] =   managerDetails3[iRow];
                //For each cell across the table row add the
                descriptions in the cellRow array
                  for (String sCell:cellRow){
                     titleVal =sCell;
                     table.addCell(sCell);
                     iCell ++;
                  }
                iCell = 0;
             iRow ++;
         }
     }
        //Add the created table to the PDF Document
        audit_document.add(table);
        //close the document
        audit_document.close();
        System.out.println("Audit Report Table Created:" + sAuditDate);
```

CHAPTER 6   PDF DOCUMENT CREATION USING JAVA

```
    } catch (Exception e) {
        e.printStackTrace();
    } finally {
//close the FileOutputStream
        try {
            if (AuditPDFfile != null) {
                AuditPDFfile.close();
            }
        } catch (IOException io) {
            //Failed to close
            System.out.println("The Audit Table PDF file failed to
            close!");
        }
    }
  }
}
```

Viewing the table shows the following output.

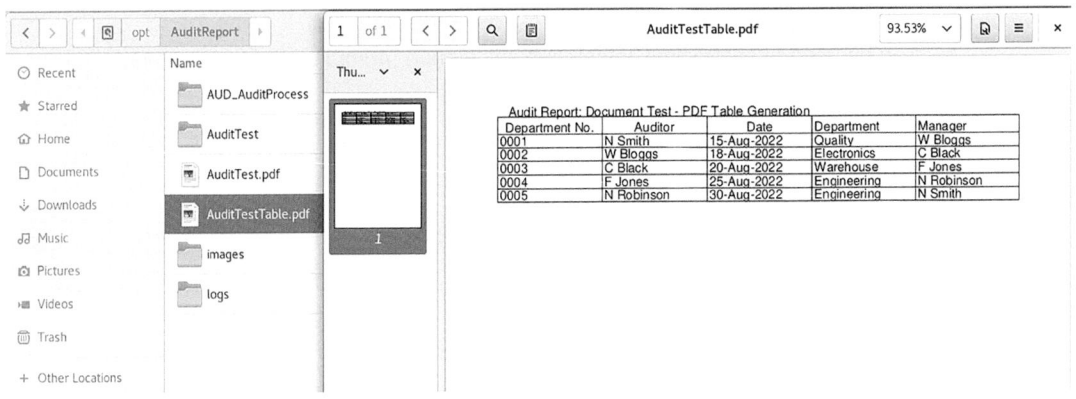

Figure 6-96. *The Test Table is output to the **AuditTestTable.pdf** file*

Chapter 6 Exercises

The following questions cover the functions using the iText pdf creation package which we covered in this chapter.

1267

CHAPTER 6   PDF DOCUMENT CREATION USING JAVA

MULTIPLE CHOICE QUESTIONS

1. The white cross on a red background icon against the IBM Case Builder Audit Master Solution panel means

    a) The Audit Master solution has new Solution edits to be saved.

    b) The Audit Master solution has been successfully deployed.

    c) The Audit Master solution failed to be deployed.

    d) The Audit Master solution has warning validation errors.

2. An abstract class, like XMLConfigReader

    a) Can be used by a subclass which can extend multiple abstract classes

    b) Can be used by a subclass which can only be used to extend the XMLConfigReader class

    c) Can be used by a subclass which could then only implement a single interface class

    d) Can be used by a subclass which can still extend multiple interface classes

3. A Java utility ResourceBundle

    a) Is often used to retrieve locale-specific String objects

    b) Is often used to retrieve a list of System properties

    c) Is often used to retrieve the Java classpath

    d) Is often used to retrieve the System Platform type

4. A PDF metadata Header is used to

    a) Hold the full layout of the pdf file as a Contents page

    b) Hold attributes such as the pdf author name, title, and a file description

    c) Hold the full security access to the pdf document

    d) Hold the list of Fonts displayed in the pdf document

CHAPTER 6   PDF DOCUMENT CREATION USING JAVA

MULTIPLE CHOICE ANSWERS

1. c) The Audit Master solution failed to be deployed.

2. d) Can be used by a subclass which can still extend multiple interface classes.

3. a) Is often used to retrieve locale-specific String objects.

4. b) Hold attributes such as the pdf author name, title, and a file description.

QUESTIONS

1. Describe the supporting libraries you need to run the Audit Report Java system.

2. What Java class object elements does the iText version 2.1.7 support for building the Audit Report pdf generation program?

3. What else could you use the pdf Table creation code for to extend the Auditing solution?

Index

A

acce web application, 212, 249, 298, 587, 684, 686, 1255
Add command, 82, 92, 95, 161, 220, 640, 746, 963
Add External JARs command, 157, 158, 719, 749, 1203, 1213
Add Library… command, 1208
add() method, 996, 1000
Apache Software Foundation (ASF), 676, 1248
Apply and Close command, 161, 407, 408, 424, 435, 715, 1205, 1215, 1217
AUDITConfig, 1046–1051
Auditing system, 996
Audit Master solution
    builder editor, 242
    document/folder event, 206, 238
    production development
        administration desktop, 98
        attributes, 107, 108, 111
        client manage roles, 113
        deployment, 113
        design object store, 99
        desktop application, 98
        download and close command, 103
        drop-down solutions, 99
        export details, 101, 102
        exported zip file, 104, 105
        export option, 100
        finish command, 109, 112
        importing file, 105–114
        import solution, 105, 106
        name/prefix values, 100
        security roles, 114
        target object store, 110
        web browser download option, 103, 104
        zip file, 102
    shell script environment, 296
    SQL database, 320
    testing
        case details, 139
        case manager desktop, 136, 137
        search command, 137, 138
        test case, 136
    testing/administration
        advanced search command, 85
        auditor role, 72
        Browse command, 94, 95
        business objects, 74
        case command button, 73
        case details, 82
        checklist items, 79, 88, 91
        command button, 92
        configuration, 91
        date/time field, 75, 76
        department case, 74
        department details, 79, 80
        document management system, 96
        documents tab, 94
        emails, 80, 81
        FileNet content, 95
        history tab, 86, 90, 96, 97
        In-basket displays, 83
        information fields, 78
        magazine view, 87
        manage roles command, 73
        manage roles menu, 71
        OK command, 81
        property entry view, 87
        required fields, 77
        search command button, 82, 83
        solution role, 70, 72
        stage details, 90
        structure creation, 84
        summary tab details, 86
        task status, 89
        users/groups command button, 71
Audit Report Java system, 1269
Audit Report program
    add **config** subfolder, 1218–1221, 1223, 1224, 1226, 1227
    Apache Commons IO jar file, 1211–1215
    Apache Commons Lang jar file, 1215–1217
    AuditMasterLogoTest.java test code, 1230–1234
    AUDITReportMain.java source, 1186–1201
    Audit_Report project classpath properties, 1202
    Audit_Report project properties, 1203, 1204
    audit review step, 1259–1263
    AuditTest class, 1228–1230
    Configuration class code, 1175, 1176
    DatabaseLoginModule class code, 1164–1170
    DatabasePrincipal class, 1171–1173
    EmailTemplate class code, 1153–1157
    IBM Case Manager Builder application, 1234–1238
    iText version 2.1.7 jar file, 1205, 1207
    Java methods, 1152
    JUnit Java test, 1208–1211
    Log4J logs subdirectory, 1244–1254
    QOperationsTest class code, 1178–1185
    TemplateContentElementDataSource.java source, 1159–1162
    testing, 1239–1244
    verifying file, 1255, 1256, 1258, 1259
AUDOperations.jar file
    AUDOperations.jardesc file, 579
    browse command, 580
    configuration.properties, 583
    exporting options, 578
    FileNet workflow system
        acce administration web application, 584
        Adapter properties, 596
        AUD_Operations queue, 597, 598, 602
        code module, 595
        CodeModules tab, 585–587
        command button, 589
        component, 584
        component tab, 596
        configuration context, 596, 597
        content button, 588
        database storage area, 590
        default values, 589
        Design Object store, 599, 600
        Document title, 586, 587
        General tab, 603
        getFolderDocuments method, 604
        load command button, 594

INDEX

AUDOperations.jar file (*cont.*)
    methods, 602
    methods list, 595
    navigation pane, 592, 593
    node system, 592
    parameters, 591, 601
    queues folder, 594
    retention period, 590
    upload option, 588
    web browser window, 587
JAR file export wizard, 577
MANIFEST.MF generation option, 581
parameters
    deployment, 606
    sendJMSMessage method, 604, 605
    templates, 606
process designer application
    action bar line commits, 608
    AUD_Operations queue, 609, 618, 620
    code module, 611
    component node, 616
    component queue configuration, 607
    component step, 613
    component step parameters, 621
    configuration menu, 607
    Design Object store, 612
    expression values, 615
    highlighted icon, 614
    initial parameter order, 619
    jar libraries, 617
    message libraries, 616
    queue properties, 608
    security JAAS credentials, 610
    sendJMSMessage method, 614
    sendJMSMessage parameters, 613
    source code, 623
    workflow installation, 617, 622
process designer workflow, 575, 576
rebuild option, 583, 584
source code, 582, 583

B

Bill of materials
    administration application, 353
    case manager system, 3–6
    command button, 339
    configuration, 360
    continuous delivery (CD), 338, 341
    downloaded packages, 340
    download status, 342
    FileNet Target Object
        administrative user, 368
        AUDQAR local queue, 370
        AUDQCF object, 378, 379
        browse message, 384, 385
        context startup, 367
        data property, 385
        highlighted menu options, 383
        initial context, 364
        JMS queue wizard, 375
        JNDI namespace, 365, 367
        LDAP server, 376, 377
        local queue, 371, 372, 375
        messaging service" "bill1, 364
        MQQueue class, 382
        object creation, 374
        properties, 380, 385
        Put message command, 383
        queue status, 381
        security configuration, 366
        server context, 369
        test message, 384
        users and groups, 369, 370
    firefox browser, 341
    hardware and software requirements, 338
    highlighted menus, 355, 356
    initial screen, 355
    installation packages, 338, 339, 342, 343, 345
    license shell script, 344
    limits.conf file, 348, 349
    link versions, 337
    Linux root user, 351, 352
    listener configuration, 360
    logging defaults, 358
    MQ-related environment variables, 349, 350
    primary installation, 348
    product details, 338
    program icon, 353
    progress status bar window, 340
    queue manager, 356–359, 362
    reconnection settings, 361
    root user file, 347, 348
    splash screen, 354
    wasadm node, 362, 363
    wasadm password, 353
    windows version, 341
Business objects
    attribute elements, 45
    audit master solution, 41–43
    AuditObj property, 43
    auditor details, 30–32, 34
    audit properties, 32, 33
    case properties, 44
    checklist item, 39, 40
    comments property, 32
    department ID property, 33
    department name/description, 35–37
    department task, 45–53
        attributes, 46
        completion date property, 47
        DepartmentName/DepartmentNumber properties, 47
        designer icon, 52
        layout container, 52, 53
        properties layout tab, 51
        properties tree menu, 46
        required option, 48, 49
        required properties, 52
        save command, 49
        search view, 53, 54
        summary case properties, 50
        type properties, 48
        views tab, 50
    description, 29, 30
    global property list, 34
    menu option, 28
    procedure name/description, 37, 38
    properties, 27, 36, 43
    search view
        audit department folder, 55
        audit department type, 62, 63
        audit master solution, 69, 70
        auditor role, 65–67
        audit solution properties, 58
        case details page, 59, 60
        components, 68
        design comments, 58, 59
        edit settings icon, 61, 62
        name/description fields, 60, 61
        OK command button, 64, 65
        page designer icon, 61
        properties, 54
        stages section, 55
        task attributes, 57

INDEX

task properties tab, 57
test environment icon, 69
workflow process, 56, 57
types, 44
unique identifier, 31
Business Process Manager (BPM), 2, 3

C

Cardinality property, 683
CaseAgeing, 929, 963, 964
CaseAnalytics, 929, 939, 962, 963, 973
Case Manager system
    administration server, 8
    audit case properties, 14
    audit master solution, 70-114
    audit system, 3
    Bill of Materials, 3-6
    BPM process, 2
    business objects (*see* Business objects)
    close command, 8, 9
    content engine web application, 9
    debugging (*see* Debugging process)
    development tool, 10
    document class
        attributes, 27
        Audit Report class, 24, 25
        creation, 23, 24
        global properties, 25, 26
    FileNet (*see* FileNet)
    FileNet databases, 9
    fix recommendation
        add-on menu option, 134, 135
        audit case details, 139-142
        base article, 131
        folder class, 131
        property definition, 133
        save and refresh command, 132
        template property, 133
        test class, 131, 132
        workplace base extensions, 131
    global properties, 21-23
    global solution properties, 13
    in-basket properties, 23
    installation document, 4
    instance administration tool, 6
    launch process, 3
    LDAP server, 6-9
    organization, 2
    overview, 1
    properties/business objects, 13-15
    property type, 14
    role solution
        auditor role, 16-18
        definition, 15
        in-baskets tab, 21
        pages tab, 18
        roles tab, 16
    solution metadata, 11-13
    video tutorial, 2
    web application server, 10
    workaround error message
        case details, 130
        choice list values, 126
        class definitions, 129
        client search, 123-125
        ContainerType property template, 125-128
        data type, 126
        initial version, 124
        reviewed properties, 127
        search command, 124, 125
        search option, 130
        single option, 127
        SQL command, 130
    work dashboard tasks, 3
CASTORE1 Database, 906, 909, 911, 915, 917, 926, 932
CAUSER Schema, 926-928, 932-934, 943, 944
CEConnection Java code, 895
CEMigrateConfig, 648
CEMigrateConfig.java Java code, 878-890
CEReplicateConfig, 648
CEReplicateConfig Java code, 760-861
Certificate Revocation Lists (CRLs)
    AUD.CLNTCONN, 457
    AUDNAMES description, 447-450
    AUDNAMES list, 445, 446
    authentication, 441, 442
    client-connection, 453, 454
    default reconnection option, 456, 457
    information menu, 440
    LDAP authentication, 440-458
    mquser1 authentication, 443
    namelist node menu, 443, 444
    OK button, 447
    pop-up message, 442
    queue manager, 454, 455
    SSL page selection, 452
    wasadm properties, 450, 451
Client application connection
    error logs
        attributes, 536, 544, 551
        AUDQAR queue, 533-535, 537-539, 555
        authorities attributes, 540
        autoreconnect/refresh options, 550
        command button, 528
        connection factories, 542, 545, 546
        connection factory wizard, 524-527
        connection options, 530
        description, 557
        description/provider version, 529
        finish command, 531
        LDAP context creation, 534
        listener option, 549
        location information, 523
        local queue, 551, 552
        logging details, 548
        message handling properties, 556
        queue creation wizard, 553, 554
        security options, 541
        subfolder, 523
        success message window, 532
        SYSTEM.DEF.CLNT.CONN channel, 558
        transactions option, 543
        transmission/dead-letter queue details, 547
    Linux, 493-495
    listener authorities
        authorities window, 482, 483
        command button, 479, 480
        existing user authorities, 479
        LISTENER.TCP profile, 481
        menu option, 482
        MQ main program, 485
        object security access, 478
        security system, 481, 483-485
    messaging creation
        CLIENT.wasadm server channel, 491, 492
        server channel, 491
        transmission queue type, 493
    missing local queue
        AUDQ1 queue creation, 517, 518
        communications tab, 522
        context queue creation, 516, 517
        drop-down list, 508, 509
        finish command button, 510, 511

1273

INDEX

Client application connection (cont.)
    LDAP server, 513, 514
    menu option, 508
    Next command, 515
    OK command button, 512
    properties, 518
    tick box creation, 510
    transmission queue, 518–521
MQSERVER environment variable, 493–495
non-privileged MQ user, 458
procedure
    accumulated authorities menu, 473
    AUD.CLNTCONN profile, 463, 464, 474
    authority records option, 461, 462
    channel authentication rule, 459
    client-connection channel, 465
    command button, 463
    initial context, 475
    LDAP root, 476
    listener program, 458
    menu option, 468
    MQSC command, 460
    queue manager, 460, 469
    queue manager authority, 470, 471
    quit command, 459
    referenced queue, 477
    security access, 466, 468
    security options, 472
    tick box selection, 467
    TOPIC Profile, 477
    user access options, 472
queue manager
    active status, 490
    configuration settings, 489
    dspmq command, 487, 488
    MQ wasadm queue, 489
    start queue manager, 490
    stopping/starting, 486
    wasadm queue, 488
    wasadm status, 487
    WebSphere MQ installation, 486
sending message, 495
testing error
    AUDD1 dead-letter queue, 504, 506
    AUDQCF queue, 503
    Dead-Letter Queue AUDD1, 496
    destination object, 502
    finish command button, 499
    LDAP server context, 501
    local queue menu, 496
    queue creation, 500, 505, 506
    queue creation process, 497, 498
    test error results, 507
    wasadm queue manager, 506, 507
close() method, 1000
Cognos Analytics web application, 915, 989
com.asb.ce.utils package, 1007, 1008
com.asb.config package, 1040–1042, 1044, 1046
com.ibm.filenet.ps.ciops.database package, 1163, 1164, 1171
com.ibm.filenet.ps.ciops package, 1058
com.ibm.filenet.ps.ciops.test package, 410, 1174, 1175, 1178, 1186
Commit My Changes command, 1237
config package, 1040
Content engine client
    administrator user password, Config.xml, 669
    config area, 669
    Config.xml, 670, 671
    files, 667, 668
    log4j.dtd, 676, 677, 679, 680, 682
    log4j.xml, 672–674

    original log4j.xml file, 674–676
    version 5.5.x jars, 667
Content engine events
    AddDocToFolder event code, 176, 177
    attributes, 148
    AuditEventHandler, 147, 149, 150
    AuditModule class package, 169–172
    CEConnection class, 152–154
    classpath node, 156
    code project, 147
    com.asb.ce package, 152
    external JARs command, 158
    jar files, 155
    jar libraries, 177, 178
    JRE versions
        AuditEventHandler project, 162
        classpath libraries, 162, 163
        command button, 160
        project classpath, 161, 162
        source code, 163–168
        system installation, 160, 161
        system library option, 159, 160
    library command, 158, 159
    open perspective command, 150
    package option, 151
    project properties, 155
    project structure, 147
    standard layout, 151
    TestHarness Java class, 173–176
    wizard types, 148
Create command, 961

D

Database Stored procedure, 141, 144, 251, 271
Datasource setup, add custom views
    CaseAgeing Data module, 964
    CaseAnalytics, 963
    CAUSER Datasource, 947, 949
    CAUSER.Dmageing_Case, 951
    CAUSER.F_Dmcaseload, 950
    CAUSER schema, 944
    Cognos Analytics, 972
    Cognos Content folder, 945
    Dashboard, 973
    Dashboard graph, 974, 976
    Dashboard panel, 961
    Data module menu item, 946
    Daysin count, 968–970
    Dmageing_Case, 955
    events server, 943
    F_Dmcaseload, 954
    graph types, 965–967
    Incoming Case Events, 967
    inner join dipslays, 958
    list of pages, 971
    New data module, 952, 953
    present data panel, 960
    relationship link procedure, 959, 960
    select sources, 949
    select tables option, 948
    Tasktypename graph, 977
    visualizations, 964
DataType property, 682
Date Created property, 696
DateDiff function, 979
DbExecute connection
    administration tool, 271
    advanced settings, 274
    auto claims solution, 273

auto insurance claims, 272
command button, 271
content navigator, 276
global toolbar settings, 274
information icon details, 275
parameters, 272
process designer desktop tab, 275
repository folder, 275
stored procedure, 304-307
    auditdb database, 305
    installation, 312-328
    parameters, 311
    procedure steps, 307
workflow step
    AuditDB connection parameters, 300-302
    auditdb database, 303
    content engine store, 299
    database connection, 300
    DbExecute connections tab, 303
    design object store, 299
    system setup, 298
workflow system
    case field properties, 310
    configuration, 307
    general system palette window, 308
    parameters, 309, 310
    process designer, 308
    task workflow map, 308-311
Debugging process
    FileNet server, 114
    logging settings, 114, 115
    save option, 116
    SystemOut.log file, 116, 117
    warning level, 115
Document class constructor, 996
Download file PDF command, 3, 4, 335, 648

E

Eclipse Java IDE program, 698
Eclipse New project
    Apache site, 716-718
    AUDIT_CEReplicate, 704, 707
    AUDIT_CEReplicate Project Classpath, 714, 715
    CEReplicate class, 713
    CEReplicate.java code, 714
    external library JAR files, 719, 720
    jar library file security, 750, 751
    Java Build Path menu, 746-750
    Java packages, 709
    Java project, 706
    Object Store security, 751, 752
    OS GUID, 758, 759
    package menu option, 710, 711
    parameters, 708
    replication program tests, running, 753-758
    supporting java projects, 726-740, 742-745
    Xerces java library, 722-726
Event Object
    action menu, 189
    AddDocToFolder, 188, 193
    AuditEventHandler code module, 191
    class handler package, 190
    creation, 188
    creation process, 193
    display name/description, 190
    handler object and links, 192
    load command button, 191
    next command, 192
    steps, 188

F

FileNet, 2
    Audit Report, 179
    bill of materials, 364
    components, 5
    content engine, 143
    content engine web application, 9
    databases, 9
    debugging process, 114
    document/folder event, 205
    fix recommendation, 131
    health status, 119
    Java content manager, 144, 145
    navigator web application, 258
    Object Store
        engine status page, 121
        status values, 120
        system information details, 123
        upgrade status, 120
        workflow process, 120-122
    target (OS2) object store, 364
    testing/administration, 95
    version details, 118
    web application, 10
    workflow system components, 584-604
Finish command, 102, 109, 112, 149, 186, 187, 198, 204, 213, 214, 368, 448, 499, 505, 511, 517, 531, 532, 537, 552, 590, 597, 713, 982

G

Getter and setter methods, 1043, 1045

H

Hamburger, 917, 925

I

IBM Case Analyzer, 903, 911, 978
IBM Case Manager Builder application, 10, 23, 211, 1234, 1237
IBM Case Manager **ViewOne** Document Viewer, 1263
IBM Cognos Analytics
    aging bands, dynamic dimensions, 979, 980
    data server connection, 990
    Datasource, 990
    development, 905
    dynamic dimensions, aging bands
        Create View statement, 913, 915
        datasource setup, 943
        logon option, 916
        query calculation, 911
        range, 911
        web application, 916
    event, 903
    exposed user **Dimension** tables, 982-985
    MS Word document, 903
    multilevel dimension procedure, 980-982
    OLAP integration, 978, 979
    release candidate builds, 904, 905
    SQL statement, DepartmentName Dimension, 985-988
    time dimension SLA calculations, 906-911
    URL, 904
    Visual Studio, downloads, 905
IBM Cognos BI Analytics, 901
IBM FileNet Content Engine Administration Console (**acce**) web application Document Viewer, 1255

INDEX

IBM FileNet object stores
    access rights, 658
    bill of materials, 648, 649
    CEConnect, 893, 894
    CEConnection Java code, 895, 896, 898, 899
    CEReplicate, 699, 701, 703, 704
    CEReplicate class, 899
    creator property, 658
    development tools used, 653, 654
    events
        definition, 684
        document class, 686
        folder class, 684, 685
    linux directory paths, 663–666
    multiplatform multilingual eAssembly, 649, 650
    nonfunctional requirements, 652
    P8 5.5.5 environment, 654–656
    PropsUtil Java code, 891–893
    replication program, 650–652
    shell script batch jobs, 658–660, 662
    unit test data, 656, 657
    unit test phase
        Java Eclipse Projects, 699
        phase 1, 687, 688, 690–694, 696
        phase 2, 697
        phase 3, 697
        phase 4, 698
Import statement, 720, 726
Initial datasource setup, custom views
    administration console, 925
    CaseLoad Data module, 940, 941
    Case Manager, 942
    CASTORE1 OLAP database, 935, 936
    CAUSER.D_Dmtasktype Dimension table, 938, 939
    CAUSER schema, 927, 928, 932, 933
    credentials, 922
    Dashboard options, 930, 931
    data connection name, 918
    data server connection, 919
    datasource connection, 926
    datasource types, 920
    graphs, 943
    loaded schema, 929
    manage menu, 917
    menu option, 923
    permissions, 924
    refresh, 937
    select tables option, 934
    signon, 921, 922
iText library, 990, 991, 999, 1000, 1228
iText methods, 992, 998

J, K

Java Archive (JAR)
    jar files, 401–403
Java Build Path Libraries, 749
Java Build Path property, 725
Java customizations, 143, 188
    auto claims solution, 251, 252
    case type workflow tasks, 252–255
    code development, 145
    components (*see* Content engine events)
    DB2 database, 312–328
    DbExecute (*see* DbExecute connection)
    DbExecute workflow, 251
    documentation, 143
    edit workflow steps option, 256
    FileNet content manager, 144, 145
    fn_eventshandler.jar
        AuditEventHandler project, 202

browse button, 203
creation, 199
export menu option, 200
file export wizard option, 201
finish command button, 204
process designer (*see* Process designer plugin)
shell script environment
    AUD_AuditProcess task workflow, 298
    audit master solution, 296
    cmd.exe command window, 290
    command line, 291
    cpetoolenv.bat, 293
    department case type, 297
    design object store, 295, 296
    edit screen, 294
    log-in command button, 293
    menu option, 295
    pedesigner.bat, 290–292
    solution definition file, 296, 297
    system error message, 292
standalone process designer
    error logs, 276
    installation window, 276, 278, 279, 281, 287–289
    older version, 283
    process designer directory, 282
    progress status bar, 278, 288
    SOAP URL link, 283, 284, 286
    software package, 277
    splash screen, 280
    windows administrator, 277
step designer application, 255–257
supporting documents, 144
task workflow page, 256
testing document/folder event
    AC.pdf document, 221–223, 228
    audit department details, 218
    audit master case, 249
    AUDIT_TEST folder, 213–215, 238
    automatic folder, 248
    case description, 216
    class properties, 220
    configurations, 225, 228
    confirmation dialog window, 245
    copy item menu, 223
    default value, 244
    department case, 215
    dependencies tab, 226
    deploy icon, 245
    display value option, 223
    documents tab, 243
    drop-down option, 219
    drop-down selection, 210
    eclipse file, 233
    file export wizard, 236
    file option, 229
    fn_eventhandlers.jar file, 235, 237, 238
    green tick icon, 246
    highlighted document, 240, 241
    jar files, 227
    local system option, 218
    log4j.xml file, 230
    pdf document, 220
    properties tab, 206, 207, 243
    redeploy all menu option, 211
    report class, 239
    report document class, 208–210, 246, 247, 250
    retention options, 213
    save/close command button, 245
    search option, 217
    solution properties tab, 207
    status window, 214

INDEX

SymbolicName properties, 207
TestHarness code, 205
TestHarness.java code, 224, 225
WARN messages, 229, 234
web application, 205, 242
XML file, 231, 232
workflow subscriptions
    AddDocToFolder, 197
    AUD_AddDocumentToFolder, 195, 199
    Audit Report class, 194
    configuration, 193-199
    enable option, 198
    event creation, 197
    finish command, 198
    object types/events, 196
    subscription, 195
    triggers screen, 197, *See also* Java Runtime Environment (JRE)
Java iText library, 990
Java Messaging Service (JMS)
    AUDOperations.jar (*see* AUDOperations.jar file)
    AUDOperationsTest.java class
        data properties, 575
        event handlers project, 570
        JUnit test output, 570-575
        properties, 574
        rebuild/deployment code, 559-570
        runmqsc command, 572
        test message, 558, 559
    bill of materials, 337-385
    client application (*see* Client application connection)
    component integration, 336
    interface development, 335
    manager installation document, 335
    operations component development
        AUD_CIOps project, 388-400
        AUDOperations component code, 390
        AUDOperationsTest class, 411-415, 417-419
        CELoginModule class, 391
        Certificate Revocation Lists (CRLs), 440-458
        CodeModule class, 386
        component integrator, 391
        component manager, 390
        configuration class, 420, 421
        configuration.properties file, 409-411
        DemoJava.zip link, 387, 388
        documentation, 390
        done command, 406
        Eclipse IDE, 386
        Eclipse project, 401
        error messages, 424-427
        GitHub, 386
        ibmorb.jar, 433
        installation screen, 405
        j2ee.jar/bootstrap.jar files, 432
        jar files, 401-403
        JRE library, 402-408
        JUnit library, 422, 424
        license agreement, 404
        MQ series channel security, 435-439
        MQ Series port, 428
        ports/library jar files, 428-435
        providerutil.jar file, 433
        sendJMSMessage method, 439
        source code, 388
        system libraries, 407, 408
        testing package/code, 410-424
    transferring workflow subscription
        Audit Master solution, 635-644
        Audit Report document, 623, 624
        AUD_JMSMessageWorkflow, 628, 630, 633

AUD_SendJMSAudit, 624
class properties, 633
command button, 640
creation, 632
department case, 636
document option, 637, 638
event action, 631
F_Comment/F_Subject fields, 644
flag property, 634
highlighted triangle icon, 635
LaunchWorkflow subscription, 631
message data, 643
message trigger, 641
MQ AUDQAR queue, 642
object creation, 635
parameters, 634
process designer, 627
property details, 639
published folder, 627
security, 629
target object store, 626
testing environment, 635
trigger creation, 625
workflow subscription, 625, 632
Java Runtime Environment (JRE)
    code module/development
        attributes, 179
        AuditEventHandler code module, 188
        browse button, 183
        CodeModules folder, 188
        command button, 184
        command selection, 182
        database storage area, 186
        document creation, 187
        document menu item, 181
        finish command, 186, 187
        fn_eventhandlers.jar file, 183
        isHiddenContainer property, 180
        mime type, 182
        property values, 185
        retention time options, 186
        system identifies, 184
        version property, 185
        wizard page, 182
    event object, 188-193
    operations component development, 402-408
    system library, 159, 160, 407, 408
Java sdk, 668
JDBC URL template, 298, 910, 921
JUnit library, 422, 1208, 1209

L

Linux CASTORE1, 910, 915
Linux commands, 262, 1213
Load metadata command, 944

M

Microsoft SQL Server Analysis Manager, 903, 904
Modify certain system properties, 752

N

New command, 271, 462, 463, 471, 479, 685, 686
New + command, 946
Next command, 101, 107, 111, 149, 150, 192, 194, 196, 102, 202, 236, 280, 283, 286, 502, 510, 515, 528, 590, 935

1277

INDEX

O

Object Authority Manager (OAM), 437
OK command, 22, 27, 51, 65, 77, 81, 404, 472, 480, 481, 512, 933, 947, 958
Online analytical processing (OLAP), 978
Open Perspective command, 150, 708
/opt/AuditReport directory, 1242
/opt/replication/libs jar library, 665

P, Q

Pdf Audit report
    audit master
        AUDITConfig.java source, 1047-1055, 1057
        AuditReportConfig.Java Class, 1009-1040
        cases, 1002
        com.asb.ce.utils Java Package, 1007, 1008
        com.ibm.filenet.ps.ciops, 1058
        config package, 1040
        ConfigVal Class, 1041, 1043, 1044
        ContentElementDataSource Java class, 1059-1062, 1068, 1069, 1071-1073
        Eclipse IDE, 1003, 1006, 1007
        JRE, 1005
        MimeType.java source, 1045, 1046
        QOperations.java source code, 1063-1152
PDF documents
    audit
        adding image, 996
        Audit Master Logo, 1000, 1002
        Header metadata settings, 994
        PdfWriter class, 997-1000
        text creation code, 994-996
    Audit Master solution, 1269
    creating auditor calendar table, 1263-1267
    Eclipse IDE, 993
    ECM Developer, 992
    import packages, 991
    iText jar library, 993
    iText methods, 992
Private gallery, 904
Process Analyzer, 905
Process designer plugin
    configuration parameters tab, 266
    connection parameters, 265
    content navigator, 258
    CPEAppletsPlugin.jar, 260, 262
    definition class option, 266
    load command, 261
    login credentials, 265
    login screen, 258
    OS2 repository, 267
    plug-in command, 259
    repositories, 264
    repository folder context menu, 268
    template menu, 269
    tracker menu actions, 269-271
PropsUtil Java code, 891

R

RandomAccessFileOrArray constructor method, 999
Real-Time Monitor, 901
Refresh command, 132, 249, 937, 951, 956, 957

S

Save command, 940
Snowflake Schema, 980, 984
SQL Server Analysis Services (SSAS), 978
SqlServer Query Analyzer, 978
Standard Operating Procedures, 647
Stored Procedure, 251, 304-307
    installation, 312-328
    Oracle version, 330, 331
    MS SQL Server, 331, 332
    parameters, 311
Structured query language (SQL)
    ADD_AUDIT_DEPT, 326
    Audit_Dept table, 312
    audit master solution, 320
    audit process task, 324
    business object values, 316
    case manager client, 323
    data fields, 317, 326, 327
    date/time string, 319
    field assignments, 315
    grant security access, 325
    hard-coded values, 328
    LaunchStep parameters, 315
    Linux server, 322
    MS SQL Server, 331, 332
    Oracle version, 330, 331
    pop-up window, 321
    properties tab, 314
    solution icon, 321
    stored procedure, 312, 313
    string variables, 327
    table creation, 312
    testing databases, 329
    test parameters, 328
    validate action menu, 318
    warning triangle, 321
    workflow validation, 319, 320

T

Target Object Store, 69, 84, 102, 110, 113, 208, 272, 323, 337, 558, 623, 626, 630, 648, 650, 651, 687, 727, 760, 878, 895, 1243

U

Update Security Event, 684, 686

V

Visual Studio, 904, 905
VMware system, 6, 982, 1206

W, X

Web Service Interface SOAP calls, 893

Y, Z

Yes command, 88, 1241

Printed by Printforce, the Netherlands